ENGINEERING THERMODYNAMICS

(In SI Units)

By

Prof. D.K. Chavan
Professor,
Mechanical Engineering Department,
Marathwada Mitra Mandal's
College of Engineering
(M.M.C.O.E.) Pune – 52

Ex. Assistant Professor
Mechanical Engineering Department,
Maharashtra Institute of Technology
M.I.T., Pune – 38

Prof. G.K. Pathak
Sr. Faculty Member,
Mechanical Engineering Department,
Maharashtra Institute of Technology
M.I.T., Pune – 38

STANDARD BOOK HOUSE

Unit of : **RAJSONS PUBLICATIONS PVT. LTD.**

1705-A, Nai Sarak, PB. No. 1074, Delhi-110006 Ph.: +91-(011)-23265506
Show Room: 4262/3, First Lane, G-Floor, Gali Punjabian, Ansari Road, Darya Ganj,
New Delhi-110002 Ph.: +91-(011)43551085 Tele Fax : +91-(011)43551185,
Fax: +91-(011)-23250212

E-mail: sbh10@hotmail.com www.standardbookhouse.com

Published by:

RAJINDER KUMAR JAIN

Standard Book House

Unit of: Rajsons Publications Pvt. Ltd.

1705-A, Nai Sarak, Delhi - 110006

Post Box: 1074

Ph.: +91-(011)-23265506 Fax: +91-(011)-23250212

Showroom:

4262/3. First Lane, G-Floor, Gali Punjabian,

Ansari Road, Darya Ganj,

New Delhi-110002

Ph.: +91-(011)-43551085, +91-(011)-43551185

E-mail: sbhl0@ hotmail.com

Web: www.standardbookhouse.com

First Edition : 2008

Second Edition : 2013

Price: **₹ 520.00**

ISBN: 978-81-89401-42-9

Typeset by:

C.S.M.S. Computers, Delhi.

Printed by:

R.K. Print Media Company New Delhi-110039

DEDICATED
To

Late Shri Kishanrao L Chavan
Late Smt. Umabai Kishanrao Chavan
Late Shri Ramkrishna I Zad

Preface

It gives us great pleasure to present a book on "Engineering Thermodynamics", for the students of Mechanical Engineering from first year to final year engineering.

In this book we will study basics of mechanical engineering, ideal real gases, compressors, gas turbines, jet propulsion, refrigeration and air-conditioning, heat transfer, compressors etc.

Every care has been taken to present the matter in precise and simple language. Self explanatory figures are included to enable the students to draw the same in the exams very easily.

From a long teaching experience, it has been observed that students commit mistakes in the units, as such, for every step in the solution of the problems, units have been made clear.

To develop more confidence in students appropriate questions from various universities are given at the end of each chapter. In the entire book SI system of units is used.

All the necessary care has been taken to avoid mistakes/misprints in the book. However it is quite likely that some mistakes/misprints might have been left unnoticed. Such mistakes and misprints of the book if brought to the notice will be gratefully acknowledged. Any suggestions to improve the usefulness of the book will be gratefully considered and appreciated.

We express our sincere thanks to, Prof. N.T. Karad, Dr. S.P. Kallurkar and also to,

(*i*) Prof. Dr. Mangesh Karad, Executive Director, MIT, Pune.
(*ii*) Prof. Dr. S.P. Kallurkar, Principal, MMCOE, Pune.
(*iii*) Prof. Dr. G.S. Tasgamkar Zeal, COE, Pune.
(*iv*) Prof. Raghubir Singh, Principal, Wingss Aeronautical College, Pune.
(*v*) Prof. Molkeri. M., HOD Mech. Engg. Deptt. REC Bhalki, Karnataka.
(*vi*) Prof. N.K. Joshi, HOD Mech. Engg. Deptt. Tolain Maritime Institute, Pune.
(*vii*) Prof. A.D. Desai, HOD; Mech. Engg. Deptt. PES, Modern COE, Pune.
(*viii*) Prof. S.M. Nakate, HOD; Mech. Engg. Deptt. MMIT, Pune.

We also express our sincere thanks to entire staff of Standard Books House New Delhi, those have taken sincere and commendable efforts to bring out this edition in very short span of time.

January 2012

Prof. D.K. Chavan
Prof. G.K. Pathak

Contents

CHAPTER 1 Basic Concepts of Thermodynamics and Zeroth Law

1.1 INTRODUCTION

In order to understand any science, everyone must get familiar with its unique terminology. This chapter is dedicated for introduction to the basic terms and concepts involved in thermodynamics. Along with the introduction to various concepts and definitions of different terms we shall also discuss the basic system of units which will be followed in the numerical part of this chapter.

1.2 THERMODYNAMIC SYSTEMS

Sometime we have to visit a general physician for certain health related problem. At that time, the physician examines our body for the accurate diagnosis of the illness. But, when we consult an ophthalmologist, he concen-

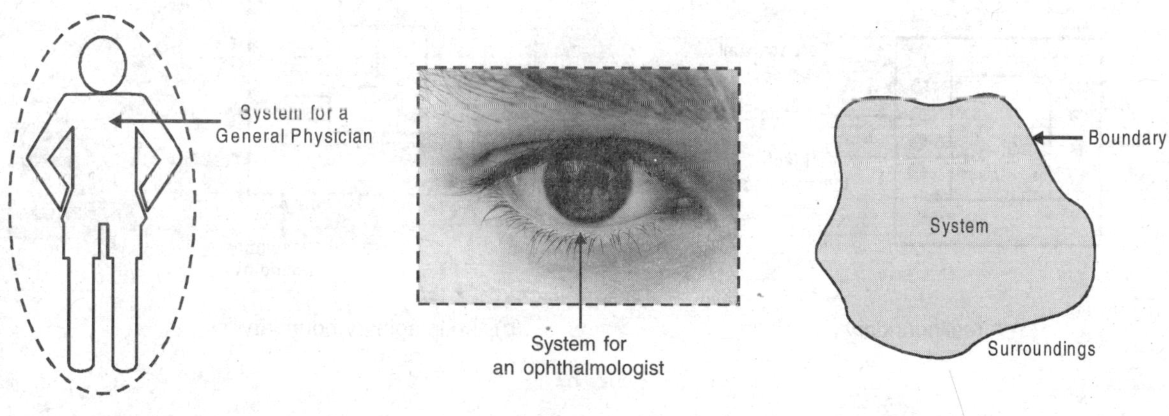

(a) Development of the concept—'System'. (b) Thermodynamic system.

Fig. 1.1

trates only a part of our body i.e., eye. Thus, every doctor decides the portion of body upon which concentration is made, according to his scope of diagnosis and treatment.

Above mentioned fact is elaborated in view of thermodynamic studies and is specified with the terms like System, Surrounding and Boundary.

A Thermodynamic System is defined as *quantity of matter or region of space upon which attention is concentrated in the analysis of the problem.*

Thermodynamic system may be defined as any part of the *Universe* which we want to study.

In order to fix the limits of a system, there must be a partition around the system. This partition is known as the *boundary of the system.*

Everything external to the boundary is called the *Surrounding.* The system along with surrounding forms an *Universe.* [Ref. Fig. 1.1(b)].

1.2.1 Types of Boundary

The working of a thermodynamic system depends on the energy interactions between the system and surrounding, which takes place across the boundary. Hence we must start our studies with the type of boundaries and its properties.

The types of boundaries on the basis of its existence are:

(i) Real Boundary. Consider an oven used for baking the food products. The wall of the oven physically encloses the heating coils and the food products. Thus it forms a real boundary. The cylinder-piston in IC engine provides a real boundary during expansion or compression of inside gases. [Refer Fig. 1.2 (a)].

(ii) Imaginary Boundary. When we wish to study the working of a tap of water, we separate the valve system from rest of the thing using an imaginary boundary. In thermodynamics the systems like nozzles, turbines, carburettor of vehicle can be defined with a concept of such an imaginary boundary. [Refer Fig. 1.2 (b)].

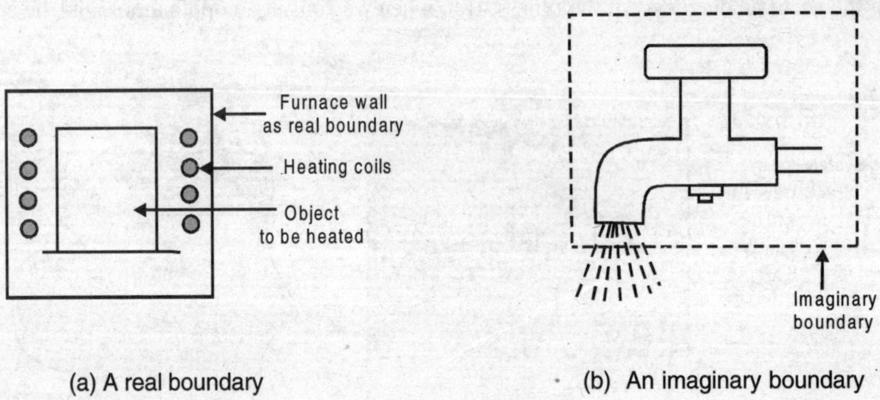

(a) A real boundary (b) An imaginary boundary

Fig. 1.2

The types of boundaries on the basis of its position are:

(i) Fixed Boundary. This type of boundary has a definite shape and encloses a fixed region of universe. A wall of a furnace is example of fixed real boundary while the system of turbine has fixed imaginary boundary.

(ii) Moving Boundary. This type of boundary may change its shape and the region of universe enclosed by it. A wall of balloon which changes its size and shape with the pressure of inside air is an example of moving real boundary.

1.2.2 Types of System

The types of systems may be viewed in a tree form as:

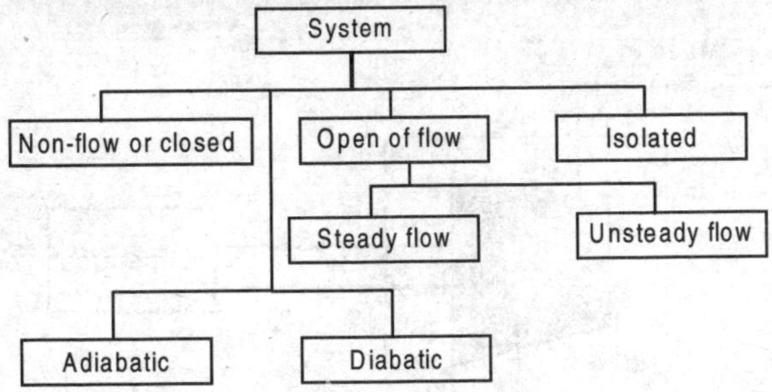

Fig. 1.3

(a) Closed or Non-flow System. It is fixed mass i.e., there is no mass transfer across the boundary. But there may be energy transfer into and out of the system as shown in Fig. 1.4 (a).

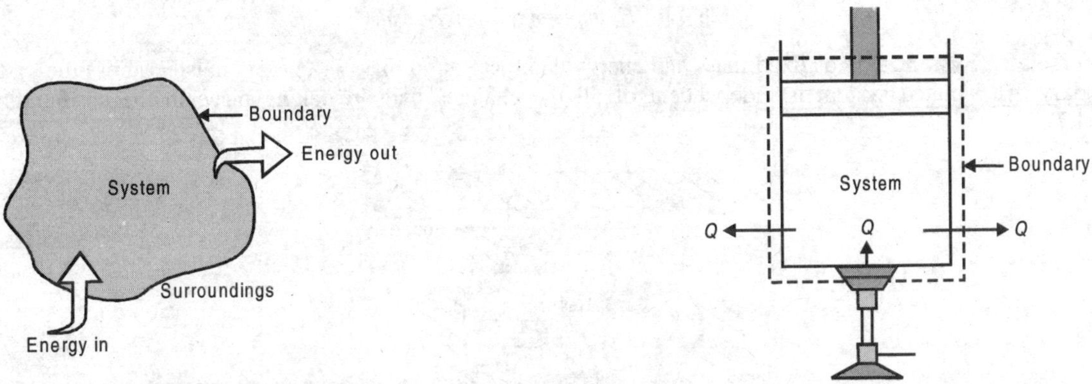

(a) A closed system

(b) Example of a closed system

Fig. 1.4

As an example, consider a system consisting of a gas, in a cylinder fitted with a piston as shown in Fig. 1.4 (b). If heat is supplied to the gas, the gas expands i.e., the volume of the gas increases so work will be done by the gas on the piston. Thus heat and work will cross the system boundary. But since no mass crosses the boundary, this system is called a *closed system.*

(b) Open System or Flow System. It is one in which both mass and energy can enter and leave the system as shown in Fig. 1.5. Most of the engineering devices are open systems.

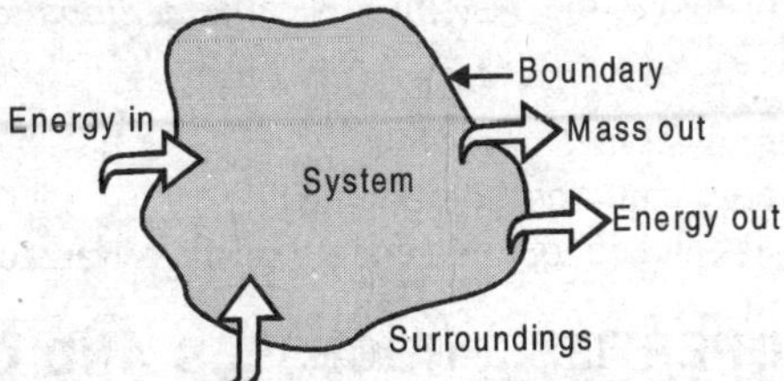

Fig. 1.5 *An open system*

For example, consider a portion of steam plant as shown in Fig. 1.6. Mass of steam enters at section (1) and after expanding into the turbine, leaves at section (2). In this case work is produced when the turbine is driven. So, as shown, the mass, work and heat are crossing the system boundary.

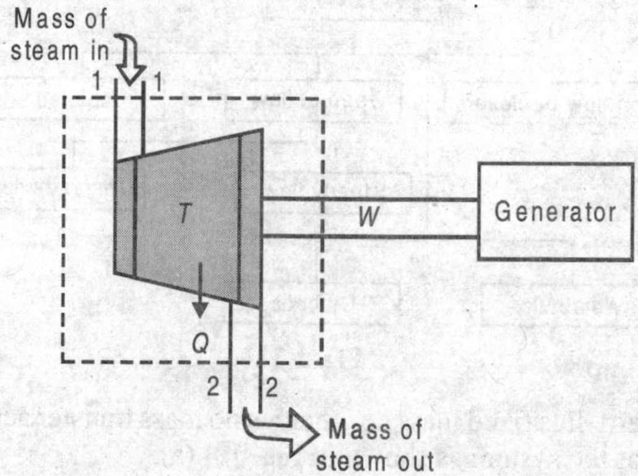

Fig. 1.6 *Example of an open system.*

(c) Isolated System. It is of fixed mass and energy and there is no mass or energy transfer across the system boundary. An isolated system is independent of all the changes that are taking place in the surroundings. (Fig. 1.7).

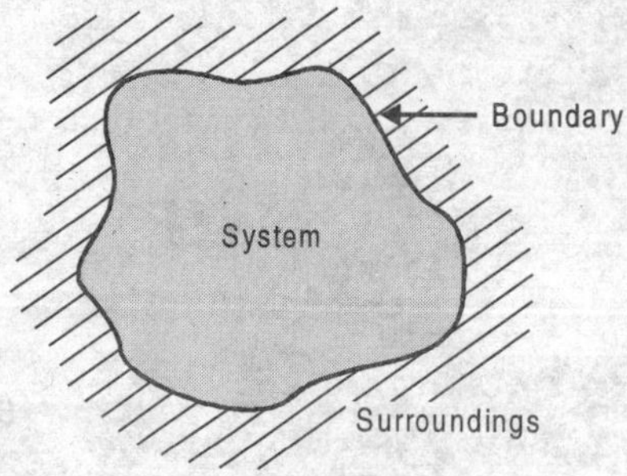

Fig. 1.7 *Isolated system (No mass or energy transfer).*

(d) Steady Flow and Unsteady Flow Systems. If the mass flow rate and energy flow rate at inlet and outlet are same, then the system is known as steady flow system.

If the rates are varying, then it is called an unsteady or transient flow system.

(e) Adiabatic and Diabatic Systems. When the system is perfectly insulated, so that neither heat enters nor it leaves the system, then it is known as *adiabatic system.*

When heat energy, either enters or leaves the system, it is called a *diabatic system.* It is not insulated.

Points to Remember :

- Closed system is of fixed mass and energy exchange.
- In open system mass and energy can cross the boundary.
- Isolated system is of fixed mass and energy. No energy or mass transfer occurs.

1.3 THERMODYNAMIC PROPERTIES, PROCESSES AND CYCLES

Every system has certain characteristics such as Pressure, Volume. Temperature, Density, Internal Energy, Enthalpy etc. by which its physical condition may be described. Such characteristics are called as *Properties* of the system.

Properties may be of two types:

1. Intensive Properties. They are independent of total mass in the system e.g., Pressure, Temperature, Density.

2. Extensive Properties. These are dependent on the total mass in the system. When the mass of the system changes, the values of the properties will change proportionately, e.g., Volume, Weight, Energy, Enthalpy, Entropy.

Properties may also be classified as —

1. Primary properties (P, V, T)
2. Secondary or derived properties (Internal energy u, Enthalpy h and Entropy s)

The properties of unit mass are known as *Specific properties* for example specific volume, specific internal energy etc. Thus all the specific properties indicate the values for unit mass and they are independent of total mass, so *all the specific properties are intensive properties.*

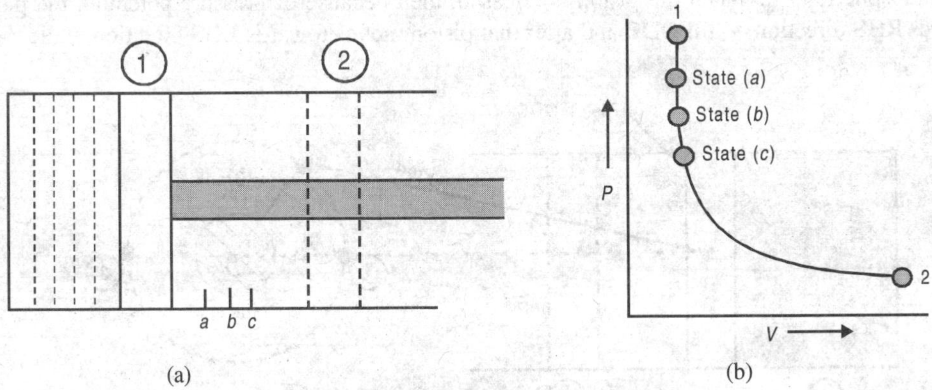

Fig. 1.8

When all the properties of a system have definite values (such as P, V, T) then the system is said to exist at a *Definite state* and the properties of the state are said to be *State functions or point functions.* Any operation in which one or more properties of the system changes, is called *Change of state,* and the series of states (and substates like a, b, c, etc. as shown in Fig. 1.9) passed during a change of state from state (1) to state (2) is called the *path* of change of state. The properties of the states and substances are called as *Path functions.*

When the path is completely specified, then the change of state is called a *Process.*

A *Process* is defined as the transformation of the system from one fixed state to another fixed state.

When any one of the properties changes, the working substance or system is said to have undergone a process.

One way of classifying a thermodynamic process is—Reversible process and Irreversible process.

(A) Reversible Process. The system will undergo a reversible process in three different conditions:

(i) Thermally Reversible Process. When the system has uniform temperature throughout the process and is in equilibrium with the surroundings, then the system is said to be in thermal equilibrium. In other words temperature gradient in the system does not exist.

Let us consider an example of working substance undergoing a process, during which it receives heat from the source. This is shown by 1–2 in Fig. 1.9.

When the working substance retraces its path in the reverse direction, then the process is called *Reversible process.* In the forward direction consider a point A where the temperature is T. At that moment the temperature of the source will be $(T + \Delta T)$, so that the heat transfer takes place from source to working substance.

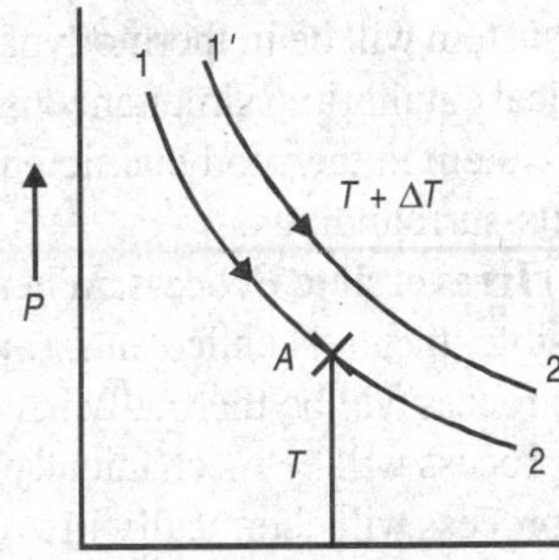

Fig. 1.9

In the reverse direction, the working substance will have to reject heat. At point A the working substance at a temperature of T has to transfer heat to source at $(T + \Delta T)$ that is not possible and the process cannot be retraced back and the process will not be reversible. To have the process reversible, the temperature gradient must be zero i.e., $\Delta T = 0$. In a thermally reversible process, therefore, temperature gradient does not exists between the surroundings (source) and the system.

(ii) Mechanically Reversible Process. A system is said to be mechanically reversible, when there are no mechanical losses either internally or externally.

Let us consider a simple piston-cylinder arrangement as shown in Fig 1.10. Position '1' represents Inner Dead Centre (IDC) and position '2' represents Outer Dead Centre position (ODC). Let space between the piston and cylinder cover be filled with a gas at a pressure of P, volume V and temperature T. And the pressure on the other side of piston be atmospheric P_{atm}. When the piston is released, then because of pressure potential, the piston starts moving towards RHS direction — till ODC and after that piston moves towards LHS direction while the crank is rotating.

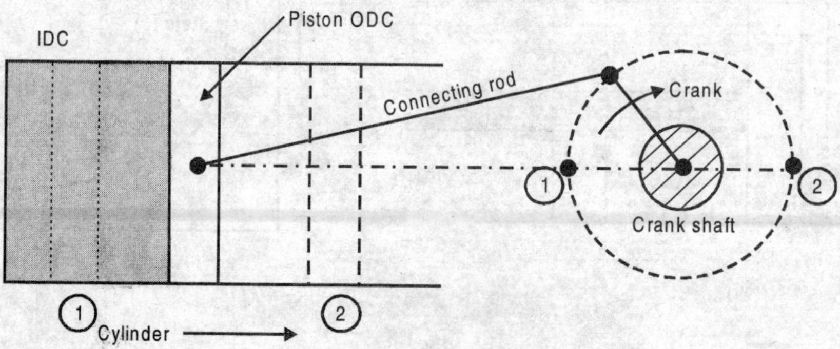

Fig. 1.10

If there are no mechanical losses the piston will reach to its original position. This type of process is called as *Mechanically Reversible Process.*

In actual practice this does not exist. Mechanical losses are mainly *frictional losses.* If the process is mechanically reversible, then the friction between the moving parts must be zero.

We can have other example as a tennis ball released from a certain height, then it will start bouncing ultimately it comes to rest. If there would not have been any mechanical losses, at the point of contact, then it would have been bouncing continuously. In practice this does not happen.

(iii) Chemically Reversible Process. The process is called chemically reversible process, when there are no chemical reactions during the process.

A system will be in thermodynamic equilibrium, when it is in thermal equilibrium, mechanical equilibrium, and chemical equilibrium simultaneously.

A system in thermodynamic equilibrium is incapable of any spontaneous change and it is in complete balance with its surroundings.

(B) Irreversible Process. When the process is not thermally reversible, mechanically reversible and chemically reversible, then it is called an *Irreversible process.*

A process will be thermally irreversible, when the heat transfer with finite temperature difference takes place.

A process will be mechanically irreversible, when the friction exists during the process.

A process will chemically irreversible, when chemical reactions take place during the process.

The various thermodynamic processes (Refer Figs. 1.11 a–c) which are used in engineering practice are:

1. Constant Pressure Process or Isobaric Process. A process during which pressure remains constant is called a constant pressure process. The law for the process is ($P = C$) and is represented on the PV-diagram by means of a horizontal line 1–2.

2. Constant Volume Process or Isochoric Process. A process during which volume remains constant is called a constant volume process. The law for the process is $V = C$ and is represented by means of a vertical line 3–4 on the PV-diagram.

3. Constant Temperature or Isothermal Process. A process during which the temperature remains constant is called a constant temperature on isothermal process. The law for the process is $PV = C$ and is represented by the curve 5–6 on the PV-diagram. Also the law of the process may be given as $T = C$.

4. Adiabatic or Isentropic Process. A process during which heat exchange with the surroundings is zero, then it is called an *Adiabatic process*. A reversible adiabatic process is called as an Isentropic process, during which entropy remains constant ($S = C$). The law for the adiabatic process is $PV^{\gamma} = C$. Irreversible adiabatic process is called a *Throttling process*, during which enthalpy remains constant. It is given by law h = constant.

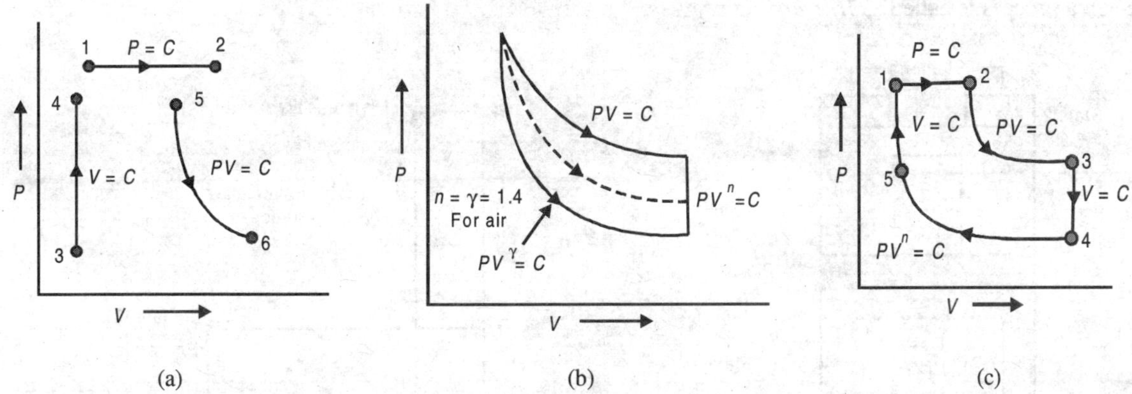

Fig. 1.11

5. General Law Process or Polytropic Process. The law for the process is $PV^n = C$. In actual practice neither we get isothermal process, nor we get adiabatic process, but what we get is the process in between these two and in known as polytropic process [Refer Fig 1.11(b)].

Thermodynamic Cycle. It is defined as the series of state changes such that the final identical with the initial state.

or

When the system undergoes the series of processes such that the end states are same, the system is said to have undergone a thermodynamic cycle.

or

Series of thermodynamic processes, the end states of which are identical, is called a *thermodynamic cycle*.

In a thermodynamic cycle, chemical composition of the working fluid during the process does not change. Thus all the properties in the initial and final states remain unchanged.

For example, water that circulates through the steam power plant. There is a change of phase during the process, but the end states are same.

Points to Remember :

- ⇨ Intensive property is independent of mass
- ⇨ Extensive property depends on mass.

1.4 QUASI-STATIC PROCESS

Quasi means nearly or almost. So quasi-static process means nearly static process or nearly stationary process, or a process which proceeds with extreme slowness.

Quasi-static process proceeds from one equilibrium state to another equilibrium state till the end of the process i.e., all the states passed during the process are all equilibrium states.

In each state the process deviates from the equilibrium state by a very small amount and immediately attains its equilibrium.

Let us consider a system consisting of a gas in a cylinder fitted with a piston as shown in Fig. 1.12 (a). Let W be the weight kept on the piston which just balances the upward force exerted by the gas. The system is initially in the equilibrium state. During initial condition, let the properties be P_I V_1 and t_1. If the weight is removed then there will be an unbalanced force between the system and the surroundings and the piston will move up till it hits the stops. Then again the system comes to another equilibrium state described by the properties P_2, V_2, t_2 but the intermediate states passed through by the system are non-equilibrium states. In Fig. 1.12 (b) points (1) and (2) are initial and final equilibrium states and since the system is passing through the non equilibrium states, it is joined by dotted line. Such a process is called as *Irreversible process.*

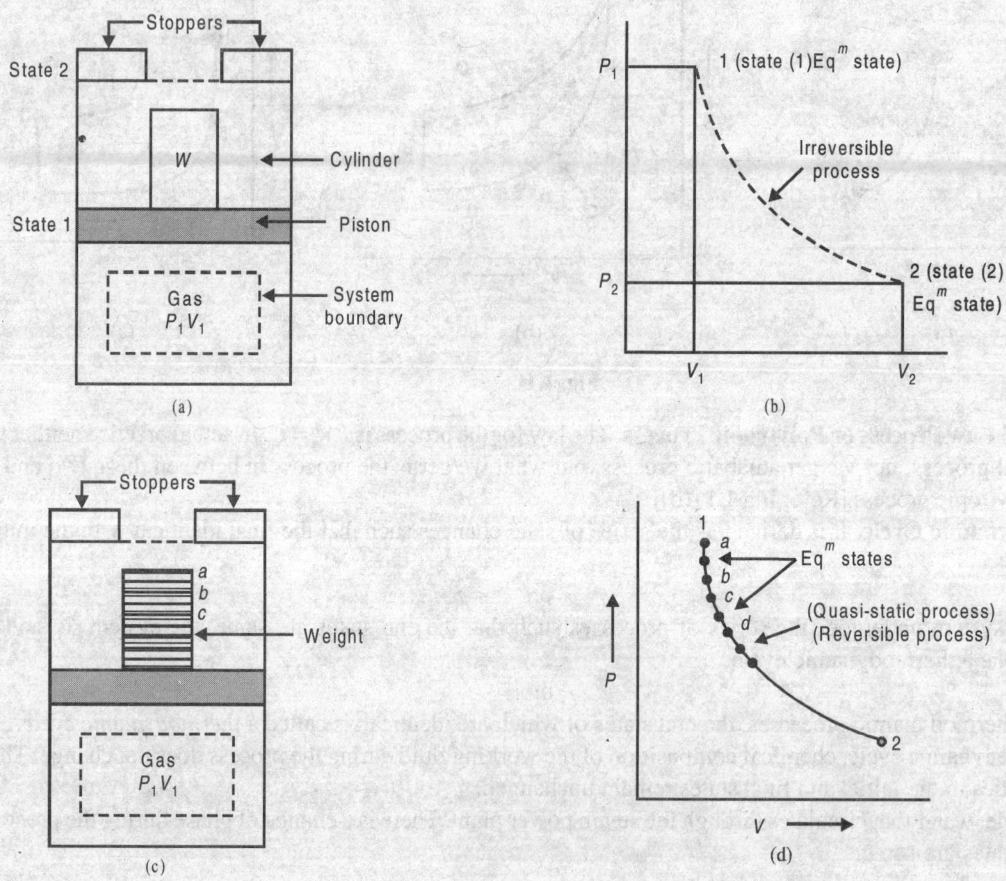

Fig. 1.12

Now if the single weight on the piston is made up of many very small pieces of weights as shown in the Fig. 1.12 (c) and these weights are removed one by one, then the departure of the state of the system from the thermodynamic equilibrium state will be very small. So, every state passed through by the system will be an equilibrium state.

A Quasi-static process is also called as *Reversible process* and is represented on the *PV*-diagram by a continuous line as shown in Fig. 1.12 (d).

In other words suppose when one weight say *a* is removed then the system will deviate from its equilibrium by a very small amount and immediately comes to an equilibrium state *a* (as shown in *PV*-diagram Fig. 1.12 (d) and when weight *b* is removed, the system again deviates by a small amount and immediately comes to another equilibrium state *b*. Thus when these small weights are removed one by one the system proceeds from one equilibrium state to another equilibrium state till end, such a process which passes through all the equilibrium states is known as quasistatic process.

1.5 WORKING SUBSTANCE

"*It is a fluid, in which energy can be stored and from which energy can be removed*".

or

"*It is a fluid which is used in thermodynamic systems, for converting one form of energy into the other*".

The working substance may be a liquid, vapour or a gas. Generally air and water are used as working substances, as they are freely available. However NH_3 and Freon-12 are also used depending upon specific applications.

1.6 UNITS AND DIMENSIONS

Nowadays generally SI (System International) system of units is used. The basic units in this system are:

System. Basic Units

Sr. No.	*Quantity*	*Unit*	*Symbols of Unit*
1.	Length	Metre	m
2.	Mass	Kilogram	kg
3.	Time	Seconds	s
4.	Temperature	Celsius/Kelvin	°C/K

The dimensions of all other quantities are derived from these basic units which are:

Sr. No.	*Quantity*	*Unit*	*Symbol of Unit*	*Alternative Unit*
1.	Force	Newton	N	—
2.	Energy (Work or Heat)	Joule	J	Nm
3.	Power	Watt	W	J/sec
4.	Pressure	Pascal	Pa	N/m²
5.	Volume	Cubic metres	m³	cu-m

A proper understanding of units of Force, Pressure, Temperature, Work, Heat and Power is necessary for solving numerical problems in Thermal Engineering.

1. Force. The unit of force in SI units is Newtons (N) A force of 1N produces an acceleration of 1 m/sec² when applied to a mass of 1 kg.

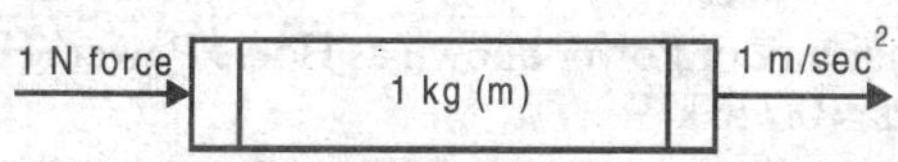

Fig. 1.13

We know that from Newton's Second law of motion $F \propto ma$

or $$F = \frac{ma}{g_c}, \quad \text{where } g_c = \text{constant of proportionality.}$$

As 1 N force produces an acceleration of 1 m/sec² in 1 kg mass, we have,

$$1 \text{ Newton} = \frac{1}{g_c} \times 1 \text{ kg} \times 1 \text{ m/sec}^2$$

$$\therefore \quad g_c = 1 \text{ in SI units and units will be}$$

$$g_c = \frac{1 \text{ kg-m}}{\text{N}-\text{sec}^2}$$

When $g_c = 1$, the product of mass and acceleration becomes the magnitude of force.

If the mass of 1 kg is allowed to fall freely under the action of standard gravitational force, it is accelerated at the rate of 9.806 m/sec² (9.81 m/sec²) and we have,

$$\text{Force} = 1 \text{ kg} \times 9.81 \text{ m/sec}^2 = 9.81 \text{ N}$$

∴ It follows that the weight of 1 kg mass equals 9.81 N and since the weight is the force,

$$\mathbf{1 \text{ kgf} = 9.81 \text{ N}}$$

2. Weight. Weight of a body is the force exerted on its mass due to gravity.

3. Density. Density is defined as mass per unit volume of the substance. It is denoted by ρ

$$\therefore \quad \rho = \frac{\text{mass}}{\text{volume}} = \frac{\text{kg}}{\text{m}^3}$$

Since unit of volume is 'm³' and mass is 'kg'.

4. Work (mechanical). When a force F acts on a body and causes displacement through a distance dl in the direction of force. then the work is said to be done. This work is equal to the product of force and distance moved.

$$W = F \times dl$$

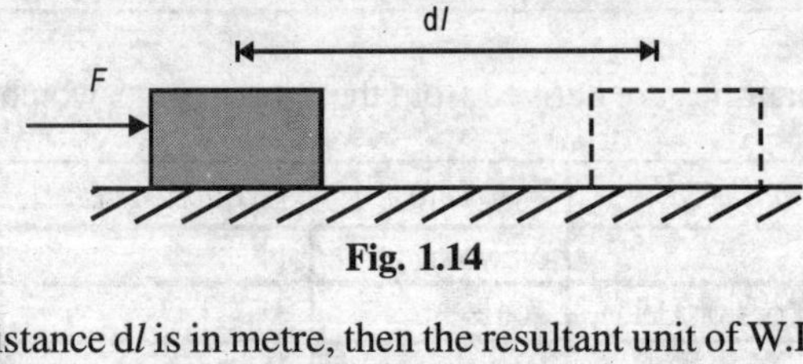

Fig. 1.14

If force F is in Newton and distance dl is in metre, then the resultant unit of W.D. will be Nm and 1 Nm = 1 joule.

5. Power. It is defined as the rate of doing work.

or $$\text{Power} = \frac{\text{W.D.}}{\text{Time taken}}$$

If the unit of W.D. is in joules and the time taken is in seconds, then the unit of Power is J/sec.
And the rate of doing work of 1 J/sec is called watt (W)

$$\therefore \quad \text{Power} = \frac{\text{J}}{\text{sec}} \text{ or W}$$

Earlier there used to be another unit of Power known as Horse Power (H.P.)

$$1 \text{ HP} = 746 \text{ watts} = 0.746 \text{ kW}$$

6. Pressure (P). Pressure is defined as the force per unit area. Thus if F is the force applied to an area A then,

$P = \dfrac{F}{A}$ and if F is in newton and A is in m² then the unit will be $\dfrac{\text{N}}{\text{m}^2}$.

This unit of pressure is called as pascal (Pa).

1.7 MECHANICAL AND THERMODYNAMIC WORK

1.7.1 Mechanical Work

$$\text{W.D.} = F \times dl$$

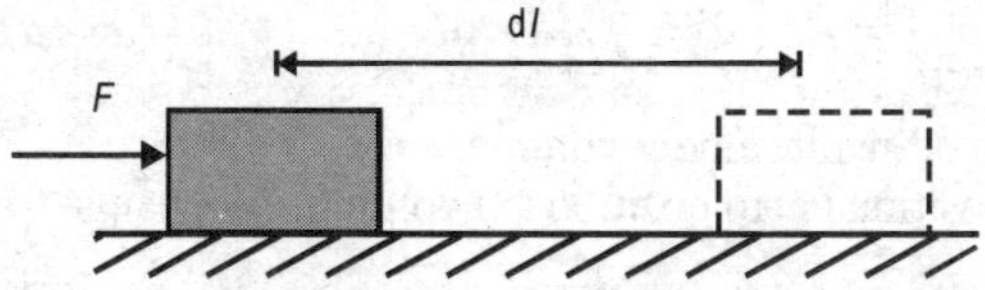

Fig. 1.15 *Mechanical work.*

When a force F acts on a body and causes a displacement through a distance in the direction of force, then the work is said to be done and this work is equal to the product of force and distance moved.

i.e. Work done $= F \times dl$

If F is in N, and dl is in m then the resultant unit will be Nm or Joule.

1.7.2 Thermodynamic Work

"It is an interaction between the system and the surroundings and the work is said to be done by the system on surroundings, if the complete external effect is lifting up of the body".

Let us consider a battery and motor system as shown in Fig. 1.16. The motor is driving a fan. The system is doing work on the surroundings.

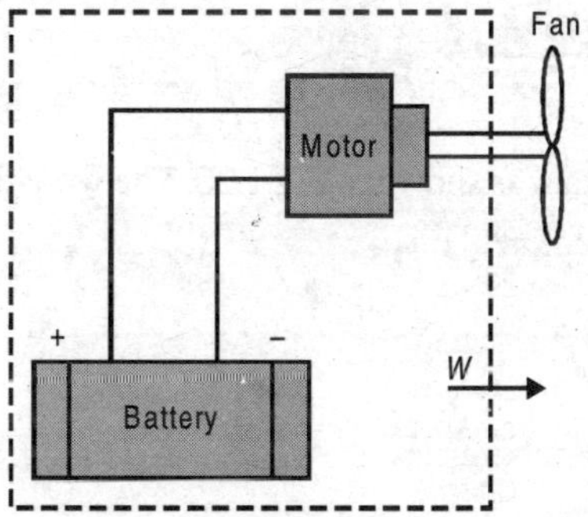

Fig. 1.16

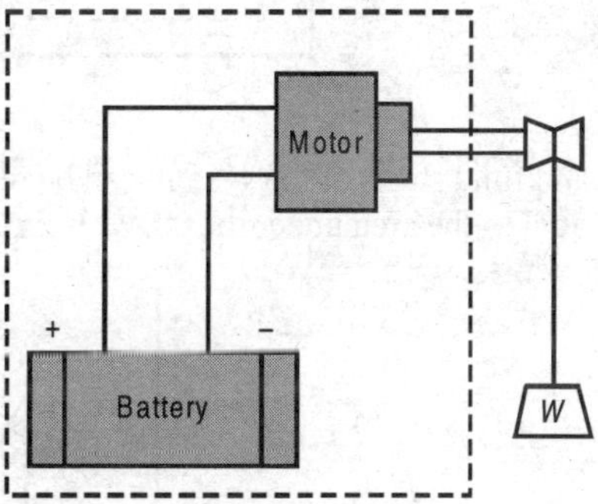

Fig. 1.17

When the fan is replaced by a pulley and weight as shown in Fig. 1.17, then the weight may be raised as the pulley is driven by the motor. Then the complete external effect is raising up the body, so the system is doing thermodynamic work.

When the work is done by the system, then it is taken as + ve, and when the work is done on the system, it is taken as –ve. (Refer Fig. 1.18)

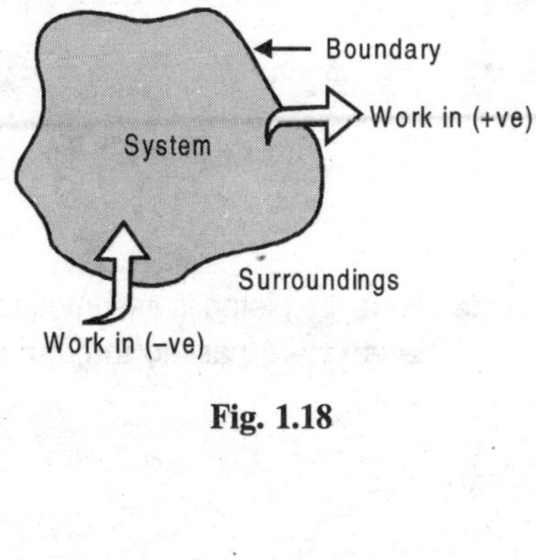

Fig. 1.18

1.8 *PDV*-WORK

Consider a system consisting of a gas in a cylinder fitted with a piston as shown in Fig. 1.19. During the initial condition of the piston i.e., when the piston is at position (1), the pressure inside the cylinder is P_1 and volume is

V_1. Let the gas expands as the piston moves to position (2). Then the pressure falls to P_2 and volume will increase to V_2.

Now, consider a small movement of the piston dl during which pressure P is assumed to be constant. If a is the cross sectional area in m^2, then the force acting on the piston is given by

$$F = P \times a$$

And the small amount of work done by the gas on the piston in causing the displacement

dl = force × displacement.

$$\delta W = F \times dl = P \times a \times dl \qquad (\because F = P \times a)$$

$$\delta W = PdV,$$

where $dV = a \times dl$ = small amount of displacement volume.

When the piston moves from position (1) to position (2) volume will change from V_1 to V_2 then the total amount of work done is given by,

$$W_{1-2} = \int_{V_1}^{V_2} PdV$$

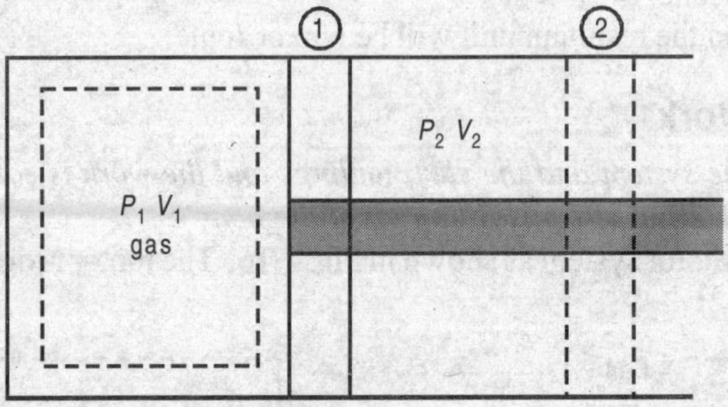

Fig. 1.19

The initial and final states of the system can be represented on PV-diagram as shown in Fig 1.20. The work done W_{1-2} will be equal to the area under the curve 1–2 on the PV-diagram.

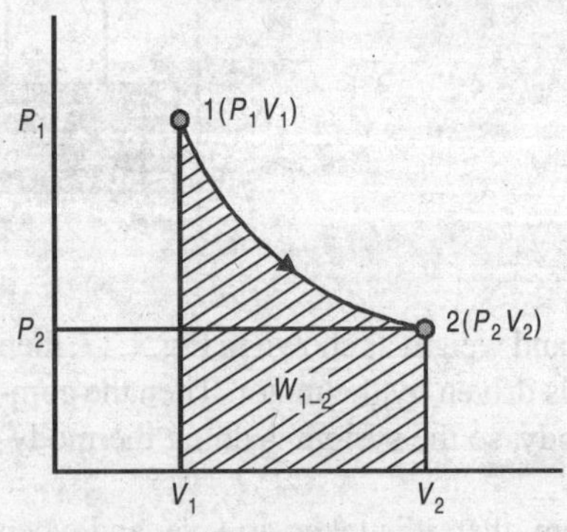

Fig. 1.20

Note: Here the piston is assumed to move very very slowly so that path 1–2 is Quasi-static (i.e., nearly static) and every state passed through is an equilibrium state.

1.9 WORK IS A PATH FUNCTION AND PROPERTIES ARE POINT FUNCTIONS

With reference to Fig. 1.20 it is possible to take a system from state (1) to state (2) by following many quasi-static paths such as *A*, *B*, *C* etc.

Now if we consider the path 1–*A*–2, then the work done during the process 1–*A*–2 will be equal to the area under the curve 1–*A*–2 and if we consider path 1–*B*–2, then the work done during the path 1–*B*–2 will be equal to the area under the curve 1–*B*–2.

So, the work done during the process directly depends upon the path and not on the end states. So, work is called Path function and δW is an inexact or imperfect differential.

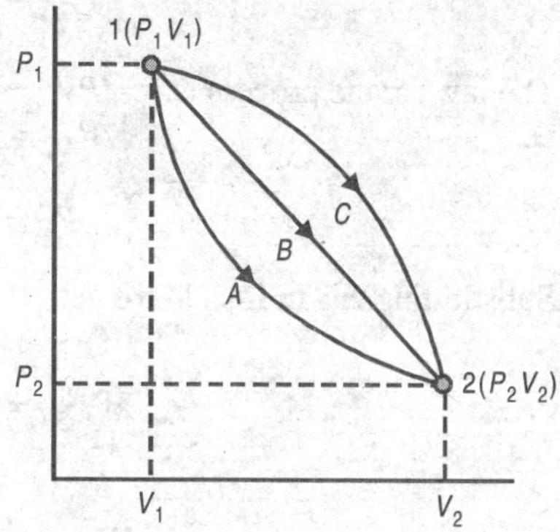

Fig. 1.21

i.e.
$$\int_1^2 \delta W \neq W_2 - W_1$$

But
$$\int_1^2 \delta W = W_{1-2} \text{ or } {}_1W_2$$

Thermodynamic properties are point functions, since for a given state, there is a definite value for each property. During change of states, the values of the properties will change and this change in values of the properties does not depend on path, but it depends upon end states and hence all the properties are point functions.

The differentials of point functions are exact or perfect differentials and the integration is simply, e.g.,

$$\int_1^2 dV = V_2 - V_1$$

Operator δ is used to denote inexact differentials, operator d is used to denote exact differentials.

1.10 EQUATIONS FOR WORK DONE IN VARIOUS PROCESSES

1.10.1 PdV Work in Various Quasi-static Processes

1. Constant pressure process (Isobaric process). Here pressure is constant.

We know that Work Done is given by,

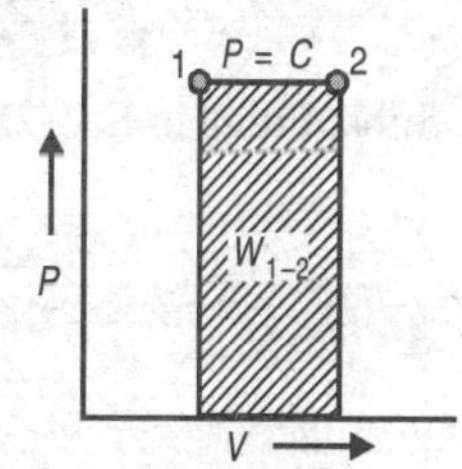

Fig. 1.22

$$W_{1-2} = \int_{V_1}^{V_2} PdV = P\int_{V_1}^{V_2} dV$$

$$= P(V_2 - V_1)\text{Nm, if } P \text{ is N/m}^2 \times V \text{ is m}^3$$

2. Constant volume process. (Isochoric process)

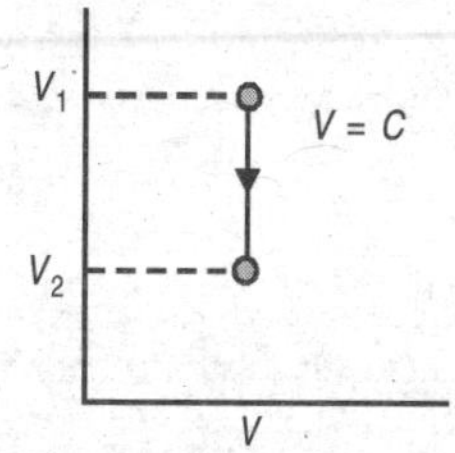

Fig. 1.23

$$W_{1-2} = \int PdV$$
$$= 0$$

During this process $V = C$ hence W.D. $= 0$

Since derivative of a constant is zero.

From *PV*-diagram area under the line 1–2 is zero.

3. Isothermal process. We know that work done in a non-flow system

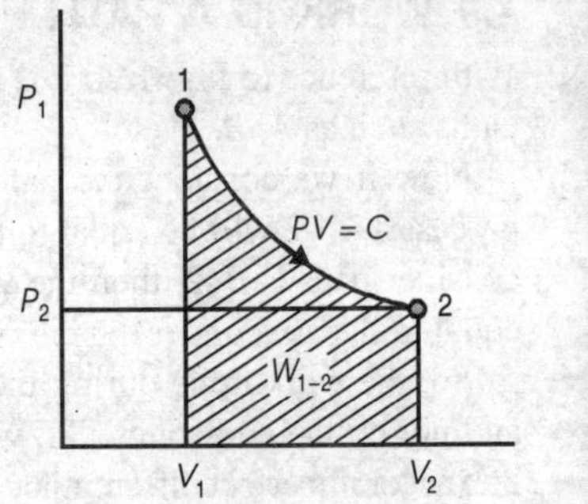

Fig. 1.24

$$W_{1-2} = \int_1^2 P.dV \quad ...(1)$$

The law for the process is, $PV = C$

i.e. $PV = P_1V_1 = C$

$$\therefore \quad P = \frac{P_1 V_1}{V}$$

Substituting this in Eq. (1) we get,

$$\therefore \quad W_{1-2} = \int_1^2 \frac{P_1 V_1}{V} \cdot dV = P_1 V_1 \int_1^2 \frac{dV}{V}$$

$$\mathbf{W_{1-2} = P_1V_1 \ln \frac{V_2}{V_1}}$$

$$\mathbf{W_{1-2} = P_1V_1 \ln \frac{P_1}{P_2}}$$

4. Adiabatic process: The law for the process is $PV^\gamma = C$ where γ is an adiabatic index. We know that, the work done in non-flow system,

$$W_{1-2} = \int_1^2 PdV \quad ...(2)$$

But for a polytropic process,

$$PV^\gamma = P_1 V_1^\gamma = P_2 V_2^\gamma = C$$

$$\therefore \quad P = \frac{C}{V^\gamma}$$

Substituting in Eq. (2), we get,

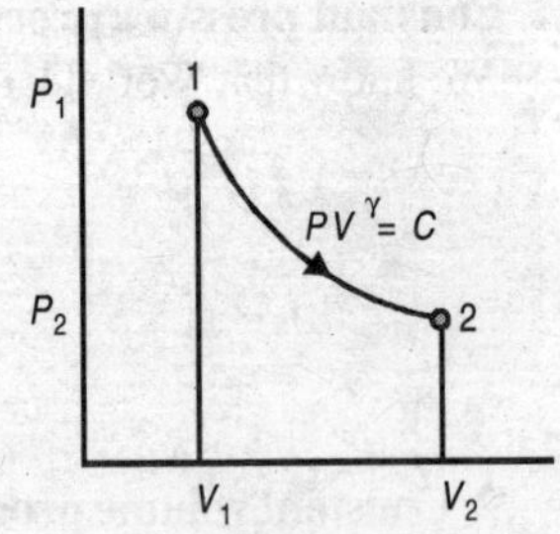

Fig. 1.25

$$W_{1-2} = \int_1^2 \frac{C}{V^\gamma} dV$$

$$W_{1-2} = \int_1^2 C V^{-\gamma} dV$$

$$= C\left[\frac{V^{-\gamma+1}}{-\gamma+1}\right]_1^2$$

$$= C\left[\frac{V_2^{-\gamma+1} - V_1^{-\gamma+1}}{-\gamma+1}\right] = \frac{CV_2^{-\gamma+1} - CV_1^{-\gamma+1}}{-\gamma+1}$$

$$= \frac{CV_2^{-\gamma+1}}{-\gamma+1} - \frac{CV_1^{-\gamma+1}}{-\gamma+1}$$

Substituting for C

$$= \frac{P_2V_2^{\gamma} \cdot V_2^{-\gamma+1}}{-\gamma+1} - \frac{P_1V_1^{\gamma} \cdot V_1^{-\gamma+1}}{-\gamma+1}$$

$$= \frac{P_2V_2^{\gamma-\gamma+1} - P_1V_1^{\gamma-\gamma+1}}{1-\gamma}$$

$$W_{1-2} = \frac{P_2V_2 - P_1V_1}{1-\gamma} \text{ or } \frac{P_1V_1 - P_2V_2}{\gamma-1} \quad \ldots(3)$$

Since $P_1V_1 = mRT_1$ and $P_2V_2 = mRT_2$

$$\mathbf{W_{1-2} = \frac{mR(T_2-T_1)}{1-r}}$$

And for unit mass $\mathbf{W_{1-2} = \frac{R(T_2-T_1)}{1-r}}$ ...(4)

5. General law process or polytropic process. The law of the process is $PV^n = C$ where n is polytropic index. The work done for this process, $\int P \cdot dV$ can be obtained by substituting γ by n.

$$\therefore \quad \mathbf{W_{1-2} = \frac{P_1V_1 - P_2V_2}{n-1}}$$

1.11 HEAT

"Heat is a form of energy, which crosses the system boundary due to the temperature difference between the system and the surroundings".

When the two bodies, one hot and the other cold are kept in contact with each other, then the hot body loses heat and becomes colder and the cold body gains heat and becomes hotter and this process continues till the thermal equilibrium is reached. At this stage the two bodies will be at the same temperature.

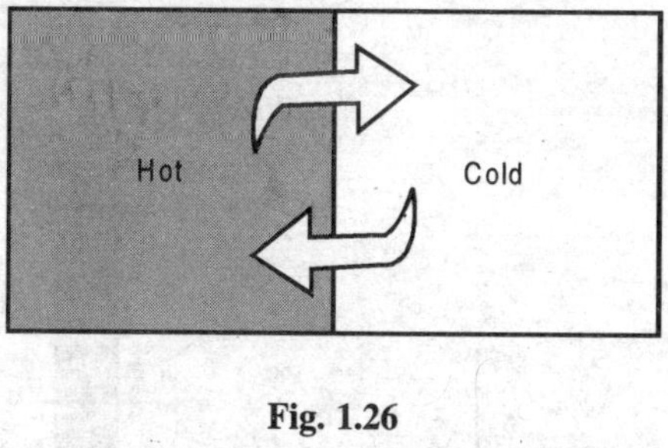

Fig. 1.26

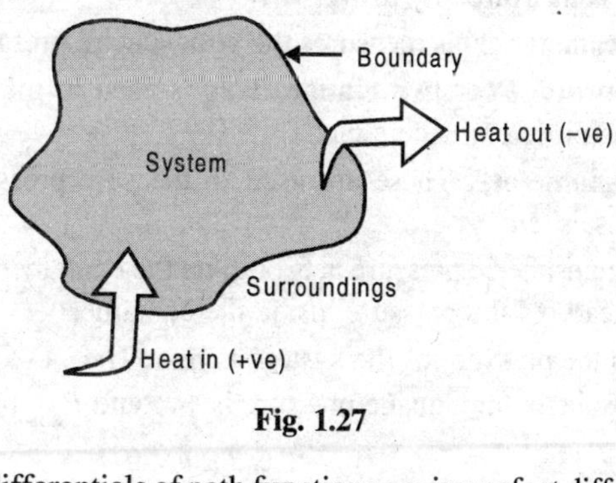

Fig. 1.27

Like work, heat is a path function and we know that the differentials of path functions are imperfect differentials. If Q is the heat transfer, then the magnitude of heat transfer during the process 1–2 is given by,

$$\int_1^2 \delta Q = Q_{1-2} \text{ or } Q_2$$

$$\int_1^2 \delta Q \neq Q_2 - Q_1$$

Note: When heat flows into the system then it is taken as +ve and when heat flows out of the system then it is taken as –ve.

1.12 PRESSURE

It is defined as force per unit area i.e.,

$$P = \frac{F}{A}$$

$$F = \text{Force in newton (N)}$$

$$A = \text{Area is a m}^2$$

Then the unit of pressure will be N/m^2, which is the basic unit of pressure, in SI units. This is also sometimes called as Pascal (Pa). Since this unit is very small, when compared to many engineering values, the units like, KPa, MPa, bar are used.

$$1 \text{ bar} = 10^5 \text{ N/m}^2 = 100 \text{ kN/m}^2 = 100 \text{ kPa}$$

Pressures are also measured in mm, or cm, of Hg or H_2O column. The pressure exerted by the atmosphere is known as atmospheric pressure and is denoted by 1 atm.

In various units, the value of 1 atm pressure are given by,

$$1 \text{ atm} = 760 \text{ mm of Hg} = 101325 \text{ N/m}^2 = 1.01325 \text{ bar.}$$

$$= 10.33 \text{ m of } H_2O = 1.033 \text{ kg/cm}^2$$

Another useful unit of pressure is 1 Torr and this is equal to pressure exerted due to 1 mm of Hg = 133.34 N/m^2.

1.13 PRESSURE MEASUREMENT

Pressure is measured by using,

1. Barometer. This measures the atmospheric pressure.

2. Bourdon Pressure Gauge. This is used to measure pressure inside the pipes/containers etc.

3. Manometers. These are used to measure pressure inside the pipes/vessels etc.

The atmospheric pressure acts both on the container and the gauge which measures the pressure inside the container.

When the pressure of the system is more (i.e., +ve pressure) than the atmosphere, then guage pressure is +ve and P_{abs} or P_{real} is given by,

$$P_{abs} = P_{atm} + P_{gauge}$$

If the pressure of the system is less (i.e., *s* –ve pressure) than the atmospheric pressure, its gauge pressure is –ve and it is known as vacuum and,

$$P_{abs} = P_{atm} - P_{vac}$$

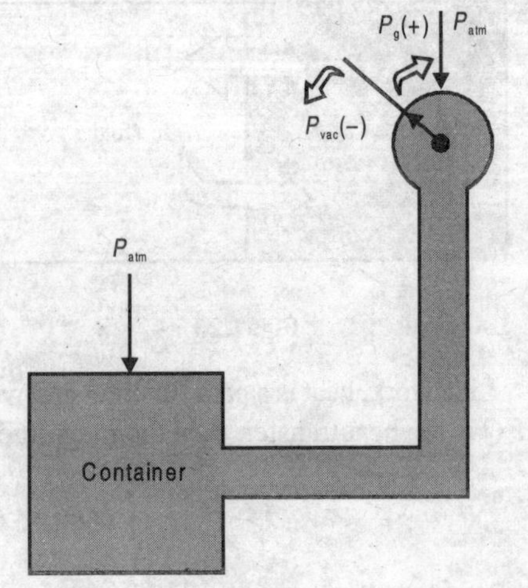

Fig. 1.28

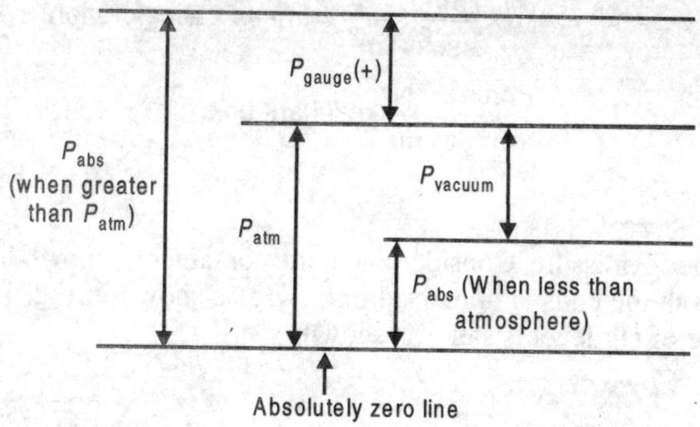

Fig. 1.29

Figure 1.29 shows the relationship between absolute and gauge pressure.

1.14 PRESSURE EXERTED DUE TO A COLUMN OF FLUID

Figure 1.30 shows a fluid of density ρ contained in container to a depth *z* or *h*. Now consider an elemental vertical cylinder of cross sectional area *a*.

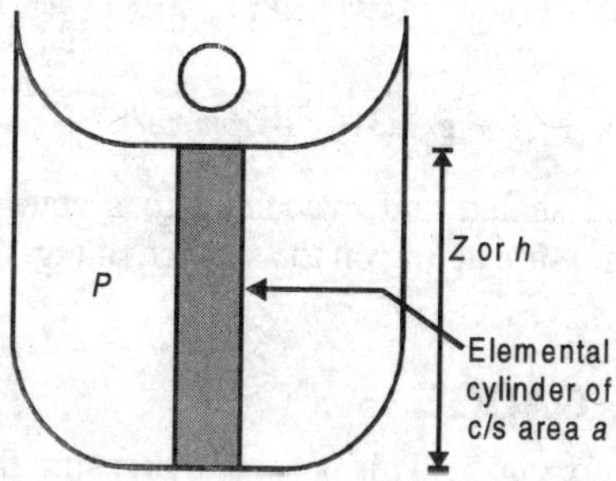

Fig. 1.30

Then we know that,

$$\text{Density} = \frac{m}{v}$$

$$\therefore \quad \text{Mass} = \text{density} \times \text{volume}$$

$$= \rho \times a \times Z$$

$$\text{But} \quad \text{Force} = \text{mass} \times \text{acceleration}$$

$$= \rho \,.\, a \,.\, Z \times g$$

But pressure exerted due to a column of fluid at a depth Z

$$= \frac{\text{Force}}{\text{Area}} = \frac{\rho \cdot a . \cdot Z \cdot g}{a}$$

$$\mathbf{P = \rho\, g\, Z}$$

Note:

Units $$P = \frac{\text{kg}}{\text{m}^3} \times \frac{\text{m}}{\text{sec}^2} \times \text{m}$$

We can write this as,
$$= \text{kg} \times \frac{\text{m}}{\text{sec}^2} \times \frac{1}{\text{m}^2} = [\text{mass} \times \text{acceleration}] \times \frac{1}{\text{m}^2}$$

$$P = \frac{\text{Force}}{\text{m}^2} = \frac{\text{N}}{\text{m}^2} \text{ resultant unit.}$$

1.15 BAROMETER

It is used to measure atmospheric pressure. Consider Hg in the container as shown in Fig. 1.31 (a) and (b). First of all, let a glass tube open at both the ends is immersed in the Hg as shown in Fig. 1.31 (a). Then the atmospheric pressure acting on the surface of Hg level is same as shown.

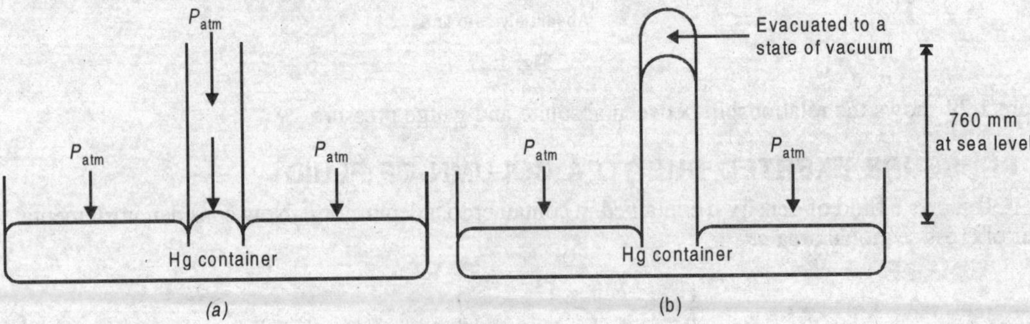

Fig. 1.31 *Barometer.*

Now if the glass tube with one end sealed and evacuated to a state of vacuum is immersed as shown in Fig. 1.31 (b) then, due to atmospheric pressure acting on the surface of Hg, the Hg will rise in the glass tube and this will be equal to 760 mm at sea level.

1.16 BOURDON PRESSURE GAUGE

This is the most common type of pressure gauge. This is used to measure the pressure inside the pipes, containers, vessels etc.

The basic element of this gauge is the tube *A*. It is elliptical in cross section and is bent into an arc of circle as shown in Fig. 1.32. End *B* of this tube is sealed, whereas open end *C* is connected to the connecting union *D* through which it can be mounted over the pipes or containers of whose pressures are to be measured.

If the pressure of the system is more than the atmosphere (i.e., +ve pressure) then the tube will tend to curl out. Conversely if the pressure of the system is less than atmosphere (i.e., –ve pressure) then the tube will tend to curl in. The change in pressure will therefore appear as the movement of the end *B*. This end *B* is connected to the quadrant gear *F* through the link *E*. The quadrant gear *F* engages with die small gear *G*, on to which a pointer *H* is attached.

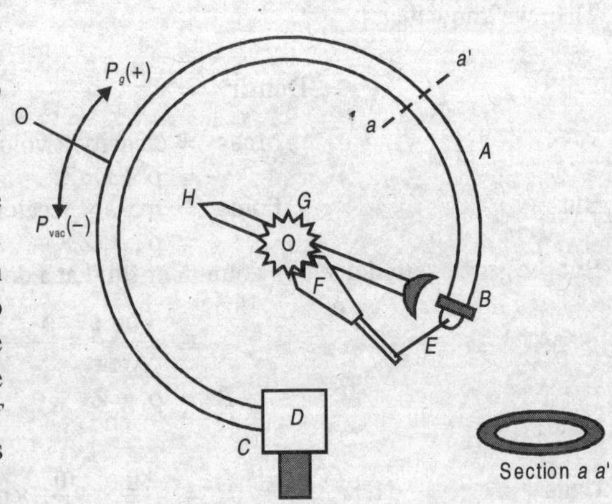

Fig. 1.32

Thus any movement due to change in pressure of the end *B* will be transmitted to the quadrant gear which will rotate the gear *G* and hence the pointer. The pointer rotates on the calibrated scale and thus gives the pressure readings directly.

1.17 MANOMETERS

These are used to measure the pressure inside the pipes, containers. It consists of a bent glass tube or a U-type glass tube which contains a liquid generally H_2O or Hg. One end of the manometer is connected to the container or pipe of whose pressure measurement is to be done and the other end is generally open to the atmosphere. The U-tube with one end closed and evacuated to state of vacuum indicates the absolute pressure directly as shown in Fig. 1.33(c).

Balancing the pressure values about the line 1–1', we get for Fig. 1.33 (a).

$$P_{abs} = P_{abs} + \rho g Z = P_{abs} + P_{gauge}$$

For Fig. 1.33 (b) $\quad P_{abs} + \rho g Z = P_{atm}$

i.e., $\quad P_{abs} = P_{atm} - \rho g Z = P_{atm} - P_{gauge} = P_{atm} - P_{vac}$

For Fig. 1.33 (c) $\quad P_{abs} = \rho g Z$

1.18 ZEROTH LAW OF THERMODYNAMICS (THERMAL EQUILIBRIUM)

When the two bodies one hot and the other cold, are placed in contact with each other, then the hot body loses heat and becomes colder and the cold body gains heat and becomes hotter, and this process continues till the thermal equilibrium is reached. In this state the two bodies will be at the same temperature.

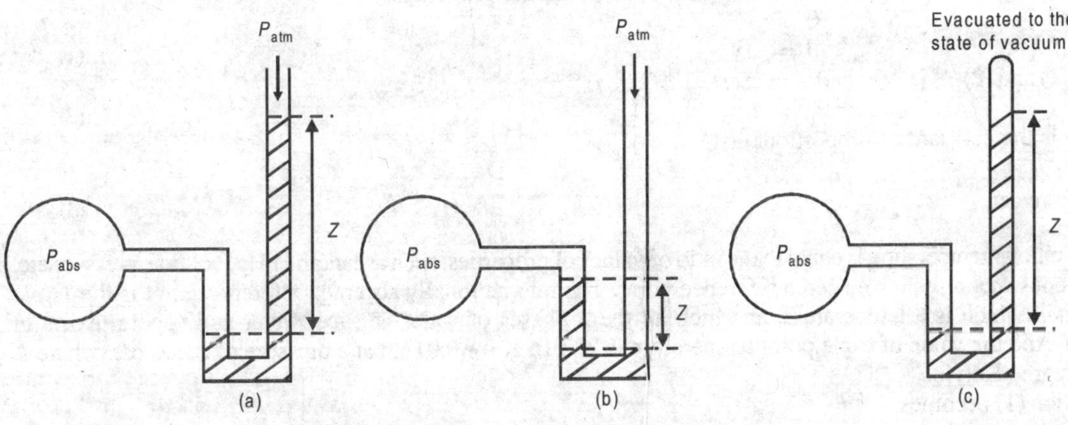

Fig. 1.33

From this a law is deduced known as *Zeroth law of thermodynamics.* It states that, "When a body *A* is in thermal equilibrium with the body *B* and also separately with the body *C*, then the bodies *B* and *C* will be in thermal equilibrium with each other. On the basis of this law, it is possible to compare the temperatures of *B* and *C* and it can be said that the temperatures of *B* and *C* are equal without actually making any contact between them. This law thus forms the basis of temperature measurement: by comparison method.

Thermal Equilibrium	**Thermal Equilibrium**	**Thermal Equilibrium**
A | *B*	*A* | *C*	*B* | *C*

If *A* is thermal equilibrium separately with *B* and *C* then *B* and *C* will also be in thermal equilibrium.

1.19 THERMOMETRIC PROPERTY AND THERMOMETERS

Temperature. It is property of the system or of a body which distinguishes a hot body from cold one.

Thermometers. In order to obtain, a quantitative measure of temperature, a reference body is used, and a certain physical characteristic of this body which changes with the temperature is selected. The selected characteristic is called the *thermometric property* and the reference body which is used in the determination of temperature is called *thermometer.*

A very common thermometer, consists of a small amount of Hg in an evacuated capillary tube. In this case the change in length of Hg is taken as the measure of temperature.

There are five different kinds of thermometers, each of them have their own thermometric properties viz.

Thermometer	*Thermometric property*	*Thermometric substance*
1. Constant volume gas Thermometer	Volume	Gas
2. Constant pressure gas Thermometer	Pressure	Gas
3. Electrical resistance Thermometer	Resistance	Wire
4. Thermocouple	Thermal emf	Wire
5. Hg in glass Thermometer	Length	Hg

1.20 PRINCIPLE OF TEMPERATURE MEASUREMENT

Let x be the value of any thermometric property as given in the table and θ be the temperature corresponding to x.

When the temperature changes, the values of the thermometric properties change proportionately. Therefore the value of temperature is proportional to the values of thermometric properties.

So we have, $\theta \propto x$

i.e., $\theta = ax$ ...(1)

For Eqs (1) and (2) $\theta_1 = ax_1$

$\theta_1 = ax_2$ ...(2)

Where a is the constant of proportionality.

Dividing we get, $$\frac{\theta_1}{\theta_1} = \frac{x_1}{x_2} \quad ...(3)$$

Thus the ratio of temperature is equal to the ratio of values of properties, such as length of Hg, volume, pressure etc.

Now, to construct a scale we need a reference point. The internationally accepted reference point is the *Triple point of water,* which is a temperature, at which all the 3-phases of water i.e., ice, water and vapour, exists in equilibrium. And the value of triple point temperature is 273.16 K (or 0.01°C) at a pressure of 0.006 bar, where *K* stands for degree Kelvin.

∴ Equation (1) becomes,

$$273.16 = ax_t \Rightarrow \frac{273.16}{x_t} = a$$

where x_t is the value of the property at triple point.

Now susbtitute this value of a in Eq. (1)

i.e. $\theta = ax$ we get, $\theta = \frac{273.16}{x_t} \cdot x$

$$\boldsymbol{\theta = 273.16 \times \frac{x}{x_t}}$$

Thus for any change in value of x (i.e., Thermometric property) the temperature θ can be found out. Thus from this equation. we can construct a scale and calibrate it to give us direct temperature readings. Here it is to be noted that x_t is the value of the property at triple point and is constant.

1.21 SCALE OF TEMPERATURE

A scale of temperature is based on some reference or fixed point to which an arbitrary value of temperature is assigned. Thus on a thermometer with some fixed or reference point, any temperature has a unique numerical value w.r.t. the fixed reference point.

1.21.1 Scales of Temperature

(1) (a) Celsius scale, (b) Fahrenheit scale.
(2) Absolute Temperature scale.
(3) Ideal Gas temperature scale.

1. (a) Celsius Scale. This scale is named after Anders the Celsius, who devised this scale. There are two fixed points on this scale. One is ice point and the other is the steam point. These two points are numbered as 0 and 100 on Celsius scale. The interval between them is divided into 100 equal parts and each division, is known as one degree Celsius and degree Celsius denoted as °C.

(b) Fahrenheit Scale. On this scale the ice point is numbered as 32 and steam point is numbered as 212. The interval between them is divided into 180 equal parts and each division, is known as 1 °F.

The temperature readings on one scale can be converted into a reading on the other scale by the following formula.

$$\frac{C}{100} = \frac{F-32}{180}$$

$$\therefore \quad C = \frac{100}{180}(F-32) = \frac{5}{9}(F-32)$$

and

$$F = \frac{9}{5}(C)+32$$

2. Absolute Temperature Scale or Kelvin Temperature Scale or Thermodynamic Temperature Scale. This is an internationally accepted scale. In this scale the temperature is measured from absolute zero temperature.

Absolute zero is the temperature at which all the vibratory, translatory and rotational motion of the molecules of a gas, is supposed to cease. A gas on cooling contracts in volume. In case of perfect gases, 'Charle's found that decreases in volume per degree centigrade decrease in temperature is 1/273. 16th of its volume at 0°C and pressure remaining constant. Thus the volume of gas will be zero at temperature –273.16°C. This temperature 273.16°C below or behind 0°C (or –273.16°C) is called the *absolute zero temperature.*

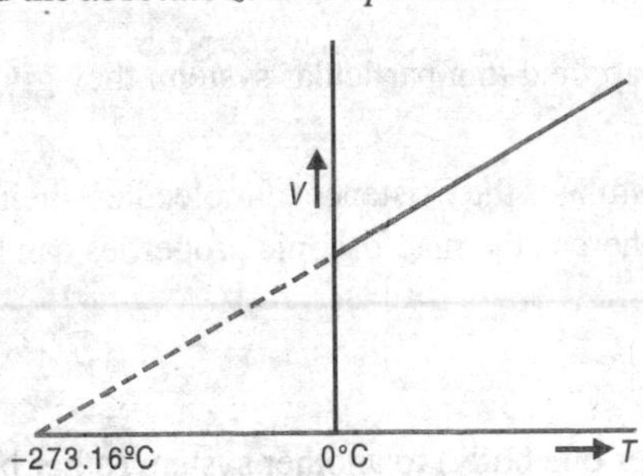

Fig. 1.34

In this scale there is one more reference point taken, the Triple point of water. This is a temperature at which all the three phases of water exists in equilibrium, and the value of triple point temperature is 0.01°C or 273.16 K at a pressure of 0.006 bar. Here it is to be noted that, the ice point temperature on this scale is 273.15 K, and steam point temperature is 373.15 K and therefore to convert the temperature given in °C to K, just add 273.15.

$K = °C + 273.15$ or in general, $K = °C + 273$. Here it is to be noted that the value of Absolute zero on this scale is –273.16°C.

3. Ideal Gas-temperature Scale. This is similar to absolute temperature scale. This scale is constructed by using constant volume gas thermometer.

In a constant volume gas thermometer, gas will be used as the working substance and as all the gases behave as ideal gases at low pressure, thus scale is known as Ideal gas temperature scale. The value of absolute zero on this scale is same as that given by Absolute temperature scale i.e., – 273.16°C.

1.22 MICROSCOPIC AND MACROSCOPIC POINT OF VIEW

There are two points of views.

1. Microscopic or Statistical Thermodynamics.
2. Macroscopic or Classical Thermodynamics, from which the behaviour of matter or working of a system can be studied.

From microscopic point of view, it is considered that, the matter is not continuous, but it is made up of a large number of identical particles called molecules. For example, consider a gas in a cylinder as a system. Then this gas is made up of number of molecules. Each molecule of a gas, at a given instant, has certain position, energy, etc., and for each molecule these changes very rapidly because of the collisions (strikings). The behaviour of the gas is described by summing up the behaviours of each molecule. And for summing up the behaviour of molecules statistical methods are employed and hence it is also called as *statistical thermodynamics.*

In macroscopic point of view a certain quantity of matter is considered, and the events that are taking place at the molecular level are not taken into account. For example, let us consider a system of an IC engine consisting of a charge in the engine cylinder. At any instant the system has certain volume depending upon the position of the piston, this volume is easily measurable.

Another quantity to describe the system is the pressure of the gas inside the cylinder. A pressure gauge can be used to measure the same. Similarly temperature, chemical composition etc. may be described. Thus, in macroscopic point of view the system will be described by large scale properties.

Though the approach based on microscopic point of view seems to be different, from that based on macroscopic point of view, but there exists a relationship between them. The relationship between macroscopic and microscopic point of view lies in the fact that the macroscopic properties are in fact the average properties of a large number of microscopic characteristics.

Hence, when both the methods are applied to a particular system, they give the same result.

Points to Remember:

⇨ The microscopic point of view postulates the existence of molecules, their movement and collisions and hence it is constantly being changed. Whereas the macroscopic properties can be measured and their existence can be felt by our senses.

1.23 ENERGY IN TRANSIT

When energy flows from one system (or one body) to another system (other body), then it is referred as energy in transit.

In general heat and work are called energies in transit. The energy transfer is due to some energy potential, e.g.,

(i) Mechanical energy or work energy are transferred due to difference in pressure potential.

(ii) Electrical energy is transferred due to difference in voltage potential.
(iii) Heat energy is transferred due to difference in temperature potential.

LIST OF FORMULAE

Note: (1) Unit of work in SI units is Nm or joules
(2) Unit of pressure in SI units is N/m² or Pa
(3) 1 bar = 10^5 N/m² = 100 kN/m² = 100 kPa
(4) 1 atm = 760 mm of Hg
= 10.33 m of water
= 101325 N/m² = 1.01325 bar
= 1.033 kg/cm²

SOLVED EXAMPLES

Example 1.1 *A fluid in a cylinder is at a pressure of 700 kN/m². It is expanded at constant pressure from a volume of 0.28 m³ to a volume of 1.68 m³. Determine the work done.*

Data: $P = 700\ kN/m^2 = 700 \times 10^3\ N/m^2$

Expanded at constant pressure,

$$V_1 = 0.28\ m^3\ V_2 = 1.68\ m^3$$

Solution

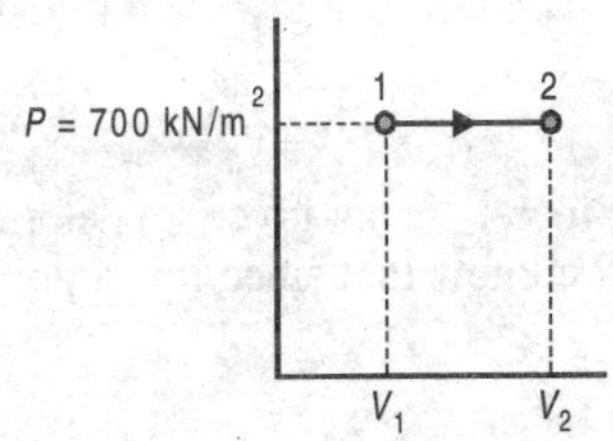

Fig. Ex. 1.1

We know that, work done $= \int_{V_1}^{V_2} PdV$

i.e., Work done $(P = C)$ $= W_{1-2} = P(V_2 - V_1)$ since

Note : Units $\frac{N}{m^2} \times m^3 = Nm$

$= 700 \times 10^3\ (1.68 - 0.28)$ Nms
$= 700 \times 10^3 \times 1.4$ Nm or Joules
$= 980000\ J = 9.8 \times 10^3\ J$

$$\mathbf{W_{1-2} = 980\ kJ = 0.98\ MJ}$$

Example 1.2 *0.112 m³ of gas has a pressure of 138 kN/m². It is compressed to 690 kN/m², according to the law $PV^{1.4}$ = C. Determine the new volume of the gas.*

Data: The gas is compressed according to the $PV^{1.4} = C$

$P_1 = 138\ kN/m^2$ $\qquad P_2 = 690\ kN/m^2$
$V_1 = 0.112\ m^3$ $\qquad V_2 = ?$

Solution

We know that polytropic process is given by $PV^n = C$

i.e., $$PV^n = P_1 V_1^n = P_2 V_2^n = C$$

$\therefore$ $$P_1 V_1^{1.4} = P_2 V_2^{1.4}$$

$$\frac{P_1}{P_2} = \left[\frac{V_2}{V_1}\right]^{1.4}$$

or $$\frac{V_2}{V_1} = \left[\frac{P_1}{P_2}\right]^{\frac{1}{1.4}}$$

or $$V_2 = V_1 \times \left[\frac{P_1}{P_2}\right]^{\frac{1}{1.4}}$$

$$= 0.112 \text{ m}^3 \times \left[\frac{138 \text{ kN/m}^2}{690 \text{ kN/m}^2}\right]^{1/1.4}$$

$$= 0.112 \text{ m}^3 \times (0.2)^{1/1.4} = 0.112 \times 0.3168$$

$$\mathbf{V_2 = 0.03555 \text{ m}^3}$$

Example 1.3 *0.014 m³ gas at a pressure of 2070 kN/m² expands to a pressure of 207 kN/m², according to the law $PV^{1.35} = C$. Determine the work done by the gas during the expansion.*

Data: Gas expands according to the law $PV^{1.35} = C$

$V_1 = 0.14 \text{ m}^3$ $\quad P_1 = 2070 \text{ kN/m}^2$

$P_2 = 207 \text{ kN/m}^2$ $\quad$ Work done = ?

Solution

The work done during a polytropic process is given by,

$$W_{1-2} = \frac{P_1V_1 - P_2V_2}{n-1} = \frac{P_2V_2 - P_1V_1}{1-n}$$

Since V_2 is not given, to determine V_2,
We know that since it is a polytropic process,

$$P_1V_1^n = P_2V_2^n$$

$$\frac{P_1}{P_2} = \left[\frac{V_2}{V_1}\right]^n$$

$$V_2 = V_1 \times \left[\frac{P_1}{P_2}\right]^{\frac{1}{n}} = 0.014 \text{ m}^3 \left[\frac{2070 \text{ kN/m}^2}{207 \text{ kN/m}^2}\right]^{\frac{1}{1.35}}$$

$$= 0.014 \times 5.5048 \text{ m}^3$$

$$\mathbf{V_2 = 0.0771 \text{ m}^3}$$

$$\text{Work done} = W_{1-2} = \frac{P_1V_1 - P_2V_2}{n-1}$$

$$= \frac{2070 \times 10^3 \text{ N/m}^2 \times 0.014 \text{ m}^3 - 207 \times 10^3 \text{ N/m}^2 \times 0.077 \text{m}^3}{1.35 - 1}$$

$$= \frac{28980 - 15959.7}{0.35} = \frac{13020.3}{0.35} = 37200.857 \text{ Nm}$$

$$= 37.2 \times 10^3 \text{ Nm or Joules}$$

$$\mathbf{W_{1-2} = 37.2 \text{ kJ}}$$

Example 1.4 *A gas is compressed hyperbolically (Isothermally) from a pressure and volume of 100 kN/m² and 0.056 m³ respectively to a volume of 0.007 m³. Determine the final pressure and the work done on the gas.*

Data: Since the gas is compressed Isothermally, PV = C i.e. $P_1V_1 = P_2V_2$

$$P_1 = 100\ kN/m^2 = 100 \times 10^3\ N/m^2$$

$$P_2 = ?$$

$$\text{Work done} = ?$$

$$V_1 = 0.056\ m^3 \qquad V_2 = 0.007\ m^3$$

Solution

Since it is an Isothermal process $P_1V_1 = P_2V_2$

$$P_2 = \frac{P_1V_1}{V_2}$$

$$P_2 = \frac{100\times1000\ \text{N/m}^2\times0.056\ \text{m}^3}{0.007\ \text{m}^3}$$

$$P_2 = \frac{5600}{0.007}$$

$$P_2 = 8\times10^5\ \text{N/m}^2$$

$$\mathbf{P_2 = 8\times10^2 kN/m^2 = 800\ kN/m^2}$$

We know that, Work done = $P_1V_1 \ln \dfrac{V_2}{V_1}$

$$= 100\times1000\times0.056\times\ln\left(\frac{0.007}{0.056}\right)$$

$$= 100\times1000\times0.056\times-2.079$$

$$W_{1-2} = -11644.973\ \text{Nm} = -11.65\ \text{kNm or kJ}$$

∴ The work done on the gas = 11.65 kJ

Example 1.5 *A non -flow reversible, quasi static process occurs for which pressure P = [V^2 + 8/V] bar. Determine the work done if volume changes from 1 m^3 to 3 m^3.*

Data:

$$P = [V_2 + 8/V]\ bar = [V^2 + 8/V]\times10^5\ N/m^2$$

$$V_1 = 1m^3,\ V_2 = 3m^2,\ \text{work done} = ?$$

Solution

We know that, work done in a closed system or displacement work done

$$W_{1-2} = \int PdV$$

$$W_{1-2} = \int_{V_1=1}^{V_2=3} PdV, \text{ subtitute for } P,$$

$$= \int_1^3\left[V^2+\frac{8}{V}\right]\times10^5\,dV = 10^5\int_1^3\left[V^2+\frac{8}{V}\right]dV$$

$$= 10^5\left[\frac{V^3}{3}+8\ln V\right]_1^3 = 10^5\left[\frac{3^3-1^3}{3}+8\ln\frac{3}{1}\right]$$

$$= 10^5\left[\frac{27-1}{3}+8\times1.0986\right]=10^5\left[\frac{26}{3}+8.789\right]$$

$$\mathbf{W_{1-2} = 1745556.5\ Nm = 17.455\times10^5\ Nm\ or\ Joules}$$

Example 1.6 *In a certain thermodynamic process of an ideal gas, the volume changes from 0.2 m³ to 0.5 m³ while pressure change occurs according to the law* $P = 1500\left[\frac{V}{100}+1\right]$. *Where P is in N/m² and V is in m³, then find out the work done by the gas in kJ.*

Data: $V_1 = 0.2\,m^3$; $V_2 = 0.5\,m^3$ *and* $P=1500\left[\frac{V}{100}+1\right] N/m^2$

Solution

We know that, work done $= W_{1-2} = \int PdV$; substitute for P

$$= \int_{0.2}^{0.5} 1500\left[\frac{V}{100}+1\right]dV = 1500\times\int_{0.2}^{0.5}\left[\frac{V}{100}+1\right]dV$$

$$= 1500\times\left[\frac{V^2}{2\times100}+V\right]_{0.2}^{0.5} = 1500\times\left[\frac{0.5^2-0.2^2}{200}+(0.5-0.2)\right]$$

$$= 1500\times(0.00105+0.3) = 451.575\ Nm$$

$$\mathbf{W_{1-2} = 0.4515\ kJ}$$

Example 1.7 *Show that for Van-der-Waals equation,* $\left[P+\frac{a}{V^2}\right](V-b) = mRT$

The work done is given by, $mRT\log_e\frac{V_2-b}{V_1-b} - a\left[\frac{1}{V_1}-\frac{1}{V_2}\right]$, *where* V_1 *and* V_2 *are initial and final volumes.*

Solution

From the given equation i.e.,

$$\left[P+\frac{a}{V^2}\right](V-b) = mRT$$

$$P = \frac{mRT}{V-b} - \frac{a}{V^2}$$

$$\therefore \quad \text{Work done} = W_{1-2} = \int PdV = \int_1^2\left[\frac{mRT}{V-b}-\frac{a}{V^2}\right]dV$$

$$= mRT\int_1^2 \frac{1}{V-b}dV - \int_1^2\frac{a}{V^2}dV$$

$$\mathbf{W_{1-2} = mRT\log_e\frac{V_2-b}{V_1-b} - a\left[\frac{1}{V_1}-\frac{1}{V_2}\right]}$$

Example 1.8 *A gas in a cylinder and piston arrangement comprise the system. Gas expands frictionlessly from 1.5 m^3 to 2 m^3 while receiving 2 Nm of work from a Paddle wheel. Pressure of gas remains constant at 6 N/m^2. Determine the net work done by the system.*

Solution

Work done by the gas on the piston,

W_{gas} = Force × Displacement = $F \times dl = P \times A \times dl = P \times dV$

The total work done by the gas, when the volume changes from V_1 to V_2 and the pressure remaining constant = $\int PdV$

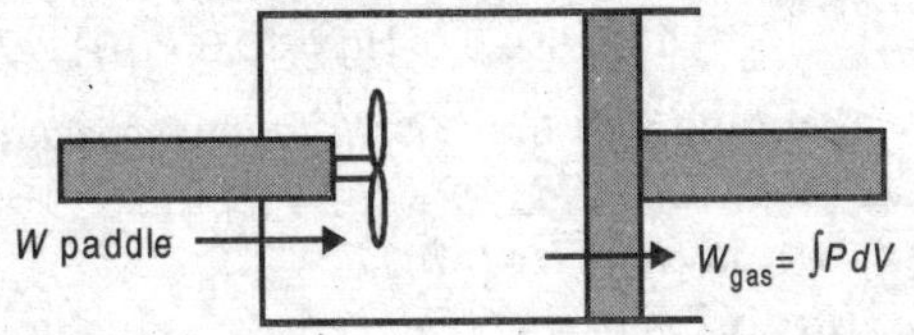

Fig. Ex. 1.8

i.e.,

$$W_{gas} = \int_1^2 PdV$$

$$= P\int_{1.5}^{2} dV \text{ since } P \text{ is constant}$$

$$= 6[V]_{1.5}^{2}$$

$$= 6[2-1.5] = 3 \text{ Nm}$$

But gas receives 2 Nm of work from the paddle wheel,

∴ **Net work done by the system = 3–2 = 1 Nm**

Example 1.9 *Convert the following readings of pressure to kPa assuming that the barometer reads 760 mm of Hg*

(a) 80 cm of Hg *(b) 40 cm of Hg vacuum*

(c) 1.5 mt of water *(d) 5.2 bar*

Solution

Let us assume ρ of Hg = 13596 kg/m^3

and g = 9.806 m/sec^2, if the values are not given

Let Z_o = Barometric pressure = 760 mm of Hg

∴ Pressure exerted due to 760 mm of Hg column in the barometer,

$$P_{atm} = \rho gh \text{ or } \rho g Z_o = 13596 \times 9.806 \times \frac{760}{1000}$$

$$= 101325.01 \text{ N/m}^2 \text{ or Pa} = 101.325 \text{ kN/m}^2 \text{ or kPa}$$

(a) Pressure of 80 cm of Hg

For 76 cm = 101.325 kPa

∴ 80 cm = ?

$$\mathbf{\frac{80}{76} \times 101.325 = 106.6578 \text{ kPa}}$$

or

We know that, pressure exerted due to a column of fluid = ρ.g.Z

$$= 13596 \times 9.806 \times 0.8 = 106657.8 \text{ N/m}^2$$

$$= 106.657 \text{ kPa}$$

(b) 40 cm of Hg vacuum

In this case, $P_{abs} = P_{atm} - P_{vac} = 76 \text{ cm} - 40 \text{ cm} = 36 \text{ cm}$

Since for 76 cm = 101.325 kPa

∴ 36 cm = ?

$$\frac{36}{76} \times 101.325 = 47.996 \text{ kPa}$$

(c) Pressure due to 1.5 mt of H_2O column on kPa

$$= \rho_w \cdot g \cdot Z_w = 1000 \times 9.806 \times 1.5$$

$$= \mathbf{14709\ N/m^2\ or\ Pa = 14709\ kPa}$$

(d)

$$5.2\text{ bar} = 5.2 \times 10^5\ N/m^2 \text{ or Pa}$$

$$= \mathbf{5.2 \times 10^2\ kPa = 520\ kPa}$$

Note: Pressure of 1 atm. equals 760 mm of Hg or 101325 N/m^2. These are standard values. To get this valve of 101325 N/m^2, we have to use ρ of Hg = 13596 kg/m^3 and g = 9.806 m/sec^2.
If we use ρ of Hg = 13.6×10^3 kg/m^3 and g = 9.81 m/sec^2 then we get P_{atm} = 101396 N/m^2.

Example 1.10 *Convert the following readings of pressure into kPa (Abs) assuming barometric pressure of 750 mm of Hg.*

(i) 55 cm of Hg (Abs)
(ii) 3.3 Atm (Gauge)
(iii) 3.2 m of H_2O (Gauge)
(iv) 4.6 bar (Abs)

Solution

For

$$P_{atm} = \rho_{Hg} \cdot g \cdot Z_o = \frac{13596 \times 9.81 \times 0.75}{1000}$$

$$= 100.32 \text{ kPa}$$

(i) 55 cm of Hg (Abs)

$$P_{abs} = 13596 \times 9.81 \times 0.55 = 73357\ N/m^2$$

$$= 73.35 \text{ kPa}$$

(ii) 3.3 atm (Gauge)

$$\text{Since } 1 \text{ atm} = 101325\ N/m^2$$

$$\therefore \quad 3.3 \text{ atm} = 3.3 \times 101325\ N/m^2$$

$$\therefore \quad P_{abs} = P_{atm} + P_{gauge} = 100.032 + \frac{3.3 \times 101325}{1000}$$

$$= 434.405 \text{ kPa}$$

(iii) 3.2 m of H_2O (Gauge)

$$P_{abs} = P_{atm} + P_{gauge} = 100.032 + \frac{100 \times 9.81 \times 3.2}{1000}$$

$$= 131.42 \text{ kPa}$$

(iv) 4.6 bar (Abs)

Since

$$1 \text{ bar} = 10^5\ N/m^2 = 10^2\ kN/m^2$$

$$\therefore \quad 4.6 \text{ bar} = 4.6 \times 100 \text{ kPa}$$

$$= \mathbf{460\ \ kPa}$$

Example 1.11 *A manometer shows a reading of 50 cm of water. If atmospheric pressure is 763 mm of Hg. Find absolute pressure in kPa. Take ρ of Hg = 13.6×10^3 kg/m^3, g = 9.81 m/sec^2.*

Solution

$$P_{abs} = P_{atm} + P_{gauge}$$

$$P_{atm} = \rho \cdot g \cdot Z_o = 13.6 \times 10^3 \frac{kg}{m^3} \times 9.81 \frac{m}{sec^2} \times 0.765 \text{ m}$$

$$= 102063.24\ N/m^2 \text{ or Pa}$$

$$= 102.063 \text{ kPa}$$

and $P_{manometter}$ or $P_{gauge} = \rho_w \cdot g \cdot Z_w = 1000 \times 9.81 \times 0.50 = 4905$ kPa $= 4.905$ kPa

$$P_{abs} = P_{atm} + P_{gauge}$$

$$P_{abs} = 102.063 + 4.905 = 106.968 \text{ kPa}$$

Example 1.12 *A vacuum gauge mounted on the condenser reads 70 cm of Hg. If atmospheric pressure is 101.325 kPa. Find absolute pressure in kPa.*

Solution

We know that $P_{abs} = P_{atm} - P_{vac}$

To find $P_{vac} = \rho \cdot g \cdot Z = 13.6 \times 10^3 \times 9.81 \times 0.70$

$= 93391.2 \text{ Pa} = 93.3912 \text{ kPa}$

$\mathbf{P_{abs} = 101.325 - 93.3912 = 7.9338 \text{ kPa}}$

Example 1.13 *A vacuum gauge connected to the tank reads 30 kPa at a location where a barometric reading is 755 mm of Hg. Determine the absolute pressure in the tank. Take density of Hg = 13590 kg/m³.*

Solution

$P_{vac} = 30 \text{ kPa}$

$Z_0 = 755 \text{ mm of Hg}$

$P_{abs} = ?$

$P_{abs} = P_{atm} - P_{vac}$ ...(1)

To find P_{atm}: $P_{atm} = \rho_{Hg} \cdot g \cdot Z_o = 13590 \times 9.81 \times 0.755$

$P_{atm} = 100655.01 \text{ N/m}^2$

$P_{atm} = 100.655 \text{ kPa}$ ...(2)

∴ From (1) $P_{abs} = P_{atm} - P_{vac} = 100.655 - 30$

$\mathbf{P_{abs} = 70.655 \text{ kPa}}$

Example 1.14 *The pressure of air in a pipe line measured with Hg manometer having one arm open to the atmosphere. The level in the open arm is 562 mm higher than the arm connected to the air pipe. Calculate the air pressure. The barometer reads 760 mm of Hg, the acceleration due to gravity is 9.81 m/sec² and the density of Hg is 13596 kg /m³.*

Data: $z = 0.560$ m $\quad Z_o = 0.760$ m

$g = 9.81$ m/sec² $\quad \rho_{Hg} = 13596$ kg/m²

Solution

Note: Since the plane 1–1' is horizontal plane, and is in the same continuous static mass of liquid, the pressure heads at the marked (*) points in figure will be equal.

Therefore the pressure along a horizontal line 1–1' is same. Balancing the pressure values about the line 1–1' we get,

$P_{abs} = P_{atm} + \rho \cdot g \cdot Z$ ...(1)

But $P_{atm} = \rho \times g \times Z_o$

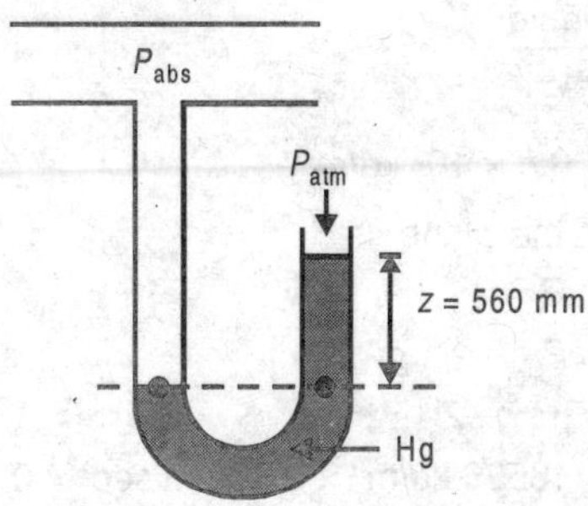

Fig. Ex. 1.14

$$= 13596 \times 9.81 \times 0.760$$
$$= 101366.3376 \text{ N/m}^2$$
$$= 101.37 \text{ kN/m}^2 = \text{kPa}$$

The pressure exerted due to the column of Hg of height $Z = \rho \times g \times Z$

$$= 13596 \times 9.81 \times 0.56 = 74690.98 \text{ N/m}^2 = \text{Pa}$$
$$= 74.69 \text{ kN/m}^2 = \text{kPa}$$

∴ From Eqs. (1), (2) and (3)

$$\mathbf{P_{abs} = 101.37 + 74.69}$$
$$\mathbf{= 176.06 \ kPa}$$

Example 1.15 *A U-tube manometer contains a liquid having a density of 800 kg/m³. When the manometer is connected to a gas pipe, the level in the open arm is 30 cm higher than the level in the column connected to the gas pipe. Find the absolute pressure of the gas in bar, if the barometric pressure is 75 cm of Hg.*

Solution

We have, Density of liquid (ρ_{liq}) = 800 kg/m³

Z = 30 cm

∴ We have, gauge pressure

$$= \rho g Z$$
$$= 800 \times 9.81 \times 0.3$$
$$= 2354.4 \text{ N/m}^2 = 0.023544 \text{ bar}$$

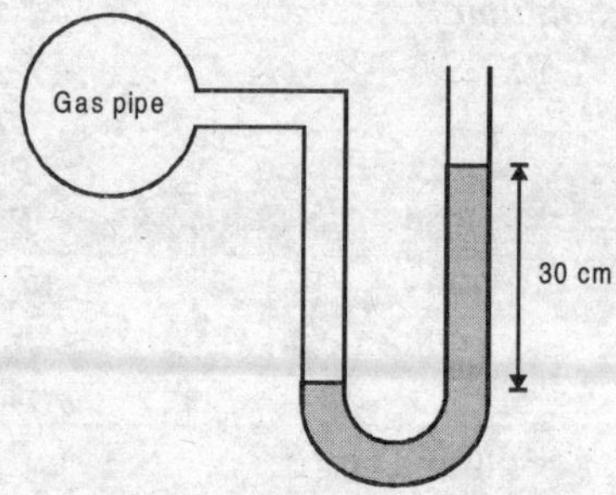

Fig. Ex. 1.15

Now $\quad P_{atm} = \rho \cdot g \cdot Z = 75$ cm of Hg

$$= 13.596 \times 10^3 \times 9.81 \times 0.75$$
$$= 99991.78 \text{ N/m}^2$$
$$= 0.9999 \text{ bar}$$

Also, $\quad P_{abs} = P_{gauge} + P_{atm} = 0.023544 + 0.9999$

$$\mathbf{P_{abs} = 1.234618 \ bar}$$

Example 1.16. A U-tube manometer is connected to a gas pipe. The level of the liquid in the manometer arm open to the atm. is 17 cm lower than the level of liquid in the arm connected to the gas pipe. The liquid in the manometer has specific gravity of 0.8. Find the abs pressure of the gas if the barometer reads760 mm of Hg. Take ρ_{Hg} = 13596 and g = 9.806.

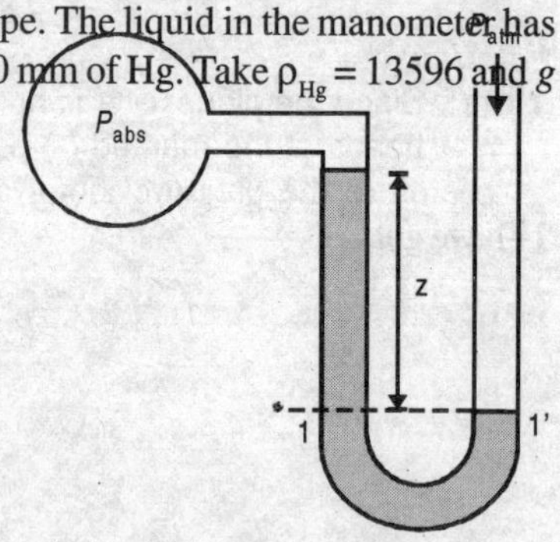

Fig. Ex. 1.16

Data: $\quad Z = 0.17 \ m \qquad P_{abs} = ?$

$\quad Z_0 = 0.76 \qquad Sp.gr = 0.8$

Solution

Note: Specific gravity = $\dfrac{\rho \text{ of given liquid}}{\rho \text{ of water}}$

∴ $\quad \rho$ of given liquid = 0.8×10^3 kg/m²

Balancing the pressure values about the plane 1–1'

$$P_{abs} + \rho gZ = P_{atm}$$
$$P_{abs} = P_{atm} - \rho \cdot g \cdot Z$$

(a) $\quad P_{atm} = \rho \cdot g \cdot Z_o \quad$ ...(1)

$$= 13596 \text{ kg/m}^3 \times 9.806 \text{ m/sec}^2 \times 0.7 \text{ m}$$
$$= 101325 \text{ N/m}^2$$

and $$\rho g Z = 0.8 \times 10^3 \times 9.806 \times 0.17 = 1333.616 \text{ N/m}^2$$

$\therefore$ $$P_{abs} = 101325 - 1333.616 = 99991.384 \text{ N/m}^2$$

$$\mathbf{P_{abs} = 0.999913 \text{ bar}}$$

Example 1.17. *A U-tube manometer with one arm open to atmosphere is used to measure the pressure of steam flowing through a pipe. Hg is used in the manometer. The height of the column open to the atm is 9.75 cm greater than the Hg level in the arm connected to the steam pipe. Steam condenses in the manometer on the steam side.*

The column of condensate is 3.4 cm. The atmospheric pressure is 76 cm of Hg. Find the absolute pressure of steam. Take,

$$g = 9.806 \text{ m/sec}^2 \quad \text{and} \quad \rho_{Hg} = 13596 \text{ kg/m}^3$$

$$P_{abs} = P \text{ of steam} = ?$$

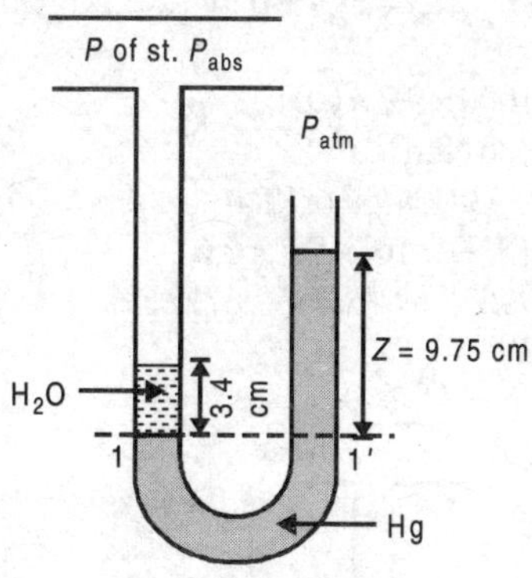

Fig. Ex. 1.17

Data: $$P_{abs} = ?$$
$$Z = 9.75 \text{ cm}$$

Solution

Balancing the pressure values about the Plane 1–1'

$$P_{abs} + \rho_w \cdot g \cdot Z_w = P_{atm} + \rho g Z$$

Now to find,

(a) $$\rho_w \cdot g \cdot Z_w = 1000 \times 9.806 \times 0.034$$
$$= 333.404 \text{ N/m}^2 \quad ...(2)$$

(b) $$P_{atm} = \rho \cdot g \cdot Z_o = 13596 \times 9.806 \times 0.76$$
$$= 101325 \text{ N/m}^2 \quad ...(3)$$

(c) $$\rho \cdot g \cdot Z = 13596 \times 9.806 \times 0.0975$$
$$= 12998.932 \text{ N/m}^2 \quad ...(4)$$

$\therefore$ Now substituting the values of Eqs (2), (3) and (4) in Eq. (1) we get,

$$P_{abs} + 333.404 = 101325 + 12998.932$$

$$P_{abs} = 114323.93 - 333.404 = 113990.53 \text{ N/m}^2$$

$$\mathbf{P_{abs} = 1.139905 \text{ bar}}$$

Example 1.18 *A Hg manometer is used to measure the pressure of steam in the mains. (i.e., main steam pipe). Its one arm having higher level of Hg is open to atmosphere at 98 kPa. The other arm is connected to steam pipe. The manometer records a level difference of 60 cm of Hg. On the lower level of Hg, 4 cm of water column accumulates due to condensation of steam. Find the absolute pressure of steam in bar.*

Take $g = 9.7 \text{ m/sec}^2$ and $\rho_{Hg} = 13.69$ gm/cc.

$\rho_w = 1000 \text{ kg/m}^3$

Data: $g = 9.7 \text{ m/sec}^2$

$\rho_{Hg} = 13.69 \text{ gm/cc} = 13.69 \times 10^3 \text{ kg/m}^3$.

$\rho_{water} = 1000 \text{ kg/m}^3$

Solution

Now balancing pressure values the plane 1–1′

$$P_{abs} + \rho_w gZ_w = P_{atm} + \rho \cdot g \cdot Z \quad ...(1)$$

Now to find,

(a) $\rho_w g Z_w = 1000 \times 9.7 \times 0.04$

$= 388 \text{ N/m}^2$

$= 0.338 \text{ kN/m}^2$ or kPa ...(2)

(b) $\rho \cdot g \cdot Z = 13.69 \times 10^3 \times 9.7 \times 0.6$

$= 79675.8 \text{ N/m}^2 = 79.6758 \text{ kPa}$

∴ From (1) $P_{abs} = P_{atm} + \rho gZ - \rho_w gZ_w$

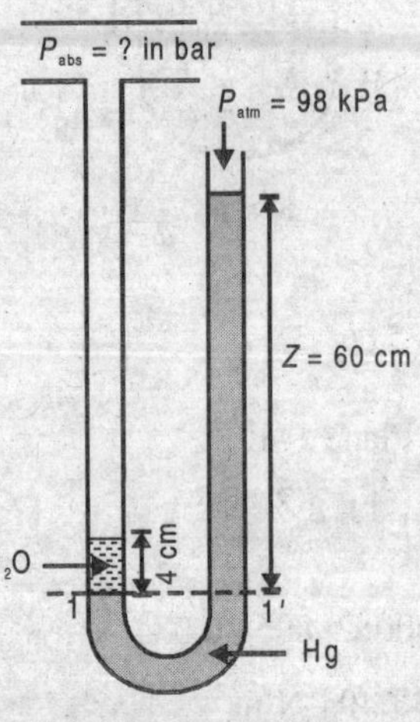

Fig. Ex. 1.18

$= 98 + 79.68 - 0.388$

$= 177.68 - 0.388$

$= 177.292 \text{ kPa}$

Since 1 bar $= 10^5 \text{ N/m}^2 = 100 \text{ kPa}$

$\mathbf{P_{abs} = 1.77292}$ **bar**

Example 1.19 *A U-tube manometer with one arm open to atmosphere is used to measure the pressure of steam flowing through a pipe. Mercury is used as a manometric fluid. The height of the column of mercury in the open end is 10 cm greater than the mercury level in the arm connected to the steam pipe. Steam condenses in the manometer on the steam side. The column of condensate is 3.5 cm in height. The atmospheric pressure is 760 mm mercury. What is the absolute pressure of steam in the pipe in bar ?*

Data: $Z_o = 0.76\ m$ P_{abs} or $P_{steam} = ?$

$g = 9.806$ $\rho_{Hg} = 13596$

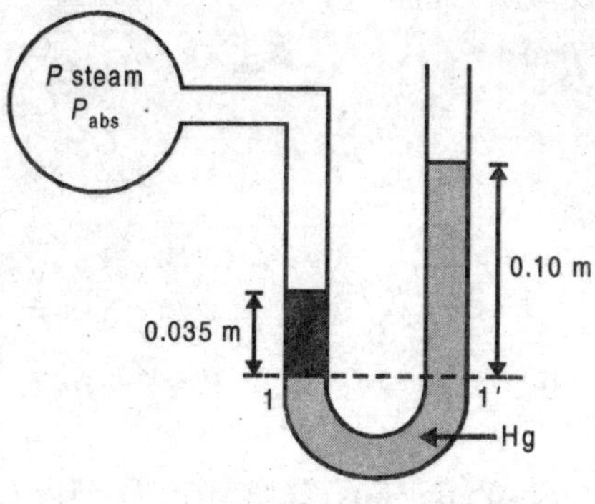

Fig. Ex. 1.19

Solution

$$P_{abs} + \rho_w g Z_w = P_{atm} + \rho_{Hg}\ g\ Z_{Hg}$$

$$P_{atm} = \rho_{Hg}\ g\ Z_o$$

$$= 13596 \times 9.806 \times 0.76$$

$$= 101325 \frac{N}{m^2}$$

$$\rho_w g Z_w = 1000 \times 9.806 \times 0.035$$

$$= 333.404\ Pa$$

$$\rho_{Hg}\ g\ Z_{Hg} = 13596 \times 9.806 \times 0.10$$

$$= 12998.93\ Pa$$

∴ From (1) $P_{abs} = P_{atm} + \rho_{Hg} g Z_{Hg} - \rho_w g Z_w$

$$= 101325 + 12998.9 - 333.404$$

[From (1), (2), (3) and (4)]

$$= 113990.52\ Pa$$

$$= 1.139\ bar$$

$$\mathbf{P_{abs} = 1.14\ bar}$$

Example 1.20 *A U-tube manometer with one arm open to atmosphere is used to measure the pressure of kerosene vapour passing in a pipe. Hg level in the open arm is 10 cm greater than that in the arm connected to the kerosene pipe. There is a condensation of kerosene vapour on the Hg column of 5.1 cm height. If the atmospheric pressure is 75.5 cm of Hg. Find the absolute pressure of vapour in kPa. Assume specific gravity of kerosene as 0.8 and that of Hg as 13.6.*

Solution

Balancing the pressure values @ the plane 1–1'

$$P_{abs\ of\ Kero.\ vap} + \rho_{kero} g Z_{kero} = P_{atm} + \rho_{Hg} g Z_{Hg} \quad ...(1)$$

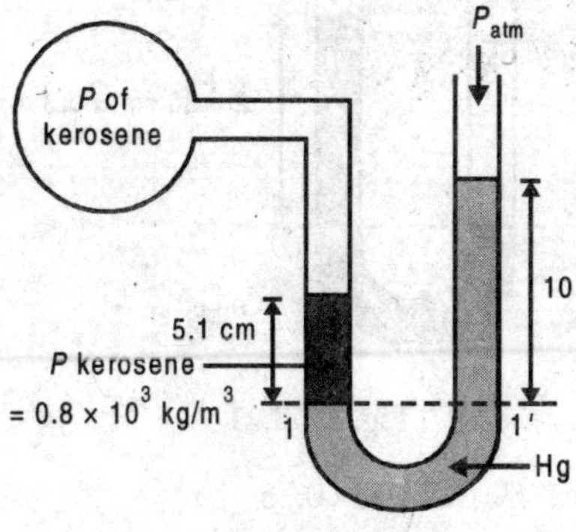

Fig. Ex. 1.20

Now to find

(a) $\rho_{kero} \cdot g \cdot Z_{kero.} = 800 \times 9.81 \times 0.051 \text{ m}$

$= 400.02 \text{ N/m}^2$ ...(2)

(b) $P_{atm} = \rho_{Hg} g Z_o$

$= 13600 \times 9.81 \times 0.755 \text{ m}$

$= 100726.08 \text{ N/m}^2$ ...(3)

(c) $\rho_{Hg} g Z_{Hg} = 13600 \times 9.81 \times 0.1 \text{ m}$

$= 13341.6 \text{ N/m}^2$ ...(4)

Sustituting Eqs (2), (3) and (4) in Eq. (1) we get,

$$P_{abs} + 400.02 = 100729.08 + 13341.6$$

$$P_{abs} = 113670.68 \text{ N/m}^2$$

$$\mathbf{P_{abs} = 113.67068 \ kP}$$

Example 1.21 *A U-tube manometer having equal arms with one arm open to atmosphere is used to measure the pressure of a gas passing through a pipe line. The fluid used in the manometer has a specific gravity of 0.8, while the height of the fluid column in the arm open to the atmosphere is 30 cm greater than the height of fluid column in the arm connected to the gas pipe line. Determine, the absolute pressure of the gas in pipe line.*

If the fluid used were Hg, what would be the difference of height of Hg-column in the 2-arms.

Take $\rho_{Hg} = 13596$ *and* $g = 9.806 \text{ m/sec}^2$

Solution

r of fluid used in the manometer $= 0.8 \times 1000 \text{ kg/m}^3$

$$P_{abs} = P_{atm} + rgZ$$

Let the barometer reads 760 cm of Hg

$$P_{atm} = rgZ_o$$

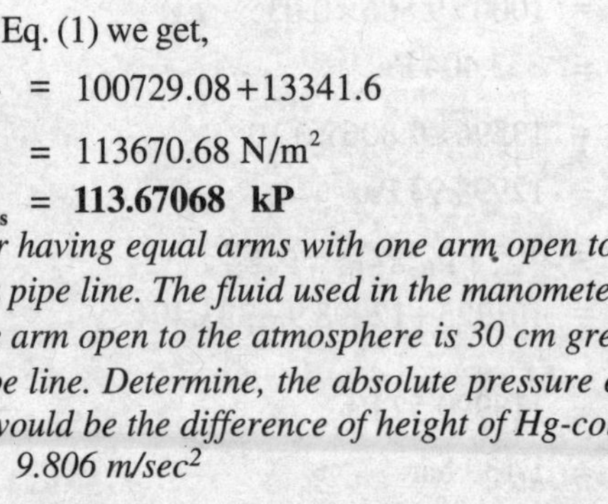

Fig. Ex. 1.21

$= 13596 \times 9.806 \times 0.76$

$= 101325 \text{ N/m}^2$

and $rgZ = 0.8 \times 10^3 \times 9.806 \times 0.3$

$= 2353.44 \text{ N/m}^2$

$$\therefore \quad P_{abs} = 101325 + 2353.44$$
$$= 1013678.44 \text{ N/m}^2$$
$$\mathbf{P_{abs} = 1.0367844 \text{ bar}}$$

If Hg is used, then to find Z,
(i.e., to produce the same absolute pressure, we have to find the equivalent column of Hg)

$$P_{abs} = P_{atm} + rgZ$$
$$103678.44 = 101325 + 13596 \times 9.806 \times Z$$
$$Z = \frac{103678.44 - 101325}{13596 \times 9.806}$$
$$= 0.01765 \text{ mt of Hg}$$
$$\mathbf{Z = 1.765 \text{ cm of Hg}}$$

Example 1.22 *A vertical composite liquid column with its upper end exposed to atmospheric pressure, comprise of 45 cms of Hg (sp. gr. 13.6); 65 cm of water and 80 cm of oil (sp. gr. 0.8). Calculate the pressure in bar,*
(i) At the bottom of the column.
(ii) At the inter surface of oil and water.
(iii) At the inter surface of water and Hg.
Assume atmospheric pressure as 100 kPa and g = 9.81 m/sec²

Solution

Since it is given P_{atm} = 100 kPa i.e., equal to 1 bar

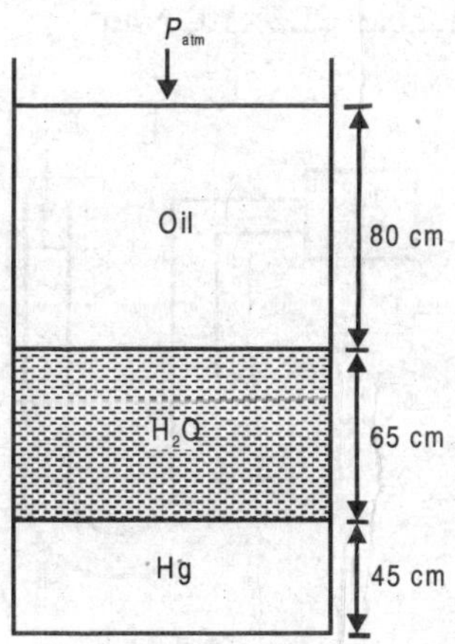

Fig. Ex. 1.22

(Since 10^5 N/m² = 1 bar = 100 kPa)

(i)
$$P_{\text{at bottom}} = P_{atm} + P_{oil} + P_{water} + P_{Hg}$$
$$= 1 + \rho_{oil} g Z_{oil} + \rho_w g Z_w + \rho g Z$$
$$= 1 + \frac{0.8 \times 1000 \times 9.81 \times 0.8}{10^5} + \frac{1000 \times 9.81 \times 0.65}{10^5} + \frac{13.6 \times 10^3 \times 9.81 \times 0.45}{10^5}$$

Note: If ρ is in kg/m³, *g* is in m/sec², *Z* is in m, then the resultant unit will be N/m².
∴ To get in bar divide by 10^5

(i) $P_{\text{at bottom}} = 1+0.062784+0.063765+0.6$

$$P_{\text{at bottom}} = \mathbf{1.72692\ bar}$$

(ii) Pressure at the interface of oil and water

$$= 1+0.062784 = 1.062784 \text{ bar}$$

(iii) Pressure at the inter surface of waterand Hg = 1 + 0.062784 + 0.063765

$$= 1.1265 \text{ bar}$$

Example 1.23 *An air tank has three manometers connected to it. The fluids in them are oil (Specific Gravity = 0.8) water and mercury (Specific Gravity = 13.6). If the absolute pressure in the tank is 1.2 bar and the barometer reads 760 mm of Hg. Estimate the height of fluid in each of manometer.*

Data: Fluids in the manometers are oil (Sp. Gr. = 0.8)

Water and mercury (Sp. Gr. = 13.6)

$$P_{abs} = 1.2 \text{ bar}$$

$$\textit{Barometric reading} = 760 \textit{ mm of Hg}$$

Height of fluid in each of manometer = ?

Solution

We know that, $P_{abs} = P_{atm} + P_{gauge}$

$$P_{atm} = \rho_{Hg}\, g Z_o = 13.6\times10^3\times9.81\times0.76$$

$$= 101396.16 \text{ N}$$

$$P_{abs} = 1.2 \text{ bar} = 1.2\times10^5 \text{ N/m}^2$$

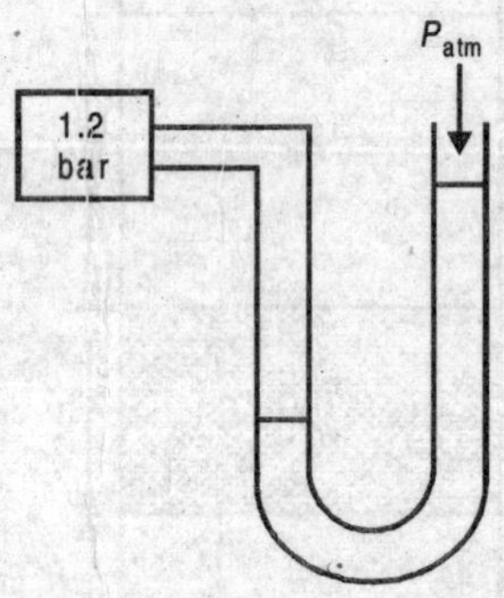

Fig. Ex. 1.23

Case I: For Oil Manometer

(a) $$\text{Sp. Gravity} = \frac{\rho \text{ of given fluid}}{\rho \text{ of water}}$$

$$0.8 = \frac{\rho_{oil}}{1000}$$

$$\therefore \quad \rho_{oil} = 800 \text{ kg/m}^3$$

$$\therefore \quad P_{gauge} = \rho_{oil}\, g Z_{oil} = 800\times9.81\times Z_{oil}$$

Subtituting these values in Eq. (1) we get,

$$1.2\times10^5 = 101396.6+800\times9.81\times Z_{oil}$$

$$\mathbf{Z_{oil} = 2.37\ m}$$

Case II: For Water Manometer

$$P_{abs} = P_{atm} + \rho_w g Z_w$$

$$1.2\times10^5 = 101396.6+1000\times9.81\times Z_w$$

$$\mathbf{Z_w = 1.89\ m}$$

Case III: For Hg Manometer

$$P_{abs} = P_{atm} + \rho_{Hg} g Z_{Hg}$$

$$1.2\times10^5 = 101396.6+13.6\times10^3\times9.81\times Z_{Hg}$$

$$\mathbf{Z_{Hg} = 0.139\ Hg}$$

Example 1.24 *A basic barometer can be used as an altitude measuring device in aeroplanes. The ground control reports a barometric reading of 753 mm of Hg, while the pilot's reading is 690 mm of Hg.*
Estimate the altitude of the plane from the ground level if the average air density is 1.25 kg/m³ and g = 9.8 m/sec².

Solution

Barometer reading at ground level $Z_{ground} = 0.753$ m

$$\text{Pilot's Barometric reading } Z_{plane} = 0.690 \text{ m}$$

$$\text{Altitude} = ?$$

$$\text{Pressure at the ground level} = \rho g Z_{ground}$$

$$= 13600\times9.8\times0.753$$

$$= 100359.84 \text{ N/m}^2 \quad ...(1)$$

$$\text{Pressure at plane level} = \rho g Z_{plane}$$

$$= 13600\times9.8\times0.690$$

$$= 91963.2 \text{ N/m}^2 \quad ...(2)$$

∴ Change of pressure at the ground level and that of plane level,

$$\Delta P = 100359.84-91963.2$$

$$= 8396.64 \text{ N/m}^2$$

∴ This is the difference of pressure, now to find altitude for this much difference of pressure,

$$\Delta P = \rho_{air} g Z_{altitude}$$

$$8396.64 = 1.25\times9.8\times Z_{altitude}$$

$$\mathbf{Z_{altitude} = 684.44\ m}$$

Example 1.25 *A manometer connected between 2 pipes as shown in Fig. Ex. 1.25. Take,*

$\rho_{water} = 1000$ kg/m³ $\rho_{Hg} = 13590$ kg/m³

Pressure at A = 400 kPa $g = 9.8$ m/sec².

Find Pressure at B = ?

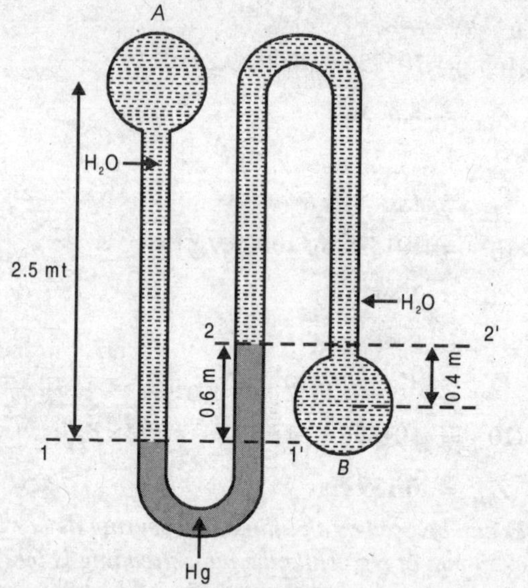

Fig. Ex. 1.25

Solution

Note. If there are more than one U-turns, then start from point and as we go downwards take it as positive, and as we go upwards take it as negative. Then equate this pressure equation to pressure at the second point.

So, writing the pressure equation,

$$P_A + P_w \ g \ Z_w - \rho_{Hg} \ g \ Z_{Hg} + \rho_w \ g \ Z_w = \ P_B$$

$$400 \times 1000 + 1000 \times 9.8 \times 2.5 - 13590 \times 9.8 \times 0.6 + 1000 \times 9.8 \times 0.4 = \ P_B$$

$$400000 + 24500 - 79409.2 + 3920 = \ P_B$$

$$\mathbf{P_B = 348510.8 \ Pascals}$$

$$\mathbf{P_B = 348.5108 \ kPa}$$

Example 1.26 *The cylinder and tubing shown in Fig. 1.26 contain oil of density 0.902 × 10³ kg/m³ for gauge reading of 2 bar. What is the total weight of piston and slab placed on it?*

Solution

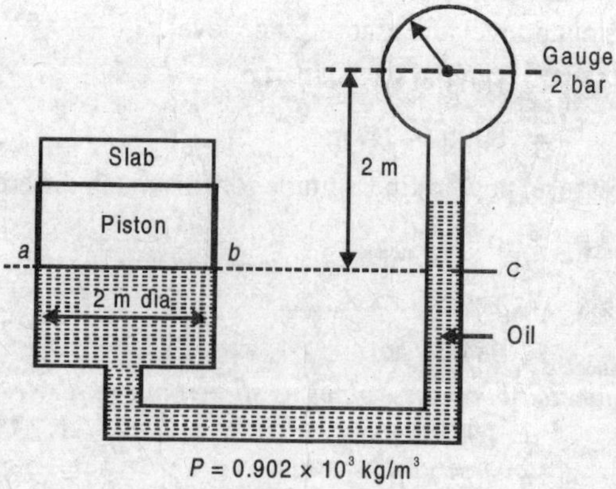

Fig. Ex. 1.26

We know that, Pressure $= \dfrac{\text{Force}}{\text{Area}} = \dfrac{\text{Weight}}{\text{Area}}$

And from Fig. Ex. 1.26 it can be seen that the pressure at $a-b-c$ level will be same.

$$\therefore \quad \frac{W}{A} = P_{gauge} + \rho_{oil}\, g h_{oil}$$

$$\therefore \quad \frac{W}{\frac{\pi}{4}(2)^2} = 2\times10^5 \frac{N}{m^2} + 0.902\times10^3\times9.81\times2\ m$$

$$\mathbf{W = 683916.05\ N\ \ (i.e.,\ weight\ of\ piston\ and\ slab)}$$

Example 1.27 *The piston shown in Fig. Ex. 1.27 is held in equilibrium by pressure of gas flowing through the pipe. The piston has a mass of 21 kg. P_I = 600 kPa; P_{II} = 170 kPa. Determine the pressure in the gas P_{III}.*

Solution

Total down ward force acting will be:

(i) Due to self weight.
(ii) Pressure P_I acting on 10 cm dia. piston.
(iii) Pressure P_{II} acting on 20 cm dia piston.

And this is balanced by pressure P_{III}.

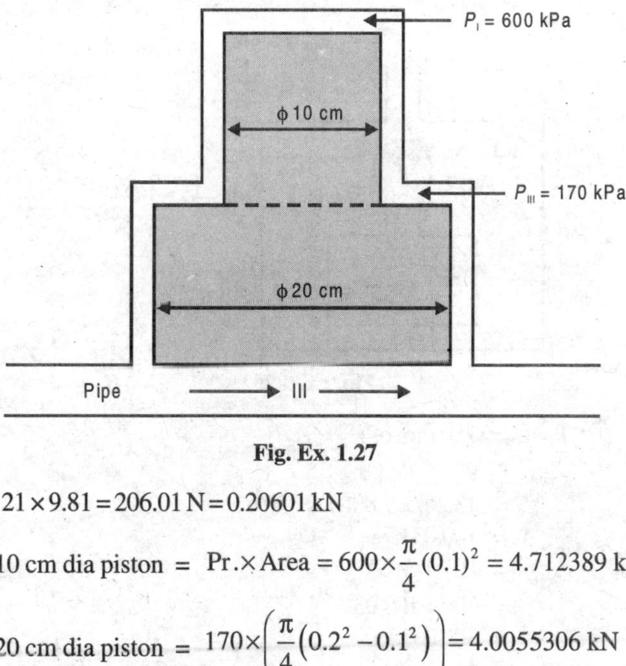

Fig. Ex. 1.27

∴ Force due to mass = 21 × 9.81 = 206.01 N = 0.20601 kN

Then force I acting on 10 cm dia piston $= \text{Pr.}\times\text{Area} = 600\times\frac{\pi}{4}(0.1)^2 = 4.712389$ kN

And force II acting on 20 cm dia piston $= 170\times\left(\frac{\pi}{4}(0.2^2-0.1^2)\right) = 4.0055306$ kN

$$\therefore \quad \text{Total downward load/force} = 0.20601 + 4.712389 + 4.005306$$
$$= 8.9239296\ \text{kN}$$

This is balanced by P_{III}

$$\therefore \quad 8.9239296 = P_{III}\times\text{Area} = P_{III}\times\frac{\pi}{4}(0.2)^2$$

$$= P_{III} \times 0.0314$$

$$\mathbf{P_{III} = 284.0575\ kPa}$$

Example 1.28 *The basic barometer can be used to measure the height of a building. If the barometric readings at top and bottom of a building are 730 and 755 mm of Hg respectively: Determine the height of the building. Assume an average air density of 1.18 kg/m³.*

Solution

The difference between pressure at top and bottom is equal to the air head of height equal to that of the building.

$$\therefore \quad P_1 - P_2 = \rho_{air} g h$$

$$(0.755 - 0.73) \times 13600 \times 9.81 = 1.18 \times 9.81 \times h$$

$$h = \frac{13600 \times 0.025}{1.18}$$

$$\mathbf{h = 288.14\ m}$$

Example 1.29 *A large container A contains another small container B, both fitted with gauges. The gauge fitted to A reads 200 kPa and gauge fitted to B reads 120 kPa.*

If barometer reads 750 mm of Hg, calculate the absolute pressure in the containers A and B.

Solution

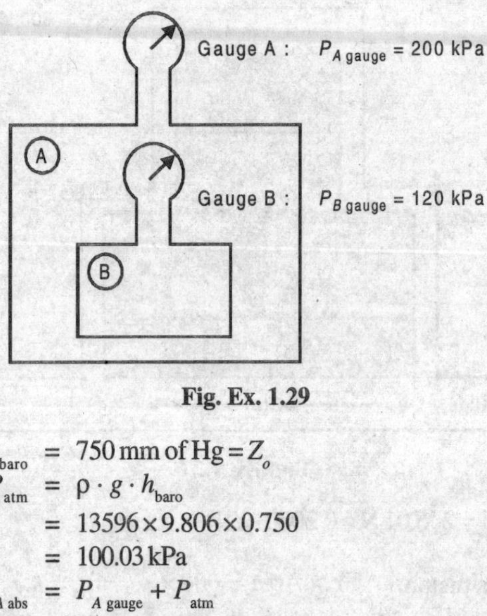

Fig. Ex. 1.29

Barometer reading

$$h_{baro} = 750 \text{ mm of Hg} = Z_o$$

$$P_{atm} = \rho \cdot g \cdot h_{baro}$$

$$= 13596 \times 9.806 \times 0.750$$

$$= 100.03 \text{ kPa}$$

$$P_{A\ abs} = P_{A\ gauge} + P_{atm}$$

$$= 200 + 100.03$$

$$= 300.03 \text{ kPa}$$

Note that gauge B reads pressure relative to its surrounding presure i.e., gauge B is surrounded by pressure of 300.03 kPa.

$$P_{B\ abs} = P_{B\ gauge} + P_{atms} \text{ of } B$$

$$= 120 + 300.03$$

$$\mathbf{P_{B\ abs} = 420.03\ kPa}$$

Example 1.30 *Convert the following temperatures into degree Fahrenheit.*

(i) 40 °C *(ii) – 20 °C*

Solution

(i) Using the relation
$$\frac{C}{100} = \frac{F-32}{180}$$
$$\frac{40}{100} = \frac{F-32}{180}$$
$$F-32 = \frac{40\times180}{100} = 72$$
$$\mathbf{F = 72+32 = 104\,°F}$$

(ii)
$$\frac{-20}{100} = \frac{F-32}{180}$$
$$F-32 = \frac{-20\times180}{100} = -36$$
$$\mathbf{F = 32-36 = -4\,°F}$$

Example 1.31 *Find the temperature which has the same value on both the centigrade and fahrenheit scales.*

Solution

Using the relation
$$\frac{C}{100} = \frac{F-32}{180}$$

Putting C = F,
$$\frac{C}{100} = \frac{C-32}{180}$$
$$\therefore \quad 180\,C = 100\,C-3200$$
$$\therefore \quad C = -40$$
$$\therefore \quad \mathbf{-40\,°C = -40\,°F}$$

Example 1.32 *A temperature scale of a certain thermometer is given by the relation t = a $\log_e$ P + b, where a and b constants and P is the thermometric property of the fluid in the thermometer. If at the ice point and steam point, thermometric properties are found to be 1.5 and 7.5 respectively. What would be the temperature corresponding to the thermometric property 3.5 on Celsius scale.*

Solution

$$t = a\log_e P+b$$
$$100 = a\log_e 7.5+b$$
$$0 = a\log_e 1.5+b$$

Solving theses two equations we get,

$\therefore \quad a = 62.1336$ and $b = -25.168$

$\therefore$ For $P = 3.5$, $\quad t = 62.1336\log_e 3.5-25.168$

$$\mathbf{t = 52.670676\,°C}$$

Example 1.33 *The temperature scale of a certain thermometer is given by t = a ln P + b, when a and b are constants and P is the thermometric property. The temperature of ice point and steam point are 100 and 300 respectively on this scale. Experiment gives values of thermometric properties of P_1 = 1.86 and P_2 = 6.81 at ice point and steam point respectively. Evaluate the temperature corresponding to a reading of thermometric property at P = 2.5.*

Solution

$$t = a \ln P + b \quad \text{...(1)}$$

$$300 = a \ln 6.8 + b$$

$$100 = a \ln 1.86 + b \quad \text{...(2)}$$

On subatracting we get, $200 = a \ln\left(\dfrac{6.81}{1.86}\right)$

$$a = \frac{200}{1.3} = 154 \quad \text{...(3)}$$

Substituting this value of a in (2) we get,

$$100 = 154 \ln 1.86 + b$$

$\therefore$ $b = 100 - 154 \ln 1.86 = 4.43$

$\therefore$ For $P = 2.5,\quad t = 154\ \ln 2.5 + 4.43$

$$\mathbf{t = 145.4°}$$

Example 1.34 *The pressure in a constant volume gas thermometer is measured as 32 mm of Hg above the atmospheric pressure at triple point of water, determine the temperature in °C, when pressure is 76 mm of Hg above atmospheric pressure. Barometric pressure is 752 mm of Hg.*

Solution

Note: In a constant volume gas thermometer, volume remains constant but as the temperature changes, pressure change occurs.

$$\frac{\theta_1}{\theta_2} = \left(\frac{P_2}{P_1}\right) \text{constant volume} \quad \text{...(1)}$$

Given that $\theta_1 = 273.16$ K (triple point of water)

$$P_1 = P_{abs} + P_{gauge} = 752 + 32$$

$$P_1 = 784 \text{ mm of Hg}$$

and $P_2 = P_{atm} + P_{gauge} = 752 + 76 = 828$ mm of Hg

Substituting these in Eq. (1) we get

$$\frac{273.16}{\theta_2} = \frac{784}{828}$$

$$\theta_2 = 288.49 \text{ K}$$

$$\boldsymbol{\theta_2 = 15.49°C}$$

Example 1.35 *The temperature T on a thermometric scale is defined in terms of a property P by the relation,*

T = a ln P + b where a and b are constants.

The temperatures of the ice point and steam points are assigned the numbers 32 and 212 respectively. Experiment gives values of P of 1.86 and 6.81 at the ice point and steam point respectively. Evaluate the temperature corresponding to a reading of P = 2.5 on the thermometer.

Solution

We have $T = a \ln P + b$

Substituting the values for ice point and steam point,

$$32 = a \ln 1.86 + b$$

and $$212 = a \ln 6.81 + b$$

Solving we get $a = 138.7,\ b = -54.07$

$\therefore$ Equation becomes $T = 138.7 \ln P - 54.07$

Now for $P = 2.5$, $T = 138.7 \ln 2.5 - 54.07$

$= \mathbf{73}$

THEORY QUESTIONS

1.1. What is Thermodynamics?

1.2. Define Thermodynamic system? Explain its types with suitable examples.

1.3. Discuss : (1) System, (2) Surrounding, (3) Boundary, (4) Universe.

1.4. What do you understand by thermodynamic system ? How these are classified ? Explain each with suitable sketch, or

Explain

1. Open system
2. Closed system
3. Isolated system

and state the examples of each.

1.5. Distinguish between: Open system and closed system.

1.6. State with reasoning whether the following systems are closed, open or isolated.

1. IC engine
2. Refrigerator
3. Pressure cooker.

1.7. Explain Thermodynamic properties, Processes and cycles.

1.8. Discuss :

1. State
2. Property

1.9. Explain : Thermodynamic properties.

1.10. Compare: Intensive and extensive properties.

1.11. What do you understand by state function and path function ?

1.12. Compare state function and path function.

1.13. Discuss – Unit extensive property is an intensive property.

1.14. Define the terms: Process.

1.15. Explain – 1. Cycle; 2. Process.

1.16. Define property and classify it with proper examples.

1.17. What is meant by Working Substance? Explain with an example.

1.18. Write a short note on Units and Dimensions.

1.19. What is Mechanical and Thermodynamic Work ? Explain.

1.20. What do you mean by thermodynamic work and mechanical work ?

1.21. What is *PdV*-work ? or Work done at the moving boundary ? or Displacement work.

1.22. Define heat and work and explain why neither is a property.

1.23. Prove that Work is a path function and Properties are point functions.

1.24. Define heat and work and explain why neither is a property.

1.25. 'Property is an exact differential'. Explain.

1.26. Derive the equations for work done in various quasi-static processes.

1.27. Write a note on Heat Energy.

1.28. Define the terms : Work and heat.

1.29. Explain the similarities and differences between heat and work.

1.30. Define the property – Pressure. State the various units used in practice.

1.31. How Pressure measurement is done ?

1.32. Write a note on Pressure exerted due to a column of fluid.

1.33. Write a short note on Barometer.

1.34. Briefly explain why you cannot use a mercury barometer to measure pressure in a gravity free environment.

1.35. Write a note on Bourdon Pressure gauge.

1.36. Explain – Bourdon's Pressure gauge with sketch.

1.37. Write a note on Manometer.

1.38. Define Thermal Equilibrium. State and explain the Zeroth law of Thermodynamics.

1.39. State and explain zeroth law of thermodynamics and explain how it leads to the concept of temperature.

1.40. What is meant by thermometric property. Enlist the temperature measuring devices.

1.41. Discuss various thermodynamic properties used for temperature measurement.

1.42. What is the principle of Temperature Measurement ?

1.43. What are the different Scales of Temperature?

1.44. Write a note on Microscopic and Macroscopic point of view.

1.45. What do you understand by microscopic and macroscopic approach of thermodynamic study ? Explain.

1.46. What is the difference between the classical and statistical approaches to Thermodynamics.

1.47. Write a note on Energy in transit.

1.48. Explain the term — Energy in transit.

1.49. What is a Quasi–Static Process? Explain.

1.50. What do you mean by Quasi-static process.

1.51. Explain why a non-quasi equilibrium compression process requires a larger work input than the corresponding quasi equilibrium one, and, why a non-quasi equilibrium expansion process delivers less work than the corresponding quasi-equilibrium one.

PROBLEMS FOR PRACTICE

1. The pressure of a fluid of specific gravity 0.8 flowing in a horizontal pipe line in determined with a simple mercury U-tube manometer. The level of Hg surface in the right limb which is open to the atmosphere is 90 mm above the centre of the pipes. The level of Hg in the left limb which is connected to the pipe is 60 mm below the centre of pipe. Determine absolute pressure of liquid in the pipe is N/m^2. Assume standard P_{atm} to be equal to 10 m of water. (*Ans.* 117641.52 N/m^2)
2. An inverted U-tube is connected across two pipes *A* and *B* carrying fluid as shown in Fig. P1.2. Calculate the pressure difference between the centre lines of the pipes.

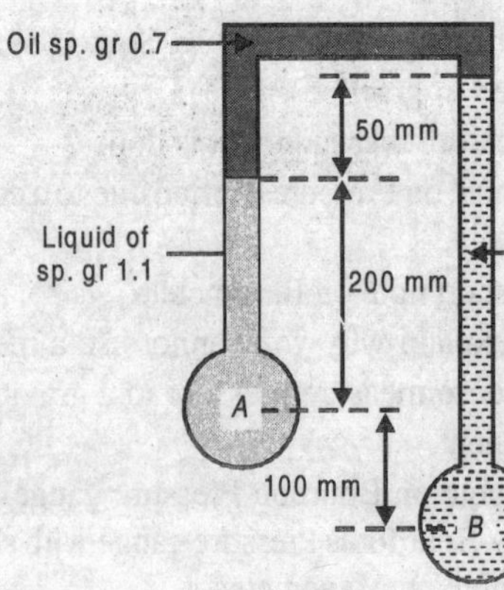

Fig. P1.2

3. In the arrangement shown in Fig. P1.3 P_3. Find the difference of pressure in the pipe *A* and *B*. (*Ans.* $P_A - P_B$ = 29135.7 N/m^2)

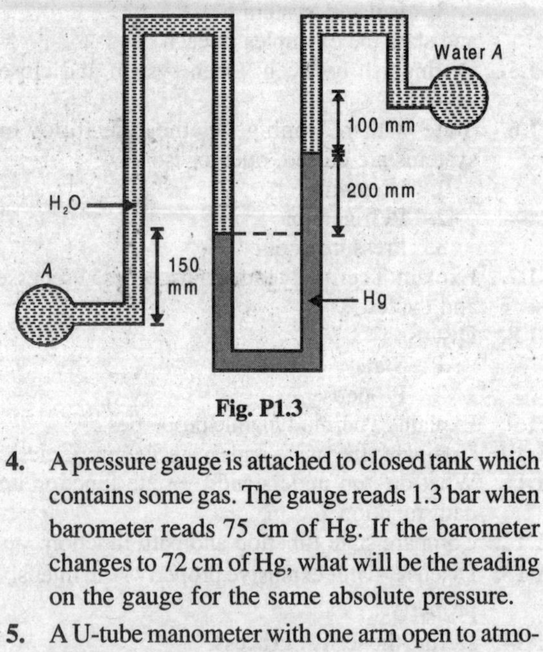

Fig. P1.3

4. A pressure gauge is attached to closed tank which contains some gas. The gauge reads 1.3 bar when barometer reads 75 cm of Hg. If the barometer changes to 72 cm of Hg, what will be the reading on the gauge for the same absolute pressure.
5. A U-tube manometer with one arm open to atmosphere is used to measure the pressure of the steam flowing through a pipe using mercury as manometric fluid. The level of the Hg column in the arm connected to the pipe is 125 mm below the level of Hg in the open arm. The column of

condensate collected is 35 mm. Taking $g = 9.81$ m/s^2, $\rho_{Hg} = 13560$ kg/m^3 and atm. pressure as 760 mm of Hg, estimate the absolute pressure of steam in the pipe in bar.

6. Both a Bourdon gauge and manometer are attached to a gas tank. If the reading on the pressure gauge is 80 kPa, determine the distance between the two fluid levels of the manometer if the fluid is (i) Hg having sp. gravity of 13.6, (ii) water.
7. A U-tube manometer with one arm open to atmosphere is used to measure the pressure of steam flowing through a pipe. The height of Hg column open to atmosphere is 15 cm greater than the Hg level in the arm connected to steam pipe. The column of condensate is 5 cm. The barometer reading is 70 cm of Hg. Find the absolute pressure of steam in kPa.
8. A water pump is required to supply water at 900 lit/min into a tank at a height of 20 m. Find the power required to drive the pump in kW.
9. Estimate the height of the atmospheric column to produce a pressure of 1 bar at the earth surface. Assume that the pressure P in Pascal and sp. volume V in m^3/kg are related according to the law $PV^{1.39} = 2.25 \times 10^5$ and $g = 9.8$ m/s^2.
10. An air tank has three manometers connected to it. The fluids in them are oil (sp. gr. = 0.85), water and Hg. If the absolute pressure in the tank is 1.24 bar and the barometric height is 75 cm of Hg, estimate the height of fluid in each manometer.
11. A gas expands in a piston—cylinder device from 0.01 to 0.03 m^3, the process being described by $P = aV^{-1} + b$ where P is in bar and V in m^3. If $a = 0.06$ bar/m^3 and work done by the gas on the piston is 10.6 kg, evaluate b.
12. In a piston cylinder arrangement the pressure is inversely proportional to the square of the volume. The initial pressure is 10 bar in the cylinder and initial volume is 0.1 m^3. The volume is now changed so that the final pressure is 2 bar. Find the work done in kJ.

CHAPTER 2

First Law of Thermodynamics

2.1 INTRODUCTION

Energy is inherent in all matters. Energy may appear in many different forms. Conversion can be made from one form of energy to another. We are unable to define the general term energy in a simple way, but we can define with precision the various forms in which it appears.

One of the consequences of Einstein's theory of Relativity is that mass may be converted into energy and energy into mass, the relation being given by the famous equation,

$$\text{Energy }(E) = m\,(\text{mass}) \times C^2 \qquad \text{(where } C - \text{velocity of light)}$$

Energy is a scalar quantity, not a vector quantity.

Except the nuclear reaction, (where mass is converted into energy) total energy of the universe is constant. For this matter, First Law Law of Thermodynamics can be expressed as—

Energy can neither be created nor destroyed (except in nuclear reactions). This is the law of conservation of energy.

2.2 JOULE'S EXPERIMENT

Joule carried out experiments in 1843, which led to the formulation of the First Law of Thermodynamics. He took a known quantity of water in a rigid vessel, which was insulated adiabatically from the surroundings. The vessel was fitted with a paddle wheel and a thermometer as shown in Fig. 2.1.

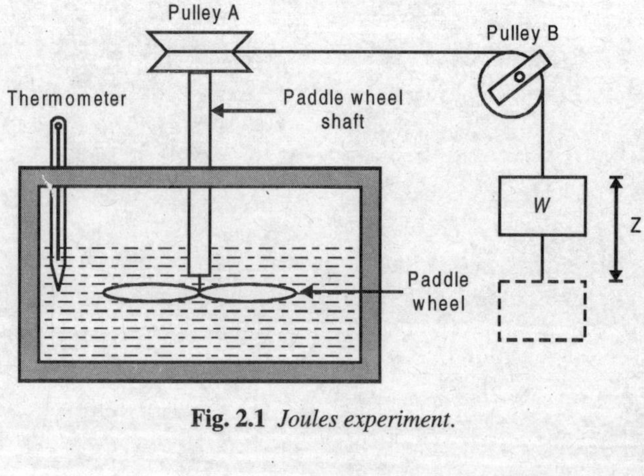

Fig. 2.1 *Joules experiment.*

Now let a certain amount of work W_{1-2} be done upon the closed system by the paddle wheel. The quantity of work can be measured by the fall of weight W which drives the paddle wheel.

Let t_1 be the initial temperature of water before work transfer and after work transfer let the temperature rise to t_2. So, the system is changing its state from state (1) to state (2) and the process 1–2 undergone by the system is shown in Fig. 2.2.

Now let the insulation be removed. Then the system will interact by heat transfer (i.e., heat will be dissipated from the system to the surroundings) till the system comes to original temperature t_1. The amount of heat transfer Q_{2-1} from the system during the process 2–1 can be estimated: ($Q_{2-1} = mC_p \Delta T$).

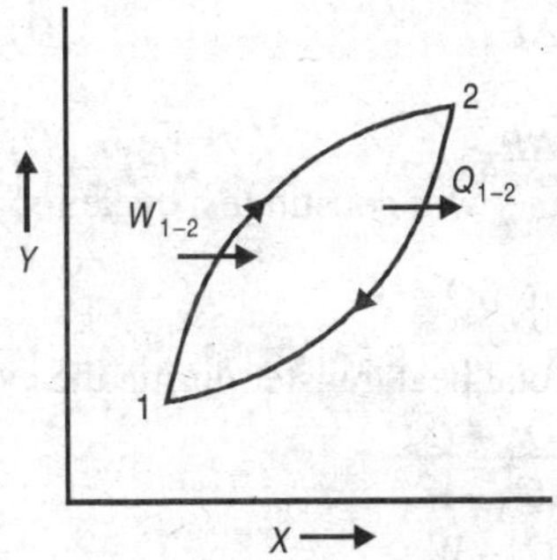

Fig. 2.2

The system thus undergoes a cycle, which consists of a definite amount of work input W_{1-2} to the system, followed by a heat transfer Q_{2-1} from the system.

Joule, repeated this experiment for different weights and different heights and in each case he found that, net work transfer during a cycle is proportional to net heat transfer,

i.e., $$\Sigma \delta W \propto \Sigma \delta Q$$

Or in other words *when a closed system undergoes any cyclic process, the cyclic integral of work is proportional to the cyclic integral of heat*. This is known as the first law of thermodynamics for a closed system undergoing a cycle.

i.e.,
$$\oint \delta W \propto \oint \delta Q$$
$$\oint \delta W = J\oint \delta Q \quad \text{Since in SI units } J = 1$$
$$\oint \delta W = \oint \delta Q$$

Thus, the I law states the general principle of conservation of energy i.e., *Energy can neither be created nor be destroyed, but energy can be converted from one form to the other.*

2.3 FIRST LAW FOR A CLOSED SYSTEM UNDERGOING A PROCESS

If a closed system undergoes a change of state or a process and during which, both work transfer and heat transfer are involved, then the net energy transfer will be stored within the system. If Q is the amount of heat transferred to the system and W is the amount of work transferred from the system, during the process as shown in Fig. 2.3, then the net energy transfer ($Q - W$) will be stored in the system.

Energy in storage is neither heat nor work, but is called as Internal energy or simply energy of the system.

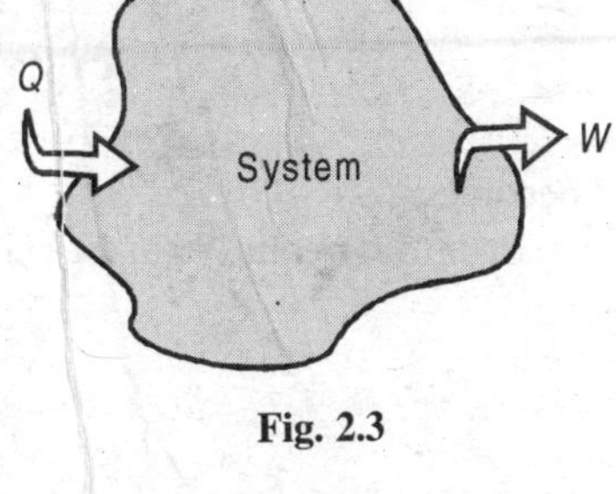

Fig. 2.3

$$\therefore \quad Q - W = \Delta E$$

or

$$Q = \Delta E + W$$

where ΔE is the change in energy. Here Q, W and ΔE all are expressed in joules.

2.4 ENERGY — A PROPERTY OF THE SYSTEM

Consider a system which changes its state from state (1) to state (2) by following the path A, and returns from state (2) to state (1) by following the path B as shown in Fig. 2.4. So the system undergoes a cycle. Now writing the I-law for the path A,

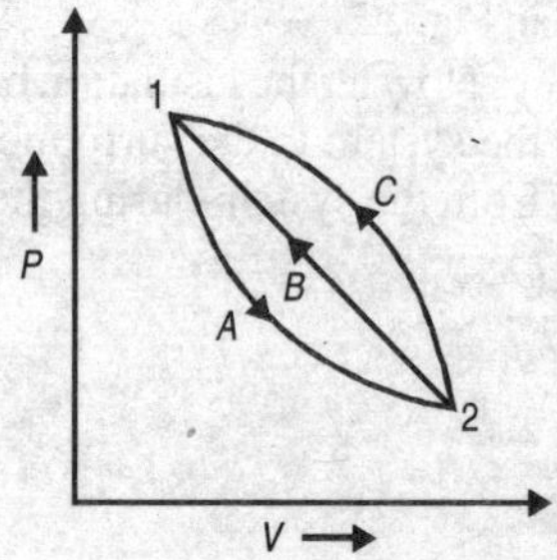

Fig. 2.4

$$Q_A - W_A = \Delta E_A \quad \text{...(1)}$$

and for path B.

$$Q_B - W_B = \Delta E_B \quad \text{...(2)}$$

The processes 1–A–2 and 2–B–1 together will constitute a cycle for which,

$$\oint \delta W = \oint \delta Q$$

i.e., Total work transfer during cycle = Total heat transfer during the cycle.

$$W_A + W_B = Q_A + Q_B$$

Rewriting

$$W_B - Q_B = Q_A - W_A \quad \text{...(3)}$$

$$\therefore \quad -(Q_B - W_B) = Q_A - W_A \quad \text{...(4)}$$

$$-\Delta E_B = \Delta E_A$$

i.e.,

$$\Delta E_A = -\Delta E_B \quad \text{...(5)}$$

Similarly if the system returns from point (2) to state point (1) by following the path C instead of path B then,

$$\Delta E_A = -\Delta E_C \quad \text{...(6)}$$

From Eqs (4) and (5),

$$\mathbf{\Delta E_B = \Delta E_C}$$

Thus the change in energy for the path B and C are same. Hence change in energy does not depend upon path, so it depends on end states. Hence, it is a point function and since properties are point functions, *Energy is a property of the system.*

2.5 DIFFERENT FORMS OF STORED ENERGIES

The total energy E is made up of Kinetic Energy (K.E.), Potential Energy (P.E.) and Internal Energy (I.E.).

$$\therefore \quad E = \text{K.E.} + \text{P.E.} + \text{I.E.}$$

Kinetic Energy is due to the motion of the fluid in the system. Potential Energy is the energy due to the gravitational force.

A part of the total energy which is stored in the molecular and atomic structure is known as *Internal energy* and is denoted by U.

In the absence of Kinetic Energy and Potential Energy i.e., when

$$\text{KE} = 0 \text{ and PE} = 0$$

$$E = U, \text{ and the I law for process}$$

$$Q = \Delta E + W$$

becomes

$$Q = \Delta U + W$$

In the differential form, Eqs (1) and (2) become

$$\delta Q = dE + \delta W$$

i.e.,

$$\delta Q = dE + PdV \left(\text{as } \delta W = PdV\right)$$

Also, $\delta Q = dU + \delta W \text{ (when } E = U\text{)}$

i.e., $\delta Q = dU + PdV$

When KE and PE are considered, then the I law for a process will be

$$\delta Q - \delta W = dU + d(\text{KE}) + d(\text{PE})$$

In the integrated form, Eq. (5) becomes

$$Q_{1-2} - W_{1-2} = U_2 - U_1 + \frac{m(C_2^2 - C_1^2)}{2} + mg(Z_2 - Z_1)$$

where C is the velocity and Z is the height.

For unit mass, Eq. (6) will be,

$$q_{1-2} - w_{1-2} = u_2 - u_1 + \frac{(C_2^2 - C_1^2)}{2} + g(Z_2 - Z_1)$$

2.6 INTERNAL ENERGY (U)

It is the energy stored in the molecular structure due to heat and work interactions.

or

In case of gases we know that the gas is made up of a number of molecules, which are moving continuously. I.E. is the energy which arises from the motion of these molecules.

If the temperature of the gas is increased, the molecular activity increases, therefore the I.E. also increases. Thus I.E. is a function of temperature and its value can be increased or decreased by adding or removing heat to or from the system.

Due to practical difficulties it is very very difficult to determine the absolute value of Internal energy. Fortunately, in most of the thermodynamic applications change in I.E. is used, since changes in the states of a system is considered. Change in I.E. is denoted by ΔU and $\Delta U = U_2 - U_1$.

For unit mass, $\Delta u = u_2 - u_1$

2.7 ENTHALPY (H)

It is an extensive property, since its value depends on mass. Enthalpy of a substance is given by the sum of Internal energy and Pressure–Volume product.

i.e., $H = U + PV$ and specific enthalpy, $h = u + Pv$

2.8 SPECIFIC HEAT AT CONSTANT PRESSURE AND SPECIFIC HEAT AT CONSTANT VOLUME

Specific heat of a gas is defined as the amount of heat required to rise the temperature of unit mass of a gas through one degree.

Let C = General specific heat

Q = Heat transfer (J)

ΔT = Change in temperature K

m = mass (kg)

Then from the definition, $$C = \frac{Q}{m \cdot \Delta T} \left[\text{units will be } \frac{\text{J}}{\text{kg-K}}\right] \quad \ldots(1)$$

or $$Q = m \cdot C \cdot \Delta T \quad \ldots(2)$$

In the differential form, $\delta Q = m \cdot C \cdot dT$ ...(3)

and for unit mass, $\delta q = C \cdot dT$ ...(4)

or $$C = \frac{\delta q}{dT}$$ ...(5)

There are 2-types of specific heats,

(1) Specific heat at constant volume C_v

(2) Specific heat at constant pressure C_p

1. Specific Heat at Constant Volume C_v

It is the amount of heat required to rise the temperature of unit mass of a gas through one degree when the volume is constant.

If a unit mass of a gas is taken in a closed vessel and is heated, the volume of the gas remains constant, but the temperature increases. As the volume remains constant, there is no external work done by the gas and as temperature of the gas increases, there is increase in internal energy of the gas.

Therefore heat supplied to the gas is completely utilised in increasing the I.E. of the gas,

and $$C_V = \left(\frac{\delta q}{dT}\right)_V = \left(\frac{du}{dT}\right)_V$$

Therefore C_v is also defined as, *the rate of change of specific internal energy with respect to temperature when the volume is kept constant.*

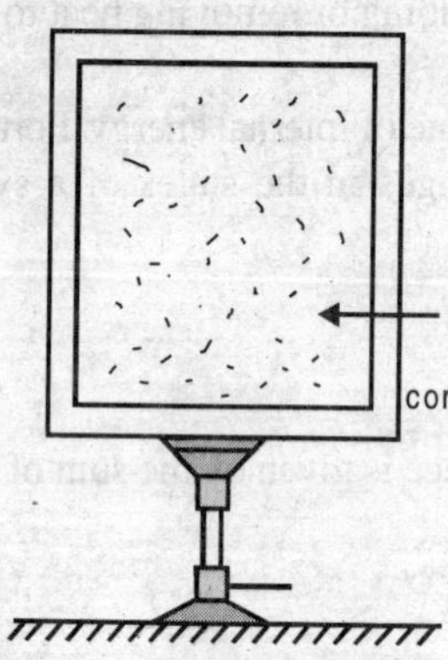

Fig. 2.5

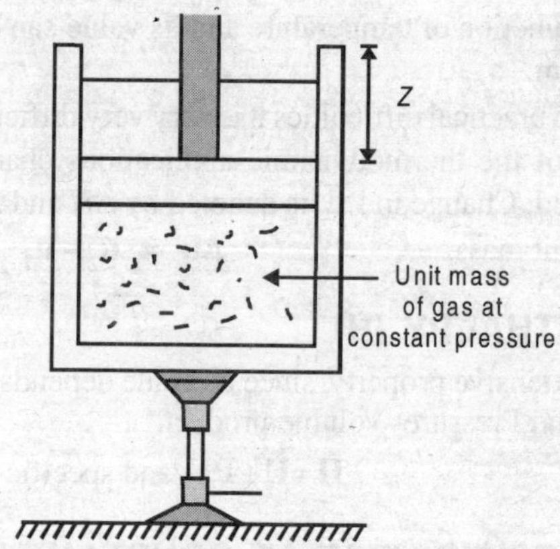

Fig. 2.6

2. Specific Heat at Constant Pressure C_p

It is the amount of heat required to rise the temperature of unit mass of a gas through one degree when the pressure is kept constant.

Consider a unit mass of a gas in a cylinder fitted with a frictionless piston as shown in the Fig. 2.6. When the gas is heated, the piston moves up, maintaining the same pressure. But the volume and temperature of the gas increases during heating. As there is increase in volume, there is external work done by the gas and as there is increase in temperature, there is increase in I.E.

Thus heat supplied when the pressure is constant, is utilised for two purposes,

(a) To do some external work.

(b) To increase the I.E. of the gas.

Whereas in case of constant volume heating, the heat supplied is completely utilised for increasing the I.E.

∴ Specific heat at constant pressure is greater than specific heat at constant volume.

i.e., $C_P > C_V$ and, $C_P = \left[\frac{\delta q}{dT}\right]_P$

Also $C_P = \left[\frac{dh}{dT}\right]_P$

Therefore, C_p is also defined as, *the rate of change of specific enthalpy with respect to temperature when the pressure is kept constant.*

2.9 ADIABATIC INDEX

It is the ratio of specific heat at constant pressure to the specific heat at constant volume and is given by,

$$K \text{ or } \gamma = \frac{C_p}{C_v}$$

Note: For air, $C_p = 1.005 \frac{kJ}{kg-K}$ **and** $C_v = 0.718 \frac{kJ}{kg-K}$ **and** $\gamma = 1.4$

2.10 PERPETUAL MOTION MACHINE OF FIRST KIND [PMM – 1]

First law states the general principle of conservation of energy i.e., energy can neither be created nor be destroyed but it can be transformed from one form to the other.

But PMM–1 is defined as a machine which will produce continuous work output without receiving any energy from any other system or the surroundings. (Ref. Fig. 2.7).

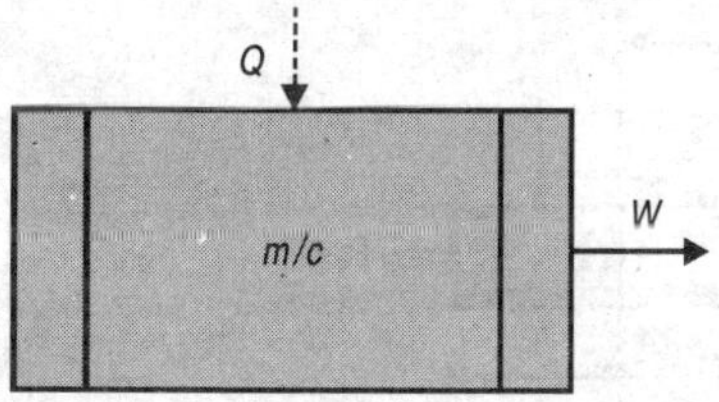

Fig. 2.7

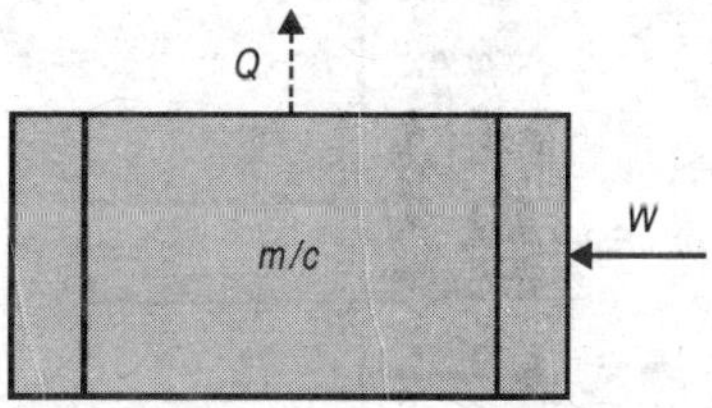

Fig. 2.8

It will create energy and thus violates I law of thermodynamics. All the attempts made so far to make such a machine have failed, thus they show the validity of I–law. Thus PMM–1 is just a conceptual machine.

Converse of PMM–1 is also true i.e., there can be no machine which would continuously consume work, without producing some other form of energy (Fig. 2.8).

2.11 FLOW PROCESS; CONTROL VOLUME; CONTROL SURFACE

Flow Process. Processes performed in the open systems are called as Flow-processes.

Figure 2.9 shows a portion of steam power plant. High pressure steam enters the steam turbines at Section (I), expands and produces work output and then it leaves the turbine at section (2).

For analysing the expansion process, 2-methods are used.

In the first method certain mass of the fluid Marked *A* is considered and this is suppose to flow through the turbine.

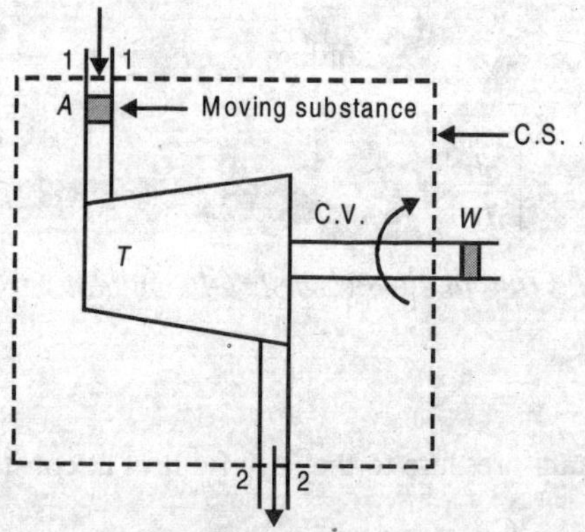

Fig. 2.9

In the second method, certain fixed volume of the system known as *control volume* (*CV*) is considered.

The moving substance flows through this control volume. The surface of the control volume is known as *control surface* (*CS*).

2.12 FLOW WORK OR FLOW ENERGY

In case of flow processes certain amount of work or energy is required to push the fluid into and out of the system. This work energy is known as *flow work or flow energy.*

Consider a flow process as shown in Fig. 2.10.

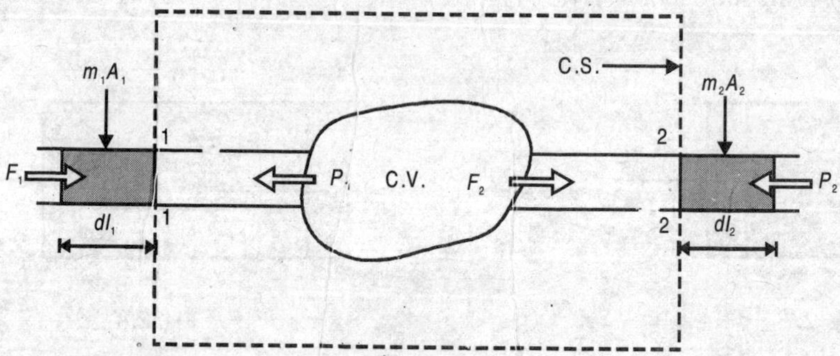

Fig. 2.10

Let F_1 be the force which forces the mass m_1 of cross sectional area A_1 into the system against the pressure P_1 of the system. Now consider a small amount of work done on the system in causing the displacement dl_1 for the mass m_1.

Thus the small amount of work done on the system in causing the displacement dl_1,

$$\delta W = -F_1 \times dl_1$$

Note : –ve sign, since the work is done on the system.

But $$F_1 = P_1 \times A_1$$

$\therefore$ $$\delta W = -P_1 \times A_1 \times dl_1$$

Since $A_1 \times dl_1 = dV_1$ = small amount of displacement volume.

$\therefore \quad \delta W = -P_1 \times dV_1$

$\therefore$ Total flow work at Sec. (1) $= -P_1 V_1$

Similarly work done by the system to force the fluid out of the system at Sec. $(2) = P_2 V_2$

$\therefore$ **Net flow work** $= P_2 V_2 - P_1 V_1$

and **for unit mass** $= P_2 v_2 - P_1 v_1$

Note: Since the flow work is entirely expressed in terms of properties of the system, the net flow work depends on the end states and it is a *Property.* Whereas the other forms of work are path functions so in problems involving flow processes, flow work is to be calculated separately.

2.13 CONDITIONS OF STEADY FLOW SYSTEM

A steady flow process should satisfy the following conditions,

(i) The mass flow rate into and out of the system are equal and do not vary with time i.e., mass in the system does not change.
(If the mass flow rate at the inlet is 10 kg/sec., then the mass flow rate at the exit will be 10 kg/sec and it does not vary with time i.e., if we note the mass flow rate say, when we come to the college and while going home also it will be same.)

(ii) The energy of the fluid at the entrance and at the exit are same and do not vary with time.

(iii) The rates of heat and work transfer into and out of the system do not vary with time.

2.14 STEADY FLOW ENERGY EQUATION (SFEE)

Steady flow energy equation is obtained by applying the first law of thermodynamics to a steady flow system.

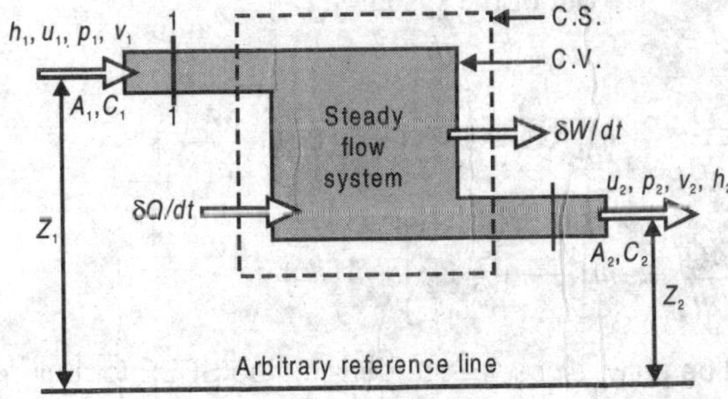

Fig. 2.11

Let,

$\dot{m}$ = Mass flow rate in kg/sec

$A_1 \, A_2$ = Cross sectional areas at sec 1–1 and 2–2 respectively in m^2

$C_1 \, C_2$ = Velocities in m/sec

$P_1 \, P_2$ = Absolute pressures in N/m^2

$v_1 \, v_2$ = Specific volumes in m^3/kg

$u_1 \, u_2$ = Specific internal energies in J/kg

$h_1 \, h_2$ = Specific enthalpies in J/kg

$Z_1 \, Z_2$ = Elevations above arbitrary reference datum line in m

Q = Net heat transfer in Joules

$$\frac{\delta Q}{dt} = \text{Net rate of heat transfer in J/sec}$$

$$\frac{\delta Q}{dm} = q = \text{Net rate of heat transfer in J/kg}$$

$$W = \text{Net work transfer in joules}$$

$$\frac{\delta W}{dt} = \text{Net rate of work transfer in J/sec}$$

$$\frac{\delta W}{dm} = w = \text{Net rate of work transfer in J/kg}$$

$$t = \text{time in seconds}$$

Figure 2.11 shows a steady flow system. Fluid entes at sec. 1–1 and leaves at sec. 2–2. Since it is a steady flow system. According to 2nd condition.

Total energy at the entrance = Total energy at the exit

∴ We can write,

$$\dot{m}\begin{bmatrix}\text{Energy carried}\\\text{into the system}\end{bmatrix} + \dot{m}\begin{bmatrix}\text{Specific enthalpy}\\\text{of entering fluid}\end{bmatrix} + \begin{bmatrix}\text{Net rate of heat}\\\text{transfer into}\\\text{system}\end{bmatrix}$$

$$= \dot{m}\begin{bmatrix}\text{Energy carried}\\\text{out of the system}\end{bmatrix} + \dot{m}\begin{bmatrix}\text{Specific enthalpy}\\\text{of the fluid}\\\text{at the exit}\end{bmatrix} + \begin{bmatrix}\text{Net rate of work}\\\text{transfer by the}\\\text{system}\end{bmatrix}$$

i.e., $$\dot{m}[\text{KE}_1 + \text{PE}_1] + \dot{m}(h_1) + \frac{\delta Q}{dt} = \dot{m}[\text{KE}_2 + \text{PE}_2] + \dot{m}(h_2) + \frac{\delta W}{dt}$$

i.e., $$\dot{m}\left[\frac{C_1^2}{2} + g\ Z_1\right] + \dot{m}(h_1) + \frac{\delta Q}{dt} = \dot{m}\left[\frac{C_2^2}{2} + g\ Z_2\right] + \dot{m}(h_2) + \frac{\delta W}{dt}$$

Note: *m* of KE and PE will be unity, since we are deriving the SFEE for unit mass.

$$\therefore \quad \dot{\mathbf{m}}\left[\frac{\mathbf{C}_1^2}{2} + \mathbf{g\ Z}_1 + \mathbf{h}_1\right] + \frac{\delta \mathbf{Q}}{\mathbf{dt}} = \dot{\mathbf{m}}\left[\frac{\mathbf{C}_2^2}{2} + \mathbf{g\ Z}_2 + \mathbf{h}_2\right] + \frac{\delta \mathbf{W}}{\mathbf{dt}} \qquad \ldots(1)$$

This is steady flow energy equation on *time basis*.

Also, since $h_1 = u_1 + P_1 v_1$ and $h_2 = u_2 + P_2 v_2$

∴ Eq. (1) becomes,

$$\therefore \quad \dot{m}\left[\frac{C_1^2}{2} + g\ Z_1 + u_1 + P_1 v_1\right] + \frac{\delta Q}{dt} = \dot{m}\left[\frac{C_2^2}{2} + g\ Z_2 + u_2 + P_2 v_2\right] + \frac{\delta W}{dt}$$

$$\text{or}\quad \dot{m}\left[\text{KE}_1+\text{PE}_1+\text{IE}_1+\text{FE}_1\right]+\frac{\delta W}{dt} = \dot{m}\left[\text{KE}_2+\text{PE}_2+\text{IE}_2+\text{FE}_2\right]+\frac{\delta W}{dt} \qquad \ldots(2)$$

2.15 STEADY FLOW ENERGY EQUATION ON MASS BASIS

For deriving this, we have to consider $\dot{m}$ = 1 kg/sec and all other quantities will be for per kg mass such as $\delta W/dm$ and $\delta Q/dm$.

∴ Eq. (1) becomes,

$$\left[\frac{C_1^2}{2}+gZ_1+h_1\right]+\frac{\delta Q}{dm} = \left[\frac{C_2^2}{2}+gZ_2+h_2\right]+\frac{\delta W}{dm}$$

$$\therefore \quad \left[\frac{C_1^2}{2}+gZ_1+h_1\right]+q = \left[\frac{C_2^2}{2}+gZ_2+h_2\right]+w.$$

This is the SFEE on mass basis.

2.16 IN A STEADY FLOW SYSTEM W.D.=$-\int v.dP$ WHEN CHANGES IN KE AND PE ARE NEGLECTED

We know that, SFEE on mass basis. From Eq. (3)

$$\left[\frac{C_1^2}{2}+gZ_1+h_1\right]+q = \left[\frac{C_2^2}{2}+gZ_2+h_2\right]+w \qquad \ldots(1)$$

$$\text{i.e.,}\quad \left[\frac{C_1^2}{2}+gZ_1+u_1+P_1v_1\right]+q = \left[\frac{C_2^2}{2}+gZ_2+u_2+P_2v_2\right]+w$$

$$\therefore \quad q = u_2-u_1+P_2v_2-P_1v_1+\frac{C_2^2-C_1^2}{2}+g(Z_2-Z_1)+w$$

$$\therefore \quad q = \Delta u+\Delta(Pv)+\Delta(\text{KE})+d(\text{PE})+w$$

In differential form, $\delta q = du+d(Pv)+d(\text{KE})+d(\text{PE})+\delta w$ ...(2)

But we also know that I law for a closed system undergoing a process is,

$$Q-W = \Delta E=\Delta U$$

For unit mass, $q-w = \Delta u$

i.e., $q = \Delta u+w$

In the differential form, $\delta q = du+\delta w$

i.e., $\delta q = du+Pdv$ ...(3)

Assuming that δq is same for closed system and for the steady flow system, equating Eqs (2) and (3)

$$du+Pdv = du+d(Pv)+d(\text{KE})+d(\text{PE})+\delta w \qquad \ldots(4)$$

$$\text{i.e.,}\quad du+Pdv = du+Pdv+vdP+d(\text{KE})+d(\text{PE})+\delta w$$

∴ We can write, $-v\,dP = \delta w+d(\text{KE})+d(\text{PE})$

∴ For total mass, $-\int vdP = w+\Delta(\text{KE})+\Delta(\text{PE})$ ...(5)

When change in KE and PE are neglected,

$$-\int vdP = w$$

This work done is given by the area behind the curve 1–2 as shown in Fig. 2.12.

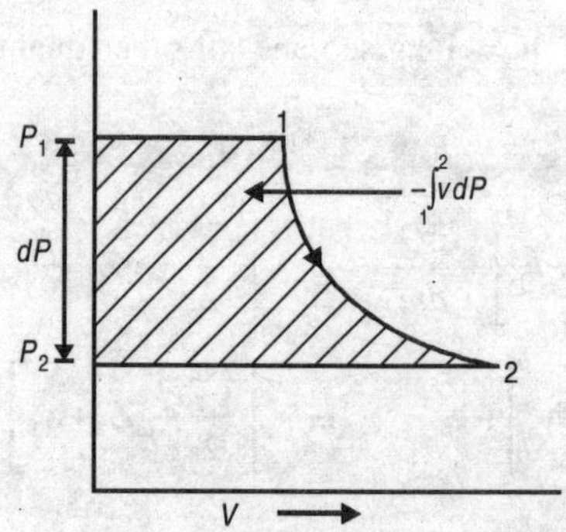

Fig. 2.12

2.17 SIGNIFICANCE OF $\int Pdv$ IN CASE OF STEADY FLOW PROCESS AND NON-FLOW PROCESS

Upto Eq. (4) derivation is same as in the above article.

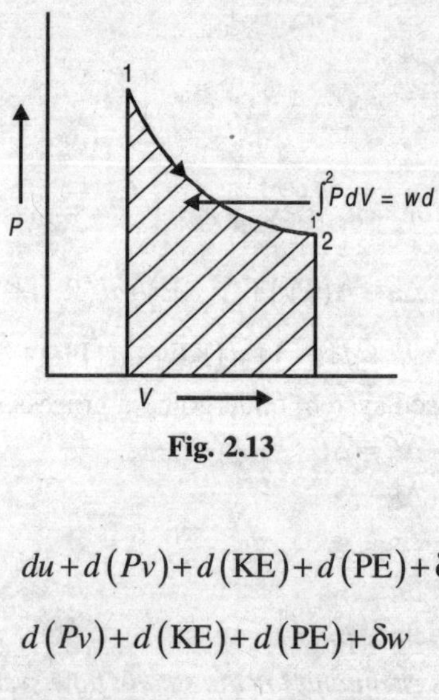

Fig. 2.13

From Eq. (4),

$$du + Pdv = du + d(Pv) + d(KE) + d(PE) + \delta w$$

$$Pdv = d(Pv) + d(KE) + d(PE) + \delta w$$

∴ For total mass, $$\int_1^2 Pdv = \Delta(Pv) + \Delta(KE) + \Delta(PE) + w$$

For a non-flow process (i.e. for a closed system),

$\int Pdv$ = work done and is given by the area under the curve 1–2 as shown in Fig. 2.13.

2.18 PROOFS

2.18.1 $-\int vdP = Q - \Delta H$

We know that, for a steady flow system,

$$Q - W = \Delta H + \Delta PE + \Delta KE \qquad \ldots(1)$$

Consider the case when the changes in PE and KE are neglected,

$$Q - W = \Delta H$$

$$\therefore \quad W = Q - \Delta H \qquad \ldots(2)$$

Also we know that, in a reversible steady flow process,

Work done $= -\int v\, dP$ when changes in KE and PE are neglected. ...(3)

Hence from (2) and (3) we can write,

$$\therefore \quad -\int \mathbf{v\,dP} = Q - \Delta H$$

2.18.2 $\int Pdv = Q - \Delta U$

We know that, for a non-flow process,

$$Q - W = \Delta U$$

$$W = Q - \Delta U \qquad \ldots(1)$$

And for a Non-flow process, $W = \int Pdv$ ...(2)

$\therefore$ From (1) and (2) $\int \mathbf{Pdv} = Q - \Delta U$

2.18.3 $-\int v\,dP - \int Pdv = -\Delta Pv$

We know that, $-\int vdP = w + \Delta KE + \Delta PE$

Also we know that, significance of $\int P\, dv$ in case of steady flow process and non-flow process,

$$\int P\, dv = w + \Delta(PE) + \Delta(KE) + \Delta Pv \qquad \ldots(3)$$

$\therefore$ Substracting Eq. (2) from (1) we get,

$$-\int \mathbf{v\,dP} - \int \mathbf{Pdv} = -\Delta \mathbf{Pv}$$

2.19 APPLICATIONS OF ENERGY EQUATIONS

(a) Boiler It is a steam generator

Characteristics of the boiler are:

(1) Change in velocity of the fluid at the entrance and exit is very small, and hence ΔKE can be neglected, i.e., $\Delta KE = 0$

(2) Change in elevation between the inlet and exit point is very small so change in PE may be neglected, i.e., $\Delta PE = 0$

(3) Since no work is done in a boiler $w = 0$

$\therefore$ From SFEE on mass basis,

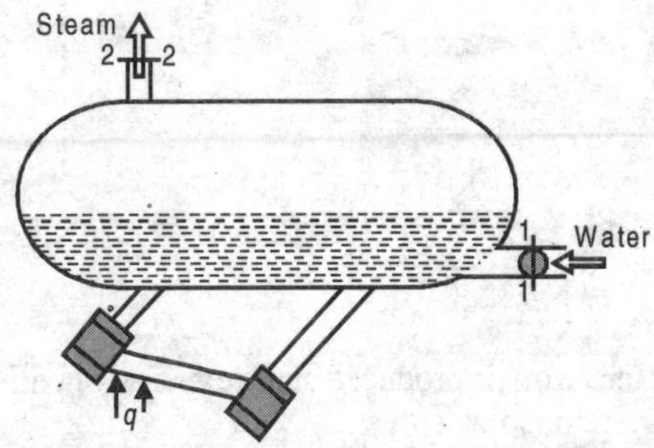

Fig. 2.14

$$\left[\frac{C_1^2}{2}+gZ_1+h_1\right]+q = \left[\frac{C_2^2}{2}+gZ_2+h_2\right]+w$$

i.e.,
$$q = \left[\frac{\left(C_2^2-C_1^2\right)}{2}+g\left(Z_2-Z_1\right)+\left(h_2-h_1\right)\right]+w \quad \ldots(1)$$

On substituting the above conditions Eq. (1) becomes,

$$q = \left[0+0+\left(h_2-h_1\right)\right]+0$$

i.e.,
$$\mathbf{q = h_2 - h_1}$$

(b) Turbine (Engine)/Compressor (or pump)

The turbines and engines are power producing devices whereas Compressors, Pumps, Blowers etc. are power consuming devices.

Charcteristic of a turbine are,

(i) Since it is insulated, negligible heat transfer.

$\therefore \quad q = 0$

(ii) Change in velocity of the fluid at the entrance and at the exit is negligible.

$\therefore \quad \Delta KE = 0$

(iii) Since change in elevation between inlet and exit point is negligible,

$\Delta PE = 0$

$\therefore$ From SFEE i.e., from Eq. (1),

$$q = \left[\frac{\left(C_2^2-C_1^2\right)}{2}+g\left(Z_2-Z_1\right)+\left(h_2-h_1\right)\right]+w$$

$\therefore$ The equation reduces to,

$$0 = \left[0+0+\left(h_2-h_1\right)\right]+w$$

or
$$\mathbf{w = h_1 - h_2}$$

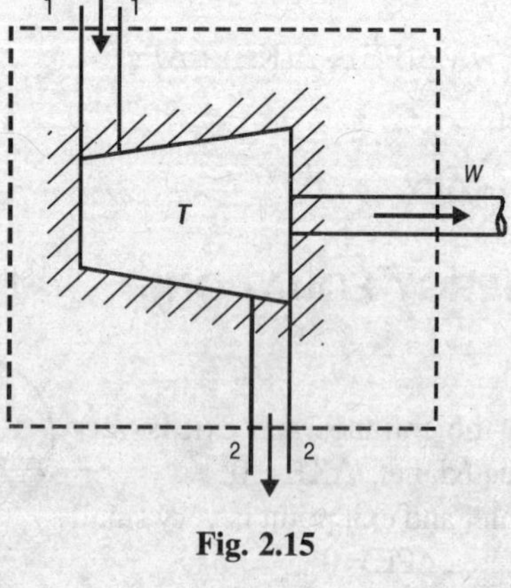

Fig. 2.15

Here work is produced at the expense of enthalpy. Similarly for an Adiabatic Pump/Compressor, work is done on the system, so w is –ve.

∴ SFEE becomes,

$$\mathbf{w = h_2 - h_1} \quad (\text{since } w \text{ is } -\text{ve})$$

(c) Throttling Process:

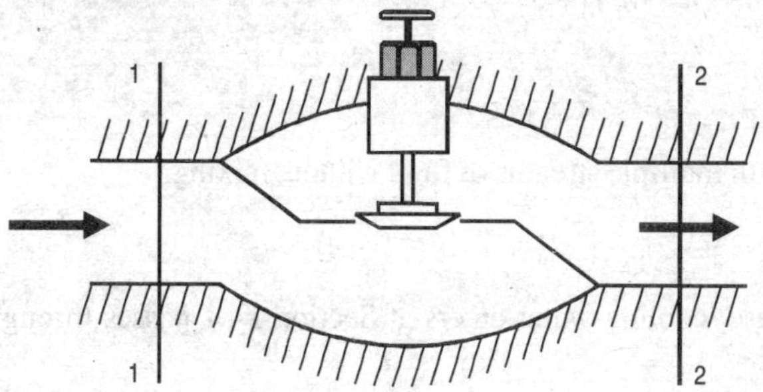

Fig. 2.16

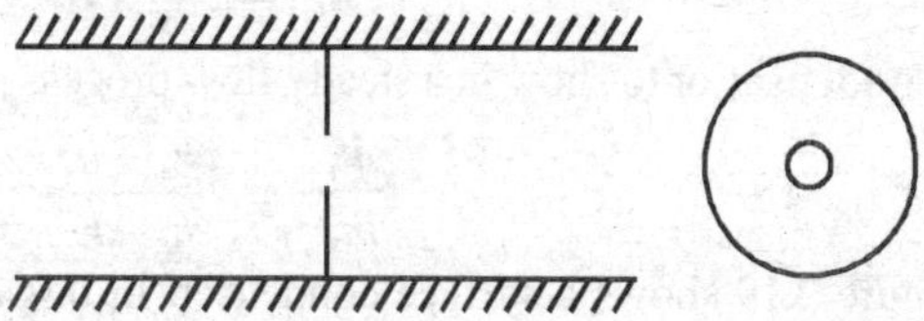

Fig. 2.17

When a fluid flows through a partially opened valve or an orifice then there is an appreciable pressure drop and the flow is said to be throttled.

Characteristics are,

(i) $q = 0$ (ii) $w = 0$ (iii) $\Delta PE = 0$

Substituting these in SFEE we get

$$q = \left[\frac{C_2^2 - C_1^2}{2} + g(Z_2 - Z_1) + (h_2 - h_1)\right] + w$$

$$0 = \left[\frac{(C_2^2 - C_1^2)}{2} + 0 + (h_2 - h_1)\right] + 0$$

i.e., $$\frac{\mathbf{C_1^2}}{\mathbf{2}} + \mathbf{h_1} = \frac{\mathbf{C_2^2}}{\mathbf{2}} + \mathbf{h_2}$$

(d) Nozzle. It is a device which is used to increase the KE of flowing fluids

(i) $q = 0$
(ii) $w = 0$
(iii) $\Delta PE = 0$

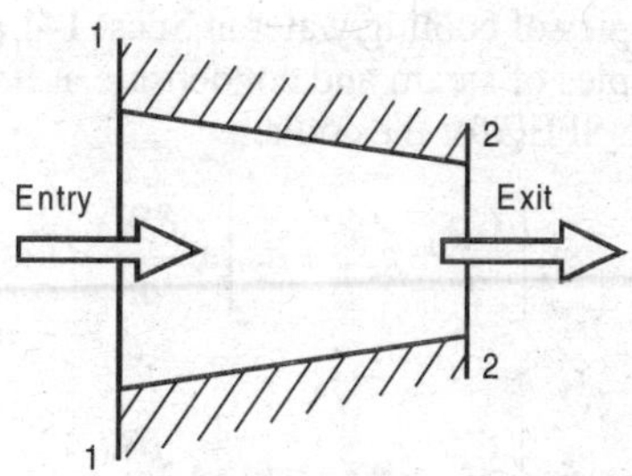

Fig. 2.18

∴ Substituting these in SFEE we get,

$$q = \left[\frac{C_2^2 - C_1^2}{2} + g(Z_2 - Z_1) + (h_2 - h_1)\right] + w$$

$$0 = \left[\frac{C_2^2 - C_1^2}{2} + 0 + (h_2 - h_1)\right] + 0$$

i.e.,
$$\mathbf{\frac{C_1^2}{2} + h_1 = \frac{C_2^2}{2} + h_2}$$

(e) Condenser or to show in a steady flow process with multiple streams of fluid without mixing.

$$\frac{\dot{m}_1}{\dot{m}_3} = \frac{h_4 - h_3}{h_1 - h_2}$$

Figure 2.19 shows a surface condenser. In this case, cooling water enters at Section 1– 1 passes through the tubes and leaves the condenser at Section 2 – 2.

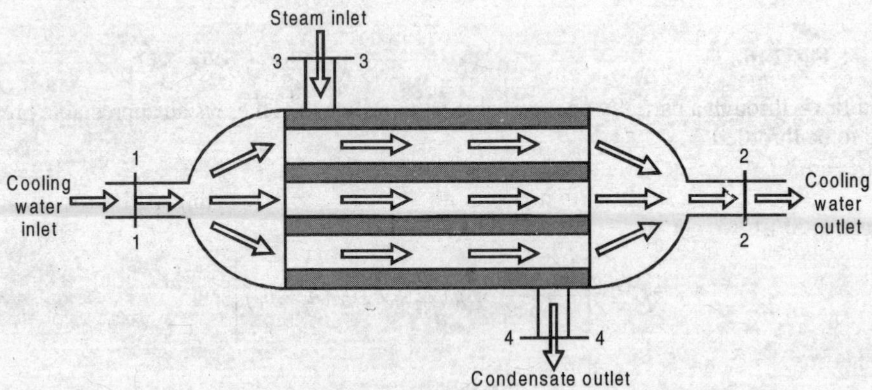

Fig. 2.19

Steam to be condensed enters the condenser at Section 3–3 and when it comes in contact with the cold surface of water tubes, it gets condensed and the condensate leaves the condenser at Section 4–4.

So, in this case working fluids are water and steam i.e., two fluids flow through the system and since they do not mix with each other, it may be considered as a steady flow system with multiple streams of fluid without mixing.

Let, $\dot{m}_1\ \dot{m}_2$ — Mass flow rates of cooling water at Secs. 1–1 and 2– 2 in kg/sec

$\dot{m}_3\ \dot{m}_4$ — Mass flow rates of steam and condenstate Secs. 3–3 and 4–4 respectively in kg/sec.

$h_1\ h_2$ — Specific enthalpies of cooling water at Secs. 1–1 asnd 2–2 respectively in J/kg.

$h_3\ h_4$ — Specific enthalpies of steam and condensate at Secs. 3–3 and 4–4 respectively in J/kg.

Since it is steady flow system, writing SFEE for the coolant.

$$\dot{m}_1\left[\frac{C_1^2}{2} + gZ_1 + h_1\right] + \frac{\delta Q}{dt} = \dot{m}_2\left[\frac{C_2^2}{2} + gZ_2 + h_2\right] + \frac{\delta W}{dt} \quad \ldots(1)$$

And for steam,

$$\dot{m}_3\left[\frac{C_3^2}{2} + gZ_3 + h_3\right] + \frac{\delta Q}{dt} = \dot{m}_4\left[\frac{C_4^2}{2} + gZ_4 + h_4\right] + \frac{\delta W}{dt} \quad \ldots(2)$$

Neglecting the changes in KE and PE, since they are very small and as the condenser is insulated $\delta Q/dt = 0$ and as no work is done in the condenser $\delta W/dt = 0$.

So, Eqs (1) and (2) reduce to,

$$\dot{m}_1 h_1 = \dot{m}_2 h_2 \quad ...(3)$$

$$\dot{m}_3 h_3 = \dot{m}_4 h_4 \quad ...(4)$$

Now since it is a steady flow system,

$$\begin{bmatrix}\text{Total energy entering the} \\ \text{condenser with steam} \\ \text{and coolant}\end{bmatrix} = \begin{bmatrix}\text{Total energy leaving the} \\ \text{condenser with coolant} \\ \text{and condensate}\end{bmatrix}$$

$$\therefore \quad \dot{m}_1 h_1 + \dot{m}_3 h_3 = \dot{m}_2 h_2 + \dot{m}_4 h_4$$

But for a steady flow system

$$\dot{m}_1 = \dot{m}_2 \text{ and } \dot{m}_3 = \dot{m}_4$$

$$\therefore \quad \dot{m}_1 (h_1 - h_2) = \dot{m}_3 (h_4 - h_3)$$

$$\therefore \quad \frac{\dot{\mathbf{m}}_1}{\dot{\mathbf{m}}_3} = \frac{\mathbf{h}_4 - \mathbf{h}_3}{\mathbf{h}_1 - \mathbf{h}_2}. \text{ Hence proved.}$$

2.20 CONTINUITY EQUATION OR LAW OF CONSERVATION OF MASS

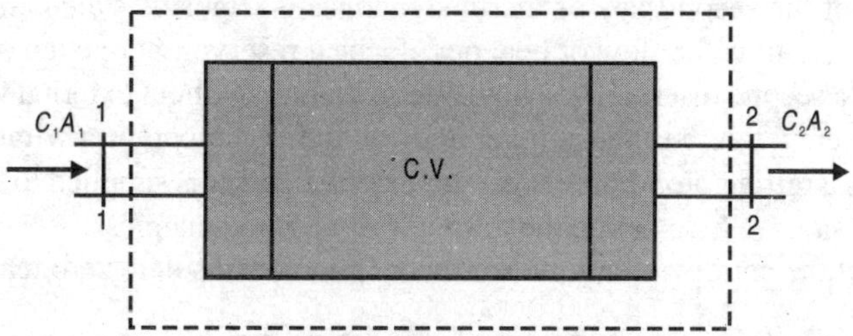

Fig. 2.20

Continuity equation is based on the principle of conservation of mass

Let
$\rho_1 \rho_2$ — Densities of the fluid at sec 1–1 and 2–2 respectively in kg/m²
$A_1 A_2$ — Cross sectional areas at sec 1–1 and 2–2 respectively in m²
$C_1 C_2$ — Velocities of the fluid at sec 1–1 and 2–2 respectively in m/sec

Then the volume of the fluid entering at sec 1–1 per unit time $= A_1 \times C_1$

$(\because A_1$ is in m^2 and C_1 is in m/sec then resultant unit will be m^3/sec$)$

$\therefore$ Mass of the fluid entering at sec 1–1 per unit time,

$$= \text{Density} \times \text{Volume}$$
$$= \rho_1 \times A_1 \times C_1$$

Simlarly mass of fluid leaving at sec 2–2 per unit $= \rho_1 A_2 C_2$

Since it is steady flow system,

Mass flow rate at sec 1–1 = Mass flow rate at sec 2–2

$$\therefore \quad \rho_1 A_1 C_1 = \rho_2 A_2 C_2$$

Since $$\rho = \frac{1}{v}$$

$$\therefore \quad \frac{A_1C_1}{v_1} = \frac{A_2C_2}{v_2}$$

or $$\rho AC = \textbf{constant}$$

$$\frac{AC}{v} = \textbf{constant}$$

These are the continuity equations.

2.21 LIMITATIONS OF FIRST LAW OF THERMODYNAMICS

We have seen that first law of thermodynamics is simply a law of conservation of energy and has its own limitations. These limitations are:

1. The law considers all forms of energies equivalent i.e., the first law of thermodynamics is a law of energy equivalence. This law is necessary condition so far as the account of energy balance is concerned with the possibility of transformation of one kind of energy to another.
2. The first law does not consider the direction of energy transformation.
3. The first law of thermodynamics does not consider the grade of the energy or energy reservoirs. It, assumes that all energy reservoirs are identical.

The second law of thermodynamics relates to the direction of energy exchange processes.

One of the statements of the second law of thermodynamics is Clausius statement and states that no cyclic process is prossible whose result is the flow of heat out of a heat reservoir at one temperature and the flow of an equal quantity of heat into a second reservoir at a higher temperature, without external work. Or, heat can not flow from one body at a lower temperature to the another body at higher temperature without any external work.

This principle is used in a refrigerator. For refrigerator external power is supplied to the compressor.

Without proof or discussion we here define the third law of thermodynamics.

The third law states that the entropy of a pure substance is zero at absolute zero temperature.

SOLVED EXAMPLES

Example 2.1 *During a cycle consisting of 4-processes, the heat transfers are 60 kJ, – 8kJ, – 34 kJ and 6 kJ. Determine the net work for the cycle.*

Solution

We know that from the first law for a closed system undergoing a cycle,

Total heat transfer in a cycle = Total work transfer in a cycle

i.e., $$\oint \delta Q = \oint \delta W$$

As given in the problem, the cycle consists of 4-processes and heat transfers for the processes are given.

$$\therefore \quad \oint \delta Q = 60\,\text{kJ} - 8\,\text{kJ} - 34\,\text{kJ} + 6\,\text{kJ}$$

$$\oint \delta Q = 24\,\text{kJ} = \oint \delta W$$

∴ **Net work transfer in a cycle = 24 kJ**

Example 2.2 *In a non-flow reversible process 300 kJ of heat leaves the system consisting of a gas. The Internal Energy of the gas remains the same. Calculate the work done.*

Data: $Q = -300$ *kJ (Since heat flows out of the system)*

$U =$ *Constant*

∴ $\Delta U = 0$

$W = ?$

Solution

We know that first law for a closed system undergoing a process $Q - W = \Delta U$. As given in problem $Q - W = 0$, since $\Delta U = 0$.

$$Q = W = -300 \text{ kJ}$$

i.e., 300 kJ of work is done on the gas.

Example 2.3 *In a non-flow reversible process, the pressure and volume are related by*

$$P = \left[\frac{3.5}{V} + 3\right] bar$$

where P is in bar and V is in m³. During the process the volume changes from 1.5 m³ to 4.5 m³ and heat added is 2000 kJ. Determine change in Internal Energy (I.E.)?

Data:

$$P = \left[\frac{3.5}{V} + 3\right] \times 10^2 \text{ kPa}$$

$$V_1 = 1.5 \text{ m}^3$$

$$V_2 = 4.5 \text{ m}^3$$

$$Q = 2000 \text{ kJ}$$

$$\Delta U = ?$$

Solution

We know that first law for a closed system undergoing a process,

$$Q - W = \Delta U$$

Q is given, ΔU is to be found

∴ To find W,

Also we know that,

$$\text{Work done} = \int P\,dV \text{ in closed system}$$

Substituting

$$P = \left[\frac{3.5}{V} + 3\right] \times 10^2 \text{ kPa}$$

We get

$$\text{Work done} = \int \left[\frac{3.5}{V} + 3\right] \times 10^2 \; dV$$

$$= 100 \int_{1.5}^{4.5} \left(\frac{3.5}{V} + 3\right) dV \text{ kJ}$$

$$= 100\left[(3.5 \ln V + 3V)\right]_{1.5}^{4.5}$$

$$= 100\left[3.5 \ln \frac{4.5}{1.5} + 3(4.5 - 1.5)\right]$$

$$= 100\left[3.5 \ln 3 + 3 \times (3)\right]$$

$$= 100[3.845 + 9] = 100[12.845]$$

$$\text{Work done} = 1284.5143 \text{ kJ}$$

$$\therefore \quad Q - W = \Delta U$$

$$2000 - 1284.5143 = \Delta U$$

$$\therefore \quad \mathbf{\Delta U = 715.4857}$$

Example 2.4 *In a boiler water enters with an enthalpy of 35 kJ/kg and steam leaves with the enthalpy of 705 kJ/kg. Find the heat transfer per kg of steam. The change in KE and PE may be neglected.*

Solution

We know that, SFEE on mass basis,

$$\left[\frac{C_1^2}{2} + gZ_1 + h_1\right] + q = \left[\frac{C_2^2}{2} + gZ_2 + h_2\right] + w$$

$$(0 + 0 + h_1) + q = (0 + 0 + h_2) + w$$

$$h_1 + q = h_2 + w$$

But in a boiler since no work is done $w = 0$

$$\therefore \quad q = h_2 - h_1$$

$$q = 705 - 35$$

$$\therefore \quad \mathbf{q = 670\ kJ/kg}$$

Example 2.5 *A piston and cylinder machine contains a fluid system which passes through a complete cycle of 4-processes. During a cycle the sum of all heat transfers is – 170 kJ. The system completes 100 cycles/min. Complete the following table showing the method for each item and calculate the net rate of work output in kW.*

Process	*Q (kJ/min)*	*W (kJ/min)*	*4E (kJ/min)*
a–b	*0*	*2170*	—
b–c	*21,000*	*0*	—
c–d	*– 2100*	—	*– 36600*
d–a	—	—	—

Solution

Process a–b

$$Q = \Delta E + W$$

$$0 = \Delta E + 2170$$

$$\mathbf{\Delta E = -2170\ kJ/min}$$

Process b–c

$$Q = \Delta E + W$$

$$21000 = \Delta E + 0$$

$$\therefore \quad \mathbf{\Delta E = 21000\ kJ/min}$$

Process c–d

$$Q = \Delta E + W$$

$$-2100 = -36600 + W$$

$$\mathbf{W = 34500\ kJ/min}$$

Process d–a

It is given that, sum of all heat transfers/cycle

i.e.,
$$\oint Q = -170\text{ kJ}$$

Secondly, the system completes 100 cycles/min

$$\therefore \quad \text{Total heat transfer/min} = -17000\text{ kJ/min}$$

$$Q_{a-b} + Q_{b-c} + Q_{c-d} + Q_{d-a} = -17000\text{ kJ/min}$$

$$0 + 21000 - 21000 + Q_{d-a} = -1700\text{ kJ/min}$$

$$\therefore \quad \mathbf{Q_{d-a} = -35900\ kJ/min}$$

While studying energy a property of the system, we have studied that change in energy for the process $1-a-2 = \Delta E_a$

And change in energy for the process,

$$2-b-1 = \Delta E_b \ (-\text{ve})$$

$\therefore$ Net change in energy for the cycle

$$= \Delta E_a - \Delta E_b = 0$$

$\therefore$ We can write cyclic integral of any property is zero or $\oint dE = 0$

So, $\Delta E_{a-b} + \Delta E_{b-c} + \Delta E_{c-d} + \Delta E_{d-a} = 0$

$$\therefore \quad -2170 + 21000 - 36600 + \Delta E_{d-a} = 0$$

$$\therefore \quad \mathbf{\Delta E_{d-a} = 17770 \ kJ/min}$$

We know that, $Q_{d-a} = \Delta E_{d-a} + W_{d-a}$

$$W_{d-a} = Q_{d-a} - \Delta E_{d-a} = -35900 - 17770$$

$$\mathbf{W_{d-a} = -53670 \ kJ/min}$$

Lastly rate of work output since

$$\oint \delta Q = \oint \delta W$$

$$= -17000 \text{ kJ/min}$$

$$\oint \delta \mathbf{W} = \textbf{Net rate of work} = \mathbf{-283.33 \ kW \ or \ kJ/sec}$$

Example 2.6 *A certain cycle consists of 3 processes. The energy transfer in each process are tabulated below.*

Process	*Q (kJ/kg)*	*w (kJ/kg)*	*Δu (kJ/kg)*
1—2	*50*	—	*20*
2—3	*– 30*	*– 40*	—
3—1	—	—	*– 30*

If the network done l kg of fluid is + 30 kJ/kg complete the table. Also find the power developed, if the mass of fluid in the cycle is 0.1 kg and the system completes 10 such cycles per sec.

Solution

Since all the equalities are given for unit mass,

For process 1–2

$$q = \Delta u + w$$

$$50 = 20 + w$$

$$\therefore \quad \mathbf{w = 30 \ kJ/kg}$$

For process 2–3

$$q = \Delta u + w$$

$$-30 = \Delta u - 40$$

$$\mathbf{\Delta u = 10 \ kJ/kg}$$

For process 3–1

It is given that, net w.d./kg $= -30$ kJ/kg

$$\therefore \quad W_{1-2} + w_{2-3} + w_{3-1} = 30 \text{ kJ/kg}$$

$$30 - 40 + w_{3-1} = 30$$

$$30 - 40 + w_{3-1} = 0$$

$$\therefore \quad \mathbf{W_{3-1} = 40 \ kJ/kg}$$

$$\therefore \quad q = \Delta u + w$$

$$q = -30 + 40$$

$$\mathbf{q = 10\ kJ/kg}$$

It is given that w.d/kg = 30 kJ/kg and mass of the fluid flowing = 0.1 kg/cycle and system completes 10 cycle per sec.

$\therefore$ $\dot{m}$ = 0.1 kg/cycle × 10 cycle/sec

$\therefore$ $\dot{m}$ = 1 kg/sec

$\therefore$ We know that w.d./kg = 30 kJ/kg

$\therefore$ Rate of w.d. or power = 30 kJ/kg × 1 kg/sec = 30 kJ/sec = 30 kW

Remember: $$\frac{\delta w}{dt} = \dot{m}\frac{\delta w}{dm}$$

Unit: $$\frac{\text{kJ}}{\text{sec}} = \frac{\text{kg}}{\text{sec}} \times \frac{\text{kJ}}{\text{kg}}$$

$$\frac{\text{kJ}}{\text{sec}} = \frac{\text{kJ}}{\text{sec}}$$

Example 2.7 *The I.E. of a certain substance is given by the following equation,*

$$u = 3.56Pv + 84$$

where u is in kJ/kg, P is in kPa, v is in m³/kg.

A system composed of 3 kg of this substance expands from an initial pressure of 500 kPa and a volume of 0.22 m³ to a final pressure of 100 kPa in a process in which pressure and volume are related by $PV^{1.2} = C$.

(i) If the expansion is quasi-static, find Q, ΔU and W for the process.

(ii) In another process the same system expands according to the same pressure-volume relation as in case (i) and from the same initial state and same final state as in case (i), but the heat transfer for this case is 30 kJ. Find the work transfer for this process.

(iii) Explain the difference in work transfer in case (i) and (ii).

Data : u is in kJ/kg

P is in kPa

$$v = m^3/kg$$
$$m = 3 kg$$
$$P_1 = 500\ kPa$$
$$V_1 = 0.22 m^3$$
$$P_2 = 100\ kPa$$
$$PV^{1.2} = C$$
$$Q = ?\ \Delta U = ?\ W = ?$$

Solution

It is given that $u = 3.56\,Pv + 84$

Change in IE for unit mass

$$\Delta u = u_2 - u_1$$
$$= 3.56\,P_2V_2 + 84 - 3.56\,P_1V_1 - 84$$
$$\Delta u = 3.56\,(P_2V_2 - P_1V_1)$$

$\therefore$ Total change in *IE* $= \Delta U = 3.56(P_2V_1 - P_1V_1)$

Note: *V* is the total volume for 3 kg mass amd it is an extensive property.

$\therefore$ It depends on mass and *P* is an intensive property.

Since in Eq. (1) V_2 is not given,

$\therefore$ $$P_1 V_1^{1.2} = P_2 V_2^{1.2}$$

$$\therefore \quad V_2 = V_1\left(\frac{P_1}{P_2}\right)^{\frac{1}{1.2}} = 0.22\left(\frac{500}{100}\right)^{\frac{1}{1.2}}$$

$$V_2 = 0.22 \times 3.83 = 0.845\ \text{m}^3$$

Now from (1)

$$\Delta U = 3.56[100 \times 0.845 - 500 \times 0.22]\ \text{kJ}$$

$$= 356\,[1 \times 0.845 - 5 \times 0.22]\ \text{kJ}$$

$$\mathbf{\Delta U = -91\ kJ}$$

For a quasi-static process, W.D. $= \int PdV$

Since it is a polytropic process,

$$W_{1-2} = \frac{P_2V_2 - P_1V_1}{1-n}$$

$$W_{1-2} = \frac{100 \times 0.845 - 500 \times 0.222}{1-1.2}$$

$$\mathbf{W_{1-2} = 127.5\ kJ}$$

$$\therefore \quad Q = \Delta U + W$$

$$Q = -91 + 127.5$$

$$\mathbf{Q = 36.5\ kJ}$$

Case (ii) Here $Q = 30$ kJ

Since the end states are same as in case (i)

$$W = Q - \Delta U = 30 - (-91)$$

$$\mathbf{W = 121\ kJ}$$

(iii) The work in case (ii) is not equal to $\int PdV$. So the process is not quasi-static.

Example 2.8 *The I.E. of a certain substance is given by the equation U = 3.5 PV + 80. Where U is in kJ, pressure P is in kPa, and volume V is in m³. A system composed of 5 kg of this substance expands from an initial pressure of 5 bar and volume of 0.22 m³ to a final pressure of 1 bar in a process in which pressure and volume are related by $PV^{1.2} = C$.*

Find for quasi-static process,

(1) Heat transfer, (2) Total change in I.E. (3) Non-flow work

Data: U in kJ ; P in kPa

V in m³;

$$m = 5\ kg$$

$$P_1 = 5\ bar = 500\ kPa$$

$$V_1 = 0.22\ m^3$$

$$P_2 = 1\ bar = 100\ kPa$$

$$P_1V_1^{1.2} = P_2V_2^{1.2} = C$$

$$\therefore \quad V_2 = V_1 \times \left(\frac{P_1}{P_2}\right)^{\frac{1}{1.2}}$$

$$V_2 = 0.22 \times (5)^{1/1.2} = 0.84\ \text{m}^3$$

Solution

Now ΔU i.e., change in I.E. will be,

$$\Delta U = U_2 - U_1 = 3.5(P_2V_2 - P_1V_1)$$
$$= 3.5[100\times0.84 - 500\times0.22]$$
$$\mathbf{\Delta U = -91\ kJ}$$
$$W_{1-2} = \frac{P_1V_1 - P_2V_2}{n-1}$$
$$= \frac{500\times0.22 - 100\times0.84}{0.2}$$
$$\mathbf{W_{1-2} = 130\ kJ}$$
$$Q = \Delta U + W = -91 + 130$$
$$\mathbf{Q = 39\ kJ}$$

Example 2.9 *A fluid is confined in a cylinder by a spring loaded frictionless piston, so that the pressure in the fluid is a linear function of the volume, P = a + bV. The internal energy of the fluid is given by the following equation U = 34 + 3.15 PV.*

Where U is in kJ, P in kPa, V in m³. If the fluid changes from an initial state of 170 kPa, 0.03 m³ to a final state of 400 kPa, 0.06 m³ with no work other than that done on the piston. Find the direction and magnitude of the work and heat transfer.

Data: $P = a + bV$; $U = 34 + 3.15\,PV$

$P_1 = 170$ kPa ; $V_1 = 0.03\ \text{m}^3$

$P_2 = 400$ kPa ; $V_2 = 0.06\ \text{m}^3$

$W = ?$ $Q = ?$

Solution

The change in I.E. of the fluid during the process is given as,

$$U_2 - U_1 = 3.15\,(P_2V_2 - P_1V_1)$$

[**Note :** Units $\frac{\text{kN}}{\text{m}^2}\times\text{m}^3 = \text{kNm} = \text{kJ}$]

$$U_2 - U_1 = 3.15\,[400\times0.06 - 1.70\times0.03]$$
$$= 100\times3.15\,[4\times0.06 - 1.70\times0.03]$$
$$U_2 - U_1 = 315\times0.189$$
$$\mathbf{U_2 - U_1 = 59.5\ kJ}$$

Now
$$P = a + bV$$
$$170 = a + b\times0.03$$
$$400 = a + b\times0.06$$

From these two equations on solving we get,

$$a = -60\ \text{kN/m}^2$$
$$b = 7667\ \text{kN/m}^5$$

We know that, W.D. during the process $W_{1-2} = \int PdV$

$$= \int_{V_1}^{V_2}(a + bV)\,dV$$

$$= a(V_2 - V_1) + b\frac{(V_2^2 - V_1^2)}{2}$$

$$= (V_2 - V_1)\left[a + \frac{b}{2}(V_1 + V_2)\right]$$

$$= 0.03\ \text{m}^3 \left[-60\frac{\text{kN}}{\text{m}^2} + \frac{7667}{2}\frac{\text{kN}}{\text{m}^5} \times 0.09\ \text{m}^3\right]$$

$$\mathbf{W_{1-2} = 8.55\ kJ}$$

Since W_{1-2} is positive, work is done by the system,

∴ Heat transfer involved is given by

$$Q = \Delta U + W$$

$$Q_{1-2} = U_2 - U_1 + W_{1-2} = 59.5 + 8.55$$

$$Q_{1-2} = 68.05\ \text{kJ}$$

Since Q_{1-2} is positive, heat flows into the system during the process.

Example 2.10 *The expression for the I.E. of a certain closed system is given by U = M + N (PV) where P is the pressure, V is the volume and M and N are constants. The system now undergoes a reversible process where Q = 0. Show that the relation between P and V is given by, $PV^K = C$ where $K = \frac{N+1}{N}$.*

Solution

As given in the problem,

$$U = M + N(P.V) \text{ differentiating we get,} \quad ...(1)$$

$$dU = O + N.P.dV + N.V.dP$$

We know that from first law for the process,

$$Q - W = \Delta U$$

$$Q = \Delta U + W$$

or

$$\delta Q = dU + \delta W$$

$$\delta Q = dU + P.dV$$

Since

$$Q = 0, 0 = dU + P.dV$$

Given,

$$dU = -P.dV \quad ...(2)$$

Equating (1) and (2)

$$-P.dV = N.P.dV + N.V\,dP$$

$$O = N.P.dV + N.V\,dP + P.dV$$

$$= (N+1)P.dV + N.V.dP$$

Dividing throughout by N, we get

$$O = \frac{N+1}{N}P.dV + V \cdot dP$$

i.e.

$$-V.dP = \frac{N+1}{N}P.dV$$

$$-VdP = K.P.dV \text{ as given } \frac{N+1}{N} = K$$

$$-\frac{dP}{P} = K.\frac{dV}{V}$$

Integrating,

$$\ln\frac{1}{P} = \ln V^K$$

or

$$\mathbf{PV^K = C}$$

where $$K = \frac{N+1}{N}$$

Example 2.11 *Following data refers to a steady flow process.*

	At entrance	At exit
Enthalpy (kJ/kg)	4000	4100
Velocity (m/sec)	50	20
Height (m)	50	10

Mass flow rate $\dot{m} = 1$ *kg/sec*

Heat transfer rate = 200 kJ/sec to the system.

Determine the power capacity of the system and state whether it is a power producing system or otherwise.

Solution

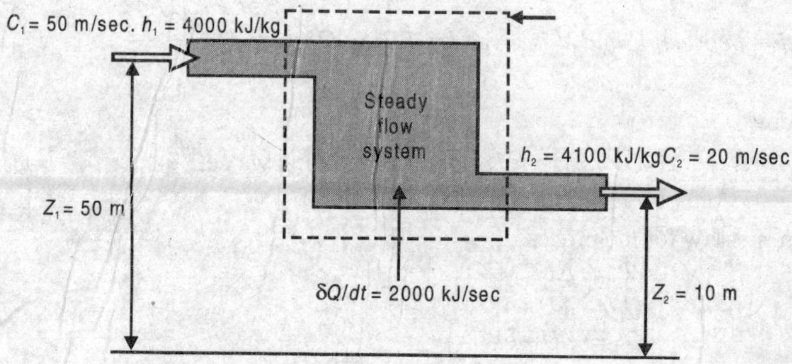

Fig. Ex. 2.11

$$\dot{m} = 1 \text{ kg/sec}$$

Since heat transfer is given in kJ/sec, we will have to use SFEE on time basis.

$$\dot{m}\left[\frac{C_1^2}{2} + gZ_1 + h_1\right] + \frac{\delta Q}{dt} = \dot{m}\left[\frac{C_2^2}{2} + gZ_2 + h_2\right] + \frac{\delta W}{dt}$$

Since power is to found,

$$\frac{\delta W}{dt} = \dot{m}\left[\frac{C_1^2 - C_2^2}{2} + g(Z_1 - Z_2) + (h_1 - h_2)\right] + \frac{\delta Q}{dt}$$

$$\frac{\delta W}{dt} = \frac{1\,\text{kg}}{\text{sec}}\left[\frac{50^2 - 20^2}{2\times1000}\,\frac{\text{kJ}}{\text{kg}} + \frac{9.81\,(40)}{1000}\,\frac{\text{kJ}}{\text{kg}} + (4000 - 4100)\right] + 200\,\frac{\text{kJ}}{\text{sec}}$$

$$\frac{\delta W}{dt} = \frac{1\,\text{kg}}{\text{sec}}\left[1.05\frac{\text{kJ}}{\text{kg}} + 0.3924\,\frac{\text{kJ}}{\text{kg}} + -100\,\frac{\text{kJ}}{\text{kg}}\right] + 200\,\frac{\text{kJ}}{\text{sec}}$$

$$\frac{\delta W}{dt} = 1.05\,\frac{\text{kJ}}{\text{sec}} + 0.3924\,\frac{\text{kJ}}{\text{sec}} - 100\,\frac{\text{kJ}}{\text{sec}} + 200\,\frac{\text{kJ}}{\text{sec}}$$

$$\boldsymbol{\frac{\delta W}{dt} = 101.4484 \text{ kJ/sec or kW}}$$

Since $\frac{\delta W}{dt}$ is + ve, it is a power producing system.

Example 2.12 *In a steady flow system 135 kJ/kg of work is done by the system. The specific volume of the fluid, pressure and velocity at the inlet are 0.37 m³/kg, 600 kPa, 16 m/sec. The inlet is 32 m above the floor level. The discharge conditions are 0.62 m³/kg, 100 kPa and 270 m/sec. The total heat loss between inlet and discharge is 9 kJ/kg. In flowing through this system, does the specific I.E. increase or decrease and by how much? The outlet is at flow level.*

Solution

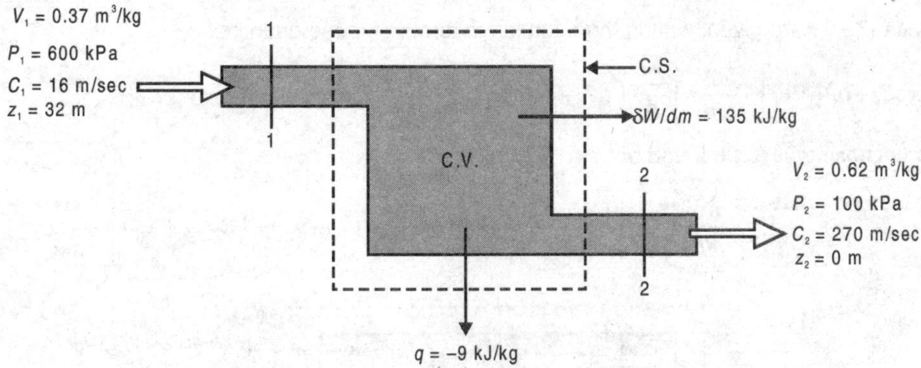

Fig. Ex. 2.12

Since work and heat transfer are given in $\frac{\text{kJ}}{\text{kg}}$

Using SFEE on mass basis,

$$\left[\frac{C_1^2}{2}+gZ_1+h_1\right]+q = \left[\frac{C_2^2}{2}+gZ_2+h_2\right]+w$$

Since pressure, specific volumes are given and change in Sp. I.E., is to be found, substitute $h_1 = u_1 + P_1v_1$ and $h_2 - u_2 + P_2v_2$ wc gct,

$$\left[\frac{C_1^2}{2}+gZ_1+u_1+P_1v_1\right]+q = \left[\frac{C_2^2}{2}+gZ_2+u_2+P_2v_2\right]+w$$

$$\therefore \quad u_1-u_2 = \frac{C_2^2-C_1^2}{2}+g(Z_2-Z_1)+(P_2v_2-P_1v_1)+w-q$$

$$u_1-u_2 = \frac{270^2-16^2}{2\times1000}\frac{\text{kJ}}{\text{kg}}+\frac{9.81(-32)}{1000}\frac{\text{kJ}}{\text{kg}}$$

$$+(100\times.62-600\times0.37)\frac{\text{kJ}}{\text{kg}}+135\frac{\text{kJ}}{\text{kg}}-\left(-9\frac{\text{kJ}}{\text{kg}}\right)$$

$$u_1-u_2 = 36.45-0.314-160+135+9$$

$$u_1-u_2 = 36.45-20.136\,\text{kJ/kg}$$

$$u_1-u_2 = 20.136\,\text{kJ/kg}$$

$$\therefore \quad \mathbf{u_2-u_1 = -20.136\ kJ/kg}$$

∴ Specific I.E. decreases by 20.236 kJ/kg

Example 2.13 *In a steady flow system, a fluid flows at the rate of 5 kg/sec. It enters at a pressure of 620 kPa, velocity of 300 m/sec. I. E., 2100 kJ/kg and specific volume 0.37 m³/kg.*

It leaves the system at a pressure of 130 kPa, a velocity of 150 m/sec; I.E. 1500 kJ/kg and specific volume 1.2 m³/kg. During its flow through system there is a heat loss of 30 kJ/kg. Determine the power capacity of the system in kW. State whether it is from or to the system. Neglect change in P.E.

Solution

Since heat transfer is given in $\frac{\text{kJ}}{\text{kg}}$, using SFEE on mass basis.

Note : Since P, u, v are given, writing the equation in terms of these quantities.

$$\left[\frac{C_1^2}{2}+gZ_1+u_1+P_1v_1\right]+\frac{\delta Q}{dm}=\left[\frac{C_2^2}{2}+gZ_2+u_2+P_2v_2\right]+\frac{\delta W}{dm}$$

Since power capacity is to be found out,

$$\frac{\delta W}{dm} = \left[\frac{C_1^2-C_2^2}{2}+g(Z_1-Z_2)+(u_1-u_2)+(P_1v_1-P_2v_2)\right]+\frac{\delta Q}{dm}$$

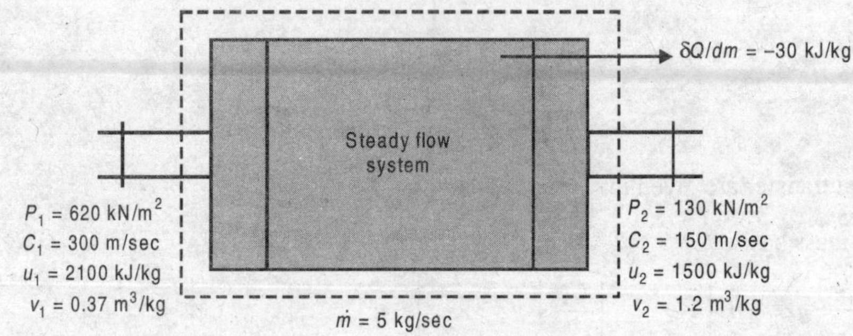

Fig. Ex. 2.13

Neglecting ΔPE as given in the problem,

$$\frac{\delta W}{dm} = \left[\frac{C_1^2-C_2^2}{2}+(u_1-u_2)+(P_1v_1-P_2v_2)\right]+\frac{\delta Q}{dm}$$

$$\frac{\delta W}{dm} = \frac{300^2-150^2}{2\times1000}\frac{\text{kJ}}{\text{kg}}+(2100-1500)+(620\times0.37-130\times1.2)-30\,\frac{\text{kJ}}{\text{kg}}$$

$$\frac{\delta W}{dm} = \left[\frac{67500}{2000}+600+73\right]\frac{\text{kJ}}{\text{kg}}-30\,\frac{\text{kJ}}{\text{kg}}$$

$$= 33.75+600+73-30$$

$$\mathbf{\frac{\delta W}{dm} = 676.75\ kJ/kg}$$

Since power is required to be expressed in kW,

Using $$\frac{\delta W}{dt} = \dot{m}\frac{\delta W}{dm}$$

Note: Mass flow rate = 5 kg/sec is given

$$\therefore \quad \text{Power} = 676.75\frac{\text{kJ}}{\text{kg}}\times 5\frac{\text{kg}}{\text{sec}}$$

$$\textbf{Power} = \mathbf{3383.75\ kW} = \frac{\delta \mathbf{W}}{\mathbf{dt}}$$

Example 2.14 *The following data is obtained from tests on an air compressor.*

	At Inlet	*At Exit*
Pressure (kPa)	*100*	*500*
Sp. volume (m^3/kg)	*0.6*	*0.15*
Sp. Internal Energy (kJ/kg)	*50*	*125*
Velocity of air (m/sec)	*8*	*4*

Rate of flow of air is 5 kg/sec.
Heat rejected to cooling water = 45 kW
Find the power required to drive the compressor in kW.

Solution

Since heat is rejected is given in kJ/sec, using SFEE on time basis,

$$\dot{m}\left[\frac{C_1^2}{2}+gZ_1+h_1\right]+\frac{\delta Q}{dt} = \dot{m}\left[\frac{C_2^2}{2}+gZ_2+h_2\right]+\frac{\delta W}{dt}$$

$$\therefore \quad \frac{\delta Q}{dt}-\frac{\delta W}{dt} = \dot{m}\left[\frac{C_2^2-C_1^2}{2}+g(Z_2-Z_1)+(P_2v_2-P_1v_1)+(u_2-u_1)\right]$$

$$-45-\frac{\delta W}{dt} = 5\left[\frac{4^2-8^2}{2\times 1000}+0+(500\times 0.15-100\times 0.6)+(125-50)\right]$$

$$-45-\frac{\delta W}{dt} = 5\left[\frac{16-64}{2000}+(75-60)+75\right]$$

$$-45-\frac{\delta W}{dt} = 5[-0.024+15+75]$$

$$\Rightarrow \quad \frac{\delta \mathbf{W}}{\mathbf{dm}} = \mathbf{-494.88\ kW}$$

∴ Power required to drive the compressor = 494.88 kW.

Example 2.15 *Steam enters a steam turbine with a velocity of 40 m/sec, enthalpy 2500 kJ/kg and leaves with a velocity of 90 m/sec and enthalpy 2030 kJ/kg. Heat losses from the turbine insulation to the surroundings are 250 kJ/min, and the steam flow rate is 5000 kg/hr. Neglect the change of P.E. Find the power developed by the turbine in kW.*

Solution

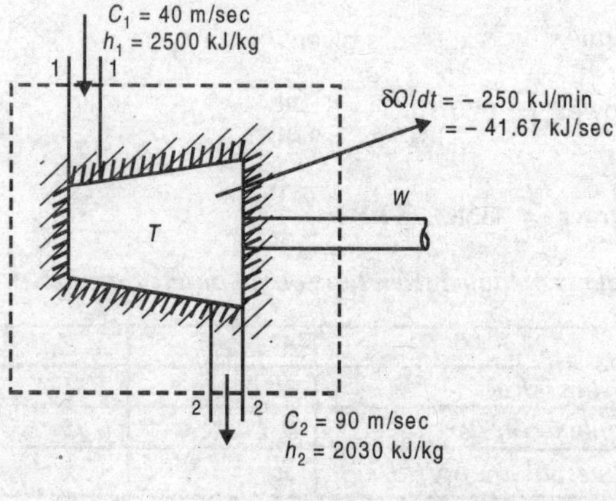

Fig. Ex. 2.15

$$\text{Steam flow rate} = 5000 \text{ kg/hr} = \frac{5000}{3600} \text{ kg/sec} = 1.389 \text{ kg/sec}$$

$$\dot{m}\left[\frac{C_1^2}{2} + gZ_1 + h_1\right] + \frac{\delta Q}{dt} = \dot{m}\left[\frac{C_2^2}{2} + gZ_2 + h_2\right] + \frac{\delta W}{dt}$$

$$\dot{m}\left[\frac{C_1^2 - C_2^2}{2} + (h_1 - h_2)\right] + \frac{\delta Q}{dt} = \frac{\delta W}{dt}$$

$$1.389 \frac{\text{kg}}{\text{sec}}\left[\frac{40^2 - 90^2}{2\times1000} \frac{\text{kJ}}{\text{kg}} + (2500 - 2030)\frac{\text{kJ}}{\text{kg}}\right] - 4.167 \frac{\text{kJ}}{\text{sec}} = \frac{\delta W}{dt}$$

$$1.389\left[-3.25 \frac{\text{kJ}}{\text{kg}} + (470)\frac{\text{kJ}}{\text{kg}}\right] - 4.167 \frac{\text{kJ}}{\text{sec}} = \frac{\delta W}{dt}$$

$$1.389 \frac{\text{kg}}{\text{sec}}\left[466.75 \frac{\text{kJ}}{\text{kg}}\right] - 4.167 \frac{\text{kJ}}{\text{sec}} = \frac{\delta W}{dt}$$

$$\frac{\delta \mathbf{W}}{\mathbf{dt}} = \mathbf{644.149 \text{ kW}}$$

Example 2.16 *Steam enters a steam turbine with a velocity of 16 m/sec and specific enthalpy 2990 kJ/kg. The steam leaves the turbine with a velocity of 37 m/sec and specific enthalpy of 2530 kJ/kg. The heat lost to the surroudings is 25 kJ/kg. The steam flow rate is 3,60,000 kg/hr. Determine the work output from the turbine in kW.*

Solution

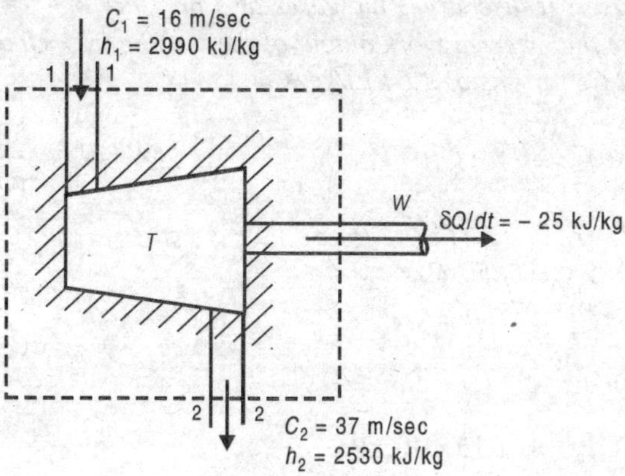

Fig. Ex. 2.16

Since heat lost is given in kJ/kg, using SFEE on mass basis,

$$\left[\frac{C_1^2}{2}+gZ_1+h_1\right]+\frac{\delta Q}{dm} = \left[\frac{C_2^2}{2}+gZ_2+h_2\right]+\frac{\delta W}{dm}$$

Since nothing is mentioned about elevation, neglecting ΔPE.

$$\therefore \quad \frac{C_1^2}{2}+h_1+\frac{\delta Q}{dm} = \frac{C_2^2}{2}+h_2+\frac{\delta W}{dm}$$

$$\therefore \quad \frac{\delta W}{dm} = \frac{\left(C_1^2-C_2^2\right)}{2}+\left(h_1-h_2\right)+\frac{\delta Q}{dm}$$

$$\frac{\delta W}{dm} = \frac{16^2-37^2}{2\times1000}\ \frac{\text{kJ}}{\text{kg}}+\left(2990-2530\right)\ \frac{\text{kJ}}{\text{kg}}-25\ \frac{\text{kJ}}{\text{kg}}$$

$$\frac{\delta W}{dm} = \frac{-1113}{2000}+460-25$$

$$\mathbf{\frac{\delta W}{dm} = 434.443\ \frac{kJ}{kg}}$$

To express power in kW, using

$$\frac{\delta W}{dt} = \dot{m}\frac{\delta W}{dm}$$

It is given that mass flow rate = 360000 kg/hr = 100 kg/sec.

$$\therefore \quad \text{Power} = 434.443\ \frac{\text{kJ}}{\text{kg}}\times100\ \frac{\text{kg}}{\text{sec}}$$

$$\mathbf{Power = 434443\ kW = \frac{\delta W}{dt}}$$

Example 2.17 *The gas turbine of a turbojet engine receives a steady flow of gases at pressure of 7.2 bar temperature of 850°C and velocity of 160 m/sec. It discharges the gases at a pressure of 1.15 bar, temperature 450°C and a velocity of 250 m/sec. Determine the external work output of the turbine in 5 kJ/kg of gas flow. The process may be assumed to be adiabatic and C_p for gas as 1.04 kJ/kg K.*

Data: $P_1 = 7.2\times10^2$ kPa $\qquad W = ?$ i.e. $\frac{\delta W}{dm} = ?$

$$Cp_{gases} = 1.04 \text{ kJ/kg-K}$$

$$T_1 = 850°C$$

$$C_1 = 160 \text{ m/sec}$$

Solution

Insulation (Adiabtic Turbine) $\therefore Q = 0$

$$P_2 = 1.15\times10^2 \text{ kPa}$$

$$T_2 = 450° \text{ C}$$

$$C_2 = 250 \text{ m/sec}$$

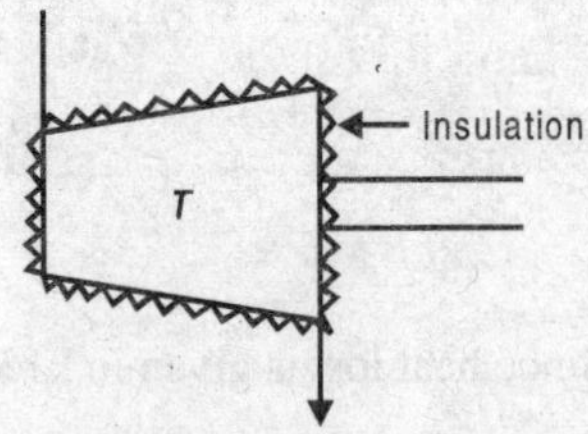

Fig. Ex. 2.17

We know that steady Flow Energy equation on mass basis is given by,

$$\left[\frac{C_1^2}{2} + gZ_1 + h_1\right] + \frac{\delta Q}{dm} = \left[\frac{C_2^2}{2} + gZ_2 + h_2\right] + \frac{\delta W}{dm} \qquad ...(1)$$

Since Adiabtic Expansion process $\frac{\delta Q}{dm} = 0$

Since they have not mentioned anything about elevation $gZ_1 = 0, gZ_2 = 0$. Then Eq. (1) becomes,

$$\left[\frac{C_1^2}{2} + h_1\right] = \left[\frac{C_2^2}{2} + h_2\right] + \frac{\delta W}{dm}$$

i.e.,

$$\frac{\left(C_1^2 - C_2^2\right)}{2} + (h_1 - h_2) = \frac{\delta W}{dm} \qquad ...(2)$$

Since $\qquad h = C_P T$

$$\therefore \quad \frac{\delta W}{dm} = (h_1 - h_2) + \frac{\left(C_1^2 - C_2^2\right)}{2}$$

$$= C_P(850 - 450) + \frac{160^2 - 250^2}{2\times1000}$$

$$= 1.04(850 - 450) + \frac{160^2 - 250^2}{2\times1000} \quad 416 + (-18.45)$$

$$\mathbf{\frac{\delta W}{dm} = 397.55 \text{ kJ/kg}}$$

Example 2.18 *The power output of an adiabatic steam turbine is 5 MW, and the inlet and exit conditions are as under,*

	Inlet	A Exit
Pressure	2 MPa	0.15 bar
Temperature (°C)	400	Dryness 0.9
Velocity (m/sec)	50	180
Elevation m	10	6

Determine the work done per unit of mass of the steam flowing through turbine and mass flow rate of the steam.

Solution

We know that, since work done per unit of mass of steam is to be found, using SFEE on mass basis.

$$\left[\frac{C_1^2}{2} + gZ_1 + h_1\right] + q = \left[\frac{C_2^2}{2} + gZ_2 + h_2\right] + w$$

or
$$w = (h_1 - h_2) + \frac{\left(C_1^2 - C_2^2\right)}{2} + g(Z_1 - Z_2)$$

Since
$$h_1 = 3248.7 \text{ kJ/kg}$$

[**Note:** From steam tables coresponding to and 20 bar (2 MPa)].

$$h_2 = h_f + x \cdot h_{fg}. \text{ Since the steam is wet at 0.15 bar from steam tables}$$

and
$$h_2 = 226 + 0.9 \times 2373.2 = 2361.88 \text{ kJ/kg}$$

$$w = (3248.7 - 2361.88) + \frac{50^2 - 180^2}{2000} + \frac{4 \times 9.81}{1000}$$

$$w = 871.91 \text{ kJ/kg} = \frac{\delta W}{dm}$$

Also, we know that,
$$\frac{\delta W}{dt} = \dot{m}\frac{\delta W}{dm}$$

$$5000\frac{\text{kJ}}{\text{sec}} = \dot{m} \times 871.91 \text{ kJ/kg}$$

$$\therefore \quad \dot{\mathbf{m}} = \mathbf{5.734 \text{ kg/sec}}$$

Example 2.19 *A turbine operating on air has inlet conditions as10 bar, 750 K and 200 m/s. While exit conditions are 1.25 bar and 40 m/sec. The mass flow rate of air is 1000 kg/hr. The flow of air is assumed to be reversible adiabatic (isentropic). Calculate,*

(i) The temperature of air at exit.

(ii) The power output of turbine.

Assume C_p = 1.053 kJ/kg K and k = 1.375 adiabatic index

Solution

$$\dot{m} = 1000 \text{ kg/hr} = 0.2777 \text{ kg/sec}$$

To find T_2, since the expansion process is adiabatic,

We know that, $$\frac{T_2}{T_1} = \left(\frac{P_2}{P_1}\right)^{\frac{k-1}{k}}$$

$$\Rightarrow \quad T_2 = T_1 \times \left(\frac{P_2}{P_1}\right)^{\frac{k-1}{k}} = 750 \times \left(\frac{1.25\times100}{10\times100}\right)^{\frac{1.375-1}{1.375}}$$

$$\mathbf{T_2 = 425.367\ K}$$

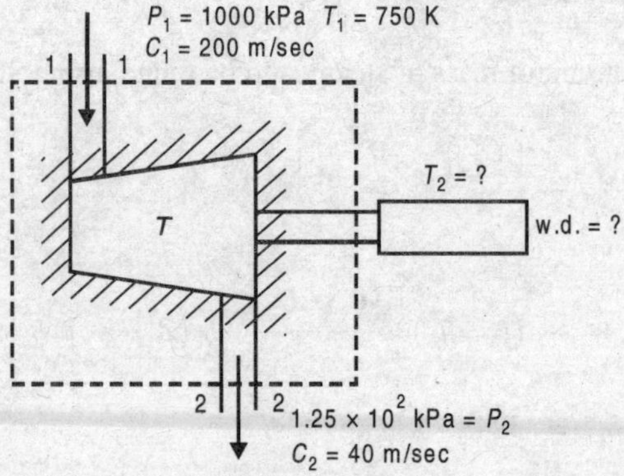

Fig. Ex. 2.19

We also know that, the SFEE is,

$$\dot{m}\left[\frac{C_1^2}{2} + gZ_1 + h_1\right] + \frac{\delta Q}{dt} = \dot{m}\left[\frac{C_2^2}{2} + gZ_2 + h_2\right] + \frac{\delta W}{dt}$$

Neglecting the ΔPE, and $\frac{\delta Q}{dt} = 0$ since the turbine is adiabatic.

$$\dot{m}\left[\frac{C_1^2}{2} + h_1\right] = \dot{m}\left[\frac{C_2^2}{2} + h_2\right] + \frac{\delta W}{dt}$$

$$-\frac{\delta W}{dt} = \dot{m}\left[\frac{\left(C_1^2 - C_2^2\right)}{2} + h_2 - h_1\right] \quad ...(1)$$

$$h_2 - h_1 = \Delta h = m \times C_p \,.\, \Delta T$$

$$= 1 \times 1.053 \times (425.367 - 750) \quad [\because m = 1 \text{ for unit mass}]$$

$$(h_2 - h_1) = -341.83855 \text{ kJ} \quad ...(2)$$

Substituting (2) in (1) we get,

$$-\frac{\delta W}{dt} = 0.277\ \frac{\text{kg}}{\text{sec}}\left[\frac{40^2 - 200^2}{2\times1000} + (-341.838)\right]$$

$$-\frac{\delta W}{dt} = 0.277[-19.2-341.838]=0.277[-361.038]$$

$$= -100.007 \text{ kW}$$

$$\frac{\delta W}{dt} = +100.007 \text{ kW.}$$

Example 2.20 *Air flows steadily at the rate of 0.5 kg/sec, through the air compressor, entering at 7 m/sec velocity, 100 kPa pressure, 0.95 m³/kg volume and leaving at 5 m/sec, 700 kPa, 0.19 m³/kg. The I.E. of the air leaving is 90 kJ/kg greater than that of air entering.*

Heat rejected to the surroundings at the rate of 58 kW.

(a) Calculate the rate of work input to the air in kW.

(b) Find the ratio of inlet pipe diameter to outlet pipe diameter.

Solution

Heat lost to the surroundings is given in kJ/sec. So using SFEE on time basis.

$$\dot{m}\left[\frac{C_1^2}{2}+gZ_1+h_1\right]+\frac{\delta Q}{dt} = \dot{m}\left[\frac{C_2^2}{2}+gZ_2+h_2\right]+\frac{\delta W}{dt}$$

Since, P, u, and v are given.

$$\dot{m}\left[\frac{C_1^2}{2}+gZ_1+u_1+P_1v_1\right]+\frac{\delta Q}{dt} = \dot{m}\left[\frac{C_2^2}{2}+gZ_2+u_2+P_2v_2\right]+\frac{\delta W}{dt}$$

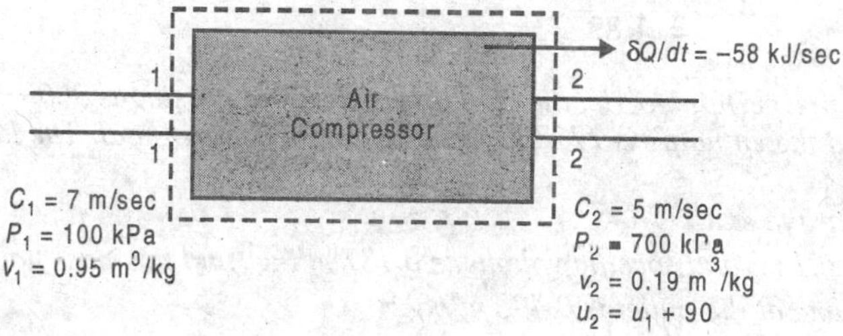

Fig. Ex. 2.20

Since power is to be found,

$$\frac{\delta W}{dt} = \dot{m}\left[\frac{C_2^2-C_1^2}{2}+g(Z_2-Z_1)+(P_2v_2-P_1v_1)+(u_2-u_1)\right]+\frac{\delta Q}{dt}$$

$$= 0.5\frac{\text{kg}}{\text{sec}}\left[\frac{5^2-7^2}{2\times1000}+0+(700\times0.19-100\times0.95)\frac{\text{kJ}}{\text{kg}}+90\frac{\text{kJ}}{\text{kg}}\right]-58\frac{\text{kJ}}{\text{sec}}$$

$$= 0.5\frac{\text{kg}}{\text{sec}}\left[(-0.012+38+90)\frac{\text{kJ}}{\text{kg}}\right]-58\frac{\text{kJ}}{\text{sec}}$$

$$= 0.5\frac{\text{kg}}{\text{sec}}\left[127.988\frac{\text{kJ}}{\text{kg}}\right]-58\frac{\text{kJ}}{\text{sec}}$$

$$\frac{\delta W}{dt} = -63.994 \frac{\text{kJ}}{\text{sec}} - 58 \frac{\text{kJ}}{\text{sec}}$$

$$\mathbf{\frac{\delta W}{dt} = -121.994\ kW}$$

(b) From continuity equation,

$$\dot{m} = \frac{A_1 C_1}{v_1} = \frac{A_2 C_2}{v_2}$$

$$\therefore \quad \frac{A_1}{A_2} = \frac{v_1}{v_2} \cdot \frac{C_2}{C_1} = \frac{0.95}{0.19} \times \frac{5}{7}$$

$$\mathbf{\frac{A_1}{A_2} = 3.57}$$

Also we know that,

$$\frac{A_1}{A_2} = \frac{\frac{\pi}{4} \cdot d_1^2}{\frac{\pi}{4} \cdot d_2^2} = 3.57$$

$$\therefore \quad \frac{d_1}{d_2} = \sqrt{3.57}$$

$$\mathbf{\frac{d_1}{d_2} = 1.89}$$

Example 2.21 *At the inlet to a certain nozzle, the enthalpy of the fluid passing is 3000 kJ/kg and the velocity is 60 m/sec. At the discharge end the enthalpy is 2757 kJ/kg. The nozzle is horizontal. The heat loss during the flow is negligible.*

(1) Find the velocity at the exit.

(2) If the inlet area is 0.1 m² and specific volume is 0.187 m³/kg, find the mass flow rate.

(3) If the specific volume at the outlet is 0.498 m³/kg.

Find the area at the exit of the nozzle.

Solution

Since the nozzle is horizontal, $\Delta PE = 0$

Since the heat loss during the flow is negligible, $q = 0$

Since no work is done in the nozle, $w = 0$

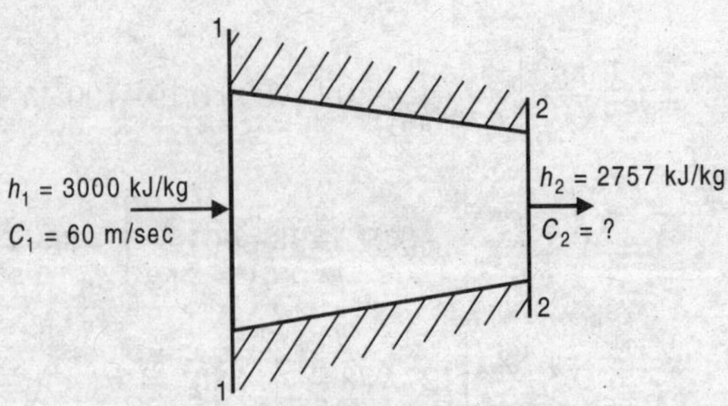

Fig. Ex. 2.21

We know that SFEE on mass basis is,

$$\left[\frac{C_1^2}{2}+gZ_1+h_1\right]+q = \left[\frac{C_2^2}{2}+gZ_2+h_2\right]+w \quad ...(1)$$

or

$$q-w = \frac{C_2^2-C_1^2}{2}+g(Z_2-Z_1)+h_2-h_1$$

On substituting the given characteristics, Eq. (1) reduces to,

$$h_1+\frac{C_1^2}{2} = h_2+\frac{C_2^2}{2}$$

$$3000\frac{\text{kJ}}{\text{kg}}+\frac{60^2}{2\times1000} = 2757\frac{\text{kJ}}{\text{kg}}+\frac{C_2^2}{2\times1000}$$

$$3000+1.8 = 2757+\frac{C_2^2}{2000}$$

$$244.8 = \frac{C_2^2}{2000}$$

$\therefore$ $$C_2^2 = 489600$$

$\therefore$ $$\mathbf{C_2 = 699.71\ m/sec}$$

(ii) Mass flow rate

$$\dot{m} = \frac{A_1C_1}{v_1} = \frac{A_2C_2}{v_2}$$

$$= \frac{0.1\,\text{m}^2\times60\,\text{m/sec}}{0.187}$$

$$\mathbf{\dot{m} = 32.08\ kg/sec}$$

(iii) Area at the exit

$$\dot{m} = \frac{A_1C_1}{v_1} = \frac{A_2C_2}{v_2}$$

$$A_2 = \frac{\dot{m}v_2}{C_2} = \frac{32.08\times0.498}{699.71}$$

$\therefore$ $$\mathbf{A_2 = 0.0228\ m^3}$$

Example 2.22 *A nozzle is a device used to increase velocity of steam flowing through it. At inlet the specific enthalpy of steam is 3000 kJ/kg. Specific volume 0.187 m³/kg and area 0.1 m² and at outlet sp. enthalpy 2762 kJ/kg, specific volume 0.498 m³/kg. There is no loss of heat, nozzle is horizontal and stream enters with the velocity of 60 m/sec.*

Find: (i) Velocity of steam of inlet, (ii) Mass flow rate of steam in kg/hr, (iii) Exit area of nozzle.

Solution

For nozzle since no work is done $w = 0$.

We know that SFEE on mass basis is,

$$\left[\frac{C_1^2}{2}+gz_1+h_1\right]+q = \left[\frac{C_2^2}{2}+gz_2+h_2\right]+w$$

On substituting characteristic, we get,

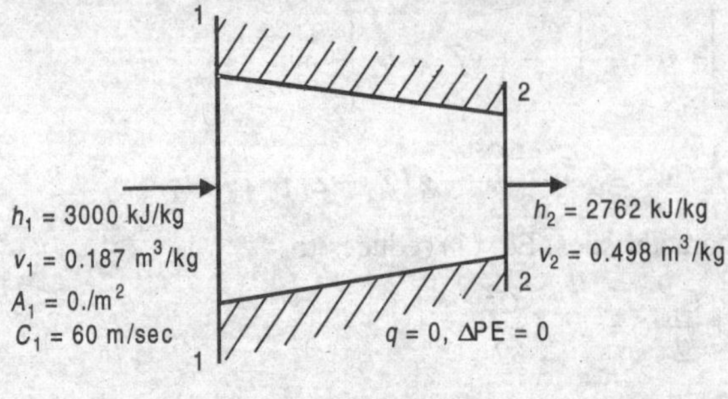

Fig. Ex. 2.22

$$h_1 + \frac{C_1^2}{2} = h_2 + \frac{C_2^2}{2}$$

$$3000 + \frac{60^2}{2\times1000} = 2762 + \frac{C_2^2}{2\times100}$$

$$\Rightarrow \qquad \mathbf{C_2 = 692.53\ m/sec}$$

(ii) Mass flow rate

$$\dot{m} = \frac{A_1C_1}{v_1} = \frac{A_2C_2}{v_2}$$

$$= \frac{0.1\times60}{0.187}$$

$$\mathbf{\dot{m} = 32.856\ kg/sec}$$

(iii) Area at the exit,

$$\dot{m} = \frac{A_1C_1}{v_1} = \frac{A_2C_2}{v_2}$$

$$A_2 = \frac{m\,.\,v_2}{C_2} = \frac{32.856\times0.4948}{692.53}$$

$$\mathbf{A_2 = 0.0231\ m^2}$$

Example 2.23 *Air enters a nozzle with a velocity of 40 m/sec. The decrease in the enthalpy in the nozzle is 180000 J/kg. Determine the exit velocity. Assume nozzle to be adiabatic.*

Solution

We know that, SFEE on mass basis is,

$$\left[\frac{C_1^2}{2} + gZ_1 + h_1\right] + q = \left[\frac{C_2^2}{2} + gZ_2 + h_2\right] + w \qquad ...(1)$$

Since for a nozzle characteristics are,

$$\Delta PE = 0,\ w = 0,\ q = 0 \text{ since adiabatic}$$

Equation (1) reduces to,

$$\frac{C_1^2}{2} + h_1 = \frac{C_2^2}{2} + h_2$$

$$\frac{C_1^2}{2} + (h_1 - h_2) = \frac{C_2^2}{2}$$

$$\frac{(40)^2}{2\times1000}+180000 = \frac{C_2^2}{2\times1000}$$

$$\therefore \quad \mathbf{C_2 = 601.33\ m/sec}$$

Example 2.24 *Air is compressed continuously by an air compressor according to the law $PV^{1.3} = C$. The pressure at the inlet is 1 bar, and at the outlet is 5 bar. The volume of the air changes from 3 m³/kg to 0.8 m³/kg and the velocity changes 25 m/sec at inlet to 130 m/sec at the delivery. The delivery connection is 12 mt above the inlet. What is the shaft power of the compressor. Is it a power absorbing or power producing system.*

Solution

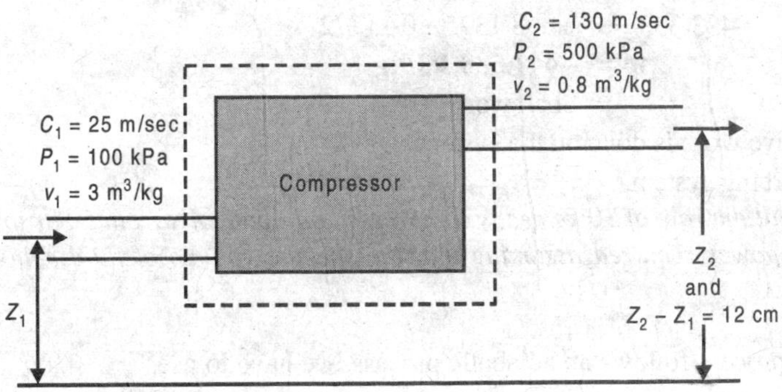

Fig. Ex. 2.24

Note. If the steady flow process or system follows. $PV^n = C$ then follow this method to solve the problem.

i.e., use $\quad -\int v\,dP = w+\Delta KE+\Delta PE$

We know that,

	Isothermal (Pv = C) (For unit mass)	*Polytropic ($Pv^n = C$)*
(i) W.D. in closed system $=\int Pdv$	$P_1v_1 \ln_e \frac{v_2}{v_1}$	$\frac{P_2v_2 - P_1v_1}{1-n}$
(ii) W.D. for open system $= -\int v\,dP$	Steady flow system, $P_1v_1 \ln_e \frac{v_2}{v_1}$ or $\frac{P_1}{P_2}$	$\frac{n}{n-1}\times P_2v_2 - P_1v_1$

For Polytropic Process,

$$-\int_1^2 vdP = \frac{n}{n-1}\times P_1v_1 - P_2v_2$$

$$= \frac{1.3}{1.3-1}\times(100\times3-500\times0.8)\frac{\text{kJ}}{\text{kg}}$$

$$= -433.333\ \text{kJ/kg}$$

and, $$\Delta KE = \frac{1}{2}\left(\frac{C_2^2-C_1^2}{1000}\right) = \frac{130^2-25^2}{2\times1000}$$

and $$\Delta KE = 8.1375 \frac{kJ}{kg}$$

$$\Delta PE = g(Z_2 - Z_1)$$

$$= \frac{9.81\times12}{1000} 0.11772 \frac{kJ}{kg}$$

Substituting these in Eq. (1) we get,

$$\therefore \quad -\int vdP = w + \Delta KE + \Delta PE$$

$$-433.333 = w + 8.1375 + 0.11772$$

$$\therefore \quad \mathbf{w = -441.588 \text{ kJ/kg}}$$

$$= \text{Shaft work}$$

Since w is (–) negative work is done on the system.

∴ It is a power absorbing system.

Example 2.25 *A fluid at the rate of 10 kg/sec is compressed adiabatically from 5 bar to 50 bar in a steady flow process. Calculate the power required assuming that the specific volume of fluid being as 0.001 m³/kg which remains almost constant.*

Solution

Since the compression process follows an adiabatic process, we have to use,

$$\text{W.D.} = -\int vdP = v.(P_2 - P_1)$$

$$\therefore \quad \text{Power required} = \dot{m}\times v\times(P_2 - P_1)$$

$$= 10 \text{ kg/sec} \times 0.001 \text{ m}^3/\text{kg } (50-5)\times100$$

$$\textbf{Power required} = \mathbf{45 \text{ kW}}$$

Example 2.26 *A blower handles 2 kg/sec of air at 20°C and consumes a power of 30 kW. The inlet and outlet velocities are 100 in/sec and 150 m/sec respectively. Estimate the exit air temperature assuming adiabatic conditions.*

Take $$R \text{ for air} = 0.287 \text{ kJ/kg-K}$$

$$\gamma = \frac{C_p}{C_v} = 1.4$$

and $$C_p = 1.005 \text{ kJ/kg-K}$$

Solution

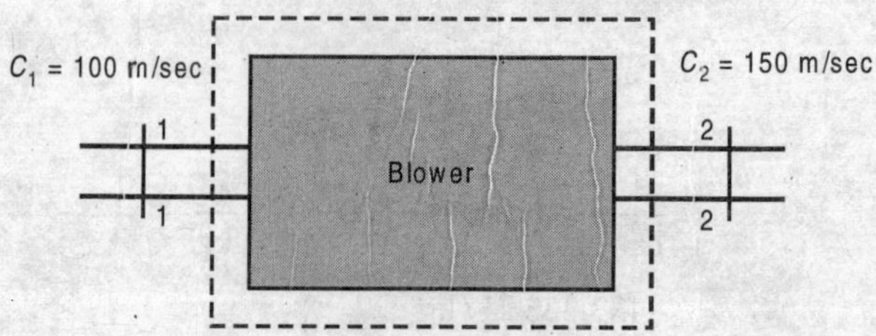

Fig. Ex. 2.26

$$\dot{m} = 2 \text{ kg/sec}$$

$$\frac{\delta W}{dt} = -30 \text{ kW} = -30 \text{ kJ/sec.}$$

$$t_2 = ?\ t_1 = 20°\text{C}$$

Since $\frac{\delta W}{dt}$ is given using the SFEE on time basis.

$$\dot{m}\left[\frac{C_1^2}{2} + gZ_1 + h_1\right] + \frac{\delta Q}{dt} = \dot{m}\left[\frac{C_2^2}{2} + gZ_2 + h_2\right] + \frac{\delta W}{dt}$$

$$\frac{\delta W}{dt} = -\dot{m}\left[\frac{C_2^2 - C_1^2}{2} + g(Z_2 - Z_1) + (h_2 - h_1)\right] + \frac{\delta Q}{dt}$$

Since adiabatic conditions are given, $\frac{\delta Q}{dt} = 0$

Neglecting the change in PE, we have,

$$\frac{\delta W}{dt} = -\dot{m}\left[\frac{C_2^2 - C_1^2}{2} + (h_2 - h_1)\right] \quad ...(1)$$

$$\frac{\delta W}{dt} = -\dot{m}\left[\frac{C_2^2 - C_1^2}{2} + (h_2 - h_1)\right]$$

$$-30\frac{\text{kJ}}{\text{sec}} = -2\frac{\text{kg}}{\text{sec}}\left[\frac{150^2 - 100^2}{2 \times 1000} + (h_2 - h_1)\right]$$

$$-30\frac{\text{kJ}}{\text{sec}} = -2\frac{\text{kg}}{\text{sec}}\left[\frac{12500}{2000} + (h_2 - h_1)\right]$$

$$-30\frac{\text{kJ}}{\text{sec}} = -2\frac{\text{kg}}{\text{sec}}\left[6.25\frac{\text{kJ}}{\text{kg}} + (h_2 - h_1)\right]$$

$$-30\frac{\text{kJ}}{\text{sec}} = -12.5\frac{\text{kJ}}{\text{sec}} - 2\frac{\text{kg}}{\text{sec}} \times (h_2 - h_1)$$

$$-30\frac{\text{kJ}}{\text{sec}} + 12.5\frac{\text{kJ}}{\text{sec}} = -2\frac{\text{kg}}{\text{sec}} \times (h_2 - h_1)$$

$$\frac{-17.5 \text{ kJ/sec}}{2 \text{ kg/sec}} = (h_2 - h_1)$$

$$\mathbf{8.75\frac{kJ}{kg} = (h_2 - h_1)}$$

We also know from the definition of C_P,

$$\left(\frac{dh}{dT}\right)_P = C_P$$

or $$dh = C_P \, . dT \ (m = 1 \text{ for unit mass})$$

or Change enthalpy $\Delta h = C_P \, \Delta T$

$$h_2 - h_1 = C_P(t_2 - t_1)$$

$$8.75 = 1.005(t_2 - 20)$$

$$\mathbf{t_2 = 28.75°C}$$

Example 2.27 *In a water cooled compressor 0.6 kg of air is compressed/sec. Power required to run the compressor is 40 kW. Heat lost to the cooling water is 30% of input and 10% of input is lost in bearings and other frictional effects. Air enters the compressor of 1 bar and 30°C. If the changes in PE and KE are neglected, estimate the exit air temperature. Take C_p = 1.005 kJ/kg-K.*

Data :

$$\dot{m} = 0.6 \text{ kg/sec.} \qquad \frac{\delta W}{dt} = 40 \text{ kW}$$

$$P_1 = 1 \text{ bar} = 10^5 \text{ N/m}^2 \qquad T_1 = 30°C$$

$$\Delta PE = 0, \ \Delta KE = 0 \qquad T_2 = ?$$

Solution

We know that,

$$\Delta H = m\, C_P(T_2 - T_1)$$

$$= 0.6 \times 1.005\,(T_2 - 30)$$

$$\Delta H = 0.603\,(T_2 - 30) \qquad ...(1)$$

Net heat lossses,

$$= 40\% \text{ of input}$$

$$= \frac{40}{100} \times 40 \text{ kW}$$

$$\frac{\delta Q}{dt} = 0.4 \times 40 = 16 \text{ kW.}$$

SFEE is given by,

$$\dot{m}\left[\frac{C_1^2}{2} + gZ_1 + h_1\right] + \frac{\delta Q}{dt} = \dot{m}\left[\frac{C_2^2}{2} + gZ_2 + h_2\right] + \frac{\delta W}{dt}$$

Since $\Delta KE = 0$ and $\Delta PE = 0$

$$\dot{m}\, h_1 + \frac{\delta Q}{dt} = \dot{m}\, h_2 + \frac{\delta W}{dt}$$

$$\therefore \quad \frac{\delta W}{dt} - \frac{\delta Q}{dt} = \dot{m}(h_1 - h_2) = \Delta H \qquad ...(2)$$

From (1) and (2)

$$40 - 16 = 0.603\,(T_2 - 30)$$

$$\mathbf{T_2 = 70° \, C}$$

Example 2.28 *A room is fitted with 2 fans each consuming 0.2 kW power. There are three lamps in the room which are provided to heat the room each consuming 200 W. Ventilation air enters the room with the enthalpy of 85 kJ/ kg and leaves the room with the enthalpy of 60 kJ/kg. The rate of air flow is 100 kg/hr. There are five persons in* *the room and heat generated by each person is 600 kJ/hr. Determine the rate at which the heat is to be removed by a room cooler, so that steady state is maintained in the room.*

Solution

$$\text{Air flow rate} = 100 \text{ kg/hr}$$

$$\text{Heat generated by each person} = 600 \text{ kJ/hr} = 0.1666 \text{ kJ/sec}$$

$$\therefore \quad \text{For 5 persons} = 0.8333 \text{ kJ/sec}$$

Since work input is given in kW using SFEE on time basis.

$$\dot{m}\left[\frac{C_1^2}{2} + gZ_1 + h_1\right] + \frac{\delta Q}{dt} = \dot{m}\left[\frac{C_1^2}{2} + gZ_2 + h_2\right] + \frac{\delta W}{dt}$$

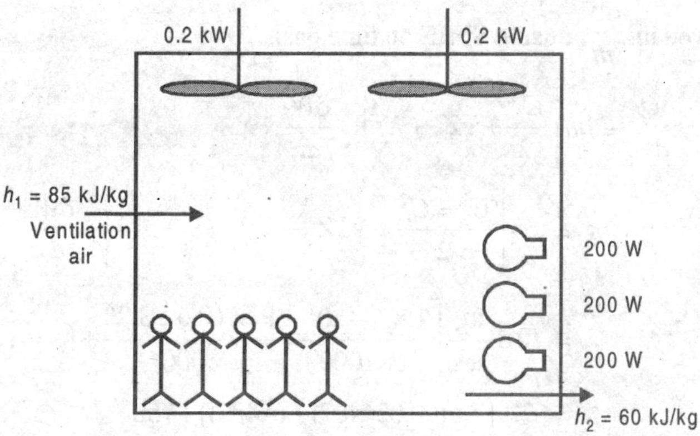

Fig. Ex. 2.28

Neglecting ΔKE and ΔPE.

$$\dot{m}\,[h_1 - h_2] + \frac{\delta Q}{dt} = \frac{\delta W}{dt}$$

$$\text{i.e., } \dot{m}\,[h_1 - h_2] + \left[\begin{pmatrix}\text{heat added due}\\ \text{to lamps}\end{pmatrix} + \begin{pmatrix}\text{heat generated}\\ \text{by 5 persons}\end{pmatrix} - \begin{pmatrix}\text{heat to be removed}\\ \text{by the cooler}\end{pmatrix}\right]$$

$$= \text{Work input to the fans (which should be } -\text{ve)}$$

$$\therefore \quad \frac{100}{3600}\frac{\text{kJ}}{\text{sec}}\left[85\frac{\text{kJ}}{\text{kg}} - 60\frac{\text{kJ}}{\text{kg}}\right] + \left[0.6\frac{\text{kJ}}{\text{sec}} + 0.833\frac{\text{kJ}}{\text{sec}} - X\right] = -0.4\frac{\text{kJ}}{\text{sec}}$$

$$0.6944 + [0.6 + 0.833 - X] = -0.4$$

$$0.6944 + 0.6 + 0.833 + 0.4$$

$$\therefore \quad \mathbf{X = 2.574 \text{ kJ/sec.}} \quad \textbf{Heat to be removed by cooler}$$

Example 2.29 *In a steady flow device, the inlet and outlet conditions are given below. Determine the heat loss/gain by the system.*

Property	*Inlet*	*Outlet*
Pressure (bar)	*10*	*0.15*
Specific volume (m^3/kg)	*0.206*	*8.93*
Specific enthalpy (kJ/kg)	*2827*	*2341*
Velocity (m/s)	*20*	*120*
Elevation m	*3.2*	*0.5*

The fluid flow rate through the device is 2.1 kg/s. The work output of the device is 750 kW.

Data: $Q = ?$

$$\frac{\delta W}{dt} = +750 \text{ kW}$$

Solution

Since work output is given in $\frac{\delta W}{dt}$, using SFEE on time basis,

$$\dot{m}\left[\frac{C_1^2}{2} + gZ_1 + h_1\right] + \frac{\delta Q}{dt} = \dot{m}\left[\frac{C_2^2}{2} + gZ_2 + h_2\right] + \frac{\delta W}{dt}$$

$$= \dot{m}\left[\frac{C_2^2 - C_1^2}{2} + g(Z_2 - Z_1) + (h_2 - h_1)\right] + \frac{\delta W}{dt}$$

$$= 2.1\frac{\text{kg}}{\text{sec}}\left[\frac{120^2 - 20^2}{2\times 1000} + \frac{9.81(0.5-3.2)}{1000} + (2341 - 2827)\right] + 750$$

$$= 2.1\,[7 + (-0.026487) + (-486)] + 750$$

$$= 2.1\,[7 - 0.025487 - 486] + 750$$

$$= 2.1\,[-479.026] + 750$$

$$= -1005.9556 + 750$$

$$\mathbf{\frac{\delta Q}{dt} = -255.95562 \text{ kW}}$$

Note: –ve sign implies that heat will flow out of system.

Example 2.30 *Air at a temperature of 15°C passes through a heat exchanger at a velocity of 30 m/c where its temperature is raised to 800°C. It then enters a turbine with the same velocity of 30 m/s and expands until the temperature falls to 650°C. On leaving the turbine, the air is taken at a velocity of 60 m/s to a nozzle where it expands until the temperature has fallen to 500° C. If the air flow rate is 2 kg/s calculate,*

(*a*) *Rate of heat transfer to the air in the heat exchanger.*

(*b*) *Power output from the turbine, assuming no heat loss, and*

(*c*) *Velocity of air at exit from nozzle, assuming no heat loss, take the specific enthalpy of air as $h = C_p T$, where $C_p = 1.005$ kJ/kg K, and T = Temperature*

Solution

For heat exchanger:

$$q_{1-2} + \dot{m}\left(h_1 + \frac{C_1^2}{2} + gz_1\right) = w_{1-2} + \dot{m}\left(h_2 + \frac{C_2^2}{2} + gz_2\right)$$

Then
$$q_{1-2} = \dot{m}(h_2 - h_1) \; (\because C_1 = C_2)$$
$$= \dot{m}\, C_P (T_2 - T_1)$$
$$q_{1-2} = 2 \times 1.005 \times (800 - 15)$$
$$\mathbf{q_{1-2} = 1577.85 \ kJ/s}$$

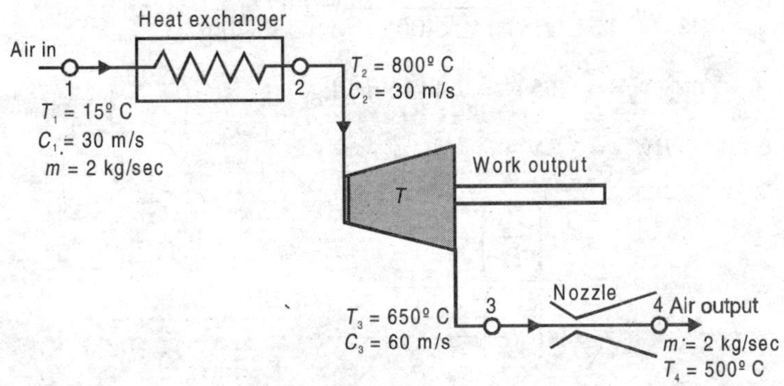

Fig. Ex. 2.30

For the turbine:

$$q_{2-3} + \dot{m}\left(h_2 + \frac{C_2^2}{2} + gz_2\right) = w_{2-3} + \dot{m}\left(h_3 + \frac{C_3^2}{2} + gz_3\right)$$

$$\therefore \quad w_{2-3} = \dot{m}\left[(h_2 - h_3) + \frac{C_2^2 - C_3^2}{2}\right] \qquad (\because q_{2-3} = 0)$$

$$= \dot{m}\left[C_P (T_2 - T_3) + \frac{C_2^2 - C_3^2}{2}\right]$$

$$= 2\left[1.005(800 - 650) + \left(\frac{30^2 - 60^2}{2 \times 1000}\right)\right]$$

$$\mathbf{W_{2-3} = 298.8 \ kW}$$

For nozzle:

$$q_{3-4} + \dot{m}\left[h_3 + \frac{C_3^2}{2} + gz_3\right] = w_{3-4} + \dot{m}\left[h_4 + \frac{C_4^2}{2} + gz_4\right]$$

$$C_4^2 - C_3^2 = 2 \times 1000 \times C_P (T_3 - T_4) \; (\because q_{3-4} = 0 \text{ and } w_{3-4} = 0)$$

$$\therefore \quad C_4^2 - C_3^2 = 2 \times 1000 \times C_P (T_3 - T_4) = 2 \times 1000 \times 1.005(650 - 500)$$

$$= 301500$$

$$\therefore \quad \mathbf{C_4 = 552.36 \ m/s}$$

Example 2.31 *Helium gas is expanded polytropically in a turbine, from 4 bar, 300°C to 1 bar such that final volume is 2.5 times the volume at inlet. Velocity of gas at exit is 50 m/s. What is the mass flow rate of gas required to produce 1 MW turbine output? How much is the heat transfer during the process ? Also determine exit area of turbine.*

Assume specific heat of helium = 5.193 kJ/kg K at constant pressure.

Solution

We know that $$R = \frac{G}{M} = \frac{8.314}{4} = 2.0785$$

$$v_1 = \frac{RT_1}{P_1} = \frac{2.0785 \times 573}{400} = 2.9774 \frac{\text{m}^3}{\text{kg}}$$

$$v_2 = 2.5\, v_1 = 7.4436 \text{ m}^3/\text{kg}$$

For Polytropic Process, $$P_1 v_1^n = P_2 v_2^n$$

$$\therefore \quad n = \frac{l_n\left(\dfrac{P_2}{P_1}\right)}{l_n\left(\dfrac{v_1}{v_2}\right)} = 1.513$$

and $$\frac{T_2}{T_1} = \left(\frac{P_2}{P_1}\right)^{\frac{n-1}{n}}$$

$$\therefore \quad T_2 = 358.11 \text{ K}$$

$$\text{Specific work} = -\int_1^2 v\, dP = \frac{n}{n-1} RT_1 \left[1 - \frac{P_2 v_2}{P_1 v_1}\right].$$

$$= \frac{1.513}{0.513} \times 2.0785 \times 573 \times \left[1 - \frac{1 \times 2.5}{4}\right]$$

$$= 1317.2 \text{ kJ/kg}$$

$$\Delta\text{KE} = \frac{1}{2} C_2^2 = 1250 \frac{\text{J}}{\text{kg}} \text{ or } 1.25 \text{ kJ/kg}$$

$$q - w_s = \Delta h + \Delta\text{KE}$$

and $$-\int v\, dP = w_s + \Delta\text{KE} \quad \text{as} \quad \Delta\text{PE} = 0$$

$$\therefore \quad \mathbf{W_s = 1317.2 - 1.25 = 1315.95 \text{ kJ/kg}}$$

and $$q = 1315.95 + CP(T_2 - T_1) + 1.25 = 1315.95 + 5.193\,(358.11 - 573) + 1.25$$

$$\mathbf{q = 201.28 \text{ kJ/kg}}$$

Then $$\mathbf{\dot{m} = \frac{100}{W_s} \text{ kJ/s} = 0.7599 \text{ kg/s}}$$

$$\therefore \quad \textbf{Exit area } \mathbf{A_2 = \frac{\dot{m}\, v_2}{C_2} = 0.1131 \text{ m}^2}$$

THEORY QUESTIONS

2.1. Write a note on 'Joule's Experiment'?

2.2. State and explain the first law of thermodynamics for a closed system undergoing a cyclic change.

2.3. Derive an equation for energy transfer for a closed system.

2.4. Explain the First Law of Thermodynamics and hence show their cyclic integral of δQ and cyclic integral of δW are equal for a closed system.

2.5. State and explain the first law of thermodynamics for a closed system undergoing a process.

2.6. Prove that 'Internal Energy' is a property of a system.

2.7. Write a note on different forms of stored energies.

2.8. Define the term 'Enthalpy' (*H*).

2.9. Write a note on'Internal Energy' (*U*).

2.10. Distinguish between heat and internal energy.

2.11. Define specific heat at constant pressure and at constant volume for a perfect gas.

2.12. Define Adiabatic Index (*g*).

2.13. What is Perpetual Motion Machine of first kind (PMM-1)?

2.14. Define a flow process and explain the terms Control Volume and control surface.

2.15. Explain the term flow work and discuss how it differs from the work done by a closed system.

2.16. Write a note on Flow work or Flow energy.

2.17. How is 'Steady flow system characterized'?

2.18. What are the conditions for a steady flow process?

2.19. State the conditions of First Law of Thermodynamics for a Steady Flow Processes.

2.20. Explain the essentials of non-flow and flow systems. Give suitable examples of each. What are advantages of steady flow system over non-flow system.

2.21. Derive the Steady Flow Energy Equation (SFEE).

2.22. Apply first law of thermodynamics to steady flow system and derive the equation of energy.

2.23. Derive Steady Flow Energy Equation on mass basis.

2.24. Explain the significance of $-\int v\,dP$

2.25. Explain the significance of $\int P dv$ in case of steady flow process and non-flow process.

2.26. Prove that, $-\int v dP = Q - \Delta H$.

2.27. Prove that, $\int P dv = Q - \Delta U$.

2.28. Obtain the energy equations for the following open systems:

(a) Boiler (b) Turbine
(c) Compressor (d) Throttling process
(e) Nozzles (f) Condenser

2.29. Derive Continuity cquation.

PROBLEMS FOR PRACTICE

1. A centrifugal air compressor delivers 15 kg of air/min. Air enters at I bar, I0 m/sec, 0.5 m^3/kg. It is discharged at 7 bar, 80 m/sec and 0.15 m^3/kg. During the process, its enthalpy increases by I60 kJ/kg and the air rejects 720 kJ/min of heat to the surroundings. Find,

(i) Power required to run the compressor.

(ii) Ratio of inlet to outlet pipe diameter. Neglect ΔPE.

2. For a steady flow system, following particulars are noted,

	Inlet	Exit
1. Pressure (bar)	1	5
2. Vol (m^3/kg)	2	0.58
3. Velocity (m/sec)	20	30
4. Elevation (m)	0	2

If the mass flow rate of the working fluid is 1 kg/sec, find the power of the system. Is it a power absorbing or power producing system? Assume that the steady flow system follows $Pv^{1.3} = C$.

3. One kg of working substance undergoes a reversible constant pressure process at 1.2 bar during which its volume changes from 1 m^3 to 1.8 m^3 and temperature changes from 50°C to 37°C. The sp. heat of a substance at constant pressure is given by $C_p = \left(1.1 + \frac{40}{t+30}\right)$ kg/kg°C where t is in °C. Find out
   (a) Heat supplied (b) Work done
   (c) Change in I.E. (d) Change in enthalpy
4. A fluid is contained in a cylinder piston arrangement. The piston is spring loaded and frictionless such that the pressure of the fluid varies with its volume according to the law $P = a + bV$. The internal energy of the fluid is given by $V = 30 + 3.2\, PV$ where U is in kJ. P in kPa, V in m^3. The fluid undergoes a change of state from 150 kPa and 0.025 m^3 to 450 kPa and 0.05 m^3. Assuming no work other than that on the piston determine the direction and magnitude of work and heat transfer.
5. What at the rate of 10 kg/sec is compressed adiabatically from 5 bar to 50 bar in a steady flow process. Calculate the power required assuming that sp. volume of water being as 0.001 m^3/kg which remains almost constant.
6. Steam enters a turbine steadily at 10 mPa and 550°C with a velocity of 60 m/s and leaves at 25 kPa with a dryness fraction of 0.95. A heat loss of 30 kJ/kg occurs during the process. The inlet area of the turbine is 150 cm^2 and exit area is 1400 cm^2. Determine : (1) mass flow rate of steam. (2) exit velocity (3) power output.
7. A closed system receives 200 kg of heat at constant volume. The system then undergoes a constant pressure process during which 150 kg of heat is rejected and 50 kJ of work is done on the system. Calculate how much of work is required to be done if the system is to be restored to its original state by means of an adiabatic process. If the initial internal energy is 50 kJ, determine the values of internal energy at the end of the constant volume and constant pressure processes.
8. In a reversible steady flow process 225 kW of heat is rejected per kg of the fluid passing through the system. During the process, pressure and volume change from 6 bar and 0.075 m^3/kg to 2 bar and 0.2 m^3/kg. Assuming that flow process follows the law PV = constant and ignoring changes in K.E. and P.E., determine the change in enthalpy of the fluid during the process.
9. Air flow steadily at the rate of 0.4 kg/sec through an air compressor, entering at 6 m/sec with pressure of 1 bar and sp. volume of 0.85 m^3/kg and leaving at 4.5 m/sec with a pressure of 6.9 bar and sp. volume of 0.16 m^3/kg. The sp. internal energy of the air leaving is 88 kJ/kg greater than that of the air entering, cooling water in the jacket surrounding the cylinder absorbs heat from the air at the rate of 59 kg/sec. Calculate the power required to drive the compressor and the inlet and outlet pipe areas.
10. In the turbine of a gas turbine unit the gases flow through the turbine at 17 kg/sec and the power developed by the turbine is 14 MW. The sp. enthalpies of the gases at inlet and outlet are 1200 kJ/kg and 360 kJ/kg respectively. Calculate the rate at which heat is rejected from the turbine. Find also the area of the inlet pipe given that sp. volume of the gases at inlet is 0.5 m^3/kg.
11. A steam turbine operates under steady flow conditions, receiving steam and leaving the steam at the following rate.

	Inlet	**Outlet**
Pressure	15 bar	130 kPa
I.E. (kJ/kg)	2500	1500
Velocity (m/sec)	300	200
Elevation (m)	3	0
Volume (m^3)	0.15	1.2 m^3 /kg

12. A closed system undergoes a cycle 1–2–3–1. If $Q_{12} = 30, Q_{23} = 10, W_{12} = 5, W_{31} = 25$ and $\Delta E_{31} = 15$, determine W_{23} and ΔE_{23}.

13. The following data is available for a test on air compressor for an air flow rate of 6 kg/sec.

	Inlet	**Outlet**
Pressure	95 kPa	7.5 bar
Sp. Internal Energy (kJ/kg)	40	150
Sp. Volume (m^3/kg)	0.95	0.162
Velocity of air (m/sec)	6	1

Estimate: (1) Power required to drive the compressor in kW. (2) Ratio of inlet pipe diameter to outlet pipe diameter.

14. A non-flow system undergoes a friction less process according to the law $P = \frac{4.5}{v} + 2$ where P is in bar, V is in m^3/kg. During this process volume changes from 0.12 m^3/kg to 0.04 m^3/kg and temperature increases by 133°C. The change in I.E. of fluid is given by $du = C_v \, dT$ where C_V = 0.71 kJ/kg K and dT is temperature change. Find out (1) Heat transfer, (2) Change in enthalpy. Assume fluid mass of 5 kg.

CHAPTER 3

Second Law of Thermodynamics

3.1 INTRODUCTION

The first law of thermodynamics deals with the conversion of energy. However it does not specify the direction of the process. The second law of thermodynamics is about the direction, quality and quantity.

3.2 LIMITATIONS OF FIRST LAW OF THERMODYNAMICS

First-law states that, when a closed system undergoes any cyclic process, then the cyclic integral of work is equal to the cyclic integral of heat

i.e. $$\oint \delta W = \oint \delta Q$$

Thus the law merely states that work transfer during a cycle is equal to the heat transfer and does not place any restriction on the direction of heat and the work transfer. It does not specify whether the process is possible in a particular direction or not at all. According to this law it can be assumed that the energy transfer can take place in either direction, since it does not specify the direction of energy transfer. But by practical experience it is observed that, even if a proposed cycle satisfies the first law, it doesn't ensure that the cycle will occur actually. This resulted in the formulation of *Second Law of Thermodynamics*. Thus a cycle will proceed only if both I and II laws of thermodynamics are satisfied.

Thus the second law involves the fact that processes will proceed in a certain direction but not in the opposite direction, for example,

(i) Consider the Joule's experiment in which the fall of weight *W* rotates the paddle wheel and increases the temperature of water. Here work is converted into heat. However reverse of this process is not possible i.e., by heating water, paddle wheel will not rotate and lift the weight. This shows that the system can operate in a cyclic process only when work is done on the system and heat is rejected out of the system. But it cannot work in a cyclic process when both work and heat transfer are positive even though such a process will not violate the first law of thermodynamics.

(ii) Consider hot tea in a cup, here heat flows from tea to the cooler surroundings. Once it is cooled, it can never be heated by the addition of heat from the cooler surroundings, without doing any work on it.

(iii) A cup of ice cream when kept in atmosphere absorbs heat and melts. But it will not solidify giving back heat without doing any work on it.

(iv) When a vehicle is stopped by applying mechanical brakes, the brakes get hot. The I.E. of the brake increases by an amount equal to the decrease in K.E. of the vehicle. According to the first law, the hot brake on cooling, should give back its increase in I.E. to the wheels of the vehicle, causing it to regain its speed. This never happens in practice.

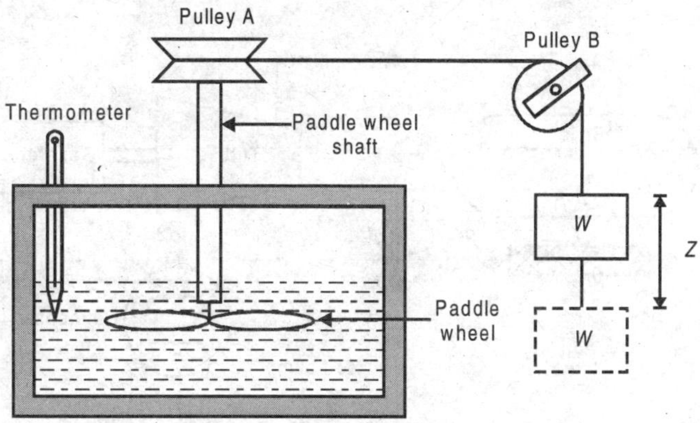

Fig. 3.1. *Joule's experiments.*

From the above examples it is obvious that, even if the first law is satisfied the processes will proceed in a particular direction while they are impossible in the opposite direction. Thus the first law of thermodynamics is necessary but not sufficient condition for the processes to take place.

It has also been practically found that all forms of energies are not equally convertible into work and the first law is silent about the extent of conversion of energy. It is necessary to study *Second Law of Thermodynamics* in this regard.

Points to Remember

- ⇨ First Law is silent about the direction.
- ⇨ It gives no indication about extent of conversion of energy.

3.3 HEAT ENGINE

Heat engine converts heat energy into mechanical energy. Heat engine works on thermodynamic cycle like Otto, Diesel, Rankine in which there is a net heat transfer to the system and a net work transfer from the system.

As we have studied, heat engines are broadly classified into two types.

(i) External combustion engines.

(ii) Internal combustion engines.

Figure 3.2 shows a simple steam power plant. In this power plant an amount of heat Q_1 is supplied from the furnace (Also called as high temperature reservoir) to the water in the boiler drum. An amount of heat Q_2 is rejected to the low temperature reservoir, like a coolant in the condenser and in doing so, an amount of work W will be produced.

The thermal efficiency η_{Th} of the heat engine is defined as,

$$\eta_{Th} = \frac{\text{Energy output}}{\text{Energy input}} \quad \text{or} \quad \eta_{Th} = \frac{\text{Energy sought for}}{\text{Energy that costs}}$$

$$\eta_{Th} = \frac{\text{Work output}}{\text{Heat supplied}} \quad \text{or} \quad \eta_{Th} = \frac{W}{Q_1} = \frac{Q_1 - Q_2}{Q_2}$$

$$\boxed{\eta_{Th} = 1 - \frac{Q_2}{Q_1}}$$

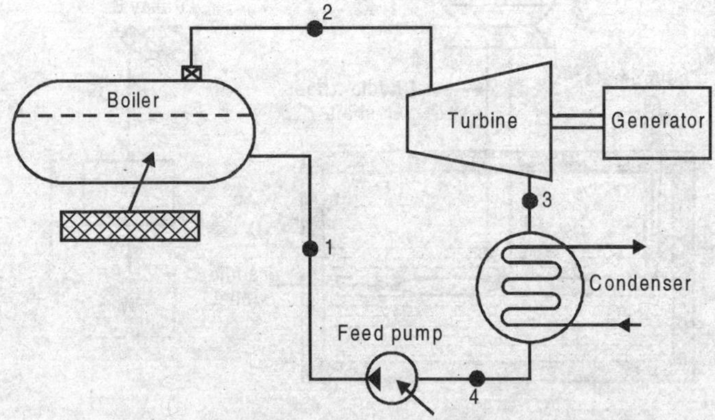

Fig. 3.2. *A simple steam power plant.*

In case of IC engines combustion of air and fuel takes place inside the engine cylinder. The products of combustion will be directly acting on the pistons of the IC engines for producing the power:

Now consider an IC engine as shown in Fig. 3.3.

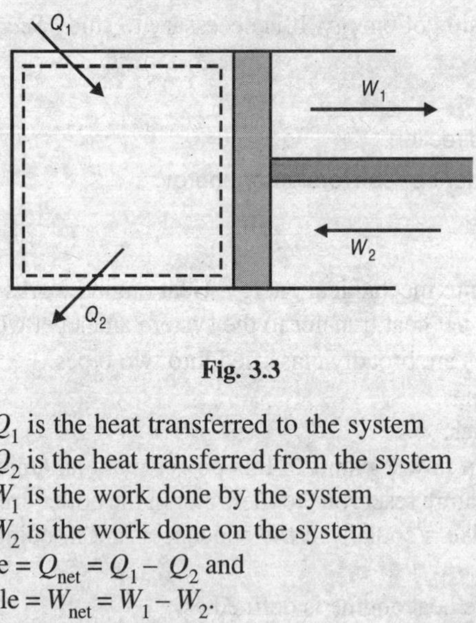

Fig. 3.3

Let Q_1 is the heat transferred to the system

Q_2 is the heat transferred from the system

W_1 is the work done by the system

W_2 is the work done on the system

Then, net heat transfer in a cycle = $Q_{net} = Q_1 - Q_2$ and

Net work transfer in a cycle = $W_{net} = W_1 - W_2$.

According to the first law of thermodynamics,

$$\oint \delta W = \oint \delta Q$$

or

$$W_{net} = Q_{net}$$

∴ We can write,

$$\mathbf{W_1 - W_2 = Q_1 - Q_2}$$

3.4 HEAT RESERVOIRS

Heat reservoir is a body of infinite heat capacity, which is capable of absorbing or rejecting any quantity of heat without suffering any change in any of its thermodynamic properties".

A heat reservoir which supplies an amount of heat Q_1 to the heat engine operating in heat engine cycle is called the High Temperature Reservoir (HTR) or Source.

The heat reservoir to which an amount of heat Q_2 is rejected by the heat engine is called the Low Temperature Reservoir (LTR) or Sink.

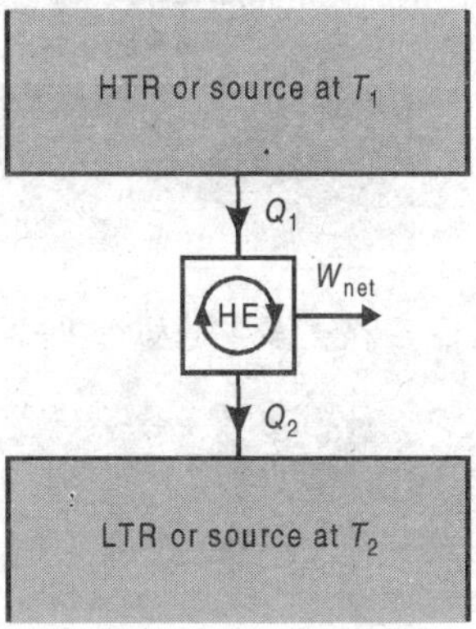

Fig. 3.4

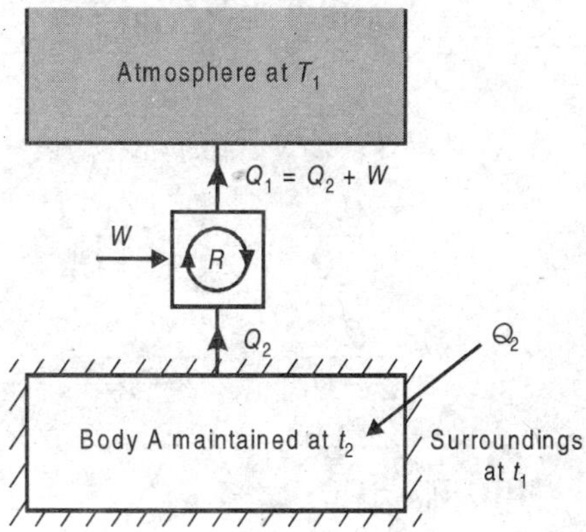

Fig. 3.5

3.5 REFRIGERATOR

"It is a device, which produces and maintains the temperature below atmospheric temperature" or

"It is a device which operating in cycle, maintains an enclosed space at a temperature lower than the temperature of the surroundings".

Let a body *A* is to be maintained at a temperature t_2, which is lower than the atmospheric temperature t_1. Even though the body, *A* is insulated, there will be a heat leakage Q_2 into the body because of the temperature difference between the body and the surroundings. In order to maintain the body *A* at constant temperature t_2, same amount of heat is to be removed at the same rate, at which it is leaking into the body.

A refrigerator is a device which operating in a cycle, removes heat Q_2 from the body a t_2 and transfers this heat to the atmosphere at t_1 by consuming an amount of work *W*.

In the refrigeration cycle, body *A*, Q_2 and *W* are of prime importance. Just like the efficiency of heat engine cycle, here a parameter called *co-efficient of performance* is important. The copies defined by,

$$[\text{COP}]_{\text{ref.}} = \frac{\text{Desired effect}}{\text{Work output}} = \frac{Q_2}{W} = \frac{Q_2}{Q_1 - Q_2} \quad ...(1)$$

3.6 HEAT PUMP

It is just opposite to the refrigerator i.e., "It is a device operating in a cycle, maintains a body say *B* at a temperature higher than the temperature of the surroundings".

Because of the temperature difference between the body and the surroundings there will be a heat leakage Q_1 from the body. The body will be maintained at constant temperature t_1, if heat is supplied into the body at the same rate at which heat leaking out of the body.

A heat pump is a device which operating in a cycle extracts heat from the atmosphere and transfers an amount of heat Q_1 to the body *B* and maintains the body *B* at constant temperature t_1.

In a heat pump, body B, Q_1 and W are of prime importance.

$$[\text{COP}]_{\text{HP}} = \frac{\text{Desired effect}}{\text{Work output}} = \frac{Q_1}{W}$$

$$= \frac{Q_1}{Q_1 - Q_2}$$

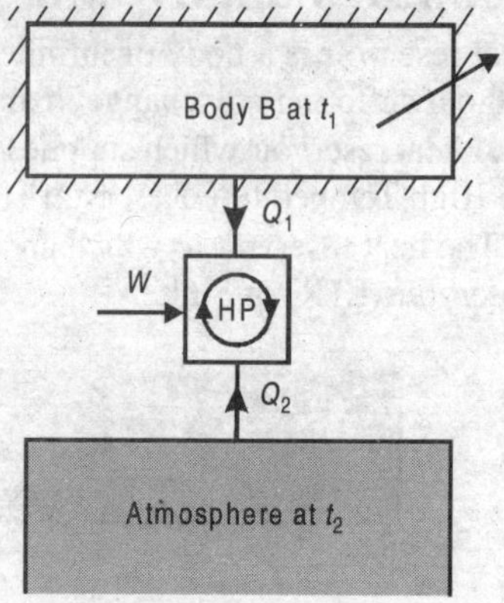

Fig. 3.5. *Heat pump.*

From Eqs (1) and (2) we can write,

$$[\text{COP}]_{\text{HP}} = [\text{COP}]_{\text{ref}} + 1$$

i.e.

$$\frac{Q_1}{Q_1 - Q_2} = \frac{Q_2}{Q_1 - Q_2} + 1$$

$$\frac{Q_1}{Q_1 - Q_2} = \frac{Q_2 + Q_1 - Q_2}{Q_1 - Q_2}$$

$$\frac{Q_1}{Q_1 - Q_2} = \frac{Q_1}{Q_1 - Q_2}$$

$$\text{LHS} = \text{RHS}$$

$$\mathbf{[COP]_{HP} = [COP]_{ref} + 1}$$

3.7 STATEMENTS OF SECOND LAW OF THERMODYNAMICS

There are two important statements of second law. They are,

(i) Kelvin Planck Statement, (ii) Clausius Statement.

(i) Kelvin Planck's Statement

"It is impossible for a heat engine to produce net work in a cycle, if it exchanges heat with a single reservoir".

or

"Heat engine cannot produce work output, if it exchanges heat with a single reservoir".

We have seen that, $$\eta_{\text{Th}} = \frac{\text{Energy output}}{\text{Energy input}} = \frac{W}{Q_1}$$

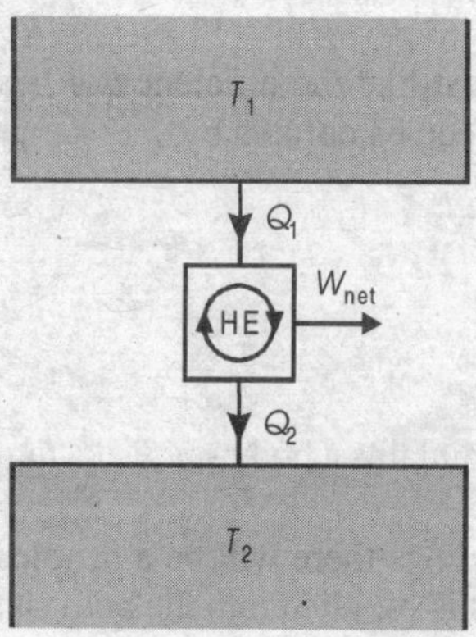

Fig. 3.7. *Heat Engine producing W_{net} by Exchanging Heat with 2 Reservoirs.*

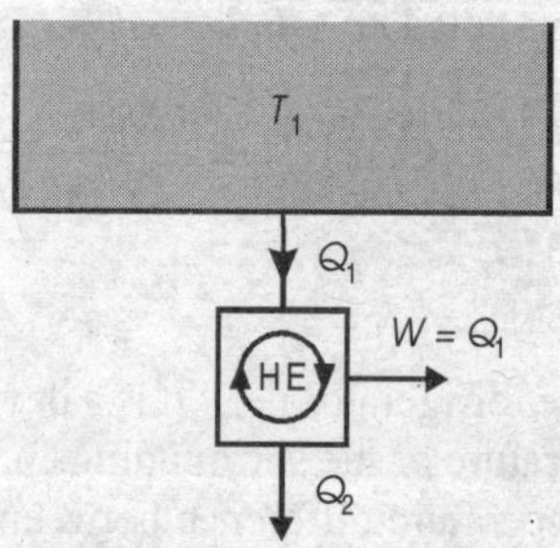

Fig. 3.8. *PMM-2.*

By practical experience, it is noted that total heat supplied cannot be converted into useful work W i.e., $Q_1 \neq W$. A heat engine can never be 100% efficient. But $W < Q_1$, therefore there has to be heat rejection i.e., $Q_2 > 0$.

Therefore for a heat engine to produce network in a cycle, it has to exchange heat with two reservoirs. (Ref. Fig. 3.7)

But if, $Q_2 = 0$, then $Q_1 = W$ or $\eta_{Th} = 100\%$, the heat engine will produce network in a complete cycle by exchanging heat with only one reservoir, thus it violates Kelvin Planck's statement. Such a heat engine is called a *Perpetual Motion Machine of the second kind. (PMM2)* Ref. Fig. 3.8. A PMM-2 is impossible and it is just a conceptual engine. All the attempts made so far to make such a machine have failed. Thus they show the validity of Kelvin Planck's, statement.

(ii) Clausius Statement

We know that, heat always flow from a hot body to a cold body. The reverse process never occurs by itself.

Clausius statement is given as, "It is impossible to construct a device, which operating in a cycle will produce no effect other than transfer of heat from a low temperature body to a high temperature body". Or

"Heat cannot flow by itself from a cold body to a hot body, in order to achieve this some work must be expended".

For example, Refrigerator—in this case heat is removed from the cold body A and transferred to the atmosphere but by consuming work input W.

3.8 EQUIVALENCE OF KELVIN PLANCK AND CLAUSIUS STATEMENTS

We have studied, the Kelvin Plank and Clausius statement as follow.

(i) Kelvin Planck's Statement. "Far a heat engine to produce W_{net} it has to exchange heat with two reservoirs".

(ii) Clausius Statement. "Heat cannot flow by itself from a cold body to a hot body. In order to achieve this some work must be expended".

At first sight these two statements will appear to be two different statements and are unconnected. But it can be easily shown that, these are the two parallel statements of second law and are equivalent in all respects.

The equivalence of the two statements will be proved if we prove that violation of one statement results into the violation of the other.

(a) Violation of Clausius statement leads into the violation of Kelvin Planck's statement. Consider a cyclic heat pump P, which transfers heat from a LTR at t_2 to HTR at t_1, without consuming work input (i.e., $W = 0$), thus violating Clausius statement.

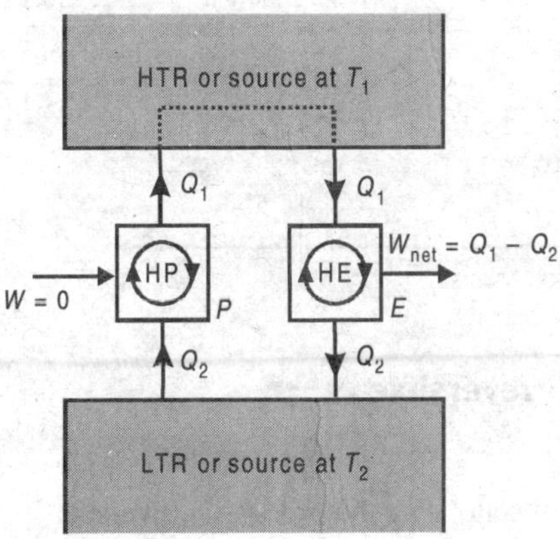

Fig. 3.9. *Violation of Clausius Statement.*

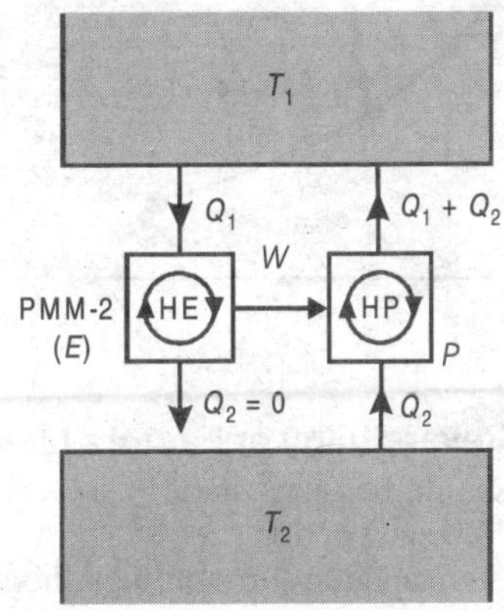

Fig. 3.10. *Violation of Kelvin–Planck Statement.*

Now, let us assume a cyclic heat engine E which also operates between the same two reservoirs at t_1 and t_2 respectively. The rate of working of the heat engine is such that it draws an amount of heat Q_1 from HTR equal to that discharged by the heat pump. Then the HTR may be eliminated and heat Q_1 discharged by the heat pump may be directly fed to the heat engine. So, the heat pump P and heat engine E acting together will form a heat engine, operating in cycles and produce net work by exchanging heat with one reservoir. This violates the Kelvin Planck's statement.

(b) Violation of Kelvin Planck Statement Leads into the Violation of Clausius Statement. Let us consider a PMM-2 (E), which produces W_{net} in a cycle by exchanging heat with only one reservoir at t_1 and thus violates Kelvin Planck's statement (ref. Fig. 3.10).

Now, let us assume a cycle heat pump P extracting heat Q_2 from LTR at t_2 and supplying heat to HTR at t_1 by consuming work W equal to that PMM-2 supplies in a complete cycle. So, E and P together will form a heat pump and producing the complete effect of transferring heat from LTR to HTR, without any external aid.

Points to Remember

⇨ At first sight the two statements appear to be different but these two are the two parallel statements and they are equivalent in all respects.

3.9 REVERSIBILITY AND IRREVERSIBILITY

Quasi—means nearly or almost. So, quasi-static process means nearly static process or a process which proceeds with extreme slowness. A quasi-static process proceeds from one equilibrium state to another equilibrium state till the end of process. All the states passed through by the system, during the process are all equilibrium states.

A quasi-static process is represented on the *PV*-diagram by a continuous curve and if we carry out the process in the reverse direction, then it is possible to retrace the same path. Hence it is known as a *Reversible process.*

Whereas in case of *Irreversible process* only end states are equilibrium states and all the intermediate states are non-equilibrium states and is represented by means of a dotted curve as shown in Fig. 3.11.

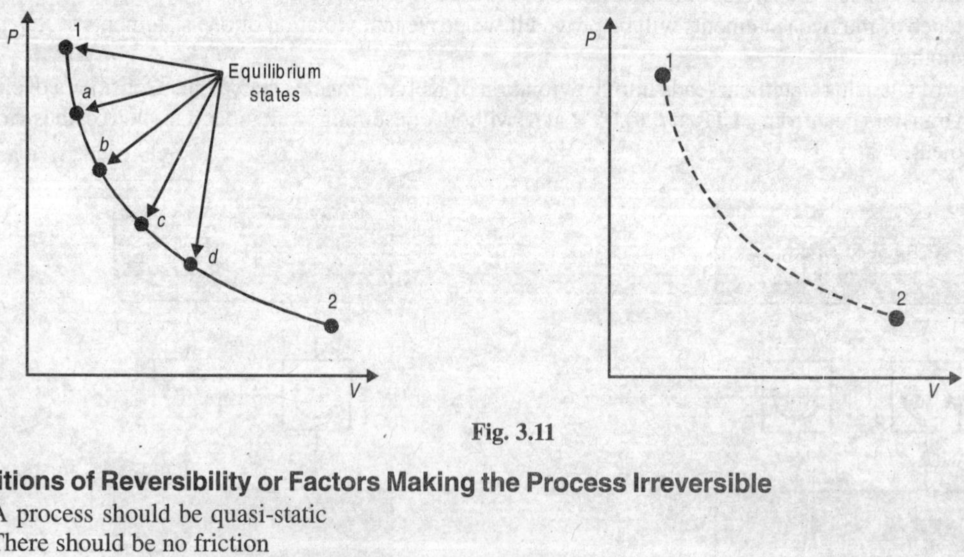

Fig. 3.11

Conditions of Reversibility or Factors Making the Process Irreversible

(i) A process should be quasi-static

(ii) There should be no friction

(iii) Both the system and the surroundings should restore their initial state after the process is reversed.

3.10 CARNOT ENGINE AND CARNOT CYCLE

Carnot, a French Engineer, was the first to introduce the concept of reversible cycle. Carnot devised an engine, which is named after him as Carnot engine. Carnot engine works between high temperature and low temperature reservoirs as shown in Fig. 3.12 .

Carnot engine works on carnot cycle. It consists of an alternate series of two reversible isothermal and two reversible adiabatic processes. Since each process is reversible, the Carnot cycle as a whole is reversible.

Carnot cycle (Figs 3.12 and 3.13) is independent of nature of working fluid. It can work with any working substance like gas, vapour or any other working substance. Let us assume that the working fluid of the engine is a gas.

Figure 3.12 shows the proposed reversible engine working between source at T_1 and sink at T_2. It consists of a cylinder, which has a piston working in it without friction. The walls of the cylinder and piston are assumed to be perfect insulators of heat. The cylinder head is so arranged that it can be a perfect conductor and perfect insulator alternatively.

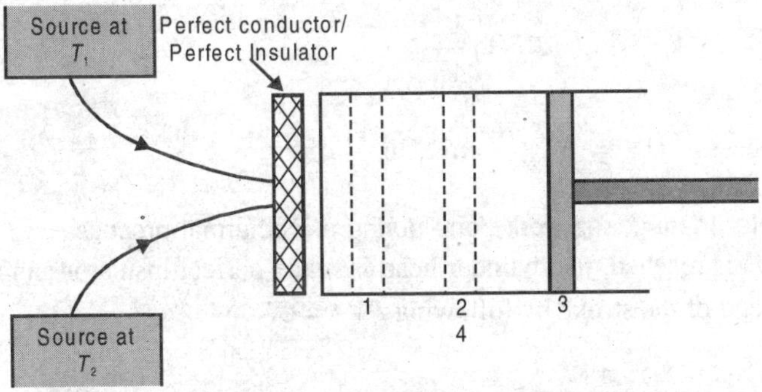

Fig. 3.12. *Carnot's engine.*

During the beginning of the stroke the cylinder head is made perfect conductor and the heat source at T_1 is brought in contact with the gas in the cylinder. The gas expands isothermally at T_1 from state point (1) to state point (2). During the expansion process, heat supplied Q_1 is used in the expansion of the gas.

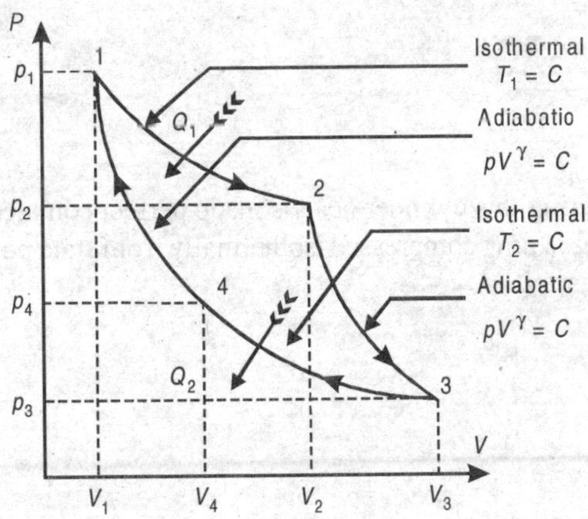

Fig. 3.13

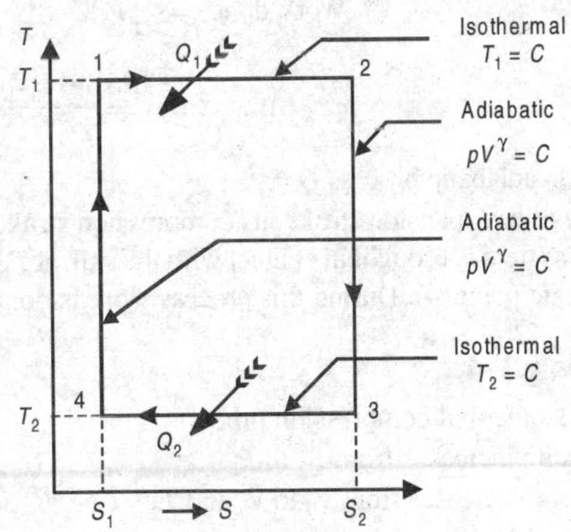

Fig. 3.14

Process 1–2

It is a Isothermal expansion process.

Law for the process is $PV = C$

Pressure falls from P_1 to P_2

Volume increases from V_1 to V_2 and, $T = C$.

$$\text{Work done} = \int_1^2 PdV$$

$$W_{1-2} = P_1V_1 \ln \frac{V_2}{V_1}$$

Since, $$P_1V_1 = mRT_1$$

$$W_{1-2} = mRT_1 \ln \frac{V_2}{V_1}$$

and For an Isothermal process, $$Q_1 = W_{1-2} = mRT_1 \ln \frac{V_2}{V_1} \quad \text{...(1)}$$

Note. For derivation refer Ideal gases work done during a Isothermal process.

When the state-point (2) is reached, the cylinder head is made perfect insulator and the gas in the cylinder is allowed to expand till the end of the stroke by following $PV^{\gamma} = C$.

Process 2–3

It is an adiabatic expansion process.

Pressure falls from P_2 to P_3

Volume increases from V_2 to V_3

Temperature falls from T_1 to T_2

$$\text{Work done} = \int PdV$$

$$W_{2-3} = \frac{P_2V_2 - P_3V_3}{\gamma - 1} = \frac{mR(T_1 - T_2)}{\gamma - 1}$$

For an adiabatic process $Q = 0$.

Now after expansion stroke, its compression stroke starts. During this cylinder head is made perfect conductor and the system is brought in contact with the sink at T_2 and the gas is now compressed isothermally from state point (3) to state point (4). During this process work is done on the gas.

Process 3–4

It is an Isothermal compression process.

Pressure increases from P_3 to P_4

Volume decreases from V_3 to V_4 and $T_2 = C$

$$\text{W.D.} = W_{3-4} = P_3V_3 \ln \frac{V_3}{V_4}$$

i.e. $$= -P_3V_3 \ln \frac{V_3}{V_4}$$

$$W_{3-4} = -mRT_2 \ln \frac{V_3}{V_4}$$

– ve sign implies work is done on the gas during compression.

For isothermal process $Q = W$

$$\therefore \quad Q_{3-4} = Q_2 = -mRT_2 \ln \frac{V_3}{V_4} \quad \text{...(2)}$$

Again the cylinder head is made perfect insulator and the gas is compressed adiabatically to state point (1).

Process 4–1

It is an adiabatic compression process.

Pressure increases from P_4 to P_1

Volume decreases from V_4 to V_1

Temperature increases from T_2 to T_1

$$\text{W.D.} = \frac{(P_4V_4 - P_1V_1)}{\gamma - 1} = \frac{-(P_1V_1 - P_4V_4)}{\gamma - 1}$$

and $Q = 0$ for an adiabatic process.

We know that, $$\eta_{Th} = \frac{\text{Work done}}{\text{Heat applied}} = \frac{W}{Q_1} = \frac{Q_1 - Q_2}{Q_1}$$

$\therefore$ From (1) and (2) $$\eta_{Th} = \frac{mRT_1 \ln \frac{V_2}{V_1} - mRT_2 \ln \frac{V_3}{V_4}}{mRT_1 \ln \frac{V_2}{V_1}}$$

Note that the ratio $$\frac{V_2}{V_1} = \frac{V_3}{V_4}$$

$$\therefore \quad \eta_{Th} = \frac{mR \ln \frac{V_2}{V_1} (T_1 \quad T_2)}{mR \ln \frac{V_2}{V_1} \cdot T_1}$$

$$\eta_{Th} \text{ or } \eta_{Carnot} = \frac{T_1 - T_2}{T_1}$$

Note. Carnot cycle is a reversible cycle. If the processes are carried out in anticlockwise direction then it will work as heat pump.

3.11 CARNOT THEOREM

Two cyclic heat engines E_A and E_B operating between the same source and sink out of which E_B is reversible (Fig. 3.15).

Carnot's theorem states that "*No engine operating between two thermal reservoirs can have an efficiency more than that of a Carnot's engine operating between the same two reservoirs*".

That is "*Carnot's engine will be more efficient than any other engine operating between the same temperature reservoirs*".

Let two heat engines E_A and E_B operate between the source at temperature T_1 and the sink at temperature T_2 as shown in Fig. 3.15.

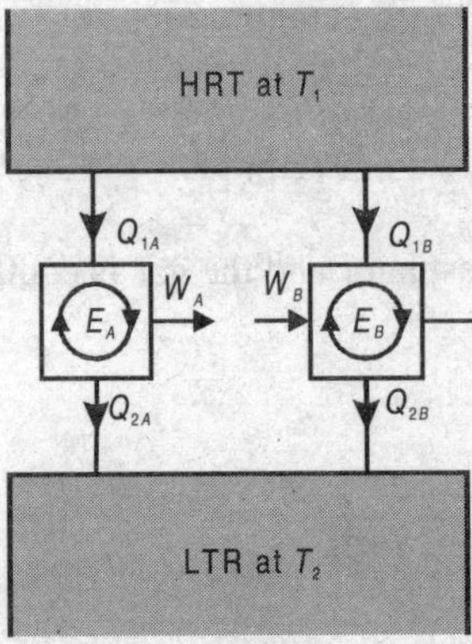

Fig. 3.15

Let E_A be any irreversible engine and E_B be any reversible heat engine. We have to prove that the efficiency of E_B is more than that of E_A. Now, let us assume that this is not true and η_A is greater than η_B. Let the rates of heat transfer are such that,

$$Q_{1A} = Q_{1B} = Q_1$$

and since

$$\eta_A > \eta_B$$

$$\frac{W_A}{Q_{1A}} > \frac{W_B}{Q_{1B}}$$

$$\therefore \quad W_A > W_B$$

Now, let E_B be reversed, since E_B is a reversible heat engine.

Note. Here the magnitudes of heat and work quantities will remain the same, but their directions will be reversed as shown in Fig. 3.16.

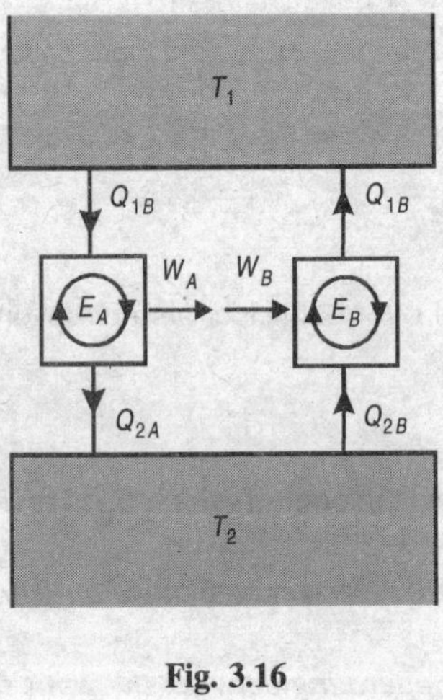

Fig. 3.16

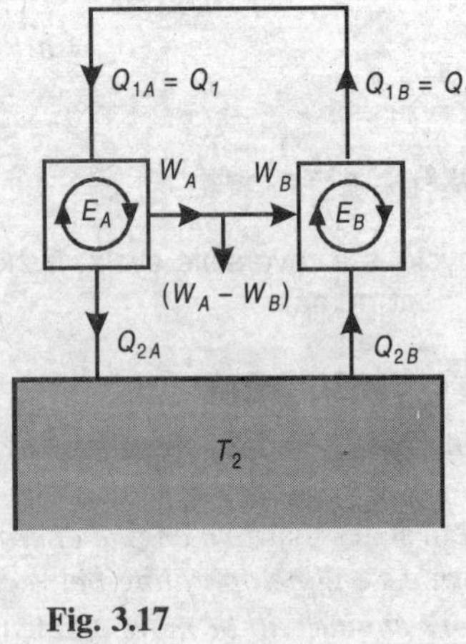

Fig. 3.17

Since $W_A > W_B$, some part of W_A (equal to W_B) may be fed to drive the reversed heat engine E_B.

Since $Q_{1A} = Q_{1B} = Q_1$ the heat discharged by E_B (reversed heat engine B) may be supplied to E_A. The source may therefore be eliminated as shown in the Fig. 3.17.

The net result is that E_A and E_B together will constitute a heat engine, which operating in a cycle, produces, network $(W_A - W_B)$ while exchanging heat with a single reservoir at T_2.

Thus violates the Kelvin Planck's statement of the second law. Hence the assumption than $h_A > h_B$ is wrong.

$$\therefore \quad \boxed{\eta_B \geq \eta_A}$$

3.12 REVERSIBLE CYCLE

If a process can occur in a reverse order, and if the initial state all energies transferred or transformed during the process can be completely restored in both system and environment, the process is called *reversible.* If a process really reversible, then no after effects or changes are evident in the system or in the environment when the process occurs in the forward and then in the reverse direction. An irreversible process that is not reversible. A quasi-static process implies an infinitely slow process since all potential differences acting on the system are infinitesimally small. Such a process may be thought of as an infinite succession of equilibrium states. It may be stopped at any time and made to proceed in the opposite direction, thereby reversing the original process in every detail and restoring the system and environment to its original state.

We know that the thermodynamic cycle is defined as a cyclic process end states of which are same. If all the processes in a cycle are reversible process, then the cycle is called a *reversible cycle.*

Reversible process and consequently reversible cycle is an ideal one. There are no losses in the form of friction, temperature differences for heat transfer to take place. Presence of fraction makes the process mechanically irreversible and the temperature difference for heat transfer during the process makes the process thermally irreversible.

Therefore reversible cycle is an ideal cycle.

Engines working on reversible cycle are called reversible engines and we have seen that:

(a) No heat engine operating in cycles between two reservoirs at different temperatures can have a greater efficiency than a reversible engine operating between the same two reservoirs and

(b) All reversible engines operating between two reservoirs at given temperatures have the same efficiency.

Carnot cycle is a reversible cycle and its efficiency depends only on the temperature.

In reversible engines, heat is received while the working substance is at the same constant temperature as the source and heat is rejected while the working substance is at the same constant temperature as the sink.

3.13 POINTS TO REMEMBER

$$\eta_{\text{Carnot}} = \frac{T_1 - T_2}{T_1}$$

From this equation we can see that, efficiency can be increased either by increasing T_1 or by decreasing T_2. But in case of IC engines since T_2 is the temperature of exhaust gases, hence it cannot be decreased less than a certain value when high temperatures are used, then the metals used in the construction of engine cylinder, cylinder head etc. will not withstand high temperature stresses and will lose some of the properties. So if we have to increase T_1, then the metals which can withstand high temperature stresses are to be used in the construction of the engine even though they are costly.

(a) It is very very difficult to construct, operate and to measure the efficiency of carrot engine. It is not possible to construct such an engine. Since during the same stroke of the piston, isothermal and adiabatic processes are to be achieved.

For isothermal process the engine has to run very slowly so that the temperature during the process is maintained constant. Such a slow speed is not desirable as the size of the engine operating at very low speed is very large. The adiabatic process needs complete insulation and the engine has to run very fast.

As in this cycle isothermal and adiabatic processes take place during the same stroke, the piston has to move very slowly for the first half stroke and very fast for the second half stroke, but this is not possible to design.

(b) The cylinder head is perfect conductor for half part of the stroke and is a perfect insulator during the remaining stroke. This is also very difficult to obtain in actual practice.

(c) It is not possible to perform frictionless reversible process.

LIST OF FORMULAE

1. Thermal efficiency of the heat engine (any engine)

$$\eta_{Th} = \frac{\text{Energy output}}{\text{Energy input}}$$

$$= \frac{\text{Work done}}{\text{Heat supplied}}$$

$$= \frac{W}{Q_1} = \frac{Q_1 - Q_2}{Q_1}$$

2. $$[\text{COP}]_{ref} = \frac{Q_2}{W} = \frac{Q_2}{Q_1 - Q_2} = \frac{T_2}{T_1 - T_2}$$

3. $$[\text{COP}]_{HP} = \frac{Q_1}{W} = \frac{Q_1}{Q_1 - Q_2} = \frac{T_1}{T_2 T_2}$$

Also, $$[\text{COP}]_{HP} = [\text{COP}]_{ref} + 1$$

4. $$\eta_{Carnot} = \frac{T_1 - T_2}{T_1} = 1 - \frac{T_2}{T_1} \quad \text{...(a)}$$

Also $$\eta_{\text{any engine}} = \frac{Q_1 - Q_2}{Q_1} = 1 - \frac{Q_2}{Q_1} \quad \text{...(b)}$$

For maximum efficiency

$$\eta_{\text{any engine}} = \eta_{Carnot}$$

$$\therefore \quad \frac{Q_2}{Q_1} = \frac{T_2}{T_1}$$

5. $$\eta_{Carnot} > \eta_{\text{any engine}}$$

SOLVED EXAMPLES

Example 3.1 *What will be the maximum efficiency of a heat engine operating between 227° C and 27° C?*

Data:

$$T_1 = 227 + 273 = 500\ K$$

$$T_2 = 27 + 273 = 300\ K$$

$$\eta_{Carnot} = ?$$

Solution

Maximum efficiency or Carnot engine is given by,

$$\eta_{Carnot} = \frac{T_1 - T_2}{T_1} = \frac{500 - 300}{500} = 0.4$$

$$\therefore \qquad \eta_{Carnot} = \mathbf{40\%}$$

Example 3.2 *What would be the COP of the carnot refrigerator and heat pump working between temperature limits of 27° C and 600 K?*

Find: COP of the carnot refrigerator and heat pump.

Data: T_1 = 600 K, T_2 = 273 + 27 = 300 K

Solution

We know that,

$$[COP]_{ref} = \frac{Q_2}{Q_1 - Q_2} = \frac{T_2}{T_1 - T_2} = \frac{300}{600 - 300}$$

$$\therefore \qquad \mathbf{[COP]_{ref} = 1}$$

$$[COP]_{HP} = \frac{Q_2}{Q_1 - Q_1} = \frac{T_2}{T_1 - T_2} = \frac{600}{600 - 300}$$

$$\therefore \qquad \mathbf{[COP]_{HP} = 2}$$

Alternatively

$$[COP]_{HP} = [COP]_{ref} + 1 = 1 + 1 = 2$$

Example 3.3 *A heat engine operates on Carnot cycle between source and sink temperatures 227°C and 27°C respectively. If the heat engine receives 400 kJ from the source, find the net work done, heat rejected to the sink and efficiency of the heat engine.*

Find: W_{net} = ?, Q_2 = ?, η_{Carnot} = ?

Data: T_1 = 227°C + 273 = 500 K, T_2 = 27°C + 273 = 300 K, Q_1 = 400 kJ

Solution

(i)
$$\eta_{Carnot} = \frac{T_1 - T_2}{T_1}$$

$$= \frac{500 - 300}{500}$$

$$\eta_{Carnot} = \mathbf{0.4 = 40\%}$$

(ii)
$$\eta = \frac{\text{Work done}}{\text{Heat supplied}} = \frac{W_{net}}{Q_1}$$

$$\therefore \qquad W_{net} = \eta_{Carnot} \times Q_1$$

$$= 0.4 \times 400 \text{ kJ}$$

$$\mathbf{W_{net} = 160 \text{ kJ}}$$

(iii)
$$W_{net} = Q_1 - Q_2$$

$$160 \text{ kJ} = 400 - Q_2$$

$$\therefore \qquad \mathbf{Q_2 = 240 \text{ kJ}}$$

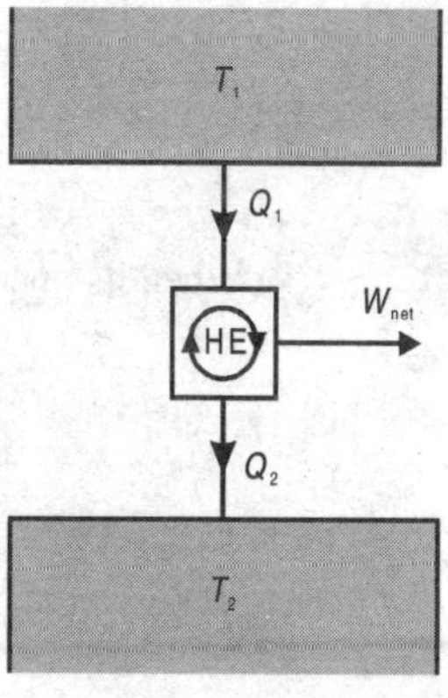

Fig. Ex. 3.3

Example 3.4 *A Carnot cycle heat engine operates between a source and sink temperatures of 227°C and 27°C. If the heat engine receives 25 kJ of heat from the source, calculate:*

1. Work done. 2. Heat rejected. 3. η_{Carnot}

What will be the COP if it operates as a heat pump and when it works as a refrigerator?

Find: $W_{net} = ?, Q_2 = ?, \eta_{Carnot} = ?, [COP]_{HP} = ?, [COP]_{ref} = ?$

Data: $T = 227 + 273 = 500\ K, T_2 = 27 + 273 = 300\ K, Q_1 = 425\ kJ$

Solution

(i)
$$\eta_{Carnot} = \frac{T_1 - T_2}{T_1} = \frac{500-300}{500}$$

$$\boldsymbol{\eta_{Carnot} = 0.4 = 40\%}$$

(ii)
$$\eta = \frac{W_{net}}{Q}$$

$$0.4 = \frac{W_{net}}{425}$$

$$\therefore \quad \mathbf{W_{net} = 170\ kJ}$$

(iii)
$$W_{net} = Q_1 - Q_2$$

$$170 = 425 - Q_2$$

$$\therefore \quad \mathbf{Q_2 = 255\ kJ}$$

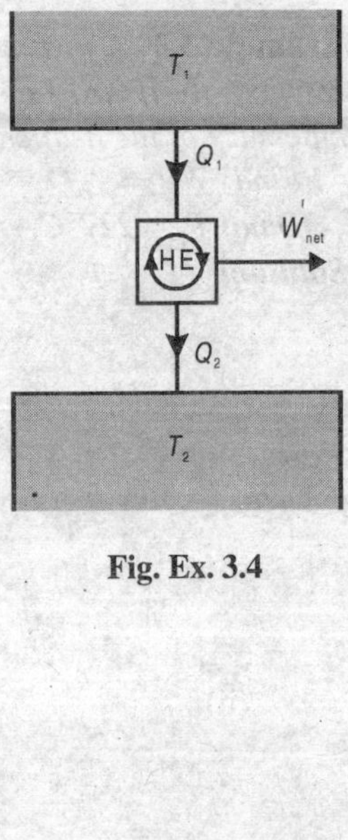

Fig. Ex. 3.4

(iv) When it operates as a heat pump:

$$[COP]_{HP} = \frac{Q_1}{W} = \frac{Q_1}{Q_1 - Q_2}$$

$$= \frac{T_1}{T_1 - T_2} = \frac{500}{500-300}$$

$$\mathbf{[COP]_{HP} = 2.5}$$

(v) When its operates as a refrigerator :

$$[COP]_{ref} = \frac{Q_2}{W} = \frac{Q_2}{Q_1 - Q_2} = \frac{T_1}{T_1 - T_2}$$

$$= \frac{300}{500-300}$$

$$\mathbf{[COP]_{ref} = 1.5}$$

Example 3.5 *A domestic refrigerator maintains a temperature of – 10ºC. The ambient (atmospheric) temperature is 35ºC. If heat leaks into the freezer at the continuous rate of 1.75 kJ/sec. What is the least power necessary to pump this heat out continuously?*

Data: Freezer temperature $T_2 = -10 + 273 = 263\ K, T_1 = 35 + 273 = 308\ K.$

Solution

For minimum power requirements or for maximum COP, we have,

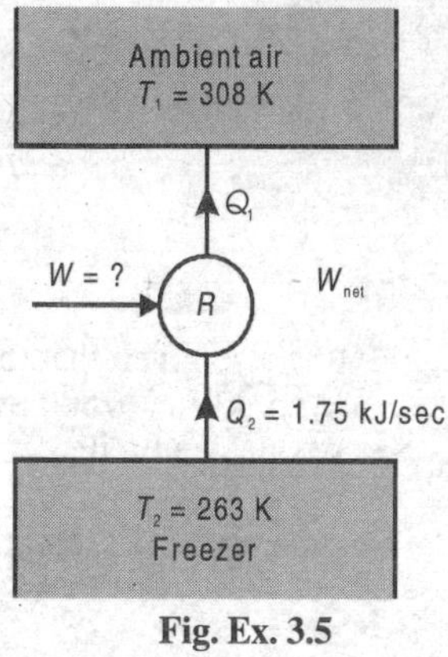

Fig. Ex. 3.5

$$\frac{Q_2}{Q_1} = \frac{T_2}{T_1}$$

$$\therefore \quad Q_1 = \frac{T_1}{T_2} \times Q_2 = \frac{308}{263} \times 1.75 \text{ kJ/sec}$$

$$Q_1 = 2.049 \text{ kJ/sec}$$

$$\therefore \quad W = Q_1 - Q_2 = 2.049 - 1.75$$

$$\mathbf{W = 0.2994 \text{ kJ/sec}}$$

Example 3.6 *A heat pump is used to maintain an auditorium hall at 25°C, when the atmospheric temperature is 10°C. The heat leakage from the hall is 1500 kJ/min. Calculate the power required to run the actual heat pump, if the COP of the actual heat pump is 30% of the COP of the Carnot heat pump working between the same temperature limits.*

Data: T_1 = 25 + 273 = 298 K, T_2 = 10 + 273 = 283 K, heat leakage Q_1 = 1500 kJ/min = 25 kJ/sec.

Solution

It is given that,

$$[\text{COP}]_{\text{actual}} = 0.3 \times [\text{COP}]_{\text{Carnot HP}}$$

We know that,

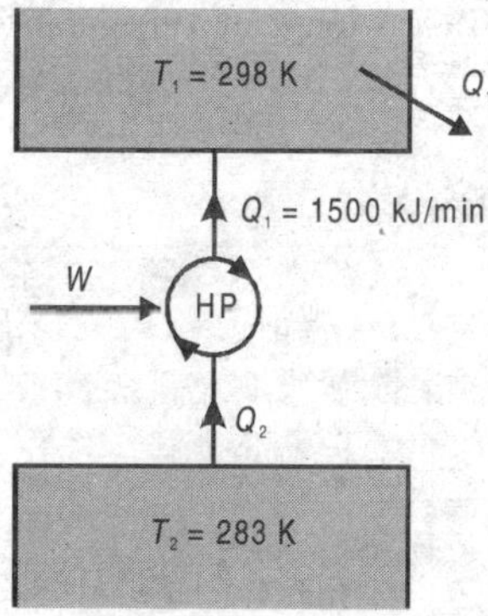

Fig. Ex. 3.6

$$[\text{COP}]_{\text{Carnot HP}} = \frac{T_1}{T_1 - T_2} = \frac{298}{298 - 283}$$

$$= 19.867$$

$$\therefore \quad [\text{COP}]_{\text{actual HP}} = 0.3 \times 19.867 = 5.96$$

Also,
$$[\text{COP}]_{\text{HP}} = \frac{Q_1}{W}$$

$$5.96 = \frac{25 \text{ kJ / sec}}{\text{W.D.}}$$

Work done or power required,

$$W = \frac{25}{5.96}$$

$$\mathbf{W = 4.195 \text{ kJ/sec}}$$

Example 3.7. *The C.O.P. of a Carnot refrigerator is 6, when it maintains the temperature of –3°C in the evaporator. Determine the condenser temperature and refrigerating effect if the power required to run the refrigerator is 7.5 kW.*

Data: T_2 = 273 – 3 = 270 K, $[COP]_{ref.}$ = 6, T_1 = ?

Solution

$$W = 7.5 \text{ kW}$$

Now
$$[\text{COP}]_{\text{ref}} = \frac{T_2}{T_1 - T_2} = \frac{\text{Evaporator temperature}}{\text{Condenser temperature} - \text{Evaporator temperature}}$$

$$6 = \frac{270}{T_1 - 270}$$

$$6(T_1 - 270) = 270$$

$$6T_1 = 270 + 6 \times 270$$

$$6T_1 = 270 + 1620$$

$$\therefore \quad T_1 = 315 \text{ K}$$

It is also given that the power required is,

= 7.5 kW = 7500 watts = 7500 J/sec.

We also know that,

$$[\text{COP}]_{\text{ref}} = \frac{Q_2}{W} = \frac{\text{Desired effect}}{\text{Work input}}$$

$$= \frac{\text{Refrigerating effect/sec}}{\text{Work supplied/sec}}$$

$$6 = \frac{\text{Refrigerating effect}}{7500}$$

$$\therefore \quad \text{Refrigerating effect} = 45000 \text{ J/sec.}$$

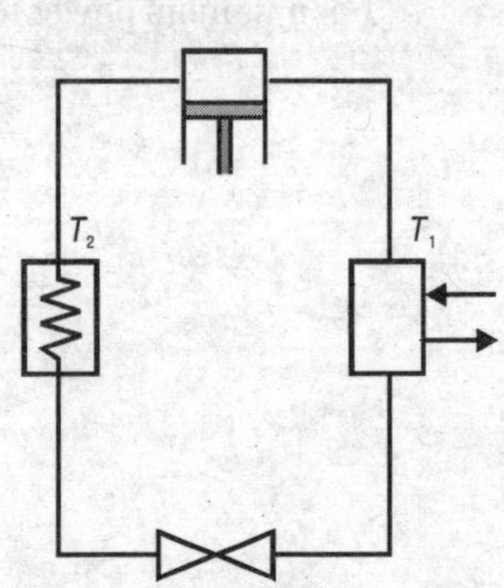

Fig. Ex. 3.7

Example 3.8 *A fish freezing plant of 100 tonnes capacity is to be maintained at – 50ºC when the outside atmospheric temperature is 40ºC. The actual C.O.P. of the refrigeration system used is 1/5 of the theoretical Carnot refrigerator working between the same temperature limits. Calculate the power required to run the plant.*

Solution

Take 1 tonne of refrigeration in SI units = 3.516 kJ/sec of heat to be removed (Q_2).

$$\therefore \quad 100 \text{ ton of capacity} = 100 \times 3.516 = 351.6 \text{ kJ/sec} = Q_2$$

$$T_1 = 40 + 273 = 313 \text{ K}$$

$$T_2 = -50 + 273 = 223 \text{ K}$$

Since $$[\text{COP}]_{\text{ref}} = \frac{1}{5}[\text{COP}]_{\text{Carnot ref.}}$$

$$\therefore \quad [\text{COP}]_{\text{Carnot}} = \frac{T_2}{T_1 - T_2} = \frac{223}{313 - 223} = 2.478$$

But $$[\text{COP}]_{\text{actual ref.}} = \frac{1}{5} \times 2.478 = 0.4956$$

Also we know that, $$[\text{COP}]_{\text{ref}} = \frac{Q_2}{W}$$

$$W = \frac{Q_2}{[\text{COP}]_{\text{ref}}}$$

$$\mathbf{W = \frac{351.6}{0.4956} = 709.44 \text{ kJ/sec.}}$$

Example 3.9 *The efficiency of a Carnot engine rejecting heat to a cooling pond at 28°C is 30%. If the cooling pond receives 1050 kJ/min. What is the power developed by the cycle in kW. Also find the temperature of the source.*

Find: $W = ?$, $T_1 = ?$

Data: $T_2 = 28 + 273 = 301$ K, $Q_0 = 1050$ kJ/min $= 17.5$ kJ/sec., $\eta_{Carnot} = 0.3$

Solution

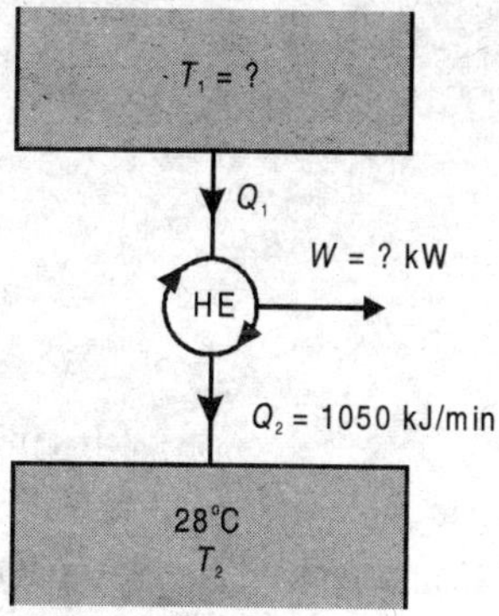

Fig. Ex. 3.9

We have,

$$\eta_{Carnot} = \frac{T_1 - T_2}{T_1}$$

$$0.3 = \frac{T_1 - 301}{T_1} = 1 - \frac{301}{T_1}$$

$$\frac{301}{T_1} = 1 - 0.3$$

$$\frac{301}{0.7} = T_1$$

$$\therefore \quad \mathbf{T_1 = 430\ K}$$

We also know that,

$$\eta = \frac{Q_1 - Q_2}{Q_1}$$

$$= 1 - \frac{Q_2}{Q_1}$$

$$0.3 = 1 - \frac{17.5}{Q_1}$$

$$\frac{17.5}{Q_1} = 1 - 0.3$$

$$\frac{17.5}{0.7} = Q_1 = 25 \text{ kJ/sec} = 25 \text{ kW}$$

$\therefore$ Power = Rate of doing work,

$$W = Q_1 - Q_2$$

$$\mathbf{W = 25 - 17.5 = 7.5\ kW}$$

Example 3.10 *300 kg of fish at 5° C is to be frozen at – 2°C. Specific heat above freezing point is 3.182 kJ/kg-K and latent heat of fusion is 234.5 (kJ/kg Freezing paint is – 2°C. A refrigerator is to reject heat to surrounding at 40°C. A refrigerator has actual COP 60% of Carnot refrigerator operating in the same temperature limits. How much power must be supplied to remove the heat in 10 hours.*

Solution

We know that,

$$[\text{COP}]_{\text{ref.}} = \frac{\text{Desired effect}}{\text{Work input}} = \frac{Q_2}{W}$$

$$= \frac{Q_2}{Q_1 - Q_2} = \frac{T_2}{T_1 - T_2} \qquad \text{...(1)}$$

Also

$$Q_2 = m[C_P \cdot \Delta t + \text{Latent heat of fusion}]$$

$$= m[C_P\, \Delta t + L_f]$$

$$= 300\left[3.187\times\left[5-(-2)\right]+234.5\right]$$

$$= 300[3.187\times 7+234.5]$$

$$Q_2 = 77032.2 \text{ kJ}$$

$$\eta_{\text{Carnot ref}} = \frac{T_2}{T_1-T_2} = \frac{271}{313-271}$$

$$= 6.452$$

$$\therefore \quad [\text{COP}]_{\text{actual ref.}} = 0.6\times 6.452 = 3.871$$

Also $$[\text{COP}]_{\text{actual}} = \frac{Q_2}{W} = \frac{77032.2}{W} = 3.871$$

$$\therefore \quad W = 19897.616 \text{ kJ}$$

$$\therefore \quad \text{Rate of work output} = \frac{19897.616}{10\times 3600}$$

Rate of work input = 0.552 kW

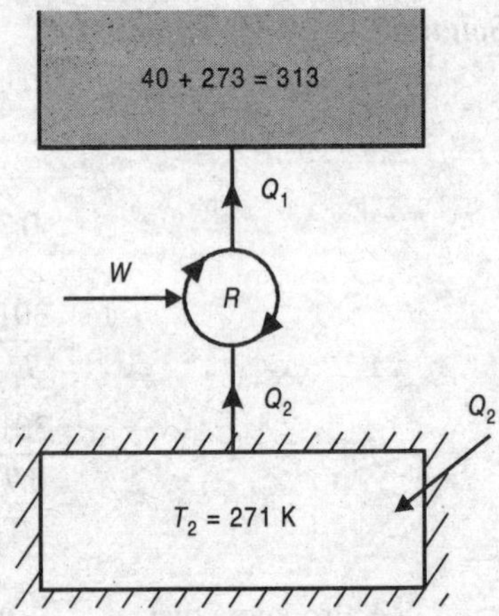

Fig. Ex. 3.10

Example 3.11 *An engineer claims his engine to develop 3.68 kW, on testing the engine consumes 0.44 kg of fuel in one hour. It has the calorific value (C.V.) of 41, 870 kJ/kg. The maximum temperature recorded in the cycle is 1400°C and minimum is 350° C. Is the engineer justified in his claim ?*

Solution

$$W = 3.68 \text{ kW} = 3.68 \text{ kJ/sec}$$

$$T_1 = 1400+273 = 1673 \text{ K}$$

$$T_2 = 350+273 = 623 \text{ K}$$

We know that, $$\eta_{\text{Carnot}} = \frac{T_1-T_2}{T_1}$$

$$= \frac{1673-623}{1673} = 0.6276 = 62.76\%$$

For actual engine, heat supplied/hr = C.V. × Mass of fuel consumption

= 41870 kJ/kg × 0.44 kg/hr = 18422.8 kJ/hr.

Work development = 3.68 kW = 3.68 kJ/sec

Work development = 13248 kJ/hr

$$\therefore \text{ Efficiency of actual engine} = \frac{\text{W.D.}}{\text{Heat supplied}} = \frac{13248}{18422.8}$$

Efficiency of actual engine = 71.91%

The actual efficiency of the engine is greater than the efficiency of carnot engine operating between the same temperature limits. So the claim of the engineer is not justified.

Example 3.12 *An inventor claims that his newly developed heat engine will produce 0.184 kW power, when the heat is supplied to engine at the rate of 70 kJ/min. The cycle operates between the maximum temperature of 1092°C and minimum temperature of 149°C. Make complete analysis of the inventors claim and clearly state whether it is worth investing any amount of money on such a project. Give reasons for your suggestions.*

Solution

$$W = 0.184 \text{ kW power}$$

$$Q = 70\,\text{kJ/min} = \frac{70}{60}\,\text{kJ/sec} = 1.17\text{ kW}$$

$$T_1 = 1092°\text{C} + 273 = 1365\text{ K}$$

$$T_2 = 149° + 273 = 422\text{ K}$$

$$\therefore \quad \eta_{Carnot} = \frac{T_1 - T_2}{T_1}$$

$$= \frac{1365 - 422}{1365} = 0.6908 = 69.08\%$$

$$\therefore \quad \eta_{actual} = \frac{\text{W.D.}}{\text{Heat supplied}} \frac{0.184}{1.17}$$

$$= 0.1577$$

$$\boldsymbol{\eta_{actual} = 15.77\%}$$

Since the efficiency of the actual engine is too low there is no point in investing any amount in such a project.

Example 3.13 *An inventor claims to have invented refrigeration machine operating between – 23°C and 27°C. It consumes 1 kW electrical power and gives 20000 kJ of refrigerating effect in one hour. Comment on his claim.*

Data: $T_1 = 300\text{ K} \qquad T_2 = 250\text{ K}$

$$W_{input} = 1\text{ kW} \qquad Q_2 = \text{Refrigerating} = \frac{20000}{3600}\text{ kJ/sec}$$

$$Q_2 = 5.55\text{ kJ/sec}$$

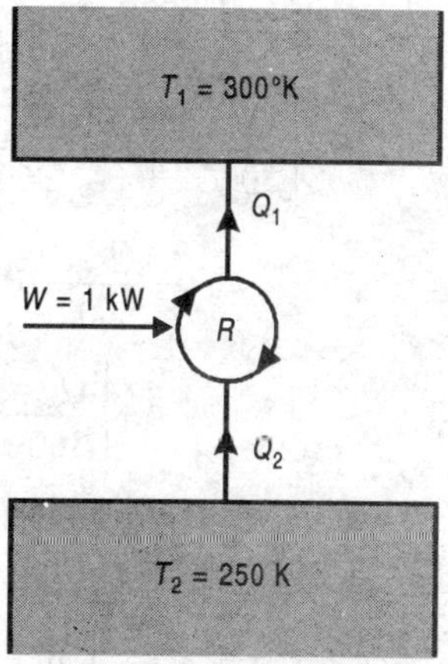

Fig. Ex. 3.13

Solution

We know that,

$$[\text{COP}]_{\text{Carnot ref}} = \frac{T_2}{T_1 - T_2} = \frac{250}{300 - 250}$$

$$[\text{COP}]_{\text{Carnot ref}} = 5 \qquad \text{...(1)}$$

$$\therefore \quad [\text{COP}]_{\text{actual ref}} = \frac{\text{Refrigerating effect}}{\text{Work input}} = \frac{5.55}{1}$$

$$= 5.55 \qquad \text{...(2)}$$

$$[\text{COP}]_{\text{actual.ref.}} > [\text{COP}]_{\text{Carnot ref.}}$$

∴ Claim of the inventor is not valid.

Example 3.14 *A reversible engine working in a cycle takes 4800 kJ of heat from a source at 800 K per minute and develops 20 kW power. The engine rejects heat to two reservoirs at 300 K and 360 K. Determine the heat rejected to each sink in kJ/min.*

Solution

It is given that,

W.D. by the reversible engine = 20 kW = 20 kJ/sec = 20 × 60 kJ/min = 1200 kJ/min.

(Since we have to find heat rejected in kJ/min)

$$Q_1 = 4800\,\text{kJ/min} \qquad \therefore\ W = Q_1 - Q_2$$

$$Q_2 = Q_1 - W$$

$$Q_2 = 4800 - 1200 = 3600\text{ kJ/min}$$

Since this heat is rejected to two sinks (*a*) and (*b*)

$$\left.\begin{aligned} Q_2 &= Q_{2a} + Q_{2b} \\ \therefore \quad Q_{2b} &= Q_2 - Q_{2a} \\ Q_{2b} &= 3600 - Q_{2a} \end{aligned}\right\} \quad ...(1)$$

We also know that, $\dfrac{Q_1}{T_1} = \dfrac{Q_2}{T_2}$

$$\therefore \quad \frac{Q_1}{T_1} - \frac{Q_2}{T_2} = 0$$

$$\therefore \quad \frac{Q_1}{T_1} - \left(\frac{Q_{2a}}{T_{2a}} + \frac{Q_{2b}}{T_{2b}}\right) = 0$$

Let Q_{2a} be the heat rejected to sink *a*

Q_{2b} be the heat rejected to sink *b*

$$\frac{4800}{800} - \left(\frac{Q_{2a}}{300} + \frac{Q_{2b}}{360}\right) = 0$$

$$\frac{4800}{800} - \left(\frac{Q_{2a}}{300} + \frac{3600 - Q_{2a}}{360}\right) = 0 \quad ...\text{(From 1)}$$

$$6 - \left[\frac{Q_{2a}}{300} + \frac{3600}{360} - \frac{Q_{2a}}{360}\right] = 0$$

i.e. $$6 - \frac{Q_{2a}}{300} - 10 + \frac{Q_{2a}}{360} = 0$$

i.e. $$-4 + \left[\frac{Q_{2a}}{360} - \frac{Q_{2a}}{300}\right] = 0$$

i.e. $$Q_{2a}\left[\frac{1}{360} - \frac{1}{300}\right] = 4$$

$$Q_{2a} \times \left[\frac{300 - 360}{360 \times 300}\right] = 4$$

$$Q_{2a} \times \left[\frac{-60}{108000}\right] = 4$$

$$Q_{2a}\left[-0.000555\right] = 4$$

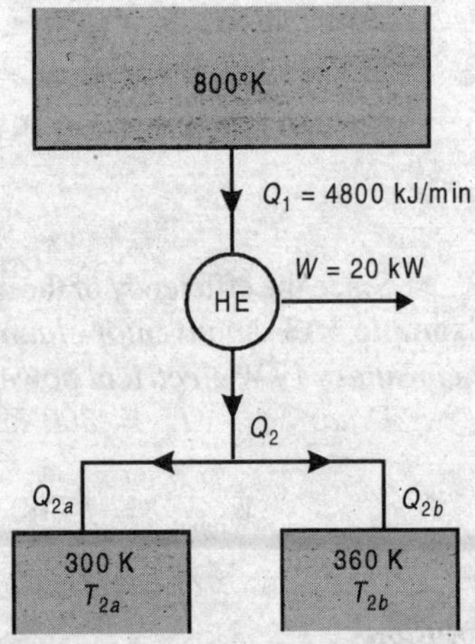

Fig. Ex. 3.14

Q_2 = −7200.00 kJ/min

$\therefore$ Since $$Q_2 = Q_{2a} + Q_{2b}$$

$$3600 = -7200 + Q_{2b}$$

$$\therefore \quad Q_{2b} = 3600 + 7200$$

Q_{2b} = 10800 kJ/min

Example 3.15 *A reversible engine working on reversible cycle is supplied with 5000 kJ/min of heat from a source at 527°C while it develops 25 kW of power. This engine rejects heat to two reservoirs at two different temperatures equal to 27°C and 97°C respectively. Determine the efficiency of the engine and heat rejected to each reservoir.*

Solution

$$T_1 = 527+273$$
$$Q_1 = 5000 \text{ kJ/min}$$
$$W = 25 \text{ kW} = 25 \text{ kJ/sec}$$
$$= 25\times60 = 1500 \text{ kJ/min}$$

and Q_2 = Heat rejected/min

We know that, $W = Q_1 - Q_2$

$\therefore$ $Q_2 = Q_1 - W$

$$Q_2 = 5000-1500=3500 \text{ kJ/min}$$

$$\left.\begin{aligned} Q_2 &= Q_{2a}+Q_{2b} \\ Q_{2a} &= Q_2 - Q_{2b} \end{aligned}\right\} \quad ...(1)$$

$$Q_{2a} = 3500-Q_{2b}$$

$$\text{Efficiency of the engine} = \frac{\text{W.D.}}{\text{Heat supplied}}$$

$$= \frac{W}{Q} = \frac{1500}{5000}\times100 = 30\%$$

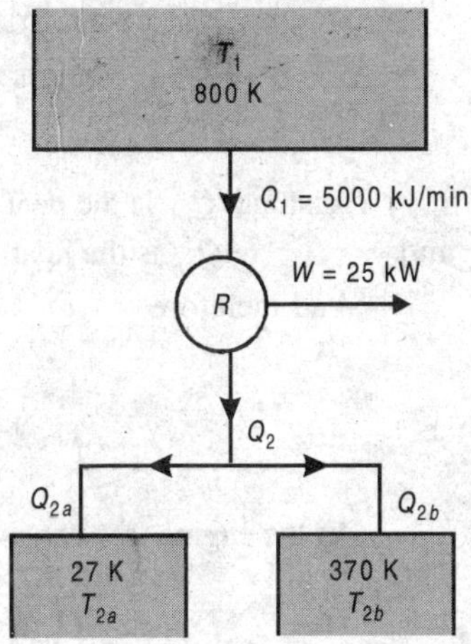

Fig. Ex. 3.15

For reversible engines, we have,

$$\frac{Q_1}{T_1} = \frac{Q_2}{T_2}$$

i.e. $$\frac{Q_1}{T_1} - \frac{Q_2}{T_2} = 0$$

i.e. $$\frac{Q_1}{T_1} - \left[\frac{Q_{2a}}{T_{2a}} + \frac{Q_{2b}}{T_{2b}}\right] = 0$$

$$\frac{5000}{800} - \left[\frac{3500-Q_{2b}}{370} + \frac{Q_{2b}}{300}\right] = 0$$

$\therefore$ We get, $\mathbf{Q_{2b} = -5090.14}$ **kJ/min**

$\therefore$ $Q_{2a} = 3500 - Q_{2b} = 3500 + 5090.14$

$\mathbf{Q_{2a} = 8590.14}$ **kJ/min**

Example 3.16 *A reversible engine is supplied with heat from two constant temperature sources at 900 K and 600 K and rejects heat to a constant temperature sink at 300 K. If the engine executes a number of complete cycles while developing 100 kW, and rejecting 3600 kJ of heat per min. Determine the heat supplied by each source per minute and efficiency of the engine.*

Solution

Now the work developed by the engine

= 100 kW

$$= 100 \text{ kJ/sec} = 100 \times 60 \text{ kJ/min}$$
$$= 6000 \text{ kJ/min}$$

We know that, $W = Q_1 - Q_2$

Given $Q_2 = 3600 \text{ kJ/min}$

$\therefore$ $W + Q_2 = Q_1$

$600 + 3600 = Q_1$

$\therefore$ $Q_1 = 9600 \text{ kJ/min}$

Assume Q_{1a} is the heat supplied by the source (*a*)

and Q_{1b} is the heat supplied by the source (*b*)

And therefore $Q_1 = Q_{1a} + Q_{1b}$

$\therefore$ $Q_{1b} = Q_1 - Q_{1a}$

i.e., $Q_{1b} = 9600 - Q_{1a}$ ...(1)

As the engine is reversible, the required condition for the engine to work is, $\dfrac{Q_1}{T_1} = \dfrac{Q_2}{T_2}$

$$\frac{Q_{1a}}{T_{1a}} + \frac{Q_{1b}}{T_{1b}} = \frac{Q_2}{T_2} \quad \text{...(2)}$$

$$\frac{Q_{1a}}{T_{1a}} + \frac{9600 - Q_{1a}}{T_{1b}} = \frac{Q_2}{T_2}$$

$$\frac{Q_{1a}}{900} + \frac{9600 - Q_{1a}}{600} = \frac{3600}{300}$$

$$\frac{Q_{1a}}{900} + \frac{9600}{600} - \frac{Q_{1a}}{600} = 12$$

$$\frac{Q_{1a}}{900} - \frac{Q_{1a}}{600} + 16 = 12$$

$$Q_{1a}\left[\frac{1}{900} - \frac{1}{600}\right] = -4$$

$$Q_{1a}\left[\frac{600 - 900}{900 \times 600}\right] = -4$$

$$Q_{1a}\left[\frac{-300}{900 \times 600}\right] = -4$$

$$Q_{1a}[-0.000555] = -4$$

$$Q_{1a} = 7200 \text{ kJ/min}$$

$\therefore$ $Q_{1b} = 9600 - 7200$

$$Q_{1b} = 2400 \text{ kJ/min}$$

i.e., Heat supplied by the source *b*

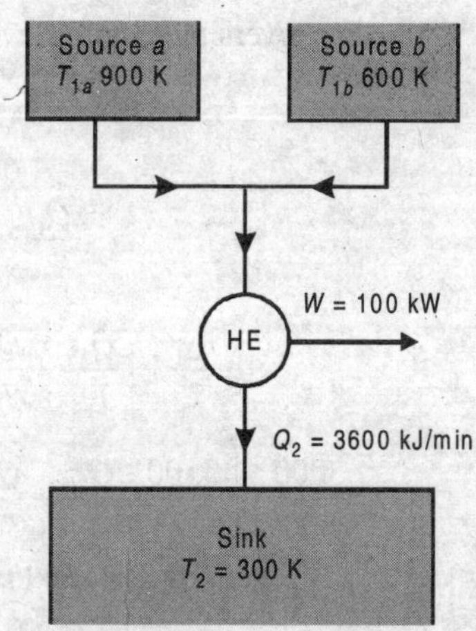

Fig. Ex. 3.16

$$\text{Now, the Engine efficiency} = \frac{\text{W.D.}}{\text{Heat supplied}}$$

$$= \frac{6000}{Q_1}$$

$$= \frac{6000}{9600} = 0.625$$

i.e., η = **62.5%**

Example 3.17 *Two reversible engines R_1 and R_2 are connected in series between a heat source S and cold body C. R_1 receives 900 kJ from S, at 540 K and R_2 rejects heat to C at 210 K. Assuming equal thermal efficiencies from R_1 and R_2, determine,*

(i) *The temperature at which heat is rejected by R_1 and received by R_2.*
(ii) *The work output W_1 and W_2 of the two engines.*
(iii) *The heat rejected by R_2 to the cold body C.*

Data: Q_1 = 900 kJ , T_1 = 540 K, T_2 = 210 K

Solution

Thermal efficiencies of R_1 and R_2 are equal to

$$\therefore \quad \text{(i)} \quad \eta_{R_1} = \frac{T_1 - T_3}{T_1}$$

$$\text{and} \quad \eta_{R_2} = \frac{T_3 - T_2}{T_3}$$

$$\text{Since} \quad \eta_{R_1} = \eta_{R_2}$$

$$\therefore \quad \frac{T_1 - T_3}{T_1} = \frac{T_3 - T_2}{T_3}$$

$$T_3T_1 - T_3^2 = T_1\,T_3 - T_1\,T_2$$

$$T_1\,T_3 - T_1\,T_3 + T_1\,T_2 = T_3^2$$

$$\therefore \quad T_3^2 = T_1\,T_2 = 540 \times 210$$

$$\therefore \quad T_3^2 = 113400$$

$$\therefore \quad T_3 = 336.75 \text{ K}$$

(ii) We know that, $W_1 = Q_1 - Q_3$ ...(1)

We also know that, $\dfrac{Q_1}{T_1} = \dfrac{Q_2}{T_2} = \dfrac{Q_3}{T_3}$

$$\therefore \quad Q_3 = \frac{T_3}{T_1} \times Q_1$$

Substituting this in Eq. (1)
We get,

$$\therefore \quad W_1 = Q_1 - Q_1\frac{T_3}{T_1}$$

$$= 900 - 900 \times \frac{336.75}{540}$$

$$\mathbf{W_1 = 338.75\ kJ}$$

Similarly $$W_2 = Q_3 - Q_2 = Q_1 \times \frac{T_3}{T_1} - Q_2$$

From $$\frac{Q_1}{T_1} = \frac{Q_2}{T_2}$$

$$Q_2 = \frac{T_2}{T_1} \times Q_1 \text{ (i.e., } Q_2 \text{ in terms of } Q_1)$$

$$W_2 = Q_1 \times \frac{T_3}{T_1} - Q_1 \times \frac{T_2}{T_1} = 900 \times \frac{336.75}{540} - 900 \times \frac{210}{540}$$

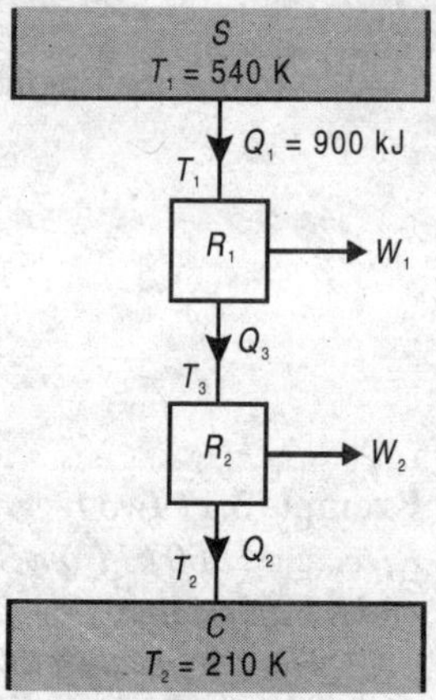

Fig. Ex. 3.17

$$\therefore \quad \mathbf{W_2 = 211.25\ kJ}$$

(iii) Heat rejected by R_2 to cold body (Q_2)

Since, $$\frac{Q_1}{T_1} = \frac{Q_2}{T_2}$$

$$Q_2 = \frac{T_2}{T_1} \times Q_1$$

$$= \frac{900 \times 210}{540}$$

$$\mathbf{Q_2 = 350\ kJ}$$

Example 3.18 *A reversible heat engine operates between two reservoirs at temperatures of 600°C and 40°C. The engine drives a reversible refrigerator, which operates between reservoirs at temperature 40°C and – 20°C. The heat transfer to the heat engine is 2000 kJ and the net work output of the combined engine refrigerator plant is 360 kJ,*

(a) *Evaluate the heat transfer to the refrigerator and the net heat transfer to the reservoir at 40°C.*

(b) *Reconsider (a) given that the efficiency of the heat engine and the C.O.P. of the refrigerator are each 40% of their maximum possible values.*

Solution

(a) Maximum efficiency of the heat engine

$$\eta_{max} = \frac{\text{W.D.}}{\text{Heat supplied}}$$

$$= \frac{W_1}{Q_1} = \frac{T_1 - T_2}{T_1}$$

i.e. $$\eta_{Carnot} \text{ or } \eta_{max} = 1 - \frac{T_2}{T_1} = 1 - \frac{313}{873} = 0.642$$

$$\therefore \quad \text{From } \eta_{max} = \frac{W_1}{Q_1} = 0.642$$

$$\therefore \quad 0.642 \times 2000 = W_1$$

$$\therefore \quad 1284 \text{ kJ} = W_1$$

Since $Q_1 - Q_2 = W_1$

$$Q_2 = Q_1 - W_1$$

$$Q_2 = 2000 - 1284$$

$$= 716 \text{ kJ}$$

Since $W_1 - W_2 = W = 360$ kJ

$$\therefore \quad W_2 = W_1 - W$$

$$= 1284 - 360$$

$$W_2 = 924 \text{ kJ}$$

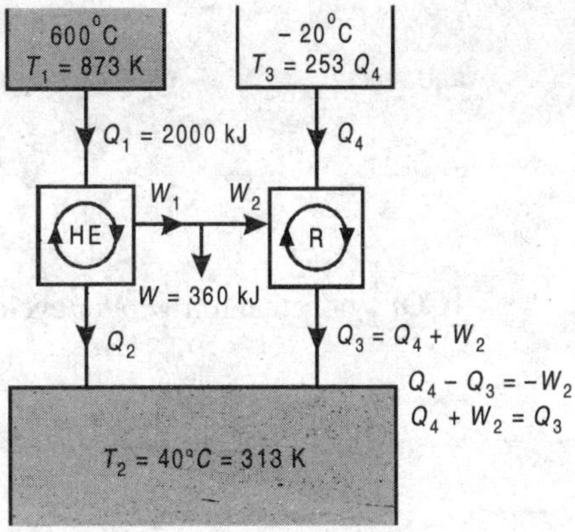

Fig. Ex. 3.18

Now we know that,

$$[\text{COP}]_{\text{ref}} = \frac{\text{Desired effect}}{\text{Work supplied}} = \frac{Q_2}{W} = \frac{Q_2}{Q_1 - Q_2} = \frac{T_2}{T_1 - T_2}$$

But from Fig. Ex. 3.18 $\quad [\text{COP}]_{\text{ref}} = \dfrac{T_3}{T_2 - T_3} = \dfrac{253}{313 - 253} = 4.22$

Also $\quad [\text{COP}]_{\text{ref}} = \dfrac{Q_4}{W_2}$ (Ref. Fig. Ex. 3.18) = 4.22

$$\therefore \quad Q_4 = W_2 \times [\text{COP}]_{\text{ref}}$$

$$Q_4 = 924 \times 4.22 = 3899 \text{ kJ}$$

$$\therefore \quad Q_3 = Q_4 + W_2 = 924 + 3899$$

$$Q_3 = 4823 \text{ kJ}$$

$\therefore$ Heat rejected to the 40°C reservoir

$$= Q_2 + Q_3$$

$$= 716 + 4823$$

$$= 5539 \text{ kJ}$$

(b) Efficiency of the actual heat engine cycle,

$$\eta_{\text{actual}} = 0.4\,\eta_{\text{max}}$$

$$= 0.4 \times 0.642$$

$$= 0.2568$$

Also since $\quad \eta_{\text{actual}} = \dfrac{W_1}{Q_1}$

$$\therefore \quad W_1 = 0.2568 \times 2000$$

$$= 513.6 \text{ kJ}$$

Since $\quad W_1 - W_2 = W$

$$\therefore \quad W_2 = W_1 - W$$

$$W_2 = 513.6 - 360$$

$$= 153.6 \text{ kJ}$$

and since $\quad Q_1 - Q_2 = W_1$

$$\therefore \quad Q_2 = Q_1 - W_1$$
$$= 2000 - 513.6$$
$$Q_2 = 1486.4 \text{ kJ}$$

$[\text{COP}]$ of actual refrigerator cycle,

$$[\text{COP}]_{\text{actual ref.}} = \frac{Q_4}{W_2}$$
$$= 0.4 \times 4.22$$
$$= 1.69$$
$$\therefore \quad Q_4 = 153.6 \times 1.69$$
$$Q_4 = 259.6 \text{ kJ}$$
$$\therefore \quad Q_2 = Q_4 + W_2$$
$$Q_2 = 159.6 + 153.6$$
$$Q_2 = 413.2 \text{ kJ}$$

$\therefore$ Heat rejected to 40°C reservoir

$$= Q_2 + Q_3$$
$$= 413.2 + 1486.4 = 1899.6 \text{ kJ.}$$

Example 3.19 *A heat pump working on the Carnot cycle takes in heat from a reservoir at 5°C and delivers heat to a reservoir at 60° C. The heat pump is driven by a reversible heat engine which takes heat from a reservoir at 840° C and rejects heat at 60°C. The reversible heat engine also drives a machine that absorbs 36 kW. If the heat pum extracts 17 kJ/sec from 5°C reservoir, determine,*

(i) The rate of heat supply from 840°C source.

(ii) The rate of heat rejection to 60°C sink.

Data: $T_1 = 60 + 273 = 333$ K, $T_2 = 5 + 273 = 278$ K, $T_3 = 840 + 273 = 1113$ K.

Solution

We know that, $\quad [\text{COP}]_{\text{HP}} = \dfrac{Q_1}{Q_1 - Q_2} = \dfrac{T_1}{T_1 - T_2}$

$$\therefore \quad [\text{COP}]_{\text{HP}} = \frac{T_1}{T_1 - T_2} = \frac{333}{333 - 278} = 6.05$$

Also, $\quad 6.05 = \dfrac{Q_1}{Q_1 - Q_2}$

$$6.05 = \frac{Q_1}{Q_1 - 17}$$

$$\Rightarrow \quad Q_1 = 20.37 \text{kW or kJ/sec.}$$

For the engine $\quad \eta_{\text{Carnot}} = \dfrac{T_3 - T_1}{T_3} = \dfrac{1113 - 333}{1113} = 0.7$

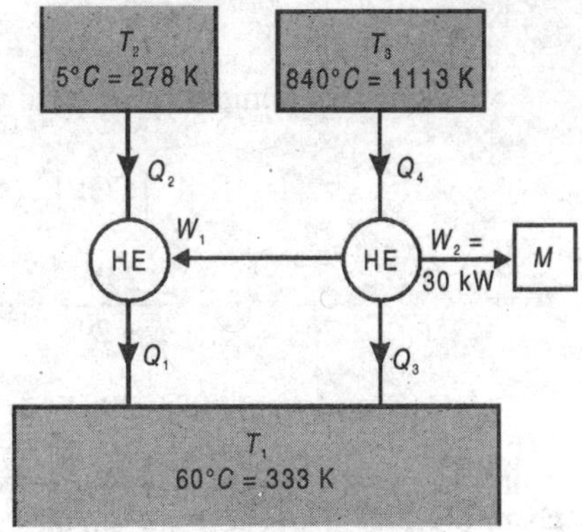

Fig. Ex. 3.19

Also, $$0.7 = \frac{W_1 + 30}{Q_4} \quad \text{...(a)}$$

We know that, $$W_1 = Q_1 - Q_2 = 20.37 - 17$$

$$W_1 = 3.37 \quad \text{... (b)}$$

From (a) and (b) we have,

$$Q_4 = 47.67 \text{ kW}$$

and $$W_E = W_1 + W_2 = 3.37 + 30 = 33.37$$

Also $$W_E = Q_4 - Q_3$$

$$\Rightarrow \quad Q_3 = Q_4 - W_E = 47.67 - 33.37$$

$$\mathbf{Q_3 = 14.3 \text{ kW or kJ/sec}}$$

∴ Heat rejected to 333 K sink $= Q_1 + Q_3 = 20.37 + 14.3 = 34.67$ kJ

Example 3.20 *A heat pump is used to heat a bunglow in the winter and then reversed to cool the bunglow in the summer. The interior temperature is to be maintained at 25°C. Heat transfer through the walls and roof is estimated to be 2800 kJ/hr°C temperature difference between the inside and outside. Estimate:*

(i) If the outside temperature in winter is 5°C, what is the minimum power required to drive the heat pump ?

(s) If the power input is the same as in part (i), what is the maximum outside temperature for which the inside temperature can be maintained at 25°C ?

Solution

(i) In winter : $$T_1 = 25 + 273 = 298 \text{ K}$$

$$T_2 = 5 + 273 = 278 \text{ K}$$

Now heat loss through walls = 2800 kJ/hr°C

∴ Net heat loss $= 2800 \times (25 - 5)$

$$Q_1 = 56000 \text{ kJ/hr}$$

For minimum power requirements of for maximum efficiency [COP] of heat pump should be maximum,

∴ Heat Pump work on reverse Carnot cycle

$$[\text{COP}]_{\text{HP}} = \frac{T_1}{T_1 - T_2} = \frac{298}{298 - 278} = 14.9$$

∴ Also, $$14.9 = \frac{Q_1}{W}$$

$$W = \frac{Q_1}{14.9} = \frac{56000}{14.9} = 3758.45 \text{ kJ/hr}$$

∴ $$\text{Minimum Power} = \frac{3758.4}{3600}$$

$$\textbf{Minimum Power} = \mathbf{1.045 \text{ kW}}$$

(ii) For summer $$T_2 = 25 + 273 = 298 \text{ K}$$

$$T_1 = \text{outside temperature} > T_2$$

$$Q_2 = \text{Heat to be removed/hr} = \text{Heat loss/hr}$$

$$Q_2 = 2800(T_1 - 298) \text{ kJ/hr}$$

As the heat pump is reversed to work as refrigerator in the summer,

$$[\text{COP}] = \frac{T_2}{T_1 - T_2} = \frac{Q_2}{W}$$

$$\frac{298}{T_1 - 298} = \frac{2800(T_1 - 298)}{3758.45}$$

$$(T_1 - 298)^2 = 460$$

$$\mathbf{T_1 = 318\ K = 45°C}$$

Example 3.21 *Three Carnot engines R_1, R_2, R_3 operate in series between two heat reservoirs which are at temperatures of 1000 K and 300 K as shown.*

Calculate intermediate temperatures if amount of work produced by these engines is in the proportions of 5 : 4 : 3.

Solution

From Fig. Ex. 3.21 by applying Carnot's Theorem we can write,

$$\frac{Q_1 - Q_2}{Q_1} = \frac{W_1}{Q_1} = \frac{T_1 - T_2}{T_1} \quad \text{...(1)}$$

and

$$\frac{Q_2 - Q_3}{Q_2} = \frac{W_2}{Q_2} = \frac{T_2 - T_3}{T_2} \quad \text{...(2)}$$

From (1) and (2) we get

$$\frac{W_1}{W_2} = T_1 - T_2 \times \frac{Q_1}{T_1} \times \frac{T_2}{Q_2} \times \frac{1}{T_2 - T_3}$$

$$= \frac{T_1 - T_2}{T_2 - T_3} \left(\text{as } \frac{Q_1}{T_1} = \frac{Q_2}{T_2} \right)$$

Therefore for Carnot engines operating in series, the temperature difference between the reservoirs, is proportional to the work ratios.

$$\therefore 1000 - T_1 : T_1 - T_2 : T_2 - 300 = 5 : 4 : 3$$

$$\therefore \quad \frac{1000 - T_1}{T_1 - T_2} = \frac{5}{4}$$

and

$$\frac{T_1 - T_2}{T_2 - 300} = \frac{4}{3}$$

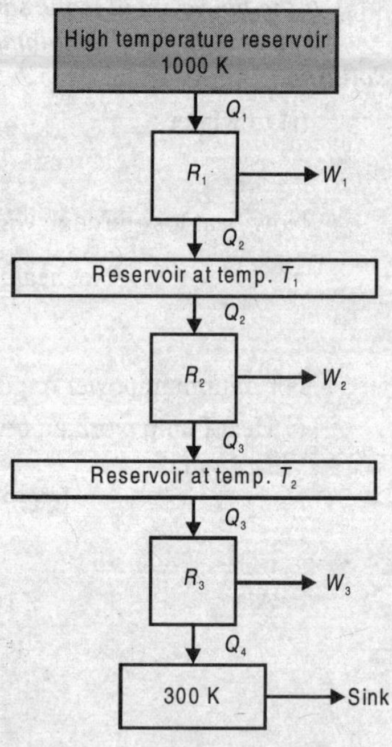

Fig. Ex. 3.21

From Eq. (1), we get,

$$4000 - 4T_1 = 5T_1 - 5T_2$$

or

$$4000 - 9T_1 = -5T_2 \quad \text{...(3)}$$

Similarly from Eq. (2), we get,

$$3T_1 - 3T_2 = 4T_2 - 1200$$

or

$$3T_1 = 7T_2 - 1200$$

Multiplying by 3 throughout we get,

or $$9T_1 = 21T_2 - 3600 \qquad ...(4)$$

Adding Eqs (3) and (4), we get,

$$4000 - 9T_1 + 9T = -5T_1 + 21T_2 - 3600$$

$$\therefore \quad 16T_2 = 7600$$

$\therefore$ $\mathbf{T_2 = 475\ K}$

Substitute $T_2 = 475$ K in Eq. (3) we get,

$$\therefore \quad 4000 - 9T_1 = (-5)(475)$$

$$\therefore \quad 9T = (4000) + (2375)$$

$\therefore$ $\mathbf{T_1 = 708.4\ K}$

Example 3.22 *Which is more effective way to increase the efficiency of Carnot engine, either to increase T_1 keeping T_2 constant; or to decrease T_2 keeping T_1 constant.*

Solution

We know that, the efficiency of Carnot cycle working between the temperature limits T_1 and T_2 is given by,

$$\eta = \frac{T_1 - T_2}{T_1}$$

Consider that the source temperature is increased by ΔT maintaining the sink temperature T_2 constant, then,

$$\eta_1 = \frac{(T_1 + \Delta T) - T_2}{T_1 + \Delta T}$$

$$= \frac{T_1 + \Delta T - T_2}{T_1 + \Delta T} \qquad ...(a)$$

Now consider that the sink temperature is decreased by ΔT maintaining source temperature T_1 constant, then,

$$\eta_2 = \frac{T_1 - (T_2 - \Delta T)}{T_1}$$

$$= \frac{T_1 + \Delta T - T_2}{T_1} \qquad ...(b)$$

Comparing Eqs. (a) and (b) it is obvious that the numerators of both the equations are same and denominator of Eq. (b) is less than the denominator of Eq. (a) as $T_1 < (T_1 + \Delta T)$

$$\therefore \quad \eta_2 > \eta_1$$

In order to increase the efficiency of Carnot cycle, decreasing to T_2 will be more effective.

Example 3.23 *A reversible engine operates between T_1 and T ($T_1 > T$). The energy rejected by this engine is received by a second reversible engine at the same temperature T. The second engine rejects the energy at temperature T_2 ($T_2 < T$). Show that*

(i) temperature T is the arithmetic mean of temperature T_1 and T_2 if the engine produces same amount of work output and

(ii) the temperature T is the geometric mean of T_1 and T_2 if the engines have the same cycle efficiency.

Solution

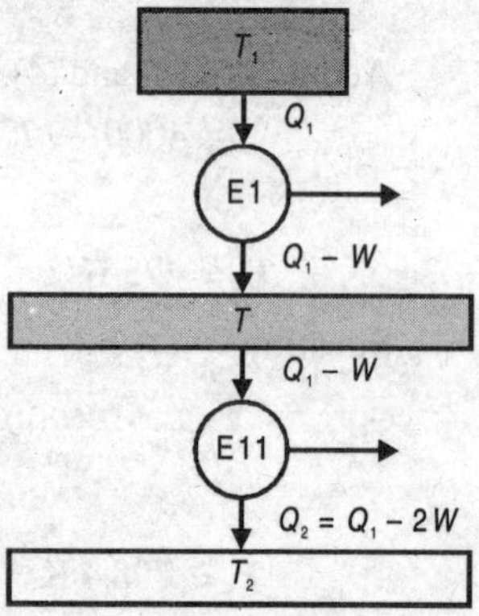

Fig. Ex. 3.23

(i)

$$\frac{Q_1}{T_1} = \frac{Q_1 - W}{T} = \frac{Q_1 - 2W}{T_2}$$

$$= C \text{ some constant}$$

$$Q_1 = CT_1, Q_1 - W = CT,$$

$$Q_{1-2} W = CT_2$$

$$\therefore \quad T = \frac{T_1 + T_2}{2} = \text{Arithmetic mean}$$

(ii)

$$\eta = \frac{T_1 - T}{T_1} = \frac{T - T_2}{T}$$

$$\therefore \quad T = \sqrt{T\,T_2} = \text{Geometric mean}$$

Example 3.24 *Two heat pumps are connected in series between two heat reservoirs at T_1 and T_2. Heat pump A pumps heat from a reservoir at T_2 and rejects heat to a reservoir at T, while the heat pump B pumps heat from reservoir at T to n reservoir at T_1. If $T_1 > T_2$, show that (a) intermediate temperature T is the arithmetic mean of temperatures T_1 and T_2 if the work input to both the pumps is equal and (b) the temperature is the geometric mean of temperature T_1 and T_2 if the heat pumps have equal C.O.P.*

Solution

(a) For equal work inputs, we can write,

$$\frac{Q_2}{T_2} = \frac{Q}{T} = \frac{Q_1}{T_1} \qquad \text{...(a)}$$

But

$$Q = Q_2 + W$$

and

$$Q_1 = Q + W = Q_2 + 2W$$

$$\therefore \quad \frac{Q_2}{T_2} = \frac{Q_2 + W}{T} = \frac{Q_2 + 2W}{T_1}$$

$$= C \text{ some constant}$$

$$\therefore \quad Q_2 = CT_2 \qquad \text{...(b)}$$

$$Q_2 + W = CT \qquad \text{...(c)}$$

$$Q_2 + 2W = CT_1 \qquad \text{...(d)}$$

Now from Eq. (c),

$$W = CT - Q_2 \quad \text{But from Eq. (b) } Q_2 = CT_2$$

$$\therefore \quad W = CT - CT_2 = C(T - T_2) \qquad \text{...(e)}$$

From Eq. (d)

$$2W = CT_1 - CT_2 \qquad \text{...(f)}$$

$$W = C\frac{(T_1 - T_2)}{2}$$

From Eqs (e) and (f)

$$C(T - T_2) = C\frac{(T_1 - T_2)}{2}$$

or $$(T - T_2) = \frac{T_1 - T_2}{2}$$

$\therefore$ $$2T - 2T_2 = T_1 - T_2$$

or $$T = \frac{T_1 + T_2}{2} \text{ which is arithmetic mean}$$

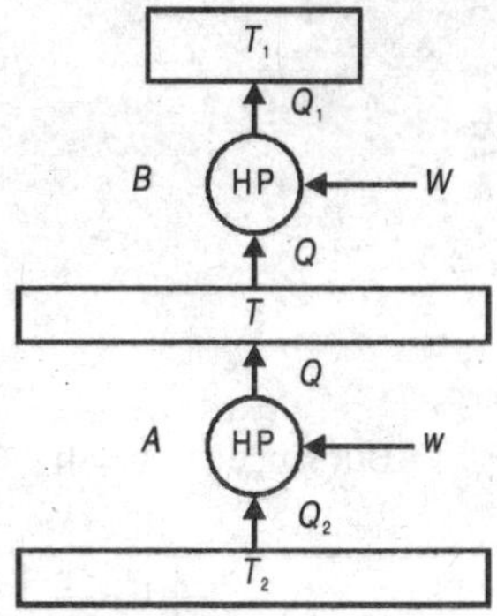

Fig. Ex. 3.24

(b) For equal C.O.P.

$$[\text{C.O.P.}]_{\text{HP(A)}} = \frac{T_1}{T_1 - T}$$

and $$[\text{C.O.P.}]_{\text{HP(B)}} = \frac{T}{T - T_2} \text{ and for equal C.O.P.}$$

$\therefore$ $$\frac{T_1}{T_1 - T} = \frac{T}{T - T_2}$$

or $$T = \sqrt{T_1 T_2} \text{ which is the geometric mean.}$$

Example 3.25 *A reversible engine works between three thermal reservoirs A, B, C. The engine absorbs an equal amount of heat from the thermal reservoirs. A and B kept at temperatures T_A and T_B respectively, and rejects heat to the thermal reservoir C kept at temperature T_C. The efficiency of the engine is μ times the efficiency of the reversible engine, which works between the two reservoirs A and C. Prove that*

$$\frac{T_A}{T_B} = (2\mu - 1) + 2(1 - 2\mu)\frac{T_A}{T_C}.$$

Solution

Given that, efficiency of the Engine = $\mu \times \eta$ of engine working between T_A and T_C

$\therefore$ Efficiency of engine A $$= \frac{T_A - T_C}{T_A}$$

and Output of Engine A $$= Q \times \frac{T_A - T_C}{T_A}$$

Similarly,

Efficiency of Engine B $$= \frac{T_B - T_C}{T_B}$$

and Output of Engine B $$= Q \times \frac{T_B - T_C}{T_B}$$

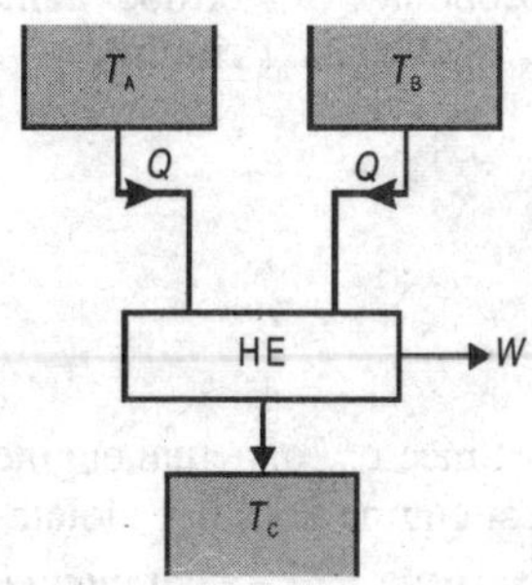

Fig. Ex. 3.25

$\therefore$ Net work output $$= Q \times \left(\frac{T_A - T_C}{T_A}\right) + Q \times \left(\frac{T_B - T_C}{T_B}\right)$$

and Net input $$= Q + Q = 2Q$$

$\therefore$ Efficiency $$= \frac{\text{Net work output}}{\text{Net input}}$$

$$= \frac{Q\left(\frac{T_A - T_C}{T_A}\right) + Q\left(\frac{T_B - T_C}{T_B}\right)}{2Q}$$

$$= \frac{1}{2}\left(1 - \frac{T_C}{T_A} + 1 - \frac{T_C}{T_B}\right)$$

But since $\eta = \mu \times \eta$ of Engine working between T_A and T_C

$$\therefore \quad \frac{1}{2}\left(1 - \frac{T_C}{T_A} + 1 - \frac{T_C}{T_B}\right) = \mu\left(1 - \frac{T_C}{T_A}\right)$$

Multiplying both sides by $\frac{T_A}{T_C}$ and rearranging terms we get the required result

$$\frac{T_A}{T_B} = (2\mu - 1) + 2(1 - 2\mu)\frac{T_A}{T_C}$$

Example 3.26 *A heat engine receives 600 kJ of heat from a high temperature source at 1000 K during a cycle. It converts 150 kJ of this heat to net work and rejects the remaining 450 kJ to a low temperature sink at 300 K. Determine if this heat engine violates the second law of thermodynamics on the basis of*

(a) the Clausius inequality and

(b) the Carnot Principle

Solution

Clausius inequality :

$$\oint \frac{\delta Q}{T} = \frac{Q_1}{T_1} - \frac{Q_2}{T_2}$$

$$= \frac{600}{1000} - \frac{450}{300}$$

$$= -0.9 \text{ kJ/K}$$

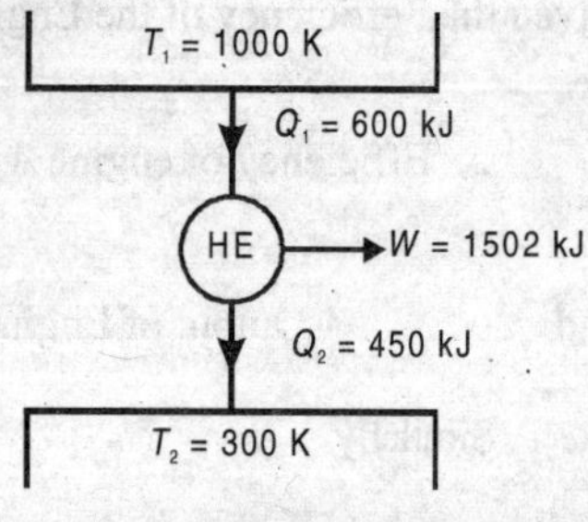

Fig. Ex. 3.26

Since $\oint \frac{\delta Q}{T}$ is negative, the heat engine does not violate second law of thermodynamics.

Carnot Principle :

$$\eta_{\text{Th (HE)}} = \frac{Q_1 - Q_2}{Q_1} = \frac{150}{600} = 25\%$$

$$\eta_{\text{Th (Carnot HE)}} = \frac{T_1 - T_2}{2T_1} = \frac{1000 - 300}{1000} = 70\%$$

Since η_{Th} of actual engine working between the same temperature limits is less than theat of Carnot engine, the heat engine does not violate second law of thermodynamics.

Example 3.27 *A heat pump is used to maintain a house at a constant temperature of 23°C. The house is losing heat to the outside air through the walls and the windows at a rate of 60,000 kJ/h while the energy generated within the house from people, lights and appliances amounts to 4000 kJ/h. For a COP of 2.5, determine the required power input to the heat pump.*

Solution

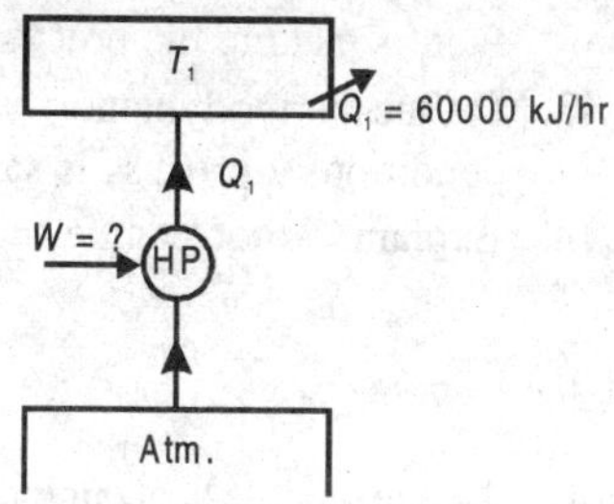

Fig. Ex. 3.27

$$[COP]_{HP} = \frac{Q_1}{W}$$

Now,
$$Q_1 = \begin{pmatrix}\text{Heat energy lost through}\\ \text{walls and windows}\end{pmatrix} - \begin{pmatrix}\text{Energy generated}\\ \text{from people, lights etc.}\end{pmatrix}$$

$$\therefore \quad Q_1 = 60{,}000 - 4{,}000$$
$$= 56{,}0000 \text{ kJ/h}$$
$$= 15.556 \text{ kJ/s}$$

$$\therefore \quad \mathbf{W} = \frac{\mathbf{15.556}}{\mathbf{2.5}} = \mathbf{6.22\ kW}$$

Example 3.28 *A heat pump is used to maintain an auditorium hall at 24°C when the atmospheric temperature is 10°C. The heat lost from the hall is 1500 kJ/min. Calculate the power required to run the heat pump if its COP is 30 percent of Carnot machine, working between the same temperature limits.*

Solution

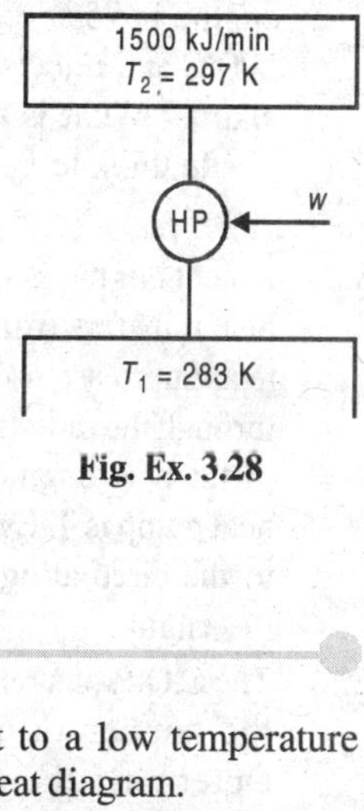

Fig. Ex. 3.28

$$COP_{HP(Carnot)} = \frac{T_2}{T_2 - T_1} = \frac{297}{297-283}$$
$$= 21.21$$

$$COP_{HP\,(Actual)} = 0.3 \times COP_{HP(Carnot)}$$
$$= 0.3 \times 21.21 = 6.36$$

$$COP = \frac{Q}{W}$$

$$\therefore \quad \mathbf{W} = \frac{\mathbf{1500/60}}{\mathbf{6.36}} = \mathbf{3.928\ kW}$$

THEORY QUESTIONS

3.1. Explain Limitation of first law of thermodynamics.

3.2. Explain Heat engine and Heat engine cycle.

3.3. What is a heat reservoir ? Explain its types.

3.4. Explain a refrigerator with a neat sketch.

3.5. Explain a heat pump with a suitable sketch.

3.6. Why we shall study Second Law of Thermodynamics? Give an example and explain the important statements of Second Law of Thermodynamics.

3.7. Explain perpetual motion machine of first and second kind.

3.8. Is it possible for a heat engine to operate without rejecting any waste heat to a low temperature reservoir? Explain with neat diagram.

3.9. In a refrigerator heat is transferred from a lower temperature medium to higher temperature one. Is this a violation of the second law of thermodynamics?

3.10. Prove that Kelvin Planck and Clausius statements are equivalent in all respects.

3.11. State and explain Kelvin–Planck constant and Clausius statements of Second Law of Thermodynamics. Prove that violation of Kelvin–Planck statement leads to violation of Clausius statement.

3.12. What is a Quasi-static process ? Explain reversibility and irreversibility. Enumerate the factors making the process irreversible.

3.13. Define thermodynamic reversibility. Under what conditions a process is said to be reversible?

3.14. Explain Carnot Engine and Carnot Cycle. Prove, that $\eta_{Carnot} = \frac{T_1 - T_2}{1}$.

3.15. State and explain Carnot's theorem.

3.16. Explain why Carnot cycle cannot be used in practical engine?

3.17. How the efficiency of IC engines is increased ?

PROBLEMS FOR PRACTICE

3.1. Two reversible heat engines *A* and *B* are operating between the same temperature limits. Engine *A* develops 37.5 kW with 30% efficiency. The engine *B* receives 7535 kJ/min of heat from source. Determine the power developed, by engine *B* and heat rejected by each engine.

(**Ans.** $W_B = 37.675$ kW, $Q_{2A} = 5250$ kJ/min $Q_{2B} = 5274.5$ kJ/min)

3.2. An inventor of a heat engine claims that his engine is 75% efficient which receives heat at 100ºC and rejects at 20ºC. How do you rate his claim ? What is the maximum efficiency you would thihk to be realistic ?

(**Ans.** Claim is wrong, 21.4%)

3.3. A heat engine is used to drive a heat pump. The heat transfers from the heat engine and from the heat pump are used to heat the water circulating through the radiators of a building. The efficiency of the heat engine is 27% and the C.O.P. of the heat pump is 4. Evaluate the Ratio of heat transfer to the circulating water to the heat transfer to the engine. (**Ans.** 1.81)

3.4. The C.O.P. of a refrigerator is 6, when it maintains the temperature of 3°C in the evaporator. Determine the condenser temperature and refrigerating effect if the power required to run the refrigerator is 7.5 kW.

3.5. An engineer claims his engine to develop 3.68 kW on testing, the engine consumes 0.44 kg of fuel in one hr. It has calorific value of 41870 kJ/kg. The maximum temperature recorded in the cycle is 1400°C and minimum is 350ºC. Is the engineer justified in his claim ?

3.6. An inventor claims that his newly developed heat engine will produce 0.184 kW power, when the heat is supplied to engine at the rate of 70 kJ/min. The cycle operates between maximum temperature of 1092°C and minimum temperature of 149°C. Make complete analysis of the inventors claim and clearly state whether it is worth investing any amount of money on such a project. Give reasons for your suggestions.

3.7. A heat engine is used to drive a heat pump. The heat transfer from the heat engine and from the heat pump are used to heat the water circulating through the radiators of a building. The efficiency of the heat engine is 27% and C.O.P. of the heat pump is 4. Evaluate the ratio of heat transfer to the circulating water to the heat transfer to the engine.

3.8. A reversible heat engine operates between two reservoirs at temperature of 650°C and 35°C. The engine drives a reversible refrigerator. Which operates between temperature of 35°C and – 20°C. The heat transfer to the engine is 1000 kJ/min. The work output (net) of the combined device is 180 kJ/min.

3.9. Estimate the refrigerating effect in kW and net heat transfer to the reservoir at 35°C. What is the power developed by the engine in kW.

3.10. A reversible heat engine used for a satellite, operates between a hot reservoir at T_1 and radiating panel at T_2. The radiation from the panel is proportional to its area and T_2^4. For a given work output and fixed T_1, show that the area of the panel will be minimum when

$$\frac{T_1}{T_2} = \frac{4}{3}$$

3.11. An inventor claims to have developed a refrigerator machine, which operates between

– 20°C and 30°C and consumes and kW of power. The machine gives refrigerating effect of 21.6 MJ/hr. Comment on the claim of the inventor.

3.12. A heat pump is used to heat at room in winter and to cool in summer. The interior temperature is to be maintained at 20°C. The heat transfer through the wall and roof is 2400 kJ/hr per degree temperature difference between inside and outside.

(i) If the outside temperature in winter is 0°C, what is the maximum power required to drive the heat pump ?

(ii) If the power input is the same as that in part (i), what is the maximum outside temperature for which inside can be maintained at 20°C.

A Carnot engine between 537°C and 37°C develops 5 kW of power. Determine,

(i) the thermal efficiency,

(ii) heat supplied per sec.

(iii) the change of entropy per second during heat rejection.

3.13. A heat pump working on Carnot cycle takes in, heat from a reservoir at 5°C and delivers heat to a reservoir at 60°C. The heat pump is driven by a reversible heat engine which takes in heat from a reservoir at 840°C and rejects heat at 60°C. The reversible heat engine also drives a machine that absorbs 30 kW. If pump extracts 17 kJ/sec from 5°C reservoir determine.

(i) The rate of heat supply from 840°C source.

(ii) The rate of heat rejection to 60°C sink.

3.14. A Carnot cycle receives heat at 527°C causing increase in entropy equal to 5 kJ/kg K. The engine delivers 2000 kJ/kg of work. Determine the efficiency of the cycle and lowest temperature in the cycle.

3.15. A small metallic object 4 kg in mass of sp. heat C_p = 0.5 kJ/kg K at a temperature of 227°C is thrown into a lake at temperature of 27°C. Calculate the change in entropy of the universe.

3.16. An inventor claims that he has designed an engine which will produce a power output of 170 kW while consuming 0.5 kg/min of fuel having C.V. of 42000 kJ/kg. He states that the upper and lower temperature of the engine are 670 K and 230 K respectively. Do you believe his claim? Make your comments.

3.17. A Carnot engine operates between the temperature limits of 180°C and 60°C and produces 200 kN m/sec of work. Determine :

(i) Heat supplied in kW

(ii) Change of entropy during heat rejection

(iii) Thermal efficiency of the engine.

3.18. Consider two Carnot heat engines operating in series. The first engine receives heat from a reservoir at 1000 K and rejects the waste heat to another reservoir at temperature *T*. The second engine receives this energy rejected by the first one, converts some of it to work and rejects the rest to a reservoir at 300 K. If the thermal efficiencies of both engines are the same, determine the temperature *T*.

3.19. A house is to be maintained at a temperature of 15°C by means of a heat pump by pumping heat from the atmosphere. Heat losses through the walls are estimated to be 0.75 kW per unit temperature difference between the inside of the house and the outside atmosphere. If the atmosphere temperature is –10°C, calculate the power required to drive the heat pump if C.O.P. of the actual heat pump is 25% of C.O.P. of Carnot's heat pump working between the same temperature limits.

3.20. A heat engine receives half of its heat at a temperature of 1000 K and the rest at 5000 K while rejecting heat to a sink at 300 K. What is the maximum possible thermal efficiency of this heat engine ?

3.21. Source *A* can supply energy at the rate of 12000 kJ /min at 320°C. A second source *B* can supply energy at the rate of 12000 kJ/min at 70°C. Which source *A* or *B* would you choose, to supply energy to an ideal reversible heat engine that is to produce large amount of flow. If the temperature of surroundings is 35°C.

3.22. A domestic refrigerator maintains a temperature of –12°C. The ambient air temperature is 35°C. If heat leaks into the freezer at the continuous rate of 2 kJ/sec, determine the least power necessary to maintain the refrigerator.

3.23. A reversible engine takes 1190 kJ/min from a reservoir at 700 K and develops 190 kJ of work/min when executing complete cycle. The engine rejects heat to two reservoirs at 600 K and 500 K. Find the heat rejected to each sink.

3.24. A system undergoes a reversible cycle during which it exchanges heat with three thermal reservoirs A, B, C at T_1, T_2 and T_3. The amount of heat exchanged is Q_1, Q_2 and Q_3. Show that

$$\frac{Q_1}{\frac{1}{T_2}-\frac{1}{T_3}}=\frac{Q_1}{\frac{1}{T_3}-\frac{1}{T_1}}=\frac{Q_1}{\frac{1}{T_1}-\frac{1}{T_2}}$$

3.25. Two Carnot heat pumps are arranged in series. Pump 1 takes 150 kJ/cycle from heat sink at 300 K. The heat output from this heat pump serves as heat input to pump 2, which delivers its output to a heat reservoir at 1200 K. If the heat pumps have same C.O.P., how much energy (kJ/cycle) must be delivered to pump 1 and to pump 2.

CHAPTER 4

Ideal Gases and Ideal Gas Processes

4.1 INTRODUCTION

An ideal gas is one, which obeys all the gas laws and the characteristic gas equation $PV = mRT$, at all temperatures and pressures.

But actually there is no perfect or Ideal gas in reality. But all the real gases like Air, O_2, H_2, N_2, He, behave as ideal gases, at very low pressures and high temperatures.

4.2 IDEAL GAS LAWS

The behaviour of ideal gas is governed by following gas laws:

1. Boyle's law 2. Charle's law 3. Avagadro's law

1. Boyle's Law. Boyle experimentally established that, the volume of a given mass of a gas is inversely proportional to the absolute pressure, when the temperature is constant.

i.e.,
$$V \propto \frac{1}{P}, \text{ for particular temperature } T$$

$$\therefore \quad V = \frac{\text{Constant of proportionality}}{P} = \frac{C}{P}$$

or
$$PV = C$$

where C is a constant.

Also
$$P_1V_1 = P_2V_2 = C$$

Consider a gas in a cylinder fitted with a frictionless piston. Let the initial properties be P_1 and V_1. Now when the gas is heated at constant temperature, then the gas expands and the pressure falls to P_2 and volume increases to V_2. This is represented on PV-diagram as shown in Fig. 4.1.

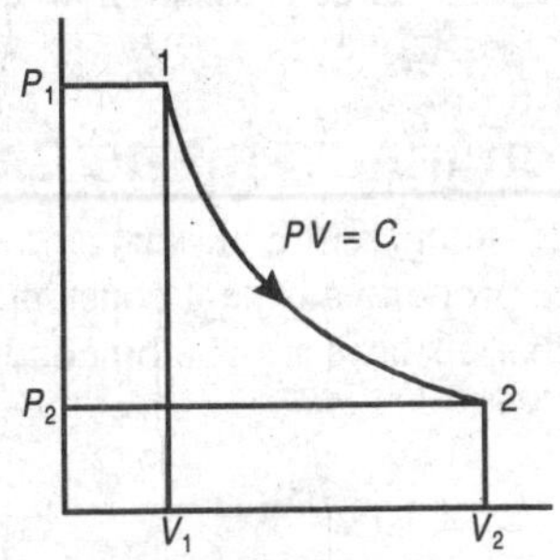

Fig. 4.1

2. Charle's Laws. (i) It states that, "when the pressure of a given mass of a gas is kept constant, then volume is directly to the absolute temperature". (Ref. Fig. 4.2)

i.e., When P = Constant

Then $V \propto T$

or V = Constant of proportionality T

or $\frac{V}{T}$ = Constant of proportionality

or $\frac{V_1}{T_1} = \frac{V_2}{T_2}$ = Constant of proportionality

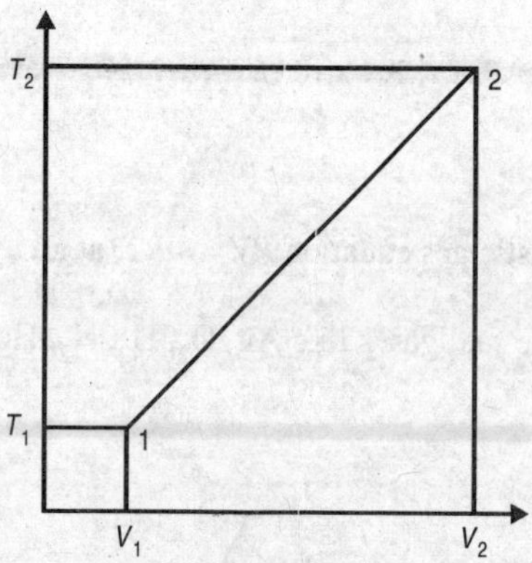

Fig. 4.2

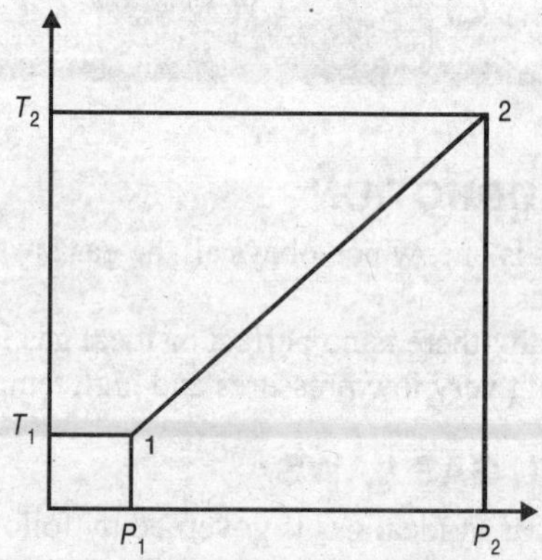

Fig. 4.3

(ii) It also states that, "when the volume of a given mass of a gas is kept constant, the pressure is directly proportional to the absolute temperature". (Ref. Fig. 4.3)

i.e. $P \propto T$

when V = Constant

or P = Constant of proportionality T

$\frac{P}{T}$ = Constant of proportionality

or $\frac{P_1}{T_1} = \frac{P_2}{T_2}$ = Constant of proportionality

3. Avagadro's Law. This law states that at the same pressure and temperature, equal volumes of different gases contain equal number of molecules.

4.3 EQUATION OF STATE OR CHARACTERISTIC GAS EQUATION

In engineering practice, pressure, volume, temperature all vary simultaneously. So, Boyle's law or Charle's law alone is not applicable, since one of three properties is kept constant. In order to establish a relationship between these three properties, Boyle's law and Charle's laws are combined together, which gives a general equation called characteristic gas equation.

According to Boyle's law,

$$V \propto \frac{1}{P} \text{ when } T = \text{Constant} \qquad \ldots(1)$$

And according to Charle's law,

$$V \propto T \text{ when } P = \text{Constant} \qquad \text{...(2)}$$

It is obvious that,

$$V \text{ is } \propto \text{ to } T \text{ and } \frac{1}{P}$$

i.e. $$V \propto \frac{T}{P} \qquad \text{...(3)}$$

or $$PV \propto T$$

or $$PV = CT \text{ where } C \text{ is a constant}$$

or $$\frac{PV}{T} = C \qquad \text{...(4)}$$

or In general $$\frac{P_1V_1}{T_1} = \frac{P_2V_2}{T_2} = \frac{P_3V_3}{T_3} = \text{........} = C$$

or Since v = specific volume in $\frac{\text{m}^3}{\text{kg}}$

Then equation will be, $\frac{Pv}{T} = C$ ...(5)

When 1 kg of gas is considered, then the constant for the Eq. (4) is written as R and is called *Characteristic gas constant*.

$$\therefore \quad \frac{Pv}{T} = R \qquad \text{...(6)}$$

Now consider m kg of gas. Multiply both sides of Eq. (6) by m

$$\frac{P\cdot(m\cdot v)}{T} = m\,.\,R$$

$$m\,.\,v = V = \text{Total volume.}$$

$$\therefore \quad \mathbf{PV = m\,R\,T}$$

This is known as characteristic equation of an ideal gas.

Note that the value of R for air = $0.287 \ \frac{\text{kJ}}{\text{kg-K}}$

4.4 UNIVERSAL GAS CONSTANT

Frequently it is required to express the equation of state, $PV = mRT$. (1) on mole basis. The mass of a substance m is equal to the product of number of moles n and the molecular weight M.

i.e., $$m = n\,.\,M \qquad \text{...(1)}$$

Hence susbtituting the value of m in Eq. (1) we get,

$$PV = n\,.\,M\,.\,RT \qquad \text{...(2)}$$

Now the molar specific volume is defined as the volume/unit mole and is denoted by $\bar{v}$,

$$\therefore \quad \bar{v} = \frac{V}{n}$$

Hence Eq. (2) becomes,

$$P\bar{v} = MRT \qquad ...(3)$$

Now if we consider two gases, *a* and *b* occupying equal molar volumes at the same temperature and pressure, Eq. (3) becomes,

$$P_a \cdot \bar{v}_a = M_a R_a T_a$$

and

$$P_b . \bar{v}_b = M_b \cdot R_b \cdot T_b$$

According to Avagardo's law, perfect gases at the same temperature and pressure, occupy the same molar volumes.

i.e.

$$\bar{v}_a = \bar{v}_b \text{ when } P_a \text{ and } T_a = T_b$$

We have,

$$\frac{P_a . \bar{v}_a}{T_a} = \frac{P_b . \bar{v}_b}{T_b}$$

$$\therefore \quad M_a . R_a = M_b . R_b = M.R$$

The product *M.R* is called the *Universal Gas Constant* and it is denoted by $\bar{R}$

$\therefore$ Eq. (3) can be written as,

$$P\bar{v} = MRT$$

$$P\bar{V} = \bar{R}T$$

and Eq. (2) can be written as,

$$P\bar{V} = nMRT$$

$$P\bar{V} = n\bar{R}T$$

and the value of $\bar{R}$ for air $= 8.3143 \dfrac{\text{kJ}}{\text{k mol K}}$

4.5 JOULE'S EXPERIMENT TO PROVE $U = f(T)$

Internal Energy-Joule's Law

Joule's Law states that the internal energy of an ideal gas depends only on the temperature of gas and is independent of changes in pressure and volume i.e., $U = f(T)$.

Joule's Experiment

The apparatus for Joule's experiment is shown in Fig. 4.4. It consists of two copper vessels *A* and *B* connected by a valve *V*. Initially *A* contained high pressure air and *B* was evacuated. Both were placed in an insulated water bath with a thermometer to measure the temperature.

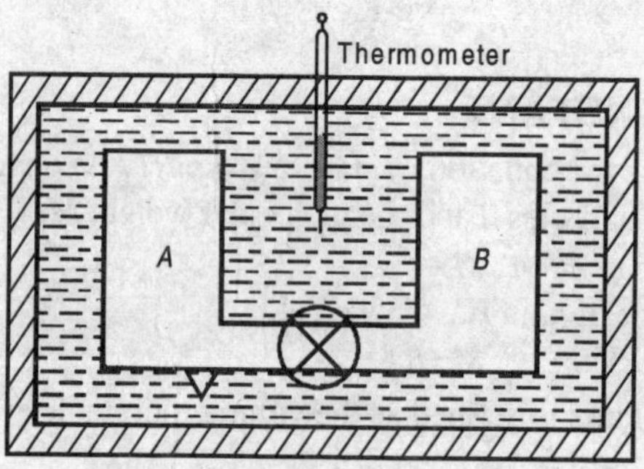

Fig. 4.4. *Joule's experiment.*

The initial temperature of water was recorded. The valve V was then opened and air is allowed to flow from vessel A to vessel B.

The operation is slow enough for air to attain equilibrium states. Eventually pressure in both the vessels will become equal and flow will be stopped. Joule observed that there was no change in the temperature of water during or after the process.

Since there was no change in temperature, there was no heat transfer between air and water. Therefore $\delta Q = 0$. Also no work is done during the process. Therefore $\delta W = 0$. By first law of thermodynamics for closed system we know that $\delta Q = dU + \delta W$.

Therefore $dU = 0$ or change in internal energy is zero. The internal energy therefore remains constant.

Again we observe that pressure and volume both changed during the process and only temperature remained constant. Joule therefore concluded that internal energy is the function of temperature only and independent of pressure and volume changes.

Thus, $$U = f(T)$$

4.6 RELATIONS BETWEEN C_p AND C_v

We have studied specific heat at constant pressure and specific heat at constant volume in the I-law of thermodynamics.

Since C_p and C_v are the properties of a system, there is a relationship between these.

We know that, specific enthalpy h is given by,

$$h = u + PV$$

$$dh = du + d(P.v) \quad \ldots(1)$$

from the definition of,

$$\mathbf{C_p = \left(\frac{\delta q}{dT}\right)_p = \left(\frac{dh}{dT}\right)_p}$$

$\therefore$ $$dh = C_p \cdot dT \quad \ldots(2)$$

and $$\mathbf{C_v = \left(\frac{\delta q}{dT}\right)_v = \left(\frac{du}{dT}\right)_v}$$

$\therefore$ $$du = C_v \cdot dT \quad \ldots(3)$$

Since, $$PV = RT$$

$$d(PV) = d(RT)$$

$$d(PV) = R \cdot dT, \text{ since } R \text{ is constant} \quad \ldots(4)$$

Substituting Eqs (2), (3) and (4) in (1) we get,

$$C_p \cdot dT = C_v\, dT + R \cdot dT$$

$\therefore$ $$C_p = C_v + R \quad \text{(Since } dT \text{ is common in both sides)}$$

or $$\mathbf{C_p - C_v = R} \quad \ldots(5)$$

We also know that the ratio $\dfrac{C_p}{C_v} = \gamma$

$\therefore$ $$C_p = \gamma \cdot C_v$$

Substituting in, $$C_p - C_v = R$$

we get, $$\gamma \cdot C_v - C_v = R$$

$$\mathbf{C_v = \frac{R}{\gamma - 1}} \quad \ldots(6)$$

and since $C_p = \gamma \cdot C_v$

$$\mathbf{C_p = \frac{\gamma . R}{\gamma - 1}} \qquad ...(7)$$

4.7 IDEAL GAS PROCESSES

4.7.1 Constant Volume Process

During this process, volume remains constant ($V = C$) and is represented on a PV-diagram by means of vertical line as shown.

When unit mass of a gas is heated in a closed vessel (i.e., $V = C$), then, since volume remains constant, no external work is done. But as temperature of the gas increases I.E. increases.

(a) Work Done. Work done in a non-flow system = $\int PdV$.

Here, $V = C \qquad \therefore \; dV = 0$

$$\therefore \; W_{1-2} = 0$$

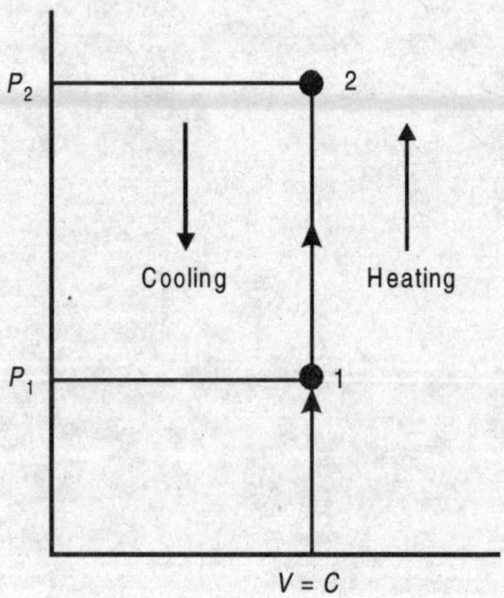

Fig. 4.5

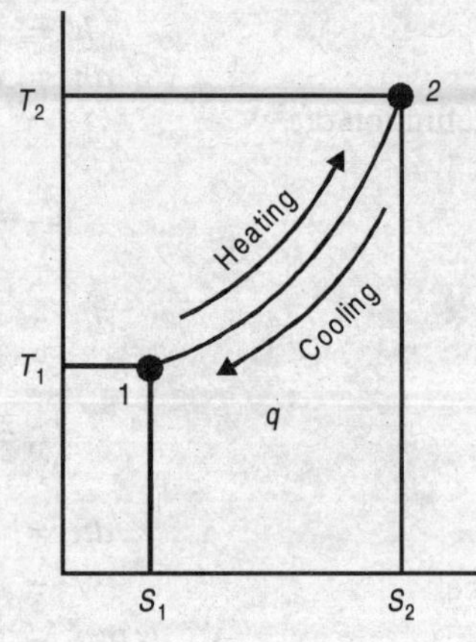

Fig. 4.6

(b) Heat Supplied. From the I–Law for a closed system undergoing a process,

$$Q = \Delta U + W \cdot \text{ In this case}$$

Since $W = 0$

$\therefore$ $Q = \Delta U$

or $q = \Delta u$ (where q is kg)

$\therefore$ $\delta q = du = C_v \cdot dT$

From the definition of C_v. For total mass,

$$\mathbf{\delta Q = dU = m \times C_v \times dT}$$

4.7.2 Constant Pressure Process

During this process pressure remains constant ($P = C$) and this process is represented by means of a horizontal line on the PV-diagram as shown.

When a unit mass of a gas is taken in a cylinder fitted with a frictionless piston and is heated. Then the piston moves up maintaining same pressure (i.e., $P = C$). As the volume of the gas increases, the work is done by the gas on the piston. As the temperature of the gas increases, Internal Energy increases. So, heat supplied in a constant pressure heating process is utilised for two purposes:

(i) For doing some external work.
(ii) For increasing the I.E. of the gas.

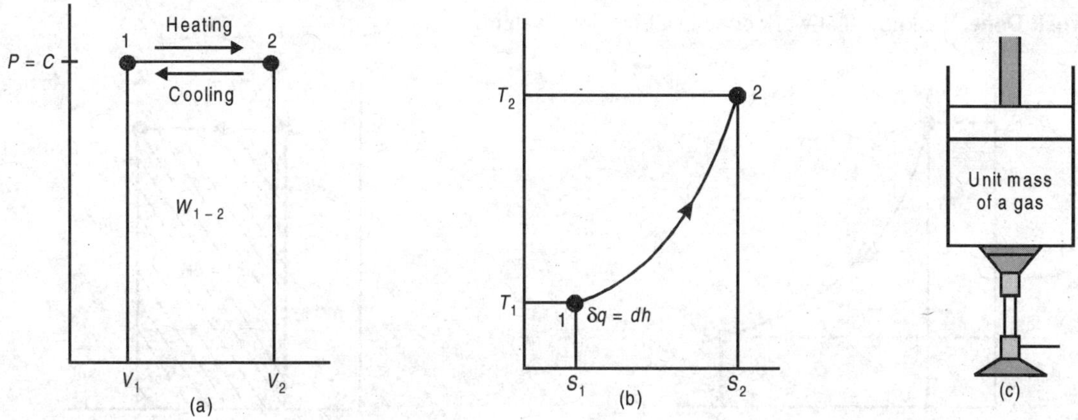

Fig. 4.7

(a) Work Done

Work done in non-flow system.

$$W_{1-2} = \int_1^2 PdV$$

Since P is constant.

$$\therefore \quad W_{1-2} = P\int_1^2 dV$$

$$W_{1-2} = P(V_2 - V_1)$$

If the pressure P is in $\frac{\text{N}}{\text{m}^2}$ and volume is in m^3 then the resultant unit of work will be N-m or J.

(b) Change in Internal Energy

$$(\Delta U) = U_2 - U_1$$

(c) Heat Transferred. From the I-law,

$$Q - W = \Delta U$$

$$Q = \Delta U + W$$

or

$$\delta Q = dU + \delta W$$

$$\delta Q = dU + P\,dV$$

$$\delta Q = dU + d(PV) \text{ since } P = C$$

$$\delta Q = d(U + PV)$$

$$U + PV = H \text{ which is enthalpy}$$

$$\delta Q = dH$$

or

$$\delta q = dh$$

where h is enthalpy/kg

i.e. heat supplied = change in enthalpy

4.7.3 Isothermal Process

During this process, temperature remains constant ($T = C$). The law for the process is $PV = C$ and is represented by means of a curve as shown on the PV-diagram. It is represented by means of a horizontal line on T–S diagram.

(a) Work Done. We know that work done in a non-flow system.

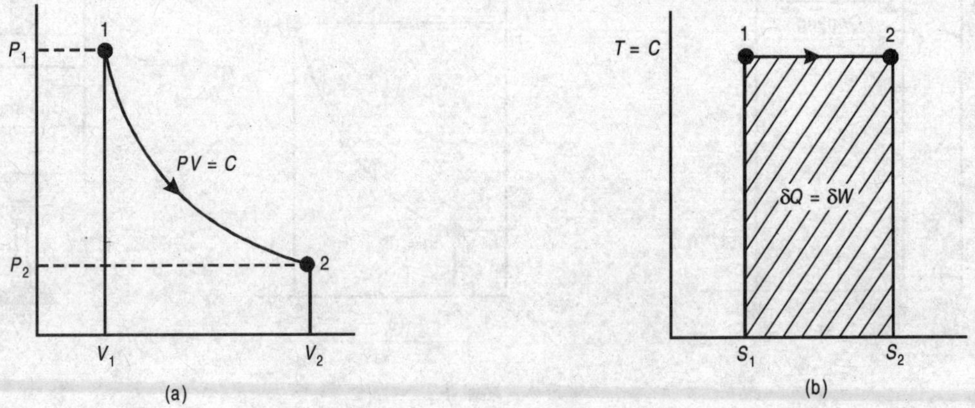

Fig. 4.8

$$W_{1-2} = \int_1^2 P.dV \qquad \ldots(1)$$

The law for the process is,

$$PV = C$$

i.e.,

$$PV = P_1 V_1 = P_2 V_2 = C$$

$$\therefore \quad P = \frac{P_1 V_1}{V} \text{ Substitute this in (1) we get,}$$

$$\mathbf{W_{1-2} = P_1 V_1 \ln \frac{V_2}{V_1}}$$

$$\mathbf{W_{1-2} = P_1 V_1 \ln \frac{P_1}{P_2} \text{ since } \frac{V_1}{V_2} = \frac{P_1}{P_2}}$$

(b) Change in IE $\quad dU = mC_v\, dT = 0$ as $T = C$, $dT = 0$

(c) Heat Transferred. From the first law for a closed system undergoing a process,

$$Q = \Delta U + W$$

$$\delta Q = dU + \delta W$$

$$\delta Q = 0 + \delta W$$

$$\mathbf{\delta Q = \delta W = P_1 V_1 \ln \frac{V_2}{V_1} = P_1 V_1 \ln \frac{P_1}{P_2}}$$

4.7.4 Reversible Adiabatic Process

A process is said to be reversible adiabatic, when the heat added or rejected during the process is zero. In this case the process should be very fast so that there is very very little time for the exchange of heat to take place.

The system for this process may be considered as piston and cylinder mechanism.

We know that first law for the process is,

$$Q - W = \Delta U$$
$$Q = \Delta U + W$$
$$\delta Q = dU + \delta W$$
$$\delta Q = dU + PdV$$
$$0 = C_v \cdot dT + PdV \quad \text{...(1)}$$

Since heat transfer in zero.

We also know that, for an ideal gas,

$$PV = RT$$

Differentiating this we get,

$$P \cdot dV + V\,dP = R \cdot dT$$
$$\frac{P \cdot dV + VdP}{R} = dT \quad \text{...(2)}$$

Substituting the value of dT in Eq. (1)

We get,

$$O = C_v \frac{(P \cdot dV + V \cdot dP)}{R} + P \cdot dV$$
$$= P \cdot dV\left(1 + \frac{C_v}{R}\right) + \frac{C_v}{R} \cdot VdP$$
$$= C_p \cdot \frac{P}{R} \cdot dV + \frac{C_v}{R} \cdot V \cdot dP$$

Dividing this equation by PV we get,

$$O = \frac{C_p}{R} \cdot \frac{dV}{V} + \frac{C_v}{R} \cdot \frac{dP}{P}$$

Now dividing by C_v throughout, we get,

$$O = \frac{C_p}{C_v R} \cdot \frac{dV}{V} + \frac{1}{R} \cdot \frac{dP}{P}$$
$$= \frac{C_p}{C_v} \cdot \frac{dV}{V} + \frac{dP}{P}$$
$$= \gamma \cdot \frac{dV}{V} + \frac{dP}{P}$$
$$\therefore \quad \frac{dP}{P} = -\gamma \cdot \frac{dV}{V} \quad \text{...(3)}$$

Integration of this equation gives

$$\int_1^2 \frac{dP}{P} = -\gamma \int_1^2 \frac{dV}{V}$$

$$\mathbf{\log_e P = -\gamma \cdot \log_e V}$$

$$\therefore \quad \left[\log_e P\right]_1^2 = -\gamma\left[\log_e V\right]_1^2$$

$$\therefore \quad \log_e P_2 - \log_e P_1 = -\gamma\left[\log_e V_2 - \log_e V_1\right]$$

$$\therefore \quad \log_e \frac{P_2}{P_1} = -\gamma \log_e \frac{V_2}{V_1}$$

$$\therefore \quad \log_e \frac{P_2}{P_1} = \log_e \left(\frac{V_2}{V_1}\right)^{-\gamma}$$

$$\therefore \quad \left(\frac{V_2}{V_1}\right)^{-\gamma} = \frac{P_2}{P_1}$$

$$\therefore \quad \left(\frac{V_1}{V_2}\right)^{\gamma} = \frac{P_2}{P_1}$$

$$\therefore \quad P_1 V_1^{\gamma} = P_2 V_2^{\gamma}$$

or
$$\mathbf{PV^{\gamma} = Constant} \qquad \ldots(4)$$

Combining this equation with ideal gas equation, we get,

$$\frac{P_1 V_1}{T_1} = \frac{P_2 V_2}{T_2}$$

$$\therefore \quad \frac{T_1}{T_2} = \frac{P_2 V_2}{P_1 V_1}$$

But
$$\frac{P_2}{P_1} = \left(\frac{V_1}{V_2}\right)^{\gamma}$$

$$\therefore \quad \frac{T_2}{T_1} = \left(\frac{V_1}{V_2}\right)^{\gamma} \cdot \frac{V_2}{V_1} = \left(\frac{V_1}{V_2}\right)^{\gamma-1}$$

$$\mathbf{\frac{T_2}{T_1} = \left(\frac{V_1}{V_2}\right)^{\gamma-1}} \qquad \ldots(5)$$

Similarly,
$$\frac{V_2}{V_1} = \left(\frac{P_1}{P_2}\right)^{\frac{1}{\gamma}}$$

$$\therefore \quad \frac{T_2}{T_1} = \frac{P_2}{P_1} \times \left(\frac{P_1}{P_2}\right)^{\frac{1}{\gamma}}$$

$$= \left(\frac{P_2}{P_1}\right)^{\frac{\gamma-1}{\gamma}}$$

$$\therefore \quad \frac{T_2}{T_1} = \left(\frac{P_2}{P_1}\right)^{\frac{\gamma-1}{\gamma}} \qquad \ldots(6)$$

$$\therefore \quad \frac{T_2}{T_1} = \left(\frac{V_1}{V_2}\right)^{\gamma-1} = \left(\frac{P_2}{P_1}\right)^{\frac{\gamma-1}{\gamma}} \qquad \ldots(7)$$

Now for Adiabatic Process ($PV^\gamma = C$) and during this process neither heat enters the system nor heat leaves the system.

(a) Work Done

Note. Derivation is just similar to work done in polytropic process, put γ in place of n.

$$\therefore \text{ Work done,} \quad W_{1-2} = \int_1^2 PdV$$

$$W_{1-2} = \frac{P_2V_2 - P_1V_1}{1-\gamma}$$

$$W_{1-2} = \frac{P_1V_1 - P_2V_2}{\gamma-1}$$

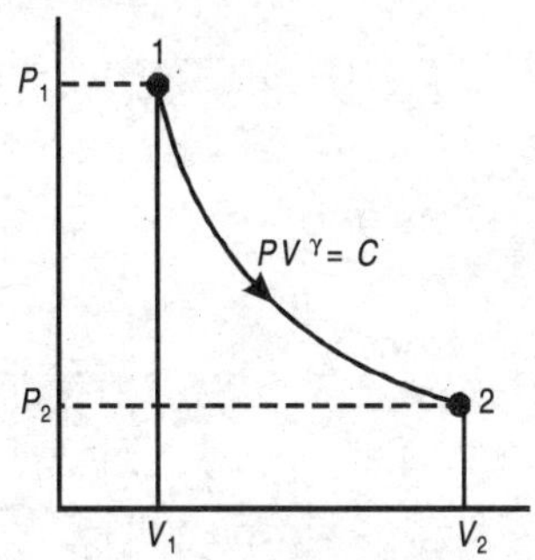

Fig. 4.9

(b) Heat supplied. Here Heat supplied = 0 (Since neither heat enters nor leaves the system during adiabatic the process).

4.7.5 Polytropic Process

Polytropic Process is given by the law ($PV^n = C$) and is represented on the PV-diagrams as shown.

(a) Work done. Work done in a non-flow process

$$W_{1-2} = \int PdV \qquad \ldots(1)$$

For a polytropic process,

$$PV^n = P_1V_1^n = P_2V^n = C$$

$$\therefore \quad P = \frac{C}{V^n}$$

Substituting in Eq. (1), we get,

$$\int Pdv = \int \frac{CdV}{V^n} = C\int V^{-n}dv$$

$$= \left[V^{1-n}\right]_1^2 = \left(V_2^{1-n} - V_1^{1-n}\right)$$

$$W_{1-2} = \frac{P_2V_2 - P_1V_1}{1-n} \text{ or } \frac{P_1V_1 - P_2V_2}{n-1} \qquad \ldots(2)$$

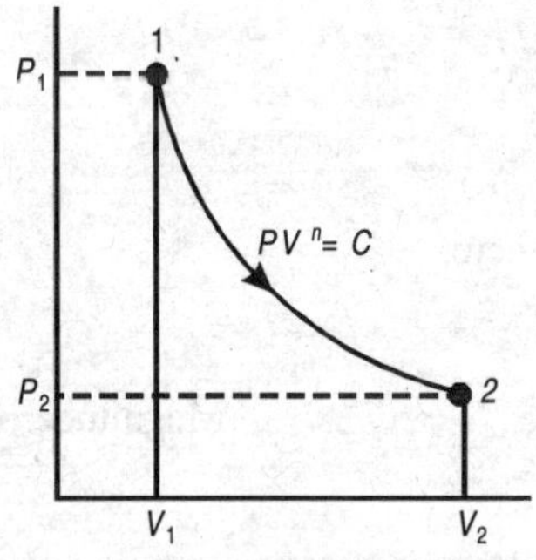

Fig. 4.10

Since, $\quad P_1V_1 = mRT_1$

and $\quad P_2V_2 = mRT_2$

$$W_{1-2} = \frac{mR(T_2 - T_1)}{1-n}$$

And for unit mass, $$W_{1-2} = \frac{R(T_2 - T_1)}{1-n}$$

(b) Heat supplied. From the I-law for closed system undergoing a process,

$$Q = \Delta U + W$$

For unit mass, $q = \Delta u + w$

In the differential form, $$\delta q = du + \delta w$$

$$\delta q = du + Pdv$$

$$q_{1-2} = C_v(T_2 - T_1) + \frac{R(T_2 - T_1)}{1-n} \quad \text{[From Eq. (3)]}$$

$$= \left[C_v + \frac{R}{1-n}\right](T_2 - T_1)$$

$$= \left[\frac{C_v - nC_v + C_p - C_v}{1-n}\right](T_2 - T_1) \text{ since } C_p - C_v = R$$

$$= \left[\frac{C_p - nC_v}{1-n}\right](T_2 - T_1)$$

Taking C_V common, we can write,

$$q_{1-2} = C_v\left[\frac{\gamma - n}{1-n}\right](T_2 - T_1), \text{ as } \frac{C_p}{C_v} = \gamma$$

$$\mathbf{q_{1-2} = C_n.(T_2 - T_1)}$$

where, $$C_n = C_v.\left[\frac{\gamma - n}{1-n}\right] \text{ or we can write as, } C_n = -C_v\left[\frac{\gamma - n}{n-1}\right]$$

i.e., $$\text{Magnitude of } C_n = C_v\left[\frac{\gamma - n}{n-1}\right] \text{ is known as Polytropic specific heat} \quad \ldots(5)$$

4.7.6 Throttling Process

Throttling is an irreversible process in which a fluid, flowing across a restriction, undergoes a drop in a total pressure, such a process occurs in the flow through a porous plug, a partially closed valve, or a small orifice. Joule and Thomson performed the basic throttling experiments in the period 1852–62, and their experiments clarified the process and let to use of throttling as a method for determining certain properties of gaseous substances.

The apparatus used in Joule Thomson experiment is shown in Fig. 4.11.

A steady stream of gas flows through a porous plug contained in a horizontal tube. This system is open, is thermally insulated ($Q = 0$) and does not exchange work with its environment ($W = 0$). At sections 1 and 2, both the

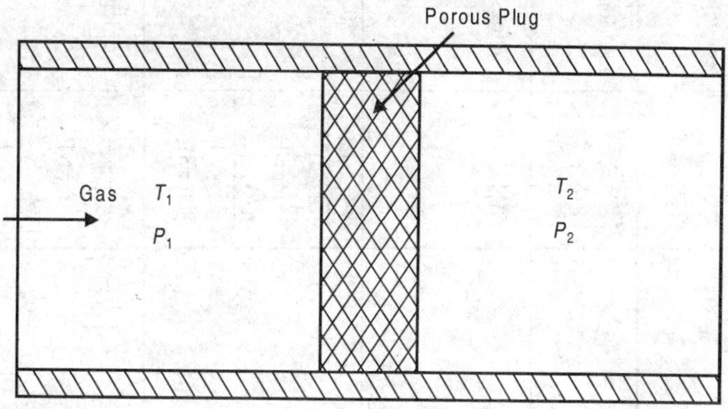

Fig. 4.11. *Throttling process, the Joule – Thomson porous plug experiment.*

temperature and the pressure are measured. If the kinetic energy does not change significantly as the fluid passes through the porous plug, the steady flow energy equation reduces to

$$Q = \Delta P + \Delta K + \Delta H + W_{sf}$$

$$\therefore \quad O = 0 + 0 + \Delta H + 0$$

or $$\Delta H = 0$$

or $$H_2 - H_1 = 0$$

or $$H_2 = H_1 \text{ or } h_1 = h_2 \text{ or } t_1 = t_2$$

Hence, in an adiabatic throttling process the enthalpy remains constant. In actual experiments with different p_2, t_2 does not remain constant but is different than t_1.

4.8 EFFECT OF VARYING '*n*'

Since it is helpful to be able to sketch curves of various thermodynamic process on *P–V* and *T–S* planes about as they should appear Fig. 4.12 will be of assistance in learning to make these sketches. It shows also the effect of varying *n*. Note that the isentropic curve on the *PV* plane is steeper than the isothermal curve and that, on *TS*-plane, the constant volume curve is steeper than the constant pressure curve when both the curves are drawn for the same temperature limits.

Expansions or compressions are imagined to take place from some common point '1'.

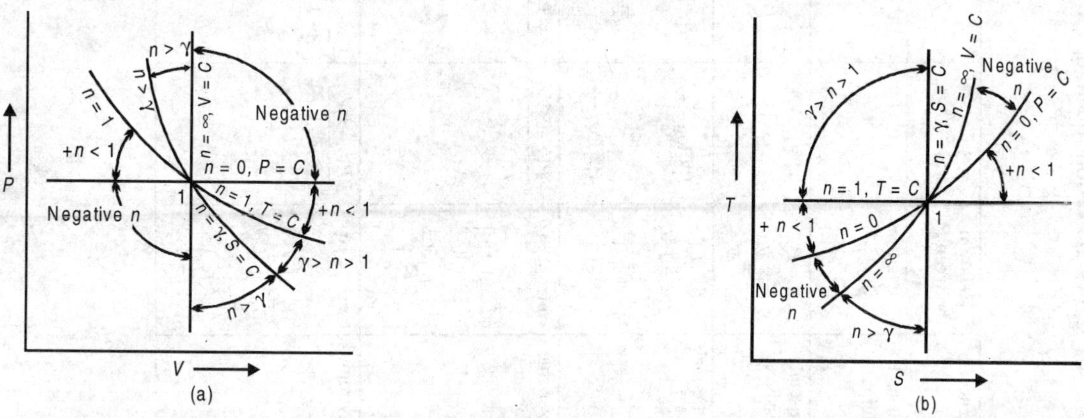

Fig. 4.12. *Effect of varying n.*

Table 4.1 shows the various quantities involved in various processes

Table 4.1

Process	Relation between P, V, T	$\int_1^2 P.dv$	$-\int_1^2 V.dP$	ΔU	ΔH	Q	Value of n	C	Work	ΔH
Constant volume	$V_1 = V_2$ $\frac{P_1}{P_2} = \frac{T_1}{T_2}$	0	$V_1(P_2 - P_1)$	$mC_v(T_2 - T_1)$	$mC_p(T_2 - T_1)$	$mC_v(T_2 - T_1)$	∞	C_v	0	$mC_v \log_e \frac{T_2}{T_1}$
Constant pressure	$P_1 = P_2$ $\frac{T_2}{T_1} = \frac{V_2}{V_1}$	$P_1(V_2 - V_1)$	0	$mC_v(T_2 - T_1)$	$mC_p(T_2 - T_1)$	$mC_p(T_2 - T_1)$	0	C_p	$mP_1(V_2 - V_1)$	$mC_p \log_e \frac{T_2}{T_1}$
Isothermal	$T_1 = T_2$ $P_1V_1 = P_2V_2$	$P_1V_1 \log_e \frac{V_2}{V_1}$ or $P_1V_1 \log_e \frac{P_1}{P_2}$	$P_1V_1 \log_e \frac{V_2}{V_1}$ or $P_1V_1 \log_e \frac{P_1}{P_2}$	0	0	$P_1V_1 \log_e \frac{V_2}{V_1}$ or $P_1V_1 \log_e \frac{P_1}{P_2}$	1	∞	$P_1V_1 \log_e \frac{V_2}{V_1}$	$R \log_e \frac{V_2}{V_1}$
Reversible adiabatic	$P_1V_1^r = P_2V_2^r$ $\frac{T_2}{T_1} = \left(\frac{V_1}{V_2}\right)^{r-1}$ $= \left(\frac{P_2}{P_1}\right)^{\frac{r-1}{r}}$	$\frac{P_1V_1 - P_2V_2}{r-1}$	$\frac{\gamma(P_1V_1 - P_2V_2)}{\gamma - 1}$	$mC_v(T_2 - T_1)$	$mC_p(T_2 - T_1)$	0	γ	0	$\frac{P_1V_1 - P_2V_2}{\gamma - 1}$ for NF $\frac{\gamma(P_1V_1 - P_2V_2)}{\gamma - 1}$ for SF	0
Polytropic	$P_1V_1^n = P_2V_2^n$ $\frac{T_2}{T_1} = \left(\frac{V_1}{V_2}\right)^{\frac{n-1}{1}}$ $= \left(\frac{P_2}{P_1}\right)^{\frac{n-1}{n}}$	$\frac{P_1V_1 - P_2V_2}{n-1}$	$\frac{n}{n-1}(P_1V_1 - P_2V_2)$	$mC_v(T_2 - T_1)$	$mC_p(T_2 - T_1)$	$mC_v.\frac{\gamma-1}{n-1}(T_2 - T_1)$ $= mC_n(T_2 - T_1)$	n	$C_v\left(\frac{\gamma-1}{n-1}\right)$ $= C_n$	$\frac{P_1V_1 - P_2V_2}{n-1}$ for NF $\frac{n(P_1V_1 - P_2V_2)}{n-1}$ for SF	$mC_n \log_e \frac{T_2}{T_1}$

SOLVED EXAMPLES

Example 4.1 *A gas whose original pressure, volume and temperature were 150 kN/m²; 0.1 m³; and 30°C is compressed such that its new pressure is 750 kN/m² and its new temperature is 70°C. Determine the new volume of the gas.*

Solution

By the characteristic equation i.e., $PV = mRT$

or $$\frac{P_1V_1}{T_1} = \frac{P_2V_2}{T_2}$$

∴ $$V_2 = \frac{P_1}{P_2}\cdot\frac{T_1}{T_2}V_1 = \frac{150}{750}\times\frac{343}{303}\times 0.1$$

$$\mathbf{V_2 = 0.0226\ m^3}$$

Note. The temperatures must always be in degrees Kelvin.

Example 4.2 *A quantity of gas has a pressure of 300 kN/m² when its volume is 0.025 m² and its temperature is 37°C. If the value of R is 0.287 kJ/kg K, determine the mass of the gas present. If the pressure of gas is now increased to 0.9 MN/m² while the volume remains constant what will be the new temperature of the gas.*

Solution

(1) By characteristic gas equation,

$$PV = mRT$$

∴ $$m = \frac{PV}{RT} = \frac{300\times10^3\times0.025}{0.287\times10^3\times310} = 0.084\text{ kg}$$

(2) We have, $$\frac{P_1V_1}{T_1} = \frac{P_2V_2}{T_2}$$ and in this case $V_1 = V_2$

∴ $$T_2 - \frac{P_2T_1}{P_1} = 310\times\frac{0.9\times10^6}{300\times10^3}$$

$$\mathbf{T_2 = 930\ K}$$

∴ $$T_2 = 930 - 273$$

$$\mathbf{T_2 = 675^\circ\ C}$$

Example 4.3 *10 kg of air is heated at constant pressure, from a temperature of 100°C to 200°C. Calculate the heat added during the process and also the change in I.E. Take gas constant* $R = 0.287\frac{kJ}{kg\text{-}K}$ *and g = 1.4 for air.*

Data: $m = 10$ kg, $T_1 = 100$°C, $T_2 = 200$°C

$$R = 0.287\frac{kJ}{kg\ K};\ \gamma = 1.4$$

Solution

We know that, heat added during constant pressure process $= m\,C_P\,(T_2 - T_1)$ ...(1)

Also, $$C_V = \frac{R}{\gamma-1} = \frac{0.287}{1.4-1}$$

$$C_V = 0.7175$$

and $$C_P - C_V = R \quad ...(2)$$

$$\Rightarrow \quad C_P = R + C_V$$

$$C_P = 0.287 + 0.7175$$

$$C_P = 1.0045 \frac{\text{kJ}}{\text{kg K}}$$

∴ Substituting (2) and (3) in (1) we get,

$$\text{Heat added} = m\, C_P (T_2 - T_1)$$
$$= 10 \times 1.0045\ (200 - 100)$$

Heat added = 1004.5 kJ

$$\text{Charge in I.E.} = \Delta U = m\, C_V (T_2 - T_1) = 10 \times 0.7175\ (200 - 100)$$

$$\mathbf{\Delta U = 717.5\ kJ}$$

Example 4.4 *A mass of 0.8 kg of air at 1 bar and 25°C is contained in a gas tight frictionless piston cylinder device. The air is now compressed to a final pressure of 5 bar. During the process heat is transferred from the air such that the temperature inside the cylinder remains constant. Calculate the heat transfer and work done during the process and direction of each in the process.*

Data : $m = 0.8$ kg, $T_1 = 298$ K, $T = C$, Isothermal process; $P_1 = 1 \text{ bar} = 10^5 \text{ N/m}^2$, $P_2 = 5 \text{ bar} = 5 \times 10^5 \text{ N/m}^2$

Solution

∴ We know that, for an isothermal process,

$$Q = W = P_1 V_1 \ln \frac{P_1}{P_2} = mRT_1 \ln \frac{P_1}{P_2}$$

$$\mathbf{Q = W = -110.11\ kJ}$$

as Q is – ve ∴ W is also – ve

Example 4.5 *1 kg of air at an initial condition of 3 bar and 175°C expands adiabatically without friction to a final condition of 1 bar. Calculate,*

(a) External work done during expansion.

(b) Change in I.E.

Take $\gamma = 1.4$ *and* $R = 0.287 \dfrac{kJ}{kg\text{-}K}$

Data : $m = 1$ kg, $T_1 = 175 + 273 = 448$ K, $\gamma = 1.4,\ R = 0.287 \dfrac{kJ}{kg\text{-}K}$; $P_1 = 3 \times 10^5 \dfrac{N}{m^2} = 3 \times 10^2 \text{ kN/m}^2$, $P_2 = 1 \times 10^2 \text{ kN/m}^2$

Solution

(a) We know that the work done in a adiabatic expansion process is given by,

$$W_{1-2} = \frac{P_1 V_1 - P_2 V_2}{\gamma - 1} \quad ...(1)$$

Here V_1 and V_2 are not given.

Since it is an adiabatic process,

$$P_1 V_1^{\gamma} = P_2 V_2^{\gamma}$$

$$\frac{P_1}{P_2} = \left(\frac{V_2}{V_1}\right)^{\gamma}$$

$$\left(\frac{P_1}{P_2}\right)^{\frac{1}{\gamma}} = \frac{V_2}{V_1}$$

$\therefore$

$$V_2 = V_1 \times \left(\frac{P_1}{P_2}\right)^{\frac{1}{\gamma}} \qquad \text{...(2)}$$

To find V_1,

$$P_1 V_1 = mRT_1$$

$$V_1 = \frac{mRT_1}{P_1} = \frac{1 \times 0.287 \times 448}{3 \times 10^2}$$

Note units : $\left[\frac{\text{kg} \frac{\text{kJ}}{\text{kg} - \text{K}} \times \text{K}}{\text{kN/m}^2} = \frac{\text{kJ}}{\text{kN/m}^2} = \frac{\text{kNm}}{\text{kN}} \times \text{m}^2 = \text{m}^3\right]$

$$V_1 = 0.429 \text{ m}^3$$

$\therefore$ From Eq. (2),

$$V_2 = 0.429 \times \left(\frac{300}{100}\right)^{\frac{1}{1.4}} = 0.429 \times (3)^{\frac{1}{1.4}}$$

$$\mathbf{V_2 = 0.940 \text{ m}^3}$$

$\therefore$ Now from Eq. (1), work done,

$$W_{1-2} = \frac{P_1 V_1 - P_2 V_2}{\gamma - 1}$$

$$= \frac{300 \times 0.429 - 100 \times 0.940}{1.4 - 1} = \frac{34.7}{0.4}$$

$$\mathbf{W_{1-2} = 86.75 \text{ kJ}}$$

(b) We also know from the I–law, that for a process, $Q = \Delta U + W$. Since it is adiabatic process $Q = 0$

$\therefore$ $$\mathbf{W = -\Delta U = 86.75 \text{ kJ} \Rightarrow \Delta U = -86.75 \text{ kJ}}$$

$\therefore$ I.E. decreases by 86.75 kJ.

Example 4.6 *Find the change of Internal energy and enthalpy of 3 kg of air, when the temperature changes from 25°C to 100°C.*

Take $C_p = 1 \frac{kJ}{kg\text{-}K}$ *and* $C_v = 0.71 \frac{kJ}{kg\text{-}K}$

Solution

In the problem, the type of process is not given, but since temprature changes, it is not an isothermal process.

$\therefore$ $$\Delta U = m \cdot C_v \cdot (T_2 - T_1)$$

$$= 3\text{ kg} \times 0.71 \frac{\text{kJ}}{\text{kg-K}} \times (373 - 298)\text{ K}$$

$$= 3 \times 0.71 \times 75$$

$$\mathbf{\Delta U = 159.75\ kJ}$$

and change of enthalpy $\Delta H = m \cdot C_P \cdot (T_2 - T_1) = 3 \times 1 \times (75)$

$$\mathbf{\Delta H = 225\ kJ}$$

Example 4.7 *0.44 kg of a gas having volume 0.28 m³ and a pressure of 1.4 bar is compressed to a pressure of 14 bar according to $PV^{1.25} = C$. Find the amount of heat transferred during the process by calculating the work done and the change of internal energy.*

Take $C_p = 1.041 \frac{kJ}{kg\text{-}K}$ and $C_v = 0.743 \frac{kJ}{kg\text{-}K}$

Data: $m = 0.44\ kg$, $V_1 = 0.28\ m^3$

$P_1 = 1.4\ bar = 140\ kPa$, $P_2 = 14\ bar = 1400\ kPa$

$PV^{1.25} = C$, $Q = ?$

$W = ?$, $\Delta U = ?$

Solution

$$C_P = 1.041 \text{ and } C_V = 0.743 \frac{\text{kJ}}{\text{kg-K}}$$

$$\text{W.D.} = \frac{P_2 V_2 - P_1 V_1}{1 - n} \quad ...(1)$$

Here only V_2 is not given and since it is a polytropic process,

$$P_1 V_1^n = P_2 V_2^n$$

$$\frac{P_1}{P_2} = \left(\frac{V_2}{V_1}\right)^n$$

i.e.

$$\frac{V_2}{V_1} = \left(\frac{P_1}{P_2}\right)^{\frac{1}{n}}$$

$$V_2 = V_1 \times \left(\frac{P_1}{P_2}\right)^{\frac{1}{n}} = 0.28 \times \left(\frac{140}{1400}\right)^{\frac{1}{1.25}}$$

$$= 0.28 \times (0.1)^{0.8}$$

$$V_2 = 0.044377\ \text{m}^3 \approx 0.044\ \text{m}^3$$

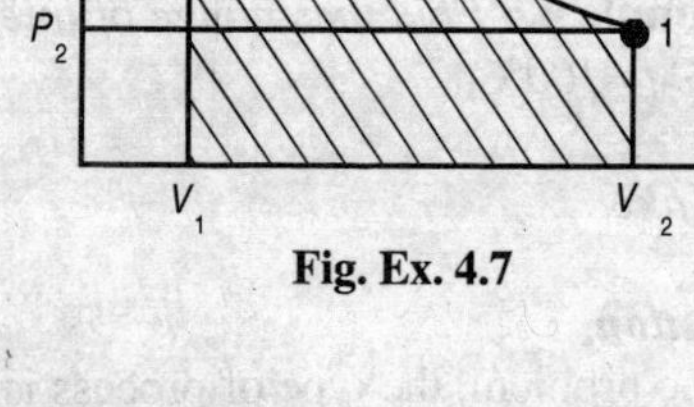

Fig. Ex. 4.7

Now work done from Eq. (1),

$$W_{1\text{-}2} = \frac{1400 \times 0.0444 - 140 \times 0.28}{1 - 1.25}$$

$$= \frac{62.16 - 39.2}{-0.25}$$

$$\mathbf{W_{1\text{-}2} = -91.84\ kJ}$$

$W_{1\text{-}2}$ is – ve, so, work is done on the gas.

We know that heat transferred in a polytropic process is given by,

We know that, $$q = C_n \cdot (T_2 - T_1)$$

$$= C_V \cdot \left(\frac{\gamma - n}{1 - n}\right)(T_2 - T_1)$$

$$= \frac{R}{\gamma - 1}\left(\frac{\gamma - n}{1 - n}\right)(T_2 - T_1)$$

$$q = \frac{R}{(1-n)}\frac{(\gamma - n)}{(\gamma - 1)} \times (T_2 - T_1)$$

$$q = \frac{(\gamma - n)}{(\gamma - 1)} \times \frac{R}{(1-n)} \times (T_2 - T_1)$$

And total heat transfer, $$Q = \frac{\gamma - n}{\gamma - 1} \times \frac{mR\,(T_2 - T_1)}{1 - n}$$

i.e. $$Q = \frac{\gamma - n}{\gamma - 1} \times \text{W.D. in a Polytropic process.}$$

Note: Remember this equation and derivation.

To find Q,

Let us find, $$\gamma = \frac{C_P}{C_V} = \frac{1.041}{0.743} = 1.401$$

$$Q = \frac{1.401 - 1.25}{1.104 - 1} \times -91.84 = \frac{0.15}{0.4} \times -91.84$$

$$\mathbf{Q = -34.583\ kJ}$$

∴ From the first law for a process,

$$Q - W = \Delta U$$

$$-34.583 + 91.84 = \Delta U$$

∴ $$\mathbf{\Delta U = 57.257\ kJ\ (Increases)}$$

Example 4.8 *A closed system containing 2 kg of working substance undergoes a reversible constant pressure process, during which the 1.E. of the system decreases by 80 kJ. If*

$C_p = 1120 \frac{kJ}{kg\text{-}K}$ *and* $C_v = 800 \frac{kJ}{kg\text{-}K}$. *Find the work done during the process.*

Data :

$$m = 2\ kg$$
$$P = C$$
$$\Delta U = -80\ kJ$$
$$C_P = 1.120 \frac{kJ}{kg\text{-}K}$$
$$C_V = 0.8 \frac{kJ}{kg\text{-}K}$$
$$W_{1-2} = ?$$

Solution

We know that work done in a constant pressure process,

$$W_{1-2} = P(V_2 - V_1) \quad ...(1)$$

Also from the I-law, $Q - W = \Delta U$ ...(2)

We know that heat supplied is utilised for,

(i) increasing I.E.,

(ii) doing external work and heat transfer,

$$\delta q = dh$$

or $$\delta Q = dH = m\, C_P\, \Delta T$$

Also, $$\Delta U = m.\, C_V.\, \Delta T \quad ...(4)$$

$$\frac{\Delta U}{m \cdot C_V} = \Delta T$$

$$\frac{-80 \text{ kJ}}{2 \text{ kg} \times 0.8 \dfrac{\text{kJ}}{\text{kg-K}}} = \Delta T$$

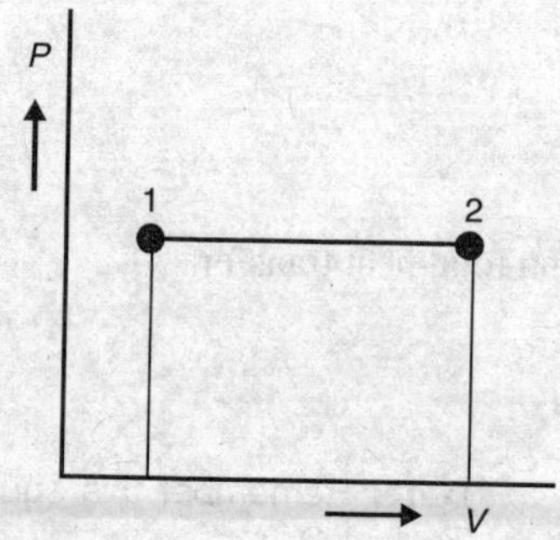

Fig. Ex. 4.8

$$\frac{-0.8}{1.6} = \Delta T$$

$$-50K = \Delta T \quad ...(5)$$

$\therefore$ From (3), $$\delta Q = m.\, C_P.\, \Delta T = 2 \times 1.12 \times (-50)$$

$$\delta Q \text{ or } Q = -112 \text{ kJ}$$

$\therefore$ From Eq. (2), $$Q - W = \Delta U$$

$$Q - \Delta U = W$$

$$-112 + 80 = W$$

$\therefore$ $$\mathbf{W = -32 \text{ kJ}}$$

Example 4.9 *A cylinder contains 0.12 m³ of air at 1 bar and 100°C. The air is compressed to 0.03 m³. The final pressure being 6 bar.*

Determine:

(i) The value of index n.

(ii) Mass of air in the cylinder.

(iii) Increase in I.E.

Take $\gamma = 1.4$, $R = 0.287$ kJ/kg-K *and* $C_v = 0.72 \dfrac{kJ}{kg\text{-}K}$

Data:

$$V_1 = 0.12 \text{ m}^3$$
$$P_1 = 100 \text{ kPa}$$
$$T_1 = 100 + 273 = 373K \qquad V_2 = 0.03 \text{ m}^3$$
$$P_2 = 600 \text{ kPa}$$

Solution

(i) To find n.

So, $$P_1 V_1^n = P_2 V_2^n$$

$$\frac{P_2}{P_1} = \left(\frac{V_1}{V_2}\right)^n$$

$$\left(\frac{600}{100}\right) = \left(\frac{0.12}{0.03}\right)^n$$

$$6 = (4)^n$$

$$\log_e 6 = n \log_e (4)$$

$$\frac{\log_e 6}{\log_e 4} = n$$

$$\frac{1.79175}{1.38629} = n$$

$$\mathbf{n = 1.29248 \approx 1.3}$$

(ii) To find mass of air in the cylinder.

We know that, $P_1V_1 = mRT_1$

$$100 \times 0.12 = m \times 0.287 \times 373$$

$\therefore$ $\mathbf{m = 0.112096\ kg}$

(iii) To find the increase in I. E.,

We have, $\Delta U = m.C_V(T_2 - T_1)$

Here T_2 is not known, so to find T_2, since it is a polytropic process,

$$\therefore \quad \frac{T_2}{T_1} = \left(\frac{V_1}{V_2}\right)^{n-1} = \left(\frac{P_2}{P_1}\right)^{n-1/n}$$

$$T_2 = T_1 \times \left(\frac{V_1}{V_2}\right)^{n-1}$$

$$= 373 \times \left(\frac{0.12}{0.03}\right)^{1.3-1}$$

$$= 373 \times (4)^{0.3} = 373 \times 1.516$$

$$T_2 = 565.36\ K$$

$$\therefore \quad \Delta U = m.C_V.(T_2 - T_1)$$

$$= 0.112\ kg \times 0.72\,(565.36 - 373)$$

$$\mathbf{\Delta U = 15.51\ kJ\ (Increases)}$$

Example 4.10 *Calculate the change of,*

(i) IE and

(ii) enthalpy of 2 kg of air when the temperature changes from 20°C to 90°C. Take

$$C_P = 1 kJ/kg\text{-}K \text{ and } C_v = 0.71 \frac{kJ}{kg\text{-}K}$$

Solution

(i) $\Delta U = m\,C_V(T_2 - T_1) = 2 \times 0.71(90 - 20)$

$$\mathbf{\Delta U = -99.4\ kJ}$$

(ii) $\Delta H = m \cdot C_V \cdot (T_2 - T_1) = 2 \times 1 \times (90 - 20)$

$$\mathbf{\Delta H = 140\ kJ}$$

Example 4.11 *10 kg of air are heated in a rigid vessel from 20°C to 100°C. If the ratio of specific heats is 1.4, estimate the value's of C_p and C_v, the change in internal energy and enthalpy.*

Data: $T_1 = 20°C, T_2 = 100°C$

Solution

$$C_V = \frac{R}{\gamma - 1}$$

and $$\gamma = 1.4 \text{ and } R = 0.287$$

$$\therefore \quad \mathbf{C_V = \frac{0.287}{0.4} = 0.7175 \frac{kJ}{kg\text{-}K}}$$

$$\mathbf{C_P = R + C_V = 0.287 + 0.7175 = 1.0045\ kJ/kg\text{-}K}$$

We know that,

Change in internal energy,

$$dU = m C_V (T_2 - T_1)$$

$$= (10)\,0.7175\,(100\text{–}20)$$

$$\mathbf{dU = 574\ kJ = 574\ kJ}$$

and change in enthalpy,

$$dH = m C_P (T_2 - T_1)$$

$$= (10)(1.0045)(100 - 80)$$

$$\mathbf{dH = 803.6\ kJ}$$

Example 4.12 *10 kg of Carbon monoxide at 40°C occupies 3 m³. Determine the gas pressure in bar. An additional mass of CO is then very slowly added to raise the tank pressure to 10 bar. Assuming that the gas temperature remains constant, how much extra mass must have been added. Assume R = 0.297 KJ/kg-K.*

Solution

We know that, $$PV = m RT$$

$$P = \frac{m RT}{V} = \frac{10 \times 0.297 \times 313}{3}$$

$$P = 309.87 \text{ kPa} = 3.0987 \text{ bar}$$

Since additional CO is added to the tank, $V = C$ and as given $T = C$.

$$\therefore \quad m \propto \text{Pressure.}$$

$$\therefore \quad \frac{m_2}{m_1} = \frac{P_2}{P_1} = \frac{10}{3.0987}$$

$$\therefore \quad m_2 = m_1 \times \frac{10}{3.0987} = 10 \times \frac{10}{3.0987}$$

$$\mathbf{m_2 = 32.2768\ kg}$$

$\therefore$ Additional mass, $$\mathbf{\Delta m = 32.2768 - 10 = 22.2768\ kg}$$

Example 4.13 *A piston and cylinder machine contains 1 kg of air, initially v = 0.8 m³/kg and T = 290 K. The air is then compressed in a slow frictionless process to a specific volume 0.2 m³/kg and a temperature 580 K. The law for compression is PV^1.5 = 0.75 with P in bar and V in m³/kg. Determine work done and heat transferred during the process. Assume for air* $C_p = 1.0 \frac{kJ}{kg\text{-}K'}$ *and* $C_v = 0.743 \frac{kJ}{kg\text{-}K}$ *and* $R = 0.287 \frac{kJ}{kg\text{-}K}$.

Data: $m = 1\ kg, v_1 = 0.8\ m^3/kg, T_1 = 290\ K, v_2 = 0.2\ m^3/kg\ \ T_2 = 580\ K$

$W = ?\ Q = ?$, *law for the Process* $= PV^{1.5} = C = 0.75$

Solution

We know that, for a Polytropic process,

$$W = \frac{P_2 V_2 - P_1 V_1}{1-n} = \frac{m\,R(T_2 - T_1)}{1-n}$$

$$= \frac{1 \times 0.287\,(580-290)}{1-1.5}$$

$$= \frac{83.23}{-0.5}\ \text{kJ}$$

$$\mathbf{W = -\ 166.46\ kJ}$$

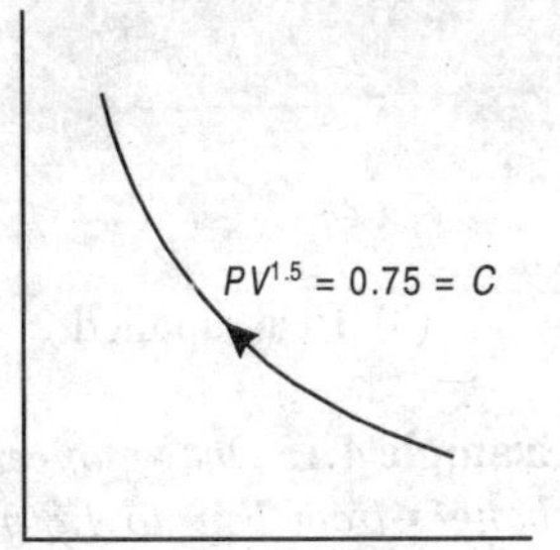

Fig. Ex. 4.13

The sign implies work is done on the air.

Also, $\Delta U = m \cdot C_V \cdot dT$

$$= 1 \times 0.743 \times (580 - 290)$$

$$\mathbf{\Delta U = 206.77\ kJ}$$

From I-law, for a process,

$$Q - W = \Delta U$$

$$Q = \Delta U + W$$

$$= 206.77 + (-\ 166.46)$$

$$\mathbf{Q = 40.31\ kJ}$$

Example 4.14 *2 kg of air at 150ºC and 3 bar expands according to law $PV^{1.2} = C$ to final pressure of 1 bar. Calculate: heat supplied, work done and change in internal energy.*

Take $R = 0.287$ kJ/kg-K and $= 1.41$.

Data: $P_1 = 3$ bar, $P_2 = 1$ bar, $T_1 = 150ºC = 423$ K

Solution

(1) Non flow work done in polytropic process

$$= \frac{mR(T_2 - T_1)}{n-1}$$

We have,

$$\frac{T_2}{T_1} = \left(\frac{P_2}{P_1}\right)^{\frac{n-1}{n}} = \left(\frac{1}{3}\right)^{\frac{0.2}{1.2}} = 0.832$$

$$T_2 = 423 \times 0.832 = 351.936\ \text{K}$$

$$\therefore \quad \text{W.D.} = \frac{0.2 \times 0.287 \times (423 - 351.936)}{1.2-1}$$

$$\textbf{W.D.} = \mathbf{203.954\ kJ}$$

(2) Heat supplied in polytropic process,

$$Q = \left(\frac{\gamma - n}{\gamma - 1}\right) \times \text{W.D. in non flow process.}$$

$$= \frac{1.41-1.2}{1.41-1} \times 203.954$$

$$\therefore \quad \mathbf{Q = 104.464\ kJ}$$

(3) Change in internal energy.

$$\Delta U = Q - W$$

$$= 104.464 - 203.954$$

$$\mathbf{\Delta U = 99.49\ kJ\ decrease}$$

Alternatively :

(2) Change in internal energy,

$$\Delta U = m\,C_V\,(T_2 - T_1)$$

$$= 2\times\frac{0.287}{1.41-1}\times(351.936-423)$$

$\therefore$ $$\mathbf{\Delta U = -99.49\ kJ}$$

(3) Heat supplied $$Q = W+\Delta U = 203.954-99.49$$

$$\mathbf{Q = 104.464\ kJ}$$

Example 4.15 *One kg of certain gas undergoes a thermodynamic constant pressure process whereby the volume changes from 1 m³ to 1.8 m³ while the temperature changes from 50°C to 450°C. The specific heat at constant pressure is given by,*

$$C_p = \left(2.5 + \frac{40}{T+20}\right) kJ/kg\ ^\circ C \text{ where } T \text{ is in } ^\circ C.$$

Find out (1) heat supplied, (2) work done, (3) change in internal energy and (4) change in enthalpy: Take $\gamma = 1.4$.

Solution

(1) Heat supplied in constant pressure process,

$$= \int m\,.\,C_P\,dT, \quad \text{As given } m = 1$$

$$= \int_{50}^{450}\left(2.5+\frac{40}{T+20}\right)dT$$

$$= \int_{50}^{450} 2.5\,dT + \int_{50}^{450}\frac{0}{T+20}\,dT$$

$$= 2.5[T]_{50}^{450} + 40[\ln(T+20)]_{50}^{450}$$

$$= 2.5(450-50) + 40\ln\left(\frac{470}{70}\right)$$

$$= 1000 + 76.169$$

Heat supplied in constant pressure process = 1076.169 kJ

(2) Change in internal energy is given by,

$$U_2 - U_1 = \int m\,C_V\,dT \text{ where } m = 1\ C_P = \gamma\,C_V$$

$$= \frac{1}{\gamma}\int C_P\,dT = \frac{1076.169}{1.4}$$

$$\mathbf{U_2 - U_1 = 768.692\ kJ}$$

(3) Work done = Heat supplied – Change in internal energy

= 1076.169 – 768.692

Work done = 307.477 kJ

(4) Change in enthalpy = Heat supplied in constant pressure process

Change in enthalpy = 1076.169 kJ

Example 4.16 *One kg of certain working substance undergoes a reversible constant pressure process at 1.2 bar during which its volume changes from 1 m³ to 2.8 m³ and temperature changes from 50°C to 370°C. The specific heat for the substance at constant pressure is given by,*

$C_p = \left(1.1 + \frac{40}{t+30}\right) kJ/kg\ °C$ *where, t is in °C. Find out:*

(i) Heat supplied *(ii) Work done*

(iii) Change in I. E. *(iv) Change in Enthalpy.*

Data:

$$P = C \qquad Mass = 1\ kg$$

$$P = 1.2\ bar = 1.2 \times 10^5 \frac{N}{m^2}$$

$$V_1 = 1\ m^3 \qquad V_2 = 1.8\ m^3$$

$$T_1 = 50°C \qquad T_2 = 370°C$$

Solution

Given,
$$C_P = \left[1.1 + \frac{40}{t+30}\right] \frac{kJ}{kg\ °C}$$

For $P = C$ process $\delta Q = dH$

$$\therefore \quad \delta Q = \text{Change in enthalpy} = \int_{T_1}^{T_2} m\,C_P\,dT$$

$$= \int_{T_1}^{T_2} 1 \times \left(1.1 + \frac{40}{t+30}\right) dT$$

$$= \int_{50}^{370} 1.1\,dT + \int_{50}^{370} \frac{40}{t+30}\,dT$$

$$= [1.1\,t]_{50}^{370} + 40\{\ln(t+30)\}_{50}^{370}$$

Change in enthalpy = 416.38 kJ

Also, for a constant pressure process,

$$\text{Heat supplied} = \text{Change in enthalpy}$$

$$\therefore \quad Q = 416.38\ \text{kJ}$$

Also,
$$\text{W.D.} = P(V_2 - V_1) = 120\,(1.8 - 1)$$

$$\mathbf{W_{1-2} = 96\ kJ}$$

According to I-law for a process,

$$Q - W = \Delta U$$

$$\Rightarrow \quad \Delta U = 416.38 - 96$$

$$\mathbf{\Delta U = 320.38\ kJ}$$

Example 4.17 *60 litres of hydrogen at 20°C and 1 bar is compressed adiabatically to 9.8 bar. It is then cooled at constant volume to a pressure P_3 such as to have temperature of 20°C an further expanded isothermally so as to reach initial state. Find:*

(i) Work done during each process.

(ii) Change in internal energy during each process. Take

$$C_p = 14.4\ kJ/kg\text{-}K$$

$$C_v = 10.28\ kJ/kg\text{-}K$$

for hydrogen and show the process on PV and T-S plane.

Data: $V_1 = 0.06\ m^3\ (H_2), T_1 = 20 + 273 = 293\ K, P_1 = 100\ kPa, PV^\gamma = c, P_2 = 980\ kPa,$

$$P_3 = T_3 = 20 + 273 = 293\ K, Pv = C, \left.\begin{matrix} C_P = 14.4\ kJ/kg\text{-}K \\ C_V = 10.28\ kJ/kg\text{-}K \end{matrix}\right\} \frac{C_P}{C_V} = \gamma = 1$$

Find : (i) W. D. during processes, (ii) during processes.

Solution

(i) Work done :

For process 1– 2, $$W_{1-2} = \frac{P_2V_2 - P_1V_1}{1-\gamma} \quad ...(1)$$

To find V_2 and since process 1 – 2 is an adiabatic process,

$$P_1V_1^\gamma = P_2V_2^\gamma$$

$$\left(\frac{P_1}{P_2}\right)^{\frac{1}{\gamma}} \times V_1 = V_2$$

$$\therefore \quad \mathbf{V_2 = 0.01175\ m^3}$$

$$\therefore \quad W_{1-2} = \frac{980 \times 0.01175 - 100 \times 0.06}{1 - 1.4}$$

$$\mathbf{W_{1-2} = -\ 13.788\ kJ}$$

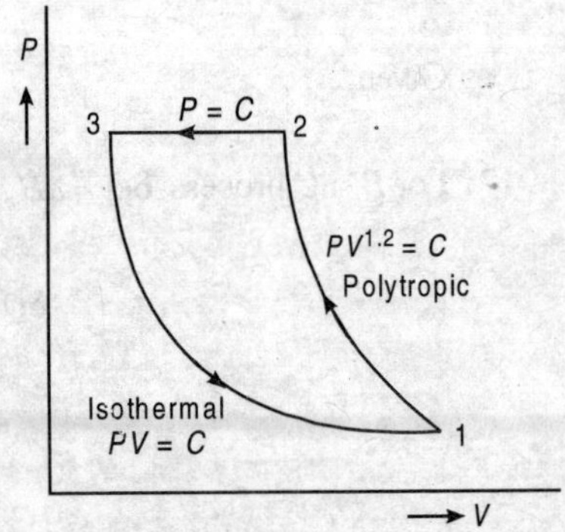

Fig. Ex. 4.17

Process 2 – 3 : Constant volume process $W_{2-3} = 0$

Process 3 – 1 : (Isothermal process) :

$$W_{3-1} = P_3\ V_3 \ln \frac{V_1}{V_3} \quad ...(2)$$

Note, $$V_2 = V_3$$

and for process 2 – 3,

$$\frac{P_2V_2}{T_2} = \frac{P_3V_3}{T_3}$$

$$\frac{P_2}{T_2} = \frac{P_3}{T_3}$$

$$\frac{980}{T_2} = \frac{P_3}{298\ (\text{i.e.}, T_1)} \quad ...(3)$$

For isothermal, $$P_1V_1 = P_3V_3$$

$$100 \times 0.06 = P_3 \times 0.01175\ (\text{as } V_2 = V_3)$$

$$\therefore \quad \mathbf{P_3 = 510.64\ kPa} \quad ...(4)$$

From (3) and (4), $$\mathbf{T_2 = 562.31\ K}$$

$$\therefore \quad W_{3-1} = P_3\ V_3 \ln \frac{V_1}{V_3}$$

$$= 510.64 \times 0.01175 \times \ln \frac{0.06}{0.01175}$$

$$\mathbf{W_{3-1} = 9.7829\ kJ}$$

(ii) Change in I. E.:

For process 1 – 2, $Q = 0$

$\therefore$ $Q - W = \Delta U$

$0 - W = \Delta U$

$\therefore$ $\mathbf{\Delta U = 13.788\ kJ\ (i.e., U_2 - U_1)}$

For process 2 – 3, $W = 0$

$$Q - W = \Delta U$$

$$Q = \Delta U = m \cdot C_V (T_3 - T_2)$$

$$P_1 V_1 = mRT_1$$

$$100 \times 0.06 = m\,(C_P = C_V)\,T_1$$

$$= m\,(14.4 - 10.28) \times 293$$

$$\mathbf{m = 0.00488\ kg}$$

$\therefore$ $Q_{2-3} = \Delta U = 0.0048 \times 10.28\,(298 - 562.3)$

$$\mathbf{\Delta U = Q_{2-3} = -13.27\ kJ}$$

for 3 – 1 $T = C$ $\Delta U = m \cdot C_V \cdot dT = 0$ as $T = C$.

Example 4.18 *3 kg of air at a pressure of 150 kPa and temperature 360 K is compressed polytropically to 750 kPa according to law $PV^{1.2}$ = constant. The air is then cooled to initial temperature at constant pressure. The air is then brought to state (1) by following PV = C. Draw the cycle on PV-diagram and determine net work and heat.*

Data: $P_1 = 150\ kPa, P_2 = 750\ kPa, P_2 = P_3, T_1 = 360\ K, V_2 = ?,\ T_3 = T_1 = 360\ K$

$V_1 = ?\ T_2 = ?\ V_3 = ?$

1 – 2 is a Polytropic process : $PV^{1.2} = C$

2 – 3 is a Constant pressure process

3 – 1 is a Isothermal process

Solution

We know that $PV = mRT$

$$P_1 V_1 = mRT_1$$

$$\Rightarrow \quad V_1 = \frac{mRT_1}{P_1} = \frac{3 \times 0.287 \times 360}{150}$$

$$\mathbf{V_1 = 2.0664\ m^3}$$

1 – 2 adiabatic process $\dfrac{T_1}{T_2} = \left(\dfrac{P_1}{P_2}\right)^{\frac{n-1}{n}} \Rightarrow \dfrac{360}{T_2}$

$$= \left(\frac{150}{750}\right)^{\frac{0.2}{1.2}} \Rightarrow T_2 = 470.76\ K$$

$$V_2 = \frac{mRT_2}{P_2} = \frac{3 \times 0.287 \times 470.76}{750}$$

$$= 0.540\ m^3$$

3 – 1 Isothermal process $PV = C$

$$P_3 V_3 = P_1 V_1 \Rightarrow V_3 = \frac{P_1 V_1}{P_3}$$

$$= \frac{150 \times 2.0664}{750}$$

$$\Rightarrow \quad V_3 = 0.413 \text{ m}^3$$

$$W_{1-2} = \frac{P_1 V_1 - P_2 V_2}{n-1} = \frac{mR(T_2 - T_1)}{1-n}$$

$$\mathbf{W_{1-2} = -476.82 \text{ kJ}}$$

$$W_{2-3} = P_2 (V_3 - V_2) = 750\ (0.413 - 0.54)$$

$$= -95.25 \text{ kJ}$$

$$W_{3-1} = P_1 V_1 \ln\left(\frac{P_3}{P_1}\right)$$

$$= 150 \times 2.0664 \ln \frac{750}{150}$$

$$\mathbf{W_{3-1} = 498.66 \text{ kJ}}$$

$$\text{Net work done} = W_{1-2} + W_{2-3} + W_{3-1}$$

$$= -73.21 \text{ kJ}$$

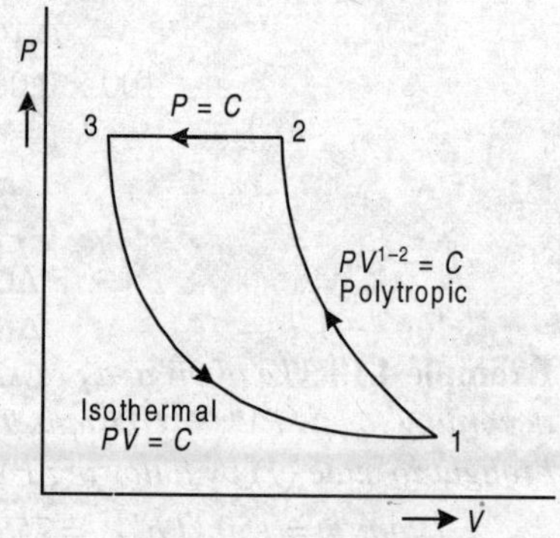

Fig. Ex. 4.18

$$Q_{1-2} = \left(\frac{\gamma - n}{\gamma - 1}\right) \times \text{Work done}$$

$$= \frac{1.4 - 1.2}{1.4 - 1.0} \times W_{1-2}$$

$$\mathbf{Q_{1-2} = -238.41 \text{ kJ}}$$

$$Q_{2-3} = \text{change in enthalpy} = m\, C_P\, \Delta t = m\, C_P (T_3 - T_2)$$

$$= \frac{m \gamma R}{\gamma - 1}(T_3 - T_2)$$

$$= \frac{3 \times 1.4 \times 0.287}{1.4 - 1}(360 - 470.76)$$

$$\mathbf{Q_{2-3} = -333.775 \text{ kJ}}$$

$$Q_{3-1} = \text{work done} = 498.86 \text{ kJ}$$

$$\therefore \quad \text{Net heat} = Q_{1-2} + Q_{2-3} + Q_{3-1}$$

$$\textbf{Net heat} = \mathbf{-73.21 \text{ kJ}}$$

$$\therefore \quad \text{Net heat} = \text{Net work done}$$

Thus the I-law for the cycle satisfies.

Example 4.19 *A certain oxygen cylinder has a capacity 250 litres and contains oxygen at a pressure of 3 MPa and a temperature of 20°C. The stop valve is opened and some oxygen is used. If the pressure and temperature of oxygen left in the cylinder falls to 1.8 MPa and 16°C respectively, determine the mass of oxygen used.*

If after the stop valve is closed the oxygen remaining in the cylinder attains its original temperature of 20°C, determine the amount of heat, transferred through the cylinder walls from the atmosphere.

Solution

Capacity of the cylinder = V = 250 litre = 0.250 m^3 as 1000 litre = 1 m^3

Density of O_2 at 0°C, 0.1013 MPa = 1.43 kg/m^3

$\therefore$ Mass of given O_2 at 0°C and 0.1013 MPa = 1.43×0.25

$$\therefore \quad m_1 = 0.3575 \text{ kg at STA}$$
$$P_2 = 3 \text{ MPA}$$
$$P_3 = 1.8 \text{ MPa}$$
$$T_2 = 20°\text{C} = 293 \text{ K}$$
$$T_3 = 16°\text{C} = 289 \text{ K}$$
$$PV = mRT$$

Since V is constant, as volume of the cylinder does not change, we have,

$$\frac{P}{T.m} = \frac{R}{V} = \text{Constant}$$

$$\therefore \quad \frac{P_1}{T_1 m_1} = \frac{P_2}{T_2 m_2}$$

$$\therefore \quad \frac{0.1013}{273 \times 0.3575} = \frac{3}{293 \times m_2}$$

$$\therefore \quad m_2 = 9.865 \text{ kg}$$

Similarly, $$\frac{P_1}{T_1 m_1} = \frac{P_3}{m_3 T_3}$$

$$= \frac{0.1013}{273 \times 0.3575} \times \frac{1.8}{289 \times m_3}$$

$$\therefore \quad m_3 = 6.001 \text{ kg}$$
$$\Delta m = m_2 - m_3$$
$$= 9.865 - 6.001$$
$$\boldsymbol{\Delta m = 3.864 \text{ kg}}$$

Also we have, $$\frac{PV}{mT} = R$$

$$\therefore \quad R = \frac{3000 \times 0.25}{9.865 \times 293} = 0.26 \text{ kJ/kg-K}$$

Heat transfer, $Q = m\,C_V(T_4 - T_3)$

{where, $T_4 = T_2$, the original temp.}

$$C_V = \frac{R}{\gamma - 1}$$

$$= \frac{R}{\left(\dfrac{C_P}{C_V} - 1\right)}$$

$$= \frac{0.26}{1.4 - 1} = 0.65 \text{ kJ/kg-K}$$

$$\therefore \quad Q = 6.001 \times 0.65\,(293 - 289)$$
$$\mathbf{Q = 15.603\ kJ}$$

Example 4.20 *A gas initially at pressure 510 kPa and volume 142 litres undergoes: a process and final pressure of 170 kPa and volume of 275 litres during which enthalpy decreased by 65 kJ. Take C_V = 0.718 kJ/kg-K. Determine:*

(i) Change in internal energy.
(ii) Specific heat at constant pressure.
(iii) Specific gas constant.

Data: Intial pressure $P_1 = 510\ kPa = 510\ kN/m^2$

Intial volume $V_1 = 142\ l = 142 \times 10^{-3}\ m^3$

Final pressure $P_2 = 170\ kPa = 170\ kN/m^2$

Final volume $V_2 = 275\ l = 275 \times 10^{-3}\ m^3$

Change in enthalpy $= -65\ kJ$ [$\because$ Enthalpy is being decreased]

$C_V = 0.718\ kJ/kg\text{-}K$

Solution

(i) We have, from first law of thermodynamics,

$$\Delta H = \Delta U + \Delta(PV)$$
$$\Delta U = \Delta H - \Delta(PV)$$
$$\Delta U = -65\times10^3 - (170\times275 - 510\times142)$$
$$= -65\times10^3 - (46750 - 72420)$$
$$= -65000 + 25670 = -39330$$
$$\therefore \quad \mathbf{\Delta U = -39.33\ kJ}$$

(ii) Also we know that,

$$\frac{\Delta H}{\Delta U} = \frac{C_P}{C_V} = \gamma$$
$$\therefore \quad C_P = C_V \times \frac{\Delta H}{\Delta U} = 0.718 \times \frac{-65}{-39.33}$$
$$\mathbf{C_P = 1.186\ kJ/kg\text{-}K}$$

(iii) Also, $C_P - C_V = R$

$$R = 1.1866 - 0.718$$
$$\mathbf{R = 0.4686\ kJ/kg\text{-}K}$$

Example 4.21 *An engine has a piston area of 0.1 m² contains gas at a pressure of 1.5 MPa. The gas expands according to a process which is represented by a straight line on P-V plane. The final pressure is 0.15 MPa. Calculate work done by gas on piston if the stroke is 0.3 m.*

Data: Area of the piston $= 0.1\ m^2$

$P_1 = 1.5\ MPa$ $\qquad P_2 = 0.15\ MPa$

Stroke $= 0.3\ m$ $\qquad W = ?$

Solution

According to the formula,

Change in volume = Area of the piston × Stroke
= 0.1 × 0.3 = 0.03 m

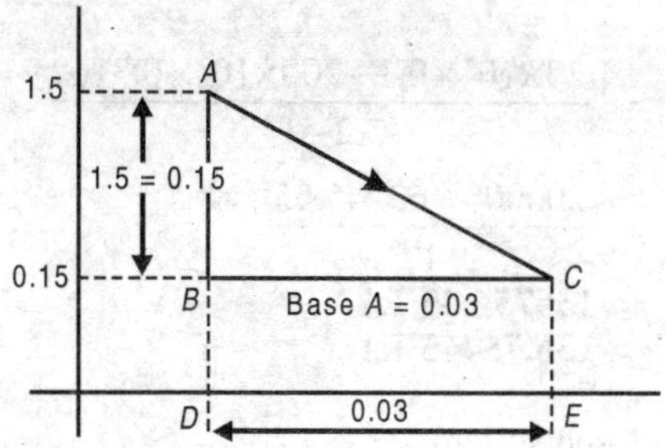

Fig. Ex. 4.21

Also we know that,

Work done by gas = Area under the curve (in this case a straight line) on the *P-V* diagram

∴ Net work done = Area of the rectangle *BCED* + Area of triangle *ABC*

$$W = \left[\{0.15\times10^6\times0.3\}+\left\{\frac{1}{2}(1.5-0.15)\times10^6\times0.03\right\}\right]$$

$$\mathbf{W = 24.75\ kJ}$$

Example 4.22 *In a piston cylinder arrangement the pressure varies inversely proportional to the square of volume during a process. The initial pressure in the cylinder is 20 bar and the initial volume is 0.1 m³. The final pressure after the expansion is 200 kPa. Estimate the work done.*

Data:

$$P \propto \frac{1}{V^2} \Rightarrow P = \frac{C}{V^2} \text{ and } PV^2 = C$$

$$P_1 = 20\ bar = 20\times10^5\ \frac{N}{m^2}$$

$$V_1 = 0.1\ m^3$$

$$P_2 = 200\ kPa = 200\times10^3\ \frac{N}{m^2}$$

$$W = ?,\ V_2 = ?$$

Solution

We know that for a Polytropic process,

$$\text{Work done} = \frac{P_1 V_1 - P_2 V_2}{n-1} \quad ...(1)$$

Since V_2 is not given,

$$P_1V_1^2 = P_2V_2^2 = C$$

$$V_1^2\times\frac{P_1}{P_2} = (V_2^2)$$

$$0.1^2\times\frac{20\times10^5}{200\times10^3} = V_2^2$$

$$0.1 = V_2^2$$

$$0.3162\ \text{m}^3 = V_2 \quad ...(2)$$

Now from Eq. (1),

$$W = \frac{20\times10^5\times0.1-200\times10^3\times0.3162}{2-1}$$

$$W = \frac{200000-63245.55}{1}$$

$$\mathbf{W = 136754.45\ kJ}$$

or $$\mathbf{W = 136.75445\ kJ}$$

Second Method :

We know that work done in a closed system,

$$= \int P\cdot dV$$

But Law for the process is,

$$PV^2 = C$$

or $$P = \frac{C}{V^2}$$

$\therefore$ From Eq. (3), $$W = \int P\cdot dV - \int_{V_1}^{V_2}\frac{C}{V^2}\cdot dV$$

$$= C\int_{V_1}^{V_2} V^{-2}\cdot dV = C\left[\frac{V^{-2+1}}{-2+1}\right]_{V_1}^{V_2} = C\left[-\frac{1}{V}\right]_{V_1}^{V_2}$$

$$W = C\left[-\frac{1}{V_2}+\frac{1}{V_1}\right] \quad ...(4)$$

Now, $$P_1V_1^2 = C$$

$\therefore$ $$20\times10^5\times0.1^2 = C = 20,000$$

Now, from Eqs (2), (4), (5),

$$W = 20000\left[-\frac{1}{0.316}+\frac{1}{0.1}\right] = 20000[-3.162+10]$$

$$= 20000[6.838]$$

$$\mathbf{W = 136754.45}$$

$$\mathbf{W = 136.754\ kJ}$$

Example 4.23 *A cylinder contains 0.12 cubic m of air at 1 bar and 90°C. It is compressed to 0.03 cubic m, the final pressure being 6 bar. Find the index of compression, increase in internal energy and heat transferred. Take:*

Characteristic gas constant = 0.287 kJ/kg-K; and

Specific heat at constant volume = 0.7176 kJ/kg-K.

Data:

$$V_1 = 0.12\ m^3 \qquad P_1 = 1\ bar = 1\times10^5\ \frac{N}{m^2}$$

$$T_1 = 90 + 273 = 363\ K \qquad V_2 = 0.03\ m^3$$

$$P_2 = 6\times10^5\ \frac{N}{m^2} \qquad n = ?$$

$$\Delta U = ? \qquad Q = ?$$

$$R = 0.287 \frac{kJ}{kg\text{-}K} \qquad C_V = 0.7176 \frac{kJ}{kg\text{-}K}$$

Solution

(i) We know that for the Polytropic process,

$$P_1V_1^n = P_2V_2^n = C$$

$$\frac{P_2}{P_1} = \left(\frac{V_1}{V_2}\right)^n$$

$$\left(\frac{6\times10^5}{1\times10^5}\right) = \left(\frac{0.12}{0.03}\right)^n$$

$$6 = (4)^n$$

$$\ln 6 = n \ln 4$$

$$\frac{\ln 6}{\ln 4} = n$$

$$\frac{1.79175}{1.38629} = n$$

$$\mathbf{n = 1.29248 = 1.3}$$

(ii) Now change in $\text{IE} = \Delta u = C_V \cdot (T_2 - T_1)$

To find T_2, since it is a polytropic process,

$$\frac{T_2}{T_1} = \left(\frac{V_1}{V_2}\right)^{n-1}$$

$$T_2 = T_1\left(\frac{V_1}{V_2}\right)^{n-1}$$

$$T_2 = 363\left(\frac{0.12}{0.03}\right)^{1.3-1}$$

$$= 363\times4^{0.3} = 363\times1.5157$$

$$\mathbf{T_2 = 550.199\ K}$$

$$\therefore \quad \Delta u = C_V\,(T_2 - T_1)\frac{\text{kJ}}{\text{kg}}$$

$$= 0.7176(550.199 - 363)$$

$$\mathbf{\Delta u = 134.334\ kJ/kg}$$

(iii) We know from the I-law,

$$q - w = \Delta u$$

$$q = \Delta u + w$$

And now to find W, $\quad W_{1-2} = \dfrac{P_2V_2 - P_1V_1}{1-n} = \dfrac{mR\,(T_2 - T_1)}{1-n}$

For unit mass,
$$W_{1-2} = \frac{R(T_2 - T_1)}{1-n}$$
$$= \frac{0.287(550.199 - 363)}{1-1.3}$$
$$= \frac{53.70}{-0.3}$$
$$\mathbf{W_{1-2} = -179.01\ kJ/kg}$$

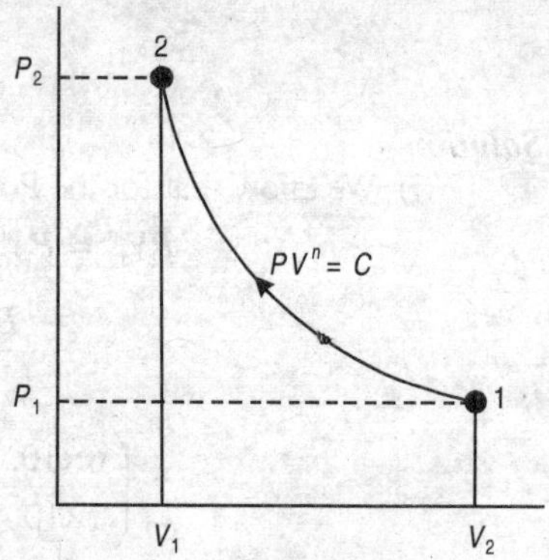

Fig. Ex. 4.23

Now heat transfer, from Eq. (1) will be,
$$q = \Delta u + W$$
$$= 134.334 + (-179.01)$$
$$\mathbf{q = -44.6765\ \frac{kJ}{kg}}$$

Example 4.24 ***Process 1:*** *Air initially at 100 kPa and 50° C undergoes reversible adiabatic compression such that its volume is reduced to 1/5th of initial volume.*

Process 2: *Then 940 kJ/kg of heat is added to this air at constant volume.*

Process 3: *Process 2 is followed by reversible adiabatic expansion up to initial volume.*

Process 4: *Finally heat is rejected at constant volume so as to reach the initial condition.*

Draw the four processes on one PV-diagram. Determine the maximum temperature, and heat rejected per kg of air. Assume adiabatic index of compression and expansion of 1.4 and constant volume specific heat as 0.717 kJ/kg-K.

Determine : (i) T_{max} = ?, (ii) Heat rejected/kg of air ?

Solution

Process (1 – 2) :
$$P_1 = 100 \text{ kPa}$$
$$PV^\gamma = C$$
$$V_2 = \frac{1}{5}V_1 = 0.2\,V_1$$
$$T_1 = 50 + 273 = 523 \text{ K}$$

Process (2 – 3) :
$$q_s = 940\ \frac{kJ}{kg}$$
$$= C_V \cdot dT$$
at $V = C$

Process (3 – 4) :
$$PV^\gamma = C$$
$$V_1 = V_4$$

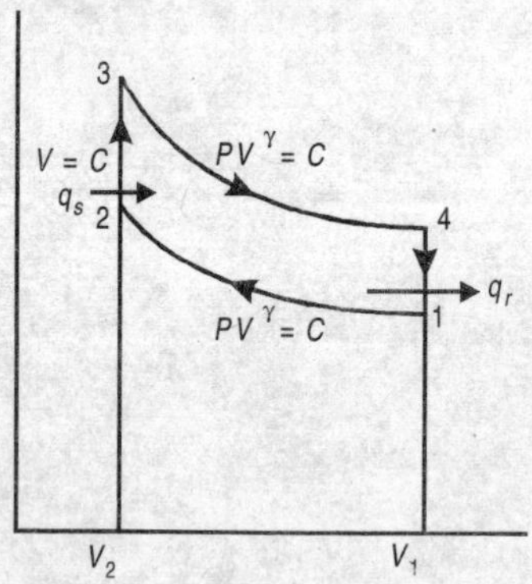

Fig. Ex. 4.24

Process (4 – 1) : $V = C$: $\gamma = 1.4$, $C_V = 0.717\ \frac{kJ}{kg\text{-}K}$

For the adiabatic 1 – 2 :
$$\frac{T_2}{T_1} = \left(\frac{V_1}{V_2}\right)^{\gamma-1}$$

$$T_2 = \left(\frac{V_1}{V_2}\right)^{\gamma-1} \times T_1$$

$$= \left(\frac{1}{0.2}\right)^{1.4-1} \times 323 \text{ as } \frac{V_2}{V_1} = 0.2$$

$$T_2 = \left(\frac{1}{0.2}\right)^{0.4} \times 323 = (5)^{0.4} \times 323$$

$$= 0.90 \times 323$$

$$\mathbf{T_2 = 614.88\ K}$$

For Process (2 – 3) : $q_s = C_V\,(T_3 - T_2)$

$$940 = 0.717(T_3 - 614.88)$$

$$\frac{940}{0.717} = T_3 - 614.88$$

$$1311.0181 = T_3 - 614.88$$

$$\mathbf{1925.898\ K = T_3}$$

This is the maximum temperature.
Now for the adiabatic 3 – 4:

$$\frac{T_4}{T_3} = \left(\frac{V_3}{V_4}\right)^{\gamma-1}$$

$$T_4 = \left(\frac{V_2}{V_1}\right)^{\gamma-1} \times T_3$$

$$= (0.2)^{1.4-1} \times 1925.9$$

$$= 0.2^{0.4} \times 1925.9$$

$$\mathbf{T_4 = 1011.686\ K}$$

Now heat rejected, $q_r = C_V\,(T_1 - T_4) = 0.717(323 - 1011.69)$

$$\mathbf{q_r = -493.7878\ \frac{kJ}{kg}}$$

– ve sign implies heat rejection.

Example 4.25 *Two vessels A and B, both containing Nitrogen are connected by a valve which is opened to allow the contents to mix and achieve an equilibrium temperature of 27°C. Before mixing, the following information is known about the gases in the two vessels.*

Vessel A: *Pressure = 1.5 MPa,*
Temperature = 50°C
and *Contents = 0.5 kg mol.*
Vessel B: *Pressure = 0.6 MPa,*
Temperature = 20°C
and *Contents = 2.5 kg*

Calculate the final equilibrium pressure, and the heat transferred to the surroundings. If the vessels had been perfectly insulated, calculate the final temperature and pressure which would have been reached. Take $\dfrac{C_p}{C_v} =$ *1.4, molecular weight = 28 kg/kg mol and R = 0.297 kJ/kg for Nitrogen.*

Solution

For the gas in vessel A

$$m_A = n_A \cdot M = 0.5 \times 28 = 14 \text{ kg}$$

$$P_A V_A = m_A RT_A$$

$$\therefore \quad V_A = \frac{14 \times 0.297 \times 323}{1.5 \times 10^3}$$

$$V_A = 0.895 \text{ m}^3$$

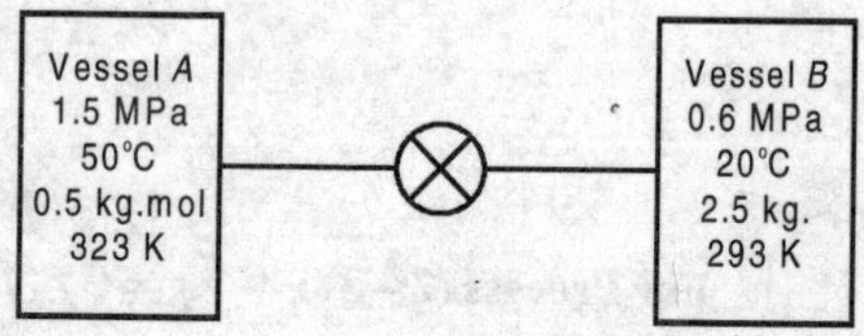

Fig. Ex. 4.25

For vessel B,
$$V_B = \frac{m_B R T_B}{P_B} = \frac{2.5 \times 0.297 \times 293}{0.6 \times 10^3} = 0.362 \text{ m}^3$$

$$\therefore \text{ Total volume of } A \text{ and } B = V_A + V_B = 1.258 \text{ m}^3$$

Also, Total mass = 16.5 kg = m

Final temprature after mixing = 300 K

For the final condition after mixing,

$$P.V. = mRT$$

$$P = \frac{16.5 \times 0.297 \times 300}{1.258} = 1168.64 \text{ kPa}$$

Since there is no work transfer,

$$\text{Heat transfer} = \text{Change in internal energy}$$
$$= U_2 - U_1$$

Taking $U = 0$ kJ at $T = 0$ K

Initial internal energy, $U_1 = m_A C_V T_A + m_B C_V T_B$

where
$$C_V = \frac{R}{\gamma - 1} = \frac{0.297}{0.4}$$

$$\mathbf{C_v = 0743 \text{ kJ/kg-K}}$$

$$\therefore \quad U_1 = [(14 \times 323) + (2.5 \times 293)] \times 0.743$$

$$\mathbf{U_1 = 3904.09 \text{ kJ}}$$

Final Internal Energy, $U_2 = m \cdot C_V \cdot T = 16.5 \times 0.743 \times 300$

$$\mathbf{U_2 = 3677.85 \text{ kJ}}$$

$$\therefore \quad \text{Heat transfer} = 3677.85 - 3904.05$$

$$\textbf{Heat transfer} = \mathbf{-226.24 \text{ kJ}}$$

If the vessels had been perfectly insulated,

$$U_1 = U_2$$

$$m_A C_V T_A + m_B C_V T_B = mC_V T$$

$$\therefore \quad T = \frac{m_A T_A + m_B T_B}{m}$$

$$= \frac{(14\times323)+(2.5\times293)}{16.5}$$

$$\mathbf{T = 318.45\ K = 45.45°C}$$

$$\text{The final pressure } P = \frac{mRT}{V}$$

$$= \frac{16.5\times0.297\times318.45}{1.258}$$

$$\therefore \quad \mathbf{P = 1240.51\ kPa}$$

THEORY QUESTIONS

4.1. Define an Ideal Gas.

4.2. Explain the concept of an Ideal Gas.

4.3. Define an Ideal gas, give the approximate conditions for the gaseouss phase of a pure substance to behave as an ideal gas.

4.4. What are the ideal gas laws ?

4.5. Write a note on Equation of State or Characteristic Gas Equation.

4.6. Write a short note on Universal Gas Constant.

4.7. Define universal gas constant. What are its units?

4.8. What is the difference between specific gas constant and universal gas constant? How are these two related ?

4.9. With the help of Joule's experiment prove that I. E. is a function of temperature.

4.10. With the help of Joule's experiment, prove that I.E. is a function of temperature.

4.11. Explain the constant internal enemy experiment of an ideal gas. Also explain upon what property is the internal energy and enthalpy of ideal gas depends ?

4.12. Prove that $C_p - C_v = R$. Also prove $C_v = \frac{R}{\gamma-1}; C_p = \frac{\gamma . R}{\gamma-1}$.

4.13. Prove that,

(a) $C_p - C_v = R$

(b) $C_v = \frac{R}{\gamma-1}$

(c) $C_p = \frac{\gamma . R}{\gamma-1}$

4.14. Derive the relation between C_p, C_v and R.

4.15. Derive the relation between C_v, γ, and R and also C_p, γ and R.

4.16. Prove that $\gamma = \frac{C_p}{C_v}$.

4.17. Explain Constant Volume Process with P–V and T–S diagrams. Also write the equation for work done and derive the equation for heat transferred and change of I.E.

4.18. Explain constant pressure process with P–V and T–S diagrams. Write the equation for work done and derive the equations for heat transferred.

4.19. What is an Isothermal process ? Represent this on P–V and T–S diagrams. Write the equation for work done, change in I.E. and heat transferred.

4.20. What is a polytropic process, show it on P–V diagram. Write the equation for work done and derive the equation for heat supplied.

4.21. Define polytropic specifc heat. What is the significance of negative sign?

4.22. Show that the polytropic sp. heat is given by

$$C_n = C_v\left(\frac{k-n}{n-1}\right)$$

where k is specific heat ratio

4.23. What is an adiabatic process? Write the equations for work done, heat supplied.

PROBLEMS FOR PRACTICE

4.1. 2 kg of air at 247°C expands isothermally from a volume of 1 m^3 to a volume of 3 m^3. Determine the work done during the process. Take:
R = 287 Nm/kg-K (**Ans.** 328 kJ)

4.2. A gas at 10 bar, 327°C occupies a volume of 0.02 m^3. This gas expands according to $PV^{14} = C$ upto a pressure of 1 bar. Find the temperature and work done during the process.
(**Ans.** T_2 = 321.16 K; 24.1 kJ)

4.3. 3 litre of air at 1200 kN/m^2 and 87°C expand polytropically according to $PV^{1.3} = C$ to a pressure of 120 kN/m^2. If for air $\gamma = 1.4$, $C_v = 0.718$ find,
(i) Polytropic specific heat.
(ii) Heat transfer per kg of air during the process and state its direction.
(iii) Change of specific IE of air.
(**Ans.** (1) – 0.239 kJ/kg-K ; 35.51 kJ/kg ; –106.27 kJ/kg)

4.4. 0.22 kg of a gas having a volume of 0.14 m^3 and a pressure of 1.4 bar is compressed to a pressure of 14 bar according to the, law $PV^{1.25} = C$. Estimate the amount of heat transferred during the process by calculating the work done and change in I.E. Take C_p = 1.041 kJ/kg-K and C_V = 0.743 kJ/kg-K.

4.5. A cylinder has a capacity of 300 litre and contains oxygen at a pressure of 3.1 MN/m^2 and temperature 18°C. The stop valve is opened and some gas is used. If the pressure and temperature of the gas left in the cylinder falls to 1.7 MN/m^2 and 15°C respectively, determine the mass of the oxygen used. If after the stop valve is closed, the oxygen remaining in the cylinder gradually attains its initial temperature of 18°C, determine the amount of heat transferred through the cylinder walls from atmosphere.
The density of oxygen at 0°C and 0.1013 MN/m^2 be taken as 1.43 kg/m^3 and $\frac{C_p}{C_v} = 1.4$.

4.6. 0.2 m^3 of mixture of fuel and air at 1.2 bar and 60°C is compressed till its pressure becomes 12 bar and temperature 270°C. Then it is ignited suddenly at constant volume and its pressure becomes twice the pressure at the end of compression. Find the maximum temperature reached and change in I.E. Also find the heat transfer during the compression process consider the mixture as perfect gas. Take
C_p = 1.005 kJ/kg-K,
and $\gamma = 1.4$

4.7. A gas initially at a pressure of 510 kPa and volume 142 litres undetgoes a process and final pressure of 170 kPa and volume of 275 litres during which enthalpy decreased by 65 kJ. Take C_v = 0.718 kJ/kg-K, determine,
(a) Change in I.E.
(b) C_p
(c) Specific gas constant

4.8. 10 kg of air is heated at constant pressure from a temperature of 100°C to 200°C. Calculate the heat added during the process and also change in I.E. Take R = 0.287 kJ/kg-K, γ = 1.4.

4.9. 50 litres of air at 1.013 bar and 100°C temperature is compressed to 28 bar. Volume of air at the end of polytropic compression is found to be 4 litres. Air is now heated at constant volume till pressure rises to 56 bar.
Assuming C_p = 1.00 kJ/kg-K and C_v = 0.71 kJ/kg-K, determine
(a) Polytropic index of compression
(b) Entropy change in each process.
Sketch the process on P–V and T–S diagram.

4.10. 3 kg of an ideal gas expanded from a pressure of 7 bar and volume 1.5 m^3 to a pressure of 1.4 bar and volume 4.5 m^3. If the change in I.E. during the process is 520 kJ. Determine,
(a) Specific gas constant
(b) Initial and final temperature
(c) Change in entropy in heat exchange. Take C_p = 1.05 kJ/kg-K

4.11. 0.01 kg of certain gas occupies a volume of 0.003 m^3 at a pressure of 7 bar and a temperature of 131°C. Calculate the molar mass of the gas. When the gas is allowed to expand until the pressure is 1 bar and final volume is 0.02 m^3, calculate final temperature.

4.12. 1 kg of perfect gas is compressed from 1.1 bar and 27°C, according. to the law, $PV^{1.3} = C$, until the pressure is 6.6 bar. Calculate the heat flow to or from the cylinder watts when the gas is argon having molecular mass of 40 kg/mol and C_p = 0.52 kJ/kg.

4.13. A certain quantity of gas has a volume of 0.11 m^3 at 0.15 MN/m^2 and 300 K. It is compressed to a pressure of 1.5 MN/m^2 according to the law $PV^{1.2}$ = constant. If the molecular mass of the gas is 27.9 and C_p = 29.04 kJ/k mol K determine heat transfer and its direction.

4.14. An ideal gas having r=1.4 and volume 1.5 m^3 at 1 bar and 27°C is compressed according to the law $PV^{1.25} = C$, to a pressure of 6 bar. Calculate volume, temperature at the end of the process. Evaluate heat transfer, work done and the change of entropy during the process.

4.15. Following pressure – volume data is available for a process:

P (bar)	20	17.5	15	12.5	10	7.5	5	2.5
V(m^3)	1.00	1.1	1.2	1.3	1.4	1.5	1.6	1.7

Estimate the work done during the process.

4.16. Air with initial condition as 27°C, 1 bar and 1.5 m^3 is compressed according to the law $PV^{1.3} = C$, to a pressure of 7.5 bar. Calculate final volume, temperature at the end of the process. Evaluate heat transfer, W.D. and change of entropy during the process. Assume r=1.4, R=0.287 kJ/kg-K.

CHAPTER 5 Entropy

5.1 INTRODUCTION

We have studied the I–law of thermodynamics. Analysis of I–law leads to the definition of a derived property known as Internal energy. Analysis of second law will lead to the definition of another derived property known as Entropy.

Clausius discovered that, when a small amount of heat δQ is supplied to a system, which is at a absolute temperature T, then it will undergo a process, and the ratio $\frac{\delta Q}{T}$ is same for all reversible processes. He assigned the value $\frac{\delta Q}{T} = dS$ and called S as entropy.

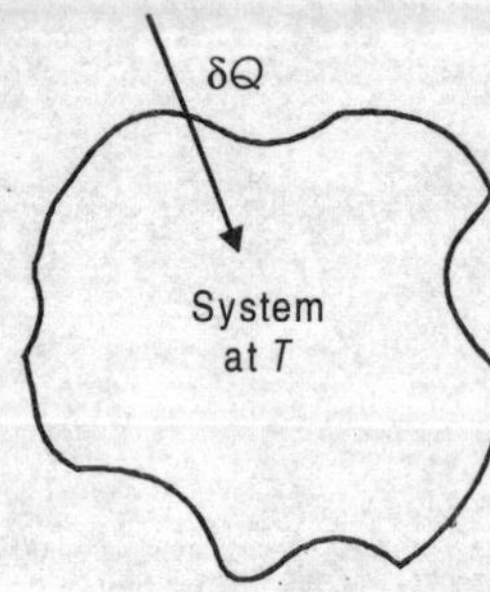

Fig. 5.1

The term entropy is taken, from Greek work 'tropee' meaning transformation. Thus, when a small amount of heat δQ is transformed to a system, the entropy changes by an amount dS. This change in entropy is regarded as the transformation content of the system.

Note that entropy is a thermodynamics property, it increases with the addition of heat and it decreases with the removal of heat.

5.2 CLAUSIUS THEOREM FOR A PROCESS AND FOR A CYCLE

Let a system be taken from an equilibrium state (1) to another equilibrium state (2) following the reversible path 1–2. Let (*a*) and (*b*) be two reversible adiabatic, which pass through the points (1) and (2) respectively. A reversible isotherm (*c*) is drawn, such that the area under 1–3–4–2 is equal to the area under the curve 1–2.

From the first law,

$$Q = \Delta U + W$$

For process 1–2, $\quad Q_{1-2} = U_2 - U_1 + W_{1-2}$ and for the process 1 – 3 – 4 – 2

$$Q_{1-3-4-2} = U_2 - U_1 + W_{1-3-4-2}$$

As we have assumed that the area under 1– 3 – 4 – 2 is equal to the area under 1–2.

i.e. $\quad W_{1-2} = W_{1-3-4-2}$

$\therefore \quad Q_{1-2} = Q_{1-3-4-2}$

$$= Q_{1-3} + Q_{3-4} + Q_{4-2}$$

But $\quad Q_{1-3} = 0$

and $$Q_{4-2} = 0 \ (\because \text{ adiabatic})$$

$\therefore$ $$Q_{1-2} = Q_{3-4}$$

Thus any reversible path can be replaced by a zig-zag path between the same end states, consisting of a reversible adiabatic process followed by a reversible isothermal process, which is again followed by a reversible adiabatic process, such that the heat transferred during the isothermal process is the same as that transferred during the original process.

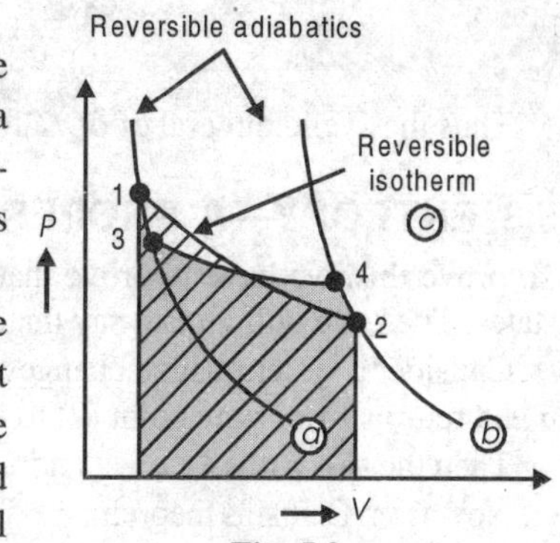

Fig. 5.2

Consider a reversible cycle as shown in Fig. 5·2. It is divided into a large number of strips by means of reversible adiabatic. Each strip may be closed at the top and bottom by means of reversible isotherms. The original closed cycle is thus replaced by a zig-zag closed path, consisting of alternate adiabatic and isothermal processes, such that the heat transferred during all the isothermal processes is equal to the heat transferred in the original cycle.

Now, for cycle 1 – 2 – 3 – 4, δQ_1 is heat absorbed reversibly at T_1 and δQ_2 is heat rejected reversibly at T_2.

Then, $$\frac{\delta Q_1}{T_1} = \frac{\delta Q_2}{T_2}$$

If heat supplied is taken as positive and heat rejected as negative, then

$$\frac{\delta Q_1}{T_1} + \frac{\delta Q_2}{T_2} = 0$$

Similarly for another elemental cycle 5 – 6 – 7 –8

$$\frac{\delta Q_3}{T_3} + \frac{\delta Q_4}{T_4} = 0$$

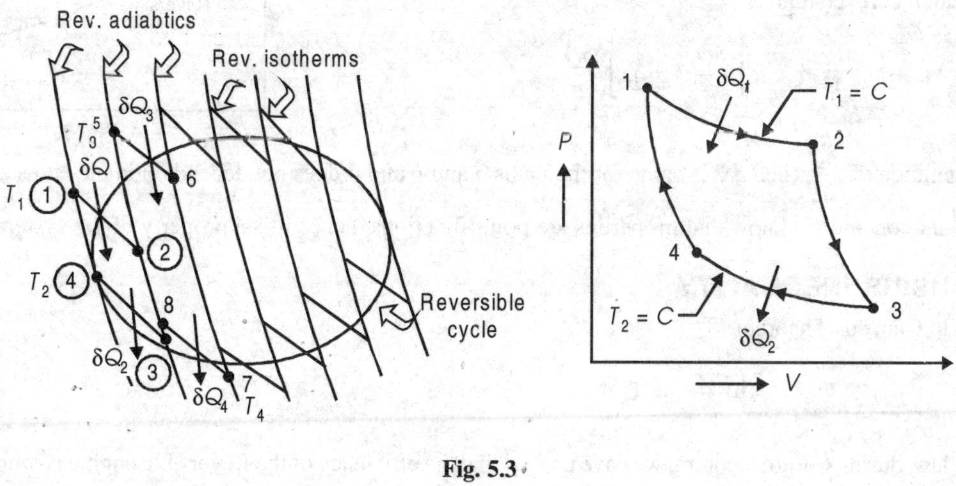

Fig. 5.3

If similar equation are written for all the elemental Carnot cycles, then for the whole (complete) original cycle.

$$\left(\frac{\delta Q_1}{T_1} + \frac{\delta Q_2}{T_2}\right) + \left(\frac{\delta Q_3}{T_3} + \frac{\delta Q_4}{T_4}\right) + \cdots = 0$$

or $$\oint_R \frac{dQ}{T} = 0$$

Thus the cyclic integral of $\delta Q/T$ for a reversible cycle is equal to zero. This is known as **Clausious Theorem**.

5.3 ENTROPY—A PROPERTY

To prove this, we have to prove that the change of entropy does not depend upon path but it depends upon end states. Then, we will able to say that, entropy is a property of the system.

Consider a system which changes its state from state point (1) to state point (2) by following the reversible path *a* and returns from state point (2) to state point (1) by following the reversible path *b*.

Then the two paths 1– *a* – 2 and 2 – *b* –1 together will form a cycle.

Now from Clausius theorem,

$$\oint_{(1-a-2-b-1)} \frac{\delta Q}{T} = 0$$

Note. May be read as integral for the reversible path 1– *a* – 2 –*b* –1.

The above integral may be replaced as the sum of two integrals one for the path *a* and other for the path *b*.

$$a\int_{R1}^{2} \frac{\delta Q}{T} + b\int_{R2}^{1} \frac{\delta Q}{T} = 0$$

Note. May be read as integral $\delta Q/T$ for the reversible path 1 – *a* – 2.

$$\therefore \quad a\int_{R1}^{2} \frac{\delta Q}{T} = -b\int_{R1}^{2} \frac{\delta Q}{T}$$

Since path *b* is reversible

$$a\int_{R1}^{2} \frac{\delta Q}{T} = b\int_{R2}^{1} \frac{\delta Q}{T}$$

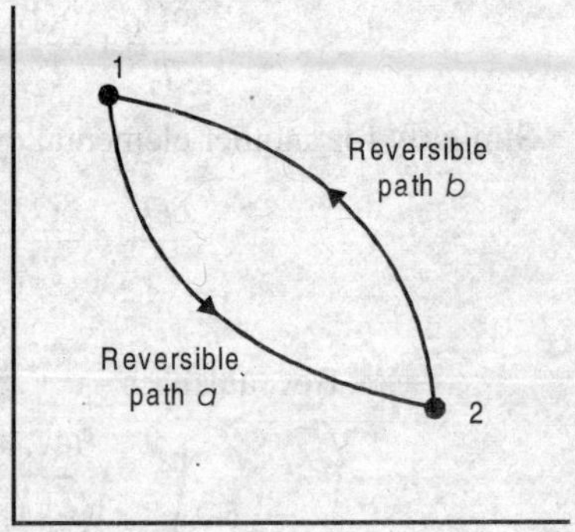

Fig. 5.4

The magnitude of $\frac{\delta Q}{T}$ (i.e., dS) is same for the paths *a* and *b* and it does not depend upon the end states, hence it is point function and we know that properties are point-functions, hence it is a property of the system.

5.4 CLAUSIUS INEQUALITY

According to Clausius Theorem,

$$\oint_R \frac{\delta Q}{T} = 0$$

In the II law, during Carnot theorem, we have proved that the efficiency of the Reversible engine is more than that of an Irreversible engine.

$$\eta_R > \eta_I$$

or $$\eta_I < \eta_R \qquad \ldots(1)$$

We also know that,

$$\eta \text{ of any engine} = \frac{\text{Heat supplied} - \text{Heat rejected}}{\text{Heat supplied}} = \frac{Q_1 - Q_2}{Q_1}$$

or For small amount of heat δQ,

$$\eta = \frac{\delta Q_1 - \delta Q_2}{\delta Q_1} = 1 - \frac{\delta Q_2}{\delta Q_1}$$

Hence the Eq. (1) will therefore be,

$$\left[1 - \frac{\delta Q_2}{\delta Q_1}\right]_I < \left[1 - \frac{\delta Q_2}{\delta Q_1}\right]_R$$

But for reversible engine,

$$\frac{Q_1}{Q_2} = \frac{T_1}{T_2}$$

$$\left[1 - \frac{\delta Q_2}{\delta Q_1}\right] = 1 - \frac{T_2}{T_1}$$

$$\therefore \quad 1 - \left(\frac{\delta Q_2}{\delta Q_1}\right)_I < 1 - \frac{T_2}{T_1}$$

$$\therefore \quad \left(\frac{\delta Q_2}{\delta Q_1}\right)_I > \frac{T_2}{T_1}$$

or opposite

$$\left(\frac{\delta Q_1}{\delta Q_2}\right) < \frac{T_1}{T_2}$$

or

$$\left(\frac{\delta Q_1}{T_1}\right)_I - \left(\frac{\delta Q_2}{T_2}\right)_I < 0$$

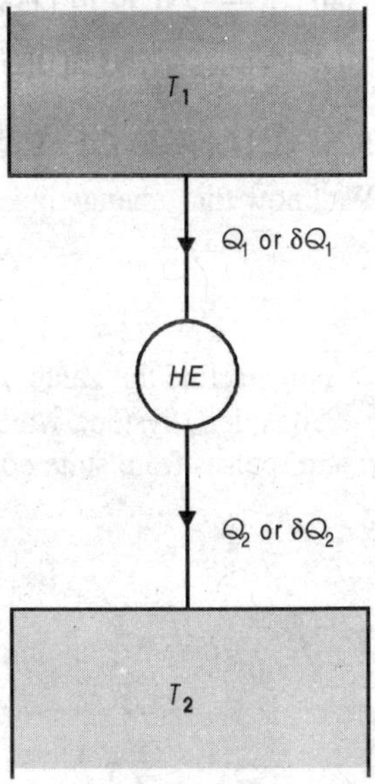

Fig. 5.5

We know that, heat added is positive. Therefore δQ_1 is positive and heat rejected is negative. Therefore δQ_2 is negative

$$\therefore \quad \left(\frac{\delta Q_1}{T_1}\right)_I + \left(\frac{\delta Q_2}{T_2}\right)_I < 0$$

i.e., The algebraic sum of $\delta Q/T$, for an irreversible cycle is always less than zero.

$$\therefore \quad \oint_R \frac{\delta Q}{T} < 0 \text{ for an Irreversible cycle.}$$

And we know that from Clausius theorem $\oint \frac{\delta Q}{T} = 0$ for a reversible cycle. Combining results for reversible and irreversible cycle, we may write,

$$\oint \frac{\delta \mathbf{Q}}{\mathbf{T}} \leq \mathbf{0}$$

Thus expression is known as Clausius Inequality. It implies whether any cyclic process is reversible or irreversible or impossible.

Note. According to Clausius Inequality

(i) $\oint \frac{dQ}{T} = 0$ for a reversible cycle

(ii) $\oint \frac{dQ}{T} < 0$ for a irreversible cycle

(iii) $\oint \frac{dQ}{T} > 0$ for a impossible cycle

(iv) All temperatures must be in K

(v) Heat received by the engine is positive

(vi) Heat rejected by the engine is negative

5.5 CHANGE OF ENTROPY IN AN IRREVERSIBLE PROCESS

We know that, change in entropy for a reversible process is given by,

$$\left(\frac{\delta Q}{T}\right)_R = (dS)_R \qquad \ldots(1)$$

Now to find the value of change in entropy in an Irreversible process.

Consider a system, which change its state from state point (1) to state point (2) by following the reversible path a and returns from state point (2) to state point (1) by following the irreversible path b as shown in Fig. 5.6.

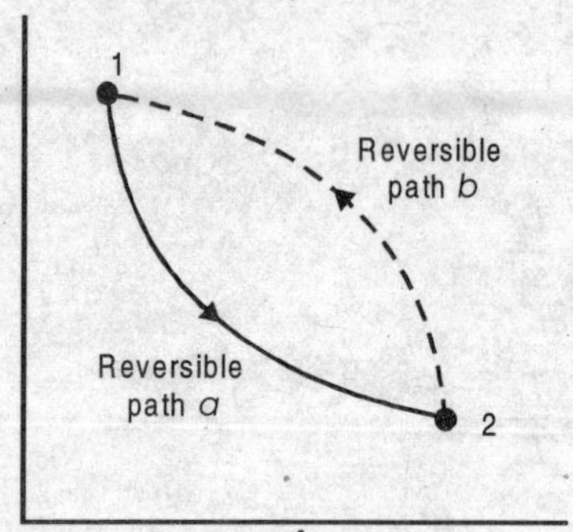

Fig. 5.6

Since cyclic integral of any property is zero and entropy is a property we can write,

$$\oint dS = a\int_1^2 (dS)_R + b\int_2^1 (dS)_I = 0 \qquad \ldots(2)$$

Now, from Eq. (1), for a reversible process we have,

$$a\int_1^2 (dS)_R = a\int_1^2 \left(\frac{\delta Q}{T}\right)_R \qquad \ldots(3)$$

Substituting this in Eq. (2) we get,

$$a\int_1^2 \left(\frac{\delta Q}{T}\right)_R + b\int_2^1 (dS)_I = 0 \qquad \ldots(4)$$

Since the processes $1 - a - 2$ and $2 - b - 1$ together will form Irreversible cycle. Applying Clausius inequality,

$$\oint_I \frac{\delta Q}{T} < 0$$

We get from Eq. (4),

$$a\int_1^2 \left(\frac{\delta Q}{T}\right)_R + b\int_2^1 \left(\frac{\delta Q}{T}\right)_I < 0 \qquad \ldots(5)$$

Subtracting Eq. (5) from Eq. (4) we get,

$$b\int_1^2 (dS)_I > a\int_2^1 \left(\frac{\delta Q}{T}\right)_I$$

In general, $$(dS)_I > \left(\frac{\delta Q}{T}\right)_I \qquad \ldots(6)$$

Combining Eqs (1) and (6), we can write,

$$\mathbf{dS \geq \frac{\delta Q}{T}}$$

where equality sign is for reversible process and inequality sign is for an Irreversible process (from Eq. 6).

Note. The effect of irreversibility is always to increase the entropy of the system.

5.6 CHANGE OF ENTROPY FOR AN ISOLATED SYSTEM

We know that, in an isolated system, matter, work or heat cannot cross the boundary of the system. Hence according to the first law of thermodynamics, the IE of the system will remain constant.

Since for an isolated system $\delta Q = 0$, from equation,

$$dS = \frac{\delta Q}{T} \qquad \ldots(1)$$

We get, $$(dS)\text{ isolated} \geq 0 \qquad \ldots(2)$$

Equation (2) states, that entropy of an isolated system either increases or remains constant and never decreases. This is known as the *Principle of increase of entropy.*

5.7 ENTROPY OF UNIVERSE

We know that all natural process are irreversible processes and during irreversible processes entropy increases and hence entropy of the universe always increases.

5.8 ENTROPY AND DISCORDER

Consider an isolated system comprising of two gases O_2 and H_2 in a separated box as shown in Fig. 5.7. When the partition or barrier is removed, the molecules of both gages get more space to move randomly and therefore collision take place between the mole molecules of the same gas as well as of both the gases, and after some time, equilibrium will be established.

If we assume that initially both gases have the same volume and temperature then after removing the partition each gas occupies double the previous volume while the temperature and the I.E. of the system remains the same, but we find that the entropy increase.

The reason is obvious, as the mixing process is irreversible and irreversible process is always associated with the increase in entropy.

Thus an irreversible process always tends to take the system (isolated) to state of greater disorder. It is a tendency on the part of nature to proceed to a state of greatest disorder. And an isolate system always tends to a state of greater entropy. So there is a close link between entropy and disorder.

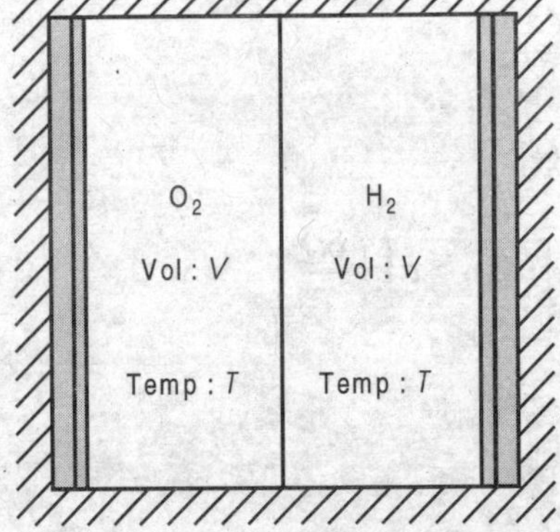

Fig. 5.7

It may be stated roughly that "**The entropy of a system is a measure of the degree of molecular disorder existing in the system**". When heat is supplied to the system, the disorderly motion of molecules increases and so the entropy of the system increases. The reverse occurs when heat is removed from the system.

5.9 PROPERTY RELATIONS FROM ENERGY EQUATIONS

For deriving the equation for change of entropy it is essential to know equation based on I and II law of thermodynamics.

We know that, change in entropy for a reversible process is given by,

$$(dS)_R = \left(\frac{\delta Q}{T}\right)_R$$

i.e., $$(\delta Q)_R = T \cdot dS \qquad \ldots(1)$$

Also we know that, the first–law applied to a closed system undergoing a process gives,

$$Q - W = \Delta U$$

or $$Q = \Delta U + W$$

In the differential form, $$\delta Q = dU + \mathrm{d}W$$

i.e., $$\delta Q = dU + PdV \qquad \ldots(2)$$

∴ From Eqs. (1) and (2) we can write,

$$T \cdot dS = dU + PdV \qquad \ldots(3)$$

This is very important equation because it gives relationship between all thermodynamic properties and does not involve any path functions.

We also know from the definition of enthalpy,

$$H = U + PV$$

or $$dH = dU + PdV + VdP$$

$$dH = T.dS + V.dP \text{ from Eq. (3)}$$

or $$T.dS = dH - V \cdot dP \qquad \ldots(4)$$

From Eq. (3) i.e.,

$$T.dS = dU + P.dV$$

$$\therefore \quad dS = \frac{dU}{T} + \frac{PdV}{T}$$

$$\therefore \quad S_2 - S_1 = \int_1^2 \frac{dU}{T} - \int_1^2 \frac{PdV}{T} \qquad \ldots(5)$$

From Eq. (4) i.e.

$$T.dS = dH - V.dP$$

$$dS = \frac{dH}{T} + \frac{V.dP}{T}$$

$$\therefore \quad S_2 - S_1 = \int_1^2 \frac{dH}{T} - \int_1^2 \frac{V \times dP}{T} \qquad \ldots(6)$$

5.10 GENERAL EQUATIONS FOR CHANGE IN ENTROPY

We know that, the firlst law applied to a closed system undergoing a process gives,

$$Q - W = \Delta U$$

Now consider a unit mass of a gas which changes its states from pressure P_1, specific volume v_1, and temperature T_1 to a new state P_2, V_2, T_2.

For unit mass $q = \Delta u + w$

In the differential form

$$\delta q = du - \delta w \qquad \ldots(1)$$

Now, from the definition of specific heat at constant volume i.e.,

$$C_v = \left(\frac{\delta q}{dT}\right)_v = \left(\frac{du}{dT}\right)_v$$

From this $\quad du = C_v \cdot dT$

$\therefore$ Eq. (1) becomes, $\quad \delta q = C_v \cdot dT + P \cdot dv \qquad \ldots(2)$

Dividing throughout by T, $\quad \dfrac{\delta q}{T} = C_v \dfrac{dT}{T} + \dfrac{P}{T} dv \qquad \ldots(3)$

But $\quad \dfrac{\delta q}{T} = ds$

$$\therefore \quad \frac{\delta q}{T} = ds = C_v \frac{dT}{T} + \frac{P}{T} dv \qquad \ldots(4)$$

From perfect gases, for unit mass, $Pv = RT$ or $\dfrac{P}{T} = \dfrac{R}{v}$

Substituting this in Eq. (4) $\quad ds = C_v \dfrac{dT}{T} + R\dfrac{dv}{v} \qquad \ldots(5)$

$$\therefore \quad \int_1^2 ds = \int_1^2 C_v \frac{dT}{T} + \int_1^2 R.\frac{dv}{v}$$

$$\mathbf{s_2 - s_1 = C_v \ ln \frac{T_2}{T_1} + R ln \frac{v_2}{v_1}} \qquad ...(6)$$

This equation gives change in entropy for unit mass, in terms of temperature ratio and volume ratio.
From perfect gases, $C_p - C_v = R$ and susbtituting this in Eq. (6) we get,

$$s_2 - s_1 = C_v \ln \frac{T_2}{T_1} + C_p \ln \frac{v_2}{v_1} - C_v \ln \frac{v_2}{v_1}$$

$$= C_p \ln \frac{v_2}{v_1} + C_v \left[\ln \frac{T_2}{T_1} - \ln \frac{v_2}{v_1} \right]$$

$$= C_p \ln \frac{v_2}{v_1} + C_v \left[\ln \frac{T_2}{T_1} \cdot \frac{v_1}{v_2} \right] \qquad ...(7)$$

From characteristic equation $P \cdot v = RT$

or
$$\frac{P_1 v_1}{T_1} = \frac{P_2 v_2}{T_2}$$

$$\therefore \quad \frac{P_2}{P_1} = \frac{T_2 v_1}{T_1 v_2} \qquad ...(8)$$

Substituting Eq. (8) in Eq. (7) we get,

$$\mathbf{s_2 - s_1 = C_p ln \frac{v_2}{v_1} - C_p \ln \frac{P_2}{P_1}} \qquad ...(9)$$

Again substituting $C_p - C_v = R$ or $C_p - R = C_v$ in Eq. (6) we get,

$$s_2 - s_1 = C_p \cdot \ln \frac{T_2}{T_1} - R \cdot \ln \frac{T_2}{T_1} + R \cdot \ln \frac{T_2}{T_1}$$

$$= C_p . \ln \frac{T_2}{T_1} - R. \left(\ln \frac{T_2}{T_1} - \ln \frac{v_2}{v_1} \right)$$

$$= C_p \cdot \ln \frac{T_2}{T_1} - \left(\ln \frac{T_2}{T_1} \cdot \frac{v_1}{v_2} \right)$$

From Eq. (8)
$$\mathbf{s_2 - s_1 = C_p . ln \frac{T_2}{T_1} - R ln \frac{P_2}{P_1}} \qquad ...(10)$$

It is to be noted that for a given change, Eqs (6), (9) and (10) will give the same result. The choice of equation is a matter of convenience.

5.11 CHANGE IN ENTROPY DURING VARIOUS PROCESSES

We have studied the equations for work done, heat transferred change in I.E. in the ideal gas chapter. Now to derive the equations for change of entropy during,

(i) Constant Volume Process

Change of entropy:

Here $v_1 = v_2$ hence $\ln \frac{v_2}{v_1} = \ln 1 = 0$

Hence from Eq. (6) (of last article)

$$s_2 - s_1 = C_v \ln \frac{T_2}{T_1} + R \ln \frac{v_2}{v_1}$$

$$\mathbf{s_2 - s_1 = C_v \ln \frac{T_2}{T_1}}$$

And from Eq. (9) i.e., $s_2 - s_1 = C_p \ln \frac{v_2}{v_1} - C_v \ln \frac{P_2}{P_1}$

$$\mathbf{s_2 - s_1 = C_v \ln \frac{P_2}{P_1}}$$

(ii) Constant Pressure Process

Change in Entropy:

Here $P_1 = P_2$

$\therefore \ln \frac{P_1}{P_2} = \ln 1 = 0$

Hence from equation which are in terms or pressure ratio (From equations derived in last article) i.e., from Eq. (9),

$$\mathbf{s_2 - s_1 = C_p \ln \frac{v_2}{v_1}}$$

And from Eq. (10), $\mathbf{s_2 - s_1 = C_p \ln \frac{T_2}{T_1}}$

(iii) Isothermal Process

Change in Entropy,

Here $T_1 = T_2$

$\therefore \ln \frac{T_1}{T_2} = \ln 1 = 0$

Hence from equations which are in terms of temperature ratio.
i.e., from Eq. (6)

$$\mathbf{s_2 - s_1 = R \ln \frac{v_2}{v_1}}$$

And from Eq. (10) $\mathbf{s_2 - s_1 = -R \ln \frac{P_2}{P_1}}$

$$\mathbf{s_2 - s_1 = R \ln \frac{P_2}{P_1}}$$

(iv) Polytropic Process

(a) We know that heat transfer in a polytropic process,

$$q = \left[\frac{\gamma - n}{1 - n}\right] C_v (T_2 - T_1)$$

In the differential form, $$\delta q = \left[\frac{\gamma - n}{1 - n}\right] C_v \cdot dT$$

Since $$\frac{\delta q}{T} = ds$$

$\therefore$ $$ds = \frac{\delta q}{T} = \left[\frac{\gamma - n}{1 - n}\right] C_v \cdot \frac{dT}{T}$$

or $$\int_1^2 ds = \int_1^2 \left[\frac{\gamma - n}{1 - n}\right] \cdot C_v \frac{dT}{T}$$

$\therefore$ $$\mathbf{s_2 - s_1 = \left[\frac{\gamma - n}{1 - n}\right] . C_v \ln \frac{T_2}{T_1}} \quad \text{...(1)}$$

This is the change in entropy, for unit mass in a polytropic process in terms of temperature ratio.

(b) *Change of entropy:* In terms of volume ratio

We know that from (1),

$$s_2 - s_1 = \left[\frac{\gamma - n}{1 - n}\right] C_v \ln \frac{T_2}{T_1}$$

But, for a polytropic process

$$\frac{T_2}{T_1} = \left(\frac{V_1}{V_2}\right)^{n-1}$$

$\therefore$ Substituting this value of $\frac{T_2}{T_1}$ in Eq. (1), we get

$$s_2 - s_1 = \left[\frac{\gamma - n}{1 - n}\right] C_v \ln \left(\frac{V_1}{V_2}\right)^{n-1}$$

$$= \frac{\gamma - n}{1 - n} C_v (1 - n) \ln \frac{V_2}{V_1}$$

$$= (\gamma - n) C_v \ln \frac{V_2}{V_1}$$

$$\mathbf{s_2 - s_1 = \frac{\gamma - n}{1 - n} R \ln \frac{V_2}{V_1}} \quad \left(\text{As } C_v = \frac{R}{\gamma - 1}\right)$$

(c) *Change of entropy polytropic process:* In terms of pressure ratio:

We know that from (1)

$$s_2 - s_1 = \frac{\gamma - n}{1 - n} C_v \ln \frac{T_2}{T_1}$$

But, for a polytropic process

$$\frac{T_2}{T_1} = \left(\frac{P_2}{P_1}\right)^{\frac{n-1}{n}}$$

Substituting this value of $\frac{T_2}{T_1}$ in Eq. (1), we get

$$s_2 - s_1 = \frac{\gamma - n}{1-n} C_v \ln\left(\frac{P_2}{P_1}\right)^{\frac{n-1}{n}}$$

$$s_2 - s_1 = \frac{\gamma - n}{1-n} \cdot \frac{R}{\gamma - 1} \cdot \frac{1-n}{n} \ln\frac{P_2}{P_1}$$

$$= \frac{\gamma - n}{1-n} \cdot \frac{R}{n} \cdot \ln\frac{P_2}{P_1}$$

$$\mathbf{s_2 - s_1 = \frac{\gamma - n}{n} C_v \ln\frac{P_2}{P_1}} \qquad \ldots(3)$$

(d) *Adiabatic Process: change of entropy for a reversible adiabatic process.* As we have studied during a reversible adiabatic process, entropy reamin constant ($S = C$) and the process is called as Isentropic process. Since $S = C$, $dS = 0$ i.e., change in entropy is zero.

5.12 REPRESENTATION OF VARIOUS PROCESSES ON T–S CHART

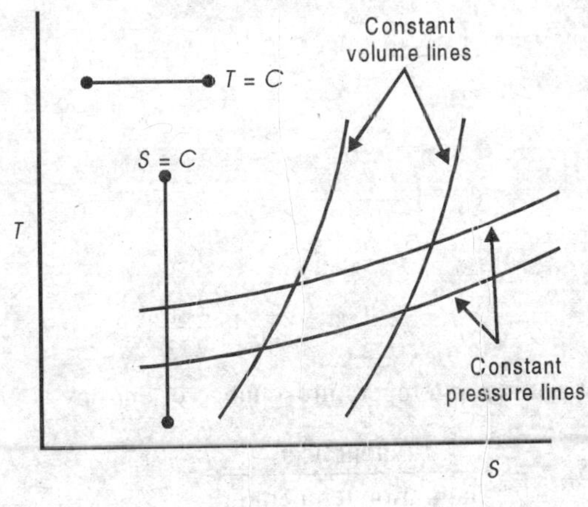

Fig. 5.8

Various processes are shown on *T–S* diagram in Fig. 5.8. Note that constant volume lines on *T–S* diagram have steep slope compared to constant pressure lines.

5.13 PURE SUBSTANCE AND CHANGE OF ENTROPY

We define a pure substance as a system which is homogeneous in composition and chemical aggregation that remains invariable with time.

In many cases we have to deal with system of fixed chemical composition with changing phases, as for example, in case of vapour power generation and referigeration cycles. In the former system water is commonly used whereas in the latter system ammonia, freon etc. are used. In these applications we find the working substance as liquid in one part of the cycle and as vapour or mixture of vapour and liquid in another part of the cycle and as vapour or mixture of vapour and liquid in another part of the cycle. The change in phase of the substance in the cycle is of physical nature only as its chemical composition remains unchanged. These fall in the category of pure substance.

We define a pure substance as a system which is homogeneous in composition and chemical aggregation than remain invariable with time.

A pure substance can exist in three different states of aggregation viz. solid, liquid and gas. It may exist in a single phase like solid, liquid, gas or more than one phase which are in equilibrium with each other. Thus a mixture of water and ice, water and steam, water and ice are all examples of pure substance.

Third law of thermodynamics states "*The entroy of a pure substance in complete thermodynamic equilibrium becomes zero at the absolute zero of temperature*".

This law enables the absolute entropies of pure substances to be calculated from the fundamental definition of entropy, with S_O set equal to zero.

$$\therefore \qquad S = \int_O^T \frac{\delta Q_{\text{rev}}}{T} \quad \text{pure substance in equilibrium}$$

Therefore, the change of entropy of a pure substance when it undergoes a reversible process, can be calculated as in case of ideal gases.

Thus, ΔS_w = Change in entropy of 1 kg water when heated at constant pressure from 0° C to saturation temperature.

$$= \int_{273}^{T_{\text{sat}}} \frac{\delta Q_{\text{rev}}}{T}$$

$$= C_{p_w} \int_{273}^{T_{\text{sat}}} \frac{dT}{T}$$

$$= C_{p_w} \log_e \frac{T_{\text{sat}}}{273} = C_{p_w} \ln \frac{T_{\text{sat}}}{273}$$

During evaporation at constant saturation temperature, change of entropy is given by

$$\Delta S_{fg} = \frac{\text{Latent heat}}{\text{Saturation temperature}} = \frac{h_{fg}}{T_{\text{sat}}}$$

Again for superheating process, the change of entropy (temperature increases from T_{sat} to T_{sup}).

$$\Delta S_{\text{sup}} = C_{p_v} \log_e \frac{T_{\text{sup}}}{T_{\text{sat}}} = C_{p_v} \ln \frac{T_{\text{sup}}}{T_{\text{sat}}}$$

5.14 THIRD LAW OF THERMODYNAMICS

Statistical analysis suggests that the entropy of a substance tends to zero as absolute zero temperature is reached.

This point has been so intensely investigated, that it is probably safe to say that at absolute zero temperature, **the entropy of a pure substance in some perfect crystalline form becomes zero, a generalisation known as the Third law.**

It is found that the specific heats approach zero as temperature tends to zero and the difference $C_p - C_v$ also approaches zero, becuase the co-efficient of thermal expansion approaches zero.

LIST OF FORMULAE

1. I–law for a closed system undergoing a cycle,

$$\oint \delta W = \oint \delta Q$$

2. I–law for a closed system undergoing a process,

$$Q - W = \Delta U$$

where Q = Heat transfer in J

W = Work transfer in J

ΔU = Change of Internal Enerrgy in J.

3. Specific heat at constant pressure,

$$C_p = \left(\frac{\delta q}{dT}\right)_P = \left(\frac{dh}{dT}\right)_P$$

$\Rightarrow$ $\delta q = C_p \cdot dT$ for unit mass

And for total mass $\delta q = m \cdot C_p \cdot dT$

4. Specific heat at constant volume

$$C_v = \left(\frac{\delta q}{dT}\right)_V = \left(\frac{du}{dT}\right)_V$$

$\delta q = C_v \cdot dT$ for unit mass

And for total mass $\delta Q = m \cdot C_v \cdot dT$

5. Adiabatic index, k or $\gamma = \frac{C_p}{C_V}$

Note for air $C_p = 1.005 \frac{\text{kJ}}{\text{kg-K}}$; $C_V = 0.718 \frac{\text{kJ}}{\text{kg-K}}$

and $\gamma = 1.4$

6. Continuity equation or law of conservation of mass

$$\rho_1 A_1 C_1 = \rho_2 A_2 C_2$$

or $\frac{A_1 C_1}{v_1} = \frac{A_2 C_2}{v_2}$ as $\rho = \frac{1}{v}$

where ρ = Density (kg/m^3)

A = Cross sectional area (m^2)

C = Velocity of fluid in (m/sec)

7. Equation of state $PT = mRT$

and R for air = 0.287 kJ/kg-K

8. Relations between C_p, C_V and R, γ

(i) $C_p - C_V = R$

(ii) $$C_V = \frac{R}{\gamma - 1}$$

(iii) $$C_P = \frac{\gamma \cdot R}{\gamma - 1}$$

9. For constant volume process ($V = C$)

(i) Work done in a closed system,

$$W_{1-2} = \int P.dV = 0 \text{ as } V = C \text{ and } dV = 0$$

(ii) From the I–law $Q - W = \Delta U$

$$Q - 0 = \Delta U = m \cdot C_V \cdot dT$$

$$\Rightarrow \quad Q = \Delta U = m \cdot C_V \cdot dT$$

10. For constant pressure process ($P = C$)

(i) Work done in a closed system,

$$W_{1-2} = \int_1^2 P.dV$$

$$W_{1-2} = P(V_2 - V_1)$$

(ii) Change in Internal energy

$$\Delta U = U_2 - U_1 = m \cdot C_V \cdot dT$$

(iii) Heat transfer $\delta Q = dH = m \cdot C_P \cdot dT$

11. For Isothermal process ($PV = C$)

(i) Work done $W_{1-2} = \int P.dV = P_1 V_1 \ln \frac{V_2}{V_1}$

or $$W_{1-2} = P_1 V_1 \ln \frac{V_2}{V_1}$$

(ii) Change in I.E. $= \Delta U = m \cdot C_V \cdot dT = 0$ as

$$T = C$$

$$\therefore \quad dT = 0$$

(iii) From I–law $Q - W = \Delta U$ as $\Delta U = 0$

$$Q = W = P_1 V_1 \ln \frac{V_2}{V_1}$$

12. Polytropic process $(PV^{\gamma}) = C$

(i) Work done $W_{1-2} = \int P.dV$

$$W_{1-2} = \frac{P_2 V_2 - P_1 V_1}{1 - n}$$

$$W_{1-2} = \frac{mR(T_2 - T_1)}{1 - n}$$

As $$PV = mRT$$

(ii) Heat transfer during polytropic process,

$$q = C_V\left[\frac{\gamma-n}{1-n}\right](T_2-T_1)$$

$$q = C_n(T_2-T_1)$$

where C_n = polytropic specific heat $= C_V\left[\frac{\gamma-n}{1-n}\right]$

Also $Q = \frac{\gamma-n}{\gamma-1} \times$ work done in a polytropic process

13. For adiabatic process $PV^\gamma = C$

(i) $$W_{1-2} = \int P.dV = \frac{P_2V_2 - P_1V_1}{1-\gamma}$$

(ii) Heat transfer $Q = 0$

14. For,

(i) Constant pressure process,
(ii) Constant volume process,
(iii) Constant temperature process.

To find the property at second state use :

$$\frac{P_1V_1}{T_1} = \frac{P_2V_2}{T_2}$$

15. For polytropic process to find the properties at the second state use :

$$\frac{T_2}{T_1} = \left(\frac{P_2}{P_1}\right)^{\frac{n-1}{n}} = \left(\frac{V_1}{V_2}\right)^{n-1}$$

and for adiabatic process, in place of n, use γ.

SOLVED EXAMPLES

Example 5.1 *A heat engine is supplied with 278 kJ/sec of heat at constant fixed temperature of 283° C and rejection takes place at 5°C. The following results were reported,*

(i) 208 kJ/sec are rejected.
(ii) 139 kJ/sec are rejected.
(iii) 70 kJ/sec are rejected.

Classify which of the results report a reversible cycle or irreversible cycle or impossible cycle.

Solution

(i) $$\oint\frac{\delta Q}{T} = \left(\frac{Q_1}{T_1}\right)+\left(\frac{Q_2}{T_2}\right)$$

$$= \left(\frac{278}{283+273}\right)+\left(\frac{-208}{5+273}\right)$$

$$= 0.5-0.748 = -2.48 < 0$$

∴ The cycle is Irreversible.

(ii)
$$\oint \frac{\delta Q}{T} = \left(\frac{278}{556}\right) + \left(-\frac{139}{278}\right)$$
$$= \frac{278}{556} - \frac{139}{278} = 0.5 - 0.5 = 0$$

$\therefore$ Cycle is reversible.

(iii)
$$\oint \frac{\delta Q}{T} = \left(\frac{278}{556}\right) + \left(-\frac{70}{278}\right)$$
$$= 0.5 - 0.252 = 0.248 > 0$$

This cycle is impossible according to Clausius inequality.

Example 5.2 *2 kg of a gas is heat at constant volume in a non flow reversible process, till its pressure is doubled. The initial temperature of gas is 50°C. Determine change in entropy.*

Take $C_v = 0.652$ kJ/kg-K and $C_p = 0.91$ kJ/kg-K.

Data: $m = 2$ kg, $P_2 = 2P_1$, or $\frac{P_2}{P_1} = 2$ (given), $T_1 = 50°C + 273$, $S_2 - S_1 = ?$, $C_v = 0.652$ kJ/kg-K, $C_p = 0.91$.

Solution

We know that change of entropy for a constant volume process is,

$$S_2 - S_1 = m \cdot C_V \cdot \ln\frac{P_2}{P_1} \qquad \left[\text{Units : kg} \times \frac{\text{kJ}}{\text{kg} - \text{K}} \times \text{ratio} = \frac{\text{kJ}}{\text{K}}\right]$$
$$= 2 \times 0.652 \times \ln = 2 \times 0.652 \times 0.6931$$
$$\mathbf{S_2 - S_1 = 0.904 \ kJ/K}$$

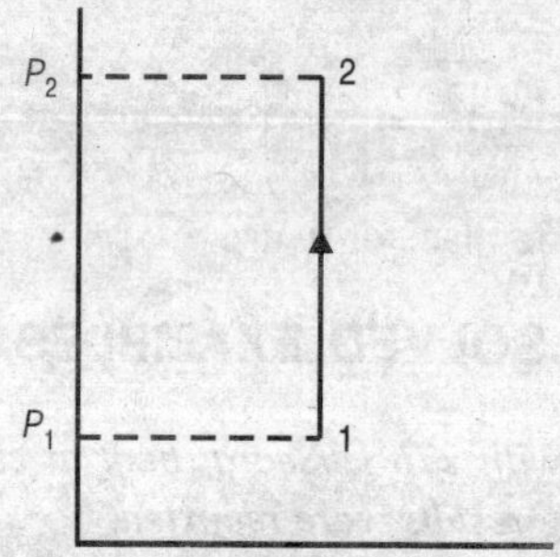

Fig. Ex. 5.2

Example 5.3 *A quantity of gas has a pressure of 8 bar and it occupies a volume of 0.015 m³ at 160°C. The gas is expand isothermally to a volume of 0.085 m³. Determine change of entropy.*

Data: $P_1 = 8$ bar $= 800$ kN/m² $\quad V_1 = 0.015$ m³

$T_1 = 160 + 273 = T_2 \quad V_2 = 0.085$ m³, $S_2 - S_1 = ?$

Solution

We know that change in entropy during isothermal process is given by,

$$S_2 - S_1 = mR \,.\, \ln\frac{v_2}{v_1}$$

$$S_2 - S_1 = mR \,.\, \ln\frac{V_2}{V_1}$$

Note. $\frac{V_2}{V_1}$ is a ratio, we can write ratio of specific volume is equal to the ratio of total volume.

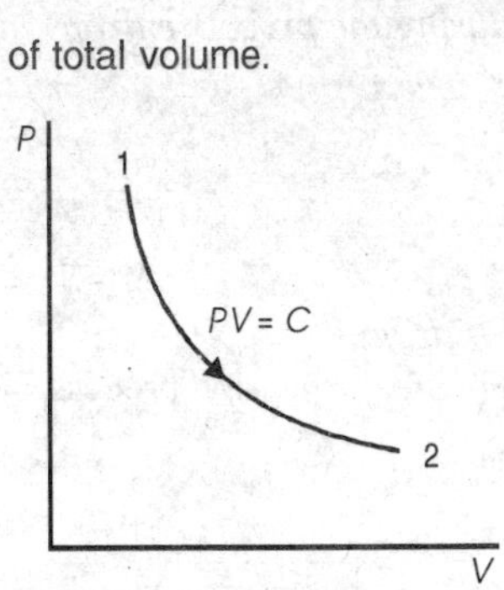

Fig. Ex. 5.3

i.e.,
$$\frac{v_2}{v_1} = \frac{V_2}{V_1}$$

We know that
$$P_1V_1 = mRT_1$$

or
$$\frac{P_1V_1}{T_1} = mR$$

Substituting this in above equation we can write,

$$S_2 - S_1 = \frac{P_1V_1}{T_1}\ln\frac{V_2}{V_1}$$

Note.
$$\frac{\text{kN/m}^2 \times \text{m}^3}{\text{K}} = \frac{\text{kNm}}{\text{K}}$$

$$= \frac{800 \times 0.015}{433} \times \ln\frac{0.0585}{0.015} = \frac{\text{kJ}}{\text{K}}$$

$$\mathbf{S_2 - S_1 = 0.048\ kJ/K}$$

Example 5.4 *3 kg of air at a pressure of 5 bar and 250°C expands reversibly in a polytropic process. The final pressure is 1.5 bar and index of expansion, is 1.25. Find the change in entropy during the process. Take C_v = 0.652 kJ/kg-K and γ = 1.4.*

Data: $m = 3$ kg $\qquad P_1 = 15$ bar $= 1500$ kN/m²

$T_1 = 250 + 273 \qquad PV^n = C$

$P_2 = 1.5$ bar $= 150$ kN/m²

$n = 125 \qquad S_2 - S_1 = ?$

Solution

Change in entropy in a polytropic process is given by (equation in terms of pressure ratio).

$$S_2 - S_1 = m\left[\frac{\gamma - n}{n} \cdot C_V \cdot \ln\frac{P_1}{P_2}\right]$$

$$= 3 \times \left[\frac{1.4 - 1.25}{1.25} \times 0.652 \times \ln\frac{15}{1.5}\right]$$

$$= 3\,[0.12 \times 0.652 \times 2.303]$$

$$S_2 - S_1 = 3\,[0.18015]$$

$$S_2 - S_1 = 0.5405\ \text{kJ/K}$$

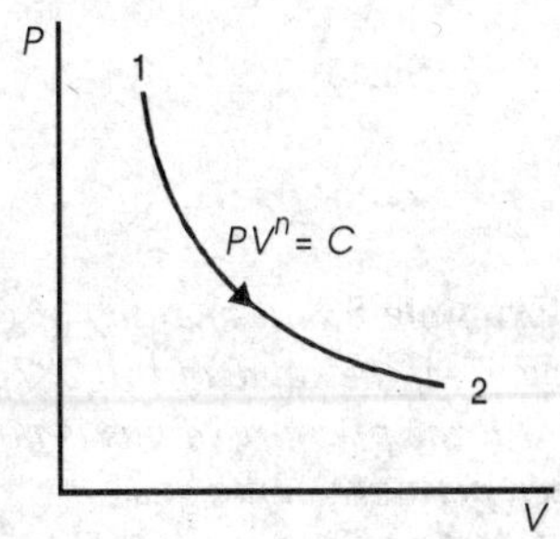

Fig. Ex. 5.4

Note. Unit of m is kg and

$$C_V = \frac{\text{kJ}}{\text{kg-K}}$$

$$\therefore \quad \text{kg} \times \frac{\text{kJ}}{\text{kg K}} = \left.\frac{\text{kJ}}{\text{K}}\right]$$

Example 5.5 *2 kg of certain gas with C_p = 0.85 kJ/kg-K and C_v = 0.70 kJ/kg-K is allowed to expand adiabatically through a partly opened valve, where by the its volume changes from V_1 = 0.0113 m³ to some higher value and* *during the process entropy increases by 0.8 kJ/K. Find the volume after expansion of gas.*

Data: $m = 0.85$ $\quad C_p = 0.85$

$C_v = 0.70$ $\quad PV = C$

$V_1 = 0.113\ m^3$ $\quad V_2 = ?$

$S_2 - S_1 = 0.8$ kJ/K

Solution

For free expansion process, $W = 0$ and given the process is adiabatic

$$Q = 0$$

∴ From I–law, $Q = \Delta U + W$

$$0 = \Delta U + 0 = \Delta U = 0$$

∴ $\text{Initial} = T_{\text{final}}$

i.e. $T_1 = T_2$

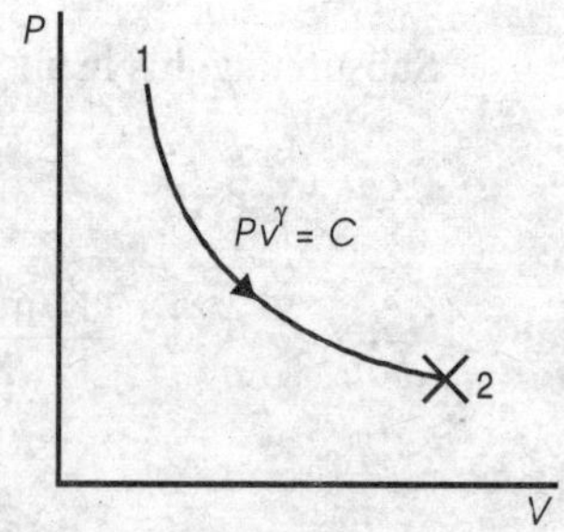

Fig. Ex. 5.5

Using the general equation for change in entropy,

$$S_2 - S_1 = m\left[C_V \ln\frac{T_2}{T_1} + R\ln\frac{v_2}{v_1}\right]$$

Since $T_1 = T_2$

$$S_2 - S_1 = mR\ln\frac{v_2}{v_1}$$

$$0.8 = 2\times[0.85 - 0.70]\ln\frac{v_2}{v_1}$$

$$\frac{0.8}{2\times 0.15} = \ln\frac{v_2}{v_1}$$

$$\ln\frac{v_2}{v_1} = 2.667$$

∴ $$\frac{v_2}{v_1} = 14.397 = \frac{V_2}{V_1}$$

Since $$\frac{v_2}{v_1} = \frac{V_2}{V_1}\text{ ratio}$$

∴ $$V_2 = 13.397\times 0.0113\ \text{m}^3$$

$$\mathbf{V_2 = 0.1627\ m^3}$$

Example 5.6 *A quantity of gas has a pressure of 300 kN/m² when its volume is 0.025 m² and its temperature is 37°C. If the value of is 0.287 kJ/kg-K determine the mass of the gas present.*

If the pressure of gas is now increased to 0.9 MN/m² while the volume remains constant what will be the new temperature of the gas.

Solution

(1) By characteristic gas equation

$$PV = mRT$$

∴ $$m = \frac{PV}{RT} = \frac{300\times 10^3\times 0.025}{0.287\times 10^3\times 310} = 0.084\ \text{kg}$$

(2) We have

$$\frac{P_1V_1}{T_1} = \frac{P_2V_2}{T_2} \text{ and in this case } V_1 = V_2$$

$$\therefore \quad T_2 = \frac{P_2T_1}{P_1} = 310 \times \frac{0.9 \times 10^6}{300 \times 10^3} = 930 \text{ K}$$

$$\therefore \quad T_2 = 930 - 273 = 657°\text{C}$$

Example 5.7 *10 kg of air is heated at constant pressure, from a temperature of 100°C to 200°C. Calculate the heat added during the process and also the change in I.E. Take gas constant R = 0.287 kJ/kg-K and γ = 1.4 for air.*

Data: $m = 10$ kg, $t_1 = 100$ °C, $t_2 = 200$°C, $T_1 = 373$ K, $T_2 = 473$ K, $R = 0.287$ kJ/kg-K, $\gamma = 1.4$

Solution

We know that, heat added during constant pressure process.

$$= mC_P(T_2 - T_1)$$

Also, $$C_V = \frac{R}{\gamma - 1} = \frac{0.287}{1.4 - 1} = 0.7175$$

and $$C_P - C_V = R$$

$$\Rightarrow \quad C_P = R + C_V$$

$$C_P = 0.287 + 0.7175$$

$$\mathbf{C_p = 1.0045 \text{ kJ/kg-K}}$$

∴ Substituting (2) and (3) in (1) we get,

$$\text{Heat added} = m\,C_P(T_2 - T_1)$$

$$= 10 \times 1.0045\,(200 - 100) = 1004.5 \text{ kJ}$$

$$\text{Charge in I.E.} = \Delta U = mC_V(T_2 - T_1) = 10 \times 0.7175\,(200 - 100)$$

$$\mathbf{\Delta U = 717.5 \text{ kJ}}$$

Example 5.8 *A mass of 0.8 kg of air at 1 bar and 25°C is contained in a gas tight frictionless piston cylinder device. The air is now compressed to a final pressure of 5 bar. During the process heat is transferred from the air such that the temperature inside the cylinder remains constant. Calculate the heat transfer and work done during the process and direction of each in the process.*

Data: $m = 0.8$ kg, $P_1 = 1$ bar $= 10^5$ N/m², $T_1 = 298$ K, $P_2 = 5$ bar $= 5 \times 10^5$ N/m², $T = C$, Isothermal process.

Solution

∴ We know that, for an isothermal process,

$$Q = W = P_1V_1 \ln\frac{P_1}{P_2} = mRT_1 \ln\frac{P_1}{P_2} = -110.11 \text{ kJ}$$

As Q is –ve ∴ W is also –ve.

Example 5.9. *1 kg of air at an initial condition of 3 bar and 175°C expands adiabatically without friction to a final condition of 1 bar. Calculate:*

(*a*) *External work done during expansion.*

(*b*) *Change in Internal Energy.*

Take γ = 1.4 and R = 0.287 kJ/kg-K

Data: $m = 1\ kg$ $P_1 = 3 \times 105\ N/m^2 = 3 \times 10^2\ kN/m^2$

$T_1 = 175 + 273 = 448\ K$ $P_2 = 1 \times 10^2\ kN/m^2$

$\gamma = 1.4$ $R = 0.287\ kJ/kg\text{-}K$

Solution

(a) We know that the work done in an adiabatic expansion process is given by,

$$W_{1-2} = \frac{P_1V_1 - P_2V_2}{\gamma - 1}$$

Here V_1 and V_2 are not given,
Since it is an adiabatic process,

$$P_1V_1^\gamma = P_2V_2^\gamma$$

$$\frac{P_1}{P_2} = \left(\frac{V_2}{V_1}\right)^\gamma$$

$$\left(\frac{P_1}{P_2}\right)^{1/\gamma} = \frac{V_2}{V_1}$$

$$\therefore \quad V_2 = V_1 \times \left(\frac{P_1}{P_2}\right)^{1/\gamma}$$

To find V_1 $\quad P_1V_1 = mRT_1$

$$V_1 = \frac{mRT_1}{P_1} = \frac{1 \times 0.287 \times 448}{3 \times 10^2}$$

$$\text{Note units : } \left[\frac{\text{kg}\dfrac{\text{kJ}}{\text{kg-K}} \times \text{K}}{\text{kN/m}^2} = \frac{\text{kJ}}{\text{kN/m}^2} = \frac{\text{kNm}}{\text{kN}} \times \text{m}^2 = \text{m}^3\right]$$

$\therefore$ From Eq. (2),

$$V_2 = 0.429 \times \left(\frac{300}{100}\right)^{1/1.4} = 0.429 \times (3)^{1/1.4}$$

$$\mathbf{V_2 = 0.940\ m^3}$$

$\therefore$ Now from Eq. (1), work done,

$$W_{1-2} = \frac{P_1V_1 - P_2V_2}{\gamma - 1}$$

$$= \frac{300 \times 0.429 - 100 \times 0.940}{1.4 - 1} = \frac{34.7}{0.4}$$

$$\mathbf{W_{1-2} = 86.75\ kJ}$$

(b) We also know from the I–law, that for a process, $Q = \Delta U + W$. Since it is adiabatic process $Q = 0$

$$\therefore \quad W = -\Delta U = 86.75\ \text{kJ} \Rightarrow \mathbf{\Delta U = -86.75\ kJ}$$

Therefore internal energy decreases.

Example 5.10 *Find the change of Internal energy and enthalpy of 3 kg of air, when temperature changes from 25°C to 100°C.*

Take C_p = 1 kJ/kg-K and C_v = 0.71 kJ/kg- K.

Solution

In the problem, the type of process is not given, but since temperature change, it is not an isothermal process.

$$\therefore \quad \Delta U = m \cdot C_V \cdot (T_2 - T_1)$$
$$= 3 \text{ kg} \times 0.71 \text{ kJ/kg-K} \times (373 - 298) \text{ K}$$
$$= 3 \times 0.71 \times 75$$
$$\mathbf{\Delta U = 159.75 \text{ kJ}}$$

$$\text{As change of enthalpy} = \Delta H = m\, C_P \cdot (T_2 - T_1)$$
$$= 3 \times 1 \times (75)$$
$$\mathbf{\Delta H = 225 \text{ kJ}}$$

Example 5.11 *0.44 kg of a gas having volume 0.28 m³ and a pressure of 1.4 bar is compressed to a pressure of 14 bar according to $PV^{1.25}$ = C. Find the amount of heat transferred during the process by calculating the work done and the change of internal energy.*

Take C_p = 1.041 kJ/kg-K and C_v = 0. 743 kJ/kg-K

Data:

$$m = 0.44 \text{ kg} \qquad V_1 = 0.28 \text{ m}^3$$
$$P_1 = 1.4 \text{ bar} = 140 \text{ kPa}$$
$$P_2 = 14 \text{ bar} = 1400 \text{ kPa}$$
$$PV^{1.25} = C \qquad Q = ? \; W = ? \; \Delta U = ?$$
$$C_p = 1.041 \text{ and } C_v = 0.743 \text{ kJ / kg-K}$$

Solution

$$\text{W.D.} = \frac{P_2V_2 - P_1V_1}{1-n} \qquad \text{...(1)}$$

Here only V_2 is not given and since it is a polytropic process,

$$P_1V_1^n = P_2V_2^n$$

$$\frac{P_1}{P_2} = \left(\frac{V_2}{V_1}\right)^n$$

i.e.,

$$\frac{V_2}{V_1} = \left(\frac{P_1}{P_2}\right)^{1/n}$$

$$V_2 = V_1 \times \left(\frac{P_1}{P_2}\right)^{1/n}$$

$$= 0.28 \times \left(\frac{140}{1400}\right)^{\frac{1}{1.25}}$$

$$= 0.28 \times (0.1)^{0.8}$$

$$V_2 = 0.044377 \text{ m}^3 = 0.0444 \text{ m}^3$$

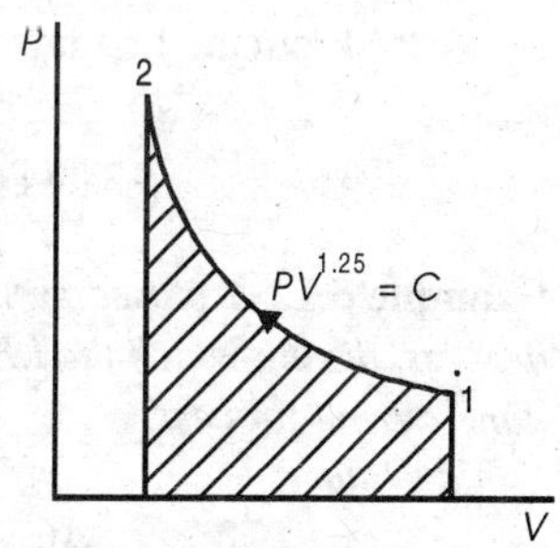

Fig. Ex. 5.11

Now work done from Eq. (1),

$$W_{1-2} = \frac{1400 \times 0.0444 - 140 \times 0.28}{1 - 1.25}$$

$$= \frac{62.16-39.2}{-0.25}$$

$$\mathbf{W_{1\text{-}2} = -\ 91.84\ kJ}$$

W_{1-2} is –ve, so work is done on the gas.

(ii) Heat transferred in a polytropic process

We know that,
$$q = C_n(T_2-T_1) = C_V \cdot \left(\frac{\gamma-n}{1-n}\right)(T_2-T_1)$$

$$= \frac{R}{\gamma-1}\left(\frac{\gamma-n}{1-n}\right)(T_2-T_1)$$

$$= \frac{R}{\gamma-1}\left(\frac{\gamma-n}{1-n}\right)\times(T_2-T_1)$$

$$q = \frac{(\gamma-n)}{\gamma-1}\times\frac{R}{1-n}\times(T_2-T_1)$$

And total heat transfer,

$$Q = \frac{\gamma-n}{\gamma-1}\times\frac{mR(T_2-T_1)}{1-n}$$

i.e.,
$$Q = \frac{\gamma-n}{\gamma-1}\times W, \text{ in a polytropic process}$$

To find Q, let us find
$$\gamma = \frac{C_P}{C_V} = \frac{1.041}{0.743} = 1.401 = 1.4$$

$$Q = \frac{1.4-1.25}{1.4-1}\times -91.84 = \frac{0.15}{0.4}\times -91.84$$

$$\mathbf{Q = -\ 34.44\ kJ}$$

∴ From the first law for a process,

$$Q - W = \Delta U$$

$$-34.44 + 91.84 = \Delta U$$

∴
$$\mathbf{\Delta U = 57.4\ kJ\ (Increase)}$$

Example 5.12 *A closed system containing 2 kg of working substance undergoes a reversible constant pressure process, during which the I.E. of the system decreases kJ. If C_p = 1120 J/kg-K and C_v = 800 J/kg-k. Find the work done during process.*

Data: $m = 2$ kg, $\Delta C = -80$ kJ, $C_v = 0.8$ kJ/kg-K, $P = C$, $C_P = 1.120$ kJ/kg-K, $W_{1-2} = ?$

Solution

We know that work done in a constant pressure process,

$$W_{1-2} = P(V_2 - V_1) \qquad \ldots(1)$$

Also from the first law, $Q - W = \Delta U$

We know that heat supplied is utilised for,

(i) Increasing I.E.

(ii) For doing external work

And heat transfer,

$$\delta q = dh$$

or $$\delta Q = dH = mC_P\Delta T$$

Also, $$\Delta U = m\, C_P\, \Delta T$$

$$\frac{\Delta U}{m \cdot C_V} = \Delta T$$

$$\frac{-80 \text{ kJ}}{2 \text{ kg} \times 0.8 \text{ kJ/kg-K}} = \Delta T \qquad \ldots(3)$$

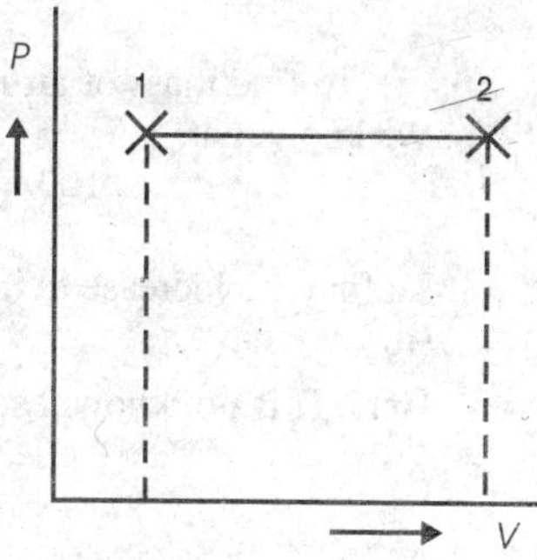

Fig. Ex. 5.12

$$= \frac{-80}{1.6} = \Delta T$$

$$-50 \text{ K} = \Delta T$$

∴ From (3) $$\delta Q = m \cdot C_P \cdot \Delta T = 1 \times 1.12 \times (-50)$$

$$\delta Q \text{ or } Q = -112 \text{kJ}$$

∴ From Eq. (2), $$Q - W = \Delta U$$

$$Q - \Delta U = W$$

$$-112 + 80 = W$$

∴ $$\mathbf{W = -32\ kJ}$$

Example 5.13 *A cylinder contains 0.12 m³ of air at 1 bar and 100°C. The air is compressed to 0.03 m³. The final being 6 bar.*

Determine,

(i) The value of index n

(ii) Mass of air in the cylinder.

(iii) Increase in internal energy .

Take $\gamma = 1.4$ $R = 0.287$ kJ/kg-K and $C_v = 0.72$ kJ/kg-K

Data: $V_1 = 0.12\ m^3$ $\quad P_1 = 100$ kPa

$T_1 - 100 + 273 = 373$ K $\quad V_2 = 0.03\ m^3$

$P_2 = 600$ kPa

Solution

So $$P_1V_1^n = P_2V_2^n$$

$$\frac{P_2}{P_1} = \left(\frac{V_1}{V_2}\right)^n$$

$$\left(\frac{600}{100}\right) = \left(\frac{0.12}{0.03}\right)^n$$

$$6 = (4)^n$$

$$\log_e 6 = n \log_e (4)$$

$$\frac{\log_e 6}{\log_e 4} = n$$

$$\frac{1.79175}{1.38629} = n$$

$$n = 1.29248$$

$$n = 1.3$$

(ii) To find mass of air in the cylinder

We know that, $P_1 V_1 = mRT_1$

$$100 \times 0.12 = m \times 0.287 \times 383$$

$$\therefore \quad \mathbf{m = 0.112096\ kg}$$

To find the increase in internal energy

(iii) We have, $\Delta U = m \cdot C_V (T_2 - T_1)$

Here T_2 is not known, so to find T_2, since it is a polytropic process,

$$\therefore \quad \frac{T_2}{T_1} = \left(\frac{V_1}{V_2}\right)^{n-1} = \left(\frac{P_2}{P_1}\right)^{\frac{n-1}{n}}$$

$$T_2 = T_1 \times \left(\frac{V_1}{V_2}\right)^{n-1} = 373 \times \left(\frac{0.12}{0.03}\right)^{1.3-1}$$

$$= 373 \times (4)^{0.3}$$

$$= 373 \times 1.516$$

$$\mathbf{T_2 = 565.36\ K}$$

$$\therefore \quad \Delta U = m \cdot C_V (T_2 - T_1)$$

$$= 0.112\ \text{kg} \times 0.72\ (565 - 373)$$

$$\mathbf{\Delta U = 15.51\ kJ\ (Increase)}$$

Example 5.14 *Calculate the change of,*

(i) Internal enemy and

(ii) Enthalpy of 2 kg of air when the temperature changes from 20°C to 90°C.

Take $C_p = 1$ kJ/kg-K and $C_v = 0.71$ kJ/kg-K

Solution

(i) $\Delta U = m \cdot C_V \cdot (T_2 - T_1) = 2 \times 0.71\ (90 - 20)$

$$\Delta U = 99.4\ \text{kJ}$$

(ii) $\Delta H = m \cdot C_P \cdot (T_2 - T_1) = 2 \times 1\ (90 - 20)$

$$\mathbf{\Delta H = 140\ kJ}$$

Example 5.15 *10 kg of air are heated in a right vessel from 20°C to 100°C. If the ratio of specific heats is 1.4, estimate the values of C_p and C_v the changes in internal energy and enthalpy.*

Solution

We have, $T_1 = 20°, \quad T_2 = 100°C$

$$C_V = \frac{R}{\gamma - 1}$$

and $\gamma = 1.4$ and $R = 0.287$

$$\therefore \quad C_V = \frac{0.287}{0.4} = 0.7175\ \text{kJ/kg-K}$$

$$C_P = R + C_V = 0.287 + 0.7175 = 1.0045\ \text{kJ/kg°K}$$

We know that,

Change in internal energy

$$dU = m\, C_V (T_2 - T_1)$$

$$= (10)\, 0.7175\, (100 - 20)$$

$$\mathbf{dU = 574\ kJ}$$

And change in enthlpy

$$dH = mC_P(T_2 - T_1) = (10)(1.0045)(100-80) = \mathbf{803.6kJ}$$

Example 5.16 *A piston and cylinder machine contains 1 kg of air, initially v = 0.8 m³ and T = 290 K. The air is then compressed in a slow frictionless process specific volume 0.2 m³/kg and a temperature 580 K. The law for compression is $PV^{1.5}$ = 0.75 with P in bar and V in m³/kg. Determine work done and heat transferred during the process. Assume for air C_p = 1.0 kJ/kg-K, C_v = 0.743 kJ/kg-K and R = 0.287 kJ/kg-K.*

Data: $m = 1$ kg, $V_1 = 0.8$ m³/kg

$T_1 = 290$ K, $V_1 = 0.2$ m³/kg

$T_2 = 580$ K, W.D. = ? Q = ?

Solution

Law for the process = $PV^{1.5} = C = 0.75$

We know that, for a polytropic process,

$$\text{W.D.} = \frac{P_2V_2 - P_1V_1}{1-n} = \frac{mR(T_2-T_1)}{1-n}$$

$$= \frac{1\times 0.287(580-290)}{1-1.5} = \frac{83.23}{-0.5}\text{ kJ}$$

$$= -166.46 \text{ kJ}$$

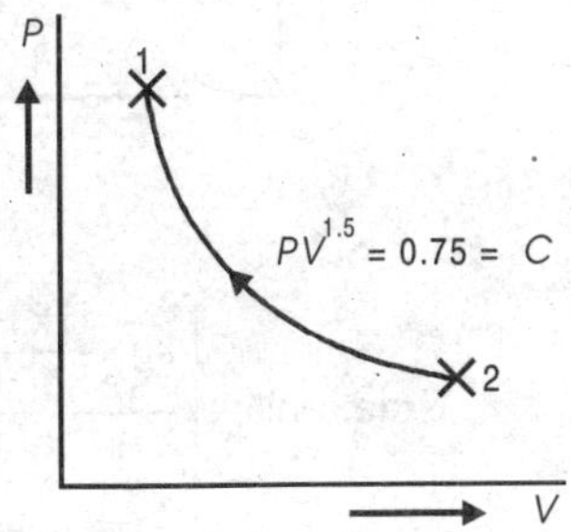

Fig. Ex. 5.16

The sign implies work is done on the air.

Also, $\Delta U = m \cdot C_V \cdot dT = 1 \times 0.743 \times (580 - 290)$

$= 206.77$ kJ

From I-law, for a process,

$$Q - W = \Delta U$$

$$Q = \Delta U + W$$

$$= 206.77 + (-166.46)$$

$$\mathbf{Q = 40.31 \text{ kJ}}$$

Example 5.17 *One kg of N_2 at a temperature of 250° C occupies a volume of 0.2 m³. The gas undergoes a constant pressure expansion without friction to a final volume of 0.36 m³. Find final pressure, final temperature, work done, change in internal energy, heat transfer and change of entropy.*

Take C_v = 0.743 kJ/kg-K and R = A297 kJ/kg-K and also represent the process on P–V and T–S diagrams.

Data: $m = 1$ kg, $T_1 = 250 + 273 = 523$ K

$V_2 = 0.36$ m³, $V_1 = 0.2$ m³.

$P_2 = ?$, $P = C$, $T_2 = ?$

$W = ?$, $\Delta U = ?$, $Q = ?$ $\Delta S = ?$

Solution

We know that,

$$\frac{P_1V_1}{T_1} = \frac{P_2V_2}{T_2}$$

Since $P = C,\ P_1 = P_2 = C$

$$\therefore \quad \frac{V_1}{T_1} = \frac{V_2}{T_2}$$

$$\therefore \quad T_2 = T_1 \times \frac{V_2}{V_1} = 523 \times \frac{0.36}{0.20}$$

$$T_2 = 941.4\text{ K}$$

Since $P_1V_1 = mRT_1$

$$P_1 = \frac{mRT_1}{V_1} = \frac{1 \times 0.297 \times 523}{0.2}$$

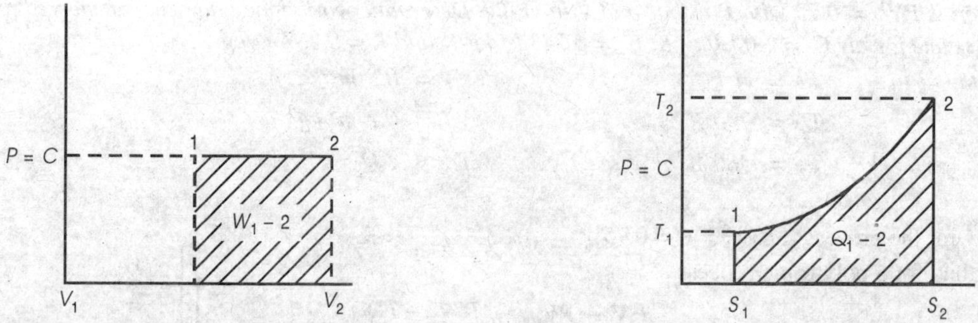

Fig. Ex. 5.17

Note. Units : $\left[\dfrac{\text{kg} \times \dfrac{\text{kJ}}{\text{kg}-\text{K}} \times \text{K}}{\text{m}^3} = \dfrac{\text{kJ}}{\text{m}^3} = \dfrac{\text{kNm}}{\text{m}^2} = \text{kPa}\right]$

$$P_1 = 776.655\text{ kPa} = P_2$$

(iii) Work done in a non-flow process = $\int PdV$

(since it is a constant pressure process).

$$= P(V_2 - V_1)$$
$$= 776.655\ (0.36 - 0.20)$$
$$W_{1-2} = 124.26\text{ kJ}$$

(iv) $\Delta U = m \cdot C_V \cdot (T_2 - T_1) = 1 \times 0.743 \times (941.4 - 523)$

$= 310.87$ kJ (+ve so I.E. increases)

Note. This $\left[\text{kg} \times \dfrac{\text{kJ}}{\text{kg}-\text{K}} \times \text{K} = \text{kJ}\right]$

(v) We also know from the I-law, for a closed system undergoing a process,

$$Q = \Delta U + W = 310.87 + 124.26$$
$$Q = 435.13\text{ kJ}$$

(vi) Change in entropy $= m \cdot C_P \cdot \ln\dfrac{T_2}{T_1}$

$= m(C_V + R)\ln\dfrac{T_2}{T_1}$ (since $C_P - C_V = R$)

$= 1 \times (0.743 + 0.297)\ln\dfrac{941.4}{523} 1 \times (1.04) \times \ln\dfrac{941.4}{523}$

Note. This $\left[\text{kg} \times \dfrac{\text{kJ}}{\text{kg}-\text{K}} \times \text{ratio} = \dfrac{\text{kJ}}{\text{K}}\right]$

$$S_2 - S_1 = 0.66112 \text{ kJ/K}$$

Example 5.18 *1 m³ of gas is filled in a closed tank. The initial condition of the gas is 3 bar and 50°C. The gas is heated until the pressure becomes 5 bar. Find the change in internal energy, work done, heat supplied, change in entropy. Take R = 0.287 kJ/kg-K and C_v = 0.711 kJ/kg-K.*

Solution

Since it is a constant volume heating process,

$$V_1 = V_2 = C = 1\text{m}^3$$
$$P_1 = 3 \text{ bar} = 3 \times 10^5 \text{ N/m}^2$$
$$T_1 = 50 + 273 = 323 \text{ K}$$
$$P_2 = 5 \text{ bar} = 5 \times 10^5 \text{ N/m}^2$$
$$R = 0.287 \text{ kJ/kg-K} = 287 \text{ J/kg-K}$$
$$C_V = 0.711 \text{ kJ/kg-K} = 711 \text{ J/kg-K}$$

(i) We know that change in IE in a constant volume process is given by,

$$\Delta U = m \cdot C_V(T_2 - T_1) \quad \ldots(1)$$

Here m and T_2 are not given

∴ To find T_2

Since $$\frac{P_1V_1}{T_1} = \frac{P_2V_2}{T_2}$$

as $$V_1 = V_2 = C$$

$$T_2 = \frac{P_2}{P_1} \times T_1$$

$$T_2 = \frac{5\times10^5}{3\times10^5} \times 323$$

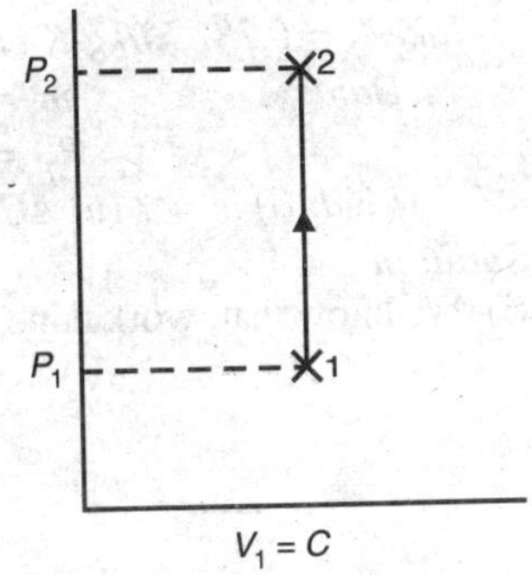

Fig. Ex. 5.18

$$T_2 = 538.33 \text{ K}$$

And since $$P_1V_1 = mRT_1$$

$$m = \frac{R_1V_1}{RT_1} = \frac{3\times10^5 \times 1}{287\times323}$$

Note units: $\left[\dfrac{\dfrac{\text{N}}{\text{m}^2}\times \text{m}^3}{\dfrac{\text{J}}{\text{kg-K}}\times \text{K}} = \dfrac{\text{Nm}}{\text{J/kg}} = \dfrac{\text{J}}{\text{J/kg}} = \text{kg}\right]$

$$m = 32362 \text{ kg}$$

∴ From Eq. (1), $$\Delta U = m \cdot C_V(T_2 - T_1)$$
$$= 3.236 \times 711 \times (538.33 - 323)$$

Note this $\left[\text{kg} \times \dfrac{\text{J}}{\text{kg-K}} \times \text{K} = \text{J}\right]$

$$\Delta U = 595462.7 \text{ J} = 495.4627 \text{ kJ}$$

(ii) Work done. In a constant volume heating process, heat supplied is completely utilised for increasing the I.E. of the gas and no external work is done since volume remains constant.

$$W_{1-2} = \int P dv = 0$$

As $$V = C,\ dV = 0$$

(iii) From first law

$$Q - W = \Delta U$$
$$Q = \Delta U + W$$
$$Q = \Delta U \text{ as } W = 0$$
$$Q = \Delta U = 495.4627 \text{ kJ}$$

(iv) $$S_2 - S_1 = mC_V \ln\frac{P_2}{P_1}$$

$$= 3.23 \times 0.711 \times \ln\frac{5\times10^5}{3\times10^5}$$

Note this $\left[\text{kg} \times \frac{\text{J}}{\text{kg-K}} \times \text{ratio} = \frac{\text{kJ}}{\text{K}} \right]$

$$S_2 - S_1 = 1.173 \text{ kJ/K}$$

Example 5.19 *2 kg of air at 150°C and 3 bar expands according to $PV^{1.2}$ = C to a final pressure of 1 bar. Find W, Q, and ΔU.*

Take R = 0.287 kJ/kg-K and γ = 1.4

Data: $m = 2$ kg $\quad T_1 = 150 + 273 = 423$ K

$P_1 = 3 \times 10^3$ N/m² $\quad n = 1.2 \quad P_2 = 1 \times 10^5$ N/m²

Find: (i) m = ? (ii) ΔU = ? Q = ? R = 0.287 kJ/kg-K, γ = 1.4

Solution

(i) We know that, work done in a polytropic process is given by,

$$W_{1-2} = \int PdV = \frac{P_1V_1 - P_2V_2}{n-1}$$

$$= \frac{mR(T_1 - T_2)}{n-1}$$

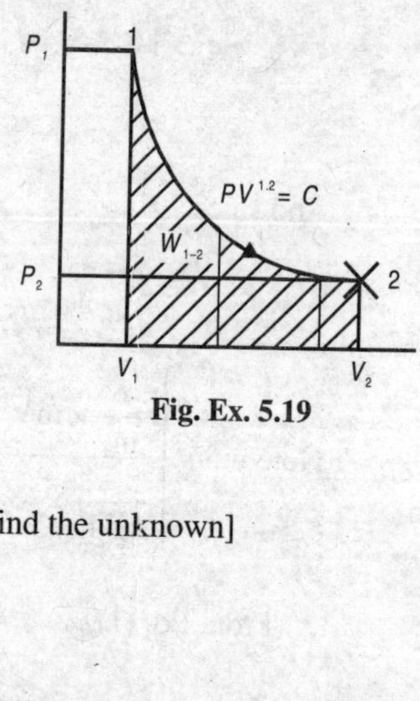

Fig. Ex. 5.19

Since T_2 is not given, to find T_2, use the relation

$$\frac{T_2}{T_1} = \left(\frac{P_2}{P_1}\right)^{\frac{n-1}{n}} \left[\frac{V_1}{V_2}\right]^{n-1}$$

Since it is a polytropic process.

[Note that if it is $P = C$; $V = C$; $T = C$ process then use $\frac{P_1V_1}{T_1} = \frac{P_2V_2}{T_2}$ to find the unknown]

$$\therefore \quad \frac{T_2}{T_1} = \left(\frac{P_2}{P_1}\right)^{\frac{n-1}{n}}$$

$$T_2 = T_1 \times \left(\frac{P_2}{P_1}\right)^{\frac{n-1}{n}} = 423 \times \left[\frac{1\times10^5}{3\times10^5}\right]^{\frac{1.2-1}{1.2}}$$

$$= 423 \times \left[\frac{1}{2}\right]^{\frac{0.1}{1.2}}$$

$$P_2 = 325.25 \text{ K}$$

W_{1-2} from Eq. (1)

$$= \frac{mR(T_1 - T_2)}{n-1} = \frac{2 \times 0.287(423 - 352.25)}{1.2 - 0.2}$$

$$W_{1-2} = 203 \text{ kJ}$$

(ii) Change in IE,

$$\Delta U = m \cdot C_V (T_2 - T_1)$$

Now,

$$\frac{C_P}{C_V} = \gamma = 1.4$$

$$C_p - C_V = T = 0.287$$

And

$$C_V = \frac{R}{\gamma - 1} = \frac{0.287}{1.4 - 1} = 0.7175 \text{ kJ/kg-K}$$

∴ From Eq. (2)

∴

$$\Delta U = 2 \times 0.7175 \times (352.25 - 423)$$

$$\mathbf{\Delta U = -101.52 \text{ kJ (Decreases)}}$$

(iii) ∴ Heat transfer

$$Q = \Delta U + W$$

$$= -102.52 + 203 \text{ kJ}$$

$$Q = 101.48 \text{ kJ}$$

(iv)

$$S_2 - S_1 = m\left[\frac{\gamma - n}{n} \times C_V \times \ln\frac{P_2}{P_1}\right]$$

$$= 2\left[\frac{1.4 - 1.2}{1.2} \times 0.7175 \times \ln\frac{3}{1}\right]$$

$$\mathbf{S_2 - S_1 = 0.26916 \text{ kJ/K}}$$

For selecting the equation to find the change of entropy, note that it is a polytropic process and pressure ratio is given, also we can select the equation in which temperature ratio is given for this problem, since temperature are given. Both the equations will give the same answer.

Example 5.20 *One kg of Nitrogen at a temperature of 150°C occupies a volume of 0.2 m³. The gas undergoes a fully restricted constant pressure expansion without friction to a final volume of – 0.36 m³. Evaluate the final pressure, final Take C_p = 743 J/kg-K, R = 0.297 kJ/kg-K.*

Solution

(a) We know that,

$$P_1 V_1 = mRT_1, \ m = 1$$

∴

$$P_1 = \frac{RT_1}{V_1} = \frac{0.297 \times 10^3 \times 423}{0.2}$$

$$= 628.155 \text{ kPa}$$

$$= P_2 = \text{final pressure}$$

(b) We know that for a constant pressure process,

$$\frac{P_1 V_1}{T_1} = \frac{P_2 V_2}{T_2} \ (P_1 = P_2)$$

∴

$$\frac{V_1}{T_1} = \frac{V_2}{T_2}$$

$$\therefore \quad T_2 = \frac{V_2}{V_1} \cdot T_1 = \frac{0.36}{0.2} \times 423$$

$$= 761.4 \text{ K or } 488.4°\text{C}$$

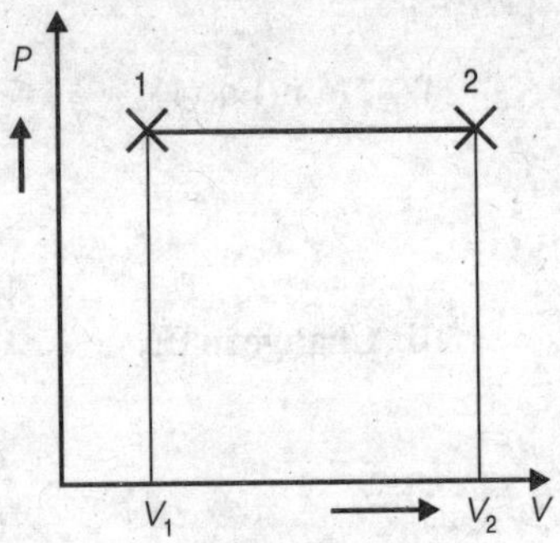

Fig. Ex. 5.20

(c) Work done in constant pressure process is given by

$$\delta Q = dH$$

$\therefore$ For unit mass, $\delta q = dh = C_P\, dT = C_P (T_2 - T_1)$

$$= 0.743 \times (761 - 423) = 251.43 \text{ kJ/kg}$$

(d) Change in entropy

$$\therefore \quad s_2 - s_1 = C_P \ln \frac{T_2}{T_1}$$

$$= 0.743 \times \ln \frac{761.4}{423}$$

$$\mathbf{s_2 - s_1 = 0.437 \text{ kJ/kg.K}}$$

Example 5.21 *1 kg of ice at –5°C is exposed to the surrounding atmosphere which is at 20°C. The ice melts and comes into thermal equilibrium with the atmosphere which is at 20°C. The ice melts and comes into thermal equilibrium with the atmosphere. Determine the entropy increase of the universe. Show the melting process on T–S diagram. Take, C_p of ice = 2.093 kJ/kg-K, Latent heat of fusion of ice = 333.3 kJ/kg.*

Solution

The melting is shown on *T–S* diagram $T_1 = 26$ K, $T_2 = T_3 = 273$ K, $T_4 = 293$ K

With reference to *T–S* diagram we have

$$S_2 - S_1 = C_P \log_e \frac{T_2}{T_1}$$

$$= 2.093 \log_e \frac{273}{268}$$

$$= 2.093 \times \log_e 1.0186$$

$$= 0.0387 \text{ kJ/kg-K}$$

$$S_3 - S_2 = h_{fg}/T_3$$

$$= \frac{33.3}{273} = 1.221 \text{ kJ/kg-K}$$

$$S_4 - S_3 = C_{Pw} \log_e \frac{T_4}{T_3}$$

$$= 4.187 \log_e \frac{293}{273}$$

$$= 0.296 \text{ kJ/kg-K}$$

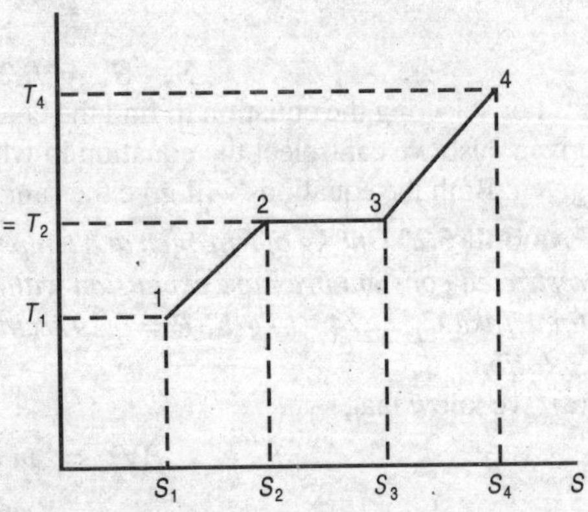

Fig. Ex. 5.21

$\therefore$ Total change in entropy of ice/water system

$$= S_4 - S_1 = 0.296 + 0.0387 + 1.221$$

$$= 1.556 \text{ kJ/kg-K}$$

Heat which is taken from surroundings

$$= 2.093\,(5) + 33.3 + 4.187\,(20)$$

$$= 10.465 + 333.3 + 83.74$$

$$= 427.505 \text{ kJ}$$

This heat is given by surrounding at constant temperature 293 K

∴ Change in entropy of the surrounding

$$= \frac{427.505}{293} = 1.4591 \text{ kJ/kg-K}$$

∴ Increase of entropy of universe

$$= \Delta S_P = 1.5560 - 1.4591$$

Increase of entropy of universe = 0.0969 kJ/kg-K

THEORY QUESTIONS

5.1. State and prove the principle of increase of entropy of the universe. Give some practical examples.

5.2. Prove that entropy is a property.

5.3. State and explain Clausius Theorem.

5.4. Explain Clausius inequality.

5.5. Write a short note on: Entropy and disorder.

5.6. Prove for polytropic process:

$$S_2 - S_1 = \frac{\gamma - n}{1-n} C_v \ln\left(\frac{T_2}{T_1}\right)$$

5.7. Explain the principle of increase of entropy. What is its significance ?

5.8. Give the criteria of reversibility, irreversibility and impossibility of a thermodynamic cycle.

5.9. State and prove the principle of increase of entropy of the universe. Give some practical examples.

5.10. Derive and express the Clausius Inequality.

5.11. Straight change of entropy during a polytropic process for an ideal gas is given by,

$$\Delta S = C_v[k-n]\ln\frac{V_2}{V_1}$$

where k = superheat ratio, n = Polytropic index.

5.12. What is the principle of increase of entropy? Explain with two practical examples.

5.13. P.T. for a poly thermo process:

$$\Delta S = \frac{\zeta - n}{\zeta - 1}.R.\ln\frac{V_2}{V_1}$$

5.14. What do you understand by the term reversibility, when need in the thermodynamic source? List to the best of your knowledge and explain all the possible irreverbilities that may be present in a system.

PROBLEMS FOR PRACTICE

5.1. 0.5 kg of air is compressed reversibly and adiabatically from 80 kPa 60°C to 0.4 MPa and is then expanded at constant pressure to the original volume. Sketch the process on P–V and T–S diagram, compute work transfer, heat transfer and change in entropy for whole path.
Take $R = 0.287$ kJ/kg-K and $\gamma = 1.4$.

(**Ans.** 93.5 kJ, 0.578 kJ /K)

5.2. A certain mass of air initially at a pressure of 480 kPa and temperature 190°C is expanded adiabatically to a pressure of 94 kPa. It is then heated at constant volume until it attains its initial temperature when the pressure is found. to be 150 kPa. State the type of compressor necessary to bring the system back to its original pressure and volume. Determine (1) the index of adiabatic expansion (2) The work done per kg of air and (3) the change in specific entropy of air.
R for air = 0.29 kJ/kg K.

(**Ans.** 1.402, –31.59 kJ/kg)

5.3. An ideal gas having γ or $K = 1.35$ and volume is 1.5 m³ at 100 kPa and 77°C. The gas is compressed according tot he law $PV^{1.25} = C$ to a pressure 3 MPa. Calculate the volume and temperature at the end of compression. Evaluate the heat transfer, work done, ΔU and change of entropy during the process.
Take $C_p = 0.98$ kJ/kg-K.

(**Ans.** –167 kJ, –584.5 kJ, 417.5 kJ, –0.322 kJ/K)

5.4. Show that when 1 kg of perfect gas is expanded through a volume ratio r the law of expansion being PV^n = constant, the change of entropy is given $(C_p - nC_v)$ Inr.

5.5. 50 litres of air at 1.013 bar and 100°C temperature is compressed to 28 bar. Volume of air at the end of polytropic compression is found to be 4 litres. Air is now heated at constant volume till pressure rises to 56 bar. Assuming $C_p = 1.00$ kJ/kg K and $C_v = 0.71$ kJ/kg K over range of working determine.

(*i*) Polytropic index of compression.

(*ii*) Entropy change in each process. Sketch the process on *P–V* and *T–S* diagram.

(**Ans.** (i) 1.31, (ii) –2.987, 10.5 kJ/K)

5.6. One kg of Nitrogen at a temperature of 150°C occupies a volume of 0.2 m^3. The gas undergoes a fully restricted constant pressure expansion without friction to a volume of 0.36 m^3. The gas is then expanded isothermally to a volume of 0.5 m^3. Represent the process on *P–V* and *T – S* diagrams and determine the change in entropy for each process and overall change in entropy. Take $C_v = 743$ J/kg-K and $R = 0.297$ kJ/kg K.

(**Ans.** 0.6089, 0.0962, 0.7051)

5.7. 4 kg of air are compressed from 40°C and 125° kPa to 250°C and 875 kPa. It is then throttled to 57 kPa. Finally it is cooled to a pressure of 125 kPa and 180°C. Calculate the overall change in entropy and also for each process.
Take $C_p = 1.055$ kJ/kg K, $C_v = 0.717$ kJ/kg K.

(**Ans.** 1.4859)

5.8. A solar thermal power plant flat plate collects or parabolic concentrators of sufficient area on which solar energy is incident are used. Flat plate collectors can give highest temperature in cycle as 90°C while parabolic concentrators give a temperature of 250°C. If the ambient temperature is 27°C. Find the minimum theoretical, area required of the solar system devéloping 10 kW power.

(i) By using flat plate collector having 60% collection efficient.

(ii) By using parabolic concentrator having 50% collection efficiency.

Assume incident plate solar energy collector, we can write. (**Ans.** (i) 96 m; (ii) 46.7 m^2)

5.9. A novel reversible heat engine plot in (*T–S*) diagram as a circle. The maximum and minimum temperatures are 1100 K and 200 K respectively and the maximum entropy change in the cycle is 2 kJ/K. Calculate the heat added to the cycle, heat rejected, the net work output and thermal efficiency of the cycle.

(**Ans.** 2006.86 kJ, 593.119, 1413.7Z kJ, 70.45)

5.10. Certain gas at a pressure of 1.4 MN/m^2 and 360° C is expanded adiabatically to a pressure of 100 kJ/m^2. The gas is then heated at constant volume until it attains 300°C when the pressure is found to be 220 kN/m^2 and finally it is compressed isothermally to the original pressure of 1.4 MN/m^2. Sketch the process on *P–V* and *T–S* diagrams. For 0.23 of gas and evaluate.

(i) ΔU during adiabatic expansion.

(ii) ΔS for each process.

Assume: $C_v = 0.705$ kJ/kg K.

(**Ans.** (i) 55.987 kJ; (ii) 0, 0.1278 kJ/K)

5.11. 60 litres of hydrogen at 20°C and 1 bar is compressed adiabatically to 9.8 bar. It is then cooled at constant volume to a pressure P_3 such as to have temperature of 20°C and further expanded isothermally so as to reach initial state: Find,

(i) Work done during each process.

(ii) Change in internal energy during each process.

Take $C_p = 14.4$ kJ/kg-K
$C_v = 10.28$ kJ/kg-K

for hydrogen and show the process on *P–V* and *T–S* plane. (**Ans.** (i) –13.788 kJ; (ii) 13.788 kJ)

5.12. 3 kg of air at a pressure of 150 kPa and temperature 360 K is compressed polytropically to 750 kPa according to law $PV^{1.2}$ = constant. The air is then cooled to initial temperature at constant pressure. The air is then heated at constant temperature till it reaches original pressure of 150 kPa. Draw the cycle on *PV* diagram and determine net work and heat.

(**Ans.** –73.21 kJ, –73.21 kJ)

5.13. Process 1: Air initially at 100 kPa and 50°C undergoes reversible adiabatic compression such that its volume is reduced to 1/5th of initial volume.

Process 2: Then 940 kJ/kg of heat is added to this air at constant volume.

Process 3: Process 2 is followed by reversible adiabatic expansion up to initial volume.

Process 4: Finally heat is rejected at constant volume so as to reach the initial condition.

Draw the four processes on one *PV* diagram. Determine the maximum temperature, and heat rejected per kg of air. Assume adiabatic index of compression and expansion of 1.4 and constant volume specific heat as 0:717 kJ.kg K.

(**Ans.** 1926 K, – 493.8 kJ/kg)

CHAPTER 6 Real Gases

6.1 INTRODUCTION

As we have studied in ideal gases that, an Ideal gas is one which obeys all the gas laws and the characteristic gas equation $PV = RT$ at all temperatures and pressures.

But actually there is no ideal or perfect gas in reality. But all the real gases like Air, O_2, N_2, H_2, He behave as ideal gases at very low pressures and high temperatures.

So a real gas does not obey gas laws $PV = RT$ for the entire range of pressure and temperature except when the pressure tends to zero and at a very high temperature. However real gas behaviour deviates from perfect gas equation at high pressures and low temperatures.

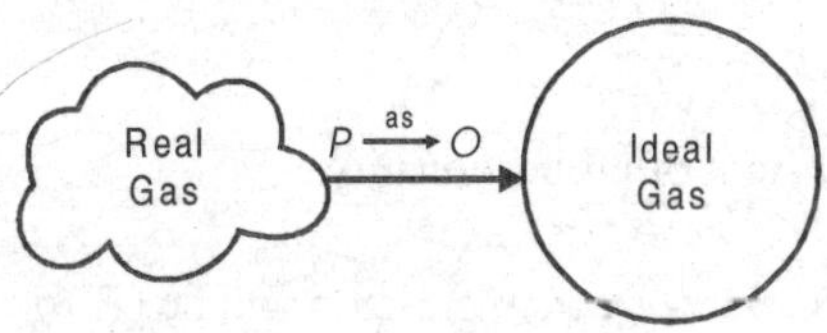

Fig. 6.1

In order to study the behaviour of real gases under all the conditions of pressures and temperatures Van der Waals' equation is most commonly used.

6.2 VAN der WAAL'S EQUATION OF STATE

The ideal gas equation $PV = RT$ is developed from kinetic theory of gases and is based on two important assumptions viz.; (i) There is very little or no attraction between the molecules of the gas (ii) The volume of the gas molecules themselves is very very small compared with the total volume of gas.

At low pressures and at very high temperatures, the intermolecular attractions and the volume of molecules compared to total volume of gas are not of much significance and the real gases obey the ideal gas equation. But when the pressure increases, the intermolecular attraction and the volume of the molecules becomes considerable when compared to total volume of gas. Then the real gases deviate from the ideal gas equation.

So, Van der Waal suggested the corrections for the ideal gas equations, so that it can be used for real gases at all pressures and temperatures.

6.2.1 Correction for Intermolecular Attractive Forces

To find a correction term for intermolecular attractive forces of the molecules, let us consider the molecule A, which is surrounded by the other molecules as shown and is attracted equally in all the directions and is in equilibrium. If we consider the molecule *B*, which is about to strike two walls of the containing vessel. Then this molecule will strike with less velocity or less momentum because of attractive force of the molecules on the inner side. When the momentum or velocity is less, then the pressure exerted by the molecules is less.

Van der Waal experimentally found that this decrease in pressure because of molecular attractive force is proportional to the square of the molal density i.e., Decrease in pressure α (Density of gas)^2.

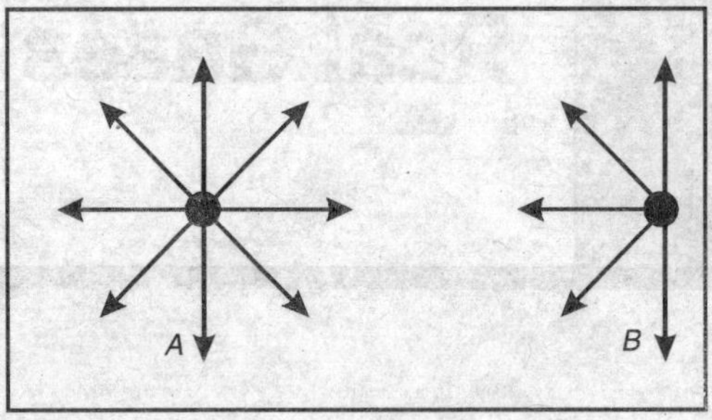

Fig. 6.2

$$p \propto (\rho_{gas})^2$$

$$p \propto \frac{1}{v^2} \quad \left(\because \rho = \frac{1}{v}\right)$$

$$p = \frac{a}{v^2}$$

where v is molal volume and a is constant of proportionality.

Now if P, is the measured pressure, the corrected, pressure will be $(P + p) = \left(P + \frac{a}{v^2}\right)$

6.2.2 Correction for Volume

This correction is to consider the self volume of the gas molecules, which is generally neglected in comparison with total volume. To correct this self volume, Van der Waal experimentally found and suggested that specific volume in ideal gas equation must be replaced by $(v - b)$. Where constant b takes care of reduction in volume due to finite size of molecules.

When we consider both these corrections, the equation of state for the Real gases can be written as,

$$\left(\mathbf{P} + \frac{\mathbf{a}}{\mathbf{v}^2}\right)(\mathbf{v} - \mathbf{b}) = \mathbf{RT} \quad \ldots(1)$$

or
$$\left(P + \frac{a}{v^2}\right) = \frac{RT}{(v-b)}$$

or
$$P = \frac{RT}{(v-b)} - \frac{a}{v^2}$$

where P is the pressure in Pa or N/m^2
R is the Gas Const. = 8314 J/kg mole K

T temp. in K
v molar volume in m^3/kg mole
b is constant in m^3/kg mole
a is constant N-m^4/(kg mole)^2.

At low pressures, b may be neglected,

$$\left(p + \frac{a}{v^2}\right)v = RT$$

$$PV + \frac{a}{v} = RT$$

$$PV = RT - \frac{a}{v} \quad \ldots(2)$$

From this equation we can see that the product of Pv is less than RT similarly at very high pressures, the term $\frac{a}{v^2}$ can be neglected when compared to the pressure P, then the Van der Waals equation can be written as,

$$P(v-b) = RT$$

$$PV - Pb = RT$$

$$PV = RT + Pb \quad \ldots(3)$$

Equation (3) is known *Clausius Equation of State* and it can be seen that the product of pressure and volume PV is more than RT.

6.3 DETERMINATION OF CONSTANTS OF VAN der WAAL'S EQUATION

Eventhough the Van der Waals equation is generally used for real gases, it has some limitations, that is the constants a and b vary with temperature and pressure. One of the methods generally used to find the value of the constants a and b is based on the fact that in case of real gases the critical isotherm on P-v diagram has zero slope at critical point. (refer Fig. 6.3).

It is to be noted that the properties at the critical state are known as critical properties such as critical pressure P_C, critical temperature T_C, critical volume v_C.

Since the critical isotherm has zero slope at critical point, we can write,

$$\left(\frac{dP}{dv}\right)_c = 0 \text{ and } \left(\frac{d^2P}{dv^2}\right)_c = 0.$$

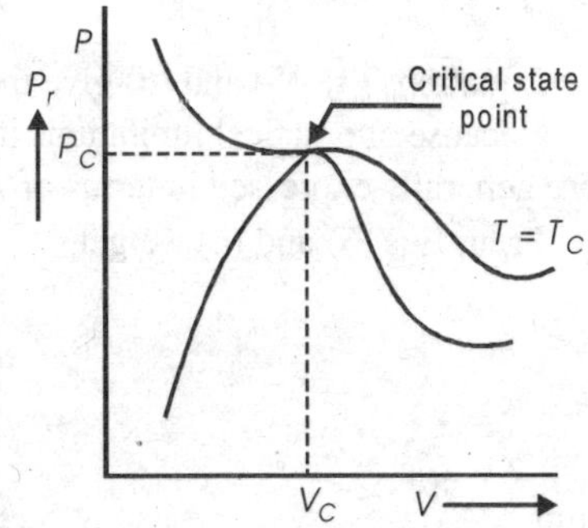

Fig. 6.3. *Critical properties on Pv diagram.*

Then from the Van der Waal's equation,

$$\left(P + \frac{a}{v^2}\right)(v-b) = RT$$

i.e.
$$P + \frac{a}{v^2} = \frac{RT}{v-b}$$

i.e.
$$P = \frac{RT}{v-b} - \frac{a}{v^2}$$

At critical point we can write, $P_c = \frac{RT_c}{v_c - b} - \frac{a}{v^2}$ ...(1)

So, differentiating this w.r.t. v_c and assuming T_c = constant we get,

$$\left(\frac{dP}{dv}\right)_c = -\frac{RT_c}{(v_c - b)^2} + \frac{2a}{v_c^3} = 0$$

or

$$\frac{-2a}{v_c^3} = \frac{RT_c}{(v_c - b)^2}$$

$$2\frac{a}{v_c^3}(v_c - b)^2 = RT_c \qquad ...(2)$$

Then

$$\left(\frac{d^2P}{dv^2}\right)_c = \frac{2RT_c}{(v_c - b)^3} - \frac{6a}{v_c^4} = 0$$

or

$$\frac{3a}{v_c^4}(v_c - b)^3 = RT_c \qquad ...(3)$$

From Eqs (2) and (3) we get, $\mathbf{v_c = 3b}$ ...(4)

Substituting v_c in Eq. (3) we get,

$$\mathbf{T_c = \frac{8a}{27Rb}} \qquad ...(5)$$

Substituting v_c and T_c in Eq. (1) we get,

$$\mathbf{P_c = \frac{a}{27b^2}} \qquad ...(6)$$

Equations (4), (5) and (6) give the values of critical properties.

Because of practical limitations it is very very difficult to measure the critical volume. So, the constants a and b are generally expressed in terms of T_c and P_c.

From Eqs (5) and (6) we get,

$$\mathbf{a = \frac{27}{64}\frac{R^2T_c^2}{P_c}} \qquad ...(7)$$

and

$$\mathbf{b = \frac{RT_c}{8P_c}} \qquad ...(8)$$

Note: When the Van der Waals equation is compared with ideal gas equation at critical state we get,

$$\mathbf{\frac{P_cV_c}{RT_c} = \frac{3}{8}} \qquad ...(9)$$

So, the ratio of $\frac{PV}{RT}$ at critical point is a constant 3/8 is same for all real gases and is unity for ideal gases.

Table below gives the constants of Van der Walls equation.

S. No.	Substance	a $Nm^4/(kmol)^2$	b $m^3/kmol$
1	Hydrogen (H_2)	25105	0.0262
2	Oxygen (O_2)	139250	0.0314
3	Carbon dioxide (CO_2)	362850	0.0423
4	Helium (He)	3417620	0.0228
5	Air	135522	0.0362
6	Water (H_2O) Vapour	551130	0.0300
7	Mercury (Hg) Vapour	2031940	0.0657

6.4 REDUCED CO-ORDINATES (VAN der WAAL'S EQUATION) IN REDUCED CO-ORDINATES

As we have studied in the previous section that the properties at the critical state are known as critical properties such as critical pressure P_c, critical temperature T_c, critical volume v_c.

Reduced co-ordinates or properties are defined as those properties which are obtained by dividing co-ordinates/properties by corresponding co-ordinates/properties at the critical state for example,

$$\text{Critical Pressure } P_r = \frac{P}{P_c} \quad \ldots(1)$$

$$\text{Critical Temperature } T_r = \frac{T}{T_c} \quad \ldots(2)$$

and

$$\text{Critical Volume } v_r = \frac{v}{v_c} \quad \ldots(3)$$

The Relationship Between Reduced Co-ordinates is known as the Law of Corresponding States:
From previous section we know from Eqs (4), (5) and (6)

$$v_c = 3b \quad \ldots(4)$$

$$T_c = \frac{8a}{27Rb} \quad \ldots(5)$$

$$P_c = \frac{a}{27 \cdot b^2} \quad \ldots(6)$$

From Eqs (1), (2) and (3) we can write,

$$P = P_r \cdot P_c = P_r \times \frac{a}{27 \cdot b^2} \quad \ldots(7)$$

$$T = T_r \cdot T_c = T_r \times \frac{8.a}{27 \cdot R \cdot b} \quad \ldots(8)$$

$$v = v_r \cdot v_c = v_r \times 3b \quad \ldots(9)$$

Substituting these values in Van der Waals equation we get,

$$\left(\mathbf{P_r} + \frac{\mathbf{3}}{\mathbf{v_r^2}}\right)(\mathbf{v_r} - \tfrac{1}{3}) = \frac{\mathbf{8}}{\mathbf{3}}\mathbf{T_r}$$

or
$$\left(P_r + \frac{3}{v_r^2}\right)(3v_r - 1) = 8T_r$$

This is the Van der Waals equation in terms of reduced co-ordinates and its is independent of *R*, *a*, and *b*.

6.5 COMPRESSIBILITY FACTOR; COMPRESSIBILITY CHART, GENERALISED COMPRESSIBILITY CHART

Compressibility Factor. The deviation of real gas from ideal gas is measured by a factor known as compressibility factor *Z*, and defined as the ratio of the actual volume to the volume predicted by the ideal gas law at the same temperature and pressure is given by $Z = \dfrac{\text{Actual volume}}{\text{Volume predicted by the ideal gas}} = \dfrac{v}{\frac{RT}{P}}$

But $\frac{RT}{P}$ is the ideal gas volume V_{ideal}. So, compressibility factor may also be defined as, the ratio of specific volume of real gas to specific volume of ideal gas i.e.,

$$\text{Compressibility Factor} = Z = \frac{v_{\text{real gas}}}{v_{\text{ideal gas}}}$$

For ideal gas
$$Z = \frac{v_{\text{ideal}}}{v_{\text{ideal}}} = 1$$

As we know all the gases behave as ideal gases at very low pressures and high temperatures. So when the pressures are lowered, then the value of *Z* tends to unity as the gas behaves as ideal.

It is to be noted that the value of *Z* may be less than unity or greater than unity depending upon the pressure and temperature. Figure 6.4 shows the compressibility factor chart which shows the values of *Z* corresponding to pressure.

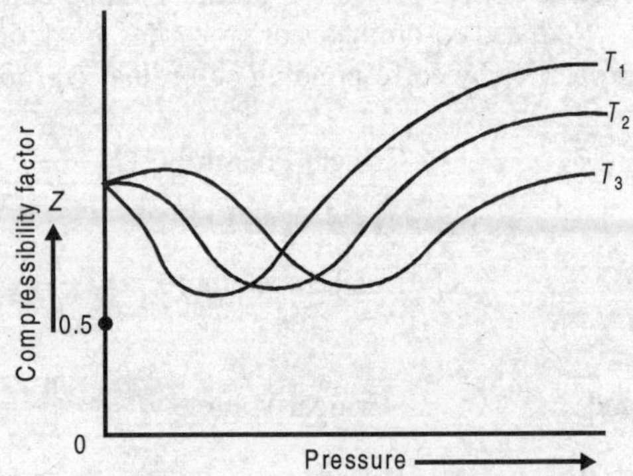

Fig. 6.4. *Compressibility factor chart.*

Note that a graph of compressibility factor Z versus P_r for different values of T_r for different gases is known as *Generalised Compressibility Chart* which is shown in Fig. 6.5.

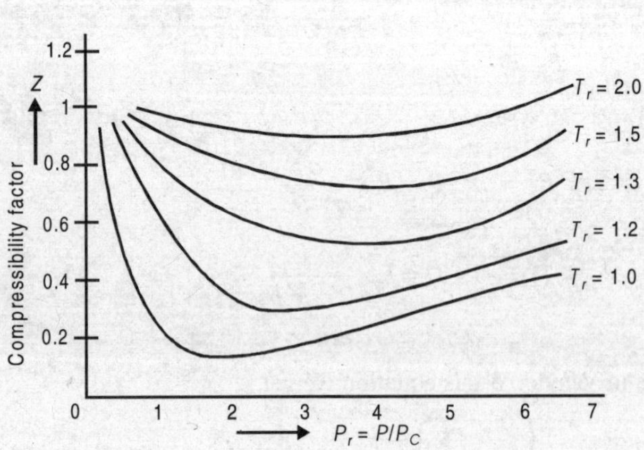

Fig. 6.5. *Generalised compressibility chart.*

6.6 VAN der WAAL'S PRINCIPLE

While studying reduced co-ordinates we have studied "that the relationship between reduced properties (i.e., $P_r T_r$ and v_r) is known as Law of corresponding states.

As we all know that all the substances exist in solid, liquid or gaseous phases and they have critical point and triple point.

And Van der Waal suggested that $P\,v\,T$ relationship can be shown by a single model and he has used critical point as reference point and critical properties as reducing properties and the principle of corresponding states may be expressed as,

$$f(P_r, v_r, T_r) = 0 \quad \text{.... (1)}$$

From the definition of compressibility factor, we know that,

$$Z = \frac{P_v}{RT} \quad \text{...(2)}$$

Since $P_r = \frac{P}{P_c}; v_r = \frac{v}{v_c}; T_r = \frac{T}{T_c}$

We can write,
$$Z = \frac{P_r \cdot P_c}{R} \cdot \frac{v_r \cdot v_c}{T_r \cdot T_c}$$

or
$$Z = \frac{P_r \cdot v_r}{T_r} \cdot \frac{P_c \cdot v_c}{R \cdot T_c}$$

$$Z = \frac{P_r \cdot v_r}{T_r} \cdot Z_c \quad \text{...(3)}$$

where $Z_c = \frac{P_c \cdot v_c}{R \cdot T_c} = \frac{3}{8}$ according to Van der Waal.

$$\therefore \quad \mathbf{Z = \frac{P_r \cdot v_r}{T_r} \times \frac{3}{8}} \quad \text{...(4)}$$

Now from Eq. (10) of reduced co-ordinates article and above Eq. (4) we get

$$\left[P_r + \frac{27P_2^r}{64T_2^r.Z^2}\right]\left(\frac{8.T_r.Z}{3P_r} - \frac{1}{3}\right) = \frac{8}{3}T_r.$$

This is another form of Van der Waals equation.

6.7 THE BEATTIE–BRIDGEMAN EQUATION OF STATE

This equation is proposed in 1928, it has five constants which have to be determined experimentally. This equation is given as below,

$$\mathbf{P = \frac{RT}{v^2}\left(1 - \frac{C}{vT^3}\right)(v + B) - \frac{A}{v^2}}$$

where $A = A_0\left(1 - \frac{a}{v}\right)$ and $B = B_0\left(1 - \frac{b}{v}\right)$

Constants of the Beattie–Bridgeman equation of state.

Gas	A_o	*a*	B_o	*b*	$C \times 10^{-4}$
Hydrogen H_2	20.0117	–0.00506	0.02096	–0.04359	0.0504
Oxygen O_2	151.0857	0.02562	0.04624	0.004208	4.8
Carbon dioxide CO_2	507.2836	0.07132	0.10476	0.07235	66.0
Helium He	2.1886	0.05984	0.01400	0.0	0.0040
Air	131.8441	0.01931	0.04611	–0.001101	4.34
Nitrogen N_2	136.2315	0.02617	0.05046	–0.00691	4.20

6.8 VIRIAL EQUATION OF STATE

We know that the equation of state of a substance is $PV = RT$ and may be expressed as,

$$P = \frac{RT}{v} + \frac{a(T)}{v^2} + \frac{b(T)}{v^3} + \frac{c(T)}{v^4} + \ldots$$

Then this equation is known as Virial Equation of state. And the co-efficients $a(T)$, $b(T)$, $c(T)$ which are the functions of temperature are called as Virial Co-efficients.

6.9 P-v-T SURFACE OF AN IDEAL GAS

We have studied the equation of state for ideal gases and real gases. These equations contain three variables P, v, T. If we plot these three variables along the three mutually perpendicular axes, then a surface is developed and is known as *P-v-T* surface.

Figure 6.6 shows the *P-v-T* surface of an ideal gas. On this figure constant pressure, constant volume, and isothermal lines are shown. It is to be noted that on this surface each point is a equilibrium state and the processes proceed from one equilibrium state to another equilibrium state till the end of the process and the processes are quasi-static.

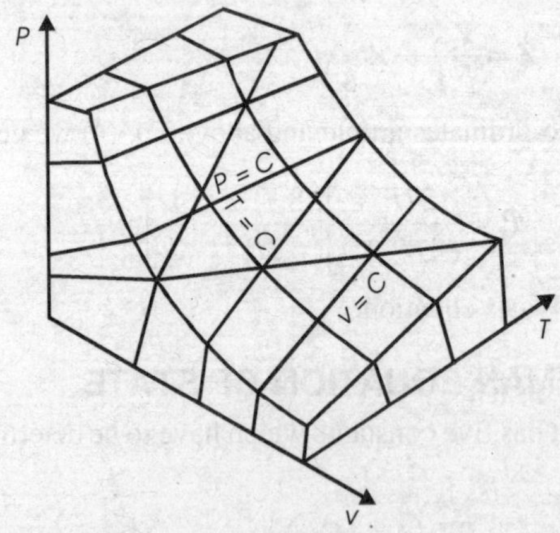

Fig. 6.6

From Fig. 6.6 we can draw *P.v* and *PT* plans for an ideal gas as shown in Fig. (i) and (ii).

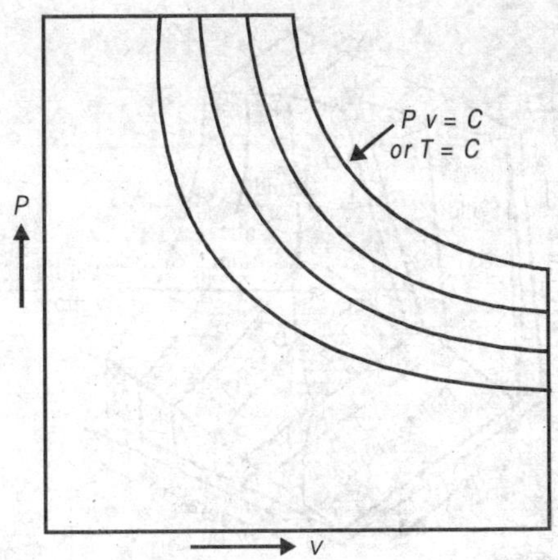

Fig. 6.7

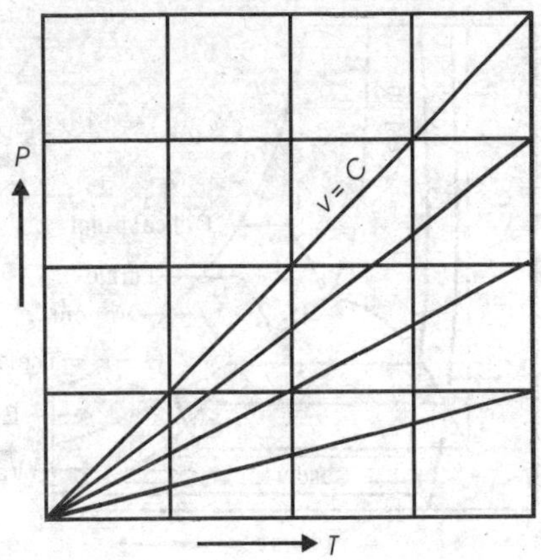

Fig. 6.8

Figure 6.9 shows the isothermal processes i.e., it graphically shows Boyle's law whereas Fig. 6.10 shows graphically Charle's law.

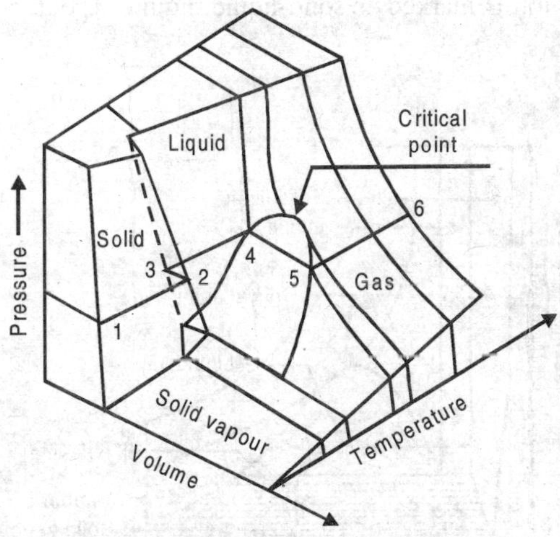

Fig. 6.9. *P-v-T surface of* H_2O.

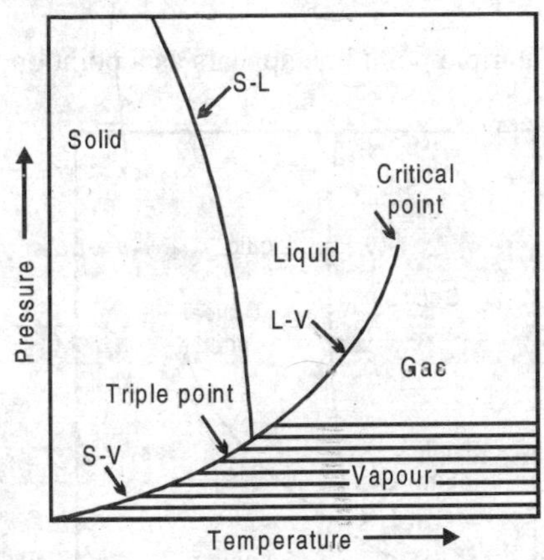

Fig. 6.10

6.10 P-v-T SURFACE FOR REAL SUBSTANCE ALONG WITH Pv AND PT DIAGRAMS FOR REAL SUBSTANCE

To study *P-v-T* surfaces of real substance, water and CO_2 are considered. The important features of the two are:

(i) Water expands on freezing

(ii) CO_2 contracts on freezing.

Figure 6.10 shows *P. v. T* surface of H_2O. Also Figs 6.10 and 6.11 show *PT* and *Pv* diagrams.

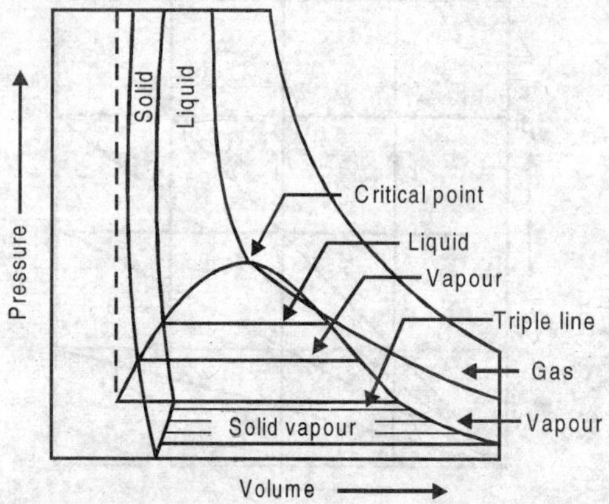

Fig. 6.11. *P–v diagram for H_2O.*

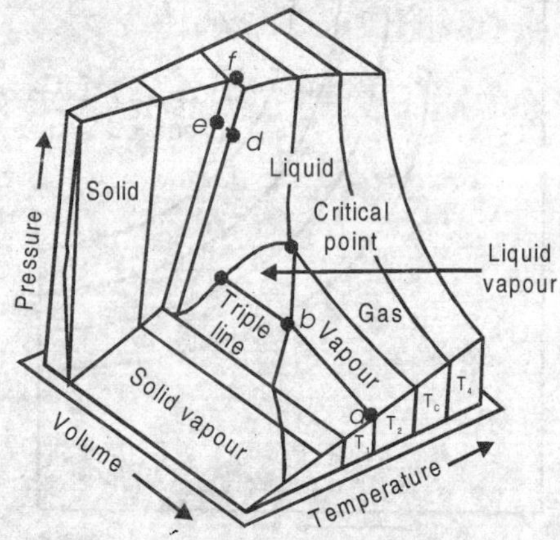

Fig. 6.12

As we have discussed in previous section that every point on the *P-v-T* surface is a equilibrium point and the processes are quasi-static.

In certain portions the substance exists in single phase i.e., it may be solid, liquid or a gas and accordingly they are marked. And if the substance has two phases, then the region is marked as solid-liquid, liquid-vapour, solid-vapour.

Note that triple point line appears as a point on *PT*-diagram.

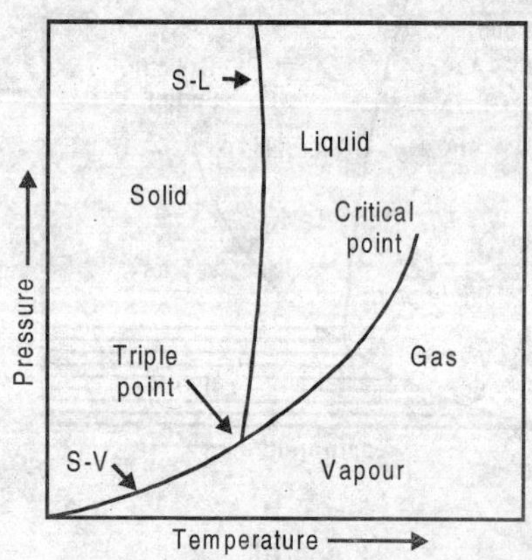

Fig. 6.13. *p-T diagram for CO_2.*

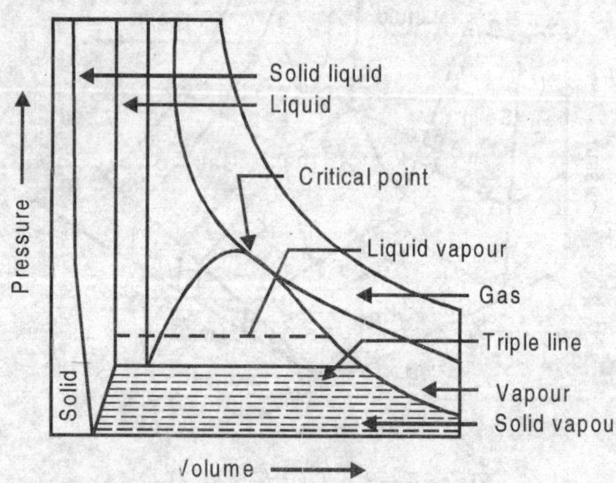

Fig. 6.14. *p-v diagram for CO_2.*

The line 1 – 2 – 3 – 4 – 5 – 6 is a constant pressure line on the *P-v-T* surface. To understand this, consider water in a closed vessel and let heat be supplied. The temperature of water will increase upto point 4 is reached. Then the steam generation starts. And the heat supplied is utilised for converting water into steam. Once the last particle of water is converted into steam the steam is known as dry saturated steam. After this point 5, if the steam is further heated, then it will be superheated along 5 – 6.

Figures 6.13, 6.14 show *P.v.T.* surface for CO_2 and *PT* and *Pv*.

SOLVED EXAMPLES

Example 6.1 *Determine pressure of 1 kg of oxygen at 100°C if the specific volume is 0.2. m^3/kg using. (i) Van der Waals equation and (ii) Ideal gas laws.*

Take for O_2,

$a = 13.93 \times 10^4$ N. m^4 (kg mole)^2; b = 0.0 314 m^3/kg mole. and R = 8314 J/kg mole K.

Data:

mass of O_2 = 1 kg

$$T = 100 + 273 = 373 \text{ K}$$

$$v = 0.2 \times 32 = 6.4 \text{ m}^3/\text{kg mole.}$$

Solution

Since *M* of Oxygen = 32

w.r.t., from Van der Waal's equation.

$$\left(P + \frac{a}{v^2}\right)(v - b) = RT$$

$$P = \frac{RT}{(v-b)} - \frac{v}{v^2}$$

$$= \frac{8314 \times 373}{6.4 - 0.0314} - \frac{13.93 \times 10^4}{(6.4)^2}$$

$$= 486939.36 - 3400.8789$$

$$= 483538.48 \text{ N/m}^2$$

$$\mathbf{P = 4.845 \text{ bar}}$$

Secondly, $Pv = RT$

$$P = \frac{RT}{v} = \frac{8314 \times 373}{6.4} = 484550.31 \text{ N/m}^2$$

$$\mathbf{P = 4.835 \text{ bar.}}$$

Example 6.2 *A certain gas obeys Van der Waals equation for which $a = 1.4 \times 10^{-1}$ N-m^4 and $b = 32 \times 10^{-3}$ m^3/kg mole. And 1 kg mole occupies 40 m^3 at 500°C calculate the pressure, by taking R = 8314.*

Solution

We know from Van der Waals equation,

$$\left(P + \frac{a}{v^2}\right)(v - b) = RT$$

$$P = \frac{RT}{v-b} - \frac{a}{v^2} = \frac{8314 \times (500 + 273)}{(40 - 32 \times 10^{-3})} - \frac{1.4 \times 10^{-1}}{(40)^2}$$

$$= 160796.69 - 8.75 \times 10^{-5}$$

$$= 160796.69 \text{ N/m}^2$$

$$P = 1.607 \text{ bar.}$$

CHAPTER

7 Non-Reactive Ideal Gase Mixtures

7.1 INTRODUCTION

The thermodynamic behaviour of a mixture of gases depends upon the individual properties of its constituent gases. Because substances in the gaseous phase are so miscible with each other, there is no limit to the number of different gaseous mixtures. Furthermore, individual constituents of gaseous mixtures often react chemically with each other, and these reactions introduce another factor that exerts a strong influence on the properties of a gas. This chapter deals only with mixtures of inert gases, in which changes in phase of constituents may take place.

A gas mixture is frequently treated as if it consisted of a single component rather than many properties of the individual constituents of an inert gas tend to be submerged, so that the gas behaves, in certain ways, as though it were a single, pure substance. The main constituents of substance are, for example, oxygen, nitrogen, argon and water vapour. But air is commonly said to have a molecular weight of 28.97 kg/K-mol, even though this figure represents a composite value based on the molecular weights and the proportions of the constituent species.

In the first case of mixtures of ideal or perfect gases, equations are derived which express the properties of mixtures in terms of properties of the constituents. Mixtures are treated in which no chemical reactions, condensation or evaporation take place. The derived expressions apply, in general to gas mixtures over a wide range of temperatures and pressures. Although the study centers about perfect gases, mixtures containing non-perfect gases often show only negligible deviation from perfect gas behaviour.

7.2 DALTON'S LAW OF PARTIAL PRESSURES

It was John Dalton (1766 – 1844) who first stated that the total pressure p_m exerted by a mixture of gases is the sum of the pressures which each gas (or vapour), would exert were it to occupy the vessel alone at the volume V_m and temperature T_m of the mixture true only of ideal gases. Thus, if p_x, p_y, p_z ...... represent respectively the individual pressures of the mixed gases x, y, z.

∴ According to Dalton's Law of partial pressures, we have

$$p_m = p_x + p_y + p_z + \ldots$$

at

$$t_m = t_x + t_y + t_z = \ldots.$$

and

$$V_m = V_x + V_y = V_z \ldots..$$

This is also a part of Gibbs – Dalton principle.

Similarly, the number of moles of the mixture is the sum of the moles of the individual components.

Thus,

$$n_m = n_x + n_y + n_z + \ldots$$

Applying the ideal or perfect gas equation to a component x and to the mixture, we get,

$$p_x V_x = p_x V_m = 8315\, n_x t_m$$

and
$$p_x V_m = 8315 n_x t_m$$

$$\therefore \quad \frac{P_x V_m}{P_m V_m} = \frac{8315\, n_x T_m}{8315\, n_m T_m}$$

or
$$\frac{P_x}{P_m} = \frac{n_x}{n_m}$$

We write this also as, $\dfrac{p_x}{p_m} = \dfrac{n_x}{n_m} = B_x$ (Volume fraction)

Thus volume fraction is the ratio of the partial pressure to the total pressure the mixture.

Figure 7.1 shows the mixtures and constituents.

Constituents

$x : m_x, p_x, V_x, T_x$

$y : m_y p_x v_y T_y$

$z : M_x p_y V_z T_z$

Mixture $\rightarrow m_m, p_m, V_m, T_m$

Fig. 7.1. *Mixture and constituents.*

Note: Here that the individual pressure exerted by a **real gas** is not necessarily given by these equations.

7.3 AMAGAT'S LAW AND VOLUMETRIC ANALYSIS

Amagat's Law states that *the total volume of a mixture of gases is equal to the sum of the volumes that would be occupied by each component at the pressure p_m and temperature T_m of the mixture.*

This law is found to be true for ideal gases only, though it is widely used for all kinds of gaseous components. Figure 7.2 shows the mixture of gases following this Law.

Gas a	Gas b	Gas c
m_a	m_b	m_c
p_m	p_m	p_m
V_a	V_b	V_c
T_m	T_m	T_m
Mixture: p_m, V_m, T_m, m_m		

$$\left.\begin{aligned} V_m &= V_a + V_b + V_c \\ p_m &= p_a = p_b = p_c \\ T_m &= T_a = T_b = T_c \\ m_m &= m_a + m_b + m_c \end{aligned}\right\} \text{Mixture}$$

Fig. 7.2. *System for Amagat's experiment.*

It may be noted here that Dalton's Law and Amagat's Law are coincident with the kinetic theory of gases. These laws assume that no intermolecular forces exist in a mixture of gases and that each constituent acts as if no other constituents were present.

When this mixture is analyzed, the perfect gas law leads to :

$$p_m V_a = n_a \bar{R} T_m$$

$$p_m V_b = n_b \bar{R} T_m$$

$$p_m V_c = n_c \bar{R} T_m$$

where n_a, n_b and n_c are the number of moles of individual gas in the mixture.

∴ Addition of these equations, we get

$$p_m (V_a + V_b + V_c) = \bar{R} T_m (n_a + n_b + n_b)$$

or $$p_m V_m = n\bar{R} T_m$$

where n is the sum of moles of the individual gases.

By dividing the perfect gas equation for each constituent by the equation for the mixture, one obtains the following:

$$\frac{V_a}{V_m} = \frac{n_a}{n}; \frac{V_b}{V_m} = \frac{n_b}{n} \text{ and } = \frac{V_c}{V_m} = \frac{n_c}{n}$$

The ratio of the partial volume of any constituent to the total volume, then, is equal to the mole fraction of that constituent.

This $\left[\frac{V_a}{V_m} = \frac{n_a}{n}\right]$ gives the volumetric analysis of the constituent gas a. Similarly, the volumetric analysis of the other gas can be determined by $\frac{n_b}{n}$ for gas b and $\frac{n_c}{n}$ for gas c. Volumetric analysis is generally given by B like B_a, B_b, B_c etc.

7.4 GRAVIMETRIC ANALYSIS

Constituents or *components* of the mixture of gases are the individual gases and vapours in a mixture. Generally, the description of a mixture of any substances like solids, liquids or gases—can be given by a *gravimetric analysis* which is the percentage or fraction by weight or mass of each constituent.

Thus if a, b and c are the constituents in a mixture and they are having weights or masses as m_a, m_b and m_c respectively , then the total mass of the mixture is

$$m_m = m_a + m_b + m_c$$

Therefore, the gravimetric analysis of the mixture is given by

% of gas a in a mixture is $G_a = \frac{m_a}{m_m} \times 100$

Similarly the percentage composition of gas b and c are given as

$$G_b = \frac{m_b}{m_m} \times 100$$

$$G_c = \frac{m_c}{m_m} \times 100$$

Here G stands for the percentage of the gas mass or for the fractional part of the said constituent.

7.5 CONVERSION BETWEEN GRAVIMETRIC AND VOLUMETRIC ANALYSIS

For converting the gravimetric analysis of the mixture of gases into volumetric analysis and vice-versa, two tables are given below.

(a) Gravimetric to Volumetric Analysis

Sr. No.	*Name of the gas*	*Molecular Weight (a)*	*Composition of gas by weight (b)*	*Relative volume* $c=\frac{b}{a}$	*% composition by volume* $=\frac{c}{\Sigma c}\times 100$
1	x	M_x	m_x	$\frac{m_x}{M_x}$	$\frac{m_x}{M_x}\times\frac{100}{\Sigma c}$
2	y	M_y	m_y	$\frac{m_y}{M_y}$	$\frac{m_y}{M_y}\times\frac{100}{\Sigma c}$
3	z	M_y	m_y	$\frac{m_z}{M_z}$	$\frac{m_z}{M_z}\times\frac{100}{\Sigma c}$
			$\Sigma b = 100$	Σc	100

(b) Volumetric to Gravimetric Analysis

Sr. No.	*Name of the gas*	Molecular weight (*a*)	% Composition of gas by weight (*b*)	*Relative volume* $c = (a)\times(b)$	*% composition by volume* $=\frac{c}{\Sigma c}\times 100$
1	x	M_x	V_x	$V_x.M_x$	$\frac{V_x.M_x}{\Sigma c}\times 100$
2	y	M_y	V_y	$V_y.M_y$	$\frac{V_y.M_y}{\Sigma c}\times 100$
3	z	M_y	V_z	$V_z.M_z$	$\frac{V_z.M_z}{\Sigma c}\times 100$
			$\Sigma b = 100$	Σc	100

The use of these tables will be best understood when the numerical problems are solved.

7.6 MOLECULAR WEIGHT AND GAS CONSTANT FOR MIXTURE

Consider the mixture of three gases *x*, *y*, and *z* having the molecular weights as M_x, M_y and M_z. Let the moles of each of these gases be $-n_x$, n_y, and n_z.

$\therefore$ Total weight or mass of the mixture can be found as

$$\text{Weight of } x \text{ gas} = m_x = n_x M_x$$
$$\text{Weight of } y \text{ gas} = m_y = n_y M_y$$
$$\text{Weight of } z \text{ gas} = m_z = n_z M_z$$

$\therefore$ Total mass of the mixture is given by

$$m_m = m_x + m_y + m_z$$

$$= n_x M_x + n_y M_y + n_z M_z$$

and the total number of moles in the mixture is

$$n_m = n_x + n_y + n_z$$

If M_m is the molecular weight of the mixture, then

$$\begin{bmatrix}\text{Total mass of} \\ \text{mixture}\end{bmatrix} = \begin{bmatrix}\text{Number of moles} \\ \text{in mixture}\end{bmatrix} = \begin{bmatrix}\text{Molecular weight} \\ \text{of mixture}\end{bmatrix}$$

$$M_m = n_m \times M_m$$

$$M_m = \text{Molecular weight of the mixture}$$

$$= \frac{m_m}{n_m}$$

$$\mathbf{M_m = \frac{m_m}{n_m}}$$

Similarly, specific gas constant for the mixture is given by

$$\bar{R} = R \times M_m$$

$$\therefore \quad \bar{R} = \text{Specific gas constant}$$

$$= \frac{\bar{R}}{M_m} \text{ kJ/kg-K}$$

$$\mathbf{R = \frac{\bar{R}}{M_m}} \text{ where } \bar{R} = 8314.3 \frac{\text{N-m}}{\text{Kmol-K}} \text{ Universal gas constant}$$

7.7 INTERNAL ENERGY, ENTHALPY, SPECIFIC HEATS AND ENTROPY OF MIXTURE OF IDEAL GASES

According to the Gibbs–Dalton Law, an extensive property of a mixture of perfect gases is the sum of the contributions of the individual constituents. Properties of a mixture can be expressed on either a mass or a mole basis and are referred to as per kg or K mol, respectively. Expressions for some of the extensive properties are presented here on a mass basis. The internal energy per unit mass of a mixture is given by

$$u = \frac{m_a U_a + m_b U_b + m_c U_c}{(m_m = 1)}$$

where m_a, m_b and m_c are the masses of gases *a*, *b* and *c* in a mixture of 1 kg and U_a, U_b and U_c are the internal energies of gases *a*, *b* and *c* in a mixture of 1 kg.

We know that internal energy is a function of temperature and for constant volume specific heats of individual gas in mixture, you can calculate the specific heat at constant volume for the mixture is given by

$$\mathbf{(C_v)_m = m_a C_{v_a} + m_b C_{v_b} + m_c C_{v_c} \text{ kJ/kg-K}}$$

on the same lines we can get the specific heat at constant pressure of a mixture containing gases *a*, *b*, and *c*.

$$\mathbf{(C_P)_m = m_a C_{p_a} + m_b C_{p_b} + m_c C_{p_c}}$$

and $$\mathbf{h_m = m_a h_a + m_b h_b + m_c h_c \text{ kJ/kg}}$$

Similarly, the entropy of the mixture is

$$S_m = m_a S_a + m_b S_b + m_c S_c \text{ kJ/kg-K}$$

7.8 PROCESSES OF GASEOUS MIXTURE

There are two basic mixing processes for ideal gases:

1. Mixing process when temperature of each gas is same. Consider a system of gases as shown in Fig. 7.3 (a) and (b).

Let the system be ideally insulated so that the mixing of two gases is *adiabatic mixing.*

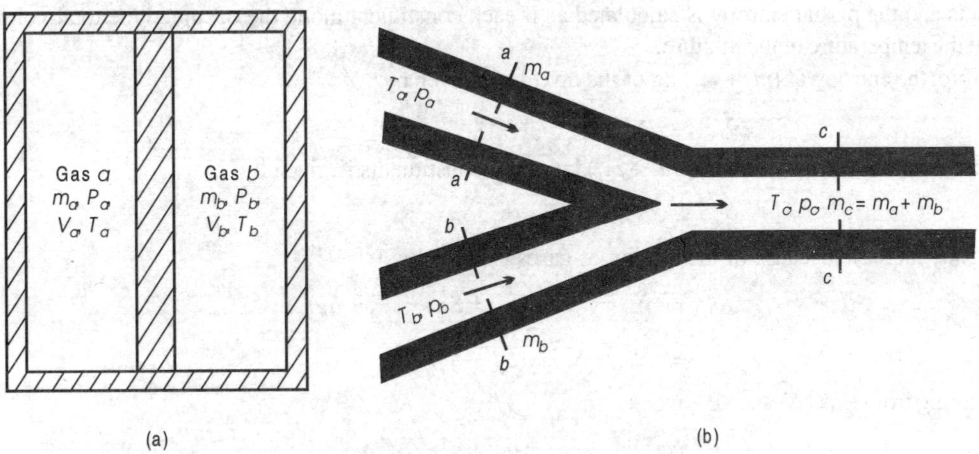

Fig. 7.3. *Adiabatic mixing.*

Let there are two gases *a* and *b* having the states as p_a, T_a, m_a and p_b, T_b, m_b as shown in Fig.7.3, Here, $T_a = T_b$. These states are shown on *T–s* diagram in Fig 7.4.

Let $p_a > p_b$ so that the pressure of the mixture is p_m and $p_a > p_m$ and $p_m > p_b$ and $t_m = t_a = t_b$.

After mixing, entropy of gas (*a*) increases and entropy of gas (*b*) decreases. Let the state of the mixture be given by the point *m*.

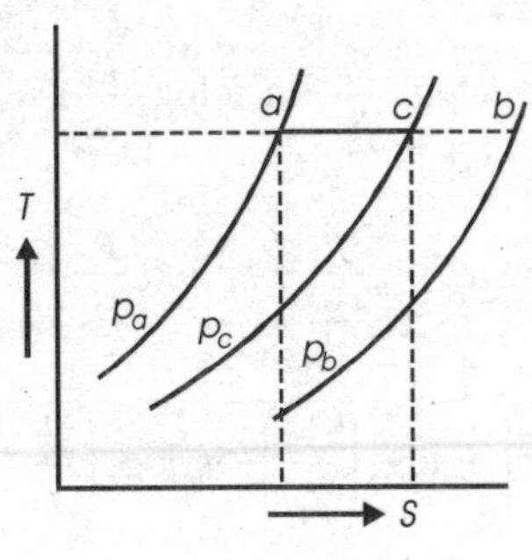

Fig. 7.4

∴ For gas *a*

$$S_a - S_{c_a} = m_a R_a \log \frac{p_a}{p_c}$$

or $$S_{c_a} - S_a = m_a R_a \log \frac{p_c}{p_a} \qquad \text{...(A)}$$

and for gas b $$S_b - S_{c_b} = m_b R_b \log_e \frac{p_b}{p_c}$$

or $$S_{c_b} - S_b = m_b R_b \log_e \frac{p_c}{p_b} \qquad \text{...(B)}$$

Then by Daltons–Gibbs law, we have the entropy of the mixture of gases is the sum of the partial entropies of the constituents and the partial entropy is calculated as if each constituent alone the occupied the total volume of the mixture at the temperature of the mixture.

Therefore, the entropy of $(m_a + m_b)$ kg of the mixture is given by

$$S_{m_2} = S_{c_a} + S_{c_b}$$

Before the mixture, the entropy of $(m_a + m_b)$ kg of the constituents is given by

$$S_{m_1} = S_a + S_b$$

∴ Entropy increase because of the mixing of gases is given by

$$\Delta S = S_{m_2} - S_{m_1} = (S_{c_a} + S_{c_b}) - (S_a + S_b)$$

$$= (S_{c_a} - S_a) + (S_{c_b} - S_b)$$

Substituting from Eqs (A) and (B), we get

$$\Delta S = m_a R_a \log_e \frac{p_c}{p_a} + m_b R_b \log_e \frac{p_c}{p_b}$$

$$= m_a \times \frac{\bar{R}}{M_a} \log_e \frac{p_c}{p_a} + m_b \frac{\bar{R}}{M_b} \log_e \frac{p_c}{p_b}$$

$$= n_a m_a \cdot \frac{\bar{R}}{M_a} \log_e \frac{p_c}{p_a} + n_b M_b \frac{\bar{R}}{M_b} \log_e \frac{p_c}{p_b}$$

$$= n_a \bar{R} \log_e \frac{p_c}{p_a} + n_b \bar{R} \log_e \frac{p_c}{p_b}$$

$$= \bar{R}\left[n_a \log_e \frac{p_c}{p_a} + n_b \log_e \frac{p_c}{p_b} \right]$$

$$= \bar{R}\left[n_a \log_e \left(\frac{1}{f_1} \right) + n_b \log_e \left(\frac{1}{f_2} \right) \right]$$

Thus we observe from this equation that the increase in entropy for a mixture depends only on the number of moles of the gases in the mixture and does not depend on the gases themselves. Mixing of gases is an irreversible process and therefore the entropy increases according to second law of thermodynamics.

Again, we observe from these equations that even if the gases approach to similarity (same gases) but the pressure are different, the change of entropy does not become zero.

2. Different gases at different initial pressures and temperatures. The system considered for this problem may be represented by the two vessels, each containing an ideal gas, at different pressures and temperatures. Let these vessels be connected to each ofter by means of pipe containing a valve as shown in Fig. 7.5.

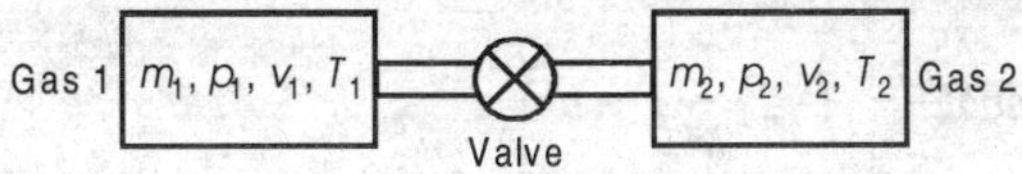

Fig. 7.5

When the valve is open, then the gases will be allowed to diffuse and the pressure and temperatures of the mixture will be in between p_1 and p_2 and T_1 and T_2.

Let m = Mass of the mixture

$$= (m_1 + m_2)$$

Internal energy is a function of the temperature and the total internal energy is the sum of the total internal energies of the individual gases.

$$\therefore \quad mu = m_1u_1 + m_2u_2$$

$$\therefore \quad u = \frac{m_1}{m}\cdot u_1 + \frac{m_2}{m}\cdot u_2$$

$$= m_{f_1}u_1 + m_{f_2}u_2$$

where m_{f_1} = mass fraction of gas 1

and m_{f_2} = mass fraction of gas 2

$$\therefore \quad u = \Sigma m_{f_1}u_1$$

$$= \Sigma\left(m_{f_1}C_{v_i}T_i\right)$$

Now we apply the specific gas equation to the individual gases and the mixture.

$$pV = mRT \text{ (for mixture)}$$

and

$$p_1V_1 = mR_1T_1$$

$$p_1V_2 = m_2R_2T_2$$

or

$$p_1V_1 + p_2V_2 = m_1R_1T_1 + m_2R_2T_2$$

or

$$(p_1 + p_2)\,V = m_1R_1T_1 + m_2R_2T_2$$

and

$$p = p_1 + p_2$$

$$\therefore \quad pV = (p_1 + p_2)\,V = mRT$$

$$= m_1R_1T_1 + m_2R_2T_2$$

$$\therefore \quad T = \text{temperature of the mixture}$$

$$= \frac{m_1R_1T_1 + m_2R_2T_2}{mR}$$

$$= \frac{\Sigma m_1R_iT_1}{mR}$$

But

$$R = \Sigma m_1R_1T_1$$

$$T = \frac{\Sigma m_1R_1T_1}{m\Sigma m_{f_1}R_i}$$

$$= \frac{\Sigma m_1 R_1 T_1}{\Sigma m_i R_i}$$

and
$$mRT = (n_m)\frac{\bar{R}}{M}T = n\bar{R}T$$

$$\therefore \quad nRT = n_1\bar{R}T_1 + n_2\bar{R}T_2$$

$$\therefore \quad T = \frac{n_1}{n}T_1 + \frac{n_2\bar{R}}{n\bar{R}}T_2$$
$$= f_1T_1 + f_2T_2$$
$$= \Sigma f_i T_i$$

Again
$$pV = mRT = m_1R_1T_1 + m_2R_2T_2$$
$$= p_1V_1 + p_2V_2$$

$$\therefore \quad p = \frac{p_1V_1 + p_2V_2}{V}$$

Note here that the final volume is the sum of the individual volumes (of the vessels)

$$\therefore \quad V = V_1 + V_2$$

The procedure followed so far for two gases can be applied to any number of gases.

Again, the change of entropy of each constituent is given by

$$\Delta S = m\left[\int_{T_1}^{T_2} C_p \frac{dT}{T} - R\log_e \frac{p_2}{p_1}\right]$$

In this equation, T_1 and p_1 and T_2 and p_2 are the temperature and pressure before mixing and after mixing. T_2 represents the mixture temperature and p_2 represents the partial pressure of the constituent in the mixture after mixing.

After calculating the changes of entropy of each constituent, addition of all changes gives the change of entropy due to mixing.

SOLVED EXAMPLES

Example 7.1 *If the analysis by volume of a product of combustion is 20% CO_2, 3.8% CO, 70% N_2 and 6.2% O_2, find the analysis by weight.*

Solution

The example can be solved by tabulating the various quantities.

Constituent	*Volume m^3 (a)*	*Mol. Weight (b)*	*Proportional weight (a) × (b)*	*% by weight*
CO_2	0.2	44	$0.2 \times 44 = 8.8$	27.984
CO	0.038	28	$0.038 \times 28 = 1.064$	3.383
N_2	0.7	28	$0.7 \times 28 = 19.6$	62.32
O_2	0.062	32	$0.062 \times 32 = 1.984$	6.31
			$\Sigma c = 31.448$	100

Thus the percentage composition of the products by weight is

$$\begin{aligned} CO_2 &= 27.984 \\ CO &= 3.383 \\ N_2 &= 62.32 \\ O_2 &= 6.31 \\ &= 100 \end{aligned}$$

Example 7.2 *A sample of water-gas contains H_2 = 38%, CO = 23%, CH_4 = 17%, C_2H_6 = 9%, H_2O = 6%, CO_2 = 4% and N_2 = 3%. Convert this to analysis by weight.*

Solution

As in Example 7.1, we will make a table for solving the problem.

S. No.	*Constituents*	*Volume m^3 of gas (a)*	*Mol. weight (b)*	*Proportional weight (a) × (b) = (c)*	*% by weight $\frac{c}{\Sigma c} \times 100$*
1	H_2	0.38	2	0.76	4.6626
2	CO	0.23	28	6.44	39.5100
3	CH_4	0.17	16	2.72	16.6871
4	C_2H_6	0.09	30	2.07	16.5644
5	H_2O	0.06	18	1.08	6.6258
6	CO_2	0.04	44	1.76	10.7975
7	N_2	0.03	28	0.84	5.1534
				$\Sigma c = 16.3$	100.00

∴ Percentage analysis of water gas, by weight, is given as

$$\left[\begin{array}{l} H_2 = 4.6626\%;\ CO = 39.51\%;\ CH_4 = 16.6871\% \\ C_2H_6 = 16.5644\%;\ H_2O - 6.6258\%; CO_2 = 10.7975\% \\ N_2 = 5.1534\% \end{array}\right]$$

Example 7.3 *The composition of a dry flue gas, as obtained by using Orsat apparatus was CO_2 = 10%, CO = 1.8%, O_2 = 7% and N_2 = 81.2% by volume. What will be the percentage composition of the flue gas by weight?*

Solution Table Calculations are as given in Table below

S. No.	*Constituents*	*Volume m^3 of gas (a)*	*Mol. weight (b)*	*Proportional weight (a) × (b) = (c)*	*% by weight $\frac{c}{\Sigma c} \times 100$*
1	CO_2	0.1	44	4.400	14.77
2	CO	0.018	28	0.504	1.687
3	O_2	0.07	32	2.240	7.497
4	N	0.812	28	22.736	76.091
				$\Sigma c = 29.88$	100.00

Example 7.4 *The diesel oil used in an engine consisted of Carbon 82%, Oxygen 7%, Hydrogen 8%, and H_2O 3% by weight. Convert this to analysis by volume.*

Solution

First we can convert weight into volume by dividing weight by the molecular weight of the constituent. We get

$$82 \text{ kg carbon} = \frac{82}{12} = 6.8333 \text{ moles}$$

$$7 \text{ kg oxygen} = \frac{7}{32} = 0.2188 \text{ moles}$$

$$8 \text{ kg hydrogen} = \frac{8}{02} = 4.0000 \text{ moles}$$

$$3 \text{ kg } H_2O = \frac{3}{18} = 0.1667 \text{ moles}$$

$\therefore$ **Total volume = 11.2188 moles**

$\therefore$ % analysis by volume is

$$\% \text{ carbon} = \frac{6.8333}{11.2188} \times 100$$

$$= \mathbf{60.9\ \%}$$

$$\% \text{ oxygen} = \frac{0.2188}{11.2188} \times 100$$

$$= \mathbf{1.95\ \%}$$

$$\% \text{ hydrogen} = \frac{4.0}{11.2188} \times 100$$

$$= \mathbf{35.65\ \%}$$

$$\% H_2O = \frac{0.1667}{11.2188} \times 100$$

$$= \mathbf{1.486\ \%}$$

Example 7.5 *In a boiler trial, the products of combustion were found to contain*

Carbon dioxide = 3.227 kg
Sulphur dioxide = 0.010 kg
Oxygen = 1.296 kg and
Nitrogen = 13.005 kg

Estimate the percentage analysis of the dry flue: (a) by weight and (b) by volume.

Solution

(a) Percentage analysis by weight. From the given data, the total weight of the dry products of combustion is

$$\Sigma p = 3.277 + 0.010 + 1.296 + 13.005$$

$$= 17.538$$

$\therefore$ Percentage analysis by weight is given as

$$\text{Carbon dioxide } \frac{3.277}{17.538} \times 100 = \mathbf{18.68\ \%}$$

Sulphur dioxide $\frac{0.01}{17.538} \times 100$ = **0.057 %**

Oxygen $\frac{1.296}{17.538} \times 100$ = **7.389 %**

Nitrogen $\frac{13.005}{17.538} \times 100$ = **74.15 %**

Total = **100.076**

Now we can convert these weight into moles, by dividing by molecular weights of the particular gases and we, then, get the relative volumes.

∴ Relative volumes of the gases are

Carbon dioxide $\frac{18.68}{44} = 0.4245$

Sulphur dioxide $\frac{0.057}{64} = 0.0009$

Oxygen $\frac{7.389}{32} = 0.2309$

Nitrogen $\frac{74.15}{28} = 2.6482$

$\Sigma = 3.3045$

∴ Percentages of volumes of the products of combustion are

Carbon dioxide $= \frac{0.4245}{3.3045} \times 100 = 12.85$

Sulphur dioxide $= \frac{0.0009}{3.3045} \times 100 = 0.027$

Oxygen $= \frac{0.2309}{3.3045} \times 100 = 6.099$

Nitrogen $= \frac{2.6482}{3.3045} \times 100 = 80.14$

Total = 99.116

Example 7.6 *Air contains 21% of O_2 and 79 % of N_2 by volume. Determine the molecular weight of air and specific gas constant.*

Solution

Molecular weight of O_2 = 32

Molecular weight of N_2 = 28

∴ Relative weights of oxygen and nitrogen in the mixture are

Oxygen $\rightarrow 21 \times 32 = 672$ kg

Nitrogen $\rightarrow 79 \times 28 = 2212$ kg

∴ Total weight of mixture = 672 + 2212 = **2884 kg**

Total number of moles are given by

$$n = n_1 + n_2 = 21 + 79 = 100$$

∴ Molecular weight of air is given by

$$nM = n_1M_1 + n_2M_2$$

$$100 \times M = 2884$$

or M = Molecule weight of air

$$= \frac{2884}{100}$$

$$= \mathbf{28.84\ kg/Kmol}$$

We know that

$$mR = \bar{R}$$

or R = specific gas constant

$$= \frac{\bar{R}}{M} = \frac{8314.3}{28.84}$$

$$= \mathbf{288.29\ N\text{-}m/kg\text{-}K}$$

Example 7.7 *10 kg of oxygen, 5 kg of* N_2 *and 15 kg of* CO_2 *are contained in a vessel where the total pressure is 300 kPa and temperature of 100°C. Calculate (a) partial pressure of each constituent in the vessel (b) the volume of the vessel and (c) if the temperature is increased to 150° C, find the total pressure.*

Solution

(a) We know that mole fraction is equal to the ratio of partial pressure to total pressure, therefore first find the mole fractions of the individual constituent of the mixture

∴ Moles of O_2 present $= \frac{10}{32} = 0.3125$

Moles of N_2 present $= \frac{5}{28} = 0.1786$

Moles of CO_2 present $= \frac{15}{44} = 0.3409$

∴ Total numbers of moles $= 0.3125 + 0.1786 + 0.3409$

$$= \mathbf{0.8320}$$

∴ Partial pressure of $O_2 = \frac{0.3125}{0.8320} \times 300$

$$= \mathbf{112.68\ kPa}$$

Similarly, partial pressure of $N_2 = \frac{0.1786}{0.8320} \times 300$

$$= \mathbf{64.399\ kPa}$$

and partial pressure of $CO_2 = \frac{0.3409}{0.8320} \times 300$

$$= \mathbf{122.92\ kPa}$$

(b) Volume of the vessel. For the mixture, we can write

$$pV = mR_m T$$

where
$$R_m = \frac{\bar{R}}{M_m}$$

where
$$M_m = \text{molecular weight of the mixture}$$

$$= \frac{\text{Total mass of mixture}}{\text{Total number of moles}}$$

$$= \frac{10+5+15}{0.8320} = \frac{30}{0.8320}$$

$$= \mathbf{36.06}$$

∴ Specific gas constant for the mixture

$$R_m = \frac{0.8320}{36.06}$$

$$= \mathbf{230.568}$$

$$V = \frac{mR_m T}{p}$$

$$= \frac{30 \times 230.568 \times 373}{300 \times 10^3}$$

$$= \mathbf{7.609\ m^3}$$

(c) Pressure after heating. Heating is at constant volume, and therefore, we have

$$\frac{p_1}{T_1} = \frac{p_2}{T_2}$$

where
$$p_1 = 300 \text{ kPa}$$
$$T_1 = 373 \text{ K}$$
$$T_2 = 150 + 273 = 423 \text{ K}$$

∴
$$p_2 = \frac{T_2}{T_1} \times p_1 = \frac{423}{373} \times 300$$

$$= \mathbf{340.21\ kPa}$$

Example 7.8 *A perfect gas mixture consists of 3 kg of nitrogen and 5 kg of carbon dioxide at a pressure of 3 bar and a temperature of 20°C. Find: (a) the mole fraction of each constituent, (b) the equivalent molecular weight of the mixture, (c) the equivalent gas constant of the mixture, (d) the partial pressures and the partial volumes, and (e) the volume and density of the mixture.*

Solution

(a) Mole fraction

Moles of nitrogen present in the mixture

$$n_{N_2} = \frac{\text{Mass}}{\text{Molecular weight}}$$

$$= \frac{3}{28}$$

$$= \mathbf{0.1071}$$

Similarly $\quad n_{CO_2}$ = Moles of carbon dioxide

$$= \frac{5}{44}$$

$$= \mathbf{0.1136}$$

$\therefore$ Total moles in the mixture

$$= 0.1071 - 0.1136$$

$$= \mathbf{0.2207}$$

$$\text{Mole fraction of } N_2 = \frac{0.1071}{0.2207}$$

$$= \mathbf{0.4853}$$

$$\text{Mole fraction of } CO_2 = \frac{0.1136}{0.2207}$$

$$= \mathbf{0.5147}$$

(b) Equivalent molecular weight of mixture

$$\text{Total moles} = 0.2207$$

$$\text{Total mass} = 3 + 5$$

$$= 8 \text{ kg}$$

$$\text{Molecular weight} = \frac{\text{Total mass}}{\text{Total number}}$$

$$= \frac{8}{0.2207}$$

$$= \mathbf{36.2483 \text{ kg m/K mol.}}$$

(c) Equivalent gas constant of mixture

$$\text{We know that } \bar{R} = MR$$

where $\bar{R}$ is the universal gas constant

$$R = \frac{\bar{R}}{M}$$

$$= \frac{8314.3}{36.2483}$$

$$= \mathbf{229.37 \text{ N-m/kg-K}}$$

(d) Partial pressures and partial volumes

$$\text{Partial pressure of } N_2 = 3 \times 0.4853$$

$$= \mathbf{1.4559 \text{ bar}}$$

$$\text{Partial pressure of } CO_2 = 3 \times 0.5147$$

$$= \mathbf{1.5441 \text{ bar}}$$

For partial volume we write specific gas equation for the gases.

$$\therefore \quad V_N = \frac{(mRT)_{N_2}}{p}$$

$$= \frac{3\times\frac{8314.3}{28}\times 293}{3\times 10^5}$$

$$= \mathbf{0.87\ m^3\ cu.m}$$

Similarly $$V_{CO_2} = \frac{5\times\frac{8314.3}{28}\times 293}{3\times 10^5}$$

$$= \mathbf{0.9228\ m^3}$$

(e) Volume and density of mixture

Volume of the mixture is given by

$$v = \left(\frac{mRT}{p}\right)_{mix}$$

$$= \frac{8\times 229.37\times 293}{3\times 10^5}$$

$$= \mathbf{1.7921\ m^3}$$

Density of mixture $$\delta_m = \frac{m_m}{V_m}$$

$$= \frac{8}{1.7921}$$

$$= \mathbf{4.467\ kg/m^3}$$

Example 7.9 *For the mixture of Example 7.8, calculate the constant volume and constant pressure specific heats. If the mixture is heated at constant volume to 35° C, find the change in internal energy, enthalpy and entropy of the mixture. Assume the specific heats remain constant and the perfect gas laws apply.*

Also, find the foregoing changes, if the heating process is preformed at constant pressure.

Solution

Constant volume process :

Taking C_v for N_2 = 0.741 kJ/kg-K

and C_v for CO_2 = 0.6615 kJ/kg-K

∴ C_v for the mixture is given by

$$c_{v_m} = \frac{3}{8}\times 0.7411 + \frac{5}{8}\times 0.6615$$

$$= 0.2779 + 0.4134$$

$$= \mathbf{0.6913\ kJ/kg\text{-}K}$$

Again taking C_p for N_2 = 1.0384 kg-K

and for CO_2 = 0.85 kJ/kg-K

C_{p_m} = specific heat of the mixture

$$= \frac{3}{8}\times 1.0384 + \frac{5}{8}\times 0.85$$

$$= 0.3894+0.5312$$
$$= \mathbf{0.9206\ kJ/kg\text{-}K}$$

For the mixture
Change in internal energy

$$(\Delta u)_m = C_v \Delta T$$
$$= 0.6913 \times (35-20)$$
$$= \mathbf{10.3695\ kJ/kg\text{-}K}$$

Total change in internal energy

$$\Delta U = 8 \times 10.3695$$
$$= \mathbf{82.956\ kJ/kg\text{-}K}$$

Change in enthalpy of the mixture

$$\Delta H = m_m C_{p_m} \Delta T$$
$$= 8 \times 0.9206 \times (35-20)$$
$$= 110.472\ \text{kJ}$$

The change in entropy of the mixture

$$\Delta S = m_m C_{pm} \log_e \frac{T_2}{T_1}$$
$$= 8 \times 0.6913 \times \log_e \left(\frac{308}{293} = 1.0512 \right)$$
$$= \mathbf{0.2761\ kJ/K}$$

Constant Pressure Process. As internal energy and enthalpy are the functions of temperatures only, the change in internal energy and change in enthalpy are the same as in constant volume process.

$$\mathbf{\Delta W = 19.812\ kJ}$$
$$\mathbf{\Delta H = 26.4\ kJ}$$

$$\Delta S = m_m C_{p_m} \log_e \frac{T_2}{T_1}$$
$$= 8 \times 0.9206 \log_e \frac{308}{293}$$
$$= \mathbf{0.3677\ kJ/K}$$

Example 7.10 *Find the increase in entropy when 2 kg of oxygen at 63°C are mixed with 6 kg of nitrogen at the same temperature. The initial pressure of each constituent is 1 bar and is the same as that of the mixture. Assume perfect gases.*

Solution

Since the mixing process is isothermal the change of entropy of the mixture is

$$\Delta S = -m_{O_2} R_{O_2} \log_e \frac{p_{O_2}}{P} - m_{N_2} R_{N_2} \log_e \frac{p_{N_2}}{P}$$

where

$$\frac{p_{O_2}}{p} = \frac{2/32}{\frac{2}{32}+\frac{6}{28}} = \frac{0.0625}{0.0625+0.2143}$$

$$= \frac{0.0625}{0.2768}$$
$$= \mathbf{0.2258}$$

Similarly

$$\frac{p_{N_2}}{p} = 1 - 0.2258$$
$$= \mathbf{0.7742}$$

$$\therefore \quad \Delta S = -2\times\frac{8314.3}{32}\log_e 0.2258$$
$$= -6\times\frac{8314.3}{28}\log_e 0.7742$$
$$= 773.285 + 455.965$$
$$= \mathbf{1229.250\ N\text{-}m/K}$$
$$= \mathbf{1.22925\ kJ/K}$$

Example 7.11 *The number of moles, the pressures and the temperatures of gases a, b, c of figure given below are given in the following table:*

Gas	*Number of moles*	*Pressure bar*	*Temperature °C*
N_2	*1*	*3*	*90*
CO	*3*	*4*	*200*
O_2	*2*	*6*	*317*

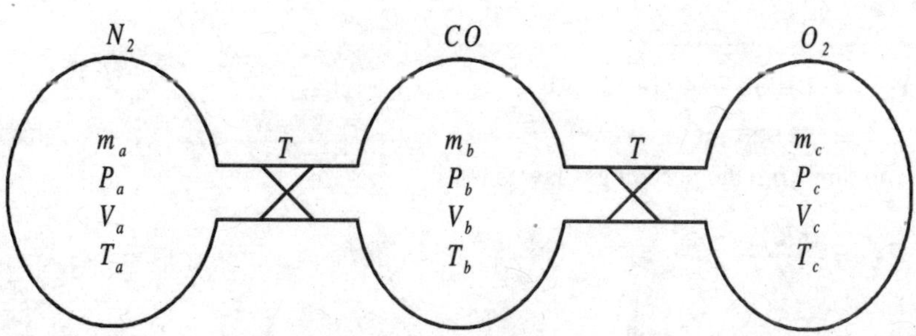

If the containers are connected, allowing the gases to mix freely, find (a) the pressure and temperature of the resulting mixture at equilibrium, (b) the change of entropy of each constituent and that of the mixture.

Solution

(a) Temperature and pressure of the mixture

Temperature of the mixture is given by

$$T_m = \frac{n_a}{n}T_a + \frac{n_b}{n}T_b + \frac{n_c}{n}T_c$$

$$= \frac{1}{6}\times 363 + \frac{3}{6}\times 473 + \frac{2}{6}\times 590$$

$$= 60.5236.5 + 196.7$$

$$= \mathbf{493.7\ K}$$

Also,

$$V_{N_2} = \frac{n_{n_2}\bar{R}T_{N_2}}{p_{N_2}}$$

$$= \frac{1\times 8314.3\times 363}{3\times 10^5}$$

$$= \mathbf{10.06\ m^3}$$

$$V_{CO} = \frac{n_{CO}\bar{R}T_{CO}}{P_{CO}}$$

$$= \frac{3\times 8314.3\times 473}{4\times 10^5}$$

$$= \mathbf{29.495\ m^3}$$

$$V_{O_2} = \frac{n_{O_2}\bar{R}\,T_{O_2}}{p_{O_2}}$$

$$= \frac{2\times 8314.3\times 590}{6\times 10^5}$$

$$= \mathbf{16.35\ m^3}$$

∴ Total volume

$$V_m = 10.06 + 29.495 + 16.350$$

$$= \mathbf{55.905\ m^3}$$

∴ Pressure of the mixture from the perfect gas law is

$$p_m = \frac{n\bar{R}T_m}{V_m}$$

$$= \frac{6\times 8314.3\times 493.7}{55.905\times 10^5}$$

$$= \mathbf{4.405\ bar}$$

(b) Change of Entropy. The partial pressures of the constituents are

$$p_{N_2} = \frac{1}{6}\times 4.405 = \mathbf{0.7341\ bar}$$

$$p_{CO} = \frac{3}{6}\times 4.405 = \mathbf{2.2025\ bar}$$

and $p_{O_2} = \frac{2}{6} \times 4.405 = \mathbf{1.4683\ bar}$

$$\Delta s_{N_2} = \left(\int_{t_1}^{t_2} C_p \frac{d_T}{T} - R \log_e \frac{p_2}{p_1} \right)_{N_2}$$

$$= 1.0384 \log_e \frac{493.7}{363} - \frac{8314.3}{28} \log_e \frac{0.7341}{3}$$

$$= 1.0384 \log_e 1.36 - 296.94 \log_e 0.2447$$

$$= 1.0384 \times 0.3075 - \frac{296.94}{1000} \times -1.4077$$

$$= \mathbf{0.3193 + 0.418}$$

$$= \mathbf{0.7373\ kJ/kg\text{-}K}$$

$$= \mathbf{0.7373 \times 28}$$

$$= \mathbf{20.644\ kJ/kg}$$

$$\Delta S_{CO} = 1.0384 \log_e \frac{493.7}{473} - \frac{8314.3}{28 \times 1000} \log_e \frac{2.2025}{4}$$

$$= 1.0384 \times 0.0429 + \frac{296.94 \log_e \times 0.309}{1000}$$

$$= 0.04455 + \frac{91.7545}{1000}$$

$$= 0.04455 + 0.0917545$$

$$= \mathbf{0.1353045\ kJ/kg\text{-}K}$$

$$= \mathbf{0.135045 \times 84}$$

$$= \mathbf{11.3652\ kJ/K}$$

$$\Delta S_{O_2} = 0.9165 \log_e \frac{493.7}{590} - \frac{8.314}{32} \log_e \frac{1.4683}{6}$$

$$= 0.9169 \times \log_e 0.8368 - 0.2598 \log_e 0.2477$$

$$= 0.9169 + 0.1782 + 0.2598 \times 1.3955$$

$$= -0.1634 + 0.3625$$

$$= \mathbf{0.1991\ kJ/kg\text{-}K}$$

$$= \mathbf{0.1991 \times 64}$$

$$= \mathbf{12.7466\ kJ/K}$$

$\therefore$ $$\Delta S_{mix} = 20.644 + 11.3652 + 12.7466$$

$$= \mathbf{44.7558\ kJ/K}$$

THEORY QUESTIONS

7.1. What is partial pressure of a gas in a mixture of gases ?

7.2. Define the terms — mass fraction and mole fraction of the gas in a mixture.

7.3. Give the procedure to find the gravimetric analysis of the mixture of gases and to convert this gravimetric analysis into volumetric analysis and vice-versa.

7.4. Explain Dalton's Law of partial pressures.

7.5. What is Amaga's Law and its application.

7.6. How will you find the molecular weight and gas constant for mixture of number of ideal non-reactive gases?

7.7. What is Gibbs–Dalton Law and its use?

PROBLEMS FOR PRACTICE

7.1. A mixture of perfect gases at 20°C has the following composition by volume :

$N_2 = 55\%, O_2 = 20\%, CH_4 = 25\%$

If the partial pressure of CH_4 = 0.5172 bar, determine:

(a) The partial pressures of N_2 and O_2

(b) The mass proportions of the mixture

(c) R_{mix} and volume per mole of mixture.

7.2. Five moles of perfect gas mixture has the following mass analysis.

$N_2 = 45\%$, He = 27%, and $C_6H_6 = 28\%$

Find

(a) the analysis by volume and the number of moles of each constituents

(b) The molecular weight of the mixture

(c) The volume of the mixture at 3 bar and 20° C

7.3. The analysis by weight of a perfect gas mixture at 20°C and 1.5 bar is 10% O_2, 70% N_2, 15% CO_2, and 5% CO. For a reference state of –15°C and atmospheric pressure, determine

(a) Partial pressures of the constituents

(b) Gas constant of the mixture

(c) Enthalpy and internal energy of the mixture

(d) Constant pressure specific heat of the mixture

(e) Entropy of the mixture.

7.4. A rigid vessel has a volume of 0.5 cu-m and contains air at bar and 20°C, Helium at 20°C is added to the vessel with the pressure is raised to 1.5 bar. Compute,

(a) the partial pressures of the final components of the mixture composed of O_2, N_2 and He.

(b) the masses of the components of the mixture.

(c) the gas constant of the mixture.

7.5. The dry products of combustion of a coal have the following volumetric composition: CO_2 11%, O_2 3.5%, N_2 85.5% . Calculate (a) the gravimetric analysis (b) the equivalent molecular weight (c) the mixture's gas constant (d) the partial pressures of each constituents for p_m = 1 bar. (e) the specific heat at constant pressure for the mixture (f) the molar specific heat using the result in (e).

CHAPTER

8 General Thermodynamic Relations

8.1 INTRODUCTION

We have studied the six thermodynamic properties of a system so far. These are, pressure (P), volume (V), temperature (T), internal energy (U), enthalpy (H) and entropy (S). In addition to the above, two more important properties, viz. Helmholtz function (A) and Gibbs function (G) will be discussed in this chapter.

Of the above eight properties, only three i.e., P, V and T can be directly measured and hence it becomes necessary to develop thermodynamic relations to express the other properties in terms of these. These combinations of properties are called 'thermodynamic gradients' and are defined as the rate of change of one property with another while a third is kept constant.

8.2 HELMHOLTZ FUNCTION OR WORK FUNCTION

Helmholtz function (A) is a combination of properties and is mathematically expressed as

$$A = U - TS \quad \text{...(1)}$$

For unit mass, specific Helmholtz function is given by,

$$a = u - TS$$

where,

u = Specific internal energy, and

s = Specific entropy

From the fist law, for a closed system, the work dW is given by

$$\delta W = \delta Q - du \quad \text{...(2)}$$

Thus, the energy converted to work is provided in part by the system, whose internal energy decreases by $-du$, and in part by the heat reservoirs with which the system is in contact and which give up a quantity of heat dQ.

The expressions for maximum amount of work can be derived when a system undergoes a process between two equilibrium states. It is assumed that system exchanges heat only with single heat reservoirs at temperature T_o.

From the principle of increase of entropy, we know that

$$dS + dS_o \geq 0 \quad \text{...(3)}$$

where,

dS = Increase in entropy of system.

dS_o = Increase in entropy of reservoir.

Now, heat dQ is absorbed by the system from the reservoir.

Hence, $$dS_0 = \frac{-dQ}{T_o} \quad \text{(Reservoir has given out heat. } \therefore dQ \text{ is –ve)}$$

Substituting above value of dS_o in Eq. (3)

$$dS_o - \frac{dQ}{T_o} \geq 0$$

$$\therefore \quad T_o dS \geq dQ \qquad \text{...(4)}$$

From Eqs (2) and (4), we get

$$dW \leq T_o dS - du \qquad \text{...(5)}$$

For a finite process between two equilibrium states, this equation can be integrated to give,

$$\int_1^2 dW \leq T_o \int_1^2 dS - \int_1^2 du$$

$$\therefore \quad W_{1-2} \leq T_o (S_2 - S_1) - (u_2 - u_1)$$

$$\therefore \quad W_{1-2} \leq (u_2 - u_1) - T_o (S_2 - S_1) \qquad \text{...(6)}$$

For a special case, where initial and final temperatures are equal and same as that of the reservoir, Eq. (6) can be written as,

$$(W_T)_{1-2} \leq (u_1 - u_2)_T - T (S_1 - S_2)_T \qquad \text{...(7)}$$

For a small change, the Helmholtz function can be written as,

$$dA = du - d(TS) \qquad \text{...(8)}$$

Integrating the above equation between two equilibrium states considering temperature (T) as constant, we get,

$$\int_1^2 dA = \int_1^2 du - T \int_1^2 dS$$

$$\therefore \quad (A_2 - A_1)_T = (u_2 - u_1) - T (S_2 - S_1)_T$$

$$\therefore \quad (A_1 - A_2)_T = (u_1 - u_2)_T - T (S_1 - S_2)_T \qquad \text{...(9)}$$

Therefore, from Eqs (7) and (9), we have,

$$(W_T)_{1-2} \leq (A_1 - A_2)_T \qquad \text{...(10)}$$

Thus, the above equation implies that in any process between two quilibrium states at the same temperature during which the sytem exchanges heat only with the reservoir, the work done is equal to or less than the decrease in Helmholtz function of the system during the process.

The maximum work is done when the process is reversible and it is less than maximum for irreversible process.

$$(W_T)_{max} = (W_T)_{rev} = (A_1 - A_2)_T$$

8.3 GIBBS FUNCTION (THERMODYNAMIC POTENTIAL)

Gibbs function (G) is a property of a system, which is given as

$$G = U - TS + PV$$

$$= (U + Pv) - TS$$

$$\therefore \quad G = H - TS$$

where,

H = Enthalpy of system

T = Temperature of system

S = Entropy of system

For unit mass, we have

$$g = h - TS$$

Consider a system that can do work in other form e.g.: electrical, magnetic works etc., in addition to PdV

Let

$$\delta W = \text{Work done by the system}$$

$$\delta W_o = \text{Form of work other than } PdV$$

We get $\delta W = Pdv + \delta W_o$

Let P = external pressure (P_o) = Constant

We now integrate the above equation from state 1 to state 2.

$$\therefore \quad \int_1^2 \delta W = P_o \int_1^2 dv + \int_1^2 \delta W_o$$

$$\therefore \quad W_{1-2} = P_o (v_2 - v_1) + W_o$$

Here, we may be electrical work (EdI) for a voltaic cell, or may be magnetic work (mdH) in case of a magnetic material.

The work done by a system exchanging heat with a reservoir at temperature T_o satisfies the following condition:

$$W_{1-2} \leq (U_1 - U_2) - T_o (S_1 - S_2) \quad \text{...(2)}$$

From Eqs (1) and (2), we get

$$P_o (v_2 - v_1) + W_o \leq (U_1 - U_2) - T_o (S_1 - S_2)$$

$$\therefore \quad W_o \leq (U_1 - U_2) - T_o (S_1 - S_2) - P_o (v_2 - v_1) \quad \text{...(3)}$$

When the initial and final equilibrium states are at same temperature and pressure, then we get

$$T_o = T \text{ and } P_o = P$$

From Eq. (3) we have $(W_o)_{P,T} \leq (U_1 - U_2)_{P,T} - T (S_1 - S_2) + P (v_1 - v_2)_{P,T}$ ...(4)

We have,

Gibbs function, $G = U - Ts + Pv$

Writing in differential form, we get

$$dG = dU - d(TS) + d(Pv)$$

Keeping P and T constant and integrating above equation from state 1 to state 2, we get:

$$\int_1^2 dG = \int_1^2 dU - T \int_1^2 dS + p \int_1^2 dV$$

$$\therefore \quad (G_2 - G_1) = (U_2 - U_1) - T (S_2 - S_1) - P (v_2 - v_1)$$

i.e., $$(G_1 - G_2)_{p,T} = (U_2 - U_1)_{p,T} - T (S_1 - S_2)_{p,T} + P (v_2 - v_1)_{p,T} \quad \text{...(5)}$$

From Eqs (4) and (5) we get,

$$(W_o)_{p,T} \leq (G_1 - G_2)_{p,T} \quad \text{...(6)}$$

It is observed that the difference between Gibbs function of a system between two equilibrium states sets the maximum limit to the work that can be performed by the system in addition to the *Pdv* work. This is observed when the two equilibrium states are at the same temperature and pressure and when the system exchanges heat with a single heart reservoir.

The work done is maximum when the process is reversible and is less for an irreversible process.

$\therefore$ For maximum work, we have

$$\left[(W_o)_{P,T}\right]_{\max} = \left[(W_o)_{P,T}\right]_{\text{rev}} = (G_1 - G_2)_{P,T}$$

Note. Gibbs function is also known as Free Energy Function.

8.4 MATHEMATICAL FUNDAMENTALS

We will now discuss briefly certain mathematical manipulations which are repeatedly useful and important from the point of view of the subject.

(*a*) *Exact Differential*: Let,

$$Z = f(x, y) \quad \ldots(1)$$

where x, y, z represent any three properties of a system.

Then,

$$dz = \left(\frac{\partial z}{\partial x}\right)_y dx + \left(\frac{\partial z}{\partial y}\right)_x dy \quad \ldots(2)$$

where,

$\left(\frac{\partial z}{\partial x}\right)_y$ = Partial derivative of z with respect to x, keeping y as constant.

$\left(\frac{\partial z}{\partial y}\right)_x$ = Partial derivative of z with respect to y, keeping x constant.

The differential dz in Eq. (2) must be an exact differential (i.e., a point function), and an equation relating point functions must meet the test for exactness. This test is that the magnitude of second mixed partial derivatives of a function are independent of order of differentiation and hence equal.

i.e.,

$$\frac{\partial^2 z}{\partial x \partial y} = \frac{\partial^2 z}{\partial y \partial x} \quad \ldots(3)$$

Now, Eq. (2) can also be written as,

$$dz = Mdx + Ndy$$

where $M = \left(\frac{\partial z}{\partial x}\right)_y$ and $N = \left(\frac{\partial z}{\partial y}\right)_x$

$$\therefore \quad \left(\frac{\partial M}{\partial y}\right)_x = \frac{\partial^2 z}{\partial x \partial y} \text{ and } \left(\frac{\partial N}{\partial x}\right)_y = \frac{\partial^2 z}{\partial x \partial y}$$

Hence, from Eq. (3), we have

$$\left(\frac{\partial M}{\partial y}\right)_x = \left(\frac{\partial N}{\partial x}\right)_y$$

(*b*) *Cyclic Relations*. We know that,

$$Z = f(x, y) \quad \ldots(1)$$

Since Z is a function of two independent variables x and y, x can similarly be expressed in terms of the other two variables i.e., y and z

$$\therefore \quad x = f(y, z) \quad \ldots(2)$$

Similarly, we can write

$$y = f(x, z) \quad \ldots(3)$$

According to the condition of exact differentials, we can write,

$$ux = \left(\frac{\partial x}{\partial y}\right)_z dy + \left(\frac{\partial x}{\partial z}\right)_y dz \quad \ldots(4)$$

and $$dy = \left(\frac{\partial y}{\partial x}\right)_z dx + \left(\frac{\partial y}{\partial z}\right)_z dz \qquad \ldots(5)$$

Substituting the value of dy from Eq. (5) in Eq. (4), we get,

$$dx = \left(\frac{\partial y}{\partial z}\right)_y dz + \left(\frac{\partial x}{\partial y}\right)_z \left[\left(\frac{\partial y}{\partial x}\right)_z dx + \left(\frac{\partial y}{\partial z}\right)_x dz\right]$$

$$= \left(\frac{\partial x}{\partial z}\right)_y dz + \left(\frac{\partial x}{\partial y}\right)_z \left(\frac{\partial y}{\partial x}\right)_z dx + \left(\frac{\partial x}{\partial y}\right)_z \left(\frac{\partial y}{\partial z}\right)_x dz$$

$$= \left[\left(\frac{\partial x}{\partial z}\right)_y + \left(\frac{\partial x}{\partial y}\right)_z \left(\frac{\partial y}{\partial z}\right)_x\right] dz + dx$$

$$\therefore \quad = \left(\frac{\partial x}{\partial y}\right)_z \left(\frac{\partial y}{\partial z}\right)_x + \left(\frac{\partial x}{\partial z}\right)_y = 0$$

$$\therefore \quad \left(\frac{\partial x}{\partial y}\right)_z \left(\frac{\partial y}{\partial z}\right)_x = -\left(\frac{\partial x}{\partial z}\right)_y$$

$$\therefore \quad \left(\frac{\partial x}{\partial y}\right)_z \left(\frac{\partial y}{\partial z}\right)_x \left(\frac{\partial z}{\partial x}\right)_y = -1$$

The above relation is called 'cyclic relation'. We can write the cyclic relation for thermodynamic properties P, V and T as

$$\left(\frac{\partial P}{\partial V}\right)_T \left(\frac{\partial V}{\partial T}\right)_P \left(\frac{\partial T}{\partial P}\right)_V = -1$$

8.5 MAXWELL'S EQUATIONS

The Maxwell relations are derived by applying the exactness test to the functions for the various properties:

(*a*) *Internal Energy*. General gas equation gives us

$$du = dQ - dW$$
$$= TdS - PdV$$

The above equation is in the form

$$dz = Mdx + Ndy$$

where $M = T, N = -P, x = S, y = V$

Hence, we can write,

$$\left(\frac{\partial T}{\partial V}\right)_S = \left(\frac{\partial P}{\partial S}\right)_V \qquad \ldots(1)$$

(*b*) *Enthalpy*. We know that,

$$dH = dU + d(pV)$$

$$\therefore \quad dH = du + PdV + VdP$$

$$= (TdS - PdV) + PdV + VdP$$

$$\therefore \quad dH = TdS + VdP$$

In the above expression, $M = T, N = V, x = S, y = P$

Hence, we can write

$$\left(\frac{\partial T}{\partial P}\right)_S = \left(\frac{\partial V}{\partial S}\right)_T \qquad \ldots(2)$$

(*c*) *Helmholtz function* (*A*). As discussed earlier,

$$dA = dU - d\,(TS)$$

$$= dU - Tds - SdT$$

$$= TdS - PdV - TdS - SdT$$

$$= -PdV - SdT$$

Thus, $M = -P, N = -S, x = V, y = T$

$$\therefore \quad -\left(\frac{\partial P}{\partial T}\right)_V = -\left(\frac{\partial S}{\partial V}\right)_T$$

i.e.

$$\left(\frac{\partial P}{\partial T}\right)_V = \left(\frac{\partial S}{\partial V}\right)_T \qquad \ldots(3)$$

(*d*) *Gibbs Function*

Now,

$$dG = dH - d(TS)$$

$$= (TdS + Vd) - TdS - SdT$$

$$= vdp - SdT$$

$$\therefore \quad M = V, N = -S, x = P, y = T$$

$$\therefore \quad \left(\frac{\partial V}{\partial T}\right)_P = \left(\frac{\partial S}{\partial P}\right)_T \qquad \ldots(4)$$

The Eqs. (1), (2), (3) and (4) are called Maxwell's equations. They apply to pure substances in process involving heat, work and internal energy.

8.6 CO-EFFICIENT OF VOLUME EXPANSION OR VOLUME EXPANSIVITY (β)

The co-efficient of volume expansion is defined as the change in volume per unit volume for a unit change of temperature at constant pressure.

It is mathematically expressed as,

$$\beta = \frac{1}{V}\left(\frac{\partial V}{\partial T}\right)_p$$

When β is positive, an isothermal increase in pressure results in rejected heat. When β is negative, an isothermal increase in pressure results in heat added.

8.7 ISOTHERMAL COMPRESSIBILITY

The isothermal compressibility or the isothermal co-efficient of compressibility is defined as the change in volume per unit volume for unit change in pressure, at constant temperature.

It is denoted by K and given by,

$$K = -\frac{1}{V}\left(\frac{\partial V}{\partial P}\right)_T$$

Note. The negative sign indicates that volume decreases as pressure increases.

8.8 ADIABATIC COMPRESSIBILITY

Adiabatic compressibility is defined as the change in volume per unit volume for unit change in pressure at constant entropy. It is denoted by K_S and given by the expression.

$$K_S = -\frac{1}{V}\left(\frac{\partial V}{\partial P}\right)_S$$

Note. Negative sign indicates that the volume decreases as pressure increases.

The ratio of isothermal and adiabatic compressibilities is equal to the ratio of specific heats.

i.e.
$$\gamma = \frac{C_P}{C_V} = \frac{K}{K_S} = \frac{-(1/V)(\partial V/\partial P)_T}{-(1/V)(\partial V/\partial P)_S} = \frac{(\partial V/\partial P)_T}{(\partial V/\partial P)_S}$$

Thus,
$$\gamma = \frac{(\partial P/\partial V)_S}{(\partial P/\partial V)_T}$$

As γ is always greater than unity, K is always greater than K_S.

8.9 RATIO OF COEFFICIENT OF VOLUME EXPANSION AND ISOTHERMAL COMPRESSIBILITY

As studied earlier, coefficient of volume expansion is given by,

$$\beta = \frac{1}{V}\left(\frac{\partial V}{\partial T}\right)_P$$

$\therefore$
$$\left(\frac{\partial V}{\partial T}\right)_P = \beta V \qquad \ldots(1)$$

Also, isothermal compressibility,

$$K = -\frac{1}{V}\left(\frac{\partial V}{\partial P}\right)_T$$

i.e.
$$\left(\frac{\partial V}{\partial P}\right)_T = -KV \qquad \ldots(2)$$

By cyclic property of differentials, we get

$$\left(\frac{\partial P}{\partial T}\right)_V\left(\frac{\partial T}{\partial V}\right)_P\left(\frac{\partial V}{\partial P}\right)_T = -1$$

Substituting Eqs (1) and (2) in above equation, we get

$$\left(\frac{\partial P}{\partial T}\right)_V\left(\frac{1}{\beta V}\right)(-KV) = -1$$

$$\therefore \quad \left(\frac{\partial P}{\partial T}\right)_V = \frac{\beta V}{KV} = \frac{\beta}{K}$$

Thus, the ratio of coefficient of volume expansion and isothermal compressibility is equal to the change in pressure with temperature at constant volume.

8.10 GENERAL EQUATION FOR CHANGE IN INTERNAL ENERGY

We will now derive certain equations for change in internal energy, in terms of pressure, volume and temperature, by considering the following conditions:

(a) Pressure (P) and volume (V) are independent

The change in internal energy in this case is given by,

$$dU = \left(\frac{\partial U}{\partial P}\right)_V dP + \left(\frac{\partial U}{\partial V}\right) dV \quad \text{...(1)}$$

Now, we can write

$$\left(\frac{\partial U}{\partial P}\right)_V = \left(\frac{\partial U}{\partial T}\right)_V + \left(\frac{\partial T}{\partial P}\right)_V = mC_V \times \frac{K}{\beta} \left\{\left(\frac{\partial U}{\partial T}\right)_V = mC_V \text{ and } \left(\frac{\partial T}{\partial P}\right)_V = \frac{K}{\beta}\right\} \quad \text{...(2)}$$

Also, we know that, $H = U + PV$ ...(3)

Taking partial derivative of above equation with respect to V keeping P constant, we get

$$\left(\frac{\partial H}{\partial v}\right)_P = \left(\frac{\partial u}{\partial v}\right)_P + \frac{\partial}{\partial v}(PV)$$

$$= \left(\frac{\partial u}{\partial v}\right)_P + P$$

$$\therefore \quad \left(\frac{\partial u}{\partial v}\right)_P = \left(\frac{\partial H}{\partial v}\right)_P - P \quad \text{...(4)}$$

We can write,

$$\left(\frac{\partial H}{\partial V}\right)_P = \left(\frac{\partial H}{\partial T}\right)_P \left(\frac{\partial T}{\partial V}\right)_P = \frac{mC_V}{\beta V} \quad \text{...(5)}$$

$$\left\{\left(\frac{\partial H}{\partial T}\right)_P = mC_P \text{ and } \left(\frac{\partial T}{\partial v}\right)_P = \frac{1}{\beta V}\right\}$$

Substituting Eq. (5) in Eq. (4), we get

$$\left(\frac{\partial U}{\partial V}\right)_P = \frac{mC_V}{\beta V} - P \quad \text{...(6)}$$

Now, substituting Eqs (2) and (6) in Eq. (1), we get,

$$dU = \left[mC_V \times \frac{K}{\beta}\right] dP + \left[\frac{mC_P}{\beta V} - P\right] dV$$

(b) Volume (V) and temperature (T) are independent

Change in internal energy is given by,

$$dU = \left(\frac{\partial U}{\partial V}\right)_T dV + \left(\frac{\partial u}{\partial T}\right)_T dT \quad \text{...(1)}$$

Substituting the value of $\left(\frac{\partial U}{\partial T}\right)_v$, as mC_V, we get

$$dU = \left(\frac{\partial U}{\partial V}\right)_T dV + mC_V dT \quad \text{...(2)}$$

From first law, we have

$$dQ = dU + dW = dU + PdV \quad \text{...(3)}$$

and from second law, we have

$$dQ = TdS \quad \text{...(4)}$$

Thus from Eqs (3) and (4),

$$TdS = dU + PdV \quad \text{...(5)}$$

Now,

$$H = U + pV$$

Differentiating above equation, we get

$$dH = dU + PdV + VdP$$

From Eq. (5), we get

$$dH = TdS + VdP$$

$$\therefore \quad TdS = dH - VdP \quad \text{...(6)}$$

The change in enthalpy, if P and T are taken as independent variables, is

$$dH = \left(\frac{\partial H}{\partial T}\right)_p dT + \left(\frac{\partial H}{\partial P}\right)_T dP$$

$$= mC_p dT + \left(\frac{\partial H}{\partial P}\right)_T dP \qquad \left\{\because \left(\frac{\partial H}{\partial T}\right)_p - mC_p\right\} \quad \text{...(7)}$$

From Eqs (5) and (6), we get

$$dU + PdV = dH - VdP$$

Substituting Eqs. (2) and (7) in above equation, we get

$$\left(\frac{\partial U}{\partial V}\right)_T dV + mC_V dT + PdV = mC_P dT + \left(\frac{\partial H}{\partial p}\right)_T dP - VdP$$

i.e. $$mC_V dT + \left[P + \left(\frac{\partial u}{\partial v}\right)_T\right] dV = mC_P dT - \left[V + \left(\frac{\partial H}{\partial P}\right)_T\right] dP$$

If pressure is assumed constant i.e., $dP = 0$, we get

$$m(C_P - C_V)(dT)_P = \left[P + \left(\frac{\partial U}{\partial V}\right)_T\right](dV)_P$$

$$\therefore \quad m(C_P - C_V) = \left[P + \left(\frac{\partial U}{\partial V}\right)_T\right]\left(\frac{\partial V}{\partial T}\right)_P$$

$$\therefore \quad m(C_P - C_V) = \left[P + \left(\frac{\partial U}{\partial V}\right)_T\right]\beta V \qquad \left\{\because \left(\frac{\partial V}{\partial T}\right)_P = \beta V\right\}$$

$$\therefore \quad \frac{m(C_P - C_V)}{\beta V} = p + \left(\frac{\partial U}{\partial V}\right)_T$$

$$\therefore \quad \left(\frac{\partial U}{\partial V}\right)_T = \frac{m(C_P - C_V)}{\beta V} - P \qquad \ldots(8)$$

Substituting Eq. (8) in Eq. (2), we get

$$dU = \left[\frac{m(C_P - C_V)}{\beta V} - P\right]dV + mC_V dT$$

(c) When temperature (T) and pressure (P) are independent
The change in internal energy is given by,

$$dU = \left(\frac{\partial U}{\partial T}\right)_P dT + \left(\frac{\partial U}{\partial P}\right)_T dP \qquad \ldots(1)$$

We know that,

$$mC_p = \left(\frac{\partial H}{\partial T}\right)_P = \left[\frac{\partial(U + PV)}{\partial T}\right]_P$$

$$= \left(\frac{\partial U}{\partial T}\right)_P + V\left(\frac{\partial P}{\partial T}\right)_P + P\left(\frac{\partial V}{\partial T}\right)_p$$

Since pressure is constant, we have

$$mC_P = \left(\frac{\partial U}{\partial T}\right)_P + P\left(\frac{\partial V}{\partial T}\right)_P$$

$$\therefore \quad \left(\frac{\partial U}{\partial T}\right)_P = mC_p - P\left(\frac{\partial V}{\partial T}\right)_P$$

$$= mC_P - P\beta V \qquad \left\{\because \left(\frac{\partial V}{\partial T}\right)_P = \beta V\right\} \qquad \ldots(2)$$

From the first law,

$$dQ = dU + dW = dU + PdV$$

Since P and T are independent variables,

$$dQ = \left(\frac{\partial U}{\partial T}\right)_P dT + \left(\frac{\partial U}{\partial P}\right)_T dp + p\left[\left(\frac{\partial V}{\partial T}\right)_P dT + \left(\frac{\partial V}{\partial P}\right)_T dP\right]$$

$$= \left[\left(\frac{\partial U}{\partial T}\right)_P + P\left(\frac{\partial V}{\partial T}\right)_P\right]dT + \left[\left(\frac{\partial U}{\partial P}\right)_T + P\left(\frac{\partial V}{\partial P}\right)_T\right]dP$$

Now, we know that $$mC_P = \left(\frac{\partial U}{\partial T}\right)_P + P\left(\frac{\partial V}{\partial T}\right)_P$$

$$\therefore \quad dQ = mC_p dT + \left[\left(\frac{\partial u}{\partial P}\right)_T + P\left(\frac{\partial V}{\partial P}\right)_T\right] dP$$

For constant volume, we have

$$(dQ)_V = mC_V (dT)_V$$

$$= mC_P (dT)_V + \left[\left(\frac{\partial U}{\partial P}\right)_T + P\left(\frac{\partial V}{\partial P}\right)_T\right] (dP)_V$$

$$\therefore \quad mC_V = mC_P + \left[\left(\frac{\partial U}{\partial P}\right)_T + P\left(\frac{\partial V}{\partial P}\right)_T\right]\left(\frac{\partial P}{\partial T}\right)_V$$

But $$\left(\frac{\partial V}{\partial P}\right)_T = -KV \text{ and } \left(\frac{\partial P}{\partial T}\right)_V = \frac{\beta}{K}$$

$$\therefore \quad mC_V = mC_P + \left[\left(\frac{\partial U}{\partial P}\right)_T - pKV\right]\frac{\beta}{K}$$

$$\therefore \quad \left(\frac{\partial U}{\partial P}\right)_T = pKV - \frac{m(C_P - C_V)K}{\beta} \qquad \ldots(3)$$

Thus, from Eqs (2) and (3) we get

$$dU = (mC_P - P\beta V)dT + \left[pKV - \frac{m(C_P - C_V)K}{\beta}\right] dP$$

8.11 TdS EQUATIONS

From the laws of thermodynamics, we have

$$dQ = dU + PdV$$

and $$dS = \frac{dQ}{T}$$

$$\therefore \quad TdS = dU + PdV$$

We will now derive some useful relations for the above equation which is sometimes known as 'general *TdS* equation'.

(a) When temperature (T) and volume (V) are independent

Change in internal energy, in this case, is given by,

$$dU = \left(\frac{\partial u}{\partial T}\right)_v dT + \left(\frac{\partial u}{\partial v}\right)_T dV \qquad \ldots(1)$$

We have $$TdS = dU + PdV$$

Substituting Eq. (1) in above equation, we get,

$$TdS = \left[\left(\frac{\partial u}{\partial T}\right)_v dT + \left(\frac{\partial u}{\partial v}\right)_T dV\right] + PdV$$

$$\therefore \quad dS = \frac{1}{T}\left[\left(\frac{\partial u}{\partial T}\right) dT + \left(\frac{\partial u}{\partial v}\right)_T dV\right] + \frac{P}{T} dV \quad \text{...(2)}$$

We can write

$$dS = \left(\frac{\partial S}{\partial T}\right)_V dT + \left(\frac{\partial S}{\partial V}\right)_T dV \quad \text{...(3)}$$

From Eqs (2) and (3), we get

$$\left(\frac{\partial S}{\partial T}\right)_V = \frac{1}{T}\left(\frac{\partial u}{\partial T}\right)_V \quad \text{...(4)}$$

and

$$\left(\frac{\partial S}{\partial V}\right)_T = \frac{1}{T}\left[P + \left(\frac{\partial u}{\partial V}\right)_T\right] \quad \text{...(5)}$$

Taking the partial derivative of Eq. (4) with respect to V, we get

$$\frac{\partial^2 S}{\partial V\,\partial T} = \frac{1}{T}\left[\frac{\partial^2 u}{\partial V \partial T}\right] \quad \text{...(6)}$$

Taking the partial derivative of Eq. (5) with respect to T, we get

$$\frac{\partial^2 S}{\partial T.\partial V} = -\frac{1}{T_2}\left[P + \left(\frac{\partial u}{\partial T}\right)_T\right] + \frac{1}{T}\left[\left(\frac{\partial P}{\partial T}\right)_V + \frac{\partial^2 u}{\partial T \partial V}\right] \quad \text{...(7)}$$

By the test of exactness, we have

$$\frac{\partial^2 S}{\partial V.\partial T} = \frac{\partial^2 S}{\partial T.\partial V}$$

i.e.

$$\frac{1}{T}\left[\frac{\partial^2 u}{\partial V\,\partial T}\right] = -\frac{1}{T_2}\left[P + \left(\frac{\partial U}{\partial V}\right)_T\right] + \frac{1}{T}\left[\left(\frac{\partial P}{\partial T}\right)_V + \frac{\partial^2 U}{\partial T \partial V}\right]$$

$$\therefore \quad \frac{1}{T_2}\left[P + \left(\frac{\partial u}{\partial V}\right)_T\right] = \frac{1}{T}\left(\frac{\partial P}{\partial T}\right)_V$$

$$\therefore \quad \frac{1}{T}\left[P + \left(\frac{\partial U}{\partial V}\right)_T\right] = \left(\frac{\partial P}{\partial T}\right)_V = \frac{\beta}{K} \cdots \quad \left\{\because \frac{\beta}{K} = \left(\frac{\partial P}{\partial T}\right)_V\right\} \quad \text{...(8)}$$

We know that

$$mC_V = T\left(\frac{\partial S}{\partial T}\right)_V$$

i.e.

$$\left(\frac{\partial S}{\partial T}\right)_V = \frac{mC_V}{T}$$

From Eq. (4), we have

$$\left(\frac{\partial S}{\partial T}\right) = \frac{1}{T}\left(\frac{\partial u}{\partial T}\right)_V$$

i.e.

$$\frac{1}{T}\left(\frac{\partial u}{\partial T}\right) = \frac{mC_V}{T} \quad \text{...(9)}$$

Substituting Eqs (8) and (9) in Eq. (2), we get,

$$dS = \frac{mC_V}{T}dT + \frac{\beta}{K}dV$$

$$\therefore \quad TdS = mC_V dT + \frac{T\beta}{K}dV \quad \text{...(10)}$$

This is known as the 'first *TdS* equation.'

(*b*) *When temperature* (*T*) *and pressure* (*P*) *are independent*

Change in internal energy,

$$dU = \left(\frac{\partial U}{\partial T}\right)_P dT + \left(\frac{\partial U}{\partial T}\right)_T dP \quad \text{...(1)}$$

Change in volume is given by,

$$dV = \left(\frac{\partial V}{\partial T}\right)_P dT + \left(\frac{\partial V}{\partial P}\right)_T dP \quad \text{...(2)}$$

For independent variables *T* and *V*, *dU* can be expressed as,

$$dU = (mC_P - P\beta V)dT + \left[pKv - \frac{m(C_P - C_V)K}{\beta}\right]dP \quad \text{...(3)}$$

Also,

$$m(C_P - C_V) = \left[P + \left(\frac{\partial u}{\partial V}\right)_T\right]\left(\frac{\partial V}{\partial T}\right)_P$$

$$= \frac{T\beta}{K}\left(\frac{\partial V}{\partial T}\right)_P$$

$$= \frac{VT\beta^2}{K} \qquad \left\{\because P + \left(\frac{\partial U}{\partial V}\right)_T = \frac{T\beta}{K}\left(\frac{\partial V}{\partial T}\right)_P = \beta V\right\} \quad \text{...(4)}$$

Substituting Eq. (4) in Eq. (3),

$$dU = (mC_P - P\beta V)dT + (PK_V - VT\beta)\,dP \quad \text{...(5)}$$

We know that,

$$\left(\frac{\partial V}{\partial T}\right)_P = \beta V \text{ and } \left(\frac{\partial V}{\partial P}\right)_T = -KV$$

Substituting above values in Eq. (2), we get

$$dV = \beta V dT - KV dP \quad \text{...(6)}$$

Now, $TdS = dU + PdV$

Substituting Eqs (5) and (6) in above equation, we get

$$TdS = (mC_P - P\beta V)dT + (PKV - VT\beta)dP + p(\beta VdT - KVdP)$$

$$\therefore \quad TdS = mC_PdT - VT\beta dP \qquad \ldots(7)$$

This is known as Second *TdS* equation.

(c) *When pressure (P) and volume (V) are independent*

The change in internal energy,

$$dU = \left(\frac{\partial U}{\partial P}\right)_V dP + \left(\frac{\partial U}{\partial V}\right)_P dV \qquad \ldots(1)$$

But as discussed earlier, for independent variables *p* and *v*, we have

$$dU = \frac{KmC_V}{\beta} dP + \left(\frac{mC_P}{\beta C_{VP}} - p\right) dV \qquad \ldots(2)$$

$$\text{Now, } TdS = dU + PdV \quad = \frac{KmC_V}{\beta} dp + \left(\frac{mC_P}{\beta V} - p\right) dV + PdV$$

$$\therefore \quad TdS = \frac{KmC_V}{\beta} dP + \frac{mC_P}{\beta V} dV \qquad \ldots(3)$$

This is known as third *TdS* equation.

8.12 SPECIFIC HEAT

Specific heat of a substance is defined as amount of heat required to raise a unit mass of a substance through unit rise in temperature. There are two processes whose specific heats are important viz. constant volume C_V and constant pressure C_P.

Specific heat at constant volume (C_V)

At constant volume, the specific heat is given by,

$$C_V = \left(\frac{\partial U}{\partial T}\right)_V = \left(\frac{\partial U}{\partial T}\right)_V \left(\frac{\partial S}{\partial T}\right)_V$$

$$= T\left(\frac{\partial S}{\partial T}\right)_V \qquad \left\{\because T = \left(\frac{\partial u}{\partial S}\right)_V\right\}$$

Specific heat at constant pressure (C_P)

At constant pressure, specific heat is given by,

$$C_P = \left(\frac{\partial h}{\partial T}\right)_p = \left(\frac{\partial h}{\partial S}\right)_p \left(\frac{\partial S}{\partial T}\right)_p$$

$$= T\left(\frac{\partial S}{\partial T}\right)_p \qquad \left\{\because T = \left(\frac{\partial h}{\partial S}\right)_p\right\}$$

8.13 SPECIFIC HEAT RELATIONS

We know that,

The first TdS equation is given by

$$TdS = mC_V dT + \frac{T\beta}{K} dV$$

$$= mC_V dT + T\left(\frac{\partial P}{\partial T}\right)_V dV \qquad \left[\because \frac{\beta}{K} = \left(\frac{\partial P}{\partial T}\right)_V\right] \qquad \text{...(1)}$$

The second TdS equation is given by

$$TdS = mC_P dT - VT\beta dP$$

$$= mC_P dT - T\left(\frac{\partial V}{\partial T}\right)_P dP \qquad \left[\because \beta V = \left(\frac{\partial P}{\partial T}\right)_P\right] \qquad \text{...(2)}$$

From Eqs (1) and (2) we get

$$mC_V dT + T\left(\frac{\partial P}{\partial T}\right)_V dV = mC_P dT - T\left(\frac{\partial V}{\partial T}\right)_P dp$$

$$\therefore \quad m(C_P - C_V)dT = T\left(\frac{\partial P}{\partial T}\right)_V dV + T\left(\frac{\partial V}{\partial T}\right)_P dP$$

$$\therefore \quad dT = \frac{T\left(\frac{\partial P}{\partial T}\right)_V}{m(C_P - C_V)} dV + \frac{T\left(\frac{\partial P}{\partial T}\right)_P}{m(C_P - C_V)} dP \qquad \text{...(3)}$$

The variables P and V are treated as independent variables. Thus, we get

$$dT = \left(\frac{\partial T}{\partial V}\right)_P dV + \left(\frac{\partial T}{\partial P}\right)_V dP \qquad \text{...(4)}$$

From Eqs (3) and (4) we get

$$\left(\frac{\partial T}{\partial V}\right)_P = \frac{T\left(\frac{\partial P}{\partial T}\right)_V}{m(C_P - C_V)} \qquad \text{...[5(a)]}$$

and

$$\left(\frac{\partial T}{\partial P}\right)_V = \frac{T\left(\frac{\partial V}{\partial T}\right)_P}{m(C_P - C_V)} \qquad \text{...[6(a)]}$$

or

$$m(C_P - C_V) = T\left(\frac{\partial P}{\partial T}\right)_V\left(\frac{\partial V}{\partial T}\right)_P \qquad \text{...[5(b)]}$$

$$m(C_P - C_V) = T\left(\frac{\partial V}{\partial T}\right)_P\left(\frac{\partial P}{\partial T}\right)_V \qquad \text{...[6(b)]}$$

From Eqs 5 (a), 5 (b) and 6 (a), 6(b), it is observed that

$$m(C_P - C_V) = T\left(\frac{\partial P}{\partial T}\right)_V \left(\frac{\partial V}{\partial T}\right)_P \quad \text{...(7)}$$

From the cyclic property of differentials, we have

$$\left(\frac{\partial P}{\partial T}\right)_V \left(\frac{\partial T}{\partial V}\right)_P \left(\frac{\partial V}{\partial P}\right)_T = -1$$

i.e.

$$\left(\frac{\partial P}{\partial T}\right)_V = \frac{1}{\left(\frac{\partial T}{\partial V}\right)_P \left(\frac{\partial V}{\partial P}\right)_T}$$

or

$$\left(\frac{\partial P}{\partial T}\right)_V = -\left(\frac{\partial V}{\partial T}\right)_P \left(\frac{\partial P}{\partial V}\right)_T$$

If we put the above value in Eq. (7), we get,

$$m(C_P - C_V) = T\left[-\left(\frac{\partial v}{\partial T}\right)_p \left(\frac{\partial p}{\partial v}\right)_T\right]\left(\frac{\partial V}{\partial T}\right)_P$$

$$= -T\left(\frac{\partial V}{\partial T}\right)_P^2 \left(\frac{\partial P}{\partial V}\right)_T \quad \text{...(8)}$$

But we know that $\left(\frac{\partial V}{\partial T}\right)_P = \beta V$ and $\left(\frac{\partial V}{\partial P}\right)_T = -K_V$

Substituting the above value in Eq. (8), we get,

$$m(C_P - C_V) = -T(\beta V)^2\left(-\frac{1}{KV}\right)$$

$$= \frac{T\beta^2 V}{K}$$

$$\therefore \quad (C_P - C_V) = \frac{T\beta^2 V}{Km}$$

But $\frac{V}{m} = V_S$ = Specific volume of gas (m³ /kg)

$$\therefore \quad C_P - C_V = \frac{T\beta^2 V_S}{Km}$$

The values of T, V_S and K are always positive. Also, even if β is negative, β^2 is always positive

$$\therefore \quad C_P - C_V > 0 \text{ i.e. } C_P > C_V$$

8.14 JOULE THOMSON COEFFICIENT

Consider a throttling process as shown in Fig. 8.1.

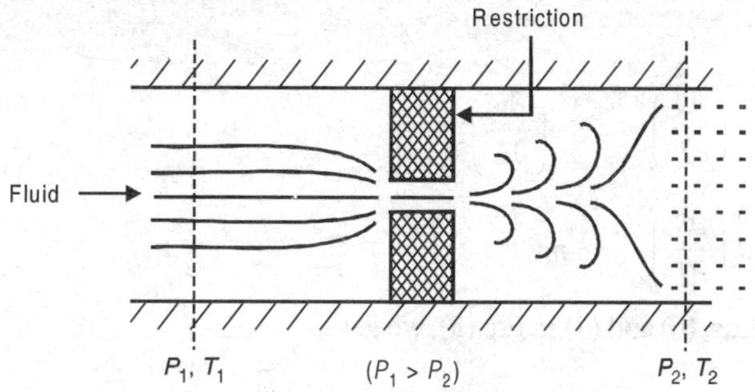

Fig. 8.1

The velocity of fluid, whose pressure is reduced by a throttling process, increases at the restriction with a consequent decrease in enthalpy. But this increase in kinetic energy is dissipated by friction, as the eddies die down after restriction.

The steady flow energy equation implies that if the flow is adiabatic and if the fluid velocity before restriction is equal to that downstream of it, the enthalpy of fluid is restored. Throttling is therefore an Isenthalpic process.

The change in temperature with drop in pressure, at constant enthalpy, is defined as the Joule–Thomson Coefficient (μ).

$$\therefore \qquad \mu = \left(\frac{\partial T}{\partial P}\right)_h \qquad \ldots(1)$$

μ varies with both temperature and pressure of the gas. The magnitude of μ is a measure of deviation of a gas from perfect gas behaviour.

For real gases, μ may be either positive or negative, depending on the thermodynamic state of the gas. If μ is greater than zero, it implies that the gas temperature decreases with decrease in pressure. If μ is less than zero, the gas temperature increases with decreases in pressure.

When $\mu = 0$, the temperature of gas remains constant with throttling. The temperature at which $\mu = 0$ is called as Inversion temperature for a given pressure.

Using usual notation, we know that enthalpy is given by

$$H = U + PV \qquad \ldots(2)$$

Writing Eq. (2) in differential form, we get

$$dH = dU + d(PV)$$

$$= dU + PdV + VdP \qquad \ldots(3)$$

We know that $\quad TdS = dU + PdV$

or $\quad dU = TdS - PdV \qquad \ldots(4)$

Substitute Eq. (4) in Eq. (3)

$$\therefore \qquad dH = TdS - PdV + PdV + VdP$$

$$\therefore \qquad dH = TdS + VdP$$

Taking P and T as independent variables, we get

$$dH = T\left[\left(\frac{\partial S}{\partial T}\right)_p dT + \left(\frac{\partial S}{\partial p}\right)_T dp\right] + Vdp$$

$$= T\left(\frac{\partial S}{\partial T}\right)_p dT + \left[T\left(\frac{\partial S}{\partial p}\right)_T + V\right] dp \qquad ...(5)$$

From Maxwell relations,

$$\left(\frac{\partial S}{\partial T}\right)_T = -\left(\frac{\partial v}{\partial T}\right)_P \qquad ...(6)$$

Also $$T\left(\frac{\partial S}{\partial T}\right)_P = mC_P \qquad ...(7)$$

Substituting value from Eqs. (6) and (7) in Eq. (5), we get

$$dH = mC_p dT + \left[V - T\left(\frac{\partial V}{\partial T}\right)_p\right] dp \qquad ...(8)$$

$$\therefore \quad dT = \frac{dH}{mC_p} + \frac{\left(T\left(\frac{\partial V}{\partial T}\right)_P - V\right)}{mC_p} dp$$

$$= \frac{dh}{C_p} + \frac{\left(T\left(\frac{\partial Vs}{\partial T}\right)_p - Vs\right)}{C_p} dp \qquad ...(9)$$

$$\left[\because dh = \frac{dH}{m} = \text{Change in sp. enthalpy (kJ/kg)}\right]$$

$$\left[\text{Also, } V_s = \frac{V}{m} = \text{Specific volume of gas (m}^3\text{/kg)}\right]$$

Assuming h and p as independent variables, we get

$$dT = \left(\frac{\partial T}{\partial h}\right)_p dh + \left(\frac{\partial T}{\partial p}\right)_h dp \qquad ...(10)$$

From Eqs (9) and (10) we get

$$\left(\frac{\partial T}{\partial h}\right)_P = \frac{1}{C_P}$$

Multiplying the numerator and denominator by T^2, we get

$$\left(\frac{\partial T}{\partial p}\right)_h = \frac{T^2}{C_P}\left[\frac{T\left(\frac{\partial V_S}{\partial T}\right)_P - V_S}{T^2}\right] = \mu \qquad \left[\because \mu = \left(\frac{\partial T}{\partial p}\right)_h\right]$$

$$\therefore \qquad \mu = \frac{T^2}{C_P}\left[\frac{\partial}{\partial T}\left(\frac{V_S}{T}\right)\right]_P$$

Note. For an ideal gas, $PV_S = RT$

$$\therefore \qquad \frac{V_S}{T} = \frac{R}{P} \quad \text{i.e.} \quad \frac{\partial}{\partial T}\left(\frac{V_S}{T}\right)_p = 0 \quad (\because\ R \text{ and } P \text{ are constants})$$

$$\therefore \qquad \mu = \frac{T^2}{C_p}\left[\frac{\partial}{\partial T}\left(\frac{V_S}{T}\right)\right]_p = 0 \text{ i.e. } \mu = 0 \text{ for ideal gas}$$

8.15 EQUATION OF STATE

The ideal gas equation of state $PV_S = RT$ can be established from the postulates of kinetic theory of gases developed by Clerk Maxwell. The two main assumptions made in deriving the perfect gas laws are:

(1) The gas moleculas are mere mass points occupying no space or negligibly small space compared to the volume of the gas.

(2) There is very little or no attraction or repulsion between the molecules.

In practice, all real gases deviate from the ideal gas laws. The deviation is small at ordinary temperatures and pressures but high at high pressure and low temperature. This is because all real gas molecules do occupy some space and to attract or repulse each other. At high pressure and low temperature, the intermolecular attraction and repulsion forces increase and also the volume of molecules becomes appreciable compared to total gas volume. Hence the ideal gas equation is not suitable for real gases at high pressure and low temperature. Some equations used to correct the ideal gas equation are given below.

(a) Van der Waal's Equation. J. D. Van der Waal, a Dutch physicist, was the first to correct the ideal gas equation $PV_S = RT$. He applied the laws of mechanics to individual molecules and introduced two correction terms in the ideal gas equation.

Van der Waal's equation is given by,

$$\left(P + \frac{a}{V_S^2}\right)(V_S - b) = RT$$

Coefficient a was introduced to account for the existence of mutual attraction between the molecules.

Coefficient b was introduced to account for the volume of molecules, and is known as co-volume.

The term $\left(a/V_S^2\right)$ is called the force of cohesion.

Real gases conform more closely with Van der Waal's equation of state than the ideal gas equation of state, particularly at high pressures. But is not obeyed by real gases in all ranges of pressures and temperatures.

(b) Clausius Equation. The Clausius equation is a modification of Van der Waal's equation and it neglects the mutual attraction between molecules. Thus, the Clausius equation is of the from:

$$P(V_S - b) = RT$$

where

b = Volume occupied by all molecules.

(c) Beattie Bridgeman Equation. In 1928, Beattie Bridgeman developed an equation of state which gives accurate results. The equation is of the form:

$$P = \frac{RT(1-\varepsilon)}{V_S}(V_S + B) - \frac{A}{V_S}$$

where,

$$A = A_O\left(1-\frac{a}{V_S}\right)$$

$$B = B_O\left(1-\frac{b}{V_S}\right)$$

$$\varepsilon = \left(\frac{C}{V_S T^3}\right)$$

The constants *Ao, Bo, a, b* and *c* have to be determined experimentally for each gas.

The Beattie–Bridgeman equation does not give satisfactory results in the critical point region, but is fairly accurate (within 2 percent) in regions where density is less than 0.8 times the critical density.

(d) Bertholet Equation. Bertholet equation is similar to the Van der Waal's equation and uses the same constants *a* as in Van der Waal's equation. The equation is of the form

$$\left(P-\frac{a}{TV_S^2}\right)(V_S-b) = RT$$

The Bertholet equation is of limited accuracy and gives properties accurate within 1 percent.

(e) Dieterici Equation. Dieterici equation of state gives very accurate properties in the neighbourhood of the critical point and on the critical isotherm. The equation is of limited accuracy and may produce large errors when used away from critical region. The equation is of the form

$$P = \frac{RT}{(V_S-b)}\cdot e^{-a/RTV_S}$$

where *a* and *b* are constants.

(f) Redlich-Kwong Equation. The equation developed by Redlich–Kwong has the following form

$$\left[P-\frac{a}{\sqrt{T}\,V_S(v_S+b)}\right](V_S-b) = RT$$

where *a* and *b* are constants.

The Redlich–Kwong equation gives good results at high pressures and is fairly accurate for temperature above the critical value.

(g) Benedict–Webb–Rubin Equation. The Benedict–Webb–Rubin equation is used for regions of high density i.e., where the density is approximately twice the critical density.

The equation has the form:

$$P = \frac{RT}{V_s}+\frac{B_oRT-A_o-C_o/T^2}{V_S^2}+\frac{bRT-a}{V_S^3}+\frac{a\alpha}{V_S^6}+\frac{C}{V_S^3T^2}\left(1+\frac{\gamma}{V_S^2}\right)e^{-\gamma/v_S^2}$$

The eight constants A_o, B_o, C_o, *a, b, c, x* and γ are to be determined experimentally.

(h) Virial Equation. The relation between PV_s and *P* may be expressed in the form of power series, as shown below.

$$PV_S = A(1+B'P+C'P^2+D'P^3+...)$$

It can be proved that for any gas,

$$\lim_{p \to 0} PV_S = RT = A$$

∴ We get, $$PV_S = RT(1+B'P+C'P^2+D'P^3+...)$$

or $$\frac{PV_S}{RT} = 1+B'P+C'P^2+D'P^3+... \quad ...(i)$$

An alternate expression for the above is

$$\frac{PV_S}{RT} = 1+\frac{B}{V_S}+\frac{C}{V_S^2}+\frac{D}{V_S^3}+... \quad ...(3)$$

Both the expressions in Eqs. (i) and (ii) are known as virial expansions or virial equations of state. These equations were first introduced by Dutch physicist Kammerlingh Onnes.

In the above equations, *B*, *C*, *D*, *B*, *C*, etc. are called virial coefficients. *B* and *B* are known as second viral coefficients, *C* and *C* are known as third virial coefficients and so on.

There exists a relation between the virial coefficients in Eqs (i) and (ii) as follows:

$$B' = \frac{B}{RT}, \quad C' = \frac{C-B^2}{(RT)^2},$$

$$D' = \frac{D-3BC+2B^3}{(RT)^3} \text{... and so on.}$$

8.16 COMPRESSIBILITY FACTOR

We know that the virial equation of state is of the form

$$\frac{PV_S}{RT} = 1+\frac{B}{V_S}+\frac{C}{V_S^2}+\frac{D}{V_S^3}+...$$

Here, the ratio (PVs/RT) is known as the compressibility factor, denoted by Z.

For ideal gas, $Z = 1$.

For real gas, Z is determined experimentally. The magnitude of Z for a particular gas at a particular pressure and temperature is a measure of deviation of the gas from ideal gas behaviours.

CHAPTER 9 Availability

9.1 INTRODUCTION

Energy exists in many forms like kinetic energy, potential energy, electrical energy, heat energy etc. While it is possible to convert most of the energies like potential energy, kinetic energy, electricity etc. into work, only some part of heat energy can be converted into work. The part of energy that can be converted into work is *Available energy* and the remainder is *Unavailable portion*.

Availability is about work potential or quality of the energy. When thermal energy is used to do work entire heat energy cannot be converted to useful work as per the second law of thermodynamics. The part that can be converted to useful work is referred to as the *Available energy* or *Availability* or *Energy*. The quality of energy is measured by Availability. As energy is used in a process it loses quality, its availability or Energy decreases.

9.2 HIGH/LOW GRADE OF ENERGY

Energies like electrical energy or potential energy which can be converted almost entirely to do work are high-grade energy. Heat on the other hand is low-grade energy since only a part of it can be converted into work. Heat is of the lowest grade in which energy exists in a very disordered state. The lower the temperature at which the heat energy exists the lower would be the grade of energy.

However low grade energy is used to produce high-grade energy. For example electricity a high-grade energy is produced in thermal or nuclear power plants using heat a low-grade energy source.

High Grade Energy Sources	**Low Grade Energy Sources**
Mechanical work	Heat
Electricity	Heat derived from combustion
Kinetic energy	Kinetic energy
Tidal power	Heat from nuclear fission

9.3 AVAILABILITY

Available energy is that of energy which is available for doing work. It is either work or that portion of energy that can be converted wholly into work. Concept of availability comes from the Second Law of Thermodynamics where some portion of heat has to be rejected to a sink to produce work. The work has to be with reference to some datum. Generally earth's atmosphere is taken as reference and is called the dead state designated by '0'. Dead state implies the temperature and pressure existing in the surroundings and are designated by T_0 and P_0.

Availability or Exergy is the maximum portion of energy that can be converted into work by ideal process that reduces the system to dead state.

9.3.1 Dead State

Any system that has temperature T and pressure P can do useful work till the temperature and pressure are reduced to T_0 and P_0. When the temperature and pressure are equal to that of the earth or dead state all transfer of energy, stops although the system contains internal energy which would be unavailable. When a system is in equilibrium with the surroundings its potential to do any work ceases. Thus when the pressure P of the system reduces to the atmospheric pressure P_0, temperature T becomes equal to T_0 and likewise the KE and PE become equal to that of the surroundings no more work can be obtained.

The properties of the system in dead state are denoted by subscript 0 i.e., as P_0, T_0, H_0, S_0, U_0, C_0 etc. Generally dead state temperature is taken at 25°C (298 K) and P_0 as 1.01325 bar unless otherwise indicated.

9.4 AVAILABILITY OF VARIOUS SYSTEMS

We can study the availability of

(a) Work Reservoir
(b) Heat Reservoir
(c) Closed System
(d) Steady Flow System

9.4.1 Work Reservoir

A work reservoir is a source of infinite work and it would not come to equilibrium with the surroundings. However the amount of W withdrawn from the work reservoir can be converted fully into useful work in the absence of any dissipative effects. Therefore Availability of work reservoir is $A = \Delta W$.

9.4.2 Heat Reservoir

(i) Infinite Heat Reservoir (at constant temperature).
(ii) Finite Heat Reservoir (where temperature changes).

9.5 IRREVERSIBILITY

Most of the real processes in nature are irreversible due to mixing, friction and heat transfers with finite temperature difference. In a reversible process there is no net increase in the entropy of the universe. In a reversible process the change in the entropy of a system equals the change in the entropy of the universe. However in the case of an irreversible process the reduction in the entropy of a system is less than the increase of entropy of the heat receiving system. This leads to loss of availability or an equivalent increase in unavailability— which however are equal. In any process, aim is always to get maximum work. In an expansion process we try to get the maximum work output. In compression process aim is to have minimum work input. Thus both in expansion as well as in compression processes aim is to maximize the work as per sign convention of work.

Maximum work (W_{rev}) during a process between two states can be obtained by reversible process. However in actual practice the actual work (W_{act}) between these two states is always less than the W_{rev}. The difference between the two is called Irreversibility.

$$I = W_{rev} - W_{act}$$

In irreversibility (I)/Exergy destruction = wasted work potential in a process. It is lost opportunity to do work. The greater the irreversibility the greater is the loss of work that could have been performed. Thus irreversibility gives the quality of the process.

Let there be two processes, one reversible and another irreversible between two states and let heat δQ_{rev} and δQ be added giving work output of W_{rev} and W_u respectively.

Thus $$\delta Q_{rev} = dE + \delta W_{rev} \quad \ldots(1)$$

and $$\delta Q = dE + \delta W \quad \ldots(2)$$

$$\delta I = (1) - (2)$$

$$\delta I = \delta Q_{rev} - \delta Q = TdS - \delta Qs$$

$$= \delta W_{rev} - \delta W$$

Integrating $I = W_{rev} - W_{act}$ = Lost work

9.6 EFFECTIVENESS

The effectiveness of a system is the ratio of the useful or actual work done to the maximum or reversible work.

Thus, Effectiveness $= \delta W/\delta W_{rev}$

9.7 SECOND LAW EFFICIENCY

So far we have been using the efficiencies and COP based on the First Law of Thermodynamics. Based on the increasing use of availability which gives indication about possibility of maximum work output or minimum work input if the process is carried out reversibly — concept of Second Law efficiency is being used these days. Second law efficiency is a measure of reversible operation.

Let us assume that an engine operating between say 1200 K and 300 K has – η_{th} of 30% (by measuring brake power and dividing by the heat input—based on First Law of Thermodynamics).

However, had the work been done in a reversible manner i.e., by Carnot Cycle.

$$\eta_{rev} = \left(1 - \frac{To}{T}\right)$$

$$= (1 - 300/200) = 0.75 = 75\%$$

The ηth is far less than the η_{rev}. We can infer that η_I is not realistic measure of the performance. To overcome we define η_{II} for various heat-work machines

$\eta_{II} = \eta_{th}/\eta_{rev}$ (For heat engines)

$\eta_{II} = W/W_{rev}$ (For power or work producing devices)

$\eta_{II} = W_{rev}/W$ (For power absorbing devices)

where W = useful work and

W_{rev} = maximum or reversible work

$\eta_{II} = COP/COP_{rev}$ (For refrigerators and heat pumps)

9.8 IMPORTANCE OF AVAILABILITY/IRREVERSIBILITY/EFFECTIVENESS/η_{II}

These concepts are based on the second law of thermodynamics and are applicable to both cycle and process. These indicate the departure of an actual process from an ideal or reversible process.

9.9 HELMONTZ AND GIBBS FUNCTIONS

Both these functions are properties (based on combination of properties just like $h = u + pv$.

Helmontz Function $F = U - TS$

or $f = u + Ts$ (per unit mass)

This function is useful for closed systems that undergo reversible, isothermal process. Decrease in the value of F equals the max work that can be done by system undergoing an isothermal process at temperature equal to that of the surroundings.

9.10 GIBBS FUNCTION

$$G = H - Ts$$

or $g = h - Ts$ (For unit mass)

9.11 SUMMARY

Availability or Exergy is the maximum useful work that can be obtained from a system under ideal conditions by reducing it to the dead state. Availability A of various systems are

(a) Heat Reservoir $\quad A = \Delta Q - T_0\,\Delta S$

(b) Closed System $\quad A = A - A_0 = (U - T_0S + P_0V) - (U_0 - T_0S_0 - P_0 + P_0V_0)$

(c) Steady Flow System $\quad A = A - A_0 = m\left\{\left(h - T_0S + \frac{C^2}{2} + gZ\right) - (h_0 - T_0S_0 + \frac{C_0^2}{2} + gZ_0)\right\}$

$$\text{Irreversibility} = W_{rev} - W = T_0 - T_0\,\Delta S$$

$$\text{Effectivness} = W/W_{rev}$$

9.12 ENTROPY CHANGE ΔS

The unavailable energy $UA = T_0\,\Delta S$. Thus change in entropy during a process is required to find out avilability and unavailability.

From Ist Law

$$\delta Q = \delta u + \delta w$$

$$\delta Q = Au + Pdv$$

Divide by T

$$\frac{\delta Q}{T} = \frac{du}{T} + Pdv$$

$$= C_v\frac{dt}{T} + R\frac{dv}{V} \quad \left(\text{Since}\,\frac{P}{T} = \frac{R}{V}\right)$$

But

$$\frac{\delta Q_{rev}}{T} = ds$$

$\therefore$

$$ds = C_V\frac{dT}{T} + R\frac{dv}{V}$$

Integrating between 1 and 2

$$s_2 - s_1 = C_V \ln\frac{T_2}{T_1} + R\frac{V_2}{V_1}$$

$$\Delta S = C_V \ln\frac{P_2V_2}{P_1V_1} + R\ln\frac{V_2}{V_1} \quad \text{...(1)}$$

$$= C_V\left(\ln\frac{P_2}{P_1} + \ln\frac{V_2}{V_1}\right) + R\ln\frac{V_2}{V_1}$$

$$= Cv\left(\ln\frac{P_2}{P_1}\right) + \ln\frac{v_2}{v_1} + (R + Cv)$$

But

$$C_v + R = C_p$$

$\therefore$

$$\Delta S = C_V \ln\frac{P_2}{P_1} + C_p\ln\frac{V_2}{V_1} \quad \text{...(2)}$$

We can also put $\frac{v_2}{v_1} = \frac{P_1T_2}{P_2T_1}$ in Eq. (1)

$$\therefore \quad \Delta S = C_V \ln\frac{T_2}{T_1} + R\ln\frac{P_1T_2}{P_2T_1}$$

$$= C_V \ln\frac{T_2}{T_1} + R\left(\ln\frac{P_1}{P_2} + \ln\frac{T_2}{T_1}\right)$$

$$= (C_V + R)\ln\frac{T_2}{T_1} + R\ln\frac{P_1}{P_2}$$

$$\Delta S = C_p \ln\frac{T_2}{T_1} + R\ln\frac{P_1}{P_2} \quad \ldots(3)$$

Entropy Changes for Various Processes are

(1) Constant volume $\quad s_2 - s_1{}^{*} = CV\ln\frac{T_2}{T_1}$

(2) Constant pressure $\quad s_2 - s_1 = C_P\ln\frac{T_2}{T_1}$

(3) Iso thermal $\quad s_2 - s_1 = R\ln\frac{V_2}{V_1}$

(4) Adiabatic $\quad s_2 - s_1 = 0$

(5) Polytropic $\quad s_2 - s_1 = C_V\left(\frac{n-r}{n-1}\right)\ln\frac{T_2}{T_1}$

9.13 AVAILABILITY OF HEAT RESERVOIR

(a) Infinite Heat Reservoir. In such a reservoir temperature remains constant when small amount of heat δQ is withdrawn from it.

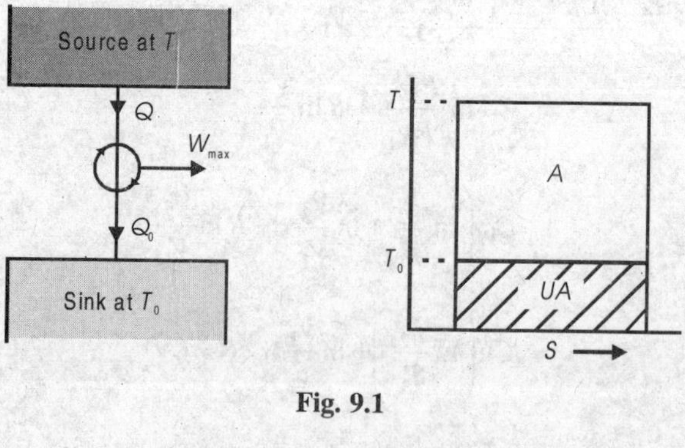

Fig. 9.1

As shown in Fig. 9.1, let a small quantity of heat δQ be withdrawn from the heat reservoir which is at average temperature T. For getting maximum work, let there be a Carnot engine operating between temperatures T and T_0 (sink temperature being of the surroundings).

$$\frac{\delta W}{\delta Q} = \left(1-\frac{T_0}{T}\right)$$

or
$$\delta W = \left(1-\frac{T_0}{T}\right)(-\delta Q) = T_0 \frac{\delta Q_{rev}}{T}$$

Integrating,
$$W_{max} = \int_{T_0}^{0} = A = -\delta Q - \int_{0}^{1} T_0 dS$$

$$= Q + T_o (s_0 - s_1)$$
$$= Q - T_o (S_1 - S_0)$$

Thus
$$W_{max} = Q - T_0 \Delta S$$
$$= \text{Heat Added} - T_0 \text{ (change in entropy)}$$

(b) Finite Heat Reservoir. In this case when heat is added or withdrawn from the reservoir which has finite capacity the temperature does not remain constant.

Consider a small quantity δ heat being withdrawn from the reservoir (1) at temperature T and producing entropy change ds as shown. Using a Carnot engine to derive maximum work between T and T_0; we have

$$\delta W'_{carnot} = \left(1-\frac{T_0}{T}\right)(-\delta Q)$$

Integrating between 0 and 1 we would get

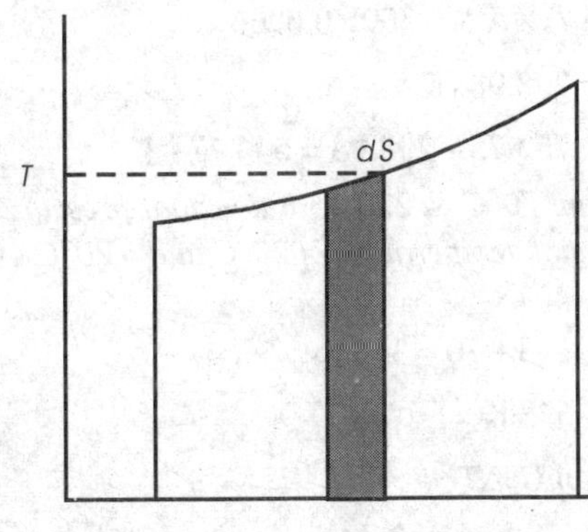

Fig. 9.2

$$W_{max} = A = \int_0^1 (-\delta Q) + \int_0^1 T_0 \frac{\delta Q_{rev}}{T}$$

$$= \int_0^1 (-\delta Q) + \int_0^1 T_0 dS$$

$$= Q - T_0 (s_1 - s_0)$$
$$= \text{Heat Added or withdrawn} - T_0 \Delta S$$

$$A = Q - T_0 \Delta S$$

$$UA = T_0(s_1 - s_0) = T_0 \Delta S$$

For the heat reservoir where temperature does not remain constant we would use appropriate equation for entropy change as applicable.

SOLVED EXAMPLES

Example 9.1 *0.5 kg of air at 500 K is heated reversibly at constant pressure to 2000 K. Find the available and unavailable energies and the heat added. Take T_0 = 300 K and C_P = 1.005 kJ/kg-K.*

Solution

$$\Delta Q = m\, C_P (T_2 - T_1)$$

$$= 0.5 \times 1.005(2000 - 500)$$

$$= 753.75 \text{ kJ}$$

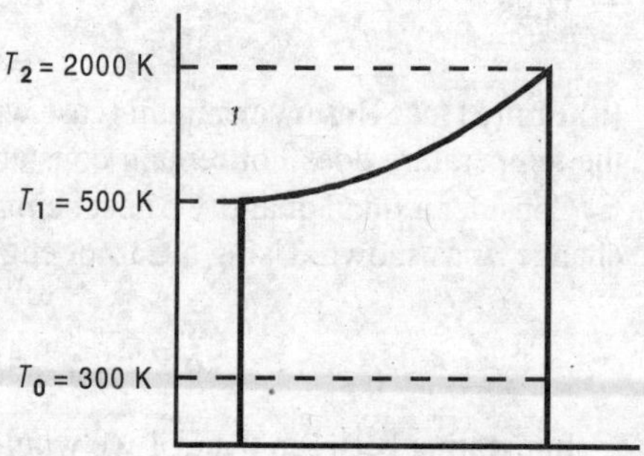

Fig. Ex. 9.1

Entropy changes at constant pressure

$$\Delta S = mC_P \ln \frac{T_2}{T_1}$$

$$= 0.5 \times 1.005 \ln \frac{2000}{500}$$

$$= 0.6966 \text{ kJ/kg-K}$$

$$UA = T_0 \times \Delta S = 300 \times 0.6966$$

$$= 208.98 \text{ kJ}$$

$$\mathbf{A = 753.75 - 208.98 = 544.77 \text{ kJ}}$$

Example 9.2 *One kg of air is heated from 70°C to 270°C at constant pressure. Determine (a) Heat supplied (b) Entropy change (c) Available energy for sink temperature of 20°C and – 20°C. Assume C_p = 1.005 kJ/kg-K for air.*

Solution

$$T_1 = 273 + 70 = 343 \text{ K}$$

$$T_2 = 543 \text{ K}$$

(a) Heat supplied $= mC_P\Delta T$

$$= 1 \times 1.005 \times (270.70) = 201 \text{ kJ}$$

(b) ΔS at const. pressure $= m\, Cp \ln \frac{T_2}{T}$

$$= 1 \times 1.005 \ln \frac{543}{343}$$

$$= 0.46167 \text{ kJ/kg.K}$$

(c) (i) Available energy for sink temperature and $20 + 273 = 293$ K

$$UA = T_0 \Delta S$$

$$= 293 \times 0.808 = 135.27 \text{ kJ}$$

$$A = \Delta Q - UA = 201 - 135.27 = 65.73 \text{ kJ}$$

(ii) Available and unavailable energy at sink temperature of

$$273 - 20 = 253 \text{ K}$$

$$T_0 = 253\text{K}$$

$$UA = T_0 \Delta S = 116.80$$

$$\mathbf{A = 201 - 116.80 = 84.2 \text{ kJ}}$$

Example 9.3 *100 kg of water at 100°C is mixed with 50 kg of water at 50°C, while the surrounding temperature is 27°C. Determine the decrease in availability due to mixing.*

Data: $T_0 = 273 + 27 = 300$ K; $m_1 = 100$ kg
$T_1 = 273 + 100 = 373$ K; $m_2 = 50$ kg
$T_2 = 273 + 50 = 323$ K

Solution

Temperature after mixing will be

$$(100 + 50)\, t_m = 100\, t_1 + 50\, t_2$$

$$= \frac{100 \times 100 + 50 \times 50}{150} = \frac{2 \times 100 + 50}{3}$$

$$t_m = \frac{250}{3} = 83.33°\text{C}$$

$$T_m = 273 + 83.33 = 356.33 \text{ K}$$

(c) Heat given by hot water at 100°C (373 K)

$$Q = m_1 \times C_P \times (100 - 83.33)$$
$$= 100 \times 4.187 \times 16.67 = 6979.73 \text{ kJ}$$

$$\text{Entropy decrease} = m_1 C_P \ln \frac{T_1}{T_m}$$

$$= 100 \times 4.187 \times \ln \frac{373}{356.33}$$

$$\Delta S = 19.1434 \text{ kJ/K}$$

$$UA_{100} = T_o \Delta S = 300 \times 19.1434 = 5743.0 \text{ kJ}$$

∴ Availability $A_{100} = \Delta Q_{100} - UA_{100}$

$$= 6979.73 - 5753.6 = 1226.13$$

(d) Heat received by cold water at 50°C

$$Q_{50} = m_2\, C_P\, \Delta T$$
$$= 50 \times 4.187 \times (83.33 - 50) = 6977.64 \text{ kJ}$$

$$\text{Entropy gain by cold water} = m_2 C_P \ln \frac{T_m}{T_2}$$

$$= 50 \times 4.187 \ln \frac{356.33}{323} = 20.56 \text{ kJ/K}$$

$$UA_{50} = T_o \Delta S = 300 \times 20.56$$
$$= 6168 \text{ kg}$$

$$A_{50} = \Delta Q_{50} - UA_{50}$$
$$= 6977.64 - 6168 = 809.64\,\text{kJ}$$

(e) Decrease in Availability = (1) – (2)

$$= 1226.13 - 809.64$$

Decrease in Availability = 416.49 kJ

Example 9.4 *Exhaust gas leave an IC engine at 827° and 1.0 atmosphere after doing 1000 kJ of work per kg of gas in the engine (C_{pg} = 1.1). The surrounding temperature is 25°C.*

(a) How much available energy is cost by throwing away exhaust gases.

(b) What is the ratio δ the lost available energy to the engine work.

Solution

$$T_1 = 827 + 273 = 1100$$
$$T_0 = 25 + 273 = 298$$

Entropy change ΔS = ?

$$= m \cdot C_{Pg} \cdot \ln \frac{T_1}{T_0}$$

$$= 1 \times 1.1 \times \ln \frac{1100}{298}$$

$$= 1.4365\ \text{kJ/kg.K}$$

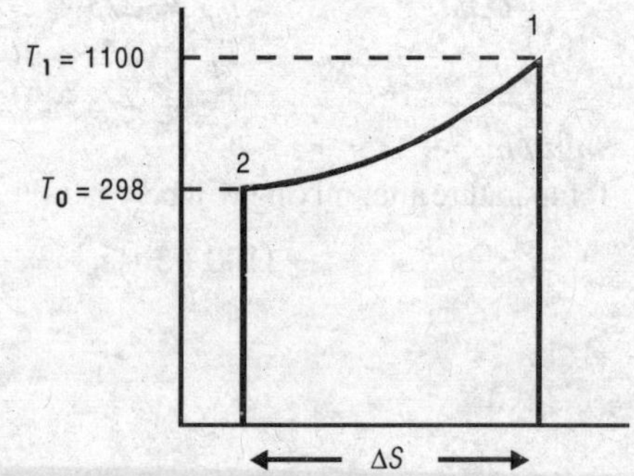

Fig. Ex. 9.4

(a)

$$UA = T_0 \times \Delta S$$
$$= 298 \times 1.4365$$
$$= 428.1\ \text{kJ/kg-K}$$

Thus lost available energy = 428 kJ/kg

(b) Ratio of lost available energy/useful work

$$= \frac{428.1}{1000} = 0.428$$

Example 9.5 *A hot spring produce water at a temperature δ 57°C, at the rate of 0.1 m³/minute. The water into a large lake with minimum temperature of 15°C. What is the rate of working of an ideal heat engine which use all the available energy.*

Solution

Now

$$T_1 = 273 + 57 = 330\ \text{K}$$
$$T_0 = 273 + 15 = 288\ \text{K}$$

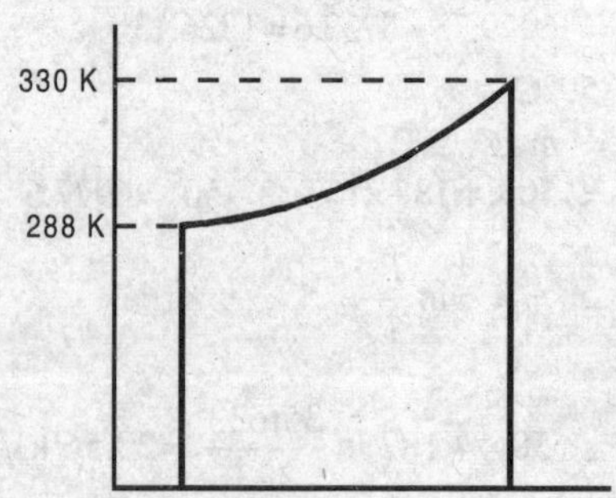

Fig. Ex. 6.5

$$\dot{m} = 0.1\ m^3/min$$

$$= \frac{0.1 \times 1000}{60} kg/sec = 1.667\ kg/s$$

(a) Heat given by the cooling water/s

$$\Delta Q = m\, C_P\,(T_1 - T_0)$$

$$= 1.6667 \times 4.187 \times (57 - 15)$$

$$= 293.09\ kJ.s$$

(b) Change of entropy of spring water

$$= \dot{m}\, C_P \ln\frac{T_1}{T_0} = 1.667 \times 4.187 \times \ln\frac{330}{288}$$

$$\Delta S = 0.94999\ kJ/K$$

$$UA = T_0\,\Delta S = 288 \times \Delta S = 273.6\ kJ$$

(c) $\text{Available energy} = Q - UA = 293.09 - 273$

$$\cong 19.5\ kJ/s = 19.5\ kW$$

Thus rate of working of an ideal heat engine which are all energy is 19.5 kW.

Example 9.6 *1 kg of air of pressure p_1 and temperature 900 K is mixed with 1 kg of air at the same pressure and 500 K. Determine the loss in availability if surrounding temperature is 27°C. Take C_P for air = 1.0 J/kg-K.*

Solution

$$T_0 = 27 + 273 = 300\ K$$

Temprature of air after mixing one kg each will be (900 + 500) / 2 = 700 K

Change in entropy.

(a) For air at 900 K $\quad \Delta S = \dot{m}\, C_P \ln\frac{T_2}{T_1}$

$$= 1 \times 1 \times \ln\frac{700}{900} = -0.2513 \qquad ...(1)$$

(b) For air at 500 K $\quad \Delta S = \dot{m} C_P \ln\frac{700}{500} = 1 \times 1 \times \ln\frac{700}{500}$

$$= 0.3365$$

(c) Net change in entropy after mixing is

$$(1) + (2) = 0.0852$$

$$\text{Loss in availability} = m \times T_0\ (\text{Net change in entropy})$$

$$= (1 + 1)\ 300 \times 0.0852$$

$$= 2 \times 300 \times 0.0852$$

Loss in Availability = 51.12 K

Example 9.7 *500 kJ of heat from a finite temperature source at 1000 K is supplied to 2 kg of gas initially at 2 bar and 350 K in a closed tank. Find the loss in available energy due to the above heat transfer. Take C_V of gas on 0.8 kJ/kg-K and temperature of surrounding as 27°C.*

Solution

With the addition of 500 kJ heat, the temperature of the gas will increase.

Thus $\quad \text{Heat added} = \dot{m}_{gas} \times C_V\,(T_2 - 350)$

where 350 K is the initial temperature

$$500 = 2 \times 0.8 \times (T_2 - 350)$$

This gives $$(T_2 - 350) = \frac{500}{2 \times 0.8} = 312.5$$

or $$T_2 = 350 + 312.5 = 662.5\text{ K}$$

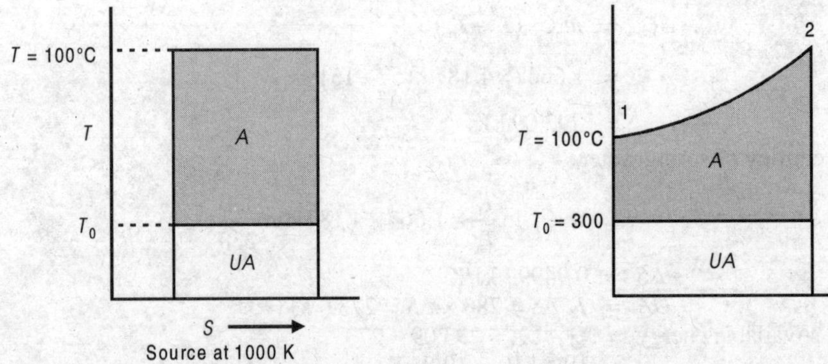

Fig. Ex. 6.7

Available UA Energy with Source

Now $$UA = T_0\,\Delta S$$
$$T_0 = 27 + 273 = 300\text{ K}$$

Change in entropy of the source at temperature of 1000 while 500 kJ is taken out is

$$\Delta S = \frac{\Delta Q}{T} = \frac{500}{1000} = 0.5\text{ kJ/kg-K}$$

$$UA = T_0 \Delta S = 300 \times 0.5$$
$$= 150\text{ kJ}$$
$$A = Q - UA = 500 - 150$$
$$= 350\text{ kJ}$$

An UA of the gas

Change in entropy of the gas at constant volume is given by

$$\Delta S = m_g\, C_V \ln \frac{T_2}{T_1}$$

$$= 2 \times 0.8 \times \ln \frac{662.5}{350}$$

$$= 1.0209398$$

$$UA = T_0 \times \Delta S = 300 \times \Delta S = 306.28\text{ K}$$

$$A = Q - UA$$
$$= 500 - 306.28 = 193.7\text{ kJ}$$

Loss in Availability due to heat transfer is

$$(1) + (2) = 350 - 193.7$$
$$\mathbf{A = 156.3\text{ kJ}}$$

Example 9.8 *Calculate the availability of a system which absorbs 1000 kJ of heat from a source at following constant temperature, while the environment temperature all three cases remain 27°C. (a) 600 K ; (b)1000 K ; (c) 1500 K.*

Solution

Now $A = Q - T_0\,\Delta S$

where $Q = 1\text{ H added} = 1000\text{ kJ}$

$\Delta S =$ Change in entropy, it will be different in all three cases.

Entropy change at constant temperature is

$$\Delta S = \int \frac{TQ}{T} = \frac{\Delta Q}{T} = \frac{1000}{T}$$

Thus

$$A = Q - T_0\,\Delta S$$

$$= 1000 - 300 \times \frac{1000}{T} = 1000\left(1 - \frac{300}{T}\right)$$

At 600 K $A_{600} = 1000\left(1 - \frac{300}{600}\right) = 1000(1 - 0.5) = 500\text{ kJ}$

At 1000 K $A_{1000} = 1000\left(1 - \frac{300}{1000}\right) = 700\text{ kJ}$

At 1500 K $A_{1500} = 1000\left(1 - \frac{300}{1500}\right) = 800\text{ kJ}$

Thus A increaseas temperature increases.

Example 9.9 *800 kJ of heat is added to a system during vaporisation process occurring at 150°C while pressure remain constant. Determine the available and unavailable energy of the surrounding temperature is 25°C.*

Solution

$UA =$ Unavailable Energy $= T_0\,\Delta S$

$A =$ Available Energy $= Q - UA$

Now $Q = 800\text{ kJ}$

$T_0 = 273 + 25 = 298\text{ K}$

$T = 273 + 150 = 423$

$$\Delta S = \int \frac{\delta Q_{rev}}{T}$$

$$= \frac{Q}{T}\text{(since temp. remain constant)}$$

$$UA = T_0 \times \Delta S$$

$$= 298 \times \Delta S = 563.59\text{ kg}$$

$$A = Q - UA = 800 - 563.59$$

$$\mathbf{A = 236.40\text{ kg}}$$

Example 9.10 *20 kg of water at 90°C is mixed with 32 kg of water at 30°C or constant pressure. Calculate the decrease in availability of surrounding air at 27°C.*

Solution

Decrease in Availability $= A)_{20} + A)_{32} - A)_{52}$

i.e., Sum of the original availability of the unmixed 20 kg and 32 kg water less the availbility of the mixture.

Temperature the mixture t_3

$$m_1 C_P t_1)_{20} + m_2 C_P t_2 = m_3 C_P t_3)_{52}$$

or $$20\times90+32\times30 = 52\times t_3$$

$$\Rightarrow \quad t_3 = 53°C\text{ (Approximately)}$$

Now $T_0 = 300$ K; $T_1 = 273 + 90 = 363$ K, $T_2 = 300$ and $T_3 = 273 + 53 = 326$ K

We have to find net change in entropy. Resulting due to 20 kg water at 363 K after mixing getting lower temperature 326

$$\Delta S_{20} = m_{20}C_P \ln\frac{363}{326} = 20\times4.187\times\ln\frac{363}{326}$$

$$= 9.0025 \quad \text{...(1)}$$

Similarly for 32 kg work at 303

$$\Delta S_{32} = m_{32}C_P \ln\frac{303}{326} = -9.8028 \quad \text{...(2)}$$

Net change in entropy = (1) + (2)

= – 0.80037

Loss in Availability = $T_0\,\Delta S = 300 \times 0.80037$

Loss in Availability = 240.112 kJ

On mixing the water at 90°C (363 K)

Example 9.11 *From a fire box of a boiler 1000 kg of heat leave hot gases at 1400°C and is given to steam at 250°C. Atmospheric temperature is 20°C. Divide the energy into available and unavailable portions (a) As it leave the hot gas (b) As it reaches the steam (c) Also show that the equation that*

Increase in unavailability = $T_0\,\Delta S$ holds good.

Solution

Now
$$T_0 = 273 + 20 = 293 \text{ K}$$
$$T_2 = 273 + 250 = 523 \text{ K}$$
$$T_1 = 273 + 1400 = 1673 \text{ K}$$

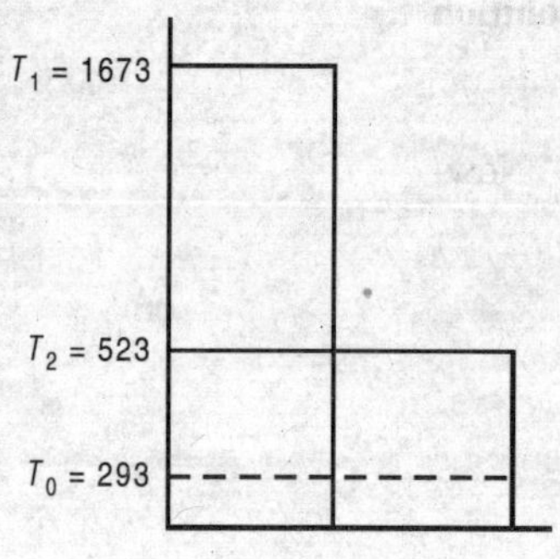

Fig. Ex. 9.11

Case (*a*) Hot Gas at 1673 K

$$UA = T_0\,\Delta S$$

Since heat is taken out at constant temperature

$$\Delta S_{Gas} = \frac{Q}{T} = \frac{1000}{1673}\text{ kJ/kg-K}$$

$$UA = 293\times\frac{1000}{1673} = 175.13\text{ kJ}$$

$$A = -1000-(-UA) = -175.13$$
$$= -824.87\text{ kg}$$

(Heat is leaving the system)

Case (*b*) Heat reaching steam at 523 K

$$\Delta S_{steam} = \frac{Q}{T} = \frac{1000}{523}$$

$$UA = T_0\,\Delta S = 293\times\frac{1000}{523} = 560.23\text{ kJ}$$

$$A = Q - T_0\Delta S = 1000 - 560.23 = 439.77\text{ kJ}$$

(c) To prove that increase in $UA = T_0\,\Delta S$, we have to find

$$\Delta S_{total} = \Delta S_{gas} + \Delta S_{steam}$$

$$= -\frac{1000}{1673} + \frac{1000}{523} = 1.3143$$

$$T_0 \Delta S = 293 \times 1.3143 = 385.1$$

Increase in unavailability $= (1) + (2)$

$$= 175.13 + 560.23$$

$$= 385.1$$

Since $(3) = (4)$, hence proved.

Example 9.12 *900 kJ of heat is added to a system during vaporisation process occurring at 200°C and at a constant pressure. Determine the available energy if the surrounding are at 17°C.*

Solution

$$\text{Available } A = Q - T_0 \Delta S$$

where

$$Q = \text{Heat added} = 900 \text{ kJ}$$

$$T_0 = 17 + 273 = 290 \text{ K}$$

$$\Delta S = \text{Entropy changes}$$

$$= \frac{\Delta Q}{T} \text{ since temperature is constant}$$

$$= \frac{900}{273+200} = \frac{900}{473} \text{ kJ/kg-K}$$

$$\therefore \quad A = 900 - 290 \times \frac{900}{473}$$

$$= 900 - 551.797$$

$$\mathbf{A = 348.203 \text{ kJ}}$$

Example 9.13 *An amount of 90 kJ of heat is added to a touch of air that is originally at 26°C. The touch contains 2.8 kg of air. The surrounding temperature is 15°C. Find available and unavailable energy.*

Solution

Here the heat is added at constant volume. C_V for air = 0.718 kJ/kg-K with the addition of 90 kJ heat, the temperature of air in the touch will increase by ΔT

Thus

$$90 = mC_V \Delta T$$

or

$$90 = 2.8 \times 0.718 \, \Delta T$$

$$\Delta T = 44.76$$

T_2 i.e., the temperature of air after heat addition

$$T_2 = 26 + 44.76 = 70.76$$

$$= 70.76 + 273 = 343.76$$

$$\cong 444 \text{ K}$$

Now change of entropy ΔS for constant volume increase

$$\Delta S = C_V \ln \frac{T_2}{T_0}$$

$$= 0.718 \ln \frac{444}{273+15}$$

$$\mathbf{\Delta S = -0.31079}$$

Example 9.14 *A temperature of 2000°C is obtained in a furnace by burning fuel in air at atmosphere pressure and ambient tempeature of 27°C. The gases can be assumed to be perfect gas with C_p = 1.0 kJ/kg-K. Determine the availability of heat in the products of combustion.*

Solution

$$T_0 = 27 + 273 = 300\text{ K}$$

For unit mass of gas –

(a) Heat addition

$$Q = mC_P \Delta T$$
$$= 1 \times 1.00 \times (2000 - 27)$$
$$= 1 \times 1983 = 1983\text{ kJ/kg}$$

(b) We have now to find change is entropy during the process

$$\Delta S = mC_P \ln \frac{T_1}{T_2}$$
$$= 1 \times 1.0 \ln \frac{2000 + 273}{27 + 273}$$
$$= \ln \frac{2773}{300} = 2.025\text{ kJ/kg-K}$$

(c)

$$UA = T_0 \Delta S = 300$$
$$= 300 \times 2.025 = 607.522\text{ kJ/kg}$$

(d)

$$A = Q - UA = 1983 - 607.522$$
$$= 1375.478\text{ kJ/kg}$$

THEORY QUESTIONS

9.1. Explain
(a) Dead state
(b) Availability
(c) Unavailable energy

9.2. Write short notes on:
(a) Irreversibility
(b) Effectiveness
(c) Second law efficiency

9.3. Comment on 'There is no energy crisis but available energy crisis'.

9.4. Derive expressions for available energies for heat source
(a) At constant temperature
(b) At variable temperature

PROBLEMS FOR PRACTICE

9.1. Determine available and unavailable apart of energy of a system which absorbs 15 MJ of heat from a source at 227°C. The surroundings are at 17°C. (**Ans.** $UA = 8.7$ MJ, $A = 6.3$ MJ)

9.2. A source at 227°C receives 7200 kJ/min from a source at a temperature of 727°C. The atmospheric temperature is 27°C. Assuming that the temperature of the system and that of the source remain constant, find the decrease in availability after the heat transfer.

(**Ans.** 2160 kJ)

9.3. A steel ball of 800 kg at 1250 K is cooled to 500K. Using steel ball as source of energy, compute available and unavailable energies. Assume specific heat of steel to be 0.5 kJ/kg-K and ambient temperature to be 27°C.

(**Ans.** $UA = 110$ MJ, $A = 190$ MJ)

9.4. Air 0.2 kg initially at a temperature of 575 K receives 300 kJ heat reversibly at constant pressure. Find available and unavailable portions of the heat added. Take C_p for air to be 1.005 kJ/kg K and ambient temperature to be 27°C.

(**Ans.** $UA = 77.16$ kJ, $A = 222.54$ kJ)

9.5. During a vaporization process at constant

pressure taking place at 180 C, 850 kJ of heat is added while ambient temperature is 299 K. Find available and unavailable energies.

(**Ans.** $UA = 55.53, A = 296.46$ kJ)

9.6. A steam stable condenses at 1000° C rejecting 3000 kJ of heat while another sample of heat condenses at 500°C rejecting the same quantity of heat. Determine the difference in the available energy between the two processes. Take sink temperature to be 27°C.

(**Ans.** $A = 2840$ kJ, $UA = 1160$ kJ)

9.7. A constant temperature source is maintained at 1000 K and the surroundings are at 290 K. If 4000 kJ heat is transferred from the source reversibly, determine available and unavailable portions of heat. (**Ans.** $A = 2840$ kJ, $UA = 1160$ kJ)

9.8. From a heat reservoir 1000 kJ of heat is withdrawn at 600 K. Find the availability and unavailability of heat if surroundings are at 290 K.

(**Ans.** $A = 516.7$ kJ, $UA = 483.3$ kJ)

9.9. 500 kJ of heat was removed at constant temperature at 527°C and supplied a system at constant temperature at 427°C. Find the loss in available energy as a result of this irreversible heat transfer. Take ambient temperature as 27°C.

(**Ans.** 26.79 kJ)

CHAPTER 10

Steam Generation and Properties, Processes of Steam

10.1 INTRODUCTION

Upto now we have considered a fluid, which under normal conditions remains in a gaseous state. Such substances are oxygen, nitrogen, air etc. Now we consider fluids, which under normal working conditions may be either in a liquid state or in a gaseous state. Such substances are known as vapours. Examples of vapours, employed in engineering practices are steam, ammonia, freon, mercury etc. The behaviour of vapours is typically represented by steam and water.

With vapours, the simple laws of perfect gas does not even approximately apply and therefore sometimes the definition of vapour is given as: "A vapour is any substance in the gaseous state which does not even approximately follow the general gas laws". The reason for this departure of the performance of vapours from the general gas law is that the vapours from the general gas law is that the vapours when heated or cooled experience some degregation work while the general gas laws are based on a perfect gas, which would experience no desegregation work when heated or cooled. It should be remembered that all known actual gases do experience a small amount of desegregation work. So we can use the formula derived for perfect gas, for vapour calculations. For this purpose we use values, which have been determined by experiments and listed in tables.

Highly superheated vapours approximately follow the perfect gas laws. Vapour obeys its own Laws.

10.2 FORMS OF MATTER

There are three forms of matter, viz. (i) Solid, (ii) Liquid and (iii) Gaseous.

In order to explain the changes in form of matter, a hypothetical experiment may be described with the help of Fig. 10.1.

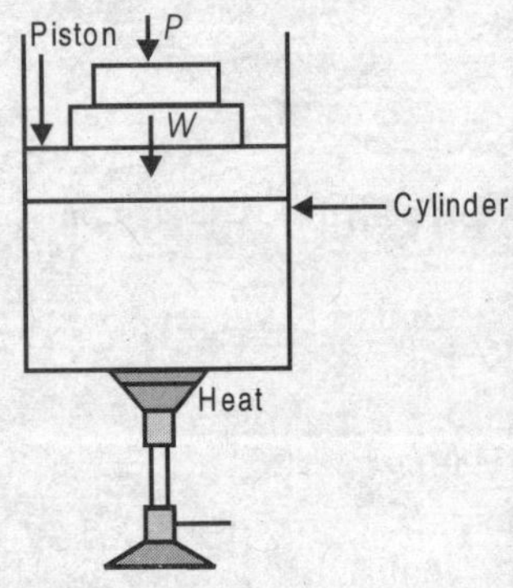

Fig. 10.1

As substance in the solid state is place in a cylinder, which is fitted with a frictionless right fitting piston being loaded by a weight as shown in Fig. 10.1. Heat is supplied to the cylinder. The substance may change in volume during the heating process but the pressure is kept constant by the piston and weights on it. The change in temperature of the substance will be recorded by a thermometer (Not shown).

At first, when the heat is added, the temperature of the solid rises and a certain value of temperature will be reached after which a considerable amount of heat is added, without a change in temperature. Through a sight window provided in the cylinder wall it can be seen that the solid is melting. When the solid has been completely transformed into liquid, further addition of heat again causes a rise in temperature. This rise in temperature continues till a point is reached where further addition of heat does not change the temperature of the substance but on observation we see that the liquid is boiling. When the liquid has been completely transformed into a vapour, the temperature again rises when the heat is added. The relation between the heat added, entropy change and the temperature of the substance is shown in Fig. 10.2.

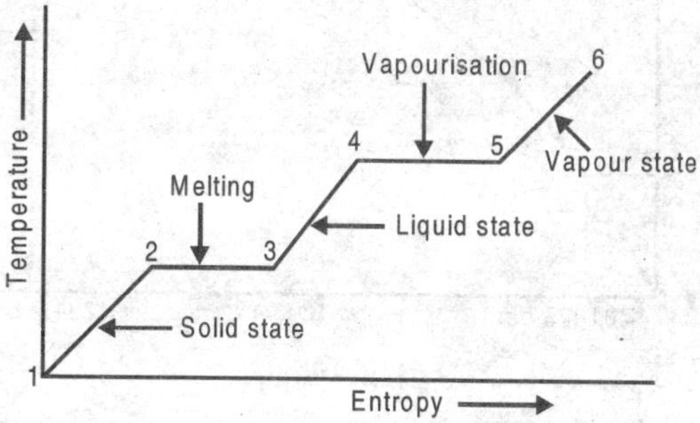

Fig. 10.2

1–2 Shows the stage during which the rise in temperature of a substance in solid state takes place.

2–3 Shows the stage during which the substance is being transformed from a solid state to a liquid state without rise in temperature.

3–4 Shows the stage during which rise in temperature of the substance in liquid state takes place.

4–5 Shows the stage during which the substance is being transformed from a liquid state to a vapour state without rise in temperature.

5–6 Shows the stage, during which rise in temperature of the substance may take place in a vapour state.

If the hot vapour is cooled the curve will be exactly retraced with all changes taking place in reverse manner. If the weight W is increased and the experiment is repeated the same general behaviour is obtained but the temperature will be higher than that in the first case. If the weight is reduced still the behaviour will be the same but temperature will be lower than that in first case.

The above hypothetical experiment brings out certain important facts.

- When a solid melts the temperature remains constant and is always the same for a given substance at a given pressure. When a liquid freezes the temperature remains constant. This temperature is called the freezing point. The quantity of heat removed or added during the change of state is constant and is known as *latent heat of fusion*. The latent heat of fusion depends upon pressure.
- If heating of water is continued after total melting, temperature increases till it starts boiling. The amount of heat added to one kg of water from freezing point to boiling point is called *sensible heat*.
- When the liquid boils the temperature remains constant and is always the same for a given substance at a given pressure. If the pressure is increased, the temperature at which the change of state takes place also increases. Similarly when the vapour condenses the temperature also remains constant. This temperature is called the *boiling point* or saturated temperature of the liquid. The definite quantity of heat removed or added during the change of state is called the *latent heat of vaporisation*. The latent heat of vaporisation depends upon pressure.

The freezing point, the latent heat of fusion, the boiling point and the latent heat of vaporisation vary widely with different substances.

When the pressure is vary low, an interesting phenomenon is noticed. A solid when heated directly transforms into a vapour without passing through an intermediate liquid state. Such a phenomenon is known as *sublimation*. A familiar example is the conversion of solid CO_2 (dry ice) directly into vapour. Figure 10.3 shows a temperature pressure curve.

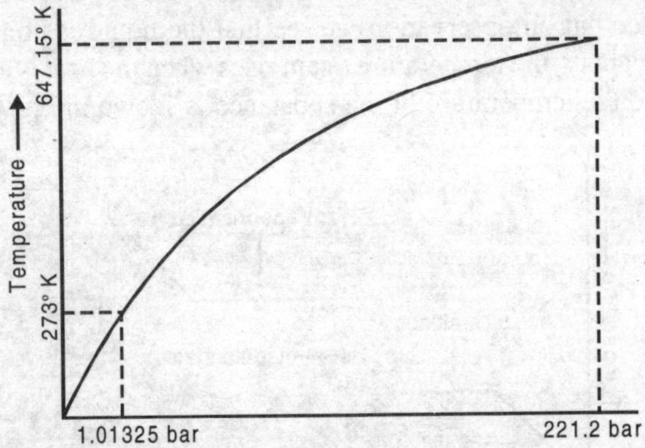

Fig. 10.3

In practice, steam is generated in a device called steam boiler.

10.3 GENERATION OF STEAM AT A GIVEN PRESSURE

Consider one kg of water at 0°C. Also let us decide upon the pressure under which one kg of water is to be heated. Consider a cylinder with a frictionless piston loaded such that the pressure *P* under consideration is acting on water. See Fig. 10.1.

Heat is added to the water so that its temperature increases from 0°C to t_{sat} (saturation temperature) at which the water boils at pressure *P*. Volume of water increases to V_f. This is shown in Fig. 10.4(a).

Heat supplied or added to 1 kg water to raise its temperature from 0°C to saturation temperature t_{sat} *is called sensible heat* because it can be detected by the sense of touch and produces a rise in temperature to be seen on a thermometer. Generally this sensible heat is denoted by h_f.

Note. V_f means volume of fluid which is taken as liquid.

Heating is continued after the water reaches saturation temperature. Evaporation of water takes place and steam is generated at pressure *P*. This heating is continued till all the water is evaporated and the temperature of the steam is saturation temperature. V_g or V_{sat} denotes the volume of steam at this condition. The external work done during evaporation is given by $W = P\,(V_{sat} - V_f)$ or $p\,(V_g - V_f)$ with proper units.

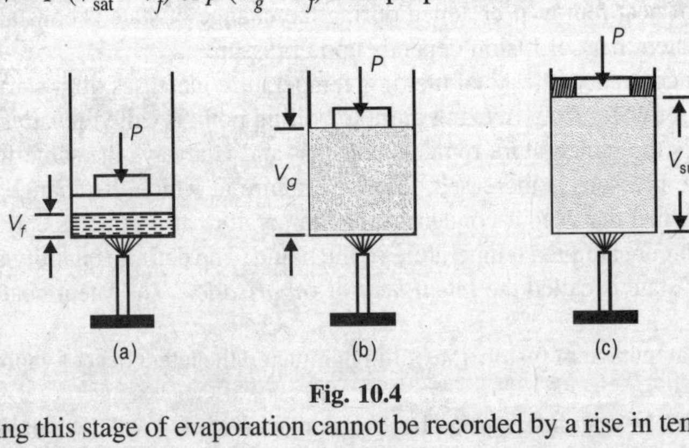

Fig. 10.4

Because the heat during this stage of evaporation cannot be recorded by a rise in temperature on the thermometer, it is called Latent heat (hidden heat).

The amount of heat required to evaporate 1 kg of water at saturation temperature to steam or vapour at saturation temperature is called Latent heat or enthalpy of evaporation and is generally denoted by h_{fg}.
This is shown in Fig. 10.4(b).

Once all the water is evaporated to saturated steam, if heat is added further, the steam behaves approximately as an ideal gas and the heating of saturated steam increases its temperature. *The steam whose temperature is greater than saturation temperature at the given pressure, is called a superheated steam*; and let its temperature be denoted by t_{sup}. Its volume will also increase to V_{sup}.

The amount of heat required to raise the temperature of l kg steam from t_{sat} *to* t_{sup} *is called superheat* and the temperature difference called *degrees of superheat.*

$$\therefore \quad \text{Superheat} = (\text{sp. heat of steam}) \times (\text{Rise in temperature})$$

$$= C_{pv}\,(t_{sup} - t_{sat})$$

$$\text{Degree of superheat} = t_{sup} - t_{sat}$$

The value of C_{pv} varies from 2.1 to 2.3 kJ/kg-K.

$$\text{Similarly,} \quad \frac{T_{sup}}{T_{sat}} = \frac{V_{sup}}{V_{sat}}$$

This is according to Charles's law of p = Const. (volumes are proportional to absolute temperatures)

$$\therefore \quad V_{sup} = V_{sat} \times \frac{T_{sup}}{T_{sat}}$$

This is shown in Fig. 10.4(c).

Note: V_{sat} and V_g also means volume of gas which is in this case, steam

$$h_f = \text{sensible heat} = C_{pw}\,(t_{sat} - 0) \text{ kJ/kg}$$

where C_{pw} = specific heat of water which is taken as 4.187 kJ/kg-K if not given.

$h_g = h_f + h_{fg}$ for 1 kg of dry and saturated steam.

Total heat or enthalpy of superheated steam

$$h = h_g + C_{pv}\,(t_{sup} - t_{sat})$$

10.4 CONDITIONS OF STEAM

Steam may occur in any one of the following three conditions:

(a) Saturated steam, which may be either dry or wet, (b) Superheated steam and

(c) Supersaturated steam

10.4.1 Saturated Steam

Saturated steam is a vapour at the temperature corresponding to the boiling point of the liquid at the given pressure. A substance, which is in the vapour state in a confined space and is in contact with some of the same substance in liquid state is always at the same temperature as the liquid and is saturated vapour.

If saturated steam does not contain any water, it is known as dry and saturated steam. It contains just sufficient heat energy to maintain all the water in a gaseous state.

If saturated steam contains liquid particles, it is known as wet steam. It does not contains sufficient heat energy to maintain all water in a gaseous state.

If some of the heat energy is absorbed from the dry saturated steam, the steam becomes wet. In practice it is difficult to get an absolutely dry saturated steam if it is produce by boiling water, because some of water particles are carried away with the steam.

10.4.2 Superheated Steam

If the temperature of the steam is greater than that of the boiling point or saturation temperature corresponding to the pressure of steam generation, the steam is known as superheated steam.

The amount of superheat in the steam is given in terms of the difference between its temperature and that of the saturated steam at the pressure of steam generation. If the temperature of the steam is 100 degrees higher than the saturation temperature corresponding to the pressure of steam generation, we say that the steam has 100 degrees of superheat.

10.4.3 Supersaturated Steam

Supersaturated steam at a particular saturation pressure has temperature less and density greater than the corresponding values given in the steam tables. This condition of steam is obtained when it is cooled by its own expansion until it contains less heat energy than the saturated steam under the same conditions. This state of steam is varying, unstable and the steam soon returns to the saturation condition. Such a state of steam occurs in expansion in a nozzle.

10.5 PROPERTIES OF STEAM

The properties of steam are interrelated. If we know certain properties, the other properties may be found out or derived. For example, if the pressure of saturated steam is observed by the pressure gauge, its temperature can be found from the steam tables in which the results of the various experiments have been tabulated.

(a) Dryness Fraction of Saturated Steam. We have seen that steam in contact with water contains liquid particles in suspension. Thus the steam consists of dry saturated and water particles in suspension.

The dryness fraction of steam is defined as the ratio of the weight of dry steam in a certain quantity of steam to the weight of total wet steam. It is generally denoted by the letter x. If W is the weight of dry steam in the total steam W_t, then $x = \frac{W_s}{W_t}$.

The dryness fraction of a wet steam may also be defined as the amount of dry steam in unit amount of the steam. When $x = 1$, the steam is said to be dry and saturated.

(b) Sensible heat (h_f): Refer Sec.10.3

(c) Latent heat or enthalpy of evaporation h_{fg} Refer Sec. 10.3

(d) Superheat and degrees of superheat: Refer Sec. 10.3

(e) Volume of wet steam: $V = xV_g$

(f) Volume of dry and saturated steam: $V = V_g$

(g) Volume of superheated steam: $V_{sup} = V_{sat} \times \frac{T_{sup}}{T_{sat}} = V_g \times \frac{T_{sup}}{T_{sat}}$

10.6 STEAM TABLES

Experimentally, accurately determined, values are tabulated in tables that are called as Steam tables. These values form the basis for many calculations concerned with steam engineering. These tables are to be used because vapours do not obey general gas laws. The values given in the tables are for one kg of dry and saturated steam but these values can also be employed for wet steam calculations. Properties of dry and saturated steam and superheated steam at various pressures and temperatures for SI units are given at the end of this book.

In order to determine the properties of steam at some pressure between those given in tables, we interpolate assuming the linear relation between these values. The method is very simple and at the same time accurate.

Steam Table: *Properties of Dry and Saturated Steam.*

Pressure bar	*Sat. temp. °C*	*Volume m³/kg*		*Enthalpy kJ/kg K*			*Entropy kJ/kg K*		
		V_f	V_g	H_f	H_{fg}	H_g	S_f	S_{fg}	S_g
10.0	179.88	0.0011274	0.19429	762.61	2013.6	2776.2	2.1382	4.4443	6.5828
10.5	182.02	0.0011303	0.18545	772.03	2005.9	2778.0	2.1588	4.4071	6.5659

Method of linear interpolation to find, say, enthalpy at 10.2 bar pressure.

Difference between two values at 10 and 10.5 is 772.03 – 762.61= 009.42 kJ / kg.

This is the difference for 0.5 bar pressure.

∴ For 0.2 bar difference, value is $\left(\frac{0.2}{0.5}\times 9.42\right) = 3.768$ kJ/kg.

Now, this is to be added to the value at 10 bar.

(This addition should be done if the value is increasing with pressure. If the value is decreasing, it should be deducted. This is seen in case of volumes.)

∴ Enthalpy at 10.2 bar pressure

$$= 762.61 + 3.768$$
$$= 766.378 \text{ kJ/kg.}$$

10.7 INTERNAL ENERGY AND ENTROPY OF STEAM

10.7.1 Internal Energy of Steams

As in case of ideal gases, internal energy of steam is given by ($u = h - pV$).

∴ For wet steam $\quad u = (h_f + xh_{fg}) - p.xVg$

For dry and saturated steam $u = h_g - p.V_g$

and for superheated steam $u = [h_g + C_{pv}(t_{sup} - t_{sat})] - \left[p.V_g.\frac{T_{sup}}{T_{sat}}\right]$

10.7.2 Entropy of Steam

The change of entropy during any reversible process is defined as the ratio of heat added or rejected to the absolute temperature at which the heat is added or rejected.

We can apply the definition for getting entropy of steam above 0°C.

Consider one kg of water at 0°C and pressure p. Let heat be added so that the temperature increases to saturation temperature corresponding to pressure p. Here the temperature changes. Let during the heating, process, T be the temperature and an infinitesimal heat δQ is added and for the process 1–2; where the temperature changes from 0ºC (273K) to T_{sat} (saturation temperature at p).

$$\therefore \quad \delta Q = Cp_w dT \text{ and } dS = \frac{\delta Q}{T}$$

$$\therefore \quad dS = \frac{Cp_w \cdot dT}{T}$$

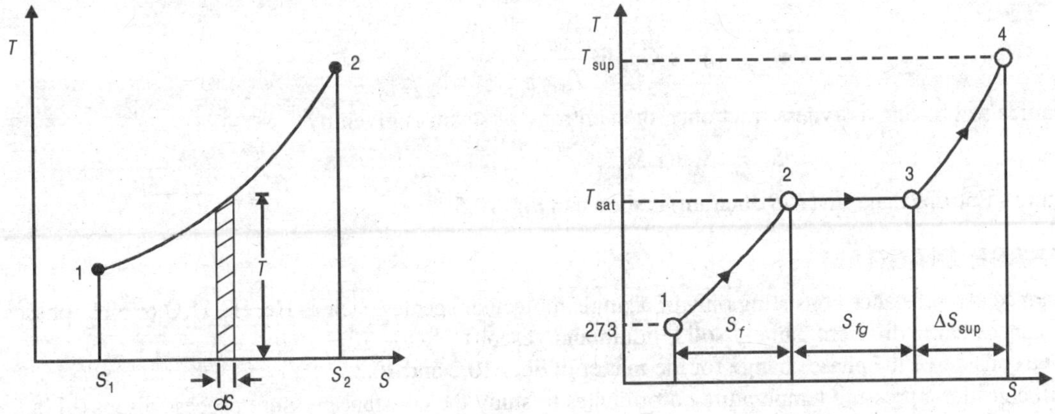

Fig. 10.5

∴ Change of entropy during sensible heating of water from 273 K to T_{sat},

$$\int_1^2 dS = S_2 - S_1 = \Delta S_{1-2} = C_{p_w} \int_{273}^{T_{sat}} \frac{dT}{T}$$

This in denoted by s_f and is given in steam tables.

∴
$$s_f = C_{p_w} \ln . \frac{T_{sat}}{273}$$

Water, at saturation temperature, if heated further, evaporation takes place. Here the heat added is latent heat or enthalpy of evaporation and temperature is saturation temperature; heat added is h_{fg}, and the temperature is constant

∴ Change of entropy during evaporation 2–3 is given by

$$\Delta S_{2-3} = \frac{h_{fg}}{T_{sat}} = s_{fg} \text{ and is given in steam tables.}$$

and the entropy of dry and saturated steam is given by

$$\Delta S_{1-3} = \left[C_{p_w} \ln \frac{T_{sat}}{273} + \frac{h_{fg}}{T_{sat}} \right] = s_g .$$

It is given in steam tables.

Similarly, if the steam is further heated so that its temperature is T_{sup} (temperature of superheated steam), then the entropy change during superheating is given by

$$\Delta S_{3-4} = Cp_v \ln . \frac{T_{sup}}{T_{sat}}$$

∴ Total change of entropy of superheated steam or simply entropy of superheated steam is given by

$$\Delta S_{1-4} = S = C_{p_w} \ln . \frac{T_{sat}}{273} + \frac{h_{fg}}{T_{sat}} + Cp_v \ln . \frac{T_{sup}}{T_{sat}} .$$

$$= S_f + S_{fg} + C_p \ln . \frac{T_{sup}}{T_{sat}} .$$

$$= S_g + C_{pv} \ln \frac{T_{sup}}{T_{sat}}$$

If the steam is wet having a dryness fraction x, then entropy of steam is given by

$$S = S_f + x . S_{fg} .$$

Temperature – Entropy diagram (T–S diagram) is shown in Fig. 10.5.

10.8 PHASE DIAGRAM

Pure substance is a substance consisting only of a single molecular species such as He, H_2, H_2O or NH_3, phases of these substances being different namely solid, liquid and gaseous.

We have considered the phase change for the matter in Secs 10.2 and 10.3.

We will consider a pressure temperature co-ordinates to study the constant pressure process discussed in Secs 10.2 and 10.3.

Figure 10.3 shows such a diagram where the phase changes are shown. Such a diagram is called a phase diagram for pure substance.

For a particular pressure, the horizontal line *a, b,c,d,e* is drawn as shown in Fig.10.6.

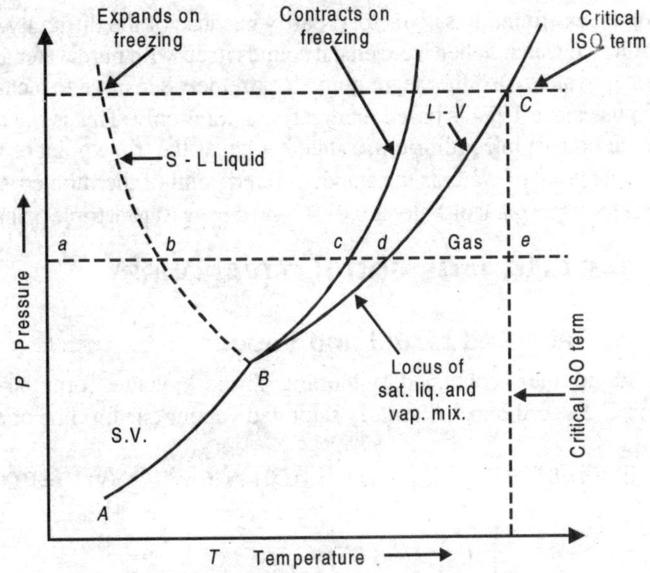

Fig. 10.6

a = Solid state
b = Melting point of ice
c = Boiling or saturation temperature point, C does not apply to substance that expand on freezing
d = dry and saturation point
dc = Supcrhcatcd vapour
bd = Liquid vapour region

This process is repeated at numerous pressure and the point analogous to b and d duly noted, say down to point B. A smooth curve through all the d points is the phase boundary between liquid and vapour; it is the locus of all points representing saturated mixtures. The smooth curve through all points b is the phase boundary for solid and liquid; points on this line represent all equilibrium mixtures of solid and liquid.

The states B and C in Fig. 10.6 are particularly significant. Point B, the junction of the three phase boundaries, is called the triple point because it represents all equilibrium mixtures of solid, liquid and vapour.

The critical point C where the curve BC ends is a state where liquid and vapour are indistinguishable. As this point is approached the latent heat of evaporation dccrcascs. At C it is zero. The temperature T_C here is called the critical temperature and the corresponding pressure is the critical pressure (P_C). The specific volume at P_C and T_C is the critical volume.

Density of water and vapour at the point C is same.

For water, P_C = 221.2 bar
T_C = 374.15°C
V_C = 0.00317 m^3/kg

10.8.1 Triple Point

We know that to determine the temperature of any system, the thermometer must be brought to thermal equilibrium with an arbitrarily chosen standard system in an easily reducible state. The temperature of the standard system in the chosen state is called the fixed point. Before 1954 there were two fixed points : (i) the ice point and (ii) steam point.

The use of two fixed points was found unsatisfactory, partly because of the difficulty of achieving equilibrium between pure ice and air-saturated water. When ice melts, it sounds itself with pure water and thus prevents intimate contact of ice and air saturated water. Also the steam point is extremely sensitive to a change in pressure.

The temperature scale in use since 1954 is based on one fixed point only. This is the temperature at which ice, liquid water, and water vapour coexist in equilibrium, a state known as the triple point of water.

The temperature of the triple point of water is the standard fixed point of thermometry.

The Celsius temperature of the triple point of water is 0.01 °C and the pressure at triple point for water is 0.006114 bar.

10.9 THE *P–V* AND *T–S* DIAGRAMS FOR H_2O (VAPOURS)

10.9.1 P–V Diagram for Saturated Liquid and Vapour

The saturated water curve (liquid line) is obtained by plotting: *P* and V_f values form the steam tables on *P–V* co-ordinates and T–S_f values on *T–S* coordinates. Similarly saturated vapour (steam) line or curves are obtained from steam tables.

(i) 221.2 bar, (ii) 374.15°C, (iii) $V_C = 0.00317\ m^3/kg$, (iv) $V_b = 0{,}0014526\ m^3/kg$, (v) $V_a = 0.0014526\ m^3/kg$.

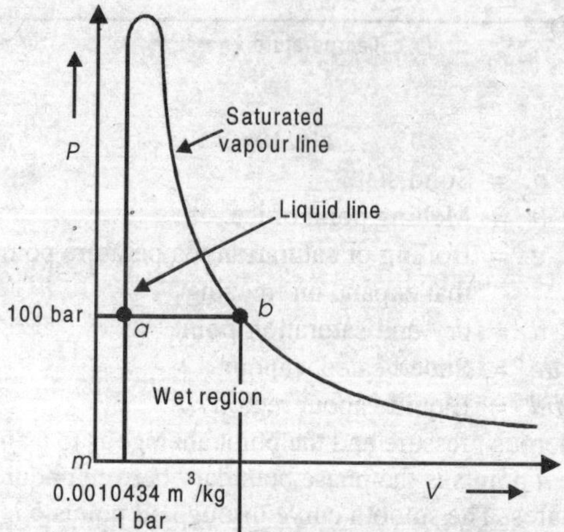

Fig. 10.7

Notice here that saturated vapour line flattens out towards the right, indicating that the volume increases at an increasing rate at low pressure.

In Fig. 10.7, *m* represents the state where $V_f = 0.0010434\ m^3/kg$ at 1 bar pressure. *C* is the critical point where pressure is 221.2 bar, saturation temperature is 374.15°C and volume is $0.317\ m^3/kg$.

This diagram is not used in practice but is useful in general to study the property relations and the vapour behaviour.

10.9.2 The Temperature–Entropy Diagram for Steam

It is possible to plot on a Temperature Entropy diagram, a chart which gives the entropy of water and steam at all temperatures and pressures. Such a chart is very convenient for solving problems containing an adiabatic (reversible) expansion of steam. A Temperature Entropy Chart (*T–S* Chart) is plotted in Fig. 10.8 with co-ordinates representing the absolute temperature and entropy. The units of entropy are reckoned from the freezing point of water.

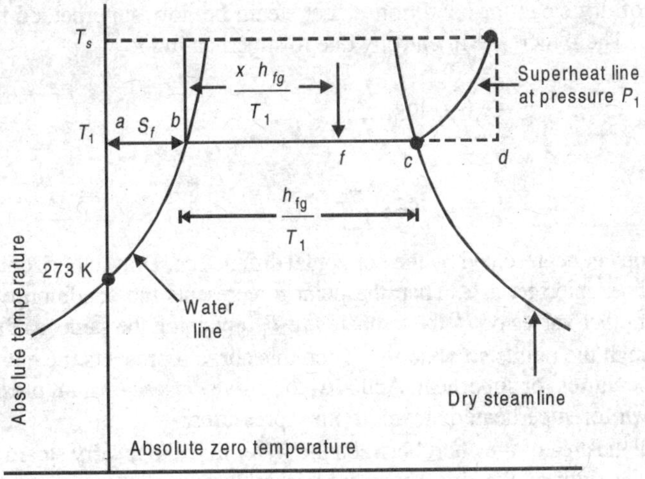

Fig. 10.8

10.9.3 Construction of T–S Diagram

Consider 1 kg steam at absolute pressure P_1 , the absolute temperature of formation at this pressure will be T_1; the value of T_1 can be obtained from the steam tables.

Then $$S_{f1} = \log_e \frac{T_1}{273}$$

$$= ab \text{ on the diagram.}$$

Let this value of S_{f1} be represent by *ab* to a convenient scale, and plot this distance on *T–S* diagram of Fig. 10.8. Repeat this for other pressure of steam and draw a curve through the points obtained. This curve is known as the saturation liquid line or simply liquid line. Sometimes it is also called water line and will give the values of S_f at all temperatures and pressure.

Again consider the steam at pressure P_1. During evaporation the change of entropy of this steam will be

$$S_{fg} = \frac{hf_g}{T_1}$$

Let this be represented by the distance *bc* and *be* plotted on the *T–S* chart. The point *C* will represent dry steam at pressure P_1. Repeat this for other pressures and temperatures and draw a curve through the points obtained. This curve is known as the dry steam line, and its horizontal distance from the vertical ordinate represents the total entropy of steam at any pressure.

If the steam at pressure P_1 had been wet, with a dryness fraction of *x*, the entropy of evaporation would be

$$\frac{x.hf_g}{T_1}$$

Let this be represented by *bf* and *be* plotted to the same scale. Then *af* is the total entropy of the wet steam, and the dryness fraction is the ratio of *bf* to *bc*.

$$\therefore \qquad x = \frac{bf}{bc}$$

Again, consider 1 kg of dry steam at condition *c*. Let steam be now superheated by increasing the absolute temperature from T_1 to T_S. Then, increase in entropy due to superheating is

$$= C_{p_w} \log_e \frac{T_s}{T_1}$$

$$= C_{p_w} \log_e \frac{T_{sup}}{T_{sat}}$$

Let this increase of entropy be represented by the horizontal distance *cd*. Draw a horizontal through T_S and a vertical through point *d* let these lines intersect at *e*. Then the point *e* represents the condition of the superheated steam at pressure P_1. Repeat this for other valves of T'_s the temperature T_1 remaining the same, as the pressure is constant.

Draw the curve *ce* through the points so obtained. Then this curve represents the entropy of superheated steam at pressure P_1, for all temperatures for superheat. Actually, the curve *ce* is a constant pressure line. Similar constant pressure lines can be drawn for superheated steam at other pressures.

It should be noticed that the area of the chart between the water line and the dry steam line represents wet steam. The area of the chart to the right of the dry steam represents superheated steam, and is known as the area of superheat, or superheat region.

Thus a complete *T–S* diagram for pressure P_1 is shown in Fig. 10.9.

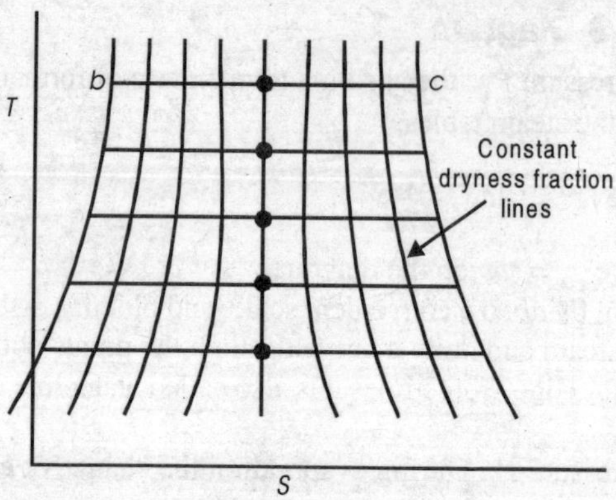

Fig. 10.9

In this diagram, the evaporation lines *bc* at all pressure can be divided into ten equal parts. Curves can then be drawn through each corresponding tenth part of these lines. These curves represent the dryness fraction of the steam at any temperature and pressure. These curves or lines are known as constant dryness fraction lines. Thus, any condition of steam can be represented by a corresponding point on the *T–S* diagram. Such a point is known as state point.

Dryness fraction lines are shown on *T–S* diagram of Fig. 10.9.

10.10 MOLLIER DIAGRAM OR TOTAL–HEAT–ENTROPY OR ENTHALPY–ENTROPY DIAGRAM (h–s DIAGRAM)

Another type of entropy diagram which is used by engineers is the Mollier diagram or Enthalpy – Entropy (*h–s*) diagram. This diagram represents the entropy and total heat or enthalpy of steam. It is shown in Fig. 10.10.

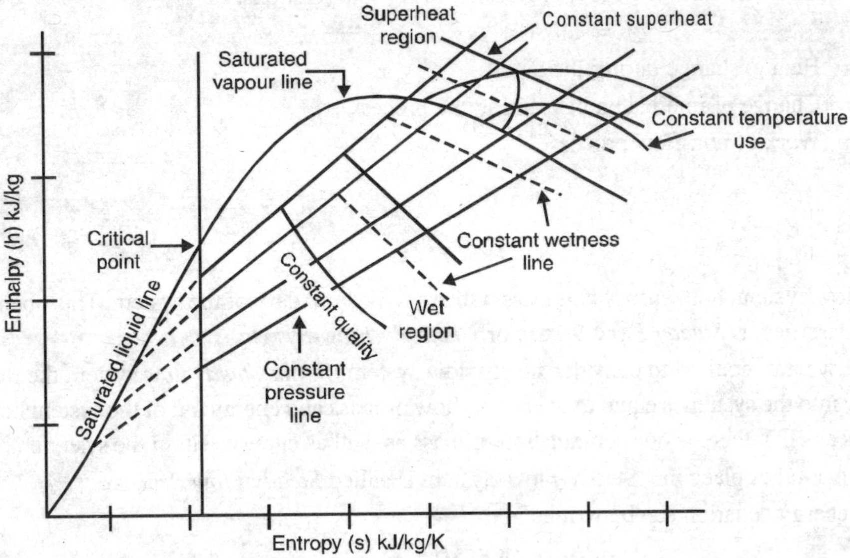

Fig. 10.10

The base of the diagram represents the entropy of the steam and the vertical ordinate represents the total heat or enthalpy of steam. The lower portion of the diagram is the wet steam area, whilst the upper portion of the diagram represents the superheated condition of the steam.

Constant pressure lines are plotted on the diagram through both the wet steam area and the superheat area. Constant pressure lines are straight lines in the wet region and curved in the superheat area. Lines of constant dryness fraction are plotted in the wet steam area, and lines of constant temperature in the superheat area.

As the base of the diagram represents entropy of the steam, an adiabatic (isentropic) expansion will be represented by a vertical line.

Again a throttling expansion will be represented by a horizontal line on this diagram, as throttling process is a constant enthalpy (isenthalpic) or total heat process.

The main use of this diagram is for finding the drop in total heat or enthalpy of steam during an isentropic expansion. The final condition of the steam after expansion can also be read off from the diagram. The drop in enthalpy during an adiabatic or isentropic expansion. The final condition of the steam after expansion can also be read off from the diagram. The drop in enthalpy during an adiabatic or isentropic expansion of steam is known as isentropic enthalpy drop.

10.11 NON-FLOW AND STEADY FLOW PROCESSES

In the previous course of thermodynamics we have studied about the different thermodynamic systems. Here we will first revise to certain extent about the system and the process.

Thermodynamic system is classified as:

1. Closed or Non-Flow system
2. Open onflow system and
3. Isolated system

In closed or non-flow system the boundary allows the energy to flow in or flow out i.e., the energy crosses the boundary of the system. Secondly, the boundary does not allow the mass to cross the boundary of the system. When the energy crosses the boundary of the system, the boundary may or may not move. The process that will take place in this closed or Non-flow system is called *Non-Flow Process*.

The energy equation for this Non-Flow Process is, with usual relations, as

$$Q = DU + W_{nf}$$

where Q = Heat exchange during the process

ΔU = Change of internal energy during the process

W_{nf} = Work of non-flow process

$$= \int_1^2 p \cdot dv$$

In open or Flow system, both energy and mass will cross the boundary of the system. This open or flow system can further be classified as *Steady-Flow System* or *Unsteady-Flow-System.*

At this stage, we are required to consider steady flow system. In this *steady flow* system the rate of mass flow and energy flow into the system is equal to the rate of flow of mass and energy out of the system i.e., $m_1 = m_2$, and $E_1 = E_2$. This means that there is no accumulation of mass as well as energy, within the system.

The process that takes place in a Steady-Flow-System is called *Steady Flow Process.*

Steady flow energy equation can be written as

$$Q = \Delta P + \Delta K + \Delta H + W_{sf}$$

where Q = Heat exchange during the process

ΔP = change in potential energy

ΔK = change in kinetic energy

ΔH = change in enthalpy

W_{sf} = work of steady flow process

$= h_1 - h_2$ for isentropic process when $\Delta K = 0$

and $h_1 - h_2 = K_1 - K_2$ for isentropic process and $W_{sf} = 0$

10.12 PROCESSES OF VAPOURS

The basic energy relations for the processes as defined for perfect gases also hold for vapours all previous equations in terms of the general symbols *W, Q, H, h, U, u, K, P* apply to any substance under the circumstances specified. The equations derived from the assumption of an ideal gas do not hold.

Remember that the areas on the *P–V* diagram under the curve at an internally reversible process represent $\int p.dv$, and that this area is the work of a non-flow process. The area behind the same curve is the $-\int v.dp$.

The vapour processes that are to be studied here are:

(i) Constant pressure process (Isobaric process)
(ii) Constant volume process (Isochoric process)
(iii) Reversible adiabatic process (Isentropic process)
(iv) Irreversible adiabatic process (Throttling process)
(v) Isothermal process
(vi) Polytropic process (Reversible)
(vii) Hyperbolic process

10.12.1 Constant Pressure Process

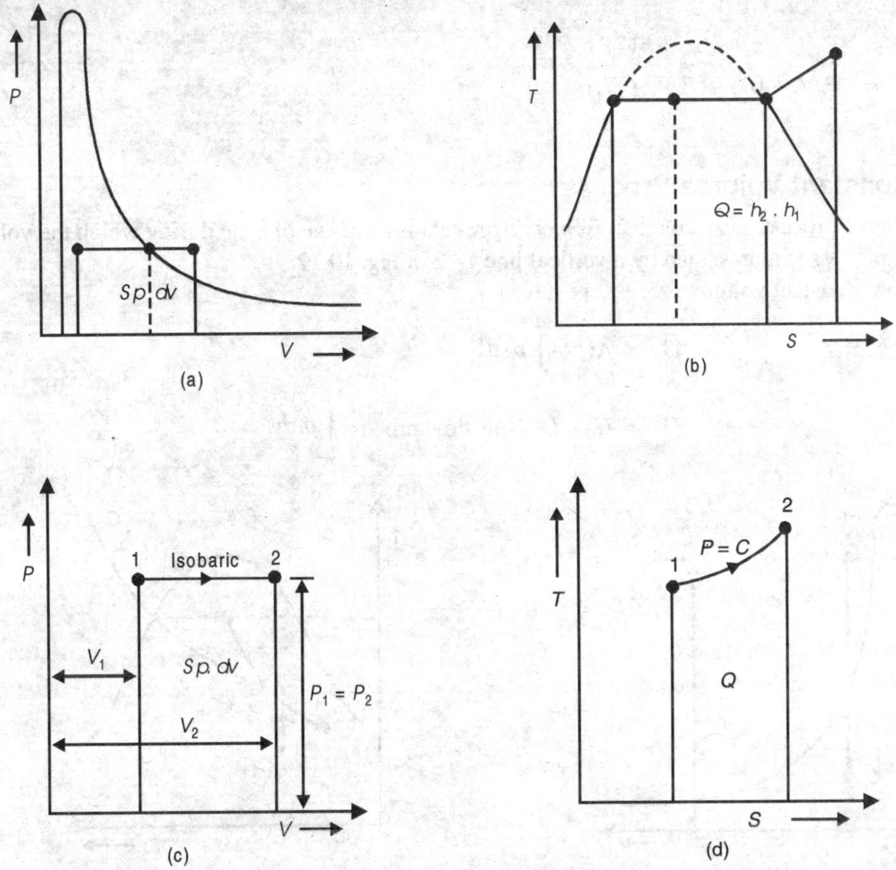

Fig. 10.11

A constant pressure, also called an isobaric process, is a change of state during which the pressure remains constant. On the *P–V* plane, Fig.10.11, the process is represented by a horizontal line, and on the *T–S* plane the process is represented by a horizontal line in the wet region and by a curve obtained from the indefinite integral of

$$ds = \frac{\delta q}{T}$$

In addition to defining relation, such as $p = c$, other quantities defining the limits of the process are needed. Suppose, in this case that point 1 is in the wet region and point 2 in the superheat region. With the pressure known, a point in the wet region is generally defined by giving its quality x or percentage moisture y. Similarly, a point 2 in the superheat region is generally, but not necessarily, defined by giving its temperature.

The work of reversible non-flow process at $p = c$ is

$$W_{nf} = \text{work of a non-flow system}$$

$$= \int p \cdot dV$$

$$= p(V_2 - V_1)$$

Also $Q = \text{Heat} = \Delta u + \int (p.dV)$

$$= u_2 - u_1 + P_2V_2$$
$$= (P_2V_2 + u_2) - (u_1 + bV_1)$$
$$= (P_2V_2 + U_2) - (P_1V_1 + U_1)$$
$$= (h_2 - h_1)$$

10.12.2 Constant Volume Process

A constant volume process, also called an *isometric process* is a change of state during which the volume remains constant. This process is represented by a vertical line 1– 2 in Fig. 10.12.

For non-flow, constant volume process we have

$$Q = \Delta U + \int p \cdot dV$$

$$Q = U_2 - U_1 \text{ Non-flow process} \int p.dv = 0$$

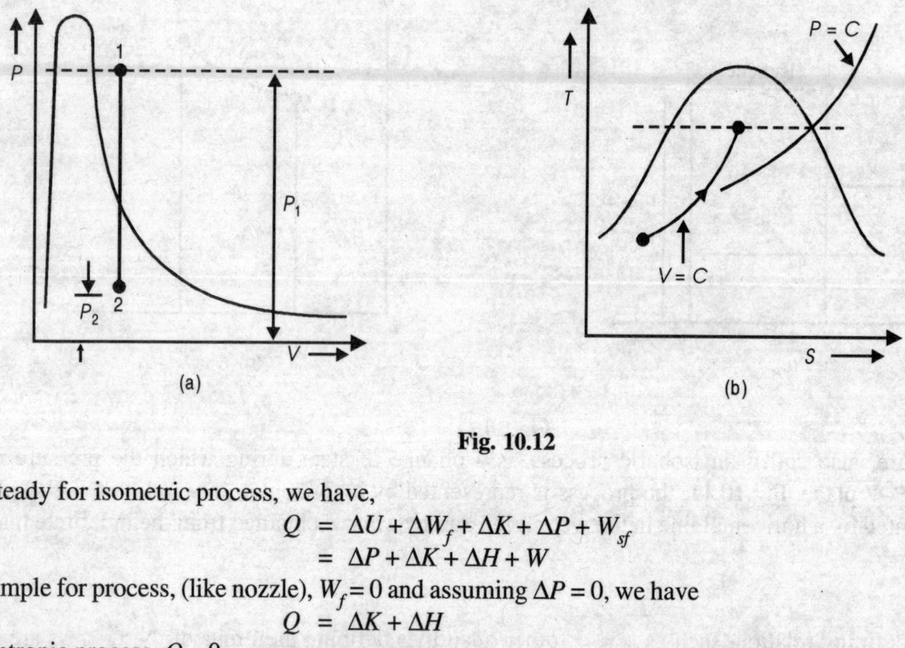

Fig. 10.12

For a steady for isometric process, we have,

$$Q = \Delta U + \Delta W_f + \Delta K + \Delta P + W_{sf}$$
$$= \Delta P + \Delta K + \Delta H + W$$

For a simple for process, (like nozzle), $W_f = 0$ and assuming $\Delta P = 0$, we have

$$Q = \Delta K + \Delta H$$

For isentropic process, $Q = 0$

$\therefore$ $$O = \Delta K + \Delta H$$

or $$\Delta H = \Delta K$$

$$H_2 - H_1 = K_1 - K_2 \text{ or } (H_1 - H_2) = (K_2 - K_1)$$
$$H = U - PV$$

10.12.3 Reversible Adiabatic Process (Isentropic Process)

Reversible adiabatic process is that process during which heat exchange is not taking place and $Q = 0$. This process is also called Isentropic process during which heat is zero and entropy change is also zero i.e., $\Delta S = 0$.

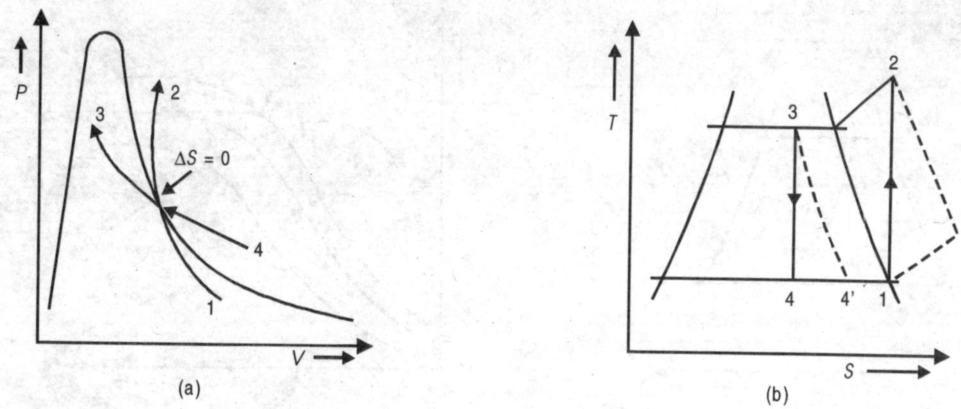

Fig. 10.13

This process is represented on *PV* plane, 1–2 Isentropic Compression, 3–4 Isentropic expansion and $T-S$ plane of Fig. 10.13.

The isentropic and irreversible adiabatic can either be non-flow or steady flow. Let the second state along isentropic be designated along an irreversible adiabatic 2'.

Then, from the non-flow energy equations with $Q = 0$ for isentropic process, we write $Q = DU + W_{nf}$

$\therefore\ 0 = \Delta U + W_{nf}$ or $W_{nf} = U_1 - U_2$ or $W = U_1 - U_2$ (Reversible) and $W = U_1 - U_2'$ (Irreversible)

For steady flow process we get, in general,

$$W = (h_3 - h_4) \text{ when } \Delta K = 0 \text{ (Reversible)}$$

or

$$W = h_3 - h_4 \text{ when } \Delta K = 0 \text{ (Irreversible)}$$

An example of steady flow to which this equation applies is a steam turbine. If the process reversible (Isentropic), $S_1 = S_2 = (S_f + xS_{fg})_2$ (Reversible non-flow or steady flow)

In which it is presumed that state 2 is in the wet region.

10.12.4 Irreversible Adiabatic or Throttling Process

This term is generally reversed for an irreversible steady flow process, but the kinetic energies are not necessarily zero. This adiabatic process ($Q = 0$) occur from the extreme of no work being done, called a throttling process. This is a simple flow process with work $W = 0$. For an ideal gas, this throttling process is at constant temperature process, therefore when $W = 0$, $\Delta P = 0$ and $Q = 0$, the energy equation is

$$Q = \Delta P + \Delta K + \Delta h + W$$

$$O = O + \Delta K + \Delta H + O$$

or

$$h_1 + K_1 = h_2 + K_2$$

If velocity changes are negligible, then $K_1 = K_2$ and $h_1 = h_2$. This is a constant enthalpy process or *Isenthalpic Process*.

Thus the throttling process is defined by $h_1 - h_2$ and it is commonly used in steam power practice to determine the quantity of steam (dryness fraction of steam). Generally this process is shown by a dotted line on *T–S* plane as shown in Fig.10.14 and by a vertical line on *h–s* diagram of Fig. 10.15.

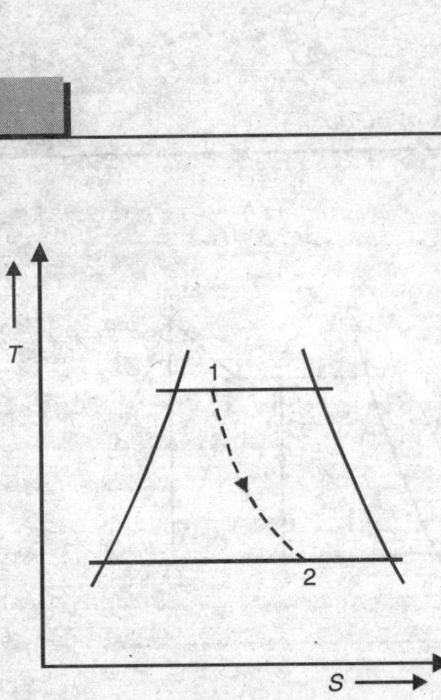

Fig. 10.14

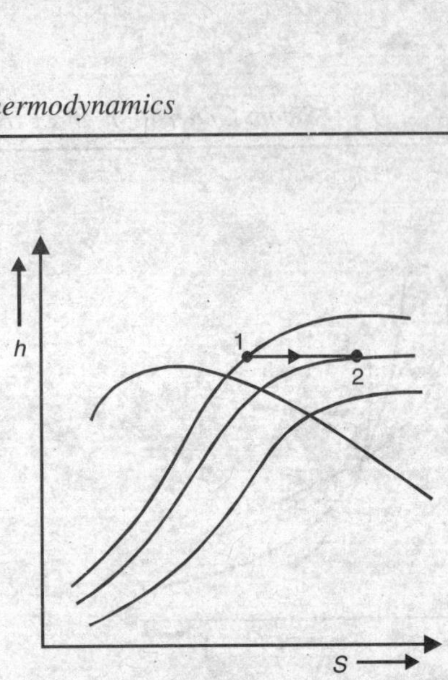

Fig. 10.15

10.12.5 Isothermal Process

An Isothermal process is one carried out at constant temperature $T = C$. Unless stated otherwise, always it will be taken as a reversible process. For an ideal gas with $T = C$, we think immediately of Boyle's Law, PV = Constant or $P_1V_1 = P_2V_2$. For steam, an 'internally reversible' isothermal process from a point in the superheat region to a point in the wet region is pictured in Fig. 10.16.

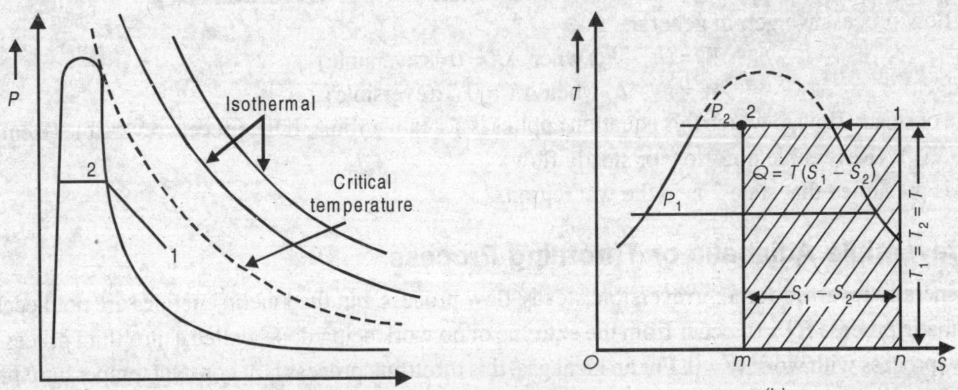

Fig. 10.16

The area m –2-1 -n, $\int T.ds = T\int ds$, represents heat

or $$Q = T(S_2 - S_1) \text{ kJ/kg}$$

where $(S_2 - S_1)$, being a negative term or number indicates that heat is passing from the substance. Then non-flow energy equation will be

$$Q = \Delta U + \int p.dv = \Delta U + W$$

or $$W = Q - \Delta U$$

$$W = Q - U_1 - U_2 \text{ kJ/kg}$$

For a steady flow isothermal process, the energy equation can be given as

$$Q = \Delta P + \Delta K + \Delta h + W$$

with $\Delta P = 0$

$$W = Q - \Delta K - \Delta h \text{ and with } \Delta K = 0$$

$$= Q - \Delta h$$

$$W = Q + h_1 - h_2$$

10.12.6 Polytropic Process

A polytropic process is an internally reversible process, which conforms to the relation $pv^n = 0$ or $P_1V_1^n = P_2V_2^n$ which n is any constant.

If a process is known or assumed to be polytropic and if two state points are defined (say, pressures and volumes are known), the value of n can be found by use of logarithms.

$$\therefore \quad P_1V_1^n = P_2V_2^n \text{ or } \frac{P_1}{P_2} = \left(\frac{V_2}{V_1}\right)^2$$

or

$$\log(P_1/P_2) = n.\log\left(\frac{V_2}{V_1}\right)$$

or

$$n = \frac{\log_e(P_1/P_2)}{\log(V_2/V_1)}$$

Using $pv^n = C$, we get,

$$\int_1^2 p.dv = \frac{P_2V_2 - P_1V_1}{(1-n)} \frac{\text{N-m}}{\text{kg}}$$

which is the work of non-flow process.

$$\therefore \quad W_{nf} = \frac{P_2V_2 - P_1V_1}{(1-n)}$$

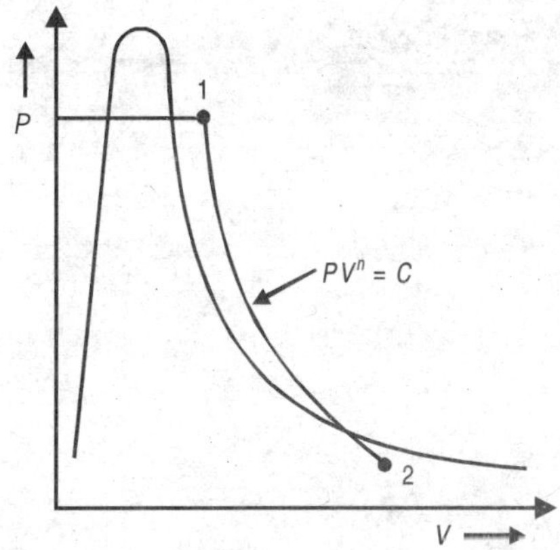

Fig. 10.17

The process is shown on P–V diagram as in Fig. 10.17.
Non-flow energy equation can be written as

$$Q = \Delta U + \int P \cdot dv$$

$$= u_2 - u_1 + \frac{P_2V_2 - P_1V_1}{1-n}$$

steady flow energy equation for the polytropic equation gives

$$Q = \Delta P + \Delta K + \Delta h + W_{sf}$$

or

$$W_{sf} = Q - \Delta P + \Delta K + \Delta h$$

when $\Delta P = 0$, $W = Q - \Delta K - \Delta h$ and again $\Delta K = 0$ gives

$$W = Q - \Delta K - \Delta h.$$

Also for flow process

$$W = -\int VdP$$

and for polytropic process $PV^n = C$

$$W = -\int_1^2 VdP = \frac{n}{n-1}(P_2V_2 - P_1V_1)$$

Besides constant volume, constant pressure, constant temperature, adiabatic (reversible and irreversible) processes the following are the other important operations:

(1) Hyperbolic and (2) Free expansion.

(1) Hyperbolic Process. When a gas or steam expands or is compressed in such a manner that the product of pressure and volume remains constant during the process of expansion or compression, the process is known as hyperbolic process because such a process on the *P–V* diagram will give a rectangular hyperbola. Hence a hyperbolic expansion or compression process follows the law,

$$\text{Pressure} \times \text{Volume} = \text{Constant}$$

$$P\,V = \text{Constant}$$

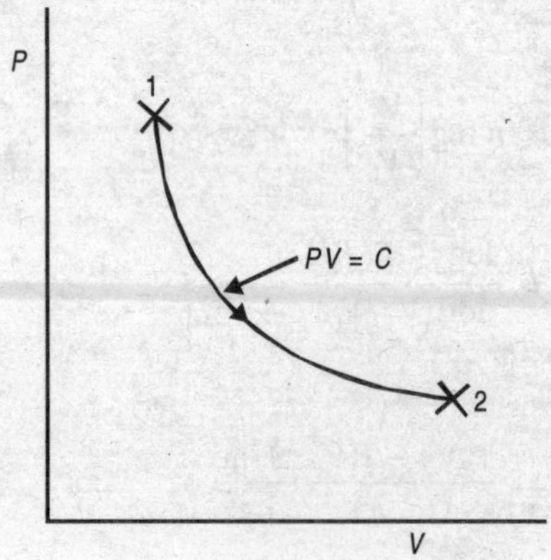

Fig. 10.18

or

$$P_1V_1 = P_2V_2$$

For ideal gas, this process is similar to isothermal process

For such a process,

$$\int_1^2 P \cdot dV = \text{work done of non-flow process}$$

$$= P_1V_1 \log_e \left(\frac{P_1}{P_2} = \frac{V_1}{V_2}\right)$$

$$-V.dp = P_1V_1 \log_e \left(\frac{P_1}{P_2} = \frac{V_1}{V_2}\right)$$

For steam, we can find the volume of steam after the process. Thus

$$P_1V_1 = P_2V_2$$

$$\therefore \quad V_2 = \frac{P_1}{P_2} - V_1$$

If

$$V_2 > V_{g2}, \text{ the steam is superheated.}$$
$$V_2 < V_{g2}, \text{ the steam is wet.}$$
$$V_2 = V_{g2},$$

the steam is dry and saturated.

when $V_2 > V_{g_2}$,

$$V_2 = V_{sat_2} \times \frac{T_{sup2}}{T_{sat2}}$$

$\therefore$ T_{sup} is determined.

$$\therefore \quad h_2 = \text{enthalpy} = h_{g2}\ {}^{T}Cpv(T_{sup} - T_{sat})$$

and $U_2 = h_2 - P_2V_2$

and hence Q heat can be calculated for non-flow process. Similarly,

when $V_2 < V_{g2}$

Then $V_2 = x_2V_{g_2}$,

$$\therefore \quad x_2 = \frac{V_2}{V_{g_2}}$$

$$\therefore \quad h_2 = h_{f2} + x_2hf_{g2}$$

$$\therefore \quad U_2 = h_2 - P_2V_2$$

and hence Q heat can be calculated. Again,

when

$$V_2 = V_{g2}$$
$$x_2 = 1$$
$$h_2 = h_{g2}$$
$$U_2 = h_{g2} - P_2V_2$$

and thus Q can be calculated.

Note: Remember we will have to calculate h_1, U_1 etc. at the condition of steam before the process commences.

(2) Free Expansion. When a fluid is allowed to expand suddenly into a vacuum chamber through an orifice of large dimensions, the free expansion of the fluid occurs. During this operation no heat has been supplied or rejected and no external work has been done; hence it follows that the total enthalpy of the fluid remains constant. This type of expansion is also called the constant total heat expansion.

After studying the different vapour process, we can conclude that for any process, it will be required to determine or calculate

(i) Heat during the process (Q)

(ii) Change in Internal energy (ΔU)

(iii) Change in Enthalpy (ΔH)

(iv) Work done (W)

(v) $\int p.dV$

(vi) $-\int V.dP$

(vii) Change in entropy (ΔS)

10.13 STEAM DRYNESS FRACTION AND ITS DETERMINATION WITH STEAM CALORIMETERS

We have defined the dryness fraction of steam as the ratio of (dry steam) to (total steam containing it) and is represented by x.

$$\therefore \quad x = \frac{\text{Dry steam (kg)}}{\text{Total steam (kg)}}$$

Dryness fraction of steam is determined with the help of steam calorimeters.

These calorimeters are as follows.

(i) Barrel Calorimeter
(ii) Separating Calorimeter
(iii) Throlling Calorimeter and
(iv) Combined separating and throtling Calorimeter

Any one combination these Calorimeters can be used for finding the dryness fraction of steam. Each of these methods will be dealt within term.

10.13.1 Barrel Calorimeter

To get the approximately dryness fraction of steam, a very simple appe called Barrel Calorimeter is used. This consists of a wooden barrel containing cold water.

Steam is taken from the pipe through which it is flowing and is condensed in the cold water.

Temperatures of water before and after condensation are recorded. Knowing the quantities of steam and water, we can very easily calculate the dryness fraction steam. The calorimeter is shown in Fig. 10.19.

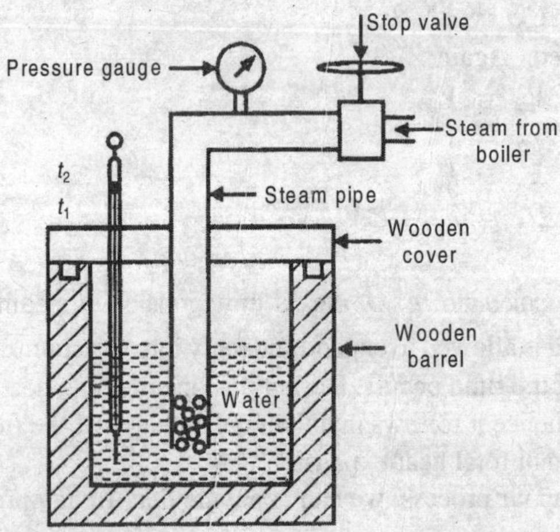

Fig. 10.19

Let,
ms = weight of steam condensed
m_w = weight of cold water in barrel
C_{pw} = sp. heat of water
x = dryness fraction of steam
h_{fg} = latent heat of the steam at pressure of steam P
t_1 = initial temperature of water before mixing steam

t_2 = final temperature of steam after mixing of steam.
t = saturation the temperature of steam at P.

Then we can find the dryness fraction of steam by using the equations given below.

[Heat lost by steam] = [Heat gained by coal water]

$$Ms\left[x.h_{fg} + C_{pw}(t - t_2)\right] = M_w C_{pw}(t_2 - t_1)$$

From this equation, we can find the dryness fraction of steam (x)

10.13.2 Separating Calorimeter

The separating calorimeter is quite simple in its action and it may be used for testing steam of practically any degree of wetness. Figure 10.20 shows a design of separating calorimeter by **Prof Carpenter.**

The sample of steam to be tested enters the calorimeter at D and passes down through the central passage DF into the perforated metal cup H. The current of steam then reverses and flows as shown by the arrows in the jacket J surrounding the inner chamber C. From the jacket the steam escapes into the atmosphere through the small orifice at the bottom and exhaust pipe E.

The water in the steam is thrown out in the cup H and collected in the chamber C. A gouge G has two scale on its dial, the inner one showing the pressure of the steam in the jacket while the outer one shows the weight of steam passing through the jacket for each pressure in ten minutes. The graduations on the outer scale are determined by experiment with steam of different pressures, the escaping steam being condensed and weighted, but in using the instrument it is only necessary to take during a ten-minute test the average reading on the outer scale to obtain the weight of steam passing through the jacket.

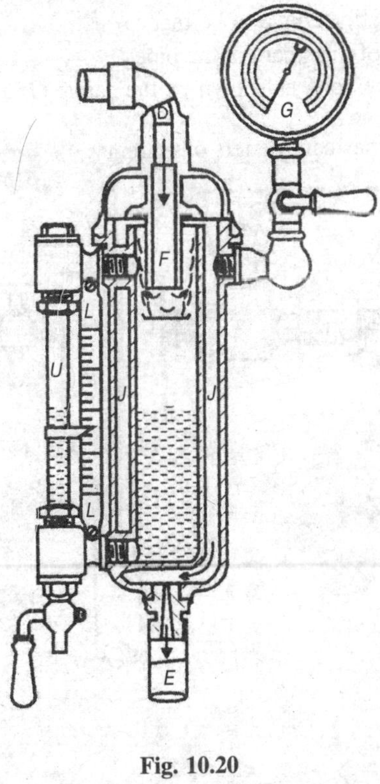

Fig. 10.20

The water separated from the steam is shown by means of the glass gauge U. A pointer P having a frictional grip of the glass tube is set to the water level at the beginning of the taste. The change in level of the separated water during a test is read off on the scale L. Which is graduated in fractions of mols.

Let W be the weight of steam passing through the jacket during a test and W is the weight of water separated in the same time, then x, the dryness fraction of the steam tested is $x = \dfrac{W}{W + W}$.

The weight of steam passing through the jacket may also be readily determined by passing the escaping steam into a bucket of water where it is condensed, the increase in the weight of the bucket and its contents giving the value of W. In that case gauge G is not required, but it used its reading may be taken as a check.

10.13.3 Throttling Calorimeter

The throttling calorimeter was first introduced by professor Peabody. Figure 10.21 shows Professor Carpenter's modification of the Peabody throttling calorimeter.

P is the pipe carrying the steam to be tested. AB is the sampling tube through which the sample of steam flows to the vessels S which is the principle part of the calometer. V is the stop valve. The sample of steam enters the vessel S through a small orifice O and expands to a pressure a little above that of atmosphere into which it escapes freely through the exhaust pipe E.

The pressure of the steam in S is measured by means of the manometer or siphon gauge G which is a glass U-tube containing mercury. G is connected to S by 2a rubber tube T. Steam to the gauge may be cut-off by means of the stop cock C. A deep pocket K, containing cylinder oil, takes the thermometer R which indicates the temperature of the steam in S. The vessel S, the stop valve V, and the exposed part of the sampling tube AB must be carefully lagged to prevent undue loss of heat by radiation.

In using this instrument four observations have to be made

(1) The pressure P_1 or temperature t_1 of the steam in the pipe P.

(2) The pressure of the steam in the vessel S as shown by the gauge G in inches/centimeters of mercury.

(3) Reading of thermometer R

(4) The height of the barometer in inches/centimeters of mercury.

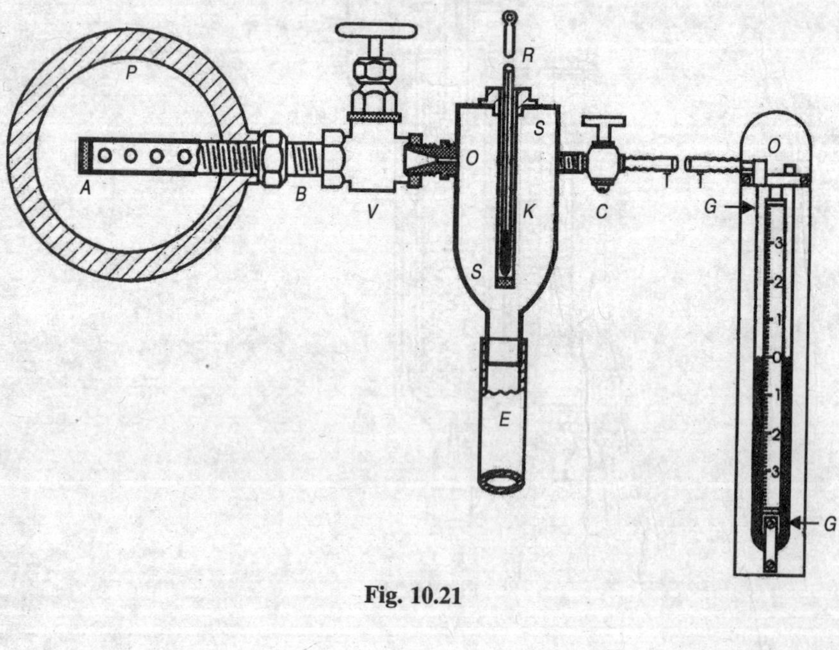

Fig. 10.21

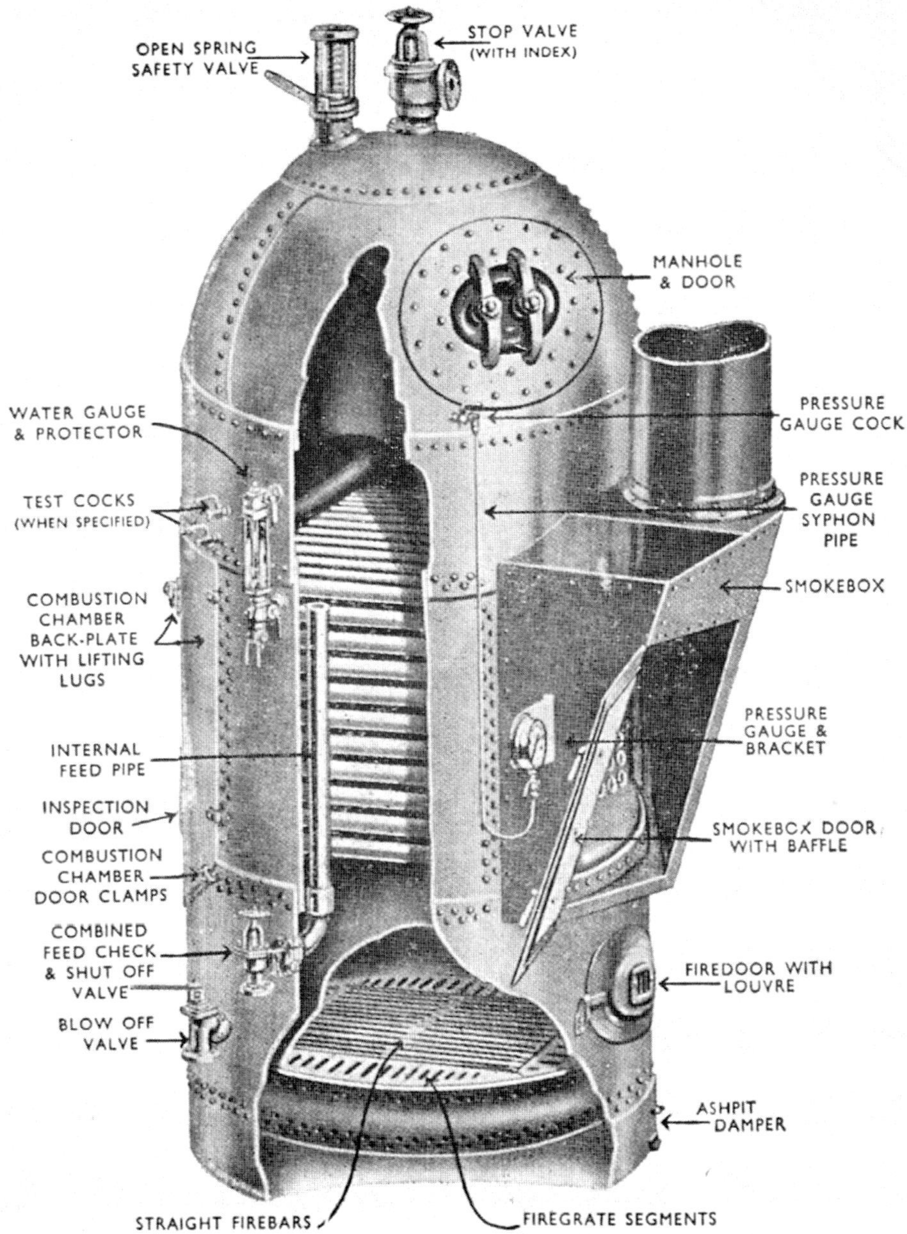

Fig : Cochran Boiler

Sterling Bent tube boiler.

Fire Tube, 6 TPH, LIPI Make Boiler

Water Tube, 90TPH, LIPI Make Boiler

Fig : Sterling Bent tube boiler.
Courtesy : LIPI BOILERS LTD. PUNE-1

The principle upon which the operation of the throttling calorimeter depends is that the total heat (enthalpy) in the steam after throttling is the same as the total heat (enthalpy) in the steam before throttling. This is given by the equation

$$h_{f1} + xh_{fg1} = h_{g2} + Cp_v(t_3 - t_2)$$

$$\therefore \quad x = \frac{h_{g_2} - h_{f1} + C_{pv}(t_3 - t_2)}{h_{fg1}}$$

where

x = dryness fraction of steam before throttling

h_{f1} = sensible heat of steam before throttling

h_{fg1} = latent heat of steam before throttling

h_{g2} = total heat or enthalpy of saturated steam at pressure after throttling

t_2 = saturated temperature of steam at pressure after throttling

t_3 = temperature of steam after throttling as indicated by thermometer R.

C_{pv} = mean specific heat of steam between t_3 and t_2 at constant.

Since the pressure after throttling is in the neighbourhood of that of the atmosphere, and since the specific heat of steam at that pressure is little affected by the temperature, C_{PV} may be taken as 2 or 2.01 kJ/kg-K.

The operation of the throttling calorimeter depends on the steam being superheated after throttling and it will fail in its purpose when the steam is so wet before throttling that it is wet after throttling. This means that t_3 must be greater than t_2. Putting $t_3 = t_2$ the value of x which may be determined by the throttling calorimeter must be greater than

$$= \frac{h_{g_2} - h_{f1}}{h_{fg1}}$$

10.14 LIMITATIONS OF CALORIMETERS

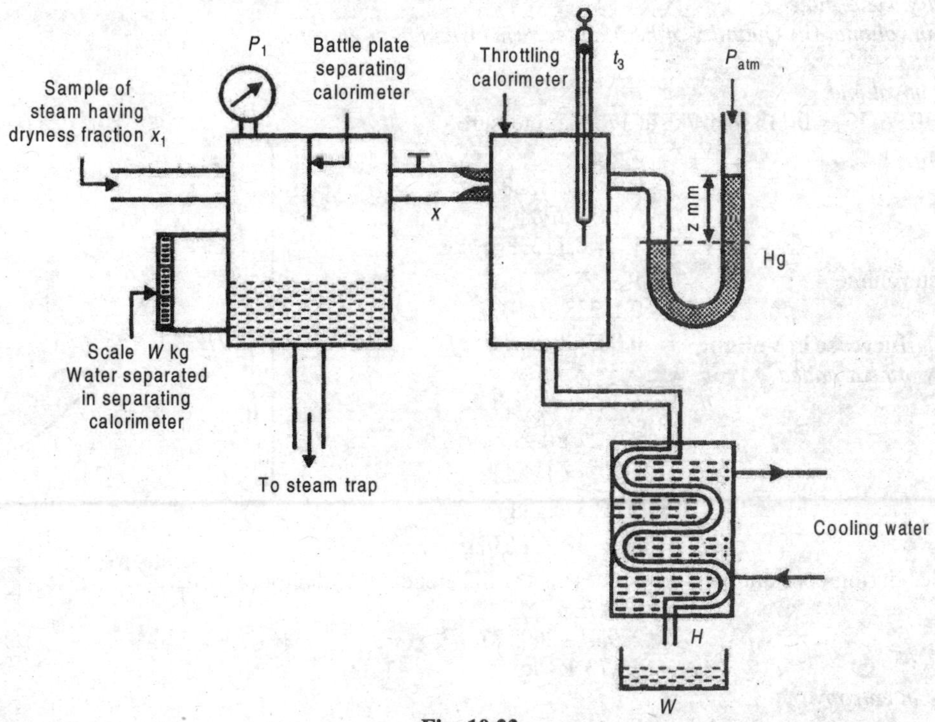

Fig. 10.22

10.14.1 Barrel Calorimeter

Basically, this calorimeter is used when we want to know the approximate dryness fraction of steam, or to have a rough idea about the dryness fraction. This calorimeter gives better results if the dryness fraction of steam is greater than 0.95.

10.14.2 Separating Calorimeter

This calorimeter will give better results when the dryness fraction of steam to be determined is above 0.95.

10.14.3 Throttling Calorimeter

Basically the condition of steam after throttling must be known so that the unknown x-dryness fraction before throttling can be calculated. For this purpose the steam after throttling must be separated and its temperature should be more than the saturated temperature of steam corresponding to the pressure after throttling . To get such a result dryness fraction before throttling should be more than 0.9.

If the dryness fraction of steam is less than 0.9, then the steam will not be superheated after throttling and the dryness before throttling cannot be found.

In such a case, a combined separating and throttling calorimeter is used. In this, steam is first admitted to separating calorimeter where some portion of the water particles are removed and the quality of steam is improved to the required level for throttling calorimeter.

A Schematic diagram of a combined separating and throttling calorimeter is shown in Fig. 10.22 and now is self explanatory.

SOLVED EXAMPLES

(I) Constant Pressure Process

Example 10.1 *One kg of steam at a pressure of 17.5 bar and dryness 0.95 is heated at constant pressure, until it is completely dry. Determine*

(i) Increase in volume. (ii) Quantity of heat added. (iii) Change in entropy.

Solution

(i) Increase in volume

We have $x_1 = 0.95$, $V_{g1} = 0.11338$ m³/kg at 17.5 bar pressure

$$\therefore \quad V_1 = x_1 V_{g1} = 0.95 \times 0.11338 = 0.1077 \text{ m}^3/\text{kg}$$

$$V_2 = V_{g2} = 0.11338 \text{ m}^3/\text{kg}$$

$$\therefore \text{ Increase in volume } = V_2 - V_1 = 0.11338 - 0.1077$$

$$\textbf{Increase in Volume} = \mathbf{0.00568 \text{ m}^3/\text{kg}}$$

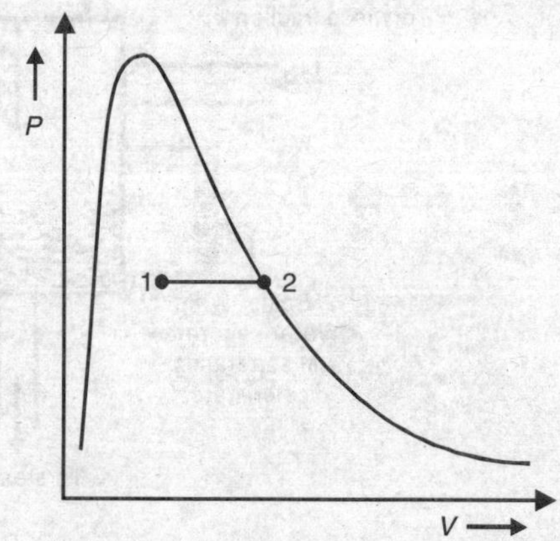

Fig. Ex. 10.1

(ii) Quantity of heat added

$$h_1 = h_{f1} + x_1 h_{fg1} = 878.27 + 0.95 \times 1915.9 = 878.27 + 1820.1 = 2698.37 \text{ kg/kg}$$

$$h_2 = h_{g2} = 2794.1 \text{ kJ/kg}$$

$\therefore$ Heat added during constant pressure process is the difference in enthalpy $(h_2 - h_1)$

$$\therefore \quad Q = h_2 - h_1 = 2794.1 - 2698.37$$

$$\mathbf{Q = 95.73 \text{ kJ/kg}}$$

(iii) Change in entropy

$$\Delta S = S_2 - S_1$$

$$= S_{g2} - (sf_1 + x_1 S_{fg1})$$
$$= (S_{f2} + S_{fg2}) - (Sf_1 + x_1 S_{fg1})$$

But $S_{f1} = S_{f2}$ and $S_{fg1} = S_{fg2}$ for $P = C$

$\therefore$ $\Delta S = S_{fg1}(1 - x_1)$
$$= 0.05 \times 4.0007$$
$$\mathbf{\Delta S = 0.200035\ kJ/kg\text{-}K.}$$

Example 10.2 *A piston cylinder arrangement is filled with a wet steam of quality 0.8 at a pressure of 0.1 MPa. Energy is added at constant pressure till the temperature of steam rises to 300°C. Calculate the heat added and the work done.*

Solution

$$0.1\ \text{MPa} = 1\ \text{bar}$$

Volume of one kg before heat is added
$$= X V_g$$
$$= 0.8 \times 1.6938$$
$$= 1.35504\ \text{m}^3/\text{kg}$$

Heat is added at constant pressure. The temperature of steam after heat added is while saturated temperature is 170.17°C

$\therefore$ Steam is superheated.

V_2 = Volume of 1 kg steam = 2.639 m^3/kg

$\therefore$ Work done/kg $= P(V_2 - V_1)$
$$= 100 \times 1\,(2.639 - 1.35504)$$
$$= 100 \times 1.28396$$

Work done/kg = 128.396 kJ/kg

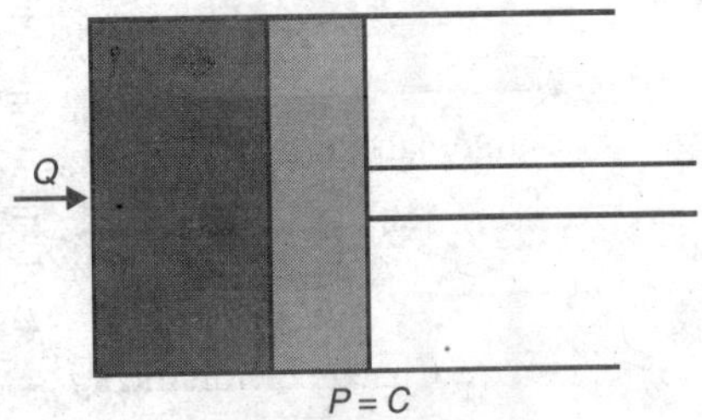

Fig. Ex. 8.2

For constant pressure process,

Q = Heat $= h_2 - h_1$ (for non-flow process)
$$= 3074.5 - (h_{f1} + x_1 h_{fg1})$$
$$= 3074.5 - (417.5 + 0.8 \times 2257.9)$$
$$= 3074.5 - 2223.82$$

Q = 850.68 kJ/kg

Example 10.3 *Steam having a dryness fraction 0.8 at a pressure of 10 bar changes its state so that the temperature of steam become 275°C. The pressure of steam remains the same. Calculate per kg of steam, the change in total heat, change in volume and change in internal energy.*

Solution

Initial Condition:

Enthalpy of steam/kg $= h_1 = h_{f1} + x_1 h_{fg1}$
$$= 762.61 + 0.8 \times 2013.6$$
$$= 762.61 + 1610.88$$
$$= 2373.59\ \text{kJ/kg}$$

Volume $V_1 = x_1 v_{g1}$
$$= 0.8 \times 0.19429$$

V_1 = 0.1554 cu.m/kg

Internal energy $u_1 = h_1 - P_1 V_1$

$$= 2373.59 - \frac{10^5}{1000} \times 10 \times 0.1554$$
$$= 2373.59 - 155.4$$
$$\mathbf{u_1 = 2218.19\ kJ/kg}$$

Final condition

$$P = 10 \text{ bar}$$
$$\text{Temperature} = 275°C$$
$$\text{Enthalpy of 1 kg steam} = hg + Cp_v\,(T_{sup} - T_{sat})$$
$$h_2 = \frac{3052.1 + 2943}{2} \quad \text{(From table Interpolation)}$$
$$= \frac{5995}{2}$$
$$= 2997.5 \text{ kJ / kg}$$
$$V_2 = \frac{0.2327 + 0.2580}{2} \quad \text{(Volume from tables)}$$
$$= \frac{0.4907}{2}$$
$$= 0.24535 \text{ m}^3$$

∴ Internal energy
$$u_2 = h_2 - P_2V_2$$
$$= 2997.5 - 100 \times 0.24535 \times 10$$
$$= 2997.5 - 245.35$$
$$= 2752.15 \text{ kJ / kg}$$
$$\therefore \text{Change in volume/kg} = V_2 - V_1$$
$$= 0.24535 - 0.1554$$

Change in volume/kg = 0.18995 m³/kg

Change in internal energy/kg
$$= 2752.15 - 2218.19$$

Change in internal energy/kg = 2533.96 kJ/kg

Example 10.4 *(a) The internal energy of 4.5 kg of wet steam at 7 bar is 11388 kJ. What is the dryness fraction of steam?*

(b) By how many heat units, the internal energy of the above steam shall increase if it is superheated at constant pressure to a temperature of 250°C ?

Assume specific heat of steam as 2.1 kJ/kg K.

Solution

(a) Dryness Fraction

Let x be the dryness fraction of steam

$$\text{Internal energy/kg} = \frac{11388}{4.5}$$
$$= 2530.67 \text{ kJ}$$
$$= h - PV$$
$$= (h_f + xh_{fg}) - p \times xvg$$

$$\therefore \quad 2530.67 = 697.06 + x \times 2064.9 - 100 \times 7 \times x \times 0.27268$$

$$\therefore \quad 2530.67 - 697.06 = 2064.9x - 190.88x$$

$$\therefore \quad 1833.61 = 1874.02x$$

$$x = \frac{1833.61}{1874.02}$$

$$\mathbf{x = 0.9784}$$

Dryness fraction of steam

(*b*) *Increase in Internal energy*

After superheating, enthalpy of 1 kg steam is given by

$$h = h_g + C_{pv}(t_{sup} - t_{sat})$$

This equation used C_p is given

$$\begin{aligned} h &= 2762.0 + 2.1(250 - 164.96) \\ &= 2762.0 + 2.1 \times 85.04 \\ &= 2762.0 + 178.584 \\ &= 2940.584 \text{ kJ/kg} \end{aligned}$$

$$\begin{aligned} V &= v_{sat} \times \frac{T_{sup}}{T_{sat}} \\ &= 0.27268 \times \frac{250 + 273}{273 + 16496} \\ &= \frac{0.27268 \times 523}{437.96} \\ &= 0.32563 \text{ m}^3/\text{kg} \end{aligned}$$

$$\begin{aligned} u &= \text{Internal energy} \\ &= h - PV \\ &= 2940.584 - 100 \times 7 \times 0.32563 \\ &= 2940.584 - 227.94 = 2712.644 \text{ kJ/kg} \end{aligned}$$

$\therefore$ Rise in heat units of Internal energy

$$\begin{aligned} &= u_2 - u_2 \\ &= 2712.644 - 2530.67 = 181.974 \text{ kJ/kg} \end{aligned}$$

Rise in heat units of Internal energy = 818.883 kJ

Example 10.5 *1.5 cu.m of steam at a pressure 9 bar and having dryness fraction 0.95 loses 9420 kJ at constant pressure. Calculate dryness fraction of the steam at the end of heat loss.*

Solution

We will have calculations on kg basis and therefore first we will find mass of steam in 1.5 cu.m at 9 bar and 0.95 dry

$$= x \,.\, Vg \qquad \text{(Volume of 1 kg steam at 9 bar and 0.95 dry)}$$

$$= 0.95 \times 0.21481$$

$$= 0.2041 \text{ cu.m}$$

$\therefore$ Mass of steam in 1.5 cu.m

$$= \frac{\text{Total volume}}{\text{Sp. volume}}$$

$$= \frac{1.5}{0.2041} = 7.35 \text{ kg}$$

$$\therefore \quad \text{1 kg steam loses} = \frac{9420}{7.35}\text{ kJ} = 1281.63 \text{ kJ at constant pressure}$$

$$\text{Initial enthalpy/kg steam} = h_1 = h_{f1} + x_1 h_{fg1}$$
$$= 742.64 + 0.95 \times 2029.5$$
$$= 742.64 + 1924 = 2670.64 \text{ kJ}$$

$$\therefore \quad \text{Final enthalpy/kg} = 2670.64 - 1281.63$$
$$= 1389.01 \text{ kJ}$$

$$\therefore \quad 1389.01 = h_{f2} + x_2 h_{fg2} = 742.64 + x_2 h_{fg2}$$
$$= 742.64 + x_2 \times 2029.5$$

$$\therefore \quad x_2 = \frac{1389.01 - 742.64}{2029.5}$$

$$\mathbf{x_2 = 0.3185}$$

Dryness fraction of steam at the end of heat loss.

Example 10.6 *A vessel contains 4 kg of steam at 10 bar and 220°C. Find the volume of the vessel. If the vessel is cooled till the steam pressure drops to 3 bar. Find the final condition of steam and the heat transfer during cooling.*

Solution

Volume of 1 kg steam at 10 bar and 220°C (from superheated steam tables and by interpolation)

$$V_1 = 0.2059 + 0.4\,(0.2328 - 0.2059)$$
$$= 0.21666 \text{ m}^3\text{/kg}$$

$$\text{Total volume of vessel} = \text{Mass of steam} \times \text{Specific volume}$$
$$= 4 \times 0.21666$$

$$\textbf{Total volume of vessel} = \mathbf{0.86664\ m^3}$$

Volume of 1 kg steam at 3 bar after cooling

$$V_{g2} = 0.60533 \text{ m}^3\text{/kg}$$

This is greater than that of V_1

∴ Steam is wet

$$\therefore \quad 0.60533\, x_2 = V_1 = 0.21666$$

$$x_2 = \frac{0.21666}{0.60533}$$

$$\mathbf{x_2 = 0.3579}$$

For constant volume process, we have

$$Q = \Delta u + \int P \cdot dv$$
$$= \Delta u \text{ as } \int P.dV = 0 \text{ (Constant volume)}$$
$$= u_2 - u_1$$

$$u_2 = h_2 - P_2 V_2$$
$$= h_{f2} + x_2 h_{fg2} - P_2 V_2$$
$$= (561.5 + 0.3579 \times 2163.2) - \frac{3 \times 10^5}{1000} \times 0.21666$$
$$= 1335.7 - 65$$

$$= 1270.7 \text{ kJ/kg}$$

$$u_1 = h_1 - p_1 v_1$$

$$= 2873.28 - \frac{10\times10^5}{1000}\times0.21666$$

$$= 2873.28 - 216.66$$

$$= 2656.62 \text{ kJ/kg}$$

$$\therefore \quad Q = u_2 - u_1$$

$$= 1270.7 - 2656.62$$

$$= -1385.92 \text{ kJ/kg}$$

∴ Total heat transfer during cooling

$$= -4\times1385.92$$

Total heat transfer during cooling = –5543.68 kJ

– ve sign indicates that the heat is rejected by the system.

Example 10.7 *1 kg of steam at initial condition of 6 bar and 0.2 dry is heated at constant volume until the pressure is 20 bar. Determine the final state of steam, heat added and change of energy.*

Solution

Initial volume of 1 kg steam

$$V_1 = x_1 V_{g1} = 0.2\times0.31546$$

$$= 0.063092 \text{ m}^3/\text{kg} = V_2$$

After cooling pressure is 20 bar

$$V_{g2} = 0.09955 \text{ and is greater than } V_1$$

∴ Steam after cooling is wet.

$$\therefore \quad 0.063092 = x_2\times0.09955$$

$$x_2 = \frac{0.063092}{0.09955}$$

$$\mathbf{x_2 = 0.634}$$

Initial entropy per kg is given by

$$S_1 = S_{f1} + x_1 S_{fg1}$$

$$= 1.931 + 0.2\times4.827$$

$$= 1.931 + 0.9654$$

$$= 2.8964 \text{ kJ/kg°K}$$

Similarly

$$S_2 = S_{f2} + x_2 S_{fg2}$$

$$= 2.447 + 0.634\times3.890$$

$$= 2.447 + 2.4663$$

$$= 4.9133 \text{ kJ/kg/K}$$

$$\Delta S = S_2 - S_1$$

= Change of entropy

$$= 4.9133 - 2.8964$$

$$\mathbf{\Delta S = 2.0169 \text{ kJ/kg-K}}$$

Now

$$u_1 = h_1 - P_1V_1 = h_{f1} + x_1 h_{fg1} - P_1V_1$$

$$= 670.4 + 0.2\times2085.1 - 100\times6\times0.063092$$

$$= 1087.42 - 37.86$$

$$= 1049.56 \text{ kJ/kg} \quad \text{...(I)}$$

$$\begin{aligned} u_2 &= h_2 - P_2V_2 \\ &= h_{f2} + x_2h_{fg2} - P_2V_2 \\ &= 908.5 + 0.634 \times 1888.7 - 100 \times 20 \times 0.063092 \\ &= 2105.94 - 126.184 \\ &= 1979.756 \text{ kJ/kg} \end{aligned} \quad \text{...(II)}$$

For constant volume process

$$\begin{aligned} Q &= \Delta u + \int P.dV = \Delta u \text{ as } \int P.dV = 0 \\ &= u_2 - u_1 + 0 \\ &= 1979.756 - 1049.56 \\ \mathbf{Q} &= \mathbf{930.196 \text{ kJ/kg}} \end{aligned}$$

Example 10.8 *A closed vessel of 0.24 cu. m capacity, constant steam at a pressure of 11 bar and at a temperature of 200°C.*

Find the quantity of steam in the vessel.

At what pressure in the vessel will the steam be dry and saturated if the vessel is cooled? The cooling is further continued till the temperature of steam in the vessel becomes 116.93°C. Find out the pressure and condition of the steam at the end of operation.

Solution

The capacity of the vesel is 0.24 m^3. Saturation temperature of steam corresponding to pressure of 11 atm. bar is 184.07°C. Therefore steam is superheated steam can be found out by Charle's Law steam behaves as a perfect gas and as the pressure of steam is constant, the volume of superheated steam can be found out by the formula

$$v_{sup} = v_{sat} \times \frac{T_{sup}}{T_{sat}}$$

$$= \frac{0.17738 \times 473}{184.07 + 273} = \frac{0.17738 \times 473}{457.07}$$

$$= 0.1836 \text{ m}^3/\text{kg}$$

As the vessel is closed, the volume of steam/kg will always remain constant whatever the pressure will be

$$\therefore \text{ Quantity of steam in vessel} = \frac{\text{Total volume}}{\text{Sp. volume}}$$

$$= \frac{0.24}{0.1836} = 1.307 \text{ kg}$$

When the steam is cooled, its quality will not be improved and the saturated steam having volume of steam = 0.1836 will have a pressure ≈10.5 bar.

Now the steam is further cooled so that its temprature is 116.93°C. The sturated pressure is 1.8 bar.

At 1.8 bar, volume of dry and saturated steam

$$= 0.97723 \text{ m}^3/\text{kg}$$

$$\therefore \quad 0.97723x = 0.1836$$

$$\therefore \quad \text{Condition of steam} = x = \frac{0.1836}{0.97723}$$

Condition of steam = 0.188

∴ Quantity steam in vessel = 1.307 kg

Pressure after cooling = 10.5 bar

Pressure after further cooling = 1.8 bar

Condition after further cooling D.F. = 0.188

Example 10.9 *Calculate the total energy and internal energy of 1 kg of steam at a pressure of bar and dryness 0.95. One cu.m of this steam is cooled at constant volume until its pressure has fallen to 5.4 bar. Determine the heat abstracted from steam.*

Solution

Total energy is enthalpy of steam

$$\therefore \quad h_1 = h_{f1} + x_1 h_{fg1}$$
$$= 742.64 + 0.95 \times 2029.5$$
$$= 742.64 + 1928.02$$
$$\mathbf{h_1 = 2670.66\ kcal/kg}$$

Internal energy kg

$$u_1 = h_1 - P_1 V_1$$
$$= 2670.66 - 100 \times 9 \times 0.95 \times 0.21481$$
$$= 2670.66 - 183.33$$
$$\mathbf{u_1 = 2497.00\ kJ/kg}$$

$$\text{Mass of one cu.m steam} = \frac{\text{Total volume}}{\text{Sp. volume}}$$
$$= \frac{1}{0.95 \times 0.21481} = 4.99 \text{ kg}$$

After cooling, the pressure is 5.4 bar at 5.4 bar, $V_{g2} = 0.34846 \text{ m}^3/\text{kg}$

$$\text{Actual volume/kg} = xv_{g2} = 0.95 \times 021481$$
$$\therefore \quad x_2 = \text{Dryness fraction of steam at 5.4 bar}$$
$$= \frac{0.95 \times 0.21481}{0.34846} = 0.586$$
$$\therefore \quad h_2 = h_{f2} + x_2 h_{fg2}$$
$$= 652.76 + 0.586 \times 2098.1$$
$$= 652.76 + 1229.5 = 1882.26 \text{ kJ/kg}$$
$$\therefore \quad u_2 = h_2 - p_2 v_2$$
$$= 1882.26 - 100 \times 5.4 \times 0.95 \times 0.21481$$
$$= 1882.26 - 110.2$$
$$= 1672.06$$
$$\therefore \quad u_2 - u_1 = \text{Change in internal energy} = \text{Heat}$$
$$\therefore \quad Q = 1672.03 - 2497.00 = -824.94 \text{ kJ/kg}$$

Total heat

$$Q = -499 \times 824.94$$
$$\mathbf{Q = -4116.45\ kJ}$$

Example 10.10 *A closed vessel of 0.2 cu.m capacity contains steam at a pressure of 8 bar and a temperature of 200°C.*

(a) *What is the weight of steam contained in the vessel ?*

(b) *The vessel is cooled and the steam becomes just dry and saturated. What will be the pressure of steam at this state? Estimate the specific entropy.*

(c) *The vessel is further cooled till the temperature drops to 160.8°C. Find the pressure and condition of steam at this state.*

Solution

(*a*) *Weight of Steam*

Specific volume of superheated steam

$$\text{Here pressure} = 8 \text{ bar}$$

$$V_{sup} = V_{sat} \times \frac{T_{sup}}{T_{sat}}$$

$$V_{sat} = 0.24026 \text{ m}^3$$

$$T_{sat} = 273 + 170.41$$

$$= 443.41°\text{K}$$

$$\therefore \quad V_{sup} = 0.24026 \times \frac{200 + 273}{170.41 + 273}$$

$$= \frac{0.24026 \times 473}{443.41} = 0.2563 \text{ m}^3/\text{kg}$$

$\therefore$ Weight of steam in the vessel

$$= \frac{\text{Actual volume}}{\text{Sp. volume}} = \frac{0.2}{0.2563}$$

Weight of steam in the vessel = 0.78 kg

(*b*) *Pressure and Entropy after cooling*

After cooling, the steam becomes dry and saturated.

$\therefore$ Volume of 1 kg dry and saturated steam = 0.2563 m^3

$\therefore$ **Saturation pressure of steam ≈ 7.5 bar**

(*c*) *Pressure and Entropy*

Saturation temperature of steam = 160.8°C

$\therefore$ Saturation pressure corresponding to this temperature ≈ 6.3 bar

$$= 6.3 \text{ bar}$$

$$V_g = \frac{0.30585 + 0.29681}{2}$$

$$= 0.30133 \text{ m}^3$$

$$\therefore \quad 0.2563 = x \times 0.30133$$

$$x = \frac{0.2563}{0.30133}$$

$$\mathbf{x = 0.85}$$

= Condition of steam (final)

(II) Isothermal Process

Example 10.11 *One kilogram of 0.3 dry steam at 10 bar is heated isothermally so as to make it dry and saturated. Calculate the mechanical work, change in internal energy, enthalpy and entropy during the process.*

Solution

Note: Here the heating process is taking place in the wet region of *T–S* diagram. During this heating process, both temperature and pressure are constant

$\therefore \quad P_1 = P_2 = 10$ bar;

$x_1 = 0.3$ and $x_2 = 1$

$\therefore \quad V_1 =$ Sp. volume of 1 kg steam before heating

$x_1 V_{g1} = 0.3 \times 0.1946$

$= 0.05838 \text{ m}^3/\text{kg}$

$\therefore$ Mechanical work available

$W = P(V_2 - V_1)$

$= 100(0.1946 - 0.05838) \times 10$

$= 1000 \times 0.13622$

$\mathbf{w = 136.22 \text{ kJ/kg}}$

$h_1 =$ Enthalpy of 1 kg

$= h_{f1} + x_1 h_{fg1}$

$= 762.2 + 0.3 \times 2013.8$

$= 1366.34 \text{ kJ/kg}$

$h_2 = h_{g2} = 2776.1 \text{ kJ/kg}$

$h_2 - h_1 =$ Change in enthalpy

$= 2776.1 - 1366.4$

$\mathbf{h_2 - h_1 = 1409.7 \text{ kJ/kg}}$

$u_1 = h_1 - P_1 V_1$

$= 1366.34 - 100 \times 10 \times 0.05838$

$= 1366.34 - 58.38$

$= 1307.96 \text{ kJ/kg}$

$u_2 = h_2 - P_2 V_2$

$= 2776.1 - 100 \times 10 \times 0.1946$

$= 2776.1 - 194.6$

$= 258.5 \text{ kJ/kg}$

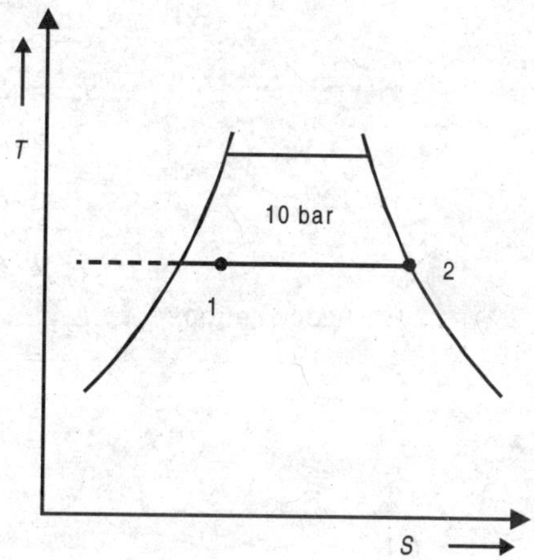

Fig. Ex. 10.11

$\therefore \quad u_2 - u_1 =$ Change in internal energy

$= 2581.5 - 1307.96$

$\mathbf{u_2 - u_1 = 1273.54 \text{ kJ/kg}}$

Again $\quad S_1 = S_{f1} + x_1 S_{fg1}$

$= 2.138 + 0.3 \times (6.582 - 2.138)$

$= 2.138 + 1.3332$

$= 3.4712 \text{ kJ/kg-K}$

$S_2 = S_{g2} = 6.582 \text{ kJ/kg-K}$

$\therefore \quad S_2 - S_1 =$ Change in entropy

$= 6.582 - 3.4712$

$\mathbf{S_2 - S_1 = 3.1108 \text{ kJ/kg-K}}$

Example 10.12 *2 kg steam at 7 bar and 190°C is compressed isothermally to 11 bar. Find the change in entropy, heat added or removed and final condition of steam. Assume* $C_p = 2.3$ *kJ/kg-K.*

Solution

Here saturation temperature at 7 bar is 164.92°C and that at 11 bar 184.06°C. Actual temperature is 190°C. Therefore, steam is superheated before and after isothermal compression. The process is shown on *T–S* diagram.

∴ Final condition of steam is superheated steam with 5.94°C of superheat

$$S_1 = S_{g1} + C_p \log_e \frac{T_{sup}}{T_{sat}}$$

$$= 6.705 + 2.3 \log_e \frac{273+190}{273+164.92}$$

$$= 6.705 + 2.3 \log_e \frac{463}{437.92}$$

$$= 6.705 + 0.1281$$

$$= 6.8330 \text{ kJ/kg-K}$$

$$S_2 = S_{g2} + C_p \log_e \frac{T_{sup}}{T_{sat}}$$

$$= 6.55 + 2.3 \log_e \frac{463}{457.06}$$

$$= 6.55 + 0.0297 = 6.5797 \text{ kJ/kg-K}$$

∴ Change in entropy

$$= S_2 - S_1 = 6.5797 - 6.8331$$

$$= -0.2534 \text{ kJ/kg-K}$$

$$= -0.5068 \text{ kJ/K.}$$

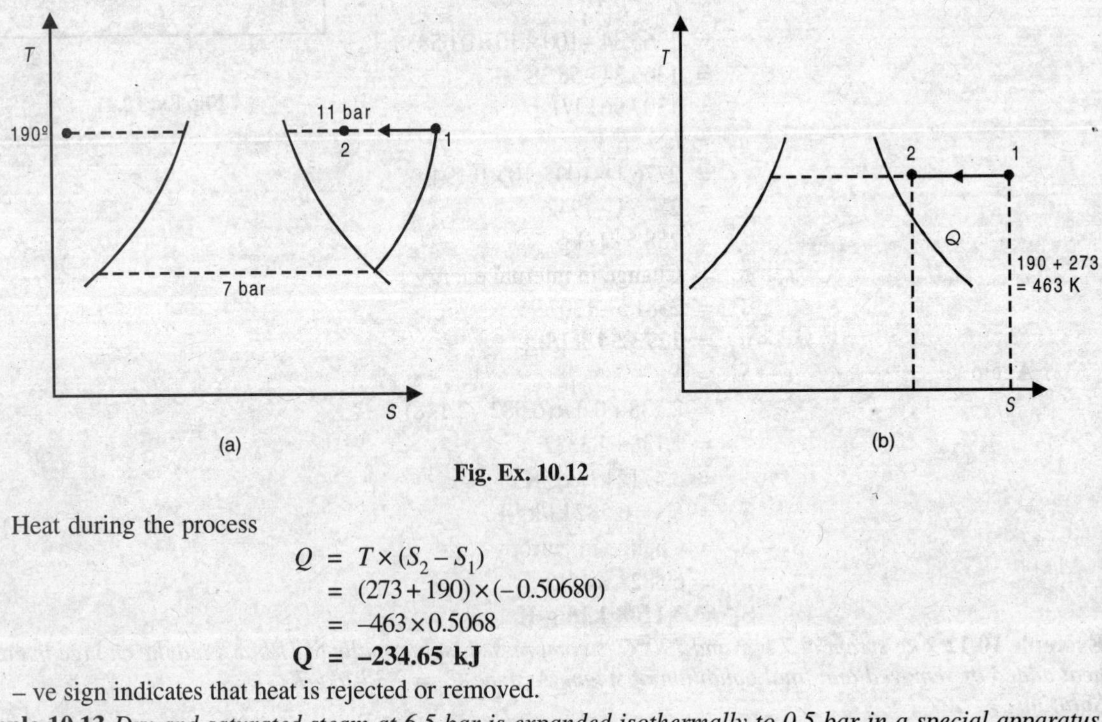

Fig. Ex. 10.12

Heat during the process

$$Q = T \times (S_2 - S_1)$$

$$= (273 + 190) \times (-0.50680)$$

$$= -463 \times 0.5068$$

$$\mathbf{Q = -234.65 \text{ kJ}}$$

– ve sign indicates that heat is rejected or removed.

Example 10.13 *Dry and saturated steam at 6.5 bar is expanded isothermally to 0.5 bar in a special apparatus. Calculate the external work available from 2 m³ of steam initially taken, as well as the change in entropy during the process.*

Solution

The process is shown in Fig. Ex. 10.13

$$T_1 = T_2 = 161.96 + 273 = 434.96\text{ K}$$

S_1 = Entropy before expansion

$$= S_{g1} = 6.730\text{ kJ/kg.K.}$$

$$S_2 = S_{g2} + C_p \ln \frac{T_{sup}}{T_{sat}}$$

$$= 7.597 + 2.3 \log_e \frac{434.96}{273 + 81.31}$$

$$= 7.597 + 2.3 \ln \frac{434.96}{354.31}$$

$$= 7.597 + 0.4717$$

$$= 8.0687\text{ kJ / Kg-K}$$

$\therefore$ Q = Heat during the expansion process

$$= T_1 \times (S_2 - S_1)$$

$$= 434.96 \times (8.0687 - 6.730)$$

$$= 434.96 \times 1.3387$$

$$= 582.28\text{ kJ / kg}$$

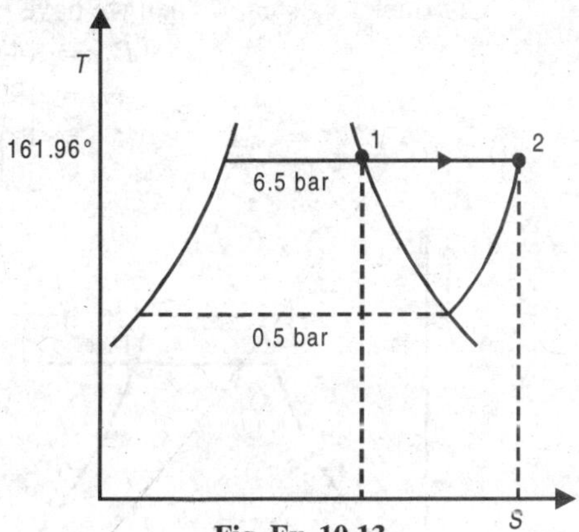

Fig. Ex. 10.13

It is positive. i.e., heat is added.

Initial volume of steam $= V_1 = 0.292847\text{ m}^3\text{/kg}$

$$\therefore \text{ Mass of steam processed} = \frac{2}{0.292847}$$

$$= 6.8295\text{ kg}$$

$\therefore$ Total heat $= 6.8295 \times 582.28$

Total Heat = 3976.68 kJ

Also change in entropy $= (S_2 - S_1) = 6.8295 \times 582.28$

$$= 9.1426\text{ kJ/K}$$

$$h_1 = h_{g1} = 2758.8\text{ kJ/kg}$$

$$\therefore \quad u_1 = h_1 - P_1V_1 = 2758.8 - 100 \times 6.5 \times 0.29284$$

$$= 2758.8 - 190.346$$

$$= 2568.454\text{ kJ/kg}$$

$$h_2 = h_{g2} + C_p (t_{sup} - t_{sat})$$

$$= 2645.3 + 2.3(161.96 - 81.31)$$

$$= 2645.3 + 185.495$$

$$= 2830.795\text{ kJ / kg}$$

$$V_2 = V_{g2} \times \frac{T_{sup}}{T_{sat}} = 3.24827 \times \frac{434.96}{354.31}$$

$$= 3.9877\text{ m}^3\text{/kg}$$

$$\therefore \quad u_2 = h_2 - P_2V_2$$

$$= 2830.795 - 100 \times 0.5 \times 3.9877$$

$$= 2830.795 - 199.383$$

$$= 2631.412 \text{ kJ/kg}$$

$$\therefore \quad u_2 - u_1 = 2631.412 - 2568.454$$

$$= 62.958 \text{ kJ/kg}$$

$$\therefore \quad Q = \Delta u + \text{Work}$$

$$3976.68 = 6.8295 \times 62.958 + \text{Work}$$

$$\therefore \quad \text{Work} = 3976.68 - 429.97$$

$$\textbf{Work} = \mathbf{3546.71 \text{ kJ}}$$

Example 10.14 *A quantity of dry and saturated steam occupies 0.2876 cu.m at 14 bar. Determine the final condition of steam if it is compressed until its volume is halved, if the compression is carried out isothermally. Determine the heat rejected during compression.*

Solution

$$\text{Mass of steam} = \frac{\text{Total steam volume}}{\text{Sp. volume}}$$

$$= \frac{0.2876}{0.14072}$$

Consider 1 kg steam. Then we have

$$P_1 = 14 \text{ bar}; \qquad V_1 = 0.14072 \text{ m}^3;$$

$$t_1 = 195.04°\text{C}; \qquad V_2 = 0.07036 \text{ m}^3;$$

$$t_2 = 195.04°\text{C}$$

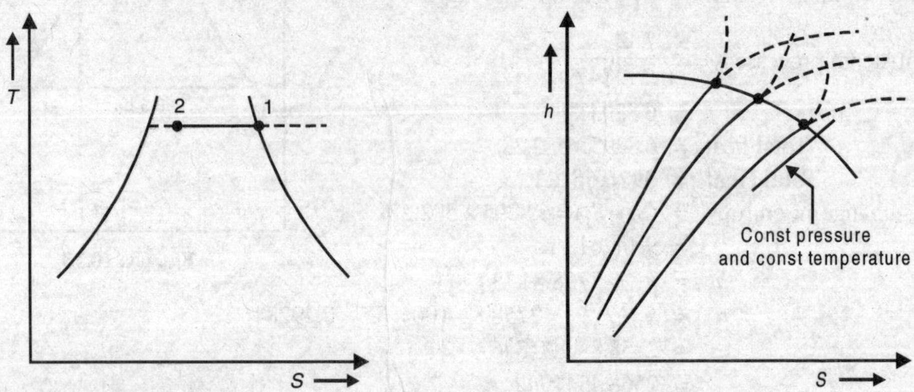

Fig. Ex. 10.14

As the temperature remains same and as the steam is initially dry and saturated, the compression of steam is at the same pressure because 195.04°C is going to be the saturation temperature at p_2 or $p_1 = p_2$.

Now as the volume is halved, the dryness fraction is given by

$$V_2 = x_2 V_{g2}$$

$$\therefore \quad 0.07036 = x_2 \times 0.14072$$

$$\mathbf{x_2 = 0.5}$$

Compression takes place in the wet region and therefore that is also a constant pressure (in addition to constant temperature process) process.

$\therefore$ Heat rejected $= (1 - x_2)\, h_{fg2}$
$= (1-0.5)\times 1957.7$
$= 0.5\times 1957.7$
$= 978.85\ \text{kJ/kg}$

$\therefore$ Total heat rejected $= 2.044\times 978.85$

Total heat rejected = 2000 kJ

(III) Isentrpoic Process

Example 10.15 *Steam at 12 bar and 100°C superheat is expanded adiabatically and reversibly till becomes dry and saturated. Calculate the lower pressure: Calculate also the work done if the process is non-flow type.*

Solution

For reversible adiabatically process, entropy before and after process is unchanged.

$$\therefore \quad S_1 = S_2$$

Assuming specific heat of steam as 2.2 kJ / kg.K

We have,

$$S_1 = S_{g1} + C_{pv} \log_e \frac{T_{\text{sup}}}{T_{\text{sat}}}$$

$$= 6.52 + 2.2 \log_e \frac{273+187.92+100}{273+187.92}$$

$$= 6.52 + 2.2 \log_e \frac{560.92}{460.92}$$

$$= 6.52 + 0.432$$

$$= 6.952\ \text{kJ/kg-K}$$

This can also be found from superheat steam tables

$$\therefore \quad S_2 = 6.952 = S_{g2}$$

as steam is dry and saturated after expansion.

$\therefore$ Pressure corresponding to $S_{g2} - 6.952$ is approximately 3.4 bar.

$$\therefore \quad \mathbf{P = 3.4\ bar}$$

For non-flow process, $Q = \Delta u + \int P.dV$

$$\therefore \quad O = \Delta u + \int P.dV$$

$\therefore$ Work done/kg $= -\Delta u = u_1 - u_2$

h_1 = Enthalpy of 1 kg steam before expansion

$= h_{g1} + C_{pv}$ (Degree of superheated)

$= 2782.6 + 2.2\times 100$

$= 2782.6 + 220$

$= 3002.6\ \text{kJ/kg}$

V_1 = Volume of 1 kg steam before expansion

$$= V_{g1} \times \frac{T_{\text{sup }1}}{T_{\text{sat }1}}$$

$$= 0.163446 \times \frac{5609.92}{460.92}$$
$$= 0.1989 \text{ m}^3\text{/kg}$$
$\therefore$ u_1 = Internal energy of 1 kg steam before expansion
$$= h_1 - P_1V_1$$
$$= 3002.6 - 100 \times 12 \times 0.1989$$
$$= 3002.6 - 238.68$$
$$= 2763 \text{ kJ/kg}$$
$\therefore$ $h_2 = h_{g2}$ = Enthalpy of 1 kg steam after expansion
$$= 2730.4 \text{ kJ/kg}$$
$$\therefore \quad u_2 = h_2 - P_2V_2$$
$$= 2730.4 - 100 \times 3.4 \times 0.539$$
$$= 2730.4 - 183.26$$
$$= 2547.14 \text{ kJ / kg}$$
$$\therefore \quad \text{Work done/kg} = u_1 - u_2$$
$$= 2763.92 - 2547.14$$

Work done/kg = 216.78 kJ/kg

Example 10.16 *Steam flows into a steam nozzle with a pressure of 14 bar when it is dry and saturated. If the expansion is isentropic to 2 bar, find the final condition of steam and change in internal energy, taking C_P = 2.2 kJ/kg.K.*

Find also the velocity of steam leaving the nozzle.

Solution

This is an example of steady flow system. Flow is isentropic.

$P_1 = 14$ bar, $P_2 = 2$ bar

$$S_1 = S_{g1} = 6.465 \text{ kJ/kg-K}$$
$$= S_2 = S_{f2} + x_2 S_{fg2}$$
$$= 1.529 + x_2 (7.127 - 1.529)$$
$$= 1.529 + 5.598\, x_2$$
$\therefore$ x_2 = Find dryness fraction of steam
$$= \frac{6.465 - 1.529}{5.598} = \frac{4.936}{5.598}$$
$$\mathbf{x_2 = 0.8817}$$
Now
$$h_1 = h_{g1} = 2787.9 \text{ kJ/kg}$$
$$h_2 = h_{f2} + x_2 h_{fg2}$$
$$= 504.5 + 0.8817 \times 2201.8$$
$$= 504.5 + 1941.33$$
$$= 244.83 \text{ kJ/kg}$$
$$V_1 = V_{g1} = 0.1409 \text{ m}^3\text{/kg}$$
$$V_2 = x_2 V_{g2} = 0.8817 \times 0.8872$$
$$= 0.7822 \text{ m}^3\text{/kg}$$
$$\therefore \quad u_1 = h_1 - P_1V_1$$
$$= 2787.9 - 100 \times 14 \times 0.1409$$
$$= 2787.9 - 197.26$$

$$= 2590.64 \text{ kJ/kg}$$

$$u_2 = h_2 - P_2V_2$$

$$= 2445.83 - 100 \times 2 \times 0.7822$$

$$= 2445.83 - 156.44$$

$$= 2289.39 \text{ kJ/kg}$$

$$\therefore \quad \Delta u = u_2 - u_1 = \text{change in internal energy}$$

$$= 2289.39 - 2590.64$$

$$\mathbf{\Delta u = -301.25 \text{ kJ/kg}}$$

Assuming change in potential energy and initial velocity, we have

$$Q = \Delta P + \Delta K + \Delta H + W_f$$

$$O = O + \Delta K + \Delta H + O$$

$$\therefore \quad \Delta K = -\Delta H = H_1 - H_2$$

$$\frac{C_2^2 - C_1^2}{2000} = H_1 - H_2$$

$$= 2787.9 - 2445.83$$

$$= 342.07$$

$$C_2 = \sqrt{2000 \times 342.07}$$

$$\mathbf{C_2 = 827.13 \text{ m/s}}$$

$$= \text{Velocity at exit}$$

Example 10.17 *Steam at 15 bar and 250°C expands reversibly and adiabatically to 3 bar in a steam turbine, find the condition of steam after expansion and the work done per kg steam admitted to the turbine. Assume ΔP and ΔK to be zero.*

Solution

This is an other example of steady flow system. Here $P_1 = 3$ bar, $T_1 = 523$ K

$$h_1 = 2923.8 \text{ kJ/kg}$$

$$S_1 = 6.711 \text{ kJ/kg.K} = S_2$$

$$\therefore \quad 6.711 = s_{f2} + x_2 s_{fg2}$$

$$\therefore \quad 6.711 = 1.671 + x_2(6.991 - 1.671)$$

$$x_2 = \frac{6.711 - 1.671}{5.320}$$

$$= \frac{5.040}{5.320}$$

$$\mathbf{x_2 = 0.9474}$$

$$= \text{Final condition of steam}$$

$$\therefore \quad h_2 = \text{Enthalpy of steam after expansion}$$

$$= h_{f2} + x_2 f_{fg2}$$

$$= 561.3 + 0.9474 \times 2163.3$$

$$= 561.3 + 2049.5$$

$$= 2610.8 \text{ kJ/kg}$$

For steady flow system, we have

$$Q = \Delta P + \Delta K + \Delta H + W_{sf}$$

$$O = O + O + \Delta H + W_f.$$

$$\therefore \quad W_{sf} = \text{work of steady flow system}$$
$$= h_1 - h_2$$
$$= 2923.8 - 2610.8$$
$$\mathbf{W_{sf} = 313.0 \text{ kJ/kg}}$$

Example 10.18 *Find the internal energy of steam 0.9 dry at 16 bar reckoned from water at 0° C. If the steam is expanded isentropically in a cylinder to 0.9 bar, find the final internal energy of the steam.*

Solution

$$h_1 = h_{f1} + x_1 f_{fg1}$$
$$= 858.3 + 0.9 \times 1933.4$$
$$= 858.3 + 1740.06$$
$$= 2598.36 \text{ kJ/kg}$$
$$v_1 = \text{Initial volume of steam}$$
$$= X_1 V_{g1} = 0.9 \times 0.123843$$
$$= 0.1115 \text{ m}^3\text{/kg}$$

Neglecting volume of water at 0°C,

$$u_1 = h_1 - P_1 V_1 = 2598.36 - 100 \times 16 \times 0.1115$$
$$= 2598.36 - 178.33$$
$$\mathbf{u_1 = 2420.03 \text{ kJ/kg-K}}$$
$$= \text{Internal energy reckoned from water at 0°C}$$

Now

$$S_1 = s_{f1} + x_1 s_{fg1} = 2.343 + 0.9(6.417 - 2.343)$$
$$= 2.343 + 3.6666$$
$$= 6.0096 \text{ kJ / kg-K}$$
$$S_2 = S_{f2} + x_2 S_{fg2} = 1.269 + x_2(7.396 - 1.269)$$
$$= 1.269 + x_2 \times 6.127$$
$$\therefore \quad S_1 = S_2 \text{ for isentropic process}$$
$$= 1.269 + 6.127$$
$$x_2 = \frac{6.0096 - 1.2690}{6.127} = \frac{4.7406}{6.127}$$
$$= 0.7737$$
$$\therefore \quad h_2 = h_{f2} + x_2 h_{fg2} = 405.1 + 0.7737 \times 2265.8$$
$$= 405.1 + 1753.1 = 2158.2 \text{ kJ / kg}$$
$$V_2 = x_2 v_{g2} = 0.7737 \times 1.87135$$
$$= 1.4479 \text{ m}^3\text{ / kg}$$
$$\therefore \quad u_2 = \text{Final internal energy above 8°C}$$
$$= h_2 - P_2 V_2$$
$$= 2158.2 - 100 \times 0.9 \times 1.4479$$
$$= 2158.2 - 130.31$$
$$\mathbf{u_2 = 2027.89 \text{ kJ/kg}}$$

Example 10.19 *Calculate the index of adiabatic expansion if steam expanded adiabatically from 6 bar and 200°C to 1 bar has dryness fraction 0.93 at the end of expansion.*

Take $C_p = 2.3$ kJ / kg. K ; $p_1 = 6$ bar; $p_2 = 1$ bar; $x_2 = 0.93$

Solution

$$V_1 = V_{g1} \times \frac{T_{sup}}{T_{sat}}$$

$$= 0.3159 \times \frac{200 \times 273}{158.81 + 273} = \frac{0.3159 \times 473}{431.81}$$
$$= 0.346 \text{ m}^3/\text{kg}$$
$$V_2 = x_2 V_{g2}$$
$$= 0.93 \times 1.6949$$
$$= 1.5762 \text{ m}^3/\text{kg}$$

Let n be the index of adiabatic expansion.

$$\therefore \quad P_1 V_1^n = P_2 V_2^n$$

or $$\left(\frac{V_2}{V_1}\right)^n = \frac{P_2}{V_2}$$

or $$n \ln \frac{V_2}{V_1} = \ln \frac{p_1}{p_2}$$

$$N = \frac{\ln P_1 / P_2}{\ln V_2 / V_1} = \frac{\ln \frac{6}{1}}{\ln . \frac{1.5762}{0.346}}$$

$$= \frac{1.7918}{1.5163}$$
$$\mathbf{n = 1.1817}$$

Example 10.20 *1 kg of steam at 100 bar and 375° C expands isentropically upto 38 bar. At the end of expansion steam is dry and saturated. For both non flow and steady flow process with $\Delta K = 0$, find the work done.*

Data: $p_1 = 100$ bar; $t_1 = 375°$ C; $p_2 = 38$ bar; $x_2 = 1$.

Solution

From superheated steam tables –

$$V_1 = \frac{0.02242 + 0.2641}{2} = \frac{0.04883}{2} = 0.024415$$

$$h_1 = \frac{2925.8 + 3099.9}{2} = \frac{6025.7}{2} = 3012.85 \text{ m}^3/\text{kg}$$

$$\mathbf{h_1 = 3012.85}$$
$$V_2 = V_{g2} = 0.052438 \text{ cu.m/kg}$$
$$h_2 = 2801.1 \text{ kJ/kg}$$

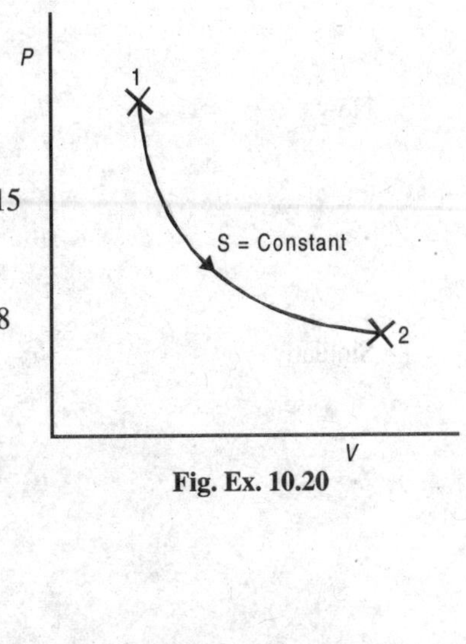

Fig. Ex. 10.20

Non-flow process :

$$u_1 = h_1 - P_1 V_1 = 3012.85 - 100 \times 0.024415$$
$$= 3012.85 - 244.15$$
$$= 2768.70 \text{ kJ/kg}$$
$$u_2 = h_2 - P_2 V_2 = 2801.1 - 100 \times 0.052438$$
$$= 2801.1 - 199.26$$
$$= 2601.84 \text{ kJ/kg}$$

$\therefore$ For non-flow process $= Q = \Delta u + \text{work}$

$$0 = (u_2 - u_1) + w_{nf}$$

$$\therefore \quad W_{nf} = \text{Work done} = u_1 - u_2$$
$$= 2768.70 + 2601.84$$
$$W_{nf} = 166.86 \text{ kJ/kg}$$

For steady flow process

with $\Delta K = 0$

We have
$$\text{Work} = -\Delta H = h_1 - h_2$$
$$= 3012.85 - 2801.1$$
$$\textbf{Work} = \textbf{277.75 kJ/kg}$$

(IV) Hyperbolic Process

Example 10.21 *Steam at 3.6 bar and 0.66 dry is compressed hyperbolically to 10 bar. Determine the heat removed from 2 m³ of steam.*

Solution

$$P_1 = 3.6 \text{ bar}; x_1 = 0.66$$
$$\therefore \quad V_1 = x_1 V_{g1} = 0.66 \times 0.5108$$
$$= 0.3371 \text{ m}^3/\text{kg}$$
$$P_2 = 10 \text{ bar}$$

Law of expansion is $PV = C$

$$\therefore \quad P_1V_1 = P_2V_2$$
$$V_2 = V_1 \times \frac{P_1}{P_2} = 0.337 \times \frac{3.6}{10}$$
$$= 0.1214 \text{ m}^3/\text{kg}$$

This is less than V_{g2}. Therefore the steam is wet after compression.

$$\therefore \quad 0.1214 = x_2 \times 0.1946$$
$$x_2 = \frac{0.1214}{0.1946}$$
$$= 0.6237$$

$$\therefore \quad \text{Work done/kg} = P_1V_1 \ln\left(\frac{V_2}{V_1} = \frac{P_1}{P_1}\right)$$
$$= -100 \times 3.6 \times 0.337 \times \ln\frac{10.0}{3.6}$$
$$= -367.8 \text{ kJ/kg}$$

Now
$$h_1 = h_{f1} + x_1 h_{fg1}$$
$$= 588.4 + 0.66 \times 2144.5$$
$$= 2003.77 \text{ kJ/kg}$$
$$\therefore \quad u_1 = h_1 - P_1V_1$$
$$= 2003.77 - 100 \times 3.6 \times 0.3371$$
$$= 2125.126 \text{ kJ/kg}$$

Similarly
$$h_2 = h_{f2} + x_2 h_{fg2}$$
$$= 762.2 + 0.6237 \times 2013.8$$
$$= 2018.2 \text{ kJ/kg}$$
$$\therefore \quad u_2 = h_2 - P_2V_2$$

$$= 2018.2 - 121.356$$
$$= 1896.844 \text{ kJ / kg}$$
$$\therefore \quad u_2 - u_1 = 1896.844 - 2125.126$$
$$= -228.282 \text{ kJ/kg}$$
$$\therefore \quad \text{Heat } Q = \Delta u + \int P \cdot dV$$
$$= -228.282 - 367.8$$
$$\textbf{Heat Q} = \mathbf{-596.082 \ kJ/kg}$$
$$= \text{Heat removed.}$$
$$= -596.082 \times \frac{2}{0.3371} = -3536.53 \text{ kJ}$$

Example 10.22 *Steam at 8 bar and 175°C is throttled down to 4 bar and then expanded hyperbolically to 1.2 bar. Determine*

(i) change in entropy during throttling
(ii) change in enthalpy during expansion and
(iii) work done

Solution

(i) Change in entropy during throttling : (1 – 2)

$$h_1 = \text{Enthalpy before throttling}$$
$$= h_{g1} + c_p (t_{sup} - t_{sat})$$
$$= 2676.3 + 2.3 (175 - 170.38)$$
$$= 2767.3 + 10.626$$
$$= 2777.923 \text{ kJ/kg}$$
$$S_1 = S_{g1} + C_p \ln \frac{T_{sup}}{T_{sat}} = 6.659 + 2.3 \ln \frac{175 + 273}{170.38 + 273}$$
$$= 6.659 + 0.0238$$
$$= 6.6828 \text{ kJ / kg.K}$$

For throttling process,

$$h_1 = h_2 > h_{g2}$$
$$\therefore \quad 2777.926 = 2737 + 2.3 (t_{sup2}) - 143.6$$
$$\therefore \quad t_{sup2} = \frac{2777.962 - 2737.7}{2.3} + 143.6$$
$$= \frac{40.226}{2.3} + 143.6$$
$$= 17.49 + 143.6 = 161.09°C$$
$$\therefore \quad S_2 = S_{g2} + C_p \ln \frac{T_{sup}}{T_{sat}}$$
$$= 6.894 + 2.3 \ln \frac{273 + 161.09}{273 + 143.6}$$
$$= 6.894 + 2.3 \ln \frac{434.09}{416.6}$$

$$= 6.894 + 0.0946$$
$$= 6.9886 \text{ kJ/kg}$$

$\therefore$ Change in entropy during throttling

$$= S_2 - S_1 = 6.9886 - 6.6828$$
$$= 0.3058 \text{ kJ/kg-K}$$

(i) Change in enthalpy during expansion (2 – 3)

$$h_2 = 2777.926 \text{ kJ/kg}$$

$$V_2 = V_{g2} \times \frac{T_{sup}}{T_{sat}}$$

$$= 0.462626 \times \frac{434.09}{416.6}$$

$$\mathbf{V_2 = 0.482 \ m^3/kg}$$

$P_2 = 4$ bar; $P_3 = 1.2$ bar

$$\therefore \quad P_2V_2 = P_3V_3$$

$$\therefore \quad V_3 = V_2 \times \frac{P_2}{P_3} = \frac{0.482 \times 4}{1.2}$$

$$= 1.607 \text{ m}^3/\text{kg}$$
$$> v_{g3}$$

$\therefore$ Steam is superheated after expansion

$$\therefore \quad 1.607 = 1.43016 \times \frac{T_{sup_3}}{104.77 + 273}$$

$$\therefore \quad T_{sup3} = \frac{1.607}{1.43016} \times \frac{377.77}{1}$$

$$= 424.436 \text{ K}$$

$$\therefore \quad h_{g3} = h_{g3} + C_p (T_{sup3} - T_{sat3})$$
$$= 2684.9 + 2.3\,(424.436 - 377.77)$$
$$= 2684.9 + 2.3 \times 46.666$$
$$= 2684.9 + 107.33$$
$$= 2792.23 \text{ kJ/kg}$$

$\therefore$ Change in enthalpy during expansion

$$h_3 - h_2 = 2792.23 - 2777.926$$
$$\mathbf{h_3 - h_2 = 14.304 \ kJ/kg}$$

(iii) Work done

$$\text{Work done} = P_2\, V_2 \ln\left(\frac{V_3}{V_2} = \frac{P_2}{P_3}\right)$$

$$= 100 \times 4 \times 0.482 \ln \frac{4}{1.2}$$

$$\textbf{Work done} = \mathbf{232.126 \ kJ/kg}$$

Example 10.23 *The steam in the steam engine cylinder expands according to the hyperbolic law from 8.5 bar and 0.85 dry to 0.7 bar.*

Find the final condition of steam and work done per kilogram assuming C_p = 2.01 kJ/kg-K.

Solution

We have $P_1 = 8.5$ bar; $P_2 = 0.7$ bar; $x_1 = 0.85$ dry

$$\therefore \quad V_1 = x_1 V_{g1} = 0.85 \times 0.227037 = 0.193 \text{ m}^3/\text{kg}$$

$$\therefore \quad P_1V_1 = P_2V_2$$

$$V_2 = \frac{P_1}{P_2} \cdot V_1 = \frac{8.5}{0.7} \times 0.193 = 2.3433 \text{ m}^3/\text{kg} < V_{g2}$$

$\therefore$ The steam is wet

$$\therefore \quad 2.3433 = x_2 \times 2.36795$$

$$\therefore \quad x_2 = \frac{2.3433}{2.36795}$$

$$\mathbf{x_2 = 0.9896}$$

$$\text{Work done/kg} = P_1 V_1 \log_e \left(\frac{V_2}{V_1} = \frac{P_1}{P_2} \right) = 100 \times 8.5 \times 0.193 \ln \frac{8.5}{0.7}$$

$$\textbf{Work done/kg} = \textbf{409.59 kJ/kg}$$

Example 10.24 *10 kg of steam at 8 bar and 0.9 dry expanded in a steam engine cylinder hyperbolically to 0.4 bar. Find the final condition of steam and the external work done during the process.*

Solution

$$P_1 = 8 \text{ bar};$$

$$P_2 = 0.4 \text{ bar};$$

$$V_1 = xV_{g1} = 0.9 \times 0.2405 = 0.21645 \text{ m}^3/\text{kg}$$

$$\therefore \quad P_1V_1 = P_2V_2$$

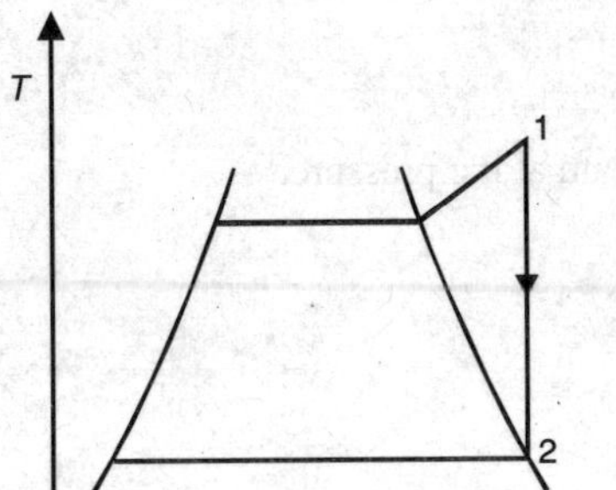

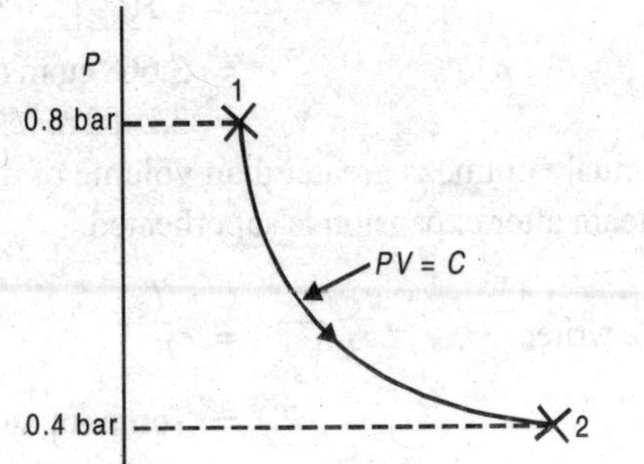

Fig. Ex. 10.24

$$\therefore \quad V_2 = \frac{P_1}{P_2} \times V_1 = \frac{8}{0.4} \times 0.21645$$

$$= 4.329 \text{ m}^3/\text{kg}$$

$$> V_{g2} (3.993)$$

$\therefore$ Steam is superheated.

$$\therefore \quad 4.329 = V_{g2} \times \frac{T_{\text{sup}\,2}}{T_{sat\,2}} = \frac{3.993 \times T_{\text{sup}\,2}}{75.9 + 273}$$

$$\therefore \quad T_{\text{sup2}} = \frac{4.329}{3.993} \times 348.9$$

$$= 378.26°\text{K}$$

$$\mathbf{T_{sup2} = 105.26° \ C}$$

$$\therefore \quad \text{Work done/kg} = P_1 V_1 \ln \frac{P_1}{P_2}$$

$$= 100 \times 8 \times 0.21645 \ln \frac{8}{0.4}$$

$$= 518.74 \text{ kJ/kg}$$

$$\textbf{Work done/kg} = \mathbf{5187.4 \ kJ}$$

Example 10.25 *Dry and saturated steam of 3.5 bar pressure is expanded hyperbolically to 0.5 bar. Using steam tables, determine*

(*i*) *Final condition of steam*

(*ii*) *Change of specific entropy during the process*

Solution

(i) Hyperbolic process is defined by PV = Const. or $P_1V_1 = P_2V_2$

Here $P_1 = 3.5$ bar and $P_2 = 0.5$ bar; $x_1 = 1$

$$\therefore \quad V_1 = V_{g1} = 0.524 \text{ m}^3/\text{kg}$$

$$\therefore \quad V_2 = \text{volume after axpansion}$$

$$= \frac{P_1 V_1}{P_2} = \frac{3.5 \times 0.524}{0.5}$$

$$= 3.668 \text{ cu.m/kg}$$

$$V_{g2} = 3.2402 \text{ m}^3/\text{kg}$$

$\therefore$ Actual volume is greater than volume of dry and saturated steam at the pressure.

$\therefore$ Steam after expansion is superheated.

$$\therefore \text{ We write,} \quad \frac{V_2}{V_{g_2}} = \frac{T_2}{T_{\text{sat}_2}}$$

or

$$T_2 = \text{Temp. of superheated steam}$$

$$= \frac{3.668}{3.2402} \times \frac{273 + 81.343}{1}$$

$$\mathbf{T_2 = 401.13 \ K}$$

(ii) Change in specific entropy:

S_2 = Entropy of 1 kg at 3.5 bar and $x_1 = 1 = S_{g1} = 6.9392$ kJ/kg-K

Entropy of 1 kg steam at 0.5 bar

$$S_2 = S_{g2} + C_{pv} \ln \frac{T_{sup}}{T_{sat}}$$

$$= 7.5947 + 2 \ln \frac{401.13}{354.343}$$

$$= 7.5947 + 2 \times 0.1240$$

$$= 7.5947 + 248$$

$$= 7.8427 \text{ kJ/kg-K}$$

$\therefore$ Change in entropy $= S_2 - S_1$

$= 7.8427 - 6.9392$

Change in entropy = 0.9035 kJ/kg-K

Example 10.26 *A quantity of dry and saturated steam occupies 0.2876 cu.m at 14 bar. Determine the final condition of the steam if it is compressed until its volume is halved, and if the compression is carried out hyperbolically.*

Determine the heat rejected during compression.

Solution

Volume of 1 kg dry and saturated steam at 14 bar = 0.14072 cu.m.

$$\therefore \quad \text{Mass of steam} = \frac{0.2876}{0.14072} = \frac{\text{Total volume}}{\text{Sp. volume}}$$

$$= 2.044 \text{ kg}$$

Hyperbolic process follows the law $PV = C$

$\therefore \quad P_1V_1 = P_2V_2$ Consider 1 kg of steam

$$\therefore \quad \frac{V_1}{V_2} = \frac{P_2}{P_1} = \frac{0.14072}{0.14072/2} = 2$$

$$\therefore \quad P_2 = P_1 \times 2 = 14 \times 2 = 28 \text{ bar}$$

$$V_2 = 0.14072/2 = 0.07036 \text{ cu.m/kg}$$

$$< V_{g2} (0.071389)$$

$\therefore$ Steam after compression is wet.

$$\therefore \quad V_2 = X_2V_{g2} = x_2 \times 0.071389$$

$\therefore \quad x_2$ = Dryness fraction after compression

$$= \frac{0.07036}{0.071389}$$

$$x_2 = 0.9856$$

Calculation for heat $\quad h_1 = 2787.8 \text{ kJ/kg}$

$$\therefore \quad u_1 = h_1 - P_1V_1 = 2787.8 - \frac{10^5 \times 14 \times 0.14072}{1000}$$

$$= 2787.8 - 197$$

$$= 2590.8 \text{ kJ/kg}$$

Similarly, $h_2 = h_{f2} + x_2 h_{fg2}$

$$= 990.48 + 0.9856 \times 1811.5$$
$$= 990.48 + 1785.41$$
$$= 2775.89 \text{ kJ / kg}$$

$\therefore$ $u_2 = h_2 - P_2V_2 (= P_1V_1)$

$$= 2775.89 - 197$$
$$= 2578.89 \text{ kJ/kg}$$

$$\text{Work done kg} = P_1 V_1 \ln \frac{V_2}{V_1}$$

$$= \frac{10^5 \times 14 \times 0.14072}{1000} \times \ln \frac{1}{2}$$

$$= 197 \ln \frac{1}{2}$$

$$= -136.55 \text{ kJ/kg}$$

$\therefore$ $Q = \Delta u + \int P \cdot dV$

$$= (u_2 - u_1) + (-136.55)$$
$$= (2578.89 - 2590.8) - 136.55$$
$$= 11.91 - 136.55$$
$$= -148.46 \text{ kJ/kg}$$

$\therefore$ Total heat rejected $= -148.46 \times 2.044$

$\therefore$ **Total heat rejected = –303.45 kJ**

(V) Polytropic Process

Example 10.27 *Steam originally at 16 bar and 0.9 dry expands in a cylinder according to law $pV^{1.1}$ = const. Until the pressure is 1.6 bar. Determine the amount of heat put into or extracted from the steam per kg during process stating the direction of heat transfer.*

Given: P_1 = 16 bar; x_1 = 0.9 dry; P_2 = 1.6 bar; $PV^{1.1}$ = constant.

Solution

From steam table V_1 = Initial volume of steam/kg

$$= x_1 V_{g1}$$
$$= 0.9 \times 0.12369$$

$$= \frac{0.1113}{n} \text{ m}^3\text{/kg}$$

We have, $P_1 V^n_{\ 1} = P_2 V_2$

$$\therefore \quad \left(\frac{V_2}{V_1}\right)^n = \frac{P_1}{P_2}$$

or $$V_2 = V_1 \cdot \left(\frac{P_1}{P_2}\right)^{1/n}$$

$$= 0.1113\left[\left(\frac{16}{1.6}\right)^{1/1.17}\right]$$

$$= 0.1113 \times (10)^{0.9091}$$

$$= 0.1113 \times 8.1115$$

$$= 0.9028 \text{ m}^3/\text{kg}$$

This is less than $V_{g2} = 1.0911$

Therefore, steam after expansion is wet and

$$x_2 = \frac{0.9028}{1.0911}$$

$$= 0.8274$$

$\therefore$ h_1 = Enthalpy of 1 kg steam before expansion

$$= h_{f1} + x_1 h_{fg2}$$

$$= 858.56 + 0.9 \times 1933.2$$

$$= 858.56 + 1739.88$$

$$= 2598.44 \text{ kJ/kg}$$

$$\therefore \quad u_1 = h_1 - P_1 V_1$$

$$= 2598.44 - 100 \times 16 \times 0.1113$$

$$= 2598.44 - 178.08$$

$$= 2420.36 \text{ kJ/kg}$$

Similarly, h_2 = Enthalpy of 1 kg steam after exp.

$$= hf_2 + x_2 h_{fg2}$$

$$= 475.38 + 0.8274 \times 2220.9$$

$$= 2302.95 \text{ kJ/kg}$$

$$\therefore \quad u_2 = h_2 - P_2 V_2$$

$$= 2302.95 - 100 \times 1.6 \times 0.9028$$

$$= 2158.50 \text{ kJ/kg}$$

$$\text{Work done/kg} = \int P.dV$$

$$= \frac{P_1 V_1 - P_2 V_2}{n-1}$$

$$= \frac{178.08 - 144.45}{1.1 - 1}$$

$$= \frac{33.63}{0.1}$$

$$= 336.3 \text{ kJ/kg}$$

Energy equation gives

$$Q = \Delta u + \int P.dV$$

$$= (u_2 - u_1) + 336.3$$
$$= (2158.5 - 2420.36) + 336.3$$
$$\mathbf{Q = 74.44\ kJ/kg}$$

Here Q is positive. This means that heat is added to the system i.e., steam.

Example 10.28 *0.9 kg steam initially at a pressure if 15 bar and temperature of 250°C expands reversibly and polytropically to 1.5 bar. Find (i) the final condition (ii) work done (iii) heat transferred and (iv) change in entropy if the index of expansion is 1.25.*

Show the process on P–V and T–S diagrams.

Given: $P_1 = 15$ *bar;* $T_1 = 250°C$; $P_2 = 1.5$ *bar;* $n = 1.25$

Solution

The process is shown on *P–V* and *T–S* diagram in Fig. Ex. 10.28.

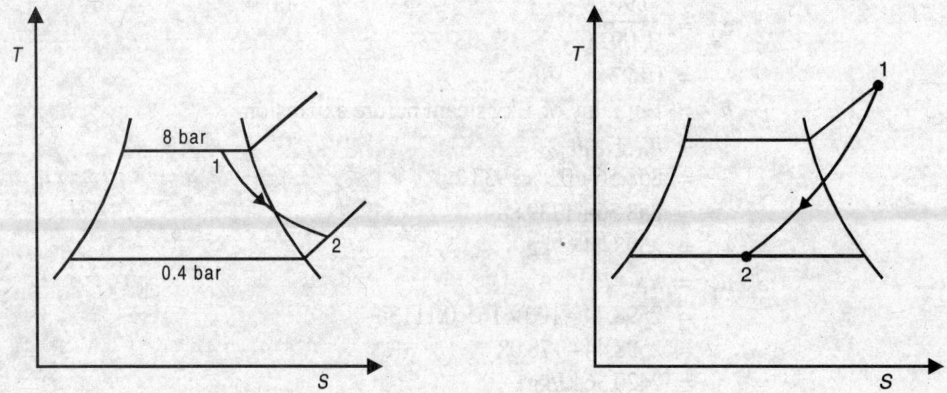

Fig. Ex. 10.28

First we will calculate on kg basis.

(*i*) *Final condition* : $V_1 = 0.1520\ m^3/kg$; $h_1 = 2923.5$ kJ/kg; $S_1 = 6.7099$ kJ/kg-K

$$\therefore \quad P_1 V^n{}_1 = P_2 V^n{}_2$$

or

$$V_2 = V_1 \times \left(\frac{P_1}{P_2}\right)^{1/n}$$
$$= 0.1520 \times \left(\frac{15}{1.5}\right)^{1/1.25}$$
$$= 0.152 \times 6.3096$$
$$= 0.9591\ m^3/kg$$
$$V_{g2} = 1.1590\ m^3/kg$$

$\therefore$ Steam is wet

$$\therefore \quad x_2 = \text{Dryness fraction of steam after expansion}$$
$$= \frac{0.9591}{1.159}$$
$$\mathbf{x_2 = 0.8275}$$

(ii) Work done :

Work done for non-flow process = $\int P \cdot dV$

$$= \frac{P_1 V_1 + P_2 V_2}{n-1}$$

$$= \frac{100}{0.25}\left[\frac{15\times0.152-1.5\times0.9591}{1}\right]$$

$$= \frac{228-143.865}{0.25}$$

$$= \frac{84.135}{0.25}$$

$$= 336.54 \text{ kJ/kg}$$

$\therefore$ Work done for 0.9 kg steam

$$= 0.9\times336.54 = 302.886 \text{ kJ.}$$

(iii) Heat transferred

$$h_1 = 2923.5 \text{ kJ/kg}$$

$$\therefore \quad u_1 = h_1 - P_1V_1 = 2923.5-228 = 2695.5 \text{ kJ/kg}$$

$$h_2 = \text{enthalpy after expansion} = h_{f2} + x_2 h_{fg2} = 467.13+0.8275\times2226.2 = 467.13+1842.18 = 2309.31 \text{ kJ/kg}$$

$$\therefore \quad u_2 = h_2 - P_2V_2 = 2309.31-143.865 = 2165.445 \text{ kJ/kg}$$

$\therefore$ For non-flow process,

$$Q = \Delta u + \int P.dV = u_2 - u_1 + \int P.dV$$

$$= (2165.445-2695.5)+336.54$$

$$= -530.055+336.54$$

$$= -193.575 \text{ kJ/kg}$$

$\therefore$ For 0.9 kg $\quad Q = -0.9\times193.515$

$$\mathbf{Q = -174.1635 \text{ kJ/kg}}$$

– ve sign indicates that heat is rejected by the steam.

(iv) Change in entropy

$$S_1 = \text{Entropy before expansion} = 6.7099 \text{ kJ/kg-K}$$

$$S_2 = \text{Entropy after expansion} = S_{f2} + x_2 S_{fg2} = 1.4336+0.8275\times5.7898 = 6.2247 \text{ kJ/kg-K}$$

∴ Change in entropy

$$\Delta S = S_2 - S_1 = 6.2247 - 6.7099$$
$$= -0.4852 \text{ kJ/kg-K}$$

∴ Total change in entropy

$$= -(0.9 \times 0.4852)$$
$$= -0.43668 \text{ kJ-K}$$

– ve sign indicates that entropy is decreasing.

Example 10.29 *1 kg of steam, initially saturated at 11 bar, expands non flow according to $pV^{1-13} = C$ to 1 bar. Determine: (a) final volume v_2 (b) x_2 (c) work (d) Δu and (e) Q. (f) If the process is steady flow which $\Delta K = +4$ kJ/kg , what is the work?*

Data: $p_1 = 11$ bar; $x_1 = 1$; $P_2 = 1$ bar; $PV^{1.3} = C$

Solution

(*a*) *Final volume V_2*

From tables,

$$V_1 = 0.17738 \text{ m}^3/\text{kg}$$
$$P_1 V^n{}_1 = P_1 V_2^n$$
$$\therefore \quad V_2 = \left(\frac{P_1}{P_2}\right)^{1/n} \cdot V_1$$

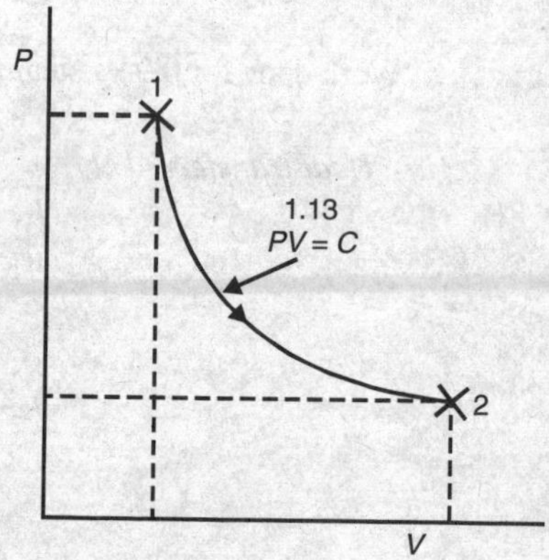

Fig. Ex. 10.29

(*b*) *Final dryness fraction x_2*

$$V_{g2} = 1.6937 \text{ m}^3/\text{kg}$$

∴ Steam is wet

$$\therefore \quad 1.481 = x_2 \times 1.6937$$
$$\therefore \quad x_2 = \frac{1.481}{1.6937}$$
$$\mathbf{x_2 = 0.8743}$$

(*c*) *Work*:

$$\text{Work of non-flow process} = \int P \cdot dV$$
$$= \frac{P_1 V_1 - P_2 V_2}{n-1}$$
$$= \frac{100}{1.13-1}[11 \times 0.17738 - 1 \times 1.481]$$
$$= \frac{100}{0.13}[1.9512 - 1.481]$$
$$= 769.231 \times 0.4702$$
$$= 361.69 \text{ kJ/kg}$$

(*d*) *Change in internal energy*

$$h_1 = h_{f1} + x_2 h_{fg1} = h_{g1}$$
$$= 2779.7 \text{ kJ/kg}$$
$$\therefore \quad u_1 = h_1 - P_1 V_1 = 2779.7 - 195.12$$
$$= 2574.58 \text{ kJ/kg}$$
$$h_2 = h_{f2} + x_2 h_{fg2} = 417.51 + 0.8743 \times 2257.9$$

$$= 417.51 + 1974.08$$
$$= 2391.59 \text{ kJ/kg}$$
$$\therefore \quad u_2 = h_2 - P_2V$$
$$= 2391.59 - 100(1.481) = 2391.59 - 148.1$$
$$= 2243.49 \text{ kJ/kg}$$
$$\therefore \quad \Delta u = u_2 - u_1 = 2243.49 - 2574.58$$
$$\mathbf{\Delta u = -331.09 \text{ kJ/kg}}$$

(e) *Work for steady flow process*

$$\text{Work done} = -\int P.dV + \Delta K = 1.13\int P.dV - 4$$
$$= 1.13 \times 361.69 - 4 = 408.71 - 4$$
$$\text{Work done} = 404.71 \text{ kJ/kg}$$

(f) *Heat Q*

For non-flow process, $Q = \Delta u + \int P.dV$

$$Q = -331.09 + 361.69$$
$$\mathbf{Q = 30.6 \text{ kJ/kg}}$$

Example 10.30 *Four kg of steam at 13 bar and 260°C expand into wet region to 95°C in a polytropic process where $PV^{1.21} = C$. Find (a) y_2, ΔH, Δu, ΔS. (b) $\int P.dV$ (c) $-\int V.dP$ (d) work for nonflow process and (e) work for steady flow process if $\Delta K = -10$ kJ/ kg (f). Find Q for both non-flow and steady flow processes.*

Solution

(a) Wetness y_2, ΔH, ΔU, ΔS :

At 95°C, saturation pressure $= P_2 = 0.8453$ bar; $P_1 = 13$ bar;

$$V_1 = 0.1769 + 0.2\,(0.1969 - 0.1769)$$
$$= 0.1769 + 0.2 \times 0.0200$$
$$= 0.1769 + 0.004$$
$$= 0.1809 \text{ cu.m/kg}$$

$$\therefore \quad V_2 = \left(\frac{P_1}{P_2}\right)^{1/n} \times V_1 = 0.1809 \times \left(\frac{13}{0.845}\right)^{1/1.20}$$
$$= 0.1809 \times (15.38)^{0.8264}$$
$$= 1.7312 \text{ m}^3\text{/kg}$$

But $V_{g2} = 1.982 \text{ m}^3\text{/kg}$

Actual volume is less than v_{g2}. Therefore steam is wet

$$\therefore \quad 1.7312 = x_2.V_{g2}$$
$$x_2 = \frac{1.7312}{1.982}$$
$$= 0.8734$$
$$\therefore \quad y_2 = \text{Wetness} = 0.1216$$
$$\therefore \quad h_1 = 2931.5 + 0.2\,(3044.3 - 2931.5)$$
$$= 2931.2 + 0.2 \times 112.8$$
$$= 2931.5 + 22.56 = 2954.06 \text{ kJ/kg}$$
$$h_2 = h_{f2} + x_2 h_{fg2} = 397.99 + 0.8734 \times 2270.2$$

$$= 397.99 + 1982.79$$
$$= 2380.78 \text{ kJ/kg}$$
$$\therefore \quad \Delta H = h_2 - h_2 = 2380.78 - 2954.06$$
$$\mathbf{\Delta H = -573.28 \text{ kJ/kg}}$$
$$S_2 = S_{f2} + x_2 S_{fg2} = 1.2501 + 0.8734 \times 6.1665$$
$$= 1.2581 + 5.3858$$
$$= 6.6359 \text{ kJ/kg-K}$$
$$S_1 = 6.7878 + 0.2\,(6.9938 - 6.7878)$$
$$= 6.7878 + 0.2 \times 0.1060 = 6.7878 + 0.02012$$
$$= 6.80792 \text{ kJ/kg-K}$$
$$\therefore \quad \Delta S = S_2 - S_1 = 6.63590 - 6.80792$$
$$\mathbf{\Delta S = 0.17202 \text{ kJ/kg-K}}$$
$$u_1 = h_1 - P_1 V_1 = 2954.06 - 100 \times 13 \times 0.1809$$
$$= 2954.06 - 235.17$$
$$= 2718.89 \text{ kJ/kg}$$
$$u_2 = h_2 - P_2 V_2 = 2380.78 - 100 \times 0.8453 \times 1.1712$$
$$= 2380.78 - 99.0$$
$$= 2281.78 \text{ kJ/kg}$$
$$\therefore \quad \Delta u = u_2 -_1 = 2281.78 - 2718.89$$
$$\mathbf{\Delta u = -437.11 \text{ kJ/kg}}$$

(b) $\int P.dV$:

$$\int P.dV = \frac{P_1 V_1 - P_2 V_2}{n-1} = \frac{100[13 \times 0.1809 - 0.8453 \times 1.7312]}{0.21}$$
$$= \frac{1}{0.21}[295.17 - 99] = \frac{136.17}{0.21}$$
$$\mathbf{\int P \cdot dV = 648.43 \text{ kJ/kg}}$$

(c) $-\int V.dP$:

$$-\int V.dP = n\int P.dV = 1.21 \times 648.43$$
$$\mathbf{\int P \cdot dV = 784.6 \text{ kJ/kg}}$$

(*d*) *Work for non-flow process*

$$W_{nf} = \int P.dV = 648.43 \text{ kJ/kg}$$

(*e*) *Work for steady flow process*

$$W_{sf} = -\int V.dP + \Delta k = 784.6 + 10 = 794.6 \text{ kJ/kg}$$

(*f*) *Q – Heat*

For non-flow process
$$Q = (u_2 - u_1) + \int P.dV$$
$$= -437.11 + 648.43 = 211.32 \text{ kJ/kg}$$

For steady flow process
$$Q = \Delta K + W_{sf} + \Delta H \,.\, 10 + 794.6 - 573.28$$
$$\mathbf{Q = 784.6 \text{ kJ/kg}}$$

Example 10.31 *One kg per second of steam expands in a polytropic proces from P_1 = 10 bar and V_1 = 0.2059 cu.m to 1 bar and 1.58 cu.m. Find (a) n in $PV^n = C$ (b) $\int P.dV$ (c) $-\int V.dP$ (d) W and Q for non-flow process (e) w and Q for steady flow process with ΔK = + 9 kJ/kg.*

Data: P_1 =10 bar; V_1= 0.2059 m³/ kg, P_2 =1 bar , V_2 = 1.58 m³/ kg

Solution

(a)

$$\text{Index } n \text{ in } PV^n = C$$

We have

$$P_1V^n_1 = PV^n_2$$

$$\therefore \quad \left(\frac{V_2}{V_1}\right)^n = \frac{P_1}{P_2} = \frac{10}{1} = 10$$

$$\therefore \quad \left(\frac{1.58}{0.2059}\right)^n = (7.6736)^n = 10$$

$$\therefore \quad n \log_{10} 7.6736 = \log_{10} 10$$

$$\therefore \quad n = \frac{1}{0.885}$$

$$\mathbf{n = 1.13}$$

(b) $\int P \cdot dV$:

$$\int P.dV = \frac{P_1V_1 - P_2V_2}{n-1}$$

$$= \frac{100}{0.13}[10\times0.2059 - 1\times1.58]$$

$$= \frac{100}{0.13}[2.059 - 1.580]$$

$$= \frac{100}{0.13}\times0.479$$

$$\int \mathbf{P \cdot dV = 368.46 \text{ kJ/kg/sec}}$$

(c) $-\int V \cdot dP$:

$$-\int V.dP = n\int P \cdot dV$$

$$= 1.13\times368.46$$

$$-\int \mathbf{V.dP = 416.36 \text{ kJ/kg/sec}}$$

(d) Work and Heat for non-flow process :

For non-flow process

$$\text{Work} = \int P \cdot dV$$

$$\textbf{Work = 368.46 kJ/kg/sec}$$

$$V_1 = 0.2059 \text{ m}^3/\text{kg}$$

$$V_{g1} = 0.19429$$

∴ Steam is superheated

$$\therefore \quad 0.2059 = 0.19429 \times \frac{T_{sup}}{T_{sat}}$$

$$= \frac{0.19429 \times T_{sup}}{179.88 + 273}$$

$$\therefore \quad T_{sup1} = \frac{0.2059}{0.19429} \times 452.88$$

$$= 480°K = 207.°C \approx 200°C$$

$$\therefore \quad h_1 = 2826.8 \text{ kJ/kg}$$

$$\therefore \quad u_1 = h_1 - P_1V_1 = 2826.8 - 100 \times 10 \times 0.2059$$

$$= 2826.8 - 205.9$$

$$= 2620.9 \text{ kJ/kg/sec}$$

$$V_2 = 1.58 \text{ m}^3\text{/kg}$$

$$V_{g2} = 1.6937$$

$$\therefore \quad x_2 = \frac{1.58}{1.6937} = 0.9329$$

$$\therefore \quad h_2 = h_{f2} + x_2 h_{fg2} = 417.51 + 0.9329 \times 2257.9$$

$$= 417.51 + 2106.32 = 2523.83 \text{ kJ/kg/sec}$$

$$\therefore \quad u_2 = h_2 - P_2V_2 = 2523.83 - 158$$

$$= 2365.83 \text{ kJ/kg/sec}$$

$$\therefore \quad Q = \Delta u + \int P \cdot dV$$

$$= (u_2 - u_1) + 368.46$$

$$= (2365.83 - 2620.9) + 368.46$$

$$\mathbf{Q = 113.39 \text{ kJ/kg/sec}}$$

(e) W and Q for steady flow process :

$$\text{Work} = -\int V \cdot dP + \Delta K$$

$$= 416.36 + 9$$

$$\textbf{Work} = \mathbf{425.36 \text{ kJ/kg/sec.}}$$

$$Q = \Delta k + \Delta H + W_{sf}$$

$$= 9 + (2523.83 - 2826.8) + 425.36$$

$$= 9 + 425.36 - 302.97$$

$$\mathbf{Q = 131.39 \text{ kJ/kg/sec}}$$

Example 10.32 *Steam at 20 bar and 300°C is expanded to four times its initial volume according to the law $PV^{1.3}$ = constant. Find the heat during the process.*

Data: P_1 = 20 bar; t_1 = 300°C ; $V_2 = 4V_1$; $PV^{1.3}$ = constant.

Solution

$$\therefore \quad P_1V_1^{1.3} = P_2V_2^{1.3}$$

$$\therefore \quad \left(\frac{V_2}{V_1}\right)^{1.3} = \frac{P_1}{P_2} = \left(\frac{4V_1}{V_1}\right)^{1.3} = 4^{1.3} = 6.063$$

$$\therefore \quad P_2 = \frac{20}{6.063} = 3.3 \text{ bar}$$

From tables, $V_1 = 0.126\ \text{m}^3/\text{kg}$

$$\therefore \quad V_2 = 4 \times 0.126 = 0.504\ \text{m}^3/\text{kg}$$

$$\therefore \quad \text{Work done/kg} = \frac{P_1 V_1 - P_2 V_2}{n-1} \qquad \text{(For non-flow)}$$

$$= \frac{100}{1.3-1}[20 \times 0.126 - 3.3 \times 0.504]$$

$$= \frac{100}{0.3}[2.52 - 1.6632]$$

$$= \frac{100 \times 0.8568}{0.3}$$

$$= 285.6\ \text{kJ/kg}$$

$$h_1 = 3025.2\ \text{kJ/kg}$$

$$u_1 = h_1 - P_1 V_1 = 3025.2 - 100 \times 20 \times 0.126$$

$$= 3025.2 - 252$$

$$= 2773.2\ \text{kJ/kg}$$

$$V_2 = 0.504\ \text{m}^3/\text{kg}$$

$$V_{g2} = \frac{0.570627 + 0.539025}{2}$$

$$= \frac{1.1099652}{2}$$

$$= 0.554826$$

$$< V_{g2}$$

$$= \frac{1.109652}{2} = 0.554826$$

$\therefore$ Steam is wet after expansion

$$\therefore \quad x_2 = \frac{0.504}{0.554826} = 0.910$$

$$\therefore \quad h_2 = h_{f2} + x_2 h_{fg2}$$

where

$$h_{f2} = \frac{570.8 + 579.8}{2}$$

$$= \frac{1150.6}{2} = 575.3\ \text{kJ/kg}$$

$$h_{fg2} = \frac{2156.9 + 2150.5}{2} = \frac{4307.4}{2}$$

$$= 2153.7\ \text{kJ/kg}$$

$$\therefore \quad h_2 = 575.3 + 0.910 \times 2153.7$$

$$= 575.3 + 1959.87$$
$$= 2535.17 \text{ kJ/kg}$$
$$\therefore \quad u_2 = h_2 - P_2V_2 = 2535.17 - 100 \times 3.3 \times 0.504$$
$$= 2535.17 - 166.32$$
$$= 2368.85 \text{ kJ/kg}$$

Applying the non-flow energy equation,

$$Q = \Delta u + \int P.dV$$
= Heat during the process
$$= (u_2 - u_1) + W_n$$
$$= (2368.85 - 2773.2) + 285.6$$
$$= -404.35 + 285.6$$
$$\mathbf{Q = -118.75 \text{ kJ/kg}}$$

– ve sign indicates that heat is rejected.

Example 10.33 *Steam expands from the condition 16 bar and 250°C to 0.9 bar according to the law $pV^{1.1}$ = constant. Find for 1 kg of steam, the expansive work and change in internal energy of steam.*

Data: P_1 = 16 bar; P_2 = 0.9 bar; T_1 = 523 K and $pV^{1.1} = C$.

Solution

V_1 = volume of 1 kg steam before expansion

$$= V_{g1}\frac{T_{sup_1}}{T_{sat_1}} = 0.123843 \times \frac{523}{474.33}$$
$$= 0.1366 \text{ m}^3\text{/kg}$$

Similarly, h_1 = Enthalpy before expansion

$$= h_{g1} + C_p(T_{sup} - T_{sat})$$
$$= 2791.7 + 2.3(250 - 201.33)$$
$$= 2791.7 + 111.941 = 2903.641 \text{ kJ/kg}$$

$\therefore$ u_1 = Internal energy

$$= h_1 - P_1V_1$$
$$= 2903.641 - 100 \times 16 \times 0.1366$$
$$= 2903.641 - 218.56$$
$$= \frac{2685.081}{1.1} \text{ kJ/kg}$$

Now we have $\quad P_1V_1 = P_1V_2$

or
$$V_2 = \left(\frac{P_1}{P_2}\right)^{1/1.1} \times V_1$$
$$= \left(\frac{16}{0.9}\right)^{1/1.1} \times 0.1366$$
$$= (17.78)^{0.9091} \times 0.1366$$
$$= 13.69 \times 0.1366$$
$$= 1.87 \text{ m}^3\text{/kg}$$
$$\approx V_{g2} \text{ approx.}$$

∴ The steam is dry and saturated after expansion.

$$\therefore \quad h_2 = h_{g2} = 2670.9 \text{ kJ/kg}$$

$$v_2 = 1.87 \text{ m}^3\text{/kg}$$

$$\therefore \quad u_2 = h_2 - P_2V_2 = 2670.9 - 100 \times 0.9 \times 1.87$$

$$= 2670.9 - 168.3 = 2502.6 \text{ kJ/kg}$$

∴ Change in internal energy/kg

$$= u_2 - u_1 = 250.2.6 - 2685.081$$

$$= -182.481 \text{ kJ/kg}$$

Work done/kg (non-flow)

$$W = \frac{P_1V_1 - P_2V_2}{n-1} = \frac{218.56 - 168.3}{0.1}$$

$$= \frac{50.26}{0.1}$$

$$\mathbf{W = 502.6 \text{ kJ/kg}}$$

Example 10.34 *At a particular instant during expansion stroke of the steam engine, the volume of the steam was 0.05 m³, the pressure and condition of steam being 8 bar and 0.9 dry respectively. Determine (i) the index of expansion (ii) the work done per kg and (iii) the change in entropy during the process.*

Data: $P_1 = 8$ bar; $P_2 = 2$ bar; $V_1 = 0.05$ m³; $V_2 = 0.15$ m³; $x_1 = 0.9$

Solution

(i) *Index of expansion*

Let n be the index of expansion.

$$\therefore \quad P_1V_1^n = P_2V_2^n$$

$$\therefore \quad \left(\frac{V_2}{V_1}\right)^n = \frac{P_1}{P_2} = \frac{8}{2} = 4$$

$$\therefore \quad \left(\frac{0.15}{0.05}\right)^n = 4 = 3^n$$

$$\therefore \quad n \log 3 = \log 4$$

$$n = \frac{\log 4}{\log 3} = \frac{0.6021}{0.4771}$$

$$\mathbf{n = 1.262}$$

(ii) *The work done* :

$$\text{Work done} = \frac{P_1V_1 - P_2V_2}{n-1} = \frac{100}{0.262}[8\times0.05 - 2\times0.15]$$

$$= \frac{40-30}{0.262} = \frac{10}{0.262} = 38.168 \text{ kJ}$$

$$\text{Specific volume} \quad V_1 = x_1V_{g1} = 0.9 \times 0.240506$$

$$= 0.2165 \text{ kJ/kg}$$

$$\therefore \quad \text{Mass of steam} = \frac{0.05}{0.2165} = 0.231 \text{ kg}$$

$$\therefore \quad \text{Work done/kg} = \frac{38.168}{0.231}$$

$$\textbf{Work done/kg} = \textbf{165.23 kJ/kg}$$

(*iii*) *Change in entropy*:

Initial entropy $S_1 = S_{f1} + x_1 S_{fg1}$

$= 2.045 + 0.9\,(6.659 - 2.045) = 6.1976$

$$\text{Final volume/kg} = \frac{0.15}{0.231}$$

$= 0.6493 \text{ m}^3\text{/kg}$

$< V_{g2}\,(0.8872)$

$$\therefore \quad x_2 = \frac{0.6493}{0.8872}$$

$= 0.732$

Final entropy $S_2 = S_{f2} + x_2 S_{fg2}$

$\therefore \quad S_2 = 1.529 + 0.732\,(7.127 - 1.529)$

$= 1.529 + 4.098$

$= 5.627 \text{ kJ/kg-K}$

$\therefore$ Change in entropy/kg $\Delta S = S_2 - S_1$

$= 5.627 - 6.1976$

$= -5706 \text{ kJ/kg-K}$

$\therefore$ Change in entropy $\Delta S = -0231 \times 0.5706 \text{ kJ/K}$

$$\boldsymbol{\Delta S = -0.1318 \text{ kJ/K}}$$

Example 10.35 *Dry and saturated steam at 20 bar passes through an expander where it is expanded to a pressure of 500 mm Hg (vacuum) (barometer reads 725 mm Hg) according to the law $pV^{1.07}$ = constant. Calculate*

(*i*) *Condition of steam after expansion*

(*ii*) $\int P.dV$

(*iii*) $-\int V.dP$

(*iv*) *Work of non-flow process*

(*v*) *Work of steady flow system with $\Delta P = \Delta K = 0$*

(*vi*) *Change in internal energy*

(*vii*) *Change in internal entropy.*

(*viii*) *Heat Q.*

Solution

(*i*) *Condition of steam expansion*:

Abs. pressure of steam after expansion

P_2 = barometer reading vacuum = 725 – 500 = 225 mm Hg

In bar $$P_2 = \frac{225 \times 1.01325}{760} = 0.3 \text{ bar}$$

$P_1 = 20 \text{ bar and } V_1 = 0.099623 \text{ m}^3\text{/kg}$

$$\therefore \quad P_1 V^n_1 = P_2 V_2^n \text{ or } V_2 = \left(\frac{P_1}{P_2}\right)^{1/n} \cdot V_1$$

$$\therefore \quad V_2 = \left(\frac{20}{0.3}\right)^{1/1-0.7} \times 0.099623$$

$$= \left(\frac{20}{0.3}\right)^{0.9346} \times 0.099623$$

$$= 5.0464 \text{ m}^3\text{/kg}$$

$$< V_{g2} (5.229)$$

$$\therefore \quad x_2 = \frac{5.0464}{5.229}$$

$$\mathbf{x_2 = 0.9651}$$

(ii) $\int P.dV$:

$$\int P.dV = \frac{P_1V_1 - P_2V_2}{n-1}$$

$$= \frac{100}{0.07}[20 \times 0.099623 - 0.3 \times 5.0464]$$

$$= \frac{199.246 - 151.392}{0.07}$$

$$= \frac{47.854}{0.07}$$

$$\mathbf{\int P.dV = 683.63 \text{ kJ/kg}}$$

(iii) $-\int V.dP$:

$$-\int V.dP = n\int P.dV$$

$$= 1.07 \times 683.63$$

$$\mathbf{-\int V.dP = 731.48 \text{ kJ/kg}}$$

(*iv*) *Work for non-flow process*

$$W_{nf} = \text{Work of non-flow process}$$

$$= \int P.dV$$

$$\mathbf{W = 683.63 \text{ kJ/kg}}$$

(*v*) *Work of steady flow process*

We have

$$W_{sf} = -\int V.dP$$

$$\mathbf{W_{sf} = 731.48 \text{ kJ/kg}}$$

(*vi*) *Change in internal energy*

$$h_1 = h_{g1} = 2797.3 \text{ kJ/kg}$$

$$\therefore \quad u_1 = h_1 - P_1V_1 = 2797.3 - 199.246$$

$$= 2598.054 \text{ kJ/kg}$$

$$h_2 = h_{f2} + x_2 h_{fg2}$$

$$= 289.3 + 0.9651 \times 2336.1$$
$$= 289.3 + 2254.57 = 2543.87 \text{ kJ/kg}$$
$$\therefore \quad u_2 = h_2 - P_2V_2$$
$$= 2543.87 - 151.392$$
$$= 2392.478 \text{ kJ/kg}$$
$$\therefore \quad \Delta u = u_2 - u_1 \text{ change in internal energy}$$
$$= 2392.478 - 2598.054$$
$$\Delta u = -205.576 \text{ kJ/kg}$$

(*vii*) *Change in entropy*

$$S_1 = S_{g1} = 6.337 \text{ kJ/kg-K}$$
$$S_2 = S_{f2} + x_2 S_{fg2}$$
$$= 0.944 + 0.9651\,(7.767 - 0.944)$$
$$= 0.944 + 0.9651 \times 6.823$$
$$= 0.944 + 6.5849$$
$$= 7.5289 \text{ kJ/kg-K}$$
$$\therefore \quad S_2 - S_2 = \text{Change in entropy}$$
$$= 7.5289 - 6.3370$$
$$\mathbf{S_2 - S_2 = 1.1919 \text{ kJ/kg-K}}$$

(*viii*) *Heat Q*

Heat during the non-flow process is given by

$$Q = \Delta u + \int p.dv$$
$$= -205.576 + 683.63$$
$$\mathbf{Q = 478.054 \text{ kJ/kg-K}}$$

For Steady Flow Process

$$Q = \Delta P + \Delta K + \Delta H + W_{sf}$$
$$= 0 + 0 + (2543.87 - 2797.3) + 731.48$$
$$\mathbf{Q = 478.05}$$

(VI) Combined Processes

Example 10.36 *Steam at 2 MN/m² and 250°C expands isentropically to 3.5 bar. It is then further expanded hyperbothically to 0.5 bar. Using steam tables, find out*

(*i*) *condition of steam after each process.*

(*ii*) *work done kg/of steam in each process.*

Data: $P_1 = 2\ MN/m^2 = 20$ *bar;* $T_1 = 250°C$*;* $P_2 = 3.5$ *bar;* $P_3 = 05$ *bar.*

Solution

The process are shown in Fig. Ex. 10.36

Isentropic Process (1 – 2) :

From steam tables : $h_1 = 2902.4$ kJ / kg.

$$S_1 = 6.5454 \text{ kJ / kg-K}$$

For isentropic process, $S_1 = S_2 = S_{f2} + x_2 S_{fg2}$

$$\therefore \quad 6.5454 = 1.7273 + x_2 \times 5.2119$$
$$\therefore \quad x_2 = \frac{6.5454 - 1.7273}{5.2119}$$

$$x_2 = 0.9244$$

$$\therefore \quad h_2 = h_{f2} + x_2 h_{fg2}$$
$$= 584.27 + 0.9244 \times 2147.4$$
$$= 2569.32 \text{ kJ/kg.}$$

Assuming non-flow process,

$$\text{Work done/kg} = u_1 - u_2$$
$$= (h_1 - P_1V_1) - (h_2 - P_2V_2)$$
$$= \left[290234 \frac{-10^5 \times 0.1114}{1000}\right] - \left[2569.32 \frac{3.5 \times 10^5 \times 0.9255 \times 0.524}{1000}\right]$$

Work done/kg = 279.81 kJ/kg

For steady flow process with $\Delta P = \Delta K = 0$

$$\text{Work done / kg} = h_1 - h_2$$

Work done/kg = 2902.4 – 2569.42 = 333.08 kJ/kg

Hyperbolic Expansion (2 – 3): The law of expansion is pV = Const.

$$\therefore \quad P_2V_2 = P_3V_3$$

where $P_2 = 3.5$ bar, $P_3 = 0.5$ bar.

$$V_2 = x_2 V_{g2} = 0.9244 \times 0.524 = 0.4844 \text{ m}^3/\text{kg}$$

$$\therefore \quad V_3 = \frac{P_2 V_2}{P_3} = \frac{3.5 \times 0.4844}{0.5} = 3.391 \text{ m}^3/\text{kg}$$

From tables, $V_3 = 3.391 > (V_{g3} = 3.2402)$

$$\therefore \quad \frac{V_3}{V_{g3}} = \frac{T_3}{T_{sat_3}}$$

$$\therefore \quad \frac{3.391}{3.2402} = \frac{T_3}{273 + 81.345}$$

$$\therefore \quad T_3 = \frac{3.391 \times 354.345}{3.2402}$$

$$\mathbf{T_3 = 370.8\ K = 97.8°C}$$

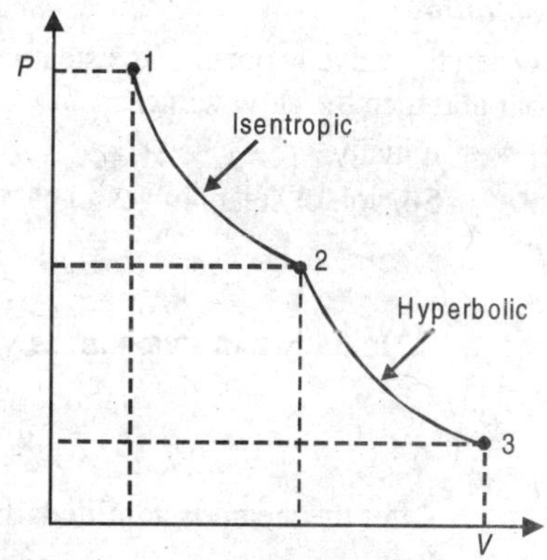

Fig. Ex. 10.36

Superheated condition

For pV = Const., the work done of non-flow process and steady flow process with $\Delta P = \Delta K = 0$, is same

$$\text{Work done/kg} = P_2 V_2 \ln \frac{P_2}{P_3}$$

$$= \frac{3.5 \times 105 \times 0.4844}{1000} \ln \frac{3.5}{0.5}$$

Work done/kg = 329.91 kJ/kg

Example 10.37 *Two boilers of equal evaporative capacities generate steam at the same pressure of 16 bar to a* *common pipe line. One boiler produces steam with 50°C superheat and other produces wet steam. If the steam in the main pipe is just dry and saturated, find the dryness fraction of wet steam.*

Solution

The flow diagram is shown above. Consider 1 kg steam from each boiler

$$h_A + h_B = 2hC$$

$$h_{gA} + cp_v\,(t_{sup} - t_{sat}) + (h_{fB} + x_B h_{fgB}) = 2h_{gc}$$

$$\therefore\ (2791.76 + 2.3 \times 50) + (858.56 + x_B \times 1933.2) = 2 \times 2791.76$$

$$2906.76 + 858.56 + 1933.2\,x_B = 5583.52$$

$$\therefore\ x_B = \frac{5583.56 - 2906.76 - 858.56}{1933.2} = \frac{1818.2}{1933.2}$$

$$\mathbf{x_B = 0.9405}$$

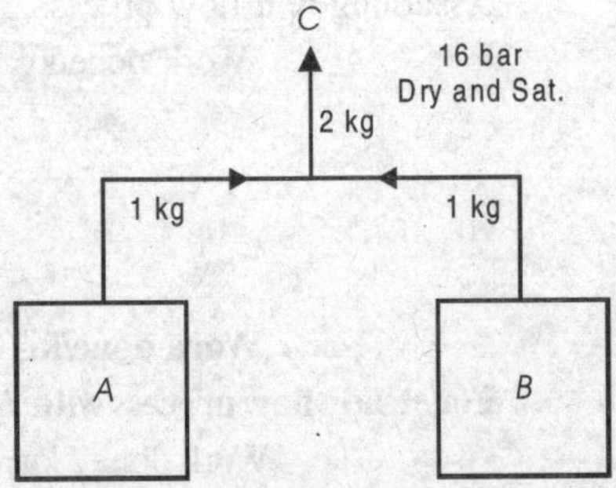

Fig. Ex. 10.37

This is Dryness fraction of wet steam.

Example 10.38 *A closed vessel of 0.3 cu.m capacity contains steam at a pressure of 1.2 bar and 0.8 dry. It is connected to a steam pipe supplying steam of 0.9 dryness fraction at a pressure of 4.5 bar. Neglecting the volume of moisture in suspension and thermal capacity of the vessel, determine, (a) the weight of steam which will pass through the steam pipe to the vessel, when the valve in the pipe is opened and (b) the final dryness of the steam in the vessel when the flow ceases.*

Solution

When the valve is opened, the steam will flow into the vessel till the pressure in the vessel pressure increases to 4.5 bar and then the flow will cease.

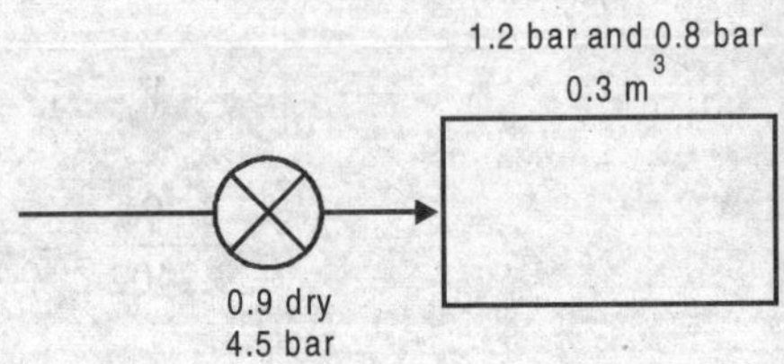

Fig. Ex. 10.38

Initially,

Sp. vol. of steam in vessel $= V_1 = x_1 V_{g1} = 0.8 \times 1.4281 = 1.14248\ \text{m}^3/\text{kg}$

$\therefore$ Initial wt. of steam in the vessel

$$W_1 = \frac{0.3}{0.14248} = 0.2626\ \text{kg}$$

After the steam is admitted, the pressure is 4.5 bar

Let W_2 = wt. of steam dryness

x_2 = Dryness fraction of steam admitted

x_3 = Dryness fraction of after mixing

Total heat of steam in vessel before mixing.

$$h_1 = 0.2626[h_{f1} + x_1 h_{fg1}] = 0.2626[439.36 + 0.8 \times 2244.1] = 586.82\ \text{kJ} \quad ...(1)$$

Total heat of steam admitted

$$h_2 = W_2\,[h_{f2} + x_2 h_{fg2}] = W_2[623.16 + 0.9 \times 2119.7] = 2530.89\,w_2\ \text{kJ} \quad ...(2)$$

Total heat of steam after mixing

$$h_3 = (W_1 + W_2)\,[h_{f3} + x_3 h_{fg3}]$$
$$= (0.2626 + W_2)\,[623.16 + x_3 \times 2119.7] \qquad ...(3)$$

$\therefore$ 1 + 2 + 3

$$586.82 + 2530.89\,W_2 = (0.2626 + W_2)(623.16 + x_3 \times 2110.7) \qquad ...(4)$$

Again volume of mixture is $0.3\ m^3$

$$\therefore \quad = (W_1 + W_2)\,x_3\,V_{g3} = 0.3$$
$$= (0.2626 + W_2)\,x_3 \times 0.41375 = 0.3$$

$$(2626 + W_2)\,x_3 = \frac{0.3}{0.41375} = 0.7251 \qquad ...(5)$$

$$\therefore \quad (2626 + W_2) = \frac{0.7251}{x_3} \qquad ...(6)$$

$$\therefore \quad W_2 = \frac{0.7251}{x_3} - 0.2626 \qquad ...(7)$$

$\therefore$ Eq. (4) gives

$$= 586.82 + 2530.89\left(\frac{0.7251}{x_3} - 0.2626\right)$$

$$= \frac{0.7251}{x_3}(623.16 + 2119.7\,x_3)$$

$$= 586.82 + \frac{1835.15}{x_3} - 664.61$$

$$= \frac{451.85}{x_3} + 1537$$

$$\frac{1835.15 - 451.85}{x_3} = 1537 + 664.61 - 586.82$$

$$\frac{1383.30}{x_3} = 2201.61 - 586.82 = 1614.79$$

$$\therefore \quad x_3 = 0.8566$$

$$\therefore \quad W_2 = \frac{0.7251}{0.8566} - 0.2626$$
$$= 0.5839\ \text{kg}$$

(a) The weight of steam admitted = 0.5839 kg

(b) Final dryness fraction = 0.8566

Example 10.39 *Steam initially at 14 bar and having specific volume of 0.12 m^3/ kg is throlled to 7 bar and then expanded hyperbolically to 1.1 bar. For each process determine change in entropy.*

Solution

Throttling Process

Volume of steam at 14 bar and dry and saturated is 0.14073 m^3/kg. Actual volume is 0.12 m^3/kg. Therefore the steam is wet before throttling.

$$\therefore \quad x_1 = \frac{0.12}{0.14073} = 0.8527$$

$$\therefore \quad h_1 = \text{Enthalpy of 1 kg steam before throttling}$$
$$= h_{f1} + x_1 h_{fg1}$$
$$= 830 + 0.8527 \times 1957.7$$
$$= 2799.32 \text{ kJ/kg}$$

This enthalpy is less than h_{g2}. Therefore steam is wet after throttling.

$$\therefore \quad h_2 = 2499.32 = h_{f2} + x_2 h_{fg2} \text{ at 7 bar}$$
$$= 697.1 + x_2 \times 2064.9$$

$$\therefore \quad x_2 = \frac{2499.32 - 697.1}{2064.9} = \frac{1802.22}{2064.9}$$
$$= 0.8728$$

Entropy of wet steam before throttling is

$$S_1 = S_{f1} + x_1 S_{fg1}$$
$$= 2.2837 + 0.8527 \times 4.1814$$
$$= 5.8492 \text{ kJ/kg-K}$$

Similarly

$$S_2 = \text{Entropy after throttling}$$
$$= S_{f2} + x_2 S_{fg2}$$
$$= 1.9918 + 0.8728 \times 4.7134$$
$$= 6.1056 \text{ kJ/kg-K}$$

∴ Change of entropy during throttling

$$\Delta S = S_2 - S_1 = 6.1056 - 5.8492$$
$$= 0.2564 \text{ kJ/kg-K}$$

Hyperbolic Process

$P_2 = 7$ bar; V_2 = volume of 1 kg steam before hyperbolic exparision

$$V_2 = x_2 V_{g2}$$
$$= 0.8728 \times 0.27268 = 0.2379 \text{ m}^3/\text{kg}$$

$$P_3 = 1.1 \text{ bar} = \text{Pressure after hyperbolic expansion}$$

$$\therefore \quad P_2 V_2 = P_3 V_3$$

$$\therefore \quad V_3 = \frac{P_2 V_2}{P_3}$$

$$= \frac{7}{1.1} \times 0.2379 = 1.5147 \text{ m}^3/\text{kg}$$

This is less than $V_{g3} = 1.5492$

∴ Steam is wet after hyperbolic expansion.

$$\therefore \quad 1.5147 = x_3 \times 1.5492$$

$$x_3 = \frac{1.5147}{1.5492}$$
$$= 0.9774 \text{ m}^3/\text{kg}$$

$$\therefore \quad S_3 = S_{f3} + x_3 S_{fg3}$$
$$= 1.3330 + 0.9774 \times 5.9947$$
$$\mathbf{S_3 = 7.1922\ kJ/kg\text{-}K}$$

Example 10.40 *Two boilers discharge equal amounts of steam into the same main. The steam from one boiler is at 17 bar and 270°C and other at 17 bar and 0.95 quality. What is (a) the equilibrium condition after mixing. ($\Delta K = 0$) (b) the change of entropy by the higher temperature steam, (c) the change of entropy of the lower temperature steam? (d) is there a net increase or decrease of entropy?*

Solution

(a) Energy diagram is shown in Fig. Ex. 10.40 $\Delta K = 0$;

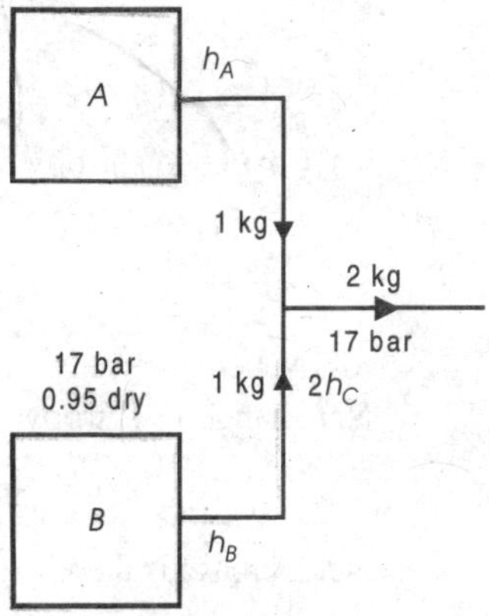

Fig. Ex. 10.40

$$h_A + h_B = 2h_c$$

From table

$$h_A = \frac{2915.3 + 3033.5}{2}$$
$$= \frac{5948.8}{2}$$
$$= 2974.4\ \text{kJ/kg}$$
$$S_{A1} = \frac{6.6398 + 6.8557}{2} = \frac{13.4955}{2}$$
$$= 6.74775\ \text{kJ/kg-K}$$
$$h_B = h_f + xh_{fg}$$
$$= 871.84 + 0.95 \times 1921.5$$
$$= 2697.26\ \text{kJ/kg}$$
$$S_{B1} = S_f + xS_{fg}$$
$$= 2.3713 + 0.95 \times 4.0244$$
$$= 6.1945\ \text{kJ/kg.K}$$

After mixing in the main pipe

$$2h_c = h_A + h_B = 2974.4 + 2697.26$$
$$= 5671.66$$
$$\therefore \quad h_C = \frac{5671.66}{2} = 2835.83\ \text{kJ/kg}$$

The steam is superheated in the main pipe.

Assuming $C_{pv} = 2.3\ \text{kJ/kg-K}$

$$2835.83 = h_g + C_{pv}(t_{sup} - t_{sat})$$
$$= 2793.4 + 2.3\,(t_{sup} - 204.31)$$
$$\therefore \quad t_{sup} = \text{Temp of steam in main pipe}$$
$$= \frac{2835.83 - 2793.4}{2.3} + 204.31$$
$$= 222.76°\text{C}$$

∴ Equilibrium condition is

$$P = 17\ \text{bar.}$$
$$t = 222.76°\text{C}$$

Entropy in main pipe is

$$S_c = S_{gc} C_{pv} \log_e \frac{T_{sup}}{T_{sat}}$$

$$= 6.3957 + 2.3 \log_e \frac{222.76 + 273}{204.31 + 273}$$

$$\mathbf{S_c = 6.48293 \text{ kJ/kg-K}}$$

(b) Change in entropy of *A* stream :

$$\Delta S_A = S_C - S_A$$
$$= 6.48293 - 6.74775$$
$$\mathbf{\Delta S_A = -0.26482 \text{ kJ/kg-K}}$$

(c) Change in entropy of *B* stream

$$\Delta S_B = S_C - S_B$$
$$= 6.48293 - 6.19450$$
$$\mathbf{\Delta S_B = 0.28843 \text{ kJ/kg-K}}$$

(d) Net change

Net change in entropy

$$= 0.28843 + (-0.26482)$$
$$= 0.02361 \text{ kJ/kg-K}$$

Net increase is there.

(VII) Separating and Throttling Calorimeter

Example 10.41 *Discuss under what circumstances, a throttling calorimeter is capable of being used as a device for measuring the steam dryness.*

Find the maximum degree of wetness that can be determined by a throttling calorimeter if the inlet pressure of steam to it is 10.5 bar and exit pressure is 1.2 bar.

Solution

Condition of the use of throttling calorimeter for finding the dryness fraction of steam is that the condition of steam after throttling must be known. For this purpose the steam should be slightly superheated so that temperature can be measured by the thermometer. In the limit, the steam should be dry and saturated after throttling.

Let x, be the dryness fraction of steam to be found.

For throttling process, enthalpies are same before and after throttling.

$$\therefore \quad h_1 = h_2$$

or

$$h_{f1} + x_1 h_{fg1} = h_{g2}$$

$$\therefore \quad 772.03 + x_1 \times 2005.9 = 2683.4$$

$$\therefore \quad x_1 = \frac{2683.4 - 772.03}{2005.9}$$

$$= \frac{1911.37}{2005.9}$$

$$= 0.9529$$

$$\therefore \quad \text{Maximum wetness} = 1 - 0.9529 = 0.0471 \text{ or } 4.71\,\%$$

Example 10.42 *The following, data were obtained with a separating and throttling calorimeter:*

Pressure in the pipe line 1.5 MPa

Condition after throttling 0.1 MPa and 110°C
Moisture collected in separating
Calorimeter/5 min 0.15 lit at 70°C
Steam condensed after throttling 3.24 kg
Find the quality of steam in the pipe.

Solution

If w is the wt. of steam separated in separating calorimeter and w the wt. of condensed steam after throttling, then dryness fraction of steam is given by

$$x = \frac{Wx_1}{W + w}$$

where x_1 is the dryness fraction of steam entering the throttling calorimeter

$$\therefore \quad x = \frac{3.24\, x_1}{3.24+0.15}$$

$$= \frac{3.24\, x_1}{3.39} \quad ...(1)$$

For throttling calorimeter,

$$[\text{Enthalpy before}] = [\text{Enthalpy after throttling}]$$

$$h_{f1} + x_1 h_{fg1} = h_{g2} + C_V(t_{sup} - t_{sat})$$

$$844.66 + x_1 \times 1945.2 = 2676.2 + 2.0[110 - 99.632]$$

$$= 2676.2 + 2 \times 10.368$$

$$= 2676.2 + 20.736$$

$$= 2696.936 \text{ kJ/kg}$$

$$\therefore \quad x_1 = \frac{2696.936 - 844.66}{1945.2}$$

$$= \frac{4852.276}{1945.2} \quad ...(2)$$

$$= 0.9522$$

$\therefore$ Dryness fraction of steam in the pipe

$$x = \frac{3.24}{3.39} \times 0.9522$$

$$\mathbf{x = 0.91}$$

Example 10.43 (*a*) *Explain the influence of pressure of the following properties of steam with the help of T–S diagram.*

(*i*) *Enthalpy of evaporation*
(*ii*) *Temperature of saturation*
(*iii*) *Enthalpy of dry saturated steam*

(*b*) *With a separating and throttling calorimeter arranged in series, the following observations, were taken. Estimate the quality of steam sampled.*

Water separated = 2 kg
Steam discharged from throttling calorimeter = 20.5 kg
Temperature of steam after throttling = 110° C

Initial pressure = 11 bar (gauge)
Barometer = 760 mm of Hg
Final pressure = 28 mm of Hg

Solution

(a) The T–S diagram for steam is shown in Fig. Ex. 10.43. Effect of pressure on

(i) Enthalpy of evaporation: Horizontal distance on T–S diagram represents to scale, the enthalpy of evaporation.
If the pressure increases, i.e., saturation temperature increases, enthalpy of evaporation decreases and vice-versa.

(ii) Temperature of saturation: As the pressure increases, saturation temperature increases and vice-versa.

(iii) Enthalpy of dry saturated steam: Enthalpy of dry saturated steam decreases with the increase in pressure and vice-versa. This is beyond certain pressure only otherwise it increases.

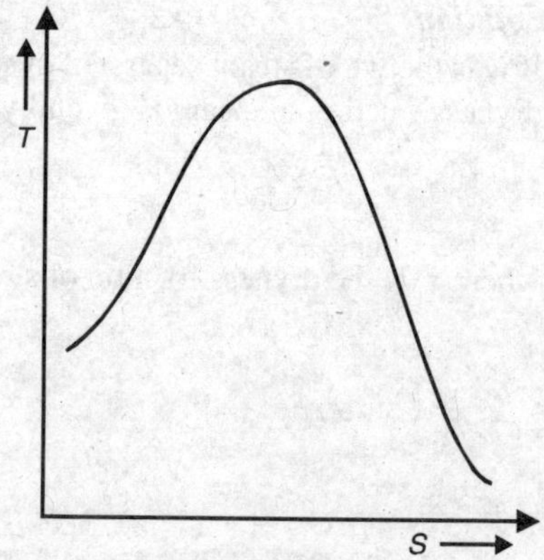

Fig. Ex. 10.43

(b) The dryness fraction of steam sampled is given by

$$x = \frac{W_1}{W_1 + W_2} . x$$

here $W_1 = 20.5$ kg, $W_2 = 2$ kg, x_1 = Dryness fraction of steam leaving the separating calorimeter and entering the throttling calorimeter Atm. pressure = 760 mm Hg = 1.01325 bar

∴ Abs. pressure of steam sampled = 11 + 1.01325 ≈ 12.01325 bar

Similarly, $28 \text{ mm Hg} = \frac{28}{700} \times 1.01325 = 0.03733 \text{ bar}$

Abs. pressure = 0.03733 + 1.01325 = 1.05058 bar, considering 1.1 bar

For throttling calorimeter, we have

$$h_1 = h_2$$

$$\therefore \quad 798.43 + x_1 \times 1984.3 = \frac{2679.6 + 2.1(110 - 102.32)}{1984.3}$$

$$= \frac{1897.298}{1984.3} = 0.9561$$

$$x = \frac{09561 \times 20.5}{22.5} = 0.871$$

PROBLEMS FOR PRACTICE

10.1. The internal energy of 2 kg of steam at a pressure of 9 bar is 4605 kJ. Calculate the dryness fraction of the steam. By how many heat units the internal energy of the above steam will increase if it is superheated, at constant pressure, to a temperature of 225°C.
Take $C_p = 2.26$ kJ/kg-K

10.2. 1 kg steam at a pressure of 10 bar and having a dryness fraction of 0.2 receives 1046 kJ heat at a constant pressure. Calculate:

(a) The dryness fraction of steam

(b) The total heat

(c) Specific volume

(d) Internal energy of steam at the end of constant pressure heat supply:

Constant Pressure Process

10.3. Two boilers one with a superheater and the other without a superheater are delivering equal quantities of steam into a common main. The pressure in the boilers and the main is 14 bar absolutely. The temperature of steam from the boiler with superheater is 325°C and that of the steam in the main is 250°C. Estimate the quality of steam supplied by other boiler.

[**Ans.** 0.9785 dry]

10.4. In a Babcock Wilcox boiler, the wet steam at 20 bar and 0.93 dry is taken, from the steam drum to the superheater, where it is superheated to 150°C superheat at constant pressure. Find the heat during the process, work done and the change of entropy during the process. Take C_P for superheated steam as 2.3 kJ / kg-K.

[**Ans.** 477.2 kJ / kg; 75.5 kJ / kg; 0.8922 kJ / kg-K]

Constant Volume Process

10.5. A pressure vessel contains 1.25 kg of steam at 5 bar and 0.7 dry. The vessel is cooled until the steam attains the final dryness fraction of 0.5. Find (a) The final pressure of steam and (b) Heat transfer during the process.

[**Ans.** (a) 2.676 bar (b) –1034.25 kJ]

10.6. A closed vessel of 0.2 m³ capacity contains steam at a pressure of 8 bar and a temperature of 200°C.

(i) What is the weight of steam contained in the vessel ?

(ii) The vessel is cooled till the steam becomes just dry and saturated. What will be the pressure of steam at this state ? Estimate the specific entropy.

(iii) The vessel is further cooled till the temperature drops; to 160.8°C. Find the pressure and condition of steam at this stage.

[**Ans.** (i) 0.78 kg (ii) 7.5 bar; 6.686 kJ/ kg-K (iii) 6.32 bar, (iv) 0.849 dry]

Isentropic Process

10.7. Steam initially at 1.5 mPa and 300°C expands reversibly and adiabatically in a steam turbine to 40°C. Determine–

(i) condition of steam after expansion

(ii) work done by the turbine per kg of steam

[**Ans.** (i) 0.818 (ii) 870 kJ / kg-K]

10.8. Steam at 2 MN/m² and 250°C is expanded isentropically to 0.35 MN/m². Using the steam tables, determine

(a) condition of steam after expansion (b) work done for non-flow process

[**Ans.** (a) 0.915 (b) 297.38 kJ/ kg]

10.9. In a steam engine cylinder steam expands hyperbolically from 8 bar pressure and dryness fraction 0.9 to 1 bar. Find the final condition of steam and change in specific volume. Find also the non-flow work during the process.

[**Ans.** (i) 107.61C; (ii) 1.51515 m³ / kg; (iii) 360 kJ / kg]

10.10. Five kilogram steam having a pressure of 16 bar and 0.9 dry is expanded in a cylinder to a pressure of 0.8 bar. Determine the final condition of steam mechanical work done and the heat transfer during the process if the process is hyperbolic.

[**Ans.** (1) 117.82C; 534.055 kJ / kg (2670.276 kJ) ; (2) 657.445 kJ]

10.11. 1 kg of wet steam at 250°C and with enthalpy of 2500 kJ is confined in a rigid vessel. Heat is supplied until the steam becomes saturated. Determine

(a) initial pressure of steam
(b) final pressure of steam
(c) heat supplied

[**Ans.** (a) 39.7894 bar (b) 48 bar (c) 262.15 kg]

10.12. A sample of steam at 15 bar has an internal energy, of 2500 kg. Find the condition of steam.

[**Ans.** $x = 0.946$]

10.13. A rigid vessel of volume 0.01 m³ contains 90% (by volume) water and 10% steam at a pressure of 2 bar. Deterrnine

(i) temperature of water
(ii) total mass of water
(iii) steam in the vessel

[**Ans.** (i) 120.2°C (ii) 9 kg (iii) .0:00113 kg]

10.14. Determine the specific volume, enthalpy of steam under the following conditions

(i) 4 MPa pressure 360°C
(ii) 150 kPa, 10% quality
(iii) 350 kPa, 10% moisture

[**Ans.** (i) 0.062 m³, 3053.31 kJ/kg (ii) 0.1159 m³, 689.6 kJ/kg (iii) 0.4717 m³;1517.2 kJ kg]

10.15. A closed vessel of 0.75 cu.m capacity contains dry and saturated steam at 5 bar. The vessel is

cold until the pressure is reduced to 2 bar. Determine,
(i) mass of steam present in the vessel
(ii) final dryness fraction
(iii) amount of heat transferred during the process.
[**Ans.** (i) 2.001 kg, (ii) 0.4232, (iii) –23.78.52 kg]

10.16. 0.125 cu.m of wet steam at 10 bar has an enthalpy of 1800 kJ/kg. Calculate mass and dryness fraction of steam.[**Ans.** $m = 1.249$ kg, $x = 0.5152$]

10.17. One kg of steam at an initial condition of 6 bar and 0.2 dry is heated at constant volume until the pressure is 20 bar. Determine the final state of steam, heat added.
[**Ans.** $x_2 = 0.635$, $Q = 932$ kJ/kg]

10.18. Three boilers *A*, *B*, *C* working at 12 bar absolute supply steam in equal quantities to a plant through a common steam pipe. The boiler *A* supplies, dry saturated steam while the steam supplied by boiler *B* is superheated to 350°C. The steam from boiler *C* is of dryness fraction 0.96. Calculate the condition of steam in common pipe. Assume that there is no loss of heat in pipes and specfic heat of superheated steam is 2.1 kJ/kg°K.
[**Ans.** 229.2 °C]

10.19. Steam is stored in a container 0.45 m³ capacity. The pressure of steam is 8 bar and temperature is 200°C. The container is cooled and pressure drops to 25 bar. Find
(i) initial internal energy in kJ
(ii) quality of steam after cooling
(iii) heat lost to atmosphere due to cooling in kJ
Neglect volume of water.
[**Ans.** (i) 4533.77 kJ, (ii) 0.3625, (iii) – 2360.42 kJ]

10.20. 1 kg of steam at 20 bar and 300°C is cooled at constant pressure till the dryness fraction of steam becomes 0.5. Find the change in enthalpy work done and the heat transfer during the process.
[**Ans.** $\Delta H = -1172$ kJ, $W = -151.5$ kJ, $Q = -1172$ kJ]

STEAM TABLES

INTRODUCTION

1. The following steam tables are based on the Mollier tables and give all the necessary data with regard to saturated and superheated steam required by Engineering students.

2. The various tables are arranged as follows:

(a) Saturated Steam

Table 1 : It provides data with reference to varying *temperatures*.

Table 2 : It provides data with reference to varying *pressures*.

(b) Superheated Steam

Table 3: It provides *Enthalpy* of steam superheated to various temperatures at different pressures.

Table 4: It provides *Entropy* of steam superheated to various temperatures at different pressures.

Table 5: It provides *Volume* of steam superheated to various temperatures at different pressures.

3. All heat energy quantities expressed in these tables are in terms of kJ/kg of steam.

4. All pressures are in terms of N/m².

IMPORTANT RELATIONSHIPS

(i) 1 J is equivalent of 1 Nm or 0.10197 m kg_f of work.

(ii) $T = t + 273$, where T = Temperature expressed in centrigrade absolute scale, (K).

(iii) Absolute pressure = Gauge pressure + Atmospheric pressure.

(iv) Volume of wet steam = $x \times V$,

where x = dryness fraction.

V_s = Specific volume of saturated Steam.

(v) Enthalpy of wet steam = Enthalpy of liquid + (Latent heat × dryness fraction).

or $$H = h_L \times x L$$

(This also applies to *Entropy* and *Internal Energy*)

(vi) 1 N/m² = 10^{-5} bar = 0.10197 kg_f/m² of pressure.

Table 1. Saturated Steam (Temperature Table)

Temperature		Pressure N/m² $p \times 10^{-5}$	Volume m²/kg		Density of Vapour kg/m²	Enthalpy kJ/kg		Latent heat L kJ/kg	Entropy kJ\kg–K		L/T kJ/kg-K
°C	K		Liquid v_i	Vapour v_i		Liquid h_i	Vapour h_s		Liquid s_i	Vapour s_v	
1	2	3	4	5	6	7	8	9	10	11	12
0	273	0.00610	0.001000	206.3	0.004847	0.00	2501.6	2501.6	0.00	9.1577	9.1577
1	274	0.00656	0.001000	192.6	0 005192	4.17	2503.4	2499.2	0.0152	9.1311	9.1159
2	275	0.00705	0.001000	179.9	0.005558	8.39	2505.2	2496.8	0.0306	9.1047	9.0741
3	276	0.00757	0.001000	168.2	0.005946	12.60	2507.1	2494.5	0.0459	9.0785	9.0326
4	277	0.00813	0.001000	157.3	0.006358	16.80	2508.9	2492.1	0.0611	9.0526	8.9915
5	278	0.00872	0.001000	147.2	0.006795	21.01	2510.7	2489.7	0.0762	9.0269	8.9507
6	279	0.00934	0.001000	137 8	0.007258	25.21	2512.6	2487.4	0.0913	9.0015	8.9102
7	280	0.01001	0.001000	129 1	0.007748	29.41	2514.4	2485.0	0.1063	8.9762	8.8699
8	281	0.01072	0.001000	121.0	0.008267	33.60	2516.2	2482.6	0.1213	8.9513	8.8300
9	282	0.01147	0.001000	113.4	0.008816	37.80	2518.1	2480.3	0.1361	8.9265	8.7904
10	283	0.01227	0.001000	106.4	0.009396	41.99	2519.9	2477.9	0.1510	8.9020	8.7510
11	284	0.01311	0.001000	99.31	0.01001	46.19	2521.7	2475.5	0.1658	8.8776	8.7118
12	285	0.01401	0.001000	93.84	0.01066	50.38	2523.6	2473.2	0.1805	8.8536	8.6731
13	286	0.01496	0.001000	88.18	0.01134	54.57	2525.4	2470.8	0.1952	8.8297	8.6345
14	287	0.01597	0.001000	82.90	0.01206	58.75	2527.2	2468.5	0.2098	8.8060	8.5962
15	288	0.01703	0.001000	77.98	0.01282	62.94	2529.1	2466.1	0.2243	8.7826	8.5583
16	289	0.01817	0.001001	73.38	0.01363	67.13	2530.9	2463.8	0.2388	8 7593	8.5205
17	290	0.01936	0.001001	69.09	0.01447	71.31	2532.7	2461.4	0.2533	8.7363	8.4830
18	291	0.02062	0.001001	65 09	0.01447	75.50	2534.5	2459.0	0.2677	8.7135	8.4453
19	292	0.02196	0.001001	61 34	0.01630	79.68	2536.4	2456.7	0.2820	8.6908	8.4088
20	293	0.02337	0.001002	57.84	0.01729	83.86	2538.2	2454.3	0.2963	8.6684	8.3721
21	294	0.02485	0.001002	54.56	0.01833	88.04	2540.0	2452.0	0.3105	8.6462	8.3357
22	295	0.02642	0.001002	51.49	0.01942	92.23	2541.8	2449.6	0.3247	8.6241	8.2994
23	296	0.02808	0.001002	48.62	0.02057	96.41	2543.6	2447.2	0.3389	8.6023	8.2634
24	297	0.02982	0.001002	45.93	0.02177	100.59	2545.5	2444.8	0.3530	8.5806	8.2276
25	298	0.03166	0.001003	43.40	0.02304	104.77	2547.3	2442.5	0.3670	8.5592	8.1922
26	299	0.03360	0.001003	41.03	0.02437	108.95	2549.1	2440.2	0.3810	8.5379	8.1569
27	300	0.03564	0.001003	38.81	0.02576	113.13	2550.9	2437.8	0.3949	8.5168	8.1219
28	301	0.03778	0.001004	36.73	0.02723	117.31	2552.7	2435.4	0.4088	8.4959	8.0871
29	302	0.04004	0.001004	34.77	0.02876	121 48	2554.5	2433.1	0.4227	8.4751	8.0524
30	303	0.04241	0.001004	32.93	0.03037	125.66	2556.4	2430.7	0.4363	8.4546	8.0181
31	304	0.04491	0.001005	31.20	0.03205	129.84	2558.2	2428.3	0.4503	8.4342	7.9339
32	305	0.04753	0.001005	29.57	0.03382	134.02	2560.0	2425.9	0.4640	8.4140	7.9500

(Contd.)

Table 1: Saturated Steam (Temperature Table) (*Contd.*)

1	2	3	4	5	6	7	8	9	10	11	12
33	306	0.05029	0.001005	28.04	0.03566	138.20	2561.8	2423.6	0.4777	8.3939	7.9162
34	307	0.05318	0.001006	26.60	0.03759	142.38	2563.6	2421.2	0.4913	8.3740	7.8827
35	308	0.05622	0.001006	25.24	0.03961	146.56	2565.4	2418.8	0.5049	8.3543	7.8494
36	309	0.05940	0.001006	23.97	0.04172	150.74	2567.2	2416.4	0.5184	8.3348	7.8164
37	310	0.06274	0.001007	22.76	0.04393	154.91	2569.0	2414.1	0.5319	8.3155	7.7835
38	311	0.06624	0.001007	21.63	0.04624	159.04	2570.8	2411.7	0.5453	8.2962	7.7509
39	312	0.06991	0.001007	20.56	0.04865	163.27	2572.6	2409.3	0.5588	8.2772	7.7184
40	313	0.07375	0.001008	19.55	0.05116	167.45	2574.4	2406.9	0.5721	8.2583	7 6862
41	314	0.07777	0.001008	18.59	0.05379	171.63	2576.2	2404.5	0.5854	8.2395	7.6541
42	315	0.08198	0.001009	17.69	0.05652	175.81	2577.9	2402.1	0.5987	8.2209	7.6222
43	316	0.08639	0.001009	16.84	0.05938	179.99	2579.7	2399.7	0.6120	8.2025	7.5305
44	317	0.09100	0.001009	16.04	0.06236	184.17	2581.5	2397.3	0.6252	8.1342	7.5590
45	318	0.09582	0.001010	15.28	0.06546	188.35	2583.3	2394.9	0.6383	8.1661	7.5278
46	319	0.10086	0.001010	14.56	0.06869	192.53	2585.1	2392.5	0.6514	3.1481	7.4967
47	320	0.10612	0.001011	13.88	0.07206	196.71	2586.9	2390.1	0.6645	3.1302	7.4737
48	321	0.11162	0.001012	13.23	0.07557	200.89	2588.6	2387.7	0.6776	8.1125	7.4349
49	322	0.11736	0.001012	12.62	0.07922	205.07	2590.4	2385.3	0.6906	8.0950	7.4044
50	323	0.12335	0.001012	12.05	0.08302	209.26	2592.2	2382.9	0.7035	8.0776	7.5741
55	328	0.15741	0.001014	9.579	0.1044	230.17	2601.0	2370.8	0.7677	7.9926	7.2249
60	333	0.1992	0.001017	7.679	0.1302	251.09	2609.7	2358.6	0.8310	7.9108	7.0798
65	338	0.2501	0.001020	6.202	0.1612	272.02	2618.4	2346.3	0.8933	7.8322	6.9389
70	343	0.3116	0.001023	5.046	0.1982	292.97	2626.9	2334.0	0.9548	7.7565	6.8017
75	348	0.3855	0.001026	4.134	0.2419	313.94	2635.4	2321.5	1.0154	7.6835	6.6681
80	353	0.4736	0.001030	3.409	0.2933	334.92	2643.8	2308.8	1.0753	7.6132	6.5379
85	358	0.5780	0.001032	2.829	0.3535	355.92	2652.0	2296.5	1.1343	7.5454	6.4111
90	363	0.7011	0.001036	2.361	0.4235	376.94	2660.1	2283.2	1.1925	7.4799	6.3074
95	368	0.8453	0.001040	1.982	0.5045	397.99	2668.1	2270.2	1.2501	7.4166	6.1665
100	373	1.0133	0.001043	1.673	0.5977	419.06	2676.0	2256.9	1.3069	7.3554	6.0485
105	378	1.2080	0.001050	1.419	0.7096	440.17	2683.7	2243.6	1.3630	7.2962	5.9332
110	383	1.4327	0.001052	1.210	0.8265	461.32	2696.3	2230.0	1.4185	7.2388	5.8203
115	388	1.6906	0.001056	1.036	0.9650	482.50	2698.7	2216.2	1.4733	7.1832	5.7099
120	393	1.9854	0.001060	0.8915	1.122	503.72	2706.0	2202.2	1.5276	7.1293	5.6017
125	398	2.3210	0.001065	0.7702	1.298	524.99	2713.0	2188.0	1.5813	7.0769	5.4956
130	403	2.7013	0.001070	0.6681	1.497	546.31	2719.9	2173.6	1.6344	7.0261	5.3917
135	408	3.131	0.001075	0.5818	1.719	567.68	2726.6	2158.9	1.6869	6.9766	5.2897
140	413	3.131	0.001080	0.5085	1.967	589.10	2733.1	2144.0	1.7390	6.9284	5.1894

(*Contd.*)

Table 1: Saturated Steam (Temperature Table) (*Contd.*)

1	2	3	4	5	6	7	8	9	10	11	12
145	418	4.155	0.001085	0.2460	2.242	610.60	2739.3	2128.7	1.7906	6.8815	5.0919
150	423	4.760	0.001090	0.3924	2.548	632.15	2745.4	2113.2	1.8416	6.8358	4.9942
155	428	5.433	0.001096	0.3464	2.886	653.78	2751.2	2097.4	1.8923	6.7911	4.8988
160	433	6.181	0.001100	0.3068	3.260	675.47	2756.7	2081.3	1.9425	6.7475	4.8050
165	438	7.008	0.001108	0.2724	3.671	697.25	2762.0	2064.8	1.9923	6.7048	4.7125
170	443	7.920	0.001110	0.2426	4.123	719.12	2767.1	2047.9	2.0416	6.6630	4.6214
175	448	8.924	0.001120	0.2165	4.618	741.07	2771.8	2030.7	2.0906	6.6221	4.4715
180	453	10.027	0.001127	0.1938	5.160	763.12	2776.3	2013.1	2.1393	6.5813	4.4426
185	458	11.233	0.001134	0.1739	5.752	785.26	2780.4	1995.2	2.1876	6.5424	4.3548
190	463	12.551	0.001141	0.1563	6.397	807.52	2784.3	1976.7	2.2356	6.5036	4.2680
195	468	13.987	0.001149	0.1408	7.100	829.88	2787.8	1957.9	2.2833	6.4654	4.1821
200	473	15.549	0.001565	0.1272	7.864	852.37	2790.9	1938.6	2.3307	6.4278	4.0971
205	478	17.243	0.001164	0.1150	8.694	874.99	2793.8	1918.8	2.3778	6.3906	4.0128
210	483	19.077	0.001173	0.1042	9.593	897.74	2796.2	1898.5	2.4247	6.3539	3.9292
215	488	21.060	0.001181	0.0946	10.57	920.63	2798.3	1877.6	2.4713	6.3176	3.8463
220	493	23.198	0.001190	0.0860	11.62	943.67	2799.9	1856.2	2.5178	6.2817	3.7639
225	498	25.501	0.001199	0.0783	12.76	966.89	2801.2	1834.3	2.5641	6.2461	3.6820
230	503	27.976	0.001209	0.0714	14.00	990.26	2802.0	1811.7	2.6102	6.2107	3.5995
235	508	30.632	0.001218	0.0652	15.33	1013.8	2802.3	1788.5	2.6562	6.1756	3.5194
240	513	33.478	0.001229	0.0596	16.76	1037.6	2802.3	1764.6	2.7020	6.1406	3.4386
245	518	36.523	0.001240	0.0546	18.31	1061.6	2801.6	1740.0	2.7478	6.1057	3.3579
250	523	39.776	0.001251	0.0500	19.99	1085.5	2800.4	1714.6	2.7935	6.0708	3.2773
255	528	43.246	0.001263	0.0459	21.79	1110.2	2798.7	1688.5	2.8392	6.0359	3.1967
260	533	46.943	0.001276	0.04213	23.73	1134.9	2796.4	1661.5	2.8848	6.0010	3.1162
265	538	50.877	0.001288	0.03871	25.83	1159.9	2793.5	1633.6	2.9306	5.9658	3.0352
270	543	55.058	0.001302	0.03559	28.10	1185.2	2789.9	1604.6	2.9763	5.9304	2.9541
275	548	59.496	0.001317	0.03274	30.55	1210.9	2785.5	1574.7	3.0223	5.8947	2.8724
280	553	64.202	0.001332	0.03013	33.19	1236.8	2780.4	1543.6	3.0683	5.8536	2.7903
285	558	69.186	0.001348	0.02773	36.06	1263.2	2774.5	1511.3	3.1146	5.8220	2 7074
290	563	74.461	0.001366	0.02554	39.16	1290.0	2767.6	1477.6	3.1611	5.7848	2.6237
295	568	80.037	0.001384	0.02351	42.53	1317.3	2759.8	1442.6	3.2079	5.7469	2.5390
300	573	85.927	0.001404	0.02165	46.19	1345.0	2751.0	1406.0	3.2552	5.7081	2.4529
305	578	92.144	0.001425	0.01993	50.18	1373.4	2741.1	1367.7	3.3029	5.6685	2.3656
310	583	98.700	0.001448	0.01833	54.54	1402.4	2730.0	1327.6	3.3512	5.6278	2.2766
315	588	105.67	0.001472	0.01686	59.33	1432.1	2717.6	1285.6	3.4002	5.5858	2.1856
320	593	112.89	0.001499	0.01548	64.60	1492.6	2703.7	1241.1	3.4500	5.5423	2.0923

(*Contd.*)

Table 1: Saturated Steam (Temperature Table) (*Contd.*)

1	2	3	4	5	6	7	8	9	10	11	12
325	598	120.56	0.001529	0.01419	70.45	1490.0	2688.0	1194.0	3.5008	5.4969	1.9961
330	603	128.63	0.001561	0.01299	76.99	1526.5	2670.2	1143.6	3.5528	4.4490	1.8962
335	608	137.12	0.001597	0.01185	84.36	1560.3	2649.7	1089.5	3.6063	5.3979	1.7916
340	613	146.05	0.001638	0.01078	92.76	1595.5	2626.2	1030.7	3.6616	5.3427	1.6811
345	618	155.45	0.001685	0.00976	102.4	1632.5	2598.9	966.4	3.7193	5.2828	1.5635
350	623	165.35	0.001741	0.00880	113.6	1671.9	2567.7	895.7	3.7800	5.2177	1.4377
355	628	175.77	0.001808	0.00786	127.2	1716.6	2530.4	813.8	3.8489	5.1442	1.2953
360	633	186.75	0.001896	0.00694	144.1	1764.2	2485.4	712.3	3.9210	5.0600	1.1390
365	638	198.33	0.002016	0.00601	166.3	1818.0	2428.0	610.0	4.0021	4.9579	0.9558
370	643	210.54	0.002136	0.00497	201.1	1890.2	2342.8	452.6	4.1108	4.8144	0.7036
374.15	647.15	221.20	0.00317	0.00317	315.5	2107.4	2107.4	0.0	4.4429	4.4429	0.0

Table 2: Saturated Steam (Pressure Table)

Pressure N/m² $p \times 10^{-5}$	Temperat-ure °C	Volume, v m²/kg		Density of Vapour, p kg/m²	Enthalpy, h, kJ/kg		Latent heat L kJ/kg	Entropy, kJ/kg–k		L/T kJ/kg-K
		Liquid v_i	Vapour v_s		Liquid h_i	Vapour h_s		Liquid s_i	Vapour s_v	
1	2	3	4	5	6	7	8	9	10	11
0.010	6.983	0.001000	129.2	0.00774	29.34	2514.4	2485.0	0.1060	8.9767	8.8707
0.015	13.036	0.001000	87.98	0.01137	54.71	2525.5	2470.7	0.1957	8.8288	8.6331
0.020	17.513	0.001001	67.01	0.01492	73.46	2533.6	2460.2	0.2607	8.7246	8.4639
0.025	21.096	0.001002	54.26	0.01843	88.45	2540.2	2451.7	0.3119	8.6440	8.3321
0.030	24.100	0.001003	45.67	0.02190	101.00	2545.6	2444.6	0.3544	8.5785	8.2241
0.035	26.694	0.001003	39.48	0.02533	111.85	2550.4	2438.5	0.3907	8.5232	8.1325
0.040	28.983	0.001004	34.80	0.02873	121.41	2554.5	2433.1	0.4225	8.4755	8.0530
0.045	31.035	0.001005	31.14	0.03211	129.99	2558.2	2428.2	0.4507	8.4335	7.9828
0.050	32.898	0.001005	28.19	0.03547	137.77	2561.6	2423.8	0.4763	8.3960	7.9197
0.055	34.605	0.001006	25.77	0.03880	144.91	2564.7	2419.8	0.4995	8.3621	7.8626
0.060	36.183	0.001006	23.74	0.04212	151.50	2567.5	2416.0	0.5209	8.3312	7.8103
0.065	37.651	0.001007	22.02	0.04542	157.64	2570.2	2412.5	0.5407	8.3029	7.7622
0.070	39.025	0.001007	20.53	0.04871	163.38	2572.6	2409.2	0.5591	8.2767	7.7176
0.075	40.316	0.001008	19.24	0.05198	168.77	2574.9	2406.2	0.5763	8.2523	7.6760
0.080	41.534	0.001008	18.10	0.05523	173.86	2577.1	2403.2	0.5925	8.2296	7.6371
0.085	42.689	0.001009	17.10	0.05848	178.69	2579.2	2400.5	0.6079	8.2082	7.6003
0.090	43.787	0.001009	16.20	0.06171	183.28	2581.1	2397.9	0.6224	8.1881	7.5657
0.095	44.833	0.001010	15.40	0.06493	187.65	2583.0	2395.3	0.6361	8.1691	7.5330
0.10	45.833	0.001010	14.67	0.06814	191.83	2584.8	2392.9	0.6493	8.1511	7.5018
0.15	53.997	0.001014	10.02	0.09977	225.97	2599.2	2373.2	0.7549	8.0093	7.2544
0.20	60.086	0.001017	7.650	0.1307	251.45	2609.0	2358.4	0.8321	7.9094	7.0773
0.25	64.992	0.001019	6.204	0.1612	271.99	2618.3	2346.4	0.8932	7.8323	6.9391
0.30	69.124	0.001023	5.229	0.1912	289.30	2625.4	2336.1	0.9441	7.7695	6.8254
0.35	72.651	0.001024	4.529	0.2210	304.20	2631.5	2327.3	0.9880	7.7168	6.7288
0.40	75.886	0.001026	3.993	0.2504	317.65	2636.9	2319.2	1.0261	7.6709	6.6448
0.45	78.743	0.001028	3.576	0.2796	329.64	2614.7	2312.0	1.0603	7.6307	6.5704
0.50	81.345	0.001030	3.240	0.3086	340.56	2646.0	2305.4	1.0912	7.5947	6.5035
0.55	83.737	0.001031	2.964	0.3374	350.61	2649.9	2299.3	1.1194	7.5623	6.4429
0.60	85.954	0.001033	2.732	0.3661	359.93	2653.6	2293.6	1.1454	7.5327	6.3873
0.65	88.021	0.001035	2.535	0.3945	368.62	2656.9	2288.3	1.1696	7.5055	6.3359
0.70	89.959	0.001036	2.365	0.4229	376.77	2660.1	2283.3	1.1921	7.4804	6.2883
0.75	91.785	0.001037	2.217	0.4511	384.45	2663.0	2278.6	1.2131	7.4570	6.2439

(Contd.)

Table 2: Saturated Steam (Pressure Table) (*Contd.*)

1	2	3	4	5	6	7	8	9	10	11
0.80	93.512	0.001038	2.087	0.4792	391.72	2665.8	2274.0	1.2330	7.4352	6.2022
0.85	95.152	0.001040	1.972	0.5071	398.63	2668.4	2269.8	1.2518	7.4147	6.1629
0.90	96.713	0.001041	1.869	0.5350	405.21	2670.9	2265.6	1.2696	7.3954	6.1258
0.95	98.204	0.001042	1.777	0.5627	411.49	2673.2	2261.7	1.2865	7.3771	6.0906
1.00	99.632	0.001043	1.694	0.5904	417.51	2675.4	2257.9	1.3027	7.3598	6.0571
1.50	111.37	0.001053	1.159	0.8628	467.13	2693.4	2226.2	1.4336	7.2234	5.7898
2.00	120.23	0.001060	0.8854	1.129	504.17	2706.3	2201.6	1.5301	7.1268	5.5967
2.50	127.43	0.001067	0.7184	1.392	535.34	2716.4	2181.0	1.6071	7.0520	5.4449
3.00	133.54	0.001073	0.6056	1.651	561.43	2724.7	2163.2	1.6716	6.9906	5.3190
3.50	138.87	0.001079	0.5240	1.908	584.27	2731.6	2147.4	1.7273	6.9392	5.2119
4.00	143.62	0.001084	0.4622	2.163	604.67	2737.6	2133.0	1.7764	6.8943	5.1179
4.50	147.92	0.001088	0.4132	2.417	623.16	2742.9	2119.7	1.8204	6.8647	5.0343
5.00	151.84	0.001092	0.3747	2.669	640.12	2747.5	2107.4	1.8604	6.8192	4.9588
5.50	155.46	0.001097	0.3425	2.920	655.66	2751.7	2096.1	1.8969	6.7869	4.8900
6.00	158.84	0.001101	0.3155	3.170	670.42	2755.5	2085.0	1.9308	6.7575	4.8267
6.50	162.00	0.001104	0.2923	3.419	684.16	2758.8	2074.7	1.9623	6.7304	4.7681
7.00	164.96	0.001108	0.2727	3.667	697.06	2762.0	2064.9	1.9918	6.7302	4.7134
7.50	167.76	0.011116	0.2554	3.916	709.30	2764.8	2055.5	2.0191	6.6812	4.8621
8.00	170.41	0.001115	0.2403	4.162	720.94	2767.5	2046.5	2.0457	6.6596	4.6139
8.50	172.95	0.001118	0.2268	4.410	732.05	2769.9	2037.9	2.0670	6.6388	4.5718
9.00	175.36	0.001121	0.2148	4.655	742.64	2772.1	2029.5	2.0941	6.6192	4.5251
9.50	177.66	0.001125	0.2040	4.890	752.82	2774.2	2021.4	2.1164	6.6078	4.4914
10.00	179.88	0.001127	0.1943	5.147	762.61	2776.2	2013.6	2.1382	6.5828	4.4446
11.00	184.07	0.001133	0.1774	5.637	781.13	2779.7	1998.5	2.1786	6.5497	4.3711
12.00	187.96	0.001138	0.1632	6.127	798.43	2782.7	1984.3	2.2161	6.5194	4.3033
13.00	191.61	0.001144	0.1511	6.617	814.70	2785.7	1970.7	2.2510	6.4913	4.2403
14.00	195.04	0.001149	0.1407	7.106	830.08	2787.8	1957.7	2.2837	6.4651	4.1814
15.00	198.29	0.001154	0.1317	7.596	844.67	2789.9	1945.7	2.3145	6.4406	4.1261
16.00	201.37	0.001159	0.1237	8.085	858.56	2791.7	1933.2	2.3436	6.4175	4.0739
17.00	204.31	0.001163	0.1166	8.575	871.84	2793.4	1921.5	2.3713	6.3957	4.0244
18.00	207.11	0.001168	0.1103	9.065	884.58	2794.8	1910.3	2.3976	6.3751	3.9775
19.00	209.80	0.001172	0.1047	9.555	896.81	2796.1	1899.3	2.4228	6.3554	3.9326
20.00	212.37	0.001176	0.0995	10.05	908.59	2797.2	1886.6	2.4469	6.3367	3.8898
21.00	214.85	0.001180	0.09489	10.54	919.96	2798.2	1878.2	2.4700	6.3187	3.8487
22.00	217.24	0.001185	0.09065	11.03	930.95	2799.1	1868.1	2.4922	6.3015	3.8093
23.00	219.55	0.001189	0.08677	11.52	941.60	2799.8	1858.2	2.5136	6.2849	3.7713
24.00	221.78	0.001193	0.0832	12.02	951.93	2800.4	1848.5	2.5343	6.2690	3.7347

(*Contd.*)

Table 2: Saturated Steam (Pressure Table) (*Contd.*)

1	2	3	4	5	6	7	8	9	10	11
25.00	223.94	0.001197	0.07991	12.51	961.96	2800.9	1839.0	2.5543	6.2536	3.6993
26.00	226.04	0.001201	0.07686	13.01	971.72	2801.4	1829.6	2.5736	6.2387	3.6651
27.00	228.07	0.001205	0.07402	13.51	981.22	2801.7	1820.5	2.5924	6.2244	3.6320
28.00	230.05	0.001209	0.07139	14.01	990.48	2802.0	1811.5	2.6106	6.2104	3.5998
29.00	231.97	0.001212	0.06893	14.51	999.53	2802.2	1802.6	2.6283	6.1969	3.5686
30.00	233.84	0.001216	0.06663	15.01	1000.84	2802.3	1793.9	2.6455	6.1837	3.5382
31.00	235.67	0.001220	0.06447	15.51	1017.0	2802.3	1785.4	2.6623	6.1709	3.5086
32.00	237.45	0.001223	0.06244	16.02	1025.4	2802.3	1776.9	2.6786	6.1585	3.4799
33.00	239.18	0.001227	0.06053	16.52	1033.7	2802.3	1768.6	2.6945	6.1463	3.4518
34.00	240.88	0.001231	0.05873	17.03	1041.8	2802.1	1760.3	2.7101	6.1344	3.4243
35.00	242.54	0.001234	0.05703	17.54	1049.8	2802.0	1752.2	2.7253	6.1228	3.3975
36.00	244.16	0.001238	0.05541	18.05	1057.6	2801.7	1744.2	2.7401	6.1115	3.3714
37.00	245.75	0.001241	0.05389	18.56	1065.2	2801.4	1736.2	2.7545	6.1004	3.3457
38.00	247.31	0.001245	0.05244	19.07	1072.7	2800.1	1728.4	2.7689	6.0896	3.3207
39.00	248.84	0.001248	0.05106	19.58	1080.1	2800.8	1720.6	2.7829	6.0789	3.2960
40.00	250.33	0.001252	0.04975	20.10	1087.4	2800.3	1712.9	2.7965	6.0885	3.2720
41.00	251.80	0.001255	0.04850	20.62	1094.6	2799.9	1705.3	2.8099	6.0583	3.2484
42.00	253.24	0.001258	0.04731	21.14	1101.6	2799.4	1697.8	2.8231	6.0482	3.2251
43.00	254.66	0.001262	0.04617	21.66	1108.5	2798.9	1690.3	2.8360	6.0863	3.2023
44.00	256.05	0.001265	0.04508	22.18	1115.4	2798.3	1682.9	2.8487	6.0286	3.1799
45.00	257.41	0.001269	0.04404	22.71	1122.1	2797.7	1675.6	2.8612	6.0191	3.1579
46.00	258.75	0.001272	0.04304	23.24	1128.8	2797.0	1668.3	2.8735	6.0097	3.1362
47.00	260.07	0.001276	0.04208	23.76	1135.3	2796.4	1661.1	2.8855	6.0004	3.1149
48.00	261.37	0.001279	0.04116	24.29	1141.8	2795.7	1653.9	2.8974	5.9913	3.0939
49.00	262.65	0.041282	0.04028	24.83	1148.2	2794.9	1646.8	2.9091	5.9824	3.0733
50.00	263.91	0.001286	0.03943	25.36	1154.5	2794.2	1639.7	2.9206	5.9735	3.0529
52.00	266.37	0.001292	0.03782	26.44	1166.8	2792.6	1625.7	2.9431	5.9561	3.0130
54.00	268.76	0.001299	0.03633	27.52	1178.9	2790.8	1611.9	2.9650	5.9392	2.9742
56.00	271.09	0.001305	0.03495	28.62	1190.8	2789.0	1598.2	2.9863	5.9227	2.9364
58.00	273.35	0.001312	0.03365	29.72	1202.3	2787.0	1584.7	3.0071	5.9066	2.8995
60.00	275.55	0.001319	0.03244	30.83	1213.7	2785.0	1571.3	3.0273	5.8908	2.8635
62.00	277.70	0.001325	0.03130	31.95	1224.8	2782.9	1558.0	3.0471	5.8753	2.8282
64.00	279.79	0.001331	0.03023	33.08	1235.7	2780.6	1544.9	3.0664	5.8601	2.8057
66.00	281.84	0.001338	0.02922	34.22	1246.5	2778.3	1531.9	3.0853	5.8452	2.7601
68.00	283.84	0.001345	0.02827	35.37	1257.0	2775.9	1518.9	3.1038	5.8306	2.7268
70.00	285.79	0.001351	0.02737	36.53	1267.4	2773.5	1506.0	3.1219	5.8162	2.6943

(*Contd.*)

Table 2: Saturated Steam (Pressure Table) (*Contd.*)

72.00	287.70	0.001358	0.02652	37.70	1277.6	2770.9	1493.3	3.1397	5.8020	2.6637
74.00	289.57	0.001364	0.02572	38.89	1287.7	2768.3	1480.5	3.1571	5.7880	2.6311
76.00	291.41	0.001371	0.02495	40.08	1297.6	2765.5	1467.9	3.1742	5.7742	2.6000
78.00	293.21	0.001377	0.02422	41.29	1307.4	2762.8	1455.3	3.1911	5.7606	2.5695
80.00	294.97	0.001384	0.02353	42.51	1317.1	2759.9	1442.8	3.2076	5.7471	2.5395
82.00	296.70	0.001390	0.02286	43.74	1326.6	2757.0	1430.3	3.2239	5.7338	2.5099
84.00	298.39	0.001397	0.02223	44.98	1336.1	2754.0	1417.9	3.2399	5.7207	2.4808
86.00	300.06	0.001404	0.02163	46.24	1345.4	2750.9	1405.5	3.2558	5.7076	2.4518
88.00	301.70	0.001411	0.02105	47.51	1354.6	2747.8	1393.2	3.2713	5.6948	2.4235
90.00	303.31	0.001418	0.02050	48.79	1363.7	2744.6	1380.9	3.2867	5.6820	2.3953
92.00	304.89	0.001425	0.01996	50.09	1372.8	2741.4	1368.6	3.3018	5.6694	2.3676
94.00	306.44	0.001432	0.01945	51.40	1381.7	2738.0	1356.3	3.3168	5.6568	2.3400
96.00	307.97	0.001438	0.01897	52.73	1390.6	2734.7	1344.1	3.3315	5.6444	2.3129
98.00	309.48	0.001445	0.01849	54.07	1399.3	2731.2	1331.9	3.3461	5.6321	2.2860
100.00	310.96	0.001453	0.01804	55.43	1408.0	2727.7	1319.7	3.3605	5.6198	2.2593
102	312.42	0.001460	0.01760	56.80	1416.7	2724.2	1307.5	3.3748	5.6076	2.2328
104	313.86	0.001467	0.01718	58.19	1425.2	2720.6	1295.3	3.3889	5.5955	2.2134
106	315.27	0.001474	0.01678	59.60	1433.7	2716.9	1283.1	3.4029	5.5835	2.1806
108	316.67	0.001481	0.01639	61.03	1442.2	2713.1	1270.9	3.4167	5.5715	2.1548
110	318.05	0.001488	0.01601	62.48	1450.6	2709.3	1258.7	3.4304	5.5595	2.1291
112	319.40	0.001496	0.01564	63.94	1458.9	2705.4	1246.5	3.4440	5.5476	2.1036
114	320.74	0.001503	0.01528	65.43	1467.2	2701.4	1234.3	3.4575	5.5358	2.0783
116	322.06	0.001511	0.01494	66.93	1475.4	2697.4	1222.0	3.4708	5.5239	2.0531
118	323.36	0.001519	0.01461	68.46	1483.6	2693.3	1209.7	3.4840	5.5121	2.0281
120	324.65	0.001526	0.01428	70.01	1491.8	2689.2	1197.4	3.4972	5.5002	2.0030
122	325.91	0.001534	0.01397	71.59	1499.9	2684.9	1185.0	3.5102	5.4484	1.9382
124	327.17	0.001542	0.01366	73.19	1508.0	2680.6	1172.6	3.5232	5.4765	1.9533
126	328.40	0.001550	0.01337	74.81	1516.0	2676.1	1160.1	3.5361	5.4646	1.9285
128	329.62	0.001559	0.01308	76.46	1524.0	2671.6	1147.6	3.5488	5.4527	1.9039
130	330.83	0.001567	0.01280	78.14	1532.0	2667.0	1135.0	3.5616	5.4408	1.8792
132	332.02	0.001576	0.01252	79.85	1540.0	2662.3	1122.3	3.5742	5.4288	1.8546
134	333.19	0.001584	0.01226	81.59	1547.9	2657.4	1109.5	3.5868	5.4168	1.8300
136	334.36	0.001593	0.01200	83.36	1555.8	2652.5	1096.7	3.5993	5.4047	1.8054
138	335.51	0.001602	0.01174	85.16	1563.7	2647.5	1083.8	3.6118	5.3925	1.7807
140	336.64	0.001610	0.01150	86.09	1571.6	2642.4	1070.7	3.6242	5.3803	1.7561
142	337.76	0.001620	0.01125	88.86	1579.5	2637.1	1057.6	3.6366	5.3679	1.7313
144	338.87	0.001630	0.01102	90.77	1587.4	2631.8	1044.4	3.6490	5.3555	1.7065

(*Contd.*)

Table 2: Saturated Steam (Pressure Table) (*Contd.*)

1	2	3	4	5	6	7	8	9	10	11
144	338.87	0.001630	0.01102	90.77	1587.4	2631.8	1044.4	3.6490	5.3555	1.7065
146	339.97	0.001638	0.01079	92.71	1595.3	2626.3	1031.0	3.6613	5.3431	1.6818
148	341.06	0.001648	0.01056	94.69	1603.1	2620.7	1017.6	3.6736	5.3305	1.6569
150	342.13	0.001658	0.01034	96.71	1611.0	2615.0	1004.0	3.6859	5.3178	1.6319
152	343.19	0.001668	0.01012	98.77	1618.9	2609.2	990.3	3.6981	5.3051	1.6070
154	344.24	0.001678	0.00991	100.9	1626.8	2603.3	976.5	3.7104	5.2922	1.5818
156	345.28	0.001688	0.00970	103.0	1634.7	2597.3	962.6	3.7226	5.2793	1.5567
158	346.31	0.001699	0.00950	105.2	1642.6	2591.1	948.5	3.7348	5.2663	1.5315
160	347.33	0.001710	0.00931	107.4	1650.5	2584.9	934.3	3.7471	5.2531	1.5060
162	348.34	0.001722	0.00912	109.7	1658.5	2578.5	920.0	3.7594	5.2399	1.4805
164	349.34	0.001733	0.00892	112.0	1666.5	2572.1	905.6	3.7717	5.2267	1.4550
166	350.32	0.001748	0.00874	114.4	1674.5	2565.5	891.0	3.7842	5.2132	1.4290
168	351.30	0.001757	0.00855	116.9	1683.0	2558.6	875.6	3.7974	5.1994	1.4020
170	352.26	0.001770	0.00837	119.5	1691.7	2551.6	859.9	3.8107	5.1855	1.3748
172	353.22	0.001782	0.00819	122.1	1700.4	2544.4	844.1	3.8240	5.1713	1.3473
174	354.17	0.001796	0.00801	124.8	1709.0	2537.1	828.1	3.8372	5.1570	1.3198
176	355.11	0.001810	0.00784	127.6	1717.6	2529.5	811.9	3.8504	5.1425	1.2921
178	356.04	0.001825	0.00767	130.4	1726.2	2521.8	795.6	3.8635	5.1278	1.2643
180	356.96	0.001840	0.00749	133.4	1734.8	2513.9	779.1	3.8765	5.1128	1.2363
182	357.87	0.001856	0.00733	136.4	1743.4	2505.8	762.3	3.8896	5.0975	1.2079
184	358.77	0.001872	0.00716	139.6	1752.1	2497.4	745.3	3.9028	5.0820	1.1792
186	359.67	0.001890	0.00700	142.8	1760.9	2488.8	727.9	3.9160	5.0661	1.1501
188	360.55	0.001907	0.00684	146.2	1769.7	2479.8	710.1	3.9294	5.0498	1.1204
190	361.43	0.001926	0.00668	149.8	1778.7	2470.6	692.0	3.9429	5.0332	1.0903
192	362.30	0.001945	0.006517	153.4	1787.8	2461.1	673.3	3.9566	5.0160	1.0594
194	363.16	0.001966	0.006358	157.3	1797.0	2451.1	654.1	3.9706	4.9983	1.0277
196	364.02	0.001988	0.006198	161.3	1806.5	2440.7	634.2	3.9849	4.9801	0.9952
198	364.86	0.002012	0.006038	165.6	1816.3	2429.8	613.5	3.9996	4.9610	0.9624
200	365.70	0.002037	0.005877	170.2	1826.5	2418.4	591.9	4.0149	4.9412	0.9263
202	366.53	0.002064	0.005714	175.0	1837.0	2406.2	569.2	4.0308	4.9204	0.8896
206	368.17	0.002125	0.005379	185.9	1859.9	2379.3	519.5	4.0651	4.8750	0.8099
210	369.78	0.002201	0.005023	199.1	1886.3	2347.6	461.3	4.1048	4.8223	0.7175
214	317.37	0.002306	0.004624	216.3	1919.0	2307.4	388.4	4.1543	4.7569	0.6026
218	379.92	0.002483	0.004115	243.0	1967.2	2248.0	280.8	4.2276	4.6622	0.4346
221.20	374.15	0.00317	0.00317	315.5	2107.4	2107.4	0.0	4.4429	4.4429	0.0

Table 3: Enthalpy of Superheated Steam

Pressure $p \times 10^{-5}$ N/m^2	Saturated Temp. °C	Saturated Enthalpy kJ/kg	Enthalpy of Superheated System at t°C										
			T = 100	150	200	250	300	350	400	450	500	550	600
1	2	3	4	5	6	7	8	9	10	11	12	13	14
0.01	69.83	2514.4	2688.6	2783.7	2808.1	2977.7	3076.8	3177.5	3279.7	3383.6	3489.2	3596.5	3705.6
0.05	32.90	2561.6	2688.1	2783.4	2879.9	2977.6	3076.7	3177.4	3279.7	3383.6	3489.2	3596.5	3705.6
0.10	45.83	2584.8	2687.5	2783.1	2879.6	2977.4	3076.6	3177.3	3279.6	3383.5	3489.1	3596.5	3705.6
0.20	60.09	2609.9	2686.3	2782.3	2879.2	2977.4	3076.4	3177.1	3279.4	3383.4	3489.0	3596.4	3705.4
0.30	69.12	2625.4	2685.1	2781.6	2878 7	2976.8	3076.1	3176.9	3279.3	3383.3	3488.9	3596.3	3705.4
0.40	75.89	2636.9	2683.8	2780.9	2878.2	2976.5	3075.9	3176.8	3279.1	3383.1	3488.8	3596.2	3705.1
0.50	81.35	2646.0	2682.6	2780.1	2877.7	2976.1	3075.7	3176.6	3279.0	3383.0	3488.7	3596.1	3705.2
0.60	85.95	2653.6	2681.3	2779.4	2877.3	2975.8	3075.4	3176.4	3278.8	3382.9	3488.6	3596.0	3705.2
0.70	89.96	2660.1	2680.0	2778.6	2876.8	2975.5	3075.2	3176.2	3278.7	3382.7	3488.5	3596.9	3705.0
0.80	93.51	2665.8	2678.8	2778.8	2876.3	2975.2	3075.0	3176.0	3278.5	3382 6	3488.4	3596.9	3705.0
0.90	96.71	2670.9	2677.5	2771.1	2875.8	2974.8	3074.7	3175.8	3278.4	3382.5	3488.3	3595.7	3704.9
1.0	99.63	2675.4	2676.2	2776.3	2875.4	2974.5	3074.5	3175.6	3278.2	3382.4	3488.1	3595.6	3704.8
1.2	104.81	2683.4	–	2774.8	2874.4	2973.9	3074.0	3175.3	3277.9	3382.1	3487.9	3595.4	3704.6
1.4	109.32	2690.3	–	2773.2	2873.4	2973.2	3073.5	3174.9	3277.6	3381.8	3487.7	3595.2	3704.5
1.6	113.32	2696.2	–	2771.7	2872.5	2972.5	3073.0	3174.5	3277.3	3381.6	3487.5	3595.0	3704.5
1.8	116.93	2701.5	–	2770.1	2871.5	2971.9	3072.6	3174.1	3777.0	3381.3	3487.3	3594.9	3704.1
2.0	120.23	2706.3	–	2768.5	2870.5	2971.2	3072.1	3173.8	3276.7	3381.1	3487.0	3594.7	3704.0
2.5	127.43	2716.4	–	2764.5	2868.0	2969.6	3070.9	3172.8	3275.9	3380.4	3486.5	3594.2	3703.6
3.0	133.54	2724.7	–	2760.4	2865.5	2967.9	3069.7	3171.9	3275.2	3379.8	3486.0	3593.7	3703.2
3.5	138.87	2731.6	–	2756.2	2862.9	2966.2	3068.5	3170.9	3274.4	3379.1	3485.4	3593.2	3702.7
4.0	143.62	2737.6	–	2752.0	2860.4	2964.5	3067.2	3170.0	3273.6	3378.5	3484.9	3592.8	3702.3
4.5	147.92	2742.9	–	2747.6	2857.8	2962.7	3065.9	3169.0	3272.8	3377.8	3484.3	3592.3	3701.9
5.0	151.84	2747.5	–	–	2855.1	2961.1	3064.8	3168.1	3272.1	3377.2	3483.8	3591.8	3701.5
5.5	155.47	2751.7	–	–	2852.5	2959.3	3063.5	3167.2	3271.3	3376.6	3483.2	3591.4	3700.1
6.0	158.84	2755.5	–	–	2849.7	2957.6	3062.3	3166.2	3270.6	3376.0	3482.7	3590.9	3700.7
6.5	161.99	2758.9	–	–	2847.0	2955.8	3061.0	3165.3	3269.8	3375.3	3482.1	3590.4	3700.3
7.0	164.96	2762.0	–	–	2844.2	2954.0	3059.8	3164.3	3269.0	3374.7	3481.6	3589.9	3699.9
7.5	167.76	2764.8	–	–	2841.4	2952.2	3058.5	3163.4	3268.3	3374.0	3481.0	3589.5	3699.5
8.0	170.41	2767.5	–		2838.6	2950.4	3057.3	3162.4	3267.5	3373.4	3480.5	3589.0	3699.1
8 5	172.94	2769.9	–	–	2835.7	2948.6	3056.0	3161.4	3266.7	3372.7	3479.9	3538.5	3698.6
9.0	175.36	2772.1	–		2832.7	2946.8	3054.7	3160.5	3266.0	3372.1	3479.4	3588.1	3698.2
9.5	177.67	2774.2	–	–	2829.8	2944.9	3053.4	3159.5	3265.2	3371.5	3478.8	3587.6	3697.8
10.0	179.88	2776.2	–	–	2826.8	2943.0	3052.1	3158.5	3264.4	3370.8	3478.3	3587.1	3697.4

(Contd.)

Table 3: Enthalpy of superheated steam (*Contd.*)

1	2	3	4	5	6	7	8	9	10	11	12	13	14
11.0	184.07	2779.7	–	–	2820.7	2939.3	3049.6	3156.6	3262.9	3369.5	3477.2	3586.2	3696.6
12.0	187.96	2782.7	–	–	2814.4	2935.4	3046.9	3154.6	3261.3	3368.2	3476.1	3585.2	3695.8
13.0	191.61	2785.4	–	–	2808.0	2931.5	3044.3	3152.7	3259.7	3366.9	3475.0	3584.3	3695.5
14.0	195.04	2787.8	–	–	2801.4	2927.6	3041.6	3150.7	3258.2	3365.6	3473.9	3383.3	3694.0
16.0	201.37	2791.7	–	–	–	2919.4	3036.2	3146.7	3255.0	3363.0	3471.7	3581.4	3692.1
18.0	207.11	2794.8	–	–	–	2911.0	3030.7	3142.7	3251.4	3360.4	3469.5	3579.5	3690.6
20.0	212.37	2797.2	–	–	–	2902.4	3025.0	3138.6	3248.7	3357.8	3467.3	3577.6	3689.2
22.0	217.24	2799.1	–	–	–	2893.4	3019.3	3134.5	3245.5	3355.2	3465.1	3575.7	3687.6
24.0	221.78	2800.4	–	–	–	2884.2	3013.4	3130.4	3242.3	3452.6	3562.9	3573.8	3685.9
26.0	226.04	2801.4	–	–	–	2874.7	3007.4	3126.1	3239.0	3349.0	3460.6	3571.9	3684.3
28.0	230.05	2802.0	–	–	–	2864.9	3001.3	3121.9	3235.8	3347.3	3458.4	3570.0	3682.5
30.0	233.84	2802.3	–	–	–	2854.8	2995.1	3117.5	3232.5	3344.6	3456.2	3568.1	3681.0
35.0	242.54	2802.0	–	–	–	2828.1	2979.0	3106.5	3224.2	3338.0	3450.6	3563.4	3676.9
40.0	250.33	2800.3	–	–	–	–	2962.0	3095.1	3215.7	3331.2	3445.0	3558.6	3672.8
45.0	257.41	2797.7	–	–	–	–	2944.2	3083.3	3324.4	3324.4	3439.3	3553.8	3668.6
50.0	263.91	2794.2	–	–	–	–	2925.5	3071.2	3198.3	3317.5	3433.7	3549.0	3664.5
55.0	269.93	2789.9	–	–	–	–	2905.7	3058.7	3189.3	3310.5	3430.3	3544.1	3660.3
60.0	275.55	2785.0	–	–	–	–	2885.0	3045.8	3180.1	3303.5	3422.2	3539.3	3656.2
65.0	280.82	2779.5	–	–	–	–	2862.9	3032.4	3170.8	3296.3	3416.4	3534.4	3652.0
70.0	285.79	2773.5	–	–	–	–	2839.4	3018.7	3161.2	3289.1	3410.6	3529.6	3647.9
75.0	290.50	2766.9	–	–	–	–	2814.0	3004.5	3151.6	3281.8	3404.7	3524.7	3643.7
80.0	294.97	2759.9	–	–	–	–	2786.8	2989.9	3141.6	3274.3	3398.8	3519.7	3639.5
85.0	299.23	2752.5	–	–	–	–	2758.2	2974.7	3131.4	3266.8	3392.8	3514.8	3635.3
90.0	303 31	2744.6	–	–	–	–	–	2959.0	8121.2	3259.2	3386.8	3509.8	3631.1
95.0	307.21	2736.4	–	–	–	–	–	2942.7	3110.7	3251.8	3380.7	3504.9	3626.9
100.0	310.96	2727.7	–	–	–	–	–	2925.8	3099.9	3243.6	3374.6	3499.8	3622.7
110.0	318.05	2709.3	–	–	–	–	–	2889.6	3077.8	3227.7	3362.2	3489.7	3614.2
			300	350	400	450	500	550	600	650	700	750	800
120.0	324.65	2689.2	–	2349.7	3054.8	3211.4	3349.6	3479.6	3605.2	3730.2	3854.3	3978.3	4102.7
130.0	330.83	2667.0	–	2305.0	3030.7	3194.6	3336.8	3469.3	3597.1	3722.9	3618.0	3972.9	4098.0
140.0	336.64	2642.4	–	2754.2	3005.6	3177.4	3323.8	3458.8	3588.5	3715.6	3841.7	3967.5	4093.3
150.0	342.13	2615.0	–	2694.8	2979.1	3159.7	3310.6	3448.3	3579.8	3708.3	3835.4	3962.1	4088.6
160.0	347.33	2584.9	–	2620.8	2951.3	3141.6	3297.1	3437.7	3571.0	3700.0	3829.1	3956.7	4084.0
170.6	352.26	2551.6	–	–	2921.7	3123.1	3283.5	3427.0	3557.8	3689.8	3819.7	3998.5	4077.0
180.0	356.96	2513.9	–	–	2890.3	3104.0	3269.6	3416.1	3553.4	3686.1	3816.5	3945.8	4074.6
190.0	361.43	2470.6	–	–	2856.7	3084.4	3255.4	3405.2	3544.5	3678.6	3810.2	3940.4	4070.0

(*Contd.*)

Table 3: Enthalpy of superheated steam (*Contd.*)

1	2	3	4	5	6	7	8	9	10	11	12	13	14
200.0	365.70	2418.4	–	–	2820.5	3064.3	3241.1	3394.1	3535.5	3671.1	3803.8	3935.0	4065.3
210.0	369.78	2347.6	–	–	2781.3	3043.6	3226.5	3382.9	3526.5	3663.6	3797.5	3925.5	4060.6
220.0	373.69	2195.6	–	–	2738.8	3022.3	3211.7	3371.6	3517.6	3656.1	3791.1	3924.1	4055.9
221.20	374.15	2107.4	–	–	2732.5	3012.3	3202.7	3363.7	3510.4	3650.2	3786.1	3920.1	4052.9
250.0	–	–	–	–	2582.0	2954.3	3165.9	3337.0	3489.9	3633.4	3771.9	3907.7	4041.9
300.0	–	–	–	–	2161.8	2825.6	3085.0	3277.4	3443.0	3595.0	3739.7	3880.3	4018.5
350.0	–	–	–	–	–	2676.4	2998.3	3215.4	3395.1	3556.1	3707.3	3852.9	3995.1
400.0	–	–	–	–	–	2515.6	2906.8	3151.6	3346.4	3517.0	3674.8	3825.5	3971.7
450.0	–	–	–	–	–	2384.2	2813.5	3086.5	3297.4	3477.8	3642.4	3798.1	3948.4
500.0	–	–	–	–	–	2293.2	2723.0	3021.1	3248.3	3438.9	3610.2	3770.9	3925.3
550.0	–	–	–	–	–	2231.4	2639.8	2956.5	3199.3	3400.2	3578.2	3744.1	3904.1
600.0	–	–	–	–	–	2187.1	2570.6	2896.2	3151.6	3362.4	3547.0	3717.4	3879.6
700.0	–	–	–	–	–	–	2476.1	2790.7	3060.4	3289.0	3480.3	3665.8	3833.3
800.0	–	–	–	–	–	–	2397.4	2708.0	2980.3	3220.3	3428.7	3616.7	3798.4
900.0	–	–	–	–	–	–	2349.9	2643.0	2913.5	3158.2	3374.6	3570.1	3752.3
1000.0	–	–	–	–	–	–	2316.1	2593.8	2857.5	3105.3	3324.4	3526.0	3714.8

Table 4: Entropy of Superheated Steam

Pressure $p \times 10^{-5}$ N/m^2	Saturated Temp. °C	Saturated Entropy kJ/kg	Enthalpy of Superheated System at t°C										
			T = 100	150	200	250	300	350	400	450	500	550	600
1	2	3	4	5	6	7	8	9	10	11	12	13	14
0.01	6.98	8.9767	9.5136	9.7527	9.9679	10.1641	10.3450	10.5133	10.6711	10.8200	10.9612	11.0957	11.2243
0.05	32.90	8.3960	8.7098	9.0094	9.2248	9.4211	9.6021	9.7704	9.9283	10.0772	10.2184	10.3529	10.4815
0.10	45.83	8.1511	8.4486	8.6888	8.9045	9.1011	9.2820	9.4504	9.6083	9.7572	9.8984	10.0329	10.1616
0.20	60.09	7.9094	8.1261	8.3676	8.5839	8.7806	8.9618	9.1303	9.2882	9.4372	9.5784	9.7130	9.8416
0.30	69.12	7.7695	7.9363	8.1791	8.3960	8.5930	8.7744	8.9430	9.1010	9.2499	9.3912	9.5257	9.6544
0.40	75.89	7.6709	7.8009	8.0450	8.2625	8.4598	8.6413	8.8100	8.9680	9.1170	9.2583	9.3929	9.5216
0.50	81.35	7.1947	7.6953	7.9406	8.1587	8.3564	8.5380	8.7068	8.8649	9.0139	9.1552	9.2898	9.4185
0.60	85.95	7.5327	7.6085	7.8551	8.0738	8.2718	8.4536	8.6224	8.7806	8.9296	9.0710	9.2056	9.3343
0.70	89.96	7.4804	7.5346	7.7825	8.0019	8.2002	8.3822	8.5511	8.7093	8.8594	8.9997	9.1344	9.2631
0.80	93.51	7.4352	7.4703	7.7195	7.9395	8.1381	8.3202	8.4892	8.6475	8.7966	8.9380	9.0727	9.2041
0.9	96.71	7.3954	7.4132	7.6638	7.8843	8.0832	8.2656	8.4347	8.5929	8.7421	8.8835	9.0182	9.1469
1.0	99.63	7.3598	7.3618	7.6137	7.8349	8.0342	8.2166	8.3858	8.5442	8.6934	8.8348	8.9695	9.1982
1.2	104.81	7.2984	–	7.5267	7.7492	7.9491	8.1319	8.3012	8.4597	8.6090	8.7505	8.8852	9.0141
1.4	109.32	7.2465	–	7.4527	7.6765	7.8770	8.0601	8.2297	8.3882	8.5376	8.6791	8.8139	8.9429
1.6	113.32	7.2017	–	7.3382	7.6133	7.8144	7.9979	8.1676	8.3263	8.4757	8.6173	8.7521	8.8809
1.8	116.93	7.1622	–	7.3309	7.5574	7.7591	7.9429	8.1128	8.2716	8.4211	8.5627	8.6975	8.8264
2.0	120.23	7.1268	–	7.2794	7.5072	7.7096	7.8937	8.0638	8.2226	8.3722	8.5139	8.6487	8.7776
2.5	127.43	7.0520	–	7.1689	7.4001	7.6042	7.7891	7.9597	8.1188	8.2686	8.4104	8.5453	8.6743
3.0	133.54	6.9909	–	7.0771	7.3119	7.5176	7.7034	7.8744	8.0388	8.1838	8.3257	8.4608	8.5898
3.5	138.87	6.9392	–	6.9993	7.2364	7.4442	7.6309	7.8024	7.9622	8.1122	8.2539	8.3890	8.5182
4.0	143.62	6.8943	–	6.9285	7.1708	7.3800	7.5675	7.7395	7.8994	8.0497	8.1919	8.3271	8.4563
4.5	147.92	6.8547	–	6.8661	7.1121	7.3232	7.5116	7.6841	7.8445	7.9945	8.1369	8.2722	8.4012
5.0	151.84	6.8192	–	–	7.5092	7.2721	7.4614	7.6343	7.7948	8.9454	8.0879	8.2233	8.3526
5.5	155.47	6.7870	–	–	7.0108	7.2256	7.4158	7.5892	7.7500	7.9008	8.0433	8.1789	8.3083
6.0	158.84	6.7575	–	–	6.9662	7.1829	7.3740	7.5479	7.7090	7.8600	8.0027	8.1383	8.2678
6.5	161.99	6.7304	–	–	6.9247	7.1434	7.3355	7.5099	7.6712	7.8224	7.9652	8.1009	8.2305
7.0	164.96	6.7052	–	–	6.8849	7.1066	7.2997	7.4745	7.6362	7.7875	7.9305	8.0663	8.1959
7.5	167.76	6.6817	–	–	6.8491	7.0721	7.2662	7.4416	7.6035	7.7550	7.8981	8.0340	8.1637
8.0	170.41	6.6596	–	–	6.8148	7.0391	7.2348	7.4107	7.5729	7.7246	7.8678	8.0038	8.1333
8.5	172.94	6.6388	–	–	6.7820	7.0091	7.2051	7.3851	7.5441	7.6960	7.8395	7.9754	8.1056
9.0	175.36	6.6192	–	–	6.7508	6.9800	7.1771	7.3540	7.5109	7.6689	7.8124	7.9486	8.0785
9.5	177.67	6.6005	–	–	6.7209	6.9523	7.1505	7.3279	7.4911	7.6433	7.7869	7.9232	8.0532

(Contd.)

Table 4: Entropy of superheated Steam (*Contd.*)

1	2	3	4	5	6	7	8	9	10	11	12	13	14
10.0	179.88	6.5828	–	–	6.6922	6.9259	7.1251	7.3031	7.4665	7.6190	7.7627	7.8991	8.0292
11.0	184.07	6.5497	–	–	6.6397	6.8764	7.0778	7.2569	7.4209	7.5737	7.7177	7.8543	7.9845
12.0	187.96	6.5193	–	–	6.5872	6.8305	7.0342	7.2144	7.3790	7.5323	7 6765	7.8132	7.9436
13.0	191.61	6.4913	–	–	6.5394	6.7878	6.9938	7.1751	7.3404	7.4940	7.6385	7.7754	7.9060
14.0	195.04	6.4651	–	–	6.4941	6.7477	6.9561	7.1386	7.3045	7.4585	7.6032	7.7404	7.7710
16.0	201.37	6.4175	–	–	–	6.6740	6.8873	7.0723	7.2394	7.3942	7.5395	7.6770	7.8080
18.0	207.11	6.3751		–	–	6.6071	6.8257	7.0131	7.1816	7.3372	7.4830	7.6290	7.7522
20.0	212.37	6.3306	–	–	–	6.5454	6.7696	6.9596	7.1296	7.2859	7.4323	7.5706	7.7022
22.0	217:24	6.3015	–	–	–	6.4879	6.7179	6.9107	7.0821	7.2394	7.3862	7.5249	7.6568
24.0	221.78	6.2690	–	–	–	6.4338	6.6699	6.8656	7.0384	7.1964	7.3439	7.4830	7.6152
26.0	226.04	6.2387	–	–	–	6.3823	6.6249	6.8236	6.9979	7.1568	7.3048	7.4443	7.5768
28.0	230.05	6.2104	–	–	–	6.3331	6.5824	6.7824	6.9601	7.1199	7.2685	7.4083	7.5412
30.0	233.84	6.1837	–	–	–	6.2857	6.5422	6.7471	6.9246	7.0854	7.2345	7.3748	7.5079
35.0	242.54	6.1228	–	–	–	6.1732	6.4491	6.6626	6.8443	7.0074	7.1580	7.2993	7.4332
40.0	250.33	6.0685	–	–	–	–	6.3642	6.5870	6.7733	6.9388	7.0909	7.2333	7.3680
45.0	257.41	6.0191	–	–	–	–	6.2852	6.5182	6.7093	6.8774	7.0311	7.1746	7.3100
50.0	263.91	5.9735	–	–	–	–	6.2105	6.4545	6.6508	6.8217	6.9770	7.1215	7.2578
55.0	269.93	5.9309	–	–	–	–	6.1388	6.3949	6.5968	6.7070	6.9276	7.0732	7.2101
60.0	275.55	5.8908	–	–	–	–	6.0692	6.3386	6.5492	6.7230	6.8818	7.0285	7.1664
65.0	280.82	5.8527	–	–	–	–	6.0008	6.2849	6.4987	6.6786	6.8390	6.9872	7.1256
70.0	285.79	5.8162	–	–	–	–	5.9327	6.2333	6.4536	6.6368	6.7993	6.9485	7.0880
75.0	290.50	5.7811	–	–	–	–	5.8641	6.1836	6.4107	6.5960	6 7615	6.9121	7.0527
80.0	294.97	5.7471	–	–	–	–	5.7942	6.1349	6.3694	6.5597	6.7262	6.8778	7.0191
85.0	299.23	5.7141	–	–	–	–	5.7211	6.0875	6.3306	6.5238	6.6924	6.8455	6.9873
90.0	303.31	5.6820	–	–	–	–	–	6.0408	6.2915	6.4894	6.6600	6.8143	6.9573
95.0	307.21	5.6506	–	–	–	–	–	5.9947	6.2537	6.4563	6.6292	6.7848	6.9287
100.0.	310.96	5.6198	–	–	–	–	–	5.9489	6.2182	6 4243	6.5994	6.7564	6.9013
110.0	318.05	5.5595						5.8571	6.1483	6.3633	6.5432	6.7031	6.8499
		t =	300	350	400	450	500	550	600	650	700	750	800
120.0	324.65	5.5002	–	5.7636	6.0810	6.3056	6.4906	6.6535	6.8022	6.9409	7.0718	7.1961	7.3147
130.0	330.83	5.4408	–	5.6661	6.0155	6.2506	6.4409	6.6069	6.7577	6 8979	7.0298	7.1550	7.2743
140.0	336.64	5.3803	–	5.5618	5.9513	6.1978	6.3937	6.5630	6.7159	6.8575	6.9906	7.1166	7.2367
150.0	342.13	5.2531	–	5.4467	5.8876	6.1468	6.3487	6.5213	6.6764	6.8195	6.9536	7.0806	7.2013
160.0	347.33	5.3178	–	5.3109	5.8240	6.0972	6.3054	6.4816	6.6389	6.7835	6.9188	7.0466	7.1681
170.0	352.26	5.1855	–	–	5.7599	6.0489	6.2636	6.4435	6.5858	6.7328	6.8698	6.9989	7.1215
180.0	356.96	5.1128	—	–	5.6947	6.0015	6.2232	6.4069	6.5688	6.7166	6.8542	6.9838	7.1067

(*Contd.*)

Table 4: Entropy of superheated Steam *Contd.*

1	2	3	4	5	6	7	8	9	10	11	12	13	14
190.0	361.43	5.0332	–	–	5.6278	5.9549	6.1839	6.3716	6.5360	6.6854	6.8241	6.9546	7.0783
200.0	365.70	4.9412			5.5585	5.9089	6.1456	6.3374	6.5043	6.6554	6.7953	6.9267	7.0511
210.0	369.78	4.8223	–		5.4863	5.8633	6.1082	6.3043	6.4737	6.6265	6.7677	6.9000	7.0251
220.0	373.69	4.5799	–	–	5.4102	5.8179	6.0716	6.2721	6.4441	6.5986	6.7410	6.8743	7.0001
221.20	374.15	4.4429	–	–	5.2955	5.7932	5.9839	6.2643	6.4042	6.5286	6.6910	6.7859	6.9823
250.0	–	–	–	–	5.1455	5.6821	5.9655	6.1801	6.3604	6.5203	6.6664	6.8025	6.9306
300.0	–	–	–	–	4.4896	5.4495	5.7972	6.0386	6.2340	6.4033	6.5560	6.6970	6.8288
350.0	–	–	–	–	4.2214	5.2031	5.6349	5.9074	6.1194	6.2988	6.4584	6.6043	6.7400
400.0	–	–	–	–	4.1190	4.9511	5.4762	5.7835	6.0135	6.2035	6.3701	6.5210	6.6606
459.0					4.0554	4.7469	5.3226	5.6654	5.9143	6.1154	6.2890	6.4451	6.5885
500.0	–	–	–	–	4.0083	4.6026	5.1782	5.5525	5.8207	6.0331	6.2138	6.3749	6.5222
550.0	–	–	–	–	3.9705	4.5000	5.0470	5.4421	5.7321	5.9558	6.1436	6.3095	6.4600
600.0	–	–	–	–	–	4.4246	4.9374	5.3463	5.6477	5.8827	6.0775	6.2483	6.4031
700.0	–	–	–	–	–	4.3182	4.7688	5.1748	5.4931	5.7480	5.9562	6.1361	6.2979
800.0	–	–	–	–	–	4.2434	4.6488	5.0382	5.3595	5.6270	5.8470	6.0354	6.2034
900.0	–	–	–	–	–	4.1852	4.5602	4.9276	5.2468	5.5195	5.7479	5.9439	6.1179
1000.0						4.1373	4.4913	4.8393	5.1505	5.4267	5.6579	5.8600	6.0397

Table 5: Volume of Superheated Steam

Pressure $p \times 10^{-5}$ N/m²	Saturated Temp. °C	Saturated Volume m³/kg	Volume of Superheated Steam at t°C										
			t = 100	150	200	250	300	350	400	450	500	550	600
1	2	3	4	5	6	7	8	9	10	11	12	13	14
0.01	6 983	129.20	172.20	195.30	218.40	241.40	264.50	287.60	310.70	333.70	356.80	379.90	403.00
0.05	32.90	28.19	34.420	39.04	43.66	48.28	52.90	57.51	62.13	66.74	71.36	75.98	80.59
0.10	45.83	14.67	17.200	19.51	21.83	24.14	26.45	28.75	31.06	33.37	35.68	37.99	40.29
0.20	60.09	7.650	8.585	9.748	10.907	12.064	13.219	14.374	15.529	16 684	17.838	18.992	20.146
0.30	69.12	5.229	5.714	6 493	7.268	8.040	8.811	9.581	10.351	11.121	11.891	12.661	13.430
0.40	75.89	3.993	4.279	4.865	5.447	6.027	6.606	7.184	7.7625	8.340	8.917	9.494	10.072
0.50	81.35	3.240	3.418	3.889	4.356	4.821	5.284	5.747	6.209	6.671	7.133	7.595	8.057
0.60	85.95	2.732	2.844	3.238	3.628	4.016	4.402	4.788	5.174	5.559	5.944	6.329	6.714
0.70	89.96	2.365	2.434	2.773	3.108	3.441	3.772	4.103	4.434	4.764	5.095	5 425	5.425
0.80	93.51	2.087	2.126	2.425	2.718	3.010	3.300	3.590	3.879	4.168	4.457	4.746	5.035
0.90	96.71	1.869	1.887	2.153	2.415	2.674	2.933	3.190	3.448	3.705	3.962	4.219	4.475
1.0	99.63	1.694	1.696	1.936	2.172	2.406	2.639	2.871	3.102	3.334	3.565	3.797	4.028
1.2	104.81	1.428	—	1.611	1.808	2.004	2.198	2.391	2.585	2.778	2.971	3.163	3.356
1.4	109.32	1.236	—	1.378	1.548	1.716	1.883	2.049	2.215	2.380	2.546	2.711	2.876
1.6	113.32	1.091	—	1.204	1.353	1.501	1.647	1.792	1.937	2.082	2.227	2.372	2.517
1.8	116.93	0.972	—	1.068	1.202	1.333	1.463	1.593	1.722	1.851	1.979	2.108	2.237
2.0	120.23	0 8854	—	0.9595	1.0804	1.1989	1.3162	1.4328	1.5492	1.6653	1.7812	1.8971	2.0129
2.5	127.43	0.7185	—	0.7641	0.8620	0.9574	1.0516	1.1452.	1.2385	1.3315	1.4244	1.5172	1.6099
3.0	133.54	0.6056	—	0.6337	0.7164	0.7964	0.8753	0.9535	1.0314	1.1090	1.1865	1.2639	1.3412
3.5	138.87	0.5240	—	0.5410	0.6120	0.6810	0.7495	0.8172	0.8848	0.9506	1.0181	1.0830	1.1504
4.0	143.62	0.4622	—	0.4707	0.5343	0.5952	0.6549	0.7139	0.7725	0:8309	0.8892	0.9474	1.0054
4.5	147.92	0.4138	—	0.4167	0.4738	0.5284	0.5816	0.6341	0.6862	0.7386	0.7908	0.8425	0 8939
5.0	151.84	0.3747	—	—	0.4250	0.4744	0.5226	0.5701	0.6172	0.6640	0.7108	0.7574	0.8039
5.5	155.47	0.3425	—	—	0.3852	0.4305	0.4745	0.5178	0.5607	0.6034	0.6459	0.6883	0.7307
6.0	158.84	0.3155	—	—	0.3520	0.3939	0.4344	0.4742	0.5136	0.5528	0.5918	0.6308	0.6696
6.5	161.99	0.2925	—	—	0.3240	0.3629	0.4005	0 4373	0.4738	0.5100	0.5461	0 5821	0.6180
7.0	164:96.	0.2727	—	—	0.2999	0.3364	0.3744	0.4057	0.4396	0.4733	0.5069	0.5403	0.5737
7:5	167.76	0.2554	—	—	0.2791	0.3134	0.3462	0.3783	0 4100	0.4416	0.4729	0.5041	0.5353
8.0	170.41	0.2403	—	—	0.2608	0.2932	0.3241	0.3543	0.3842	0 4137	0.4432	0 4725	0.5017
8 5	172.94	0.2268	—	—	0.2447	0.2754	0.3047	0.3332	0.3613	0.3892	0.4169	0.4446	0.4721
9.0	175.36	0.2148	—	—	0.2303	0.2596	0.2874	0.3144	0.3410	0.3674	0.3936	0.4197	0.4458
9.5	177.67	0.2040	—	—	0.2175	0.2455	0.2719	0.2976	0 3228	0.3479	0.3727	0.3975	0.4222
10.0	179.94	0.1943	—	—	0.2059	0.2327	0.2580	0.2824	0.3065	0.3303	0.3540	0.3775	0.4010
11.0	184.07	0.1774	—	—	0.1859	0.2107	0.2339	0.2563	0.2782	0.3000	0.3215	0.3430	0.3643
12.0	—	0.1632	—	—	0.1692	0.1924	0.2139	0.2345	0.2547	0.2747	0.2945	0.3142	0.3338
13.0	191.61	0.1511	—	—	0.1551	0.1769	0.1969	0.2160	0.2348	0.2533	0.2716	0.2898	0.3080
14 0	195.06	0.1407	—	—	0.1429	0.1635	0.1823	0.2002	0.2177	0.2349	0.2520	0.2690	0.2859
16.0	201.37	0.1237	—	—	—	0.1419	0.1587	0.1745	0.1900	0.2051	0.2202	0.2351	0.2499
18.0	207.11	0.1103	—	—	—	0.1250	0.1402	0.1546	0.1684	0.1820	0.1954	0.2087	0.2219
20.0	212.37	0.0995	—	—	—	0.1114	0.1255	0.1386	0.1511	0.1634	0.1756	0.1876	0.1995
22 0	217.24	0.0906	—	—	—	0.10035	0.11343	0.12547	0.13700	0.14825	0.15934	0.17030	0.18119
24.0	221.78	0.08320	—	—	—	0.09107	0.10336	0.11455	0.12522	013561	0.14582	0.15591	0.18991
26.0	226.04	0.07680	—	—	—	0.08320	0.09483	0.10532	0.11526	0.12491	0.13438	0.14374	0.15301

(Contd.)

Table 5: Volume of superheated Steam (*Contd.*)

1	2	3	4	5	6	7	8	9	10	11	12	13	14
28.0	230.05	0.07139	—	—	—	0.07644	0.08751	0.09740	0.10671	0.11574	0.12458	0.13330	0.14194
30.0	233.84	0.06663	—	—	—	0.07055	0.08116	0.09053	0.09931	0.10779	0.11608	0.12426	0.13234
35.0	242.54	0.05703	—	—	—	0.05869	0.06842	0.07678	0.08449	0.09189	0.09909	0.10617	0.11315
40.0	250.33	0.04975	—	—	—	—	0.05883	0.06645	0.07338	0.07996	0.08634	0.09260	0.09876
45.0	257.41	0.04404	—	—	—	—	0.05134	0.05840	0.06472	0.07068	0.07643	0.08204	0.08757
50.0	263.91	0.03943	—	—	—	—	0.04530	0.05194	0.05779	0.06325	0.06342	0.07360	0.07862
55.0	269.93	0.03563	—	—	—	—	0.04033	0.04665	0.05212	0.05719	0.06202	0:06666	0.07131
60.0	275.55	0.03244	—	—	—	—	0.03614	0.04222	0.04738	0.05210	0.05659	0.06094	0.06518
65.0	280.82	0.02972	—	—	—	—	0.03225	0.03850	0.04340	0.04783	0.05203	0.05598	0.06012
70.0	285.79	0.02737	—	—	—	—	0.02946	0.03523	0.03992	0.04413	0.04809	0.05189	0.05559
75.0	290.50	0.02533	—	—	——	—	0.02672	0.03242	0.03694	0.04095	0.04469	0.04825	0.05323
80.0	294.97	0.02353	—	—	—	—	0.02426	0.02995	0.03431	0.03431	0.03814	0.04170	0.04510
85.0	299.23	0.02193	—	—	—	—	0.02200	0:02774	0.03199	0.03567	0.03909	0.04232	0.04544
90.0	303.31	0.02050	—	—	—	—	—	0.02579	0.02993	0.03348	0.03674	0.03982	0.04280
95.0	307.21	0.01921	—	—	—	—	—	0.02402	0.02805	0.03151	0.03465	0.03759	0.04045
100.0	310.96	0.01804	—	—	—	—	—	0.02244	0.02641	0.02974	0.03276	0.03560	0.03832
110.0	318.05	0.01601	—	—	—		—	0.01961	0.02351	0.02668	0.02950	0.03214	0.03466
				$t = 350$	**400**	**450**	**500**	**550**	**600**	**650**	**700**	**750**	**800**
120.0	324.65	0.01428	—	0.01721	0.02108	0.02412	0.02679	0.02926	0.03160	0.03387	0.03607	0:03822	0.04033
130.0	330.83	0.01280	—	0.01510	0.01902	0.02194	0.02440	0.02682	0.02902	0.03114	0 03319	0.03519	0.03716
140.0	336.64	0.01150	—	0.01321	0.01723	0.02008	0.02251	0.02472	0.02680	0.02880	0.03072	0.03260	0.03444
150.0	342.13	0.01034	—	0.01146	0.01566	0.01845	0.02080	0.02291	0.02488	0.02677	0.02859	0.03036	0.03209
160.0	347.33	0.009308	—	0.009764	0.01427	0.01703	0.01929	0.02132	0.02320	0.02499	0.02672	0.02839	0.03002
170.0	352.26	0.008371	—	—	0.01303	0.01576	0.01797	0.01992	0.02104	0.02271	0.02431	0.02587	0.02738
180.0	356.96	0.007498	—	—	0.01191	0.01464	0.01678	0.01867	0.02040	0.02204	0.02360	0.02512	0.02659
190.0	361.43	0.006677	—	—	0.01089	0.01362	0.01573	0.01755	0.01922	0.02079	0.02229	0.02374	0:02515
200.0	365.70	0.005877	—	—	0.009947	0.01271	0.01477	0.01655	0.01816	0.01967	0.02111	0.02250	0.02385
210.0	369.78	0.005023	—	—	0.009071	0.01187	0.01391	0.01564	0.01720	0.01866	0.02004	0.02138	0.02267
220.0	373 69	0.003728	—	—	0:003251	0.0111	0.01312	0.01481	0.01633	0.01774	0.01907	0.02036	0.02160
221.20	374.15	0.00317	—	—	0.003041	0.009878	0.01212	0.01395	0.01553	0.01708	0.01847	0.0200	0.01987
250.0	—	—	—	—	0.005014	0.009171	0:01113	0.01272	0.01413	0.01542	0.01663	0.01779	0.01891
300.0	—	—	—	—	0.002831	0.006735	0.008681	0.01017	0:01144	0.01258	0.01365	0.01465	0.01562
350.0	—	—	—	—	0.002179	0.004950	0.006925	0.008364	0.009519	0.01056	0.01152	0.01242	0.01327
400.0	—	—	—	—	0.001909	0.003675	0.005616	0.006982	0.008088	0.009053	0.009935	0.01075	0.01152
450.0	—	—	—	—	0.001801	0.002913	0.004625	0.005934	0.006984	0.007886	0.008699	0.009452	0.01016
500.0	—	—	—	—	0 001729	0 002492	0 003882	0.005113	0.006111	0.006960	0.007720	0.008420	0.00907
550.0	—	—	—	—	0.001672	0.002246	0.003320	0.004484	0.005454	0 006200	0.006925	0.007581	0.008195
600.0	—	—	—	—	0.001632	0.002084	0.002952	0.003947	0.004835	0.005596	0.006269	0:006885	0.007460
700.0	—	—	—	—	—	0.001890	0.002467	0.003922	0.003972	0.004652	0.005257	0.005808	0 006321
800.0	—	—	—	—	—	0.001772	0.002188	0.002764	0.003379	0.003974	0.004519	0.005017	0.005481
900.0	—	—	—	—		0.001691	0.002013	0.002458	0.002967	0.003476	0.003964	0.004419	0.004841
1000.0	—	—	—	—		0.001629	0.001893	0.002246	0.002668	0.003106	0.003536	0.003952	0.004341

CHAPTER 11

Simple Vapour Power Cycles

11.1 INTRODUCTION OR IDEAL STEAM PLANT CYCLES

Figure 11.1 shows the simple steam power plant. The heat energy released due to the combustion of fuel will be utilised in the boilers for converting water into steam (i.e., vapour) and this steam is then expanded into the steam engine/steam turbines to obtain useful work. The steam after producing work output, is generally condensed in the condensers. If the condensate is pure, then it is pumped as feed water to the boiler by means of a feed pump.

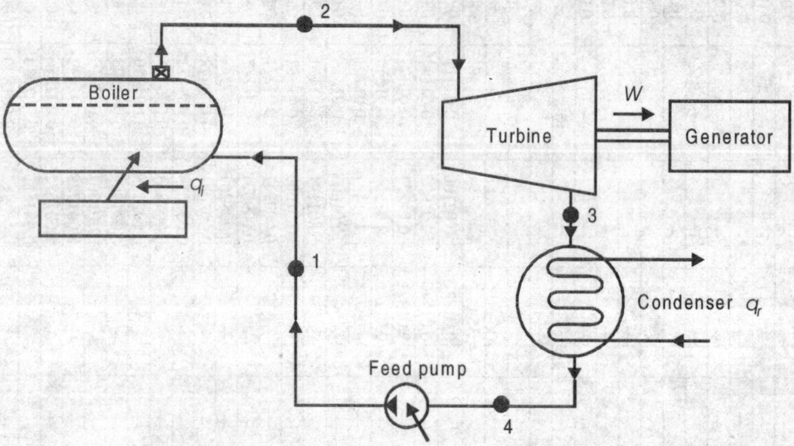

Fig. 11.1. *General arrangement of steam power plant.*

11.2 CARNOT CYCLE AND STEAM PLANT

As we know, the Carnot cycle has the greatest possible efficiency between any two given temperature limits T_1 and T_2 and is given by $\eta_{th} = \dfrac{T_1 - T_2}{T_1}$. It is important therefore to see whether Carnot cycle can be successfully applied to the steam plant.

Referring the T–S diagram as shown in Fig. 11.2.

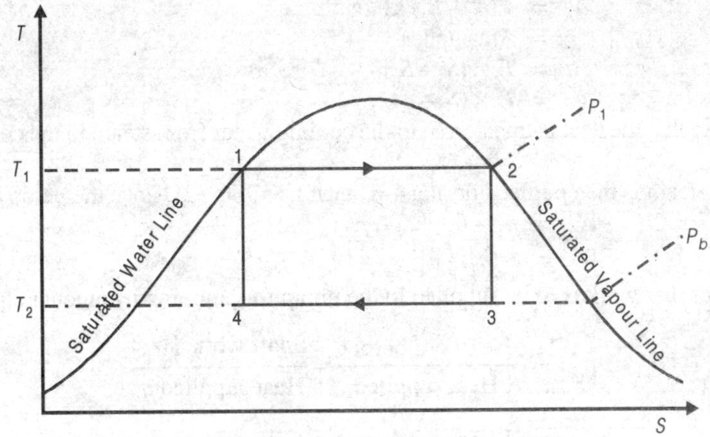

Fig. 11.2. *Carnot cycle on T–S diagram.*

Process 1–2:
This is an isothermal heat addition process and during this, water in the boiler gets converted into steam at constant temperature and pressure.

At point 2, steam is dry saturated.

Process 2–3:
The steam is expanded adiabatically isentropically in the steam turbine or steam engine from pressure P_1 to P_b.

Process 3–4:
The steam after producing work output in the turbine enters the condenser. During this process isothermal heat rejection q_r takes place and results into condensation of steam.

Process 4–1:
Partially condensed steam at 4 is fed into the feed pump, which pumps/compresses the mixture of wet steam and condensate isentropically to boiler pressure.

Since all the devices in Steam Power Plant are steady flow ones, from steady flow energy equation (SFEE) for unit mass,

$$q - w = \Delta h + \Delta KE + \Delta PE \quad ...(1)$$

Since the change in KE and PE are small, on neglecting them Eq. (1) reduces to

$$q - w = \Delta h \quad ...(2)$$

We can write the energy equation for all the devices of Steam Power Plant considering that, boiler produces no work, turbine is adiabatic, condenser produces no work and pump is adiabatic.

∴ Heat added in the boiler q_1 during 1 – 2

$$q_i = h_2 - h_1 \quad (\because W = 0 \text{ for boiler})$$

i.e.
$$q_i = h_2 - h_1 \text{ kJ/kg}$$
$$= \text{Area under } 1 - 2$$
$$= T_1 \times (S_2 - S_1) \quad ...(3)$$

Work producėd by the turbine W_T during the expansion process 2–3 and since expansion adiabatic.

$$q = 0$$

∴
$$0 - W_T = h_3 - h_2$$
$$W_T = h_2 - h_3 \text{ kJ/kg} \quad ...(4)$$

During the process 3–4 isothermal heat rejection takes place and results into condensation of steam (since $W =$ 0)

$$\begin{aligned} q_r &= h_4 - h_3 \\ q_r &= -(h_3 - h_4) \text{ kJ/kg} \qquad \text{...(5)} \\ &= \text{Area under } 3-4 \\ &= T_2 \times (S_3 - S_4) \\ &= T_2 \times (S_2 - S_1) \end{aligned}$$

Negative sign implies that the heat is transferred to the cooling water from steam in the condenser by an amount equal to $(h_3 - h_4)$

Partially condensed steam is then pumped during 4–1, then from Eq. (2), since the pump is adiabatic $q = 0$

$$\therefore \quad -W_p = h_1 - h_4$$

or

$$W_P = -(h_1 - h_4)$$

Negative sign implies that work is to be supplied to the pump for pumping by amount equal to $(h_1 - h_4)$

$$\eta_{th} = \frac{\text{Net work output}}{\text{Heat supplied}} = \frac{\text{Shaft work } (W_s)}{\text{Heat supplied } q_i}$$

$$\eta_{th} = \frac{\text{Difference between turbine and pump work}}{q_i}$$

$$= \frac{W_T - W_P}{q_i}$$

$$\eta_{th} = \frac{[h_2 - h_3] - [h_1 - h_4]}{[h_2 - h_1]} \qquad \text{...(7)}$$

$$= \frac{(h_2 - h_1) - (h_3 - h_4)}{h_2 - h_1}$$

$$= \frac{T_1(S_1 - S_2) - T_2(S_1 - S_1)}{T_1(S_2 - S_1)}$$

$$= \frac{T_1 - T_2}{T_1} \text{ as for ideal gas.}$$

Even though, the Carnot efficiency is maximum for a given value T_1 and T_2. But because of some inherent practical difficulties, as discussed below, Carnot cycle cannot be applied for the steam plant.

Carnot cycle when applied to S.P.P is practical upto a point that, Isothermal heat addition and the adiabatic expansion of steam in the turbine, which is quite reasonable.

The impractical part is in handling of steam in the condenser and pump. In the condenser, the steam is only partially condensed and the condensation must be stopped at 4. Also the feed pump must be capable of handling both wet steam and condensate.

A slight modification is made to this Carnot cycle which however produces a more practical cycle, called Rankine cycle. This cycle is accepted as, the practical but ideal cycle for the steam plant even though it has reduced thermal efficiency.

11.3 RANKINE CYCLE

To obtain Rankine cycle, modification made to Carnot cycle, is that, instead of stopping the condensation at some intermediate condition, the condensation is completed as shown in Fig. 11.3.

On *T–S* diagram Carnot cycle, would be 1'–2–3–4'. For Rankine cycle, however, condensation is continued until it is complete at 4. At this point there is all water. This can be successfully dealt with by the feed pump.

Process 4–1:

Pumping process, where water of condenser pressure P_b is raised to boiler pressure P_1. The pumping compression process is isentropic. At point (1) temperature is lower than T_{sat}. Subcooled state:

Process 1–2: (1 – 1' and 1' – 2)

During 1 – 1' feed water is heated to T_{sat} (may occur in economiser/boiler itself.) During 1' – 2 heat added q_t at constant temperature and pressure P_1 convert water into steam in the boiler.

Process 2–3:

Isentropic expansion of steam in the turbine from boiler pressure P_1 to back pressure P_b, which results into turbine work W_T

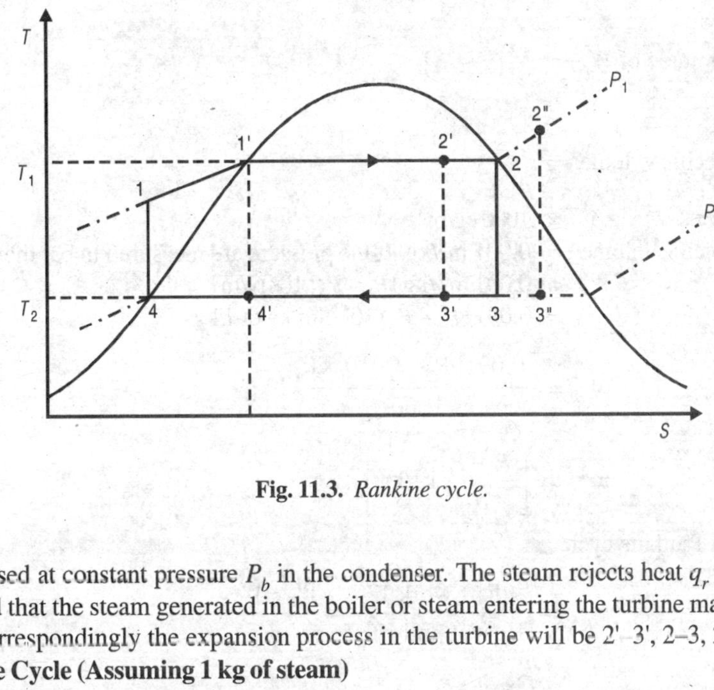

Fig. 11.3. *Rankine cycle.*

Process 3–4:

The steam is condensed at constant pressure P_b in the condenser. The steam rejects heat q_r to the cooling water.

It should be noted that the steam generated in the boiler or steam entering the turbine may be wet, dry-saturated and superheated. Correspondingly the expansion process in the turbine will be 2'–3', 2–3, 2"–3" respectively.

Analysis of Rankine Cycle (Assuming 1 kg of steam)

From Eq. (5) we get $q - w = \Delta h$

Process 1–2:

In the boiler heat q_i converts water into steam

$$q_i = h_2 - h_1 \text{ (since } q = 0)$$

Process 2–3:

In the turbine, when the steam expands, work will be produced.

$$W_T = h_2 - h_3 \text{ kJ/kg (since } q = 0)$$

Process 3–4:

In the condenser condensation will be completed and an amount of heat q_r is rejected to the cooling water.

$$q_r = h_4 - h_3$$

i.e.,

$$q_r = -[h_3 - h_4] \text{ Negative sign implies heat rejection.}$$

Process 4–1:

Feed pump pumps the water and increases the pressure of water from P_b to P_1.

$$0 - W_p = h_1 - h_4 \text{ (Since } q = 0)$$

or $$W_P = -(h_1 - h_4)$$

Negative sign implies work to be supplied to pump.

Also since pump is a steady flow equipment, we know that work done in a steady flow system.

$$= \int v dP$$

$$\therefore \quad W_P = -\int_{P_b}^{P_1} v dP$$

The specific volume of water remains constant since the effect of pressure is negligible on liquids.

$$\therefore \quad W_P = -V\int_{P_b}^{P_1} dP$$

$$\therefore \quad \text{Magnitude of } W_P = V[P_1 - P_b]$$

We also know that

$$\text{Specific volume} = \frac{1}{\rho}$$

and for water $$\rho_w = 1000 \text{ kg/m}^3$$

$\therefore$ Specific volume $= 0.001$ m^3/kg. If the pressure are measured in bar then,

$$W_P = 0.001 \text{ m}^3\text{/kg } [P_1 - P_b]\, 10^5 \text{ N/m}^2$$

$$= 0.001\,(P_1 - P_b)\, 10^5 \text{ Nm/kg or J/kg}$$

$$= \frac{0.001(P_1 - P_b)10^5 \text{ kJ}}{1000} \text{ kJ/kg}$$

$$W_P = \left[\frac{P_1 - P_b}{10} \text{ kJ/kg}\right] \quad ...(8)$$

And the efficiency of Rankine cycle,

$$\eta_R = \frac{\text{Shaft work } W_s}{\text{Heat supplied } q_i} = \frac{W_T - W_p}{q_i}$$

$$\therefore \quad \eta_R = \frac{[h_2 - h_3] - [h_1 - h_4]}{[h_2 - h_1]}$$

Also we can write this as,

$$\therefore \quad \eta_R = \frac{[h_2 - h_3] - [h_1 - h_4]}{[h_2 - h_4] - [h_1 - h_4]}$$

Since the pump work $W_P = h_1 - h_4$ is very small quantity, it can be neglected,

$$\therefore \quad \eta_{th} = \frac{[h_2 - h_3]}{[h_2 - h_4]}$$

Note: For solving the problems we can read the value of $h_4 = h_f$ from steam tables at back pressure P_b, h_2 and h_3 can be determined with the help of Mollier diagram or by calculations from steam tables. Since $W_P = h_1 - h_4$ the volume of h_1 can be determined by using Eq. (8) i.e., by finding W_p.

It should be noted that $h_R < \eta_{carnot}$ because the mean temperature of heat addition process in Rankine cycle is lower than the mean temperature of heat addition process in Carnot cycle. Whereas the mean temperature of heat rejection process in both the cycles is same.

In order to improve the efficiency of Rankine cycle, superheated steam has to be used, as represented by the process 2–2". It improves the cycle efficiency, as it increases the mean temperature of heat addition process in the boiler. Also when the superheated steam is used, it reduces the corrosion of turbines.

The Rankine efficiency has the same form as before, from Fig. 11.3 on similar lines

$$\eta_R = \frac{h_2'' - h_3''}{h_2'' - h_4''}$$

and when the steam is wet, then the cycle would be 1 – 1' – 2' – 3' – 4 –1

and $$\eta_R = \frac{h_2' - h_3'}{h_2' - h_4'}$$

11.4 EFFECT OF INLET PRESSURE (BOILER PRESSURE) AND BACK PRESSURE (EXHAUST PRESSURE) AND SUPERHEAT ON RANKINE EFFICIENCY

The effect of inlet pressure, back pressure and the superheat on the performance of Rankine cycle can be studied with the help of *T–S* diagram as shown in Fig. 11.4.

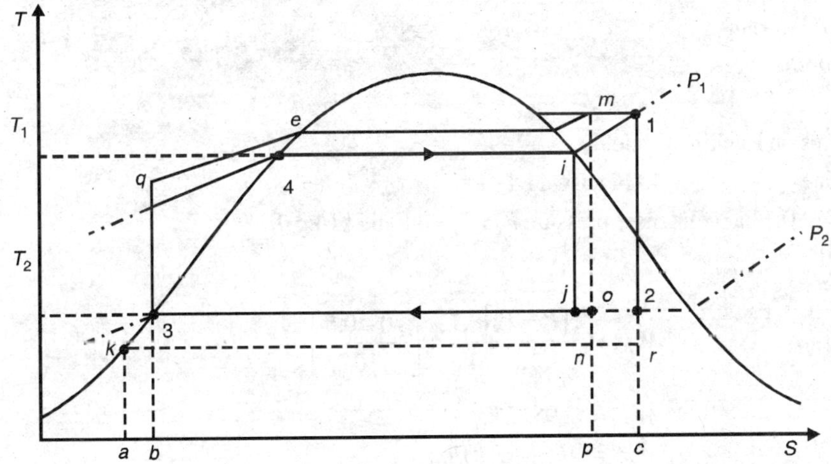

Fig. 11.4. *Rankine cylce varying working conditions.*

11.4.1 Boiler or Inlet Pressure

If the boiler pressure is raised, the evaporation line is raised as the saturation temperature is increased from 4–1 to *em*. The maximum temperature t_1 is kept constant i.e., t_m; $= t_1$. 1 – 2 – 3 – 4 –1 is the basic cycle taken for comparison.

With increased boiler pressure, keeping the condenser pressure and temperature constant, the new cycle will be m 03 *qem*. The amount of heat rejected is decreased given by C20 pc and difference in the heat added –*b*341*cb* and *bqempb* is small and the thermal efficiency increases. It is experimentally found that at $p_1 = 170$ bar and $p_2 = 0.14$ bar, the efficiency is maximum and decrease with increase in p_1.

11.4.2 Condenser Pressure and Temperature is Lowered

The new cycle wilt be 1*dk* 41. Work done is increased and heat rejected will be less so that the efficiency will be increased. Nature, however, preçisely defines the limit of improvement that may be obtained by this means. Atmospheric temperature and therefore temperature of the cooling water puts the limits of the steam temperature and pressure.

11.4.3 Superheat

If the steam is dry and saturated at the inlet of turbine, and if the steam is superheated then the work is increased and the proportional rise of work is more than the heat supplied so that the efficiency of the Rankine cycle increases.

It is interesting to note that condensing plants commonly operate more efficiently in the winter than in summer because of the lower cooling water temperature in the winter.

SOLVED EXAMPLES

Example 11.1 *A steam turbine operates on the Rankine cycle. It receives steam from the boiler at 30 bar, 250°C and it is exhausted into a condenser at 0.5 bar. Condensate is returned to the boiler by a feed pump. Calculate,*

(1) Heat supplied in the boiler drum.

(2) Dryness fraction of steam entering the condenser.

(3) Work done.

(4) Steam rate/kWh and

(5) Efficiency of the cycle.

Solution

Since it operates on Rankine Cycle,

Refer to Rankine cycle 1–2–3–4, (Fig. Ex. 11.1)

h_4 from tables at 0.5 bar condenser pressure $h_4 = h_f = 340.56$ kJ/kg.

We know that,

$$W_P = \frac{[P_1 - P_b]}{10}\frac{\text{kJ}}{\text{kg}} = \left[\frac{30-0.5}{10}\right]\frac{\text{kJ}}{\text{kg}}$$

$$W_P = 2.95 \text{ kJ/kg}$$

Also,

$$W_P = [h_1 - h_4] \text{ kJ/kg}$$

or

$$h_1 = W_p + h_4 = 2.95 + 340.56 \text{ kJ/kg}$$

$$h_1 = 343.51 \text{ kJ/kg}$$

From steam tables, h_2 at 30 bar and 250°C

$$h_2 = 2855 \text{ kJ/kg } (2854.8)$$

and

$$s_2 = 6.28 \text{ kJ/kg-K}$$

For isentropic process (2–3)

Entropy before expansion s_2 = Entropy after expansion s_3

$$6.28 = s_{f3} + x_3.s_{fg3} = 1.0912 + x_3.(6.5035)$$

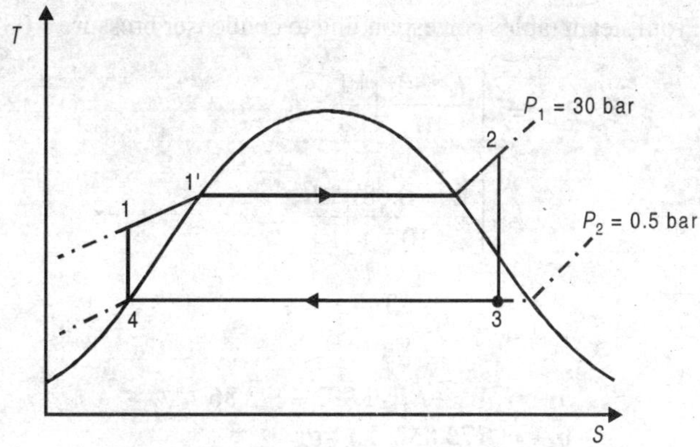

Fig. Ex. 11.1

∴ Dryness fraction of steam entering the condenser

$$x_3 = 0.797 \quad \text{...(A)}$$

∴ $h_3 = h_{f3} + x_3 . h_{fg3}$ corresponding to condenser pressure

$$= 340.56 + 0.797 \times 2305.4$$

$$h_3 = 2177.9 \text{ kJ/kg}$$

Now,

(1) Heat supplied in the boiler drum

$$q_i = h_2 - h_1 = 2855 - 343.51$$

$$q_i = 2511.49 \text{ kJ / kg}$$

(2) Dryness fraction of steam entering condenser

$$x_3 = 0.797 \text{ as above}$$

(3) W.D. or shaft work

$$W_s = W_T - W_P = [h_2 - h_3] - [h_1 - h_4]$$

$$= (2855 - 2177.9) - 2.95 = 674.15 \text{ kJ/kg}$$

(4) Since 1 kWh = 3600 kJ

$$\text{Steam rate} = \frac{3600}{W_s \text{ (kJ/kg)}} = \frac{3600}{674.15}$$

$$= 5.34 \frac{\text{kg}}{\text{kWh}}$$

(5) $$\eta_R = \frac{W_s}{q_1} = \frac{674.15}{2511.49}$$

$$\boldsymbol{\eta_R = 26.84 \ \%}$$

Example 11.2 *Steam at 20 bar and 360°C expands in a steam turbine to 0.08 bar. It is then condensed in a condenser to saturated water. The pump feeds back the water to the boiler. Assume ideal Rankine cycle and determine.*

(i) The net work done/kg of steam

(ii) η_R

Solution

Refer Fig. Ex. 11.2. From steam tables corresponding to condenser pressure 0.08 bar $h_4 = h_f = 173.86$ kJ/kg.

We know that, $$W_P = \left[\frac{P_1 - P_b}{10}\right]\frac{\text{kJ}}{\text{kg}} = h_1 - h_4$$

$$W_P = \left[\frac{20 - 0.08}{10}\right]\frac{\text{kJ}}{\text{kg}}$$

$$W_P = 1.992\frac{\text{kJ}}{\text{kg}} = h_1 - h_4$$

or $$h_1 = W_P + h_4 = 1.992 + 173.86$$

$$\mathbf{h_1 = 175.852\ kJ/kg}$$

or corresponding to 20 bar and 360° C, $h_2 = 3160.62$

At 350° C, $h = 3138.6$ kJ/kg and 400° C $= h$ 3248.7 kJ/kg

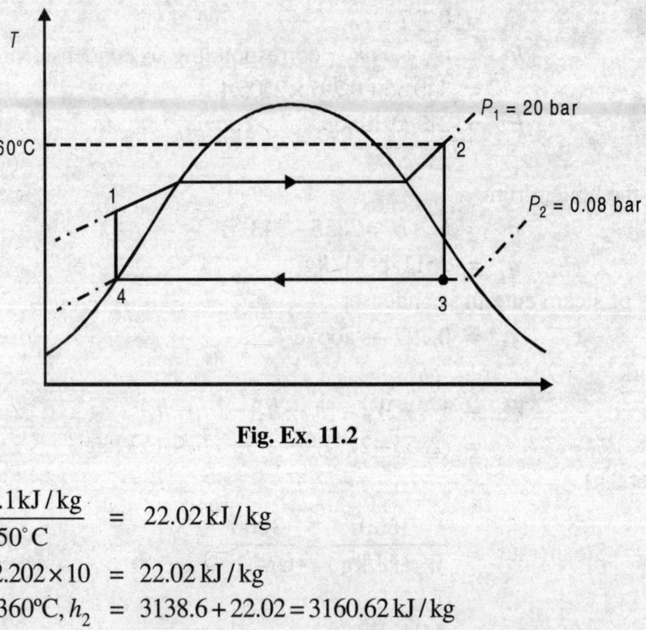

Fig. Ex. 11.2

$$\text{Difference is } \frac{110.1\,\text{kJ/kg}}{50^\circ\text{C}} = 22.02\,\text{kJ/kg}$$

$$\therefore \quad 2.202 \times 10 = 22.02\,\text{kJ/kg}$$

$$\therefore \quad \text{At } 360^\circ\text{C}, h_2 = 3138.6 + 22.02 = 3160.62\,\text{kJ/kg}$$

This method of finding the values at intermediate points is known as Interpolation.

To find h_s:

For isentropic process 2–3

Entropy before expansion s_2 = Entropy after expansion s_3

To find $s_2 = 7.1296$ at 400° C

6.9596 at 350° C

$$0.17/50 = 0.0034\,\text{kJ/kg-K}$$

$$0.0034 \times 10 = 0.034\,\text{kJ/kg-K}$$

$\therefore \quad 6.9596 + 0.034 = 6.9936 \text{ kJ/kg-K}$

$$6.9936 = f_{f3} + x_3.s_{fg3} \text{ at condenser pressure}$$
$$= 0.5925 + x_3.7.6371$$

$\therefore \quad x_3 = 0.8382$

$$h_3 = h_{f3} + x_3\ h_{fg3} \text{ at condenser pressure}$$
$$= 173.86 + 0.8382 \times 2403.2$$
$$h_3 = 2188.15 \text{ kJ/kg}$$

Now, $\quad \text{Work done} = W_s = W_T - W_P = (h_2 - h_3) - W_P$

$$= (3160.62 - 2188.15) - 1.992$$
$$= 970.478 \text{ kJ/kg}$$

and $\quad \eta_R = \dfrac{W_S}{q_i} = \dfrac{970.478}{h_2 - h_1} = \dfrac{970.478}{3160.62 - 175.852}$

$$\boldsymbol{\eta_R = 32.51\%}$$

Example 11.3 *A steam turbine operates on Rankine cycle. Steam entering the turbine has a pressure and temperature of 50 bar and 500°C, expands to a condenser pressure of 0.05 bar. Steam is condensed completely. Find heat supplied, turbine work, pump work, dryness fraction of steam entering the condenser and Rankine cycle efficiency.*

Solution

Refer Fig. Ex. 11.3. Since the steam is superheated,

From steam tables corresponding to 0.05 bar.

$$h_4 = h_f = 137.8 \text{ kJ/kg}$$

Also pump work $\quad W_P = \dfrac{P_1 - P_b}{10} = \dfrac{50 - 0.05}{10}$

$$W_P = 4.995 \text{ kJ/kg} = h_1 - h_4$$

$\therefore \quad h_1 = 4.995 + 137.8$

$$h_1 = 142.795 \text{ kJ / kg}$$

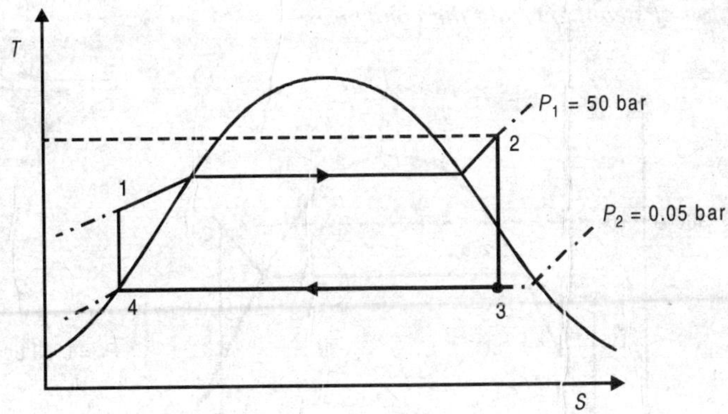

Fig. Ex. 11.3

Now, h_2 corresponding to 50 bar and 500° C from superheated steam tables,

$$h_2 = 3433.7 \text{ kJ/kg}$$

To find h_3; from the isentropic process 2–3

Entropy before expansion = Entropy after expansion.

$$s_2 = s_3$$

$$s_2 = s_{f3} + x_3 \cdot s_{fg3} \quad \text{...(A)}$$

Now s_2 at 50 bar and 500° C = 6.977 kJ/kg-K

∴ From (A), $6.977 = 0.4763 + x_3 \times (7.92)$

(At condenser pressure 0.05 bar)

$$\Rightarrow \quad x_3 = 0.8208 = 0.82$$

$$h_3 = h_{f3} + x_3 . h_{fg3} = 137.77 + 0.82 \times 2423.8$$

$$\mathbf{h_3 = 2125.286 \ kJ/kg}$$

Now,

(i) Heat supplied, $q_i = h_2 - h_1 = 3433.7 - 142.795$

$$\mathbf{q_i = 3290.905 \ kJ/kg}$$

(ii) Turbine work $W_T = h_2 - h_3 = 3433.7 - 2125.286$

$$\mathbf{W_T = 1308.414 \ kJ/kg}$$

(iii) Pump work W_P = 4.995 kJ/kg, already found in the beginning.

(iv) Dryness fraction of steam entering the condenser.

$$x_3 = 0.8208 \text{ already found}$$

(v) Rankine cycle efficiency

$$\eta_R = \frac{W_s}{q_i}$$

$$\eta_R = \frac{W_T - W_p}{q_i} = \frac{1308.414 - 4.995}{3290.905}$$

$$\boldsymbol{\eta_R = 39.606 \ \%}$$

Example 11.4 *Steam at 20 bar and 360°C is expanded in a steam turbine to 0.08 bar. It then enters a condenser where it is condensed to saturated liquid. It is then fed back to the boiler. Determine,*

(i) *Net work (shaft work) per kg of steam.*

(ii) *Dryness fraction of steam entering the condenser.*

Solution

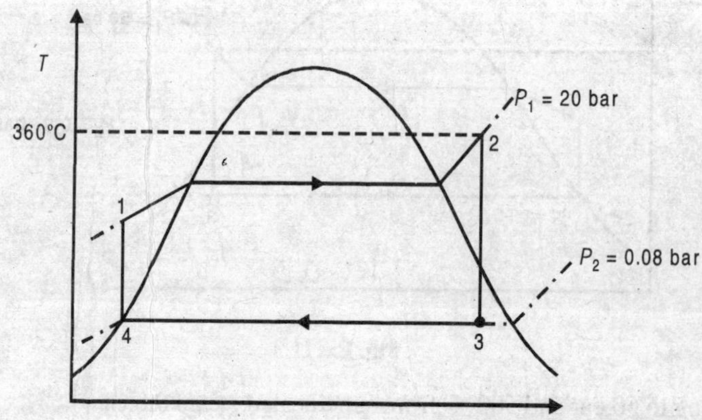

Fig. Ex. 11.4

From steam tables at 0.08 bar

$$h_4 = h_f = 173.9 \text{ kJ/kg}$$

And pump work $$W_P = \left[\frac{P_1 - P_b}{10}\right] = \left[\frac{20-0.08}{10}\right]$$

$$W_P = 1.992 \text{ kJ/kg} = h_1 - h_4$$

$$h_1 = 1.992 + h_4$$

$$h_1 = 1.992 + 173.9$$

$$h_1 = 175.892 \text{ kJ/kg}$$

h_2 corresponding to 20 bar and 360° C from superheated steam tables = 3160.62 kJ/kg (By interpolation)

To find h_3 from the isentropic process 2–3

Entropy before expansion = Entropy after expansion

$$s_2 = s_3$$

$$s_2 = s_{f3} + x_3 \cdot s_{fg3} \qquad \text{...(A)}$$

Now s_2 at 20 bar and 360° C = $6.994 \dfrac{\text{kJ}}{\text{kg K}}$ (By interpolation)

Now from (A) $$6.994 = 0.593 + x_3 \times 7.637 \text{ at } 0.08 \text{ bar}$$

$$6.994 - 0.593 = 7.637 x_3$$

$$0.8381 = x_3$$

$\therefore$ $$h_3 = h_{f3} + x_3 \cdot h_{fg3}$$

at $$0.08 \text{ bar} = 173.9 + 08381 \times 2403.2$$

$$\mathbf{h_3 = 2188.157 \ kJ/kg}$$

(i) $\therefore$ Net work or shaft work = $W_s = W_T - W_P$

$$W_s = [h_2 - h_3] - [h_1 - h_4]$$

$$= (3160.62 - 2188.157) - (175.892 - 173.9)$$

$$= (972.4626) - (1.992)$$

$$\mathbf{W_s = 970.4706 \ kJ/kg}$$

(ii) Dryness fraction of steam entering the condenser $x_3 = 0.8381$ as found earlier.

Example 11.5 *A steam power plant separating on Rankine cycle, receives steam from boiler at 3.5 MPa and 350°C. It is exhausted to condenser at 10 kPa. Calculate,*

(i) Energy supplied per kg of steam generated in boiler.

(ii) Quality of steam entering the condenser.

(iii) Rankine cycle efficiency considering feed pump work.

(iv) Specific steam consumption.

Data: P = 3.5 MPs = 35 bar, P_b = 10 kPa = 0.1 bar

As $100 \dfrac{kN}{m^2} = 1 bar$

(i) $q_i = ?$ *(ii)* $x_3 = ?$

(iii) $\eta_R = ?$ *(iv)* $SR = ?$

Solution

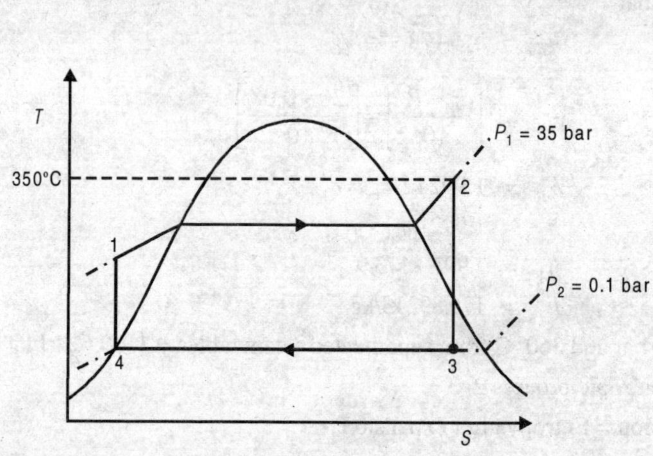

Fig. Ex. 11.5

From steam tables corresponding to 0.1 bar,

$$h_4 = h_f = 191.8 \text{ kJ/kg}$$

Also pump work $\quad W_p = \dfrac{P_1 - P_b}{10}$

$$= \frac{35 - 0.1}{10} = 3.490 \text{ kJ/kg} = h_1 - h_4$$

i.e.

$$3.499 + h_4 = h_1$$

$$3.490 + 191.8 = h_1$$

$$195.29 = h_1$$

And h_2 corresponding to 35 bar and 350° C from superheated steam tables = 3106.45 kJ/kg by interpolation.

To find h_3 from the isentropic process 2–3

$$s_2 = s_3$$

$$s_2 = s_{f3} + x_3 \cdot s_{fg3} \quad \text{...(A)}$$

Now s_2 at 35 bar and 350° C = 6.663 $\dfrac{\text{kJ}}{\text{kg-K}}$ by interpolation.

∴ From (A) $6.663 = 0.649 + x_3 \times 7.502$ at 0.1 bar

$$\frac{6.663 - 0.649}{7.502} = x_3$$

$$0.8016 = x_3$$

∴

$$h_3 = h_{f3} + x_3.h_{fg3} \text{ at 0.1 bar}$$

$$= 191.8 + 0.8016 + 2392.9$$

$$h_3 = 2110.075 \text{ kJ/kg}$$

Now,

(i) Heat supplied $\quad q_i = h_2 - h_1 = 3106.45 - 195.29$

$$q_i = 2911.16 \text{ kJ/kg}$$

(ii) $x_3 = 0.8016$ dryness fraction of steam which is already found.

(iii) Rankine cycle efficiency η_R considering feed pump work,

$$\eta_R = \frac{W_s}{q_i} = \frac{W_T - W_P}{q_i}$$

$$= \frac{[h_2 - h_3] - 4.90}{2911.16}$$

$$= \frac{(3106.45 - 2110.0) - 4.9}{2911.16}$$

$$\eta_R = 0.34226$$

$$\eta_R = 34.226\%$$

(iv) $$\text{Steam Rate} = \frac{3600}{W_S} = \frac{3600}{[W_T - W_P]}$$

$$= \frac{3600}{[h_2 - h_3] - 4.9} = \frac{3600}{996.375 - 4.9}$$

$$= \frac{3600}{991.475}$$

$$\textbf{S.R.} = \textbf{3.630 kg/kWh}$$

Example 11.6 *Find heat supplied, turbine work, pump work, shaft work, efficiencies, steam rate of Carnot and Rankine cycle using steam between 30 bar and 0.04 bar.*

Steam is dry and saturated at 30 bar. Compare the results.

Data: $q_i = ?$ $W_T = ?$

$W_P = ?$ $W_S = ?$

$\eta_c = ?$ $\eta_R = ?$

$S.R. = ?$

$P_1 = 30$ bar $P_b = 0.04$ bar.

Solution

Refer Fig. Ex. 11.6

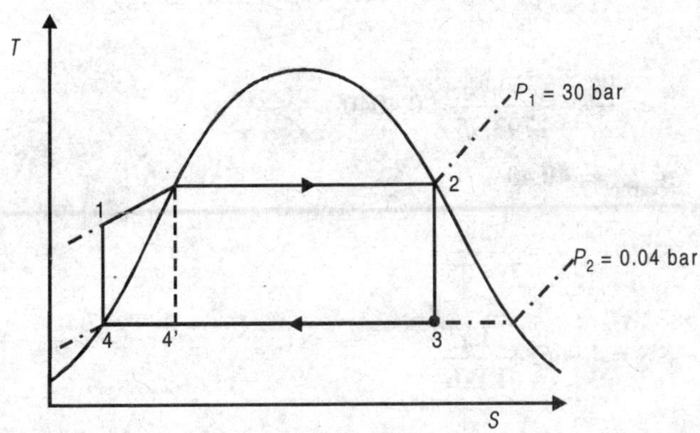

Fig. Ex. 11.6

First consider the Carnot cycle 1' – 2 – 3 – 4'

From steam tables at 30 bar,

$$h_{1'} = h_f = 1008.35 \text{ kJ/kg}$$

$$s_{1'} = s_f = 2.6455 \text{ kJ/kg-K}$$

and

$$h_2 = h_g \text{ at 30 bar} = 2802.3 \text{ kJ/kg}$$

$$s_2 = s_g = 6.1837 \text{ kJ/kg-K}$$

For the isentropic process 2–3

$$s_2 = s_{f3} + x_3 . s_{fg3} \text{ at 0.04 bar}$$

$$6.1837 = 0.423 + x_3 \times 8.053$$

$$x_3 = 0.7154$$

$$\therefore \quad h_3 = h_{f3} + x_3 h_{fg3}$$

$$= 121.41 + 0.7154 \times 2433.1$$

$$h_3 = 1862.04 \text{ kJ/kg}$$

Now for Isentropic Compression process 4'–1'

$$s_1' = s_4' = S'_{f_4} + x_4' s'_{fg_4}$$

$$2.6455 = 0.4225 + x_4' \times 8.0530$$

$$x_4' = 0.276 \qquad \text{Nearer to liquid line}$$

$$\therefore \quad h_4' = h'_{f_4} + x'_4 h'_{fg_4}$$

$$= 121.41 + 0.276 \times 2433.1$$

$$h_4' = 792.94 \text{ kJ/kg}$$

Now,

(i) Heat supplied $\quad q_i = h_2 - h_{1'}$

$$= 2802.3 - 1008.35$$

$$q_i = 1793.95 \text{ kJ/kg}$$

(ii) Turbine work $\quad = W_T = h_2 - h_3 = 2802.3 - 1862.04$

$$W_T = 940.26 \text{ kJ/kg}$$

(iii) Pump work $\quad W_P = h_{1'} - h_{4'} = 1008.35 - 792.94 = 215.41 \text{ kJ/kg}$

(iv) Shaft work $\quad W_s = W_T - W_P = (940.26) - (215.41) = 724.85 \text{ kJ/kg}$

(v) Carnot cycle efficiency

$$\eta_{\text{carnot}} = \frac{W_s}{q_i} = \frac{724.85}{1793.95} = 0.4040$$

$$\boldsymbol{\eta_{\text{carnot}} = 40.40\%}$$

(iii) $\quad$ Steam Rate $= \dfrac{3600}{W_s} = \dfrac{3600}{724.85}$

$$\mathbf{SR = 4.96654 \frac{kg}{kWh}}$$

II. Rankine Cycle 1 – 2 – 3 – 4

From Fig. Ex. 11.6 it is to be noted that

h_2 and h_3 will remain same.

$$\therefore \quad h_2 = 2802.3\ \text{kJ/kg}$$
$$h_3 = 1862.04\ \text{kJ/kg}$$

and
$$h_4 = h_f \text{ at } 0.04\ \text{bar} = 121.41\frac{\text{kJ}}{\text{kg}}$$

Now to find h_1,

We also know that, pump work

$$W_P = \frac{P_1 - P_b}{10} = \frac{30 - 0.04}{10} = 2.996$$
$$\cong 3.0\ \text{kJ/kg}$$

Also,
$$W_p = h_1 - h_4$$
$$W_p + h_4 = h_1$$
$$3 + 121.41 = h_1$$
$$\Rightarrow \quad h_1 = 124.41\ \text{kJ/kg}\ 124.41\ \text{kJ/kg}$$

Now,

(a) Heat supplied
$$q_i = h_2 - h_1 = 2802.3 - 124.41$$
$$q_i = 2677.89\ \text{kJ/kg}$$

(b) Turbine work
$$W_T = h_2 - h_3 = 940.26\frac{\text{kJ}}{\text{kg}} \cdot \text{It will remain same.}$$

(c) Pump work
$$W_P = h_1 - h_4 = 124.41 — 121.41$$
$$\mathbf{W_p = 3\ kJ/kg}$$

(d) Shaft Work
$$W_S = W_T = h_2 - h_3 = 940.26 - 3 = 937.26\ \text{kJ/kg}$$

(e) Rankine Cycle Efficiency

$$\eta_R = \frac{W_S}{q_i} = \frac{937.26}{2677.89}$$
$$\eta_R = 0.34999$$
$$\eta_R = 34.999\%$$

(f) Steam Rate,

$$\text{SR} = \frac{3600}{W_s} = \frac{3600}{937.26}$$
$$\mathbf{SR = 3.8409\ kg/kWh}$$

Comparison:

(1) Pump work for Carnot Cycle is more, since it has to handle large amount of mixture of wet steam and condensate.

(2) Turbine work in both the cases is same as the state points are same.

(3) Carnot cycle efficiency is more than Rankine cycle efficiency.

Example 11.7 *Calculate the cycle efficiency and steam consumption in kg/kWh for Carnot cycle and Rankine cycle using steam between two pressure of 20 bar and 0.1 bar. The steam is dry and saturated at 20 bar.*

Find: $\eta_{carnot} = ?\quad \eta_R = ?\quad SR = ?$

Solution

Refer Fig. Ex. 11.7

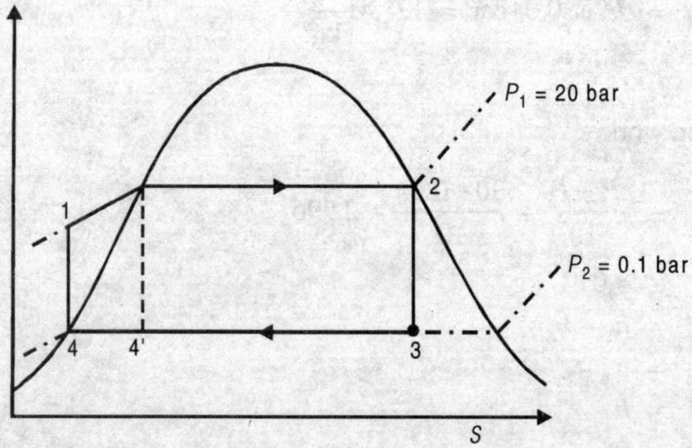

Fig. Ex. 11.7

(1) Consider carnot Cycle 1' – 2 – 3 – 4'

$$h_{1'} = h_f \text{ at 20 bar} = 908.6 \frac{\text{kJ}}{\text{kg}}$$

$$s_{1'} = s_f = 2.447 \frac{\text{kJ}}{\text{kg-K}}$$

and
$$h_2 = h_g \text{ at 20 bar} = 2797.2 \frac{\text{kJ}}{\text{kg}}$$

$$s_2 = s_g \text{ at 20 bar} = 6.337 \frac{\text{kJ}}{\text{kg-K}}$$

For Isentropic Process 2–3

$$s_2 = s_3$$
$$s_2 = s_{f3} + x_3 \cdot s_{fg3}$$
$$6.337 = 0.649 + x_3 \times 7.502$$
$$x_3 = 0.75819 \cong 0.76$$
$$\therefore \quad h_3 = h_{f3} + x_3 . s_{fg3} = 191.8 + 0.76 \times 2392.9$$
$$h_3 = 2006.09 \text{ kJ/kg}$$

Now for Isentropic Compression process 4' – 1'

$$s_{1'} = s_{4'} = s_{f4'} + x_{4'} \times s_{fg4'}$$

$$2.447 = 0.649 + x_{4'}\, 7.502 \text{ at } 0.1 \text{ bar}$$

$$0.2397 = x_{4'}$$

$$h_{4'} = h_{f_{4'}} + x_{4'} \times h_{fg_{4'}}$$

$$h_{4'} = 191.8 + 0.2397 \times 2392.9$$

$$h_{4'} = 765.305 \text{ kJ/kg}$$

Now
$$\eta_{carnot} = \frac{W_s}{q_i} = \frac{W_T - W_P}{q_i}$$

$$\eta_{carnot} = \frac{[h_2 - h_3] - [h_{1'} - h_{4'}]}{[h_2 - h_{1'}]}$$

$$= \frac{(2797.2 - 2006.09) - (908.6 - 765.3)}{2797.2 - 908.6}$$

$$= \frac{797.11 - 143.3}{1888.6} = 034618$$

$$\boldsymbol{\eta_{carnot} = 34.618\ \%}$$

And Steam Rate
$$= \frac{3600}{W_S} = \frac{3600}{W_T - W_P}$$

$$= \frac{3600}{[h_2 - h_3] - [h_{1'} - h'_{4'}]}$$

$$\text{S.R} = \frac{3600}{653.81} = 5.506 \text{ kg/kWh}$$

(II) Now Consider Rankine Cycle = 1 – 2 – 3 – 4

$$h_4 = h_f \text{ at } 0.1 \text{ bar} = 191.8 \frac{\text{kJ}}{\text{kg}}$$

From Fig. Ex. 11.7 h_2 and h_3 will remain same.

i.e.
$$h_2 = 2797.2 \text{ kJ/kg}$$
$$h_3 = 2006.09 \text{ kJ/kg}$$

Now to find $h_{1'}$, we also know that,

$$W_P = \frac{P_1 - P_b}{10} = \frac{20 - 0.1}{10}$$

$$W_P = 1.99 \text{ kJ/kg} = h_1 - h_4$$

$$\therefore \quad h_1 = 1.99 + 191.8$$

$$h_1 = 193.79 \text{ kJ/kg}$$

$$\text{Now, } \eta_g = \frac{W_s}{q_i} = \frac{W_T - W_P}{q_i}$$

$$\eta_R = \frac{[h_2 - h_3] - W_P}{[h_2 - h_1]}$$

$$= \frac{(2797.2-2006.09)-1.99}{(2797.2-193.79)}$$

$$= \frac{791.11-1.99}{2603.41} = \frac{789.12(\text{i.e. } W_s)}{2603.41}$$

$$= 0.30311$$

$$\eta_R = 30.311\,\%$$

and Steam Rate $= \dfrac{3600}{W_s} = \dfrac{3600}{[h_2 - h_3] - W_p} = \dfrac{3600}{789.12}$

S.R = 4.562 kg/kWh

Example 11.8 *A steam plant operating on Rankine cycle generates superheated steam at 10 bar and 380°C. Condensation occurs at 0.06 bar. Calculate the thermal efficiency of the plant and compare this with the thermal efficiency of Carnot cycle working between the same temperature limits.*

Find: $\eta_R = ?$ $\eta_{carnot} = ?$

Solution

Refer Fig. Ex. 11.8.

(I) Rankine Cycle (1 – 2 – 3 – 4)

From steam tables, at 0.06 bar,

$$h_4 = h_f = 151.5 \text{ kJ/kg}$$

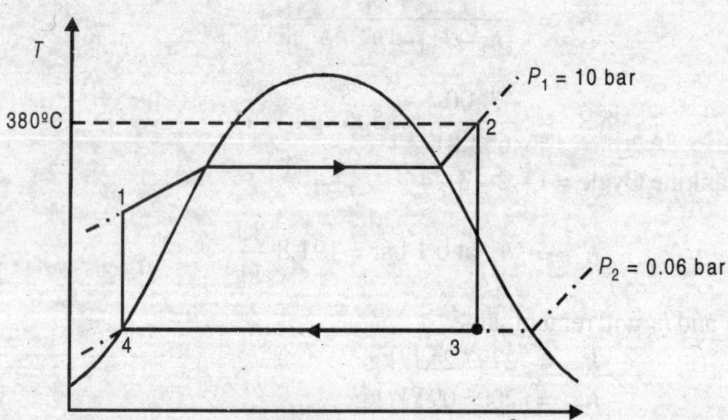

Fig. Ex. 11.8

Also Pump work

$$W_P = \frac{P_1 - P_b}{10} = \frac{10-0.06}{10}$$

$$W_P = 0.994 \text{ kJ/kg} = h_1 - h_4$$

$$0.994 + 151.5 = h_1$$

$$h_1 = 152.494 \text{ kJ/kg}$$

And h_2 at 10 bar and 380°C from superheated steam tables = 3222.04 kJ/kg by interpolation.

To find h_3; from the Isentropic process 2–3

$$s_2 = s_3$$

$$s_2 = s_{f3} + x_3 s_{fg3} \quad ...(A)$$

Now s_2 at 10 bar and 380°C = 7.4.014 kJ/kg-K

Now from Eq. (A),

$$7.4014 = 0.521 + x_3 \times 7.81 \text{ at } 0.06 \text{ bar}$$

$$\frac{7.4014 - 0.521}{7.81} = x_3$$

$$0.8809 = x_3$$

Now

$$h_3 = h_{f3} + x_3 \,.\, h_{fg3} \text{ at } 0.06 \text{ bar}$$

$$= 151.5 + 0.8809 \times 2416$$

$$h_3 = 2279.93 \text{ kJ/kg}$$

Now, (i)

$$\eta_R = \frac{W_S}{q_i}$$

$$\eta_R = \frac{W_T - W_p}{q_i}$$

$$\eta_R = \frac{[h_2 - h_3] - W_p}{[h_2 - h_1]}$$

$$\eta_R = \frac{(3222.04 - 2279.93) - (0.994)}{(3222.04 - 152.494)}$$

$$\eta_R = \frac{(942.108 - 0.994)}{3069.546}$$

$$\eta_R = \frac{941.1149}{3069.546}$$

$$\eta_R = 0.30659.$$

$$\boldsymbol{\eta_R = 30.659\ \%}$$

(II) Carnot Cycle 1'–2–3–4'

From steam tables at 10 bar

$$h_{1'} = h_f = 762.6 \text{ kJ/kg}$$

$$s_1 = s_f = 2.138 \text{ kJ/kg-K}$$

And $h_2 = 3222.04$ and $h_3 = 2279.93$

These will remain same.

Now for Isentropic process 4'–1'

$$s_{1'} = s_{4'} = s_{f4'} + x_{4'} \times s_{fg4'}$$

$$2.138 = 0.521 + x \times 7.81 \text{ at } 0.06 \text{ bar}$$

$$\frac{2.138 - 0.521}{7.81} = x_{4'}$$

$$0.2070 = x_{4'}$$

$$h_{4'} = h_{f_{4'}} + x_{4'} \times h_{fg_{4'}} \text{ at } 0.06 \text{ bar}$$

$$= 151.5 + 0.2070 \times 2416$$

$$h_{4'} = 651.714 \text{ kJ/kg}$$

Now,

$$\eta_{\text{carnot}} = \frac{W_s}{q_i} = \frac{W_T - W_P}{q_i}$$

$$= \frac{[h_2 - h_3] - [h_{1'} - h_{4'}]}{h_2 - h_{1'}}$$

$$= \frac{(3222.04 - 2279.93) - (762.6 - 651.714)}{3222.04 - 762.6}$$

$$= \frac{(942.11) - (110.886)}{(2459.44)}$$

$$= 0.33797$$

$$\eta_{\text{carnot}} = \mathbf{33.797\%}$$

Hence Carnot Cycle efficiency is more than Rankine cycle for an engine working between same temperature limits.

THEORY QUESTIONS

11.1. Derive the expression for efficiency of Carnot cycle.

11.2. Draw Carrot cycle on P–V and T–S diagram.

11.3. Explain Rankine cycle with P–V and T–S diagram.

11.4. Derive an expression for efficiency of Rankine cycle.

11.5. Compare Carnot cycle with Rankine cycle.

11.6. Define the terms : (i) Work ratio (ii) SSC.

11.7. Write a short note on factors affecting performance of Rankine cycle.

11.8. Explain why Rankine cycle and not Carnot cycle is considered as the thermodynamic cycle for the steam power plant.

11.9. Discuss the effects of the following parameters on the performance of Rankine cycle:

(i) Steam pressure at turbine inlet.

(ii) Steam temperature at turbine inlet.

(iii) Steam pressure at turbine exhaust.

11.10. Draw a neat sketch of Rankine cycle of steam plant, label the parts and describe the working of each component, giving the formula for energy exchange for the same.

11.11. Enumerate the advantages of Rankine cycle over the Cannot cycle for steam power plant.

11.12. Explain the drawbacks of Cannot cycle for a steam power plant. How are they overcome in a Rankine cycle ?

11.13. In what respect does the Rankine cycle differ from the Carnot cycle. Draw Carnot and Rankine cycle on T–S plane.

11.14. Enumerate the advantages of Rankine cycle over the Carnot cycle for steam power plant.

11.15. Why is Carnot cycle not practicable for a steam power plant.

11.16. Draw a neat sketch of Rankine cycle steam plant, label the part and describe the working of each component, giving formula for energy exchange for the same.

11.17. Explain clearly how the following affects the performance of a Rankine cycle.

(i) Lowering the condensation pressure.

(ii) Superheating the steam to high temperature.

(iii) Increasing the boiler pressure.

11.18. Parameters affecting the performance of Rankine cycle.

11.19. Explain the factors affecting the efficiency of Rankine cycle.

11.20. Explain with *T–S* diagram, factors affecting performance of Rankine cycle.

11.21. Explain the effect of following on the efficiency of Rankine cycle.

(i) Increasing the average temperature at which heat is supplied.

(ii) Decreasing the temperature at which heat is rejected.

11.22. Show the Rankine cycle on

(i) Wet

(ii) Dry saturated

(iii) Superheated

PROBLEMS FOR PRACTICE

11.1. A steam power plant operating on Rankine cycle recieves steam from boiler at 35 bar and 350°C. It is exhausted to condenser at 0.1 bar, calculate

(i) Energy supplied/kg of steam generated in boiler

(ii) Quality of steam entering the condenser

(iii) Rankine cycle efficiency considering feed pump work.

[**Ans.** (i) 2911.16 kJ/kg, (ii) 0.8016, (iii) 34.226%]

11.2. Consider a steam power plant operating on the ideal Rankine cycle. The steam enters the turbine at 3 MPa and 350°C and is condensed in the condenser at a pressure of 10 kPa. Determine

(i) The thermal efficiency of this power plant.

(ii) The thermal efficiency if the steam is superheated to 600°C instead of 350°C.

(iii) The thermal efficiency if the boiler pressure is raised to 15 MPa while the turbine inlet temperature is maintained at 600° C.

CHAPTER 12

Improved Rankine Cycle

12.1 INTRODUCTION

Carnot cycle is the most efficient thermodynamic cycle for all the heat engines utilising the heat energy and producing mechanical energy. We have also observed that there are limitations for the use of this cycle in practice.

The ideal practical cycles for IC Engines are Otto cycle (for Petrol and gas engines) and Diesel cycle (for Diesel or oil engines). Similarly the ideal practical cycle for steam—is the Rankine Cycle.

12.2 DIFFERENCE IN CARNOT AND RANKINE CYCLE

Figure 12.1 shows saturated cycle for both Carnot and Rankine cycle. Carnot cycle is 1'–2–3–4' and Rankine cycle 1–2–3–4.

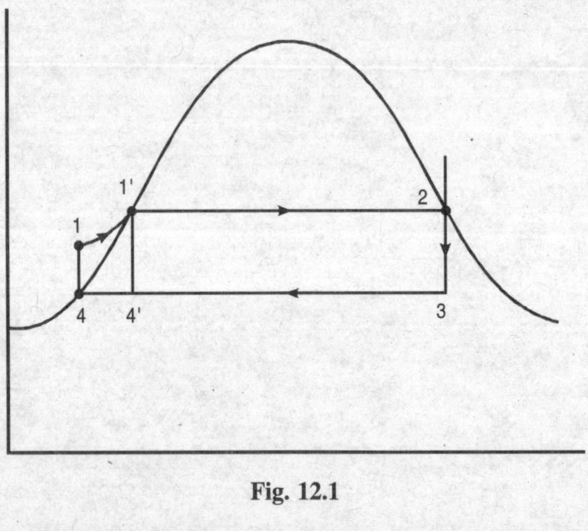

Fig. 12.1

Comparing these two cycles we observe that all the processes except water heating 1–1' are reversible processes. Rankine cycle efficiency is less than Carnot efficiency because of only one irreversible process 1–1'. If we could eliminate this irreversibility, we can get the Rankine cycle efficiency same as that of Carnot efficiency.

To eliminate this irreversible process, we have to heat water from 1–1' by some internal sources only. This type of the heating of water within the cycle is called Regenerative heating of water.

12.3 REGENERATIVE FEED HEATING CYCLE

Regenerative feed water heating can be achieved with the help of steam extracted or bled from the turbine, in the device called feed water heaters.

This irreversibility can be eliminated by the process of regeneration i.e., by internal and reversible heat exchange between the steam expanding in the turbine and water in the economizer. Figure 12.2 shows a flow diagram and a *T-s* diagram of the process for a saturated Rankine cycle.

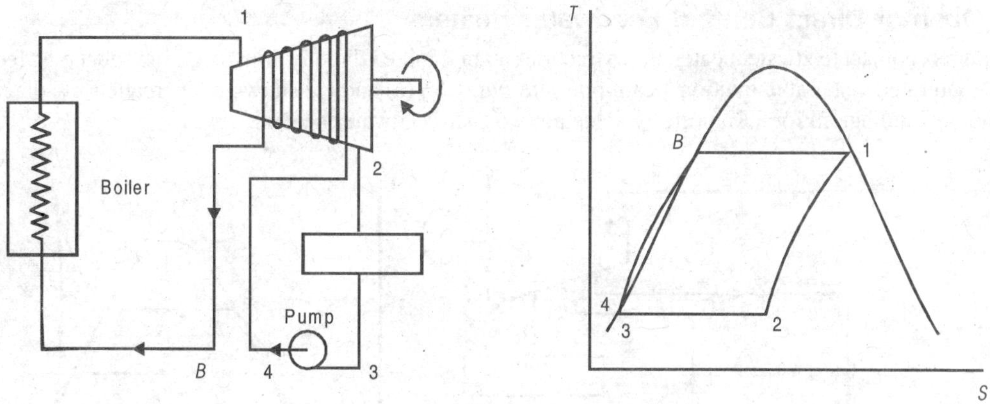

Fig. 12.2. *Ideal regeneration of rankine cycle.*

The feed water after the condenser is carefully passed over the turbine, where it receives heat from the expanding steam reversibly at all times (i.e., with zero temperature difference). The water is progressively heated until it enters the steam generator at point *B*. The economizer and the irreversibilities introduced along with it, are thus eliminated. The cycle would receive and reject heat at constant temperature and would have same efficiency as that of Carnot cycle, working between same temperature limits.

This heating of feed water is called *ideal regenerative feed water heating*. It is however not employed as it has some serious limitations. They are as follows:

Limitations of Ideal Regeneration

1. Water is flowing outside the turbine in tubes while the steam is expanding over turbine blades placed inside. In this situation, no adequate surface is available for effective heat transfer.
2. The mass flow rates of water and steam are very large. Hence effectiveness of heat exchange is very low.
3. The final dryness of steam leaving the turbine is very low and amount of moisture content is unacceptable for proper turbine operation and efficiency.

12.4 PRACTICAL–REGENERATIVE FEED WATER HEATING

In practical method of feed water heating, economizer is not eliminated totally but its size is reduced. The steam expands adiabatically through the turbine as usual. The feed water at 4 (Refer Fig. 12.2) is heated in steps and not continuously, by the steam bled from the turbine at selected stages. Heating of liquid takes place in heat exchangers, called *feed water heaters*.

Feedwater heating was introduced in early 1920's and since then has become an important and essential component of large thermal power plants, which use between five to eight such feed water heaters.

Feedwater heating is done in steps and not continuously. The feedwater therefore enters the boiler below point *B*. Also, feedwater heaters themselves are heat exchangers (They are practically very much like condensers) and have their own irreversibilities and losses. Thus we cannot attain an ideal situation of Fig. 12.2 and Rankine cycle cannot attain Carnot efficiency. A well designed Rankine cycle is however a closest practical cycle to Carnot and hence is widely accepted for most power plants.

12.5 TYPES OF FEEDWATER HEATERS

There are three types of feedwater heaters in use. They are,

1. Open or direct contact type
2. Closed type with drains cascaded backward
3. Closed type with drains pumped forward

These types are now discussed and analysed in detail.

12.5.1 Open or Direct Contact Feedwater Heaters

In open or direct contact feedwater heater, the extraction steam is mixed directly with incoming subcooled feedwater; to produce saturated water at extraction steam pressure Fig. 12.3 (a) and (b) shows a schematic flow diagram and corresponding *T-s* diagram for a Rankine cycle using two such feedwater heaters.

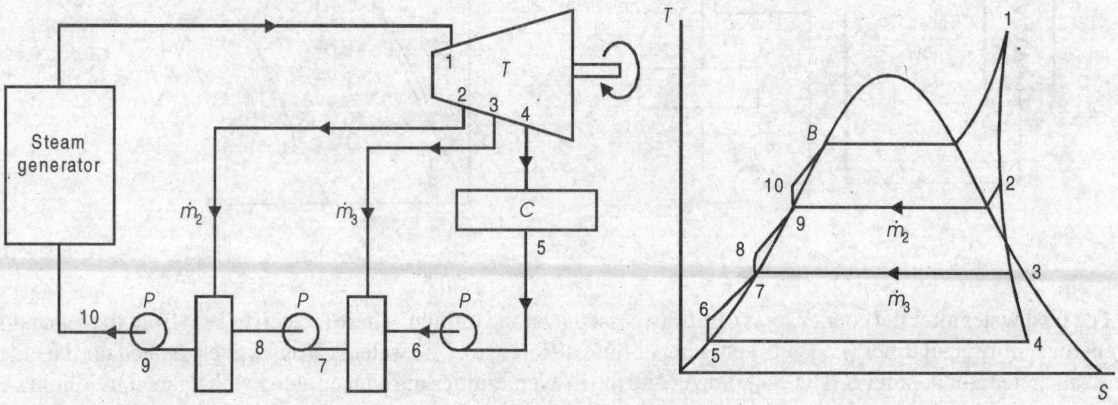

Fig. 12.3. *Schematic flow and T-s diagrams of a non-ideal superheat rankine cycle with two open-type feedwater heaters.*

Condensate water leaves the condenser at 5 and is pumped to 6 to a pressure equal to that of the extraction steam at 3. The now-subcooled water at 6 and wet steam at 3 mix in the low pressure feed water heater to produce saturated water at 7.

Amount of bled steam $\dot{m}_3$ should essentially be such that subcooled water at 6, should be raised close to its saturation point. If it were much less, the water will be heated to below its saturation temperature and advantage of feedwater heating would be partially lost. If it were more there will be surplus steam at point 7. This would mean loss of turbine work and presence of steam-water mixture at 7 which is difficult to pump.

The pressure at 6-7 should be less than extraction steam pressure at 3 or else reverse flow of condensate water would enter the turbine at 3. The second pump pressurises saturated water at 7 to a subcooled condition at 8, which is at the pressure of extraction steam at 2. In the high pressure feed water heater superheated steam at 2, mixes with subcooled water at 8 to produce saturated water at 9. This is pressurised to 10 by a pump to enter the steam generator at its working pressure.

Steam at 2 or 3 loses its latent heat of vaporisation which is much larger than the gain of sensible heat by water at 8 or 6. Therefore the quantity of extracted steam $\dot{m}_2$ and $\dot{m}_3$ is only a small fraction of the steam passing through the turbine.

In addition to a condensate pump, one pump for each open feedwater heater is required. This is a disadvantage.

Open feed water heaters because of vigorous action of mixing remove non-condensible gases like Air, O_2, H_2, CO_2 from the feed water. They are therefore known as *deareating heaters* or *DA*.

12.5.2 Mass Flow Rates and Energy Balance

This is done on the basis of unit mass flow rate at the turbine inlet (point 1). The mass balance therefore is as follows:

$$\begin{aligned}
\text{Mass flow between points 1 and 2} &= 1 \text{ kg/s}\\
\text{Mass flow between points 2 and 9} &= \dot{m}_2\\
\text{Mass flow between points 2 and 3} &= 1-\dot{m}_2\\
\text{Mass flow between points 3 and 7} &= \dot{m}_3\\
\text{Mass flow between points 3 and 4} &= 1-\dot{m}_2-\dot{m}_3\\
\text{Mass flow between points 4 and 7} &= 1-\dot{m}_2-\dot{m}_3\\
\text{Mass flow between points 6 and 7} &= 1-\dot{m}_2-\dot{m}_3\\
\text{Mass flow between points 7 and 9} &= 1-\dot{m}_2\\
\text{Mass flow between points 9 and 1} &= 1
\end{aligned} \quad \text{...(12.1)}$$

where $\dot{m}_2$ and $\dot{m}_3$ are fraction of 1.

For energy balance, consider the low and high pressure feed water heaters. At low pressure feed water heater, heat lost by condensing steam equals heat gained by feed water heater, or,

$$\dot{m}_2(h_2-h_9) = (1-\dot{m}_2)(h_9-h_8) \quad \text{...(12.2)}$$

Similarly for high pressure feed water heaters,

$$\dot{m}_3(h_3-h_7) = (1-\dot{m}_2-\dot{m}_3)(h_7-h_6) \quad \text{...(12.3)}$$

where h_2, h_3, h_6, h_7, h_3, h_9 etc. are specific enthalpies at respective points.

Since admission, extraction and condenser pressures are known, all the enthalpy values are known. For two Eqs. (12.2) and (12.3), therefore, there are only two unknowns $\dot{m}_2$ and $\dot{m}_3$. For any number of feedwater heaters there will be as many equations as there are unknowns, so solutions are always possible.

The Cycle Analysis

Based on the mass flow rates, cycle analysis can now proceed as follows:

1. Heat added $\quad q_s = v(h_1-h_{10})$ kJ/kg

2. Turbine work $\quad W_T = 1(h_1-h_2)+(1-\dot{m}_2)(h_2-h_3)+(1-\dot{m}_2-\dot{m}_3)(h_3-h_4)$ kJ/kg

3. Pump work

$$\left|\sum W_P\right| = (1-\dot{m}_2-\dot{m}_3)(h_6-h_5)+(1-\dot{m}_2)(h_8-h_7)+1(h_{10}-h_9) \text{ kJ/kg} \quad \text{...(12.4)}$$

$$= (1-\dot{m}_2-\dot{m}_3)v_5(p_6-p_5)+(1-\dot{m}_2)v_7(p_8-p_7)+1\times v_9(p_{10}-p_9) \text{ kJ/kg}$$

4. Heat rejected $\quad |q_r| = (1-m_2-m_3)(h_4-h_5)$ kJ/kg

5. Net cycle work $\quad = \Delta W_{|net|} = |W_T - W_P|$ kJ/kg

6. Cycle thermal efficiency $= \eta_{th} = \dfrac{\Delta W_{net}}{q_s}$

7. $\quad \text{W.R.} = \dfrac{\Delta W_{net}}{W_T}$

8. Specific steam consumption, SSC $= \dfrac{3600}{\Delta W_{net}}$ kg/k Wh

If turbine and pump efficiencies are considered then,

$$W_T\text{. actual} = \eta_{th} \times W_T$$

where η_t = turbine efficiency and η_p = pump efficiency

and $$W_P\text{. actual} = \frac{W_P}{\eta_P}$$

12.5.3 Closed Type Feed Water Heaters with Drains Cascaded Backward

This is the simplest and most commonly used type in power plants. It is a shell and tube type heat exchanger. The feedwater passes through the tubes, and the bled steam, on the shell side transfers its heat energy to it and condenses. Thus, they are like surface condensers, operating at higher pressures than those in the main condenser. Since feedwater does not mix with the steam and passes through the tubes of successive close feed water heaters, only one condensate pump is sufficient to pressurise the water from condenser to the boiler. However, to reduce the pressure rise in one pump, one condensate pump before the feed water heaters and one feed water pump after the feed water heaters are used. If deareater heater is used, one boiler feed pump is automatically required after the same.

Figure 12.4 shows a simplified flow diagram and corresponding *T-s* diagram of a non-ideal superheat Rankine cycle. There are two feedwater heaters. One pump 5–6 pressurises the condensate, sufficient to pass through the two feedwater heaters and enters the steam generator at 8.

The bled steam condensed at each fwh is fed back to the next lower pressure fwh. The condensate from the last fwh is led back to the main condenser. One can imagine, then, a cascade from high pressure to lower-pressure heaters. Hence, the name.

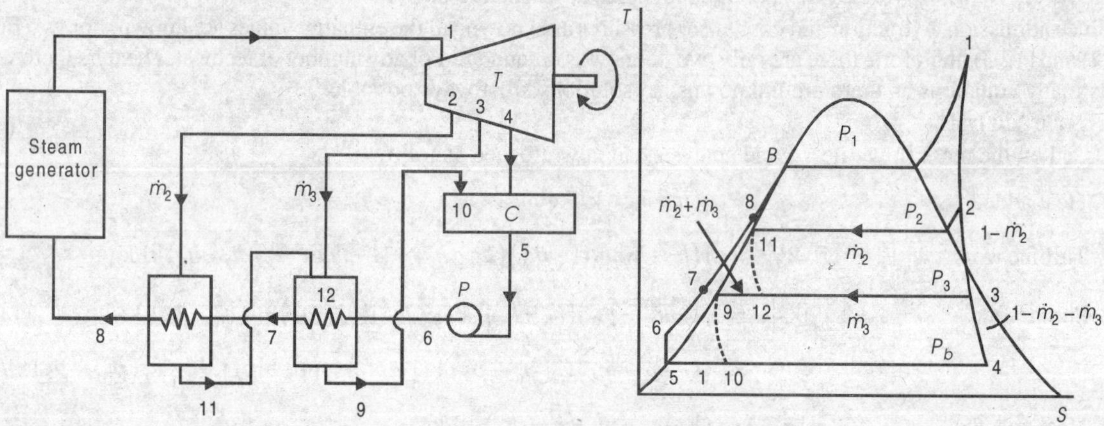

Fig. 12.4. *Schematic flow and T-s diagrams of a non-ideal superheat rankine cycle with two closed-type feed water heaters with drains cascaded backward.*

12.5.4 Operation

Wet steam bled at 3 at the rate of $\dot{m}_3$ enters the low pressure fwh and heats the subcooled water at 6. The temperature of water at 7 is less than saturation temperature at 3. This difference is called the *terminal temperature difference* (TTD) and is defined for all closed fwhs.

TTD or TD = saturation temperature of bled steam – exit water temperature

The value of TTD for low pressure heaters is positive and is of the order of 3°C. For high pressure heaters it may be negative and lies between 0 and – 3°C. In case no data is available, temperature at 7 is taken equal to saturation temperature at 3.

The drain from low pressure fwh is led to the main condenser and enters it as two phase mixture at 10. The process 9 – 10 is throttling process from the pressure corresponding to 9 and the main condenser pressure. Due to throttling there is some loss of availability. Also enthalpy at 9 and 10 are equal, as the process is constant enthalpy one.

The high pressure fwh receives superheated steam bled from the turbine at 2 that flows on the shell side at the rate of $\dot{m}_2$ and heats the water entering the tubes at 7. The inlet steam at 2 is superheated, hence temperature at 8 may be higher than saturation temperature of that steam and TTD may be negative. The drain at 11 may be slightly subcooled; and is throttled to the low pressure fwh which it enters at 12 as a two phase mixture. There it joins with the steam bled at 3 and aids the heating of water in low pressure fwh. The drain from low pressure fwh is thus $\dot{m}_2 + \dot{m}_3$, which is throttled to the main condenser at 10. The higher pressure exit water at 8 is led to the steam generator.

12.5.5 Mass Flow Rate and Energy Balance

A mass balance based on unit flow rate at turbine inlet point 1, is given clockwise by

$$\begin{aligned} \text{Mass flow between 1 and 2} &= 1 \\ \text{2 and 3} &= 1-\dot{m}_2 \\ \text{3 and 10} &= 1-\dot{m}_2-\dot{m}_3 \\ \text{10 and 1} &= 1 \\ \text{2 and 12} &= \dot{m}_2 \\ \text{3 and 12} &= \dot{m}_3 \\ \text{12 and 10} &= \dot{m}_2+\dot{m}_3 \end{aligned} \quad \text{...(12.5)}$$

The energy balances on the high and low pressure heaters are given respectively by,

$$\dot{m}_2\,(h_2 - h_{11}) = h_8 - h_7 \quad \text{...(12.6)}$$

$$\dot{m}_3\,(h_3 - h_9) + \dot{m}_2\,(h_{12} - h_9) = h_7 - h_6 \quad \text{...(12.7)}$$

For throttling processes, we have, $h_{12} = h_{12}$ and $h_{10} = h_9$.

All the relevant pressures and the enthalpies of the Eqs. (12.6) and (12.7) are known. Since there are only two unknowns $\dot{m}_2$ and $\dot{m}_3$, the equations can be solved.

12.5.6 Cycle Analysis

From mass flow rates, energy balance equations and enthalpy values, we can determine the amounts of steam extracted i.e. $\dot{m}_2$ and $\dot{m}_3$ and then the pertinent cycle parameters. We assume mass flow rate of 1 kg/s at turbine inlet (point 1).

1. Heat added $q_s = h_1 - h_8$
2. Heat rejected $q_r = (1-\dot{m}_2-\dot{m}_3)\,(h_4 - h_5) + (\dot{m}_2 + \dot{m}_3)\,(h_{10} - h_5)$
3. Turbine work $W_T = (h_1 - h_2) + (1 - \dot{m}_2)\,(h_2 - h_3) + (1 - \dot{m}_2 - \dot{m}_3)\,(h_3 - h_4)$
4. Pump work $|W_p| = h_6 - h_6 = v_5\,(p_6 - p_5)$
5. Net cycle work $\Delta W_{net} = W_T - |W_P|$ ...(12.8)
6. Cycle thermal efficiency $= \eta_{th} = \dfrac{\Delta W_{net}}{q_s}$
7. Work ratio = W. R. $= \dfrac{\Delta W_{net}}{W_T}$
8. Specific steam consumption, SSC $= \dfrac{3600}{\Delta W_{net}}$

12.5.7 Closed Type Feed Water Heaters with Drains Pumped Forward

This is a second closed type fwh. It avoids throttling and the irreversibilities therein but has to use a small additional pump. Like previous closed type fwh, it is also a shell and tube type heat exchanger. The feedwater passes through the tubes, the steam passes over them; transfers heat and condenses. They do not mix and only one pump is sufficient to carry water from the condenser to the boiler. If, however, *DA* is used one more pump would be necessary.

The drain from these fwhs is pumped forward into the main feedwater line. Figure 12.5 shows a simplified flow diagram and corresponding *T-s* diagram for a non-ideal superheat Rankine cycle with two such fwhs. Like open fwh these heaters also require an additional pump per heater. But the pump is small and carry only fractional flows corresponding to $\dot{m}_2$ and $\dot{m}_3$ while pumps in open fwh carry full steam and water flow.

12.5.8 Operation

The drain of low pressure fwh at 13 is pumped forward to the main feedwater line and enters it at 14. It mixes with exit water from l.p. heater at 7 and forms mixture at 8. Point 8 is closer to 7 than 1.4 on the *T-S* diagram because flow at 7 is much greater than the drain flow $\dot{m}_3$.

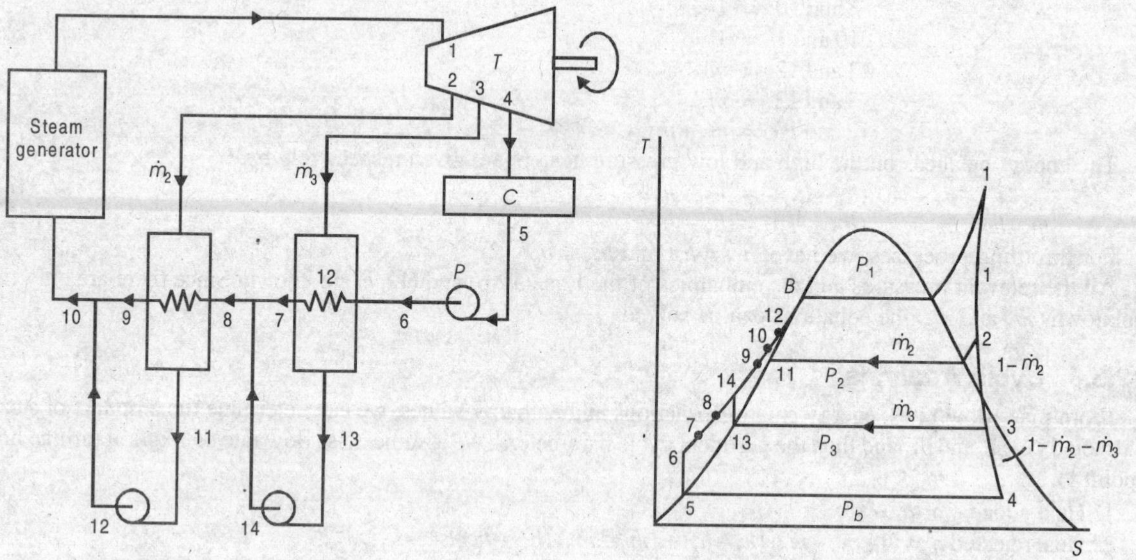

Fig. 12.5. *Schematic flow and T-s diagrams of non-ideal superheat Rankine cycle with two closed-type feedwater heaters with drains pumped forward.*

The water at 8 enters h.p. heater and is heated to 9. The drain leaves the heater at 11, is pumped to 12. It mixes with the feedwater at 9 to form full feedwater flow at 10, which crow goes to boiler.

12.5.9 Mass Balance and Energy Balance

A mass balance based on a mass flow rate of 1 kg/s at turbine inlet (point 1) is given clockwise on *T-S* diagram by,

$$\begin{aligned} \text{Mass flow between 1 and 2} &= 1 \\ \text{2 and 12} &= \dot{m}_2 \\ \text{2 and 3} &= 1-\dot{m}_2 \\ \text{3 and 14} &= \dot{m}_3 \\ \text{3 and 7} &= 1-\dot{m}_2-\dot{m}_3 \\ \text{at 14} &= \dot{m}_3 \end{aligned} \qquad \text{...(12.9)}$$

$$8 \text{ and } 9 = 1-\dot{m}_2$$
$$\text{at } 12 = \dot{m}_2$$
$$10 \text{ and } 1 = 1$$

The energy balance equations are,

h.p. heater $\dot{m}_2 (h_2 - h_{11}) = (1-\dot{m}_2)(h_9 - h_8)$ ...(12.10)

and l.p. heater $\dot{m}_3 (h_3 - h_{13}) = (1-\dot{m}_2-\dot{m}_3)(h_7 - h_6)$ ...(12.11)

The values of h_9 and h_7 are obtained from the temperatures t_9 and t_7 which are equal to the saturation temperature of the steam at each heater minus its terminal temperature difference, or

$$t_9 = t_{11} - \text{TTD} \quad \text{h. p. heater} \qquad \text{...(12.12a)}$$

and
$$t_7 = t_{13} - \text{TTD} \quad \text{I. P. heater} \qquad \text{...(12.12b)}$$

h_{10} needed for q_A and h_8 to be used in Eq. (12.10) are obtained from h_{11} and h_{14} respectively, or

$$h_{12} = h_{11} + v_{11}\frac{p_{12}-p_{11}}{\eta_p} \qquad \text{...(12.13a)}$$

and
$$h_{14} = h_{13} + v_{13}\frac{p_{14}-p_{13}}{\eta_P} \qquad \text{...(12.13b)}$$

where η_P is pump efficiency.

Thus

$$h_{10} = m_2 h_{12} + (1-m_2) h_9 \qquad \text{...(12.14a)}$$

and
$$(1-m_2) h_8 = m_3 h_{14} + (1-m_2-m_3) h_7 \qquad \text{...(12.14b)}$$

12.5.10 Cycle Analysis

1. Heat added, $q_s = h_1 - h_{10}$
2. The turbine work, $W_T = (h_1 - h_2) + (1-\dot{m}_2)(h_2 - h_3) + (1-\dot{m}_2-\dot{m}_3)(h_3 - h_4)$
3. Pump work, $W_P = (1-\dot{m}_2-\dot{m}_3)(h_6 - h_5) + \dot{m}_3 (h_{14} - h_{13}) + \dot{m}_2 (h_{12} - h_{11})$
4. Net work $\Delta W_{\text{net}} = |W_T - W_P|$ ...(12.15)
5. Thermal Efiiciency, $\eta_{\text{th}} = \dfrac{\Delta W_{\text{net}}}{q_s}$
6. Work Ratio W. R. $= \dfrac{\Delta W_{\text{net}}}{W_T}$
7. Specific steam consumption, SSC $= \dfrac{3600}{\Delta W_{\text{net}}}$

12.6 COMPARISON OF CLOSED FEED WATER HEATERS

Table 21.1 comparison of closed feed water heaters.

Drains cascaded backward	*Drains pumped forward*
1. There is a loss of availability due to throttling.	1. No throttling, no loss of availability.
2. Number of pumps are less-one or two.	2. Number of pumps are more; one for each 3. feedwater heater.
3. Simple and less expensive system.	4. Complex and expensive system.
4. Drain is slightly subcooled. This imparts more energy to the water and reduces loss of availability.	5. Is used as the lowest pressure feed water heater in an otherwise all cascaded system. This prevents throttling of the combined cascaded flows to the condenser.

12.7 THE PLACEMENT OF FEEDWATER HEATERS

When number of fwhs are used, a natural question arises, as to where to place these fwhs in the cycle or what are the pressures at which steam is to be bled from the turbine, that will result in the maximum increase in the efficiency?

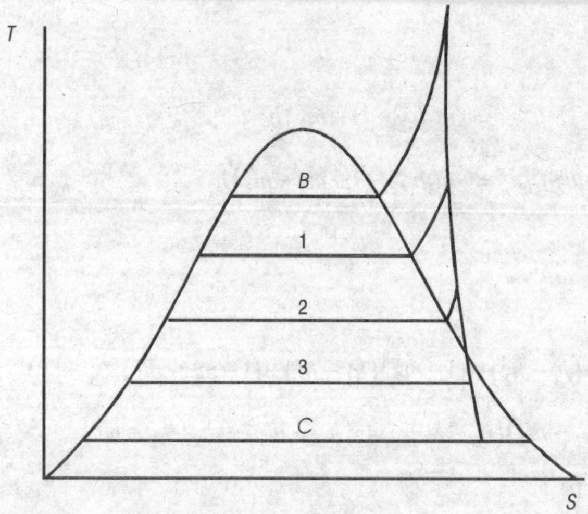

Fig. 12.6. *One feedwater heater in three possible positions.*

To answer this let us consider for the sake of simplicity that only one fwh is used. The aim of using any feedwater heater is to heat the feedwater as close to the boiler temperature as possible. We may now place the feedwater heater at position 1, 2 or 3. (Fig. 12.6) T_B and T_C are saturation temperatures at boiler and condenser pressures respectively. Now at position 1 the heat transfers are because of temperature differences $T_B - T_1$ and $T_1 - T_C$. At position 3 the heat transfers are because of $T_B - T_3$ and $T_3 - T_C$. In either case one of the temperature differences is very large. To minimise the temperature differences, we have to select position 2 which is in the middle i.e. $T_B - T_2 = T_2 - T_C$. Thus we find the temperature which is *halfway between* T_B and T_C and then obtain the saturation pressure at that temperature. Feed water placed at this pressure gives us optimum position of maximum efficiency. Note that the temperature at which steam is actually bled from the turbine may be superheated.

If two fwhs are used, the temperature range $T_B - T_C$ is divided into three equal parts and for three fwhs it is divided into 4 equal parts.

Thus, in general, for *n* fwhs the optimum temperature rise per heater would be given by

$$\Delta T_{opt} = \frac{T_B - T_C}{n+1} \quad ...(12.14)$$

Consider for example a cycle where boiler pressure is 100 bar and condenser pressure is 0.08 bar and three fwhs are to be used. We find the pressures at which they are placed. From steam tables we get saturation temperatures at 100 bar and 0.08 bar. They are $T_B = 310.96°C$ and $T_C = 41.534°C$ respectively. From above Eq. (12.14) we have,

$$\Delta T_{opt} = \frac{310.96 - 41.534}{4} = 67.36$$

Thus the saturation temperatures of fwhs will be found by adding ΔT_{opt} successively from $T_C = 41.534$ onwards and then finding the corresponding saturation pressures from the steam tables. Table 12.1 summarises the results.

Table 12.2

Component	*Saturation temperature*	*Saturation pressures at saturated temperature*
The boiler	310.96°C	100 bar
The high pressure fwh	240.60°C	34 bar
The intermediate, pressure fwh	176.24°C	9.2 bar
The low pressure fwh	108.88°C	1.35 bar
The condenser	41.534°C	0.08 bar

Thus, the fwhs will be placed at 34 bar, 9.2 bar and 1.35 bar respectively.

12.8 IRREVERSIBILITIES IN RANKINE CYCLE

All ideal cycles whether gas cycles or vapour cycles are assumed to be made up of reversible processes only. In the actual cycles, however, irreversibilities are introduced because of factors like heat transfer through finite temperature difference. friction, mixing etc.

In Rankine cycle, we will discuss two kinds of irreversibilities:

1. **External irreversibilities.** Introduced as a result of temperature differences between combustion gases in the furnace and feed water or steam in the boiler tubes and the temperature differences between the condensing steam and condenser cooling water.
2. **Internal irreversibility.** Introduced because of fluid friction and throttling in the turbine and the pump and mixing.

12.8.1 External Irreversibilities

In Fig. 12.4 (a), line *ab* represents combustion gases in a counter-flow furnace and 4 – B – 1 represents water in a saturated Rankine cycle. Line *cd* represents condenser cooling water and 2 – 3 represents condensing steam.

The temperature differences between line *ab* and 4 – B – 1 – 1' and between 2 – 3 and line *cd* are not constant.

Figure 12.7 also shows temperature *vs* heat exchanger path length diagrams for (b) parallel flow and (c) counter flow steam generators and effect of flow directions in the steam generators. *L* – 1 and *e* – *B* represent minimum approach point called *pinch point*; and is finite. For very small pinch point temperature difference, the overall temperature differences are low and hence reversibilities are lower but the size and cost of steam generators are

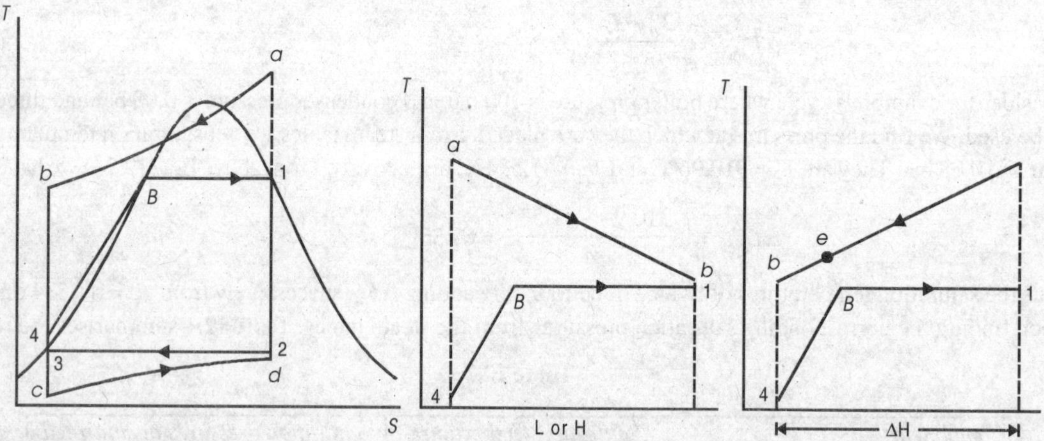

(a) External irreversibility with Rankine cycle

(b) Effect of flow direction on external irreversibility: (i) parallel flow, (ii) counterflow

Fig. 12.7

high. For very large pinch point temperature differences, size and cost of steam generators are low but overall temperature differences and irreversibilities are higher and plant efficiencies are lower. The most economical pinch point temperature difference is obtained by optimization of fixed cost (based on capital costs), and operating costs (based on fuel costs).

Figure 12.7 also shows that overall temperature differences between the combustion gases and water and steam are greater for parallel flow than counter flow heat exchange in steam generators. Heat transfer considerations also favour counter flow as it gives higher overall heat transfer coefficients. The counter flow is thus favoured over parallel flow.

12.8.2 Internal Irreversibilities

They are result of fluid friction, throttling and mixing in turbines, pumps, and pressure losses in pipes, bends, valves etc.

The actual expansion of steam in the turbine is adiabatic but not adiabatic reversible (isentropic) and during expansion entropy increases. Refer Fig. 12.8. The ideal expansion, which is isentropic is 1 – 2s but the actual expansion is 1 – 2.

The pump pressure P_4 has to be more than turbine inlet pressure P_1 because of pressure drop in heat exchangers, feed water heaters, pipes, bends and valves etc. Thus P_5 is the pressure of steam leaving the steam generator, while P_5, is the pressure at turbine inlet. Heat loss in the turbine decreases entropy of steam from 5' to 1.

As a result, dryness and therefore enthalpy at 2 is more than that at 2s. The irreversible losses in the turbine are represented by *turbine isentropic efficiency* and is given by the ratio of turbine actual work to the ideal isentropic work. Hence,

$$\eta_T = \frac{h_1 - h_2}{h_1 - h_{2s}} \qquad ...(12.15)$$

Well designed turbines give efficiencies of around 90%. η_T given above is not to be confused with *cycle thermal efficiency*. It is also overall isentropic efficiency. The individual turbine stages may have different efficiencies.

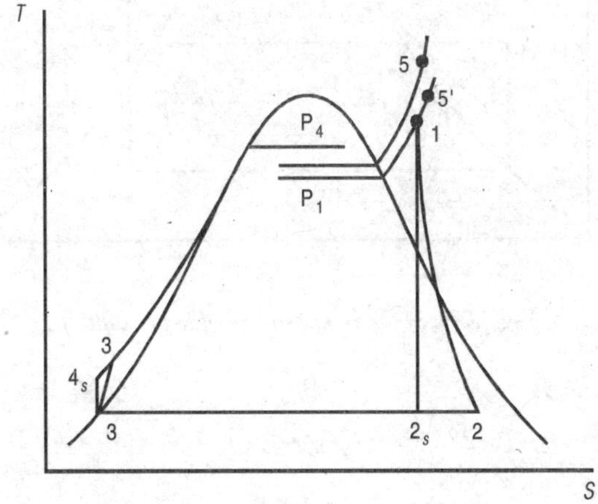

Fig. 12.8. *A T-s diagram of an internally irreversible superheat rankine cycle.*

The compression of water in the pumps is also adiabatic but not isentropic. The entropy increases. The ideal compression is 3 – 4s while the actual one is 3 – 4. The losses are represented by *pump isentropic efficiency* which is the ratio of ideal isentropic work to the actual work taken by the pump. Hence

$$\eta_P = \frac{h_3 - h_{4s}}{h_3 - h_4} \qquad ...(12.16)$$

In both Eqs. (12.14) and (12.16), the smaller quantity is in the numerator. The actual pump work can be obtained by

$$|W_p| = \frac{h_{4s} - h_3}{\eta_P} = \frac{v_3(p_4 - p_3)}{\eta_P}$$

Thus we pay penalty for the turbine and pump irreversibilities. The turbine produces less work and the pump absorbs more work.

12.9 MEAN TEMPERATURE OF HEAT ADDITION

In Rankine cycle, heat is added at constant pressure but not at constant temperature. Refer Fig. 12.9. There is a large difference between temperature of water at the entry of the economiser at point 4 and that of steam at the exit of the superheater at point 1. To find mean temperature of heat addition T_m, construct a rectangle 6–2–3–2–5 on the base 2–3 such that area 6–2–3–5 and area 1–2–3–4–B-1 are equal. Recall a similar construction of a rectangle on indicator diagram of an IC engine to find the mean effective pressure P_m.

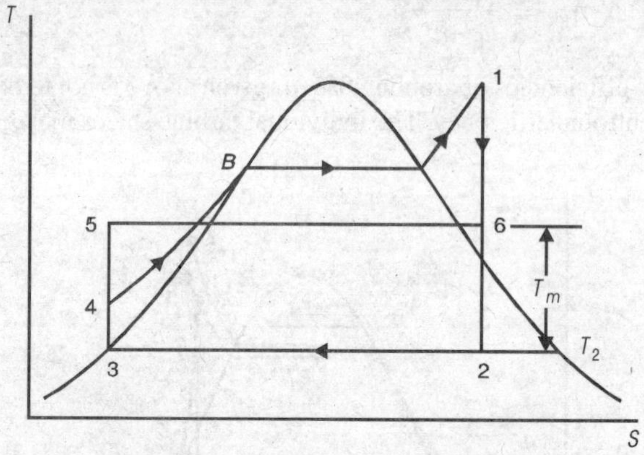

Fig. 12.9. *Mean temperature of heat addition.*

Now heat supplied is,

$$q_s = h_1 - h_4 = T_m(s_1 - s_4)$$

$$\therefore \quad T_m = \frac{q_s}{s_1 - s_4}\,^\circ\text{C} = \frac{h_1 - h_4}{s_1 - s_4}$$

Now heat rejected is

$$q_r = T_2(s_1 - s_4)$$

$$\therefore \quad \eta_{\text{Rankine}} = 1 - \frac{q_r}{q_s} = 1 - \frac{T_2}{T_m}$$

Since T_2 is fixed by condenser pressure for which very little improvement is possible over present practices, the efficiency largely depends on T_m. Larger the value of T_m greater is the cycle efficiency.

It will be realised now that superheating of steam will raise the value of T_m over the saturated cycle and hence will increase the cycle efficiency.

12.10 REHEAT CYCLE

Raising the turbine inlet pressure i.e., boiler outlet pressure or boiler working pressure as well as raising the temperature of superheat improves the thermal efficiency of the Rankine cycle. Higher working pressures and temperatures increase mean temperature T_m of heat addition; and takes T_m closer to gas line *ab* reducing the irreversibilities. Refer Fig. 12.7 (b).

Increasing the inlet pressure from P_1 to P_2 however results in the lowering of the final dryness fraction from 4 to 4' and all the stages after point *d* on the expansion line get progressively wet steam. The moisture present in the steam, in the form of suspended water particles hit the high speed turbine blades with a large relative velocities and erode them.

To improve the thermal efficiency of the cycle over and above that of superheat cycle as well as to improve the final dryness of steam, reheating of steam is used.

Figure 12.10. (a) and (b) show a schematic flow diagram and *T-s* diagram of Rankine cycle that uses superheat and resuperheat or reheat.

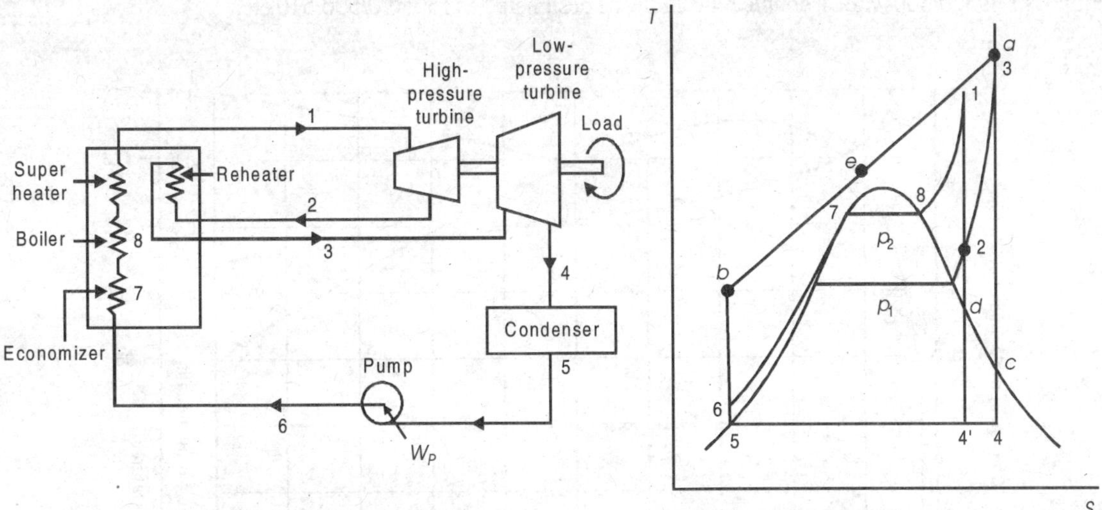

(a) Schematic of a Rankine cycle with superheat and reheat

(b) *T-S* diagram of a Rankine cycle of Fig. (a)

Fig. 12.10

Reheating is resuperheating of steam at point 2 after it expands in high pressure section of turbine. The steam after expansion is sent to the steam generator where it is reheated (resuperheated) to the original temperature with the help of hot flue gases. The reheated steam now expands in the low pressure section of the turbine to the condenser pressure.

12.10.1 Advantages of Reheat Cycle

There are following advantages of reheat cycle is,

1. In reheat cycle, heat is added twice; from 6 to 1 and from 2 to 3. This increases the value of T_m—the temperature of heat addition and keeps the evaporator, superheat reheat portion from 7 to 3 closer to the flue gas line *ae*. This results in higher cycle efficiency.
2. It improves the final dryness of steam (from 4' to 4) and reduces turbine blade erosion.
3. It increases cycle work by increasing the cycle area. This increases the turbine power.

Modern fossil fuel power plants employ superheat and at least one stage of reheat. Some employ two. More than two reheats however, result in cycle complications and increased capital costs, that are not justified by improvements in efficiency.

12.10.2 Effect of Reheat Pressure Ratio on Cycle Efficiency

The pressure p_2 at which the steam is reheated affects the cycle efficiency. Figure 12.11 shows the relation between increase in the efficiency $\Delta\eta$ percent and the ratio of reheat pressure to initial pressure.

At $p_2/p_1 = 1$ there is no reheat and $\Delta\eta = 0$. As p_2 is lowered the efficiency improves and reaches the maximum at p_2 equal to about 20 percent of p_1. Lowering p_2 further decreases the efficiency and at p_2/p_1 less than say 0.025 the efficiency decreases and $\Delta\eta$ becomes negative. The optimum pressure ratio is between 0.2 and 0.25.

Figure 12.8 is plotted for initial pressure of 170 bar and initial and reheat temperature of 500° C, but the optimum value of p_2/p_1 holds good for most modern power plants. The graph shows variation of T_2 and X_4 also. Note that reheating improves quality of steam.

The superheat reheat plant is often designated by $p_1/T_1/T_3$ in bar and degrees centigrade. The above case for example is 170/500/500 while a double reheat may be designated as 150/490/500/510 etc.

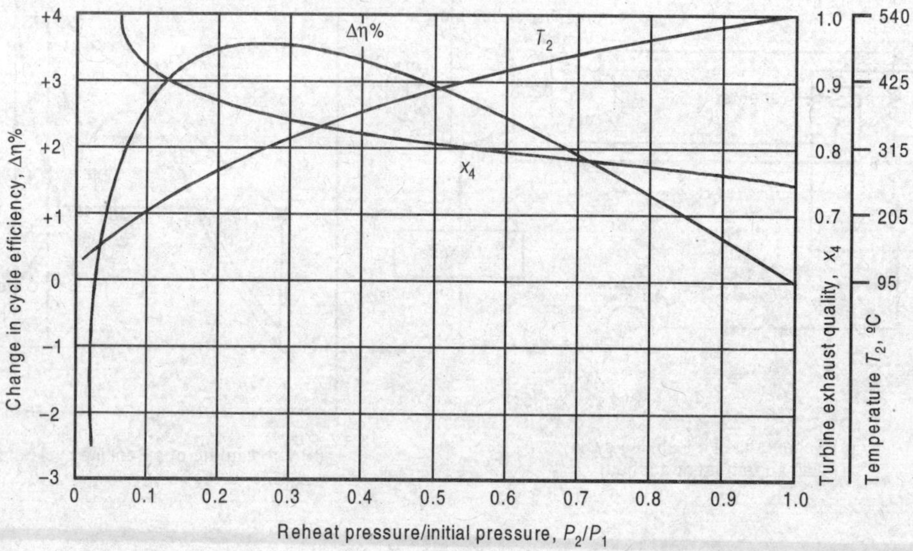

Fig. 12.11. *Effect of reheat-to-initial pressure ratio on efficiency, high-pressure turbine exit temperature, and low-pressure turbine exit quality.*

12.11 TYPICAL LAYOUT OF A STEAM POWER PLANT

Figure 12.12 shows a typical layout of a steam power plant. It shows steam and water circuits of the plant which includes reheaters and feedwater heaters. Note the positions of four closed feedwater heaters and one open feedwater heater (deaerator). Note also that there are two *hp*, two *ip* and one *lp* feedwater heater.

12.12 COGENERATION

Cogeneration is the simultaneous generation of electricity and steam (or heat) in a single power plant. It has long been used by certain industries which need large quantities of steam or heating as well as electrical power. Examples are cement factories, paper mills, chemical industries, textile plants, fertilizer plants, extraction plants, edible oil industries, refineries etc. Cogeneration is used if the electricity thus produced is cheaper, than that purchased from the state electricity boards.

In a cogeneration plant if,

E = electrical energy generated

ΔH_s = heat energy generated or heat in the process steam

and Q_s = total heat added to the plant

then the cogeneration plant efficiency is given by

$$\eta_{co} = \frac{E + \Delta H_s}{Q_A} \qquad \text{...(12.17)}$$

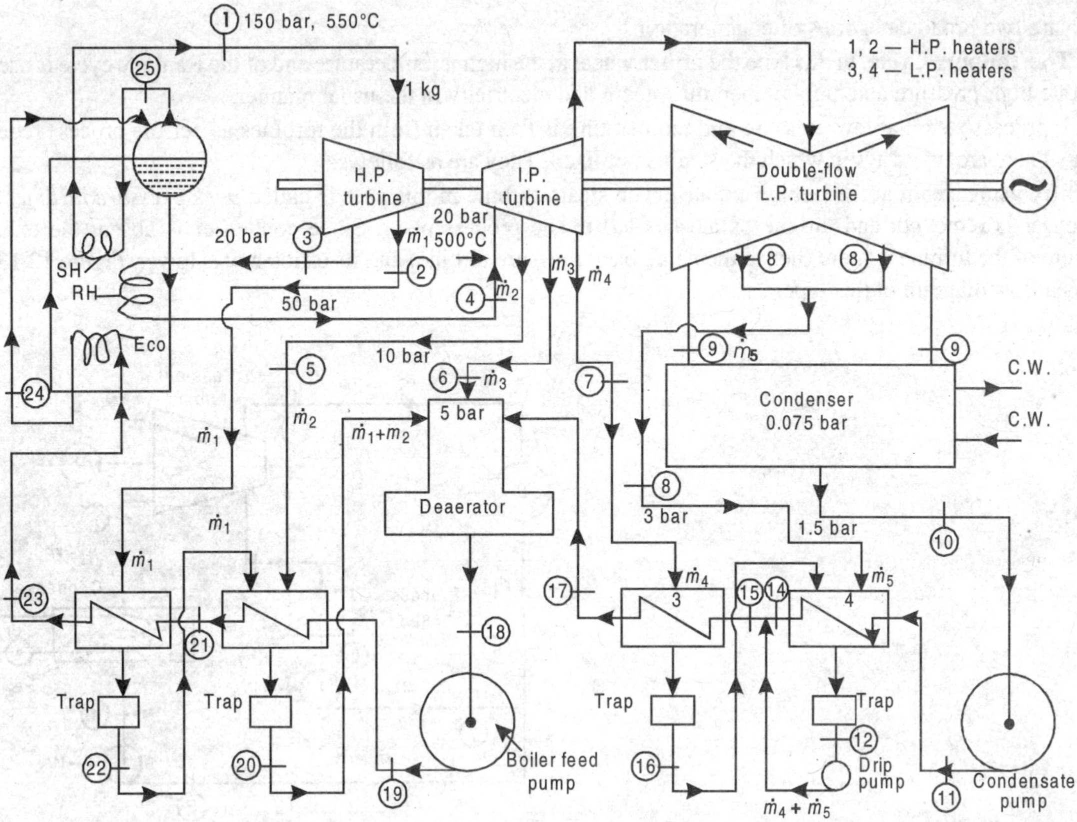

Fig. 12.12. *Schematic flow diagram of a modern steam power plant.*

Now consider the generation of electricity and steam separately.

Total energy output of the plant = $E + \Delta H_s$ as stated earlier. Let e be the electrical fraction of total energy output,

or
$$e = \frac{E}{E + \Delta H_s}$$

Then the heat added per unit total energy output is,

$$\frac{e}{\eta_e} + \frac{1-e}{\eta_h}$$

where η_e = electric plant efficiency and η_h = steam (or heat) generation efficiency. The *combined efficiency* η_c *for separate generation* is therefore given by,

$$\eta_c = \frac{1}{\left(\dfrac{e}{\eta_e} + \dfrac{1-e}{\eta_h}\right)}$$

and cogeneration is beneficial if $\eta_{co} > \eta_c$.

12.13 TYPES OF COGENERATION

There are two broad categories of cogeneration.

1. The Topping Cycle. In this type the primary heat at the higher temperature end of the Rankine cycle is used to generate high-pressure and high temperature steam and electricity in the usual manner.

The process steam at low pressure and temperature is then taken from the turbines as per the process requirements. There are two ways in which the steam is utilized. They are as follows:

(a) By using steam at the turbine exhaust. The steam turbine in this case is called *back pressure turbine*. The condenser is removed, and turbine exhaust is led to the process plant. Since condenser is absent the exhaust pressure of the turbine is more (hence the name back pressure turbine) and its efficiency is lower. Figure 12.13 (a) shows a flow diagram of this type.

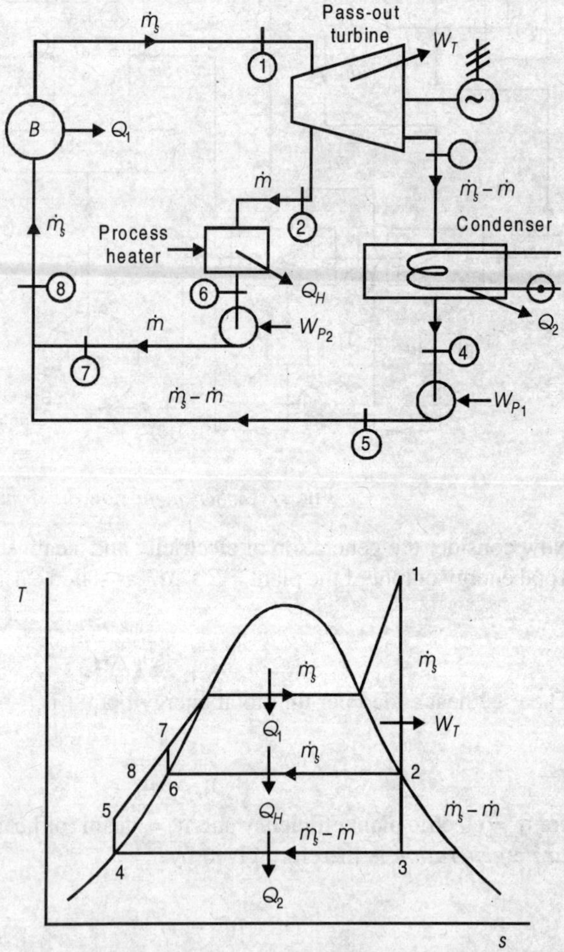

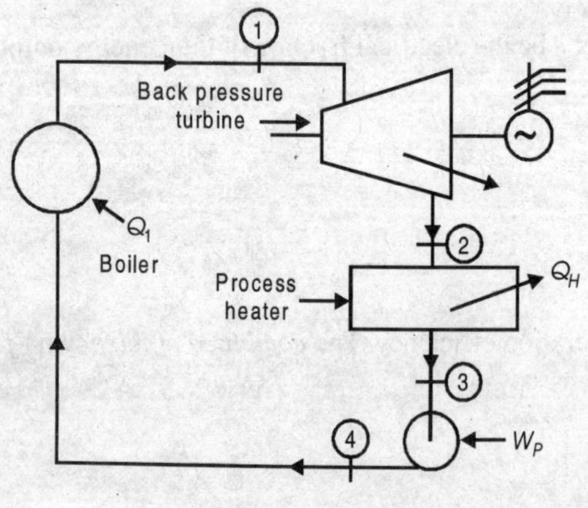

(a) Congeneration plant with a back pressure turbine

(b) Congeneration plant with an extraction turbine

Fig. 12.13

(b) By extracting steam from the turbine at an intermediate state; much as for feedwater heating. The turbine of this type is known as *extraction turbine* or *passout turbine*. Since condenser is used the efficiency of the cycle is higher; and steam extracted has much higher pressure and temperature. Figure 12.13 (b) shows a flow diagram of this type.

The main advantage of this cycle is its ability to use wide range of fuels including wood, coal, petroleum, coke, refuse etc. and hence is used where low quality refuse and byproduct fuels are readily available.

This type however is seldom used because of its high initial cost and low power-to-heat ratio; especially at high process-steam pressures.

2. Bottoming Cycle. Here first the primary heat at high temperature is directly used for process requirements. The low grade (low temperature and availability) waste heat from process plant is used to generate electricity. Figure 12.14. The efficiency of the cycle is obviously low, –10 percent for low and upto 20 percent at high temperature applications. Given that the waste heat is free, the cycle economics may be favourable.

The main application of bottoming cycles are in the processes of the chemical and material processing industries such as cement, limestone and steel.

Since the temperature of waste heat is low, hydrocarbon fluids are often used as working medium.

12.13.1 Types of Topping Cycle

Since the bottoming cycle has very low combined efficiency; it is of little thermodynamic or economic interest and only the topping cycle therefore can provide true savings in primary energy. In addition, most process applications require a low grade (low temperature and availability) steam which can be conveniently produced in a topping cycle.

Some of the arrangements for cogeneration in topping cycle are as under.

(a) Steam turbine power plant with a back pressure turbine—refer section 12.13.
(b) Steam turbine power plant with steam extraction from condensing turbine—refer section 12.13.
(c) Gas turbine power plant with a heat recovery boiler—using gas turbine exhaust to generate steam.
(d) Combined steam gas turbine power plant—The steam turbine is either of the back pressure type (a) or the extraction-condensing type (b) above.

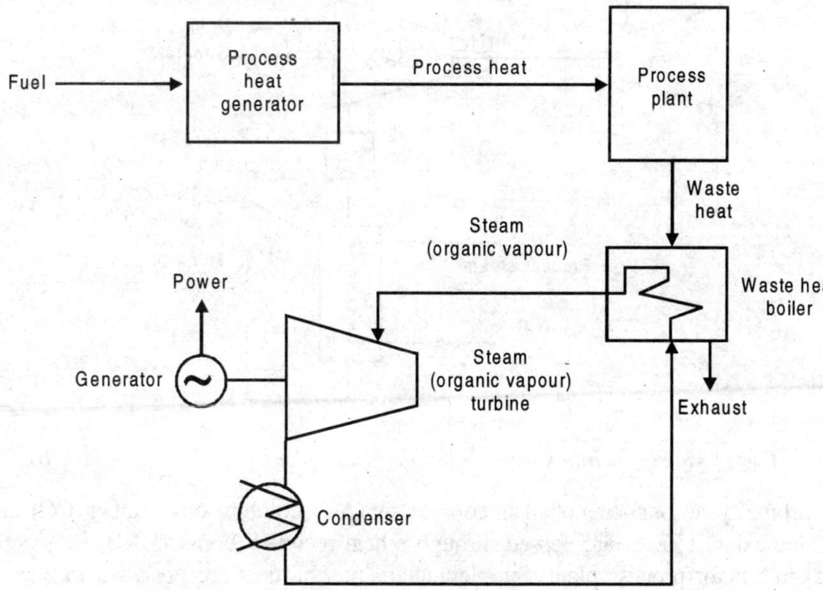

Fig. 12.14. *Cogeneration plant using a steam or organic bottoming cycle.*

The back pressure steam-turbine plant (a) is most suitable only when electricity demand is low compared with the heat demand. The combined cycle plant (d) above is most suitable only when the electric demand is high, about comparable to heat demand or higher. The gas turbine cycle (c) lies in between. Only the extraction condensing plant (b) above is suitable over a wide range of ratios.

12.14 COGENERATION WITH A GAS TURBINE PLANT

Gas turbines are used by themselves in wide range of services. Some of them are:

1. As a powering unit for all types of air-crafts.
2. For driving mechanical equipment such as pumps, compressors in industrial plants.
3. As a powering unit for railways and racing cars.
4. To drive electric generators in electric power plants — To produce power for peak loads.

There is also growing interest in using gas turbines for cogeneration and in combined cycle plants. The working temperatures of Gas Turbine are quite high (Max. temperature about 1200 – 1300°C) and hot exhaust gases leaving the turbine at about 500°C, still have a great deal of heat energy stored in them. This heat energy can be used for various purposes such as,

1. Drying and direct heating—In food industry.
2. Heating of fluids in process industries—to heat air, oil, water which is a working medium in a process plant.
3. Producing steam by using a waste recovery boiler—The steam produced may be used for process application or for generation of electricity by using steam turbines. This is known as "cogeneration" for the former and "combined cycle" for the latter.

Figure 12.15 shows a schematic flow diagram of a gas turbine plant used for cogeneration, with a heat recovery boiler.

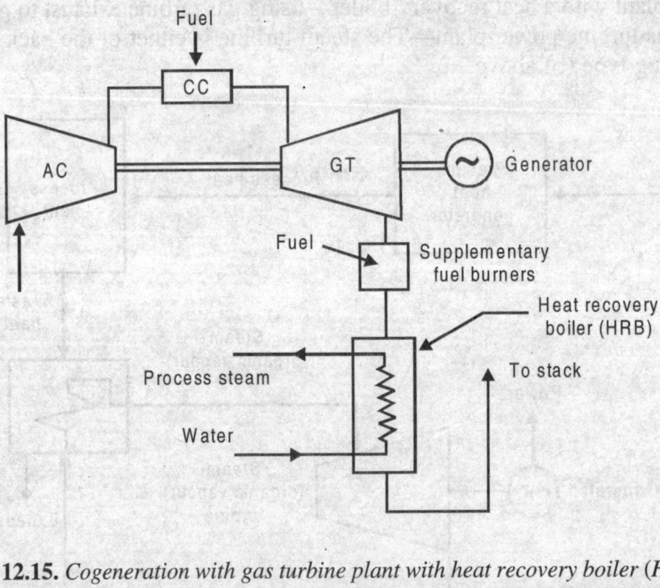

Fig. 12.15. *Cogeneration with gas turbine plant with heat recovery boiler (HRB).*

A simple gas turbine cycle consisting of an air compressor (AC), combustion chamber (CC) and gas turbine (GT) is used. The turbine exhaust gases are passed through a heat recovery boiler (HRB) to generate process steam which is then taken out for process plant. Supplementary fuel burners are provided to burn additional fuel to increase the mass flow rate of process steam for a short period, if required.

12.15 COMBINED CYCLE POWER PLANTS

In *combined cycle* power plants both gas and steam turbines supply electrical power to the grid. The idea of combined cycle has grown out of the need to improve the simple Brayton-cycle efficiency by utilising the waste heat in the turbine exhaust gases. This can be done by regeneration also. The regeneration however has some serious drawbacks as given below.

1. Regeneration does not increase output. In fact regeneration reduces turbine pressure ratio and hence the net plant output by a few percent.
2. Because of large heat transfer surfaces and large gas and air piping; regeneration makes the plant costly.
3. With regeneration the optimum pressure ratio for maximum efficiency reduces sharply (from 20 – 30 to 7 – 10). This reduces the power output.

Thus raising the efficiency of a gas turbine plant by regeneration is costly. A means was saught whereby both efficiency and power are increased. A solution was found in using the large quantity of energy in the turbine exhaust, to generate steam for a steam turbine plant. This is a natural solution. Gas turbine is a high temperature machine (1100°C – 1625°C) while steam turbine is a low temperature machine (550 – 650°C). The joint operation with gas turbine at "hot end" and a steam turbine at "cold end" is called a *combined-cycle power plant.*

12.15.1 Advantages of Combined Cycle Power Plant

1. High overall plant efficiency—over 50% can be attained.
2. High total power–combined power of GT and ST is available.
3. Quick starting of plant–GT can be started quickly.
4. Low water requirement–GT needs no cooling. Only water required is for the condenser circulation. Sometimes condensers also are air cooled. [e.g. Uran combined-cycle plant]. Then water requirement is very low. Considering huge quantities of water required for condensers, this is a distinct advantage.
5. Phased installation—GT plant which takes less time to be installed, is installed first and stars producing power, while ST plant is being constructed.
6. Low pollution level—In the CC plant operation, formation of NO_x and CO_2 is much lower than that in coal fired plant.
7. Can be used for both peak load and base load operations.

12.16 COMBINED CYCLE POWER PLANT WITH HEAT RECOVERY BOILER

Figure 21.16 shows a schematic flow diagram of such a combined cycle. A gas turbine plant consists of air compressor (AC), combustion chamber (CC) and gas turbine (GT). The exhaust of the gas turbine goes to the heat recovery boiler (HRB) to produce superheated steam. The steam is used in a steam turbine plant which consists of steam turbine (ST), condenser (C), pumps (CP) and (BFP), feed water heater (FWH) and deareator (DA). The HRB consists of economizer (EC), boiler (B), drum (SD) and superheater (SU). The gas goes to stack after HRB. Both steam and gas turbine drive electric generators (G).

The gas turbine is operated with a high air to fuel ratio (400 : 1) to ensure sufficient air in the gas turbine exhaust for further combustion. Plant output can be increased for short periods during load peaks by burning, additional full in supplementary fuel burners. HRB has no refractory lining nor water walls. Hence, its temperature is limited to 750°C.

In small plants steam turbine output is less than the gas turbine output by as much as 500 percent. In large plants, steam turbine output is greater than the gas turbine output by upto 8 : 1.

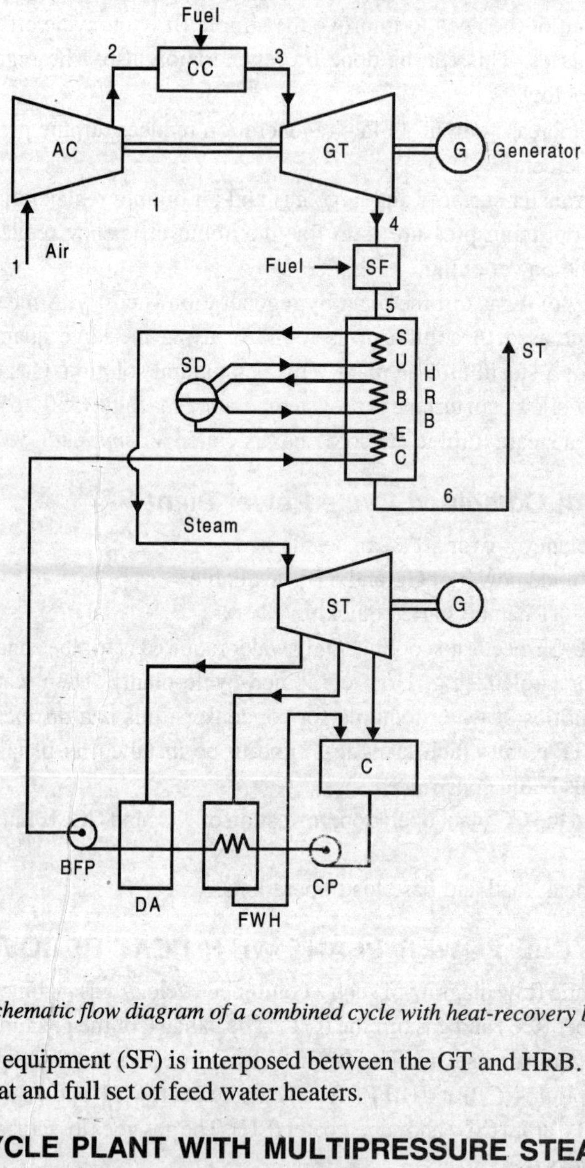

Fig. 12.16. *Schematic flow diagram of a combined cycle with heat-recovery boiler (HRB).*

The supplementary firing equipment (SF) is interposed between the GT and HRB. The steam cycle is designed for high efficiency with reheat and full set of feed water heaters.

12.17 COMBINED CYCLE PLANT WITH MULTIPRESSURE STEAM

This plant allows the gas from MR to leave at a lower temperature and increases the efficiency of the plant.

Figure 12.17(a) shows a schematic flow diagram of the plant. The "dual-pressure cycle" has a HRB which has two steam circuits. The high pressure circuit supplies steam to the high pressure stages of the turbine and low pressure circuit supplies steam to the low pressure stages of the same turbine.

Figure 12.17(b) shows the temperature-enthalpy diagram of both gas and steam circuits in the HRB.

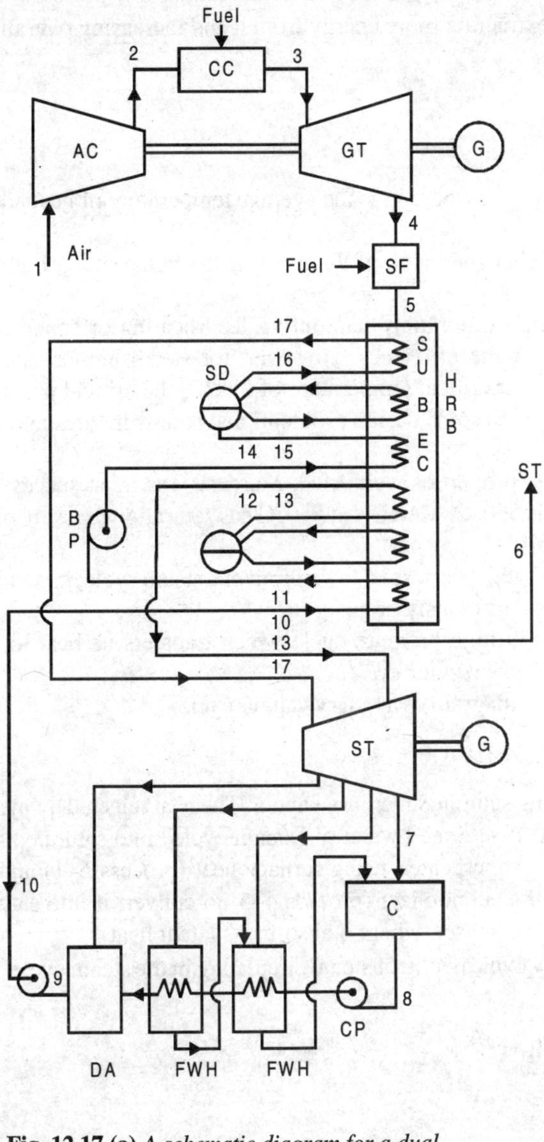

Fig. 12.17 (a) *A schematic diagram for a dual-pressure combined cycle.*

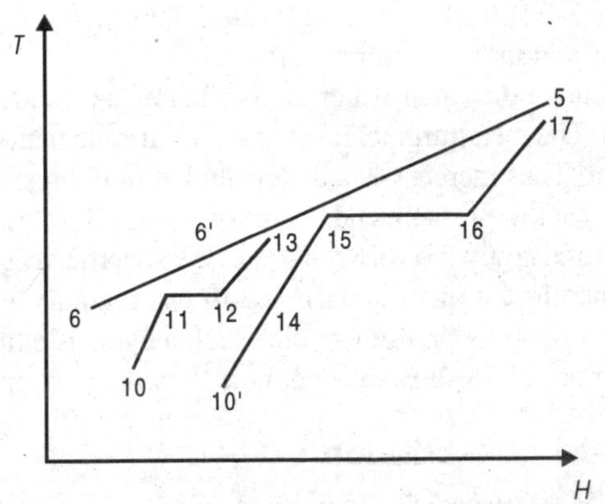

Fig. 12.17 (b) *Temperature-enthalpy (T-H) diagram of the heat recovery boiler of the dual-pressure combined cycle.*

Exhaust gas leaving the gas turbine enters the supplementary firing (SF) at 4 and the HRB at 5, leaving it to the stack at 6. Condensate leaving the condenser at 8 enters the low pressure economizer at 10 via condensate pump, two feedwater heaters, one DA and boiler feed pump. Process 10 – 11 is feed heating of low pressure circuit; followed by evaporation to 12 and superheat to 13. Superheated steam at 13 enters the steam turbine at low pressure stage.

Water from 11 is pumped by booster pump (BP) to 14 and goes to the high pressure economizer. Evaporation occurs from 15 to 16 and superheat to 17. High pressure superheated steam at 17 enters the steam turbine first stage.

In *T-H* diagram we see that low pressure steam boils at a lower temperature (12) below that of high pressure steam (15). Single high pressure circuit is shown by 10'–15–16–17 with gas leaving at 6'. Adding low pressure circuit allows the gas to leave at a lower temperature (6), thus extracting more energy from it and increasing overall cycle efficiency.

12.18 BINARY VAPOUR CYCLE

The efficiency of a reversible cycle is given by $\eta_{rev.} = 1 - \frac{T_1}{T_2}$, where T_1 is the average temperature of heat addition and T_2 is the average temperature of heat rejection. Higher the value of T_1 or lower the value of T_2, higher the efficiency.

The irreversibilities of Rankine cycle are least during evaporation than during feed heating or superheating. Thus, higher the saturation temperature of steam, higher the efficiency. However, for steam higher saturation temperature is accompanied by high pressure. Recall that at critical temperature of 374°C, the critical pressure of steam is 221 bar—a very high value. This high pressure poses some design problems and results in thick and heavy boiler and turbine components.

Fluids other than water having better thermodynamic properties were tried. Mercury was most successful of them. The metallurgical limit of steam turbine materials is 560°C. Mercury at 560°C has saturation pressure of only 12 bar. Thus mercury is a better fluid at high pressure end.

But at low pressure end, mercury is a bad fluid. At 30°C which is normal lower limit of a steam cycle, its saturation pressure is only 3.6×10^{-6} bar; and its specific volume is enormously large.

Therefore a plant is designed to use mercury vapour in high pressure cycle which transfers its heat to water vapour in a low pressure cycle. Such a cycle is called *Binary vapour cycle*.

Figure 21.18 shows a schematic flow diagram and a *T-s* diagram of a binary vapour cycle.

12.18.1 Description

The "Mercury cycle" *abcde* is a simple Rankine cycle using saturated mercury vapour. The heat rejected by mercury during condensation (process *b–c*) is transferred to convert saturated water of "steam cycle" into saturated steam (process 5–6). The saturated steam is superheated in the superheater, using furnace heat (process 6–1) and then admitted to steam turbine. The condensate is heated in the economizer (process 4–5) to convert it into saturated liquid which is then admitted to steam boiler—mercury condenser, where it absorbs its latent heat.

Let $\dot{m}$ represent the flow rate of mercury in the mercury cycle per kg of steam circulating in the steam cycle. Then for 1 kg of steam,

1. Heat supplied, $q_s = \dot{m}(h_a - h_d) + (h_1 - h_6) + (h_5 - h_4)$ kJ/kg
2. Heat rejected, $q_r = (h_2 - h_3)$ kJ/kg
3. Turbine work, $W_T = \dot{m}(h_a - h_b) + (h_1 - h_2)$ kJ/kg
4. Pump work, $W_P = \dot{m}(h_d - h_c) + (h_4 - h_3)$
5. Cycle efficiency,

$$\eta_{cycle} = \frac{q_s - q_r}{q_s} = \frac{W_T - W_P}{q_s} = \frac{W_{net}}{q_s} \qquad \ldots(12.14)$$

6. Specific steam consumption,

$$\text{S.S.C.} = \frac{3600}{W_{net}} = \text{kg/kWh}$$

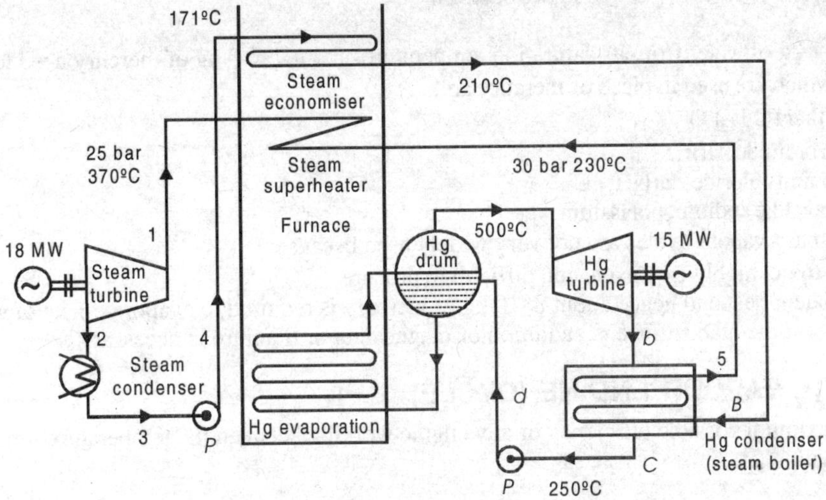

(a) Binary vapour cycle.

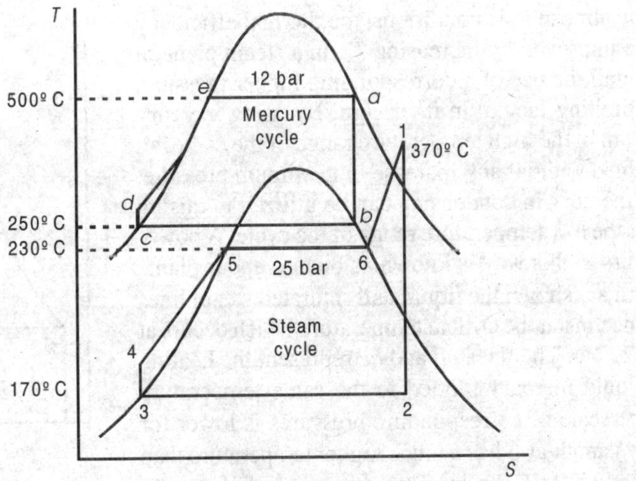

(b) *T-s* diagram of binary vapour cycle.

Fig. 12.18

7. Quantity of mercury per kg of steam.
   The energy balance equation for mercury condenser steam boiler is,
   $\dot{m}(h_b - h_c) = (h_6 - h_5)$

$$\dot{m} = \frac{h_6 - h_5}{h_b - h_c} \text{ kg/kg of steam}$$

To vapourise 1 kg of water from its saturation temperature, about 7 to 8 kg of mercury need to be condensed. Other fluids which are used in place of mercury are:

1. Diphenyl ether $(C_6H_5)_2O$
2. Aluminium bromide $AlBr_3$.
3. Zinc ammoniun chloride $Zn(NH_3)_2Cl$.
4. Liquid metals like sodium, potassium.

However, mercury vapour cycles are not very widely used because:

1. Mercury is toxic, highly expensive and difficult to procure.
2. It has low latent heat and hence about 8–10 kg of mercury is required to evaporate 1 kg of water.
3. It does not wet the tube surface, so addition of magnesium or titanium is necessary.

12.19 BINARY VAPOUR ENGINE (CYCLE)

We know that maximum possible efficiency of any engine working between the temperature limits of T_2 and T_1 is given by the equation .

$$\eta = \frac{T_1 - T_2}{T_1}$$

where T_1 = Higher temperature
T_2 = Lower temperature (Atmospheric temperature)

Since T_2 is fixed by atmospheric conditions, the thermal efficiency of a plant can only be improved by increasing T_1. In a steam plant an increase of T_1 may entail the use of a corresponding higher pressure, which is one of the limiting factors in its design. By using mercury vapour in place of steam in the high temperature range of the cycle, an increase in T_1 is obtained without any increase in maximum pressure. The heat rejected by mercury in condensing can be utilized in raising superheated steam for the low temperature range of the cycle. A power plant using two vapours in this way is known as binaryvapour plant.

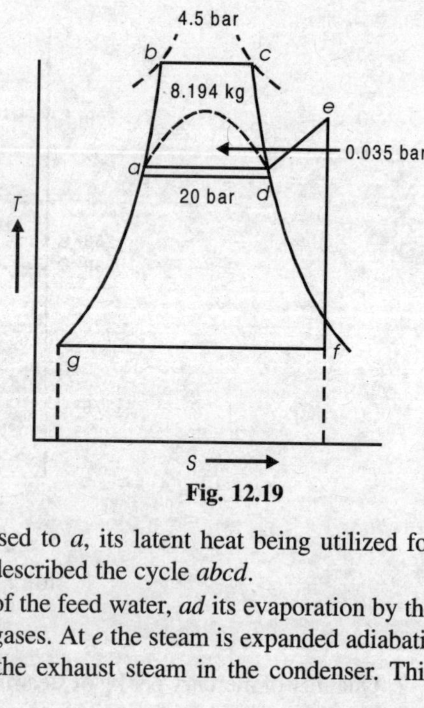

Fig. 12.19

On the *T-s* diagram; are shown the liquid and saturated steam lines for water and steam, meeting at the critical temperature which occurs at a steam pressure of 221 bar. The lines *ab* and *bc* represent the heating and evaporation of liquid mercury plotted to the same temperature scale as the steam, the scale of corresponding pressures is lower for mercury. The mercury vapour at *C* has a much higher temperature than steam at the same pressure. At *C* the mercury expanded adiabatically through a separate mercury vapour turbine to *d*; from *d* it is condensed to *a*, its latent heat being utilized for evaporating a corresponding amount of steam. The mercury has thus described the cycle *abcd*.

The steam cycle is represented by *gadef*, *ga* represents the heating of the feed water, *ad* its evaporation by the condensing mercury, and *de* the superheating of the steam by the flue gases. At *e* the steam is expanded adiabatically through a steam turbine to *f*; *fg* represents the condensing of the exhaust steam in the condenser. This completes the steam cycle.

To obtain the correct amount of heat from the condensing mercury for evaporating the steam, it is found that 8.196 (8.2 approx.) kg of mercury are required per kg of steam. It will be seen from the areas of *T-s* diagram that the use of mercury for the high temperature range of the cycle gives a higher efficiency than that obtainable from steam with the same addition of heat. Using pressure limits of 24 bar and. 0.07 bar, a steam plant gives an efficiency of 24–25% whilst that of the binary vapour plant is 50%.

A diagrammatic view of a binary vapour plant is shown below. The hot gases from the furnace pass through the mercury boiler, through the liquid mercury heater, then through the steam superheater and finally through the feed water economiser, after which they are exhausted to the chimney.

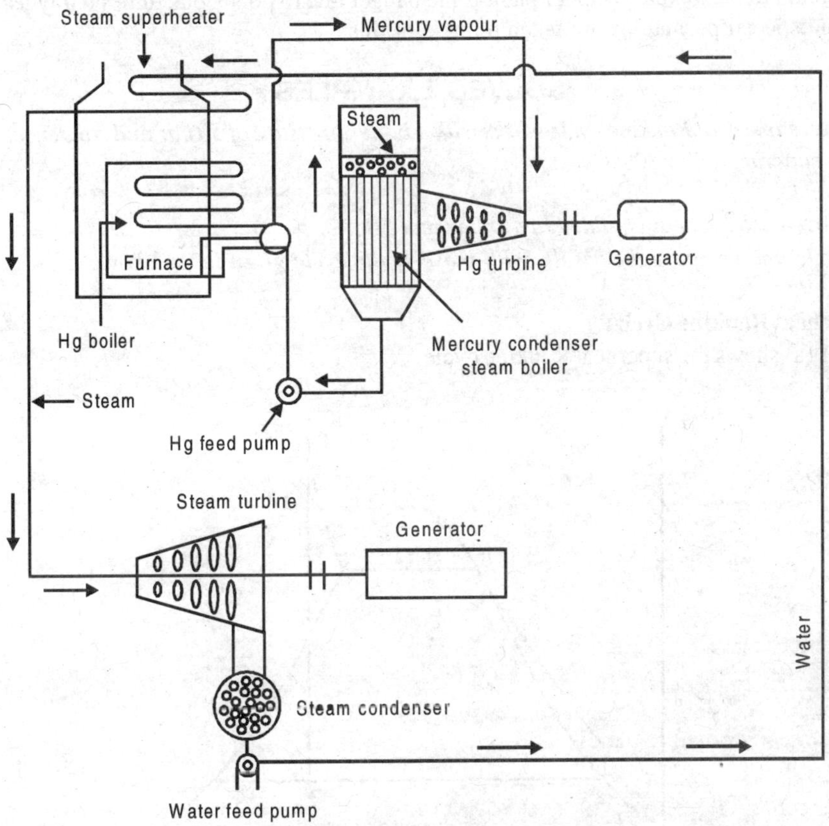

Fig. 12.20

The liquid mercury passes from the mercury liquid heater to the mercury vapour boiler where it is evaporated. It then flows to the mercury turbine through which it is expanded to its low pressure limit. From here it exhausts to the mercury condenser steam boiler where its latent heat is given upto the hot feed water; this operation condenses the mercury whilst at the same time the feed water is evaporated to steam. The mercury is then returned to the mercury liquid boiler, thus completing its cyclc.

Referring to figure, the feed water from the economiser (not shown) is evaporated to steam in the mercury condenser steam boiler and then passes to the superheater where it is superheated by the hot flue gases. From the superheater the superheated steam passes to the steam turbine in which it is expanded down to the steam condenser pressure, it is then condensed in the steam condenser and passes back to the economiser, thus completing the cycle of the water and steam.

At the Hartford-USA–binary vapour plant, erected by the General Electric Co., the weight of mercury in use in the plant is 60 tonnes. A higher efficiency is obtained than is possible from a steam plant of the same output. Daily saving of 150 T of coal compared with the consumption of an equivalent steam plant. Five binary vapour plants have been erected by the General Electric Co.

In the early stages of the development of the mercury boiler, it was found that the liquid mercury did not wet the heating surface of the tubes. There was a tendency for a film of mercury vapour to form at the tube surface which prevented a good heat transfer to the liquid. The difficulty was overcome by a 0.002 percent solution of magensium and titanium being used in the boiler as a surface wetting agent.

The chief objection to the mercury vapour plant is the danger due to poisonous fumes if any leakage of mercury vapour occurs, and special pecautions are taken to prevent this.

SOLVED EXAMPLES

Example 12.1 (a) *A superheat Rankine cycle works with a boiler pressure of 50 bar and condenser pressure of 0.08 bar. The steam is superheated to 500°C.*

Analyse the cycle.

(b) For the above cycle, reheat is added at 1.0 bar and 500°C.

Analyse the cycle and compare the results with those of the cycle given in (a) above.

Solution

Case (a) Superheat Rankine Cycle:

Figure Ex. 12.1 (a) shows the superheat Rankine cycle.

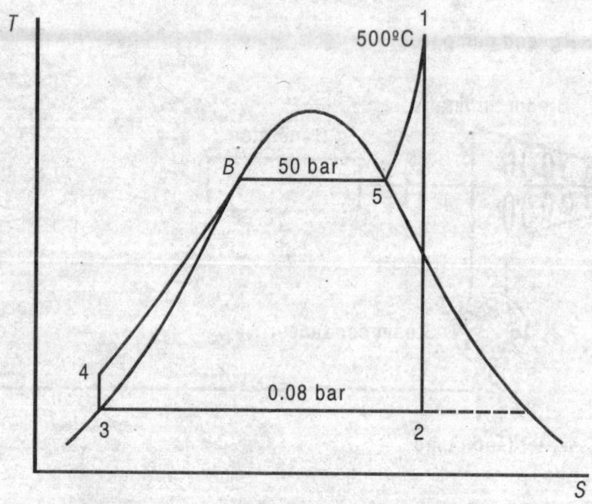

Fig. Ex. 12.1(a) *Superheat rankine cycle.*

Our first aim is to determine enthalpies at various salient points of the cycle. From the steam tables we have,

(i) at 50 bar 500°C

$$h_1 = 3433.7 \text{ kJ/kg}$$
$$s_1 = 6.977 \text{ kJ/kg-K}$$

(ii) at 0.08 bar

$$s_{f3} = 0.5925 \text{ kJ/kg-K}$$
$$s_{fg2} = 7.6371 \text{ kJ/kg-K}$$
$$h_{f3} = 173.86 \text{ kJ/kg}$$
$$h_{fg2} = 2403.2 \text{ kJ/kg}$$

1. To find X_2 equate entropies at point 1 and point 2.

$$6.977 = 0.5925 + x_2(7.6371)$$

$$\therefore \quad x_2 = 0.386$$

$$\therefore \quad h_2 = 173.86 + 0.836 \times 2403.2 = 2182.93 \text{ kJ/kg}$$

2. Turbine work = $W_T = h_1 - h_2 = 3433.7 - 2182.93 = 1250.76$ kJ/kg
3. Pump work = $W_P = 0.001\,[50 - 0.08] \times 100 = 5$ kJ/kg
4. Net work = $W_{net} = W_T - W_P = 1250.76 - 5.0 = 1245.76$ kJ/kg
5. Heat supplied = $q_s = h_1 - h_4 = h_1 - (h_3 - W_P)$

$$= 3433.7 - 178.76 = 3254.84 \text{ kJ/kg}$$

6. Rankine efficiency = $h_r = \dfrac{W_{net}}{q_s} = \dfrac{1250.76}{3254.84} = 0.3843$ i.e. 38.43%

7. Work ratio = $\dfrac{W_{net}}{W_T} = \dfrac{1245.76}{1250.76} = 0.996$

8. Specific steam consumption = $\dfrac{3600}{1250.76} = 2.878$ kg/kWh

Case (b) Superheat Reheat Cycle: Figure Ex. 12.1 (b) shows the cycle.

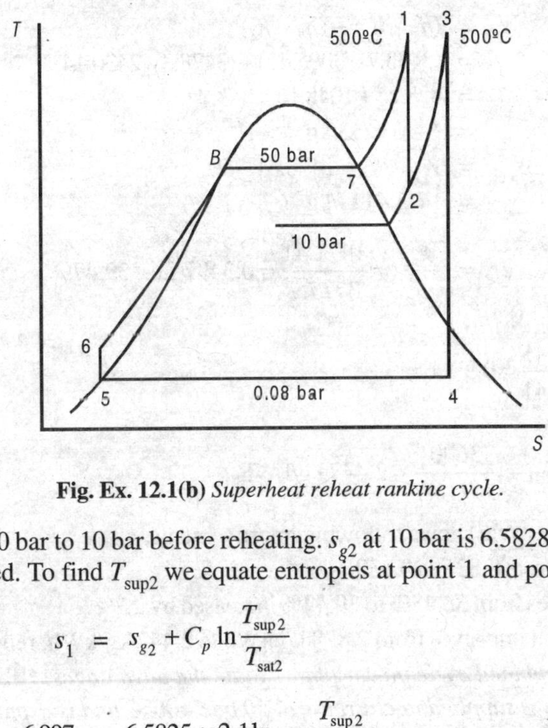

Fig. Ex. 12.1(b) *Superheat reheat rankine cycle.*

The steam expands from 50 bar to 10 bar before reheating. s_{g2} at 10 bar is 6.5828. Since $s_1 > s_{g2}$ the condition of steam at point 2 is superheated. To find T_{sup2} we equate entropies at point 1 and point 2.

or
$$s_1 = s_{g2} + C_p \ln \frac{T_{sup2}}{T_{sat2}}$$

$$\therefore \quad 6.997 = 6.5825 + 2.1 \ln \frac{T_{sup2}}{273 + 179.88}$$

$$\therefore \quad T_{sup2} = 551.62 \text{ K} = 278.62\,°\text{C}$$

$\therefore$ At 10 bar and 278.62°C, we have from steam tables (by interpolation)

$$h_2 = 3005.45 \text{ kJ/kg}$$

At 10 bar and 500 °C, we have from steam tables,

$$h_3 = 3478.3 \text{ kJ/kg and } s_3 = 7.7627$$

To find x_4 equate entropies at 3 and 4.

$$\therefore \quad s_3 = s_4$$

$$\therefore \quad 7.7627 = 0.5925 + x_4 7.6371$$

$$x_4 = 0.939$$

$$\therefore \quad h_4 = 173.86 + 0.939 \times 2403.2 = 2430.14 \text{ kJ/kg}$$

Let us analyse the cycle.

1. Heat supplied $= q_{s1} = h_1 - h_6 = 3433.7 - (h_5 + W_P)$ $\quad [W_p = 5 \text{ kJ/kg}]$

$$= 3433.7 - 178.86$$

$$= 3254.84 \text{ kJ/kg}$$

also heat supplied under reheat

$$= q_{s2} = h_3 - h_2$$

$$= (3478.3 - 3005.45)$$

$$= 472.85 \text{ kJ/kg}$$

$$\therefore \quad \text{Total heat supplied} = q_s = q_{s1} + q_{s2}$$

$$= 3254.84 + 472.85$$

$$= 3727.7 \text{ kJ/kg}$$

2. Turbine work $\quad W_T = (h_1 - h_2) + (h_3 - h_4)$

$$= (3433.7 - 3005.45) + (3478.3 - 2430.14)$$

$$= 428.25 + 1048.16$$

$$= 1476.41 \text{ kJ/kg}$$

3. Net work $= \quad W_{net} = W_T - W_P = 1476.41 - 50$

$$= 1471.41 \text{ kJ/kg}$$

4. Cycle efficiency $\quad h_r = \dfrac{W_{net}}{q_s} = \dfrac{1471.41}{3727.7} = 0.3497$ i.e., 39.47%

5. Work ratio $= \dfrac{W_{net}}{W_T} = \dfrac{1471.41}{1476.41} = 0.9966$

6. Specific steam consumption $= \dfrac{3600}{1471.41} = 2.447$ kg/kWh

Comments: Superheat reheat cycle shows following improvements,

1. Final dryness is improved from 0.836 to 0.939.
2. Rankine efficiency improves from 38.43% to 39.47% increased by 2.7%.
3. Specific steam consumption improves from 2.878 kg/kWh to 2.447 kg/kWh; reduction by 15%.

Example 12.2 (a) *What is a combined cycle power plant? State the advantages.*

(b) In a Rankine cycle, steam is supplied to a turbine at 50 bar. 400°C and is expanded to a pressure of 0.2 bar. Find the thermal efficiency of the plant if the reheat pressure is 7 bar and reheated steam temperature 400°C. Assume isentropic efficiency of the turbine as 80% throughout.

Solution

The cycle is shown on *T–S* diagram.

Given: $\quad P_1 = 50$ bar

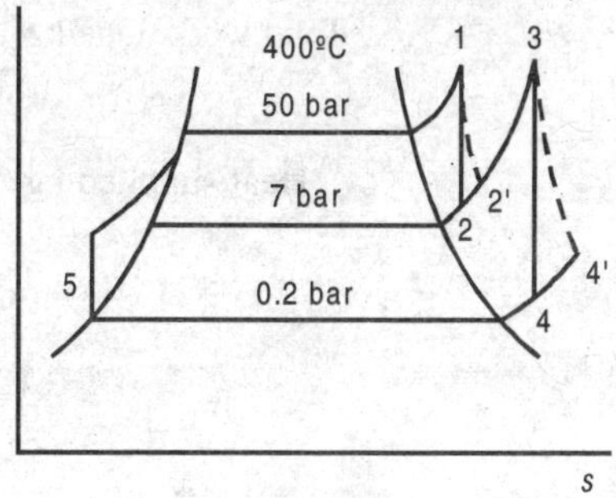

Fig. Ex. 12.2

$$T_1 = 673\text{ K} = T_3$$

$$P_3 = 7\text{ bar}$$

$$P_4 = 0.2\text{ bar}$$

$$\eta_t = 80\,\%$$

From Steam tables,

$$h_1 = 3198.3\text{ kJ/kg};\ h_3 = 3269\text{ kJ/kg}$$

$$h_{f_5} = 251.45\text{ kJ/kg}$$

$$s_1 = 6.6508;\ s_3 = 7.6362$$

$$s_2 = 1.9918 + x_2 \times 4.7134 \qquad \because\ s_1 = s_2$$

$$x_2 = \frac{6.6508 - 1.9918}{4.7134} = \frac{4.6590}{4.7134}$$

$$= 0.9885$$

$$h_2 = 697.06 + 0.9885 \times 2064.9$$

$$= 697.06 + 2041.07$$

$$= 2738.13\ h_2 = 2738.13$$

From *T–s* diagram,

$$\eta_t = \frac{h_1 - h_{2'}}{h_1 - h_2} \qquad \therefore\ 0.8 = \frac{3198.3 - h_{2'}}{3198.3 - 2738.13}$$

$$\therefore\quad 0.8 \times 460.17 = \mathbf{368.14\ kJ/kg}$$

= work of turbine I

$$\therefore\quad h_{2'} = 3198.3 - 368.14 = 2830.16\text{ kJ/kg}$$

Similarly, for 2nd stage turbine (after reheating)

$$s_3 = s_4 \qquad \therefore\ 7.6362 = 0.8321 + x_4 \times 7.0773$$

$$\therefore\quad x_4 = \frac{7.6362 - 0.8321}{7.0773} = \frac{6.8041}{7.0773}$$

$$= \mathbf{0.9614}$$

$$\therefore\quad h_4 = 251.45 + 0.9614 \times 2358.4$$

$$= 251.45 + 2267.36$$

$$= \mathbf{2518.81\ kJ/kg}$$

$$\therefore\quad 0.8 = \frac{h_3 - h_{4'}}{h_3 - h_4} = \frac{3269 - h_{4'}}{3269 - 2518.81}$$

$$\therefore\quad h_{4'} = 3269 - 0.8 \times 750.19$$

$$= 3269 - 600.15$$

$$= \mathbf{2668.85\ kJ/kg}$$

$$W_{t2} = 600.15\text{ kJ/kg}$$

$$\therefore \quad \text{Total work done/kg} = 368.14 + 600.15 - W_p$$
$$= 968.15 - W_p$$
$$\text{Heat supplied/kg} = (h_1 - h_{f_5}) + (h_3 - h_{2'}) - W_p$$
$$= (3198.3 - 251.45) + (3269 - 2830.16) - W_p$$
$$= 2946.85 + 439.84 - W_p$$
$$= 3389.69 - W_p$$
$$\therefore \quad W_p = \frac{p_1 - p_4}{10} = \frac{50 - 0.2}{10} = \frac{49.8}{10} = \mathbf{4.98\ kJ/kg}$$
$$W = (968.29 - 4.98) = \mathbf{963.31\ kJ/kg}$$
$$Q = 3386.69 - 4.98 = \mathbf{3381.71\ kJ/kg}$$
$$\therefore \quad \text{Thermal efficiency} = \frac{W}{Q} = \frac{963.31}{3381.71}$$
$$= \mathbf{0.285} \text{ or } \mathbf{28.5\%}$$

Example 12.3 *In a two stage steam turbine working on ideal Rankine cycle, steam enters the first stage at 90 bar and 550°C temperature. Second stage entry temperature after reheating is 550°C. Steam is never wet during expansion in both sides. Find the cycle efficiency and specific steam consumption. Also find the steam flow rate in the turbine in kg/sec if the turbine is directly connected to an alternator of 150 MW output. Assume 0.9 as alternate efficiency. Draw T-s diagram for the cycle. Neglect pump work.*

Solution

Assume dry and saturated steam after expansion.

$$h_1 = 3518.95$$
$$s_1 = 6.8087$$
$$P_2 = 5.5 \text{ bar}$$
$$h_2 = 2750$$
$$h_3 = 3600$$
$$h_4 = 2590$$

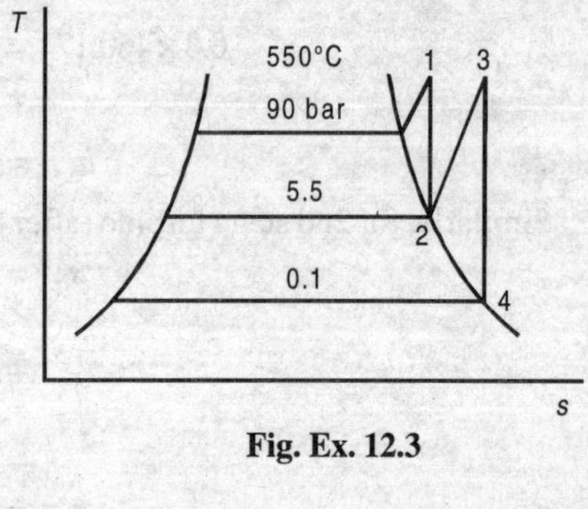

Fig. Ex. 12.3

$p_1 = 90$ bar, $p_2 = 5.5$ bar $= p_3$; $p_4 = 0.1$ bar.

$$\text{Work done/kg} = (h_1 - h_2) + (h_3 - h_4)$$
$$= (3518.95 - 2750) + (3600 - 2590)$$
$$= 768.95 + 1010$$
$$= \mathbf{1778.95\ \ kJ/kg}$$
$$\text{Heat supplied/kg} = (h_1 - h_{f4}) + (h_3 - h_2)$$
$$= (3518.95 - 191.83) + (3600 - 2750)$$
$$= 3397.12 + 850$$
$$= \mathbf{4247.12\ \ kJ/kg}$$
$$\therefore \quad \text{Cycle efficiency} = \frac{W}{Q_A} = \frac{1778.95}{4247.12}$$
$$= 0.4189$$
$$= \mathbf{41.89\%}$$

$$\text{Specific steam consumption} = \frac{3600}{\text{work/kg}}$$

$$= \frac{3600}{1778.95}$$

$$= \mathbf{2.027\ kg/kWh}$$

$$\text{Alternator output} = 150\ \text{MW}$$

$$\therefore \quad \text{Alternator input} = \frac{150}{0.9} = 166.67\ \text{MW} = \mathbf{166670\ kW(kJ/s)}$$

$$\text{Steam flow rate} = \frac{166670}{1778.95} = \mathbf{93.69\ kg/s}$$

Example 12.4. (*a*) *Write a short note on cogeneration.*

(*b*) *In a single feed water, regenerative cycle, the steam enters the turbine at a pressure of 30 bar and 400°C. The exhaust pressure of steam is 0.1 bar. The exhaust pressure of steam is 0.1 bar. The feed water heater is direct-contact type which operates at a pressure of 5 bar. Find the thermal efficiency of the cycle and steam rate. Show the flow diagram, this regenerative cycle on h-s and T-s diagram. (Neglect pump work).*

Solution

The flow diagram and *T-s, h-s*, diagrams are shown. Given data:

$p_1 = 30$ bar, $p_2 = 5$ bar $\qquad p_3 = 0.1$ bar, $T_1 = 400$°C.

For steam table, $\qquad h_{f2} = 640.12$ kJ, $h_{f4} = 191.83$ kJ

From chart, $\qquad h_1 = 3225$ kJ, $p_2 = 2780$ kJ, $h_3 = 2200$ kJ

Consider the heater:

$$\text{Heat given by system} = \text{Heat gained by water}$$

$$\dot{m}_2\left(h_2 - h_{f_2}\right) = \left(1-\dot{m}_2\right)\left(h_{f_2} - h_{f_3}\right)$$

$$\dot{m}_2(2780 - 640.12) = (1-\dot{m}_2)(640.12 - 181.83)$$

$$\dot{m}_2(2139.88) = (1-\dot{m}_2)(458.29)$$

$$\frac{2139.88}{458.29} = 4.6693 = \frac{1-\dot{m}_2}{\dot{m}_2} = \frac{1}{\dot{m}_2} - 1$$

$$\frac{1}{\dot{m}_2} = 5.6693 \text{ or } \dot{m}_2 = 0.1764$$

$$\dot{m}_2 = \mathbf{0.1764}$$

Neglecting pump work,

$$\text{Work done/kg} = 1(h_1 - h_2) + (1-\dot{m}_2)(h_2 - h_3)$$

$$= (3232 - 2780) + (1 - 0.1764) \times (2780 - 2200)$$

$$= 452 + 0.8236 \times 580$$

$$= 452 + 477.69$$

$$= \mathbf{929.69\ kJ/kg}$$

$$\text{Heat supplied/kg} = h_1 - h_{f2}$$
$$= 3232 - 640.12$$
$$= \mathbf{2591.88\ kJ/kg}$$

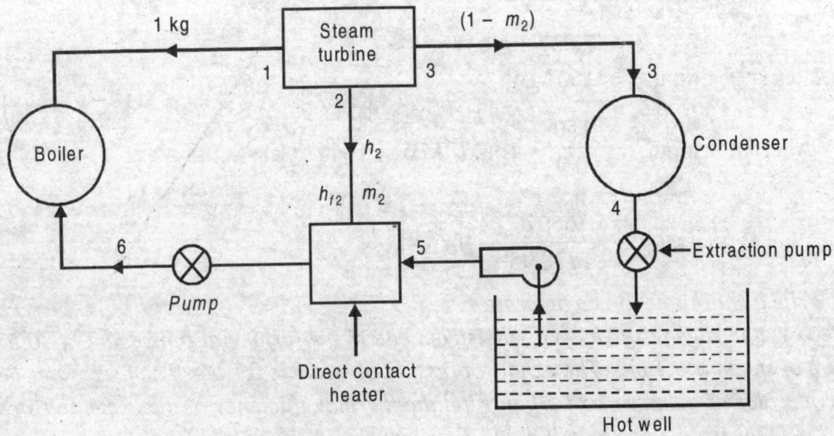

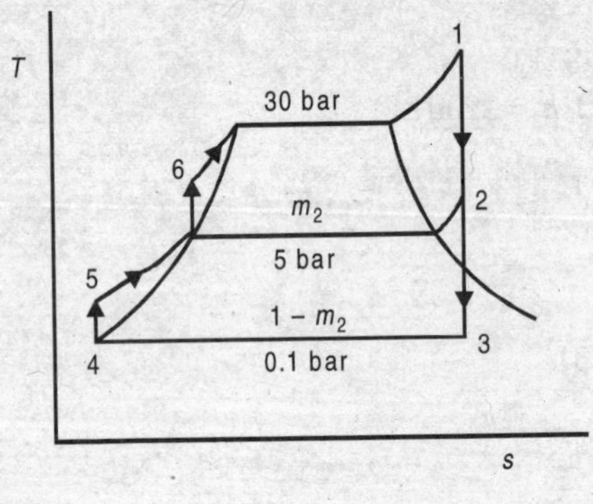

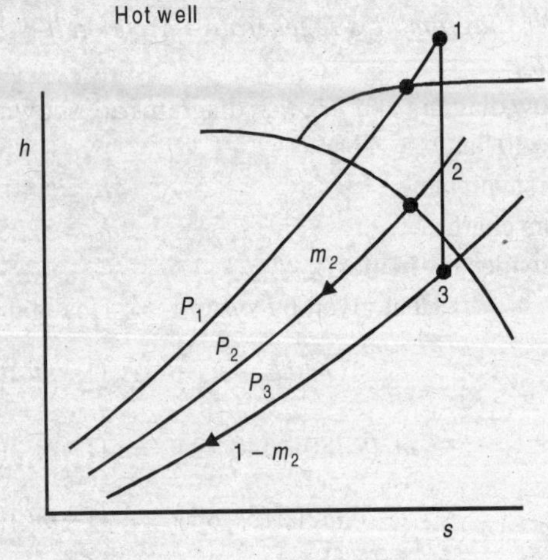

Fig. Ex. 12.4

∴ Thermal Efficiency of the cycle

$$= \frac{\text{Work done/kg}}{\text{Heat added/kg}}$$

$$= \frac{929.69}{2591.88}$$

$$= \mathbf{0.3587}$$
$$= \mathbf{35.87\%}$$

$$\text{Steam rate} = \frac{3600}{\text{Work done/kg}}$$

$$= \frac{3600}{929.69} = \mathbf{3.8722\ kg/kWh}$$

Example 12.5 (*a*) *Explain briefly how a steam turbine plant under variable load is governed.*

(*b*) *A steam power plant equipped with a generative and reheat systems is supplied with steam to HP turbine at 80 bar and 470°C. The part of the steam is extracted at 7 bar for feed water heating and the remaining is reheated to 350°C. It is then expanded in the L.P. turbine to 0.035 bar. Determine the following:*

(*i*) *Power generated by the plant for a steam flow rate of 50 kg/s.*

(*ii*) *Mass of steam bled off per hour for feed heating and the percentage of steam supplied, bled.*

(*iii*) *Efficiency of the plant.*

Draw the plant schematically and represent the cycle on h-s diagram.

Solution

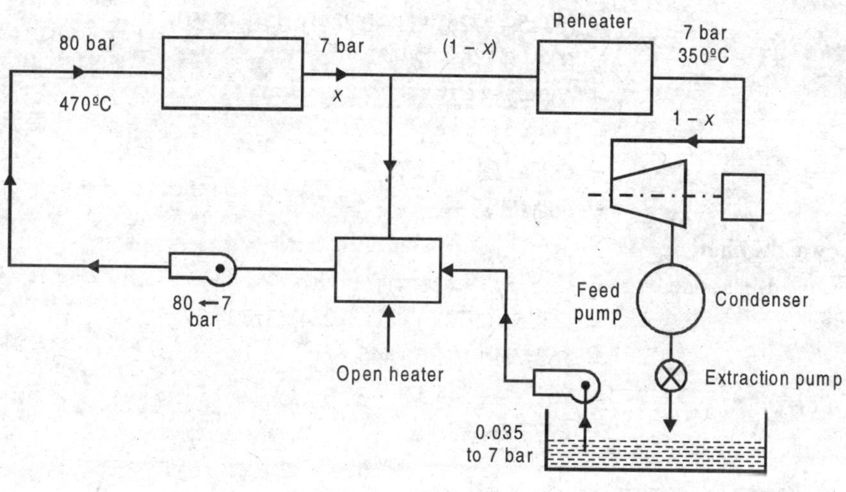

Fig. Ex. 12.5 (a)

Plant in equation is shown on a schematic diagram.

$$p_1 = 80 \text{ bar}$$
$$T_1 = 470°\text{C}$$
$$p_2 = 7 \text{ bar}$$
$$T_3 = 350°\text{C}$$
$$p_4 = 0.035 \text{ bar}$$
$$\text{Steam flow rate} = 50 \text{ kg/s}$$

(i) Power generated : (ii) Steam bled

$h_1 = 3325$ kJ/kg $\quad h_2 = 2725$ kJ/kg

$h_3 = 3170$ kJ/kg $\quad h_4 = 2250$ kJ/kg

$h_{f4} = 111.85$ kJ/kg

$h_{f2} = 697.06$ kJ/kg

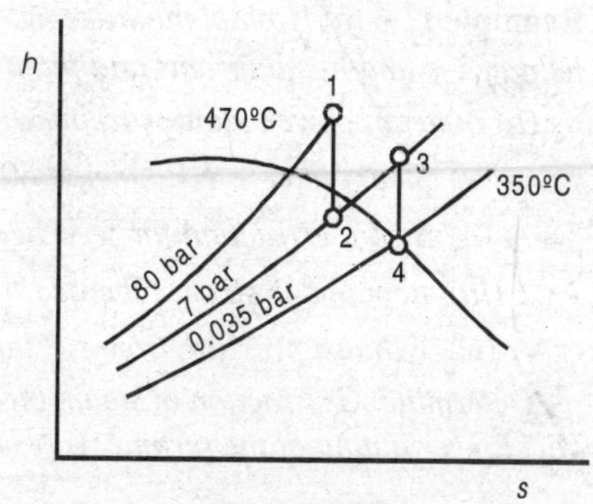

Fig. Ex. 12.5 (b)

Assume 1 kg steam supplied to H.P. turbine and x kg is bled for feed water heating. Again assume a open (direct) feed heater (steam and water are mixed together).

For the feed heater

$$\text{Heat lost by steam} = \text{Heat gained by feed water}$$

$$x(h_2 - h_{f2}) = (1-x)(h_{f2} - h_{f4})$$

$$\therefore \quad (2725 - 697.06)x = (1-x)(697.06 - 111.85)$$

$$2027.94x = (1-x)585.21$$

$$= 585.21 - 585.21x$$

$$\therefore \quad x(2027.94 + 585.21) = 585.21$$

$$x = \frac{585.21}{2613.15}$$

$$= \mathbf{0.224\ kg/kg\ steam = 22.4\%}$$

$$\text{Power generated} = (h_1 - h_2) + (1-x)(h_3 - h_4)$$

$$= (3325 - 2725) + (1 - 0.224)(3170 - 2250)$$

$$= 600 + 0.776 \times 920$$

$$= 600 + 713.92$$

$$= \mathbf{1313.92\ kJ/kg}$$

$$= \mathbf{65696\ kJ/s = kW}$$

$$= \mathbf{66\ MW}$$

(iii) Efficiency of the plant

$$\text{Heat added/kg} = (h_1 - h_{f2}) + (1-x)(h_3 - h_2)$$

$$= (3325 - 697.06) + (1 - 0.224)(3170 - 2725)$$

$$= 2627.94 + 0.776 \times 445$$

$$= 2627.94 + 345.32$$

$$= \mathbf{2973.26\ kJ/kg}$$

$$\therefore \quad \text{Efficiency of the plant} = \frac{W}{Q_A} = \frac{1313.92}{2973.26}$$

$$= \mathbf{0.4419}$$

$$= \mathbf{44.19\%}$$

Example 12.6 (*a*) *Explain the working of a modern reheat regenerative cycle used for steam power plants with the help of T-s and h-s diagrams and plant schematic.*

(*b*) *A steam power plant working on reheat regenerative cycle has the following data:*

(*i*) *Steam inlet to HP turbine 80 bar 470°C.*

(*ii*) *Steam extraction for feed heating at 7 bar.*

(*iii*) *Remainder steam reheated to 350°C in reheater.*

(*iv*) *Exhaust pressure from LP turbine = 0.035 bar.*

Determine: (*i*) *Fraction of steam bled off for reheating.* (*ii*) *Heat supplied/kg of steam in boiler and in reheater.* (*iii*) *Power output of the turbine.* (*iv*) *Thermal efficiency.*

Solution

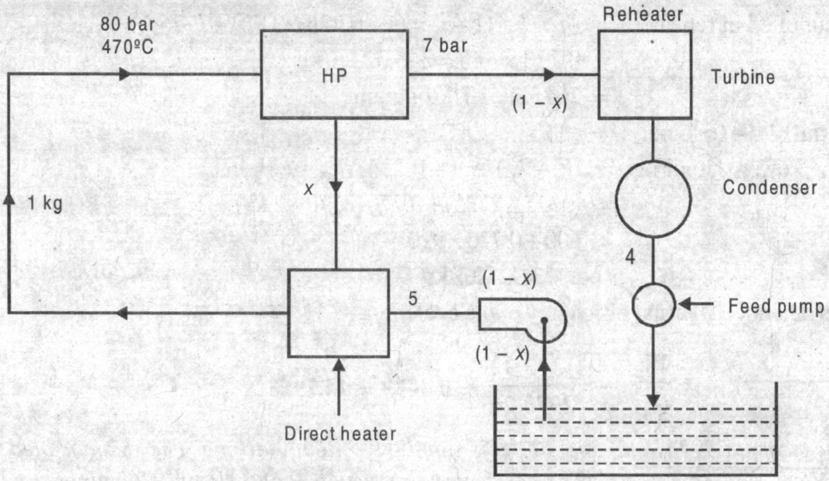

Fig. Ex. 12.6(a)

Assume direct heater.

Referring to Mollier chart,

$h_1 = 3325$ kJ/kg, $h_2 = 2725$, $h_{f2} = 697.06$, $h_{f3} = 3170$, $h_4 = 2250$, $h_{f4} = 111.85$

(i) Fraction of steam bled off or for feed heating: Let x be the steam bled per kg of steam supplied to turbine.

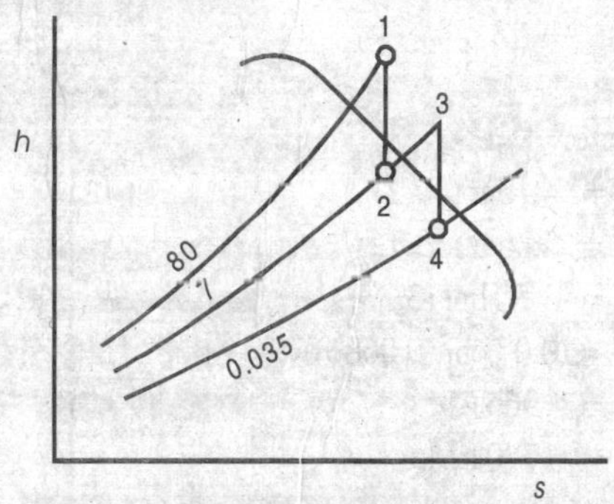

Fig. Ex. 12.6(b)

$\therefore$ Heat given by steam = Heat gained by water

$$x[h_1 - h_{f2}] = (1-x)(h_{f2} - h_{f4})$$

$$\therefore \quad x(2725 - 697.06) = (1-x)(697.06 - 111.85)$$

$$2027.94x = 585.21 - 585.21x$$

$$x = \frac{585.21}{2613.15} = \mathbf{0.224}$$

x = **0.224 kJ/kg steam supplied to turbine**

i.e., **22.4%**

(ii) Heat supplied: Heat supplied in boiler/kg

$$= h_1 - h_{f2} = 3325 - 697.06$$

$$= \mathbf{2627.94\ kJ/kg}$$

Heat supplied in reheater $= (h_3 - h_2)(1 - x) = (1 - 0.224)(3170 - 2725)$

$= 0.776 \times 445 = \mathbf{345.32\ kJ/kg}$

$= \mathbf{2973.26\ kJ/kg\ steam}$

(iii) Power putput of the turbine:

Power output/kg $= (h_1 - h_2) + (1 - 0.224)(h_3 - h_4)$

$= (3325 - 2725) + 0.776(3170 - 2250)$

$= 600 + 0.776 \times 920$

$= \mathbf{1313.92\ kJ/kg}$

(iv) Thermal efficiency: Thermal efficiency of the cycle

$$= \frac{W}{Q_A} = \frac{1313.92}{2973.26} = \mathbf{0.4419 = 44.19\%}$$

Example 12.7 *The steam at 392 N/cm² and 300°C is supplied to steam turbine. The steam is bled for regenerative heating at 137.2 N/cm² and 33.3 N/cm². The condenser pressure is 0.686 N/cm². Calculate*

(i) At each stage the optimum weight of bled beam per kg of steam supplied to the turbine and

(ii) the cycle efficiency.

Assume turbine efficiency of 0.8 for each portion of the expansion and that the feed water leaving each heater is raised to the steam temperature entering the heater.

The heaters used are indirect heaters and the condensate coming out from the first heater is passed through the second heater is passed to the hot well. Neglect pump work.

Solution

Indirect heater

∴ Steam does not mix with feed water

$p_1 = 40$ bar (39.2)

$p_2 = 14$ bar (13.72)

$p_3 = 3.50$ bar (3.33)

$p_4 = 0.07$ bar (0.0686)

$h_1 = 2965$ kJ/kg

$h_a = 1740$ kJ/kg

$h_{2'} = 2740 + (2964 - 2740) \times 0.2$

$= 2740 + 45 = \mathbf{2785\ kJ/kg}$

$h_b = 2540$ kJ/kg

$h_3 \doteq (2785 - 2540) \times 0.2 + 2540$

$= 2540 + 0.2 \times 245 = 2540 + 49 = \mathbf{2589}$

$h_{f3} = 584$ kJ/kg

$h_c = 2330$ kJ/kg

$h_4 = 2330 + 0.2(2589 - 2330)$

$= 2330 + 51.8$

$= \mathbf{2381\ kJ/kg}$

$h_{f4} = 160$ kJ/kg

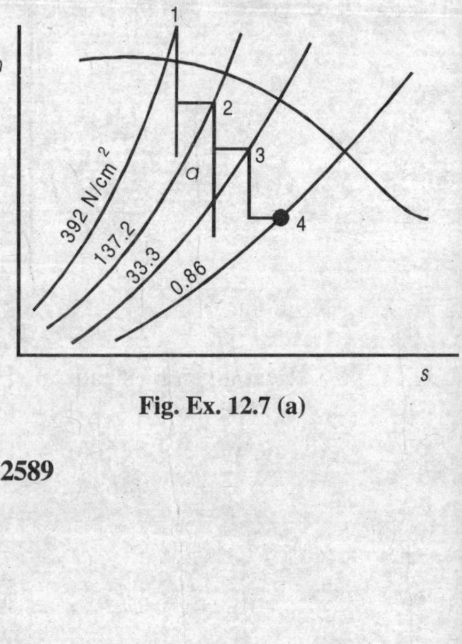

Fig. Ex. 12.7 (a)

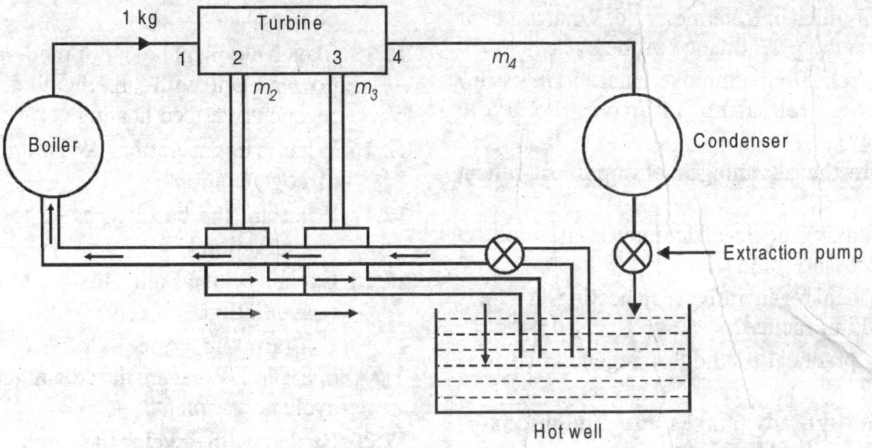

Fig. Ex. 12.7 (b)

(i) Heater at 2

$$\dot{m}_2(h_2 - h_{f_3}) = 1(h_{f2} - h_{f3}) = 830 - 584 = 246$$

$$\therefore \quad \dot{m}_2 = \frac{246}{2785-584} = \frac{246}{2201} = \mathbf{0.1118\,kg}$$

Heater at 3

$$\dot{m}_3(h_3 - h_{f_4}) + \dot{m}_2(h_{f_3} - h_{f_4}) = 1(h_{f3} - h_{f4})$$

$$\dot{m}_3\,(2589-160) + 0.1118\,(584-160) = 584 - 160 = \mathbf{424}$$

$$\therefore \quad \dot{m}_3 = \frac{424-47.4}{2429} = \mathbf{0.155\ kg}$$

(ii) Cycle efficiency

$$\text{Work done/kg supplied} = 1(h_1 - h_2) + (1-\dot{m}_2)(h_2 - h_3) + (1 - \dot{m}_2 - \dot{m}_3)(h_3 - h_4)$$

$$= (2965-2785) + (1-0.1118)\,(2785-2589) + (1-0.1118-0.155)\,(2589-2381.8)$$

$$= 506\text{ kJ/kg}$$

$$\text{Heat input/kg} = 1(h_1 - h_{f2}) = 2965 - 830 = 2135\text{ kJ/kg}$$

$$\therefore \quad \text{Cycle efficiency} = \frac{\text{Work output}}{\text{Heat input}} = \frac{506}{2135} = 0.237$$

$$= \mathbf{23.7\,\%}$$

THEORY QUESTIONS

12.1. Explain. Why Rankine cycle rather than Carnot cycle is used as a standard cycle for steam power plants.

12.2. What is work ratio? What is its importance in power cycles?

12.3. Explain the effect of following parameters on the performance of Rankine cycle:
- (a) admission pressure,
- (b) inlet temperature,
- (c) exhaust pressure.

12.4. Explain the "external" and "internal" irreversibilities of Rankine cycle. What are their effects on the performance of the cycle?

12.5. What is 'reheating' employed in Rankine cycle? How does reheating improve the cycle efficiency?

12.6. What are the advantages of superheat-reheat cycle?

12.7. How Rankine cycle efficiency is affected by reheat pressure ratio?

12.8. Explain ideal regenerative feed heating. Why it is not used in practice?

12.9. What is practical method of regenerative feed heating?

12.10. What are the advantages and limitations of regenerative feed heating?

12.11. What are different types of feed water heaters?

12.12. What is DA? Where and why it is used in power plants?

12.13. Compare following closed feed water heaters.
- (i) With drains cascaded backwards
- (ii) With drains pumped forward.

12.14. How feed water heaters are placed ideally in a power plant?

12.15. Sketch a typical layout of modern power steam power plant with superheating, repeating and regenerative feed heating.

12.16. What is cogeneration? What are the categories of cogeneration?

12.17. What are the basic types of topping cycles in cogeneration?

12.18. Explain with a sketch the gas turbine plant with cogeneration.

12.19. Explain the concept of a combined cycle operation. What are the advantages of combined cycle power plant?

12.20. Explain with a sketch the combined cycle power plant with a heat recovery boiler.

12.21. What is a multipressure combined cycle plant? What are its advantages?

12.22. What is binary vapour cycle ? Explain with a sketch the working of binary vapour power plant. What are its limitations?

PROBLEMS FOR PRACTICE

12.1. Find the ideal cycle efficiency and steam consumption of reheat cycle operating between pressure of 30 and 0.04 bar with a superheat temperature of 450°C. Assume that the first expansion is carried out to the point where the steam is dry saturated and that the steam is reheated to the original temperature. The feed pump work may be neglected.

[**Ans.** $\eta = 0.387$, s.s.c. = 2.39 kg/kWh]

12.2. Steam is supplied to a two stage turbine at 40 bar and 350°C. It expands in the first turbine until it is just dry saturated, then it is reheated to 350°C and expanded through the second stage turbine. The condenser pressure is 0.035 bar. Calculate the work output and the heat supplied per kg of steam for the plant. Calculate also the specific steam consumption and the cycle efficiency.

[**Ans.** (i) 1290 kJ, (ii) 3362 kJ, (iii) 2.79 kg/kWh, (iv) 38.4 %]

12.3. In a reheat cycle, the initial steam pressure and the maximum temperature are 150 bar and 550°C respectively. If the condenser pressure is 0.1 bar and the moisture at the condenser inlet is 15%, and assuming ideal processes, determine (a) the reheat pressure, (b) the cycle efficiency and (c) the steam rate.

[**Ans.** (a) 13.5 bar, (b) 43.6 %, (c) 2.05 kg/kWh]

12.4. In a reheat steam cycle, the maximum steam temperature is limited to 500°C. The condenser pressure is 0.1 bar and the quality at turbine exhaust is 0.8778. Had there been no reheat the exhaust quality would have been 0.7592. Assuming ideal processes, determine (a) the reheat pressure, (b) the boiler pressure, (c) the cycle efficiency, and (d) the steam rate.

[**Ans.** (a) 30 bar, (b) 150 bar, (c) 50.51 %, (d) 1.9412 kg/kWh]

12.5. Find the cycle efficiency and steam consumption of regenerative cycle with one open feed heater. The steam leaves the boiler at 30 bar and superheated to 450°C. The bleed pressure is so that the temperature difference is divided into approximately equal steps.

[**Ans.** $\eta = 0.373$, s.s.c. = 4.29 kg/kWh]

12.6. Calculate the efficiency of regenerative cycle using a single feed water heater, neglecting the feed pump work. The steam leaves the boiler at 30 bar and superheated to 450°C and the condenser pressure is 0.035 bar. The heater is of closed type and bled steam is flashed into the condenser. [**Ans.** Bleeding pressure = 2.7 bar, Steam extracted = 0.165 kg, Efficiency = 0.401]

12.7. A steam turbine is supplied with steam at 40 bar and 450°C and exhausted at 0.035 bar. Neglecting the feed pump work, find the ideal Rankine cycle efficiency of the plant and compare it with the ideal efficiency, which could be obtained if regenerative feed heating were employed. Assume the use of open feed water heater at one bleeding point at 3 bar. Also calculate the amount of steam extracted per kg of steam supplied.

[**Ans.** Ideal efficiency = 0.391, Ideal efficiency with feed water heater = 0.415, Steam extracted = 0.173 kg of steam supplied]

12.8. In a single-heater regenerative cycle the steam enters the turbine at 30 bar, 400°C and the exhaust pressure is 0.10 bar. The feedwater heater is a direct contact type which operates at 5 bar. Find (a) the efficiency and the steam rate of the cycle, and (b) the increase in mean temperature of heat addition, efficiency and steam rate, as compared to the Rankine cycle (without regeneration). Neglect pump work.

[**Ans.** (a) 35.36 %, 3.93 kg/kWh, (b) 27.4°C, 1.18%, 0.47 kg/kWh]

12.9. A regenerative steam cycle employing three closed heaters the steam is supplied to the turbine at 42 bar and 500°C and is exhausted to the condenser at 0.035 bar. The bled steam for heating is taken at pressure of 15.4 and 0.5 bar. Assuming the ideal processes and neglecting pump work, calculate:

(i) the fraction of steam bled at each stage
(ii) power output per kg of steam flow rate
(iii) the cycle efficiency.

[**Ans.** (1) 0.0969, 0.0902, (ii) 1133.6 kW/kg, (iii) 43.6%]

12.10. A simple steam power cycle uses solar energy for the heat input. Water in the cycle enters the pump as a saturated liquid at 40°C, and is pumped to 2 bar. It then evaporates in the boiler at this pressure, and enters the turbine as saturated vapour. At the turbine exhaust the conditions are 40°C and 10% moisture. The flow rate is 150 kg/h. Determine (a) the turbine isentropic efficiency, (b) the net work output, (c) the cycle efficiency, and (d) the area of the solar collector needed if the collectors pick up 0.58 kW/m^2.

[**Ans.** (a) 76.7 %, (b) 15.5 kW, (c) 12.78 %, (d) 182 m^2]

12.11. The net power output of an ideal reheat regenerative steam cycle is 80 MW. Steam enters the h.p. turbine at 80 bar, 50W and expands till it becomes saturated vapour. Some of the steam then goes to an open feedwater heater and the balance is reheated to 400°C, after which it expands in an L.P. turbine to 0.07 bar. Compute: (a) the reheat pressure, (b) the steam flow rate to the h.p. turbine, (c) the cycle efficiency, and (d) the rate of flow of cooling water in the condenser if the temperature rise of water is 8°C. (e) If the velocity of steam flowing from the turbine to the condenser is limited to 130 in/s, find the diameter of the connecting pipe.

[**Ans.** (a) 6.5 bar, (b) 58.4 kg/s, (c) 43.7%, (d) 3146.5 kg/s, (e) 2.97 m]

12.12. A binary vapour cycle operates on mercury and steam. Saturated mercury vapour at 4.5 bar is supplied to the mercury turbine, from which it exhausts at 0.04 bar. The mercury condenser generates saturated steam at 15 bar which is expanded in a steam turbine to 0.04 bar. (a) Find the overall efficiency of the cycle. (b) If 50,000 kg/h of steam flows through the steam turbine, what is the flow through the mercury turbine? (c) Assuming that all processes are reversible, what is the useful work done in the binary vapour cycle for the specified steam flow? (d) If the steam leaving the mercury condenser is superheated to a temperature of 300°C in a superheater located in the mercury boiler, and if the internal efficiencies of the mercury and steam turbines are 0.85 and 0.87 respectively, calculate the overall efficiency of the cycle. The properties of saturated mercury are given below:

$p(bar)$	$t(°C)$	h_f (kJ/kg)	h_g	s_f (kJ/kg–°K)	s_g	v_f (m³/kg)	v_g
4.5	450	62.93	355.98	0.1352	0.5397	79.9×10^{-6}	0068
0.04	216.9	29.98	329.85	0.0808	0.6925	76.5×10^{-6}	5.178

[**Ans.** (a) 53%, (b) 59.35×10^4 kg/h, (c) 28.5 MW, (d) 46.2%]

12.13. In a combined cycle, the gas turbine portion has a pressure ratio of 16. Air enters the compressor at 300 °K at a rate of 14 kg/s and heated to 1500 °K in the combustion chamber. The combustion gases leaving the gas turbine are used to heatsteam to 460°C at 100 bar in a heat exchanger. The combustion gases leave the heat exchanger at 420°C. The steam leaving the turbine is condensed at 15 kPa. Assuming all the compression and expansion processes to be isentropic, determine (1) the mass flow rate of the steam, (2) the net power output and (3) the thermal efficiency of the combined cycle. Assume constant specific heat for air at room temperature.

[**Ans.** (1) 1.275 kg/s, (2) 7881 kW, (3) 66.3%]

12.14. A pass-out two-stage turbine receives steam at 50 bar, 350°C. At 1.5 bar, the h.p. steam exhausts and 12,000 kg of steam per hour are taken at this stage for process purposes. The remainder is reheated at 1.5 bar to 250°C and then expanded through the i.p. turbine to a condenser pressure of 0.05 bar. The power output from the turbine unit is 3750 kW. Take isentropic efficiency for h.p. and i.p. stages as 0.84 and 0.81 respectively.

Calculate the boiler capacity required.

[**Ans.** 18.5 t/h]

12.15. Steam is generated at 70 bar, 500°C and expands in a turbine to 30 bar with an isentropic efficiency of 77%. At this condition, it is mixed with steam twice its mass at 30 bar, 400°C. The mixture then expands with an isentropic efficiency of 80% to 0.06 bar. At a point in the expansion where the pressure is 5 bar, steam is bled for feedwater heating in a direct contact heater, which raises the feedwater to the saturation temperature of the bled steam. Calculate the mass of steam bled per kg of high pressure steam and the cycle efficiency. Assume that the L.P. expansion condition line is straight.

[**Ans.** 0.53 kg, 31.9 %]

CHAPTER 13 Air Standard Cycles

13.1 INTRODUCTION

In all heat engines — particularly Internal combustion Engines — the common working substance is air. In addition to air, these different engines use different fuels and in different quantities. To avoid the effect of different properties of fuels used, and because the quantity of fuel is very very small as compared to air, (generally air to fuel ratio is 12 to 16) we generally neglect the fuel quantity and hence the effect of fuel, and the study is theoretical only. Therefore, the working substance in the study is assumed to be air only. The cycle of working of the engine is called as theoretical or standard cycle and as the working substance is air, the cycle is called *Air Standard Cycle*.

We will study in this chapter four different air standard cycles viz.

(i) Otto cycle; (ii) Diesel cycle; (iii) Combination of Otto and Diesel Cycles; (iv) Stirling cycle.

13.2 ASSUMPTIONS MADE FOR AIR STANDARD CYCLES

The analysis of all air standard cycles is based upon the following assumptions:

(1) The working substance used in the engine is air.
(2) Air is assumed to behave as an ideal gas and therefore obeys all gas laws.
(3) The physical properties of air in the cylinder remain unchanged for the pressures and temperatures encountered in practice i.e., specific heats (C_P and C_V) do not very with the temperature. Also specific heat ratio $\gamma = \dfrac{C_P}{C_V}$ is constant.
(4) Compression and expansion process are isentropic (reversible and adiabatic).
(5) No chemical reaction takes place in the cylinder. Heat is supplied or rejected by bringing a hot body or a cold body in contact with the cylinder cover at the appropriate points during the cycle.
(6) The cycle is considered or assumed to be closed cycle, with the same air always remaining in the cylinder to repeat the cycle.

Because of many simplifying assumptions, it is clear that the air standard cycle approximations do not closely represent the conditions within the cylinder of an actual engine. Although the quantities calculated from this approximations are considerably in error, the trends shown are usually correct, so that the general effect on the efficiency of such variables as inlet pressure or compression ratio may be calculated. Because of the simplicity of air standard cycle calculations, it is often used to obtain rough answers to complex engine problems.

13.3 OTTO – CYCLE: (CONSTANT VOLUME CYCLE)

To avoid high values of pressures and compression ratios, a practical cycle was introduced by a German Scientist

Dr. A .N. Otto in 1876 and it was successfully applied for the working of petrol and gas engines. This cycle consists of two isentropic processes and two constant volume processes as shown in Fig. 13.1.

The sequences of operations is as follows: air as the working substance in the cylinder is at a pressure P_1, volume V_1 and temperature T_1 and is shown by a point 1 on both *P–V* and *T–S* diagram in Fig. 22.1 (a) and (b).

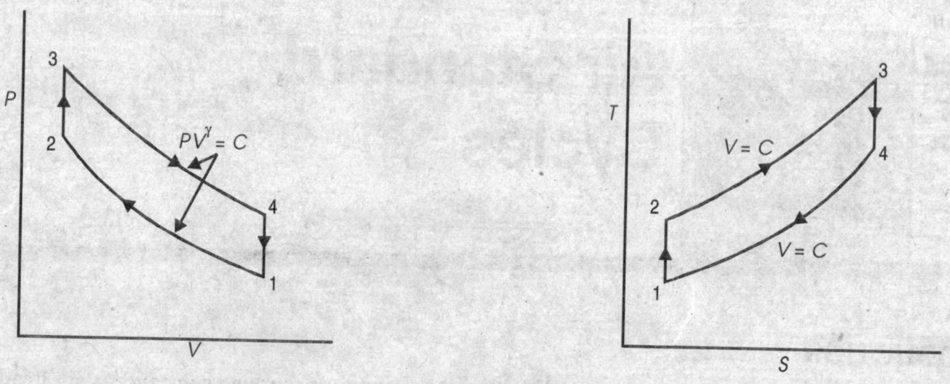

Fig. 13.1

Process 1–2. Air is compressed isentropically to 2. Ratio V_1 to V_2 is called compression ratio donated by r_k. Cover is adiabatic. At the end of compression, diabatic cover replaces adiabatic cover.

Process 2–3. At the end of compression, heat is added at constant volume by bringing a hot body near the cylinder diabatic cover. Heat is added at constant volume and hence pressure and temperature increase. Because the heat is added at constant volume, the cycle is called as *Constant Volume Cycle.*

Process 3–4. After heat is added, air expands isentropically to state 4. Here the expansion ratio is given by $r_e = \dfrac{V_4}{V_3}$. For Otto cycle $r_k = r_e$ Adiabatic cover is kept during expansion.

Process 4–1. At the end of isentropic expansion, adiabatic cover is replaced by diabatic cover and the cold body is brought near the cover. Heat is rejected at constant volume thus decreasing the pressure and temperature to initial values.

Thus the cycle is completed.

(a) Thermal Analysis of Cycle:

Heat added/kg = heat added during 2–3

$$Q_A = C_V(T_3 - T_2) \text{ kJ/kg}$$

Heat rejected/kg = heat rejected during 4–1

$$Q_R = C_V(T_4 - T_1) \text{ kJ/kg}$$

Because $Q = 0$ during 1–2 and 3–4

$$\text{Work done/kg} = Q_A - Q_R$$

$$= C_V(T_3 - T_2) - C_V(T_4 - T_1)$$

or Work done /kg = Area of $P - V$ diagram

= (Area under 3–4) – (Area under 2–1)

$$= \frac{p_3v_3 - p_4v_4}{\gamma - 1} - \frac{p_2v_2 - p_1v_1}{\gamma - 1}$$

We can replace *pv* product by *RT* for unit mass.

$\therefore$ Work done /kg $= \dfrac{R}{\gamma-1}[T_3 - T_4 - T_2 + T_1]$

$$= C_V[(T_3 - T_2) - (T_4 + T_1)]$$

$$= C_V(T_3 - T_2) - (T_4 + T_1)$$

$$= Q_A - Q_R \text{ as before.}$$

$\therefore$ Thermal efficiency of the cycle = Air standard cycle efficiency shortly given by ASE

$$\text{ASE} = \frac{\text{Work done/kg}}{\text{Heat supplied/kg}}$$

$$= \frac{Q_A - Q_R}{Q_A}$$

$$= 1 - \frac{Q_R}{Q_A}$$

$$= 1 - \frac{C_V(T_4 - T_1)}{C_V(T_3 - T_2)}$$

$$\mathbf{ASE = 1 - \frac{T_4 - T_1}{T_3 - T_2}} \quad ...(1)$$

This is the efficiency given in terms of temperatures.

For compression process, we have

$$\frac{\text{Higher temperature}}{\text{Lower temperature}} = (\text{Compression ratio})^{\gamma-1}$$

$$\therefore \quad \frac{T_2}{T_1} = r_k^{\gamma-1}$$

$$\therefore \quad T_2 = T_1 r_k^{\gamma-1}$$

Similarly, for expansion process 3–4

We have $\dfrac{T_3}{T_4} = r_e^{\gamma-1} = r_k^{\gamma-1}$

$$T_3 = T_4 \cdot r_k^{\gamma-1}$$

Substituting these values of T_2 and T_3 in Eq. (1) we get

$$\text{ASE} = 1 - \frac{T_4 - T_1}{T_4 \cdot r_k^{\gamma-1} - T_1 \cdot r_k^{\gamma-1}}$$

$$= 1 - \frac{T_4 - T_1}{r_k^{\gamma-1}(T_4 - T_1)}$$

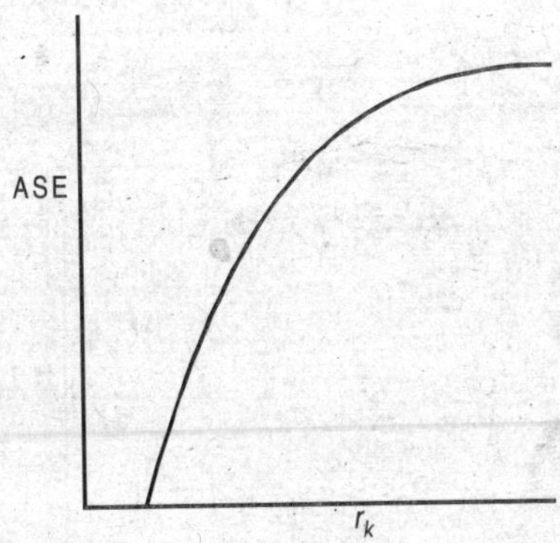

Fig. 13.2

$$\mathbf{ASE = 1 - \frac{1}{r_k^{\gamma-1}}} \quad ...(2)$$

We note from these equations that the air-standard efficiency of Otto–Cycle depends upon the compression ratio only and it increases with the increase in compression ratio. The variation of ASE with compression ratio r_k is shown in Fig. 13.2. For $r_k = 1$, efficiency is zero.

The compression ratio used in actual engine working on this cycle lies between 5 and 9 and depends upon the type of the fuel used.

Example. *Show that the compression ratio for the maximum work to be done per kg of air in an Otto cycle is given by*

$$r_k = \sqrt{\left(\frac{T_3}{T_1}\right)^{1/\gamma-1}}$$

where T_3 and T_1 are the upper and lower limits of absolute temperatures in the cycle.

Solution

$$\begin{aligned}\text{We have, Work done/kg} &= Q_A - Q_R \\ &= C_V [(T_3 - T_2) - C_V (T_4 - T_1)] \\ &= C_V (T_3 - T_2 - T_4 + T_1)\end{aligned}$$

We get $T_2 = T_1 r^{\gamma-1}{}_k$ and $T_3 = T_4 r_k^{\gamma-1}$

$$\therefore \qquad T_4 = \frac{T_3}{r_k^{\gamma-1}}$$

$$\therefore \qquad \text{Work done/kg} = C_V\left(T_3 - T_1 r_k^{\gamma-1} - \frac{T_3}{r_k^{\gamma-1}} + T_1\right)$$

$\therefore$ Work done is a function of r_k only as T_3 and T_1 are constants

$\therefore$ Work is maximum when $\dfrac{dw}{dr_k} = 0$

$$\therefore \qquad 0 = 0 - (\gamma-1)T_1 \cdot r_k^{\gamma-1} - 1 + T_3 \frac{\gamma-1}{T_k^{(\gamma-1)^2}} \cdot r_k^{\gamma-2}$$

$$= -T_1(\gamma-1) + T_3 \cdot \frac{\gamma-1}{T_k^{(\gamma-1)^2}}$$

$$\therefore \qquad T_1 = T_3 . \frac{1}{r_k^{(\gamma-1)^2}}$$

$$\text{or} \qquad r_k^{(\gamma-1)^2} = \frac{T_3}{T_1}$$

$$\mathbf{r_k} = \sqrt{\left(\frac{\mathbf{T_3}}{\mathbf{T_1}}\right)^{\mathbf{1/\gamma-1}}}$$

$$= \left(\frac{T_3}{T_1}\right)^{\frac{1}{2(\gamma-1)}}$$

$$= \left(\frac{T_3}{T_1}\right)^{\frac{1}{2\times 0.4}}$$

$$= \left(\frac{T_3}{T_1}\right)^{1.25}$$

$$\therefore \qquad \text{Max. Eff.} = 1-\frac{1}{r_k^{\gamma-1}}$$

$$= 1-\frac{1}{\sqrt{\frac{T_3}{T_1}}}$$

$$\eta_{max} = 1-\frac{1}{\sqrt{\frac{T_1}{T_3}}} = 1-\sqrt{\frac{T_3}{T_1}}$$

Example. *An engine working on Otto–cycle in which the salient points 1, 2, 3, 4 has upper and lower temperature limits T_3 and T_1. If the maximum work per kg air is to be done, then show that the intermediate temperature is given by*

$$T_2 = T_4 = \sqrt{T_1 T_3}$$

Solution

$$\text{Work done/kg} = Q_A - Q_R$$

$$= C_V(T_3 - T_2) - C_V(T_4 - T_1)$$

$$= C_V(T_3 - T_2 - T_4 + T_1)$$

As $\qquad r_k = r_e$ then $\dfrac{T_2}{T_1} = \dfrac{T_3}{T_4}$

or $\qquad T_4 = \dfrac{T_1 T_3}{T_2}$

Substituting this value of T_4 in the equation for work done, we get

$$\dot{W} = C_V\left[T_3 - T_2 - \frac{T_1 T_3}{T_2} + T_1\right]$$

Here T_3 and T_1 are the upper and lower limits and are constants.

$\therefore$ Work done is a function of T_2 only

$\therefore$ Work is maximum when $\dfrac{dw}{dT_2} = 0$

$$\therefore \qquad \frac{dw}{dt_2} = 0 = 0 - 1 + \frac{T_1 T_3}{T_2^2} + 0$$

$$= -1+\frac{T_1T_3}{T_2^{\,2}}$$

$$\therefore \quad T_1T_3 = T_2^2$$

$$\mathbf{T_2} = \sqrt{\mathbf{T_1T_3}}$$

Similarly $$T_4 = \frac{T_1T_3}{T_2} = \frac{T_1T_3}{\sqrt{T_1T_3}}$$

$$\mathbf{T_4} = \sqrt{\mathbf{T_1T_3}}$$

$$\therefore \quad \mathbf{T_2} = \mathbf{T_4} = \sqrt{\mathbf{T_1T_3}}$$

(b) Mean Effective Pressure–MEP

Mean effective pressure is defined as that constant pressure, acting on the piston throughout the stroke, which will produce the work equal to that of the cycle.

If d and L are the diameter and stroke of the piston, then the work produced by the mean effective pressure is given by

$$= P_m \times A \times L = P_m \times \text{stroke volume}$$
$$= P_m \times V_s$$

This is equal to the work of the cycle.

$$\therefore \quad P_m = \text{Mean effective pressure}$$
$$= \frac{\text{Work done during cycle}}{\text{Stroke volume}}$$

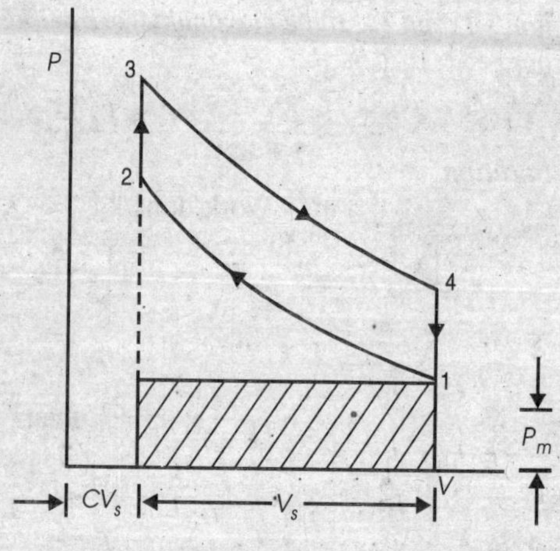

Fig. 13.3

Now stroke volume $= V_s$ from $P–V$ diagram

$$= V_1 - V_2$$

$$= V_1\left[1-\frac{V_2}{V_1}\right]$$

$$= V_1\left[1-\frac{1}{r_k}\right]$$

$$= V_1\left[\frac{r_k-1}{r_k}\right]$$

Also work done during cycle $=$ Area of $P–\dot{V}$ diagram 1–2–3–4–1

$=$ Area under $(3-4)$ – Area under $(2-1)$

$$= \frac{P_3V_3 - P_4V_4}{\gamma-1} - \frac{P_2V_2 - P_1V_1}{\gamma-1}$$

We will write all pressures in terms of p_1 and volumes in terms of V_1

First consider volumes,

$$\therefore \quad V_1 = V_4 \text{ and } V_2 = V_3$$

$$r_k = \frac{V_1}{V_2} \quad \therefore \quad V_2 = V_3 = \frac{V_1}{r_k}$$

Now consider pressures

$$\text{For process } 1-2\ P_1V_1^{\gamma} = P_2V_2^{\gamma}$$

$$\therefore \quad P_2 = P_1 \times \frac{V_1^{\gamma}}{V_2^{\gamma}} = P_1 \times r_k^{\gamma}$$

Then for process 3 – 4

$\frac{P_3}{P_2}$ is the explosion ratio say, denoted by $\propto$,

$$\therefore \quad \frac{P_3}{P_2} = \propto \text{ and } P_3 = P_2 \propto$$

$$= P_1 \cdot r_k^{\gamma} \cdot \propto$$

$$\text{For process } 3-4\ p_3v_3^{\gamma} = p_4v_4^{\gamma}$$

$$\text{or} \quad p_4 = p_3 \times \left(\frac{V_3}{V_4}\right)^{\gamma} = p_3 \times \left(\frac{V_2}{V_1}\right)^{\gamma} = p_3 \times \frac{1}{r_k^{\gamma}}$$

$$\therefore \quad = \frac{p_1 \times r_k^{\gamma}.\alpha}{r_k^{\gamma}}$$

$$= p_1\alpha$$

Substituting these values in the equation for p_m, we get

$$p_mV_1\left[\frac{r_k-1}{r_k}\right] = \frac{1}{\gamma-1}\left[p_1\alpha r_k^{\gamma}\frac{V_1}{r_k} - p_1\alpha V_1 - p_1\frac{V_1}{r_k}r_k^{\gamma} + p_1V_1\right]$$

$$p_m = \frac{r_k \cdot p_1}{(r_k-1)(\gamma-1)}\left[\alpha.r_k^{\gamma-1} - \alpha - r_k^{\gamma-1} + 1\right]$$

$$= p_1 \cdot \frac{r_k}{(r_k-1)(\gamma-1)} \cdot \left[r_k^{\gamma-1}(\alpha-1) - (\alpha-1)\right]$$

$$= p_1 \cdot \frac{r_k}{(r_k-1)(\gamma-1)} \cdot \left[(\alpha-1)\left(r_k^{\gamma-1}-1\right)\right]$$

Alternative method

This method considers temperatures instead of pressures and volumes

$$\text{Work done/kg} = C_v(T_3 - T_2) - C_v(T_4 - T_1)$$

We have, $T_2 = T_1 - r_k^{\gamma-1}, T_3 = \alpha T_2 = \alpha.T_1 r_k^{\gamma-1}$

$$T_4 = \frac{T_3}{r_k^{\gamma-1}} = \frac{\alpha.T_1.r_k^{\gamma-1}}{r_k^{\gamma-1}} = \alpha T_1$$

$$\therefore \qquad p_m \times v_1 . \frac{r_k - 1}{r_k} = \frac{R}{\gamma - 1}\left[\alpha . T_1 r_k^{\gamma-1} - T_1 r_k^{\gamma-1} - \alpha T_1 + T_1\right]$$

$$V_1 = \text{Volume of kg} = \frac{RT_1}{p_1}$$

$$\therefore \qquad p_m \times \frac{RT_1}{p_1} . \frac{r_k - 1}{r_k} = \frac{RT_1}{\gamma - 1}\left[r_k^{\gamma-1}(\alpha - 1) - (\alpha - 1)\right]$$

$$\mathbf{p_m} = \frac{\mathbf{p_1 . r_k}}{(\gamma - 1)(\mathbf{r_k} - 1)}\left[(\alpha - 1)(\mathbf{r_k}^{\gamma-1} - 1)\right]$$

Same as before

$$\therefore \qquad \mathbf{p_m} = \mathbf{p_1} . \frac{\mathbf{r_k}(\mathbf{r_k}^{\gamma-1} - 1)}{(\mathbf{r_k} - 1)} \times \frac{(\alpha - 1)}{(\gamma - 1)}$$

If area of p–v diagram and length of diagram are given, then mean effective pressure can be found by

$$p_m = \frac{\text{Area of diagram (mm)}^2}{\text{Length of diagram (mm)}} \times \text{Pressure scale}$$

The unit for pressure scale is bar/mm

So that unit of p_m is bar.

If the scale is N/m²/mm, then p_m-N/m² and so on.

13.4 DIESEL CYCLE: (CONSTANT PRESSURE CYCLE)

This cycle was introduced by Rudolph Diesel in 1893 and it is successfully used in all Diesel engines or oil engines. This cycle consists of two isentropics, one constant pressure and one constant volume processes.

The cycle is shown on P–V and T–S diagrams as shown in Fig. 13.4.

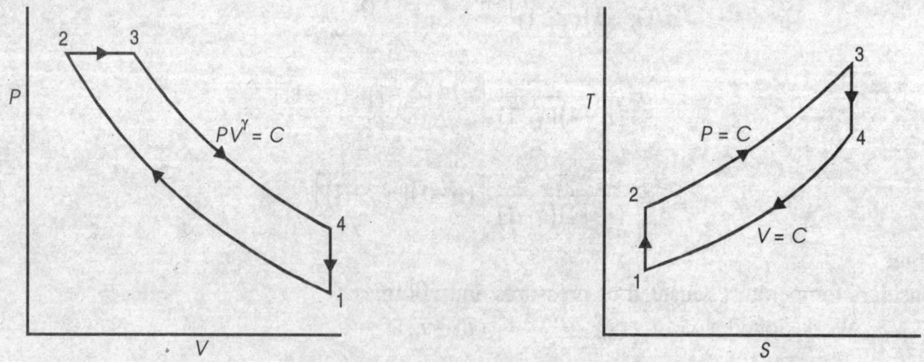

Fig. 13.4

The sequence of operations is as follows:

As before, air as the working substance in the cylinder is at pressure P_1, volume V_1 and temperature T_1 and is shown by state point 1 on both P–V and T–S diagram of Fig. 13.4 (a) and 13.4 (b).

1–2 : Air is compressed isentropically to 2. Ratio V_1 to V_2 is called compression ratio denoted by r_k. Cylinder cover is adiabatic.

2–3: At the end of compression, diabetic cover is placed and heat is added at constant pressure by bringing a hot body near the cylinder cover. Heat is added for a part of the stroke. At the point 3, the heat addition is stopped or cut-off. The ratio of the volume of cut-off point to the initial volume before heat addition is called the cut-off ratio and is donated by r_c. Therefore, $r_c = \frac{V_3}{V_2}$. Because the heat is added at constant pressure, the cycle is also called as constant pressure cycle.

3–4: After the heat is added, air expands isentropically to state 4. Here the expansion ratio is given by $r_e = \frac{V_4}{V_3} = \frac{V_1}{V_3}$.

Here the cover is adiabatic.

$$\therefore \quad r_e = \frac{V_1}{V_3} = \frac{V_1}{V_2} \times \frac{V_2}{V_3} = r_k / r_e$$

or

$$\mathbf{r_k = r_e \cdot r_c}$$

4 – 1: Adiabatic cover is replaced by diabatic cover at the end of isentropic expansion, the cold body is brought near the cover. Heat is rejected at constant volume thus decreasing the pressure and temperature to initial values. Thus the cycle is completed.

13.4.1 Thermal Analysis of Cycle

$$\text{Heat added kg} = Q_A = C_p(T_3 - T_2)\,\text{kJ/kg}$$

$$\text{Heat rejected kg} = Q_R = C_v(T_4 - T_1)\,\text{kJ/kg}$$

$$\therefore \quad \text{Work done/kg} = W = Q_A - Q_R$$

$$= C_p(T_3 - T_2) - C_V(T_4 - T_1)$$

$\therefore$ Air standard efficiency

$$\text{ASE} = \frac{\text{Work done/kg}}{\text{Heat added/kg}}$$

$$= \frac{Q_A - Q_R}{Q_A} = 1 - \frac{Q_R}{Q_A}$$

$$= 1 - \frac{C_p(T_3 - T_2) - C_V(T_4 - T_1)}{C_P(T_3 - T_2)}$$

$$= 1 - \frac{C_V(T_4 - T_1)}{C_P(T_3 - T_2)}$$

$$= 1 - \frac{(T_4 - T_1)}{\gamma(T_3 - T_2)}$$

Now we have

$$\frac{T_2}{T_1} = r_k^{\gamma-1} \quad \therefore \quad T_2 = T_1 . r_k^{\gamma-1}$$

$$\frac{T_3}{T_2} = \frac{V_3}{V_2} = r_c \text{ for constant pressure process.}$$

$$\therefore \quad T_3 = T_2 \cdot r_c = T_1 \cdot r_k^{\gamma-1} \cdot r_c$$

Again
$$\frac{T_3}{T_4} = r_e^{\gamma-1} = \left(\frac{r_k}{r_c}\right)^{\gamma-1}$$

$$\therefore \quad T_4 = T_3 \times \left(\frac{r_k}{r_c}\right)^{\gamma-1}$$

$$T_4 = T_1 \cdot r_k^{\gamma-1} \cdot r_c \times \frac{r_c^{\gamma-1}}{r_k^{\gamma-1}}$$

$$= T_1 \cdot r_c^{\gamma-1}$$

Substituting these values of temperatures in the equation for ASE we get

$$\text{ASE} = 1 - \frac{T_1 r_c^{\gamma-1} - T_1}{\gamma\left[T_1 \cdot r_k^{\gamma-1} \cdot r_c - T_1 \cdot r_k^{\gamma-1}\right]}$$

$$= 1 - \frac{r_c^{\gamma} - 1}{\gamma \cdot r_k^{\gamma-1}(r_c - 1)}$$

$$\therefore \quad \textbf{ASE} = \mathbf{1 - \frac{r_c^{\gamma} - 1}{\gamma \times r_k^{\gamma-1}(r_c - 1)}}$$

From this equation we see that the efficiency decreases as cut-off ratio increases.

Again as the compression ratio is equal to that for Otto-cycle, ASE of Otto-cycle is more than ASE of Diesel cycle.

In pratice, the compression ratio of Diesel engines ranges from 12 to 22.

13.4.2 Mean Effective Pressure of Diesel Cycle

Work done = Area of P–V diagram

= Area 1–2–2–3–4–1

= Area under (2–3–4) Area under (2–1)

$$= p_2(V_3 - V_2) + \frac{p_3V_3 - p_4V_4}{\gamma - 1} - \frac{p_2V_2 - p_1V_1}{\gamma - 1}$$

As before, we will write all pressure in terms of P_1 and all volumes in terms of V_1.

First consider volumes

$$V_2 = \frac{V_1}{r_k}, \frac{V_3}{V_2} - r_c$$

$$\therefore \quad V_3 = V_2 r_c = \frac{V_1 \cdot r_c}{r_k}$$

$$V_4 = V_1$$

Again for pressures

$$p_2 = p_1 \cdot r_k^{\gamma} = p_3$$

$$\frac{p_3}{p_4} = r_e^{\gamma} = \left(\frac{r_k}{r_c}\right)^{\gamma}$$

$$\therefore \quad p_4 = \frac{p_3 \cdot r_c^{\gamma}}{r_k^{\gamma}} = \frac{p_1 r_k^{\gamma} \cdot r_c^{\gamma}}{r_k^{\gamma}} = p_1 r_c^{\gamma}$$

Substituting these values we get,

$$p_m \times V_1 \cdot \frac{r_k - 1}{r_k} = \left(\frac{V_1 r_c}{r_k} - \frac{V_1}{r_k}\right) p_1 r_k^{\gamma} + \left(\frac{p_1 r_k^{\gamma} \cdot \frac{V_1 r_c}{r_k} - p_1 r_c^{\gamma} \cdot V_1}{\gamma_k}\right) - \left(\frac{p_1 r_k^{\gamma} \cdot \frac{V_1}{r_k} - p_1 V_1}{\gamma - 1}\right)$$

$$p_m \times \frac{r_k - 1}{r_k} = \left[r_k^{\gamma-1}(r_c - 1) + \frac{r_k^{\gamma-1} . r_c - r_c^{\gamma}}{\gamma - 1} - \frac{r_k^{\gamma-1} - 1}{\gamma - 1}\right] p_1$$

$$= \frac{p_1}{\gamma - 1}\left[r_k^{\gamma-1}(r_c - 1)(\gamma - 1) + \left(r_k^{\gamma-1} - r_c^{\gamma-1}\right) - r_c\left(r_k^{\gamma-1} - 1\right)\right]$$

$$= \frac{p_1 \cdot r_k^{\gamma-1}}{\gamma - 1}\left[(r_c - 1)(\gamma - 1) + (r_c - 1) - \left(\frac{r_c^{\gamma} - 1}{r_k^{\gamma-1}}\right)\right]$$

$$p_m = \frac{p_1 \cdot r_k^{\gamma-1} \cdot r_k}{(\gamma - 1)(r_k - 1)}\left[(r_c - 1)(\gamma - 1 + 1) - \frac{r_c^{\gamma} - 1}{r_k^{\gamma-1}}\right]$$

$$= \frac{p_1 r_k^{\gamma-1}}{(\gamma - 1)(r_k - 1)}\left[\gamma(r_c - 1) - \frac{r_c^{\gamma-1} - 1}{r_k^{\gamma-1}}\right]$$

$$\mathbf{p_m} = \frac{\mathbf{p_1}}{(\gamma - 1)(\mathbf{r_k} - 1)}\left[\mathbf{r_k}^{\gamma-1} \times \gamma(\mathbf{r_c} - 1) - \mathbf{r_k}\left(\mathbf{r_c}^{\gamma} - 1\right)\right]$$

13.4.3 Alternative Method

$$\text{Work done/kg} = Q_A - Q_R$$

$$= C_P(T_3 - T_2) - C_V(T_4 - T_1)$$

Here

$$C_p = \frac{R\gamma}{\gamma - 1},$$

$$C_v = \frac{R}{\gamma - 1}$$

$$\frac{T_2}{T_1} = r_k^{\gamma-1} \qquad \therefore T_2 = T_1 \cdot r_k^{\gamma-1}$$

$$\frac{T_3}{T_4} = r_e^{\gamma-1} = \left(\frac{r_k}{r_c}\right)^{\gamma-1}$$

$$\therefore \quad T_4 = \frac{T_3 \times r_c^{\gamma-1}}{r_k^{\gamma-1}} = T_1 \cdot r_c^{\gamma-1}$$

Substituting these values in the equation for work done, we get

$$p_m \cdot V_1 \cdot \frac{r_k - 1}{r_k} = p_m \cdot \frac{RT_1}{p_1} \cdot \frac{r_k - 1}{r_k}$$

$$= \frac{\gamma R}{\gamma - 1}\left[T_1 r_c \cdot r_k^{\gamma-1} - T_1 r_k^{\gamma-1}\right] - \frac{R}{\gamma - 1}\left[T_1 r_c^{\gamma} - T_1\right]$$

$$\therefore \quad P_m \cdot \frac{r_k - 1}{r_k} \times \frac{1}{P_1} = \frac{\gamma}{\gamma - 1}\left[r_k^{\gamma-1}(r_c - 1)\right] - \frac{1}{\gamma - 1}\left[r_c^{\gamma} - 1\right]$$

$$\therefore \quad p_m = \frac{p_1 . r_k}{(r_k - 1)(\gamma - 1)}\left[\gamma . r_k^{\gamma-1}(r_c - 1) - \left(r_c^{\gamma} - 1\right)\right] \text{ as before}$$

13.5 DUAL COMBUSTION OR SEMI-DIESEL CYCLE

From the expression for the air standard efficiency of the Diesel cycle, it is seen that the efficiency decreases as the value of cut-off ratio (r_c) increase and therefore, it is desirable to keep the value of r_c small as possible to increase the efficiency.

This can be done by supplying the heat partly at constant pressure. This type of the cycle is called Dual Combustion cycle or semi-Diesel cycle. The cycle is represented on *P–V* and *T–S* diagram as in Fig. 13.5.

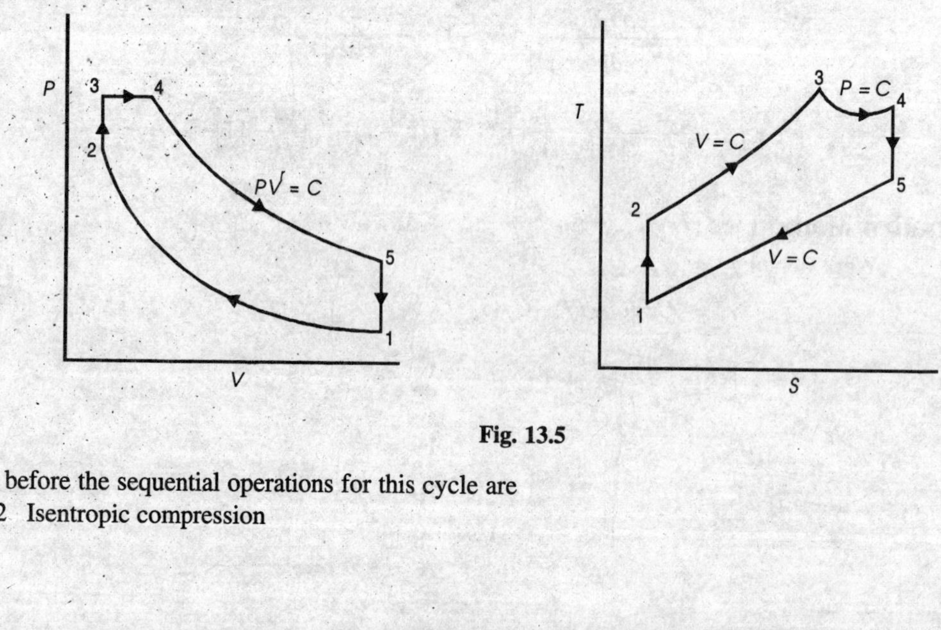

Fig. 13.5

As before the sequential operations for this cycle are

1–2 Isentropic compression

2–3 Constant volume heat addition
3–4 Constant pressure heat addition
4–5 Isentropic expansion
5–1 Constant volume heat rejection

$$Q_A = \text{Heat added/kg}$$
$$= C_V(T_3 - T_2) + C_P(T_4 - T_3)$$
$$Q_R = \text{Heat rejected/kg}$$
$$= C_V(T_5 - T_1)$$

$\therefore$

$$W = \text{Work done/kg}$$
$$= Q_A - Q_R$$
$$= C_V(T_3 - T_2) + C_P(T_4 - T_3) - C_V(T_5 - T_1)$$

$\therefore$ Air standard efficiency is given by

$$\text{ASE} = \frac{W}{Q_A} = \frac{C_V(T_3 - T_2) + C_P(T_4 - T_3) - C_V(T_5 - T_1)}{C_V(T_3 - T_2) + C_P(T_4 - T_3)}$$

$$= 1 - \frac{C_V(T_5 - T_1)}{(T_3 - T_2) + C_P(T_4 - T_3)}$$

As before, all the temperatures in the above equation are to be expressed in terms of T_1.

Students are to derive the expression of Air Standard Efficiency (ASE) terms of compression ratio r_k cut-off ratio r_c explosion ratio α and the specific heat ratio γ and get,

$$\mathbf{ASE} = 1 - \frac{1}{r_k^{\gamma-1}}\left[\frac{\alpha \times r_c^{\gamma} - 1}{(\alpha - 1) + \gamma - \alpha(r_c - 1)}\right]$$

It is obvious from this equation that the efficiency of the dual-combustion cycle can be increased by increasing the value of explosion ratio (i.e., more heat is added at constant volume) and decreasing the value of r_C (less heat is added during constant pressure process) keeping the total heat added constant.

The Otto and Diesel cycles are special cases of Dual cycle. If $r_c = 1$, then $V_3 = V_4$ and the cycle becomes Otto-cycle. Again if $\alpha = 1$, then $P_2 = P_3$ and the cycle becomes Diesel cycle.

Again applying the same procedures used in Otto-cycle and Diesel cycles, the mean effective pressure of the dual combustion can be calculated (students have to derive this relation as an exercise).

$$\mathbf{p_m} = \frac{\mathbf{p_1}}{(r_k - 1)(\gamma - 1)}\left[r_k^{\gamma}.\alpha\gamma(r_c - 1) + r_k^{\gamma} - 1\right]$$

13.6 STIRLING CYCLE (REGENERATIVE CYCLE)

Stirling constructed a hot air engine in 1845 which works on stirling cycle. Theoretically, the efficiency of the Stirling cycle is equal to Carnot cycle. The constructuion of the generator and its function is explained below:

Regenerator: The construction of the regenerator is shown in Fig. 13.6 which alternately stores and rejects heat in a manner which theoretically, is thermodynamically reversible.

The air near the lower end of the filling (important part of the generator) is maintained at a higher temperature T_2 by supplying the heat from external heat source. The air near the upper end of the wire mesh is maintained at a lower temperature T_1 with the help of external heat sink (circulation of water as shown). The wire mesh filling is used to store the heat. The temperature of the wire gauge varies uniformly from T_1 to T_2 as shown. The heat storing capacity of the wire mesh should be large.

Stirling cycle: The stirling cycle is represented on P–V and T–S diagrams as shown.

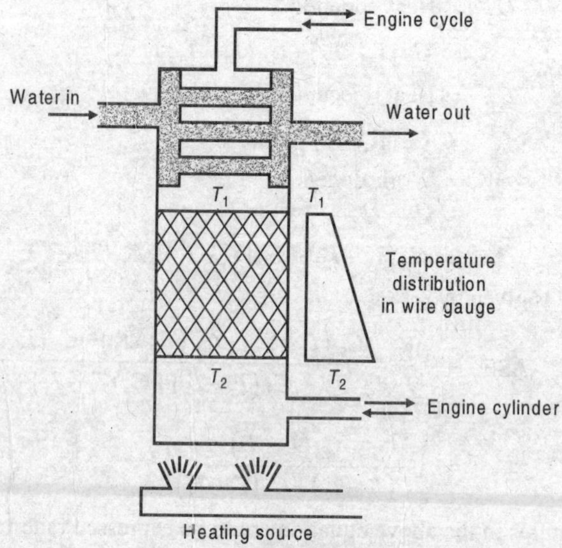

Fig. 13.6

The sequence of operations is given below:

(1) The air is compressed isothermally from 1–2.

(2) The air state 2 is passed into the regenerators from the top at a temperature T_1. The air passing through the matrix gets heated from T_1 to T_2 at constant volume. This heat transfer from the matrix to air is considered thermodynamically reversible because in each element of the matrix heat is transfer due to infinitesimal difference in temperature.

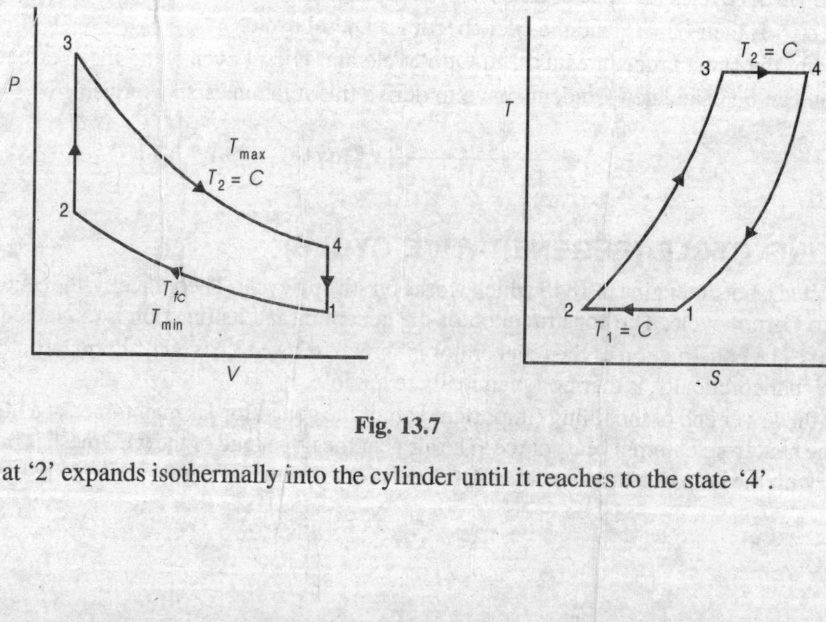

Fig. 13.7

(3) Then the air at '2' expands isothermally into the cylinder until it reaches to the state '4'.

(4) The air coming out of the engine at temperature T_2 (at 4) enters the generator from the bottom and gets cooled while passing through the generator at constant volume and it comes out at a temperatures T_1 and the cycle is repeated. This heat transfer during 4–1 processes is also considered thermodynamically reversible.

The work done by the cycle per kg air and air standard efficiency of the cycle can be calculated as follows:

$$Q_A = \text{Heat supplied during isothermal expansion process } 3-4$$

$$= p_3 v_3 \log_e \frac{V_4}{V_3} = RT_2 \log_e \left(\frac{V_4}{V_3} = \frac{V_1}{V_2} \right)$$

$$= RT_1 \log_e r_k$$

$$Q_A = \text{Heat rejected during isothermal expansion compression } 1-2$$

$$= RT_1 \log_e r_k$$

$$\therefore \quad \text{Work done} = Q_A - Q_R = r \log_e r_k [T_2 - T_1]$$

$$\text{ASE} = \frac{W}{Q_A} = \frac{R(T_2 - T_1)\log_e r_k}{RT_1 \log_e r_k}$$

$$= \frac{T_2 - T_1}{T_1}$$

Thus the efficiency of the Stiriling cycle is equal to that of Carnot cycle when both are working within same temperature limits.

Heat transferred during 2–3 and 4–1 are equal.

In practice, 10–20% heat losses occur in the generator and the efficiency of the generator lies between 90% and 80%.

If the generator efficiency (η_r) is considered then

$$Q_A = RT_2 \log_e r_k + (1 - \eta_r) C_V (T_2 - T_1)$$

$$Q_R = RT_1 \log_e r_k + (1 - \eta_r) C_V (T_2 - T_1)$$

$$\text{Efficiency} = \frac{R(T_2 - T_1)\log_e r_k}{RT_1 \log_e r_k + (1 - \eta_r) C_v (T_2 - T_1)} \qquad \text{...(4)}$$

$$= \frac{T_2 - T_1}{T_2} \text{ when } \eta_r = 100\%$$

The cycle has not become very successful even though its ideal efficiency is sufficiently high because the regenerator volume has to be very large compared with the engine size.

$$p_1 V_2 = p_2 V_2$$

$$p_2 = p_1 \cdot \frac{V_1}{V_2} = p_1 \cdot r_k$$

$$\frac{p_3}{p_2} = \alpha = \text{pressure ratio}$$

$$= \text{expansion ratio}$$

$$\therefore \quad p_3 = \alpha \cdot p_2 = \alpha \cdot p_1 r_k$$

$$p_m \times V_s = p_m (V_1 - V_2) = p_m V_1 \left[\frac{r_k - 1}{r_k}\right]$$

$$= p_3 V_3 \log_e \frac{V_4}{V_3} - p_2 V_2 \log_e \frac{V_1}{V_2}$$

$$= \log_e r_k . [p_3 - p_2] V_2$$

$$= \log_e r_k . [p_3 - p_2] \frac{V_1}{r_k} = \frac{V_1}{r_k} \log_e r_k [\alpha p_1 r_k - p_1 r_k]$$

$$= V_1 \log_e r_k . p_1 (\alpha - 1)$$

$$p_m = p_1 \cdot \frac{r_k}{r_k - 1} (\alpha - 1) \log_e r_k$$

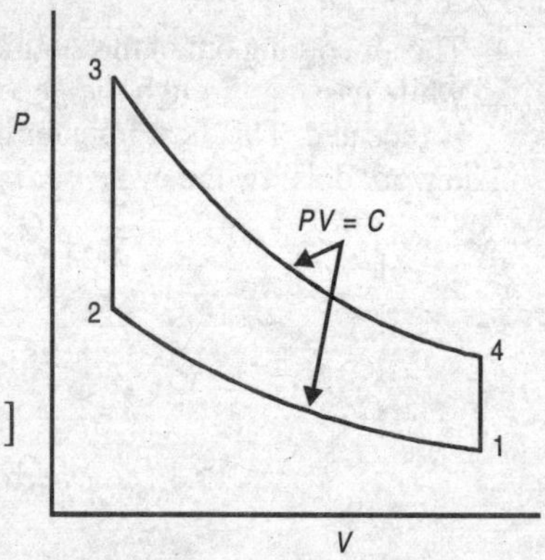

Fig. 13.8

13.7 THE ERICSSON CYCLE

John Ericsson (1803–1889), been a Swede, methodomical stubborn, irascibe, designer and inventor extraordinary, had a passion for machinery knowing that a reversible cycle is the most efficient, Ericsson set out to design an air engine to operate on such a reversible cycle.

In addition to Carnot cycle, there are few reversible cycles one of which is the Ericsson cycle. The reason for doing this is that the Ericsson cycle incorporates a feature of particular importance in modern power plants and other cycles, a *'Regenerative Effect'*.

As in the Carnot cycle, heat is added and rejected at constant temperature, so that the transfer of heat can theoretically occur with a temperature difference ΔT which approaches zero and reversibility. In Fig. 13.9 below, heat is received from an external source at T_1 along *ab*.

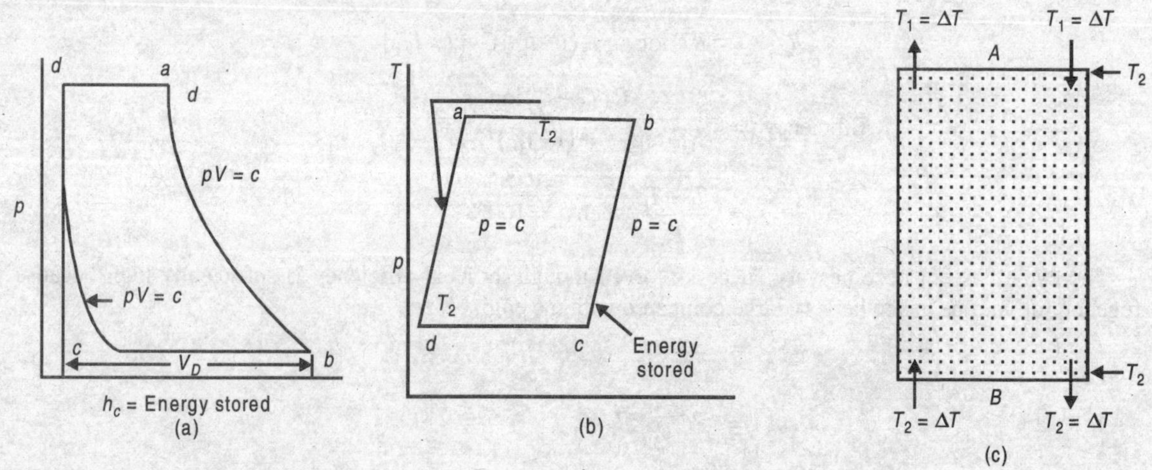

Fig. 13.9

13.8 ERICSSON CYCLE

This cycle consists of two isothermal processes *ab* and *cd*, and two constant pressure processes *bc* and *da*. The hot air then leaves the cylinder and passes through a *regenerator* while the pressure remains constant, *bc*. The regenerator is a chamber in which energy can be transferred from the gas and stored. In practice, a regenerator of the kind intended have may contain a labyrinth of brick-work or it may be filled with mesh wire. One end of the regenerator is at a temperature T_1 and the other and at T_2, and there is a gradual temperature gradient from one to the other.

To get the energy into storage reversibly, we imagine the gas entering the regenerator at a temperature $T_1 + \Delta T$, Fig. 13.9 (c), so that heat begins to be transferred to the contents of the regenerator at A, as the gas passes through the generator, its temperature drops gradually to $T_2 + \Delta T$ at the exits B_1 at all points in the regenerator, the gas temperature is higher than the adjacent contents temperature by an amount ΔT, which is zero in the ideal reversible case.

After this storage operation, the gas rejects that at constantly temperature to an external receiver, the sink; that is there is an isothermal compression of the gas *cd*. Then this gas reenters the regenerators at B, Fig. 13.9 (c), at a temperature $T_2 - \Delta T$, and being slightly cooler than the contents, it receivers heat and continues to receive heat until the gas leavers at a temperature $T_1 - \Delta T$ at A, as in the previous passage in the opposite direction, there is always a temperature difference of ΔT between the gas the adjacent contents. In the limit, a ΔT becomes zero, the whole operturs becomes reversible. It is essential to observe that the heat the gas in moving from B to A is the same as the energy stored from the gas passed from A to B. *This heat does not involve and external source*, it is an interchange of heat within the system. Now, if the source of heat and the sink are temperature T_1 and T_2 respectively, all processes are *reversible externally and internally*.

Thus with a system consisting of the gas and the regenerator, no matter crosses the boundaries, and the energy crossing consists of Q_A, Q_R and W.

$$Q_A = RT_1 \log_e \frac{V_b}{V_a} \text{ kJ along } ab/\text{kg}$$

$$= RT_1 \ln \frac{V_b}{V_a}$$

$$Q_R = RT_1 \log_e \frac{V_d}{V_c} = -RT_2 \log_e \frac{V_c}{V_d}$$

$$= -RT_2 \ln \frac{V_c}{V_d} \text{ kJ along } cd/\text{kg}$$

From Charle's low, we have

$$\frac{V_c}{V_b} = \frac{T_2}{T_1} \text{ and } \frac{V_d}{V_a} = \frac{T_2}{T_1} \text{ or } \frac{V_c}{V_d} = \frac{V_b}{V_a}$$

Using this relation, we find the work to be

$$W = \Sigma Q = (T_1 - T_2) R \ln \frac{V_b}{V_a} \text{ kJ/kg}$$

And the thermal efficiency is

$$\eta_{Th} = \frac{W}{Q_A}$$

$$= \frac{T_1 - T_2}{T_1}$$

Which is the same as that of the Carnot cycle, the highest possible efficiency for any working substance operating between temperatures T_1 and T_2, the work may also be taken from *pv* plane by algebraically summing the ∫p.d *v* quantities for the processes of the cycle i.e. $\oint p.dv$.

$$W = p_a V_a \left| \eta \frac{V_b}{V_a} + p_b (V_c - V_b) + p_c V_c \ln \frac{V_d}{V_a} + p_d (V_a - V_d) \right.$$

$$\text{kg} - \text{m} = (T_1 - T_2) \ln \frac{v_b}{v_a}. \text{ As before.}$$

13.9 BRAYTON CYCLE

George Brayton was a contemporary of Otto and was an engineer from Boston. Joule and Brayton independently proposed the cycle that is the ideal prototype of the actual unit. An unsuccessful turbine unit was built as for back as 1872 and 1906 a unit that produced net power had been built. Metallurgical developments are raising the highest temperatures–say 800 to 1100°C – and more if high cost and or if short life is accepted. A better knowledge of aerodynamics has been partly responsible for improving efficiency of both the centrifuge and axial – flow compressors.

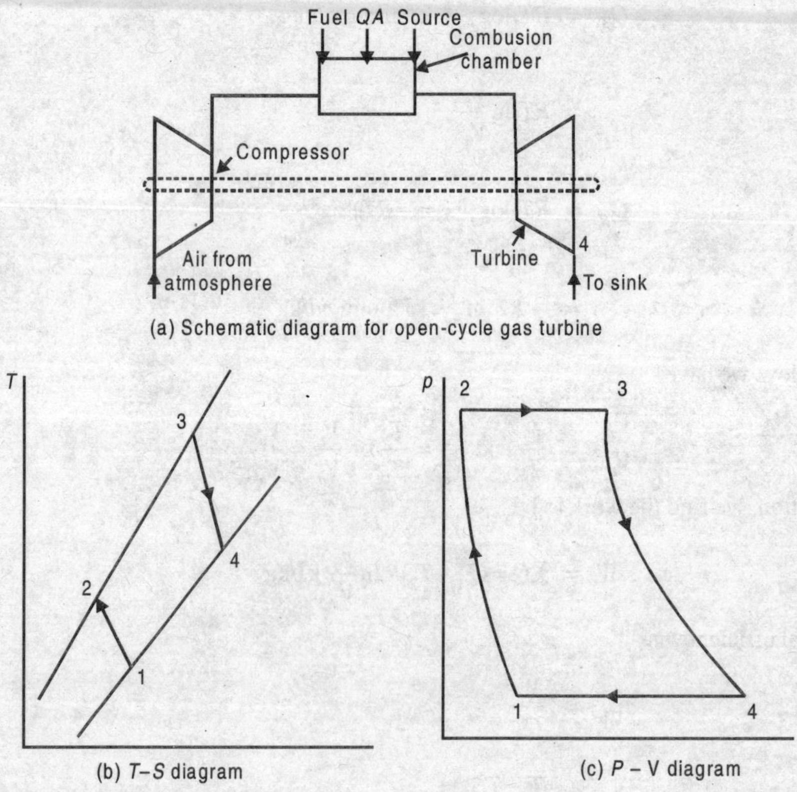

Fig. 13.10 *(a) Schematic diagram for open-cycle gas turbine and (b) and (c) are T–S and p–v diagrams for the schematic diagram.*

Gas turbine working on Brayton cycle consist of the following process :

1–2 Isentropic compression – $W_c = -c_p (T_1 - T_2)$ kJ/kg

2–3 Constant pressure heat addition $Q_s = C_p (T_3 - T_4)$ kJ/kg

3–4 Isentropic expansion in turbine $W_t = C_p(T_3 - T_4)$ kJ/kg

4–1 constant pressure heat rejection $Q_R = C_p(T_4 - T_1)$ kJ/kg

Here atmosphere acts as a sink.

Net work of the cycle
$$= W_\eta = W_t + W_c$$
$$= C_p (T_3 - T_4) - C_p (T_1 - T_2)$$
$$= C_p (T_3 - T_4) - C_p (T_4 - T_1)$$
$$= Q_A - Q_R$$

Heat added/kg, $Q_A = C_p (T_3 - T_2)$

∴ Thermal efficiency of the cycle

$$\eta_{Th} = \frac{W_n}{Q_A} = \frac{Q_A - Q_R}{Q_A}$$

$$= 1 - \frac{Q_R}{Q_A}$$

$$= 1 - \frac{C_p (T_4 - T_1)}{C_p (T_3 - T_2)}$$

$$= 1 - \frac{T_4 - T_1}{T_3 - T_2}$$

From the $T - S$ or p–V diagrams we have

$$\frac{T_2}{T_1} = \left(\frac{p_2}{p_{1_{\gamma-1}}}\right)$$

$$= r_p \frac{p_2}{p_1} = \text{pressure ratio}$$

$$= r\beta$$

$$\therefore \quad \frac{T_2}{T_1} = T_1 \left(r_{p_{\gamma-1\gamma}}\right) = \frac{p_3}{p_4}$$

Similarly
$$\frac{T_3}{T_4} = T_1 \left(r_p\right)^{\gamma-1\gamma} = \left(\frac{p_3}{p_4}\right)^{\gamma-1\gamma}$$

$$\therefore \quad T_3 = T_4 \left(r_p\right)^{\gamma-1\gamma}$$

Substituting these values of T_2 and T_3 in the relation for is η_{Th}, we get

$$\eta_{Th} = 1 - \frac{T_4 - T_1}{r_p^{\gamma-1\gamma}}$$

$$= 1 - \frac{1}{\left(r_p\right)^{\gamma-1\gamma}}$$

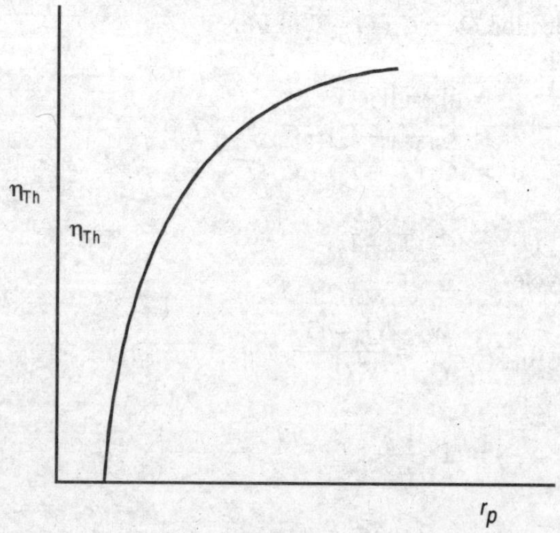

Fig. 13.11 *Graph showing relation between η_{Th} and r_p.*

From this relation we say that the thermal efficiency of the Brayton cycle is a function of (r_p) only. As r_p, the pressure ratio increase thermal efficiency increases. This is shown in Fig. 13.11.

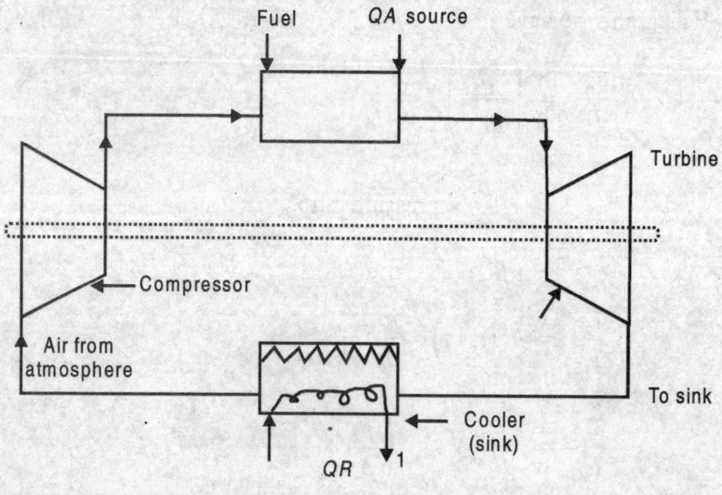

Fig. 13.12 *Benton closed cycle*

Figure 13.12 shows the schematic diagram of closed-cycle, and is self-explanatory.
Analysis of this closed cycle is similar to open-cycle.
The *T-S* and *p-v* diagrams are also same those for Open-cycle.

13.10 COMPARISON OF OTTO, DIESEL AND DUAL COMBUSTION CYCLE

The important parameters in cycle analysis are compression ratio, maximum pressure, maximum temperature, heat input, output etc. The cycle can be analysed for the different parameters.

(a) For same compression ratio and same heat input. Figure 13.13 shows the three cycles on *P–V* and *T–S* diagrams. All the cycles start from the same initial state 1 and air is compressed isentropically to the same point 2 as compression ratio is same.

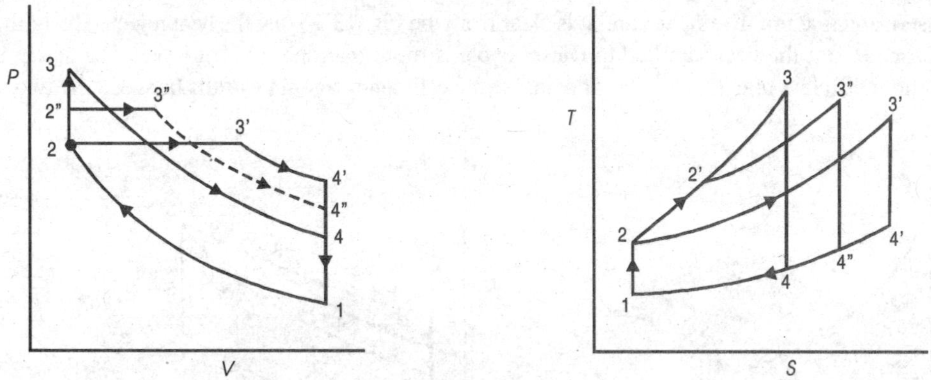

Fig. 13.13

It is seen that adding the heat at constant volume resulting the highest maximum temperature and pressures (Otto–cycle). Adding the heat at constant pressure results in the lowest maximum temperatures (Diesel cycle) while the values for the limited pressure cycle (Dual cycle) lie between Otto and Diesel cycles.

For the same heat input and compression ratio, minimum heat is rejected by the Otto-cycle, maximum heat is rejected by Diesel cycle and the heat rejected is in between Otto and Diesel in case of dual cycle. Consequently, work output and efficiency is maximum in Otto-cycle and the least work done and efficiency in case of Diesel cycle. Dual cycle work and efficiency lie in between the two–Otto and Diesel cycles.

(b) For constant maximum pressure and same heat input

Otto-cycle 1–2–3–4–1

Diesel cycle 1–2'–3'–4'–1

Dual cycle 1–2"–3"–5"–1

Figure 13.14 compares the three cycles on the basis of same maximum pressure and same heat input.

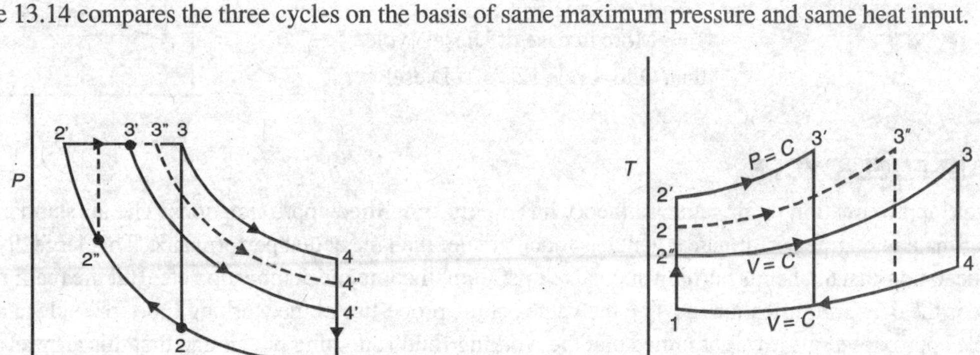

Fig. 13.14

For the same maximum pressure the points 2, 3', and 3", must be on the same pressure line and the heat input, area 2356 (Otto) = area 2'3'5'6 = area 2". 3" and 5"6. From the diagram, heat rejected its maximum in Otto-cycle, minimum in Diesel cycle and intermediate in dual cycle.

Therefore, for the same heat input, Diesel cycle produces more work, Otto-cycle less work and dual cycle intermediate work. Therefore, for maximum pressure and same heat input. Diesel cycle is the most efficient, dual cycle come the next and otto-cycle has the least efficiency.

(c) For the same maximum pressure and temperature. Figure 13.15 compares Otto -Diesel cycles for the same maximum pressure and maximum temperature. It is clear from the Fig. 13.15 that the heat rejected by both Otto and Diesel cycles is same. But the heat supplied to Diesel cycle is more than that for Otto-cycle. Hence in this case, Diesel cycle is more efficient than Otto cycle. The dual cycle efficiency again lies/falls between the two to others.

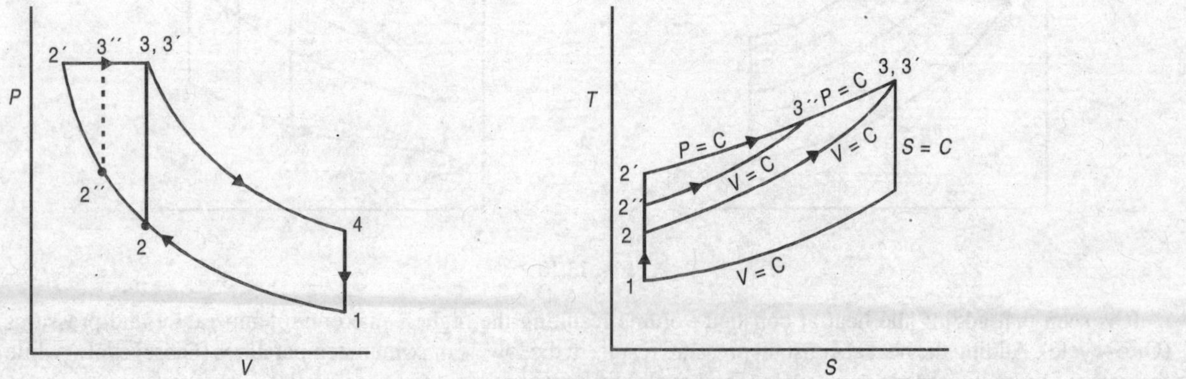

Fig. 13.15

(d) For the same maximum pressure and output. For same work output, point 3' will be on left side of 3. Therefore heat rejected in Diesel is less.

$$\therefore \qquad \eta = \frac{\text{Work output}}{\text{Work output + Heat rejected}}$$

= 12342 – Otto

= More in case of Diesel cycle than Otto -cycle 12'3'4'1–Diesel

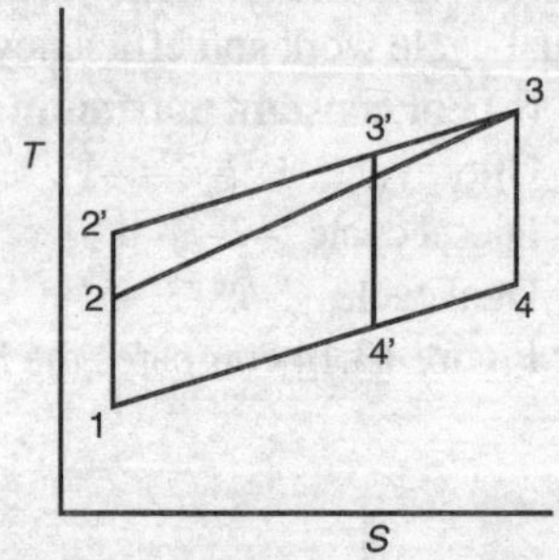

Fig. 13.16

13.11 AIR-FUEL CYCLE

The air cycle approximation or air standard theory has highly simplified approximations. The air standard theory gives an estimate of engine performance which is much greater than the actual performance. This large divergence is partly due to non-instantaneous burning and valve operation, incomplete combustion etc. But the main reason of divergence is the over simplification in using the values of the properties of the working fluid for cycle analysis. In the air cycle approximations it was assumed that the working fluid is nothing but air and that this air was a perfect gas and had a constant specific heats. In actual engine the working fluid is not air but a mixture of air, fuel and residual gases. Furthermore, the specific heats of the working substance are not constant but increase as temperature rises and finally, the products of combustion are subjected to dissociation at high temperatures.

If the actual physical properties of the cylinder gases before and after burning are taken into account a much closer approach to actual performance figures is achieved.

The theoretical cycle based on the actual properties of the cylinder gases on the actual properties of the cylinder gases is called the fuel-air cycle and represents a nearly attainable ideal for comparison with actual performance.

For fuel-air cycle calculations take into consideration the following :

1. The actual composition of the cylinder gases i.e., fuel + air + combustion gases. The fuel air ratio is changed during the operation of the engine which changes the relative amounts of CO_2, CO, water vapour etc.
2. The variation in the specific heat of these gases with temperature. Specific heat increases with temperature except the monoatomic gases.
3. The fact that the fuel and air do not completely combine chemically at high temperatures (above 1600 K), CO, H_2, H and O_2 may be present at equilibrium conditions.
4. The variation in the number of molecules present in the cylinder as the pressure and temperature change. The number of molecules present after combustion depends on the fuel-air ratio and the pressure and temperature after the combustion.

The following assumptions are commonly made for fuel-air cycle analysis :

1. There is no chemical change in either fuel or air prior to combustion.
2. Subsequent to combustion, the charge is always in chemical equilibrium.
3. There is no heat exchange between the gases and the cylinder walls in any process i.e., they are adiabatic. Also expansion and compression processes are frictionless.
4. In reciprocation engines the velocities are negligibly small.

13.12 VARIABLE SPECIFIC HEATS

All gases, except monoatomic gases, show an increase in specific heat at temperatures. The increase in specific heat does not follow any particular law. However, over the temperature range in general use for gases in heat engines (300 K to 1300 K) the specific heat curve is nearly a straight line which may be approximately expressed in the form,

$$C_p = a + KT; C_v = b + KT$$

where a, b and K are constants.

$$\therefore \quad R = C_p - C_v = (a + KT) - (b + KT)$$

$$= (a - b)$$

Above 1500 K, the specific heat increases much more rapidly and may be approximately expressed in the form

$$C_p = a + K_1T + K_2T^2$$

$$C_v = b + K_1T + K_2T^2$$

The physical explanation of increases in specific heat is that as the temperature is raised, larger and larger fractions of the heat input go to produce motion of the atoms within the molecules. Since temperature is the result of motion of the molecules as a whole the energy which goes into moving the atoms does not contribute to the temperature rise. Hence more heat is required to raise the temperature of unit mass through one degree. This heat—by definition—is specific heat.

For air C_p = 1.005 at 0°C and 1.2645 at 2000°C.

SOLVED EXAMPLES
(CONSTANT VOLUME CYCLE (OTTO-CYCLE))

Example 13.1 *An engine is operating on ideal Otto cycle for which the following information is available.*

Maximum temperature = 1277 ºC
Exhaust temperature = 447 ºC
Ambient condition = 1 bar and 37 ºC
Air consumption = 2 kg/min

Estimate: (i) Compresion ratio (ii) Air Standard efficiency (iii) Power output.

Assume : $C_p = 1.005$ *kJ/kg-K,* $C_v = 0.718$ *kJ/kg.K*

Solution

Ideal Otto-cycle is shown on *P–V* diagram given Fig. Ex. 13.1

Referring to Fig. Ex. 13.1

$$T_1 = 37 + 273 = 310\,\text{K}$$
$$p_1 = 1\,\text{bar}$$
$$T_3 = \text{Max. temp.} = 1277 + 273 = 1550\,\text{K}$$
$$T_4 = \text{Exhaust temperature} = 447 + 273 = 720\,\text{K}$$
$$C_p = 1.005\,\text{kJ/kg.K}$$
$$C_V = 0.718\,\text{kJ/kg.K}$$

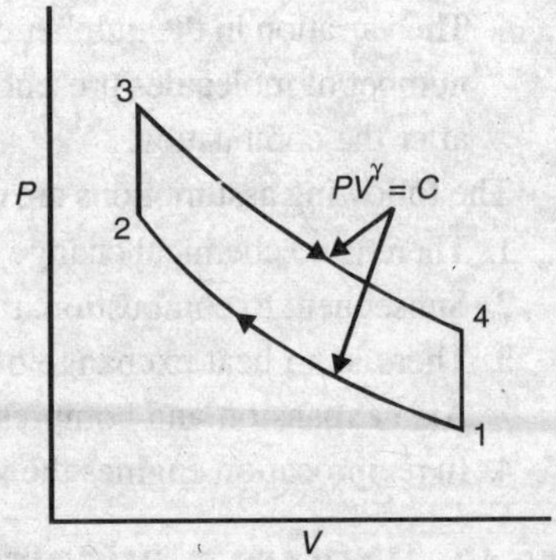

Fig. Ex. 13.1

$$\therefore \quad \gamma = \frac{C_p}{C_V} = \frac{1.005}{0.718} = 1.4$$

(i) Compression ratio :

For Otto-cycle $\quad r_k = \text{Compression ratio} = r_e$

$\therefore$ For expansion process

$$\frac{T_3}{T_4} = r_e^{\gamma-1} = r_k^{\gamma-1}$$
$$= \frac{1550}{720} = 2.153$$
$$\therefore \quad r_k^{0.4} = 2.153$$
$$r_k = (2.153)^{1/0.4} = (2.153)^{2.5}$$
$$\mathbf{r_k = 6.8}$$

(ii) Air standard efficiency

$$\text{ASE} = 1 - \frac{1}{r_k^{\gamma-1}}$$
$$= 1 - \frac{1}{2.153}$$
$$= 1 - 0.4645$$
$$\textbf{ASE} = \mathbf{0.5355 = 53.55\%}$$

(iii) Power output

$$\text{Work done/kg} = Q_A - Q_R$$
$$= C_V(T_3 - T_2) - C_V(T_4 - T_1)$$

We have
$$\frac{T_3}{T_2} = \frac{T_4}{T_1} \text{ or } \frac{T_3}{T_2} = \frac{T_2}{T_1}$$

$$\therefore \quad T_2 = T_1 \times \frac{T_3}{T_4} = \frac{310 \times 1550}{720}$$

$$\mathbf{T_2 = 667.36\ K}$$

$$\therefore \quad \text{Work done/kg} = 0.718(1550 - 667.36) - 0.718(720 - 310)$$
$$= 0.718 \times 882.64 - 0.718 \times 410$$
$$= 633.74 - 294.38$$

$$\textbf{Work done/kg} = \mathbf{339.36\ kJ/kg}$$

$$\therefore \quad \text{Power output} = \text{mass of air/sec} \times \text{work per kg}$$
$$= \frac{2}{60} \times 339.36 = 11.312 \text{ kJ/sec}$$

$$\textbf{Power output} = \mathbf{11.312\ kW}$$

Example 13.2 *A four stroke engine working on Otto-cycle has a swept volume of 0.1 m³. The compression ratio is 7. The condition at the start of the cycle is pressure 1 bar and temperature 90°C. The heat addition at constant volume is 100 kJ/cycle. Find ideal efficiency, mean effective pressure and temperature at key points in the cycle. Assume air as working substance with C_v = 0.718 kJ/kg.K and γ = 1.4.*

Solution

The cycle is shown on *P–V* diagram of Fig. Ex. 13.2 given

We have
$$P_1 = 1 \text{ bar}$$
$$T_1 = 273 + 90 = 363 \text{ K}$$
$$Q_A = 100 \text{ kJ/cycle}$$
$$r_k = 7$$

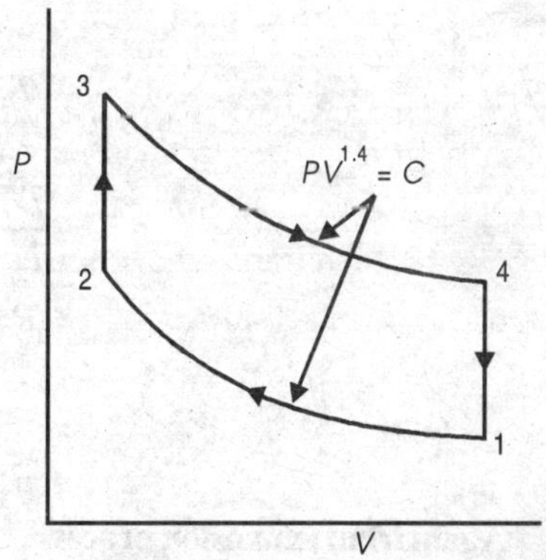

Fig. Ex. 13.2

(i) Ideal efficiency :

$$\text{Ideal efficiency} = \text{Air standard efficiency}$$
$$= 1 - \frac{1}{r_k^{\gamma-1}}$$
$$= 1 - \frac{1}{7^{0.4}}$$
$$= 1 - \frac{1}{2.1780}$$
$$= 1 - 0.459 = 0.541$$

$$\textbf{Ideal efficiency} = \mathbf{54.1\%}$$

(ii) Pressure and temperature at key points :

$$\frac{T_2}{T_1} = r_k^{\gamma-1} = 7^{0.4} = 2.178$$

$$T_2 = 363 \times 2.178$$

$$\mathbf{T_2 = 790.61\ K}$$

$$\frac{p_2}{p_1} = \left(\frac{V_1}{V_2}\right) = 7^{1.4} = 15.245$$

$$\mathbf{p_2 = 1 \times 15.245 = 15.245\ bar}$$

$$Q_A = mC_V(T_3 - T_2) = \frac{p_2 v_2}{RT_2} C_V (T_3 - T_2)$$

Now compression ratio $= \frac{V_1}{V_2} = \frac{V_2 + V_S}{V_2}$

$$= 1 + \frac{V_S}{V_2}$$

$\therefore$ $$\frac{V_S}{V_2} = 7 - 1 = 6. V_S = 0.1 \text{ cu.m}$$

$\therefore$ $$V_2 = \frac{V_S}{6} = \frac{0.1}{6} = 0.0167 \text{ m}^3$$

$\therefore$ $$100 = \frac{10^5 \times 15.245 \times 0.0167}{287 \times 790.61} \times 0.718(T_3 - T_2)$$

$$100 = 0.1122 \times 0.718(T_3 - T_2)$$

$$T_3 - T_2 = \frac{100}{0.1122 \times 0.718}$$

$$= 1241.32$$

$$T_3 = 1241.32 + 790.61$$

$$\mathbf{T_3 = 2031.93\ K}$$

$\therefore$ $$\frac{T_3}{T_2} = \frac{p_3}{p_2} \text{ for constant volume heating}$$

or $$P_3 = \text{pressure after heat addition}$$

$$= \frac{2031.92}{790.61} \times 15.245$$

$$\mathbf{P_3 = 39.18\ bar}$$

Again from expansion process

$$\frac{p_3}{p_4} = \left(\frac{V_4}{V_3}\right)^{\gamma} = 7^{1.4} = 15.243$$

$\therefore$ $$p_4 = \frac{p_3}{15.243} = \frac{29.18}{15.243}$$

$$\mathbf{p_4 = 2.57}$$

$\therefore$ $$T_4 = T_3 \times \left(\frac{1}{7}\right)^{0.4} = \frac{2031.93}{2.178} = \mathbf{932.93\ K}$$

(iii) Mean effective pressure

$$\text{Heat rejected/cycle} = mC_v(T_4 - T_1)$$
$$= 0.11220 \times 0.178(932.92 - 363)$$
$$= 45.91 \text{ kJ}$$
$$\therefore \quad \text{Work done} = Q_A - Q_R$$
$$= 100 - 45.91$$
$$= 54.09 \text{ kJ}$$

$$\therefore \quad 54.09 = p_m \times V_1 \cdot \left(\frac{r_k - 1}{r_k}\right)$$
$$= p_m \times 0.1167 \times \frac{7-1}{7}$$
$$= p_m \times 0.7/7$$

If p_m is bar, them

$$54.09 = 100 p_m \times 0.1167 \times \frac{6}{7}$$
$$p_m = \frac{54.09 \times 7}{100 \times 6 \times 0.1167}$$
$$\mathbf{p_m = 5.41 \text{ bar}}$$

II. CONSTANT PRESSURE CYCLE (DIESEL CYCLE)

Example 13.3 *A Diesel cycle operates at a pressure of 1 bar at the beginning of compression and the volume is compressed to $\frac{1}{16}$ of the initial volume. Heat is supplied until volume is twice that of the clearance volume. Calculate mean effective pressure of the cycle. Take K= 1.4.*

Solution

With reference to P–V diagram shown

$$\text{Compression ratio} = r_k = 16$$
$$\text{Cut-off ratio} = \frac{V_3}{V_2} = 2$$

Work done/Cycle/kg air

$$= p_2(V_3 - V_2)$$
$$= \frac{p_3V_3 - p_4V_4}{\gamma - 1} - \frac{p_2V_2 - p_1V_1}{\gamma - 1}$$

$$\text{We have } \frac{p_2}{p_1} = r_k^{\gamma} \therefore p_2 p_1 r_k^{\gamma} = p_3$$

$$\frac{p_3}{p_4} = \left(\frac{V_4}{V_3}\right)^{\gamma} = r_e^{\gamma} = \left(\frac{r_k}{r_e}\right)^{\gamma}$$

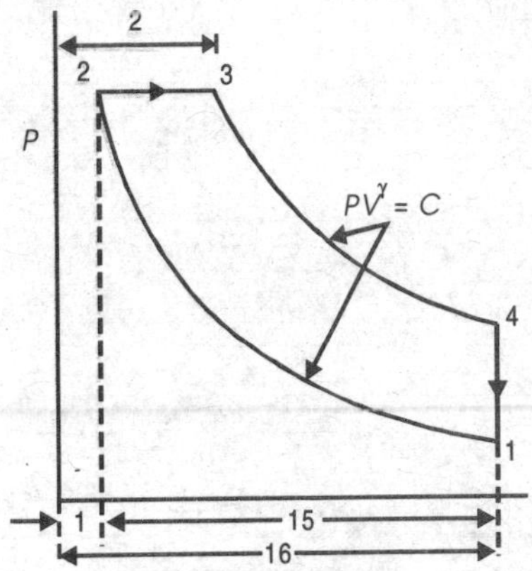

Fig. Ex. 13.3

$$p_4 = \frac{p_3 - r_c^{\gamma}}{r_k^{\gamma}} = \frac{p_1 \times r_k^{\gamma} r_c^{\gamma}}{r_k^{\gamma}} = p_1 r_e^{\gamma}$$

Similarly

$$V_2 = \frac{V_1}{r_k}, \frac{V_3}{V_2} = r_c$$

$\therefore$

$$V_3 = V_2 r_c = \frac{V_1 r_c}{r_k}; V_4 = V_1$$

Substituting these values in the equation of work we get

$$\text{Work done} = P_1 r^k K\left(\frac{V_1 r_c}{r_k} - \frac{V_1}{r_k}\right) + \left(\frac{p_1 r^k K \frac{V_1 r_c}{r_k} - \frac{P_1 r^k{}_c V_1}{1}}{\gamma - 1}\right) - \left(\frac{p_1 r^k K \frac{V_1}{r_k} - p_1 V_1}{\gamma - 1}\right)$$

$$= p_1\left[\frac{r^{\gamma}{}_K . V_1}{r_k}(r_c - 1). + \frac{V_1}{\gamma - 1}\left(\frac{r^{\gamma}{}_K r_c}{r_k} - r^{\gamma}{}_e\right) - \frac{V_1}{K - 1}\left(\frac{r^{\gamma}{}_K}{r_k} - 1\right)\right]$$

$$= p_m \times (V_1 - V_2) = p_m \times \left(V_1 - \frac{V_1}{r_k}\right) = p_m V_1\left(\frac{r^{\gamma}{}_K}{r_k} - 1\right)$$

$$p_m \times \frac{r^{\gamma}{}_K - 1}{r_k} = \frac{p_1}{(K-1)}\left[\frac{(\gamma-1)(r_c - 1)r^{\gamma}{}_K}{r_k} + \frac{r^{\gamma}{}_c r_c - r_k r^{\gamma}{}_c}{\gamma - 1} - \frac{r^{\gamma}{}_K - r_k}{r_k}\right]$$

$$p_m = \frac{P_1}{(r_k - 1)(\gamma - 1)}\left[(\gamma-1)(r_c - 1)r^{\gamma}{}_k + r^{\gamma}{}_k (r_c - 1) - r_k (r^{\gamma}{}_c - 1)\right]$$

Here $r_k = 16, r_c = 2, \gamma = 1.4, P_1 = 1$

$$p_m = \frac{1}{(16-1)(1.4-1)}\left[0.4 \times 1 \times 16^{1.4} + 16^{1.4} \times 1 - 16(2^{1.4} - 1)\right]$$

$$= \frac{1}{15 \times 0.4}\left[0.4 \times 16^{1.4} + 16^{1.4} - 16(2^{1.4} - 1)\right]$$

$$= \frac{1}{6}\left[0.4 \times 4.851 + 4.57 - 16(2.636 - 1)\right]$$

$$= \frac{1}{6}[19.404 + 48.51 - 16.1636] = \frac{1}{6} p[67.914 - 26.176]$$

$$= \frac{1}{6} \times 41.738$$

$$\mathbf{p_m = 6.96\ bar}$$

Example 13.4 *In an air standard Diesel cycle, compression begins at 103 kPa and 300°K. After compression heat addition is of 545 kJ/kg of air, the peak pressure reached in the cycle is 4.7 MPa. Calculate*

(i) *fuel cut-off ratio*
(ii) *compression ratio*
(iii) *maximum temperature in the cycle*
(iv) *air standard efficiency*
Assume $\gamma = 1.4$ and $C_p = 1.004$ kJ/kg.

Solution

$$P_1 = 1.03 \text{ bar}$$
$$Q_A = 545 \text{ kJ/kg}$$
$$P_2 = p_3 = 47 \text{ bar}$$

(i) With reference to *P–V* diagram

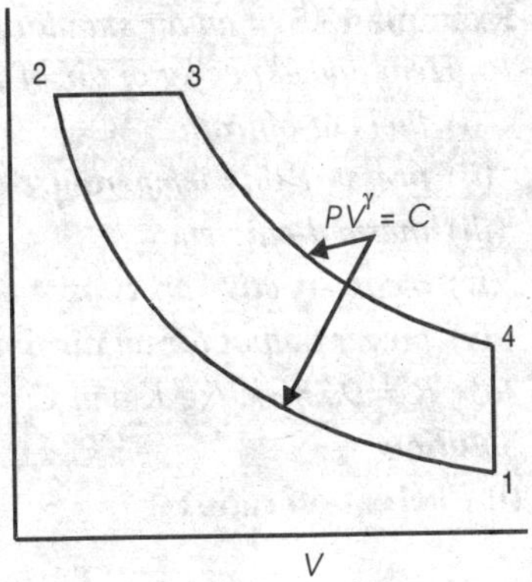

Fig. Ex. 13.4

$$\frac{P_2}{P_1} = \left(\frac{V_1}{V_2}\right)^{\gamma} = r_k^{1.4} = \frac{47}{1.03} = 45.63$$

$$r_k = (45.63)^{1/1.4} = 45.63^{0.7143}$$
$$= 15.32$$
$$\mathbf{r_k = \text{Compression ratio} = 15.32}$$

(ii) Fuel cut off ratio

$$\frac{T_2}{T_1} = \left(\frac{P_2}{P_1}\right)^{\frac{\gamma-1}{\gamma}} = (45.63)0^{04/1.4}(45.63)^{0.286}$$
$$= 2.983$$

$\therefore$ $$T_2 = 300 \times 2.983$$
$$= 894.9 \text{ K}$$

$\therefore$ $$Q_A = 545 = C_P(T_3 - T_2) = 1.004\,(T_3 - 894.9)$$

$$T_3 = \frac{545}{1.004} + 894.9 = 542.83 + 894.9$$
$$= 1437.73 \text{ K}$$

$\therefore$ $$\frac{T_3}{T_2} = \frac{V_3}{V_2} = r_c \text{ (cut-off ratio)} = \frac{1437}{894.9} = 1.61$$

$$\mathbf{r_c = \text{cut-off ratio} = 1.61}$$

(iii) Maximum temperature in the cycle

$$T_3 = \text{Maximum. temp. in the cycle}$$
$$= \mathbf{1437.73 \text{ K} = 1164.73°C}$$

(iv) Air standard efficiency

$$r_e = r_k / r_c = \frac{15.32}{1.61} = 9.52$$

$\therefore$ $$(T_3/T_4) = (r_e)^{\gamma-1} = (9.52)^{0.4} = 2.462$$

$\therefore$ $$T_4 = \frac{1437.73}{2.462} = 584 \text{ K}$$

$\therefore$ $$Q_R = \text{Heat rejected} = 1.004\,(584 - 300)$$

$$= 1.004 \times 284 = 285 \text{ kJ/kg}$$

$$\therefore \quad \text{Work done/kg} = Q_A - Q_R = 545 - 285$$

$$= 260 \text{ kJ/kg}$$

$$\therefore \quad \text{Thermal efficiency} = \text{ASE} = \frac{W}{Q_A} = \frac{260}{545} = 0.477$$

$$= 47.7\%$$

$$\textbf{ASE} = \textbf{47.7 \%}$$

Example 13.5 *In an air standard Diesel Cycle, compression begins at 1 bar and 300°K. The compression ratio is 16. Heat added per kg of air of 1000 kJ/kg. Determine:*

(i) fuel cut-off ratio

(ii) pressure and temperature at the end of expansion

(iii) thermal efficiency

(iv) mean effective pressure

(v) power output for an air flow rate of 0.2 kg/sec

Take R = 0.286 kJ/kg K and C_p = 1.0 kJ/kg K.

Solution

(i) Fuel cut-off ratio :

$$\frac{T_2}{T_1} = r_k^{\gamma-1} = 16^{0.4} \qquad R = C_p - C_V$$

$$= 3.031 \qquad \therefore \quad C_V = C_P - R = 1.0.286$$

$$\therefore \quad T_2 = 300 \times 3.031 \quad \therefore \quad \gamma = \frac{1}{0.714} = 1.4$$

$$= 909.3 \text{ K}$$

Now

$$Q_A = C_p(T_3 - T_2)$$

$$\therefore \quad 1000 = 1(T_3 - 909.3)$$

$$\therefore \quad T_3 = 1000 + 909.3 = 1909.3 \text{ K}$$

$$\therefore \quad \frac{T_3}{T_2} = \frac{V_3}{V_2}$$

$$\therefore \quad r_c = \text{Cut-off ratio} = 1909.3/909.3 = 2.1$$

$$\therefore \quad \mathbf{r_c} = \textbf{Cut-off ratio} = \mathbf{2.1}$$

(ii) Pressure and temperatures at the end of expansion:

$$\frac{P_3}{P_4} = \left(\frac{V_4}{V_3}\right)^{\gamma} = (r_e)^{\gamma} = \left(\frac{r_k}{r_c}\right)^{\gamma} = \left(\frac{16}{2.1}\right)^{1.4} = (7.619)^{1.4} = 17.17$$

Also

$$\frac{P_2}{P_1} = r_k^{1.4} = 16^{1.4} = 48.51$$

$\therefore \quad P_2 = 1 \times 4851 = 48.51 \text{ bar} = P_3$

$\therefore \quad P_4 = \dfrac{P_3}{17.17} = \dfrac{48.51}{17.17} = 2.825 \text{ bar}$

$\dfrac{T_3}{T_4} = (r_e)^{\gamma-1} = (7.619)^{1.4} = 2.253$

$\therefore \quad T_4 = \dfrac{T_3}{2.253} = \dfrac{1909.3}{2.253} = 847.4 \text{ K}$

$\therefore \quad \mathbf{P_4 = 2.285 \text{ bar}}$

$\mathbf{T_4 = 847.4 \text{ K}}$

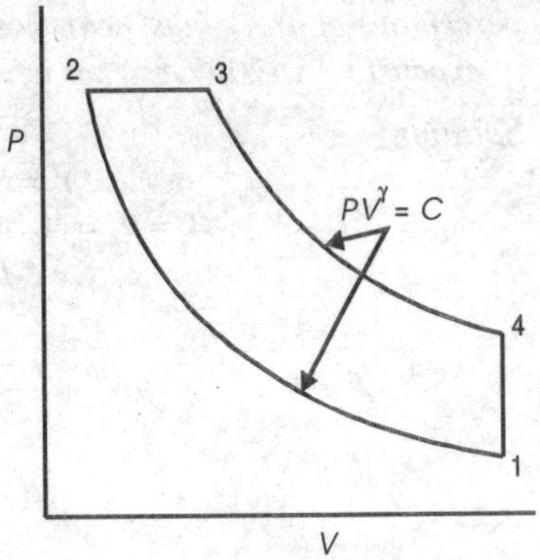

Fig. Ex. 13.5

(iii) Thermal efficiency :

$Q_R = \text{Heat rejected} = C_V(T_4 - T_1) = 0.714\,(847.4 - 300)$

$= 0.714 \times 574.4 = 390.84 \text{ kJ/kg}$

$\therefore \quad \text{Thermal efficiency:} = \dfrac{Q_A - Q_R}{Q_A} = \dfrac{1000 - 390.84}{1000} = \dfrac{609}{1000} = 0.60916$

$= 60.916\,\%$

ASE = 60.916 %

(iv) MEP

$\text{Work done/kg} = 608.16 \text{ kJ} = P_m \times V_s = P_m\,(V_1 - V_2)$

But $\quad V_1 - V_2 = V_1 \dfrac{V_1}{r_k} \text{ as } \dfrac{V_1}{V_2} = r_k$

$= V_1 \dfrac{(r_k - 1)}{r_k} = \dfrac{RT_1}{P_1} \cdot \dfrac{15}{16} = \dfrac{286 \times 300}{10^5} \times \dfrac{15}{16}$

$\therefore P_m \times \dfrac{286 \times 300}{10^5 \times 1000} \times \dfrac{15}{16} = 609.16$

$\therefore \quad P_m = \dfrac{609.16 \times 16}{15} \times \dfrac{1000 \times 10^5}{286 \times 300} = 10^5 \times 7.573 \text{ N/m}^2$

$= 7.573 \text{ bar}$

$\therefore \quad \mathbf{p_m = 7.573 \text{ bar}}$

(v) Power output

$\text{Work done/kg} = Q_A - Q_R = 1000 - 390.84 = 609.16 \text{ kJ}$

Flow rate of air = 0.2 kg/sec

power developed = 0.2×609.16

= **121.832 (kJ/sec = kW)**

III. DUAL COMBUSTION CYCLE

Example 13.6 *An oil engine works on the ideal dual combustion cycle with a volume compression ratio of 10. The temperature at the beginning of compression is 80°C and the maximum pressure of the cycle is 40 bar. At what* *percentage of stroke must heat reception of constant pressure cease in order that the temperature of the beginning of expansion is 1700°C? What is the ratio of the work done during expansion to that done during compression?*

Solution

$$r_k = 10 \qquad T_1 = 80 + 273 = 353\text{ K}$$

$$P_3 = P_4 = 40\text{ bar} \qquad T_4 = 1700 + 273 = 1973\text{ K}$$

Let $P_1 = 1$ bar

We have, $\dfrac{T_2}{T_1} = r_k^{\gamma-1} = 10^{0.4}$

$= 25.12$ bar

$P_3 = 40$ bar

$\therefore \quad \dfrac{P_3}{P_2} = \dfrac{T_3}{T_2} = \dfrac{40}{25.12}$

$\therefore \quad T_3 = \dfrac{40}{25.12} \times 886.7$

$= 1411.94$ K

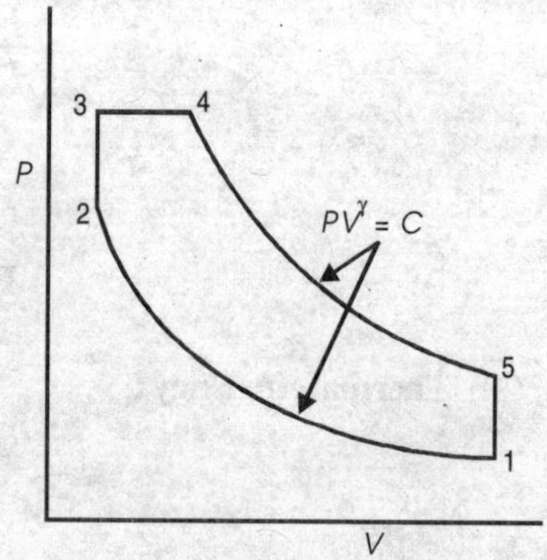

Fig. Ex. 13.6

$\therefore$ Cut-off ratio $r_c = \dfrac{T_4}{T_3} = \dfrac{1973}{1411.94}$

$= 1.3974$

Now $r_k = 10 = \dfrac{V_1}{V_2}$

$\therefore \quad V_1 = 10V_2$

$\therefore \quad V_s = V_1 - V_2 = 10V_2 - V_2 = 9V_2$

Again $r_c = \dfrac{V_4}{V_3} = 1.3974V_2$

$\therefore \quad V_4 - V_3 = V_4 - V_2 = 1.3974V_2 - V_2$

$= 0.3874V_2$

% of stroke at which heat addition is stopped

$= \dfrac{V_4 - V_3}{V_1 - V_2} \times 100$

$= \dfrac{0.3974V_2}{9V_2} \times 100$

$= 0.04416 \times 100 = 4.416\%$

Work done during expansion

$$W_e = \frac{P_4V_4 - P_SV_S}{\gamma - 1} = \frac{R}{\gamma - 1}[T_4 - T_5]$$

For expansion $r_e = \dfrac{r_k}{r_c} = \dfrac{10}{1.3974}$

$$= 7.156$$

$$\therefore \quad \frac{T_4}{T_5} = (7.156)^{0.4} = 2.1972$$

$$\therefore \quad T_s = \frac{1973}{2.1972} = 897.96 \text{ K}$$

$$\therefore \quad W_e = \frac{R}{\gamma - 1}[T_4 - T_5] = C_v(T_4 - T_5)$$

$$= 0.718[1973 - 897.96]$$

$$= 0.718 \times 1076.04 = 772.6 \text{ kJ/kg}$$

Work during compression

$$W_e = \frac{P_2V_2 - P_1V_1}{\gamma - 1} = \frac{R}{\gamma - 1}(T_2 - T_1)$$

$$= 0.718(886.7 - 353)$$

$$= 0.718 \times 533.7$$

$$= 383.197 \text{ kJ/kg}$$

$$\therefore \quad \textbf{Ratio } \frac{\mathbf{W_e}}{\mathbf{W_c}} = \frac{\mathbf{772.6}}{\mathbf{383.197}} = \mathbf{2.019}$$

Example 13.7 *The following data refer to an oil engine operating on dual cycle i.e., with heat reception partly at constant volume and partly at constant pressure and with heat rejection at constant volume. Compression ratio 11.6, pressure and temeparture at the beginning of compression 1 bar and 320°K, percentage increase of pressure during constant volume burning 53% and percentage volume increase during constant pressure during 38%. If it is assumed that $C_p = 1.089$ and $C_v = 0.795$ and that the compression and expansion curved are isentropic, find MEP of the cycle.*

Solution

Here
$$\gamma = \frac{C_p}{C_v} = \frac{1.089}{0.795}$$

$$= 1.37$$

Given:
$$r_k = \frac{V_1}{V_2} = 11.6$$

$$P_1 = 1 \text{ bar}$$

$$T_1 = 320 \text{ K}$$

$$\frac{P_3 - P_2}{P_2} \times 100 = 53 = \frac{P_3}{P_2} - 1 = 0.53$$

$$\therefore \quad \frac{P_3}{P_2} = 1.53$$

$$\frac{V_4 - V_3}{V_3} \times 100 = 38$$

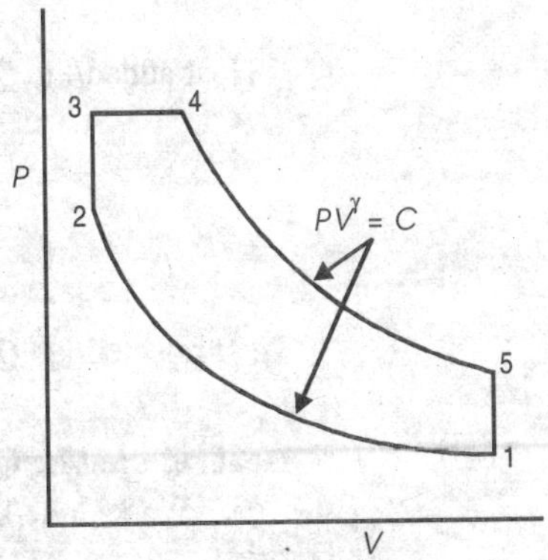

Fig. Ex. 13.7

$$\frac{V_4}{V_3} - 1.38 = r_c$$

$$\therefore \quad P_2 = P_1 \times r_k^{1.37} = 1 \times 28.73 \text{ bar} = 2873 \text{ bar}$$

$$\therefore \quad P_3 = 28.73 \times 1.53 = 43.954 \text{ bar} = P_4$$

$$\therefore \quad r_e = \text{expansion ratio} = \frac{r_k}{r_c} = \frac{11.6}{1.38}$$

$$= 8.406$$

$$\therefore \quad \frac{P_4}{P_5} = r_e^{\gamma} = (8.406)^{1.37}$$

$$= 18.48 \text{ bar}$$

$$\therefore \quad P_5 = \frac{43.954}{18.48} = 2.378 \text{ bar}$$

Also

$$\frac{T_4}{T_5} = r_e^{\gamma-1} = (8.406)^{0.37}$$

$$= 2.1984$$

and

$$\frac{T_2}{T_1} = r_e^{\gamma-1} = 11.6^{0.37} = 2.4766$$

$$\therefore \quad T_2 = 320 \times 2.4766 = 792.5 \text{ K}$$

$$T_3 = 792.5 \times 1.53$$

$$= 1212.5255 \text{ K}$$

$$T_4 = 1212.5255 \text{ K}$$

$$T_5 = \frac{1673.28}{2.1984}$$

$$= 761.1377 \text{ K}$$

$$\text{Heat added/kg } Q_A = C_V(T_3 - T_2) + C_P(T_4 - T_3)$$

$$= 0.795(1212.5255 - 792.5) + 1.089(1673.28 - 792.5)$$

$$= 0.795 \times 420.0255 + 1.089 \times 460.7545$$

$$= 333.92 + 501.76$$

$$= 835.68 \text{ kJ/kg}$$

$$\text{Heat rejected/kg } Q_R = C_V(T_5 - T_1)$$

$$= 0.795 \times 441.1377$$

Heat rejected/kg Q_R = 350.70 kJ/kg

$$\therefore \quad \text{Work done/kg} = Q_A - Q_R$$

$$= 835.68 - 350.7$$

Work done/kg = 484.98 kJ/kg

$$\therefore \quad P_m V_1 \frac{r_k - 1}{r_k} = 484.98$$

$$P_m \times \frac{287 \times 320}{1000} \times \frac{11.6-1}{11.6} = 484.98$$

$$\mathbf{P_m = 5.78\ bar}$$

IV. STIRLING CYCLE

Example 13.8 *A striling regenerative air engine works between temperatures of 400°C and 15°C, the ratio of isothermal expansion is 3. Calculate the ideal efficiency when (a) the engine is fitted with a regenerator (b) when the efficiency of the regenerator is 0.8. Take C_p = 1 kJ/kg-K and C_v = 0.718 kJ/kg-K.*

Solution

(a) When regenerator is 100% efficient.

In this case, the efficiency is the same as that for Carnot cycle

$$\therefore \quad \eta = \frac{T_1 - T_2}{T_1} = 1 - \frac{T_2}{T_1}$$

where $T_1 = 400 + 273 = 673\text{ K}$

$\therefore$ $T_2 = 15 + 273 = 288\text{ K}$

$$\therefore \quad \eta = \frac{673-288}{673}$$

$$= \frac{386}{673}$$

$$= 0.5721$$

$$\eta = \mathbf{57.21\%}$$

(b) When regenerator efficiency is 80%

When the regenerator is not 100% efficient, more heat is taken by the working fluid and more heat is rejected by the working fluid. Thc nct output remains the same.

Heat added during the cycle

$$Q_A = RT_2 \log_e r_e + (1-\eta_r) C_v (T_2 - T_1)$$

where r_e = Isothermal expansion ratio = 3

η_r = Regenerator efficiency

= 80% or 0.8

$$\text{Work done/cycle} = (T_1 - T_2) R \log_e r_e$$

$$= (673-288)(1-0.718)\log_e 3$$

$$= 385 \times 0.282 \times 1.0986$$

$$= 119.276\text{ kJ/kg}$$

$$\text{Heat supplied during cycle/kg} = T_1 R \log_e r_e + (1-\eta_r) C_v (T_1 - T_2)$$

$$= 673 \times 0.282 \times 1.0986 + (1-0.8) \times 0.718 \times (673-288)$$

$$= 208.5 + 0.2 \times 0.718 \times 385$$

$$= 208.5 + 55.286$$

∴ Efficiency of the engine

$$= \frac{\text{Work done}}{\text{Heat supplied}}$$

$$= \frac{119.276}{263.786}$$

$$= 0.4522$$

Efficiency of the engine = 45.22%

Example 13.9 *A stirling cycle has a temperature ratio of 3 and volumetric compression ratio of 80% and has only 75% effectiveness in heat exchnages. Calculate the efficiency of the cycle.*

Solution

$$\frac{T_1}{T_2} = 3 \; r_k = 8$$

$$\eta_r = 0.75$$

Work done/kg = Area under (3 – 4) – Area under (2 – 1)

$$= P_3V_3 \log_e \frac{V_4}{V_3} - P_2V_2 \log_e \frac{V_1}{V_2}$$

$$= RT_1 \log_e 8 - RT_2 \log_e 8$$

$$= R(T_1 - T_2)\log_e 8$$

$$= RT_2\left(\frac{T_1}{T_2} - 1\right)\log_e 8$$

$$= 0.287T_1(3-1) \times 2.0794$$

$$= 2 \times 0.287 \times 0.0794T_2$$

Work done/kg = 1.1936 T_2 kJ/kg

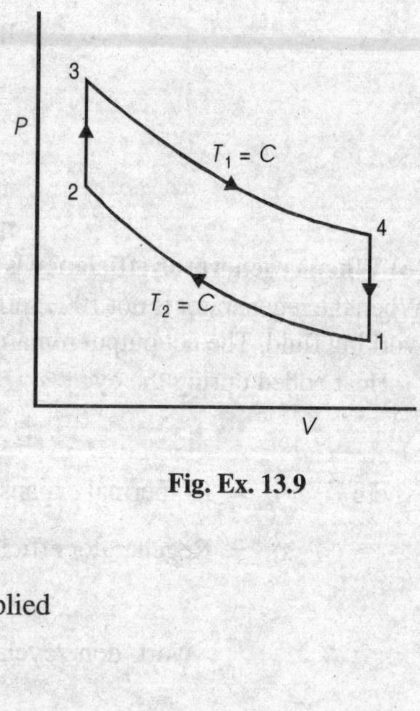

Fig. Ex. 13.9

Heat supplied/kg when η_r= 100%

$$= P_3V_3 \log \frac{V_4}{V_3} RT_1 \log_e 8$$

When regenerator effectiveness is 0.75, additional heat required to be supplied

$$= (1-\eta_r)C_V(T_1 - T_2)$$

$$= 0.25 \times 0.718T_2(3-1)$$

$$= 2 \times 0.25 \times 0.718T_2$$

$$= 0.359\,T_2$$

∴ Total heat supplied/kg

$$= 0.287 \times T_2 \log_e 8 + 0.359T_2$$

$$= 0.287 \times 3T_2 \times 2.0794 + 0.359T_2$$

$$= 1.7904T_2 + 0.359T_2$$

Total heat supplied/kg = 1.8263 T_2 kJ/kg

∴ Efficiency of the cycle

$$ASE = \frac{W}{Q_A} = \frac{1.1936T_2}{1.8263T_2}$$

$$= 0.6536$$

$$\textbf{ASE} = \textbf{65.36\%}$$

ADDITIONAL SOLVED EXAMPLES

Example 13.10 (*a*) *Show that for maximum work, the intermediate temperature in Otto-cycle is given by* $T_2 = T_4 = \sqrt{T_1T_3}$ *where* T_1 *and* T_3 *are initial and maximum temperature respectively. Also find out maximum efficiency of Otto cycle having maximum temperature 1000°C and minimum temperature 15°C. Take* $\gamma = 1.4$.

(*b*) *A four stroke engine working on Otto-cycle has a swept volume of 0.1 m³. The compression ratio is 7. The condition at the start of the cycle is 1 bar and temperature 90°C. The heat addition at constant volume is 100 kJ/cycle. Find ideal efficiency, mean effective pressure and temperature at the key points in the cycle. Assume air at working substance with* $C_v = 0.718$ *kJ/kg.K and* $\gamma = 1.4$.

Solution

(a) For derivation of $T_2 = T_4 = \sqrt{T_1T_3}$ refer theory

$$T_1 = 15+273=288 \text{ K},$$

$$T_3 = 1000+273=1272$$

$$\therefore \quad T_2 = \sqrt{1273\times 288} = 605.5 \text{ K}$$

$$\therefore \quad \text{Max. Eff.} = 1-\frac{T_4-T_1}{T_3-T_2} = 1-\frac{605-288}{1273-605}$$

$$= 1-\frac{317}{668} = 1-0.4746$$

$$\textbf{Max. Eff.} = \textbf{52.54\%}$$

$$\therefore \quad \frac{T_2}{T_1} = (r_k)^{\gamma-1} = \frac{605}{288} = 2.1$$

$$\therefore \quad ASE = 1-\frac{1}{2.1} = 1-0.4762$$

$$= 0.5238 = 52.38\%$$

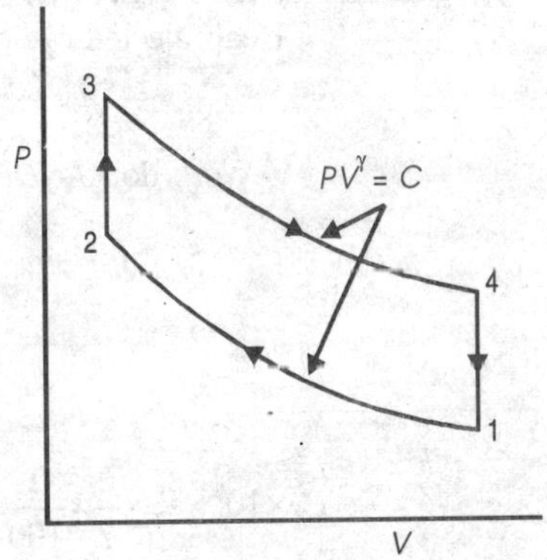

Fig. Ex. 13.10

$$T_1 = 273+90=363\text{K}$$

$$P_1 = 1 \text{ bar}$$

$$r_k = 7$$

$$\therefore \quad T_2 = T_1 r_k^{\gamma-1} = 363\times 7^{0.4}$$

$$= 363\times 2.1779$$

$$= 7906\text{K}$$

$$P_2 = P_1.r_k^{\tau} = 1\times 7^{1.4} = 15.24 \text{ bar}$$

Mass of air in the cycle

$$m = \frac{P_1V_2}{RT_1} = \frac{1\times10^5\times0.1}{287\times363} = 0.096 \text{ kg}$$

$$\therefore \quad \text{Heat added/cycle} = 100 \text{ kg} = m(T_3 - T_2)$$

$$= 0.096(T_3 - 790.6)$$

$$\therefore \quad T_3 = \frac{100}{0.096} + 790.6 = 1041.67 + 790.36$$

$$= 1832.27 \text{ K}$$

$$\therefore \quad P_3 = P_2\times\frac{T_3}{T_2} = 15.245\times\frac{1832.27}{790.6}$$

$$\mathbf{P_3 = 35.33 \text{ bar}}$$

Again

$$P_4 = \frac{P_3}{15.245} = \frac{35.33}{15.245}$$

$$\mathbf{P_4 = 2.1376 \text{ bar}}$$

$$T_4 = \frac{T_3}{2.1779} = \frac{1832.27}{2.1779}$$

$$\mathbf{T_4 = 841.3 \text{ K}}$$

(c) Mean Effective Pressure (MEP)

$$\text{Heat rejected/cycle} = mC_4(T_4 - T_1)$$

$$= 0.096\times0.718(841.3 - 363)$$

$$= 32.97 \text{ kJ/cycle}$$

$$\therefore \quad \text{Work done/cycle} = Q_A - Q_R$$

$$= 100 - 32.97$$

$$= 67.07 \text{ kJ}$$

$$= P_m\times V_1.\frac{r_k - 1}{r_k}$$

$$\therefore \quad P_m\times10^5\times V_1\times\frac{6}{7}\times\frac{1}{100} = 67.03$$

$$P_m = \frac{67.03\times7}{0.6\times100}$$

$$\mathbf{P_m = 7.82 \text{ bar}}$$

Example 13.11 *(a) With the help of P–V and T–S diagrams compare Otto, Diesel and Dual combustion cycles on the basis of same compression ratio and same heat added to each cycle.*

(b) In an engine working on Diesel cycle inlet pressure and temperature are 1 bar and 17°C respectively. Pressure at the end of adiabatic compression is 35 bar. The ratio of expansion i.e., after constant pressure heat addition is 5. Calculate:

(i) Heat addition
(ii) Heat rejection
(iii) Efficiency of the cycle

Take $\gamma = 1.4$*,* $C_p = 1.004$*,* $C_v = 0.717$ *kJ/kgK*

Data

$$P_1 = 1 \text{ bar}, \quad T_2 = 17 + 273 = 290 \text{ K}$$
$$P_2 = P_3 = 35 \text{ bar} \quad r_e = 5$$

(*b*) *The cycle is shown on P–V diagram.*

First we will find the temperatures at the salient points

Solution

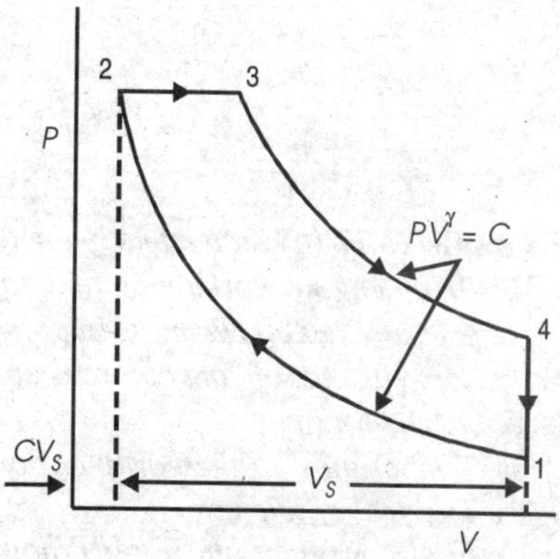

Fig. Ex. 7.11

We have $P_1V_1^\gamma = P_2V_2^\gamma$

$$\therefore \quad \left(\frac{V_1}{V_2}\right)^\gamma = \left(\frac{P_2}{P_1}\right) = 35$$

$$\therefore \quad \frac{V_2}{V_1} = 35^{1/14} = 35^{0.7143}$$
$$= 12.674 \ r_k$$
$$= \text{compression ratio } r_k$$

$$\therefore \quad T_2 = T_1 \times (35)^{0.286} = 2.7644$$
$$= 290 \times 2.7644 = 301.68 \text{ K}$$

How, $r_k = r_c \times r_e$

$$\therefore \quad r_c = \text{cut-off ratio}$$
$$= \frac{12.674}{5}$$
$$= 2.5348$$

$$\therefore \quad r_c = \frac{V_3}{V_2} = \frac{T_3}{T_4}$$
$$T_3 = 2.5348 \times T_2 = 2.5348 \times 801.68$$
$$= 2032.1 \text{ K}$$

Again, $\frac{T_3}{T_4} = r_e^{\gamma-1} = 5^{0.4}$
$$= 1.9037$$
$$T_4 = \frac{2032.1}{1.9037}$$
$$= 1067.45 \text{ K}$$

(i) Heat addition:

$$\text{Heat added/kg} = Q_A = C_p(T_4 - T_1)$$
$$= 1.004(2032.1 - 801.68)$$
$$= 1.004 \times 1230.42$$

Heat added/kg = 1235.34 kJ/kg

(ii) Heat rejected:

$$\text{Heat rejected/kg} = C_v(T_4 - T_1)$$

$$= 0.717\,(106.45 - 290)$$
$$= 0.717 \times 777.45$$

Heat rejected/kg = 557.43 kJ/kg

(iii) Efficiency of cycle:

Thermal efficiency of the cycle

$$= \eta_{th} = 1 - \frac{Q_R}{Q_A}$$

$$= 1 - \frac{557.43}{1235.34} = 1 - 0.4512$$

$$= 0.5488 \text{ or } 54.88\,\%$$

Example 13.12 (*a*) *State clearly the assumptions of air standard cycle analysis.*

(*b*) *In an engine working on limited pressure at the end of adiabatic compression is 164.5 N/cm². The maximum cycle pressure is 300 N/cm². Compression begins at 1 bar and 27°C. The amount of heat added at constant volume is the same as that of constant pressure. Assume heating volume of the fuel as 41840 kJ/kg, and relative efficiency as 55%. Determine*

(*i*) *Temperature at the cardinal points of the cycle*
(*ii*) *Thermal efficiency*
(*iii*) *Fuel consumption in kg/IP/hour*
(*iv*) *Power developed per cu-m, displacement at engine speed of 1800 rpm*

The engine works on four stroke cycle.

Solution

Given:

$$P_1 = 1 \text{ bar}$$
$$T_1 = 27 + 273 = 300°\text{K}$$
$$P_2 = 164.5 \text{ N/cm} = 164.5 \times 10^4 \text{ N/m}^2 = 16.45 \text{ bar}$$
$$P_3 = 300 \text{ N/cm}^2 = 30 \text{ bar}$$

Cal. value of fuel used = 41840 kJ/kg and Rel. eff = 55%

$$\therefore \quad T_2 = T_1 \times \left(\frac{16.45}{1}\right)^{0.4/1.4} = 300 \times 2.2275$$

$$T_2 = 668.26°\text{K}$$

$$\therefore \quad r_k = (2.2275)^{3.5} = 16.5$$

Again,

$$\frac{P_3}{P_2} = \frac{T_3}{T_2} = 16.5$$

$$T_3 = T_2 \times \frac{30}{16.45} = 668.26 \times \frac{30}{16.45}$$

$$\mathbf{T_3 = 1278.71\ K}$$

$$C_v(T_3 - T_2) = 0.718\,(1278.71 - 668.26)$$
$$= 0.718 \times 610.45$$
$$= 438.3 \text{ kJ/kg}$$
$$= Q_{A_P} = C_p\,(T_4 - T_3) = 1.005[T_4 - 1278.71]$$

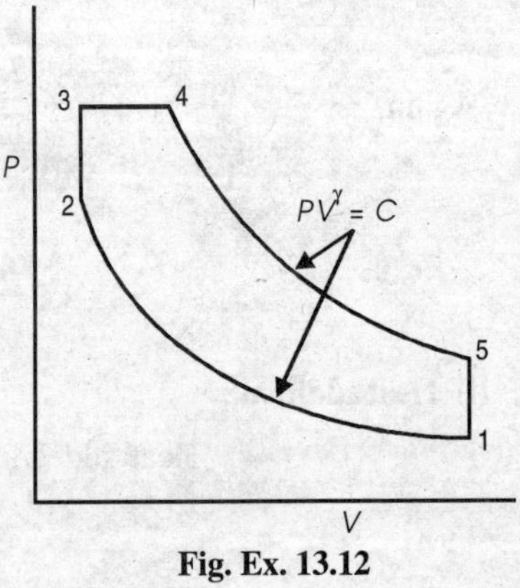

Fig. Ex. 13.12

$$T_4 = \frac{438.3}{1.005} + 1278.71 = 436.12 + 1278.71$$

$$\mathbf{T_4 = 1714.83\ K}$$

$$\frac{T_4}{T_3} = \frac{V_4}{V_3} = r_c = \frac{1714.83}{1278.71}$$

$$= 1.341$$

$$r_e = r_k / r_c = \frac{16.5}{1.341} = 12.3$$

$$\frac{T_4}{T_5} = (r_e)^{0.4} = (12.3)^{0.4} = 2.729$$

$$T_5 = (r_e)^{0.4} = (12.3)^{0.4} = 2.729$$

$$\mathbf{T_5 = \frac{1714.83}{2.279} = 628.37\ K}$$

(ii) Thermal efficiency

Heat added/kg = $2 \times 438.3 = 876.6$ kJ/kg

Heat rejected/kg = $1.005\,(628.37 - 300)$

= 330 kJ/kg

Work done/kg = $876.6 - 330$

= 546.6 kJ/kg

$$\text{ASE} = \frac{546.6}{876.6} = 0.6235$$

ASE = 62.35%

Actual thermal efficiency = $0.55 \times 0.6235 = 0.343$ or 34.3%

(iii) Fuel consumption

Indicated thermal efficiency

= 0.343

$$= \frac{\text{IP} \times 3600}{\text{Fuel/hr} \times \text{cal. value}}$$

$$= \frac{\text{IP} \times 3600}{\text{Fuel/hr} \times 41840}$$

$\therefore$ Fuel consumption $= \dfrac{3600}{41840 \times 0.343}$

Fuel consumption = 0.2508 kg/IP/hour

(iv) Power developed per m³ displacement

Actual work/done kg = 546.6×0.55

$$= \frac{300.60}{P_m \times V_s} \text{ kJ/kg}$$

$$\therefore \quad 300.6 = P_m \times V_1 \frac{r_k - 1}{r_k}$$

$$= P_m \times \frac{10^5}{1000} \times \frac{1 \times 287 \times 300}{1 \times 10^5} \times \frac{1.2275}{2.2275}$$

$$= P_m \times 47.45$$

$$\therefore \quad P_m = \frac{300.6}{47.45} = 6.335 \text{ bar}$$

∴ Work done/m³ displacement

$$\doteq P_m \times \frac{\pi}{4} \frac{d^2 \times L \times 1800/2 \times 60}{1000}$$

$$= \frac{6.335 \times 10^5 \times 1 \times 1800/2 \times 60}{1000}$$

$$= \frac{6.335 \times \frac{\pi}{4} d^2 \times L \times 1800/2 \times 60}{1000}$$

$$= 6.335 \times 100 \times 15 \text{ J/S}$$

Work done/m³ displacement = 9.502 kW

Example 13.13 (*a*) *Compare air standard Otto and Diesel cycle for:*

(*i*) *Same compression ratio and same heat input.*

(*ii*) *Same maximum pressure and same heat input*

(*b*) *Explain in brief why actual cycle deviates from theoretical cycle*

(*c*) *What will be loss in the ideal efficiency of a Diesel engine with compression ratio 16 if the fuel cut-off is delayed from 6 to 8%.*

Solution

First consider the case when fuel cut-off takes place at 6% stroke

$$V_3 - V_2 = 0.06\, V_s$$

$$\therefore \quad r_c = \frac{CV_s + 0.06V_s}{CV_s}$$

$$= \frac{C + 0.06}{C}$$

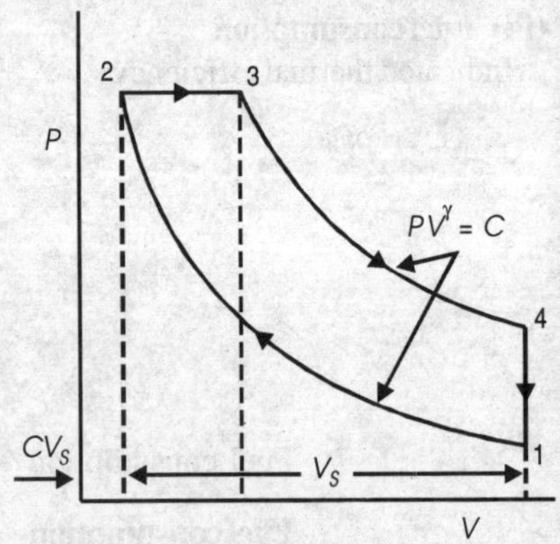

Fig. Ex. 13.13

and
$$\frac{V_1}{V_2} = r_k = 16 = \frac{CV_s + V_s}{CV_s}$$

$$= \frac{C+1}{C} = 1 + \frac{1}{C}$$

$$\therefore \quad \frac{1}{C} = 15$$

or
$$C = \frac{1}{15}$$

$$r_c = \frac{1/15 + 0.06}{1/15}$$

$$= 1+15\times0.06$$
$$= 1+0.90=1.9$$

$$\frac{V_3}{V_2} = \text{cut-off ratio}$$

$$(\text{ASE})_1 = 1-\frac{1}{\gamma\cdot r_k^{\gamma-1}}\left[\frac{r_c^{\gamma}-1}{r_c-1}\right]$$

$$= 1-\frac{1}{\gamma\cdot r_k^{\gamma-1}}\left[\frac{1.9^{1.4}-1}{19-1}\right]$$

$$= 1-\frac{1}{1.4\times16^{0.4}}\left[\frac{2.456-1}{0.9}\right]$$

$$= 1-\frac{1}{1.4\times3.031}\left[\frac{1.456}{0.9}\right]$$

$$= 1-0.3812$$
$$= 0.6188 \text{ or } 61.88\%$$

When cut off is delayed to 8%

$$r_c = 1+(r_k-1)\,0.08$$
$$= 1+1.20=2.2$$

$$\text{ASE}_2 = 1-\frac{1}{1.4\times3.031}\left[\frac{2.2^{1.4}}{2.2-1}\right]$$

$$= 1-\frac{1}{4.2434}\times\frac{3.016}{1.2}$$

$$= 1-0.5922$$
$$= 0.4078 \text{ or } 40.78\%$$

∴ Loss of ideal efficiency

$$= 61.88-40.78$$
$$= 21.1\%$$

$$\%\text{ of loss} = \frac{21.1}{61.88}\times100$$

$$\textbf{\% of loss} = \textbf{34.1\%}$$

Example 13.14 (*a*) *State the assumptions made for Air Standard Cycles.*

(*b*) *An oil engine with 20 cm cylinder diameter and 30 cm stroke works on theoretical Diesel cycle. The initial pressure and temperature of the air used are 1 bar and 30°C. Cut-off ratio is 2.6. Assume compression ratio as 16 and working fluid is air. Find:*

(*i*) *Pressure and temperature at all salient points*

(*ii*) *Theoretical air standard efficiency*

(*iii*) *Mean Effective pressure*

(*iv*) *Power developed by the engine if the working cycle per minute are 400. Take ideal conditions.*

Solution

$$P_1 = 1 \text{ bar}$$

$$T_1 = 303°K$$
$$r_k = 16$$
$$r_c = 2.6$$
$$r_e = \frac{16}{2.6} = 6.1538$$
$$P_2 = P_1 \times r_k^{\gamma} = 1 \times 16^{1.4}$$
$$= 48.503 \text{ bar}$$
$$T_2 = T_1 \times r_k^{\gamma-1} = 303 \times 16^{0.4}$$
$$= 303 \times 3.031 = 918.52$$
$$P_2 = P_3 = 48.503 \text{ bar}$$
$$\frac{V_3}{V_2} = r_c = \frac{T_3}{T_2}$$
$$\mathbf{T_3 = 2.6 \times 918.52 = 2388.152\ K}$$
$$\frac{P_3}{P_4} = r_e^{\gamma} = (6.01538)^{1.4} = 12.7292$$
$$P_4 = \frac{48.503}{12.7292} = 3.8104 \text{ bar}$$

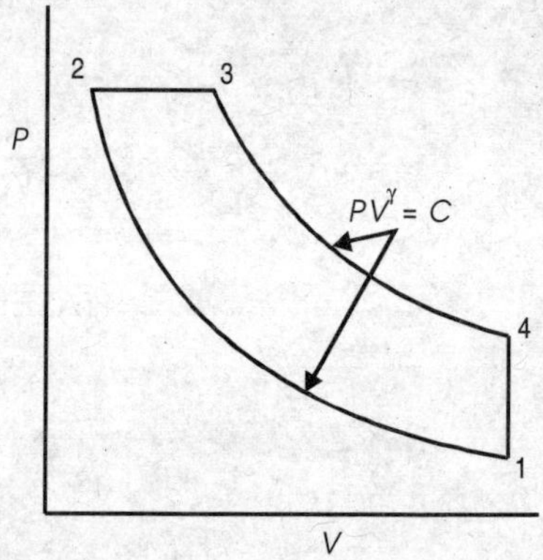

Fig. Ex. 13.14

Also

$$\frac{T_4}{T_1} = \frac{P_4}{P_1} = 3.8104$$
$$T_4 = 303 \times 3.8104 = 1145.55 \text{ K}$$

(iii) Theoretical ASE

$$Q_A = \text{Heat rejected/kg} = C_p (T_3 - T_2)$$
$$= 1.005(2388.152 - 918.52)$$
$$= 1.005 \times 1469.532 = 1476.88 \text{ kJ}$$
$$\mathbf{Q_R = Heat\ rejected/kg = C_v\ (T_4 - T_1)}$$
$$= 0.718\,(1154.55 - 303)$$
$$\mathbf{Q_R = 609.26\ kJ}$$
$$\text{ASE} = 1 - \frac{Q_R}{Q_A}$$
$$= 1 - \frac{609.26}{1426.88}$$
$$= 1 - 0.4125$$
$$= \mathbf{0.5875} \text{ or } \mathbf{58.75\%}$$

(iii) Mear effective pressure

$$\text{Work done/kg} = 1476.88 - 609.26$$
$$= 867.62$$
$$= P_m \times V_s$$
$$= P_m \cdot V_1 \cdot \frac{r_k - 1}{r_k}$$

$$= P_m \times \frac{RT_1}{P_1} \times \frac{r_k - 1}{r_k}$$

$$= P_m \times \frac{287 \times 10^5 \times 303}{1000 \times 10^5} \times \frac{15}{16}$$

$$P_m = \frac{867.62 \times 16}{15} \times \frac{1}{303 \times 0.287}$$

$$\mathbf{P_m = 10.64\ bar}$$

(iv) Power developed

$$\text{Mass of air/second} = \frac{P_1 V_1}{RT_1}$$

$$= \frac{1 \times 10^5}{287 \times 303} \times \frac{\pi d^2 L}{4} \times \frac{400}{60}$$

$$= 1.15 \times \frac{\pi}{4} \times (0.2 \times 0.2) \times 0.3 \times \frac{400}{60}$$

$$= 1.15 \times \frac{\pi}{4} \times 0.04 \times 0.3 \times \frac{400}{60}$$

$$= 0.07226 \text{ kg/sec}$$

$$\therefore \quad \text{Work/sec} = 867.62 \times 0.07226$$

$$\textbf{Work/sec} = \mathbf{62.7\ kJ/sec = kW}$$

Example 13.15 *(a) Derive the expression for thermal efficiency for air standard Otto cycle and prove that, for maximum work, compression rate should be*

$$r_k = \left(\frac{T_3}{T_1}\right)^{1.25}$$

where T_3 and T_1 are upper and lower and temperature limits of cycle respectively.

Take adiabatic index = 1.4

(b) Find air standard efficiency of 4-stroke engine working on otto-cycle for given data:

Bore diameter = 13.7 cm

Stroke length = 13.0 cm

Clearance vol. = 280 cm³

(c) Compare Otto, Diesel and Dual air standard cycle for some comp. ratio, same heat input. Explain with neat P–V and T–S diagram.

Solution

$$\text{Work done/kg} = Q_A - Q_R$$

$$= C_V(T_3 - T_2) - C_V(T_3 - T_2) - C_V(T_4 - T_1)$$

$$\text{Heat supplied/kg} = Q_A = C_V(T_3 - T_2)$$

$$\therefore \quad \text{Thermal efficiency} = \frac{Q_A - Q_R}{Q_A}$$

$$= 1-\frac{Q_R}{Q_A}=1-\frac{T_4-T_1}{T_3-T_2}$$

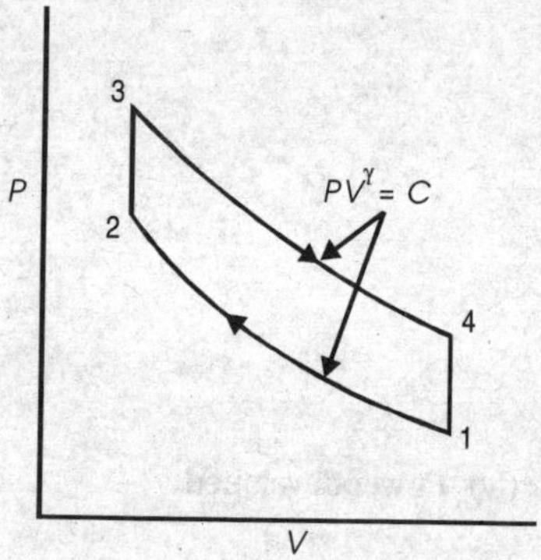

Fig. Ex. 13.15

We have $\frac{T_2}{T_1} = r_k^{\gamma-1}$

$$\frac{T_3}{T_4} = (r_e)^{\gamma-1}=r_k^{\gamma-1}$$

$$T_3 = T_4 r_c^{\gamma-1} \text{ and } T_2 = T_1 r_k^{\gamma-1}$$

Thermal efficiency $= 1-\frac{T_4-T_1}{T_4 r_k^{\gamma-1}-T_1 r_k^{\gamma-1}}$

$$= \frac{1}{r_k^{\gamma-1}}$$

Work done $= C_V\left[T_3-T_2-T_4+T_1\right]$

From above $T_2 = T_1.r_k^{\gamma-1}$

and $T_4 = \frac{T_3}{r_k^{\gamma-1}}$

$$C_V\left[T_3-T_1 r_k^{\gamma-1}-\frac{T_3}{r_k^{\gamma-1}}+T_1\right]=W$$

Work done is maximum when $\frac{dW}{dr_k}=0$

$$-(\gamma-1)T_1 r_c^{\gamma-1-1}+T_3(\gamma-1)^{(-\gamma+1-1)r_c}=0$$

$$T_1 r_k^{\gamma-1} = T_3 r_k^{-\gamma}$$

$$\therefore \quad \frac{T_3}{T_1} = \frac{r_k^{\gamma-2}}{r_k^{-\gamma}}=r_k^{\gamma-2+\gamma}=r_k^{2(\gamma-1)}$$

$$\therefore \quad r_c = \left(\frac{T_3}{T_1}\right)^{1/2(\gamma-1)}=\left(\frac{T_3}{T_1}\right)^{1/2(1.4-1)}$$

$$= \left(\frac{T_3}{T_1}\right)^{1/0.8}=\left(\frac{T_3}{T_1}\right)^{1.25}$$

$$\mathbf{r_k} = \left(\frac{\mathbf{T_3}}{\mathbf{T_1}}\right)^{\mathbf{1.25}}$$

(b) Air standard efficiency = $1-\dfrac{1}{r_k^{\gamma-1}}$

$$r_k = \frac{V_s + \text{cl. vol}}{\text{cl. vol}}$$

where V_s = Stroke volume

$$= \frac{\pi}{4} d_2 \times l$$

$$= \frac{\pi}{4} \times (13.7)^2 \times 3$$

$$= 1916.348 \text{ cm}^3$$

$$\therefore \quad r_k = \frac{1916.348 + 280}{280}$$

$$= \frac{2196.348}{280}$$

$$= 7.844$$

$$\eta_{th} = 1 - \frac{1}{(7.844)^{0.4}} = 1 - \frac{1}{2.2794}$$

$$= 1 - 0.4387$$

$$\boldsymbol{\eta_{th} = 0.5613 \text{ or } 56.13\%}$$

(ii) Heat supplied/kg air

$$\text{Heat supplied/kg} = C_p(T_3 - T_2)$$

$$= 1.005[1753 - 873.05]$$

$$= 1.005 \times 879.95$$

$$\textbf{Heat supplied/kg} = \mathbf{884.35 \text{ kJ/kg}}$$

(iii) $$\frac{T_3}{T_4} = r_e^{\gamma-1} = \left(\frac{r_k}{r_c}\right)^{\gamma-1}$$

$$= \left(\frac{16}{2.0079}\right) = (7.9685)^{0.4} = 2.294$$

$$\therefore \quad T_4 = \frac{1753}{2.294} = 764.17 \text{ K}$$

$$\therefore \quad Q_R = \text{Heat rejected} = C_V (T_4 - T_1)$$

$$= 0.718(764.17 - 288) = 0.718 \times 476.17$$

$$= 341.89 \text{ kJ/kg}$$

$$\therefore \quad \text{Work done/kg} = Q_A - Q_R = (884.35 - 341.89)$$

$$= 542.46 \text{ kJ/kg}$$

The cycle efficiency

$$= 1 - \frac{Q_R}{Q_A} = 1 - \frac{341.89}{884.35} = 1 - 0.3866$$

The cycle efficiency = 0.6134 i.e., 61.344%

(iv) MEP

$$\text{Work done} = P_m \times \text{Stroke volume}$$

$$\text{Stroke volume} = V_s = V_1 - V_2 = V_1\left[1 - \frac{V_2}{V_1}\right]$$

$$= V_1\left[1 - \frac{1}{r_k}\right] = V_1 \times \frac{r_k - 1}{r_k}$$

$$\therefore \quad 542.46 \times 1000 = P_m \times 10^5 \times \frac{16-1}{16}\left[\frac{RT_1}{P_1}\right]$$

$$= P_m \times 10^5 \times \frac{15}{16}\left[\frac{287 \times 288}{1 \times 10^5}\right]$$

$$= P_m \times 10^5 = 0.7749$$

$$P_m = \frac{542.46 \times 1000}{10^5 \times 0.7749} = \frac{5.4246}{0.7749}$$

$$\mathbf{P_m = 7}$$

Example 13.16 (*a*) *Compare the limitations of air standard cycle. What are the assumptions of fuel air cycle?*

(*b*) *The parameters of the initial stage of 1 kg of air in the cycle of an IC Engine are 0.95 bar, 65°C. The compression ratio is 11. Compare the values of ideal thermal efficiency for*

(*i*) *Constant volume cycle*

(*ii*) *Constant pressure cycle*

(*iii*) *Dual combustion cycle with equal heat supplied in constant pressure and constant volume heat supply process.*

Assume total heat supplied in the cycle 800 kJ.

Solution

(i) Constant volume cycle:

$$\text{Thermal efficiency} = 1 - \frac{1}{r_k^{\gamma-1}}$$

$$= 1 - \frac{1}{11^{0.4}} = 1 - \frac{1}{2.6095}$$

$$= 1 - 0.3832$$

Thermal efficiency = 0.6168 or 61.68%

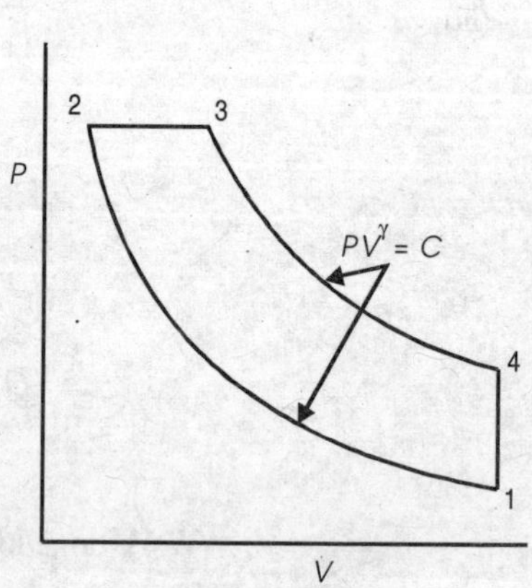

Fig. Ex. 13.16 (a)

(ii) Constant pressure cycle

$$P_1 = 0.95 \text{ bar}$$

$$T_1 = 273 + 65 = 338 \text{ K}$$

$$\frac{T_2}{T_1} = r_k^{\gamma-1} = 11^{0.4} = 2.6095$$

$$T_2 = 338 \times 2.6095$$

$$= 882.01 \text{ K}$$

$$\eta = \frac{W}{Q_A} = \frac{Q_A - Q_R}{Q_A}$$

$$= \frac{800-314.81}{800} = \frac{485.19}{800}$$

$$= \mathbf{0.6065 \text{ or } 60.65\%}$$

$$mC_p(T_3 - T_2) = 1.005(T_3 - 882.011) = 880$$

$$T_3 - 882.011 = \frac{800}{1.005} = 796.02$$

$$T_3 = 796.02 + 882.01$$
$$= 1678.031 \text{ K}$$

$$\frac{V_3}{V_2} = \frac{T_3}{T_4} = r_e = \frac{1678.031}{882.011} = 1.9025$$

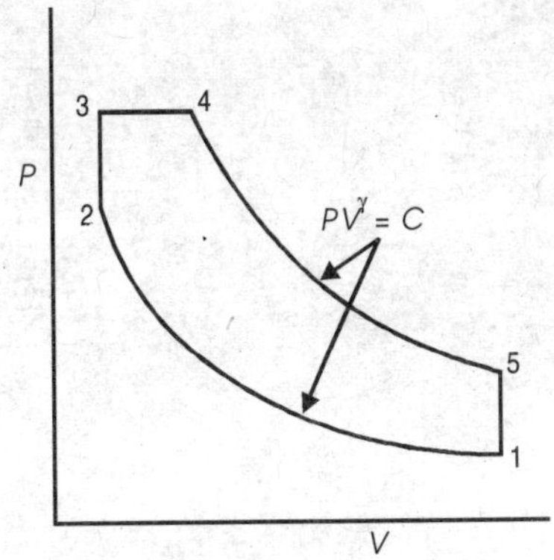

Fig. Ex. 13.16 (b)

Expansion ratio $r_e = \frac{r_k}{r_c} = \frac{11}{1.9025}$

$$= 5.782$$

$\therefore$ $$\frac{T_3}{T_4} = r_e^{\gamma-1} = 5.782^{1.4-1} = 2.0176$$

$\therefore$ $$T_4 = \frac{T_3}{2.0176} = \frac{1678.031}{2.0176} = 831.7 \text{ K}$$

$$Q_R = C_V(T_4 - T_1) = 0.718\,[831.7 - 338]$$
$$= 0.781 \times 493.7$$
$$= 354.48 \text{ kJ}$$

$\therefore$ $$\eta = 1 - \frac{Q_R}{Q_A} = 1 - \frac{354.48}{800} = 1 - 0.4431$$

$$\eta = \mathbf{0.5569 \text{ or } 55.69\%}$$

(iii) Dual combustion cycle:

$$P_2 = 0.95 \times 11^{1.4} = 0.95 \times 28.7$$
$$= 27.27 \text{ bar}$$
$$T_2 = 882.11 \text{ K}$$
$$Q_{AV} = 400\text{kJ} = (T_3 - T_2) \times C_V$$
$$= 0.718\,(T_3 - 882.011)$$

$$T_3 = \frac{400}{0.718} + 882.011$$
$$= 557.1 + 882.01 = 1439.11 \text{ K}$$

Again $$Q_{Ap} = 400 = 1.005\,[T_4 - T_3]$$
$$= 1.005\,[T_4 - 1439.11]$$

$$T_4 = \frac{400}{1.005} + 1439.11 = 398 + 1439.11$$
$$= 1837.11 \text{ K}$$

$$R_c = \frac{11}{1.277} = 9.614$$

$$\frac{T_4}{T_5} = 8.164^{0.4} = 2.366$$

$$T_5 = \frac{1837.11}{2.366} = 776.46°C$$

$$Q_R = 0.718\,[776.46 - 338] = 0.718 \times 438.46 = 314.81 \text{ kJ}$$

$\therefore$

$$\eta = \frac{W}{Q_A} = \frac{Q_A - Q_R}{Q_A}$$

$$= \frac{800 - 314.81}{800} = \frac{485.19}{800}$$

$$= \mathbf{0.6065 \text{ or } 60.65\%}$$

EFFECT OF VARIABLE SPECIFIC HEAT ON OTTO-CYCLE EFFICIENCY

Example 13.17 *What is the effect of percentage change in the efficiency of Otto-cycle having a compression ratio of 7, if the specific heat at constant volume increases by 1%?*

Solution

The otto-cycle efficiency is given by

$$\eta = 1 - \frac{1}{r_K^{\gamma-1}}$$

We have $C_p - C_v = R$ and $\gamma - 1 = \frac{R}{C_v}$ as $C_v = \frac{R}{\gamma - 1}$

$\therefore$

$$\eta = 1 - \frac{1}{r_K^{R/C_v}}$$

$\therefore$

$$1 - \eta = r_K^{R/C_v} \text{ or } \log_e(1-\eta) = -\frac{R}{C_v} \cdot \log_e r_K$$

Differentiating,

$$-\frac{1}{1-\eta} \cdot d\eta = \frac{R}{C_v^2} \cdot \log_e r_K \cdot dC_v$$

or

$$d\eta = -(1-\eta)\frac{R}{C_v^2} \cdot \log_e r_K \cdot dC_v$$

or

$$\frac{d\eta}{\eta} = -\frac{(1-\eta)}{\eta} \cdot (\gamma - 1) \cdot \log_e r_K \cdot \frac{dC_v}{C_v} \qquad \text{....(1)}$$

$$\eta = 1 - \frac{1}{r_K^{\gamma-1}} = 1 - \frac{1}{T^{1.4-1}} = 0.54 \text{ or } 54\ \%$$

$\therefore$

$$\frac{d\eta}{\eta} = -\frac{1 - 0.54}{0.54} \times (1.4 - 1)\log_e 7 \times \frac{1}{100}$$

$$= -\frac{0.46\times0.4}{54}\times[\log_e 7 = 1.946] = -0.00663$$

$$= \mathbf{-0.663\ \%\ (approx.)}$$

OTTO-CYCLE: EFFECT OF VARIABLE SPECIFIC HEAT ON MAXIMUM PRESSURE

Example 13.18 *A petrol engine of compression ratio 6 uses a fuel of calorific value* 44000 kJ/kg. *The air to fuel ratio is 15 : 1. The pressure and temperature of the charge at the end of suction stroke are 1 bar and 60°C respectively. Determine the maximum pressure in the cylinder if the index of compression is 1.32 and the specific heat at constant volume is expressed by the relation* $C_v = \frac{0.712}{0.712} + \frac{20.94T^{-5}}{20.94}$ *where T is the temperature in K. Compare this value with that of constant specific heat* $C_v = (0.712\ kJ/kg°K)$.

Solution

We have $p_2 = p_1,\ r_K^{n-1} = 1\times6^{1.32}$

$= 10.64$ bar

$T_2 = T_1\times r_K^{n-1} = 333\times6^{0.32}$

$= 333\times1.772$

$= \mathbf{590\ K}$

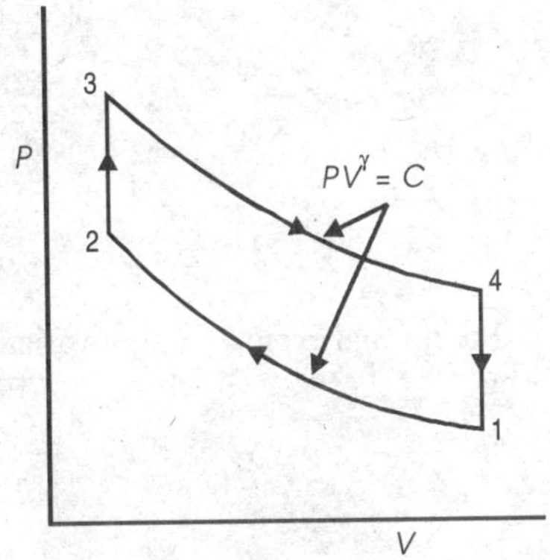

Fig. Ex. 13.18

Assume 1 kg air : Fuel $= \frac{1}{15}$ kg

$\therefore$ Heat supplied $= \frac{44000}{15} = \mathbf{2933.33\ kJ}$

Mass of charge $= 1+\frac{1}{15} = \frac{16}{15}$ kg

$\therefore$ Q = Heat added $= \text{Mass}\times\text{specific heat}\times(T_3 - T_2)$

But we have $dQ = m\cdot C_v\cdot dT = m\cdot(0.712 + 20.94\,T^{-5})\cdot dT$

$\therefore$ $Q = m\int(0.712+20.94\,T^{-5})\cdot dT$

$$\therefore\quad 2933.33 = \frac{16}{15}\left[0.712\,T - \frac{20.94\times T^{-4}}{4}\right]_{T_2}^{T_3}$$

$$\therefore\quad \frac{2933.33\times15}{16} = 0.712(T_3 - 590) - 5.235(T_3^{-4} - T_2^{-4} - 590)$$

We can solve this by another method:

If T_3 is the maximum temperature after heat addition — then mean temperature during heating is $\frac{T_3+T_2}{2}$.

$$\therefore \quad \frac{2933.33\times 15}{16} = \left[0.712+20.94\times 10^{-5}\left(\frac{T_3+590}{2}\right)\right]\times (T_3-590)$$

$$2750 = \left(1.424+20.94\,T_3\times 10^{-5}+12354.6\times 10^{-5}\right)(T_3-590)$$

$$2750 = 10^{-5}\left(\frac{1.424}{10^{-5}}+20.94\,T_3+12354.6\right)(T_3-590)$$

$$10^5\times 2750 = (142400+20.94\,T_3+12354.6)\,(T_3-590)$$

$$= (20.94\,T_3+154754.6)\,(T_3-590)$$

$$= 20.94\,T_3^2+154754.6\,T_3-91305214-12354.6\,T_3$$

$$= 20.94\,T_3^2+142400\,T_3-91305014$$

$$10^5\times 131.3 = T_3^2+6800\,T_3-4.36\times 10^6$$

$$0 = T_3^2+6800\,T_3-10^6\,(4.36+13.13)$$

$$= T_3^2+6800\,T_3-17.49\times 10^6$$

$$T_3 = 3095 \text{ K by trial and error method}$$

$$\therefore \quad p_3 = 10.64\times \frac{3095}{590}$$

For normal cycle, (without variation of specific heat)

$$2750 = 0.712\,(T_3-590)$$

$$\therefore \quad T_3 = \frac{2750}{0.712}+590$$

$$= 3862.36+590$$

$$= 4452.36\,°\text{K}$$

$$\therefore \quad \frac{p_3}{p_2} = \frac{T_3}{T_2}$$

or

$$p_3 = p_2\times \frac{T_3}{T_2} = 10.64\times \frac{4452.36}{590}$$

$$= \mathbf{80.3\ bar}$$

$$\frac{2933.33\times 15}{16} = \left[0.712+20.94\times 10^{-5}\left(\frac{T_3+590}{2}\right)\right](T_3-590)$$

$$2750\times 10^5 = \left[71200+20.46\left(\frac{T_3+590}{2}\right)\right](T_3-590)$$

$$= [71200+10.46\,T_3+6171.4]\,(T_3-590)+6171.4$$

$$= [77371.4+10.46\,T_3]\,(T_3-590)$$

$$= 77371.4\,T_3+10.46\,T_3^2-45.65\times 10^6-6171.4\,T_3$$

$$= 71200\,T_3 + 10.46\,T_3^2 - 45.65 \times 10^6$$

$$= 71200\,T_3 + 10.46\,T_3^2 - (45.65 + 275)10^6$$

$$= 71200\,T_3 + 10.46\,T_3^2 - 320.65 \times 10^6$$

$$= 6807\,T_3 + T_3^2 - 30.65 \times 10^6$$

$$= T_3^2 + 6807\,T_3 - 30.65 \times 10^6$$

By trial and error method

$$T_3 = 3095 \text{ K}$$

$$\therefore \quad p_2 = p_1 \times \frac{3095}{590}$$

$$= \frac{10.64 \times 3095}{590}$$

$$= \mathbf{55.8\ bar}$$

DIESEL ENGINE – % STROKE IN CONE

Example 13.19 *If a Diesel engine combustion is assumed to begin at IDC and to be at constant pressure. The air to fuel ratio is 28 to 1, the calorific value of the fuel is 42000 kJ/kg and the specific heat of the products of combustion is given by –*

$$C_v = 0.712 + 18.85 \times 10^{-5} T$$

$$R \text{ for the products} = 287 \text{ N-m/kg/K}$$

If the compression ratio is 14 : 1, and the temperature at the end of compression is 800 K. What percentage of the stroke during combustion is completed.

Solution

For one kg fuel the charge is (28 + 1 = 29 kg) and the heating value is 42000 kJ/kg.

$$\therefore \quad dQ = \int_2^3 C_p \cdot dT$$

But

$$C_p = R + C_v$$

$$= 0.287 + 0.712 + 18.82 \times 10^{-5}\,T$$

$$= 0.999 + 18.85 \times 10^{-5}\,T$$

$$\therefore \quad \frac{42000}{29} = \int_2^3 \left(0.999 + 18.85 \times 10^{-5}\,T\right) dT$$

$$1448.3 = \left[0.999\,T + \frac{18.85 \times 10^{-5}\,T^2}{2}\right]_{800}^{T_3}$$

$$= 0.999\,(T_3 - 800) + 9.425 \times 10^{-5}\,(T_3^2 - 800^2)$$

$$= 0.999\,T_3 - 800 + 9.425 \times 10^{-5}\,T_3^2 - 60.32$$

$$0 = 9.425 \times 10^{-5}\,T_3^2 + 0.999\,T_3 - (860.32 + 1448.3)$$

$$= 9.425 \times 10^{-5}\,T_3^2 + T_3 - 2308.62$$

$$= T_3^2 + 10610\,T_3 - 244.95 \times 10^5$$

$$= T_3^2 + 10610\, T_3 - 24.495 \times 10^6$$

For $T_3 = 2000$, $\quad 0 = 4 \times 10^6 + 21.22 \times 10^6 - 24.495 \times 10^6$

$= 25.22 \times 10^6 - \mathbf{24.495 \times 10^6}$

For $T_3 = 1990 \rightarrow \quad 0 = (3.96 \times 10^6 + 21.11 \times 10^6)$

$= 25.07 \times 10^6$

$1980 \rightarrow \quad = (3.92 + 21.01 = 24.93) \times 10^6$

$1975 \rightarrow \quad = 3.90 + 20.95 = 24.85 \times 10^6$

$1960 \rightarrow \quad = (3.84 + 20.8 = 24.64) \times 10^6$

$1950 \rightarrow \quad = 3.80 + 20.69 = \mathbf{24.49 \times 10^6}$

$\therefore \quad \mathbf{T_3 = 1950^\circ}$

$\therefore$ Cut off ratio $= \dfrac{V_3}{V_2} = \dfrac{T_3}{T_2} = \dfrac{1950}{800}$

$= \mathbf{2.4375}$

$\therefore$ Combustion occupies

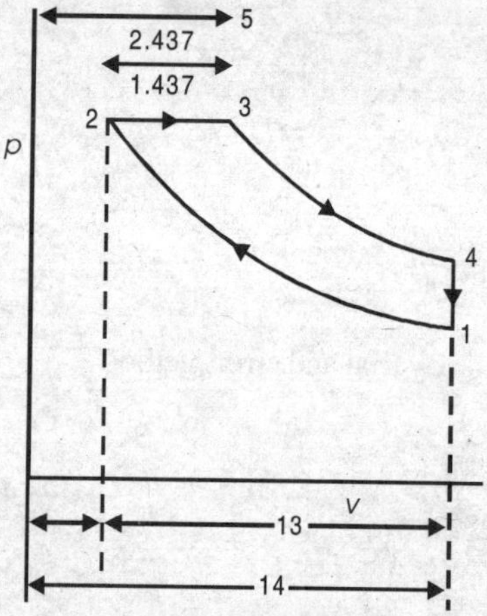

Fig. Ex. 13.19

$$= \frac{1.4375}{13} \times 100$$

$= \mathbf{11.06\ \%\ stroke}$

ACTUAL CYCLE

Actual cycle efficiency is much lower than the air standard efficiency due to various losses occurring in the actual engine operation. These losses are —

(1) Losses due to variation of specific heats with temperature.
(2) Chemical equilibrium losses or dissociation losses.
(3) Time losses.
(4) Losses due to incomplete combustion.
(5) Direct heat losses.
(6) Exhaust blow-down losses.
(7) Pumping losses.

ASE > Fuel-Air Cycle efficiency > Actual cycle efficiency

Air Cycle → Correction for the characteristics of fuel-air medium becomes → Fuel-Air Cycle

i.e. Air Cycle → Variable specific heat losses + Dissociation losses → Fuel-Air cycle

Fuel-air cycle → Losses (listed above) → Actual Cycle

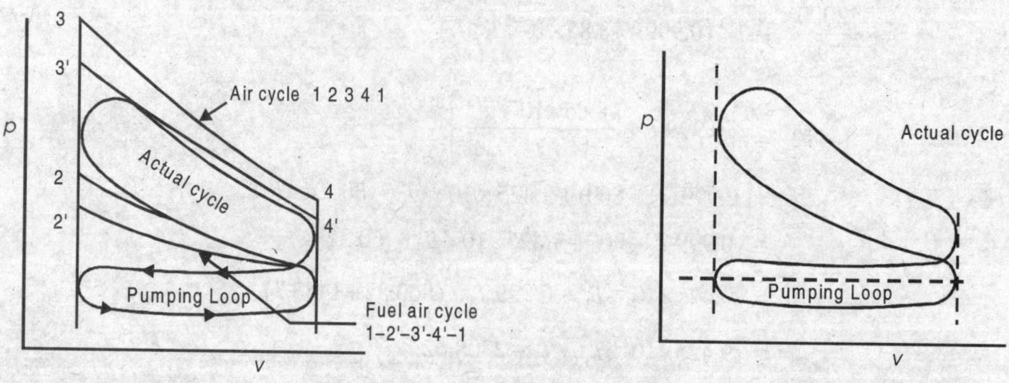

Fig. 13.17

PROBLEMS FOR PRACTICE

13.1. An air engine working on Otto cycle the compression ratio raised from 5 to 6. Compare the change in efficiency due to the rise in compression ratio. [**Ans.** 8.12% rise]

13.2. Calculate the thermal efficiency of an engine working on the Otto cycle. The bore and stroke of the cylinder are 17 cm and 30 cm respectively. The clearance volume is 2025 cm^3.
[**Ans.** 44.45%]

13.3. An engine working on the Otto cycle has a suction pressure of 1 bar and a pressure of 13.5 bar at the end of compression. Find (a) compression ratio, (b) the percentage clearance, (c) thermal efficiency, (d) if the initial temperature is 27°C, find the temperature at the end of compression, (e) if the pressure at the end of combustion is 23 bar, what is the mean effective pressure?
[**Ans.** (a) 6.65%, (c) 53.18%, (d) 639 K, (e) 2.385 bar]

13.4. The maximum and minimum pressure in an Otto cycle are 21 bar and 1 bar. With thermal efficiency of 47.5 %, determine (a) compression ratio, (b) mean effective pressure. The working substance has $C_p = 1.055$, $C_V = 0.754$.
[**Ans.** (a) 5, (b) 3.4 bar]

13.5. Prove that for Otto cycle the optimum compression ratio for maximum work done is given by

$$r_k = \left(\frac{T_3}{T_1}\right)^{1.25}$$

where T_3 = Max. temperature

T_1 = Min. temperature

13.6. In an air standard Diesel cycle, compression ratio is 16 and compression begins at 1 bar and 300. Heat added pre kg is 1500 kJ/kg. Find:

(a) Cut-off ratio
(b) Presure and temperature at the end of expansion
(c) Thermal efficiency
(d) Mean effective pressure
(e) Power output if flow rate of air is 0.1 kg/sec.

[**Ans.** (a) 0.1832, (b) 1905.86 K, 6.36 bar (c) 35.77%, (d) 11.11 bar, (e) 89.42 kW]

13.7. A diesel engine has a cylinder of 25 cm diameter and 40 cm of stroke. Cut-off takes place at 5% of the stroke. Estimate (a) Air standard efficiency, (b) Mean effective pressure. Assume clearance volume 1.2 litres and pressure at the start of suction 1 bar. For 4 stroke operation at the start of suction 1 bar. For 4 stroke operation, determine power developed by the engine at 5 rev/sec.
[**Ans.** (a) 63.7% ,(b) 6.05 bar, (c) 29.7 kW]

13.8. In an engine working on air standard Diesel cycle, the compression ratio is 16. At the beginning of compression (isentropic), the conditions are pressure 1 bar and temperature 15°C. Heat is added at constant pressure till temperature is 1480°C. Calculate:
(i) Cut-off ratio, (ii) Heat supplied, (iii) Cycle efficiency, (iv) Mean effective pressure.
[**Ans.** (i) 2.008, (ii) 884.35 kJ/kg, (iii) 61.33%, (iv) 7 bar]

13.9. A gas engine working on Otto cycle has 200 mm bore and stroke 300 mm. Clearance volume of 1.78 litres. Engine when running at 300 rpm the number of firing cycles per min. are 135. Indicated mep is 518 kPa. Gas consumption is 8.8 m^3/hour and C_v of gas = 16350 kJ/m^3. Determine:

(i) Air standard efficiency
(ii) Indicated power and indicated thermal efficiency.

[**Ans.** (i) 52 %, (ii) 12.21 kW, 30.56 5]

13.10. An air standard Otto-cycle with a compression ratio of 3.55. At the beginning of the compression the air is at 98 kPa and 32°C. During the heat addition process the pressure is trippled. The cylinder volume is 2800 cm^3. Determine:

(i) The thermal efficiency of the cycle by computing heat added and heat rejected.
(ii) Mean effective pressure of the cycle.

Take for air C_P = 1.005 kJ/kg–°K
C_V = 0.718 kJ/kg–°K

[**Ans.** (i) 57.63%, (ii) 7.551 bar]

13.11. An air standard dual cycle has a compression ratio of 15:1. The intake conditions of air are

0.95 bar and 17°C. The pressure ratio during constant volume heating is 15:1 and the volume ratio during the constant pressure part is 1.8:1. Calculate:

(i) Maximum pressure and maximum temperature attained in the cycle.

(ii) The heat input and heat rejection

(iii) The thermal efficiency of the cycle per kg per cycle.

Assume of air $\gamma = 1.4, R = 0.287$ kJ/kg-°K

[**Ans.** (i) 63.146 bar, 2313.2 K, (ii) 1337.4 kJ/kg, (iii) 62.81%]

13.12. A dual combustion cycle has compression ratio of 15 to 1 with air suction at 1 bar, 25°C. The maximum pressure of the cycle is limited to 55 bar. If the heat supplied at constant volume is twice that supplied at constant pressure, Determine:

(i) Explosion ratio

(ii) Cut-off ratio and

(iii) Thermal efficiency

[**Ans.** (i) 1.24, (ii) 65.95%]

13.13. The following data refers to an air standard Diesel cycle –

Maximum temperature 1600 K

Temperature after expansion 700 °K

Condition of air before compression 90 kPa, 300°K

Determine:

(i) Ratio of compression

(ii) Air standard efficiency

[**Ans.** (i) 14.44, (ii) 61.9%]

13.14. An engine develops 50 kW shaft output at full load with mechanical eff. 75%. If mechanical loss remain constant, determine its mechanical efficiency at half load. [**Ans.** 60%]

13.15. A stirling cycle works between temperature of 500°C and 50°C, the ratio of isothermal expansion being 401. Calculate the cycle efficiency if regenerator returns only 85% of heat supplied to it. [**Ans.** 58.21%]

13.16. In the analysis of Diesel cycle the temperatures recorded at the salient points are as follows.

$T_1 = 300$ °K, $T_2 = 909$ °K, $T_3 = 1500$ °K, $T_4 = 604.6$

(i) With cut-off ratio held fixed, to what value must the compression ratio be changed if the thermodynamic efficiency of the cycle is to be increased by 5%?

(ii) By how much [kJ/kg] is heat added per cycle changed?

Assume for air $C_p = 1.005$ kJ/kg.°K, $\gamma = 1.4$

[**Ans.** (i) 20.13, (ii) 53.34 kJ/kg]

13.17. The mean effective pressure of an ideal Diesel cycle is 8 barand the ratio of compression is 12. If suction pressure is 100 kPa, determine the cut-off ratio.

Derive the formula you may use. [**Ans.** 2.42]

13.18. The initial conditions for an Otto cycle operating with a compression ratio of 8 are 1 bar, 27°C. At the beginning of compression stroke, the volume of air is 3.8 litres and 5.7 kJ of heat energy is supplied to the air during constant volume heating process.

Calculate the pressure and temperature at the end of each process and determine mean effective pressure of the cycle.

[**Ans.** 18.38 bar, 690K, 66.33 bar, 2490 K, 3.61 bar, 1082K]

13.19. In an ideal Diesel cycle the pressure and temperature at the beginning of the adiabatic compression are 98.5 KN/m^2 and 60°C respectively. The maximum pressure attained during the cycle is 4.5 MN/m^2 and the heat received during the cycle is 580 kJ/kg gas. Determine, for the cycle,

(i) Volume ratio of compression.

(ii) The temperature at the end of compression.

(iii) The temperature at the end of heat reception.

(iv) The temperature at the end of adiabatic expansion.

(v) The net work done/kg of gas.

(vi) The thermal efficiency.

Take $g = 1.4$, $C_P = 1.004$

[**Ans.** (i) 15.33, (ii) 992.4, (iii) 1577, (iv) 635.2 k, (v) 363.3 kJ/kg, (vi) 62.698]

13.20. (a) Derive an expression for thermal efficiency of an ideal dual combustion cycle in terms of compression ratio, explosion ratio, fuel cut-off ratio and specific heat ratio. Hence obtain an expression for ASE of Otto-Cycle and that of diesel cycle.

(b) An engine works on the constant volume cycle. It has a bore, of 80 mm and a stroke of 85 mm. Clearance volume of the engine is 0.06 litre. The actual thermal efficiency of the engine is 22 %.

Determine the relative efficiency of the engine. Take $\gamma = 1.4$ **[Ans.** 38.8 %]

13.21. An air standard dual cycle has a compression ratio 16 and compression begins at 1 bar, 50°C. The maximum pressure is 70 bar. The heat transferred to air at constant pressure is equal to that at constant volume.

Estimate:

(i) The pressure and temperature at the points of the cycle.

(ii) The cycle efficiency.

(iii) The mean effective pressure of the cycle.

Take $C_V = 0.718$, $C_P = 1.005$kJ/kg °K.

[**Ans.** (i) 48.5 bar, 979 K, 70 bar, 1413 K, 70, 1723 K, 1.91 bar, 615 K, (ii) 66.32 %, (iii) 4.458 bar]

CHAPTER 14 Gas Turbines

14.1 INTRODUCTION

A gas turbine is a turbo-machine and basically similar to steam turbine regarding its working principle.

The first turbine to produce useful work was probably a windmill, where no compression and no combustion exist. The gas turbine as we know today include a compression process and a heat addition or combustion process. Joule and Braylon independently proposed the cycle that is the ideal prototype of the actual unit. In 1906, a unit that produced net-power had been built.

There were two main reasons or obstacles to be overcome as revealed by thermodynamic analysis. To produce and deliver a sizeable amount of power, it is noted that—

- the temperature at the beginning of expansion must be high and the lack of materials to withstand these high temperatures.
- the compressor and turbine must operate at a high efficiency.
- for high efficiency, the lack of knowledge of aerodynamic and thermodynamics of flow mechanism.

Metallurgical developments are raising the highest permissible temperatures of about 800°C to 1100°C.

A better knowledge of aerodynamics has been responsible for improving the efficiency of both the centrifugal and axial compressors.

The gas turbines are also driven by the exhaust of internal combustion engines. Such turbines are used for supercharging IC Engines.

14.2 CLASSIFICATION OF GAS TURBINES

Basically, gas turbines are classified according to three main events. These are—

1. Combustion process
2. Path of working substance
3. Action of combustion gases in turbine.

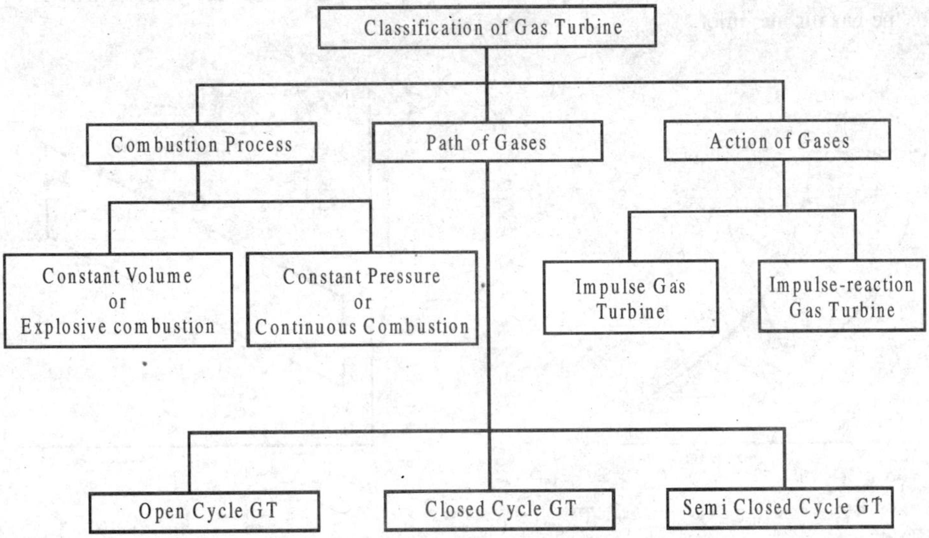

14.3 WORKING OF SIMPLE, CONSTANT PRESSURE, OPEN-CYCLE GAS TURBINE

The schematic diagram of a simple, Constant Pressure Cycle is shown in Fig. 14.1 .

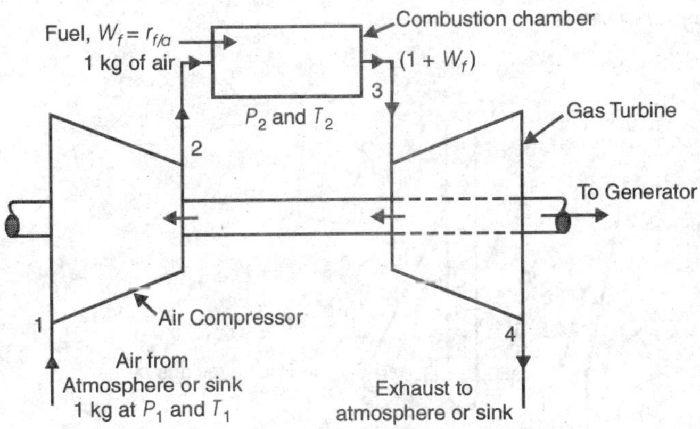

Fig. 14.1

This gas turbine works on Joule or Brayton cycle. Constant Pressure Cycle is Joule cycle. Generally, standard cycle for gas turbine working on constant pressure cycle is Joule cycle and here mass of fuel is generally neglected.

In Brayton cycle analysis, mass of fuel is taken into consideration.

Many times, Joules cycle and Brayton cycle are taken as identical.

Air continuously enters the compressor from atmosphere at P_1, v_1, and T_1. This state is shown on diagram by 1. The compressor may be of axial or centrifugal type. Air is compressed isentropically to state 2 given by P_2, v_2, T_2.

This process of compression is shown by 1–2 on *P–V* and *T–S* diagrams, as shown in Fig. 14.2 (a) and (b).

After compression, air enters the combustion chamber in the open-cycle. Some of this air goes around the outside of the combustor proper and some supplies oxygen for burning the fuel which is continuously injected as shown in Fig. 14.3. Both these streams mix together so that the temperature of the gases formed is within the limits required for the gas turbine inlet.

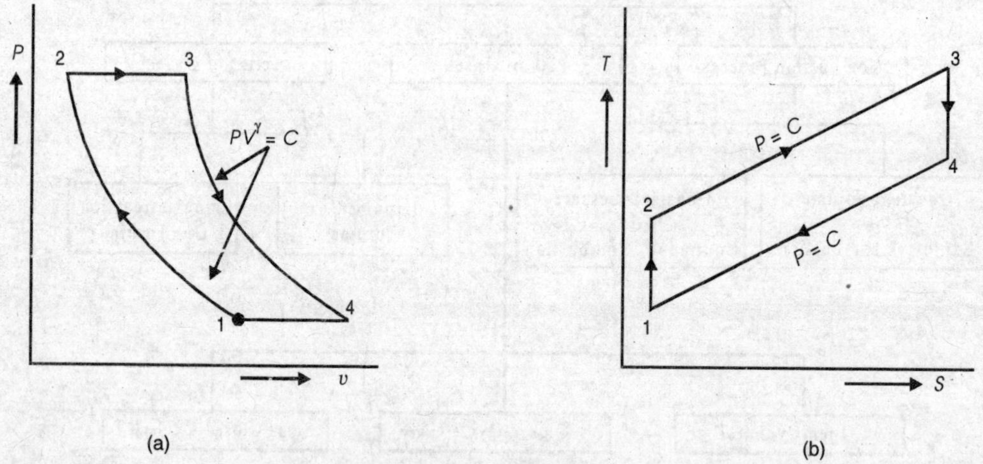

Fig. 14.2

Here the combustion is continuous and takes place at constant pressure. This is shown by 2–3 in Fig. 14.2 (a) and (b). Temperature of the gases increases and the temperature of these gases entering the turbine is the maximum temperature that can be sustained by the turbine blades. This maximum temperature that is acceptable to the blade material is known as *ceiling temperature.*

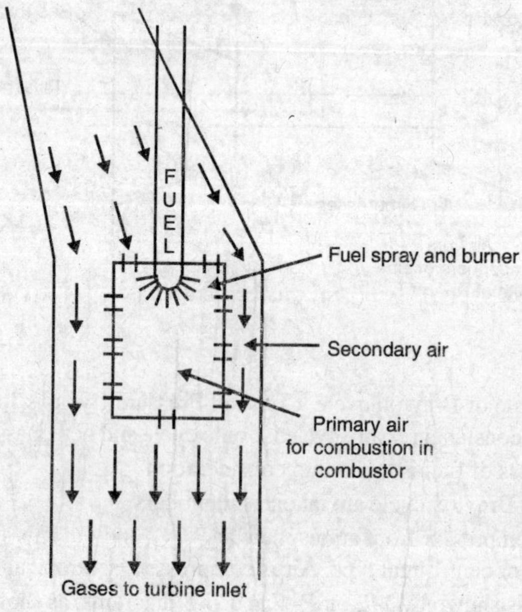

Fig. 14.3. *Combustion chamber*

Gas at the state 3 expands isentropically through the turbine and this expansion is represented by 3–4 as shown in Fig. 14.2. This expansion is complete expansion upto atmospheric pressure as against an expansion in an IC Engine and is an incomplete expansion.

After the expansion, gases will be exhausted—in open cycle—to the atmosphere in condition 4. During this process, the heat is rejected to the atmosphere. This rejection of heat takes place at constant pressure. This process is represented by 4–1 on *P–v* and *T–S* diagrams of Fig. 14.2. Thus the cycle is complete.

You will note that all the processes are continuous. Thus we say that

1–2 is isentropic compression.

2–3 is constant pressure heat addition

3–4 is isentropic expansion and

4–1 is constant pressure heat rejection.

Generally, turbine is coupled to compressor and generator. Turbine supplies power required to drive the compressor and the remaining power to generator or to the aircraft (jet or propeller or both).

It is observed that 60–70 % of the power developed by the turbine is absorbed by the compressor and the remaining power is used to generate power or power required to operate the aircraft.

14.4 ANALYSIS OF THE SIMPLE JOULE OR BRAYTON CYCLE

To analyse the simple cycle, we have to assume certain facts. These assumptions are

(a) Compression and expansion are isentropic.

(b) Heat addition process in the combustion chamber and the heat rejection process are assumed to take place at constant pressure.

(c) There is no pressure loss in the combustion chamber, connecting pipes, and heat exchangers.

(d) The working fluid is assumed to be perfect or ideal gas and that the composition of the fluid does not change during its passage through the cycle.

(e) The flow of fluid is steady flow and the mass flow is constant throughout the cycle is constant.

(f) Other losses during the operation of the machine/turbine are absent. (Frictional loses at the bearings etc. are assumed to be absent).

(g) The velocities of the gas flowing between the two components remain constant.

With these assumptions we will analyse the Joule/Brayton cycle. Consider 1 kg of fluid.

Work of compressor is given by

$$W_C = \frac{P_1 v_1 - P_2 v_2}{\gamma - 1}$$

$$= \frac{R}{\gamma - 1}[T_1 - T_2]$$

$$= C_V [T_1 - T_2] \text{ and is –ve.}$$

$$= C_V (T_2 - T_1) \text{ positive quantity.}$$

Similarly, work of expansion in turbine is given by

$$W_t = \frac{P_3 V_3 - P_4 V_4}{\gamma - 1}$$

$$= \frac{R}{\gamma - 1}[T_3 - T_4] \text{ and is positive.}$$

Flow work during 2–3 is given by

$$W_{f_{2-3}} = (P_3 v_3 - P_2 v_2) = R[T_3 - T_2]$$

and during 4–1,

$$W_{f_{4-1}} = (P_4 v_4 - P_1 v_1) = R[T_4 - T_1]$$

∴ The net work done/kg

$$\begin{aligned} W_n &= C_V(T_3 - T_4) - C_V(T_2 - T_1) + R(T_3 - T_2) - R(T_4 - T_1) \\ &= C_V[T_3 - T_2 - T_4 + T_1] + R[T_3 - T_2 - T_4 + T_1] \\ &= (C_V + R)(T_3 - T_2 - T_4 + T_1) \\ &= C_P(T_3 - T_2 - T_4 + T_1) \\ &= C_P(T_3 - T_2) - C_p(T_4 - T_1) \\ &= Q_A - Q_R \end{aligned} \quad \text{...(1)}$$

where Q_A = Heat added/kg

Q_R = Heat rejected/kg

OR

Work of an open cycle or flow cycle 1–2–3–4 is given by the area of P–v diagram or area 1–2–3–4–1.

∴ W_n = Area 1–2–3–4–1

$$\begin{aligned} &= \begin{bmatrix}\text{Area behind the}\\ \text{curve } 3-4\end{bmatrix} - \begin{bmatrix}\text{Area behind the}\\ \text{curve } 2-1\end{bmatrix} \\ &= \frac{\gamma}{\gamma-1}[P_3 v_3 - P_4 v_4] - \frac{\gamma}{\gamma-1}[P_2 v_2 - P_1 v_1] \\ &= \frac{\gamma R}{\gamma-1}[T_3 - T_4] - \frac{\gamma R}{\gamma-1}[T_2 - T_1] \\ &= C_P(T_3 - T_4) - C_p(T_2 - T_1) \\ &= C_P[(T_3 - T_2) - (T_4 - T_1)] \\ &= C_P(T_3 - T_2) - C_p(T_4 - T_1) \text{ as before} \\ &= Q_A - Q_R \quad \text{...(2)} \\ &= (h_3 - h_4) - (h_2 - h_1) \quad \text{...(3)} \\ &= \text{Turbine work} - \text{Compressor work} \\ &= W_t - W_c \end{aligned}$$

∴ Net work of the cycle/kg

$$\begin{aligned} W_n &= \text{Work of turbine} - \text{Work of compressor} \\ &= W_t - W_c \end{aligned} \quad \text{...(4)}$$

Heat added during the cycle/kg of fluid or air is given by

$$Q_A = C_P(T_3 - T_2) \quad \text{...(5)}$$

∴ Thermal efficiency of Joule cycle or Brayton cycle is given by

$$\eta_{th} = \frac{\text{Net work of cycle}}{\text{Heat supplied}}$$

$$= \frac{W_t - W_c}{Q_A} = \frac{Q_A - Q_R}{Q_A}$$

$$= 1 - \frac{Q_R}{Q_A}$$

$$= 1 - \frac{C_P(T_4 - T_1)}{C_P(T_3 - T_2)}$$

$$\eta_{th} = 1 - \frac{T_4 - T_1}{T_3 - T_2} \qquad \text{...(6)}$$

This efficiency is given in terms of temperatures.
Referring to P–v diagram of Fig. 14.2, we have

$$\frac{T_2}{T_1} = \left(\frac{P_2}{P_1}\right)^{\frac{\gamma-1}{\gamma}}$$

and

$$\frac{T_3}{T_4} = \left(\frac{P_3}{P_4}\right)^{\frac{\gamma-1}{\gamma}}$$

But

$$\frac{P_2}{P_1} = \frac{P_3}{P_4} = \text{Pressure ratio}$$

$$= r_p$$

∴

$$\frac{T_2}{T_1} = (r_p)^{\frac{\gamma-1}{\gamma}} \text{ and } \frac{T_3}{T_4} = (r_p)^{\frac{\gamma-1}{\gamma}}$$

∴

$$T_2 = T_1 - r_p^{\frac{\gamma-1}{\gamma}} \qquad \text{...(7)}$$

$$T_3 = T_4 \cdot r_p^{\frac{\gamma-1}{\gamma}} \qquad \text{...(8)}$$

Substituting these values in Eq. 6 we get

$$\eta_{th} = 1 - \frac{T_4 - T_1}{T_4 \cdot r_p^{\frac{\gamma-1}{\gamma}} - T_1 \cdot r_p^{\frac{\gamma-1}{\gamma}}}$$

$$= 1 - \frac{T_4 - T_1}{r_p^{\frac{\gamma-1}{\gamma}}(T_4 - T_1)}$$

$$= 1-\frac{1}{r_p^{\frac{\gamma-1}{\gamma}}}$$

$$\eta_{th} = 1-\frac{1}{r_p^{\frac{\gamma-1}{\gamma}}} \quad \ldots(9)$$

It is observed from Eq. (9) that the thermal efficiency of Joule or Brayton cycle depends only on r_p, the pressure ratio. As the pressure ratio increases, thermal efficiency of the Joule cycle increases and this variation is shown by a curve given in Fig. 14.4.

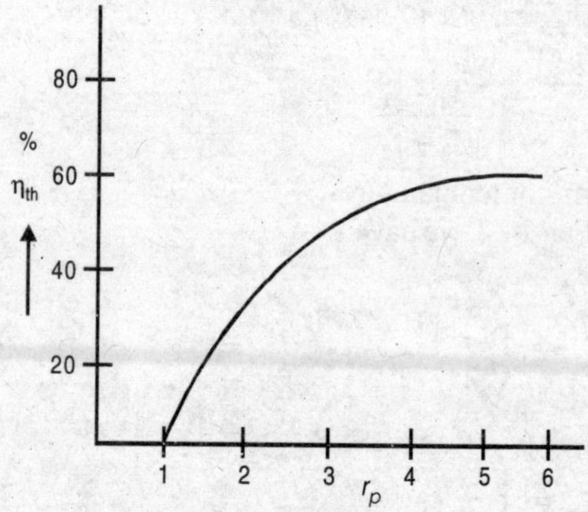

Fig. 14.4

At $r_p = 1$, $\eta_{th} = 0$

Ceiling temperature limits the pressure ratio.

We have, from Eq. 2

$$W_n = C_p\left[T_3 - T_4 - T_2 + T_1\right]$$

and
$$\frac{T_2}{T_1} = \frac{T_3}{T_4} = r_p^{\frac{\gamma-1}{\gamma}}$$

$\therefore$
$$T_4 = \frac{T_3 \times T_1}{T_2}$$

Now work is a function of T_2 that is work depends on T_2 only as

T_1 = Initial temperature and is constant.

T_3 = Ceiling temperature and is constant.

$\therefore$
$$\text{Work } W_n = C_p\left(T_3 - \frac{T_1 \times T_3}{T_2} - T_2 + T_1\right) \quad \ldots(10)$$

This work is maximum when $\frac{dW_n}{dT_2} = 0$

$\therefore$ Differentiating Eq. 10 and equating it to zero, we get

$$\frac{dW_n}{dT_2} = 0 = \frac{T_1 \times T_3}{T_2^2} - 1$$

or

$$T_2^2 = T_1 \times T_3$$

$$\mathbf{T_2 = \sqrt{T_1\, T_3}} \qquad \ldots(11)$$

14.5 WORK RATIO

To know the extent to which the work developed by the turbine is required or supplied to run the compressor, another ratio called *Work Ratio* is introduced in the study. Work ratio is defined as the ratio of the net work $(W_t - W_c)$ to the turbine work developed (W_t). It is denoted by *WR*.

$$\therefore \quad WR = \frac{\text{Net work}}{\text{Turbine work}}$$

$$= \frac{W_t - W_c}{W_t}$$

$$= 1 - \frac{W_c}{W_t}$$

$$= 1 - \frac{C_p(T_2 - T_1)}{C_p(T_3 - T_4)}$$

$$= 1 - \frac{T_1\left(\frac{T_2}{T_1} - 1\right)}{T_3\left(1 - \frac{T_4}{T_3}\right)} \qquad \ldots(12)$$

But

$$\frac{T_2}{T_1} = r_p^{\frac{\gamma-1}{\gamma}}$$

$$\frac{T_3}{T_4} = r_p^{\frac{\gamma-1}{\gamma}}$$

Substituting these ratios in the $WR = 1 - \dfrac{T_1\left(r_p^{\frac{\gamma-1}{\gamma}} - 1\right)}{T_3\left(1 - \dfrac{1}{r_p^{\frac{\gamma-1}{\gamma}}}\right)}$

$$= 1-\frac{T_1}{T_3}\left[\frac{r_p^{\frac{\gamma-1}{\gamma}}\left(r_p^{\frac{\gamma-1}{\gamma}}-1\right)}{\left(r_p^{\frac{\gamma-1}{\gamma}}-1\right)}\right]$$

$$\mathbf{WR} = 1-\frac{T_1}{T_3}\times r_p^{\frac{\gamma-1}{\gamma}} \qquad \ldots(13)$$

From this relation we note that, for the given temperatures T_1 and T_3, *WR* is maximum when $r_p^{\frac{\gamma-1}{\gamma}}$ is minimum.

Similarly, for the given pressure ratio r_p, again *WR* is maximum when $\frac{T_1}{T_3}$ is minimum, i.e., T_1 is minimum and T_3 is maximum.

Maximum T_3 is according to metallurgical considerations and varies between 800°C to 1000°C.

Similarly T_1 will be depending on the atmospheric conditions and may be taken as varying between 15°C and 30°C.

14.6 CONCEPT OF MAXIMUM PRESSURE RATIO AND OPTIMUM PRESSURE RATIO

We are having, now, two quantities that will be considered. One is the net work (W_{net}) and the work ratio (*WR*).

$$\text{Net work} = W_t - W_c$$

$$= C_{p_g}(T_3-T_4)-C_{p_g}(T_2-T_1)$$

$$= C_{p_g} T_3\left(1-\frac{T_4}{T_3}\right)-C_{p_g} T_1\left(\frac{T_2}{T_1}-1\right)$$

$$= C_{p_g} T_3\left(1-\frac{1}{r_p^{\frac{\gamma-1}{\gamma}}}\right)-C_{p_g} T_1\left(r_p^{\frac{\gamma-1}{\gamma}}-1\right)$$

$$= C_{p_g}\left[T_3\left(\frac{r_p^{\frac{\gamma-1}{\gamma}}-1}{r_p^{\frac{\gamma-1}{\gamma}}}\right)-C_{p_g} T_1\left(r_p^{\frac{\gamma-1}{\gamma}}-1\right)\right]$$

Therefore, net work is a function of r_p only for constant T_3 (ceiling temperature) and T_1 (atmospheric temperature).

Similarly, work ratio which is given by

$$WR = 1-\frac{T_1}{T_3}\cdot r_p^{\frac{\gamma-1}{\gamma}}$$

is also a function r_p — the pressure ratio and temperature ratio.

Thus we note that the efficiency of a theoretical cycle is a function of pressure ratio while the performance of the cycle is a function of both the pressure ratio and temperature ratio.

Consider now the different cycles with different r_p and the same maximum cycle temperature and the same initial temperature and pressure.

Consider a new-cycle 1– 2"– 3"– 4"– 1 (Fig. 14.5) where the pressure is increased such that the temperature is very near the ceiling temperature T_3. In this case, heat added is less. This heat addition will be zero in case the temperature after compression is same as ceiling temperature. Thus in the limit when $T_{2''} = T_{3''} = T_{max}$, no heat being added in the combustion chamber and the cycle plots as a single vertical line and no network is available ($W_t = W_c$). The maximum pressure ratio can be obtained equating net output equal to zero.

i.e.
$$W_{net} = W_t - W_c = 0$$

$$\therefore \quad 0 = C_p(T_3 - T_4) - (T_2 - T_1)C_p$$

$$\therefore \quad C_p T_3 \left(1 - \frac{1}{r_p^{\frac{\gamma-1}{\gamma}}}\right) = C_p \left(r_p^{\frac{\gamma-1}{\gamma}} - 1\right) T_1$$

or
$$\frac{T_3}{T_1} = \frac{\left(r_p^{\frac{\gamma-1}{\gamma}} - 1\right) \cdot r_p^{\frac{\gamma-1}{\gamma}}}{r_p^{\frac{\gamma-1}{\gamma}} - 1}$$

$$= r_p^{\frac{\gamma-1}{\gamma}}$$

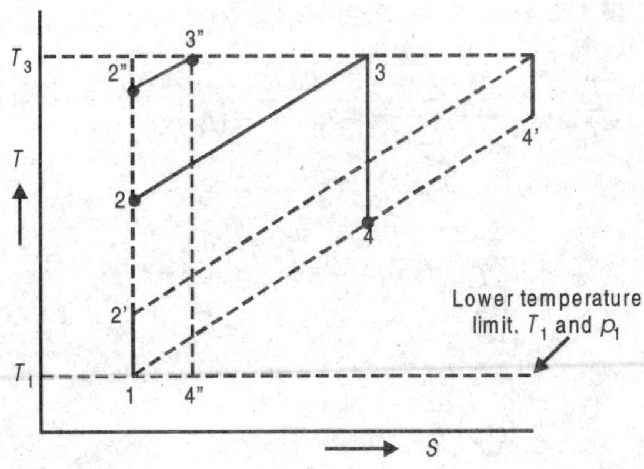

Fig. 14.5

This pressure ratio is maximum pressure ratio written as $(r_p)_{max}$ and is given by

$$(r_p)_{max} = \left(\frac{T_3}{T_1}\right)^{\frac{\gamma}{\gamma-1}} \quad ...(14)$$

We will now consider another case where the lowest possible pressure is considered. Consider another cycle 1–2'–3'–4'–1 where some low pressure ratio is considered, and the thermal efficiency will be very poor. In this cycle, it will be noticed that the heat added in the combustion chamber is very large as compared with the work obtained. Also average temperature at which heat is added and rejected is very small. In the limit, the pressure ratio will be one (1) and the heat addition and rejection lines are same on *T–S* diagram and net work is gain zero.

Between the extreme values of pressure ratios of $r_p = 1$ and $r_p = (r_p)_{max}$, there lies an optimum value for which the specific output, the net output per unit mass flow is maximum. Thus taking T_1 and T_3 as constant temperatures, differentiate expression for W_{net} with respect to r_p and equating it to zero, we get an optimum pressure ratio $(r_p)_{opt}$.

Consider, therefore, a cycle 1–2–3–4–1, Fig. 14.5.

$$\text{Net work/kg} = W_n = W_t - W_c$$

$$= C_p(T_3 - T_4) - C_p(T_2 - T_1)$$

$$= C_p\, T_3\left(1 - \frac{1}{r_p^{\frac{\gamma-1}{\gamma}}}\right) - C_p\, T_1\left(r_p^{\frac{\gamma-1}{\gamma}} - 1\right)$$

Differentiating this equation w.r.t. r_p and equating it to zero, we get

$$O = \frac{d\,W_n}{d\,r_p} = C_p\, T_3\left[\frac{\frac{\gamma-1}{\gamma}\cdot r_p^{\frac{\gamma-1}{\gamma}-1}}{r_p^{\frac{2(\gamma-1)}{\gamma}}}\right] - C_p\, T_1\left[\frac{\gamma-1}{\gamma}\cdot r_p^{\frac{\gamma-1}{\gamma}-1}\right]$$

$$\therefore \quad \frac{T_3}{T_1} = \frac{r_p^{-\frac{1}{\gamma}} \cdot r_p^{\frac{2(\gamma-1)}{\gamma}}}{r_p^{-\frac{1}{\gamma}}} = r_p^{\frac{2(\gamma-1)}{\gamma}} = (r_p)_{opt}^{\frac{2(\gamma-1)}{\gamma}}$$

$$\therefore \quad (r_p)_{opt} = \left(\frac{T_3}{T_1}\right)^{\frac{\gamma}{2(\gamma-1)}}$$

But

$$\frac{T_3}{T_1} = (r_p)_{max}^{\frac{\gamma-1}{\gamma}}$$

$$\therefore \quad (r_p)_{opt} = (r_p)_{max}^{\frac{\gamma-1}{\gamma}\times\frac{\gamma}{2(\gamma-1)}}$$

$$= \left(r_p\right)_{max}^{\frac{1}{2}}$$

$$\therefore \qquad \left(\mathbf{r_p}\right)_{opt} = \sqrt{\left(\mathbf{r_p}\right)_{max}} \qquad \ldots(15)$$

Figure 14.6 shows the variation of work ratio and efficiency w.r.t. r_p — the pressure ratio.

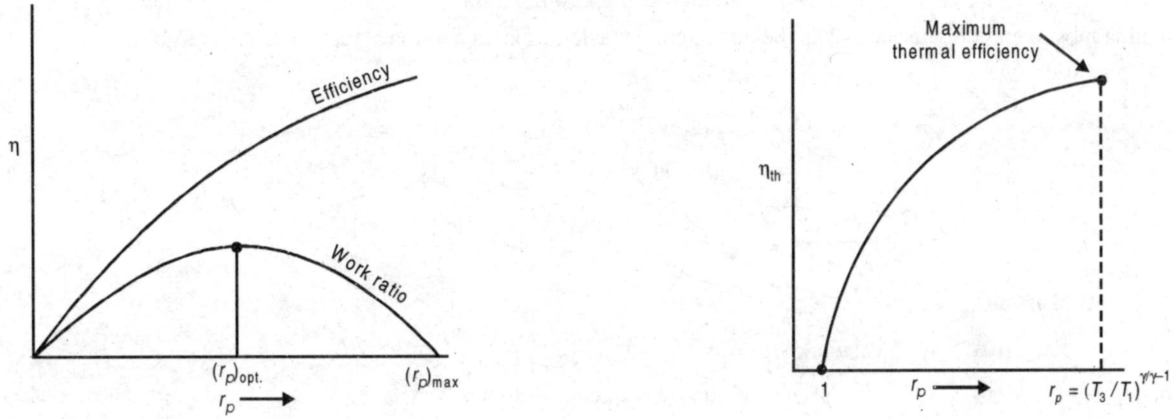

Fig. 14.6

14.7 ACTUAL GAS-TURBINE CYCLE WITH FLUID FRICTION

As you know, the efficiency of an actual cycle is the actual net work delivered, divided by the actual energy chargeable against the cycle. However, since three or more different pieces of equipment are needed in order for the cycle to operate, and since the efficiency of each is significant, it is customary to study each separately. As steady flow machines with $\Delta K = 0$, the works of the fluid are adiabatic flow are

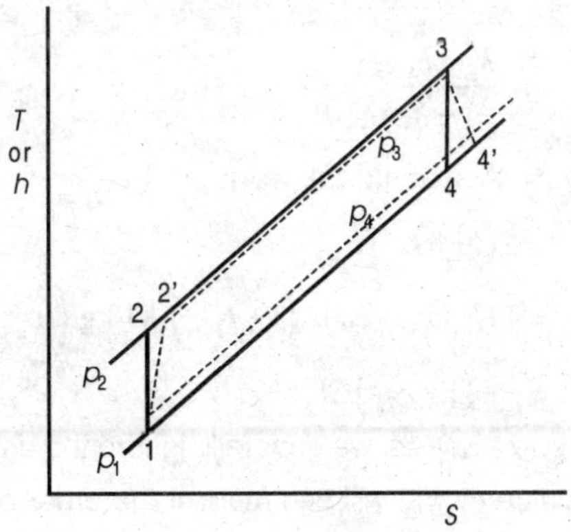

Fig. 14.7. *Brayton cycle with fluid frcition.*

I. Compressor

$$\text{Ideal work} = W_c = h_1 - h_2$$

$$= C_p(T_1 - T_2)$$

$$= -\Delta h_s \text{ kJ/kg air}$$

But actual work is more than this ideal or isentropic compression work because of the fluid friction. Compressor isentropic efficiency takes into cosideration this fluid friction and is given by

$$\eta_c = \frac{\text{Isentropic work/kg}}{\text{Actual work of compressor}}$$

and is most commonly used. 1–2' is the actual compression process and the actual work is given by

$$\text{Actual work} = W_{c'} = h_1 - h_{2'}$$

$$= C_p(T_1 - T_{2'})$$

$$= -\Delta h_{\text{act}} \text{ kJ/kg}$$

$$\therefore \quad \eta_c = \frac{C_p(T_2 - T_1)}{C_p(T_{2'} - 1)} = \frac{T_2 - T_1}{T_{2'} - 1}$$

II. Turbine

$$\text{Ideal work} = W_t = (h_3 - h_4)(1 + r_{f/a})$$

where $r_{f/a}$ = Fuel to air ratio

$$\therefore \quad W_t = (1 + r_{f/a})C_p(T_3 - T_4)$$

Again, actual work is less than this isentropic work because of the fluid friction. Turbine isentropic efficiency takes into consideration this fluid friction and is given by

$$\eta_t = \frac{\text{Actual work}}{\text{Isentropic work}}$$

$$= \frac{C_p(T_3 - T_{4'}) \times (1 + r_{f/a})}{C_p(T_3 - T_4) \times (1 + r_{f/a})}$$

$$= \frac{h_3 - h_{4'}}{h_3 - h_4}$$

$$\therefore \quad W_{t'} = \text{Actual turbine work}$$

$$= (h_3 - h_{4'})$$

$$= C_p(T_3 - T_{4'}) \times (1 + r_{f/a}) \text{ kJ/kg air}$$

$$= -(1 + r_{f/a})\Delta h_{t'}$$

Note. In making an *air standard cycle analysis*, we generally ignore the amount of fuel ($r_{f/a} = 0$). With the current temperature ceilings, the maximum value of $r_{f/a}$ will be small, of the order of 0.02 or less.

$$\therefore \quad \text{Net actual work} = (W_n)_{\text{actual}}$$

$$= W_{t'} - W_{c'}$$

$$= C_p(T_3 - T_{4'}) - C_p(T_{2'} - T_1). \text{ Neglecting } r_{f/a}$$

$$= C_p\,\eta_t\,T_3\left(1-\frac{T_4}{T_3}\right)-\frac{C_p\,T_1}{\eta_c}\left(\frac{T_2}{T_1}-1\right)$$

But $$\frac{T_2}{T_1} = \left(r_p\right)^{\frac{\gamma-1}{\gamma}} = \frac{T_3}{T_4} \text{ neglecting pressure drop.}$$

$\therefore$ $$\text{Net actual work/kg} = C_p\,\eta_t\,T_3\left[1-\frac{1}{r_p^{\frac{\gamma-1}{\gamma}}}\right]-\frac{C_p\,T_1}{\eta_c}\left(r_p^{\frac{\gamma-1}{\gamma}}-1\right)$$

Also, $$\text{heat supplied/kg} = C_p\left(T_3-T_{2'}\right)$$
$$\approx C_p\left(T_3-T_2\right)$$

This approximation is only for simplification of the problem without any appreciable error.

Substitute x for $\left(r_p\right)^{\frac{\gamma-1}{\gamma}}$ and $\frac{T_3}{T_1} = y$

$\therefore$ Heat supplied/kg air

$$= Q_A = C_p\left(T_3-T_2\right)$$
$$= C_p\left(y\cdot T_1-T_1\cdot x\right)-C_pT_1\left(y-x\right)$$

Therefore, the actual thermal efficiency of the cycle is given by

$$(\eta_{th})_{act.} = \frac{\text{Actual net work}}{\text{Heat supplied}}$$

$$= \frac{\eta_t\,T_3\left(1-\frac{1}{x}\right)-\frac{T_1}{\eta_c}(x-1)}{T_1\left(y-x\right)}$$

$$= \frac{\eta_t\,T_1\,y\left(1-\frac{1}{x}\right)-\frac{T_1}{\eta_c}(x-1)}{T_1\left(y\ \ x\right)}$$

$$= \frac{\eta_t\,y\left(1-\frac{1}{x}\right)-\frac{(x-1)}{\eta_c}}{\left(y-x\right)}$$

For maximum thermal efficiency, we have

$$\frac{d\eta_{th}}{dx} = 0 = \frac{\left\{\left(y-x\right)\left[\eta_t\,y\times\frac{1}{x^2}-\frac{1}{\eta_c}\right]+\left[\eta_t\,y\left(1-\frac{1}{x}\right)-\frac{x-1}{\eta_c}\right]\right\}}{\left(y-x\right)^2}$$

$$= \left[\frac{y\left(y-x\right)}{x^2}\eta_t-\frac{y-x}{\eta_c}\right]+\left[\frac{\eta_t y\left(x-1\right)}{x}-\frac{x-1}{\eta_c}\right]$$

$$= \left[\frac{\eta_c\ \eta_t\ y(y-x)-x^2(y-x)}{\eta_c\ x^2}\right]+\left[\frac{\eta_t\ \eta_c\ y(x-1)-x(x-1)}{x\,\eta_c}\right]$$

$$= (y-x)\left[\frac{\eta_c\ \eta_t\ y-x^2}{x^2\ \eta_c}\right]+(x-1)\left[\frac{\eta_t\ \eta_c\ y-x}{\eta_c\ x_2}\right]$$

$$= \frac{(y-x)\left(\eta_c\ \eta_t\ y-x^2\right)+x(x-1)(\eta_t\ \eta_c\ y-x)}{\eta_c\ x^2}$$

$$= (y-x)\left(\eta_c\ \eta_t\ y-x^2\right)+x(x-1)(\eta_t\ \eta_c\ y-x)$$

$$= y^2\eta_c\eta_t - x^2y - xy\,\eta_c\eta_t + x^3 + \left(x^2-x\right)\left(\eta_c\eta_t y - x\right)$$

$$= y^2\eta_c\eta_t - x^2y - xy\eta_c\eta_t + x^3 + x^2y\eta_c\eta_t - x^3 - xy\eta_c\eta_t + x^2$$

$$= x^2y(\eta_c\eta_t - 1) - 2xy\eta_c\eta_t + y^2\eta_c\eta_t + x^2$$

$$= x^2\left[y(\eta_c\eta_t-1)+1\right] - 2xy\eta_c\eta_t + y^2\eta_c\eta_t$$

$$= \left[y(\eta_c\eta_t-1)+1\right] - \frac{2y}{x}\eta_c\eta_t + \frac{y^2}{x^2}\eta_c\eta_t$$

$$O = \frac{1}{x^2}\left(y^2\eta_c\eta_t\right) - \frac{1}{x}(2y\eta_c\eta_t) + \left[y(\eta_c\eta_t-1)+1\right]$$

This is a quadratic in x.

$$\therefore \qquad \frac{1}{x} = \frac{2y\eta_c\eta_t \pm \sqrt{4y^2\eta_c^2\eta_t^2 - 4\left(y^2\eta_c\eta_t\right)\left[y(\eta_c\eta_t-1)+1\right]}}{2y^2\eta_c\eta_t}$$

Taking + ve root only, we have

$$\frac{1}{x} = \frac{1+\sqrt{(1-y)+\dfrac{y-1}{\eta_c\eta_t}}}{y}$$

$$\therefore \qquad x = \frac{y}{1+\sqrt{(1-y)+\dfrac{y-1}{\eta_c\eta_t}}}$$

$$= \frac{y}{1+\sqrt{(y-1)\left[\dfrac{1}{\eta_c\eta_t}-1\right]}}$$

Substituting the values of y we get

$$x = (r_p)_{opt} = \left\{ \frac{\frac{T_3}{T_1}}{1+\sqrt{\left(\frac{T_3}{T_1}-1\right)\left[\frac{1}{\eta_c \eta_t}-1\right]}} \right\}^{\frac{\gamma}{\gamma-1}}$$

Following the same procedure for the condition for actual optimum pressure ratio for specific output, we get

$$(r_p)_{opt} = \left((r_p)_{opt} \frac{T_3}{T_1}\right)^{\frac{\gamma}{2(\gamma-1)}}$$

14.8 AIR RATE

Generally we use symbol m_f to designate the mass of fuel used. Its units are defined by the context, usually either m_f kg fuel/kg air or m_f kg fuel/unit of work say kg fuel/kWh. In this case it is generally called the *Specific Fuel Consumption.*

Similarly, symbol m_a is used to designate the mass of air used. Its units are defined by the context, usually either m_a kg air/kg fuel or m_a kg air/unit of work say kg air/kWh.

∴ Air rate of an engine or turbine is given by the ratio.

$$\dot{m}_a\,(\text{kg/kWh}) = \frac{3600\,(\text{kJ/hour})}{\text{Work done/kg air}}$$

$$\dot{m}_a = \textbf{Air rate} = \frac{\textbf{3600}}{\textbf{Work done / kg air}}$$

SOLVED EXAMPLES

Example 14.1 *The air enters the compressor of an open cycle constant pressure gas turbine at 1 bar, 20°C. The pressure of air at the end of compression is 4 bar. The isentropic efficiency of compressor and turbine are 80% and 85% respectively. The air to fuel ratio is 90. If the air low rate is 3 kg/s, find*

(i) Power developed, (ii) Thermal efficiency.

(Given: C_p of air = 1.0 kJ/kg-K, γ = 1.4 for air, Heat value of fuel = 41800 kJ/kg)

Solution

The gas turbine cycle is shown on *T–S* diagram given below.

Given:

$$P_1 = 1 \text{ bar} = P_4$$

$$T_1 = 20 + 273 = 293 \text{ K}$$

$$P_2 = 4 \text{ bar} = P_3$$

$$\frac{\text{Air}}{\text{Fuel}} = \frac{90}{1}$$

$$\eta_c = 80\%, \eta_t = 85\%$$

$$\text{C. V. of fuel} = 41800 \text{ kJ/kg}$$

First we will calculate all the temperatures at the salient points.

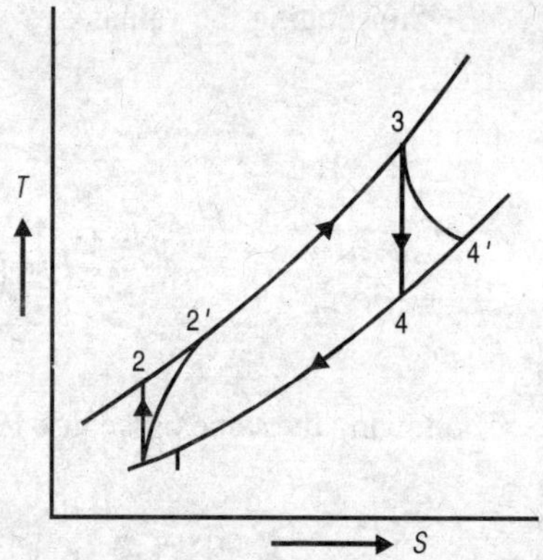

Fig. Ex. 14.1

$$\frac{T_2}{T_1} = \left(\frac{P_2}{P_1}\right)^{\frac{\gamma-1}{\gamma}} = \left(\frac{4}{1}\right)^{\frac{1.4-1}{1.4}}$$

$$= (4)^{0.2857} = 1.486$$

$$\therefore \quad T_2 = 293 \times 1.486 = \mathbf{435.4\ K}$$

Again, $$\eta_c = \frac{T_2 - T_1}{T_{2'} - T_1}$$

$$\therefore \quad 0.8 = \frac{435.4 - 293}{T_{2'} - 293} = \frac{142.4}{T_{2'} - 293}$$

$$\therefore \quad T_{2'} = 293 + \frac{142.4}{0.8}$$

$$= 293 + 178 = \mathbf{471\ K}$$

$$\text{Fuel supplied/kg air} = \frac{1}{90} \text{ kg}$$

$$\therefore \quad \text{Heat supplied/kg air} = \frac{1}{90} \times 41800 = \mathbf{464.44\ kJ/kg}$$

$$\therefore \quad 464.44 = 1 \times 1 \times (T_3 - T_{2'})$$

$$\therefore \quad T_3 = 464.44 + 471 = \mathbf{935.44\ K}$$

$$\therefore \quad T_4 = \text{Temperature after expansion}$$

$$= \frac{T_3}{1.486}$$

$$= \frac{935.44}{1.486}$$

$$= \mathbf{629.5\ K}$$

Again $$\eta_t = \frac{T_3 - T_{4'}}{T_3 - T_4}$$

$$= \frac{935.44 - T_{4'}}{935.44 - 629.5}$$

$$\therefore \quad 0.85 = \frac{935.44 - T_{4'}}{305.94}$$

$$\therefore \quad 0.85 \times 305.94 = 935.44 - T_{4'}$$

$$\therefore \quad T_{4'} = 935.44 - 260.05$$

$$= \mathbf{675.39\ K}$$

(i) Power Developed

$$\text{Compressor work/sec} = mC_p (T_{2'} - T_1)$$

$$= 3 \times 1 \times (471 - 293)$$

$$= 3 \times 178$$

$$= 534 \text{ kJ/sec}$$

$$= \mathbf{534\ kW}$$

$$\text{Turbine work/sec} = m_g C_p (T_3 - T_4)$$

$$= \left(3 + \frac{3}{90}\right) \times 1 \times (935.44 - 675.39)$$

$$= \frac{91}{30} \times 1 \times 260.05$$

$$= \mathbf{788.82 \ kJ/s = kW}$$

$$\therefore \quad \text{Net power developed} = W_t - W_c$$

$$= 788.82 - 534$$

$$= \mathbf{254.82 \ kW}$$

(ii) Thermal Efficiency of Cycle

$$\text{Thermal efficiency} = \frac{\text{Work developed}}{\text{Heat supplied}}$$

$$= \frac{254.82}{464.44} = \mathbf{0.5486 = 54.86\ \%}$$

Example 14.2 *(a) Describe with neat sketch and T–S diagram an open cycle and closed cycle gas turbine. List the advantages of closed cycle gas turbine over open cycle gas turbine.*

(b) In a gas turbine plant air compressor takes in air at a temperature of 15°C and compresses it to 4 times the initial pressure with an isentropic efficiency of 82%. The air is then passed through a regenerator before reaching the combustion chamber. The effectiveness of the regenerator is 78%. The maximum temperature of the cycle is 600°C and the isentropic efficiency of the turbine is 70%. Find the thermal efficiency of the plant. Take R = 0.287 kJ/kg-K.

Solution

Given:

$$T_1 = 15 + 273 = 288 \text{ K}$$

$$T_3 = 600 + 273 = 873 \text{ K}$$

$$r_p = 4, \eta_c = 82\%$$

$$\eta_t = 70\%, \eta_R = 78\%$$

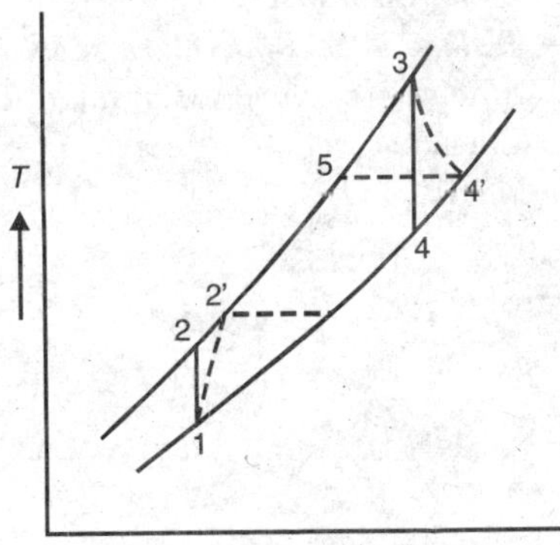

Fig. Ex. 14.2

$$\frac{T_2}{T_1} = \frac{T_3}{T_4} = (r_p)^{(\gamma-1)/\gamma}$$

$$= (4)^{0.2857} = 1.486$$

$$\therefore \quad T_2 = 1.486 \times 288 = \mathbf{427.96 \ K}$$

$$T_4 = \frac{873}{1.486} = \mathbf{587.48 \ K}$$

For compressor,

$$\eta_c = \frac{T_2 - T_1}{T_{2'} - T_1} = \frac{427.96 - 288}{T_{2'} - 288}$$

$$\therefore \quad T_{2'} = 288 + \frac{139.96}{0.82} = 288 + 170.68 = \mathbf{458.68 \ K}$$

Similarly for turbine,

$$\eta_t = \frac{T_3 - T_{4'}}{T_3 - T_4} = \frac{873 - T_{4'}}{873 - 587.48}$$

$$0.7 \times 285.52 = 199.86 = 873 - T_{4'}$$

$$\therefore \quad T_{4'} = 873 - 199.86 = \mathbf{673.14\ K}$$

$$\therefore \quad \text{Net work/kg} = W_{tur} - W_{com} = 1.005 \times (199.86 - 170.68)$$
$$= 1.005 \times 29.18$$
$$= \mathbf{29.33\ \ kJ/kg}$$

For generator,
$$\eta_R = \frac{T_5 - T_{2'}}{T_{4'} - T_{2'}}$$

$$\therefore \quad 0.78 = \frac{T_5 - 458.68}{673.14 - 458.68}$$

$$\therefore \quad T_5 = 458.68 + 0.78 \times 214.46$$
$$= 458.68 + 167.28$$
$$= \mathbf{625.96\ \ K}$$

∴ Heat added in combustion chamber/kg

$$= C_p (T_3 - T_5) = 1.005 \times (873 - 625.96)$$
$$= 1.005 \times 247.04$$
$$= \mathbf{248.27\ \ kJ/kg}$$

∴ Thermal efficiency of the plant is given by

$$\eta_{th} = \frac{W_n}{Q} = \frac{29.33}{248.27} = \mathbf{0.1182\ or\ 11.82\ \%}$$

Example 14.3 (*a*) *Explain the processes of multistaging, intercooling and reheating as applied to gas turbine cycle. What are their advantages ?*

(*b*) *The following particulars refer to a gas turbine cycle.*

Air entry at 0.98 bar, 20°C

Compressor discharge pressure = 4.2 bar

Turbine inlet temperature = 850°C

Efficiency of compressor and turbine = 0.8 each

Calculate thermal efficiency of actual and theoretical cycles.

Solution

(b)
$$T_1 = 293\ \text{K}$$
$$P_1 = 0.98\ \text{bar}$$
$$P_2 = 4.2\ \text{bar}$$
$$T_3 = 850 + 273 = \mathbf{1123\ K}$$
$$\eta_c = \eta_t = 0.8$$

$$\frac{T_2}{T_1} = \left(\frac{4.2}{0.98}\right)^{\frac{\gamma-1}{\gamma}} = (1.2857)^{0.286} = 1.5162$$

$$\therefore \quad T_2 = 293 \times 1.5162 = \mathbf{444.25\ K}$$

Similarly,
$$T_4 = \frac{1123}{1.5162} = \mathbf{740.67\ K}$$

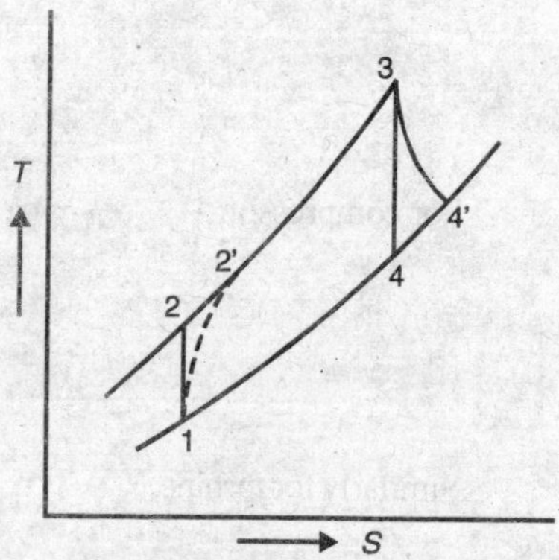

Fig. Ex. 14.3

Theoretical efficiency,

$$\eta_{theo} = 1 - \frac{T_4 - T_1}{T_3 - T_2} = 1 - \frac{740 - 293}{1123 - 444.25}$$

$$\eta_{theo} = 1-\frac{447.67}{678.75}=1-0.6596$$
$$= \mathbf{0.3404 \text{ or } 34.04\ \%}$$

$$\eta_c = \frac{T_2 - T_1}{T_{2'} - T_1}$$

$$\therefore \quad 0.8 = \frac{444.25-293}{T_{2'}-293}$$

$$\therefore \quad T_{2'} = 293+\frac{444.25-293}{0.8}$$

$$\therefore \quad T_{2'} = 293+\frac{151.25}{0.8} = \mathbf{482.1\ K}$$

$$\therefore \quad Q_{A'} = T_3 - T_{2'} = 1123 - 482.1 = \mathbf{640.9\ kJ/kg}$$

$$\eta_t = \frac{T_3 - T_{4'}}{T_3 - T_4}$$

$$0.8 = \frac{1123-T_{4'}}{1123-740.67}$$

$$382.33 \times 0.8 = 1123 - T_{4'}$$

$$T_{4'} = 1123 - 305.864 = 817.136\ K$$

$$W_{t'} = 1.005 \times (1123 - 817.136)$$
$$= \mathbf{307.393\ \ kJ/kg}$$

$$W_{c'} = C_p (T_{2'} - T_1) = (482.1 - 293) \times 1.005$$
$$= 1.005 \times 189.1$$
$$= 190.045\ kJ/kg$$

$$\therefore \quad \text{Net work} = W_{t'} - W_{c'}$$
$$= 307.393 - 190.045 = \mathbf{117.348\ kJ/kg}$$

∴ Actual efficiency of cycle,

$$\eta_{act} = \frac{W_{n'}}{Q_{A'}} = \frac{117.348}{640.9} = \mathbf{0.183 = 18.3\ \%}$$

THEORY QUESTIONS

14.1. What is a Gas turbine? In what respect it is different from IC Engines?

14.2. What are the assumptions made while analysing a Joule or Brayton cycle?

14.3. Derive the condition for maximum efficiency and maximum network from Gas turbine.

14.4. What is work ratio? In what way maximum pressure ratio is different than optimum pressure ratio? Derive the relation between them.

14.5. What are the methods used for improving the performance of a simple gas turbine plant? Explain them briefly.

14.6. How will you compare gas turbine with IC engines and steam turbine.

14.7. Explain with neat sketch the working of a semi-closed cycle gas-turbine. What are the advantages and disadvantages of semi-closed cycle gas turbine?

PROBLEMS FOR PRACTICE

1. The following data refer to a gas turbine plant.

Compressor inlet pressure	= 1 bar
Compressor inlet temperature	= 37 °C
Compressor efficiency	= 75 %
Pressure ratio	= 4
Turbine inlet temperature	= 1113 K
Turbine efficiency	= 83 %
Calorific value of fuel	= 44 MJ/kg
Air flow rate	= 75 kg/min.

Find :

(a) Air-fuel ratio

(b) Net Power output

(c) Overall efficiency of plant

Assume : $C_{pa} = 0.996$, $C_{va} = 0.71$, $C_{pg} = 1.046$, $C_{vg} = 0.744$

[**Ans.** (a) 114.5, (b) 35.18 kW, (c) 7.32 %]

2. The following refers to a closed-cycle gas turbine plant using helium as working fluid and incorporation two-stage compression with intercooling and two stage expansion with reheating, temperature at entry to each compression stage is 270°C; pressure at entry to each compression stage is 270°C, pressure at entry to first stage compression stage and exit from the second turbine stage is 1 bar; first compression stage pressure ratio is 6 and isentropic efficiency is 85% temperature at inlet to each expansion stage is 1150 °C and isentropic efficiency of each expansion stage is 0.9; reheat pressure is 6 bar; for helium adiabatic index is 1.24 and cycle $\gamma_p = 10.05$. Calculate the cycle thermal efficiency.

[**Ans.** 22.31%]

3. In a gas turbine plant the air pressure and temperature before compression are 1 bar and 28°C. Pressure ratio = 3.5 and the temperature of gas before expansion is 700°C. If the isentropic efficiencies of the compressor and turbine are 80 % and 85 % respectively, calculate

(a) The efficiency of the cycle

(b) The turbine exhaust temperature

(c) % change in efficiency if the regenerator of 0.7 effectiveness is added to the plant.

[**Ans.** (a) 17.14 %, (b) 724.157 K, (c) 56.12 %]

4. A gas turbine operates on Brayton cycle. The temperature range is 1000 K and 288 K. Find pressure ratio for maximum power output. Also determine thermal efficiency, work ratio, specific power output. Take $C_p = 1.005$ kJ/kg.K and $\gamma = 1.4$. Expansion and compression is isentropic.

[**Ans.** $(r_p)_{opt.} = 8.831$, $\eta_{th} = 46.34$ %, WR = 46.34 %, sp. output = 215.83]

5. An open circuit, continuous combustion, constant pressure gas turbine takes in air at 1.01 bar and 15°C. The air is compressed through a pressure ratio of 6. The air then passes at constant pressure through a heat exchanger, the effectiveness of which is 65 %. From the heat exchanger air passes at constant pressure through a combustion chamber in which the temperature is raised to 870°C. From the combustion chamber the air passes through a gas turbine in which it is expanded to a pressure of 1.01 bar and it then passes through the heat exchanger to exhaust. The isentropic efficiency of the compressor is 85 % while that of turbine is 80 %. Neglect the mass of fuel and take air mass flow rate as 4 kg/sec. Take $\gamma = 1.4$ and $C_p = 1.005$ kJ/kg.K.

Determine :

(a) The net power output of the plant.

(b) The exhaust temperature from the heat exchanger.

(c) The thermal efficiency of the plant.

(d) The thermal efficiency of the plant if there was no heat exchanger.

[**Ans.** (a) 612 kW, (b) 600 K, (c) 33.5 % (d) 24.1 %]

CHAPTER 15

Jet Propulsion and Rocket Propulsion

JET PROPULSION

15.1 HISTORY

Since the beginning of aviation in the early 1900s, the speed and flight endurance of powered aircraft have been determined by the power and efficiency of their propulsion systems—piston engines at first and now powerful jet engines.

Jet engines are the most widely used form of propulsion on commercial and military aircraft. They generate power when a mixture of compressed air and fuel is ignited and the resulting hot gases escape through an exhaust nozzle. The reaction of the hot expanding gases exiting the exhaust nozzle produces thrust in the opposite direction and moves the aircraft forward.

Most jet engines today fall into three categories—turbojet, turbofan and turboprop. Turbojets incorporate a turbine-driven compressor to pull air into the engine and compress it before fuel is injected into the combustion chamber and ignited. Turbofan engines are turbojets in which additional power is generated by compressor blades that extend beyond the exterior of the main engine casing. Turboprop engines are turbojets that have propellers which provide extra thrust.

Today's jet engines have succeeded in flying faster than speed of sound, making best use of sophisticated intake systems that minimize drag even at hypersonic speeds, efficient fuel control systems, newly invented materials that can sustain very high temperatures, auxiliary power units that ensure starting in worst conditions and vectored thrust nozzle that can direct thrust in required direction.

Here, we focus on modern jet engines which are used for military and civil purposes, their major components, and systems. Type comparison helps to identify the best engine for a particular application and set of situations. It also focuses on advanced designs like Ramjet, Scramjet, etc.

15.2 HISTORICAL EVOLUTION

Jet engines can be dated back to the first century AD, when Hero of Alexandria invented the aeolipile. This used steam power directed through two jet nozzles so as to cause a sphere to spin rapidly on its axis. So far as is known, it was never used for supplying mechanical power, and the potential practical applications of Hero's invention of the jet engine were not recognised. It was simply considered a curiosity.

Jet propulsion only literally and figuratively took off with the invention of the rocket by the Chinese in the 11th century. Rocket exhaust was initially used in a modest way for fireworks but gradually progressed to propel some quite fearsome weaponry; and there the technology stalled for hundreds of years.

The problem was that rockets are simply too inefficient to be useful for general aviation. Instead, by the 1930s, the piston engine in its many different forms (rotary and static radial, air-cooled and liquid-cooled inline) was the only type of powerplant available to aircraft designers. This was acceptable as long as only low performance aircraft were required, and indeed all that were available.

However, engineers were beginning to realise conceptually that the piston engine was self-limiting in terms of the maximum performance which could be attained; the limit was essentially one of the propeller efficiency. This seemed to peak as blade tips approached the speed of sound. If engine, and thus aircraft, performance were ever to increase beyond such a barrier, a way would have to be found to radically improve the design of the piston engine, or a wholly new type of powerplant would have to be developed. This was the motivation behind the development of the gas turbine engine, commonly called a "jet" engine, which would become almost as revolutionary to aviation as the Wright brothers' first flight.

The earliest attempts at jet engines were hybrid designs in which an external power source supplied the compression. In this system (called a thermojet by Secondo Campini) the air is first compressed by a fan driven by a conventional piston engine, then it is mixed with fuel and burned for jet thrust. The examples of this type of design were the Henri Coandâ's Coandâ-1910 aircraft, and the much later Campini Caproni CC.2, and the Japanese Tsu-11 engine intended to power Ohka kamikaze planes towards the end of World War II. None were entirely successful and the CC.2 ended up being slower than the same design with a traditional engine and propeller combination.

The key to a practical jet engine was the gas turbine, used to extract energy to drive the compressor from the engine itself. The gas turbine was not an idea developed in the 1930s: the patent for a stationary turbine was granted to John Barber in England in 1791. The first gas turbine to successfully run self-sustaining was built in 1903 by Norwegian engineer Ægidius Elling. The first patents for jet propulsion were issued in 1917. Limitations in design and practical engineering and metallurgy prevented such engines reaching manufacture. The main problems were safety, reliability, weight and, especially, sustained operation.

15.3 JET ENGINES

15.3.1 Introduction

A jet engine is an engine that discharges a fast moving jet of fluid to generate thrust in accordance with Newton's third law of motion. This broad definition of jet engines includes turbojets, turbofans, rockets and ramjets and water jets, but in common usage, the term generally refers to a **gas turbine Brayton cycle engine** used to produce a jet of high speed exhaust gases for special propulsive purposes in Illustration 1.

The principle of jet propulsion is based on Newton's second and third laws of motion. Momentum is imported to a mass of fluid in such a manner that the reaction of the imparted momentum gives a propulsive force. This is done by expanding the gas which is at high pressure and temperature through a nozzle so that the gas with markedly increased velocity in the form of jet comes out into the atmosphere and its reaction in the opposite direction gives the propulsive force. The open cycle gas turbine is most suited for jet propulsion. If the working fluid is expanded in the turbine such that the power developed is only sufficient to drive the compressor and accessories and the rest expansion is achieved in a nozzle which is placed just after the turbine. This unit will serve as a jet propulsion system. The gas from the turbine while passing through the nozzle will be accelerated and will come out in the form of jet with a tremendous velocity. The reaction of this jet propels the unit in the forward direction (i.e., opposite direction of the jet).

Illustration 1. *Jet Engine*

The jet propulsion engines are classified as follows:

1. Atmospheric jet engines.
2. Rocket engines.

1. Atmospheric jet engines require oxygen from atmospheric air for combustion of the fuel. As a result their performance depends to a great extent on the forward speed of the engine and on the atmospheric pressure and temperature.

2. The rocket engine carries its own oxidizer for the combustion of the fuel and is, therefore, independent of atmospheric air. It requires maximum specific fuel consumption among all types of propulsion systems.

Atmospheric jet engines are further classified as follows.

(a) Steady Combustion Systems (Continuous Air Flow)

(i) Turbo jet

(ii) Turbo jet with after burner (also known as turboram jet, turbojet with tail-pipe burning and turbojet with reheater).

(iii) Turboprop (also known as prop-jet)

(iv) Ram jet (also known as athodyds and Lorin tube)

(b) Intermittent Combustion System-Intermittent air flow.

(i) Pulse jet (also known as aeropulse, resojet, schmidt tube and intermittent jet).

The turbojet, the turbojet with after burner and turbo-prop are all modified forms of simple open-cycle gas turbine. The ram jet and pulse jet are athodyds (aero-thermodynamic ducts) i.e., straight duct-type of jet engines without compressor and turbine wheels.

Rocket engines are further classified as: (i) Liquid-propellent and (ii) solid propellent.

Rocket propulsion should be regarded as a source of power for achieving objectives obtainable by other methods. Some of the rocket applications are

(a) artillery barrage rockets.
(b) anti tank 'bazcoka' rockets
(c) all types of guided missiles
(d) aircraft launched rockets
(e) jet assisted take-off for air planes
(f) engines for long range, high speed guided missiles and pilot less aircraft
(g) transonic airplanes such as D-558.

15.4 THE TURBOJET

The turbojet engine consists of a diffuser at the entrance which slows down the entrance air and thereby compresses it, called the raming effect; a simple open gas turbine cycle and an exit nozzle which expands the gas and converts the thermal energy of the exit gas into kinetic energy which is emitted out in the form of a jet. The increased velocity of air thereby produces thrust. A schematic arrangement of a turbo-jet engine is shown in Fig. 15.1. A corresponding *T–S* diagram is shown. The basic cycle for the turbojet engine is the joule or Brayton cycle. In process 0 to 1, the air entering from the atmosphere is diffused isentropically from velocity V_a down to zero i.e., $V_1 = 0$. This means that the diffuser has an efficiency of 100 percent. This is termed as **ram-compression**. Process 0–1' is the actual compression (ram). Processs 1'–2 shows the isentropic compression of air and 1'–2' shows the actual compression of air. 2'–3 shows the heat addition at constant pressure $p_2 = p_3$. Process 3–4 shows isentropic expansion of gas in the turbine and 3–4' shows the actual expansion in the turbine. Process 4'–5 shows the isentropic expansion of the gas in the nozzle and 4'–5' shows the actual expansion of gas in the nozzle. Consider 1 kg of working fluid flowing through the system.

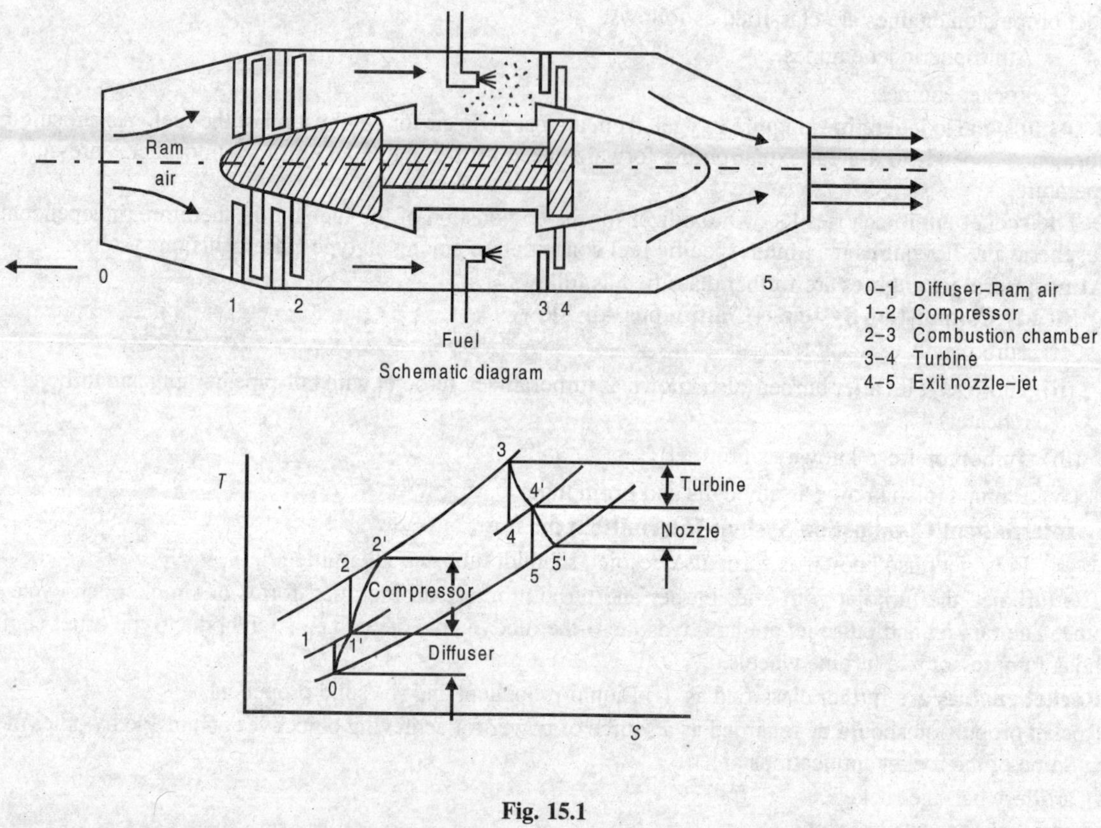

Fig. 15.1

Diffuser: Energy equation between 0 and 1 gives

$$\frac{V_a^2}{2} + h_0 + Q_{01} = \frac{V_1^2}{2} + h_1 + W_{01}$$

V_a = Velocity of air entering from atmosphere.
In ideal diffuser, $V_1 = 0$, $Q_{01} = 0$, $W_{01} = 0$

$$\therefore \quad \frac{V_a^2}{2} + h_0 = h_1$$

or

$$C_p T_1 = C_p T_o + \frac{V_a^2}{2}$$

$$T_1 = T_o + \frac{V_a^2}{2C_p} \quad \ldots(1)$$

Process 0–1' shows the actual process in the diffuser.

$$\text{Diffuser efficiency} = \eta_d = \frac{T_1 - T_0}{T' - T_0} = \frac{h_1 - h_0}{h_{1'} - h_0}$$

$$\therefore \quad h_{1'} = h_0 + \frac{V_a^2}{2\,\eta_d}$$

$$\therefore \quad T_{1'} = T_0 + \frac{V_a^2}{2\,\eta_d \cdot C_p} \quad \ldots(2)$$

Compressor
Using energy equation,

$$h_1 + \frac{V_1^2}{2} + Q_{1-2} + W_c = h_2 + \frac{V_2^2}{2}$$

Neglecting the kinetic energy changes,

$$W_c = \text{Ideal work of compressor}$$
$$= h_2 - h_1 - C_p (T_2 \quad T_1)$$

$$\text{The actual compressor work} = h_{2'} - h_1 = \frac{h_2 - h_1}{\eta_C} = \frac{C_p (T_2 - T_1)}{\eta_C}$$

This compressor work has to be supplied by the turbine.

Combustion Chamber

$$\text{The ideal heat supplied} = C_p (T_3 - T_2) = h_3 - h_2$$

$$\text{The actual heat supplied} = (1+F) h_3 - h_{2'}$$
$$= C_{pg} (1+F) T_3 - C_{pa} T_{2'}$$

Turbine
Using energy equation between 3–4, we get

$$h_3 + \frac{V_3^2}{2} + Q_{3-4} = h_4 + \frac{V_4^2}{2} + W_t \quad Q_{3-4} = 0$$

then
$$W_t = \text{turbine work}$$
$$= (h_3 - h_4) + \frac{V_3^2 - V_4^2}{2}$$
$$= h_3 - h_4 \text{ neglecting } \Delta\text{KE}$$
$$= C_p (T_3 - T_4)$$
$$\text{Actual turbine work} = h_3 - h_{4'} = C_p (T_3 - T_{4'})$$
$$= C_p (T_3 - T_4)\eta_t$$

For simplification, for turbo-jet engine,
$$W_c = W_t$$

or
$$C_p (T_3 - T_{4'}) = C_p (T_3 - T_4)\eta_t = \frac{(T_2 - T_1) C_p}{\eta_c}$$

or
$$T_{4'} = T_3 - (T_3 - T_4)\eta_t = T_3 - \frac{T_2 - T_1}{\eta_c}$$

Jet-nozzle

Using energy equation between states 4 and 5,
$$h_4 + \frac{V_4^2}{2} = h_5 + \frac{V_5^2}{2}$$

In actual case
$$h_{4'} + \frac{V_{4'}^2}{2} = h_{5'} + \frac{V_{5'}^2}{2}$$

$$\therefore \quad \frac{V_{5'}^2}{2} = h_{4'} - h_{5'} \text{ as } V_4 << V_5$$

$$\therefore \quad V_{5'} = \sqrt{2C_p (T_{4'} - T_{5'})}$$
$$= \sqrt{2C_p \eta_n (T_{4'} - T_5)} \qquad \ldots(3)$$

$$\therefore \quad \text{Thermal efficiency} = \frac{(h_3 - h_{5'}) - (h_{2'} - h_0)}{h_3 - h_{2'}}$$
$$= \frac{(T_3 - T_{5'}) - (T_{2'} - T_0)}{T_3 - T_{2'}}$$

15.4.1 Nozzles

The primary object of a nozzle is to expand the exhaust stream to atmospheric pressure, thereby producing a high velocity jet, relative to the vehicle. If the fully expanded jet velocity exceeds the flight velocity, there will be a forward thrust on the airframe.

15.4.2 Convergent Nozzles

Simple convergent nozzles are used on many jet engines. If the nozzle pressure ratio is above the critical value (about 1.8:1) a convergent nozzle will choke, resulting in some of the expansion to atmospheric pressure taking place downstream of the throat (*i.e.* smallest flow area), in the jet wake. Increasing the nozzle pressure ratio further will not increase the throat Mach number beyond unity. Downstream (*i.e.* external to the nozzle) the flow is free to expand to supersonic velocities.

15.4.3 Convergent-divergent Nozzle (De Laval Nozzle)

At high nozzle pressure ratios, much of the expansion will take place downstream of a convergent nozzle, which is somewhat inefficient. Consequently, some jet engines incorporate a convergent-divergent nozzle, to allow most of the expansion to take place within the nozzle to maximise thrust. However, unlike the con-di nozzle used on a conventional rocket motor, when such a device is used on a jet engine it has to be a complex variable geometry device, to cope with the wide variation in nozzle pressure ratio encountered in flight and engine throttling. This further increases the weight and cost of such an installation.

15.4.4 Ejector Nozzle

The simpler of the two is the *ejector nozzle*, which creates an effective nozzle through a secondary airflow and springloaded petals. At subsonic speeds, the airflow constricts the exhaust to a convergent shape. As the aircraft speeds up, the two nozzles dilate, which allows the exhaust to form a convergent-divergent shape, speeding the exhaust gasses past Mach 1. More complex engines can actually use a tertiary airflow to reduce exit area at very low speeds. Advantages of the ejector nozzle are relative simplicity and reliability. Disadvantages are average performance (compared to the other nozzle type) and relatively high drag due to the secondary airflow. Notable aircraft to have utilized this type of nozzle include the SR-71, Concorde, F-111, and Saab Viggen.

15.4.5 Iris Vectored Thrust Nozzle

For higher performance, it is necessary to use an *Iris nozzle*. This type uses overlapping "petals" which mechanically adjusts the petals with hydraulics. Although more complex than the ejector nozzle, it has significantly higher performance and smoother airflow. As such, it is employed primarily on high-performance fighters such as the F-14, F-15, F-16, though is also used in high-speed bombers such as the B-1B. Some modem iris nozzlc additionally have the ability to change the angle of the thrust.

Illustration 2. *Iris Vectored thrust nozzle*

Rocket motors also employ convergent-divergent nozzles, but these are usually of fixed geometry, to minimise weight. Because of the much higher nozzle pressure ratios experienced, rocket motor con-di nozzles have a much greater area ratio (exit/throat) than those fitted to jet engines.

15.4.6 Afterburner (Thrust Booster or Augmentation)/Tail Bunching

Many military combat engines incorporate an afterburner (or reheat) in the engine exhaust system. Its purpose is to provide a temporary increase in thrust for situations such as take-off, or in military aircraft, combat or supersonic flight. This is achieved by injecting additional fuel into the jet pipe downstream of the turbine. This fuel is ignited by the hot exhaust gasses and adds greatly to the thrust of the engine. The advantage of afterburning is significantly increased thrust, the disadvantage of afterburning is its very high fuel consumption and inefficiency but this is acceptable for the short periods in which reheat is usually used.

Jet engines are referred to as operating wet when reheat is being used, and dry when the engine is used without afterburner.

A jet engine afterburner is an extended exhaust section containing extra fuel injectors, and since the jet engine upstream (*i.e.*, before the turbine) will use little of the oxygen it ingests, the afterburner is, at its simplest, a type of ramjet. When the afterburner is turned on, fuel is injected, which ignites readily, owing to the relatively high temperature of the incoming gases. The resulting combustion process increases the afterburner exit (nozzle entry) temperature significantly, resulting in a steep increase in engine net thrust. As well as an increase in afterburner exit stagnation temperature, there is also an increase in nozzle mass flow (*i.e.* afterburner entry mass flow plus the effective afterburner fuel flow), but a decrease in afterburner exit stagnation pressure owing to a fundamental loss due to heating plus friction and turbulence losses.

The resulting increase in afterburner exit volume flow is accommodated by increasing the throat area of the propulsion nozzle. Otherwise, the upstream turbomachinery rematches (probably causing fan surge in a turbofan application).

15.4.7 Limitations

Due to their high fuel consumption, afterburners are not used for extended periods. They are only used when it is important to have as much thrust as possible. This includes takeoffs from short runways (as on an aircraft carrier) and air combat situations.

15.4.8 Efficiency

Since the exhaust gas already has reduced oxygen due to previous combustion, and since the fuel is not burning in a highly compressed air column, the afterburner is generally inefficient compared with the main combustor. Afterburner efficiency also declines significantly if, as is usually the case, the tailpipe pressure decreases with increasing altitude.

Illustration 2. *A statically mounted Pratt and Whitney J58 engine with full after burner on disposing of the last of the SR-71 fuel prior to program termination.*

Afterburners do produce markedly enhanced thrust as well as (typically) a very large, impressive flame at the back of the engine. This exhaust flame may show *shock-diamonds*, which are caused by shock waves being formed due to slight differences between ambient pressure and the exhaust pressure. These imbalances cause oscillations in the exhaust jet diameter over distance and cause the visible banding where the pressure and temperature is highest.

15.5 TURBOJET ENGINE

Turbojets are the simplest and oldest kind of general purpose jet engine. Two different engineers, Frank Whittle in Britain and Hans von Ohain in Germany, developed the concept independently during the late 1930s. On 27 August 1939 the Heinkel He 178 became the world's first aircraft to fly under turbojet power, thus becoming the first practical jet plane.

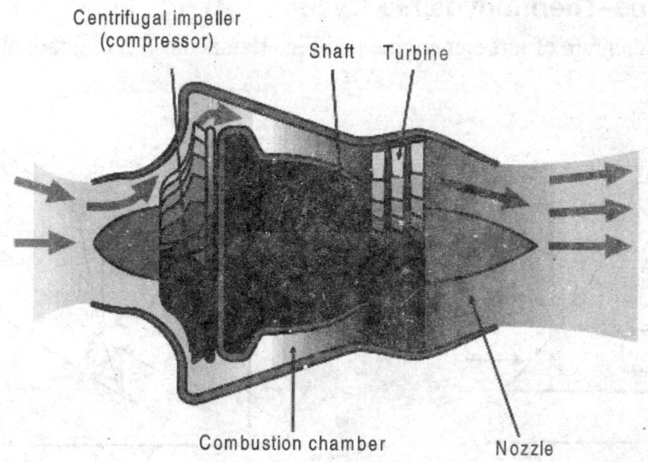

Fig. 15.2. *Schematic diagram showing the operation of a centrifugal flow turbojet engine. The compressor is driven via the turbine stage and throws air outwards, requiring it to be redirected parallel to the axis of thrust.*

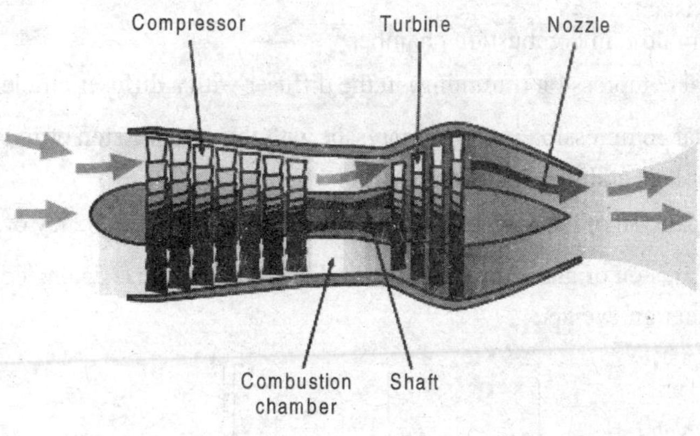

Fig. 15.3. *Schematic diagram showing the operation of an axial flow turbojet engine. Here, the compressor is again driven by the turbine, but the air flow remains parallel to the axis of thrust.*

A turbojet engine is used primarily to propel aircraft. Air is drawn into the rotating compressor via the intake and is compressed to a higher pressure before entering the combustion chamber. Fuel is mixed with the compressed air and ignited by flame in the eddy of a flame holder. This combustion process significantly raises the temperature of the gas. Hot combustion products leaving the combustor expand through the turbine, where power is extracted to drive the compressor. Although this expansion process reduces the turbine exit gas temperature and pressure, both parameters are usually still well above ambient conditions. The gas stream exiting the turbine expands to ambient pressure via the propelling nozzle, producing a high velocity jet in the exhaust plume. If the momentum of the exhaust stream exceeds the momentum of the intake stream, the impulse is positive, thus, there is a net forward thrust upon the airframe.

15.5.1 Turbojet Engine–Thermodynamic Cycle

Figure 15.4 shows the *P–V* diagram of turbojet engine and Fig. 15.5 is the *T–S* diagram of the turbojet engine.

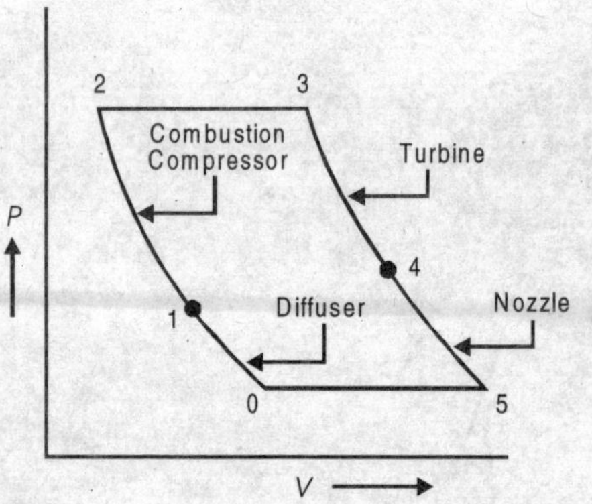

Fig. 15.4

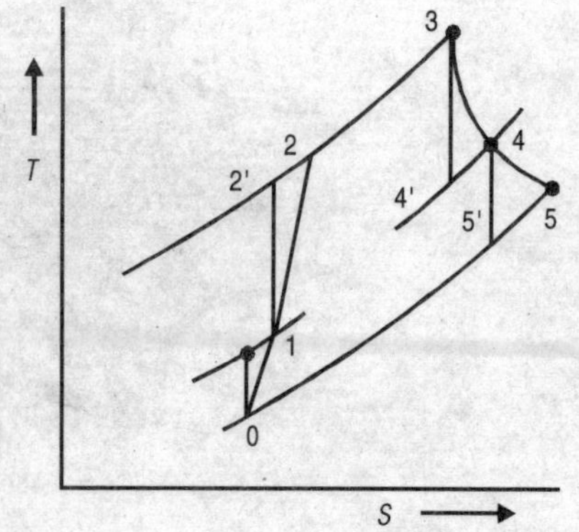

Fig. 15.5

While analysing the cycle we assume that

1. Specific heat is constant.
2. There is no pressure drop in combustion chamber.

0 – 1 is the actual compression (ramming) in the diffuser with a diffuser efficiency of η_d.

1 – 2 is the actual compression in the compressor with the compression efficiency η_c.

2 – 3 Heating or heat addition in the combustion chamber.

3 – 4 Actual expansion of gases in gas-turbine with an isentropic efficiency of η_t.

4 – 5 Actual expansion of gases through jet-nozzle with an nozzle efficiency of η_n.

After solving this *T–S* diagram, we get,

$$\text{Exit Velocity} = \left\{2\eta_n \cdot C_p \cdot T_{o_4}\left[1-\left(\frac{P_5}{P_{o_4}}\right)^{\frac{\gamma-1}{\gamma}}\right]\right\}^{1/2} \quad \text{....(1)}$$

Let

m_a = mass of air inducted in kg/sec

m_f = mass of fuel injected in kg/sec

$$\frac{F}{A} = \text{Fuel to air ratio}$$

$$V_p = V_a = \text{Velocity of incoming air at diffuser inlet}$$

$\therefore$ Initial momentum of air $= m_a \times V_a$
and final or momentum at exit

$$= (m_a + m_f)V_e$$

$\therefore$ Force = Thrust developed

$$T = (m_a + m_f)V_e - m_a \times V_a$$

$$= m_a\left[\left(1 + \frac{m_f}{m_a}\right)V_e - V_a\right] \quad \text{...(2)}$$

Note that velocity of plane is equal to the velocity of incoming air and sometimes the inlet velocity is also denoted by V_p.

$$\therefore \quad T = m_a\left[\left(1 + \frac{m_f}{m_a}\right)V_e - V_a\right]$$

Here, $(m_a + m_f)V_e$ is known as gross thrust and $m_a V_a$ is known as inlet drag.

If we neglect the fuel per kg of air or fuel to air ratio is very very small and many times it is neglected.

$$\therefore \quad \text{Thrust} = m_a V_e - m_a V_a$$

$$= m_a (V_e - V_a)$$

$$= m_a \cdot V_a \left[\frac{V_e}{V_a} - 1\right]$$

Here $\frac{V_a}{V_e}$ = Effective speed ratio = α

$$\therefore \quad \text{Thrust } T = m_a V_a \left[\frac{1}{\alpha} - 1\right]$$

$$\mathbf{T = m_a\ V_a \left[\frac{1}{\alpha} - 1\right]} \quad \text{...(3)}$$

The thrust developed by the engine should overcome the drag on the air-craft and in doing so develops power, which is called *thrust power* and is given by

$$P_T = \text{Thrust power}$$

$$= \text{Thrust} \times \text{Velocity at inlet}$$

$$= T \times V_a$$

$$= m_a V_a [V_e - V_a]$$

$$\mathbf{P_T = m_a\ V_a [V_e - V_a]} \quad \text{...(4)}$$

Specific thrust is the thrust produced kg of air flow. It is also sometimes called as *specific impulse*.

$$\therefore \quad \text{Specific thrust} = \frac{T}{m_a}$$

$$= \frac{m_a(V_e - V_a)}{m_a}$$

$$= V_e - V_a \qquad ...(5)$$

$$= V_e\left(1 - \frac{V_a}{V_e}\right)$$

$$= V_e(1 - \alpha) \qquad ...(6)$$

Specific thrust is a criterion of the size of the engine required to produce a given total thrust.

From the equation for the thrust (2a) we observe that if the effective jet velocity V_e is kept constant, then the specific thrust decreases with increase in plane velocity $V_a = V_p$ or flight speed. specific thrust will be maximum when $\alpha = 0$ i.e. $V_a = 0$ or flight velocity is zero.

15.6 THERMAL EFFICIENCY OF A TURBOJET ENGINE

Thermal efficiency is defined as the ratio of propulsive power developed at the exhaust nozzle to the heat supplied by the fuel. This ratio indicates the degree of utilisation of fuel in accelerating the fluid flow.

∴ Thermal efficiency is given by

$$\eta_{th} = \frac{\text{Propulsive power}}{\text{Heat supplied}}$$

Here, the propulsive power is given by the net increase in energy of the working fluid between inlet and exit.

∴ Propulsive power P is given by

$$P = \frac{1}{2}\left[(m_a + m_f)V_e^2 - m_a V_a^2\right]$$

$$= \frac{1}{2}\left[m_a V_e^2 - m_a V_a^2\right] \text{ if } m_f \text{ is neglected}$$

$$= \frac{m_a}{2}\left(V_e^2 - V_a^2\right) \qquad ...(7)$$

$$\therefore \quad \eta_{th} = \frac{P}{\text{Heat supplied}}$$

$$= \frac{\frac{m_a V_e^2}{2}\left[1 - \left(\frac{V_i}{V_e}\right)^2\right]}{m_f \times \text{C.V.}}$$

$$= \frac{\frac{m_a V_e^2}{2}\left[1 - \alpha^2\right]}{m_f \times \text{C.V.}}$$

$$= \frac{\frac{V_e^2}{2}\left(1 - \alpha^2\right)}{\frac{m_f}{m_a} \times \text{C.V.}}$$

$$= \frac{V_e^2\left(1-\alpha^2\right)}{2\cdot\frac{F}{A}\cdot \text{C.V.}}$$

$$= \frac{V_e^2\left(1-\alpha^2\right)}{2\cdot f\cdot \text{C.V.}} \quad \text{where } f = \frac{m_f}{m_a}$$

$$\therefore \qquad \eta_{th} = \frac{V_e^2\left(1-\alpha^2\right)}{2f\times \text{C.V.}} \qquad \ldots(8)$$

15.7 PROPULSIVE EFFICIENCY

Propulsive efficiency is defined as the ratio of useful propulsive power or thrust power to the sum of thrust power and unused kinetic energy of the jet. It is generally denoted by η_p. Here the kinetic energy of the jet is relative to the earth viz. $\frac{m(V_e - V_a)^2}{2}$.

$$\therefore \qquad \eta_p = \frac{m_a(V_e - V_a)V_a}{m_a\left[V_a(V_e - V_a) + \frac{(V_e - V_a)^2}{2}\right]}$$

$$= \frac{V_a(V_e - V_a)}{(V_e - V_a)\left[V_a + \frac{(V_e - V_a)}{2}\right]}$$

$$= \frac{V_a}{V_a + \frac{V_e - V_a}{2}}$$

$$= \frac{2V_a}{2V_a + V_e - V_a}$$

$$= \frac{2V_a}{V_a + V_e}$$

$$= \frac{2}{1 + \frac{V_e}{V_a}}$$

$$= \frac{2V_a}{V_e\left(1 + \frac{V_a}{V_e}\right)}$$

$$= \frac{2\alpha}{(1+\alpha)} \quad \text{...(9)}$$

For maximum propulsive efficiency, we differentiate Eq. (9) with respect to α and equating to zero.

$$\frac{d\eta_p}{d\alpha} = 0$$

or when $\alpha = 1$

$$\eta_p = 1 \text{ i.e., } 100\ \%$$

$\therefore$ $V_e = V_a$

and in this case no thrust is available or produced and thus no useful work output can be obtained.

∴ To get the thrust power, $V_e > V_a$ or $V_e > V_p$

∴ From Eq. (4) we have

$$\text{Thrust power} = TP = m_a\,V_a\,(V_e - V_a)$$

and

$$\alpha = \frac{V_a}{V_e}$$

$\therefore$

$$V_a = V_e \cdot \propto$$

$$TP = m_a\,V_e\alpha(V_e - V_a)$$

$$= m_a\,V_e^2\alpha(1-\alpha)$$

Differentiating this w.r.t. α, and equating it to zero, we get the condition for maximum thrust power.

$$\therefore \quad \frac{d(TP)}{d\alpha} = 0 = m_a\,V_e^2 - m_a\,V_e^2 \times 2\alpha$$

$$\therefore \quad 0 = m_a\,V_e^2\,[1-2\alpha]$$

or

$$\alpha = \frac{1}{2} = 0.5 \quad \text{...(10)}$$

∴ For this condition, the propulsive efficiency is given by

$$\eta_p = \frac{2\alpha}{1+\alpha}$$

$$= \frac{2\times 0.5}{1+0.5} = \frac{1}{1.5}$$

$$= \mathbf{0.667} \text{ or } \mathbf{66.7\ \%}$$

15.8 OVERALL EFFICIENCY OF PROPULSIVE SYSTEM

Overall efficiency of the propulsive system gives the performance of the propulsive efficiency. It indicates the extent to which the system utilizers the energy supplied. Also, the overall efficiency is defined as the ratio of the rate of doing useful propulsive work and the rate at which energy is supplied to the system. It is indicated by η_o.

$$\therefore \quad \eta_o = \text{Overall efficiency}$$

$$= \frac{\text{Useful propulsive work}}{\text{Energy supplied}}$$

$$= \frac{\text{System output}}{\text{Energy input}} \times \frac{\text{Transmission output}}{\text{System output}} \times \frac{\text{Useful propulsive work}}{\text{Transmission output}}$$

$$= \eta_{th} \times \eta_{tr} \times \eta_p$$

$$\eta_o = \eta_{th} \times \eta_{tr} \times \eta_p \quad \text{...(11)}$$

15.9 THRUST SPECIFIC FUEL CONSUMPTION (TSFC)

Similar to engines and turbines, the specific fuel consumption in case of jet propulsive system is on the basis of thrust produced.

Hence, the thrust specific fuel consumption (TSFC) is defined as the amount of fuel required to produced unit thrust per hour. Thrust take here is Newton. Therefore,

$$\text{TSFC} = \frac{\text{Mass of fuel/hour}}{\text{Thrust}}$$

$$= \frac{m_f \times 3600}{F} \text{ kg/Nh} \quad \text{...(12)}$$

Here m_f = Mass of fuel/sec

and F = Thrust in Newtons

Again,

$$\text{TSFC} = \frac{m_f \times 3600}{F}$$

$$= \frac{\frac{m_f}{m_a} \times 3600}{\frac{F}{m_a}}$$

Here

$$\frac{m_f}{m_a} = \frac{\text{Fuel}}{\text{Air}} \text{ Ratio} = f$$

$\therefore$

$$\text{TSFC} = \frac{3600 f}{\frac{F}{m_a}} \quad \text{...(13)}$$

Sometimes, specific fuel consumption is based on thrust power. Then, Thrust Power Specific Fuel Consumption (TPSFC) is defined as the ratio of fuel consumed per hour per unit thrust power.

$\therefore$

$$\text{TPSFC} = \frac{m_f \times 3600}{TP} \; \frac{\text{kg of fuel}}{\text{thrust kWH}} \quad \text{...(14)}$$

The nature of the variation curves for thrust, thrust power, TSFC and TPSFC against air or flight speed are shown in Fig. 15.6.

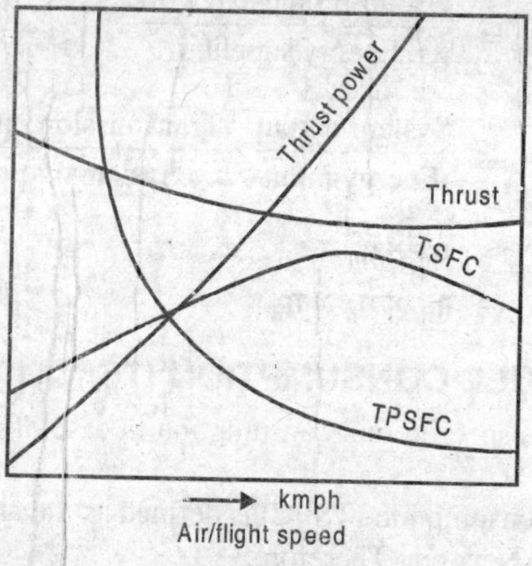

Fig. 15.6. *Performance curves for jet engine.*

The performance of the jet propulsion system is affected mainly by two parameters as
1. Forward speed of the air-craft and
2. Altitude of the air-craft.

15.10 ADVANTAGES AND DISADVANTAGES OF JET PROPULSION OVER THE OTHER SYSTEM

15.10.1 Advantages

Following are the advantages of jet propulsion :

1. **Low-specific weight.** The specific weight of the jet propulsion is one-fourth to one-half of the reciprocating engine.
2. **No unbalance force.** There are no reciprocating parts and so jet propulsion is free from unbalanced forces. Greater reliability is thus achieved.
3. **Small frontal area.** The frontal area of jet propulsion is less than one-fourth the frontal area of the reciprocating engines which decreases drag greatly and hence makes available greater power, particularly at high loads. This also reduces the air cooling problem.
4. **No restriction in power output.** Compared to reciprocating engine the jet propulsion with greatly increased power output can be built because the power is not limited by detonation. The unit can work/operate over a large range of mixture strength.
5. **High-speed.** The speed of jet propulsion is not limited by propeller. High speed can be obtained.
6. **Neither lubrication nor radiators.** Jet propulsion requires neither internal lubrication nor radiators as reciprocating engines require.
7. At high speed greater than 900 kmph and at an altitude greater than 10,000 metres, the efficiency of the jet is much higher than that of a propeller.
8. Combustion and delivery of power is continuous, whilst peak and fluctuating pressures do not occur.
9. There is no ship stream loss, the drag is reduced, and warm compressed air is available for cabin heating.
10. The unit permits of a better position of the pilot whilst the absence of a propeller permits a smaller under carriage.

15.10.2 Disadvantages

The disadvantages from which jet propulsion suffers are as follow :

1. Particularly at low pressure, the thermal efficiency is lower. At low altitude and speed upto 150 m/sec/540 kmph, the fuel consumption is 2 to 3 times that of a reciprocating engine.
2. The plant is very noisy, materials costly and life short.
3. The compression-pressure ratio is not constant as in the reciprocator, but varies approximately with the square of the speed.
4. There are certain difficulties which are encountered in the running of the propulsive unit.

SOLVED EXAMPLES

Example 15.1 *A jet propelled engine having two jets and working on turbojet system has a velocity of 210 m/s, when flying at an altitude of 12000 m, the density of air at this altitude being 0.172 kg/m³. The resistance or drag of the plane is 6670.8 N and propulsion efficiency of the jet is 50%. Calculate*

(i) *Absolute velocity of the jet.*
(ii) *Quantity of air compressed/minute.*
(iii) *Diameter of the jet.*
(iv) *Net output of gas turbine plant.*
(v) *TSFC.*
(vi) *Air-fuel ratio.*

Take $\eta_{overall}$ = *18%, CV = 4.895 × 10⁴ kJ/kg.*

Solution

Given : V_i = 210 m/s, CV = 4.894 × 10⁴ kJ/kg, ρ_{air} = 0.172 kg/m². Thrust, F = 6670.8 N.

(i) Propulsion efficiency,
$$\eta_p = \frac{2\alpha}{1+\alpha};\ \alpha = \frac{V_i}{V_e}$$

$$= \frac{2V_i}{V_e + V_i}$$

$$\therefore \quad 0.5 = \frac{2\times 210}{V_e + 210}$$

$$\therefore \quad V_e = 630 \text{ m/s}$$

But absolute velocity of jet,
$$V = V_e - V_i = 630 - 210 = \mathbf{420\ m/s}$$

(ii) Propulsive thrust,
$$F = \dot{m}_i (V_e - V_i)$$

$$6670.8 = \dot{m}_i (420)$$

$$\therefore \quad \dot{m}_i = 15.88 \text{ kg/s}$$

$$\rho_{air} = 0.172 \text{ kg/m}^3$$

$$\therefore \text{ Volume of air compressed/min} = \frac{15.88}{0.172}\times 60 = \mathbf{5539.8\ m^3/min}$$

(iii) Let a_e be the area of one jet. Neglecting mass of fuel,

$$\text{Volume compressed/min} = 2(a_e V_e) \times 60$$

$$\therefore \quad a_e = \frac{5539.8}{2 \times 60 \times 630} = 0.0733 \text{ m}^2$$

But

$$a_e = \frac{\pi}{4} d_e^2 = 0.0733$$

$$\therefore \quad d_e = 0.305 \text{ m}$$

or

$$d_e = 30.5 \text{ cm}$$

(iv)

$$\eta_p = \frac{\text{Thrust power}}{\text{Net output of turbine}}$$

$$\therefore \text{ Net output of turbine plane} = \frac{6670.8 \times 210}{1000 \times 0.5} = \mathbf{2801.7\ kW}$$

(v) Let $\dot{m}_f$ be fuel consumed/hour.

$$\eta_{overall} = \frac{\text{Propulsive work}}{\text{Heat supplied (by fuel)}}$$

$$\therefore \quad 0.18 = \frac{2801.7 \times 3600}{\dot{m}_f \times 48950}$$

$$\therefore \quad \dot{m}_f = 1144.7 \text{ kg/h}$$

$$\text{TSFC} = \frac{\dot{m}_f}{\text{Thrust}} = \frac{1144.7}{6670.8 \times V_i}$$

$$= \frac{1144.7}{\frac{6670.8}{1000} \times 210} = \mathbf{0.817\ kg/kWh}$$

(vi) Air-fuel ratio,

$$\text{A/S} = \frac{\text{Mass of air}}{\text{Mass of fuel}}$$

$$= \frac{15.88 \times 60 \times 60}{1144.7} = \mathbf{49.94}$$

Example 15.2 *A German Jumo turbojet plant uses petrol having a calorific value of 4.184 × 10⁴ kJ/kg. The fuel consumption is 0.15 kg/Nh of thrust when the thrust is 8829 N. The air-craft velocity is 425 m/s. The mass of air passing through the compressor is 19.8 kg/s. Calculate the air-fuel ratio and overall efficiency.*

Solution

Given : CV = 41840 kJ/kg, TSFC = 0.156 kg/Nh, F = 8829 N,

V_i = 425 m/s, $\dot{m}_i$ = 19.8 kg/s

To find : η_o, A/F

$$\dot{m}_f = \frac{0.15}{3600} \times 8829 = 0.3678 \text{ kg/s}$$

$$A/F = \frac{\text{Mass flow rate of air}}{\text{Mass flow rate of fuel}}$$

$$= \frac{19.8}{0.3678} = \mathbf{53.8:1}$$

Thrust power, $T_p = F \times V_i$

$= 8829 \times 425 = 3752.3$ kW

Heat input, $Q_i = \dot{m}_f \times \text{CV}$

$= 0.3678 \times 41840$

$= 15388.7$ kW

Overall efficiency, $\eta_o = \dfrac{T_p}{Q_i} \times 100$

$$= \frac{3752.3}{15388.7} \times 100$$

$= \mathbf{24.4\ \%}$

Example 15.3 *A turbojet engine consumes air at the rate of 60 kg/s when flying at a speed of 950 km/h. Calculate*

(i) *Exit velocity of jet when the enthalpy change for the nozzle is 210 kJ/kg.*
(ii) *Fuel flow rate in kg/s if A/F = 65.*
(iii) *TSFC.*
(iv) *Thermal efficiency of the plant when combustion efficiency is 93% land CV of fuel is 42500 kJ/kg.*
(v) *Propulsive power.*
(vi) *Propulsive efficiency.*
(vii) *Overall efficiency.*

Solution

Given : $\dot{m}_i = 60$ kg/s, $\Delta h = 210$ kJ/kg, $A/F = 65:1$, $V_i = 950$ km/h,

$\eta_{\text{comb}} = 0.93$, CV $= 42500$ kJ/kg

$$V_i = \frac{950 \times 1000}{3600} = 263.9 \text{ m/s}$$

(i) Exit velocity of jet, $V_e = \sqrt{2\,\Delta h \times 1000}$

$= \sqrt{2 \times 210 \times 1000}$

$= \mathbf{648.1\ m/s}$

(ii) Fuel consumption, $\dot{m}_f = \dfrac{\text{Air consumption}}{(A/F)}$

$$= \frac{60}{65} = \mathbf{0.923\ kg/s}$$

(iii) Thrust produced or propulsive thrust,

$$F = m_i (V_e - V_i)$$

$= 60 \times (648.1 - 263.9)$

$$= 23052 \text{ N}$$

Thrust specific fuel consumption,

$$\text{TSFC} = \frac{\dot{m}_f}{T_p}$$

$$= \frac{0.923}{23052}$$

$$= \mathbf{4 \times 10^{-5} \text{ kg/N thrust/s}}$$

(iv)
$$\eta_{th} = \frac{\text{Work output}}{\text{Heat supplied}}$$

$$= \frac{\text{Gain of KE/kg}}{\dot{m}_f/\text{kg of air} \times \text{CV} \times \eta_{comb} \times 1000}$$

$$= \frac{\dfrac{(648.1)^2 - (263.9)^2}{2}}{\dfrac{1}{65} \times 42500 \times 0.93 \times 1000} = \mathbf{28.81\ \%}$$

(v)
$$\text{Propulsive power} = m_i \left(\frac{V_e^2 - V_i^2}{2} \right) \times \frac{1}{1000}$$

$$= \frac{60 \times \left[(648.1)^2 - (263.9)^2 \right]}{2 \times 1000}$$

$$= \mathbf{10511.7 \text{ kW}}$$

(vi) Propulsive efficiency,
$$\eta_p = \frac{2V_e}{(V_i + V_e)}$$

$$= \frac{2\alpha}{(1+\alpha)} \text{ where } \alpha = \frac{V_i}{V_e} = 0.4072$$

$$= \frac{2 \times 0.4072}{(1 + 0.4072)} = \mathbf{57.87\ \%}$$

(vii) Overall efficiency,
$$\eta_o = \frac{(V_e - V_i)V_i}{10^3 \times (F/A) \times \text{CV} \times \eta_{comb}}$$

$$= \frac{(648.1 - 263.9) \times 263.9}{10^3 \times (1/65) \times 42500 \times 0.93}$$

$$= \mathbf{15.9\ \%}$$

THEORY QUESTIONS

15.1. Compare air breathing engines and rocket engines.
15.2. How rocket engines are classified?
15.3. Writ e a note on 'rocket propellants'.
15.4. Explain 'solar rocket propulsion'.
15.5. What is staging of rockets? What are the advantages and disadvantages of staging?
15.6. Give applications of rocket engines.

PROBLEMS FOR PRACTICE

1. What is meant by thrust ? Derive the thrust equation for a general propulsion system.
2. Explain clearly the various efficiencies associated with a propulsion device.
3. What do you mean by thrust augmentation ? Explain how it is effected.
4. What is meant by jet propulsion ? What are the basic differences between jet propulsion cycle and shaft power cycle ?
5. A naval air-craft is powered by a turbojet engine, with provision for flap blowing. When landing at 55 m/s, 15% of the compressor delivery is bled off for flap blowing and it can be assumed to be discharged perpendicular to the direction of flight. If a propelling nozzle of area 0.13 m^2 is used, calculate the net thrust during landing. Given that the engine operating conditions are as follows:

Compressor pressure ratio	= 9.0
Compressor isentropic efficiency	= 0.82
Turbine inlet temperature	= 1275 K
Turbine isentropic efficiency	= 0.87
Combustion pressure loss	= 0.45 bar
Nozzle isentropic efficiency	= 0.95
Mechanical efficiency	= 0.98
Ambient conditions	= 1 bar, 288 K

   The ram pressure and temperature rise can be regarded as negligible. [**Ans.** 20794 N]
6. A turbojet engine flying at a speed of 960 km/hr consumes air at the rate of 54.5 kg/s. Calculate:
   (i) Exit velocity of jet when the enthalpy change for the nozzle is 200 kJ/kg and velocity coefficient is 0.97.
   (ii) Fuel flow rate in kg/s when air-fuel ratio is 75 : 1.
   (iii) Thrust specific fuel consumption.
   (iv) Thermal efficiency of the plant when the combustion efficiency is 93% and calorific value of the fuel is 45000 kJ/kg.
   (v) Propulsive power.
   (vi) Propulsive efficiency and
   (vii) Overall efficiency.
   [**Ans.** 613.5 m/s, 0.727 kg/s, 4.3 × 10^{-5} kg/N/s, 25.4%, 8318 kW, 60.6%,16.6%]
7. A turbo-jet has a speed of 750 km/hr while flying at an altitude of 10000 m. The propulsive efficiency of the jet is 50% and overall efficiency of the turbine plant is 16%. The density of air at 10000 m altitude is 0.173 kg/m^3. The drag on the plant is 6250 N. The calorific value of the fuel is 48000 kJ/kg. Calculate
   (i) Absolute velocity of the jet
   (ii) Volume of air compressed/min
   (iii) Diameter of the jet
   (iv) Power output in kW and
   (v) Air-fuel ratio.
   [**Ans.** 417.3 m/s, 5194 m^3/min, 415 mm, 2500 kW, 46%]

CHAPTER 16

Refrigeration

16.1 INTRODUCTION

The term *refrigeration* is used to denote the maintenance of a system or body at a lower temperature than of its *surroundings*. The system maintained at the lower temperature is known as *refrigerated system*.

A machine that produces temperature less than atmospheric temperature is called a *refrigerator*.

Generally any reversible heat engine will act as refrigerator if it is run in the reversed direction by means of external power.

16.2 METHODS OF REFRIGERATION

There are different methods of refrigeration. They are

1. Refrigeration by non-cyclic methods
   (a) Refrigeration by using ice or snow
   (b) Refrigeration by evaporation
2. Refrigeration by cyclic methods or by means of mechanical or artificial systems as
   (a) by using non-condensable gases such as air, CO_2 and SO_2
   (b) by using condensable vapours
      (i) vapour compression refrigeration system
      (ii) vapour absorption refrigeration system
3. Other methods of refrigeration.
   (a) Magnetic refrigeration system
   (b) Vortex tube system
   (c) Thermo-electric refrigeration system
   (d) Steam-jet refrigeration system
   (e) Ultra-sound refrigeration.

16.2.1 Natural Methods

(a) By using ice or snow. The natural methods include the utilization of ice or snow produced naturally in cold climate. Ice melts at 0°C, so when it is placed in a space or system warmer than 0°C heat flows into ice and the space is cooled. The ice in turn melts into water by absorbing its latent heat at the rate of 335 kJ/kg. The river water, if it is quite cold, may also be used for cooling certain systems, but the removal of heat will not be as fast and satisfactory as in the case when ice or snow is used.

The natural methods of refrigeration were actually used in early days. With the present developments in science and technology and the subsequent rise in the standard of leaving the refrigeration requirements have become so large that the natural methods have become inadequate and therefore obsolete.

When ice is to be used for refrigeration, winter ice can be stored by packing it in straw and dried weeds, for use in the summer months.

(b) Evaporative cooling process or Adiabatic cooling process. In the old days, it was observed by the people that evaporation of water naturally causes cooling effect and hence they used to put their wine or other liquids into porus earthenware containers placed in the open atmosphere such as roof.

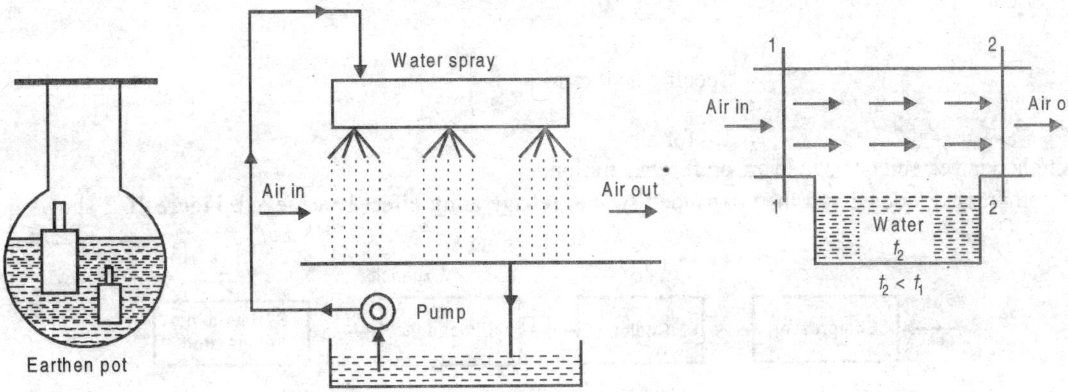

Fig. 16.1. *Examples of evaporative cooling.*

We have every day experience that water kept in an earthen pot becomes cooled because of the evaporation of water absorbing latent heat of evaporation from water in the pot and coming out through the pores. This effect is experienced very well in summer.

Air flowing over the surface of water whose temperature is less than the incoming air, then the air is cooled because of the evaporation during which latent heat is taken from air and water sump.

Same effect is obtained when air is flowing across the spray of water as shown in Fig. 16.1. The cooling is taking place without exchanging the heat with the surrounding and this type of the cooling is called as *Adiabatic cooling*.

Evaporation of sweat on our body thereby producing cooling effect is the best example of evaporative cooling.

16.2.2 Mechanical Refrigeration

Artificial or *mechanical* methods of producing *cold* have been used since 1850. Methods of mechanical refrigeration include the use of non-condensable gases and condensable gases or vapours. When non-condensable gases are used, refrigeration can be achieved by two methods such as

(a) reversible or isentropic expansion of gases against the restraining force such as a piston-cylinder mechanism.

(b) irreversible adiabatic expansion of gases from the high pressure to a low pressure through a very minute orifice or similar restriction.

(a) Isentropic expansion of gases. If we expand high-pressure air or gas in a piston-cylinder mechanism, then the temperature of gas decreases depending on the extend of expansion and the temperature of gas becomes less than atmosphere. This cold gas can be used to cool the object or the system.

We know that if,

T_1 = Initial temperature of gas (K) before expansion

T_2 = Final temperature of gas (K) after expansion

P_1 = Pressure of gas before expansion (bar)

P_2 = Pressure of gas after expansion (bar)

then,

$$\frac{T_1}{T_2} = \left(\frac{P_1}{P_2}\right)^{\frac{\gamma-1}{\gamma}}$$

and

$$T_2 = \frac{T_1}{\left(\frac{P_1}{P_2}\right)^{\frac{\gamma-1}{\gamma}}}$$

where

$$\gamma = \text{Specific heat ratio } \left(\frac{C_P}{C_V}\right)$$
$$= 1.4 \text{ for air}$$

This method is very much used in air-craft refrigeration.

Air is compressed, cooled and then expanded so that refrigerating effect is achieved. Figure 16.2 shows this process.

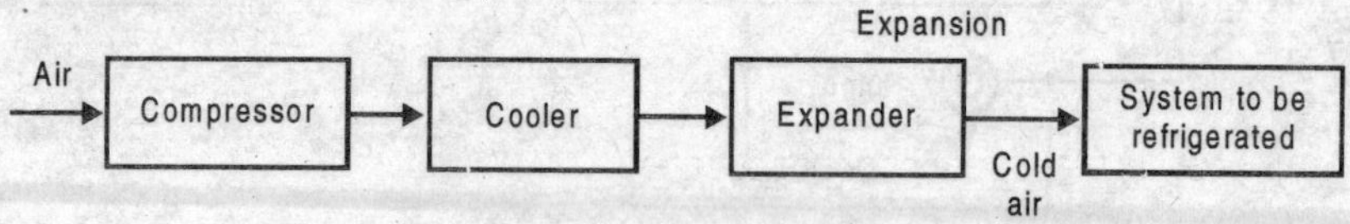

Fig. 16.2. *Refrigeration by expansion of air.*

Advantage of such a method is that the gas (air) is available free but the major disadvantage will be the lubrication difficulties for the piston cylinder mechanism.

(b) Refrigeration by irreversible adiabatic expansion of gases (Throttling) or Joule–Thomson Effect. Throttling is an irreversible process in which a fluid, flowing across a restriction, undergoes a drop in total pressure. Such a process occurs in the flow through a porous plug, a partially closed valve, or a small orifice. Joule and Thomson performed the basic throttling experiments in the period 1852–62, and their experiments clarified the process and led to use of throttling as a method for determining certain properties of gaseous substances.

The apparatus used in the Joule–Thomson experiment is shown in Fig. 16.3.

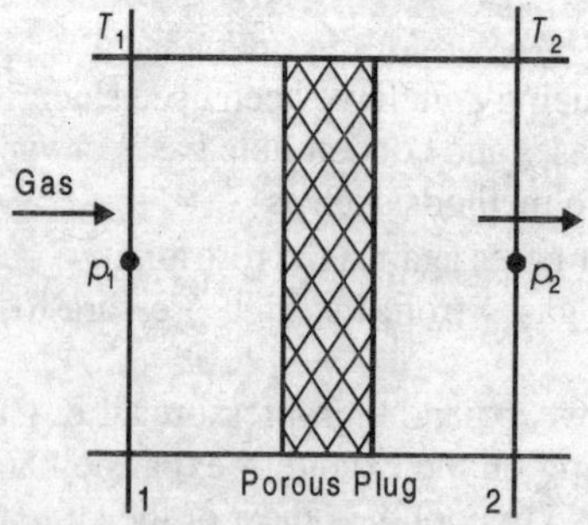

Fig. 16.3. *Throttling process. The Joule-thomson porous-plug experiment.*

A steady stream of gas flows through a porous plug contained in a horizontal tube. This system is open, is thermally insulated ($Q = 0$) and does not exchange work with its environment ($W = 0$). At sections 1 and 2 both the temperature and the pressure are measured. If the kinetic energy does not change significantly as the fluid passes through the porous plug, the steady flow energy equation reduces to

$$h_1 = h_2 \qquad ...(1)$$

Hence, in an adiabatic throttling process the enthalpy remains constant. When a series of Joule–Thomson experiments is performed at the same initial temperature t_1 and pressure p_1 but with different downstream pressure, it is found that the temperature t_2 changes. The results from these experiments can be plotted as a constant enthalpy curve on *T–P* plane. If several different rates of flow are established at each condition of t_1 and p_1, a series of isenthalpic curves is obtained. These are shown in Fig. 16.4. The maximum point on each curve is called the *inversion point* and the locus of the inversion points is called the inversion curve. The slope of an isenthalpic curve is called the Joule–Thomson coefficient μ_h and is expressed by

$$\mu_h = \left(\frac{\partial T}{\partial p}\right)_h \qquad ...(2)$$

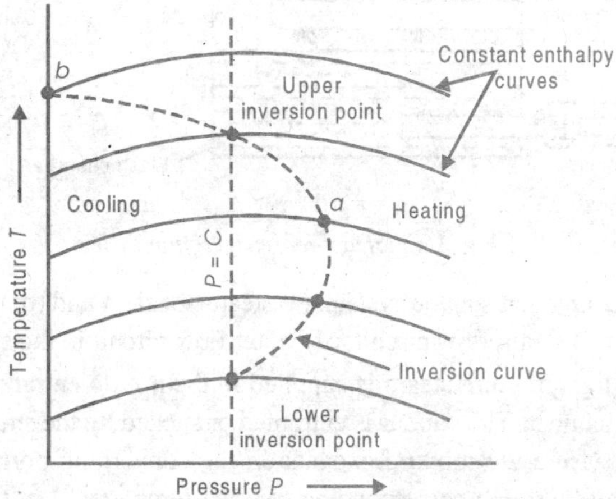

Fig. 16.4. *Inversion and constant enthalpy curves.*

At the left side of an inversion point, μ_h is positive; at the right side of an inversion point, μ_h is negative; at the inversion point μ_h is zero. Since there is always a pressure drop in the throttling process, ΔP is always negative. Therefore when μ_h is positive, the temperature change is negative, and throttling produces cooling. Similarly, when μ_h is negative, the change is positive, and throttling produces a rise in temperature, even though the process is adiabatic. At any particular pressure, μ_h is positive only within a certain temperature range. These temperatures are called the upper and lower inversion points and are indicated in Fig. 16.4. Between these two temperatures, throttling causes a drop in the temperature of the gas; outside of this temperature range, throttling results in a temperature rise. Above a certain pressure, throttling can cause only a cooling effect. This is point *a* in Fig. 16.4. Similarly, above a certain temperature throttling can cause only a heating effect. This is point *b* in Fig. 16.4.

In case of a perfect gas, enthalpy is a function of temperature alone, and therefore the temperature of the gas remains constant so that $\mu_h = 0$. In case of a real gas the temperature generally does change with pressure and μ_h is a function of both temperature and pressure.

16.2.3 Steam-Jet Refrigeration

This system is one of the methods employed by chemical agent machines for achieving mechanical refrigeration. Evidently water is used here as refrigerant.

Steam–Jet system is primarily used in installations where chilled water is required or the systems like cold storage where the temperatures required are more than 5°C as water starts changing its volume at 4°C and starts freezing at 0°C. The principle of flash cooling is employed in achieving the objective. A typical refrigeration system using steam-jet is shown diagrammatically in Fig. 16.5.

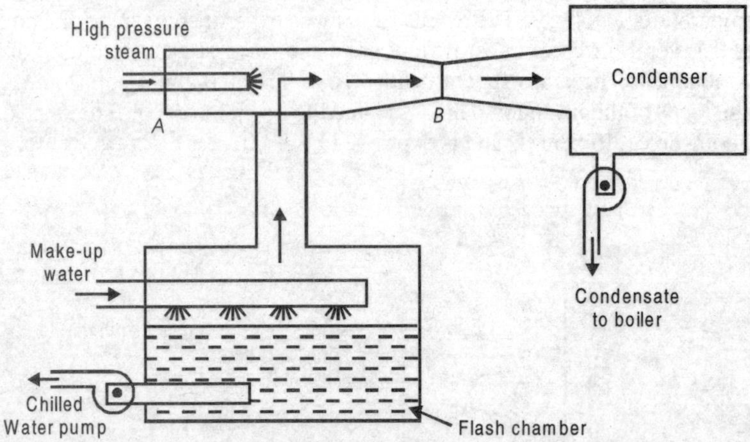

Fig. 16.5. *Steam-jet refrigeration system.*

There are two important components in the system, (a) steam nozzle *A* and (b) the flash chamber. Also, there are two distinct circuits of fluids (i) steam flow circuit, (ii) water flow circuit in the system.

(i) Steam flow circuit. High pressure steam is supplied to the nozzle entrance at *A*. Here it takes with it any vapour flashed in the flash chamber. The vapour is entrained or sucked by the suction created by the high velocity of steam flow in the nozzle. Here the mixture expands in the converging portion upto the throat *B*. It is then compressed to a pressure say 5 cm Hg corresponding to 35°C temperature in the diverging diffuser. The compressed fluid passes on to the condenser. The condensate which consists of the flashed vapour and steam from supply source is pumped back to the boiler.

(ii) Water flow circuit. The chilled water required is circulated from the flash chamber, by a pump on to the point of consumption, and the balance if any is returned back to the chamber as spray.

In the *flash chamber* low pressure of the order of 6 mm Hg is maintained due to the suction of vapour by steam flow at nozzle. The flash point is therefore low corresponding to saturation temperature at that pressure say 5°C. The water is chilled due to absorption from it of the latent heat required for the evaporation of a certain fraction of water during flashing. The amount of water flashed in the chamber plus any other quantity of chilled water drawn from the chamber (say for drinking or other purposes) is made up by an equivalent amount of fresh supply through the spray.

It will be noted that for obtaining water at 5°C, with a condensing temperature say 35°C, the operating pressures are 6 mm Hg and 5 mm Hg respectively, which are extremely low values. The compression ratio of approximately 8 also approaches the limit of efficiency operation. Note, if temperatures below 0°C are required, antifreez should be added to water.

16.2.4 Refrigeration by using Liquid-Gases

We know that a liquid can be vaporized at any desired temperature by changing its pressure. Further, heat is required to be added to the liquid during vaporization when the liquid phase changes to the gaseous phase. Accordingly, a vapourizing liquid can be used to produce refrigeration at any temperature. For instance, at a pressure of about 1 atm. ammonia may be made to boil at –33°C or at a pressure of about 5 atm. Freon –22 boils at 0°C.

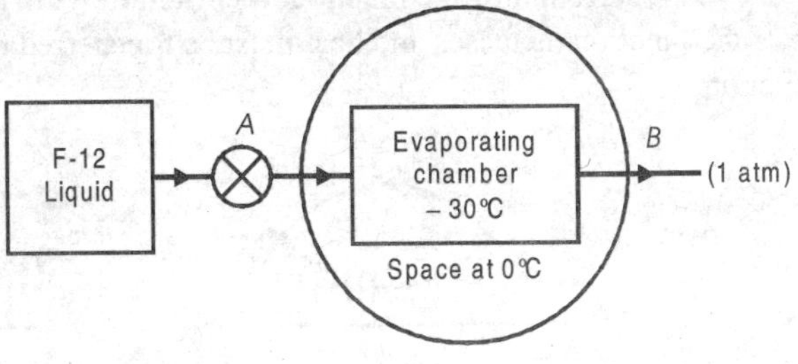

Fig. 16.6. *Refrigeration by using liquid-gas.*

Hence in an arrangement in which a system containing some refrigerant in liquid form at a certain pressure is exposed to another system at temperature greater than refrigerating temperature.

A simple scheme of such an arrangement is shown in Fig. 16.5. Liquid Freon F-12 is supplied to the evaporating chamber through a valve at a pressure of about 1 atm. Since the boiling temperature of F-12 at this pressure is – 30°C, heat flows from the surround space at 0°C and makes the F-12 boil. Thus the space will be cooled as long as there is a supply of liquid F-12 to the evaporating chamber.

The arrangement of Fig. 16.6, however, suffers from two important drawbacks.

(a) The cost of replacing Freon-12 (an expensive refrigerant) will be prohibitive as the evaporated vapour leaks out to the atmosphere.

(b) If the system were to use some other refrigerant like ammonia, it may become a hazard to life due to its discharge to atmosphere since it is a highly toxic and irritating fluid.

To eliminate these drawbacks, it will be necessary to let the refrigerant fluid work in a *closed loop* and be used again and again.

16.2.5 Thermo-Electric Refrigeration

Thermoelectric effects refer to phenomenon involving the interchange of heat and electrical energy. One familiar example is the heating effect due to the flow of an electric current in a resistor. The electric power dissipation depends on the current flow I and the potential V according to the equation $P = VI$ or $P = VI = I^2R$. This type of energy conversion is irreversible, although some thermoelectric effects are reversible.

There are three thermoelectric effects. They are: (i) The Seebeck effect, (ii) Peltier effect, and (iii) Thomson effect.

(i) The Seebeck Effect

In 1833 Seebeck held the junctions of two dissimilar electrical conductors at different temperatures and noted that an open-circuit voltage was generated that was proportional to the difference in temperature between the two junctions. A familiar example which illustrates the Seebeck effect is the ***thermocouple***. The heat supplied at one of the junctions is converted into electrical energy, which is then measured.

(ii) The Peltier Effect

In the Peltier effect, which is the inverse of the Seebeck effect, an electric current is sent through a junction of two dissimilar materials A and B, and this results in either a heating or cooling of the junction. The rate of heating or cooling is proportional to the current flow, and is given by the equation

$$\dot{Q} = \pi_{AB} I$$

where the coefficient π_{AB} is a function of temperature called the ***Peltier coefficient***. The subscript AB indicates that the Peltier coefficient also depends on the two materials A and B. A typical unit of π_{AB} is the volt. If current flows in the direction of the potential drop characteristic of the junction, then the temperature of the junction decreases, and a refrigerating effect occurs in the space surrounding the junction. If current flows in the direction of the junction's potential rise, the temperature of the junction increases and heat must be transferred from the junction to maintain constant temperature at the junction.

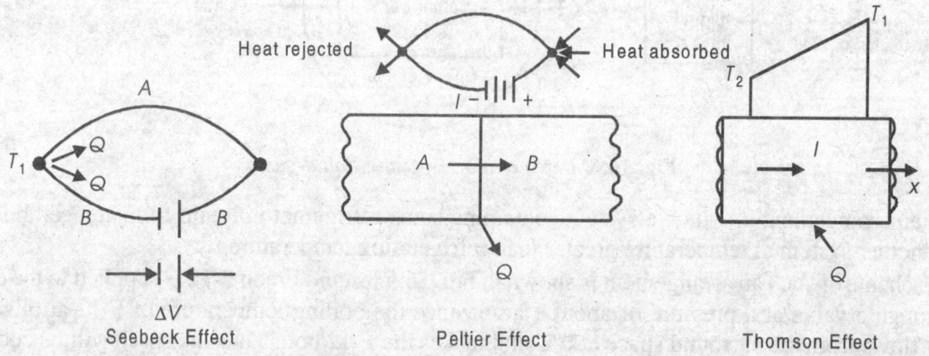

Fig. 16.7 *Thermoelectric effects.*

(iii) The Thomson Effect

When an electric current flows through a homogeneous material, heat is exchanged between the environment and the material, and the rate of transfer of heat depends on both the quantity of current that flows through the material and the temperature gradient that exists in the material. The Thomson coefficient τ is defined as

$$\tau_A = \frac{d\dot{Q}/dx}{I(dT/dx)}$$

where $d\dot{Q}/dx$ is the rate of heat interaction per unit length of the conductor and dT/dx is the temperature gradient along the length of the conductor. Like the Seebeck and Peltier effects, the Thomson effect is reversible; on the other hand, it involves only a single substance.

The Seebeck, Peltier and Thomson effects occur in both metals and semiconductors, but *these effects are more pronounced in semiconductors*. The three effects are shown in Fig. 16.7.

If Seebeck would probably have tried reversing the direction of flow of electrons in the thermoelectric circuit (by externally applying a potential difference in the reverse direction), he would have found a refrigerating effect. But this very important effect (refrigeration) was discovered by Mr. Jean Charles Athanase Peltier, in 1834. He observed during his experiments that when a small current was passed through the junction of two dissimilar wires, the junction was cooled as shown in Fig. 16.7.

Peltier effect is the basis for *thermoelectric refrigeration*. A practical thermoelectric refrigeration circuit using semiconductor materials is shown in Fig. 16.8.

Thermoelectric refrigerators cannot compete with the vapour compression refrigeration systems because of its low coefficient of performance, small refrigerating effect. They are preferred in some specific applications because of their small size, simplicity, quietness and reliability.

Mention may be made here of certain non-conventional refrigeration system which are used for specific applications only and will not be discussed here.

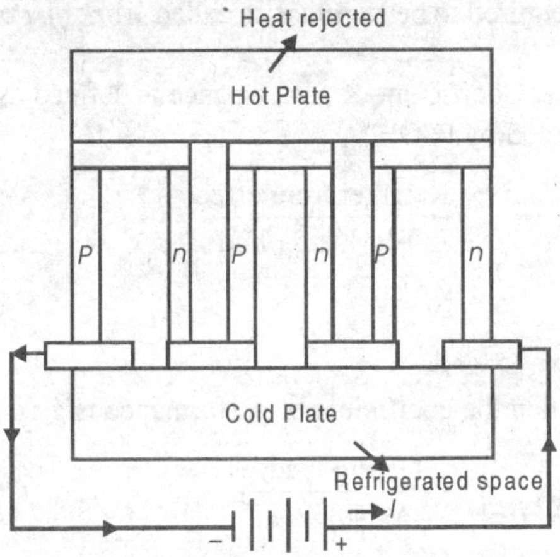

Fig. 16.8. *Thermoelectric refrigerator.*

Non-conventional refrigeration systems are,

(i) Magnetic refrigeration systems
(ii) Vortex-tube system
(iii) Pulse jet refrigeration system
(iv) Ultrasound refrigeration system

16.3 AIR REFRIGERATION SYSTEM

The virtues of air refrigeration, including cheapness, availability and safety of the refrigerant used made it quite appealing to the early inventors of refrigeration machines. Air refrigeration dates back to the middle of nineteenth century, but it was not until 1877 that a reliable unit was developed. It was used mostly for cold-storage of meat on ships. However, the equipment was bulky and inefficient. The bulkiness was due to the reciprocating compressor and expanders and low speed of rotation. Rapid developments of the more efficient and less bulky vapour compression systems made air-refrigeration rather short-lived.

Recent developments in the rotory compressors and gas turbines have renewed interest in air refrigeration systems. Rotory machines caused great reduction in weight per unit of refrigeration.

16.3.1 Terms and Definitions used in Refrigeration

For further studies of refrigeration following terms and definitions will be of importance.

(i) Refrigeration. Refrigeration is defined as the process (cyclic or non-cyclic) by means of which temperatures of the particular body or space, less than that of surroundings or atmosphere, are produced.

(ii) Refrigerants. The working substance used in the refrigeration system is called Refrigerants. It may be air, gas or chemical compounds.

(iii) Mechanical Refrigeration. When mechanical components such as compressors and electric motor are used for refrigeration, the system is called as *Mechanical refrigeration system.*

(iv) Refrigerating Effect. The amount of heat extracted from a body or space for cooling purpose is called *Refrigeration Effect*. It is denoted by *N*.

(v) Heating Load. The amount of heat energy supplied to the body or space, for heating purposes, by a heat pump is called *Heating load* for the system.

(vi) Work or Power Supplied. To run the system, energy in the form of Heat energy directly or to the electric motor to drive the compressor is required to be supplied, is called *Work or Power Supplied.* Generally it is denoted by letter *W*.

(vii) Coefficient of Performance. Coefficient of performance is defined as the ratio of the refrigerating effect to the work supplied. It is denoted by $[\text{COP}]_{\text{Ref.}}$.

$$\therefore \qquad (\text{COP})_{\text{Ref.}} = \frac{\text{Refrigerating effect}}{\text{Work supplied}}$$

$$= \frac{N}{W}$$

When heat pump is used, then the coefficient of performance is given by

$$(\text{COP})_H = \frac{\text{Heating load}}{\text{Work supplied}}$$

$$= \frac{N+W}{W}$$

$$= 1+\frac{N}{W}$$

$$= 1 + (\text{COP})_{\text{Ref.}}$$

COP is, therefore, the extent to which the work or energy supplied is utilized in producing the desired effect. The properties of the refrigerant should be such that it will give highest COP.

(viii) Unit of Refrigeration. (*Rating for refrigeration*) Unit of refrigeration is Tonne Refrigeration shortly written as TR.

Fig. 16.9. *Schematics for refrigeration and heat pump.*

One TR is defined as the heat extracted from 1 tonne water at 0°C to produce ice at 0°C in one day. One tonne in refrigeration practice is the smaller tonne and is corresponding to 2000 lbs (pounds). Latent heat of ice is taken as 335 kJ.

$$\therefore \qquad 1\text{ TR} = \frac{2000}{2.216} \times \frac{335}{24 \times 3600}$$

$$\simeq 3.516 \text{ (kJ/sec} = \text{kW)}$$
$$\simeq \mathbf{3.52\ kW}$$

(ix) Subcooling or Undercooling. In case of vapour compression cycle, vapours after compression are passed through a condenser where the heat is rejected by the refrigerant to the cooling medium. If the temperature of the condensed liquid refrigerant coming out of the condenser is less than the saturation temperature of the refrigeraut liquid at the pressure of compressor outlet, the liquid is said to be *undercooled* or *subcooled*.

(x) EER (Energy Efficiency Ratio) and SEER (Seasonal Energy Efficiency Ratio). We have defined the above term $(\text{COP})_R$ for the refrigerating machine as

$$(\text{COP})_R = \frac{\text{Refrigerating effect}}{\text{Energy supplied}}$$

Here the units for numerator and denominator are same so that COP is a pure number.

The other way to expressing the performance of the machine is the termed Energy Efficiency Ratio (EER). At present this term is used in United States of America and is defined as the ratio of the amount of heat energy removed per watt-hour of electrical energy consumed. It is noted here that the two terms in the numerator and denominator are same as COP but their units are different.

$$\therefore \quad \text{EER} = \frac{\text{Cooling capacity in kJ/hour}}{\text{Power input in Watt}}$$

$$= \frac{\text{kJ}}{\text{Wh}}$$

$$\textbf{EER} = \frac{\textbf{kJ}}{\textbf{Wh}}$$

The anticipated performance of a refrigerating device, over an average season is sometimes called the *Seasonal Energy Efficiency Ratio* (SEER) and it is also expressed in kJ/Wh

$$\therefore \quad \textbf{SEER} = \frac{\textbf{kJ}}{\textbf{Wh}}.$$

16.3.2 Refrigeration Load

As the temperature of the body or space is less than that of the surroundings, it is but natural that heat will flow into the space by the combined effect of conduction, convection and radiation. In addition to this heat flow, there is always some air leakage and ventilation and hence additional heat removal is required. Products when placed fresh, then the temperature of the products or bodies is more than the surroundings. And, therefore, to maintain the temperature constant in the refrigerator, all heat coming in, should be removed. This total heat to be removed is called the *refrigeration load*. This is the actual *refrigerating effect*.

Hence the heat sources in a refrigerator are as follows:

1. Heat transmission/transfer
2. Product heat
3. Air leakage and ventilation
4. Miscellaneous sources

16.3.3 Heat Pump

A heat pump is a thermodynamic system operating in a cycle which removes heat from a low temperature body and delivers heat to a high temperature body. In accomplishing this, the heat pump receives external energy in the form of work. The system or working substance undergoes a series of processes which constitute the heat pump cycle.

The heat pump may be used as a refrigerator where the primary function is the transfer of heat from a low temperature system. In a heating system, utilising a heat pump, the primary function is the transfer of heat to a high temperature system.

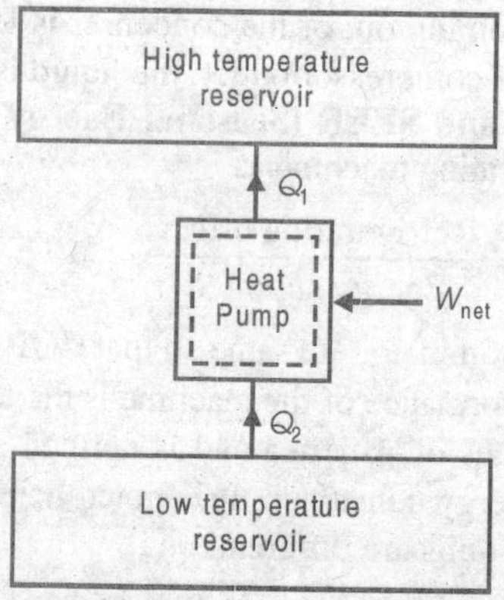

Fig. 16.10. *Principle of heat pump.*

An index of performance of a heat pump is the coefficient of performance, COP.
Referring to Fig. 16.10, let

$$Q_1 = \text{heat rejected by system}$$
$$Q_2 = \text{Heat transferred to system}$$
$$W_{Net} = \text{Net work required by system}$$

Applying the first law of thermodynamics to the cycle gives

$$\oint dQ + \oint \partial W = 0$$

or

$$Q_1 - Q_2 = W_{Net}$$

∴ COP for a heat pump is

$$(COP)_{Heat\ pump} = \frac{Q_1}{Q_1 - Q_2}$$
$$= \frac{Q_2 + W_{Net}}{W_{Net}}$$
$$= 1 + \frac{Q_2}{W_{Net}}$$
$$= 1 + (COP)$$

16.3.4 Carnot Cycle for Refrigeration

A refrigeration system removes heat from a low-temperature region and transfers heat to a high-temperature region.

The Carnot cycle can serve as the model of the ideal refrigeration cycle. Considering all refrigerators operating between the same two thermal reservoirs, the maximum coefficient of performance is attained by the Carnot refrigeration cycle.

Figure 16.11 shows the Carnot refrigeration cycle on T–S and P–V diagram. The working substance is a condensable vapour or it may be any gas or air (non-condensable). Starting at state 1 and following the arrows, the four processes of the cycle are

1–2 reversible adiabatic compression
2–3 reversible isothermal heat rejection
3–4 reversible adiabatic expansion
4–1 reversible isothermal heat absorption

During the heat rejection process (2–3), condensation takes place at a constant temperature T_h and the area *a*–3–2–*b*–*a* in the *T–S* diagram represents the heat removed Q_h. Similarly, during the heat absorption process 4–1, evaporation takes place as heat is absorbed from the surrounding space at a temperature T_c, and the area *a*–4–1–*b*–*a* in the *T–S* diagram represents the refrigeration effect Q_c. The coefficient of performance of the Carnot refrigeration cycle is given by

$$(\text{COP})_{\text{Ref.}} = \frac{Q_c}{Q_h - Q_c}$$

$$= \frac{T_c}{T_h - T_c} \qquad \ldots(3)$$

It is not practical, however, to operate a refrigeration unit according to the Carnot cycle requirements. The main difficulty lies in the isothermal heat interaction processes. These processes proceed at a very low rate so that very large heat transfer surfaces are needed. Furthermore, the presence of the liquid phase of the working fluid in the compressor creates problems, for it causes severe erosion of the compressor. This difficulty is avoided in the Rankine refrigeration cycle. When air or gas are used for Carnot cycle, Coefficient of Performance is given by

$$(\text{COP})_{\text{Ref.}} = \frac{T_c}{T_h - T_c}$$

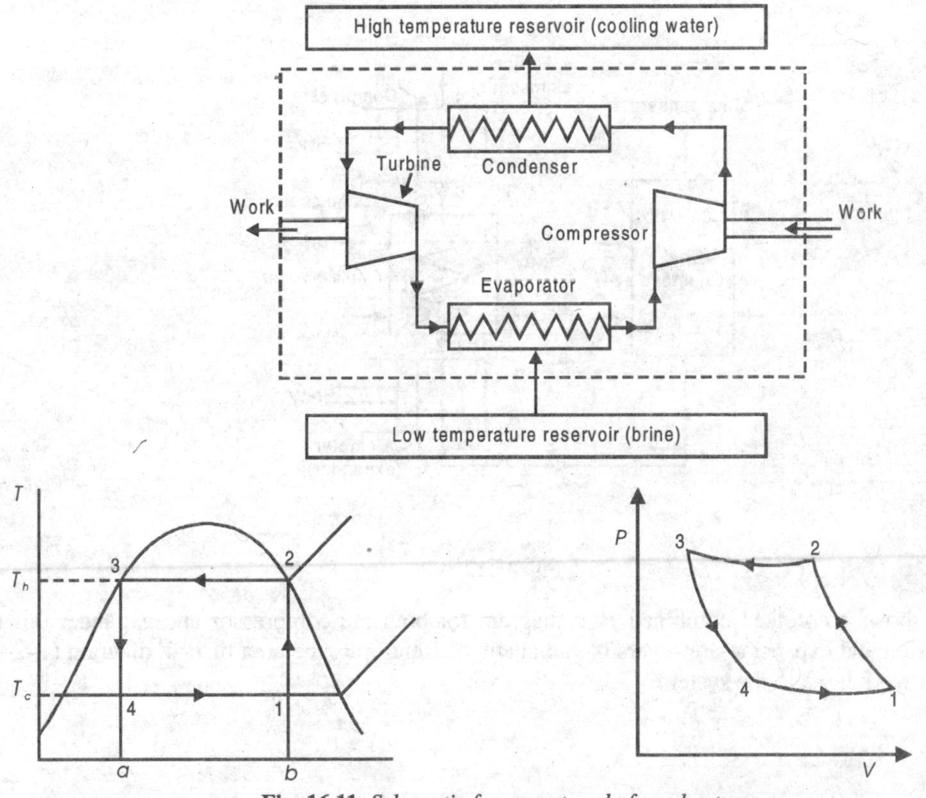

Fig. 16.11. *Schematic for carnot cycle for a heat pump.*

The ideal gas refrigeration cycle operates on a *reversed Brayton* or *Joule cycle*. This reversed Joule cycle is called *Bell-Coleman Cycle*.

16.3.5 Practical Cycle for Air Refrigeration–Bell-Coleman Cycle

Modification of the ideal reversed Carnot cycle so as to make it practicable has resulted in Bell-Coleman Cycle.

The distinguishing features of this system as compared to vapour system is that the *refrigerant air* remains in the gaseous phase throughout the cycle and does not undergo liquification and evaporation as a vapour refrigeration system.

Air refrigeration cycle may be further classified as (i) Open cycle and (ii) Closed or dense air cycle.

The isothermal operations are replaced by constant pressure operations.

(i) Open Cycle. Figure 16.12 shows a flow diagram of the open air system. The name is applied because air is discharged into cold chamber at atmospheric pressure and allowed to come in direct contact with the cold body. The compressor draws the air from the cold chamber, compresses it, and delivers it to the cooler. As air flows through the cooler, heat is removed from it causing a great decrease in its volume. In the ideal case, the pressure remains constant during cooling; in the actual case there is slight pressure drop. The high pressure air is then allowed to expand and made to do work; this work will be done at the expense of its enthalpy (total heat) and there will be a big drop in its temperature. This expansion of air is carried out in the expansion cylinder (air motor). The low temperature air leaving the expansion cylinder then enters the cold chamber and abstracts heat from the bodies kept in the chamber and in turn lowers temperature in the cold chambers.

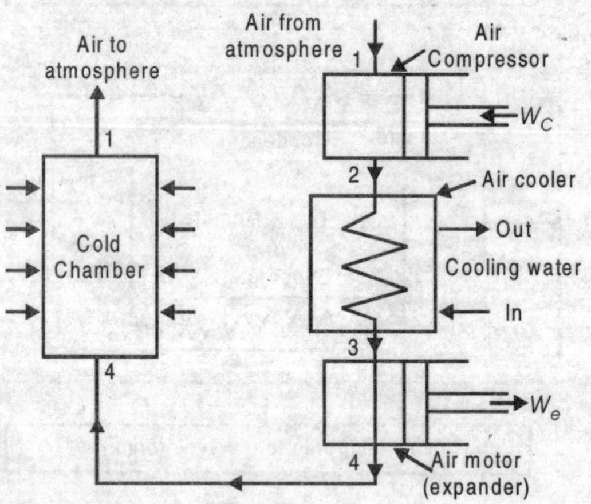

Fig. 16.12

Figure 16.13 shows theoretical combined P–V diagram for both the compressor and expander (air motor). Ideally, compression and expansion are reversible adiabatic or isentropic, the area of P–V diagram (1–2–3–4–1) represents the net work input to the system.

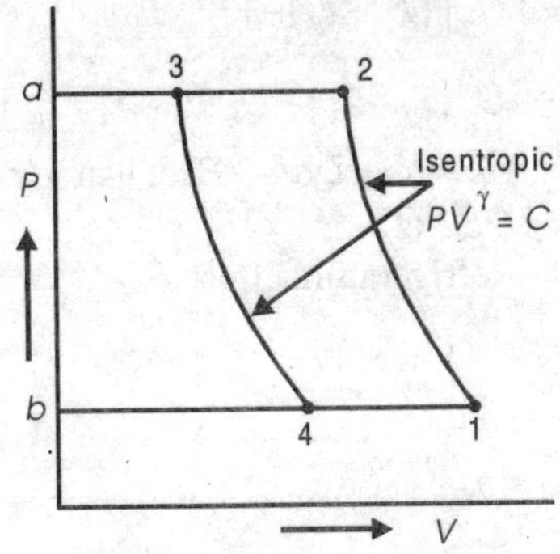

Fig. 16.13. *Combined P–V diagram for Bell-Coleman.*

Bell-Coleman cycle is also shown on *T–S* diagram of Fig. 16.14.

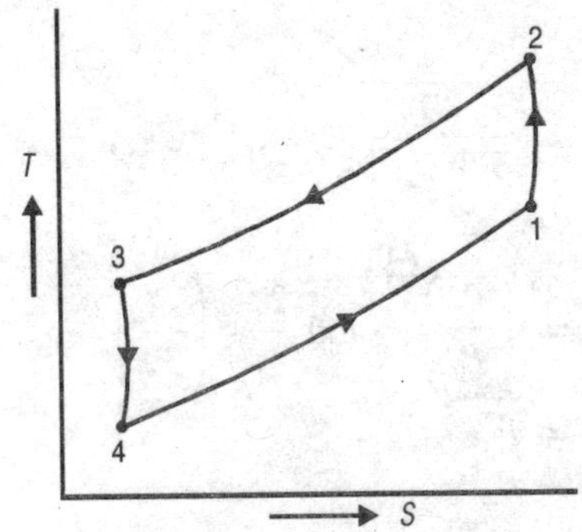

Fig. 16.14. *Bell–Coleman cycle on T–S diagram.*

1–2 Isentropic compression
2–3 Constant pressure heat-rejection
3–4 Isentropic expansion in expander
4–1 Constant pressure heat-addition

Let T_1, T_2, T_3, T_4 be the absolute temperatures at the points 1, 2, 3 and 4 respectively.

From the *P–V* diagram, we get

$$\begin{aligned}
\text{Work done} &= \text{Area } 1\text{–}2\text{–}3\text{–}4\text{–}1 \\
&= \text{Area } (1\text{–}2\text{–}a\text{–}b\text{–}1) - \text{Area } (4\text{–}3\text{–}a\text{–}b\text{–}4) \\
&= \frac{(P_1V_1 - P_2V_2)\gamma}{\gamma - 1} - \frac{(P_3V_3 - P_4V_4)\gamma}{\gamma - 1} \\
&= \frac{\gamma R}{\gamma - 1}[T_2 - T_1] - \frac{\gamma R}{\gamma - 1}[T_3 - T_4] \\
&= \frac{\gamma R}{\gamma - 1}[T_2 - T_1 - T_3 + T_4]
\end{aligned}$$

$$= C_p\left[(T_2 - T_3) - (T_1 - T_4)\right]$$

$$= C_p(T_2 - T_3) - C_p(T_1 - T_4)$$

= Heat rejected – Heat extracted

$$= Q_R - Q_A$$

Heat extracted = Refrigerating effect

$$= C_p(T_1 - T_4)$$

∴ Coefficient of performance is given by

$$\text{COP} = \frac{N}{W_{\text{Net}}}$$

$$= \frac{C_p(T_1 - T_4)}{C_p(T_2 - T_3) - C_p(T_1 - T_4)}$$

$$= \frac{1}{\dfrac{T_2 - T_3}{T_1 - T_4} - 1}$$

But $\dfrac{T_2}{T_1} = \left(\dfrac{P_2}{P_1}\right)^{\frac{\gamma-1}{\gamma}}$ and $\dfrac{T_3}{T_4} = \left(\dfrac{P_3}{P_4}\right)^{\frac{\gamma-1}{\gamma}}$

But $P_2 = P_3$

and $P_1 = P_4$

∴ $\dfrac{T_2}{T_1} = \dfrac{T_3}{T_4}$

∴ $$\text{COP} = \frac{1}{\dfrac{T_1(r_p)^{\frac{\gamma-1}{\gamma}} - T_4 \cdot r_p^{\frac{\gamma-1}{\gamma}}}{T_1 - T_4} - 1}$$

$$= \frac{1}{\dfrac{r_p^{\frac{\gamma-1}{\gamma}}(T_1 - T_4)}{T_1 - T_4} - 1}$$

$$= \frac{1}{r_p^{\frac{\gamma-1}{\gamma}} - 1}$$

$$= \frac{1}{\frac{T_3}{T_4} - 1}$$

$$= \frac{T_4}{T_3 - T_4}$$

$$\textbf{COP} = \frac{T_4}{T_3 - T_4} \qquad \ldots(5)$$

If the expansion and compression are not isentropic, then expansion and compression may be considered as polytropic and $PV^n = C$ is the law of compression and expansion.

$$\therefore \quad \text{Work done/kg} = W_{\text{Net}} = \frac{n(P_2V_2 - P_1V_1)}{n-1} - \frac{n}{n-1}(P_3V_3 - P_4V_4)$$

$$= \frac{nR}{n-1}[T_2 - T_1] - \frac{nR}{n-1}[T_3 - T_4]$$

Note here that, $W_{\text{Net}} \neq Q_R - Q_A$
but heat during compression and expansion are to be taken into consideration.

(ii) Closed or Dense Air Cycle. The basic air standard cycle is shown in Fig. 16.15. Air is contained in the system and is compressed from B by a reciprocating or rotory *compressor*. The compressed air at C is passed on to the *air cooler* where it cools down at constant pressure, to stare D.

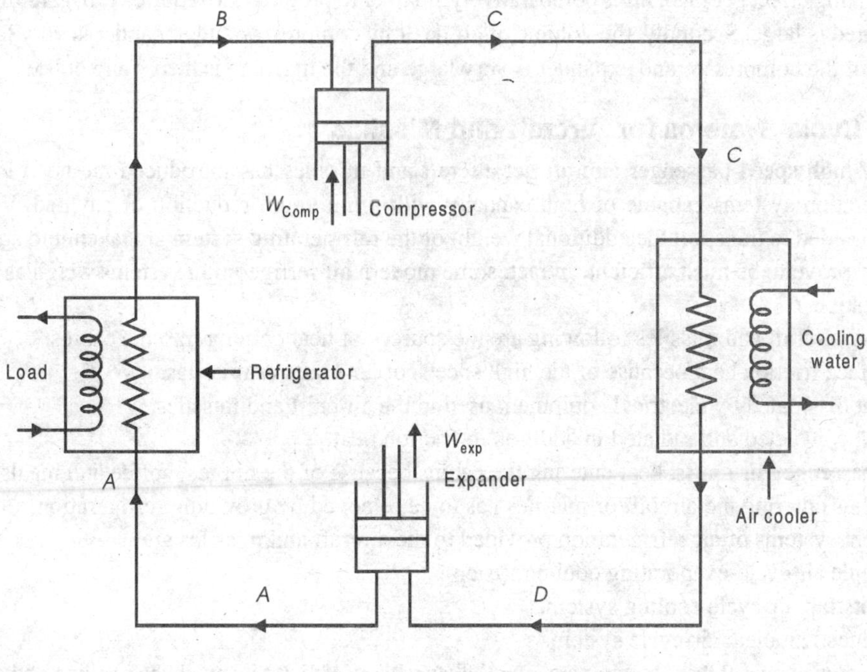

Fig. 16.15. *Dense or closed air cycle.*

The high pressure gas at D is then passed on to an expansion cylinder. Here the air expands upto A, delivers work and gets cooled. The work delivered may be fed back to the compressor by mounting the compressor and the expander on the same shaft. The cooler air at A is passed through the refrigerator. The heat picked up by the air in the refrigerator causes its state to change to point B and the cycle repeats.

The cycle discussed above is known as *closed or dense air system.* If the cool air is used as such for air conditioning or other purposes and is not taken back to the compressor, then the fresh charge of air is drawn by the compressor and the system is called *Open Air System* as discussed in the previous cycle or system.

The open air system is inherently limited to operation at atmospheric pressure only whereas in the dense air system the pressure after expander may be even lower than atmospheric, hence lower temperatures may be attained.

The $P–V$ and $T–S$ diagrams will be same as for open cycle and the numerical analysis of the cycle or system is also same.

16.3.6 Advantages and Disadvantages of Air Refrigeration

The main advantage of air refrigeration is that the working substance or air refrigerant is available always in the atmosphere. Further it is free or no cost is involved. The open air cycle system is simple. Thus we can say that

1. Availability of air is there
2. Air is free
3. System is very simple

The disadvantages are

1. Freezing of the moisture in the air used in the system. This is particularly for open system. A deposit of snow is liable to choke up the valves and ducts carrying air.
2. Coefficient of performance (COP) of the sytem is very low.
3. Refrigerating effect per kg of air is comparatively small or to produce the required refrigerating effect mass of air required is large. Secondly, the volume of air dealt by compressor and expander is very large. Therefore, the size of the compressor and expander is very large and the machine is heavy and bulky.

16.3.7 Air Cycle Systems for Aircraft and Missiles

The advent of high speed passenger aircraft, jet aircraft and missiles has introduced the need for compact and simple refrigeration systems capable of high capacity with a minimum reduction of payload. When the power requirements needed to transport the additional weight of the refrigerating system are taken into account, air cycle system usually prove to be most efficient. In fact, some modern air refrigeration systems weigh as low as 2 kg per ton of refrigeration.

In passenger aircraft and missiles following are the sources of heat coming into the cabins

(i) Surface friction heat because of the high speeds of air-crafts or missiles.
(ii) Heat dissipated by electrical equipment used in the air craft and missiles.
(iii) Heat convected and radiated in addition to friction heat.
(iv) In passenger air-crafts, heat entering the cabins because of the hot tea, coffee and meals served.

The heat, thus entering the aircraft or missiles has to be removed by providing refrigeration.

The different systems of air refrigeration provided to the aircraft and missiles are

(i) Simple air cycle-evaporating cooling system.
(ii) Bootstrap air-cycle cooling system.
(iii) Reduced ambient air-cycle system.

The schematic diagrams of these systems are given below without elaborating much the working of these systems.

(i) Simple air Cycle-Evaporating Cooling System

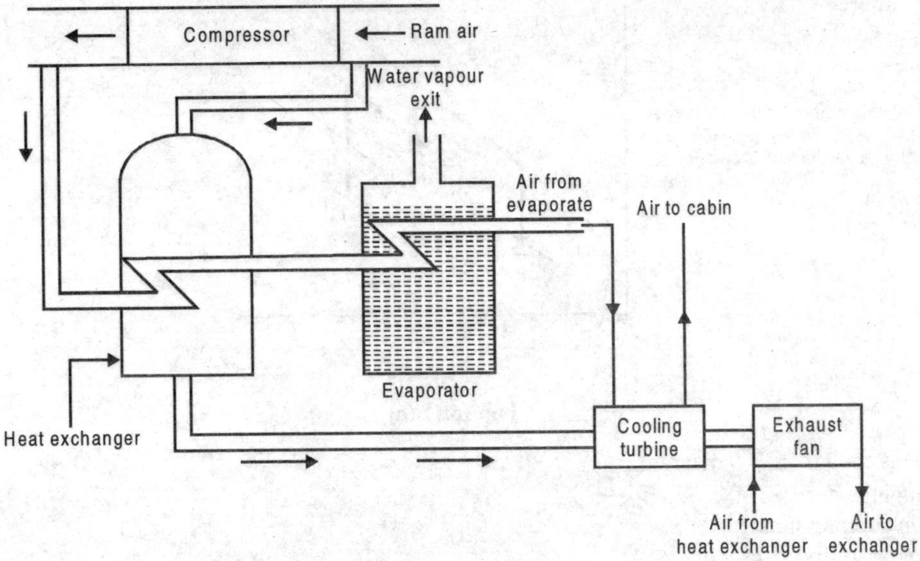

Fig. 16.16. *Simple air cycle—Evaporating cooling system.*

Ram air is at lower temperature than compressed air and hence compressed air is cooled by ram air in heat exchanger. Ram air is inducted by a cooling fan. This fan is driven by the cooling turbine as shown in Fig. 16.16. Compressed air leaves the heat exchanger and is passed through the evaporator where water is evaporated and the heat required is taken from the air and is cooled. This air is expanded through a cooling turbine and this air is supplied to the cabin.

(ii) Bootstrap Air-Cycle Cooling System. This system is shown in Fig. 16.17 (a) and (b). The cycle is shown on *T–S* diagram.

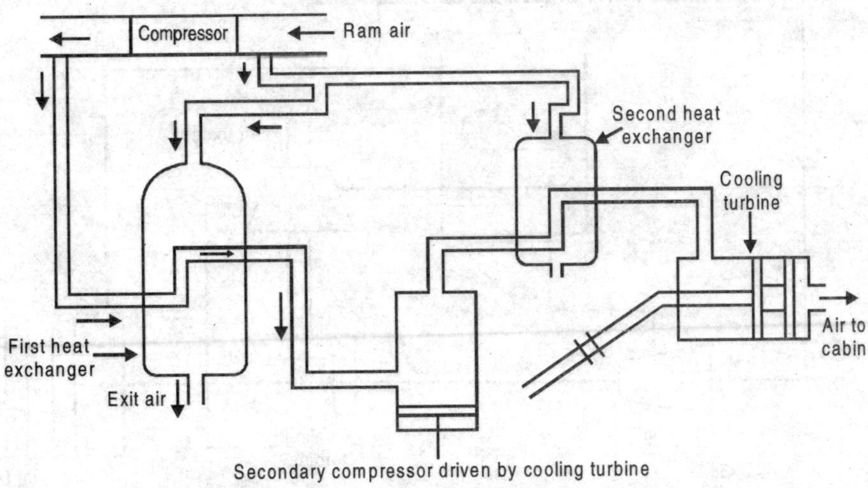

Fig. 16.17 (a) *Bootstrap air cooling cycle.*

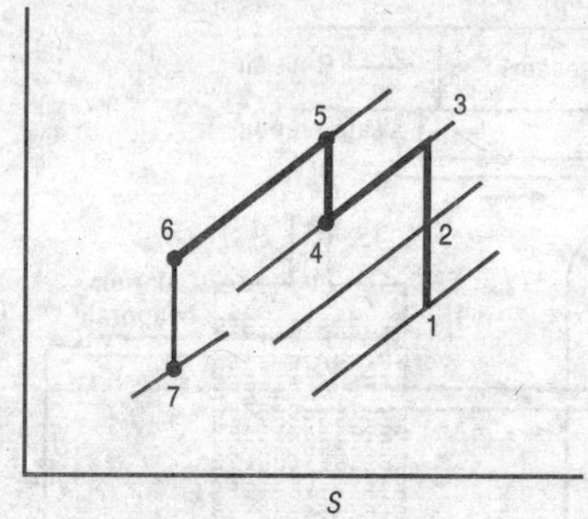

Fig. 16.17 (b)

The cycle is shown on *T–S* diagram

1–2 Ramming effect.

2–3 Main compression.

3–4 Cooling in first *H-E*.

4–5 Compression in secondary compressor.

5–6 Cooling second *HE*.

6–7 Expansion in cooling turbine.

(iii) Reduced Ambient Air-Cycle Cooling System. This system is diagrammatically shown in Fig. 16.18 (a) and *T–S* diagram of Fig. 16.18 (b).

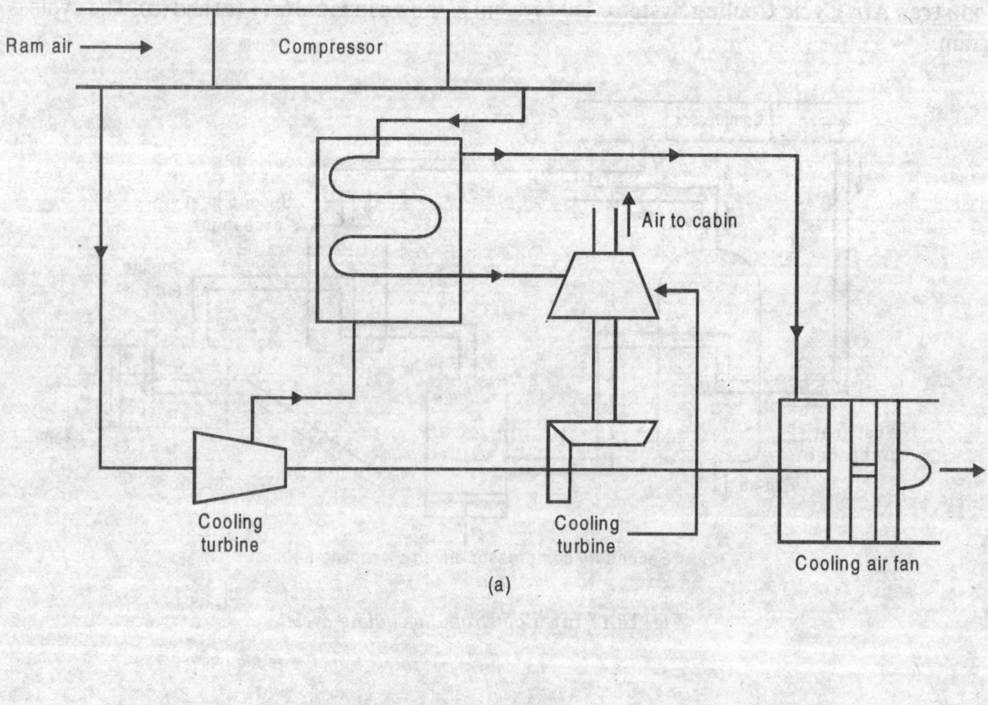

(a)

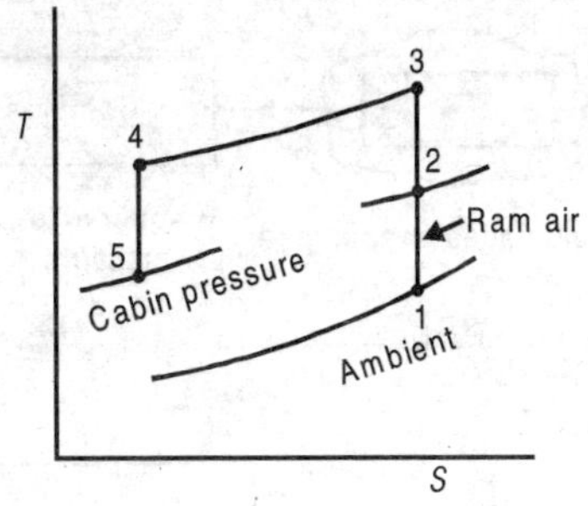

Fig. 16.18. *Reduced ambient air cycle.*

1–2 Ramming effect.
2–3 Compression in main compressor.
3–4 Cooling in heat exchanger.
4–5 Expansion in the cooling turbine.

16.4 SIMPLE VAPOUR COMPRESSION REFRIGERATION SYSTEM

A refrigerator may be defined as a machine for producing cold. Refrigerators are used for the manufacture of ice and for the cooling of storage chambers in which perishable food is stored. Theoretically, any reversible heat engine would act as a refrigerator when run in a reversed direction by means of external power; the engine then becomes a heat pump, which pumps heat from a cold body and delivers it to a hot body. The cold body usually consists of a tank of brine which is kept at a low temperature by the refrigerator; the cold brine is circulated through pipes, around the storage chamber which is to be cooled. The hot body is a supply of water which abstracts the heat from the working substance of the refrigerator. The net action of the refrigerator is to pump heat from the brine and to deliver this heat to the cooling water. By doing this it is transferring heat from a cold body to a hot body; this operation, according to the *Second Law of Thermodynamics*, can only be performed by the aid of external work. Hence, it is necessary to supply power from an external source in order to drive the refrigerator.

16.4.1 Vapour Compression System

For a closed cycle of refrigeration employing the condensable refrigerant vapour, the following processes are required.

(i) Compression of the vapour, thereby increasing pressure.
(ii) Condensing these vapours and rejecting heating to the cooling medium (usually water or atmospheric air).
(iii) Expanding the condensed liquid refrigerant thereby lowering the pressure and corresponding saturation temperature.
(iv) Evaporating the liquid refrigerant thereby absorbing heat from the body or space to be cooled or refrigerated.

It is due to this requirements of compression that the system is called *Vapour Compression System* and the cycle of operation is called *Vapour Compression Cycle of Refrigeration.*

This cycle, incorporating the compressor and condenser is shown in Fig. 16.19. Here the liquid at state *D*, the discharge of the condenser, is still at the same pressure of compressor discharge. If it were allowed to boil as it is, it will do so at the saturation temperature and will require heat to be supplied to it. Heat could possibly flow only from some temperature higher than this saturation temperature. But the requirement is to achieve a low temperature.

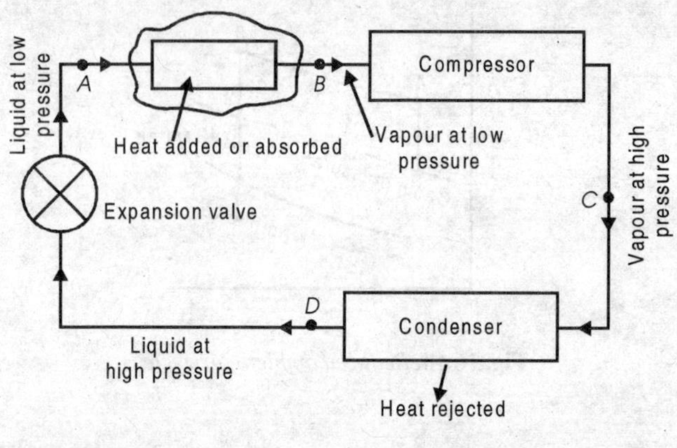

Fig. 16.19. *Schematic of vapour compression cycle.*

In order to produce the temperature less than the temperature of the body, say –10ºC, the liquid at state *D* should be throttled down to the pressure which is the saturation pressure corresponding to –10ºC. For this purpose an expansion device is provided at point between condenser and evaporator. The state of the liquid at *A*, is low pressure liquid having the saturation temperature of –10ºC. Hence, the system once charged with Freon or Ammonia will continuously work, where the refrigerant will liquify and evaporate alternatively, producing refrigeration effect at evaporator and heat rejection at condenser. Neither the cylinder of Freon nor any fresh supply of the same is required.

A simple vapour compression system of refrigeration, consists of the equipments as follows.

(a) Compressor
(b) Condenser
(c) Expansion valve
(d) Evaporator

The schematic diagram of the arrangement is shown in Fig. 16.20. The low temperature, low pressure vapour at state *B* is compressed by the *compressor*. This vapour is condensed into high pressure liquid at state *D* in the *condenser* and then passes through the *expansion valve*. Here the liquid is throttled down to a lower pressure and passes on to the evaporator, where it absorbs heat from the surroundings or the circulating fluid (being refrigerated) and vaporises to low pressure vapour of state *B*. The cycle then repeats. The exchange of energy is as follows.

(a) Compressor requires work *W*. The work energy is supplied by the system from the surroundings.

(b) During condensation heat Q_1, equivalent of latent heat of condensation etc. is lost. This heat energy flows out *from the system to the surroundings*.

(c) During evaporation heat Q_2, equivalent of latent heat of vaporisation is absorbed by the refrigerant. This heat is again an inlow of energy *into the system*.

(d) There is *no exchange of energy* during the throttling process through the expansion valve, as this process occurs at constant enthalpy.

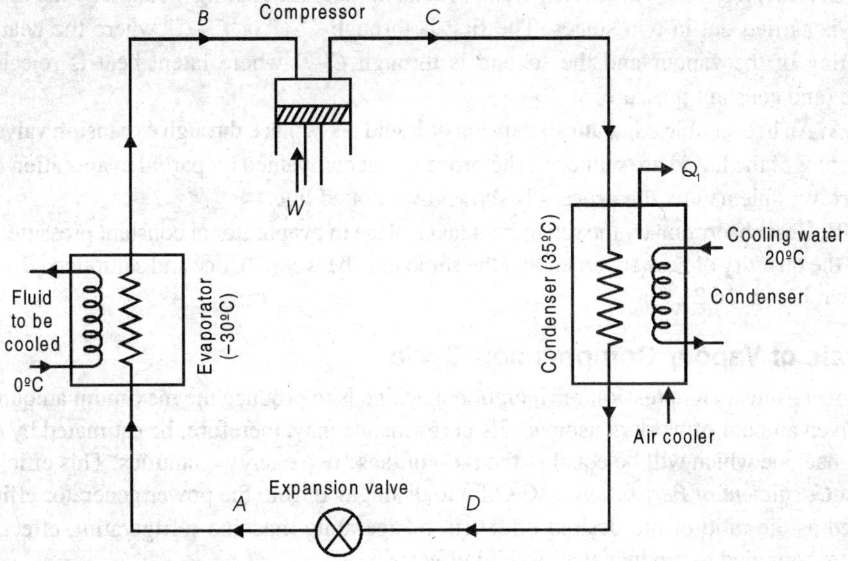

Fig. 16.20

Hence, by making an energy balance we get,

$$Q_1 = Q_2 + W$$

Figure 16.21 shows on *T–S* diagram for the equipment arrangement of Fig. 16.20. The cycle works between the temperatures T_1 and T_2 representing the condenser and evaporator temperature respectively.

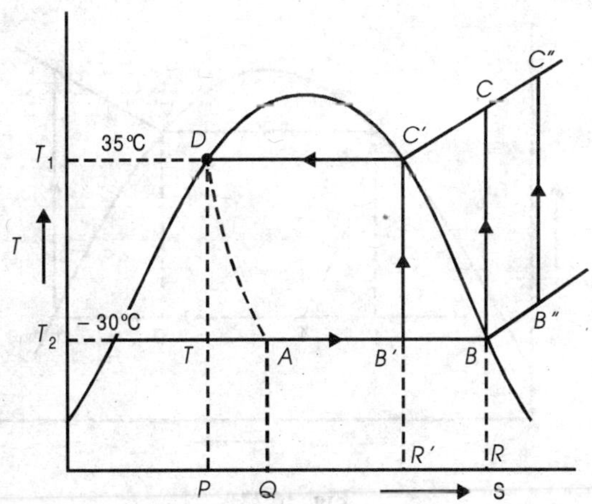

Fig. 16.21

The various processes of the cycle *A–B–C–D* are as given below:

(i) Process B–C. Isentropic compression of the vapour from state is *B* to *C*. If the vapour state, *B* or superheated, (*B*"), the compression is called *Dry Compression*. If initial state is wet, (*B*'), the compression is called *wet compression* as represented by *B'–C'*.

(ii) **Process C–D.** Heat is rejected by the refrigerant, in condenser, to the cooling medium (water ar air), at constant pressure. It is carried out in two stages. The first is through *C–C′* or *C′–C″* where the heat is rejected by desuperheating of the vapour and the second is through *C′–D* where latent heat is rejected at constant temperature (and constant pressure).

(iii) **Process D–A.** An irreversible adiabatic expansion of liquid takes place through expansion valve. The pressure and temperature of the liquid are reduced. The process is accompanied by partial evaporation of some liquid. Due to its irreversible nature, the process is shown by a dotted line.

(iv) **Process A–B.** Heat absorption by the refrigerant takes place in evaporator at constant pressure. The final state depends on the quantity of heat absorbed and the same may be wet (*B′*), dry and saturated (*B*), or superheated (*B″*) as shown in Fig. 16.21.

16.4.2 Analysis of Vapour Compression Cycle

The purpose of a mechanical compression refrigeration machine is to produce the maximum amount of refrigerating effect for a given amount of work consumed. Its performance may, therefore, be estimated by calculating the efficiency of the machine which will be equal to the ratio of these two energy quantities. This efficiency in refrigeration is called a *Coefficient of Performance* (C.O.P.) to distinguish from the power generator efficiency.

COP is defined as the ratio of the desired effect (in refrigerating machine refrigerating effect is the desired effect) to the energy supplied to produce that desired effect.

$$\therefore \qquad \text{COP} = \frac{\text{Refrigerating effect}}{\text{Energy supplied}}$$

Work of compression/kg is given by $W = h_2 - h_1$ and is supplied. (Fig. 16.22)

Similarly, the refrigerating effect is the heat absorbed by the refrigerant in the evaporator at constant temperature and pressure and is equal to the change in enhalpy during the process.

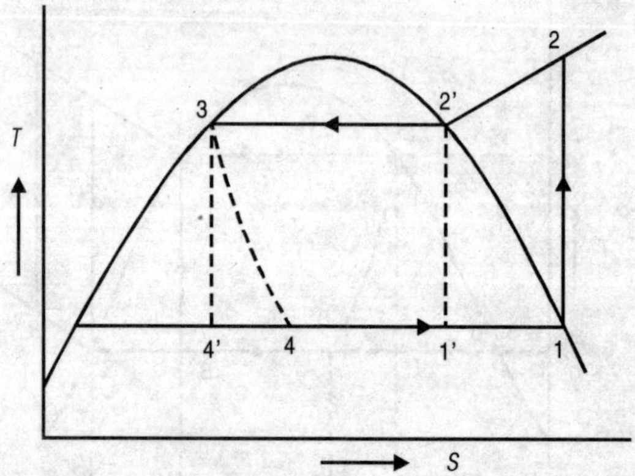

Fig. 16.22

$$\therefore \qquad \text{Refrigerating effect/kg} = h_1 - h_4 \text{ kJ/kg}$$

Letters *A*, *B*, *C*, *D* are replaced by 4,1, 2 and 3 and shown *T–S* diagram of Fig. 16.22.

$$\therefore \qquad \text{COP} = \frac{h_1 - h_4}{h_2 - h_1}$$

But $\qquad h_3 = h_4$ (isenthalpic process)

$$\therefore \quad \text{COP} = \frac{h_1 - h_3}{h_2 - h_1}$$

Instead of throttle valve or expansion valve, if an expander (piston-cylinder arrangement) is used, the expansion is taken as isentropic as compression. This isentropic expansion is shown on *T–S* diagram of Fig. 16.22 by a vertical line 3–4'. Then the expansion work is to be taken into consideration.

$$\text{Expansion work/kg} = h_3 - h_{4'}$$

and for cycle analysis, it is supplied to compressor.

∴ Net work supplied from outside/kg

$$W = \text{Work of compression} - \text{Work of expansion}$$

$$= (h_2 - h_1) - (h_3 - h_{4'})$$

Also, refrigerating effect/kg is given by

$$N = h_1 - h_{4'}$$

and

$$\text{COP} = \frac{N}{W} = \frac{h_1 - h_{4'}}{(h_2 - h_1) - (h_3 - h_{4'})}$$

$$\therefore \quad \mathbf{(COP)_{Ref.}} = \frac{\mathbf{h_1 - h_4}}{\mathbf{h_2 - h_1}} \quad \text{with expansion valve}$$

$$\mathbf{(COP)_{Ref.}} = \frac{\mathbf{h_1 - h_{4'}}}{\mathbf{(h_2 - h_1) - (h_3 - h_{4'})}} \quad \text{with expansion cylinder}$$

If we know the total capacity of the refrigerating system say *x TR* then, mass flow rate of refrigerant can be determined as

Mass flow rate of refrigerant (kg/sec)

$$= \frac{3.516\ TR}{\text{Ref. effect/kg}} = \frac{3.516x}{\text{Ref. effect/kg}}$$

16.4.3 Pressure-Enthalpy (Total heat) Chart

The pressure-total heat, or pressure-enthalpy chart is probably the most convenient chart for refrigeration calculations; this is the chart recommended by the refrigeration sub-committee of IME. Such a chart giving common features is shown in Fig. 16.23. Horizontal lines represent constant pressure lines, vertical lines are constant enthalpy lines, while lines of constant temperature and of constant entropy are also plotted and are shown in Fig. 16.23.

For different refrigerants, different *P–h* charts are available.

(a) Constant pressure process is shown by the horizontal line.
(b) Constant enthalpy lines represent throttling or isenthalpic process.
(c) Constant entropy line represents isentropic or adiabatic (reversible) process.
(d) Constant temperature lines represent isothermal processes.

16.4.4 Simple Vapour Compression Cycle on P-h Chart

A simple vapour compression cycle is shown by 1–2–3–4–1 on *P–h* chart of Fig. 16.24.

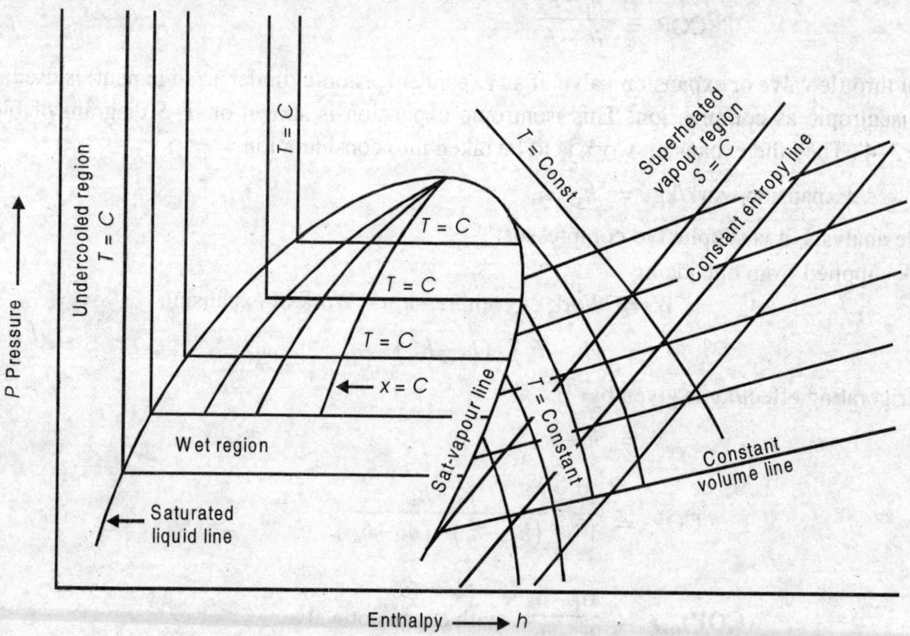

Fig. 16.23. *P-h chart.*

1–2 Isentropic compression in compressor.
2–3 Constant pressure cooling (Heat rejection).
3–4 Isenthalpic expansion through expansion valve.
4–1 Constant pressure heat absorption.

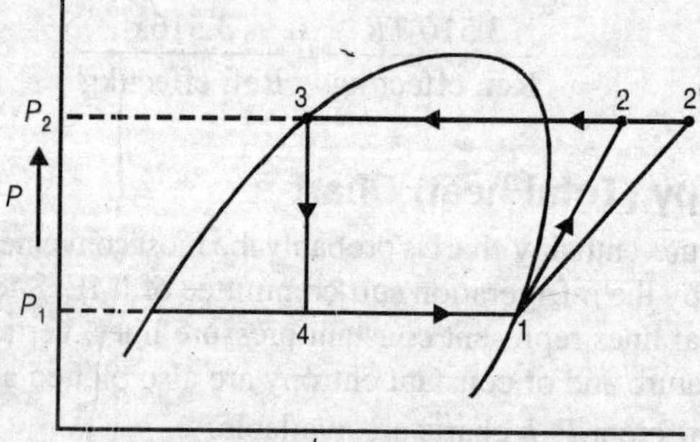

Fig. 16.24

(a) Mass of Refrigerant in Circulation

$$m_r = \frac{\text{Total refrigerating effect/sec}}{\text{Refrigerating effect/kg}}$$

$$= \frac{3.516 \times \text{TR}}{\text{Ref. effect/kg}}$$

$$= \frac{3.156(\text{TR})}{\text{Ref. effect/kg}} \text{ kg/sec}$$

(b) Piston Displacement

Let the specific volume of the vapour at 1 i.e., at the suction to the compressor, be V_1 and let the total volumetric efficiency of the compressor be $\eta_{vol.}$, then the piston displacement per min. $= \frac{V_1}{\eta_{Vol.}} \times m_r$ where m_r is the mass of refrigerant per minute.

(c) Power Required

(i) If the compression is isentropic, then work of compression $= h_2 - h_1$ kJ/kg

$$\text{Hence the power required} = \frac{m_r(h_2 - h_1)}{60} \text{ kJ/sec or kW}$$

(ii) If the actual compression is not isentropic, but it is according to any other law (actual) as shown by line 1–2', then work of compression is given by $(h_{2'} - h_1)$ kJ/kg .

$$\therefore \quad \text{Actual power required} = \frac{m_r(h_{2'} - h_1)}{60} \text{ kW}$$

(d) Heat Rejected in Condenser

Heat rejection process is 2–3 and is at constant pressure.

$$\therefore \quad \text{Heat rejected/kg} = h_2 - h_3 \text{ kJ/kg}$$

$$\text{or} \quad \text{Heat rejected/min} = m_r(h_2 - h_3) \text{ kJ/min}$$

$$\text{or} \quad \text{Heat rejected/sec} = \frac{m_r(h_2 - h_3)}{60} \text{ kW}$$

16.4.5 Actual Vapour Compression Refrigeration Cycle

In the preceding sections we have discussed the theoretical vapour compression cycle. The cycle as applied in practice, however, differs considerably from the theoretical cycle.

An actual vapour compression cycle is shown in Fig. 16.25 on *T–S* diagram. A few of the important points of deviations of this cycle as compared to the ideal theoretical cycle are explained below.

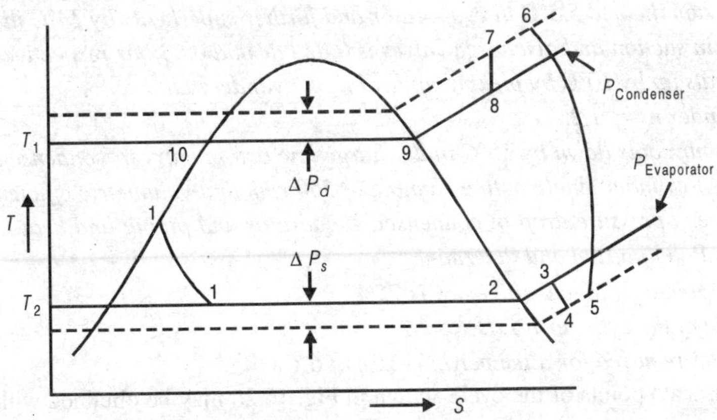

Fig. 16.25 *Actual vapour compression cycle.*

(i) Process 1–2–3. This process represents the passage of refrigerant through the evaporator coil at $P_{evaporator}$ and T_2 represents the entry of vapour to compressor in superheated condition. Superheating upto this point may be either due to larger removal of heat from the evaporator; (this is an advantage) or due to heat picked up by the vapour in suction piping which is a disadvantage.

(ii) Process 3–4–5–6–7–8. This process represents the passage of vapour through the compressor, and shows several departures from the theoretical process.

(a) Both the suction and discharge valves of the compressor are actuated by pressure difference. Hence the actual suction pressure p_s in the cylinder is lower than the evaporator pressure (p_{evap}). Similarly the actual discharge pressure (p_d) will be higher than the condenser pressure (p_{cond}). There is thus *throttling effect* in both the suction and discharge valves. Pressure 3–4 and 7–8 represent this throttling effect through Δp_s and Δp_d respectively.

(b) As soon as the cold vapour enters the cylinder it gets heated by coming in contact with the hot cylinder walls. Hence the temperature of the vapour increases. This heating is represented by process 4–5. Similarly there is a cooling effect at the discharge which is represented by process 6–7.

(c) States 5 and 6, therefore, represent the initial and final condition of vapour during actual compression. The compression may neither be isentropic nor polytropic. It may be assumed that the heat absorbed by the gas during the first part of compression is equal to the heat rejected during the latter part. The actual compression process is as shown by process 5–6 in Fig. 16.25.

(iii) Process 8–9–10–11. This process represents the passage of refrigerant through the condenser at pressure (p_{Cond}). Process 8–9 is actually the removal of superheat, 9–10 is the removal of latent heat and 10–11 represents the subcooling of refrigerant liquid.

(iv) Process 11. This process represents the throttling of subcooled liquid from state 11 so as to result in state 1 at the entry to the evaporator, after expansion.

The following example illustrates the method of considering most of the departures of actual cycle from the theoretical cycle discussed above.

Illustrative Example

An ammonia compression machine is required to cater for a load of 100 tons. The cooling water temperature requires the condenser to work at 35°C and the brine temperature requires evaporator to work at –30°C. The other data is given below :

(i) Temperature at entry to expansion valve 30°C.

(ii) The vapour is superheated 5.5°C in evaporator and further superheats by 14°C in suction line.

(iii) Pressure drop in suction and discharge valves is 0.02 bar and 0.03 bar respectively.

(iv) The vapour heats up by 10°C by picking up heat from cylinder walls.

(v) Compression index n = 1.2.

(vi) Discharge vapour cools down by 55°C in discharge line before entry to condenser.

(vii) Compressor is 2-cylinder single acting running at 560 rpm and volumetric efficiency is 0.57.

Ignore the losses e.g. pressure drop in condenser, evaporator and piping and heat transfer in liquid line etc. Draw the cycle on a P–h diagram and determine :

(a) B.P. of the compressor if Mech. efficiency is 75%

(b) Bore and stroke of the cylinder if L/D = 1.25

(c) Condenser water required for a temperature rise of 6°C.

Solution The various state points of the cycle shown in Fig. 16.26 may be obtained with the help of P-h chart as given below :

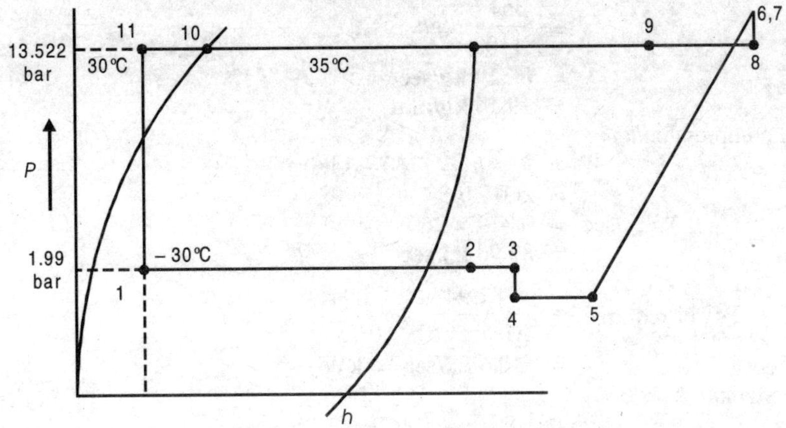

Fig. 16.26

$$P_1 = P_2 = P_3 = 1.199 \text{ bar}$$
$$t_2 = -30 + 5.5 = -24.5°C, \quad h_2 = 1440 \text{ kJ/kg}$$
$$t_3 = -24.5 + 14 = -10.5°C, \quad h_3 = 1456 \text{ kJ/kg}$$
$$v_3 = 1.04 \text{ m}^3\text{/kg}$$
and
$$P_4 = P_5 = 1.199 - 0.02 = 1.179 \text{ bar}, \quad h_3 = h_4 = 1456 \text{ kJ/kg}$$
$$t_5 = -10.5 + 10 = 0.5°C, \quad h_5 = 1480 \text{ kJ/kg}$$
$$V_5 = 1.3 \text{ m}^3\text{/kg}$$
$$P_8 = P_9 = P_{10} = P_{11} = 13.522 \text{ bar}$$
$$P_6 = 13.522 + 0.03 = 13.552 \text{ bar}$$

For compression, $P_5 v_5^{1.2} = P_6 v_6^{1.2}$

$$\therefore \quad \left(\frac{V_5}{V_6}\right)^{1.2} = \frac{P_6}{P_5} = \frac{13.552}{1.179} = 11.494$$

$$\therefore \quad \frac{V_5}{V_6} = (11.494)^{\frac{1}{1.2}} = (11.494)^{0.8333} = 7.651$$

$$\therefore \quad V_6 = \frac{1.3}{7.651} = 0.1699 \text{ m}^3\text{/kg}$$

and
$$h_6 = h_8 = 1740 \text{ kJ/kg}$$
$$t_8 = 138°C$$
$$\therefore \quad t_9 = 138 - 55 = 83°C$$
$$\therefore \quad h_9 = 1630 \text{ kJ/kg}$$
$$h_{11} = 338 \text{ kJ/kg}, \ t_{11} = 30°C$$
$$= h_1$$

Now we can find the required quantities.

(a) The mass of NH_3 in circulation :

$$m_r = \text{Mass of refrigerant kg/sec}$$
$$= \frac{\text{TR} \times 3.516}{h_2 - h_1} = \frac{3.516 \times 100}{1440 - 338}$$

$$= \frac{351.6}{1102} \text{ kg/s}$$
$$= \mathbf{0.319 \text{ kg/sec}}$$
$$= \mathbf{19.14 \text{ kg/min}}$$

Work of compression/kg

$$W = h_6 - h_5 = 1740 - 1480$$
$$= 260 \text{ kJ/kg}$$

$\therefore$ Work/sec $= 0.319 \times 260$
$$= \mathbf{82.94 \text{ kJ/sec}}$$

$\therefore$ Power required $= \dfrac{82.94}{\text{Mech. eff.}} = \dfrac{82.94}{0.75}$
$$= \mathbf{110.6 \text{ kJ/sec} = kW}$$

(b) Bore and Stroke :

Total displacement/min $= \dfrac{m_r \cdot V_3}{\eta_{\text{Vol.}}}$

$$= \frac{19.14 \times 1.04}{0.57}$$
$$= \mathbf{34.922 \text{ m}^3\text{/min}}$$
$$= 2 \times \frac{\pi d^2}{4} \times L \times N$$
$$= \frac{2 \times \pi \times d^2 \times 1.25d}{4} \times 560$$
$$= \mathbf{1100\, d^3}$$

$\therefore$ $d^3 = \dfrac{34.922}{1100} = 0.03174$

$\therefore$ $d = 0.3166 \text{ m} = \mathbf{31.66 \text{ cm}}$

$\therefore$ $L = 1.25 \times 31.66 = \mathbf{39.58 \text{ cm}}$

(c) Heat Rejected in Condenser and Water Required :

Heat rejected/min $= m_r (h_g - h_{11})$

$$= 19.14\,(1630 - 338)$$
$$= 19.14 \times 1292$$
$$= \mathbf{24729 \text{ kJ/min}}$$

$\therefore$ Water required/min $= \dfrac{24729}{6 \times 4.187}$

$$= \mathbf{984.35 \text{ kg/min}}$$

16.4.6 Under Cooling or Subcooling

Undercooling or subcooling of the liquid takes place when the liquid is cooled below the saturation temperature corresponding to condenser pressure, before admitted to the throttle valve or expansion valve. Undercooling of the liquid is generally along the constant pressure line. This is represented by line 3–3' both on *T–S* and *P–h* diagrams of Fig. 16.27. Many times this undercooling is shown wrongly along the liquid line of *T–S* diagram (3–3").

Undercooling is brought about by circulating greater quantity of cooling water through the condenser.

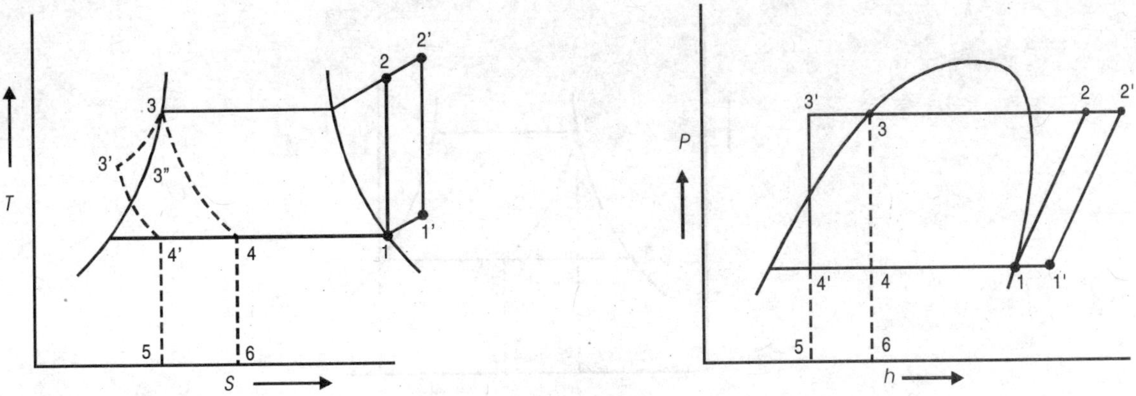

Fig. 16.27

Many times undercooling of liquid refrigerant coming out of the condenser is brought about by the vapours coming from the evaporator thus vapours are superheated to some extent. This arrangement is shown in Fig. 16.28.

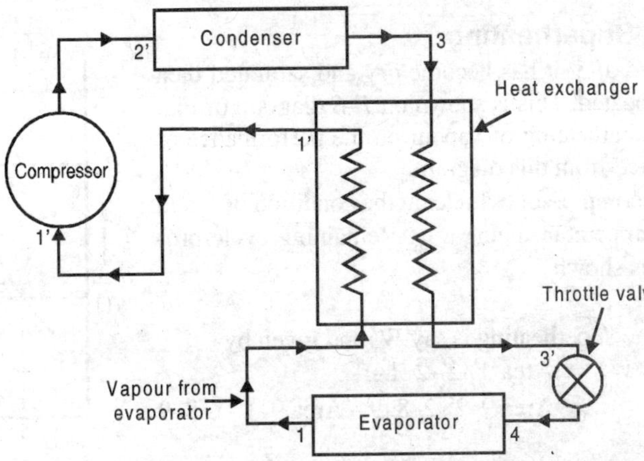

Fig. 16.28. *Schematic showing arrangement for undercooling liquid.*

16.4.7 Effect of Undercooling

The coefficient of performance COP may be increased by what is known as precooling or undercooling. It will be noticed from the diagrams of Fig. 16.27 that the net refrigerating effect has been increased by the area 4'–4–6–5'–4'. The work done in compressed is not increased in that proportion and hence the coefficient of performance is increased by undercooling the liquid.

16.4.8 Effect of an Expansion Cylinder

If, instead of expanding the high-pressure liquid at 3 (coming from the condenser) through an expansion valve, it is expanded in an expansion cylinder or expander, thus driving the piston, work may be obtained which could be utilized in helping to drive the compression cylinder or compressor. By this method less external power would be required to drive or run the refrigerator. The work saved by the expansion cylinder is represented by the area 3–3'–4'. The high pressure liquid at 3 is expanded isentropically to 4' and the wet vapour is then forced through the brine tank or refrigerator. The net refrigerating effect is now increased by the area 4'–4–5–6–4' and the work required to drive the compression is less and hence the coefficient of performance is increased to a large extent.

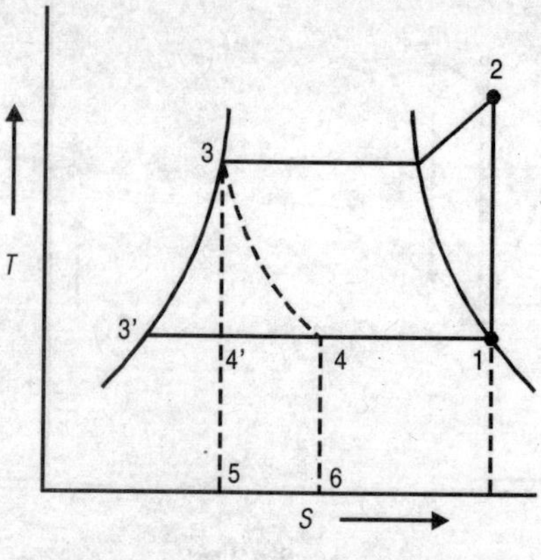

Fig. 16.29. *Effect of expansion cylinder.*

The expansion cylinder is not used in practice, as the work saved is insufficient to overcome the friction of the necessary mechanism.

16.4.9 The Effect of Superheating

If the vapour is compressed after it has become dry and saturated then the vapour becomes superheated. This is shown on *T–S* diagram of Fig. 16.30. The effect of this superheating of vapour on the performance of the cycle COP can be studied from this diagram.

The vapour is drawn in the compressor cylinder at the condition of 1' and is compressed to 2', the temperature being T_{sup}. Remaining cycle processes remain unchanged as shown.

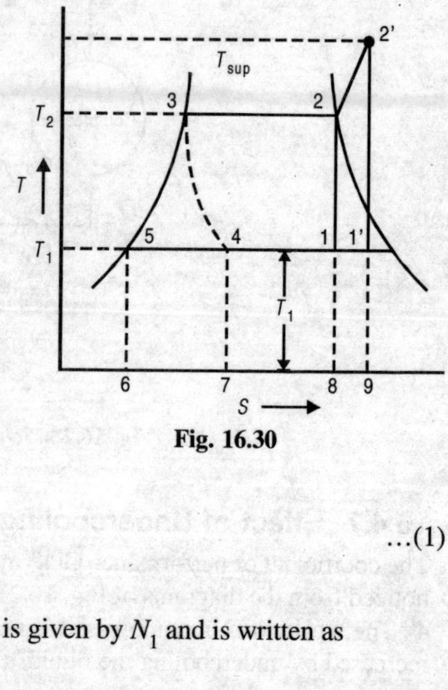

Fig. 16.30

Referring to Fig. 16.30,

Additional work done due to superheating is say W_1 and given by

$$W_1 = \text{Area } 1'\text{–}2'\text{–}2\text{–}1\text{–}1'$$

$$= \text{Area } 9\text{–}2'\text{–}2\text{–}8\text{–}9 - \text{Area } 9\text{–}1'\text{–}1\text{–}8\text{–}9$$

$$= \text{Superheat} - T_1 \times C_{pv} \log_e \frac{T_{sup}}{T_2}$$

$$= C_{PV}\left(T_{sup} - T_2\right) - T_1 \times C_{PV} \log_e \frac{T_{sup}}{T_2} \quad \ldots(1)$$

where C_{pv} is the specific heat of the superheated vapour.

Similarly, addition heat abstracted from brine due to superheating is given by N_1 and is written as

$$N_1 = \text{Addition refrigerating effect}$$

$$= \text{Area } 9\text{–}1'\text{–}1\text{–}8\text{–}9$$

$$= T_1 \times C_{pv} \log_e \frac{T_{sup}}{T_2} \quad \ldots(2)$$

Now, if W = Work without superheating

= Area 1–2–3–4–1

and N = Refrigerating effect

= Heat abstracted without superheating

then, COP = Coefficient of performance with superheating

$$= \frac{N + N_1}{W + W_1}$$

From the diagram it is noticed that the ratio of $\frac{W_1}{W}$ is much greater than the ratio of $\frac{N_1}{N}$. The effect of this that the coefficient of performace is lowered or decreased by superheating.

In practice it is not usual to superheat.

16.4.10 Coefficient of Performance when T_1, T_2, Specific Heat of Liquid and Latent Heat at Higher Temperature Given

Let Fig. 16.31 represents the T–S diagram of the cycle and in this case let the vapour have a dryness fraction x at the end of compression.

From the diagram of Fig. 16.31, we have

$$x = \frac{3\text{–}2}{3\text{–}11}$$

and $\quad$ 3–11 = Change in entropy during evaporation

$$= \frac{h_{fg_2}}{T_2}$$

$$\therefore \quad 3\text{–}2 = \frac{x \cdot h_{fg_2}}{T_2}$$

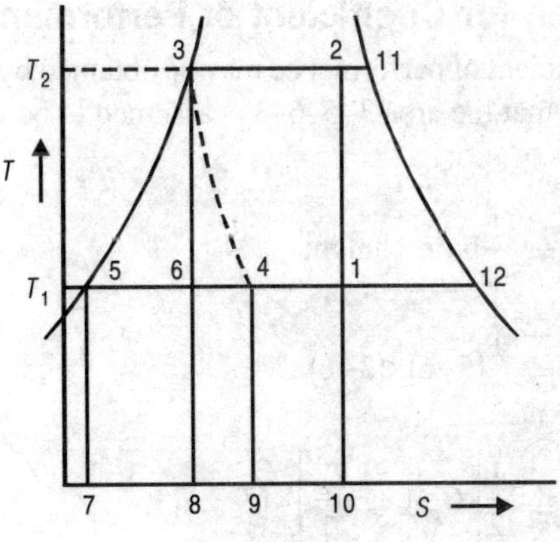

Fig. 16.31

Also, $\quad$ area 3–5–6–3 = Liquid heat at T_2 above that of T_1.

$$= C_{pi}\,(T_2 - T_1)$$

Similarly, $\quad$ 5–6 = Entropy change during liquid heating

$$= C_{p_i} \log_e \frac{T_2}{T_1}$$

Also, as 3–4 is an isenthalpic process,

$$\text{Area } 3\text{–}5\text{–}6\text{–}3 = \text{Area } 4\text{–}6\text{–}8\text{–}9\text{–}4$$

From these equation we get

$$C_{p_l}(T_2 - T_1) - C_{p_l}\log_e \frac{T_2}{T_1} = (6-4)\times T_1$$

from which (6–4) can be calculated.

Then for the performance of coefficient,

$$\begin{aligned} W &= \text{Work done} \\ &= \text{Area } 1\text{–}2\text{–}3\text{–}4\text{–}1 \\ &= \text{Area } 3\text{–}5\text{–}6\text{–}3 + \text{Area } 1\text{–}2\text{–}3\text{–}6\text{–}1 \\ &= T_1\times(6-4) + x\cdot h_{fg_2}\times(T_2 - T_1) \end{aligned}$$

and

$$\begin{aligned} N &= \text{Refrigerating effect} \\ &= \text{Area } 1\text{–}4\text{–}9\text{–}10\text{–}1 \\ &= \{(2-3)-(6-4)\}\times T_1 \\ &= \left\{\frac{x\cdot h_{fg_2}}{T_2} - (6-4)\right\}\times T_1 \end{aligned}$$

and

$$\text{COP} = \frac{N}{W} \text{ can be calculated.}$$

16.4.11 Approximate Solution for Coefficient of Performance

An approximate solution for the coefficient of performance may be obtained by assuming the liquid line (3–5) to be a straight line Fig. 16.30. This means that the area 3–5–6–3 is assumed to be a triangle.

Then triangle 3–5–6–3

$$\begin{aligned} &= \frac{1}{2}\text{base}\times\text{height} \\ &= \frac{1}{2}(5-6)\times(3-6) \\ &= \frac{1}{2}\left[C_{p_l}\log_e\frac{T_2}{T_1}\right]\times(T_2 - T_1) \end{aligned}$$

$$\begin{aligned} \text{Also triangle } 3\text{–}5\text{–}6\text{–}3 &= \text{rectangle } 4\text{–}6\text{–}8\text{–}9 \\ &= T_1\times(4-6) \end{aligned}$$

From these two equations,

$$(4-6)\times T_1 = \frac{1}{2}\left[C_{p_l}\log_e\frac{T_2}{T_1}\right]\times(T_2 - T_1)$$

$$\therefore \qquad (4\text{–}6) = \frac{T_2 - T_1}{T_1} \times \frac{1}{2} C_{p_i} \log_e \frac{T_2}{T_1}$$

By using this value of (4–6), we can calculate work done *W* and refrigerating effect *N* by using the value of (4–6). Note here that for CO_2 as a refrigerant liquid line 3–5 cannot be regarded as a straight line.

16.4.12 Limitations of Air Refrigeration System

After noting down the working of Bell–Coleman air refrigeration systems, we observe certain limitations in using air-refrigeration systems. These limitations may be enlisted as follows.

1. The power required to get unit of refrigeration is comparatively large and hence the performance is very poor. It is observed that approximately 4 kW are required to produce 1 TR. For vapour compression refrigeration system, power requirement is approximately 1 kW only for every TR capacity.
   Therefore, the air refrigeration system is to be used when other systems can not be used.
2. When the air refrigeration plant is to be stationary, *then atmosphere should not be dusty*. In spite of the dusty atmosphere, if air refrigeration system is to be used, then dust-separators must be provided, otherwise wear and tear of the components will result.
3. Moisture contained in the air is a big problem as moisture condenses and even freezes choking the valve and creating problems for the running of this air system.
4. To have the required capacity of the air refrigerating system, the quantity of air required is very large and the volume of air to be handled is very large as compared to vapour compression system. In this case the size and weight of the compressor is very large and the compressor is heavy and bulky.

Thus when the air refrigerating system is absolutely necessary, then and then only this system is used.

16.5 VAPOUR ABSORPTION REFRIGERATION SYSTEM

We now come to the discussion of the vapour absorption refrigeration system. The power utilized in vapour compression system is high grade energy i.e., *electric power* for running the compressor motor. In absorption refrigeration system, however, heat is directly utilized as source of energy. Of course, it may be preferable to utilize heat as such as it avoids undergoing through the various energy transformations required in the generation of electrical energy from heat energy.

It may be recalled that in the compression system the vapour was compressed by undergoing a great change in volume during the compression process. Accordingly the major part of the power was consumed in the process. If means were available for rising this pressure of the refrigerant without appreciably altering its volume, the work requirements will be enormously reduced (by about 95% or so). This may possibly be done by dissolving the refrigerant in some absorbent and supplying the heat to the solution for compression purposes. The absorption cycle achieves this objective by placing the refrigerant in solution before the so called compression process and by removing from the solution immediately after the process. The absorption of the vapour is governed by **Raoult's law.**

The basic difference between vapour compression and vapour absorption cycles will thus be to replace the compressor of the vapour compression cycle by a set of equipment which fulfils the objective discussed above. The other important element i.e condenser, expansion device and evaporator will exist in both systems.

16.5.1 Simple Vapour Absorption Cycle

Figure 16.32 illustrates the simplest scheme of equipment required for the replacement of the compressor.

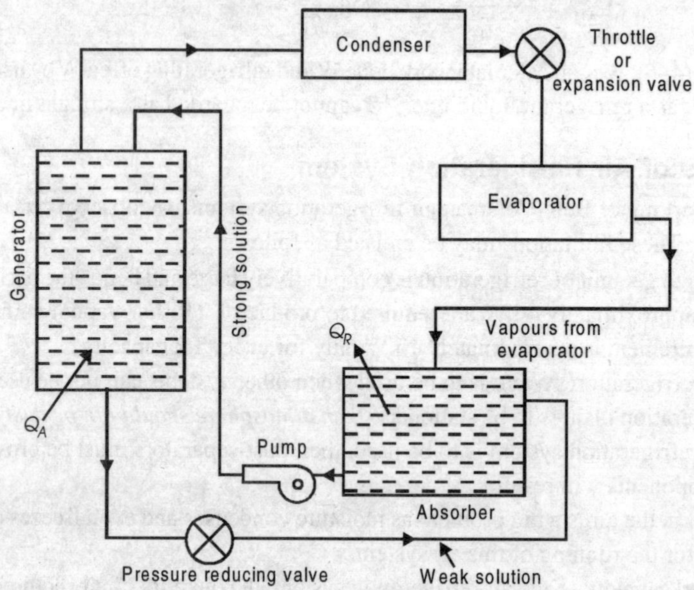

Fig. 16.31. *Equipment replacing the compressor (Simple vapour absorption system of refrigeration).*

Q_A = Heat added or supplied
Q_R = Heat rejected to coolant.

Figure 16.32 equipment replacing the compressor (Simple vapour absorption system of refrigeration).

The most commonly used fluids in the absorption system are ***water*** as absorbent and ***ammonia*** as refrigerant. The vapour from the evaporator is allowed to be mixed and *absorbed* in the absorber. The heat of absorption generated in the process is rejected from the absorber to the circulating cold water in a heat exchanger dipped in the solution contained in the absorber. The strong acqua-ammonia solution from the absorber is pumped upto the condenser pressure and fed to the generator which is the main energy consuming element of the system. Heat is supplied to the generator. The boiling point of refrigerant NH_3, is lower than that of the absorbing liquid H_2O, hence the vapours leaving the generator are predominantly those of refrigerant. These vapours then pass on to the condenser. The liquid refrigerant from the condenser, then, passes through an expansion valve or throttle valve to the evaporator where it absorbs heat from the substances or bodies to be refrigerated. Liquid refrigerant is then evaporated and the vapours enter the absorber completing the cycle.

The weak acqua-ammonia solution in the generator left due to separation of refrigerant vapour is drained back to the absorber for repeating the cycle.

The weak acqua-ammonia solution leaving the generator is at high pressure and the pressure in the absorber is the evaporator pressure which is less than the generator or condenser pressure, and hence a pressure reducing valve is provided in the weak solution line to the absorber.

The energy requirements of the system are at the generator and at the pump as compared to those at compressor in the vapour compression system. Since the volume of liquid handled by the pump is too small, the power required here is almost negligible as compared to that by the generator.

16.5.2 Practical Absorption Refrigration Cycle

The replacement of the compressor by the simple arrangement of Fig. 16.33 is not very economical in practice. In order to make improvements certain additional auxiliary items are provided in the system. They include, ***analyzer***, a ***rectifier***, and two ***heat exchangers***. The practical absorption cycles as developed after incorporating these auxiliaries is shown in Fig. 16.33.

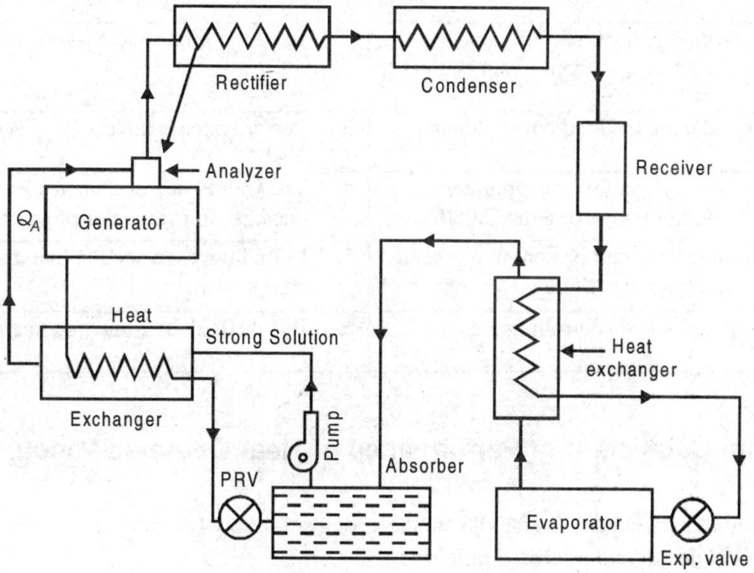

Fig. 16.33. *Practical absorption system.*

(a) Analyzer. The ammonia vapours leaving the generator may contain certain moisture, and therefore it should be freed from any trace of water vapour before passing on to the condenser and then to the expansion valve, otherwise the water vapour is likely to freeze in the small valve passage and choke the flow.

The function of the analyzer is to remove the moisture as far as possible. It is an open types of cooler and forms an integral part of the generator, mounted on its top. Both the strong acqua-ammonia solution from the absorber and the condensate removed in rectifier are introduced from the top and flow downwards. The hot rising vapour of ammonia therefore comes in contact with the same and gets cooled. Thus most of the water vapour is condensed and drips back into the generator. This helps in salvaging a certain portion of heat in outgoing vapour which would otherwise have been rejected out through the condenser.

(b) Rectifier. It is a closed type of cooler and is actually a miniature condenser where any traces of water vapour left in the ammonia vapour, are *removed* by condensation .The cooling is achieved by circulating water as is done in an ordinary condenser .The condensed acqua is drained back to the generator through the analyzer.

(c) Heat Exchangers. Two heat exchangers are provided to internally exchange heat from the higher temperature fluid to the lower temperature fluid so that one is cooled and the other is heated.

One heat exchanger is provided between liquid receiver and evaporator so that the liquid is *subcooled* and vapour is heated up. Another heat exchanger is located between generator and absorber so that the strong acqua is *heated up* before going on to the analyzer and weak acqua is cooled before entering the absorber.

16.5.3 Comparison between Vapour Compression and Absorption System

Some of the important points of difference between the vapour absorption are listed below.

Absorption	*Compression System*
1. Moving parts are only in the pump, which is a small element of the system. Hence quiet in operation.	1. Moving parts are in the compressor. More wear and tear and noisy operation.
2. Use low grade energy heat therefore may be worked on exhaust system from steam engines etc.	2. Use high grade energy–mechanical work.
3. No effect of reducing the load on performance.	3. Performance is adversely affected at partial loads.
4. The system can work on lower evaporator pressure also, without, lowering the C.O.P.	4. The C.O.P. decreases considerably with decrease in evaporator pressure.
5. Liquid traces of refrigerant present in piping at the exit of evaporator constitute no danger.	5. Liquid traces in suction line may damage the compressor.
6. Automatic operation for controlling the capacity.	6. It is difficult as compared to absorption unit.

16.5.4 Maximum Coefficient of Performance of Heat Operated Vapour Absorption System

We know that refrigerant vapours are liberated from the strong solution when heated in a generator. This type of the machine or system is called a ***heat operated machine***.

Let T_h be the temperature at which heat is supplied to the strong solution.

We will think of a system shown in Fig. 16.34.

T_k is the temperature at which heat is rejected to atmosphere; T_o is the temperature of the body to be refrigerated.

Thermal efficiency of the engine is given by

$$\eta_{th} = \frac{\text{Work}}{\text{Heat supplied}}$$

$$= \frac{\text{Work}}{Q_h}$$

$$\text{Work of the engine} = \eta_{th} \times Q_h$$

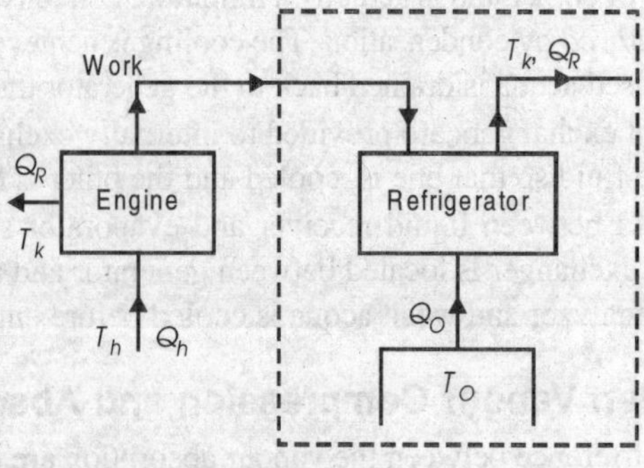

Fig. 16.34

This work, it is assumed, is used for refrigeration system.

For refrigeration system, T_o is the temperature at which heat is given to refrigerant (actually this is the heat abstracted from the body to be refrigerated) and the heat is rejected the atmosphere at T_k temperature.

To get the maximum COP, we have to consider the Carnot reversed engine.

$$(\text{COP})_{\text{carnot}} = \frac{T_o}{T_k - T_o} = \frac{Q_o}{Q_R - Q_o}$$

Similarly, we consider the engine as a Carnot engine and

$$(\eta_{\text{th}})_{\max} = \frac{T_h - T_k}{T_h} = \frac{\text{Work}}{Q_h}$$

$$Q_o = (\text{COP})_{\text{carnot}} \times \text{Work}$$

and
$$Q_h = \frac{\text{Work}}{(\eta_{\text{th}})_{\text{carnot}}}$$

∴ COP of the cycle

$$= \frac{Q_o}{Q_h} = \frac{\text{Work} \times \eta_{\text{carnot}}}{\text{Work}/\eta_{\text{th}}}$$

$$= \eta_{\text{th}} \times \eta_{\text{carnot}}$$

This is maximum when η_{th} is maximum which is the Carnot thermal efficiency

$$(\textbf{COP})_{\max} = (\eta)_{\text{carnot}} \times (\textbf{COP})_{\text{carnot}}$$

Note here that this maximum C.O.P. is not to be compared with COP of vapour compression system which is $\frac{Q_o}{W}$.

∴
$$(\textbf{COP})_{\max} = \frac{T_h - T_k}{T_h} \times \frac{T_o}{T_k - T_o}$$

In this case

T_h = Temperature of the source should be as high as possible (source temperature)

T_k = Sink temperature which should be as possible.

and T_o = Refrigerator temperature should be as possible.

16.5.5 Single-Effect and Double-Effect Cycles

In vapour absorption refrigeration system, the temperature of the heat source play an important role. The heat given by the source may be used in one single stage or in more than one stage. Accordingly the system is called as *single stage* or *single effect cycle* of operation and if the heat is supplied in two stages then the system is called *two stage* or *double effect cycle* of operation.

It is found that when the temperature of the heat source is upto 105°C, the heat is utilized in a single stage and the cycle is a *single effect cycle.* When the temperature of heat source is greater than 105°C, then generally the heat is utilized in two stages and the system becomes a *double-effect cycle.*

1. **Single Effect Cycle.** A simple and practical absorption system using ammonia as refrigerant and water as absorbent described in the previous articles is an example of single-effect cycle system for vapour absorption refrigeration system.

2. **Double Effect Cycle.** The schematic sketch of double-effect system is shown in Fig. 16.35. Vapours of refrigerants are generated in two stages as shown. For this purpose two heat exchangers are used.

Fig. 16.35 for double-effect system is self explanatory.

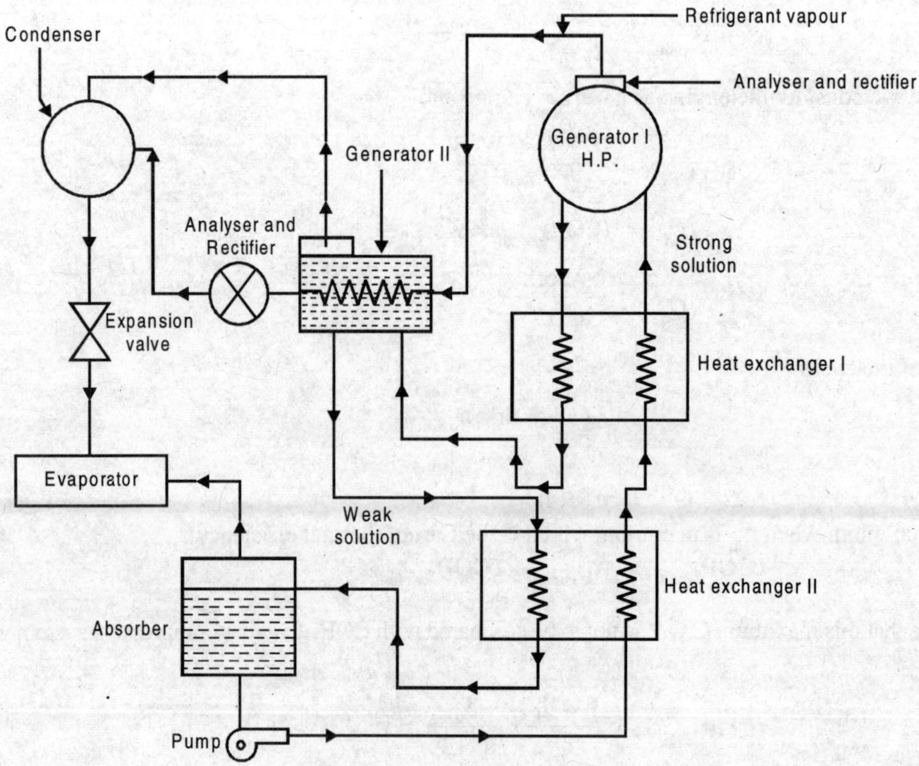

Fig. 16.35. *Double effect cycle.*

16.5.6 Electrolux Refrigrator (The Ammonia – Hydrogen Refrigerator)

This type of refrigerator takes advantage of the fact that the liquid ammonia evaporators very readily in hydrogen. For this refrigerator, no pump is required as the circulation is maintained by gravity and the heat is supplied by a small gas jet or electric heater.

Figure 16.36 shows a diagrammatic view of Electrolux refrigerating system. This system consist of an absorber *A* which contains a strong solution of ammonia dissolved in distilled water. When a gas burner *B* is lighted the circulation of the system commences because of the warming of ammonia solution in the boiler or heater *D*. The strong ammonia solution now flows from *A* through the intercharger *C*, where it is heated by the hot work solution returning from *D*. It then passes around the heater *D*, which causes the ammonia vapour to be driven out of the solution. Now the released ammonia vapour passes through a rectifier, after which it enters the condenser *E*. The object of the rectifier is to drain off any particles of water coming with the ammonia vapours. The hot weak solution left behind *D* drains back into the absorber *A*. In doing so it passes through the intercharger *C*, where it gives its heat to the strong solution flowing from *A*.

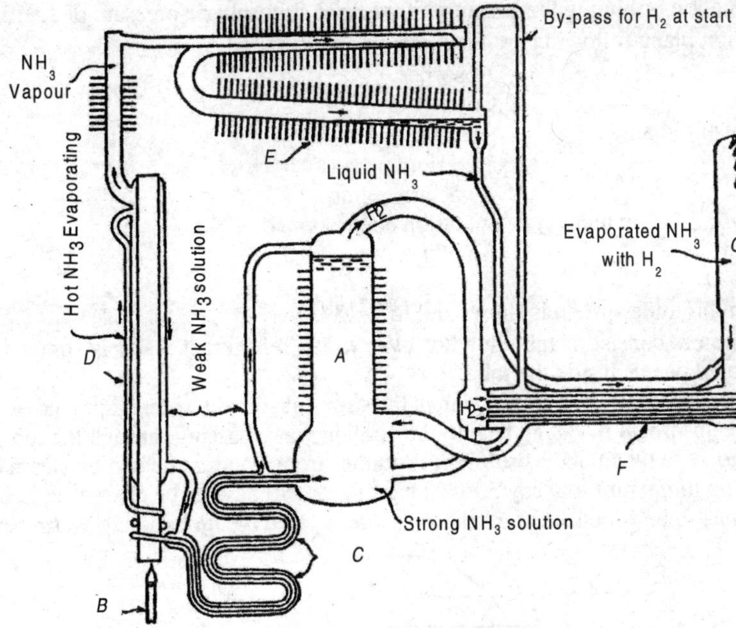

Fig. 16.36. *Electrolux refrigerator.*

After condensation in the condenser, ammonia vapours forms the liquid and this ammonia liquid flows by gravity into the evaporator *G*. This is full of hydrogen into which the liquid ammonia evaporates. Liquid ammonia absorbs latent heat producing intense cold around *G*, which is situated in the food cabinet or any space which is to be refrigerated and thus the cabinet or space is cooled and the objects place in the cabinet or space are cooled. The mixture of hydrogen and ammonia vapour now flows through the gas heat exchanger *F*, where it cools the fresh supply of hydrogen flowing into *G*. From *F* the NH_3 ammonia vapour and hydrogen gas flows into the absorber *A*, where the hydrogen rises to the top and flows back to the evaporator *G*. Thus, the cycle is complete.

The main advantage of this Electrolux refrigerator is that there is no moving parts like compressor, pump or fan and therefore there is no noise and no machinery to give mechanical troubles.

The coefficient of performance of this refrigerator is given by the equation

$$COP = \frac{\text{Heat absorbed by evaporator } G}{\text{Heat supplied by } B}$$

Total pressure in the system is about 15 bar. H_2 gas is present in the evaporator at about 11–12 bar pressure. The quantity of H_2 is so proportioned that the practical pressure of NH_3 within evaporator will be only a smaller portion of the total pressure for the pressure of NH_3 vapours in the evaporator will only 4–3 bar. This low pressure causes NH_3 to boil or evaporator at low temperatures. Here the Dalton's law of partial pressure is applied.

Other substances that can be used in vapour absorption system are

(i) Water as the refrigerant and lithium bromide salt solution as absorbant.

(ii) Water are refrigerant and lithium chloride brine as the absorbant.

(iii) Methylene chloride as refrigerant and dimethyl ether of tatraethylene glycol as absorbant.

16.5.7 Lithium Bromide-water Absorption Refrigeration System

We have seen that the situation or boiling temperature of water at atmospheric pressure of 1.01325 bar is 100°C.

As usual for a absorption plant, following are the main components

1. Evaporator
2. Absorber
3. Generator
4. Condenser

Additional Components used are

1. A heat exchanger
2. Two fluid pump
3. A purge unit
4. A vacuum pump
5. An automatic decrystallization unit
6. A slution control valve
7. A steam or hot water valve
8. An educator and
9. A control centre

A typical water-lithium bromide system is shown in Fig. 16.37.

1. **Evaporator.** Evaporators consist of tube bundles over which refrigerant water is sprayed and evaporated. The liquid to be cooled passes inside the tube.
2. **Absorbers.** Absorbers are tube bundles over which the strong absorbent water is sprayed. Refrigerant vapour is condenser into the absorbent releasing heat to the cooling water passing through the tubes.
3. **Generator.** Generators are tube bundles submerged in the strong solution heated by steam or hot water. Thus to have a boiling temperature of say 30°C, the pressure will have to be decreased to nearly 0–042 bar. Therefore,vacuum has to be produced if all the pressure is to be mentioned below atmospheric pressure.

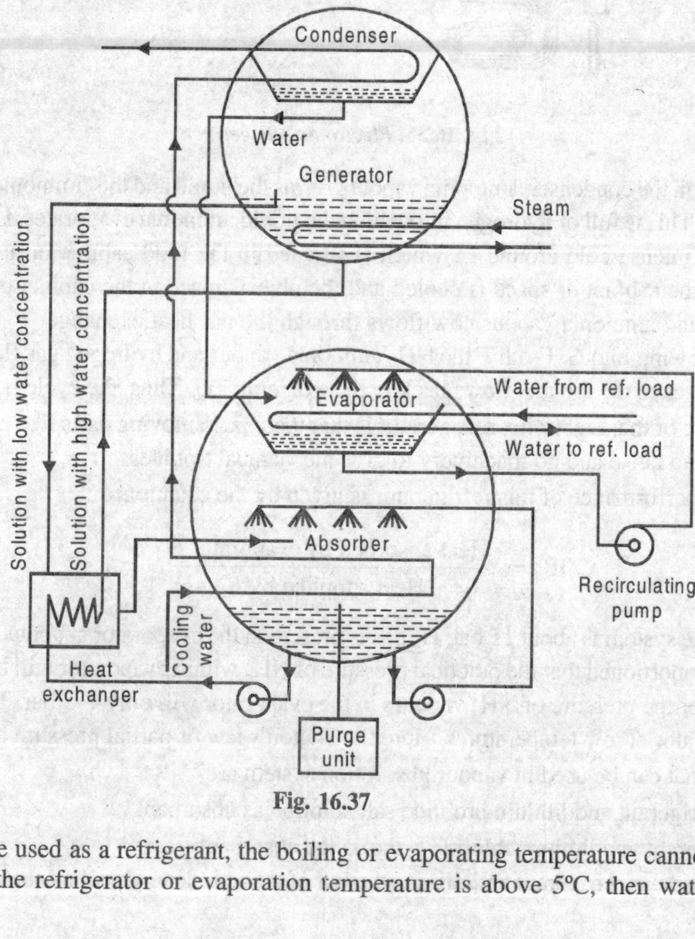

Fig. 16.37

When water is to be used as a refrigerant, the boiling or evaporating temperature cannot be reduced below 5°C. Therefore, if the refrigerator or evaporation temperature is above 5°C, then water can be used as a refrigerant. Then the absorbent for water must have a great affinity for water. The substances which can be used as absorbent are common salt (sodium chloride), calcium chloride, lithium bromide etc.

According to the tests made on these different salts, *lithium bromide* was found to possess the best overall characteristics required for an absorbent to be used in large capacity absorption chillers.

Lithium bromide is available in crystals and these crystals are dissolved in water and the mixture is called *Lithium Bromide Solution* which is used in the absorption chiller. The amounts of lithium bromide and water in a solution are measured by weights and not by volume 40% concentration means that 40 % of the total weight is lithium bromide.

4. **Condensers.** Condensers are tube bundles located over which vapours are present coming from the generators and the cooling water flows through the tubes. This cooling water first passes through the absorber.
5. **Heat Exchangers.** These are of steel shelf and tube type and tube type heat exchangers.
6. **Purge Unit.** This is used to remove non-condensable gases. These gases are present in small quantities and will rise the total pressure in the absorber sufficiently and the evaporator pressure changes thus changing the evaporating temperature. A very small pressure rise can cause appreciable change in the evaporating temperatures.
7. **Expansion Device.** Normal mechanical expansion valves are not used in absorption units. The flow of refrigerant liquid to evaporator is controlled by an orifice or other fixed restriction between the condenser and the evaporator (just like capillary tube in small Household refrigerator).
8. **Pumps.** Pumps which are used for recirculation are shown in Fig. 16.37.

16.6 REFRIGERANTS

A refrigerant is a working substance which is used in refrigerating machine. It is a working fluid through which heat is transferred from the cold body to the hot body. Formerly the number of useful refrigerants was very limited. Air, ammonia (NH_3), carbon dioxide (CO_2), sulphur dioxide (SO_2) were the only four refrigerants which has survived the adolescent stage. Somewhat later methyl chloride (CH_3Cl) came in general use for domestic and commercial units but since the development of the fluorinated hydrocarbons, a great variety of refrigerants has been put at the disposal of the refrigerating engineer. They are largely used in modern plants. The most common Freon refrigerants in use are

Freon-11 ($CHCl_3$), Freon-12 (CF_2Cl_2)
Freon-13 (CF_3Cl), Freon-21 ($CHFCl_2$)
Freon-22 (CHF_2Cl), Freon-113 ($C_2F_3Cl_3$)
and Freon-114 ($C_2F_4Cl_2$)

Curiously enough the use of air as a refrigerant is both older and more recent than the other vapour refrigerants.

16.6.1 Desirable Properties of Refrigrant

Desirable refrigerant possess such chemical, physical and thermodynamic properties which permit their practical application in refrigerating machines in an efficient way. No refrigerant may be ideal for all jobs. The best suited refrigerant may be selected on the basis on its characteristics and the type of service under consideration. The best suited refrigerant may be selected on the basis of its characteristics and the type of service under consideration. The referigerator selected for service should possess as many qualities as practicable out of the following:

1. The condensing pressure should not be excessive.
2. The boiling temperature should be low.
3. The critical temperature should be high.
4. The latent heat of vaporisation should be high.
5. The specific heat of liquid should be low.
6. The specific volume of vapour should be low.
7. It should be non-corrosive to metals and inert to oils.
8. The refrigerant should be chemically stable.

9. The refrigerant should be non-inflammable and non-explosive.
10. It should be non-toxic and harmless to human beings or living cells.
11. It should be easily detectable for leaks.
12. It should be easily available and have low cost.
13. It should have satisfactory heat transfer and viscosity coefficient.

16.6.2 Classification of Refrigerants

As refrigeration is a process of heat removal, for this purpose a medium for heat transfer is necessary. Such a medium of heat transfer, which absorbs heat by evaporating at low temperature and gives up heat by condensing at high temperature and pressure is termed as *refrigerant.* In the broadest sense the term refrigerant is also applied to the cooling medium like cold water, brine solution and antifreezes. Thus the refrigerant may be classified as ***Primary*** refrigerants and ***Secondary*** refrigerants.

Primary refrigerants as commonly interpreted include only those working medium which actually pass through the cycle of evaporation, compression, condensation and liquefaction e.g., NH_3, CO_2, SO_2, Freon group etc.

Secondary refrigerants include those working medium which transfer heat from a remote point to the evaporator of the refrigeration system. Cooling medium such as ice and solid CO_2 are therefore not primary refrigerants but secondary refrigerants. Desirable properties of secondary refrigerants mainly are its specific gravity, freezing point, thermal conductivity, viscosity and specific heat.

Advantages of the secondary refrigerants over the direct cooling system are,

1. Easy to handle and easy to control its flow.
2. Temperatures can be controlled by controlling the flow.
3. A toxic refrigerant, such as NH_3, will be away from the refrigeration load.
4. Lengths of the primary refrigerants pipe or tubing can be eliminated thereby pressure drop and leakages of the primary refrigerants are eliminated to a large extent.

Nowadays number of refrigerants are commercially available and these are classified into several groups as under:

I. Halocarbon Compounds. This group is a chemical compound formed from the base methane CH_4 and ethane C_2H_6 and include refrigerants which contain one or more of the three halogens i.e., Chlorine, Fluorine and Bromide. These refrigerants are:

F-11 Trichloromonefuoro methane CCl_3F
F-12 Dichlorodifluoro methane CCl_2F_2
F-22 Monochlorodifuloro methane $CHClF_2$
F-114 Dichlorotetrafluoro ethane $CClF_2CClF_2$

II. Azeotropes. This group consist of mixtures of refrigerants which do not separate into their compounds with pressure or temperature changes and have fixed thermodynamic properties. e.g., F-500(F-12-.73.8%+ F-152, 26.2%).

III. Hydrocarbons. This group contain organic compounds, and many hydrocarbons are used as refrigerant in industrial installations. They are highly flammable but possess satisfactory thermodynamic properties.

Examples are: CH_4 (methane), C_2H_6 (ethane) etc.

IV. Inorganic Compound. These refrigerants are,

711 Ammonia NH_3
718 Water H_2O
729 Air
744 Carbon dioxide CO_2
764 Sulphur dioxides SO_2

V. Oxygen Compounds. This group contains oxygen element e.g., ethyl-ether $C_2H_5OC_2H_6$

VI. Nitrogens Compounds. This group contains nitrogen element e.g., methyl-amine CH_3NH_2.

VII. Unsaturated Organic Compounds. This group is mainly hydrocarbon group of base ethylene C_2H_4 and propylene C_3H_6. There are thus number of refrigerants commercially available and are sold under different trade names.

16.6.3 Properties of Important Refrigerants

Although a refrigerant has not been found yet which has all the desirable properties, Freon-12 carbon dioxide and ammonia are commonly used for very low temperature work because of their low boiling points. But for the units used in household refrigerants, there is a wide range of available refrigerants each of which in comparison with others has certain desirable qualities.

The properties of some of the common refrigerants in use are given below:

1. Ammonia (NH_3). This primary refrigerant has long been used as refrigerant in large ice-making and cold storage plants. At atmospheric pressure it boils at –33°C and is therefore suitable for low temperature work. It is much less expensive than some other refrigerants. It is well suited for one or two stage reciprocating compressors. It is regarded as the most useful refrigerants because of its high latent heat, moderate working pressures, high critical temperature and small sensible heat compared with the latent heat. It requires low volume of gas to be circulated per ton of refrigeration. Its heat transfer properties are excellent. It is very soluble in water forming ammonium hydroxide. When heat is applied to ammonia hydroxide, ammonia is released in vapour form. It is used in operating on absorption system of refrigeration. It is not miscible with oil. Ammonia is almost never used in comfort cooling for public places.

It is toxic and has extremely offensive and pungent (irritaing) odour. Leaks can be easily detected on account of its pungent odour. It attacks many non-ferrous metals like copper, zinc, brass etc. in presence of humidity. It has no corrosive effect on iron and steel. The exact location of leak is shown by the dense white fumes produced by burning sulphur taper or by use of test papers.

The density of ammonia is smaller than that of air and lubricating oil.

2. Sulphur-dioxide (SO_2). It is one of the refrigerants commonly used in household refrigerators because of the very low working pressure. Its boiling point at atmospheric pressure is –10°C. It has a fairly large latent heat and high critical temperature. It is non-inflammable and non-explosive. It is highly toxic. SO_2 when used in a hermatically sealed compressor within domestic refrigerating units, does not present a serious problem even though it is toxic. It has a very pungent and suffocating odour. Leakage is easily detected on account of its pungent odour and the white smoke which forms when strong ammonia water (28% solution) is brought in the presence of gas.

If is very corrosive when in contact with moisture and therefore extreme care should be taken to prevent even the slightest bit of moisture entering to system to avoid the formation of sulphurous (H_2SO_3) acid which is corrosive in effect.

3. Carbon Dioxide (CO_2). Formerly this refrigerants was prepared for marine work and in dairies. The heat extracted per unit volume is the highest known for many refrigerants, but the high working pressure (75 ata), the low critical temperature (31°C) the low COP and the excessive power required per ton of refrigeration has restricted its use. Now its use as a refrigerant in the proper sense of the word is on the decline but it is increasingly used as dry ice. It is used for reciprocating compressor only. It is non-toxic but will suffocate in high concentration. Leaks of CO_2 gas are difficult to locate for the reason that the gas is colourless and odourless. It is non-inflammable. BP at atm-pressure = – 78.5°C. It has no corrosive effect on copper, copper alloys, iron or steel. It has no effect on oil and greases. If is miscible with oil. Due to high working pressure, the compressors, condenser and pipe connecting them must be designed for greater strength.

Dry ice is very well suited for cooling of railway refrigerator vans, lorries etc. as well as for ice-cream during distribution and sales etc. It is readily available and its cost is low.

4. Methyl Chloride (CH_3Cl). Methyl chloride is suitable for small refrigerators and domestic units. It has low working pressure, fairly high latent heat and high critical temperature. Its boiling point is –23.7°C at atmospheric pressure. It is inflammable but does not been reality. It is colourless and has sweet odour. Its odour resembles that of chloroform and hence leaks can be easily detected. It is somewhat toxic. Large concentration of vapour in air may produce unconciousness or even death but when well diluted with air it has little effect on human body.

In the presence of moisture it is very corrosive to zinc magnetism and aluminium and it produces solvent action on rubber It does affect copper, iron or steel-methyl chloride dissolves practically all types of oil and glycerine.

When the fluorinated hydrocarbons gained foothold as refrigerants methyl chloride was abandoned in most countries and in some the use is now restricted by the authorities.

5. Fluorinated Hydrocarbons. These refrigerants are halogen derivaties of saturated hydrocarbon i.e., of methane, ethane etc. Fluorine and chlorine are substituted for one or more of the hydrogen atoms and investigations have shown that an increasing number of fluorine atoms reduces health hazard and flammability and increases stability.

The different types in use cover a large range of normal boiling points and molecular weights. Thus some are not suitable for reciprocating and some for centrifugal compressors and some are well suited for low temperature work and some for heat pumps.

They are all more or less miscible with oils. They are all safe as long as they do not come into contact with red hot surfaces or open fire. In these cases they decompose forming poisonous products, viz. hydrochloric acid (HCl), Hydrofluric acid (HF) and phosgene (CCL_2O). They are, however, inflammable.

Like most other organic compounds their heat transfer properties are not so good as those of water and ammonia.

In order to avoid confusion an international agreement has been proposed to use letter F (Freon).

F-11 (Monofluotrichloro Methane). This refrigerant is low pressure type, well suitable for centrifugal compressors. High molecular weight. Reduces the number of stages for a given pressure ratio. F-11 is the least stable refrigerant of Freon group.

F-12 (Difluorodichloro Methane). The development of the fluorinated refrigerants with the manufacture of F-12 which rapidly gained a widespread application in household units and commercial plants. BP at atm =29.8°C.

It is well suited for reciprocating and rotary compressors. However, at normal temperatures the swept volume necessary for a certain output is about 70% larger than when ammonia is used.

F-12 attacks magnesium and zinc, but copper and its alloys can be safely used so that pipe lines are easily laid. F-12 attacks natural rubber.

The discharge pressure are lower for F-12 than for ammonia at the same temperatures. Mass of F-12 is much more for a given output than ammonia plants.

F-21 (Monofluorodichloro Methane). This is a low pressure refrigerant. In certain countries it has been applied in household refrigerators. With rotory compressors and in absorption units. It has been proposed as a refrigerant in plants working with centrifugal compressors.

The properties of F-21 are similar to those of other fluorinated refrigerants.

F-22 (Difluromonochloro Methane).

BP at atm. pressure = – 40.77°C
Critical temperature = – 96.2°C
Corrosive action = Nil

Swept volume necessary is very much less than that for F-12.

F-22 has been increasingly used in recent years as a substitute for F-12. The range of application goes from small commercial units to medium size industrial units. It is mostly used for reciprocating compressors. With regard to miscibility with oil F-22 presents the peculiarity that is fully miscible with oil at the temperatures in the high side of the plant, whereas it is only partially miscible at usual evaporating temperatures. For this an effective oil separator should always be used.

Water (H_2O). Water freezes at 0° C at atmospheric pressure and therefore the applications requiring operating temperatures above this point, water has been used as a working fluid. The specific volume of water vapour at atmospheric temperatures is very great and hence reciprocating compressors can not be used. But centrifugal compressors are used with success.

16.6.4 Refrigerant Drying Agents

To remove moisture from the refrigerating systems the refrigerant drying agents are frequently used. These agents commonly used are

1. Silica gel (silicon dioxide)
2. Activated alumina (Aluminium oxide)
3. Anhydrous calcium sulphate

First two are available in granular form and are of absorption type. The last is available in the granular form and cast sticks and is of absorption type.

16.6.5 CFCs and Non-CFCs

From the properties of the halocarbon compounds containing one or more of the three halogens viz. Chlorine, Fluorine and Bromine, reduces health hazards and flammability. It is also observed that the stability increases as the number of fluorine atoms increases. That way these refrigerants are more or less ideal refrigerants with few exceptional qualities. The referigerants which do not contain hydrogen atoms, containing only carbon, fluorine and chlorine are known as CFCs (carbon-fluorine compounds) and are very much detrimental for the ozone layer round the earth.

If these compounds contain hydrogen atoms in addition to carbon, fluorine and chlorine, they are called as HCFs. Thus we have CFCs as CFCs— (1) F-12, (2) F-11 and many others.

Similarly HFCs F-22, 9($CHCF_2$) F-123 ($CHCl_2$ CF_3) and many others.

Refrigerants which do not contain any halogens i.e., chlorine, fluorine and bromine are termed as **non-CFCs.**

16.7 APPLICATIONS OF REFRIGERATION

Mechanical refrigeration serves human beings throughout their life. It has found tremendous applications in manufacture of ice, preservation of food stuff and in air conditioning. An analysis of the various applications of refrigeration indicates that they may be classified into three general groups:

1. The industrial process.
2. Preservation of perishable goods.
3. Providing comfortable environments.

Mechanical refrigeration provides aid in oil refining and rubber manufacturing, in processing food stuff and beverages and in treatment of metals or concrete for dams. It has helped in saving large quantities of food from being spoiled, by the development of frozen foods industry where the perishable food can be stored at low temperatures for a considerable amount of time without loosing its nutritive value or flavour. Further it has made the tropical regions more comfortable to live in by summer air conditioning of buildings. There are just a few examples wherein the science of refrigeration has helped us.

Specifically speaking the various applications of refrigeration may be listed as follows:

I. (a) Processing of foods products.
(b) Processing of farm crops.
(c) Processing of textiles, printing work, photographic materials etc.

(d) Cooling of concrete for dams.
(e) Treatment of air for blast furnaces
(f) Processing of tobacco, petroleum and other chemical products.

II. (a) Manufacture of ice.
(b) Freezing or chilling, storage and transportation of food stuff including beverages, meat, poultry products, dairy products, fish, fruits, vegetables, fruit juices etc.

III. (a) Industrial air conditioning
(b) Comfort air conditioning of hospitals, residences, hotels, restaurant, theaters, offices etc.

SOLVED EXAMPLES (BELL–COLEMAN CYCLE)

Example 16.1 *In a Bell–Coleman cycle, air is drawn into the cylinder of a compressor at a pressure of 1 bar and –5°C. It is compressed adiabatically to 5 bar at which pressure it is cooled to 15°C. It is then expanded in the expansion cylinder to atmospheric pressure and discharged into the refrigerator chamber. Law of expansion is $PV^{1.2}$ = Constant.*

Find the work done on the air and the COP of the refrigeration plant and also the capacity of the plant in TR for a mass of flow of 10 kg/min.

Solution

The cycle for this problem is given in Fig. Ex. 16.1

For diagram shown we have,

$$P_1 = 1 \text{ bar},\ T_1 = 273 - 5 = 268 \text{ K},\ P_4 = 1 \text{ bar}$$
$$P_2 = 5 \text{ bar},\ T_3 = 273 + 15 = 288 \text{ K},\ P_3 = 5 \text{ bar}$$

We require to calculate all temperatures at the salient points.

We have

$$\frac{T_2}{T_1} = \left(\frac{P_2}{P_1}\right)^{\frac{\gamma-1}{\gamma}} = (5)^{\frac{0.4}{1.4}}$$
$$= 5^{0.286}$$
$$= \mathbf{1.5845}$$
$$T_2 = 268 \times 1.5845$$
$$= \mathbf{424.65\ K}$$

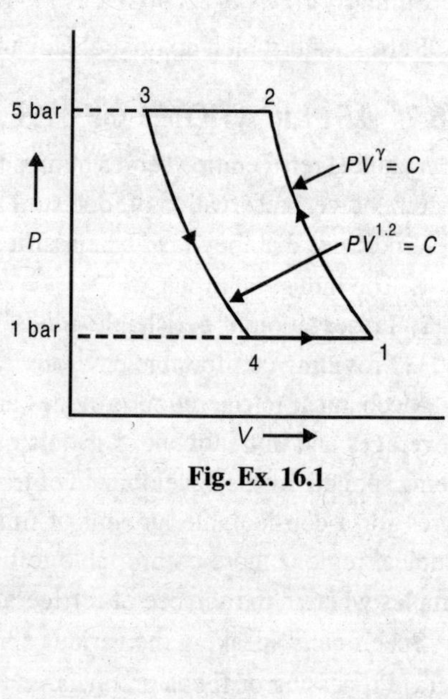

Fig. Ex. 16.1

Similarly,

$$\frac{T_3}{T_4} = \left(\frac{P_3}{P_4}\right)^{\frac{n-1}{n}} = 5^{\frac{1.2-1}{1.2}}$$
$$= (5)^{\frac{1}{6}} = \mathbf{1.3084}$$
$$T_4 = \frac{T_3}{1.3084} = \frac{288}{1.3084}$$
$$= \mathbf{220.12\ K}$$

$\therefore$ Work done/kg = Area of *P–V* diagram

$$= \frac{\gamma}{\gamma-1}[P_2V_2 - P_1V_1] - \frac{n}{n-1}[P_3V_3 - P_4V_4]$$

$$= \frac{1.4}{0.4} \cdot R[T_2 - T_1] - \frac{1.2}{0.2} \cdot R[T_3 - T_4]$$

$$= 3.5 \times 0.287\,(424.65 - 268) - 6 \times 0.287\,(288 - 220.12)$$

$$= 157.355 - 116.89$$

$$\mathbf{W = 40.465 \text{ kJ/kg}}$$

$$\text{Refrigerating effect/kg} = C_p\,(T_1 - T_4)$$

$$N = 1.005\,(268 - 220.12)$$

$$= \mathbf{48.12 \text{ kJ/kg}}$$

$$\therefore \quad \text{COP} = \frac{N}{W}$$

$$= \frac{48.12}{40.465}$$

$$= \mathbf{1.19}$$

Total refrigerating effect/sec

= (Mass flow of air/sec) × (Refrigerating effect/kg)

$$= \frac{10}{60} \times 48.12$$

= **8.02 kJ/s**

or = **8.02 kW**

∴ Capacity of the plant in TR

$$= \frac{\text{Total refrigerating effect/sec}}{3.516}$$

$$= \frac{8.02}{3.516}$$

= **2.281 TR**

Example 16.2 *In a refrigerator working on Bell–Coleman cycle, air is drawn into the compressor from the cold chamber at a pressure of 0.95 bar and temperature of -5°C. After compression (isentropic) to 5 bar, air is cooled at constant pressure to temperature of 15°C. The polytropic expansion is according to $PV^{1.3} = C$ and air is expanded to 1 bar to the chamber. Calculate (a) COP, (b) TR if $m_a = 1.5$ kg/min.*

Solution

The process taking place in a refrigerator are shown on *P–V* diagram of Fig. Ex. 16.2.

We have $P_1 = 0.95$ bar, $P_4 = 1$ bar

$$T_1 = -5 + 273 = 268 \text{ K}$$

$$T_3 = 15 + 273 = 288 \text{ K}$$

$$P_2 = P_3 = 5 \text{ bar}$$

We have

$$\frac{T_2}{T_1} = \left(\frac{P_2}{P_1}\right)^{\frac{\gamma-1}{\gamma}}$$

$$= \left(\frac{5}{0.95}\right)^{0.286}$$

$$= \mathbf{1.608}$$

$$\therefore \quad T_2 = 1.608 \times 268 = \mathbf{431\ K}$$

Similarly,

$$\frac{T_3}{T_4} = \left(\frac{P_3}{P_4}\right)^{\frac{n-1}{n}} = \left(\frac{5}{1}\right)^{\frac{1.3-1}{1.3}}$$

$$= (5)^{\frac{0.3}{1.3}} = \mathbf{1.45}$$

$$\therefore \quad T_4 = \frac{T_3}{1.45} = \frac{288}{1.45}$$

$$= \mathbf{198.62\ K}$$

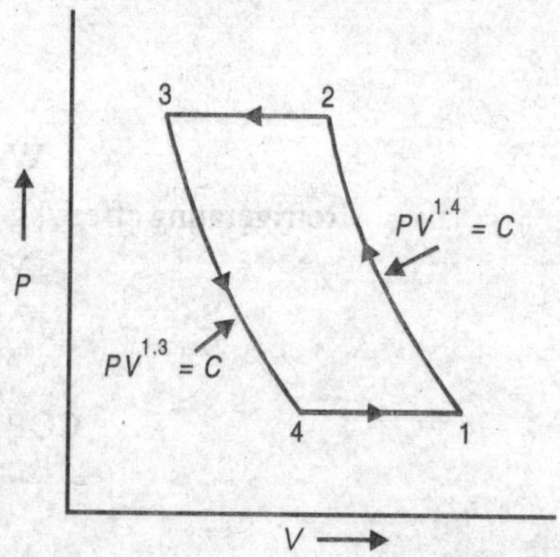

Fig. Ex. 16.2

(a) COP

Refrigerating effect/kg

$$N = C_p (T_1 - T_4)$$

$$= 1.005\ (268 - 198.48)$$

$$= \mathbf{69.87\ kJ/kg}$$

$$\text{Work done/kg} = C_{pa}(T_2 - T_1) - \frac{n}{n-1} \cdot R(T_3 - T_4)$$

$$W = 1.005 \times (431 - 268) - \frac{1.3 \times 0.287}{0.3} \times (288 - 198.62)$$

$$= 1.005 \times 163 - 1.213 \times 89.38$$

$$= 163.81 - 108.45$$

$$= 55.36 \text{ kJ/kg}$$

$$\therefore \quad \text{COP} = \frac{N}{W} = \frac{69.87}{55.36}$$

$$= \mathbf{1.2621}$$

(b) Capacity in TR

$$\text{Refrigerating effect/sec} = \frac{\text{Mass of air}}{\text{sec}} \times \frac{\text{Refrigerating effect}}{\text{kg}}$$

$$= \frac{1.5}{60} \times 69.87$$

$$= \mathbf{1.7467\ kJ/sec}$$

$$= \frac{1.7467}{3.516} \text{ TR}$$

$$= \mathbf{0.497}$$

Example 16.3 *A refrigerator working on Bell–Coleman cycle operates between the pressure limts of 1.05 bar and 8.5 bar. Air drawn from the cold chamber at 10°C is compressed and then it is cooled to 30°C before entering the expansion cylinder. Expansion and compression follows the law $PV^{1.35} = C$. Determine*

(i) Theoretical COP

(ii) Mass flow rate of air require l in kg/min for a refrigerating capacity of 10 TR if the actual COP is 0.5 of the theoretical.

(iii) Power required to run the plant.

Solution

P–V diagram for the example data is shown in Fig. Ex. 16.3.

For calculating the temperatures of the salient points we have,

$$\frac{T_2}{T_1} = \left(\frac{P_2}{P_1}\right)^{\frac{n-1}{n}} = \left(\frac{P_3}{P_4}\right)^{\frac{n-1}{n}} = \frac{T_3}{T_4}$$

$$\therefore \quad \frac{T_2}{T_1} = \left(\frac{8.5}{1.05}\right)^{\frac{0.35}{1.35}} = \mathbf{1.7199}$$

$$\therefore \quad T_2 = 283 \times 1.7199 = \mathbf{486.73\ K}$$

$$T_4 = \frac{T_3}{1.7199} = \frac{303}{1.7199} = \mathbf{176.2\ K}$$

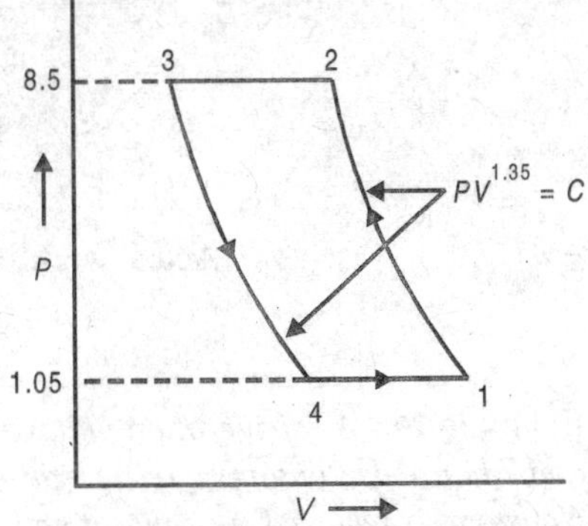

Fig. Ex. 16.3

Now we calculate the required facts.

(i) Theoretical COP

$$\text{Work done/kg} = \text{Area of } P\text{–}V \text{ diagram}$$

$$= \frac{n}{n-1}(P_2V_2 - P_1V_1) - \frac{n}{n-1}(P_3V_3 - P_4V_4)$$

$$= \frac{1.35}{0.35} \cdot R(T_2 - T_1) - \frac{1.35}{0.35} \cdot R(T_3 - T_4)$$

$$= \frac{1.35}{0.35} \times 0.287[486.73 - 283 - 303 + 107.334]$$

$$= \mathbf{85.16\ kJ/kg}$$

Refrigerating effect/kg

$$N = C_{pa}(T_1 - T_4) = 1.005(283 - 176.2)$$

$$= \mathbf{107.334\ kJ/kg}$$

$$\therefore \quad \text{COP} = \frac{N}{W} = \frac{107.334}{85.16}$$

$$= \mathbf{1.26}$$

(ii) Air Flow Rate

$$\text{Actual refrigerating effect} = 10\ \text{TR}$$

$$= 10 \times 3.516$$

$$= \mathbf{35.16\ kW} = \text{kJ/sec}$$

$$\text{Refrigerating effect/kg} = 107.334\ \text{kJ/kg}$$

$$\therefore \quad \text{Actual mass flow rate} = \frac{35.16}{107.334}\ \text{kg/sec}$$

$\therefore$ Required mass flow rate in kg/min

$$= \frac{35.16 \times 60}{107.334}$$

$$= \mathbf{19.65\ kg/min}$$

(iii) Power Required

$$\text{Actual COP} = 0.5 \times 1.26 = 0.63$$

$$= \frac{\text{Actual refrigerating effect}}{\text{Actual work}}$$

$$= \frac{35.16}{\text{Actual work}}$$

$$\therefore \quad \text{Actual work} = \frac{35.16}{0.63}$$

$$= 55.81 \text{ kJ/sec}$$

$$= \mathbf{55.81 \text{ kW}}$$

Example 16.4 *A refrigeration unit working on Bell–Coleman cycle operates between the pressure limits of 1 bar and 7 bar and maintains a refrigeration temperature of –13°C. Heat from the air compressed according to $PV^{1.3}$ = Constant, is removed such that it gets cooled to 37°C before expansion. Index of expansion is 1.35. Calculate:*

(i) Theoretical COP

(ii) Mass flow rate of air required to manufacture of ice, at 0°C from water at 30°C, at the rate of 5 tonnes per day.

(iii) Power required to run the plant.

Take C_{pa} = 1 kJ/kg-K

γ = 1.4

Latent heat of fusion of ice at 0°C = 336 kJ/kg

Specific heat of water = 4.187 kJ/kg-K

Solution

The $P–V$ diagram is shown in Fig. Ex. 16.4. We have,

$$T_1 = -13 + 273 = \mathbf{260 \text{ K}}$$

$$T_3 = 307 + 273 = \mathbf{310 \text{ K}}$$

$$P_1 = P_4 = 1 \text{ bar}$$

$$P_2 = P_3 = 7 \text{ bar}$$

We calculate all temperatures

$$\frac{T_2}{T_1} = \left(\frac{P_2}{P_1}\right)^{\frac{n-1}{n}} = \left(\frac{7}{1}\right)^{\frac{1.3-1}{1.3}} = 1.567$$

$$\therefore \quad T_2 = 360 \times 1.567$$

$$= \mathbf{407.42 \text{ K}}$$

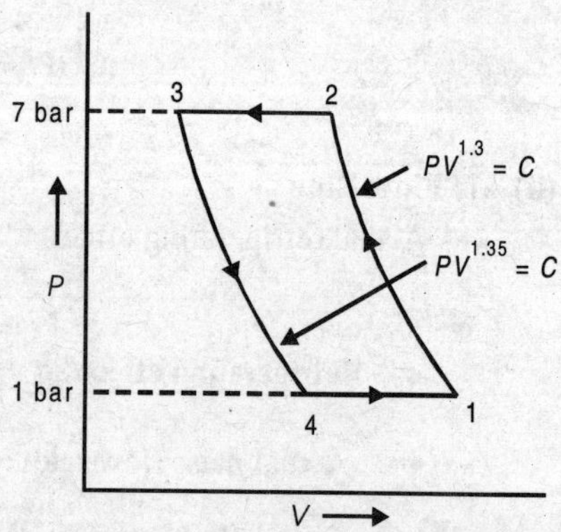

Fig. Ex. 16.4

Similarly,

$$\frac{T_3}{T_4} = \left(\frac{P_3}{P_4}\right)^{\frac{n-1}{n}} = \left(\frac{7}{1}\right)^{\frac{1.35-1}{1.35}}$$

$$= \mathbf{1.6563}$$

$$\therefore \quad T_4 = \frac{311}{1.6563}$$

$$= \mathbf{187.76 \text{ K}}$$

(i) Theoretical COP

$$\text{COP} = \frac{\text{Refrigerating effect}}{\text{Work supplied}}$$

$$\text{Refrigerating effect/kg} = C_{pa}(T_1 - T_4)$$

$$N = 1 \times (260 - 187.76)$$

$$= \mathbf{72.24\ kJ/kg}$$

$$\text{Work done/kg} = W$$

$$= \frac{1.3}{0.3} \cdot R[T_2 - T_1] - \frac{1.35}{0.35} \cdot R[T_3 - T_4]$$

$$= \frac{1.3 \times 0.287}{0.3}[407.42 - 260] - \frac{1.35 \times 0.287}{0.35}[310 - 187.76]$$

$$= 183.34 - 136.43$$

$$= \mathbf{46.91\ kJ/kg}$$

$$\therefore \quad \text{COP} = \frac{72.24}{46.91}$$

$$= \mathbf{1.54}$$

(ii) Air flow Rate

Actual refrigerating effect/min

= Actual heat extracted from water to produce ice at 0°C/min

= Sensible heat + Latent heat

= $m_w C_{pw} (t_w - 0) + m_w \times$ Latent heat

$$= \frac{5 \times 1000}{24 \times 60}\{4.187 \times 30 + 336\}$$

$$= \frac{5000}{1440} \times 461.61$$

$$= \mathbf{1598.75\ kJ/min}$$

∴ Mass flow rate of air

$$= \frac{1598.75}{72.24}$$

$$= \mathbf{22.131\ kg/min}$$

(iii) Power Required

$$\text{Work done per kg} = 46.91\ \text{kJ/kg}$$

$$\text{Mass flow rate/sec} = \frac{22.131}{60}\ \text{kg/sec}$$

$$\therefore \quad \text{Work done/sec} = 46.91 \times \frac{22.131}{60}\ \text{kJ/sec}$$

$$\therefore \quad \text{Power required} = \mathbf{17.3\ kW}$$

Example 16.5 *A refrigerating machine of 8 TR capacity works on ideal Bell–Coleman cycle. The pressure of air entering the compressor is 1 bar and corresponding temperature is 20°C. The highest pressure reached in the cycle is 6 bar. The air after compression gets cooled to 40°C before entering the expansion cylinder. Assuming expansion and compression to be adiabatic with* $\gamma = 1.4$, *calculate*

(i) *COP of the cycle*

(ii) *Mass flow rate of refrigerant*

(*iii*) *Power required*
(*iv*) *Piston displacement of compressor*
(*v*) *Bore and stroke of compressor cylinder if compressor used is double acting and runs at 300 rpm (Given: Bore = Stroke)*

Solution

The cycle processes are shown on *P–V* diagram of Fig. Ex. 16.5.

Given :

$$P_1 = P_4 = 1 \text{ bar}$$
$$P_2 = P_3 = 6 \text{ bar}$$
$$T_3 = 40 + 273 = 311 \text{ K}$$
$$T_1 = 20 + 273 = 293 \text{ K}$$
$$\text{Capacity} = 8 \text{ TR}$$

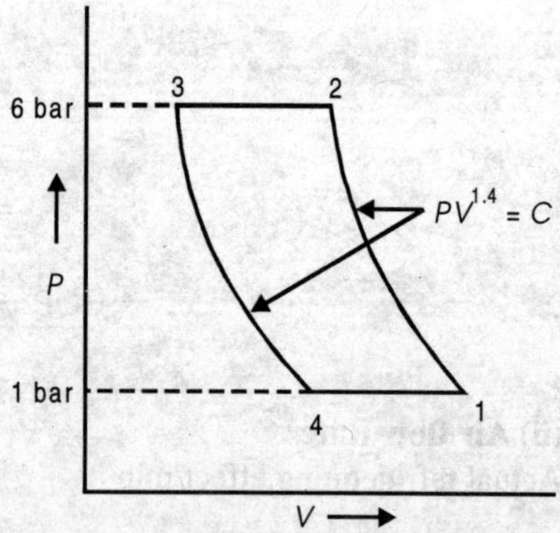

Fig. Ex. 16.5

For temperatures T_2 and T_4, we have

$$\frac{T_2}{T_1} = \frac{T_3}{T_4} = \left(\frac{P_2}{P_1}\right)^{\frac{\gamma-1}{\gamma}} = (6)^{\frac{0.4}{1.4}}$$
$$= \mathbf{1.6694}$$

$\therefore$ $T_2 = 293 \times 1.6694 = \mathbf{489.12\ K}$

and $T_4 = \dfrac{311}{1.6694} = \mathbf{187.5\ K}$

(i) COP of Cycle

$$\text{Work done/kg} = Q_{2\text{–}3} - Q_{1\text{–}4}$$
$$= C_{pa}(T_2 - T_3) - C_{pa}(T_1 - T_4)$$
$$= 1.005\ (489.12 - 311 - 293 + 187.5)$$
$$= \mathbf{70.95\ kJ/kg}$$

Note: As compression and expansion are adiabatic, $W = Q_R - Q_A$.

$$\text{Refrigerating effect/kg} = C_{pa}(T_1 - T_4)$$
$$= 1.005\ (293 - 187.5)$$
$$= \mathbf{106.03\ kJ/kg}$$

$\therefore$ COP of the cycle is given by

$$\text{COP} = \frac{N}{W} = \frac{106.03}{70.95}$$
$$= \mathbf{1.494}$$

(ii) Mass Flow Rate of Refrigerant

$$\text{Capacity of machine} = 8 \text{ TR} = 8 \times 3.516$$
$$= 28.128 \text{ kW}$$
$$= \mathbf{28.128\ kJ/sec}$$

But refrigerating effect/kg = 106.03 kJ/kg

$\therefore$ Mass flow rate of refrigerant

$$= \frac{28.128}{106.03} = \mathbf{0.265\ kg/s}$$
$$= \mathbf{15.91\ kg/min}$$

(iii) Power Required

$$\text{Work done/sec} = \text{Mass flow rate/sec} \times \text{Work/kg}$$
$$= 0.265 \times 70.95$$
$$= 18.8 \text{ kJ/sec}$$
$$= \mathbf{18.8 \text{ kW}}$$

(iv) Piston Displacement of Compressor. Neglecting clearance – V_1 is the piston displacement/sec

$$= \text{Volume of air in flow/sec}$$

$$= \frac{mRT_1}{P_1}$$

$$= \frac{0.265 \times 287 \times 293}{1 \times 10^5}$$

$$= \mathbf{0.22284 \text{ m}^3\text{/sec}}$$

(v) Bore and Stroke of Compressor Cylinder. As comperssor is double acting,

$$V_1 = 2 \times \frac{\pi d^2}{4} \times L \times \frac{\text{RPM}}{60}$$

$$\therefore \quad 0.22284 = 2 \times \frac{\pi d^3}{4} \times \frac{300}{60}$$

$$\therefore \quad d^3 = 0.0284$$
$$d = L = 0.305 \text{ m}$$
$$= \mathbf{30.5 \text{ cm}}$$

Example 16.6 *A vapour compression refrigeration plant operates between evaporation and condensation temperatures of –10°C and 45°C respectively. The refrigerant is dry and saturated vapour at the entry to the compressor. It is discharged at 102°C from the compressor. The bore and stroke of the compressor are 80 mm each. It runs at 720 rpm with a volumetric efficiency of 80%. The liquid refrigerant enters the expansion valve at 35°C. Determine*

(1) COP

(2) Mass low rate of refrigerant

(3) Capacity of the plant in TR

Take specific heat of liquid refrigerant =1.62 kJ/kg-K. You may use the properties of refrigerant tabulated below.

Sat. temperature	V_g	h_f kJ/kg	h_g kJ/kg	S_f kJ/kg-K	S_g kJ/kg-K
–10°C	*0.233*	*45.4*	*460.7*	*0.183*	*1.762*
45°C	*0.046*	*133.0*	*488.6*	*0.485*	*1.587*

Solution

(i) COP

The cycle of operation is shown on *T–S* diagram in Fig. Ex. 16.6.

In Fig. Ex. 16.6, 1–2 is an isentropic process

$$\therefore \quad S_1 = S_2$$

$$\therefore \quad S_{g_1} = S_{g_2} + C_{p_v} \log_e \frac{T_2}{T_{sat2}}$$

$$\therefore \quad 1.762 = 1.587 + C_{p_v} \log_e \frac{102+273}{45+273}$$

$$\therefore \quad 1.762 - 1.587 = C_{p_v} \log_e \frac{375}{318} = \mathbf{0.175}$$

$$\therefore \quad C_{p_v} = \frac{0.175}{\log_e 1.1792}$$

$$= \mathbf{1.061\ kJ/kg\text{-}K}$$

Again, Work done/kg $= h_2 - h_1$

where $h_1 = 460.7$ kJ/kg

$$h_2 = h_{g_2} + C_{p_v}(T_2 - T_{sat2})$$

$$= 488.6 + 1.061\,(102 - 45)$$

$$= \mathbf{549.08\ kJ/kg}$$

$\therefore$ Work done/kg $= 549.08 - 460.7$

$$= \mathbf{88.38\ kJ/kg}$$

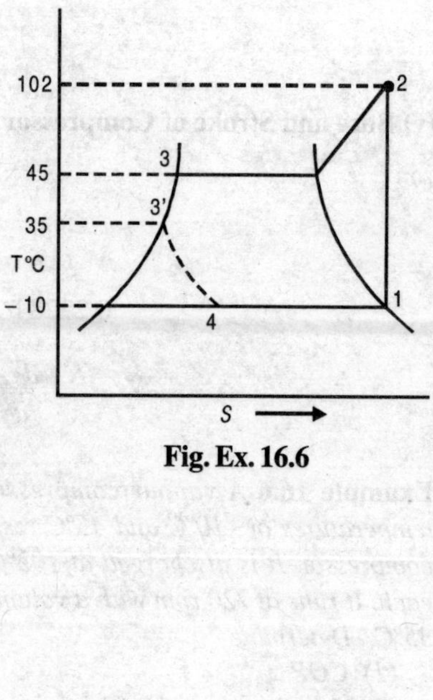

Fig. Ex. 16.6

Also, refrigerating effect/kg

$$N = h_1 - h_4 = h_1 - h_{3'}$$

where

$$h_{3'} = h_3 - C_{p_a}(45 - 35)$$

$$= 133 - 1.62 \times 10 = 133 - 16.2$$

$$= \mathbf{116.8\ kJ/kg}$$

$$\therefore \quad N = 460.7 - 116.8$$

$$= 343.9 \text{ kJ/kg}$$

$$\therefore \quad \text{COP} = \frac{N}{W} = \frac{343.9}{88.38}$$

$$= \mathbf{3.89}$$

(ii) Mass Flow Rate of Refrigerant

Theoretical volume taken in the compressor/min

= Displacement volume of compressor/min

$$= \frac{\pi d^2}{4} \times L \times N$$

$$= \frac{\pi}{4} \times \left(\frac{80}{1000}\right)^2 \times \frac{80}{1000} \times 720 = \mathbf{0.2895\ m^3/min}$$

$$= \frac{\pi}{4} \times \left(\frac{80}{1000}\right)^2 \times \frac{80}{1000} \times 720 = \mathbf{0.2895\ m^3/min}$$

$\therefore$ Actual volume of refrigerant vapour/min

= Theoretical volume × Volumetric efficiency

$$= 0.2895 \times 0.8$$

$= \mathbf{0.2316\ m^3/min}$

But specific volume of vapour $= \mathbf{0.233\ m^2/kg}$

$\therefore$ Mass of refrigerant/min $= \dfrac{0.2316}{0.233}$

$= \mathbf{0.994\ kg/min}$

(iii) Capacity of the Plant in TR

Refrigerant flow/sec $= \dfrac{0.9941}{60}$ kg/s

Refrigerating effect/kg $=$ 343.9 kJ/kg

$\therefore$Refrigerating effect/sec $= \dfrac{343.9 \times 0.9941}{60}$

$= \mathbf{5.698\ kJ/sec = kW}$

$\therefore$ TR capacity $= \dfrac{5.698}{3.576}$

$= \mathbf{1.62\ TR}$

Example 16.7 *28 tonnes of ice from and at 0°C is produced per day in an ammonia refrigeration plant. The temperature range in the compressor is from 25°C to –15°C. The refrigerant is dry and saturated at the end of compression. If actual COP is 60% of the theoretical COP, calculate the power supplied or required to drive the compressor. Assume latent heat of ice =335kJ/kg. Use properties of refrigerant given below:*

Temperature °C	h_f *kJ/kg*	h_g *kJ/kg*	S_f *kJ/kg-K*	S_g *KJ/kg-K*
+25	*+100.04*	*1319.22*	*+0.3473*	*4.4852*
–15	*–54.56*	*13.04.99*	*–2.1338*	*5.0585*

Solution

T–S diagram for the cycle is shown in Fig. Ex. 16.7

(i) Theoretical C.O.P.

$$\text{Theoretical C.O.P} = \frac{h_1 - h_4}{h_2 - h_1} = \frac{h_1 - h_3}{h_2 - h_1}$$

First we have to find all the enthalpies.

$h_2 =$ Enthalpy of vapour after compression

$= \mathbf{1319.22\ kJ/kg}$

3–4 is isenthalpic process

$\therefore \quad h_3 = h_4$

or $\quad h_{f_3} = h_4 = \mathbf{100.04\ kJ/kg}$

To find h_1,we will require to know the quality of vapour at compressor inlet. For this – $S_2 = S_1$

$$\therefore \quad 4.4852 = S_{f_1} + x_1 S_{f_{g1}}$$

$$= -2.1338 + x_1 \left(S_{g_1} - S_{f_1}\right)$$

$$= -2.1338 + x_1 (5.0585 + 2.1338)$$

$$\therefore \quad x_1 = \frac{4.4852 + 2.1338}{7.1923} = \frac{6.6190}{7.1923}$$

$$= \mathbf{0.9203}$$

$$\therefore \quad h_1 = h_{f_1} + x_1 h_{fg_1}$$

$$= -54.56 + 0.9203\,(1304.99 + 54.56)$$

$$= -54.56 + 0.9203 \times 1359.55$$

$$= \mathbf{1196.62 \ kJ/kg}$$

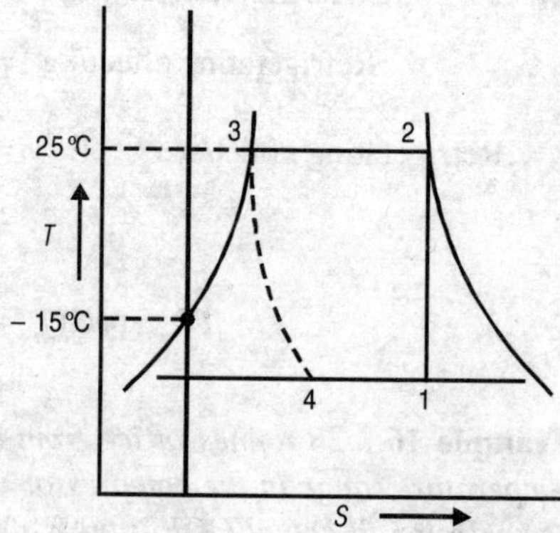

Fig. Ex. 16.7

∴ Theoretical COP

$$\frac{h_1 - h_3}{h_2 - h_1} = \frac{1196.62 - 100.04}{1319.22 - 1196.62}$$

$$= \mathbf{8.9444}$$

$$\therefore \quad \text{Actual COP} = 0.6 \times 8.9444$$

$$= \mathbf{5.3664}$$

Again actual refrigerating effect /sec

$$= m_w \left[C_{p_w} (t_w - 0) + \text{Latent heat} \right]$$

$$= \frac{28 \times 1000}{24 \times 3600} [0 + 335]$$

$$= \mathbf{108.565 \ kJ/kg}$$

$$\therefore \quad \text{Actual COP} = \frac{\text{Actual referigerating effect/sec}}{\text{Power required}}$$

$$= \frac{108.56\,(\text{kW})}{\text{Power}\,(\text{kW})}$$

$$\therefore \quad \text{Power (kW)} = \frac{108.565}{5.3664}$$

$$= \mathbf{20.23 \ kW}$$

Example 16.8 *An NH_3 refrigeration plant has a capacity of 20 TR. The condensation and evaporation temperatures are 35°C and –20°C respectively. Refrigerant is dry and saturated at the entry to the compressor. There is no under cooling of the liquid refrigerant. The actual COP is 0.7 times the theoretical COP, determine,*

(i) Mass flow rate of refrigerant

(ii) Power required to drive compressor

(iii) Diameter and stroke of the compressor running at 4 RPS and L= D. Its volumetric efficiency is 80% and is single acting sketch the cycle on P–h diagram.

Solution

P–h diagram for th example is shown in Fig. Ex. 16.8.

We can use the *P–h* chart for NH_3 and find the enthalpies from the chart.

We get these enthalpies from the chart.

$$h_1 = 1260 \text{ kJ/kg} \qquad h_2 = 1488 \text{ kJ/kg}$$

$$h_3 = h_4 = 365 \text{ kJ/kg}$$

$$\therefore \quad \text{C.O.P.} = \frac{h_1 - h_4}{h_2 - h_1} = \frac{1260 - 365}{1488 - 1260}$$

$$= \frac{895}{228}$$

$$= \mathbf{3.925}$$

But actual C. O. P. $= 0.7 \times 3.925$

$= \mathbf{2.7475}$

Actual refrigerating effect $= 20$ TR

$= 20 \times 3.516$ kJ/sec

$=$ **70.32 kJ/sec**

But refrigerating effect/kg $= 895$ kJ/kg

$$\therefore \quad \text{Mass of refrigerant kg/s} = \frac{70.32}{895}$$

$=$ **0.0787 kg/s**

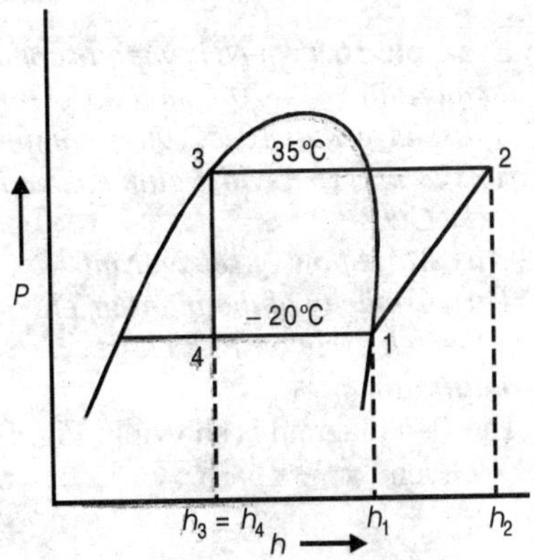

Fig. Ex. 16.8

$$\text{Now we have, Actual COP} = \frac{\text{Actual ref effe. (kW)}}{\text{Actual power (kW)}}$$

$$\therefore \quad 2.7475 = \frac{70.32 \text{ (kW)}}{\text{Actual power (kW)}}$$

$$\therefore \quad \text{Actual power} = \frac{70.32}{2.7475}$$

$=$ **25.6 kW**

Power required to drive the compressor

$=$ **25.6 kW**

(iii) Diameter and Stroke of Compressor

Actual volume of refrigerant at the compressor inlet/kg $= 0.54$ m^3/kg

$\therefore$ Volume/sec $= 0.0787 \times 0.54$

$=$ **0.0425 m³/sec**

$\therefore$ Theoretical Swept volume/sec

$$= \frac{0.0425}{0.8}$$

$=$ **0.05312 m³/sec**

$$\text{But swept volume/sec} = \frac{\pi d^2}{4} \times L \times \frac{N}{60}$$

But $D = L$

$$\therefore \quad 0.05312 = \frac{\pi D^3}{4} \times \frac{N}{60}$$

$$= \frac{\pi D^3}{4} \times 4 = \pi D^3$$

$$\therefore \quad D^3 = \frac{0.05312}{\pi} = 0.0169$$

$\therefore \quad D$ = Diameter of cyclinder

$= (0.0169)^{1/3}$

$= 0.2565$ m

= **25.65 cm**

= Stroke of piston

Example 16.9 *An NH_3 vapour compressor cycle refrigerator operates between an evaporation and condensation temperature of – 10°C and 35°C respectively. The refrigerant is dry and saturated at entry to the compressor and is discharged at 102°C after compression. The single acting compressor has a base 80 mm and stroke 80 mm. It runs at 480 rpm with volumetric efficiency of 80%. Assuming liquid subcooling of 10°C, calculate:*

(i) COP

(ii) Mass flow of refrigerant

(iii) Capacity of the plant in TR

Sketch cycle on P–h chart

Solution

The *P–h* diagram is shown in Fig. Ex. 16.9.

Given

t_2 = 102°C

t_3 = 35–10

= 25°C

t_1 = 10°C

η_{vol} = 80%

D = L = 80 mm, rpm = 480

t_a = 35°C undercooling by 10°C

From *P–h* chart for NH_3, we get the following enthalpies

h_1 = 1450 kJ/kg, h_2 = 1680 kJ/kg

h_3 = h_4 = 325 kJ/kg and v_3 = 0.4175 m³

Now we calculate the required quantities.

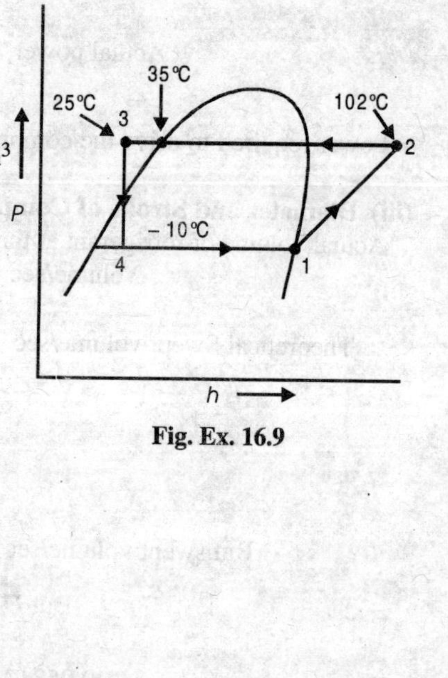

Fig. Ex. 16.9

(i) COP

Refrigerating effect/kg $= h_1 - h_3 = 1450 - 325$

= **1125 kJ/kg**

Work supplied/kg $= h_1 - h_2 = 1680 - 1450$

= **230 kJ/kg**

$$\therefore \quad \text{COP} = \frac{\text{Ref. effect/kg}}{\text{Work supplied/kg}}$$

$$= \frac{1125}{230} = \mathbf{4.891}$$

(ii) Mass Flow Rate of Refrigerant

Stroke volume/min. = Theoretical volume/min.

$$= \frac{\pi D^2}{4} \times L \times N$$

$$= \frac{\pi}{4} \times \left(\frac{80}{1000}\right)^3 \times 480$$

∴ Actual volume of refrigerant taken in/min

$$= \frac{\pi}{4} \times \left(\frac{80}{1000}\right)^3 \times \frac{480}{1} \times 0.8$$

$$= 0.618 \text{ m}^3/\text{min}$$

$$\text{Specific volume} = 0.4175 \text{ m}^3/\text{kg}$$

$$\text{Mass of refrigerant/min} = \frac{\text{Volume/min}}{\text{Specific volume}}$$

$$= \frac{0.618}{0.4175}$$

$$= \mathbf{1.48 \text{ kg/min.}}$$

(iii) Capacity of Plant in TR

$$\text{Refrigerating effect/sec} = \frac{1.48}{60} \times 1125 \text{ kJ/s}$$

and $1 \text{ TR} = 3.516 \text{ kJ/sec}$

∴ TR capacity of the plant

$$= \frac{1.48 \times 1125}{60 \times 3.516}$$

$$= \mathbf{7.8925 \text{ TR}}$$

Example 16.10 *A food storage chamber requires a refrigeration system of 12 TR capacity when an evaporator temperature of –8°C and condenser temperature of 30°C are maintained. The system uses F-12 as refrigerant (R-12) with following properties. It is subcooled by 5°C before entering throttle valve and vapour is supplied heated by 6°C before entering the compressor. The specific heats of refrigerant are C_{P_v} = 1.235 and C_P (liquid) = 0.7327 kJ/kg-K. Determine:*

(i) Mass flow rate of refrigerant
(ii) Refrigerating effect/kg and
(iii) COP of cycle

Also draw the cycle on T–S chart assuming isentropic compressor properties of R-12

Temperature °C	*Specific enthalpy*		*Specific entropy*	
	Liquid	*Vapour*	*Liquid*	*Vapour*
–8	*411.3*	*569.8*	*4.1618*	*4.7564*
30	*447.9*	*586.5*	*4.2874*	*4.7438*

Solution

The *T–S* diagram for the cycle is shown in Fig. Ex. 16.10

Here 3–3' is under cooling

First find the enthalpies at the salient points.

$$h_1 = h_{g_1} + C_{p_v} \times (\text{degrees of superheat})$$

$$= 569.8 + 1.235 \times 6$$

$$= 569.8 + 7.310$$

$$= \mathbf{577.11 \text{ kJ/kg}}$$

For isentropic compression

$$S_1 = S_2$$

$$\therefore\ 4.7564 + C_{P_V} \log_e \frac{273-2}{265} = 4.7564 + 1.235 \log_e \times \log_e \frac{273}{265}$$

$$= 4.784 \text{ kJ/kg-K}$$

$$S_2 = S_{g_2} + C_{P_V} \log_e \frac{T_2}{303} = 4.7438 + 1.235 \log_e \frac{T_2}{303}$$

$$S_1 = S_2, T_2 = \mathbf{306.11\ K}$$

$$\therefore\ h_2 = h_{g_2} + C_{P_V}(t_{sup} - t_{sat})$$

$$= 586.5 + 1.235\,(306.11 - 303)$$

$$= \mathbf{590.34\ kJ/kg}$$

Also

$$h_3 = h_3 - c_{p_v} \times \text{degrees of undercooling}$$

$$= 447.9 - 0.7327 \times 5$$

$$= \mathbf{444.24\ kJ/kg}$$

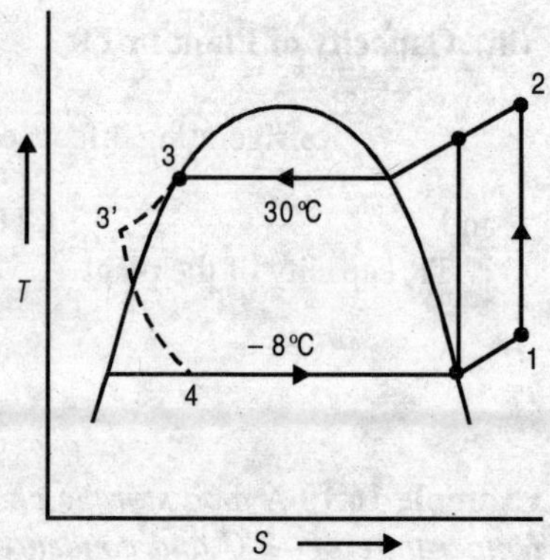

Fig. Ex. 16.10

(i) Refrigerator Effect/kg

$$\text{Ref. effect/kg} = h_1 - h_4 = h_1 - h_{3'}$$

$$= 577.11 - 444.24$$

$$= \mathbf{132.87\ kJ/kg}$$

(ii) Mass Flow of Refrigerant

$$\text{Actual refrigerator effect} = 12 \text{ TR}$$

$$= 12 \times 3 - 516 \text{ kW} = \text{kJ/sec}$$

$\therefore$ Mass flow rate of refrigerant kg/sec

$$= \frac{12 \times 3.516}{132.87}$$

$$= \mathbf{0.3175\ kg/s}$$

$$= \mathbf{19.05\ kg/min}$$

(iii) COP

$$\text{C. O. P.} = \frac{\text{Ref. effect}}{\text{Work supplied}}$$

$$= \frac{h_1 - h_{3^1}}{h_2 - h_1}$$

$$= \frac{132.87}{590.34 - 577.11}$$

$$= \frac{132.87}{13.23}$$

$$= \mathbf{10.04}$$

THEORY QUESTIONS

16.1. What is refrigeration? Enumerate the applications of refrigeration.

16.2. What is the unit of refrigeration?

16.3. What do you understand under cooling or sub-cooling?

16.4. What is heat-pump?

16.5. Distinguish clearly between ideal refrigeration cycle and the Carnot reversed cycle.

16.6. Why do you require refrigeration in a aeroplane flying in a sub-zero temperature atmosphere at an altitude?

16.7. What is a regenerative system used in air refrigeration? Explain with neat sketch.

16.8. What are the limitations of air refrigeration system?

16.9. Explain the working of a simple vapour compression cycle machine. Derive the relation between work required, refrigerating effect and COP.

16.10. Draw a *P–h* chart and show on that the simple theoretical vapour compression cycle and actual cycle. Show the cycle with undercooling on *P–h* chart.

16.11. Explain with the neat sketch (*T–S* diagram) the effect of superheat, undercooling and condenser pressure on the performance of the working of vapour compression cycle.

16.12. What is the principle on which the vapour absorption cycle works?

16.13. Explain with the neat sketch a simple vapour absorption cycle working.

16.14. What modifications are required in the simple cycle to become a practical cycle.

16.15. What are the combinations of refrigerant and absorbent?

16.16. Explain with the neat sketch the working of Electrolux refrigerating system.

16.17. Compare vapour compression and vapour absorption cycles.

16.18. Explain with neat sketch the Lithium-bromide-water absorption system.

16.19. What is the single effect and double effect refrigeration systems. Explain.

16.20. Derive the relation for the maximum C.O.P. of a Heat operated vapour absorption cycle.

PROBLEMS FOR PRACTICE

16.1. An air refrigeration plant working on Carnot cycle between the temperatures limits of 34°C and –1°C has a capacity of 10 tons. Determine the power input to the plant. [**Ans.** 4.52 kW]

16.2. An air refrigeration plant working on Carnot cycle produces refrigeration capacity equivalent of production of tons of ice in 24 hours at 0°C from water at 20°C. The compressor temperature limits are 30°C and –15°C. Determine the power input to the plant. The latent heat of ice is 335 kJ/kg.

[**Ans.** 8.45 kW]

16.3. In a refrigerator plant working on the Bell–Coleman cycle the air is drawn in from the cold chamber at –5°C and 1.03 bar. The same air is compressed adiabatically to 5.25 bar before being cooled in the heat exchanger to 15°C. After the exchanger the air is expanded adiabatically and reversibly to cold changer state. Assume C_p = 1.004 kJ/kg-K and C_v = 0.717 kJ/kg/K for air. Determine

(i) work done on the cycle per kg of air.

(ii) the refrigeration effect produced in cold chamber.

(iii) C.O.P.

[**Ans.** (i) 51.85 kJ/kg, (ii) 87.51 kJ/kg air, (iii) 1.69]

16.4. In a Bell–Coleman cycle, air is drawn from cold chamber at a pressure of 1.03 bar and –10°C, compressed isentropically to 0.18 bar. Cooled to 25°C. It is expanded with the law of $PV^{1.3}$ =Constant. C_p = 1.004 kJ/kg/K, C_v = 1.717 kJ/kg-K. Determine

(a) Work input/kg of air

(b) Refrigeration produced in cold chamber

(c) COP

[**Ans.** (a) 51.03 kJ/kg, (b) 66.16 kJ/kg, (c) 1.3]

16.5. Following data is related to a dense air refrigeration cycle. Capacity = 1 TR, pressure and temperature at compressor suction = 5

bar and –18°C, pressure and temperature at the inlet of expander = 20 bar and 16°C. Determine
(a) COP
(b) Heat removed in cooler
(c) Piston displacement of compressor.
[**Ans.** (a) 2.06, (b) 5.22 kJ/sec, (c) 0.32 m³/min]

16.6. A simple saturated ammonia vapour compression plant operates between 40°C and –20°C. Calculate
(a) Refrigerating effect/kg
(b) Work input/kg
(c) Heat rejected from condenser/kg
(d) C.O.P
(e) Mass flow rate in kg/sec per TR
Compare the above mentioned parameters for dry and saturated vapours at the suction with altering other conditions. Take C_{pv} = 3.52 kJ/kg-K.
[**Ans.** (a) 856.66 (b) 242.12 kJ/kg (c) 1098.8 (d) 3.54 (e) 0.0041] [(a) 1046.5 (b) 313.93 (c) 1360.43 (d) 3.33 (e) 0.00336]

Temperature °C	Pressure bar	h_f kJ/kg	h_g kJ/kg	S_f kJ/kg-K	S_g kJ/kg-K
– 14	1.7	108.7	1437.6	0.65	5.9
+ 40	15.6	391.1	1489.9	1.65	5.15

16.7. A F-12 refrigeration system operates between temperature limits of –6°C and 36°C. The vapour is 95 % dry at the beginning of isentropic compression and the liquid leaving the condenser is at 30°C. Assume actual COP as 50 % of theoretical COP, calculate the amount of ice produced in kg/kWh at 0°C from water at 30°C. Latent heat of ice = 335 kJ/kg. Specific heat of refrigerant in liquid stage is 1.24 kJ/kg-K. [**Ans.** 21.657 kg/kWh]

Temperature °C	h_f kJ/kg	h_g kJ/kg	S_f kJ/kg-K	S_g kJ/kg-K
– 6	30.53	184.94	0.1217	0.6996
+ 36	70.55	201.80	0.2591	0.6836

16.8. Compare the COP of an ammonia refrigeration cycle which uses wet compression with one which uses dry compression and operates between –20°C and 30°C. Assume compression to be isentropic and no subcooling of liquid refrigerant. In wet compression, vapour leaving the compressor is dry and saturated whereas in dry compression, vapour entering the compressor is dry and saturated.
[**Ans.** Wet compression COP = 1.025, Dry compression COP = 4.113]

Sat. Temperature °C	h_f kJ/kg	h_{fg} kJ/kg	h_g kJ/kg	S_f kJ/kg-K	S_g kJ/kg-K
– 20	89.8	1330	1420	0.3684	5.6244
+ 30	323.1	1145.8	1468	1.2037	4.9842

16.9. 30,000 kg of ice per day is produced from and at 0°C in an ammonia ice plant working on vapour compression cycle. The temperature range of the cycle is –15°C to 25°C. The vapour at the end of compression dry saturated. The relative COP of the system is 65 %. Find the power required to drive the plant. Latent heat of ice = 335 kJ/kg. [**Ans.** 5.303 kW]

16.10. A refrigeration system using NH_3 works on standard vapour compression cycle. The condensing temperature is 30°C. Investigate the performance of the system based on

(a) C O P
(b) Mass of refrigerant handled by compressor.
(c) Volume of refrigerant handled by compressor.
(d) Temperature of refrigerant at compressor delivery.

The system operates at temperature in the evaporator of –10°C and 16°C respectively. Capacity of the system is 10 TR.

Sat. Temperature °C	Sp. Volume V_g m^3/kg	Sp. Enthalpy kJ/kg			Specific entropy kJ/kg-K	
		h_f	h_{fg}	h_g	S_f	S_g
– 10	0.4189	112.4	1314.2	1426.6	0.4572	5.5789
+ 30	0.1107	323.1	1145.8	1468.9	1.2037	4.9842

[**Ans.** (a) 4.125, (b) 1.697 kg/min, (c) 0.7811 m^3/min, (d) 116.2°C]

Table 16.1: Properties of Refrigeration R-116

Saturation temp. in °C (t)	Saturation pressure in bar (p)	Specific volume in m^3/kg		Specific enthalpy in kJ/kg			Specific entropy in kJ/kg K	
		Liquid (v_f)	Vapour (v_g)	Liquid (h_f)	Vapour (h_g)	Latent (h_{fg})	Liquid (s_f)	Vapour (s_g)
–40	0.05096	0.000617	2.7625	0.00	203.48	203.48	0.0000	0.8730
–38	0.05942	0.000619	2.5360	1.65	202.83	204.48	0.0068	0.8701
–36	6.06787	0.000621	2.3094	3.30	202.18	205.48	0.0137	0.8671
–34	0.07633	0.000623	2.0827	4.95	201.53	206.48	0.0206	0.8642
–32	0.08478	0.000624	1.8562	6.60	200.88	207.48	0.0274	0.8613
–30	9.09323	0.000626	1.6296	8.25	200.23	208.48	0.0343	0.8583
–28	0.10345	0.000627	1.4448	9.90	199.58	209.48	0.0410	0.8554
–26	0.11586	0.000629	1.3123	11.55	198.94	210.49	0.0477	0.8533
– 24	0.12828	0.000631	1.1798	13.20	198.29	211.49	0.0544	0.8508
–22	0.14292	0.000633	1.0722	14.85	197.65	212.50	0.0611	0.8483
– 20	0.15869	0.000634	0.9770	16.51	196.99	213.50	0.0678	0.8462
– 18	0.17414	0.000636	0.8793	18.19	196.33	214.52	0.0741	0.8441
– 16	0.19280	0.000638	0.8045	19.82	195.71	215.53	0.0808	0.8420
– 14	0.21260	0.000639	0.7345	21.49	195.05	216.54	0.0875	0.8403
– 12	0.23390	0.000641	0.6713	23.17	194.38	217.55	0.0938	0.8387
10	0.25848	0.000643	0.6191	24.83	193.73	218.56	0.1009	0.8366
– 8	0.28306	0.000645	0.5671	26.50	193.07	219.57	0.1063	0.8349
– 6	0.30280	0.000647	0.5194	28.17	192.42	220.59	0.1126	0.8332
– 4	0.33967	0.000649	0.4801	29.85	191.76	221.61	0.1189	0.8315
– 2	0.36983	0.000651	0.4409	31.54	191.09	222.63	0.1252	0.8303
0	0.40360	0.000653	0.4069	33.22	190.42	223.64	0.1315	0.8296
1	0.42190	0.000653	0.3920	34.04	190.08	224.14	0.1344	0.8280
2	0.44021	0.000654	0.3770	34.90	189.75	224.65	0.1374	0.8274
3	0.45852	0.000655	0.3620	35.74	189.41	225.15	0.1405	0.8267
4	0.47683	0.000656	0.3471	36.59	189.06	225.65	0.1436	0.8261
5	0.49720	0.000657	0.3340	37.43	188.73	226.16	0.1465	0.8254
6	0.51917	0.000658	0.3225	38.28	188.38	226.66	0.1498	0.8248
7	0.54118	0.000659	0.3109	39.13	188.04	227.17	0.1526	0.8242
8	0.56317	0.000660	0.2994	39.98	187.70	227.68	0.1562	0.8236
9	0.58517	0.000661	0.2878	40.84	187.35	228.18	0.1586	0.8229
10	0.60717	0.000662	0.2763	41.68	187.01	228.69	0.1616	0.8223
11	0.63320	0.000663	0.2673	42.53	186.67	229.20	0.1646	0.8219
12	0.65921	0.000664	0.2584	43.39	186.32	229.71	0.1675	0.8215
13	0.68523	0.000665	0.2495	44.24	185.97	230.21	0.1706	0.8210

(Contd.)

Saturation temp. in °C (t)	Saturation pressure in bar (p)	Specific volume in m³/kg		Specific enthalpy in kJ/kg			Specific entropy in kJ/kg K	
		Liquid (v_f)	Vapour (v_g)	Liquid (h_f)	Vapour (h_g)	Latent (h_{fg})	Liquid (s_f)	Vapour (s_g)
14	0.71124	0.000666	0.2405	45.10	185.62	230.72	0.1738	0.8206
15	0.73727	0.000667	0.2316	45.95	185.27	231.22	0.1767	0.8200
16	0.76552	0.000669	0.2235	46.81	184.92	231.73	0.1796	0.8194
17	0.79655	0.000670	0.2163	47.67	184.56	232.23	0.1825	0.8170
18	0.82759	0.000671	0.2092	48.54	184.19	232.73	0.1855	0.8186
19	0.85862	0.000672	0.2021	49.40	183.84	233.24	0.1885	0.8180
20	0.88966	0.000673	0.1950	50.26	183.48	233.74	0.1913	0.8173
21	0.92070	0.000674	0.1878	51.12	183.12	234.24	0.1943	0.8169
22	0.95625	0.000675	0.1821	51.99	182.75	234.74	0.1972	0.8165
23	0.99237	0.000676	0.1764	52.85	182.39	235.24	0.2003	0.8163
24	1.02496	0.000677	0.1708	53.72	182.03	235.75	0.2031	0.8160
25	1.06117	0.000678	0.1651	54.59	181.66	236.25	0.2061	0.8157
26	1.10074	0.000679	0.1595	55.46	181.29	236.75	0.2089	0.8152
27	1.13880	0.000680	0.1542	56.32	180.93	237.25	0.2120	0.8150
28	1.17917	0.000681	0.1495	57.20	180.56	237.76	0.2148	0.8148
29	1.22273	0.000682	0.1452	58.07	180.18	238.25	0.2177	0.8144
30	1.26070	0.000684	0.1401	58.94	179.82	238.76	0.2206	0.8140
31	1.30665	0.000685	0.1362	59.82	179.43	239.25	0.2234	0.8135
32	1.34821	0.000686	0.1316	60.70	179.05	239.75	0.2261	0.8131
33	1.39568	0.000687	0.1278	61.57	178.67	240.24	0.2290	0.8128
34	1.44414	0.000689	0.1242	62.45	178.28	240.73	0.2320	0.8126
35	1.49276	0.000690	0.1205	63.32	177.91	241.23	0.2349	0.8123
36	1.54130	0.000691	0.1168	64.20	177.52	241.72	0.2378	0.8119
37	1.58983	0.000692	0.1131	65.07	177.15	242.22	0.2406	0.8116
38	1.64000	0.000693	0.1096	65.95	176.77	242.72	0.2433	0.8114
39	1.69571	0.000694	0.1066	66.84	176.66	243.20	0.2462	0.8112
40	1.75145	0.000696	0.1037	67.74	175.95	243.69	0.2490	0.8110
41	1.80718	0.000697	0.1007	68.63	175.55	244.18	0.2518	0.8108
42	1.86292	0.000698	0.0977	69.52	175.15	244.67	0.2546	0.8106
43	1.91866	0.000699	0.0947	70.41	174.75	245.16	0.2574	0.8104
44	1.97952	0.000700	0.0921	71.31	175.34	245.65	0.2600	0.8102
45	2.04300	0.000701	0.0896	72.20	173.94	246.14	0.2029	0.8100
46	2.10641	0.000702	0.0871	73.08	173.55	243.63	0.2658	0.8098
47	2.16980	0.000704	0.0847	73.98	173.14	247.12	0.2686	0.8096
48	2.23310	0.000705	0.0823	74.87	172.73	247.60	0.2713	0.8094
49	2.29670	0.000706	0.0798	75.76	172.31	248.07	0.2741	0.8092

Table 16.2: Properties of Refrigeration R-12

Saturation temp. in °C (t)	Saturation pressure in bar (p)	Specific volume in m³/kg		Specific enthalpy in kJ/kg			Specific entropy in kJ/kg K	
		Liquid (v_f)	Vapour (v_g)	Liquid (h_f)	Vapour (h_g)	Latent (h_{fg})	Liquid (s_f)	Vapour (s_g)
–100	0.01185	0.000600	10.1951	– 51.84	193.84	142.00	– 0.2567	0.8628
– 95	0.01864	0.000604	6.6231	– 47.56	191.78	144.22	– 0.2323	0.8442
– 90	0.02843	0.000608	4.4206	– 43.28	189.74	146.46	– 0.2086	0.8273
– 85	0.04254	0.000613	3.0531	– 39.00	187.73	148.73	– 0.1856	0.8122
– 80	0.06200	0.000617	2.1519	– 34.72	185.74	151.02	– 0.1631	0.7985
– 75	0.08826	0.000622	1.5462	– 30.42	183.74	153.32	– 0.1412	0.7861
–70	0.12298	0.000627	1.1314	– 26.12	181.75	155.63	– 0.1198	0.7665
– 65	0.16807	0.000632	0.8421	– 21.81	179.77	157.96	– 0.0988	0.7648
– 60	0.22665	0.000637	0.6401	– 17.48	177.77	160.29	– 0.0783	0.7558
– 55	0.30052	0.000643	0.4930	– 13.14	175.76	162.62	– 0.0581	0.7475
– 50	0.39237	0.000648	0.3845	– 8.78	173.73	164.95	– 0.0394	0.7401
– 45	0.50512	0.000654	0.3035	– 4.39	171.66	167.27	– 0.0190	0.7334
– 40	0.64190	0.000660	0.2422	0.00	169.60	169.60	0.0000	0.7274
– 38	0.70460	0.000663	0.2221	1.76	168.76	170.52	0.0075	0.7251
– 36	0.77196	0.000665	0.2040	3.53	167.91	171.44	0.0149	0.7230
– 34	0.84421	0.000667	0:1877	5.31	167.05	172.36	0.0224	0.7209
– 32	0.92776	0.000670	0.1729	7.08	166.20	173.28	0.0298	0.7190
– 30	1.00441	0.000673	0.1596	8.86	165.34	174.20	0.0371	0.7171
– 28	1.09311	0.000676	0.1475	10.64	164.47	175.11	0.0444	0.7153
– 26	1.18778	0.000678	0.1364	12.43	163.59	176.02	0.0516	0.7135
– 24	1.28858	0.000681	0.1265	14.22	162.71	176.93	0.0588	0.7118
– 22	1.39581	0.000683	0.1173	16.02	161.81	177:83	0.0660	0.7102
– 20	1.50972	0.000686	0.1090	17.82	160.91	178.73	0.0731	0.7087
– 18	1.63104	0.000689	0.1014	19.62	160.01	179.63	0.0801	0.7073
– 16	1.75963	0.000692	0.0944	21.43	159.10	180.53	0.0871	0.7059
– 14	1.89575	0.000695	0.0880	23.23	159.19	181.42	0.0941	0.7045
– 12	2.04605	0.000698	0.0821	25.05	157.26	182.31	0.1010	0.7032
– 10	2.19172	0.000701	0.0767	26.87	156.32	183.19	0.1080	0.7019
– 8	2.35272	0.000704	0.0717	28.70	155.36	184.06	0.1148	0.7007
– 6	2.52244	0.000707	0.0672	30.53	154.41	184.94	0.1217	0.6996
– 4	2.70116	0.000710	0.0630	32.37	153.43	185.80	0.1285	0.6986
– 2	2.88921	0.000713	0.0591	34.20	152.47	186.67	0.1352	0.6975
0	3.08690	0.000717	0.0555	36.05	151.48	187.53	0.1420	0.6965
1	3.18974	0.000719	0.0538	36.98	150.97	187.95	0.1453	0.6961

(Contd.)

Saturation temp. in °C (t)	Saturation pressure in bar (p)	Specific volume in m³/kg		Specific enthalpy in kJ/kg			Specific entropy in kJ/kg K	
		Liquid (v_f)	Vapour (v_g)	Liquid (h_f)	Vapour (h_g)	Latent (h_{fg})	Liquid (s_f)	Vapour (s_g)
2	3.29513	0.000720	0.0521	37.90	150.48	188.38	0.1489	0.6956
3	3.40310	0.000721	0.0505	38.83	149.98	188.81	0.1521	0.6951
4	3.51367	0.000723	0.0490	39.76	149.47	189.23	0.1553	0.6947
5	3.62690	0.000725	0.0475	40.69	148.96	189.65	0.1568	0.6943
6	3.74280	0.000727	0.0461	190.07	41.62	148.45	0.1620	0.6938
7	3.86141	0.000729	0.0447	190.49	42.56	147.93	0.1653	0.6933
8	3.98283	0.000730	0.0434	190.91	43.50	147.41	0.1686	0.6929
9	4.10702	0.000732	0.0422	191.32	44.43	146.89	0.1719	0.6925
10	4.23407	0.000734	0.0410	191.74	45.37	146.37	0.1752	0.6921
11	4.36442	0.000736	0.0398	192.15	46.31	145.84	0.1784	0.6917
12	4.49763	0.000738	0.0386	192.56	47.26	145.30	0.1817	0.6913
13	4.63386	0.000739	0.0375	192.97	48.20	144.77	0.1850	0.6909
14	4.77312	0.000741	0.0365	193.38	49.15	144.23	0.1883	0.6906
15	4.91545	0.000743	0.0355	193.79	50.10	143.69	0.1915	0.6902
16	5.06087	0.000745	0.0345	194.19	51.05	143.14	0.1948	0.6898
17	5.20942	0.000747	0.0335	194.59	52.00	142.59	0.1981	0.6894
18	5.36117	0.000749	0.0326	194.99	52.95	142.04	0.2013	0.6891
19	5.51614	0.000751	0.0317	195.38	53.91	141.47	0.2046	0.6888
20	5.67441	0.000753	0.0308	195.78	54.87	140.91	0.2078	0.6884
21	5.83635	0.000756	0.0300	196.17	55.83	140.34	0.2110	0.6881
22	6.00171	0.000758	0.0292	196.56	56.79	139.77	0.2143	0.6878
23	6.17050	0.000759	0.0284	196.96	57.75	139.21	0.2174	0.6875
24	6.34269	0.000761	0.0276	197.34	58.73	138.61	0.2207	0.6872
25	6.51840	0.000764	0.0269	197.73	59.70	138.03	0.2239	0.6868
26	6.69765	0.000766	0.0262	198.11	60.67	137.44	0.2271	0.6865
27	6.88048	0.000768	0.0255	198.50	61.65	136.85	0.2303	0.6862
28	7.06704	0.000770	0.0248	198.87	62.63	136.24	0.2335	0.6859
29	7.25738	0.000772	0.0241	199.25	63.61	135.64	0.2368	0.6856
30	7.45103	0.000775	0.0235	199.62	64.59	135.03	0.2400	0.6853
31	7.64903	0.000777	0.0230	199.99	65.58	134.41	0.2431	0.6850
32	7.85089	0.000779	0.0225	200.36	66.57	133.79	0.2463	0.6847
33	8.05662	0.000782	0.0218	200.73	67.56	133.17	0.2495	0.6845
34	8.26621	0.000784	0.0212	201.09	68.56	132.53	0.2527	0.6842
35	8.48000	0.000786	0.0207	201.45	69.56	131.89	0.2559	0.6839

Saturation temp. in °C (t)	Saturation pressure in bar (p)	Specific volume in m^3/kg		Specific enthalpy in kJ/kg			Specific entropy in kJ/kg K	
		Liquid (v_f)	Vapour (v_g)	Liquid (h_f)	Vapour (h_g)	Latent (h_{fg})	Liquid (s_f)	Vapour (s_g)
36	8.69766	0.000789	0.0202	201.80	70.55	131.25	0.2591	0.6836
37	8.91904	0.000792	0.0196	202.16	71.55	130.61	0.2623	0.6833
38	9.14483	0.000794	0.0191	202.51	72.56	129.95	0.2654	0.6830
39	9.37497	0.000796	0.0186	202.86	73.57	129.29	0.2685	0.6828
40	9.60897	0.000799	0.0182	203.20	74.59	128.61	0.2718	0.6825
41	9.84793	0.000802	0.0177	203.54	75.61	127.93	0.2750	0.6822
42	10.09131	0.000804	0.0173	203.87	76.62	127.25	0.2782	0.6820
43	10.33862	0.000807	0.0168	204.21	77.65	126.56	0.2814	0.6817
44	10.59021	0.000810	0.0164	204.55	78.68	125.87	0.2846	0.6814
45	10.84655	0.000813	0.0160	204.87	79.71	125.16	0.2878	0.6811
46	11.10758	0.000815	0.0156	205.19	80.75	124.44	0.2909	0.6808
47	11.37304	0.000818	0.0153	205.51	81.79	123.72	0.2941	0.6805
48	11.64290	0.000821	0.0149	205.83	82.83	123.00	0.2973	0.6802
49	11.91724	0.000824	0.0146	206.14	83.88	122.26	0.3005	0.6800
50	12.19655	0.000827	0.0142	206.45	84.94	121.51	0.3037	0.6797

Chart I

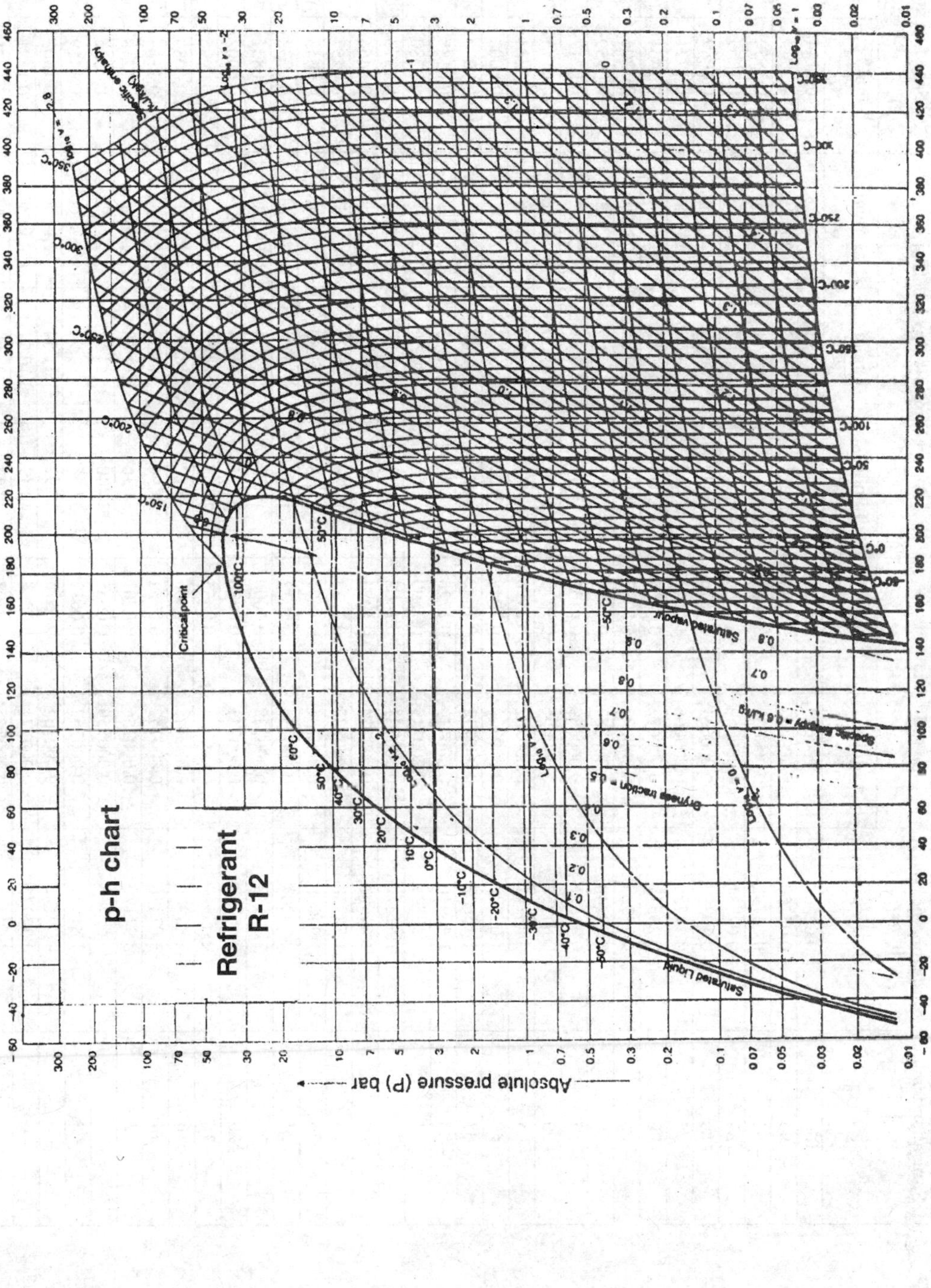
p-h chart
Refrigerant R-12
Absolute pressure (P) bar
Specific enthalpy kJ/kg
Critical point
Saturated Liquid
Saturated vapour

Table 16.3: Properties of Refrigeration R-22

Saturation temp. in °C	Saturation pressure in bar	Specific volume in m^3/kg		Specific enthalpy in kJ/kg			Specific entropy in kJ/kg K	
		Liquid (v_f)	Vapour (v_g)	Liquid (h_f)	Vapour (h_g)	Latent (h_{fg})	Liquid (s_f)	Vapour (s_g)
100	0.02009	0.000643	8.3412	– 63.45	267.18	203.73	–0.3144	1.2293
– 95	0.03150	0.000647	5.4344	– 58.14	264.33	206.19	–0.2843	1.2004
– 90	0.04792	0.000652	3.6381	– 52.87	261.51	208.64	–0.2550	1.1736
– 85	0.07731	0.000656	2.5204	– 47.61	258.72	211.11	–0.2269	1.1489
– 80	0.10393	0.000661	1.7816	– 42.40	256.00	213.60	–0.1989	1.1267
– 75	0.14759	0.000666	1.2842	– 37.17	253.28	216.11	–0.1721	1.1066
– 70	0.20517	0.000672	0.9420	– 31.93	250.55	218.62	–0.1461	1.0874
– 65	0.27965	0.000677	0.7037	– 26.68	247.84	221.16	–0.1206	1.0702
– 60	0.37448	0.000683	0.5351	– 21.42	245.09	223.67	–0.0959	1.0543
– 55	0.49621	0.000689	0.4131	– 16.13	242.31	226.18	–0.0712	1.0396
– 50	0.64758	0.000696	0.3229	– 10.81	239.50	228.69	–0.0473	1.0262
– 45	0.83241	0.000702	0.2556	– 5.40	236.60	231.20	–0.0234	1.0137
– 40	1.05586	0.000709	0.2049	0.00	233.67	233.67	0.0000	1.0024
– 38	1.15852	0.000712	0.1882	2.20	232.45	234.65	0.0096	0.9982
– 36	1.26910	0.000715	0.1728	4.40	231.25	235.65	0.0188	0.9940
– 34	1.38731	0.000719	0.1590	6.58	230.02	236.60	0.0280	0.9902
– 32	1.51324	0.000721	0.1465	8.83	228.77	'237.60.	0.0373	0.9860
– 30	1.64690	0.000724	0.1353	11.05	227.50	238.55	0.0464	0.9283
– 28	1.78938	0.000727	0.1253	13.29	226.23	239.52	0.0557	0.9785
– 26	1.94069	0.000731	0.1161	15.58	224.90	240.48	0.0645	0.9747
– 24	2.10207	.0.000734	0.1077	17.77	223.64	241.41	0.0733	0.9710
– 22	2.27448	0.000738	0.1000	19.99	222.34	242.33	0.0821	0.9676
– 20	2.45793	0.000741	0.0930	22.21	221.04	243.25	0.0908	0.9638
– 18	2.65310	0.000744	0.0865	24.47	219.70	244.17	0.0996	0.9605
– 16	2.85903	0.000748	0.0806	26.72	218.36	245.08	0.1080	0.9576
– 14	3.07876	0.000752	0.0752	28.94	217.02	245.96	0.1164	0.9538
– 12	3.31172	0.000756	0.0701	31.16	215.68	246.84	0.1248	0.9508
– 10	3.55793	0.000759	0.0655	33.40	214.32	247.72	0.1336	0.9479
– 8	3.81321	0.000763	0.0612	35.66	212.94	248.60	0.1419	0.9454
– 6	4.09172	0.000767	0.0573	37.92	211.54	249.46	0.1503	0.9425
– 4	4.37972	0.000771	0.0536	40.19	210.11	250.30	0.1591	0.9396
– 2	4.68317	0.000775	0.0503	42.52	208.62	251.14	0.1675	0.9370

(*Contd.*)

Saturation temp. in °C	Saturation pressure in bar	Specific volume in m^3/kg		Specific enthalpy in kJ/kg			Specific entropy in kJ/kg K	
		Liquid (v_f)	Vapour (v_g)	Liquid (h_f)	Vapour (h_g)	Latent (h_{fg})	Liquid (s_f)	Vapour (s_g)
0	5.00207	0.000779	0.0471	44.94	207.03	251.97	0.1763	0.9345
1	5.16841	0.000781	0.0457	46.16	206.21	252.37	0.1809	0.9333
2	5.33917	0.000783	0.0443	47.38	205.39	252.77	0.1855	0.9320
3	5.51434	0.000785	0.0429	48.64	204.53	253.17	0.1901	0.9303
4	5.69352	0.000787	0.0416	49.91	203.65	253.56	0.1943	0.9291
5	5.87621	0.000790	0.0403	51.16	202.79	253.95	0.1989	0.9278
6	6.06276	0.000792	0.0391	52.41	201.92	254.33	0.2035	0.9266
7	6.25393	0.000794	0.0379	53.68	201.03	254.71	0.2077	0.9253
8	6.44993	0.000797	0.0368	54.94	200.14	255.08	0.2123	0.9241
9	6.65034	0.000799	0.0357	56.22	199.22	255.44	0.2169	0.9228
10	6.85517	0.000801	0.0346	57.52	198.29	255.81	0.2211	0.9216
11	7.06621	0.000803	0.0336	58.80	197.35	256.15	0.2257	0.9203
12	7.27724	0.000805	0.0326	60.07	196.41	256.48	0.2303	0.9190
13	7.49793	0.000808	0.0316	61.38,	195.45	256.83	0.2345	0.9178
14	7.72552	0.000810	0.0307	62.72	194.45	257.17	0.2391	0.9165
15	7.95517	0.000813	0.0298	64.02	193.48	257.50	0.2437	0.9152
16	8.18758	0.000815	0.0289	65.32	192.48	257.81	0.2483	0.9140
17	8.42552	0.000817	0.0281	66.63	191.50	258.13	0.2529	0.9127
18	8.63241	0.000820	0.0273	67.95	190.49	258.44	0.2571	0.9115
19	8.91793	0.000822	0.0266	69.27	189.46	258.73	0.2617	0.9102
20	9.17241	0.000825	0.0258	70.59	188.41	259.00	0.2663	0.9089
21	9.43310	0.000827	0.0251	71.93	187.36	259.29	0.2709	0.9077
22	9.69931	0.000830	0.0244	73.31	186.27	259.58	0.2755	0.9065
23	9.97103	0.000833	0.0237	74.66	185.19	259.85	0.2801	0.9052
24	10.24828	0.000835	0.0230	76.04	184.07	260.11	0.2847	0.9039
25	10.53103	0.000838	0.0224	77.39	182.99	260.38	0.2889	0.9027
26	10.81931	0.000841	0.0218	78.79	181.85	260.64	0.2935	0.9014
27	11.11310	0.000844	0.0212	80.16	180.73	260.89	0.2981	0.9002
28	11.41241	0.000846	0.0206	81.54	179.58	261.12	0.3023	0.8989
29	11.69034	0.000849	0.0200	82.96	178.41	261.37	0.3069	0.8977
30	12.03448	0.000852	0.0194	84.38	177.22	261.60	0.3115	0.8964
31	12.35103	0.000855	0.0189	85.77	176.04	261.81	0.3161	0.8948
32	12.67310	0.000858	0.0184	87.17	174.85	262.02	0.3207	0.8935
33	13.00070	0.000861	0.0179	88.57	173.65	262.22	0.3249	0.8922

(Contd.)

Saturation temp. in °C	Saturation pressure in bar	Specific volume in m³/kg		Specific enthalpy in kJ/kg			Specific entropy in kJ/kg K	
		Liquid (v_f)	Vapour (v_g)	Liquid (h_f)	Vapour (h_g)	Latent (h_{fg})	Liquid (s_f)	Vapour (s_g)
34	13.33793	0.000864	0.0174	90.00	172.40	262.40	0.3295	0.8909
35	13.68276	0.000867	0.0169	91.43	171.15	262.58	0.3337	0.8893
36	14.03034	0.000870	0.0165	92.85	169.89	262.74	0.3383	0.8880
37	14.38207	0.000874	0.0161	94.24	168.64	262.88	0.3429	0.8868
38	14.74345	0.000877	0.0156	95.63	167.37	263.00	0.3471	0.8851
39	15.11103	0.000881	0.0152	97.03	166.10	263.13	0.3517	0.8838
40	15.48965	0.000884	0.0148	98.44	164.77	263.21	0.3563	0.8822
41	15.86827	0.000887	0.0144	99.82	163.47	263.29	0.3601	0.8809
42	16.25793	0.000891	0.0140	101.24	162.14	263.38	0.3651	0.8797
43	16.65380	0.000894	0.0137	102.68	160.80	263.48	0.3693	0.8780
44	17.05517	0.000898	0.0133	104.12	159.46	263.58	0.3739	0.8767
45	17.46552	0.000902	0.0130	105.58	158.09	263.67	0.3785	0.8755
46	17.88414	0.000906	0.0126	107.04	156.72	263.76	0.3827	0.8738
47	18.30827	0.000910	0.0123	108.51	155.36	263.87	0.3871	0.8723
48	18.73793	0.000914	0.0119	109.98	153.95	263.93	0.3915	0.8709
49	19.12414	0.000918	0.0116	111.42	152.57	263.99	0.3959	0.8694
50	19.61380	0.000922	0.0113	112.86	151.19	264.05	0.4003	0.8680

Chart II

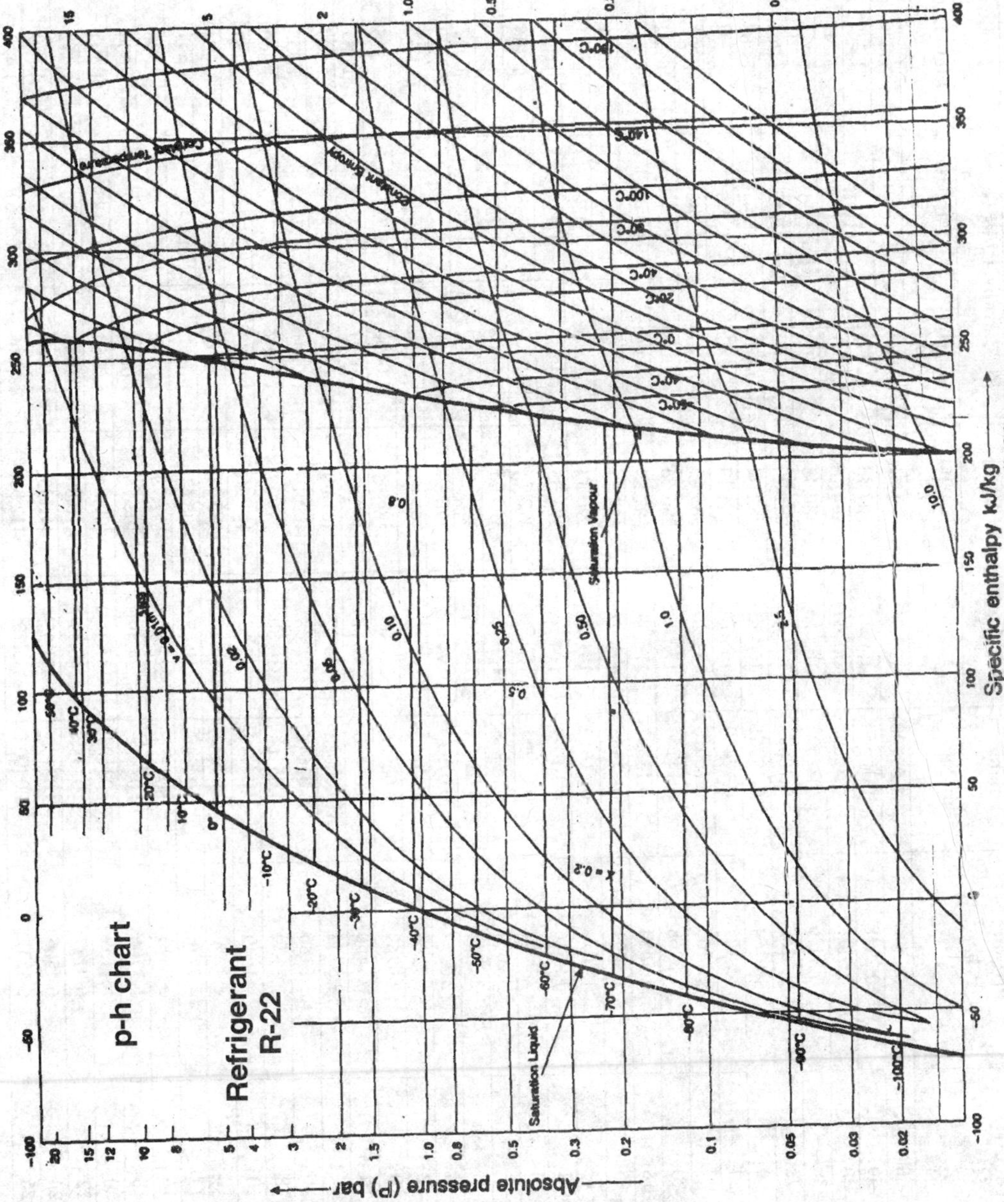

p-h chart for refrigerant R-22

Table 16.4: Properties of Refrigerant R-134a

Temp. T °C	Pressure P, bar	Volume V_g m^3/kg	Enthalpy, kJ/kg		Entropy, kJ/kg K		Sp Heat, kJ/kg K	
			Liquid h_f	Vapour h_g	Liquid s_f	Vapour s_g	Liquid C_{pf}	Vapour C_{pg}
–60	0.1594	1.07700	123.96	361.51	0.6871	1.8016	1.220	0.685
–50	0.2948	0.60560	136.21	367.83	0.7432	1.7°?2	1.229	0.712
–40	0.5122	0.36095	148.57	374.16	0.7973	1.7649	1.243	0.740
– 30	0.8436	0.22596	161.10	380.45	0.8498	1.7519	1.260	0.771
– 28	0.9268	0.20682	163.62	381.70	0.8601	1.7497	1.264	0.778
–26	1.0164	0.18961	166.16	382.94	0.8704	1.7476	1.268	0.785
–24	1.1127	0.17410	168.70	384.19	0.8806	1.7455	1.273	0.791
– 22	1.2160	0.16010	171.26	385.43	0.8908	1.7436	1.277	0.798
–20	1.3268	0.14744	173.82	386.66	0.9009	1.7417	1.282	0.805
– 18	1.4454	0.13597	176.39	387.89	0.9110	1.7399	1.286	0.812
– 16	1.5721	0.12556	178.97	389.11	0.9211	1.7383	1.291	0.820
– 14	1.7074	0.11610	181.56	390.33	0.9311	1.7367	1.296	0.827
– 12	1.8516	0.10749	184.16	391.55	0.9410	1.7351	1.301	0.835
– 10	2.0052	0.09963	186.78	392.75	0.9509	1.7337	1.306	0.842
– 8	2.1684	0.09246	189.40	393.95	0.9608	1.7323	1.312	0.850
– 6	2.3418	0.08591	192.03	395.15	0.9707	1.7310	1.317	0.858
– 4	2.5257	0.07991	194.68	396.33	0.9805	1.7297	1.323	0.866
– 2	2.7206	0.07440	197.33	397.51	0.9903	1.7285	1.329	0.875
0	2.9269	0.06935	200.00	398.68	1.0000	1.7274	1.335	0.883
2	3.1450	0.06470	202.68	399.84	1.0097	1.7263	1.341	0.892
4	3.3755	0.06042	205.37	401.00	1.0194	1.7252	1.347	0.901
6	3.6186	0.05648	208.08	402.14	1.0291	1.7242	1.353	0.910
8	3.8749	0.05284	210.80	403.27	1.0387	1.7233	1.360	0.920
10	4.1449	0.04948	213.53	404.40	1.0483	1.7224	1.367	0.930
12	4.4289	0.04636	216.27	405.51	1.0579	1.7215	1.374	0.939
14	4.7276	0.04348	219.03	406.61	1.0674	1.7207	1.381	0.950
16	5.0413	0.04081	221.80	407.70	1.0770	1.7199	1.388	0.960
18	5.3706	0.03833	224.59	408.78	1.0865	1.7191	1.396	0.971
20	5.7159	0.03603	227.40	409.84	1.0960	1.7183	1.404	0.982
22	6.0777	0.03388	230.31	410.89	1.1055	1.7176	1.412	0.994
24	6.4566	0.03189	233.05	411.93	1.1149	1.7169	1.420	1.006
26	6.8531	0.03003	235.90	412.95	1.1244	1.7162	1.429	1.018

(Contd.)

Temp. T °C	Pressure P, bar	Volume V_g m³/kg	Enthalpy, kJ/kg		Entropy, kJ/kg K		Sp Heat, kJ/kg K	
			Liquid h_f	Vapour h_g	Liquid s_f	Vapour s_g	Liquid C_{pf}	Vapour C_{pg}
28	7.2676	0.02829	238.77	413.95	1.1338	1.7155	1.438	1.031
30	7.7008	0.02667	241.65	414.94	1.1432	1.7149	1.447	1.044
32	8.1530	0.02516	244.55	415.90	1.1527	1.7142	1.457	1.058
34	8.6250	0.02374	247.47	416.85	1.1621	1.7135	1.467	1.073
36	9.1172	0.02241	250.41	417.78	1.1715	1.7129	1.478	1.088
38	9.6301	0.02116	253.37	418.69	1.1809	1.7122	1.489	1.104
40	10.165	0.01999	256.35	419.58	1.1903	1.7115	1.500	1.120
42	10.721	0.01890	259.35	420.44	1.1997	1.7108	1.513	1.138
44	11.300	0.01786	262.38	421.28	1.2091	1.7101	1.525	1.156
46	11.901	0.01689	265.42	422.09	1.2185	1.7094	1.539	1.175
48	12.527	0.01598	268.49	422.88	1.2279	1.7086	1.553	1.196
50	13.177	0.01511	271.59	423.63	1.2373	1.7078	1.569	1.218
52	13.852	0.01430	274.71	424.35	1.2468	1.7070	1.585	1.241
54	14.553	0.01353	277.86	425.03	1.2562	1.7061	1.602	1.266
56	15.280	0.01280	281.04	425.68	1.2657	1.7051	1.621	1.293
58	16.033	0.01212	284.25	426.29	1.2752	1.7041	1.641	1.322
60	16.815	0.01146	287.49	426.86	1.2847	1.7031	1.663	1.354
62	17.625	0.01085	290.77	427.37	1.2943	1.7019	1.686	1.388
64	18.464	0.01026	294.08	427.84	1.3039	1.7007	1.712	1.426
66	19.344	0.00970	297.44	428.25	1.3136	1.6993	1.740	1.468
68	20.234	0.00917	300.84	428.61	1.3234	1.6979	1.772	1.515
70	21.165	0.00867	304.29	428.89	1.3332	1.6963	1.806	1.567
72	22.130	0.00818	307.79	429.10	1.3430	1.6945	1.846	1.626
74	23.127	0.00772	311.34	429.23	1.3530	1.6926	1.890	1.693
76	24.159	0.00728	314.96	429.27	1.3631	1.6905	1.941	1.770
78	25.227	0.00686	318.65	429.20	1.3733	1.6881	2.000	1.861
80	26.331	0.00646	322.41	429.02	1.3837	1.6855	2.069	1.967

Chart III

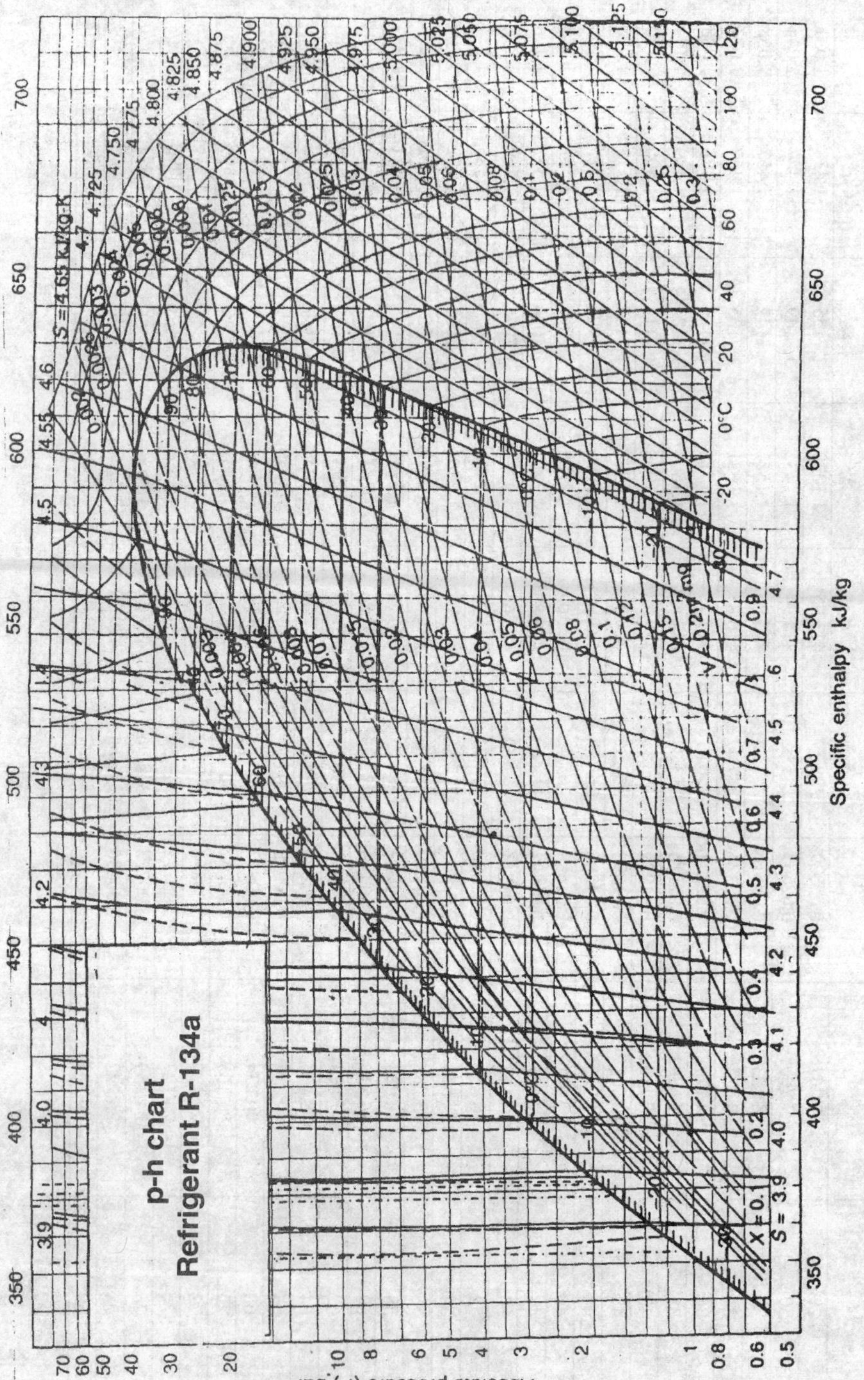
p-h chart
Refrigerant R-134a
Absolute pressure (P), bar
Specific enthalpy kJ/kg

Table 16.5: Properties of Ammonia

Saturation temp. in °C (t)	Saturation pressure in bar (p)	Specific volume in m^3/kg		Specific enthalpy in kJ/kg			Specific entropy in kJ/kg K	
		Liquid (v_f)	Vapour (v_g)	Liquid (h_f)	Vapour (h_g)	Latent (h_{fg})	Liquid (s_f)	Vapour (s_g)
– 50	0.40896	0.001426	2.6281	– 44.43	1417.70	1373.27	– 0.1943	6.1603
– 48	0.45972	0.001431	2.3565	– 35.44	1412.24	1376.80	– 0.1551	6.1192
– 46	0.51600	0.001436	2.1177	– 26.60	1406.80	1380.20	– 0.1157	6.0789
– 44	0.57710	0.001441	1.9062	– 17.81	1401.12	1383.31	– 0.0769	6.0394
– 42	0.64455	0.001446	1.7196	– 8.97	1395.64	1386.67	– 0.0384	6.0008
– 40	0.71793	0.001451	1.5537	0.00	1390.02	1390.02	0.0000	5.9631
– 38	0.79384	0.001456	1.4077	8.97	1384.16	1393.13	0.0381	5.9262
– 36	0.88607	0.001462	1.2775	17.81	1378.54	1396.35	0.0758	5.8900
– 34	0.98096	0.01467	1.1614	26.84	1372.67	1399.51	0.1131	5.8545
– 32	1.08165	0.001472	1.0574	35.68	1366.80	1402.48	0.1506	5.8198
– 30	1.19586	0.001477	0.9644	44.66	1360.94	1405.60	0.1876	5.7856
– 28	1.31724	0.001483	0.8820	53.68	1354.85	1408.53	0.2242	5.7521
– 26	1.44790	0.001488	0.8069	62.61	1348.84	1411.45	0.2607	5.7195
– 24	1.58841	0.001494	0.7397	71.73	1342.66	1414.39	0.2970	5.6872
– 22	1.73986	0.001500	0.6793	80.76	1336.52	1417.28	0.3330	5.6556
– 20	1.90276	0.001505	0.6244	89.78	1330.24	1420.02	0.3684	5.6244
– 18	2.07807	0.001511	0.5750	98.76	1323.96	1422.72	0.4043	5.5939
– 16	2.26551	0.001517	0.5303	107.83	1317.45	1425.28	0.4397	5.5639
– 14	2.46634	0.001523	0.4896	116.95	1310.93	1427.88	0.4747	5.5356
– 12	2.68069	0.001529	0.4526	126.16	1304.38	1430.54	0.5096	5.5055
– 10	2.90896	0.001536	0.4189	135.37	1297.68	1433.05	0.5443	5.4770
– 8	3.15365	0.001541	0.3884	144.35	1290.98	1435.33	0.5789	5.4487
– 6	3.41380	0.001548	0.3604	153.56	1284.37	1437.97	0.6139	5.4210
– 4	3.69069	0.001554	0.3348	162.77	1277.25	1449.02	0.6473	5.3940
– 2	3.98427	0.001561	0.3113	171.98	1270.19	1442.17	0.6812	5.3670
0	4.29586	0.001567	0.2.898	181.20	1263.25	1444.45	0.7151	5.3405
1	4.45848	0.001571	0.2798	185.80	1259.69	1445.49	0.7321	5.3277
2	4.62662	0.001574	0.2702	190.40	1256.14	1446.54	0.7487	5.3145
3	4.79931	0.001578	0.2610	195.17	1252.42	1447.59	0.7653	5.3017
4	4.97682	0.001582	0.2521	199.85	1248.78	1448.63	0.7818	5.2888
5	5.15862	0.001585	0.2436	204.46	1245.10	1449.59	0.7989	5.2765
6	5.34745	0.001589	0.2354	209.06	1241.43	1450,40	0.8154	5.2638

(Contd.)

Saturation temp. in °C (t)	Saturation pressure in bar (p)	Specific volume in m³/kg		Specific enthalpy in kJ/kg			Specific entropy in kJ/kg K	
		Liquid (v_f)	Vapour (v_g)	Liquid (h_f)	Vapour (h_g)	Latent (h_{fg})	Liquid (s_f)	Vapour (s_g)
8	5.73820	0.001595	0.2201	218.50	1234.04	1452.54	0.8487	5.2389
9	5.94186	0.0015 98	0.2128	223.11	1230.28	1453.39	0.8652	5.2266
10	6.15103	0.001603	0.2060	227.72	1226.50	1454.22	0.8814	5.2141
11	6.36641	0.001607	0.1992	232.53	1222.77	1455.30	0.8979	5.2020
12	6.58731	0.001610	0.1928	237.15	1218.96	1456.11	0.9142	5.1900
13	6.81420	0.001614	0.1866	241.92	1215.04	1456.96	0.9307	5.1780
14	7.04717	0.001617	0.1807	246.60	1211.20	1457.80	0.9470	5.1659
15	7.28276	0.001621	0.1751	251.44	1207.19	1458.63	0.9634	5.1542
16	7.53104	0.001624	0.1696	256.14	1203.33	1459.47	0.9794	5.1421
17	7.78138	0.001628	0.1643	259.88	1200.36	1460.24	0.9956	5.1302
18	8.03862	0.001633	0.1592	265.54	1195.38	1460.92	1.0118	5.1186
19	8.30551	0.001637	0.1543	270.35	1191.40	1461.75	1.0280	5.1073
20	8.57241	0.001641	0.1496	275.16	1187.44	1462.60	1.0442	5.0956
21	8.85172	0.001645	0.1450	279.77	1183.44	1463.21	1.0604	5.0843
22	9.13655	0.001648	0.1407	284.56	1179.28	1463.84	1.0763	5.0729
23	9.42690	0.001652	0.1365	289.37	1175.26	1464.63	1.0924	5.0616
24	9.72690	0.001656	0.1324	294.19	1171.14	1465.33	1.1083	5.0503
25	10.02760	0.001661	0.1284	298.90	1166.94	1465.84	1.1242	5.0391
26	10.34069	0.001665	0.1246	303.82	1162.77	1466.59	1.1402	5.0279
27	10.66137	0.001669	0.1210	308.63	1158.59	1467.22	1.1563	5.0170
28	10.99172	0.001673	0.1174	313.45	1154.40	1467.85	1.1721	5.0061
29	11.32758	0.001678	0.1140	318.26	1150.19	1468.45	1.1879	4.9951
30	11.66896	0.001682	0.1107	323.08	1145.79	1468.87	1.2037	4.9842
31	12.01655	0.001686	0.1075	327.89	1141.61	1469.50	1.2195	4.9733
32	12.37517	0.001691	0.1045	332.71	1137.23	1469.94	1.2350	4.9624
33	12.74482	0.001695	0.1015	337.52	1132.84	1470.36	1.2508	4.9517
34	13.12137	0.001700	0.0987	342.48	1128.44	1470.92	1.2664	4.9409
35	13.50345	0.001704	0.0960	347.50	1123.93	1471.43	1.2821	4.9302
36	13.89379	0.001709	0.0932	352.29	1119.41	1471.70	1.2978	4.9196
37	14.29517	0.001713	0.0907	357.25	1114.94	1472.19	1.3135	4.9091
38	14.70620	0.001718	0.0882	362.12	1110.33	1472.45	1.3290	4.8985
39	15.12276	0.001723	0.0857	367.11	1105.76	1472.87	1.3445	4.8885
40	15.54483	0.001727	0.0834	371.93	1101.37	1473.30	1.3600	4.8774

(*Contd.*)

Saturation temp. in °C (t)	Saturation pressure in bar (p)	Specific volume in m^3/kg		Specific enthalpy in kJ/kg			Specific entropy in kJ/kg K	
		Liquid (v_f)	Vapour (v_g)	Liquid (h_f)	Vapour (h_g)	Latent (h_{fg})	Liquid (s_f)	Vapour (s_g)
41	15.98551	0.001732	.0.0811	376.95	1096.55	1473.50	1.3754.	4.8669
42	16.43172	0.001738	0.0789	381.79	1091.91	1473.70	1.3908	4.8563
43	16.88344	0.001742	0.0768	386.76	1087.15	1473.91	1.4065	4.8461
44	17.34482	0.001747	0.0747	391.79	1082.33	1474.12	1.4221	4.8357
45	17.81724	0.001752	0.0727	396.81	1077.52	1474.33	1.4374	4.8251
46	18.30070	0.001757	0.0708	401.84	1072.70	1474.54	1.4528	4.8147
47	18.79518	0.001763	0.0689	406.86	1067.82	1474.68	1.4683	4.8045
48	19.29931	0.001768	0.0670	411.89	1062.87	1474.76	1.4835	4.7938
49	19.80965	0:001773	0.0652	416.91	1057.93	1474.84	1.4990	4.7834
50	20.33103	0.001779	0.0636	421.94	1052.98	1474.92	1.5148	4.7732

Chart IV

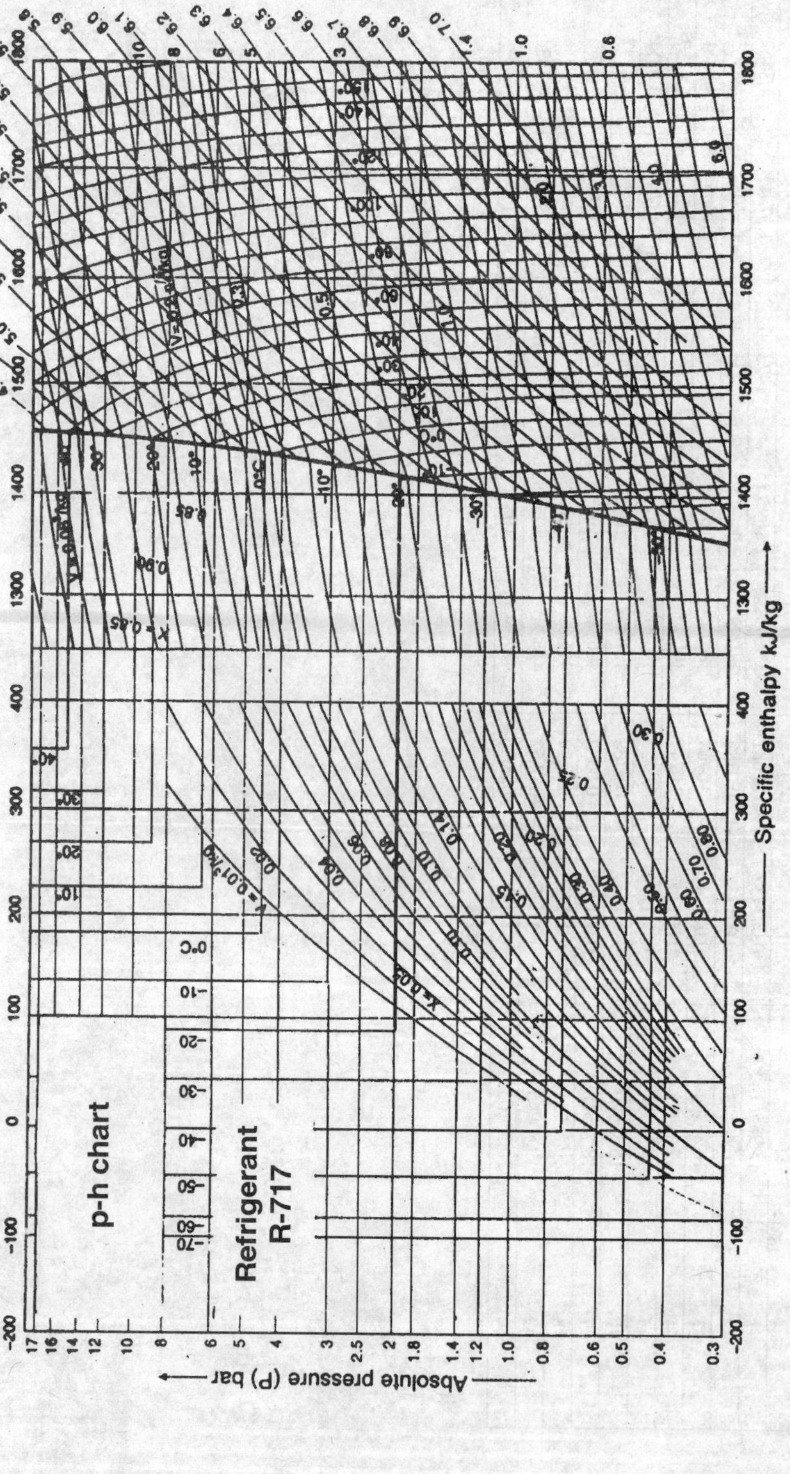
p-h chart
Refrigerant R-717
Specific enthalpy kJ/kg
Absolute pressure (P) bar

CHAPTER 17 Air Conditioning

17.1 INTRODUCTION

We have seen the behaviour of the mixture of ideal gases. In addition to mixtures consisting of ideal gases, we may frequently come across such mixtures where one of the constituents is a vapour. The *vapour* in these mixtures is at such a low partial pressure that we can treat it as an *ideal gas*. Hence the general rules, applicable to ideal gas mixtures apply to these mixtures as well.

A significant difference in the behaviour of *gas-vapour mixtures* as compared to *ideal-gas mixtures*, however, lies in the fact that the composition of the former may change on heating or cooling while that of the later does not change normally. When the temperature of the gas-vapour mixture is lowered below a certain limit, it may condense or solidify and thus the composition of the mixture may change.

Normal atmospheric air is the most familiar example of a gas-vapour mixture. Since air has a great affinity for water vapour it is very difficult to find absolutely dry air. We usually come across *moist air* which is actually a mixture of dry air and water vapour. The detailed study of the properties of moist air has been important due to its wide application in *air-conditioning*, *cooling towers*, *evapourative condensers* and *weather forecasting* etc.

The science which deals with the properties of air-water vapour mixtures has been given the name of *psychometrics*. We shall, therefore, deal with psychometrics in this chapter.

17.2 PSYCHOMETRIC TERMS AND DEFINITIONS

Following are some important terms and definitions used with reference to *air-water vapour* mixtures.

1. Moist-Air. Ordinarily available atmospheric air is the *moist air*. This is not necessarily at a particular pressure. It is a natural atmosphere.

The name is given to differentiate it from *dry air*. The latter is free from any moisture and technically speaking has the following volumetric composition.

Oxygen O_2 = 20.29%, Nitrogen N_2 = 78.03%, Argon = 0.94%, Carbon dioxide = 0.03% plus traces of rare gases.

For all practical purposes we take air contains — Oxygen 21% and Nitrogen 79% by volume. Similarly, dry air is assumed to contain oxygen O_2 23% and nitrogen N_2 77% by weight.

Moist air, as already mentioned, is the mixture of dry air with water vapour unless specifically stated otherwise, the total pressure of the mixture is assumed to be 1.01325 bar or 1 atm.

The relative amount of water vapour in the atmosphere or the moist air is very small, say about 1 to 2%. Hence the partial pressure of the water vapour in the atmosphere is also very small. At ordinary temperature, therefore, the moisture in atmosphere will be present in the *superheated state*.

2. Dry Bulb Temperature (DBT). The temperature of a mixture of gases as recorded or indicated by an ordinary accurate thermometer is called the *dry bulb temperature*, abbreviated as DBT or t_{db} or t_{d}.

3. Dew Point Temperature (DPT). If a sample of moist air is cooled at constant pressure then the temperature of both the dry air and water vapour part are reduced simultaneously. Dry air being a gas mixture remain unchanged in composition. The water vapour, however, may change its phase and therefore cause a change in the composition of the mixture. Figure 17.1 represents a *T–S* diagram of steam (for that purpose moisture vapour in air), show the process of cooling of water-vapour part of the mixture.

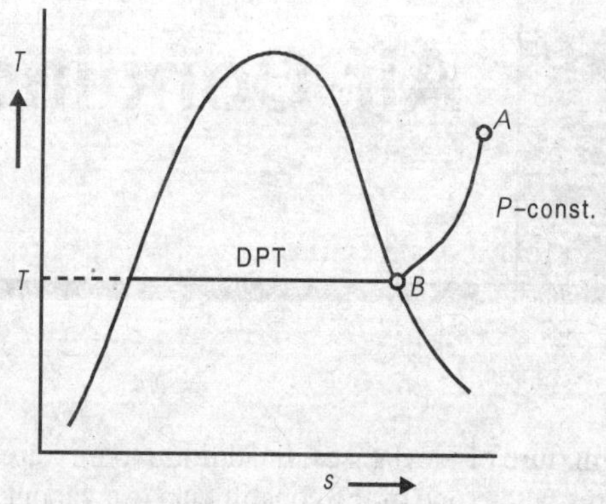

Fig. 17.1. *T–S diagram of water vapour part.*

Point *A* represents the original state of water vapour in the mixture. This vapour is in a superheated state. The pressure at point *A* equals the partial pressure of the vapour and the temperature equals the temperature of the mixture. Now if we cool the mixture at constant pressure, the water vapour will be cooled at constant partial pressure as long as the composition of the mixture does not change.

Curve *AB* represents the cooling at constant partial pressure until the vapour reaches the saturated state at point *B*. Any further cooling will result in the *condensation* of vapour i.e., release of moisture. The temperature at *B* is called the *dew-point-temperature*. This temperature, therefore, corresponds to the *saturation temperature* of steam at the *partial pressure* of water vapour in the original mixture.

Dew Point Temperature of the given air is that temperature at which the given moisture or water vapour in air saturates the air. At this temperature, the moisture is on the verge of condensing if the air temperature is reduced below this temperature. Dew Point Temperature is denoted by DPT or t_{dpt}.

4. Saturated Air. It is the mixture of dry air and saturated water vapour at the dry-air temperature. In other words, moist air in which the partial pressure of water vapour equals the saturation pressure of steam at the temperature of air is known as *Saturated Air*. State *B* in Fig. 17.1 indicates the water vapour part of the saturated air at temperature *T*. It may be noted that that it is not the air which is saturated but it is actually water vapour is a saturated vapour at its partial pressure.

Air is saturated with water vapour means that the vapour holding capacity of air at its temperature is complete and maximum and at that temperature it cannot hold any more water vapour.

5. Unsaturated Air. Moist air at a particular temperature which is not saturated is known as *unsaturated air*. It has a capacity to hold or to absorb still more water vapour. Water vapour in an unsaturated air is in a superheated condition.

6. Specific Humidity or Humidity Ratio (W_H). Specific humidity is the ratio of the mass of water vapour m_v to the mass of the associated dry air in a sample of moist air. It is denoted by *W*.

$$\therefore \qquad W_H = \frac{\text{Mass of water vapour}}{\text{Mass of air (dry)}}$$

$$= \frac{m_v}{m_a}$$

where $$m_v = \frac{p_v V}{R_v T}$$

and $$m_a = \frac{p_a V}{R_a T}$$

Here p_v and p_a are the partial pressures of water vapour and dry air respectively in the sample of moist air at the temperature T and volume V.

$$\therefore \quad W_H = \frac{m_v}{m_a}$$

$$= \frac{p_v V / R_v T}{p_a V / R_a T}$$

$$= \frac{p_v \times R_a}{p_a \times R_v}$$

$$= \frac{p_v}{p_a} \times \frac{287}{8314.3/14}$$

$$= \frac{0.622 p_v}{p_a}$$

But If p = total pressure of air = barometric pressure

$$= p_v + p_a$$

$$\therefore \quad p_a = p - p_v$$

$$\therefore \quad W_H = \frac{0.622 p_v}{p - p_v} \quad \text{...(1)}$$

7. Degree of Saturation (μ). Degree of saturation is defined as the ratio of the actual mass of water vapour present in the moist air to the maximum mass of water vapour in the air at the same dry bulb temperature. It is denoted by μ.

With maximum water vapour in the air, the air becomes saturated and the water vapour is m_s (mass of water vapour at saturation condition).

$$\therefore \quad \mu = \text{Degree of saturation}$$

$$= \frac{m_v}{m_s} \text{ or } = \frac{W_v}{W_s} \quad \text{...(2)}$$

$$= \frac{0.622\ p_v /(p - p_v)}{0.622\ p_s / p - p_s}$$

$$= \frac{p_v}{p_s} \times \frac{p - p_s}{p - p_v} \quad \text{...(3)}$$

8. Relative Humidity. Water vapour contained in the air is called *humidity*. The *Relative Humidity* is defined as the ratio of the actual partial pressure of water vapour in the moist air to the saturation pressure p_s of the vapour at the same dry-bulb temperature of air. It is denoted by symbol ϕ or RH

$$RH = \phi = \frac{m_v}{m_s}$$

$$= \frac{p_v V / R_v T}{p_s V / R_v T}$$

$$= \frac{p_v}{p_s}$$

$$\boldsymbol{\phi} = \frac{\mathbf{p_v}}{\mathbf{p_s}} \quad \ldots(4)$$

$$\therefore \quad \mu = \frac{p_v}{p_s} \times \frac{p - p_s}{p - p_v} = \frac{p_v}{p_s} \times \frac{p\left[1 - \left(p_s / p\right)\right]}{p\left[1 - \left(p_v / p\right)\right]}$$

$$= \phi . \frac{1 - \dfrac{p_s}{p}}{1 - \dfrac{p_s \phi}{p}}$$

or

$$\phi = \frac{\mu}{1 - (1 - \mu)(p_s / p)} \quad \ldots(5)$$

Equation (5) gives a relationship between the relative humidity and degree of saturation.

It may be pointed out that there is no convenient method to measure ϕ, μ or W_H. They must therefore, be calculated by suitable means. All the equations noted above require the values of p_s and p_v. The values of p_s may be obtained from the steam tables. There are two methods by which p_v can be calculated.

1. p_s is the saturation pressure at the dry bulb temperature of air. This is found from steam tables.

If we get or know the dew point temperature of air, then the saturation pressure corresponding to the dew-point temperature is the vapour pressure p_v. From Fig. 17.1, *B* corresponds to the dew point temperature and the pressure at T_B is the partial pressure of water vapour at *A*.

The accurate measurement of DPT, however, is difficult.

2. The second method involves the determination of wet bulb temperature which is discussed below.

9. Wet-Bulb Temperature (WBT) or t_{wb} or t_w. The temperature recorded by a thermometer whose bulb is covered by a wet fibrous cloth is known as the *wet-bulb temperature*. It is denoted either by WBT or t_{wb} or t_w. Frequently a wet bulb thermometer and a dry bulb thermometer are mounted together, and the combined instrument is known as *psychometer*.

Figure 17.2 shows a sling psychometer consisting of two thermometers side by side on a holder. A handle is provided so that the whole device may be whirled through the air. The vigorous rotation of the psychometer helps to get quick and dependable readings on the wet bulb thermometer.

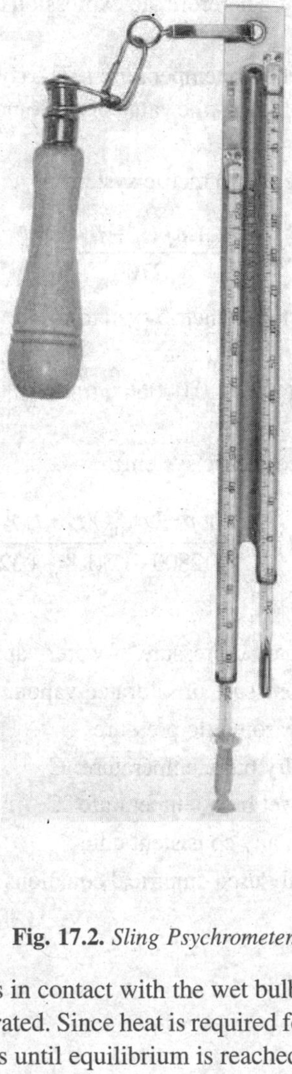

Fig. 17.2. *Sling Psychrometer.*

When the surrounding moist air comes in contact with the wet bulb, vapourisation of water takes place on the bulb till the surrounding air becomes saturated. Since heat is required for vapourisation the temperature of both the air and that of water on the bulb decreases until equilibrium is reached. The equilibrium temperature is known as the *wet-bulb temperature*, abbreviated as WBT or t_{wb} or t_e.

Evidently, the less the relative humidity of the ambient air the greater will be its affinity to absorb more moisture, hence greater will be the evaporation of water from the bulb surface and therefore the greater will be its wet-bulb depression $(t - t')$ below dry bulb temperature value. If the ambient air is *saturated*, there will be no evapouration of the bulb and hence dry bulb and wet bulb temperatures will be the same.

$\therefore$ Wet bulb depression $= t_d - t_w = t - t'$

$\phi = 1, \quad t_d = t_w = t'$

or for Wet bulb depression = 0,

$$\phi = 1$$

It may be noted that the wet bulb temperature is an indirect measure of the moisture content of the air. Due to various inaccuracies in the measurement (which is not carried on under ideal conditions), the temperature t_w can not be considered as a thermodynamic property. Therefore, no expression can be found analytically to obtain the value of p_v in terms of $t_w = t'$.

However, since the measurement of wet bulb temperature t_w is easy, certain empirical relations exist which are used to obtain the value of partial pressure p_v from the value of t_w. Some of the empirical equations involving p_v and t_w are,

(i) Modified Apjohn Equation (as converted to metric system)

$$p_v = (p_s)_{t_w} - \frac{1.8p(t_d - t_w)}{2700} \qquad \text{...(6)}$$

(ii) Modified Ferret Equation (as converted to metric system).

$$p_v = (p_s)_{t_w} - 0.00066\ p(t_d - t_w)\left(1 + \frac{1.8t_d}{1571}\right) \qquad \text{...(7)}$$

(iii) The Carrier Equation (as converted to metric system)

$$p_v = (p_s)_{t_w} - \frac{(p - (p_s)_{t_w})(t_d - t_w)1.8}{2800 - 1.3(1.8t_d + 32)} \qquad \text{...(8)}$$

where

p_v = partial pressure of water vapour

$(p_s)_{t_w}$ = pressure of saturated vapour at t_w

P = barometric pressure

t_d = dry bulb temperature °C

t_w = wet bulb temperature °C

The pressures P, $(P_s)_{tw}$ and P_v may be in any consistent units.

In addition to these, two more commonly used empirical equations to find p_v, may be considered.

(iv) $$p_v = (p_s)_{t_w} - \frac{\left[p - (p_s)_{t_w}\right](t_d - t_w)}{1527.4 - 1.3t_w} \qquad \text{...(9)}$$

(v) $$p_v = (p_s)_{t_w} - \frac{\left[p - (p_s)_{t_w}\right](t_d - t_w)}{1544 - 1.44t_w} \qquad \text{...(10)}$$

10. Thermodynamic Wet Bulb Temperature or Temperature of Adiabatic Saturation. If the saturation process as discussed above for the measurement of wet bulb temperature is carried on under ideal conditions (no radiation and other heat losses), then the temperature so recorded will be a thermodynamic property and is known as thermodynamic wet bulb temperature or temperature of adiabatic saturation. It is denoted as t^*. The concept of this temperature is given below.

A mixture of dry air and water vapour at temperature t_a flows through an insulated duct, Fig. 17.3, pass over a surface of water. It is assumed that (i) the flow is steady (ii) there is no work done and (iii) the process is adiabatic.

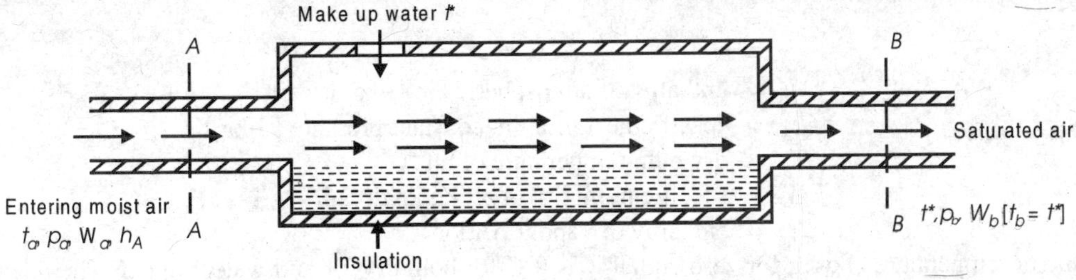

Fig. 17.3. *Adiabatic saturation process.*

Since the air entering the duct is unsaturated, as it passes over the water surface some of the water is evapourated. Both air and water are cooled due to evapouration. Eventually, the water attains a *steady state* temperature when the heat energy transferred to water from air is exactly equal to the heat energy required to vapourise the water. When this state of thermal equilibrium is reached, the latent heat of vapourisation water is supplied completely by the air and no further drop in water temperature occurs. This *steady state temperature* of water is known as the adiabatic saturation temperature t^* of entering air. In order to maintain the thermal equilibrium the make up water must be supplied at the *adiabatic saturation temperature* t^*.

If the insulated duct is sufficiently long, the air leaving the duct will be saturated air and its temperature will be equal to water temperature t^*.

A simple energy balance for the adiabatic process described above will show that, for a given initial state of air and for complete saturation of air at exit, the *exit air temperature* t^* can have but one value. Thus t^* is a property of the air-vapour mixture.

Note that the state of air continuously changes from the inlet to the exit of the duct. But for all these different states of air, the temperature of adiabatic saturation is the same because the steady-state water temperature has the same value all along the duct. Thus the above process occurs at a constant temperature of adiabatic saturation.

The process of adiabatic saturation may be traced on a *T–S* diagram by curve *AB* in Fig. 17.4. The partial pressure of water vapour increases during the process from p_a to p_b while the total pre-pressure *p* of the mixture remains the same.

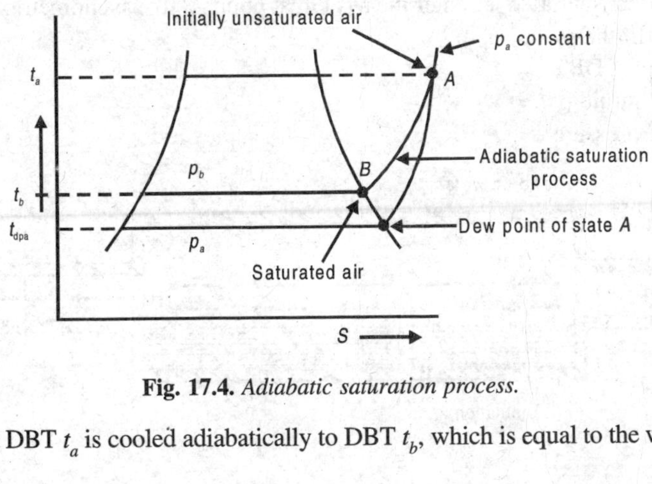

Fig. 17.4. *Adiabatic saturation process.*

The vapour initially at DBT t_a is cooled adiabatically to DBT t_b, which is equal to the wet bulb temperature t^*.

11. Enthalpy of Moist Air. The *enthalpy* of a mixture of dry air and water vapour is equal to the sum of the enthalpy of (i) 1 kg of dry air and (ii) the associated water vapour.

$$\therefore \quad h = h_a + Wh_v \qquad \text{...(11)}$$
$$= C_{pa} t_d + wh_v$$

where

h = enthalpy of atmospheric air kJ/kg dry air
C_{pa} = specific heat of air at constant pressure kJ/kg-K
T_d = dry bulb temperature of air °C
W = specific humidity of air kg/kg dry air
h_v = enthalpy of vapour per kg

The value of temperature chosen for zero enthalpy is 0°C for both dry air and water vapour. The relationship between the enthalpy of dry air and its temperature is not quite linear and values taken from NBS Circular No. 564, for the standard atmospheric pressure of 101.325 kPa and suitably modified for the chosen zero, for the basis of the tables of properties of humid air. An approximate equation for the enthalpy of dry air over the range of 0 °C to 60 °C is

$$h_a = 1.007\,t - 0.026 \qquad \text{...(12)}$$

and for lower temperature down to – 10 °C, the approximate equation is

$$h_a = 1.005\,t \qquad \text{...(13)}$$

The enthalpy of vapour over ice, however, is based on the information in the ASHRAE Handbook of Fundamentals (1967).

For the purpose of approximate calculations, without recourse to the psychrometric tables, we may assume that, in the range of 0°C to 60°C, the vapour is generated from water at 0°C and that the specific heat of superheated steam is constant. The following equation can then be used for the enthalpy of water vapour

$$h_v = 2501 + 1.84\,t \qquad \text{...(13-a)}$$

These equations can be combined to give us an approximate expression for the enthalpy of humid air at a barometric pressure of 101.325 kPa.

$$\therefore \quad h = h_a + wh_v$$
$$= (1.007\,t - 0.026) + W(2501 + 1.84t)$$
$$= 1.005\,t_{db} + W(2501 + 1.84\,t_{db})$$
$$h = 1.005\,t_{db} + W(2501 + 1.84\,t_{db}) \qquad \text{...(14)}$$

We can find specific humidity from adiabatic saturation system.

Referring to Fig. 17.3, let unsaturated air enter the system at point A. Its psychometric properties are,

Enthalpy = h_a
DBT = t_a
Specific humidity = w_a
Total pressure = P

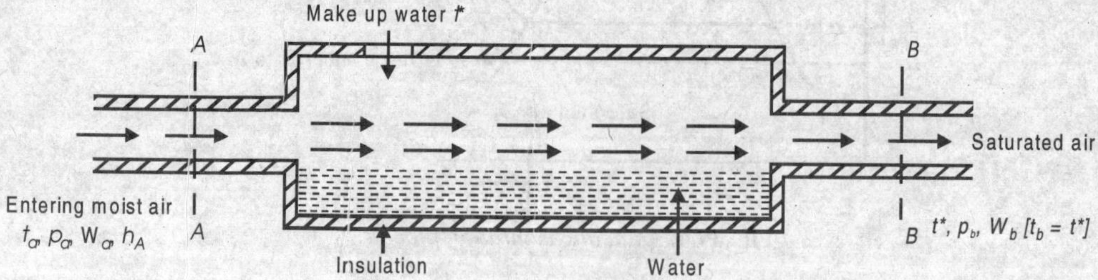

Let saturated air leave the system at *B*. Its corresponding properties are

$$\text{Enthalpy} = h_B = h_{sb}$$
$$\text{DBT} = \text{WBT} = t^* = t_{wb}$$
$$\text{Specific humidity} = w_{bs}$$
$$\text{Total pressure} = P$$

Water at t_{wb} which is supplied, has liquid enthalpy h_{cb}

System is adiabatic.

$$\therefore \quad h_{sb} = h_a + (w_{bs} - w_a) h_{lb}$$

$$\therefore \quad h_a - w_a \cdot h_{lb} = h_{sb} - w_{bs} h_{lb} \quad \ldots(15)$$

But $h_a = (h_{\text{dry air}})_a + w_a$ (enthalpy of superheated vapour at t_a per kg)

$$= h_{\text{dry air}} + w_a h_{vb}$$

h_{sb} = enthalpy of 1 kg air at $t_{ws} + w_{bs}$ (enthalpy of saturated vapour at t_{wb} per kg of vapour)

$$= (h_{\text{dry air}})_b + w_{bs} h_{vb}$$

Knowing $t_{db} = t_{wb}$, and substituting these values in Eq. 15, we get,

$$\mathbf{w_a} = \frac{\mathbf{W_{bs} L_{bs} - 1.005(t_{da} - t_{db})}}{\mathbf{h_{va} - h_{wb}}} \quad \ldots(16)$$

By applying fundamental procedures of statistical mechanics, Goff and Gratch calculated accurate thermodynamic properties of moist air for standard sea level conditions (total pressure 760 mm Hg).

Calculations for specific volume, specific enthalpy and specific entropy of *unsaturated moist air* are approximately given closely by the following relations

Volume $\rightarrow v = v_a + \mu\,(v_{as} - v_a)$

Enthalpy $\rightarrow h = h_a + \mu\,(h_{as} - h_a)$ Refer table on next page 17.10   ...(17)

Entropy $\rightarrow s = s_a + \mu\,(s_{as} - s_a)$

17.3 PSYCHROMETRIC TABLE

Psychrometric properties of air at standard Barometric Pressure of 760 mm Hg or 1.01325 bar are tabulated for temperatures from – 10°C to 60°C.

The Psychrometric Tables are given on next page.

17.4 PSYCHROMETRIC CHART

In the preceding sections we considered various thermodynamic properties of moist air, and the equations relating them, with the help of which we accurately obtain the solution of problems concerning moist air. However, calculations in some cases is tedious and time consuming. A graphical plot of the properties of moist air may reduce the calculations as the various processes may directly be shown on such a graph and the required results obtained. Such a monograph is called a *psychrometric chart*. It is shown in Fig. 17.5.

If the pressure and two other independent properties of moist air are known, the thermodynamic state of moist air is uniquely fixed. A particular chart is therefore constructed for a specified value of barometric pressure usually 76 mm Hg. The choice of coordinates is arbitrary. However, most of the charts employ DBT and specific humidity as basic co-ordinates.

Psychometric Properties of Air at Standard Barometric Pressure of 76 mm Hg or 1.01325 bar

	Properties of water and steam					Properties of dry air at 760 mm of Hg		Properties of mixture of dry air and saturated steam i.e. (1+ w) kg. at 760 mm Hg		
Tem pera ture °C	Saturated pressure of water and steam p_s		Enthalpy of saturated water h_f or h_w	Enthalpy of saturated steam h_g	Specific volume of saturated steam V_g	Specific volume V_a	Enthalpy h_a	Volume of mixture (1 + w) kg of dry air V_{as}	Enthalpy of mixture (1 + w) kg per kg of dry air h_{as}	Specific humidity per kg of dry air W_s
°C	mm Hg	Bar	kJ/kg	kJ/kg	m³/kg	m³/kg	kJ/kg	m³/kg	kJ/kg	kg/kg
0	4.580	0.006108	0.00	2501.6	206.31	0.7734	0.000	0.7781	9.473	0.003789
1	4.924	0.006566	4.17	2503.4	192.61	0.7762	1.006	0.7813	11.204	0.004076
2	5.291	0.007055	8.39	2505.2	179.92	0.7791	2.012	0.7845	12.982	0.004381
3	5.681	0.007575	12.60	12507.1	168.17	0.7819	3.018	0.7878	14.811	0.004707
4	6.097	0.008129	16.80	2508.9	157.27	0.7848	4.024	0.7911	16.696	0.005054
5	6.539	0.008718	21.01	2510.7	147.16	6.7876	5.030	0.7944	18.639	0.005424
6	7.010	0.009345	25.21	2512.6	137.79	0.7904	6.036	0.7978	20.644	0.005818
7	7.510	0.010012	29.41	2514.4	129.06	0.7933	7.041	0.8012	22.713	0.006237
8	8.041	0.010720	33.60	2516.2	120.97	0.7961	8.047	0.8046	24.852	0.006683
9	8.605	0.011472	37.80	2518.1	113.44	0.7990	9.053	0.8081	27.064	0.007157
10	9.204	0.012270	41.99	12519.9	106.43	0.8018	10.059	0.8116	29.352	0.007661
11	9.839	0.013116	46.19	2521.7	99.909	0.8046	11.065	0.8152	31.724	0.008197
12	10.512	0.014014	50.38	2523.6	93.835	0.8075	12.071	0:8188	34.179	0.008766
13	11.225	0.014965	54.57	2525.4	88.176	0.8103	13.077	0.8225	36.726	0.009370
14	11.982	0.015973	58.75	2527.2	82.900	0.8132	14.084	0.8262	39.370	0.010012
15	12.781	0.017039	62.94	2529.1	77.978	0.8160	15.090	0.8300	42.113	0.010692
16	13.628	0.018168	67.13	2530.9	73.384	0.8188	16.096	0.8338	44.963	0.011413
17	14.524	0.019362	71.31	2532.7	69095	0.8217	17,103	0.8377	47.926	0.012178
18	15.470	0.020624	75.50	2534.5	65.087	0.8245	18.108	0.8414	51.008	0.012989
19	16.470	0.021957	79.68	2536.4	61.341	0.8274	19.114	0.8457	54.216	0.013848
20	17.53	0.023366	83.86	2538.2	57.838	0.8302	20.121	0.8498	57.555	0.014758
21	18.64'	0.024853	88.04	2540.0	54.561	0.8330	21.127	0.8540	61.035	0.015721
22	19.82	0.026422	92.23	2541.8	51.492	0.8359	22.133	0.8583	64.660	0.016741
23	21.06	0.028076	96.41	2543.6	48.619	0.8387	23.140	0.8627	68.440	0.017821

Properties of water and steam						Properties of dry air at 760 mm of Hg		Properties of mixture of dry air and saturated steam i.e. (1+ w) kg. at 760 mm Hg		
Tem pera ture °C	Saturated pressure of water and steam p_s		Enthalpy of saturated water h_f or h_w	Enthalpy of saturated steam h_g	Specific volume of saturated steam V_g	Specific volume V_a	Enthalpy h_a	Volume of mixture (1 + w) kg of dry air V_{as}	Enthalpy of mixture (1 + w) kg per kg of dry air h_{as}	Specific humidity per kg of dry air W_s
°C	mm Hg	Bar	kJ/kg	kJ/kg	m³/kg	m³/kg	kJ/kg	m³/kg	kJ/kg	kg/kg
24	22.37	0.029821	100.59	2545.5	45.926	0.8416	24.146	0.8671	72.385	0.018963
25	23.75	0.031660	104.77	2547.3	43.402	0.8444	25.153	0.8717	76.500	0.020170
26	25.50	0.033597	108.95	2549.1	41.034	0.8472	26.159	0.8764	80.798	0.021448
27	26.73	0.035636	113.13	2550.9	38.813	0.8501	27.165	0.8811	85.285	0.022798
28	28.34	0.037782	117.31	2552.7	36.728	0.8529	28.172	0.8860	89.976	0.024226
29	30.03	0.040040	121.48	2554.5	34.769	0.8558	29.179	0.8910	94.878	0.025735
30	31.81	0.042415	125.66	2556.4	32.929	0.8586	30.185	0.8962	100.006	0.027329
31	33.69	0.044911	129,84	2558.2	31.199	0.8614	31.192	0.9015	105.369	0.029014
32	35.65	0.047534	134.02	2560.0	29.572	0.8643	32.198	0.9069	110.979	0.030793
33	37.72	0.050288	138.20	2561.8	28.042	0.8671	33.205	0.9125	116.857	0.032674
34	39.89	0.053180	142.38	2563.6	26.601	0.8700	34.212	0.9183	123.011	0.034660
35	42.17	0.056216	146.56	2565.4	25.245	0.8728	35.219	0.9242	129.455	0.036756
36	44.55	0.059400	150.74	2567.2	23.967	0.8756	36.226	0.9303	136.209	0.038971
37	47.06	0.062739	154.92	2569.0	22.763	0.8785	3 7.233	0.9366	143.290	0.041309
38	49.69	0.066240	159.09	2570.8	21.627	0.8813	38.239	0.9431	150.713	0.043778
39	52.44	0.069908	163.27	2572.6	20557	0.8842	39.246	0.9498	158.504	0.046386
40	55.31	0.73750	167.45	2574.4	19.546	0.8870	40.253	0.9568	166.683	0.049141
41	58.33	0.677773	171.63	2576.2	18592	0.8898	41.261	0.9640	175.265	0.052049
42	61.49	0.081986	175.81	2577.9	17.692	0.8927	42.268	0.9714	184.275	0.055119
43	64.80	0.086391	179.99	2579.7	16.841	0.8955	43.275	0.9792	193.749	0.058365
44	68.25	0.091001	184.17	2581.5	16.036	0.8983	44.282	0.9872	203.699	0.061791

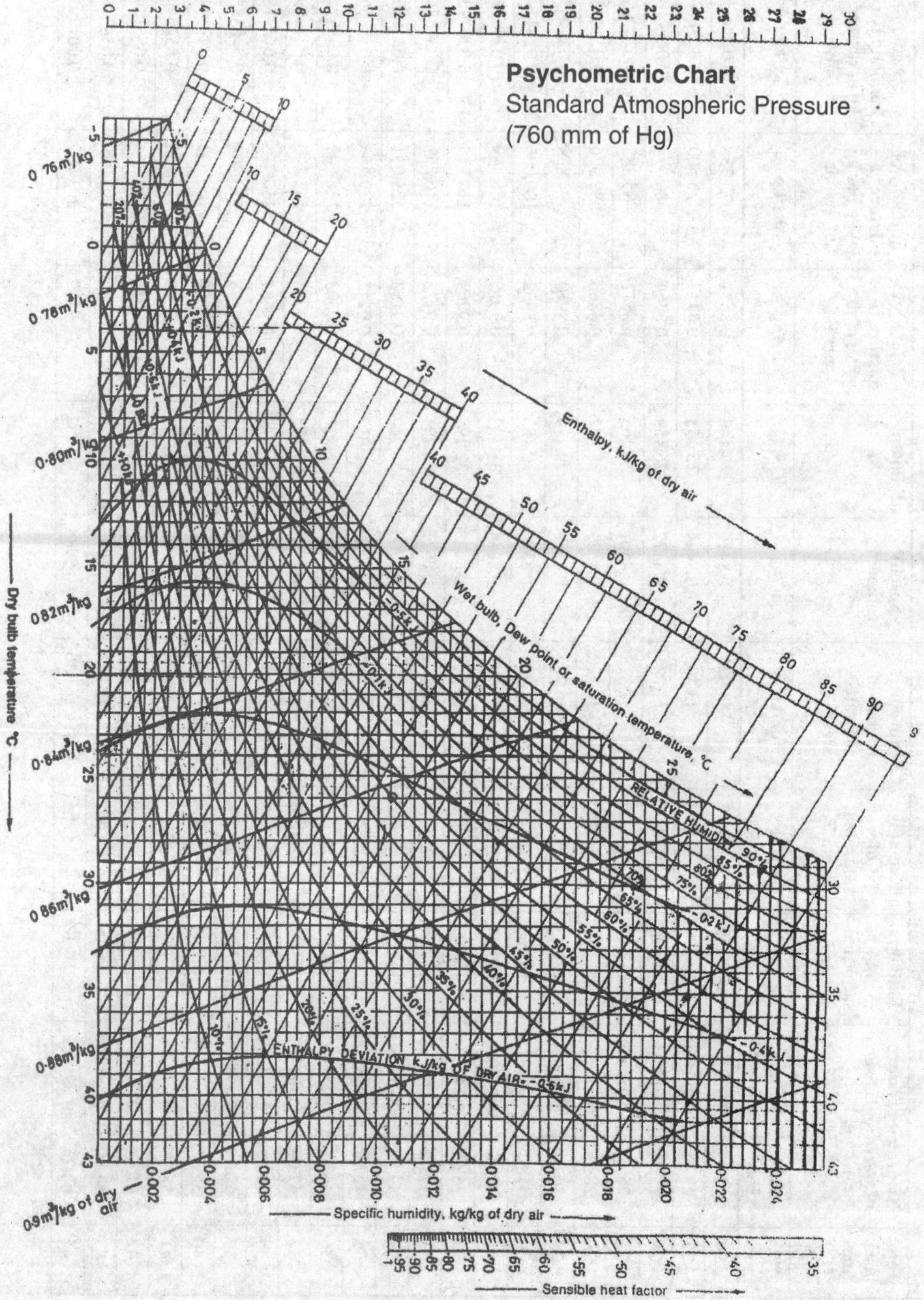

Fig. 17.5. *Psychometric chart.*

In 1923 Mollier introduced a chart using enthalpy and specific enthalpy as oblique co-ordinates, but the inclination of enthalpy lines was so chosen that specific humidity lines were horizontal. His chart, therefore, had the appearance similar to that of Fig. 17.5. Equation (1) indicates that the vapour pressure is approximately a linear function of the specific humidity; hence pressure scale will be an auxiliary ordinate scale parallel to the specific humidity scale.

The mass basis of the chart is 1 kg dry air plus associated water vapour. The various lines drawn on the psychrometric chart are discussed now with reference to Fig. 17.5.

(a) Constant Dry Bulb Temperature Lines. All vertical lines perpendicular to the horizontal d.b.t. scale are constant temperature lines. (–5°C to 43°C). Fig. 17.5 (a).

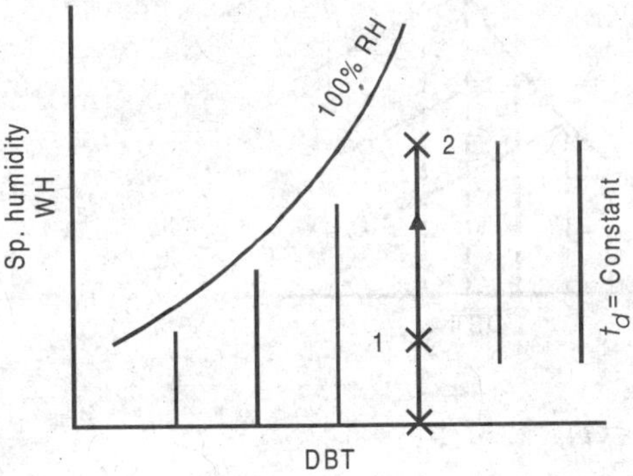

Fig. 17.5. (a)

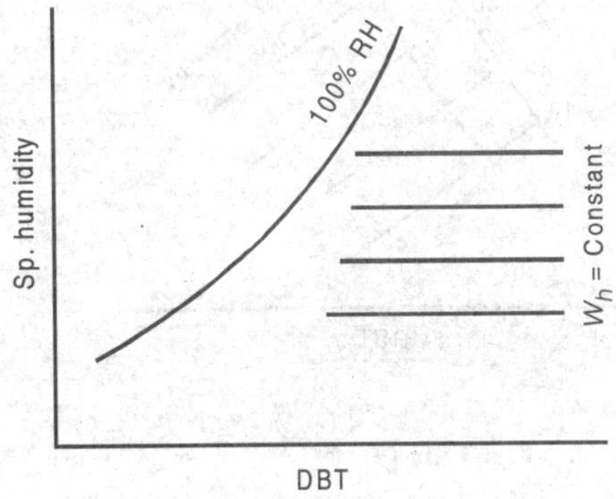

Fig. 17.5. (b)

(b) Constant Specific Humidity Lines. All horizontal lines perpendicular to the vertical specific humidity scale are constant specific humidity lines. $(kg)_{vapour}$ per $(kg)_{dry\ air}$ [0 to 0.026]. Fig. 17.5(b).

(c) Saturation Curve. The extreme left hand curve is the saturation curve which represents state of saturated air at different values of d.b.t. (DBT). It is a100% relative humidity line or a curve. Fig. 17.5 (c).

(d) Constant R.H. Lines. Suppose we want to construct a constant RH line of $\phi = 50\%$. We can select dry bulb temperatures t_1, t_2, t_3 etc. and find out the corresponding saturation pressures p_{s1}, p_{s2}, p_{s3} etc. from the steam tables. Since $p_v = \phi . p_s = 0.5\ p_s$, the values of p_{v1}, p_{v2}, p_{v3} etc. are readily calculated. Since $w = 0.622\ p_v$, we can find out w_1, w_2, w_3 etc. for $\phi = 50\%$. The $P - P_v$ curve passing through (t_1, w_1), (t_2, w_2), (t_3, w_3) is the required $\phi = 50\%$ line or curve. Fig. 17.5 (d).

(e) Constant Enthalpy Lines. The enthalpy lines are inclined straight lines and uniformly spaced as shown in Fig. 17.5 (e). These lines are parallel to wet bulb temperature lines and are drawn up to saturation curve. Most of these lines coincide with bulb temperature lines. Fig. 17.5 (e).

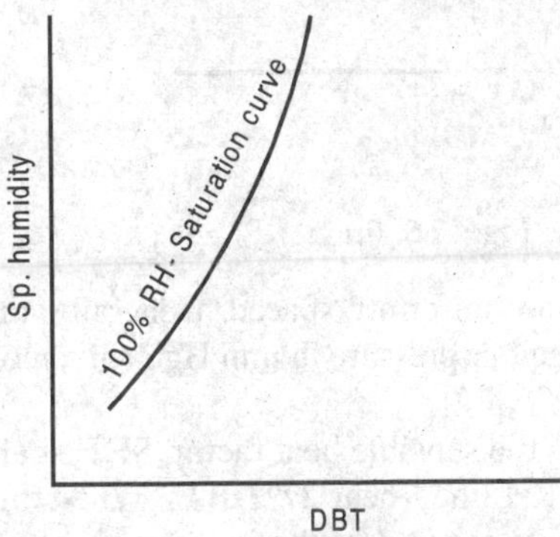

Fig. 17.5. (c)

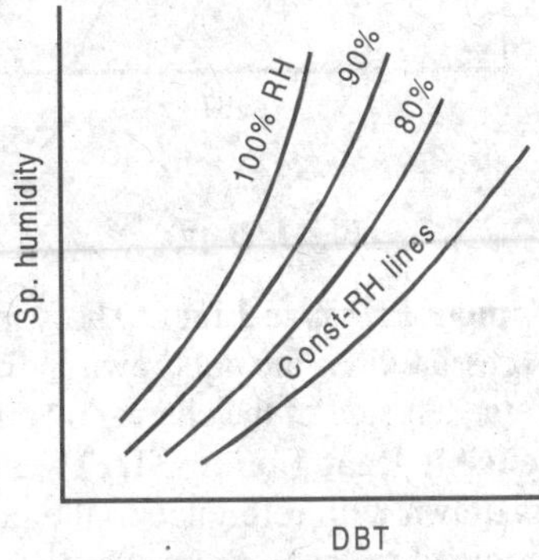

Fig. 17.5. (d)

(f) Constant Wet Bulb Temperature Lines. The wet-bulb temperature lines are inclined lines and non-uniformly spaced as shown in Fig. 17.5 (f). At any point on the saturation curve, the dry bulb and wet-bulb temperature are equal. The values of wet bulb temperature are generally given along the saturation curve of the chart Fig. 17.5 (f).

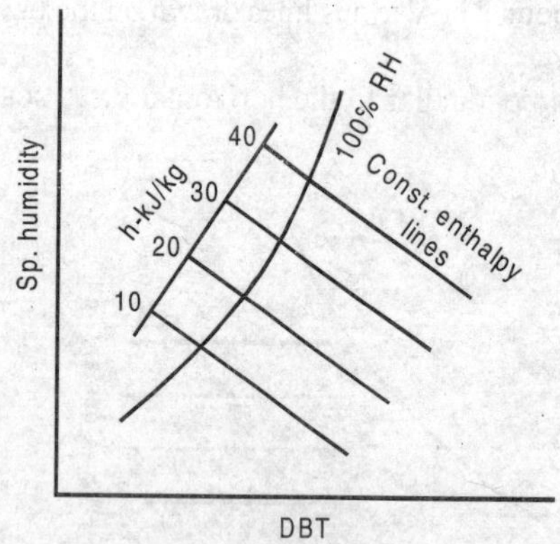

Fig. 17.5. (e)

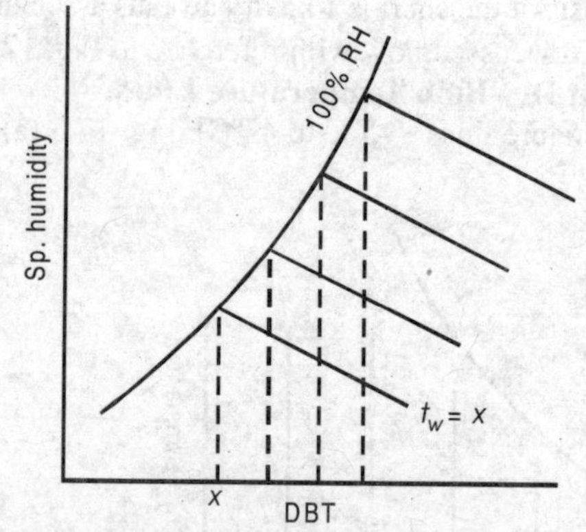

Fig. 17.5. (f)

(g) Constant Specific Volume Lines. The constant specific volume lines are inclined straight lines and are uniformly spaced as shown in Fig. 17.5 (g). These lines are drawn up to saturation curve. The value of specific volume represented by them are generally given at the base of the chart. Fig. 17.5 (g).

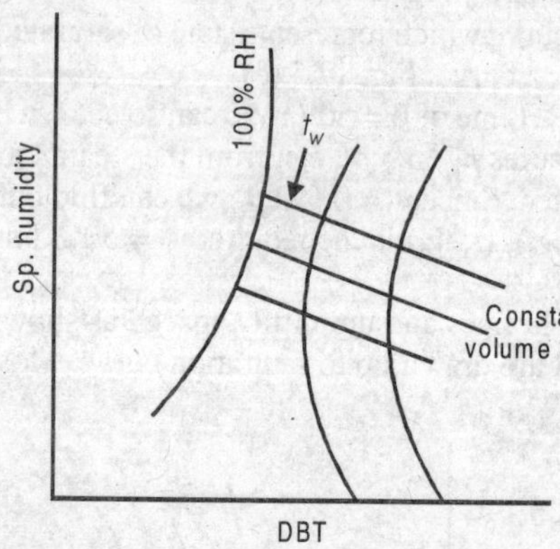

Fig. 17.5. (g)

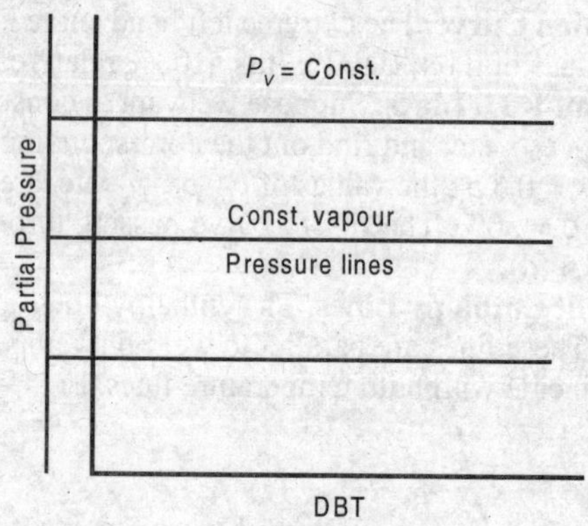

Fig. 17.5. (h)

(h)Vapour Pressure Lines. The vapour pressure lines are horizontal and uniformly spaced. Generally, the vapour pressure lines are not drawn in the main chart. But a scale showing vapour pressure in mm Hg is given on the extreme left side of the chart shown in Fig. 17.5 (h).

(i) Sensible Heat Factor (SHF) Scale. The scale on the extreme right is the sensible heat factor, SHF, scale which is drawn with reference to the point shown as a dark circle at 50% RH line. (near 27° DBT). When this point is joined to the appropriate value on the SHF scale, the resulting line represents the inclination of all heating or cooling processes of moist air which are carried on according to that particular value of SHF.

The sensible heat factor (SHF) is defined as the ratio of the sensible heat (added or removed) to the total heat in the process.

$$\therefore \qquad \text{SHF} = \frac{\text{Sensible heat}}{\text{Total heat}}$$

$$= \frac{\text{Sensible heat}}{\text{Sensible heat + Latent heat}}$$

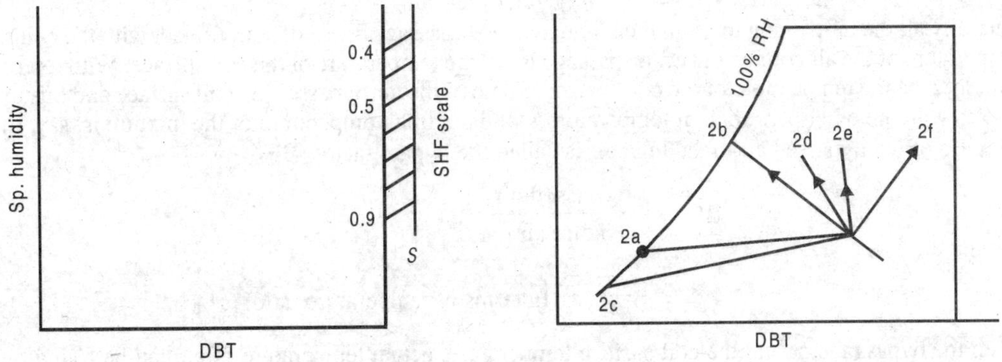

Fig. 17.5. (i)

17.5 PSYCHROMETRIC PROCESSES

The common *Psychrometric processes* which are usually encountered in practice and which may be easily solved with the help of a psychrometric chart are given below and are indicated in Fig. 17.6.

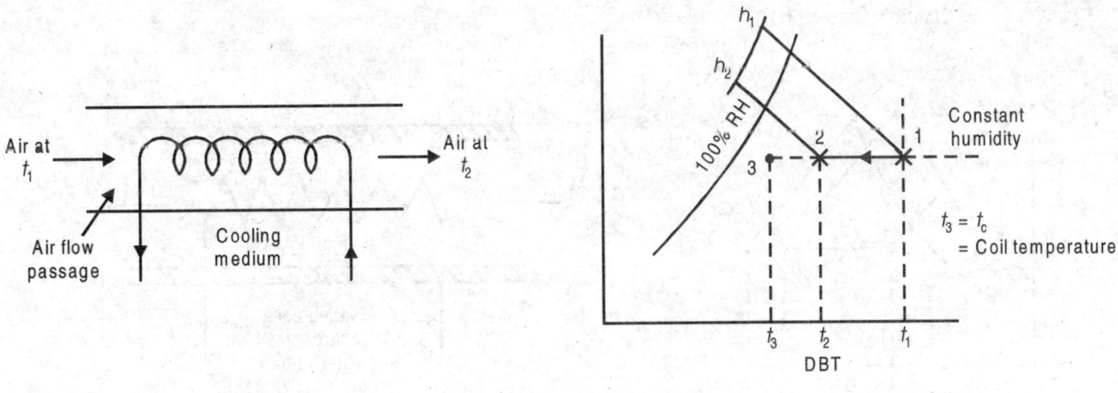

Fig. 17.6. (a)

1. Simple heat transfer process:

(a) Sensible cooling process

(b) Sensible heating process.

2. Simple Humidification process at same DBT.

3. Simple dehumidification process at same DBT.

4. Heat transfer processes with change in humidity.

(a) Cooling and dehumidification
(b) Heating and humidification
(c) Heating and dehumidification
(d) Cooling and humidification

5. Adiabatic mixing of two samples of air. These processes will now be explained individually.

1. (a) Sensible Cooling. This process involves cooling of air at constant moisture content. Process 1–2 is a horizontal line from right to left. Evidently, such cooling of air has to be carried on by bringing it into contact with a body or a surface temperature of which is not below the dew point temperature of air at 1.

$$\text{Heat removed} = h_1 - h_2 \text{ kJ/kg of dry air.}$$

Generally all the air passing through the air duct or flow passage does come in contact with the coil surface and the temperature of the air coming out of the passage is not the temperature of the coil surface. With reference to Fig. 17.6 (a), let t_c be the temperature of the coil surface. Some of the air bypasses the coil surface and this bypassed air at t_1 mixes with the other part of air at temperature t_c and the final temperature of the mixture is, say, t_2.

The ratio of the bypassed air to the total air is called the bypass factor (BF).

$$\therefore \quad \text{BF} = \frac{\text{Bypassed air}}{\text{Total air}}$$

$$= \frac{t_2 - t_c}{t_1 - t_c} \text{ (In terms of temperature ratio)} \quad \ldots(18)$$

Given the bypass factor and the coil surface temperature, actual temperature of air can be found by using the equation.

(b) Sensible Heating Process. This process involves of heating of air without the change of humidity or humidity content.

Heating process line is a straight line from left to right Fig. 17.6 (b)

Heat supplied/kg of air = $h_2 - h_1$ kJ/kg dry air.

For this heating process,

$$\text{Bypass factor (BF)} = \frac{t_c - t_2}{t_c - t_1} \quad \ldots(19)$$

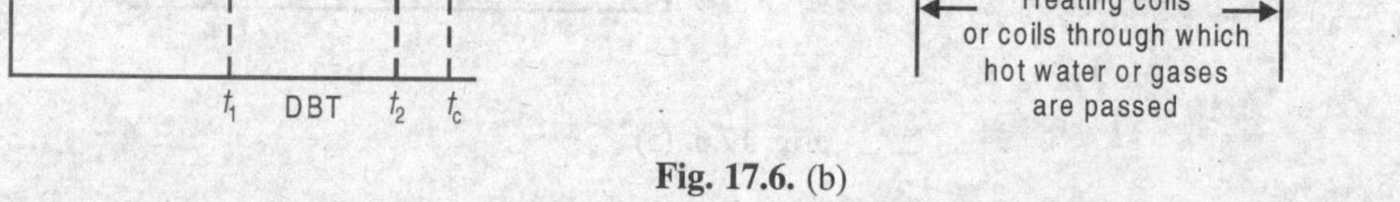

Fig. 17.6. (b)

The amount of bypass depends upon
(a) Velocity of air over the coil.
(b) Number of rows of coil pipes in the direction of flow.
(c) In case of a coil with fins attached, the bypass factor depends on the number of fins per unit length on the coil.

2. Simple Humidification Process at the same DBT. Adding moisture at constant dry bulb temperature can be represented by a vertical line on a psychrometric chart.

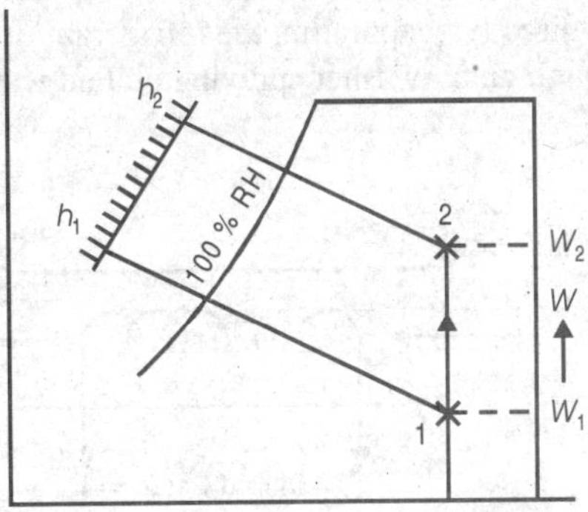

Fig. 17.6. (c)

Pure humidifying process is possible only when heat of vaporisation is supplied from a source other that air without affecting the dry bulb temperature. This happens when water at the dry-bulb temperature of the entering air is sprayed in the chamber whose temperature is maintained at the air DB temperature.

$$\text{Heat supplied} = h_2 - h_1 \text{ kJ/kg dry air} = (w_2 - w_1) \times \text{Latent heat}$$

$$\text{Moisture added} = w_2 - w_1 \text{ kg/kg dry air.}$$

3. Simple Dehumidification Process at same DBT. Pure dehumidification occurs without the change of dry-bulb temperature of air, when air is passed over chemical absorbing moisture and heat evolved, when moisture is absorbed, is removed immediately by the surrounded walls. These walls are maintained at the dry-bulb temperature of air. The chemical is also maintained at the dry-bulb temperature of air. The process of pure dehumidification is shown in Fig. 17.6(d).

$$\text{Heat removed} = h_1 - h_2 = (w_1 - w_2) \times \text{Latent heat.}$$

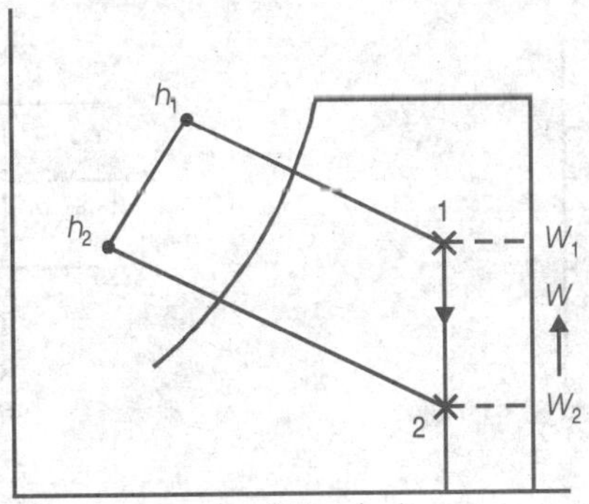

Fig. 17.6. (d)

Pure or simple humidification and dehumidification processes are not found in practice and are always accompanied by heating and cooling.

4. (a) Cooling and Dehumidification Process. This process involves lowering of both temperature and specific humidity, and may be carried on by bringing the moist air into contact with a chilled body or surface having its temperature lower than the dew point temperature of the air.

The cooling and dehumidification process of moist air may be carried on by any one of the methods given below:

1. By causing the air to flow over a *chilled surface*, below dew point temperature, which may be cooled by circulating chilled water or by directly evaporating the refrigerant of a refrigeration system, within the coil.
2. By causing the air to flow through an air washer spraying chilled water below dew point temperature.

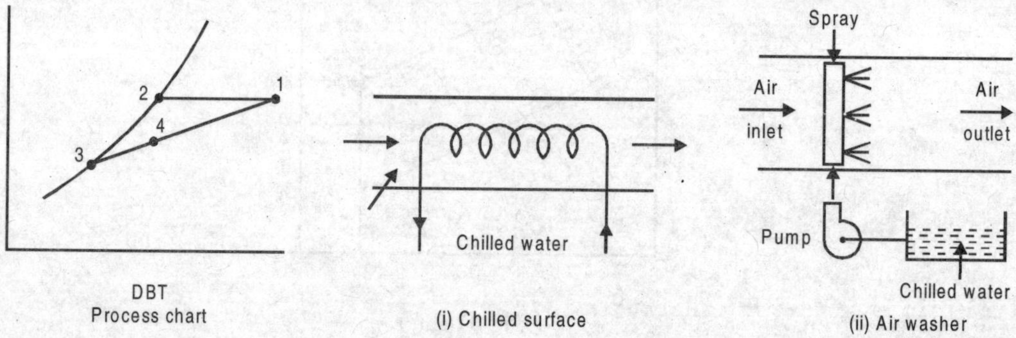

Fig. 17.6. (e)

These processes have practical applications in summer air conditioning. Since the actual process is difficult to trace on the chart, only end states are shown distinctly and joined by dotted curve. But possible processes can be shown as – (1–2) is the sensible cooling up to the dew point temperature and then cooling of air below dew point temperature as shown by (2–3). t_3 is the coil surface temperature and if the coil is having a bypass factor then the actual state of air may be given by point (4). t_3 is the apparatus dew point temperature ADP and the bypass factor is given by

$$BF = \frac{t_4 - t_3}{t_1 - t_3} = \frac{t_4 - ADP}{t_1 - ADP}$$

(b) Heating and Humidification Process. This process is the reverse of that discussed in 4 (a) above. In this process the specific humidity and the dry bulb temperature of air are increased.

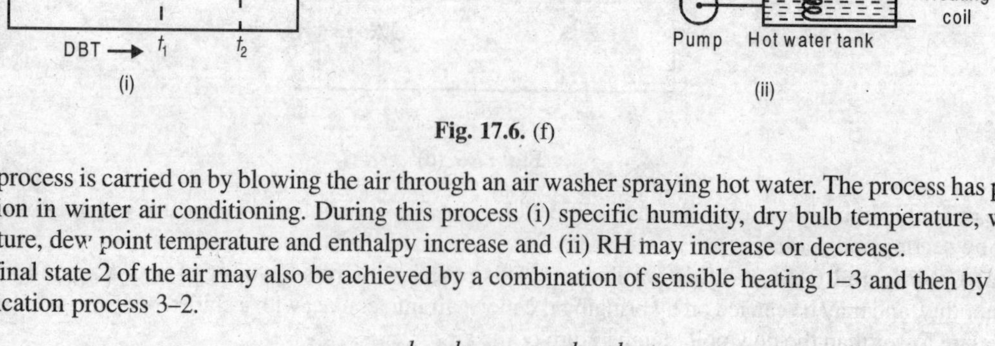

Fig. 17.6. (f)

This process is carried on by blowing the air through an air washer spraying hot water. The process has practical application in winter air conditioning. During this process (i) specific humidity, dry bulb temperature, wet bulb temperature, dew point temperature and enthalpy increase and (ii) RH may increase or decrease.

The final state 2 of the air may also be achieved by a combination of sensible heating 1–3 and then by sensible humidification process 3–2.

$$SHF = \frac{h_3 - h_1}{(h_3 - h_1) + (h_2 - h_3)} = \frac{h_3 - h_1}{h_2 - h_1}$$

Heating and humidification can also be achieved by injecting into air, the steam. This process is generally used in textile industry where high humidity is to be maintained. Rise in dry bulb temperature is generally very little. Instead of hot water, steam will be injected.

(c) Heating and Dehumidification Process. There are certain chemical substances like *silica gel* and *activated alumina* which have an affinity for moisture. If air is passed over such chemicals it may be dehumidified. Usually in doing so the moisture is condensed and gives up its latent heat, raising the dry bulb temperature of air.

This process is reverse of adiabatic saturation process.

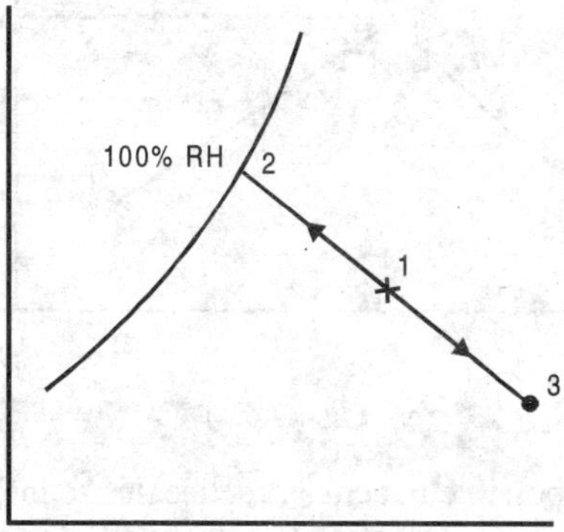

Fig. 17.6. (g)

This is shown in Fig. 17.6(g). In figure, 1–2 adiabatic saturation process 1–3 adiabatic dehumidification process.

The path followed during this (adiabatic dehumidification) process is along the constant wet bulb temperature line or constant enthalpy line.

Actually there are two types of chemicals used for chemical dehumification. The are

1. *Absorbents.* They include water solutions or brines of calcium chloride, lithium chloride, lithium bromide and ethylene glycol. Generally these are used in absorption refrigeration systems.
2. *Adsorbents.* These are the substances in solid state and while absorbing moisture from air, they do not change chemically and physically. These are silica gel and activate alumina (porous amorphous form of aluminium dioxide)

(d) Cooling and Humidification (adiabatic humidification). Air passing through a spray chamber with recirculated water follows the process line 1–2 in Fig. 17.6(g). Whereby it cools due to the evaporation of water. Such cooling of air shown schematically in Fig. 17.6(h), will accompany humidification following only one path i.e., constant wet-bulb temperature line.

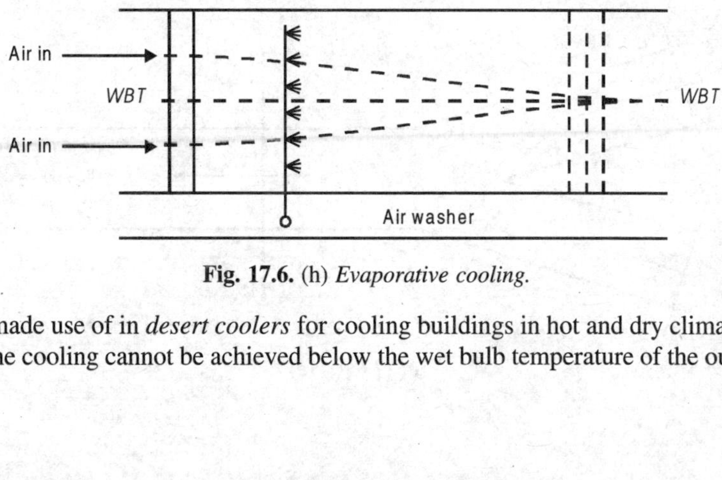

Fig. 17.6. (h) *Evaporative cooling.*

This principle is made use of in *desert coolers* for cooling buildings in hot and dry climate so as to make them comfortable. Since the cooling cannot be achieved below the wet bulb temperature of the outside or incoming air, at which temperature the leaving air will be saturated, this technique of cooling has only limited use. Hence the system becomes ineffective in monsoon season.

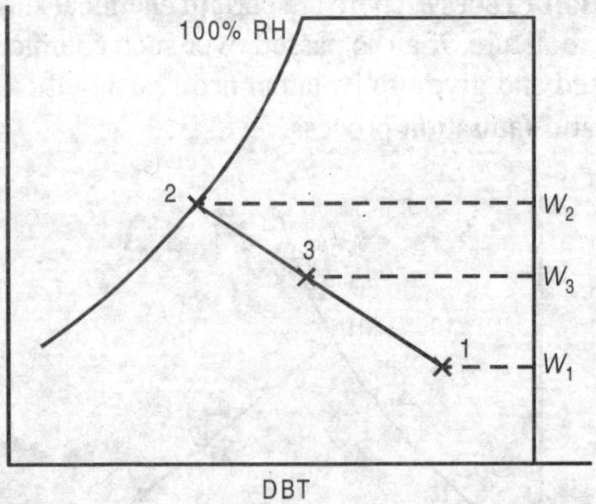

Fig. 17.6. (i)

Perfect humidification of air is not possible in actual practice due to inefficiency of spray chamber. Therefore, the final condition of air at outlet is represented by state 3 on the line 1–2 as shown in Fig. 17.6(i). The effectiveness or the humidifying efficiency of the spray chamber is given by

$$\eta_H = \frac{\text{Actual drop in DBT}}{\text{Ideal drop in DBT}} = \frac{t_1 - t_3}{t_1 - t_2}$$

$$= \frac{\text{Actual increase in humidity}}{\text{Ideal increase in humidity}} = \frac{w_3 - w_1}{w_2 - w_1}$$

5. Adiabatic Mixing of Air Streams. Very often air having different conditions are mixed in the air conditioning system, such as the mixing of fresh air and return or recirculated air before it enters the process plant. The psychrometric chart can be used to find the resulting condition of the mixture.

Figure 17.6 (j) shows the schematic diagram showing two streams of air mixing together.

Let the two streams of masses m_1 and m_2, humidity rations w_1 and w_2 and enthalpies h_1 and h_2 respectively, mix together without any heat exchange with surroundings and also exchange of moisture to give a stream of mass $m_3 = m_1 + m_2$, humidity $w_3 = w_1 + w_2$ and enthalpy $h_3 = h_1 + h_2$.

$$\therefore \quad m_3 = m_1 + m_2$$

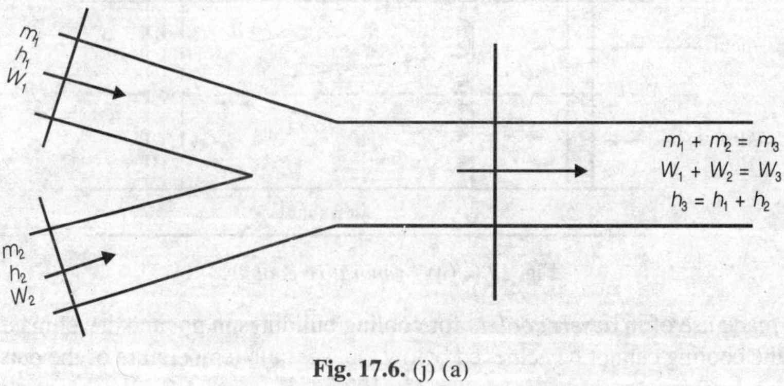

Fig. 17.6. (j) (a)

$$m_3 w_3 = m_1 w_1 + m_2 w_2$$
$$m_3 h_3 = m_1 h_1 + m_2 h_2$$

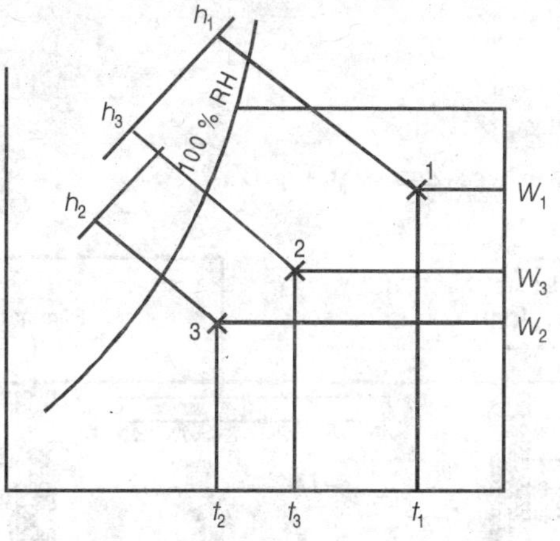

Fig. 17.6. (j) (b)

Eliminating m_3 and rearranging the above equations, we get

$$\frac{m_1}{m_2} = \frac{w_3 - w_2}{w_1 - w_3} = \frac{h_3 - h_2}{h_1 - h_3}$$

The final condition of the mixture can be found with the help of chart as follows.

First determine the percentage of air in each stream, in the mixture. Let in stream one, this is x%. Therefore, $(100 - x)$ is the percentage of air mass in other stream.

Then locate the two points 1 and 2 from the given conditions of two streams. Join 1 and 2. Next step is very important. Multiply the dry bulb temperature of each air by its percentage in the mixture. Let this be $x.t_1$. This is the contribution of stream (1) DB degrees to the mixture. Similarly for second stream $(100 - x)\, t_2$ will be DB degrees contributing in the mixture.

$\therefore \quad x\, t_1 + (100 - x)\, t_2 = 100 \times t_3$

$\therefore \quad t_3$ can be found.

Draw a vertical at t_3 to meet the line 1–2 in 3. Then all other properties can be found from the chart.

17.6 EVAPORATIVE COOLING

We know that liquid will evaporate if there is a difference between the vapour pressures in the air and the saturation pressure of liquid (say water) at the water temperature. The larger the difference between the vapour pressure and saturation pressure, the rate of evaporations larger or higher. The evaporation of water will have a cooling effect on water and thus reduce its temperature. Then the saturation pressure will reduce and the rate of evapouration will reduce. This will continue till some kind of quasi-steady operation is reached. This phenomenon, thus, explains why the water temperature is always less than the surrounding air particularly in dry climate like summer. It is therefore said that the air at the water surface will always be saturated because of the direct contact with water, and thus the vapour pressure.

Then, the natural tendency of water to evaporate in order to achieve pressure equilibrium, forms the basis for the operation of the *evaporative cooler*. Evaporative coolers are commonly used in dry climates (tropical regions) and provide effective cooling.

17.7 AIR WASHERS

Air washer is a device that may be used for

(a) Cooling and dehumidification
(b) Cooling and humidification
(c) Heating and humidification
(d) Humidification
(e) Adiabatic saturation
(f) Cooling.

The schematic diagram for air washer is shown in Fig. 17.7.

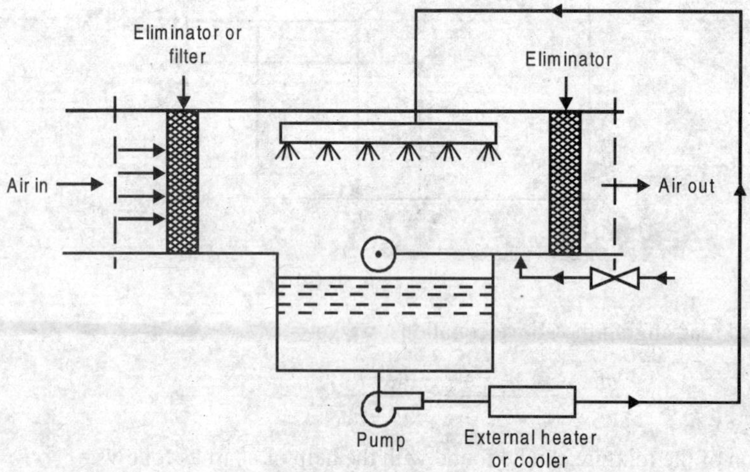

Fig. 17.7. *Air washer.*

Following are the processes that can be carried out with the help of washer shown in Fig. 17.7 and they are shown on psychrometric chart of Fig. 17.8.

1–2a Sensible cooling. Water temperature is equal to dew-point temperature of air.

1–2b Adiabatic saturation. It takes place along the wet bulb temperature line. Water is at WBT

1–2c Cooling and dehumidification. Temperature of water is below the dew point temperature of air.

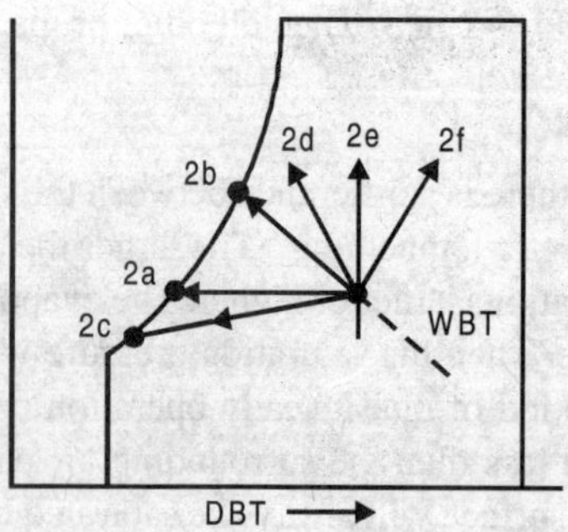

Fig. 17.8. *Various processes on psychrometric chart.*

1–2d Cooling and humidification. When wet bulb temperature is less than water spray temperature and this spray water temperature is less than the dry bulb temperature of air, this cooling and humidification process takes place. The water is required to be heated externally.

1–2e Humidification. Water spray temperature is equal to the dry bulb temperature of air. The water externally heated. (Latent heat of evaporation)

1–2f Heating and Humidification. If the spray water is heated and maintained at temperature higher than the dry bulb temperature of air entering the washer. In this case, the dry bulb temperature of air leaving the washer is more than the inlet temperature of air. Wet bulb is higher than the initial WBT and higher humidity ratio than the entering air.

In comparison with the cooling coil, the air washer has a number of disadvantages:

(a) It is more bulky.

(b) Cost wise it is more expensive as far as maintenance is concerned.

(c) Water is required in greater quantity.

(d) Corrosion is a greater risk.

(e) Rusting, deposition of scales in the chilling plant is also a great problem. Because of rusting and deposition of scales, heat trouser efficiency is greatly affected and cost of refrigeration plant is increased.

The air washers allows humidification of air and therefore we can think of steam injection for humidification and it is safe.

There are three different types of air washers. They are:

(i) Spray type air washer.

(ii) High velocity sprays type air washer.

(iii) Cell-type air washers.

1. Spray Type Air Washer. There is no standardisation for air washers. The essential requirements in the air washer operations are

(a) uniform distribution of the air across the spray chamber.

(b) an adequate amount of spray water broken up into fine droplets.

(c) good spray distribution across the air stream.

(d) sufficient length of travel through the spray and the wetted surfaces.

(e) the elimination of free moisture from the outlet air.

Refer Fig. 17.7 for the simple air-washer used in practice. A pumps recirculates water at a rate greater than the evapouration rate. Washers are commonly available from 0.8 to 118 m^3\sec. capacity. This type of washer is used primarily as an evaporative cooler or humidifier.

2. High-Velocity Spray Type Air Washer.

Air velocity range is 6 to 9 m/s –12 m/s.

Maximum capacity up to 70 m^3/s.

These are generally rectangular in cross-section.

3. Cell-Type Air Washer. These washers obtain intimate air-washer contact by passing the air through cells packed with glass, metal or fiber screens. Water passes over the cells arranged in tires. Behind the cells are blade-type or glass-mat eliminates. Most cell-type washers are arranged for concurrent air and water flow. They are also constructed for special duty with counter.

Current flow characteristics or in a combination of both arrangements, cell washers come to many sizes of insulated or uninsulated construction and standard washers are available up to 10 cells high by 12 cells wide with a capacity of up to 100 m^3/s.

Atomisation of the spray water is not required in cell type washers, but good water distribution over the face of the cell essential. A saturation effectiveness of 90 to 97% is possible.

17.8 COMFORT – PHYSIOLOGICAL CONSIDERATIONS

Man and many other living creatures may be compared in some respects with an automatically controlled furnace. Food consisting mainly of carbon, hydrogen, oxygen, nitrogen and certain minerals air taken periodically by a man and likewise coal containing the more or less the same elements is fed to the boiler. Air is continually drawn in and the oxygen combines with carbon and hydrogen in coal and food producing CO_2 which is exhaled by man and leaves with the gases in boiler. O_2 combines with hydrogen to form water vapour which is also exhausted. Water leaves the man body through the sweat glands and evaporates from the skin surface. Human body has a marvelous control system that maintains our body temperature at about 37°C (or 98.6°F).

Heat evolved from this oxidation process is termed *metabolism* similar to combustion process in the furnace.

The amount of heat generated and dissipated by the human body varies considerably with *activity* and *age* as well as *size and gender*. The human body has a complex regulating system acting to maintain the deep body temperature of about 37% (98.6°F) regardless of the environmental conditions. A normal, healthy person generally feels most comfortable when the environment is maintained at conditions where the body can easily maintain a thermal balance with the surroundings. ASHRAE specifies conditions in which 80% or more of the occupants will find the environment thermally acceptable.

The environmental factors that affect a person's thermal balance and therefore which influence thermal comfort are

(a) the dry bulb temperature
(b) the humidity of the surrounding air
(c) the relative velocity, of the surrounding air
(d) The temperature of any surface that can directly view any part of the body and thus exchange radiation.

In addition, the personal variables that influence thermal comfort are *activity* and *clothing*.

The basic mechanisms that the body uses to control body temperatures are metabolism, blood circulation near the surface of the skin (cutaneous blood circulation), respiration and sweating. *Metabolism* determines the rate at which energy is converted from chemical to thermal form within the body, and *blood circulation* controls the rate at which the thermal energy is carried to the surface of the skin. In *respiration*, air is taken in at ambient conditions but leaves saturated with moisture and very near the body temperature. *Sweating* has a significant effect on the rate at which energy can be carried away from the skin by heat and mass transfer.

Heat is exchanged between the body and its environment by four modes:

1. Evapouration (E)
2. Radiation (R)
3. Convection (C)
4. Positive or negative heat (S).

Storage of that would cause the deep tissue temperature to rise or fall. For a normally clothed, healthy human being in a comfortable condition or environment and engaged in a non-strenuous activity, $S = 0$ and the thermoregulatory system of the body is able to modify the losses by radition and convection to maintain a stable, satisfactory temperature. Evaporative losses occur in three ways: by exhalation of saturated water vapour from the lungs, by a continual normal process of insensible perspiration, and by an energy mechanism of sweating.

Insensible perspiration results from the body fluids oozing through skin under osmatic pressure and forming microscopic droplets on the surface which, because of their small size, evapourate virtually instantaneously, not being felt or seen and hence termed insensible.

If the body temperature tends to rise, the thermoregulatory system increases the evaporative loss by operating sweat-glands selectively and flooding strategic surfaces. In extreme cases the body is entirely covered with sweat that must evaporate on the skin to give a cooling effect. If the sweat rolls off or is absorbed by clothing, the cooling effect will be reduced.

Then we write the thermal energy balance in the equation form as:

$$M - W = E + R + C + S$$

where M = Thermal energy generated by metabolism

W = Work or energy spend in doing the activity

17.9 HUMAN REQUIREMENTS OF COMFORT

For a common person, the following conditions are desirable

1. Dry bulb temperature should be between 22°C and 26°C
2. The relative humidity should be between 20% and 60%
3. The average velocity of air in the room should preferably not to exceed 0.15 m/s
4. The carbon dioxide content should not exceed about 0.1%
5. The temperature difference between the feet and the head level should be as small as possible, normally not exceeding 1.5°C and should never exceed 3°C.

Not all these variables are directly amenable to regulation and no air-conditioning system is able to achieve control over all of them. The two most important are dry bulb temperature and air velocity.

17.10 HUMAN COMFORT AND EFFECTIVE TEMPERATURE

It is now logical to ask the desired temperature–humidity relationship is. People react in widely different ways to atmospheric conditions around them. A comfortable atmospheric condition for one person may be chilly to another or too warm for the third person. Some people may like the moist climate of the sea coast; others may prefer the dry weather. Dry bulb temperature and relative humidity are, therefore, not the indication for the comfort for the individual. Air movement is another factor that decides the comfort conditions for the human beings.

The ASHRAE has conducted research over many years involving hundreds of persons in different localities to find the combination of the three factors dry bulb temperature, relative humidity and air movements so that the people are comfortable. From these results of tests, a concept of effective temperature is developed.

Effective temperature is a measure of comfort which involves the combined effects of dry bulb temperature, relative humidity and air movement.

It was found that for a given air velocity, there are number of different combinations of dry bulb temperature, relative humidity which would give the same feeling of comfort to a high percentage of the people. *All these combinations were said to have the same effective temperature.*

Effective temperature scale is not present as is the case of dry-bulb temperature.

Yaglau and others using a survey of the responses of the people to a relatively short term exposure in different environments found the scale of *Effective Temperature*, defined as the temperature of still, and saturated air that gives a feeling of comfort similar to that of another combination of three relevant environmental variables.

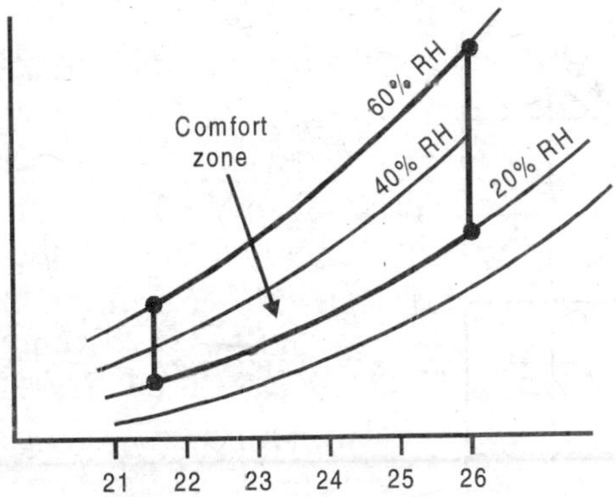

Fig. 17.9. *Normal Comfort Zone.*

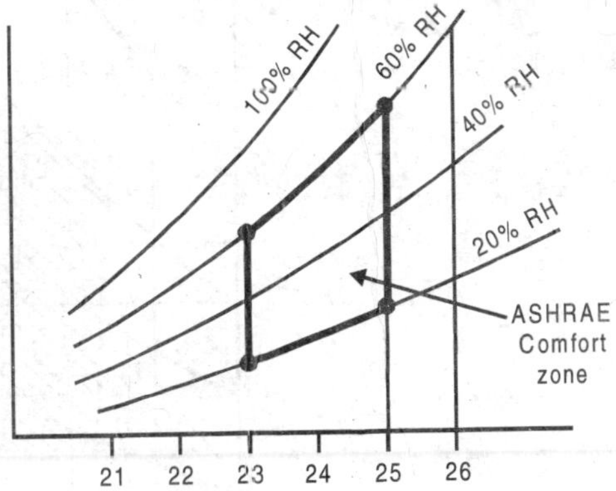

Fig. 17.10. *ASHRAE Comfort Zone.*

Figure 17.9 shows normal comfort zone where DBT varies between 22°C, and 26°C, relative efficiency varies between 20% and 60% while air velocity varies from 0.1 to 0.15 m/s.

Figure 17.10 shows the new comfort zone as recommended by ASHRAE comfort standard. The comfort zone lies within about 23°C and 25°C and between 20% RH and 60% RH.

Thus we conclude that *effective temperature is the index of sensations of warmth produced* by the combined effect of air temperature, humidity and air motion.

Effective temperature cannot be measured directly by dry-bulb thermometer but from the effective temperature chart.

Other forms of comfort zone and comfort charts are shown in Figs 17.11 and 17.12.

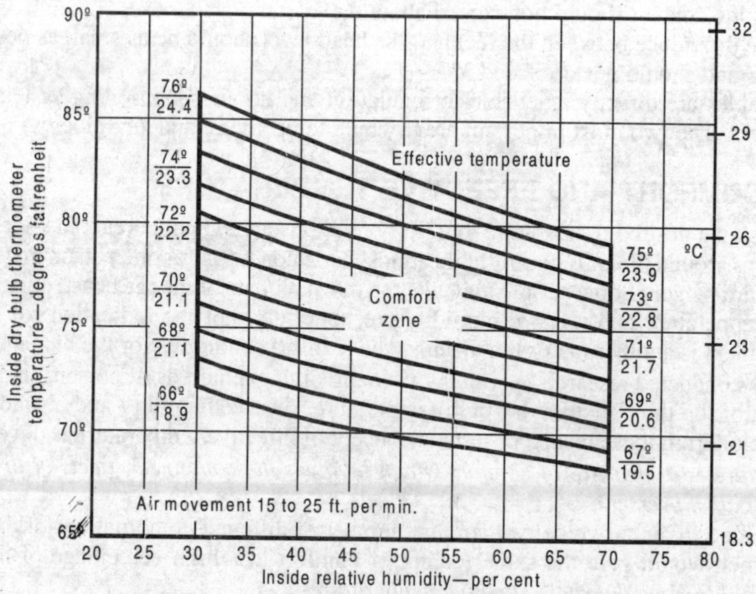

Fig. 17.11. *Comfort zone and chart.*

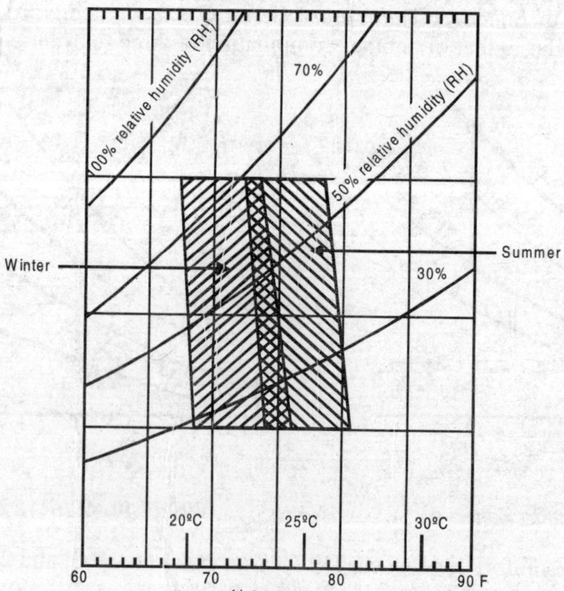

Fig. 17.12. *Other form of comfort-zone chart.*

17.11 AIR CONDITIONING SYSTEM CLASSIFICATION

Air conditioning systems can be classified in a number of ways such as

I. The cooling or heating fluid that is used.

Under this category, there are three possible groups in regard to the fluid used:

(*a*) *All air systems:* These systems use only air for cooling and heating.

(*b*) *All water* (*hydronic*) *systems:* These systems use only water for cooling and heating.

(*c*) *Air-water combination systems:* These systems use both water and air for cooling and heating.

II. Unitary and Central Systems

(*a*) *A unitary system* uses packaged equipment. This is most, if not all, of the system components (fan, coils, refrigeration equipment) are furnished as an assembled package from the manufacturer.

(*b*) *A central or built system* is one whose components are furnished separately and installed and assembled by the contractor.

III. Single-zone or Multiple Zone Systems

Out of these three ways of classification, here we will study about the unitary and central systems.

17.12 APPLICATIONS OF AIR CONDITIONING

Air conditioning is basically divided in two types:

1. Comfort air conditioning
2. Industrial air conditioning

We will enumerate the applications accordingly.

I. Comfort Air Conditioning

Following are the applications for comfort air conditioning.

1. Comfort air conditioning had its first major use in motion picture theaters (early 1920s).
2. Multiform office buildings
3. At the end of 1920s came the introduction of the first self contained room air conditioners
4. Apartments
5. Hotels, restaurants and night clubs
6. Hospitals
7. Departmental stores
8. Drug stores
9. Dress shops
10. Grocery shops
11. Banks
12. Dance halls and skating rinks
13. Transportation: trains, air planes, buses, automobiles, trollys and many more applications can be listed.

II. Industrial Air Conditioning Applications

In this case, human comfort element is not considered. Following are some of the applications of industrial air conditioning:

1. Textile industry
2. Printing industry
3. Drugs and chemicals
4. Libraries and museums
5. Metal working
6. Laboratories and many more.

SOLVED EXAMPLES

Example 17.1 *Atmospheric air at 35°C DBT and 60% RH is conditioned to 22°C DBT and 55% RH. This is done first by cooling and dehumidifying and then by heating. If the quantity of air flow is 60 m³/min, find:*

(*i*) *Weight of water drained*

(*ii*) *Capacity of cooling coil*

(*iii*) *Capacity of heating coil.*

Solution

Initial condition of air, then process and final conditions of air are represented on psychrometric chart as shown in Fig. Ex. 17.1.

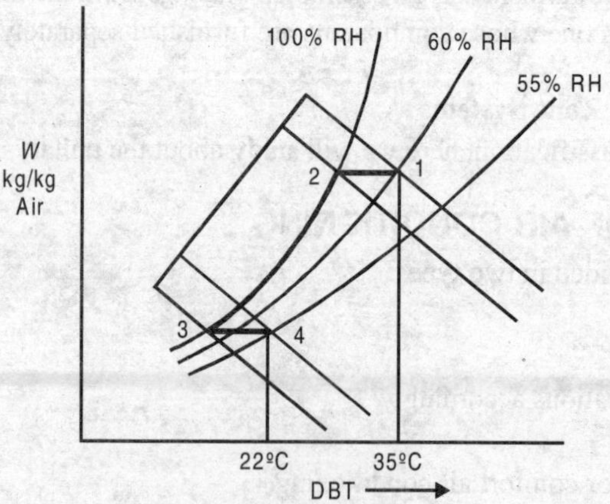

Fig. Ex. 17.1

1. Weight of Water Drained

$$p_{s_1} = 42.17 \text{ mm of Hg}$$

$$\text{RH} = \frac{p_{v_1}}{p_{s_1}}$$

$$\therefore \quad p_{v_1} = 42.17 \times 0.6$$

$$= 25.302 \text{ mm of Hg}$$

$$\therefore \quad w_1 = \frac{0.622 p_{v_1}}{\left(p - p_{v_1}\right)} = \frac{0.622 \times 25.302}{760 - 25.302}$$

$$= \frac{0.622 \times 25.302}{734.698}$$

$$= \mathbf{0.0214208 \text{ kg/kg of dry air}}$$

$$\text{and} \quad w_{s_1} = \mathbf{0.03663 \text{ kg/kg of dry air}}$$

$$\therefore \quad \mu_1 = \frac{w_1}{w_{s_1}}$$

$$= \frac{0.02142}{0.03663}$$

$= \mathbf{0.5848}$

Similarly, at the final state,

$$p_{s_4} = 19.82 \text{ mm of Hg}$$

$$RH_4 = \frac{p_{v_4}}{p_{s_4}}$$

$$\therefore \quad p_{v_4} = 0.55 \times 19.82$$

$= \mathbf{10.901 \text{ mm of Hg}}$

$$\therefore \quad w_4 = \frac{0.622 \times 10.901}{760 - 10.901} = \frac{0.622 \times 10.901}{749.099}$$

$= \mathbf{0.00905 \text{ kg/kg of dry air}}$

$$w_{s_4} = 0.01668 \text{ kg/kg of dry air}$$

$$\therefore \quad \mu_4 = \frac{w_4}{w_{s_4}}$$

$$= \frac{0.00905}{0.01668}$$

$= \mathbf{0.5426}$

∴ Weight of water drained/kg of dry air

$= w_1 - w_4$

$= 0.0214208 - 0.0090514$

$= \mathbf{0.0123694 \text{ kg/kg of dry air}}$

Initial volume of 1 kg air

$= v_a + \mu_1 (v_{as} - v_a)$

$= 0.8728 + 0.5848 \times (0.9239 - 0.8728)$

$= 0.8728 + 0.5848 \times 0.0511$

$= \mathbf{0.9027 \text{ cu.m/kg}}$

∴ Weight of air flowing per minute

$$= \frac{60}{0.9027} = \mathbf{66.47 \text{ kg/min.}}$$

∴ Weight of water drained/min.

$= 0.0123694 \times 66.47$

$= \mathbf{0.8222 \text{ kg/min.}}$

2. Capacity of Cooling Coil

Enthalpy of 1 kg at the initial condition

$$h_1 = h_{a_1} + \mu_1 \left(h_{as_1} - h_{a_1}\right)$$

$= 35.25 + 0.5848 \times (129.17 - 35.25)$

$= 35.25 + 54.92$

$= \mathbf{90.17 \text{ kJ/kg of dry air}}$

Dew point temperature of air at 4 is approximately 12.5°C

$\therefore \quad h_3 = 35.445$ kJ/kg of dry air

and
$$h_4 = h_{a_4} + \mu_4 (64.522 - 22.149)$$
$$= 22.149 + 0.5426 \times (64.522 - 22.149)$$
$$= 22.149 + 22.992$$
$$= \mathbf{45.141 \text{ kJ/kg of dry air}}$$

$\therefore$ Capacity of cooling air
$$= h_1 - h_3$$
$$= 90.17 - 35.445$$
$$= 54.725 \text{ kJ/kg}$$
$$= 54.725 \times 66.47 \text{ kJ/min}$$
$$= 3637.57 \text{ kJ/min}$$
$$= \mathbf{60.63 \text{ kW}}$$

3. Capacity of Heating Coil

Capacity of Heating Coil $= h_4 - h_3$
$$= 45.141 - 35.445$$
$$= 9.696 \text{ kJ/kg of dry air}$$
$$= 9.696 \times 66.47 \text{ kJ/min}$$
$$= \mathbf{644.493 \text{ kJ/min}}$$
$$= \mathbf{10.74 \text{ kJ/sec or kW}}$$

Example 17.2 *(a) Obtain all psychrometric properties of moist air having 35°C DBT and 14°C DPT, (b) 200 c mm air at 30°C DBT and 90% RH is mixed with 300 cm of air at 10° C DBT and 100% RH. Determine the final condition of air after mixing. Also find the moisture separated if any.*

Solution

(a) DBT is 35°C $\quad \therefore \quad p_s = 42.17$ mm of Hg

DPT is 14°C $\quad \therefore \quad p_v$ = vapour pressure

w = specific humidity at 14°C DPT

$= \mathbf{0.010012}$ **kg/kg of dry air.**

$$= \frac{0.622\, p_v}{760 - p_v} = \frac{0.622}{\dfrac{760}{p_v} - 1}$$

$$\therefore \quad \frac{760}{p_v} = 1 + \frac{0.622}{0.010012}$$
$$= 1 + 62.125 = 63.125$$

$$\therefore \quad p_v = \frac{760}{63.125} = \mathbf{12.04 \text{ mm of Hg}}$$

$\therefore \quad \phi$ = Relative humidity
$$= \frac{p_v}{p_s} = \frac{12.04}{42.17}$$
$$= \mathbf{0.2855 \text{ or } 28.55\ \%}$$

μ = Degree of saturation
$$= \frac{w}{w_s} = \frac{0.010012}{0.036756}$$
$$= \mathbf{0.2724 \text{ or } 27.24\ \%}$$

h = Enthalpy of 1 kg air

$$= h_a + \mu(h_{as} - h_a)$$
$= 35.219 + 0.2724 \times (129.455 - 35.219)$
$= 35.219 + 0.2724 \times 94.236$
$=$ **60.89 kJ/kg of dry air**

t_{wb} = Wet bulb temperature
$=$ **20.9583°C**

V = Volume of 1 kg dry air

$$= v_a + \mu(V_{as} - V_a)$$
$= 0.8728 + 0.2724 \times (0.9242 - 0.8728)$
$= 0.8728 + 0.2724 \times 0.0514$
$=$ **0.8868 cu-m/kg of dry air**

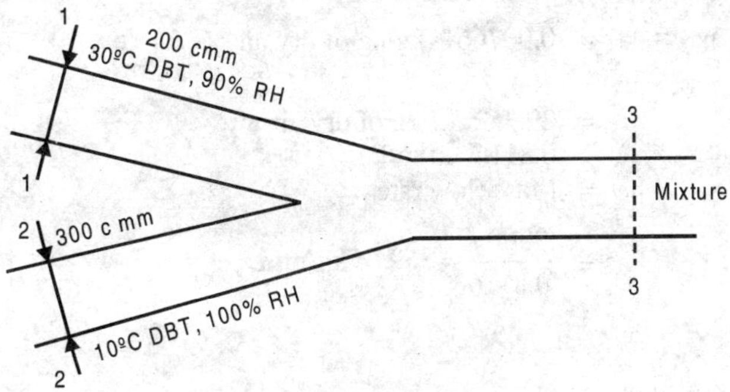

Fig. Ex. 17.2

Stream 1:

p_{s_1} = 31.81 mm of Hg

$\therefore$ $$\phi_1 = \frac{p_v}{p_s} = \frac{p_v}{31.81} = 0.9$$

$\therefore$ $p_{v_1} = 0.9 \times 31.81 =$ **28.629 mm of Hg**

$\therefore$ w_1 = Specific humidity

$$= \frac{0.622\,p_v}{760 - p_v} = \frac{0.622 \times 28.629}{760 - 28.629}$$

$$= \frac{17.807}{731.371} = \mathbf{0.02435\ kg/kg\ of\ dry\ air}$$

w_{s_1} = 0.027329 kg/kg of dry air

$\therefore$ μ_1 = Degree of saturation

$$= \frac{W}{W_s} = \frac{0.02435}{0.027329} = \mathbf{0.891}$$

$\therefore$ $$h_1 = h_{a_1} + \mu_1\left(h_{as_1} - h_{a_1}\right)$$

$$= 30.185 + 0.891\ (100.006 - 30.185)$$
$$= 30.185 + 0.891 \times 69.821$$
$$= \textbf{92.396 kJ/kg of dry air}$$

$$V_1 = V_{a_1} + \mu_1 \left(V_{as_1} - V_{a_1}\right)$$
$$= 0.8586 + 0.891\ (0.8962 - 0.8586)$$
$$= 0.8586 + 0.891 \times 0.0376$$
$$= \textbf{0.8921 cu-m/kg of dry air}$$

$\therefore$ m_1 = Air flow/min

$$= \frac{200}{0.8921} = \textbf{224.19 kg/min}$$

Stream 2:

$$p_s = p_v = 9.204 \text{ mm of Hg}$$
$$w_{s_2} = w_2 = 0.007661 \text{ kg/kg of dry air}$$
$$\phi_2 = \mu_2 = \textbf{100 \%}$$
$$h_2 = 29.352 \text{ kJ/kg of dry air}$$
$$V_2 = 0.8116 \text{ cu-m/kg}$$

$\therefore$ m_2 = Mass flow rate

$$= \frac{300}{0.8116} = \textbf{369.64 kg/min.}$$

Stream 3:

We can write

$$m_1 w_1 + m_2 w_2 = m_3 w_3$$
$$224.19 \times 0.02435 + 369.64 \times 0.007661 = [(224.19 + 369.64) \times w_3]$$

$$\therefore \quad w_3 = \frac{5.459 + 2.8318}{593.83}$$
$$= \textbf{0.01396 kg/kg of dry air}$$

Similarly,

$$m_1 h_1 + m_2 h_2 = m_3 h_3$$
$$224.19 \times 92.396 + 369.64 \times 29.352 = 593.83\ h_3$$

$$\therefore \quad h_3 = \frac{20714.259 + 10849.673}{593.83}$$
$$= \textbf{53.153 kJ/kg of dry air}$$

Referring the tables we get wet bulb temperature as **18.699°C.**

Again enthalpy of air is given by

$$h_3 = 53.153 = t_{d_3} + w_3\left[2501 + 1.82\, t_{d_3}\right] \text{ assuming } C_{pa} = 1.$$
$$= t_{d_3} + 0.01396 \times 2501 + 0.01396 \times 1.82\, t_{d_3}$$
$$= t_{d_3} + 34.92 + 0.02541\, t_{d_3}$$

$\therefore$ t_{d_3} = Dry bulb temperature of the mixture

$$= \frac{53.153 - 34.92}{1.02541} = \frac{18.233}{1.02541} = \textbf{17.78°C}$$

Example 17.3 (*a*) *Define and explain the following terms used in air conditioning.* (*1*) *Psychrometry* (*2*) *DPT,* (*3*) *WBT,* (*4*) *RH.*

(b) What is Comfort Air Conditioning? Recommend suitable psychrometric processes for following atmospheric conditions. Show the processes on handmade psychrometric chart.

(1) 35°C DBT, 65% RH, *(2) 40°C DBT 10% RH.*

(c) Atmospheric air at 30°C DBT and 60% RH is passed over a cooling coil with bypass factor of 0.15 at the rate of 200 m³/min. The coil surface temperature is 14°C. Find supply air DBT, cooling coil capacity in TR and amount of moisture separated per hour.

Solution

(b) Assume comfort conditioning requirements as–

DBT = 22°C; RH = 40%

1. 1 is atmospheric condition of air. 2 is required condition of air. Processes recommended: 1–a–sensible cooling. *a–b*: conndensation or separation of moisture and *b*–2 is sensible heating.

2. 1 is atmospheric condition.

2 is required condition.

To achieve 2, we proceed as–

1–*a* sensible cooling

a–b moisture removal or condensation.

b–2 sensible heating or use a cooling coil having surface temperature t_c having a bypass factor BF = $\frac{2-c}{1-c}$.

(c) Figure Ex. 17.3 shows the process.

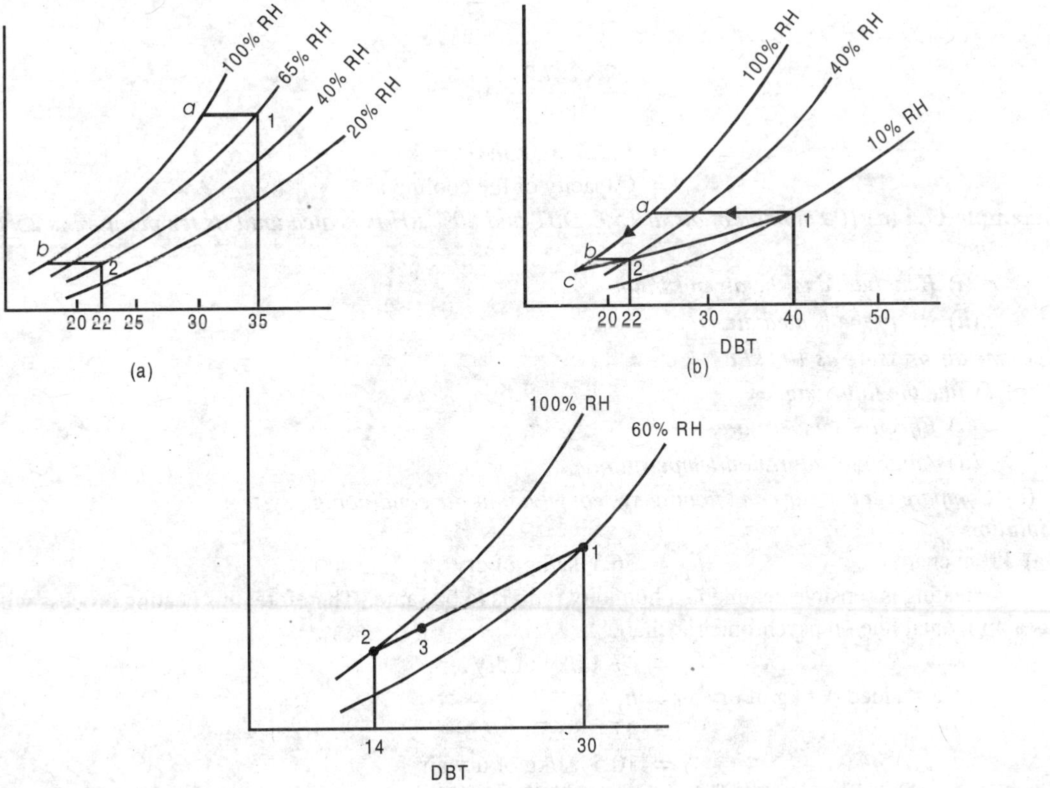

Fig. Ex. 17.3

1 is the condition of atmospheric air. 2 is the condition if the air is passed over the cooling coil having 100% efficiency or zero bypass factor.

1–2 is measured from chart. This is $2\frac{3''}{8}$ or $\frac{19''}{8}$ or 6 cm approximately. Cooling coil has BF = 0.15. Then point 3 is located on the chart and from chart DBT of the supply air is found to be 16.4°C approximately.

Specific volume of atmospheric air is 0.88 cu-m/kg (from chart).

$$\therefore \quad \text{Mass of air flowing} = m_a = \frac{200}{0.88} = \mathbf{227.273\ kg/min.}$$

$$\text{From chart, } h_1 = 71\text{ kJ/kg.; } W_{H_1} = 0.016\text{ kg/kg of dry air}$$

$$h_3 = 44\text{ kJ/kg.; } W_{H_3} = 0.0108\text{ kg/kg of dry air}$$

$$\therefore \quad \text{Moisture separated/hour} = \left(W_{H_1} - W_{H_3}\right) \times \text{Mass of air/hr}$$

$$= (0.016 - 0.0108) \times 227.273 \times 60$$

$$= 0.0052 \times 227.273 \times 60$$

$$= \mathbf{70.91\ kg/hour}$$

$$\text{Also heat removed from air/sec.} = (h_1 - h_3) \times \text{Mass of air/sec.}$$

$$= (71-44)\times\frac{227.273}{60}$$

$$= \frac{27\times227.273}{60}$$

$$= \mathbf{102.273\ (kJ/sec = kW)}$$

= Capacity of the cooling air.

Example 17.4 (*a*) *100 m³/min. of air at 15°C DBT and 80% RH is heated until its temperature is 25°C. Find the following*

(*i*) *Heat added to the air in kJ/min.*

(*ii*) *RH of the heated air.*

Assume air pressure as 100 kPa.

(*b*) *Define the following:*

(*i*) *Effective temperature.*

(*ii*) *Adiabatic saturation temperature.*

(*c*) *Compare central air conditioning system with unit air conditioning system.*

Solution

(a) From chart, $h_1 = 36.5$ kJ/kg of dry air

Heating is sensible heating i.e., humidity ratio is to be same. Therefore, this heating process will be shown as a horizontal line on psychrometric chart.

$$h_2 = 47\text{ kJ/kg of dry air}$$

$$\therefore \text{Heat added per kg of dry air} = h_2 - h_1$$

$$= 47 - 36.5$$

$$= 10.5\text{ kJ/kg of dry air}$$

$$\text{Specific volume of air} = v_1 = 0.8282\text{ cu-m/kg}$$

$$\therefore \quad \text{Mass of air} = \frac{100}{0.8282} = \mathbf{120.744\ kg/min.}$$

$$\therefore \quad \text{Heat added per minute} = 120.744 \times 10.5$$

$$= \mathbf{1267.81\ \ kJ/min}$$

From chart, RH of heated air = **43%**

Example 17.5 *50 m³ of air at 38°C and 60% RH is cooled to 32°C maintaining its specific humidity constant. Without using psychrometric chart, determine*

(i) Relative humidity of cooled air. (ii) Heat removed from air.

Use the following formula to calculate specific enthalpy of moist air.

h = 1.005 (DBT) + w [2500 + 1.88 (DBT – DPT))

where DBT = Dry bulb temperature and DPT = Dew point temperature.

Solution

(i) **RH of cooled air :** Given : 50 m³ of air at DBT = 38°C and RH = 60 %.

$$(\text{DBT})_2 = 32\ °\text{C}$$

Initial air (before cooling):

$$p_s = 49.69\ \text{mm Hg}$$

$$\phi = \text{RH} = 60\ \%$$

$$\therefore \quad \phi = 0.6 = \frac{p_v}{p_s} = \frac{p_v}{49.69}$$

$$\therefore \quad p_v = 0.6 \times 49.69 = 29.814\ \text{mm of Hg}$$

$$\therefore \quad w = \frac{0.622 p_v}{p - p_v} = \frac{0.622 \times 29.814}{760 - 29.816} = \frac{0.622 \times 29.814}{730.184}$$

$$= 0.0254\ \text{kg/kg of air}$$

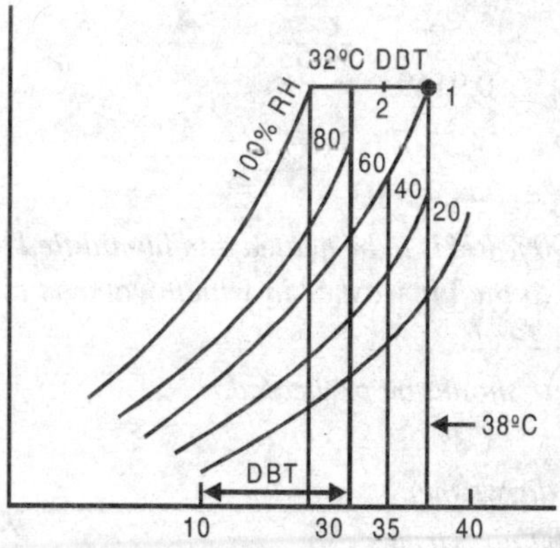

Fig. Ex. 17.5

Dew point temperature of air is that temperature at which the w = 0.0254 kg/kg of dry air, saturates air. From table,

$$\text{DPT} = 29 - 0.222$$

$$= \mathbf{28.778\ °C}$$

After sensible cooling, temperature is 32°C p_s at 32°C = 35.65 mm of Hg.

$$p_v = 29.814 \text{ mm of Hg}$$

$$\therefore \quad \text{RH at } 32°\text{C} = \frac{29.814}{35.65} = 83.63\,\%$$

(ii) Heat removed from air :

$$\begin{aligned} h_1 &= \text{Specific enthalpy before cooling} \\ &= 1.005 \times 38 + 0.0254 \times [2500 + 1.88\,(38 - 28.778)] \\ &= 38.19 + 0.0254 \times [2500 + 1.88 \times 9.222] \\ &= 38.19 + 0.0254 \times [2500 + 17.34] \\ &= 38.19 + 63.94 \\ &= \mathbf{102.13 \text{ kJ/kg of dry air}} \end{aligned}$$

$$\begin{aligned} h_2 &= \text{Specific enthalpy after cooling (32° DBT)} \\ &= 1.005 \times 32 + 0.0254 \times [2500 + 1.88\,(32 - 28.778)] \\ &= 32.16 + 0.0254 \times [2500 + 6.057] \\ &= 32.16 + 63.654 \\ &= \mathbf{95.814 \text{ kJ/kg}} \end{aligned}$$

$$\begin{aligned} \therefore \text{ Heat removed/kg of dry air} &= 102.13 - 95.814 \\ &= \mathbf{6.316 \text{ kJ/kg of dry air}} \end{aligned}$$

Initially, specific volume of air

$$v_a = \frac{m_a R_a T_a}{p_a} = \frac{1 \times 287 \times 311}{\frac{730.184}{760} \times 1.01325 \times 10^5} = \frac{287 \times 311}{0.9735 \times 10^5}$$

$$= \mathbf{0.9169 \text{ m}^3\text{/kg}}$$

$$\therefore \quad \text{Mass of air} = \frac{50}{0.9169} = 54.53 \text{ kg}$$

$$\begin{aligned} \therefore \quad \text{Total heat removed} &= 54.53 \times 6.316 \\ &= \mathbf{344.42 \text{ kJ}} \end{aligned}$$

Example 17.6 *Air at 10°C DBT and 90% RH is to be heated and humidified to 35°C DBT and 22.5°C WBT. The air is preheated sensibly before passing to the air washer in which water is circulated. The relative humidity of air coming out of the air washer is 90%. Find:*

(1) The temperature to which the air should be preheated.

(2) The total heating required.

(3) Make up water required in the air washer.

(4) The humidifying efficiency of the air washer.

(Use of psychrometric chart is expected while solving the problem).

Solution

1–2 Sensible heating $h_1 = 27.2$, $h_2 = 42$ kJ/kg

2–3 Sensible humidifying $h_3 = 66.5$ kJ/kg

3–4 Heating and dehumidifying $h_4 = 66.5$ kJ/kg

$W_1 = W_2 = 0.00712$, $W_3 = 0.0168$, $W_1 = 0.0122$

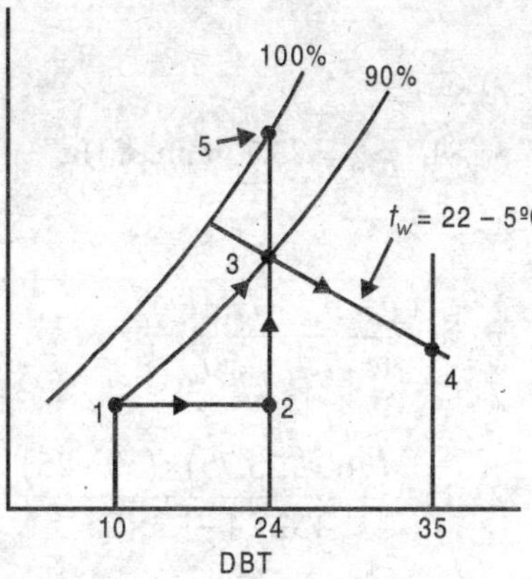

Fig. Ex. 17.6

Referring to the diagram,

(1) The temperature to which the air should be preheated $= T_2 =$ **24°C**

(2) The total heating required

$$= h_3 - h_1 = 66.5 - 27.2$$
$$= \mathbf{39.3\ kJ/kg}$$

(3) Make up water in the air washer

$$= w_3 - w_2$$
$$= 0.01680 - 0.00712$$
$$= \mathbf{0.00968\ kg/kg\ of\ dry\ air}$$

(4) Humidifying efficiency of the washer

$$\eta_H = \frac{w_3 - w_2}{w_5 - w_2}$$
$$= \frac{0.01680 - 0.00712}{0.01840 - 0.00712}$$
$$= \frac{0.00968}{0.01128}$$
$$= 0.8581$$
$$= \mathbf{85.81\ \%}$$

Example 17.7 (*a*) *Derive the expression to calculate the enthalpy of moist air.*

(*b*) *Define the terms*

(*1*) *Evapourative cooling*

(*2*) *Comfort chart.*

(*c*) *In a laboratory test, a sling psychrometer recorded 30°C DBT and 25°C WBT. Calculate*

(*1*) *Vapour pressure*

(*2*) *Relative humidity*

(*3*) *Specific humidity*

(*4*) *DPT (Dew point temperature)*

(*5*) *Degree of saturation.*

Solution

(c)
$$t_d = 30°C, t_w = 25°C$$
$$p_{sd} = 31.81,\ p_{sw} = 23.75 \text{ mm of Hg}$$

(i) Vapour pressure:

$$p_v = p_{sw} - \frac{(p - p_{sw})(t_d - t_w)}{1527.4 - 1.3\,t_w}$$

$$= 23.75 - \frac{(760 - 23.75)\times(30-25)}{1527.4 - 1.3\times 25}$$

$$= 23.75 - \frac{736.25\times 5}{(1527.4 - 32.5)}$$

$$= 23.75 - 2.4625$$

$$= \mathbf{21.2875 \text{ mm of Hg}}$$

(ii) Relative humidity: Partial presure of vapour at saturation at 30ºC is $p_s = 31.81$

$$\therefore \quad \text{Relative humidity} = \frac{p_v}{p_s} = \frac{21.2875}{31.81}$$

$$= \mathbf{0.6692} \text{ or } \mathbf{66.92\ \%}$$

(iii) Specific humidity:

$$w = \frac{0.622\,p_v}{(p - p_v)} = \frac{0.622\times 21.2875}{760 - 21.2875}$$

$$= \frac{0.622\times 21.2875}{738.7125}$$

$$= \mathbf{0.0179 \text{ kg/kg of dry air}}$$

(iv) Dew point temperature: Dew point temperature is the temperature at which the given moisture saturates the air.

From table, $t_{dp} = \mathbf{24\ °C}$

(v) Degree of saturation:

$$w = 0.0179$$
$$w_s = 0.027329$$
$$\therefore \quad \mu = \text{Degree of saturation}$$

$$= \frac{w}{w_s} = \frac{0.0179}{0.027329}$$

$$= \mathbf{0.655} \text{ or } \mathbf{65.5\ \%}$$

THEORY QUESTIONS

17.1. Explain the difference between comfort and industrial air conditioning.

17.2. Define the term 'Human Comfort'. What factors are involved in Humån comfort air conditioning? Explain.

17.3. What do you understand by effective temperature? In what way it is connected with comfort air conditioning.

17.4. In what way the human body compensates and reacts automatically to the warm of hot environment? Explain.

17.5. Explain how does the human body attempt to compensate for a cool environment which tends to lower the internal temperature.

17.6. Draw a 'hand-drawn' psychrometric chart and show on it the comfort zone for both winter and summer air conditioning.

17.7. What is an air conditioning system? In what way it is different from refrigeration system?

17.8. Classify the air conditioning systems.

17.9. Distinguish clearly between unitary and central air conditioning systems.

17.10. What is the difference between winter and summer air conditioning systems? What different equipments are required?

PROBLEMS FOR PRACTICE

17.1. What do you understand by the term "Air Conditioning"? How will you classify the different air conditioning systems?

17.2. Distinguish clearly between comfort air conditioning and industrial air conditioning.

17.3. What are the applications of air conditioning?

17.4. In what way an industrial air conditioning is applied in practice? Give examples.

17.5. What are the factors on which comfort air conditioning depends? Explain.

17.6. How the human body responds to the following different environmental conditions?

(1) Dry hot weather
(2) Humid and hot weather and
(3) Cold weather.

17.7. Define the following terms as applied in air conditioning language:

(a) Dry bulb temperature (DBT)
(b) Wet bulb temperature (WBT)
(c) Dew point temperature (DPT)
(d) Humidity, specific humidity and absolute humidity
(e) Relative humidity
(f) Degree of saturation
(g) Saturated air
(h) Enthalpy of air.

17.8. What do you understand by the terms "Adiabatic saturation temperature" and "Thermodynamic Wet Bulb Temperature"?

17.9. Are the dry bulb temperature and dew point temperature are same? When they are same?

17.10. When dry bulb temperature, wet bulb temperature and dew point temperature are numerically same?

17.11. Define and explain the terms "Effective temperature" and "Comfort zone".

17.12. Will "Cooling of air" mean air conditioning? Explain.

17.13. What is evaporative cooling? Under what conditions it will take place?

17.14. What is best example of evaporative cooling in air conditioning systems?

17.15. What are psychrometric tables and psychrometric charts?

17.16. How will you construct psychrometric chart?

17.17. Enumerate the different psychrometric processes that are taking place.

17.18. Explain with neat sketch/diagram each of these processes and show them on the hand-drawn psychrometric chart.

17.19. For the given state of air, how will you find dew-point temperature with the help of psychrometric chart and psychrometric table?

17.20. What is a by-pass factor? Explain and define.

Where this factor is considered?

17.21. Distinguish clearly the central air conditioning system and unitary air conditioning system.

17.22. Explain with a neat sketch the working of "Window Air Conditioner". Can this be called as unitary air conditioning system?

17.23. What do you understand by 'Split Unit' air conditioner? What are the advantages and disadvantages of this unit?

17.24. What is DX type central air conditioning system? Explain.

17.25. What is an air washer? Explain the different types of air washers.

17.26. Where chilled water air washer will be used? Explain.

17.27. Explain the 'recirculation of air' in air conditioning system. Is there any advantage for this recirculation?

17.28. Find the relative humidity and specific humidity for air at 30°C and having dew point temperature of 15°C.

[**Ans.** RH = 40.169%, w_H = 0.010783 kg /kg of dry air]

17.29. Air is supplied to a room at 17°C and 60% RH. If the barometer reads 1.01325 bar, calculate the specific humidity and dew point of the air.

[**Ans.** w_H = 0.007213 kg/kg of dry air, DPT = 9.18°C]

17.30. One kg of air at 40°C DBT and 50% RH is mixed with 2 kg of air at 20°C DBT and 20° DPT. Calculate temperature and specific humidity of the mixture.

[**Ans.** T_3 = 270 C, w_3 = 0.0179]

17.31. Outside air at 760 mm of Hg pressure has its DBT and WBTs 300 °C and 200° C respectively. Determine its specific humidity, relative humidity, vapour density, DPT and enthalpy of air.

[**Ans.** w = 0.0 1048, f = 39.56%, p_a = 0.01199 kg/m^3; DPT = 14.75°C, h = 56.9 kJ/kg of dry air]

17.32. Outside air flowing at the rate of 2 kg/sec at 30°C DBT, 21°C WBT mixes adiabatically with return air stream flowing at the rate of 3 kg/sec at 21°C DBT and 50% RH. Determine the DBT, WBT, w and h of the mixed stream.

[**Ans.** h = 49.5 kJ/kg; 0.0092; DBT = 24.6°C; WBT = 17.6°C]

17.33. Air at 30°C DBT and 60% RH is passed through a cooling coil at the rate of 200 kg/min. The surface temperature of the coil is 14°C and by-pass factor is 0.12. Find the DBT, cooling coil capacity in TR, moisture condensed.

[**Ans.** 15.95°C; TR = 52.3; 1.08 kg/min]

17.34. 45 m^3 /min of air at 20°C and 75% RH is heated till its DBT becomes 30°C. Calculate the mass flow rate of air, RH and heat added per hour.

[**Ans.** 53.3 kg/min; ϕ = 41%; Q = 33.42 kJ/hour]

17.35. A small room requires 6 m^3/min of conditioned air at 20°C DBT and 60% RH, while the surrounding conditions in winter are 12°C DBT and 10°C WBT. Conditioned air is obtained by first heating the air-heating coil having a by-pass factor 0.4 and then by adiabatic humidifying. Determine the surface temperature of coil and the capacity of the humidifier.

[**Ans.** ADP = 32.75°C; 0.76 kg/hr]

17.36. 50 cu-m/min. of air at 15°C DBT and 80% RH is heated until its temperature becomes 22°C. Find,

(1) Heat added per hour

(2) RH of the heated air.

Assume barometer reads 760 mm of Hg.

[**Ans.** (1) 26126.9 kJ/hr; (2) RH = 51.7%]

17.37. Outside air at 12°C and 75% RH is to be conditioned to a temperature of 22°C DBT and 60% RH. Amount of air supplied is 100 m^3/min.

The required condition is achieved first by heating and then by adiabatic humidifying. Find the following

(1) Quantity of steam required per hour. Steam at 2 bar and 0.96 dry.

(2) Quantity of water required per hour in the humidifier. You can use chart.

[**Ans.** (1) 156.4 kg/hr, (2) 251 cg/hr]

17.38. Atmospheric air – 760 mm of Hg—at 40°C

DBT and 60% RH is passed through the conditioning system wherein 10 gm of water per kg of dry air are removed. Find:

(a) Relative humidity

(b) Wet bulb temperature and

(c) Dew point temperature.

[**Ans.** (a) 69%, (b) 25.2°C, (c) 23.6°C)]

17.39. Outside air – 760 mm of Hg—at 30°C DBT and 25°C WBT is heated to 40°C. If the air supply is 150 m^3 /min, find the amount of heat added and RH, WBT of the air.

[**Ans.** 1762.73 kJ/min; 36.8%; 27.75 °C]

17.40. Cold air at 10° C and at 60% RH is required for an industrial processes at the rate of 100 cu-m/ hr. When the outside conditions are 41°C DBT and 25°C WBT, the required conditions are achieved first by cooling and dehumidifying and then by heating. Determine

(a) Heat removed from the air per minute,

(b) Relative humidity and wet bulb temperature.

CHAPTER 18

Introduction to Heat Transfer

18.1 DEFINITION

We have repeatedly used the term heat (Q) and heat exchanges between bodies without discussing the details of the phenomena by which heat is transferred. The purpose of this chapter is to give the student a general idea of the nature of the problems and to provide an introductory background for further study.

Heat transfer is a name of general process applied to any device that effects a transfer of heat from one substance to another.

The various devices that employ the principles of heat transfer are: (i) steam boilers (ii) surface condensers, evaporators, closed feed water heaters and the automobile radiator etc.

Whenever a temperature gradient exists within a system, or when two systems at different temperatures are brought into contact, energy is transferred. The process by which the heat energy transport takes place is known as heat transfer.

18.2 MODES OF HEAT TRANSFER

Heat transfer can be defined as the transmission of energy from one region to another as a result of a temperature difference between them. Since temperature differences exist all over the universe, the phenomena of heat flow over the universe, the phenomena of heat flow are as universal as those associated with gravitational attractions. Unlike gravity, however, heat flow is governed not by a unique relationship, but rather by a combination of various independent laws of physics.

The literature of heat transfer generally recognises three distinct modes of heat transmission:

(i) Conduction
(ii) Radiation and
(iii) Convection

Strictly speaking, only conduction and radiation should be classified as heat transfer processes, because only these two mechanisms depend for their operation on the mere existence a temperature difference. The last of the three, convection, does not strictly comply with the definition of heat transfer because it depends for its operation on mechanical mass transport also. But since convection also accomplishes transmission of energy from regions of higher temperatures to regions of lower temperatures, the term "heat transfer by convection", has been generally accepted.

18.3 FOURIER'S LAW OF CONDUCTION

The basic relation for heat transfer by conduction was proposed by French scientist JBJ Fourier in 1882. It states that q_k the rate of heat flow by conduction in a material, is equal to the product of the following three quantities:

(i) K – the thermal conductivity of the material

(ii) A – the area of the section through which heat flows by conduction and

(iii) $\frac{dT}{dx}$ – the temperature gradient at the section i.e., the rate of change of temperature T with respect to distance in the direction of heat flow or

$$q_k = -K.A\frac{dT}{dx}$$

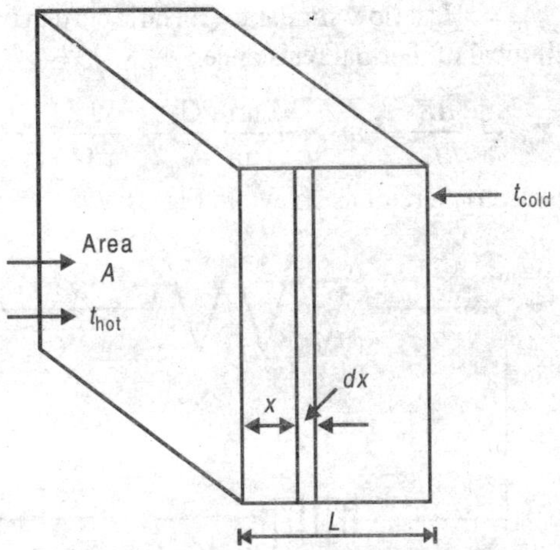

Fig. 18.1 *One dimensional heat flow.*

$$q_k \to \text{kJ/hr}; A \to m^2; \frac{dT}{dx} \to \text{°C/m}$$

$$K \to \frac{q_k}{A \cdot dT/dx} \to \frac{\text{kJ}}{\text{hr}} \times \frac{1}{\text{m}^2} \times \frac{1}{\text{°C/m}} \to \frac{\text{kJ}}{\text{hr.m°C}} \text{ units of K}$$

Refer Fig. 18.1.

Consider a wall or slab of thickness L, let A be the area of heat flow across the wall. Let the temperatures on both sides of the wall be T_{hot} and T_{cold}. Consider a section at a distance x from hot surface and let dx and dT be the thickness of the strip and temperature drop across the strip as shown. Therefore, applying Fouriers Law, we get

$$q_k = \frac{-KA.dT}{dx}$$

We have

$$\frac{q_k}{A} = -\int K.\frac{dT}{dx}$$

$$\therefore \quad \frac{q_k}{A}\int_O^l dx = -K\int_{Hot}^{Cold} dT \quad K.\int_{Cold}^{Hot} dT$$

$$\therefore \quad \frac{q_k}{A}.L = K(T_{hot} - T_{cold})$$

or

$$q_k = \frac{AK}{L}(T_{hot} - T_{cold})$$

We can write this also as,

$$q_k = \frac{AK}{L} \cdot \Delta T = \frac{\Delta T}{L/AK}$$

Here we can compare with simple electric circuit as shown in Fig.17.2.

Here ΔT = Driving force for heat transfer similar to (Voltage)

$\frac{L}{AK}$ = Thermal resistance R_k similar to (Electric Resistance)

q_k = Heat flow similar to (Electric current)

Thermal conductance is the reciprocal of thermal resistance

$$K_k = \frac{AK}{L} \rightarrow m^2 \times \frac{\text{kJ/hrm°C}}{\text{m}} \rightarrow \frac{\text{kJ}}{\text{hr°C}}$$

∴ This can be looked upon as an electric circuit as shown in Fig. 18.2.

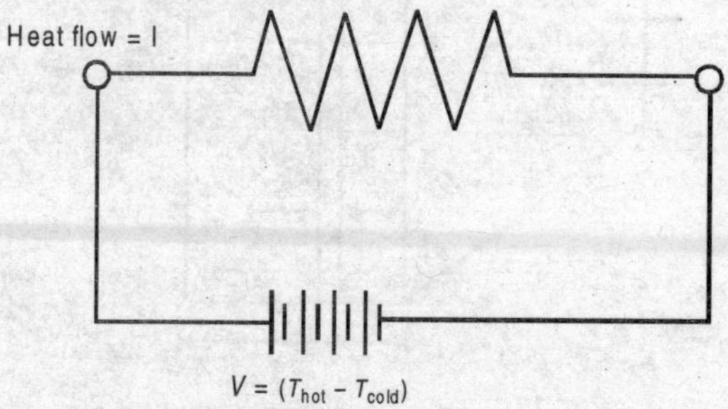

Fig. 18.2. *Electric analog.*

18.3.1 Composite Walls

A composite wall, typical of the type used in a large furnace, is shown in Fig.18.3. The inner layer, which is exposed to the high temperature gases, is made of firebrick. The intermediate layer consists of an insulating brick and is followed by an outer layer of ordinary red bricks.

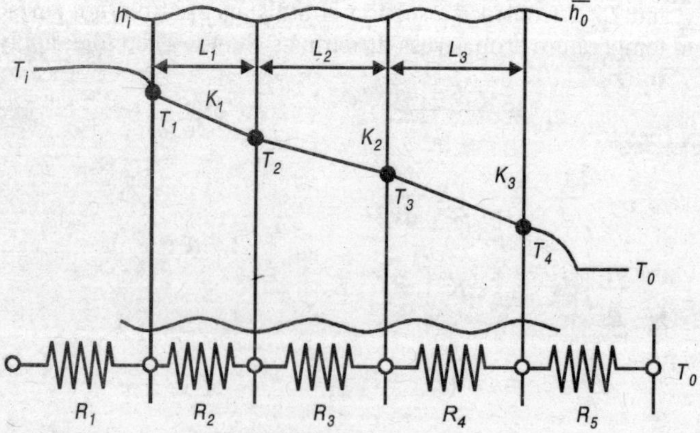

Fig. 18.3

The temperature of the hot gases is T_i and the unit surface conductance over the interior surface is $\bar{h}_i$. The atmosphere surrounding the furnace is at a temperature t_0 and the unit surface conductance over the exterior surface is $\bar{h}_0$.

Under these conditions there will be a continuous heat flow from the hot gases through the walls to the surroundings. Since the heat flow through a given area A is the same for any section, we obtain

$$q = \bar{h}_i A(T_i - T_1) = \frac{K_1 A}{L_1}(T_1 - T_2) = \frac{K_2 A}{L_2}(T_2 - T_3)$$

This equation can be written in terms of the thermal resistance of the various section as

$$q = \frac{T_i - T_1}{R_1} = \frac{T_1 - T_2}{R_2} = \frac{T_2 - T_3}{R_3} = \frac{T_3 - T_4}{R_4} = \frac{T_4 - T_0}{R_5}$$

where the resistances may be determined. Solving for the various temperature differences, we have

$$\begin{aligned} T_i - T_1 &= qR_1 \\ T_1 - T_2 &= qR_2 \\ T_2 - T_3 &= qR_3 \\ T_3 - T_0 &= qR_4 \\ T_4 - T_0 &= qR_5 \end{aligned}$$

Adding, we get

$$T_i - T_0 = q(R_1 + R_2 + R_3 + R_4 + R_5)$$

$$\therefore \quad q = \frac{T_i - T_0}{R_1 + R_2 + R_3 + R_4 + R_5}$$

$$\therefore \quad \text{Heat flow} = \frac{\text{Overlap temp. potential}}{\text{Sum of resistance in the path}}$$

18.3.2 Conduction of Heat through Hollow Cylinder

The heat flow through a cylinder is considered along radial direction. Therefore the Fourier equation of conduction for cylinder can be written as

$$Q = -KA\frac{dT}{dr}$$

Figure 18.4 (a) shows hollow pipe having inside radius R_1 and outside radius R_2. Let the inside temperature be T_1 and outside temperature T_2. The conductivity of the pipe material is say K.

Consider an elementary ring at a radius of r and having a thickness dr as shown in Fig. 18.4(a). The heat flow equation through an element dr under steady state condition can be written as

$$Q = -K.(2\pi r)\frac{dT}{dr}$$

where L is the length of pipe and $A = 2\pi rL$. Integrating this equation between the limits of R_1 and R_2 and T_1 to T_2 as the temperature changes from T_1 and T_2 through the thickness of the cylinder ($R_2 - R_1$),

$$\int_{R_1}^{R_2} Q.\frac{dr}{r} = \int_{T_1}^{T_2} -K2\pi L.dT$$

$$\therefore \quad Q.\log_e \frac{R_2}{R_1} = -2\pi LK(T_2 - T_1)$$

$$= 2\pi LK(T_1 - T_2)$$

$$\therefore \quad Q = \frac{2\pi LK(T_2 - T_1)}{\log_e \frac{R_2}{R_1}}$$

$$= \frac{T_1 - T_2}{\frac{1}{2\pi LK}\log_e \frac{R_2}{R_1}}$$

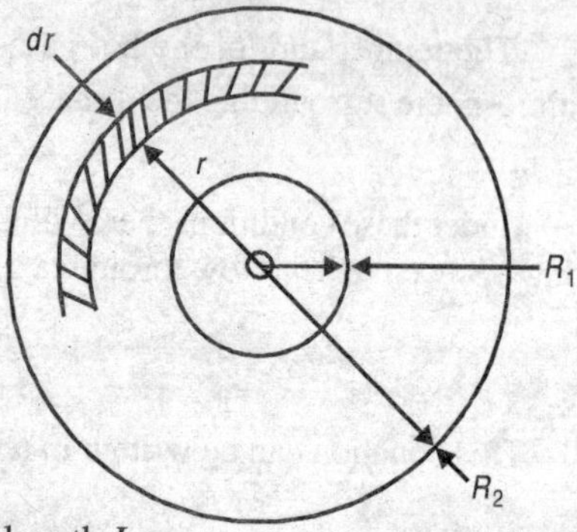

Fig. 18.4

As before $\left[\frac{1}{2\pi LK}\right]\log_e \frac{R_2}{R_1}$ is known as thermal resistance of the pipe of length L.

18.4 CONVECTION HEAT TRANSFER

Convection heat transfer is defined as the transfer of heat through the agency of particles of fluid which receive heat from a high temperature source and move to the locality of a lower temperature sink to reject heat. The sink may be some body exposed to the fluid, or it may be cooler particles of fluid. It should be noted that convection heat transfer occurs by means of fluid motion, in contrast to conduction heat transfer, which takes place entirely by means of intermolecular energy transfers.

Convection heat transfer may be further classified according to the mode of motivating the flow. If the convection currents are caused by density differences brought about by temperature gradients within the fluid, the convection is said to be *natural*. If the flow is aided by some mechanical device such as a pump (fan) the resulting convection currents are forced.

18.5 NEWTON'S LAW OF COOLING

The rate of heat transfer by convection between a surface and a fluid may be computed by the relation

$$q_c = \bar{h}CA.\Delta T$$

where q_c = rate of heat transfer by convection (kJ/hr)

A = heat transfer area (m^2)

ΔT = difference between the surface temperature T_S and the temperature of the fluid $T\infty$ at some specified location (°C).

$\bar{h}_c$ = average unit thermal convective conductance often called the surface coefficient of heat transfer or the convective heat transfer coefficient kJ/hr.m^2.°C.

Thermal conductance K_c for convective heat transfer $K_c = \bar{h}_c . A$ and thermal resistance

$$R_c = \frac{1}{\bar{h}_c A}$$

The relation expressed by this equation was originally proposed by the British scientist, Issac Newton in 1701. This is also known as Newton's law of cooling. The evaluation of the convective heat transfer coefficient is difficult because convection is a very complex phenomenon.

18.6 FILM COEFFICIENT OF HEAT TRANSFER

In heat exchangers, we are mostly concerned with walls separating liquids or gases from each other. In these cases we do not know the temperatures of both surfaces of the separating walls, but only the temperatures of the liquids or fluids on both sides of the walls. These temperatures are indicated as t_1 and t_2 as shown in Fig. 18.5. By measuring the temperature field in the liquids one obtains the curves shown.

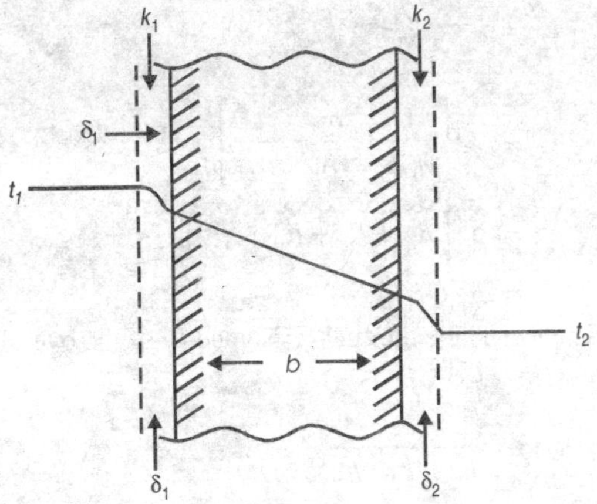

Fig. 18.5

The temperature gradient is confined to a relatively narrow layer of thickness δ quite close to the wall, whereas at a grater distance from the wall in most cases only small temperature differences exist. Simplified, the temperature curve can be replaced by the dashed broken line. This can be explained by assuming that a thin film of liquid (of the thickness δ') adheres to (i) the wall whereas outside thin film, temperature differences vanish as a result of mixing motions of the liquid. This picture over simplifies the actual process considerably as we shall see later, but it brings out the salient features. Within the film, the heat transfer takes place by conduction, as in a solid wall. The temperature in the film, therefore, is again linear, and the flow of heat follows the equation

$$Q = \frac{K}{b}.A.(t_1 - t_2)$$

into which the thermal conductivity K of the liquid or of the gas and the thickness of the film δ' are to be inserted.

Thus, for the heat flow to the wall, the equation $Q = \frac{K}{\delta'}A(t - t_w)$ is obtained.

From this equation the rate of heat flow Q can be calculated if the film thickness δ' is known. The latter, however, depends to a great extent upon the external flow conditions, for instance upon the velocity with which the liquid flows along the wall, upon the shape of the wall, upon the structure of the wall surface and similar factors. In engineering practice it has become customary to calculate with the expression K/δ, rather (2) than directly with the film thickness δ'. This value is called film heat transfer coefficient and is designated by the letter h. Thus the expression

$$Q = hA\,(t - t_w)$$

as formulated by Issac Newton is obtained. The value $(1/hA)$ is called thermal resistance R_t, of the film heat transfer process

$$R_t = 1/hA$$

From Fig. 18.3 we have,

$$t_1 - t_{w1} = \frac{1}{h_1 A}.Q$$

$$t_{w1} - t_{w2} = \frac{b}{kA}.Q$$

$$t_{w2} - t_2 = \frac{1}{h_2 A} \cdot Q$$

∴ Adding, we get
$$t_1 - t_2 = Q\left[\frac{1}{h_1 A} + \frac{b}{kA} + \frac{1}{h_2 A}\right]$$
$$= Q\left[R_{t_1} + R_c + R_{t_2}\right]$$
$$= R_0 Q$$

The sum of the individual resistances is the thermal resistance R_0 of the overall heat transfer process. We have

$$Q = UA\,(t_1 - t_2)$$

where
$$U = \frac{1}{h_1} + \frac{b}{k} + \frac{1}{h_2}$$

The value U is called the overall heat transfer coefficient.

18.7 THERMAL RADIATION

It will be recalled that conduction and convection heat transfer depend on the existance of a medium to convey heat. The third mode of heat transfer, *radiation*, differs from the other two in that a medium is not required. Radiation is the term applied to the process by which energy is continually emitted from the surfaces of all bodies. This energy is called *radiant energy* and is in the form of electromagnetic waves. These waves travel with the velocity of light and are transmitted through a vacuum as well as through substances which are transparent to them. In fact, transmission through a vacuum is better, since intervening media absorb some, if not all, of the radiant energy.

When the electromagnetic waves fall on the body which is not completely transparent to them, they are absorbed and there energy is converted to heat. The process by which radiant energy is converted to heat is not completely understood. However, it is known that all substances emit and absorb radiant energy at a rate which depends on the absolute temperature and physical properties of the substance.

The waves incident upon the surface of the substance may be partly absorbed, partly reflected and partly transmitted through the substance. The fraction of the radiant energy which is absorbed is called *absorptivity* (α). The fraction reflected is termed the *reflectivity* ρ and the fraction transmitted through the substance is known as the *transmissivity* τ. The relationship between the absorptivity, reflectivity and transmissivity is written as

$$\alpha + \rho + \tau = 1$$

When the transmissivity is zero, as it is for most solids *opaque* to light, the substance is said to be *opaque* to radiation. Conversely, if transmissivity is unit, the substance is transparent to radiation. No substance is perfectly transparent but the less dense fluids, such as gases, have a high transmissivity.

An ideal *reflector* is a body whose surface reflects all radiant energy incident upon it. Highly polished surfaces are good approximations to an ideal reflector. An *ideal absorber* absorbs all radiant energy incident upon its surface and hence its absorptivity is unity.

Only the radiant energy absorbed can contribute to increasing the internal energy of the substance, and since the rate at which radiant energy is emitted depends on the temperature of the emitter, a good absorbed is also a good emitter. Therefore, we define an *ideal radiator* as a body which absorbs all radiant energy incident upon its surface. We see that the characteristics of an ideal radiator are an absorptivity of unity, and reflectivity and transmissivity of zero.

The rate at which energy is radiated at a wavelength λ, per unit area of surface is called *monochromatic emissive power*, E_λ. The *total emissive power*, E, is the energy radiated per unit time and per unit area of surface at all wavelengths.

The total emissive power of an ideal radiator may be found by integrating the monochromatic emissive power over all wavelengths.

or
$$E = \int_{\lambda=0}^{\lambda=\infty} E_\lambda d\lambda.$$

18.8 KIRCHHOFF'S RADIATION LAW

No true ideal radiator exist in nature, although some substances, such as lampblack and platinum black, are good approximations. Kirchhoff conceived of a method of producing a close approximations to an ideal radiator through the use of a hollow enclosure, as shown in Fig. 18.6.

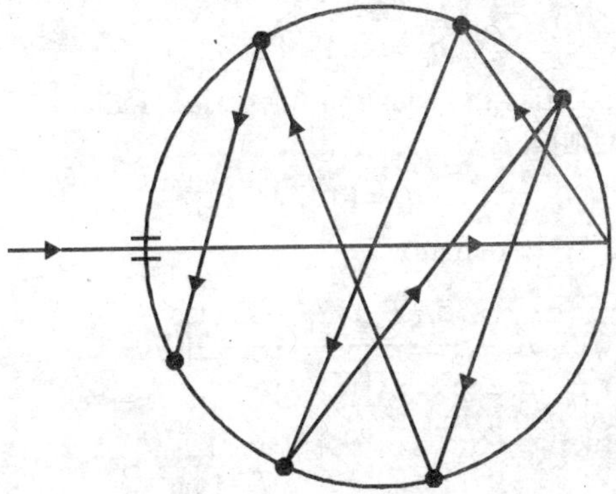

Fig.18.6. *Hollow sphere for producing black-body radiation.*

A small hole is cut in a large hollow enclosure, such as a sphere, and the radiant energy incident upon this hole is partially absorbed by the inside surface of the enclosure. The rest of the radiant energy is diffused within the enclosure. The radiant energy enetring through the hole continues to be absorbed reflected and reradiated, but only a very small amount finds its way out through the hole. Hence, the hole behaves very much like an ideal radiator.

18.9 STEFAN BOLTZMANN'S LAW OF RADIATION

The quantity of energy leaving a surface as radiant heat depends upon the absolute temperature and the nature of the surface. A perfect radiator or black-body emits radiant energy from its surface at a rate q_r given by

$$q_r = \sigma A_1 T_1^4 \text{ kJ/hr}$$

where A_1 is the surface area in sq.m. T_1 is the surface temperature in degrees Kelvin and σ is a dimensional constant and is known as Stefan–Boltzmann constant after two Austrian scientists, J. Stefan who in 1879 found the above equation experimentally and L. Boltzmann, who in 1884 derived it theoretically. Here 6 = 20.42×10^{-8} kJ/hr. m^2 T^4 = 5.671 W/m^2.T^4.

Stefan–Boltzmmnn Law of radiation states that any black body surface above a temperature of absolute zero radiates heat at a rate proportional to the fourth power of the absolute temperature.

While the rate of emission is independent of the conditions of the surroundings, a net transfer of radiant heat requires a difference in the surface temperature of any two bodies between which the exchange is taking place. If the black-body radiates to an enclosure which completely surrounds it and whose surface is also black i.e. absorbs all the radiant energy incident upon it, the net rate of radiant heat transfer is given by,

$$q_r = \sigma A_1 \left(T_1^4 - T_2^4\right)$$

where T_2 is the surface temperature of the enclosure in K.

Real bodies do not meet the specifications of an ideal radiator but emit radiation at a lower rate than the black-bodies. If they emit, at a temperature equal to that of a black-body, a constant fraction of the black-body emission at each wavelength, they are called *Gray Bodies*. The net rate of heat transfer from a gray body at T_1 to a black surrounding body at T_2 is

$$q_r = \sigma A_1 \varepsilon_1 \left(T_1^4 - T_2^4\right)$$

where ε is the emissivity of the gray surface and is equal to the ratio of emission from the gray surface to the emission from a perfect radiator at the same temperature.

If neither of the two bodies is a perfect radiator and if two bodies possess a given geometrical relationships to each other, the net heat transfer by radiation between them is given by

$$q_r = \sigma A_1 F_{1-2} \left(T_1^4 - T_2^4\right)$$

where F_{1-2} is a modulus which modifies the equation for perfect radiators to account for the emissivities and relative geometries of the actual bodies.

$$q_r = K_r \left(T_1^4 - T_2^4\right)$$

By comparison the conductance K_r for radiation is

$$K_r = \frac{\sigma A_1 F_{1-2}\left(T_1^4 - T_2^4\right)}{T_1 - T_2} \text{ kJ/hr°C}$$

Unit thermal conductance for radiation $\bar{h}_r$ is

$$\bar{h}_r = \frac{K_r}{A_1} = \frac{\sigma F_{1-2}\left(T_1^4 - T_2^4\right)}{T_1 - T_2} \frac{\text{kJ}}{\text{hr}} \text{m}^2/°\text{C or } \frac{\text{kJ}}{\text{hr.m}^2\,°\text{C}}$$

Similarly, the thermal resistance for radiation is given by

$$R_r = \frac{T_1 - T_2}{\sigma A_1 F_{1-2}\left(T_1^4 - T_2^4\right)}$$

18.10 HEAT EXCHANGER

A heat exchanger is an apparatus in which the processes of heating or cooling occur, i.e., heat is transferred from one substance to another. There is a large number of different heat exchangers or heat exchange apparatuses varying both in application and design. In respect to the principle of operation, heat exchangers may be divided in three categories as follows.

18.10.1 Direct Contact Heat Exchanger

The simplest type of heat exchanger is a container in which a hot and a cold fluids are mixed directly. In such a system both fluids will reach the same final temperature and the amount of heat transferred can be estimated by equating the energy lost by the hotter fluid to the energy gained by the hotter fluid to the energy gained by the cooler one. Open feed water heaters, desuperheaters, water cooling towers, scrubbers and jet condensers are the examples of heat transfer equipment employing direct mixing of fluids.

18.10.2 Recuperative Heat Exchangers

In heat exchangers of the recuperative variety, hot and cold fluids flow simultaneously through the heat exchanger and the heat is transferred through a wall separating the fluids. This group unites such heat transfer equipments such as steam boilers, water heaters, condensers etc. There are many forms of such equipment ranging from a simple pipe—within a pipe with a few sq.m. of heat transfer surface upto complex surface condensers and evaporators with many hundreds of sq.m. of heat transfers surface. In between these extremes is a broad field of common shell and tube exchangers. These units are widely used because they can be constructed with large heat transfer surfaces in a relatively small volume, can be frabricated from alloys to resist corrosion, and are suitable for heating, cooling, evaporating or condensing all kinds of fluids.

18.10.3 Regenerative Heat Exchanger

A regenerative heat exchanger is an apparatus in which one and the same heating surface is alternately exposed to the hot and cold fluids. The heat carried by the hot fluid is taken away by and accumulated in the walls of the apparatus and is then transferred to the cold fluid flowing through the heat exchanger. Regenerators of open hearth and glass-melting furnaces, air preheaters of blast furnaces are some of the specimens of regenerative heat-exchange equipment.

The process of heat transfer in recuperative and regenerative heat exchangers is invariably bound with the surface of solid. That is why they are known as *surface exchanger.*

The special names given to heat exchanger are usually determined by their application and designation for example, steam boilers, furnaces, water heaters, evaporators, super-heaters, condensers etc. In spite of the great variety of shapes, layouts, principles of operation and working media, heat exchange apparatuses—ultimately serve one and the same purpose—transfer of heat from one, hot fluid to another cold fluid. The design fundamentals are, therefore common to all.

The complete design of a heat exchanger can be broken down into three major phases:

1. The thermal analysis
2. The preliminary mechanical design
3. Design for manufacture.

Here the emphasis will be on the thermal design. This phase of the design is primarily concerned with the determination of the heat-transfer surface area required to transfer heat at a specified rate for given flow rates and temperatures of the fluids.

The mechanical design involves considerations of the operating temperatures and pressures, the corrosive characteristics of one or both fluid, the relative thermal expansions and accompanying thermal stresses, and the relation of the heat exchanger to other equipment concerned.

The design for manufacture requires the translation of the physical characteristics and dimensions into a unit, which can be built at a low cost. Selection of materials, seals, enclosure, and the optimum mechanical arrangement have to be made and the manufacturing procedures must be specified.

18.11 BASIC TYPES OF HEAT-EXCHANGERS

The simple type of shell and tube heat exchanger is shown in Fig. 18.7. It consists of a tube or a pipe located concentricity inside another tube which forms the shell for this arrangement. One of the fluids flows through the inner tube and the other through the annulus formed between the inner and the outer tube. Since both fluid streams traverse the exchanger only once, this arrangement is called *single pass* heat exchanger. If both fluids flow in the same direction, the heat exchanger is a *parallel flow type*; if the fluids move in the opposite direction, the exchanger is of the counter flow type.

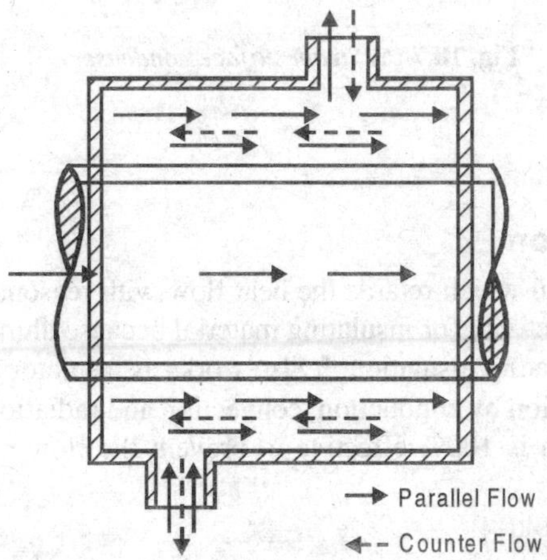

Fig. 18.7

The temperature difference between the hot and the cold fluid is, in general, not constant along tube, and the rate of heat flow vary from section to section. To determine the rate of heat flow one must therefore use an appropriate mean temperature difference.

When the two fluids flowing along the heat transfer surface move at right angles to each other, the heat exchanger is of the *cross flow type*. Three separate arrangements of this type of exchanger are possible. In the first case, each of the fluids is unmixed as it passes through the exchanger and therefore, the temperatures of the fluids leaving the heater section are not uniform, being hotter one side than on the other. A flat-type heater, a design used for turbine regenerators to claim the energy of the exhaust gases, or an automobile radiator approximates this type of exchanger.

In the second case, one of the fluids is unmixed and the other is perfectly mixed as it flows through the exchanger. The temperature of the mixed fluids will be uniform across any section and will vary in the direction of flow. An example of this type is the cross flow air heaters. The air flowing over the bank of tubes is mixed, while the hot gases inside the tubes are confined and therefore do not mix.

In the third case, both of the fluids are mixed as they flow through exchanger, that is the temperature of both fluids will be uniform across the section, and will vary only in the direction of flow. Thus type of arrangement is less important than the other two.

In order to increase the effective heat transfer, optimum heat transfer surface area per unit volume, is provided. Most commercial heat exchangers provide for more than a single pass through the tubes, and the fluid flowing outside the tubes in the shell is routed back and forth by means of baffles. Such an exchanger is known as "*multi-pass*" or *multiflow heat exchanger*. Figure 18.7(a) shows a three pass surface condenser.

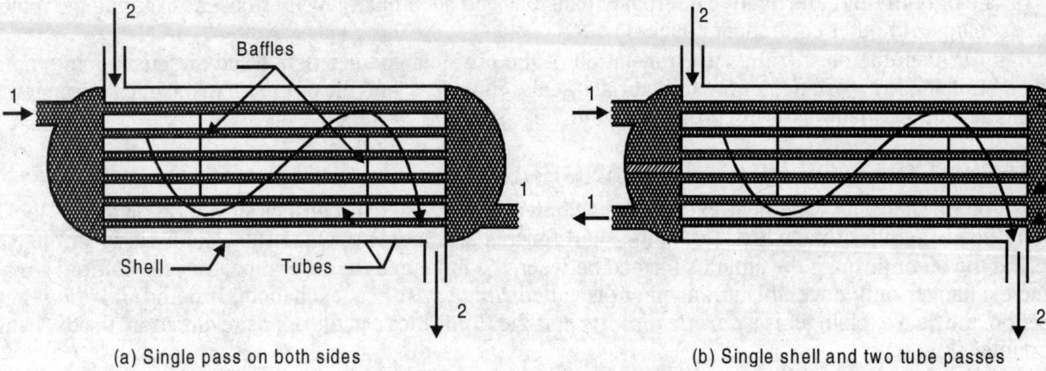

Fig. 18.7 (a) *Steam surface condenser.*

18.12 INSULATION

18.12.1 Purpose of Insulation

The insulation is defined as a material which retards the heat flow with reasonable effectiveness. It is not always necessary to have low thermal conductivity for insulating material because aluminum foil has five thousand times greater conductivity than low temperature insulation. It also works as insulator when used with air spaces.

Heat is transferred through insulation by conduction, convection and radiation or by combination of these three modes. There is no insulation which is 100% effective to prevent the flow of heat transfer under temperature gradient.

The purpose of insulation is two fold

(a) to prevent the flow of heat from the system to the surroundings as in the case of steam and hot water pipes which are used for air-conditioning in winter.

(b) to prevent the flow of heat from the surroundings to the system as in the case of brine pipes which are used for air conditioning in summer or refrigerator.

The insulations are commonly used for the following industrial purposes.

(i) Air-conditioning systems
(ii) Refrigerators and food preserving halls
(iii) Preservation of liquid gases
(iv) Boilers and steam pipes
(v) Insulating bricks in all types of furnaces.

18.12.2 Required Properties of Insulating Material

The desired properties of insulators are:

1. Thermal conductivity. Some insulators have higher thermal conductivity than other. It is best to choose insulator having the lowest conductivity. The choice of the insulating material also depends upon outer practical requirements.

2. Structural strength. Some insulating materials tend themselves readily to various types of construction and easily handled under all circumstances. Structural strength generally is obtained by the use of wood or steel frame work. For curved surfaces, material such as ground cork or mineral wool has been used extensively.

3. Weight. The weight of insulation for use in household cabinets is factor of small importance but when insulation under consideration is for use in truck, marine work, rail road, refrigerated car, aeroplane and similar application, weight often becomes a deciding factor.

4. Origin of insulation. Insulators derived from vegetable or animal sources are always subject to decomposition over a period of time. Such materials may develop odours unless special consideration is given to the method of applications.

5. Resistance to vermin. Insulators of the vegetable origin are regarded as food by certain forms of vermin. In tropical and semitropical countries, this difficulty may be troublesome on account of the presence of roaches, ants, mites and other forms of vermin, so that the vegetable type insulators need special protection.

6. Freedom from odour. Many foods absorb odours when they are exposed to odours. For this reason, the insulation used in refrigerator cabinets should be odourless when it is dry. In the presence of moisture, certain forms of vegetable insulation favour the growth of mold which produces odours and soon destroys the insulation. Insulation of this type when used in cabinets must be kept dry to avoid such difficulties.

7. Resistance to moisture absorption. An insulator which absorbs moisture deteriorates rapidly, especially if used on low temperature equipments. The rate of deterioration depends on the following factors:

(a) the thickness of insulation
(b) the temperature and humidity of atmosphere
(c) method of application of the insulation
(d) the temperature difference between interior and exterior faces of the insulation.

8. Resistance to fire. A few insulators are extensively resistant to combustion and some burn very readily. Resistance to fire is not deciding factor, but in some application, such as, fire proof warehouse, is important.

18.13 TYPES OF INSULATIONS

Insulating materials are frequently classified in accordance with the temperatures at which they are used, such as low temperature (<100°C), moderate or steam temperature (100–600°C) and high temperature (> 600°C). This classification is not entirely satisfactory because many insulating materials can be properly employed in two or more of these temperature ranges. It is better to classify these insulators according to physical structure.

1. Loose Fill Insulators. This type of material is generally made in powered, granulated, cellular or fibrous form: (a) Mineral wool (b) Diatomaceous Earth (c) Venniculate (d) Silica Aerogal (e) Crushed insulating brick, asbestos fibres, granulated cork (f) Shredded redwood bark (g) Batts or Flexible Insulation.

2. Slab Insulation. Slab insulation consists usually of wood or sugar cane fibres, processed with an adhesive: (a) Cork board slabs (b) Rock cork (c) Balsa wood (d) Rubatex (e) Foam glass (f) Celotex (g) Hair felts (h) Kapok (i) Palco Bark (j) Insulating papers.

3. Insulating Fire-Brick. The insulating fire bricks are generally made by mixing an organic material with the refractory before moulding and the organic material is subsequently burned during firing leaving voids. The moulded brick is dried, fired. Standard size of brick is $6 \times 12 \times 25$ cm.

4. Reflective Insulation. The effectiveness of metallic surface for retarding heat transfer was well-known to physicists early in the nineteenth century. Conduction and convection are practically eliminated by maintaining a high vacuum in the space.

(a) Aluminum Foil (Alfol)
(b) Low temperature reflective insulation
(c) Lead Tin alloy coated steel sheets.

18.14 EXTENDED SURFACES (FINS)

The equations of conduction heat transfer presented so far have not included the influence of heat losses or gains at the surfaces of the conducting body. When such concurrent heat flow must be considered, the surface coefficient of heat transfer h enters the basic equations. Also it is seen that the overall heat transfer resistance of a plane wall is determined principally by the greatest single resistance. If this resistance is one of the convection resistances, the heat flow through the wall can be increased by putting fins (extended surfaces) on the surface where this large resistance occurs. Such extended surfaces are widely used, for example, in economizers in steam power plants, convectors for steam and hot water heating systems, or electrical transformers, for the cylinders of air-craft engines, two wheeler engines etc.

There are many forms and shapes of these extended surfaces. Some of these are as follows :

1. A thin rod. This is transferring heat at its surface to a surrounding fluid and which is connected at its base to a heated wall. This system is shown in Fig. 18.8.

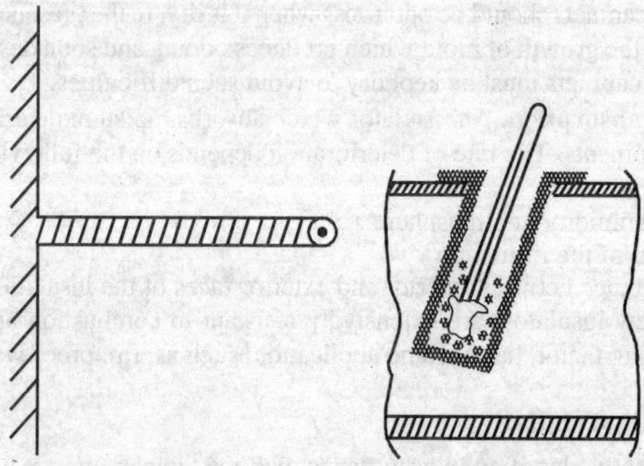

Fig. 18.8. *Thermometer packet or well.*

2. The Rectangular Fin. This is the simplest case for the plane finned surface. As long as the height of the fins on a tube is comparatively small with respect to the tube diameter, the formula derived from the plane wall can also be used for the tube. This system is shown in Fig. 18.9.

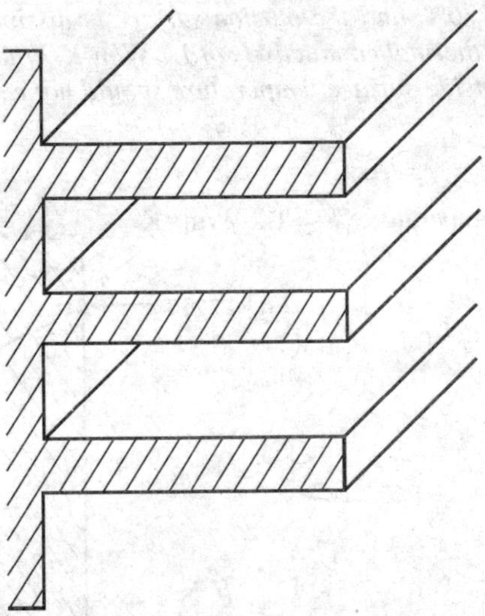

Fig. 18.9. *Heating surface with rectangular fins.*

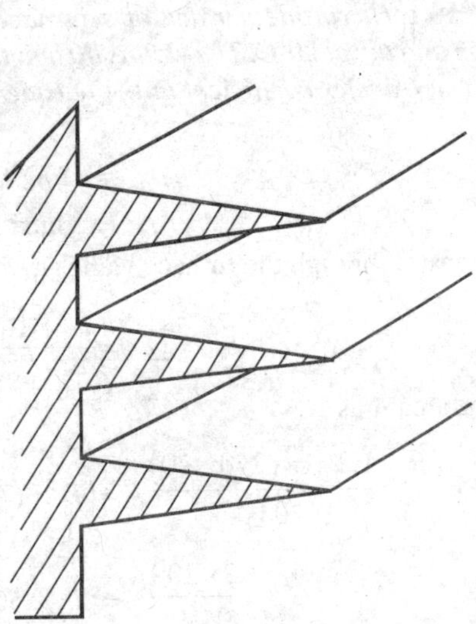

Fig. 18.10. *Heating surface with traingular fins.*

3. Straight Fins of Triangular Profile. In determining the optimum fin, the question arises as to whether or not weight advantage can be gained by using a shape other that rectangular for the fin cross-section. Generally, a straight fin of triangular cross-section is considered. Such a fin is shown in Fig. 18.10.

The mathematical treatment in this case is similar to the case of the fin of rectangular cross-section except that the area nornal to the heat flow is a function of the distance along the fin, decreasing as the fin length increases.

4. Cylindrical Fins. Fins which are arranged around tubes are called cylindrical fins and are quite important from an engineering point of view.

Such a fin system is shown in Fig. 18.11. Here again the treatment is substantially the same as for the rectangular fin except that the area must be allowed to vary with the radius.

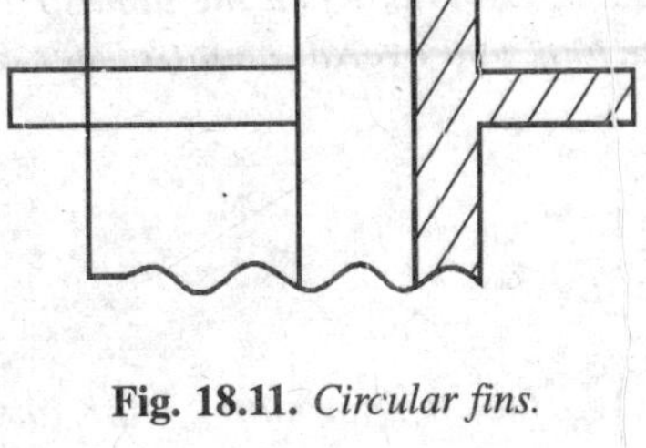

Fig. 18.11. *Circular fins.*

SOLVED EXAMPLES

Example 18.1 *The inside surface of a furnace wall is at 270°C and the outside surface is dissipating heat by convection into air at 20°C. The wall is 40 mm thick and has thermal conductivity of 1.2 W/m-K. What is minimum value of heat transfer coefficient at the outside surface if outside surface temperature should not exceed 70°C.*

Solution

Given:

$$t_1 = 270°C = t_2 = 70°C = t_3 = 20°C$$

$$t_2 = \text{Outside Surface Temperature}, K = 1.2 \text{ W/m}\circ\text{K}$$

Heat transfer through the furnace wall is given by

$$Q = \frac{KA(t_1 - t_2)}{L} = h_c A(t_2 - t_3)$$

Assume area A as 1

$$\therefore \quad \frac{1.2 \times 1 \times (270 - 70)}{0.04} = h_c \times (70 - 20)$$

$$\frac{1.2 \times 200}{0.04} = h_c \times 50$$

$$h_c = \text{heat transfer coefficient}$$

$$= \frac{1.2 \times 200}{50 \times 0.04}$$

$$= \mathbf{120\ W/m^2K}$$

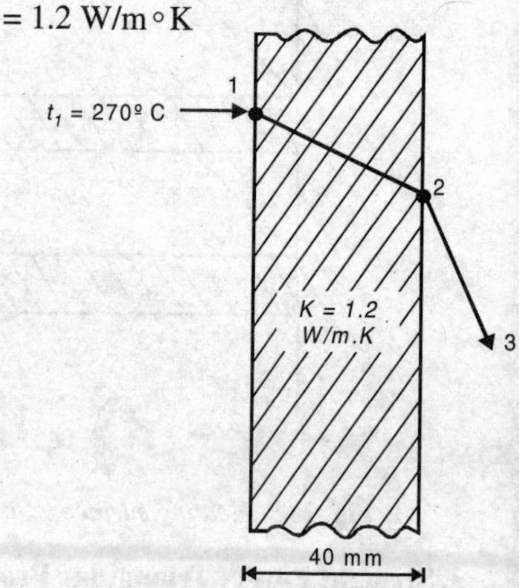

Fig. Ex. 18.1

Example 18.2 *A black body at 30°C is heated to 100°C. Calculate the change in its emissive power.*

Solution

(c) Emissive power of block body $= \sigma A T^4$

where $\sigma = 20.52 \times 10^{-8}$ kJ/hr K^4

$\therefore$ For unit area, $A = 1\ m^2$

Increase of emmissive power

$$= \sigma(T_2^4 - T_1^4)$$

$$= 20.52 \times 10^{-8} (100 + 273)^4 - (30 + 273)^4$$

$$= 20.52 \times 10^{-8} \left[373^4 - 303^4\right]$$

$$= 20.52 \times 109.28$$

$$= \mathbf{2242.4256\ kJ/hr.}$$

Example 18.3 (*a*) *A mild steel tank of wall thickness 12 mm contains water at 100°C. Calculate the rate of heat loss per square meter area of the surface of the tank when the atmospheric temperature is 20°C. Also calculate temperature of the outside surface of the tank and overall heat transfer coefficient.*

Assume the following :

Conductivity of mild steel 50 W/m-K

Convection heat transfer coefficient

(*i*) *on water side 2850 W/m² L*

(*ii*) *on air side 10W/m² K*

(*b*) *A 60 W incandescent lamp has coil surface temperature 2500 K and room temperature 300 K. Estimate the surface area of the coil in mm².*

(c) *Derive an equation of conduction heat transfer rate in terms of temperatures dimensions and thermal conductivity for a thick, hollow cylinder.*

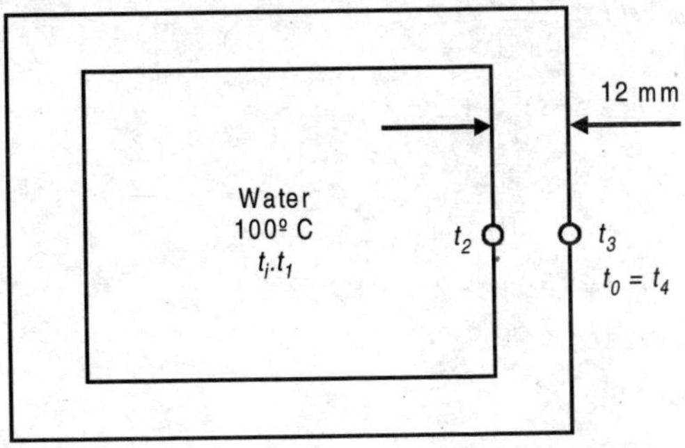

Fig. Ex. 18.3

Solution

Consider 1 sq. m of tank surface. The system is given in Fig. 18.13. Let Q be the heat rate

$$Q = h_i A(t_1 - t_2) \text{ (Water to inner surface)}$$

$$= \frac{KA(t_2 - t_3)}{x} \text{ (Inner to outer surface)}$$

$$= h_0 A(t_3 - t_4) \text{ (Outer surface to air)}$$

$$\therefore \quad t_1 - t_2 = \frac{Q}{h_i} \text{ for unit area}$$

$$t_2 - t_3 = \frac{Q.x}{K} \text{ for unit area}$$

$$t_3 - t_4 = \frac{Q}{h_0} \text{ for unit area}$$

$$t_1 - t_4 = Q\left[\frac{1}{h_i} + \frac{x}{K} + \frac{1}{h_0}\right]$$

$$100 - 20 = Q\left[\frac{1}{2850} + \frac{12/1000}{50} + \frac{1}{10}\right]$$

$$80 = Q[0.0003509 + 0.00024 + 0.1] = Q \times 0.1006$$

$$Q = \text{Rate of heat transfer/m}^2$$

$$= \frac{\mathbf{80}}{\mathbf{0.1005989}} = \mathbf{795.3\ W}$$

Again $$795.3 = h_0 A(t_3 - t_4) = 10[t_3 - 20]$$

$$t_3 - 20 = \frac{795.3}{10}$$

$$t_3 = 79.5 + 20$$

$$t_3 - 20 = \frac{795.3}{10} = 79.53$$

$$t_3 = 79.5 + 20 = 99.53°C$$

Temperature of the outside surface = 99.53°C

For overall heat transfer coefficient

We have $$Q = UA\,(t_i - t_0)$$

$$795.3 = U\,(100 - 20)$$

U = overall heat transfer coefficient

$$= \frac{795.3}{80} = 9.94 \text{ W/m}^2\text{/K}$$

(b) Heat transfer from the incandescent lamp is 60 W.

∴ $$60 = \sigma\left(T_2^4 - T_1^4\right)$$

where σ = 5.671×10^{-8} W/m²/K⁴

A = Area sq. m.

T_1 = Lamp surface temperature

T_2 = Room temperature

∴ $$60 = 5.671 \times 10^{-8} \times A \times (2500^4 - 300^4)$$

$$A = \frac{60\times10^8}{5.671\times\left(2500^4 - 300^4\right)}$$

$$= \frac{60\times10^8}{5.671\times10^8\left(25^4 - 3^4\right)}\text{m}^2 = \frac{10.58\times10^6}{(625-9)(625+9)}\text{mm}^2$$

$$= \mathbf{27.091\ mm^2}$$

= Coil/surface area

(c) For derivation of the required equation, **Refer 18.5.**

Example 18.4 (*a*) *Determine the rate of heat loss per hour through a 3.5 m² boiler plate if it is 20 mm thick with thermal conductivity 50 W/m-K the inner surface of the wall is covered with a layer of scale and deposits 2 mm thickness having a conductivity of 1 W/m-K. The hot gas temperature of the plate is 250°C while that on the cold air (outside) is 200°C. Also determine interface temperature.*

(*b*) *determine the loss by radiation from a steel tube 70 mm diameter and 3 m long at a temperature of 227°C the tube is located in a huge room having wall temperature 27°C.*

Solution

The plate with scale and deposit is shown in Fig. Ex. 18.4

With reference to this figure,

We write $$Q = \frac{K_1A_1(t_1-t_2)}{x_1} = \frac{K_2A_2(t_2-t_3)}{x_2}$$

$$\therefore \quad t_1 - t_2 = \frac{Q.x_1}{K_1A_1}$$

$$t_2 - t_3 = \frac{Q.x_2}{K_2A_2}$$

$$t_1 - t_3 = \frac{Q}{A_1}\left[\frac{X_1}{K_1}+\frac{X_2}{K_2}\right]$$

$$250-200 = \frac{Q}{3.5}\left[\frac{2}{1000\times1}+\frac{20}{1000\times50}\right]$$

$$= \frac{Q}{3.5\times1000}[2+0.4]$$

$$Q = \frac{50\times3.5\times1000}{2.4} = 72916.7 \text{ W}$$

= Rate of heat loss

= **72.92 kJ/s**

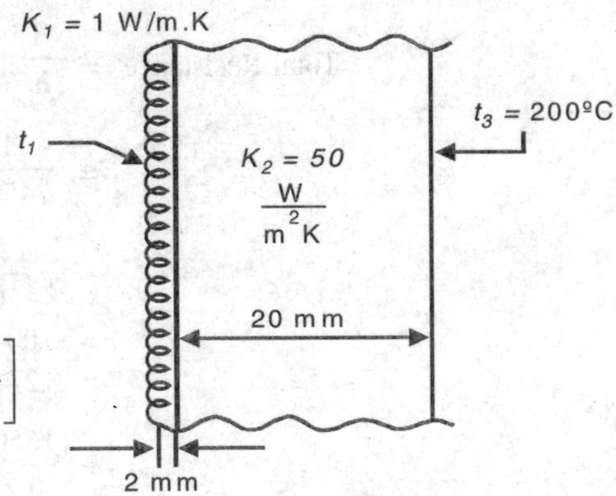

Fig. Ex. 18.4

Rate of heat per hour

$= 72.92 \times 3600$ kJ/hr $=$ 26512 kJ/hr

$=$ **26.25 × 10⁴ kJ/hr**

Example 18.5 *A furnace wall consists of two lawyer 22.5 cm of brick (K = 4.984 kJ /hr.m °C/m) and 12.5 cm of insulating brick (K= 0.623 kJ/hr.m °C/m) the temperature inside the furnace is 1650 °C and the unit surface conductance at the inside wall is 245.28 kJ/hr.m °C the temperature of the surrounding atmosphere is 27 °C and the unit surface conductance at the outer wall is 40.88 kJ /hr. m² °C. Neglecting the thermal resistance of the morter joints. Estimate (a) the rate of heat loss per sq.m of wall and the temperatures at the (b) intersurface and (c) outersurface.*

Solution

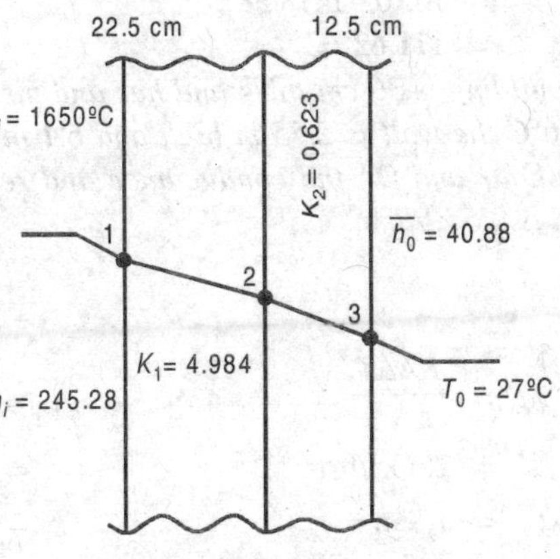

Fig. Ex. 18.5

(a) The rate of heat flow is given by

$$\frac{q}{A} = \frac{T_I - T_O}{\text{Total Resistance}}$$

$$\text{Total Resistance} = \frac{1}{h_I} + \frac{L_1}{K_1} + \frac{L_2}{K_2} + \frac{1}{h_O}$$

$$= \frac{1}{245.28} + \frac{22.5/100}{4.984} + \frac{12.5/100}{0.623} + \frac{1}{40.88}$$

$$= 0.00408 + 0.04514 + 0.20064 + 0.02446$$

$$= 0.27432$$

$$\therefore \quad q = \frac{1650-27}{0.27432} = \frac{1623}{0.27432}$$

$$= \mathbf{5916.9\ kJ/hr.m^2}$$

(b) We have $\frac{q}{A} = 5916.9\ \frac{T_i - T_1}{R_1} = \frac{1650 - T_1}{0.00408}$

$$\therefore \quad T_1 = \text{Inner wall temperature}$$

$$= 1650 - 5916.9 \times 0.00408 = 1650 - 24.14$$

$$= \mathbf{1625.86 \circ C}$$

(c) Again for outer surface temperatures

We have $\frac{q}{A} = \frac{T_i - T_3}{R_1 + R_2 + R_3} = \frac{1650 - T_3}{0.00408 + 0.04514 + 0.20064}$

$$5916.9 = \frac{1650 - T_3}{0.24986}$$

$$\therefore \quad T_3 = 1650 - 5916.9 \times 0.24986$$

$$= 1650 - 1478.38$$

$$= \mathbf{171.62\ °C}$$

Example 18.6 *A brick wall of a building is 30 cm thick and has and inside surface temperature of 24°C and an outside surface temperature of –6°C the wall is 2.75 m high and 6.1 m long. Estimate (a) the heat transfer by conduction through the wall per hour and (b) the conductance and resistance of the wall. The coefficient of conductivity of the brick material is 2.6 kJ/hr.m °C.*

Solution

(a) Heat transfer by conduction

We have $q = \frac{KA\Delta T}{x}$ $\quad A$ = Area of the wall

Where K = 2.6 kJ/hrm °C $\quad$ = 2.75 × 6.1 = 16.8 m^2

$\Delta T = t_2 - t_1$ $\quad x$ = Thickness of wall in m

$= -(t_1 - t_2)$ $\quad = \frac{30}{100} = 0.3$ m

$$= -(24+6) = -30\,°C$$

Substituting these values, we get

$$q = \frac{2.6 \times 16.8 \times 30}{0.3}$$

$$= \mathbf{4368\ kJ/hr}$$

(b) Conductance and resistance of the wall

$$\text{Resistance} = \frac{t_1 - t_2}{q} = \frac{30}{4368}$$

$$= \mathbf{0.00687\ C.hr/kJ}$$

$$\text{Conductance} = \frac{1}{\text{Resistance}}$$

$$= \frac{4368}{30}$$

$$= \mathbf{145.6\ kJ/m\ °C}$$

Example 18.7 *A temperature of the inner surface of a furnace wall is 450°C and that of the outer surface 200°C. The ambient temperature is 50°C. It is required to reduce the heat loss from the outer surface by doubling the thickness of the brickwork. Assuming constant values convective heat transfer coefficient of air and thermal conductivity of wall material calculate percent decrease in heat loss.*

Solution

Let K be the conductivity of the wall material and $\bar{x}$ be the thickness.

Let T_0 be outside temperature and $\bar{h}_0$ be the unit coefficient of heat transfer for outer surface.

∴ Heat transfer

$$q = \frac{450 - T_0}{\frac{x}{K} + \frac{1}{h_0}} \text{ per unit area} \quad \ldots\text{(i)}$$

When the wall is doubled in thickness, heat transfer.

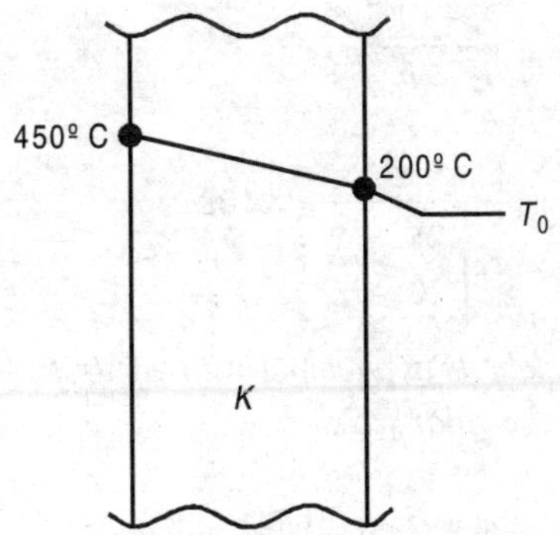

Fig. Ex. 18.7

$$q = \frac{450 - T_0}{\frac{2x}{K} + \frac{1}{h_0}} \quad \text{...(ii)}$$

$$\text{Decrease in heat loss} = (\text{i} - \text{ii})$$

$$= (450 - T_0)\left\{\frac{450 - T_0}{\frac{x}{K} + \frac{1}{h_0}} - \frac{450 - T_0}{\frac{2x}{K} + \frac{1}{h_0}}\right\}$$

$$\text{Percentage decrease in heat lose} = \frac{(450 - T_0)\left\{\frac{1}{\frac{x}{K} + \frac{1}{h_0}} - \frac{1}{\frac{2x}{K} + \frac{1}{h_0}}\right\}}{(450 - T_0)\left\{\frac{1}{\frac{x}{K} + \frac{1}{h_0}}\right\}}$$

$$= 1 - \frac{\frac{1}{2x/K} + \frac{1}{h_0}}{\frac{1}{x/K} + \frac{1}{h_0}} = 1 - \frac{\frac{x}{K} + \frac{1}{h_0}}{\frac{2x}{K} + \frac{1}{h_0}}$$

$$= \frac{\frac{2x}{K} + \frac{1}{h_0} - \frac{x}{K} - \frac{1}{h_0}}{\frac{2x}{K} + \frac{1}{h_0}}$$

$$= \frac{\frac{x}{K}}{\frac{2x}{K} + \frac{1}{h_0}}$$

$$= \frac{\mathbf{x}}{\mathbf{K}\left[\frac{\mathbf{2x}}{\mathbf{K}} + \frac{\mathbf{1}}{\mathbf{h_0}}\right]} \times \mathbf{100}$$

Example 18.8 *Assuming earth as a block body at a temperature of 300K. Estimate the energy received by unit area of the earth's surface perpendicular to the solar rays.*

Solution

According to Stefan–Boltzamn's law heat absorbed = Heat radiated

For a black-body

∴ Heat received by unit area of the earth's surface perpendicular to the solar ray

$$Q = \sigma T^4 = 20.52 \times 10^{-8} \times (300)^4$$

$$= 20.52 \times 10^{-8} \times 10^{8} \times 3^{4}$$
$$= 20.52 \times 81$$
$$= \mathbf{1662.12\ kJ/hr}$$

Example 18.9 *Determine the loss of heat by radiation from a steel tube 70 mm diameter and 3 m long at a temp. of 227 °C the tube is located in a large room having wall temperature 27°C.*

Solution

$$Q = A.\sigma\left(T_H^4 - T_C^4\right)$$

$$A = 2\pi DL = 2\pi \times \frac{70}{1000} \times 3\,\text{m}^3$$

$$= \frac{42\pi}{100} = 0.42\,\pi\text{m}^2$$

$$= \frac{22}{7} \times \frac{42}{100} = \frac{132}{100} = 1.32$$

$$= 1.32 \times 20.52 \times 10^{-8}\,[\,500^4 - 300^4]$$
$$= 1.32 \times 20.52\,[5^4 - 3^4\,]$$
$$= 1.32 \times 20.52\,[\,625 - 81\,]$$
$$= 1.32 \times 20.25 \times 544$$
$$= 14730 \text{ kJ/hr}$$
$$= \mathbf{4094\ W} \text{ or } \mathbf{4.094\ kW}$$

Example 18.10 *The inside temperature of refrigerator is required to be 7 °C the walls are constructed with two mild steel sheets 3 mm thick with 5 cm of glass wool insulation between them. The heat transfer coefficient on inner and outer surface of refrigerator are 10W/m°C and 12.5 W/m²C. Find the rate of heat removed from the refrigerator when it is kept in a kitchen room. The temperature in kitchen is 28°C and K for mild steel is 40W/°C and glass wool is 0.04 W/m°C.*

Solution

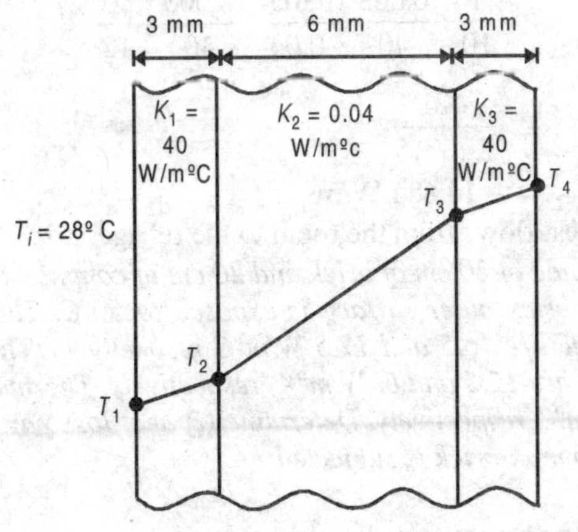

Fig. Ex. 18.10

This is a problem of composit walls and the wall in problem is shown in above Fig. Ex. 18.10.

$T_1 = 7°C$, $T_0 = 28°C$, $K_1 = 40$ W.m°C K_3; $x_1 = 3$ mm $= x_3$ and $x_2 = 50$ mm, $h_i = 10$ W/m²°C and $h_0 = 12.5$ W/m²/°C. Consider unit area and we write

$$Q = h_i A\,(T_i - T_1) \quad \therefore \quad T_i - T_1 = \frac{Q}{h_i A}$$

$$= \frac{K_1 A_1 (T_1 - T_2)}{x_1} \quad T_1 - T_2 = \frac{Q \times x_1}{K_1 A_1}$$

$$= \frac{K_2 A_2 (T_2 - T_3)}{x_2} \quad \therefore \quad T_2 - T_3 = \frac{Q \times x_2}{K_2 A_2}$$

$$= \frac{K_3 A_3 (T_3 - T_4)}{x_3} \quad \therefore \quad T_3 - T_4 = \frac{Q \times x_3}{K_3 A_3}$$

$$= h_0 A_0 (T_4 - T_0) \quad \therefore \quad T_4 - T_0 = \frac{Q}{h_0 A_0}$$

Adding these we get $T_i - T_0 = Q\left[\dfrac{1}{h_i A_i} + \dfrac{1}{\dfrac{K_1 A_1}{x_1}} + \dfrac{1}{\dfrac{K_2 A_2}{x_2}} + \dfrac{1}{\dfrac{K_3 A_3}{x_3}} + \dfrac{1}{h_0 A_0}\right]$

Assuming unit area of heat transfer

$$Q = \frac{T_i - T_0}{\dfrac{1}{h_i} + \dfrac{x_1}{K_1} + \dfrac{x_2}{K_2} + \dfrac{x_3}{K_3} + \dfrac{1}{h_0}}$$

$$= \frac{7 - 28}{\dfrac{1}{10} + \dfrac{0.003}{40} + \dfrac{0.05}{0.04} + \dfrac{0.003}{40} + \dfrac{1}{12.5}}$$

$$= \frac{-21}{1.43015}$$

$$= \mathbf{-14.684\ W/m^2}$$

The –ve sign indicates that the heat flows from the room to the refrigerator.

Example 18.11 *A furnace wall is made of 20 cm of brick and 20 cm of common brick. The fire brick is exposed to hot gases at 1400°C and common brick outer surface is exposed to 50°C. The convection and radiation heat transfer coefficient towards gas side are 16.5 and 17.5 W/m°C respectively. The convection end radiation heat transfer coefficient towards outside are 12.5 and 6. W/m°K respectively. Thermal conductivity's of fire brick and common brick are 4 and 0.065 W/m°C respectively. Determine (i) heat loss par m² area of the furnace wall, (ii) maximum temperature to which common brick is subjected.*

Solution

1. The heat flow through the furnace composite wall per unit area is given by

$$Q = h_i\,(T_i - T_1)$$

$$= \frac{K_1 \cdot (T_1 - T_2)}{x_1}$$

$$= \frac{K_2.(T_2 - T_3)}{x_3}$$

$$= h_0 \ (T_3 - T_0)$$

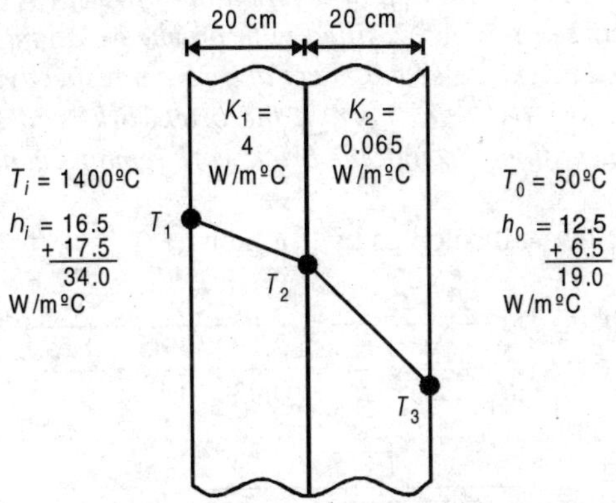

Fig. Ex. 18.11

$$\therefore \qquad Q = \frac{T_i - T_0}{\frac{1}{h_i} + \frac{x_1}{K_1} + \frac{x_2}{K_2} + \frac{1}{h_0}}$$

We take here $h_i = 34 = \overline{h}_c + \overline{h}_r$

$$= 16.5 + 17.5 = 34$$

$$= \frac{1400 - 50}{\frac{1}{34} + \frac{0.2}{4} + \frac{0.2}{0.65} + \frac{1}{19}} = 1350$$

$$= \frac{1350}{0.0294 + 0.05 + 0.3077 + 0.0526}$$

$$= \frac{1350}{0.4397} = 3070 \ \text{W/m}^2$$

$$= \mathbf{3.07 \ kW/m^2}$$

2. Maximum temperature to which common brick is subjected is T_2

$$\therefore \qquad Q = h_i \,(T_i - T_1)$$

$$\therefore \qquad 3070 = 34 \,(1400 - T_1)$$

$$\therefore \qquad \frac{3070}{34} = 90.294 = 1400 - T_1$$

$$\therefore \qquad T_1 = 1400 - 90.294$$

$$= 1309.706 \ ^\circ\text{C}$$

Similarly

$$Q = \frac{K_1 (T_1 - T_2)}{x_1}$$

$$\therefore \quad T_1 - T_2 = \frac{3070 \times 0.2}{4} = 153.5$$

$$T_2 = 1309.706 - 153.5$$

$$= \mathbf{1156.206\ °C}$$

Example18.12 *A brick work of a furnace is built up of layers laid of fire clay (K_1= 0.93 W/m °C) and red brick (K_3= 0.7W/m °C and the space between the two is filled with crushed diatomite brick (K= 0.13W/m °C). The thickness of fireday, diatomobile and brick are 12 cm, 5 cm and 25 cm respectively.*

What should be the thickness of the red brick layer if the brickwork is to be laid without diatomite filling between the two layers so that the heat flow through the brick work remains constant.

Solution

With the diatomite filling, the heat flow will be given by (for unit area)

$$Q = \frac{T_1 - T_4}{\frac{x_1}{K_1} + \frac{x_2}{K_2} + \frac{x_3}{K_3}} = \frac{T_1 - T_4}{\frac{0.12}{0.93} + \frac{0.05}{0.13} + \frac{0.25}{0.7}}$$

$$= \frac{T_1 - T_4}{0.1290 + 0.3846 + 0.3571}$$

$$= \frac{T_1 - T_4}{0.8708} \text{ W/m}^2 \qquad \ldots(i)$$

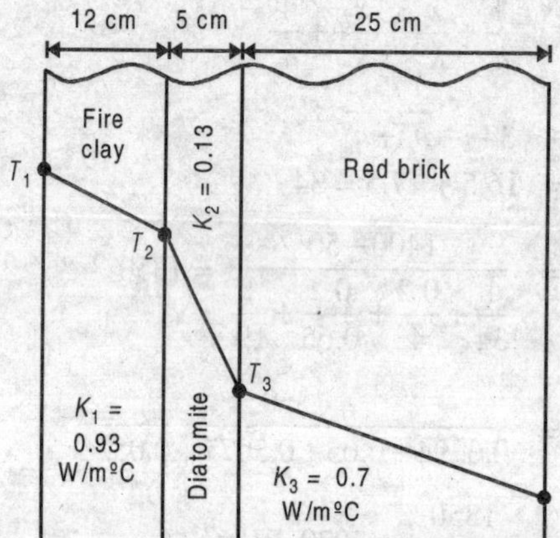

Fig. Ex.18.12

For the same Q-heat flow and without diatomite fillings, we have

$$Q = \frac{T_1 - T_4}{\frac{x_1}{K_1} + \frac{x_3}{K_3}} = \frac{T_1 - T_4}{\frac{0.12}{0.93} + \frac{x_3}{0.7}}$$

$$\therefore \quad \frac{T_1 - T_4}{0.8708} = \frac{T_1 - T_4}{0.1290 + \frac{x_3}{0.7}}$$

$$\therefore \quad \frac{1}{0.8708} = \frac{1}{0.1290+\frac{x_3}{0.7}}$$

$$\therefore \quad 0.1290+\frac{x_3}{0.7} = 0.8707$$

$$\therefore \quad \frac{x_3}{0.7} = 0.8707 - 0.1290$$

$$= 0.7417$$

$\therefore \quad x_3$ = Thickness of the red brick layer

$$= 0.7417 \times 0.7$$

$$= 0.51919 \text{ m}$$

$$= \mathbf{51.919 \text{ cm}}$$

Example 18.13 *Hot air at a temperature of 656°C is flowing through a steel pipe of 120 mm diameter. The pipe is covered with two layers of different insulating materials of thickness 60 mm and 40 mm and their corresponding thermal conductivities are 0.24 and 0.4 W/m°C the inside and outside heat transfer coefficients are 60 and 12 W/m² °C. The atmosphere is at 20 °C. Find the rate of heat loss from a 60 m length of pipe. Neglect the resistance of the steel pipe.*

Solution

The pipe with insulations is shown in Fig. Ex. 18.13 we have

$$R_1 = 60 \text{ mm} = 0.06 \text{ m}$$

$$R_2 = 60 + 60 = 120 \text{ mm} = 0.12 \text{ m}$$

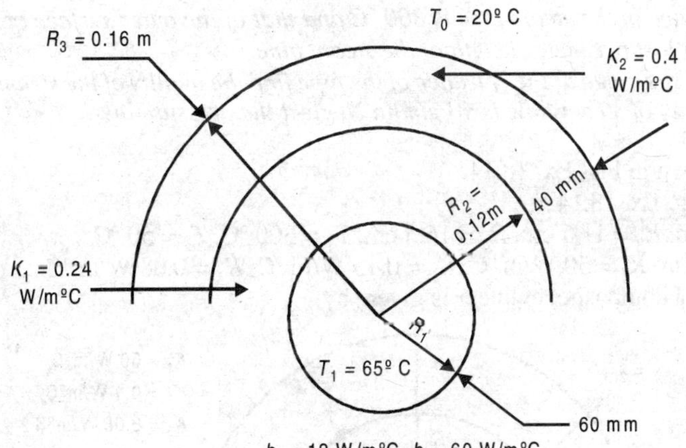

Fig. Ex.18.13

$$R_3 = 60 + 60 + 40 = 160 \text{ mm} = 0.16 \text{ m}$$

K_1 = 0.24 W/m °C $\qquad$ K_2 = 0.4 W/m °C

h_i = 60 W/m °C $\qquad$ h_0 = 12 W/m °C

Length of the pipe to be considered is 60 m

Heat transfer through a circular, hollow pipe with raddi r_1 and r_2 is given by

$$Q = \frac{K.2\pi(t_1 - t_2)}{\log_e \frac{r_2}{r_1}} = \frac{t_1 - t_2}{\frac{1}{2\pi L}.\frac{1}{K}\log_e \frac{r_2}{r_1}}$$

Applying this equation to the problem, we have

$$Q = \frac{t_i - t_o}{\frac{1}{2\pi L}\left[\frac{1}{h_i R_1} + \frac{1}{K_1}\log_e\frac{R_2}{R_1} + \frac{1}{K_2}\log_e\frac{R_3}{R_2} + \frac{1}{h_0 R_3}\right]}$$

$$= \frac{65-20}{\frac{1}{2\pi L}\left[\frac{1}{60\times0.06} + \frac{1}{0.24}\log_e\frac{0.12}{0.06} + \frac{1}{0.4}\log_e\frac{0.16}{0.12} + \frac{1}{12\times0.16}\right]}$$

$$= \frac{45}{0.002653[0.2778+2.888+0.7192+0.5208]}$$

$$= \frac{45}{0.002653\times4.4058}$$

$$= \frac{45}{0.01169}$$

$$= \mathbf{3.8494\ kW}$$

Example 18.14 *A steam pipe of 16 cm inside diameter and 17 cm of outside diameter is covered with two layers of insulation. The thickness of first and second layers are 3 cm and 5 cm respectively and their corresponding conductivities are 0.15 and 0.08 W/m °C respectively. The conductivity of the pipe material is 50 W/m °C. The temperature of inner surface of the steam pipe is 300 °C and that of the outer surface of insulation is 50 °C. Find*

(i) the quantity of heat lost per meter length of the steam pipe.

(ii) if the steam is fully saturated at the entrance of the pipe find the quality of the steam coming out of one meter length of pipe. The quantity of steam flow is 40 gm/hr. Neglect the pressure loss.

Solution

The insulated pipe is shown in Fig. Ex. 18.14.

(i) Referring to this Fig. Ex. 18.14.

R_1 = 8 cm, R_2 = 8.5 cm, R_3 = 11.5 cm, R_4 = 16.5 cm, T_1 = 300 °C, T_4 = 50 °C

Length of pipe = L = 1m. K_1 = 50 W/m °C, K_2 = 0.15 W/m °C, K_3 = 0.08 W/m °C.

The heat loss through a composite, cylinder is given by

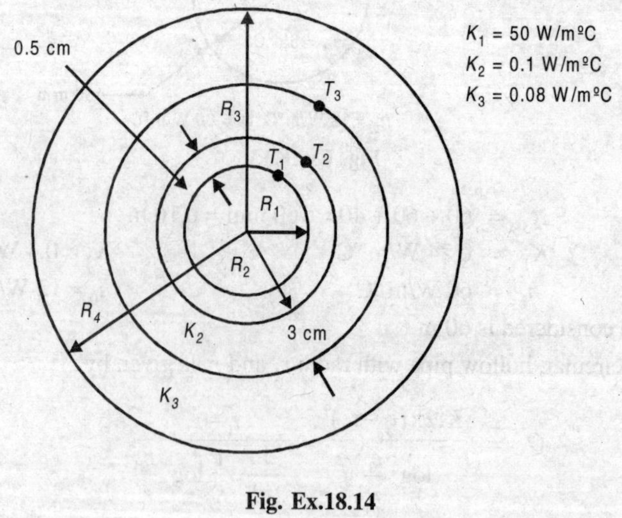

Fig. Ex.18.14

$$Q = \frac{2\pi L(T_1 - T_4)}{\frac{1}{K_1}\log_e \frac{R_2}{R_1} + \frac{1}{K_2}\log_e \frac{R_3}{R_2} + \frac{1}{K_3}\log_e \frac{R_4}{R_3}}$$

$$= \frac{2\pi \times 1(300 - 50)}{\frac{1}{50}\log_e \frac{8.5}{8} + \frac{1}{0.15}\log_e \frac{11.5}{8.5} + \frac{1}{0.08}\log_e \frac{16.5}{11.5}}$$

$$= \frac{1570.8}{0.001212 + 2.0152 + 4.513}$$

$$= \frac{1570.8}{6.5291} = 240.58 \text{ W/m}$$

$= 0.24058$ kW = **866 kJ/hr**

(ii) In enthalpy or total heat of dry and saturated steam at 300 °C is obtained form steam tables and is 2751 kJ/kg total heat of steam at 1 m ahead of the entrance is given by

$$2751 - \frac{866}{40} = 2751 - 21.65$$

$= 2729.35$ kJ/kg

$= h_f + xhfg$

$= 1345 + x \times 1406$

$\therefore$ x = Dryness fraction of steam

$$= \frac{2729.35 - 1345}{1406} = \frac{1384.35}{1406}$$

= **0.9846**

Example 18.15 *A brickwall of a building is 30 cm thick and has an inside surface temperature of 24 °C, an outside surface temperature of –6 °C. The wall is 2.75 metres high and 6.1 metres long. Estimate (a) the heat transfer by conduction through the wall per hour and (b) the conductance and resistance of the wall.*

The coefficient of conductivity of the brick material is 2.6 kJ/m.hr °C.

Solution

Refer Fig. Ex. 18.15 from the problem.

(a) Heat transfer: We have a relation between the various quantities as

$$Q = \frac{-K.A\Delta T}{x}$$

where K = 2.6 kJ/m.hr °C

A = Area of wall in sq.m

$= 2.75 \times 6.1$

$= 16.8$ m^2

ΔT = Temperature difference across the wall

$= 24\ (-6) = 24 + 6$

$= 30$ °C

x = Thickness if wall in metres

$$= \frac{30}{100} \text{ m}$$

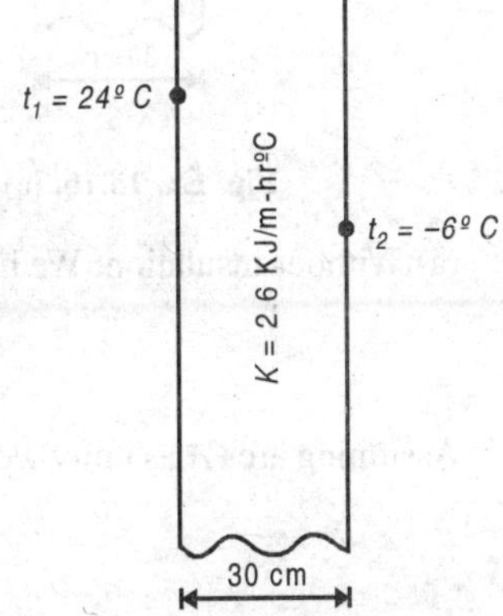

Fig. Ex. 18.15

Substituting these values in the above equation, we get

$$Q = \frac{2.6 \times 16.8 \times 30}{0.3} = 62 \times 16.8$$

$$= \mathbf{4368\ kJ/hr}$$

(b) Conductance and resistance of the wall:

$$\text{Resistance} = \frac{t_1 - t_2}{Q} = \frac{30}{1068}$$

$$= 0.00687 \text{ hr.C/kJ}$$

$$\text{Conductance} = \frac{1}{\text{Resistance}} = \frac{4368}{30}$$

$$= \mathbf{145.6\ kJ/hr\ ^\circ C}$$

Example 18.16 *The inside and outside surface of a furnace wall 30 cm thick of refractory bricks (K = 5.66 kJ/hr.m°C) are at 1650°C and 320°C respectively. Find the reduction in heat loss through the wall to be obtained by adding 30 cm thickness of insulating bricks for which K = 1.26 kh/hr m°C. Assuming the inside surface temperature of refractory bricks to remain fixed at 1650°C. The temperature of outside surface of the brick may be taken as 27°C.*

Solution

Arrangement are shown in Fig. Ex. 18.16 (a) and (b)

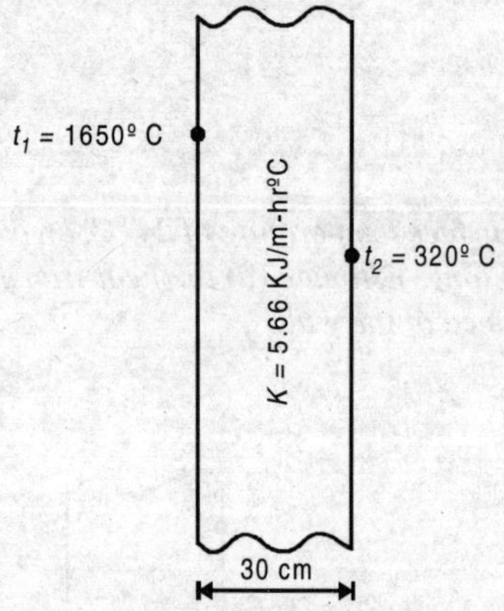

Fig. Ex. 18.16. (a)

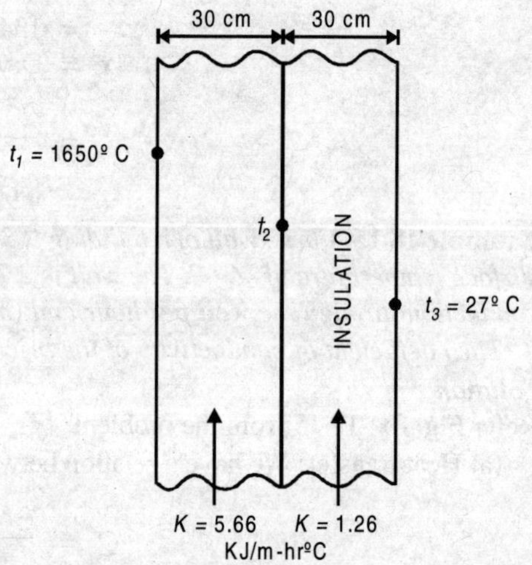

Fig. Ex. 18.16. (b)

(a) Without insulation: We have

$$Q = \frac{-K A \Delta T}{x}$$

Assuming area A as unity we get

$$Q = \frac{566 \times 1 \times (320 - 1650)}{0.3}$$

$$= 25093 \text{ kJ/hr/m}^2$$

(b) With insulation: For the refractory brick wall, we have

$$Q = -\frac{K_1 A \Delta T}{x_1} = \frac{5.66\times(t_1 - t_2)}{0.3}$$

$$= \frac{5.66\times(t_1 - t_2)}{0.3} = 18.87\ (t_1 - t_2)$$

or $$t_1 - t_2 = \frac{Q}{18.87} \quad \text{...(i)}$$

Similarly for insulating brick wall,

$$Q = \frac{-K_2 A(t_3 - t_2)}{x_2} = \frac{1.26\times 1\times t_2 - t_3}{0.3}$$

$$= 4.2\ (t_2 - t_3)$$

$\therefore$ $$t_2 - t_3 = \frac{Q}{4.2} \quad \text{...(ii)}$$

Adding Eqs (i) and (ii) we get,

$$t_1 - t_3 = Q\left[\frac{1}{18.87} + \frac{1}{4.2}\right] = Q[0.05299 + 0.2381]$$

$$= Q\times 0.29109$$

$\therefore$ $$Q = \frac{t_1 - t_3}{0.29109} = \frac{1650 - 27}{0.29109}$$

$$= 5575.6\ \text{kJ/hr/m}^2$$

$\therefore$ Reduction in heat loss through the wall

$$= 25093 - 5575.6 = \mathbf{1951.4\ kJ/hr/m^2}$$

Example 18.17 *A cold room has one of the walls which measures 4.6 cm × 2.3 m constructed of bricks, 11.5 cm thick, insulated externally by cork slabbing 7.5 cm thick. The cork is protected externally by wood 2.5 cm-thick.*

Estimate the leakage through the wall per 24 hours if the interior temperature is –20°C and the exterior 18°C. The thermal conductivities of brick, cork and wood are 41, 1.9 and 7.5 kJ/hr.m°C respectively.

What will be the temperatures at the interfaces ?

Solution

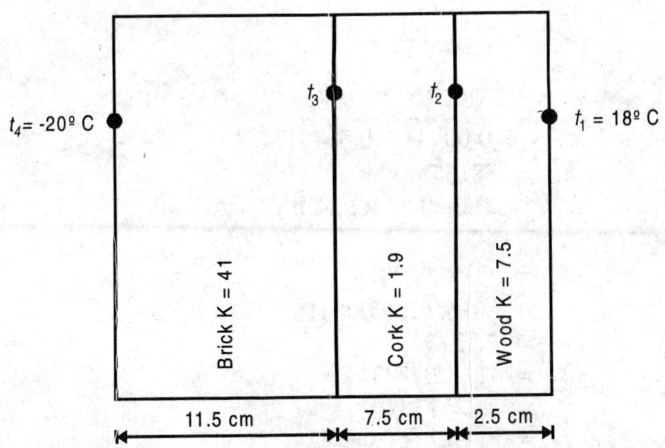

Fig. Ex. 18.17

Figure Ex. 18.17 shows the arrangement of the various materials in the example.

For wood we have
$$Q = \frac{KA(t_2 - t_1)}{x}$$
$$= 7.5 \times \frac{(4.6 \times 2.3)}{2.5/100}(t_1 - t_2)$$
$$= 3174\ (t_2 - t_1)$$
$$\therefore \quad t_2 - t_1 = \frac{Q}{3174}$$
$$= Q \times 0.000315$$
$$t_2 - t_3 = Q \times 0.003731$$

Similarly for cork we have
$$Q = \frac{1.9 \times 4.6 \times 2.3(t_2 - t_3)}{7.5/100}$$
$$= 268.03\ (t_3 - t_4)$$
$$\therefore \quad t_3 - t_4 = Q \times 0.00373 \qquad \text{...(ii)}$$

And for bricks, we have
$$Q = \frac{41 \times 4.6 \times 2.3(t_3 - t_4)}{11.5/100}$$
$$= 3772\ (t_3 - t_4)$$
$$\therefore \quad t_3 - t_4 = \frac{Q}{3772}$$
$$= Q \times 0.000265 \qquad \text{...(iii)}$$

Adding these three equations, we get
$$t_2 - t_4 = Q[0.000315 + 0.003731 + 0.000265]$$
$$= Q \times 0.00431$$
$$\therefore \quad Q = \frac{t_1 - t_4}{0.00431}$$
$$= \frac{18 + 20}{0.00431} = \frac{38}{0.00431}$$
$$= 8816.7 \text{ kJ/hr}$$
$$= \mathbf{211600.93 \text{ kJ/24 hr.}}$$

For interface temperatures
$$t_1 - t_2 = Q \times 0.000315$$
$$= 8816.7 \times 0.000315$$
$$= 2.7773$$
$$\therefore \quad t_2 = 18 - 2.7773$$
$$= \mathbf{15.2277\ °C}$$

Similarly
$$t_2 - t_3 = \mathbf{8816.7 \times 0.00373}$$
$$= 32.89$$

$$\therefore \quad t_3 = 15.2227 - 32.89$$
$$= \mathbf{-17.667\ °C}$$

Example 18.18 *Prove that temperature distribution. Along the thickness of slab is uniform that is linear, if the thermal conductivity of the material of the slab is constant.*

Compute the heat loss per sq. metre of surface area for a furnace wall, 30 cm thick. The inner outer surface temperatures are 320°C and 38°C respectively. The variation of the thermal conductivity with the temperature is given by the following the relation

$$K = \left(0.01256T - 4.2 \times 10^{-6} T^2\right) kJ/hr.m°C.$$

Solution

Consider a wall having a thickness as L and consider a plane surface at a distance x from the left face of the wall. Let the temperature at this surface be t and let dx be the elementary thickness of the slab at x. This is shown in Fig. Ex. 18.18.

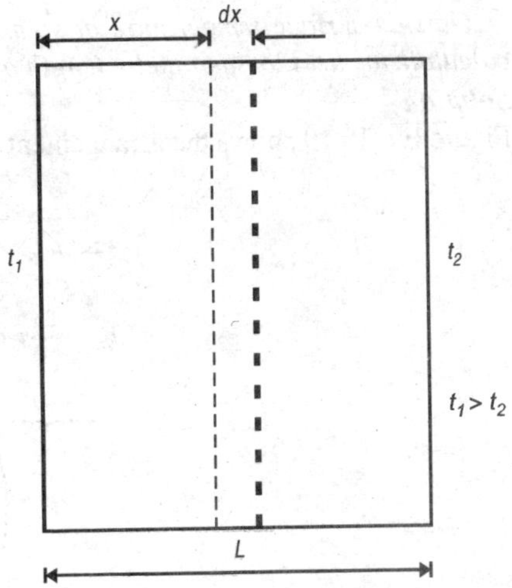

Fig. Ex. 18.18

$$Q = -\frac{KA.dt}{dx}$$

$$Q.dx = -KA.dt$$

Integrating this equation,

$$Q\int_0^L dx = -KA\int_{t_1}^{t} dt$$

$$\therefore \quad QL = -KA(t - t_1)$$

$$\therefore \quad Q = -\frac{KA(t - t_1)}{L}$$

This shows that the temperature variation is linear one or uniform.

$$\frac{QL}{AK} = t_1 - t$$

or
$$\mathbf{t = t_1 - \frac{QL}{AK}}$$

For the problem given, thermal conductivity of the material is varying.

$$\therefore \quad \text{We have } Q = \frac{AK.dt}{dx}$$

$$= -A\left[0.01256T - 4.2T_2 \times 10^{-6}\right]\frac{dT}{dx}$$

Integrating, we get
$$Q'L = A\left[\frac{0.01256T^2}{2} . \frac{4.2T^3}{3} \times 10^{-6}\right]_{38}^{320}$$

$$QL = A\left[0.00628\left(320^2 - 38^2\right) - \frac{4.2}{3} \times 10^{-6}\left(320^3 - 38^3\right)\right]$$

$$= 1\left[0.00628(358 \times 282) - 1.4 \times 10^{-6} \times 10^{6}\,(32.768 - 0.055)\right]$$

$$= 634 - 1.4 \times 32.713$$
$$= 634 - 45.78$$
$$= 588.2 \text{ kJ/hr/m}^2$$

$$\therefore \quad Q = \frac{588.2}{0.3}$$
$$= \mathbf{1960\ kJ/hr/m^2}$$

Example 18.19 *A steam pipe line 11.5 cm outside diameter, is covered with two layers of different materials. The first layer is 5 cm thick and a thermal conductivity of 0.222 kJ/hr.m °C. The second layer is 3 cm thick and as a conductivity of 3.14 kJ/hr.m °C.*

Outside surface temperature of steam pipe is 235 °C and that of the outer surface of lagging is nm 38 °C calculate the heat loss per metre length of pipe per hour, and the temperature between the two layers of insulation.

Solution

Figure Ex. 18.19 shows the arrangement of the pipe with layers of different materials.

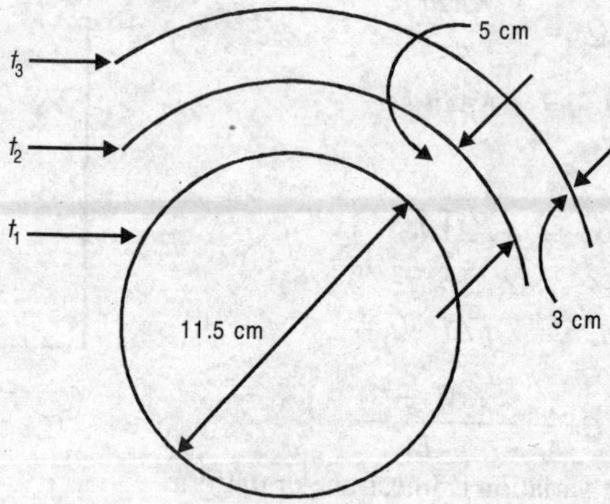

Fig. Ex. 18.19

We have for cylindrical pipes,

$$Q = \frac{2\pi l \times K(t_1 - t_2)}{\log_e \dfrac{D}{d}}$$

For the inner layer. $\log_e \dfrac{D}{d} = \log_e \dfrac{11.5+10}{11.5}$

$$= \log_e 1.87$$
$$= 0.6257$$

Similarly for outer layer

$$\log_e \frac{D}{d} = \log_e \frac{21.5+6}{21.5}$$
$$= \log_e 1.28$$
$$= 0.2409$$

Taking the length of the pipes as 1 metre,

$$t_1 - t_2 = \frac{Q \times 0.6257}{2\pi \times 0.22}$$
$$= 0.4486\ Q \quad \ldots(i)$$

Similarly,
$$t_2 - t_3 = \frac{Q \times 0.2569}{2\pi \times 3.14}$$
$$= 0.0125\ Q \quad \ldots(ii)$$

Adding Eqs (i) and (ii), we get,

$$t_1 - t_3 = Q \times [0.4486 + 0.0125]$$
$$= Q \times 0.4611$$
$$\therefore \quad 235 - 38 = 0.4611\ Q$$
$$\therefore \quad \mathbf{Q} = \frac{\mathbf{197}}{\mathbf{0.4611}}$$
$$= \mathbf{427.24\ kJ/hr}$$

From Eq. (i) we have,

$$235 - t_2 = 0.4486 \times 427.24$$
$$= 191.66$$
$$t_2 = 235 - 191.66$$
$$= \mathbf{43.34\ ^\circ C}$$
= Temperature between the two layers of insulation.

Example 18.20. *A wall of refrigerated van of 1.5 mm of steel sheet at outer surface, 10 mm plywood at a inner surface and 2 cm of glass wool in between. Calculate the rate of heat flow if the temperature at the inside and outside surface are –15 and 24 °C. Take*

K (steel) = 23.2 W/m °C
K (glasswood) = 0.014 W/m °C
K (plywood) = 0.052 W/m °C

Solution

This is shown in Fig. Ex. 18.20 consider unit area of flow of heat (1 m^2)

With refrence to Fig. Ex. 18.20, we have

$$Q = \frac{KA(t_1 - t_2)}{x_1}$$
$$= \frac{23.2 \times 1(24 - t_2)}{1.5/1000}$$
$$\therefore \quad (24 - t_2) = \frac{Q \times 1.5}{1000 \times 23.2} = \frac{0.0646Q}{1000}$$
$$t_2 - t_3 = \frac{Q \times 20}{1000} \times \frac{1}{0.014}$$
$$t_3 - t_4 = \frac{Q \times 10}{1000} \times \frac{1}{0.052}$$
$$= \frac{Q \times 192.31}{1000}$$

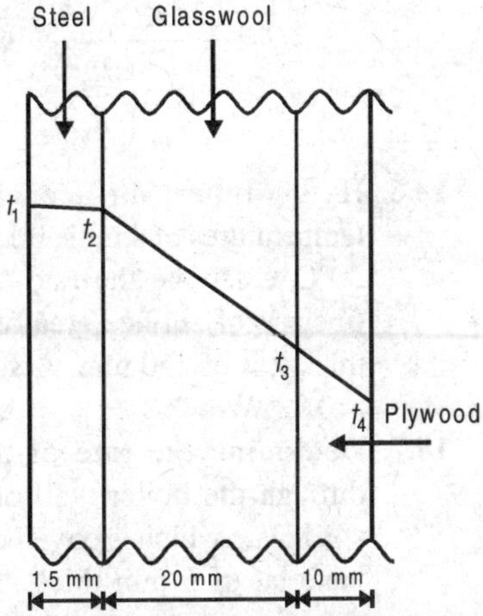

Fig. Ex. 18.20

Adding we get

$$24 + 15 = \frac{Q}{1000}[0.0646+1428.57+192.31]$$

$$39 = \frac{Q}{1000}1620.9446$$

$$Q = \frac{39000}{1620.9446} = \mathbf{24.06\ W/m^2}$$

PROBLEMS FOR PRACTICE

18.1. What are the modes of heat transfer. Explain very briefly each one mentioned.

18.2. State the following laws

(i) Fouriers' law

(ii) Newton' law of cooling and

(iii) Stefa–Boltzmann law of radiation.

18.3. Explain the terms (i) convective heat transfer coefficient and (ii) overall heat transfer coefficient.

18.4. Derive an expression for the heat loss in W/m^2 through a composite wall of *n* layers.

(a) without considering convective heat transfer coefficients and

(b) with considering the convective heat transfer coefficient.

18.5. Prove that the heat flow through the hollow cylinder is given by

$$Q = \frac{2\pi LK(T_1 - T_2)}{\log_e\left(\frac{R_2}{R_1}\right)} \text{ where } T_1 > T_2$$

18.6. If the inner surface and outer surface temperatures of simple brickwall are 40ºC and 20ºC, calculate the rate of heat transfer per (meter)^2 of surface area of the wall having a thickness of 250 mm. Assume *K* for the brick is 0.52 W/m °C. (**Ans.** 41.6 W/m^2)

18.7. Determine the rate of heat transfer/flow through the boiler wall of 20 nun thickness steel plate which is covered with an insulating material of 5 mm thick. The temperature at the inner and outer surfaces of the all are 300ºC and 50°C respectively. Assume *K* for steel = 58 W/m°C and for insulation as 0.116 W/m.K. (**Ans.** 5.8 kW/m^2)

18.8. A steel tank of wall thickness 10 mm contains water at 90ºC. Calculate the rate of heat loss per unit area of tank surface when the atmospheric temperature is 15ºC. *L* for steel is 50 W/m°C and the heat transfer coefficients for inside and outside the tank are 2800 and 11W/m^{2}K respectively. Calculate also the temperature of the outside surface of the tank. (**Ans.** 820 W/m^2, 89.6ºC)

18.9. Sheets of brass and steel, each 10 mm thick, are placed in contact with each other. The outer surface of brass is at 100°C and the outer surface of steel is at 0°C. What is the temperature of the common interface? Assume that the conductivities of brass and steel are in the ratio of 2 : 1. (**Ans.** 66.7ºC)

18.10. The wall of the furnace is made up of 250 mm of fire brick, *K* = 1.05 W/m°C; 120 mm of insulation brick, *K* = 0.85 W/mºC, and 200 mm of red brick, *K* = 0.85 W/m°C. The inner and outer surface temperatures walls are 850ºC and 65ºC respectively. Calculate the temperatures at the contact surfaces. (**Ans.** 703ºC, 210ºC)

18.11. Explain the following

(i) Fouriers' law of heat conduction

(ii) Newton's law of cooling

(iii) Stephan–Boltzman law of radiation heat transfer.

18.12. State different types of heat exchangers.

18.13. Explain the phenomenon of heat transfer by convection. State the Newton's Law of cooling.

18.14. State any three exchangers normally observed in thermal power plant, mentioning clearly the type of heat exchange.

(**Ans.** The heat exchangers that are normally used in thermal power plant are

1. Air preheater—Multipass, shell and tube type.
2. Economiser—Multipass, tube assembly kept in the flue gas passage.
3. Superheater—Steam carrying tubes kept in the flow gas passages)

18.15. Steam Generator:

(i) Fire tube boiler furnace tubes through which gases flow and tubes are surrounded by water

(ii) Water tube boiler-water tubes carrying water are placed in gas passage. Gases and water both are flowing.

18.16. Explain the following:

(a) Conduction heat transfer and its associated property

(b) Convection heat transfer and its associated property.

Table 18.1. Properties of Metals at 20°C

Metal	Density	Thermal	Specific Heat	Thermal
	ρ kg/m³	α m²/s × 10⁶	c J/kg-K	k W/mK
Aluminium, Pure	2707	94.44	896	204.2
Steel 0. 5% carbon	7833	14.72	465	53.6
Nickel Steel 20% Ni	7983	5.28	461	19.1
Chrome Steel 20% Cr	7683	6.67	461	22.5
Constantan 60% Cu	8922	6.11	410	22.7
Magnesium, pure	1746	96.94	1013	171.3
Nickel, pure	8906	22.78	444	90.0
Nickel Chrome	8666	4.44	44	17.2
Tungsten Steel 10% W	8313	13.61	419	48.5
Opper, pure	8954	12.22	381	386.0
Bronze	8666	8.61	343	25.9
Brass	8522	33.89	385	110.7
Silver, pure	10524	165.56	235	406.8
Tungsten	19350	62.78	134	162.7
Zinc, pure	7144	41.11	385	112.1
Tin, pure	7304	38.61	226	4.1
Cr Ni steel 15% Cr 10% Ni	7865	5.28	461	19.1

1W/mK = 0.86 kcal/m hr°C 1 J/kg-K = 238.9 10^{-6} kcal/kg°C

Table 18.2. Properties of Metals at 20°C

Metal	Density	Thermal Diffusivity	Specific Heat	Thermal Conductivity
	ρ kg/m³	α m²/s × 10⁶	c J/kgK	k W/mK
Berylium	1840	50.97	1675	157
B	2500	10.90	1047	28.6
Cadmium	8660	46.67	230	93
Carbon (graphite)	1700–2300	122.22	670	116.3–174.5
Chromium	7150	218.06	448	69.8
Cobalt	8800	17.64	448	69.8
Gold	19300	127.00	129	317.0
Lithium	5340	40.28	3308	68.6
Molybdenum	10200	54.44	253	140.7
Platinum	21460	24.58	132	69.8
Potassium	870	155.56	737	100.0
Rhodium	12450	48.6	248	150.0
Sodium	975	94.44	1197	109.3
Silicon	2330	93.4	703	153
Thorium	11700	38.17	118	54.0
Uranium	19100	12.70	113	27.4
Vanadium	5900	11.94	496	34.9
Titanium	4540	6.22	532	15.12
Zirconium	6570	12.50	278	22.7

1W/mK=0.86 kcal/m hr°C 1 J/kg-K = 238.9 × 10^{-6} kcal/kg°C

Table 18.3. Properties of Insulating, Building, and Other Materials

Substance	Temperature °C	k, W/m°C	ρ, kg/m³	C kJ/kg°C	α, m²/s × 10⁷
		Structural and heat resistant materials			
Asphalt	20–55	0.74–0.76			
Brick:					
Building brick, common	20	0.69	1600	0.84	5.2
Face		1.32	2000		
Carborundum brick	600	18.5			
	1400	11.1			
Chrome brick	200	2.32	3000	0.84	9.2
	550	2.47			9.8
	900	1.99			7.9
Diatomaceous earth,					
moulded and fired	200	0.24			
	870	0.31			
Fireclay brick, burnt	500	1.04	2000	0.96	5.4
1330°C	800	1.07			
	1100	1.09			
Burnt 1450°C	500	1.28	2300	0.96	5.8
	800	1.37			
	1100	1.40			
Missouri	200	1.00	2600	0.96	4.0
	600	1.47			
	1400	1.77			
Magnesite	200	3.81		1.13	
	650	2.77			
	1200	1.90			
Cement, portland		0.29	1500		
Mortar	23	1.16			
Concrete, cinder	23	0.76			
Stone 1–2–4 mix	20	1.37	1900–2300	0.88	8.2–6.8
Glass, window	20	0.78 (avg)	2700	0.84	3.4
Corosilicate	30–75	1.09	2200		
Plaster, gypsum	20	0.48	1440	0.84	4.0
Metal lath	20	0.47			
Wood lath	20	0.28			
Stone					
Granite		1.73–3.98	2640	0.82	8–18
Limestone	100–300	1.26–1.33	2500	0.90	5.6–5.9
Marble		2.07–2.94	2500–2700	0.80	10–13.6
Sandstone	40	1.83	2160–2300	0.71	11.2–11.9
Wood: (across the grain)					
Balsa,	30	0.055	140		
Cypress	30	0.097	460		
Fir	23	0.11	420	2.72	0.96
Maple or oak	30	0.166	540	2.4	1.28
Yellow pine	23	0.147	640	2.8	0.82
White pine	30	0.112	430		

Table 18.3. *Contd.*

Substance	Temperature °C	k, W/m°C	ρ, kg/m³	C kJ/kg°C	α, m²/s × 10⁷
			Insulating material		
Asbestos:					
Loosely packed	–45	0.149			
	0	0.154	470–570	0.816	3.3–4
	100	0.161			
Asbestos–cement boards	20	0.74			
Sheets	51	0.166			
Felt, 16 laminations/cm	38	0.057			
	150	0.069			
	260	0.083			
8 laminations/cm	38	0.078			
	150	0.095			
	260	0.112			
Corrugated, 150 plies/m	38	0.087			
	93	0.100			
	150	0.119			
Asbestos cement	–	2.08			
Balsam wool,	32	0.04	35		
Cardboard, corrugated	–	0.064			
Celotex	32	0.048			
Corkboard,	30	0.043	160		
Cork, regranulated	32	0.045	45–120	1.88	2–5.3
Ground	32	0.043	150		
Diatomaceous earth	0	0.061	320		
(Sil–o–cel)					
Felt, hair	30	0.036	130–200		
Wool	30	0.052	330		
Fiber, insulating board	20	0.048	240		
Glass wool,	23	0.038	24	0.7	22.6
Insulex, dry	32	0.064			
		0.144			
Kapok .	30				
Magnesia, 85%	38	0.067	270		
	93	0.071			
	150	0.074			
	204	0.080			
Rock wool,	32	0.040	160		
Loosely packed	150	0.067	64		
	260	0.087			
Sawdust	23	0.059			

CHAPTER 19

Reciprocating Air Compressors

19.1 INTRODUCTION

Air compressor is a power absorbing machine which provides high pressure air. It takes the air from the atmosphere, compresses it to a high pressure and high pressure air will be stored in a storage vessel (reservoir) from where it can be taken out for use.

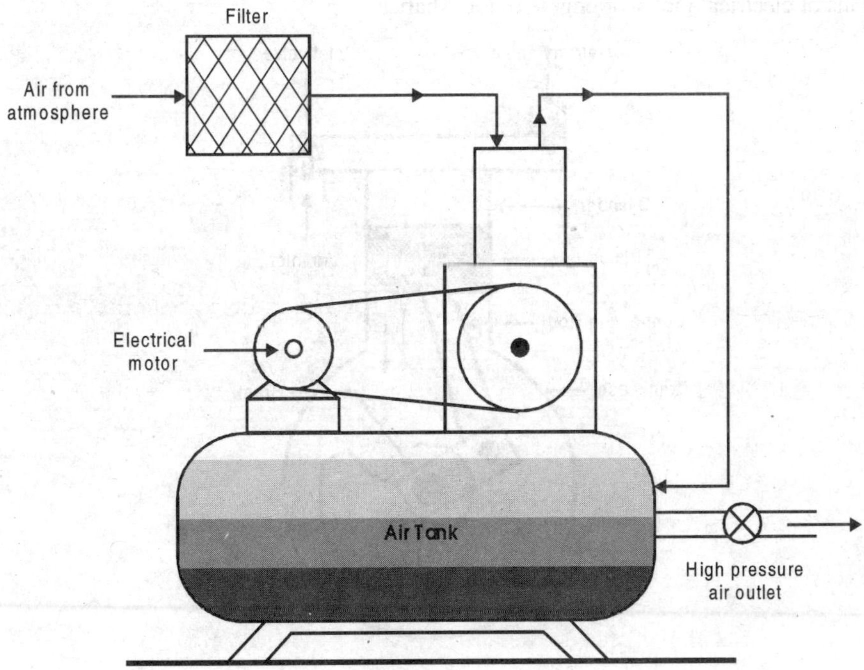

Fig. 19.1

There are many uses of high pressure air in industry. The main uses of compressed air are,

1. For inflating automobile tyres.
2. To clean workshop machines, generators etc.

3. To operate air operated drills, hammers (Pneumatic tools).
4. To inject fuel in the diesel engine cylinder.
5. In spray painting.
6. To operate air brakes in automobiles.
7. To operate compressed air engines/air motors in mines.

(In mines IC engines and electricity are not used because of fire risks. So high pressure air operated machines are used).

19.2 CLASSIFICATION

Air compressors may be broadly classified as given below: (i) Reciprocating and (ii) Rotary.

The principle parts of a reciprocating air compressor are same as that of a reciprocating I.C. engine. The air compressor may also be as follows,

(i) Single Acting or Double Acting. The air is admitted to one side of the piston or air is admitted to either side of the piston alternately.

(ii) Single Stage/Multistage. In case of single stage air is compressors, air is compressed in a single cylinder completely, whereas in case of multistage air compressor, air where air is compressed in several stages. And each stage in pressure of air increases.

19.3 SINGLE STAGE/SINGLE ACTING RECIPROCATING AIR COMPRESSOR

It can be seen that the construction is just similar to reciprocating I.C. engine. In this case crank of the compressor is driven by means of electrical motor or engine output shaft.

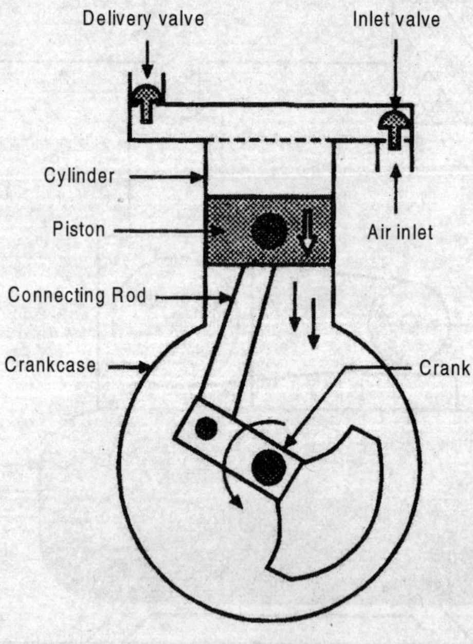

Fig. 19.2

During suction stroke, piston will be moving downwards and the pressure in the cylinder falls below the atmospheric pressure. As a result of which inlet valve opens and the air is drawn from the atmosphere. Then during compression stroke air will be compressed by means of upward moving piston and delivered through the delivery valve, when the pressure in the cylinder increases beyond the prespecified value of delivry valve.

19.3.1 Compressor Valves and Types

Compressor valves are automatic i.e., there is no valve operating mechanism provided, but they are operated because of the pressure difference.

Types of compressor valves:

(i) Reed types
(ii) Plate type
(iii) Poppet type,
(iv) Thimble valves

(i) Reed Type Valves. Figure 19.3 shows the reed type valve during compression the pressure of air increases and the pressure devleoped at (1) increases beyond the delivery chamber pressure, (2) then the valve plate (which has 6–8 reeds) lifts and the air is delivered.

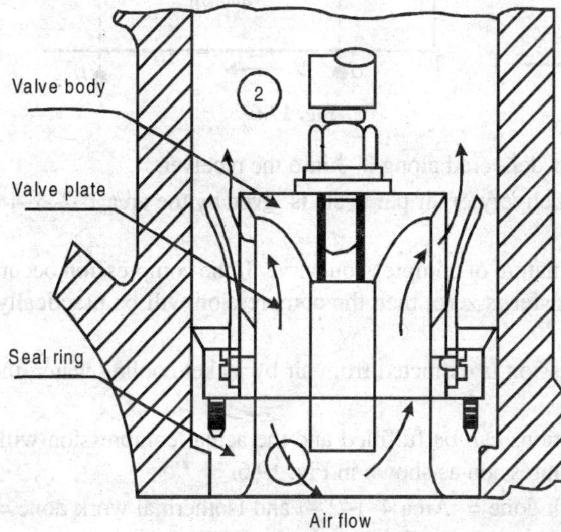

Fig. 19.3

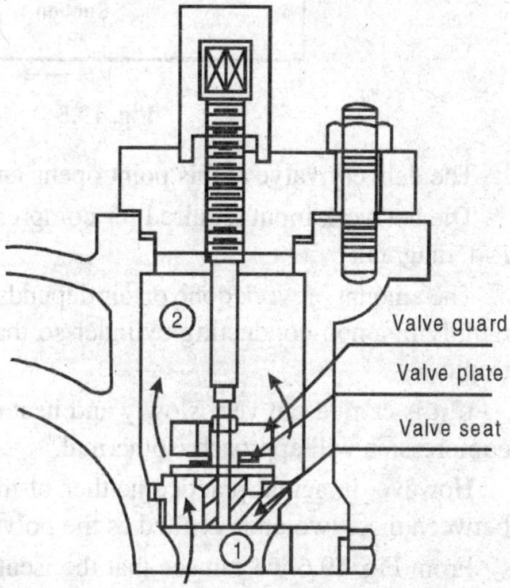

Fig. 19.4

(ii) Plate Type Valves. As shown in Fig. 19.4 valve plate is held between valve seat and valve gaurd. In between valve plate and valve gaurd spring is provided. When the air pressure increases at (1), then the valve plate is lifted against then compression of spring and high pressure air is delivered to side (2).

19.4 INDICATOR DIAGRAM

The events described for the single stage air compressor can be represented on P–V diagram as shown in Fig. 19.5. The diagram is drawn for the compressor without clearance volume.

Clearance Volume. It is the volume of space provided between piston head and cylinder head when the piston is at TDC. This is provided in order to avoid striking of piston heat with cylinder head. Also in this clearance volume, during the delivery stroke some volume of air will be trapped, when the piston comes to TDC. This volume of air during the downward stroke expands and produces the necessary vacuum.

During suction stroke, the charge of air is drawn along 4–1 at constant pressure P_1, which is slightly below P_{atm}. At point 1 piston completes suction stroke and starts compression stroke. During starting both the valves being closed air will be compressed along 1–2. At point 2 pressure P_2 is reached which is slightly higher than the pressure in the receiver.

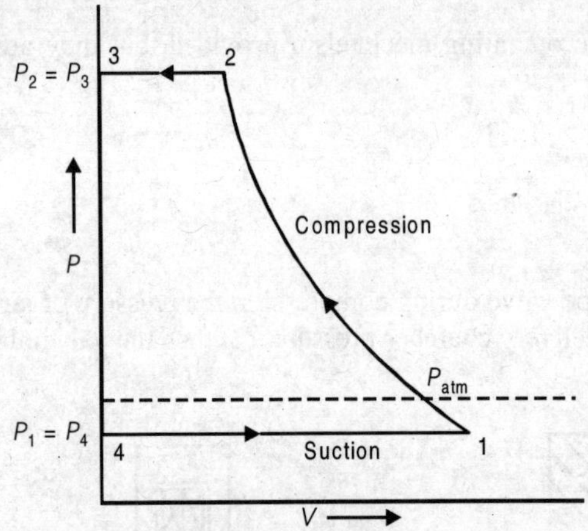

Fig. 19.5

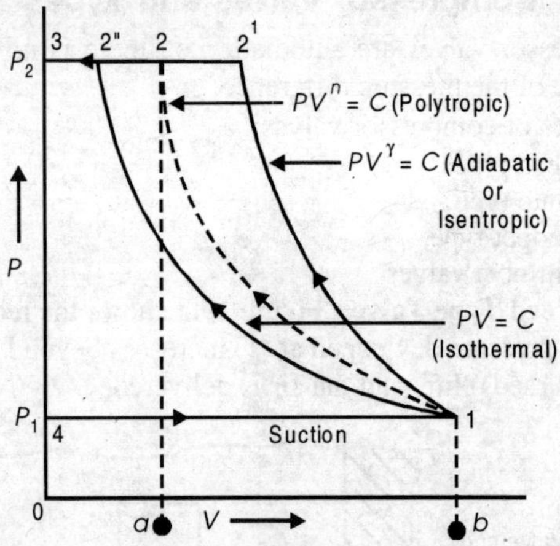

Fig. 19.6

The delivery valve at this point opens and the air is delivered along 2–3 into the receiver.

The net work input required for compression and delivery of air per cycle is given by the area 1–2–3–4 on the *P–V* diagram.

The amount of work done on air depends upon the nature of compression curve. If the compression occurs very rapidly in a non-conducting cylinder so that heat transfer is zero, then the compression will be practically isentropic.

If it is carried out very slowly and heat of compression is extracted from air by jacket cooling water, then the compression will approach isothermal.

However in actual practice neither of these conditions can be fulfilled and the actual compression will be in between these two and is called as the polytropic compression as shown in Fig. 19.6.

From Fig. 19.6 we can see that the isentropic work done = Area 4–1–2'–3 and Isothermal work done = Area 4–1–2"–3. So, if we achieve isothermal compression curve, the work input required will be minimum.

Note: For isothermal compression process $PV = C$, and $n = 1$

For isentropic compression process $PV^\gamma = C$ and $\gamma = 1.4$ for air.

In actual practice we achieve the process in between these two and for water cooled cylinders the value of n is 1.3.

19.5 WORK INPUT FOR COMPRESSOR

(i) If the Compression Process Follows $PV^n = C$

As shown in Fig. 19.4 work done during polytropic compression and delivery is given by the area 1–2–3–4 on the *P–V* diagram.

Now

$$\text{Area } 1-2-3-4 = [\text{Area } 2-3-0-a] + [\text{Area } 1-2-a-b] - [\text{Area } 4-1-0-b]$$

$$= P_2(V_2V_3) + \frac{P_2V_2 - P_1V_1}{n-1} - P_1(V_1 - V_4)$$

As $\quad V_C = V_3 = V_4 = 0$

$$= P_2V_2 + \frac{P_2V_2 - P_1V_1}{n-1} - P_1V_1$$

$$\text{W. D.} = \frac{(n-1)P_2V_2 + P_2V_2 - P_1V_1 - (n-1)(P_1V_1)}{n-1}$$

$$= \frac{n.P_2V_2 - P_2V_2 + P_2V_2 - P_1V_1 - nP_1V_1 + P_1V_1}{n-1}$$

$$= \frac{n(P_2V_2 - P_1V_1)}{n-1}$$

$$= \frac{n.P_1V_1\left\{\dfrac{P_2V_2}{P_1V_1} - 1\right\}}{n-1}$$

But as $PV^n = C; \quad = \dfrac{V_2}{V_1} = \left(\dfrac{P_1}{P_2}\right)^{1/n} = \left(\dfrac{P_2}{P_1}\right)^{-1/n}$

$$\therefore \quad \text{W.D.} = \frac{n}{n-1}P_1V_1\left\{\frac{P_2}{P_1}\left(\frac{P_2}{P_1}\right)^{-1/n} - 1\right\}$$

Work Done or Indicated Power Required to Drive Compressors

$$\mathbf{W.D\ or\ I.P.} = \frac{\mathbf{n}}{\mathbf{n-1}}\mathbf{P_1 . V_1}\left\{\left(\frac{\mathbf{P_2}}{\mathbf{P_1}}\right)^{\frac{\mathbf{n-1}}{\mathbf{n}}} - \mathbf{1}\right\}$$

Note generally,

W = in kW $\qquad$ n = index of compression no units

P_1 = kN/m^2 $\qquad$ V_1 = m^3/sec

P_2 = kN/m^2 (delivery pressure)

Also $$\mathbf{W.D.} = \frac{\mathbf{n}}{\mathbf{n-1}}\mathbf{mRT_1}\left\{\left(\frac{\mathbf{P_2}}{\mathbf{P_1}}\right)^{\frac{\mathbf{n-1}}{\mathbf{n}}} - \mathbf{1}\right\}$$

As $P_1V_1 = mRT_1$ and as air is the ideal gas.

(ii) If the Compression Process follows $PV^\gamma = C$ (i.e., Adiabatic Process). The replace n by γ and the derivation is same as above and

$$\text{W.D.} = \frac{\gamma}{\gamma-1}P_1V_1\left[\left(\frac{P_2}{P_1}\right)^{\frac{\gamma-1}{\gamma}} - 1\right]\text{kW}$$

(iii) If the Compression is Isothermal i.e., it follows PV = const. Then W.D = Area 1 – 2 – 3 – 4 = Area under 2 – 3 + Area under 1 –2 – Area under 4 – 1

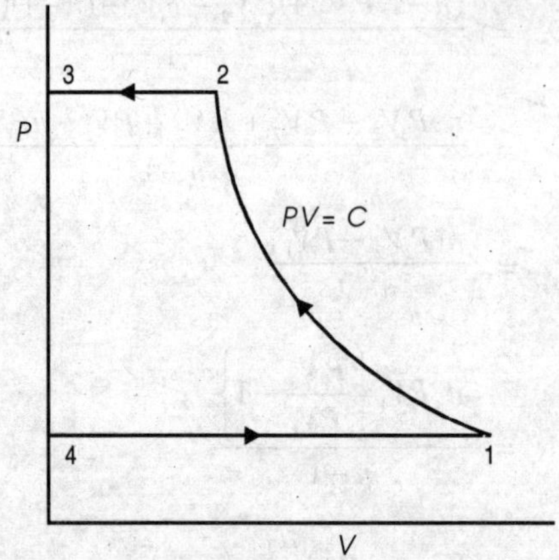

Fig. 19.7

$$= P_2V_2 + P_1V_1 \ln \frac{V_1}{V_2} - P_1V_1$$

$$\textbf{W.D.} = \mathbf{P_1V_1 \ln \frac{V_1}{V_2}}$$

or

$$\textbf{W.D.} = \mathbf{P_1V_1 \ln \frac{P_2}{P_1}}$$

As $P_1V_1 = P_2V_2 = C$

Note. All the expressions are arranged to give positive value, since we are interested in the magnitude of work input.

19.6 ISOTHERMAL EFFICIENCY

It is defined as the ratio of,

$$\eta_{iso} = \frac{\text{Isothermal work of compressor}}{\text{Actual work input}}$$

And Isothermal work input $= P_1V_1 \ln \frac{P_2}{P_1}$ or $= mRT_1 \ln \frac{P_2}{P_1}$ as we have studied in earlier article.

Note. η_{iso} gives an indication of closeliness of the compression process to isothermal compression i.e., lower the value of n higher is η_{iso}.

19.7 MEASURES OR METHODS TO IMPROVE ISOTHERMAL EFFICIENCY

(i) Effective Cooling System. This can be provided by providing fins on the cylinder and cylinder head and air blown on the fins. When the air is blown, it removes heat from the compressed air and dissipates it to the atmosphere.

Secondly cooling water jackets may also be provided around the cylinder and cylinder head for large sized compressors.

(ii) Multistaging with Intercoolers. In this method, the compression of air is carried out in two or more stages i.e., two or more cylinders. The pressure of air is increased in each cylinder. It is a common practice to provide inter-coolers between the cylinders of multistage compressor for the purpose of cooling the compressed air to intake (or atmosphere) air temperature before it enters the next cylinder. It is this cooling between the cylinders that keeps the compression very near to isothermal.

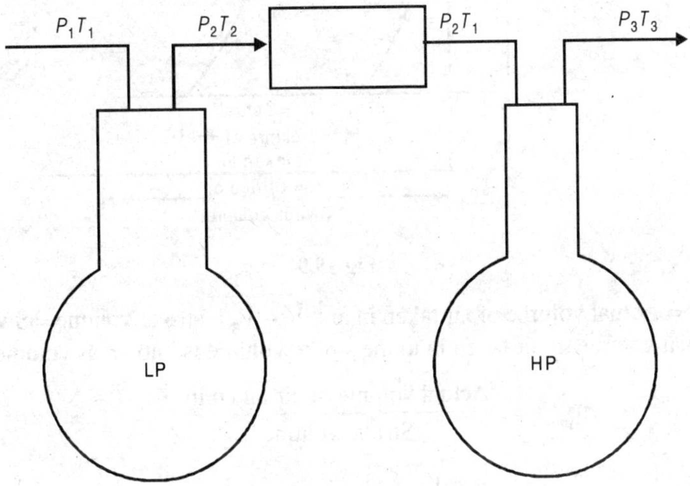

Fig. 19.8

(iii) Spray Injection. In this method a small amount of water is injected or sprayed into the compressed air. It immediately vapourises by absorbing heat from air and thereby keeps the temperature of air low.

(iv) Suitable Cylinder Dimensions. To have effective heat transfer, surface to volume ratio should be higher i.e., a cylinder with large dia. and short stroke should be selected. But this has the limitation since increased dia increases clearance volume and hence reduces volumetric efficiency.

19.8 EFFECT OF CLEARANCE ON VOLUMETRIC EFFICIENCY

The volume of space provided between piston heat and cylinder head when the piston is at TDC is known as clearance volume. In actual compressors it is provided to safeguard the piston from striking the cylinder head.

The events taking place in a compressor with clearance are same as those taking place in a compressor without clearance.

Figure 19.9 shows the *P–V* diagram for a single stage/single acting reciprocating air compressor with clearances volume. In this case both compression and expansion curves are assumed to be polytropic following the law $PV^n = C$.

As shown in Fig. 19.9, 1– 2 is polytropic compression, 2 – 3 is delivery stroke. At point 3, piston reaches TDC and delivery of high pressure air stops and some air is trapped in the clearance volume at the delivery pressure P_2.

As the piston moves downwards this air which is trapped in the clearance volume expands according to $PV^n = C$ and produces the necessary vacuum. At point 4 the inlet valve opens and actual volume of air taken is along 4–1.

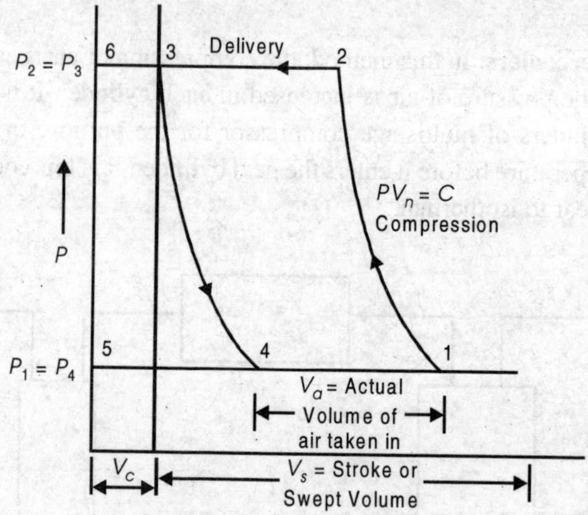

Fig 19.9

Hence $V_1 - V_4 = V_a$ = Actual volume of air taken in and $V_1 - V_3$ = stroke volume = Swept volume. And the ratio of actual volume of air taken in to the stroke volume is known as volumetric efficiency i.e.,

$$\eta_{vol.} = \frac{\text{Actual volume of air taken in}}{\text{Stroke volume}}$$

$$= \frac{V_1 - V_4}{V_1 - V_3} = \frac{V_a}{V_s}$$

i.e.,

$$\boldsymbol{\eta}_{\textbf{vol.}} = \frac{\mathbf{V_a}}{\mathbf{V_s}} \qquad ...(1)$$

For the process 3–4

$$P_3V_3^n = P_4V_4^n$$

$$\left(\frac{P_3}{P_4}\right)^{1/n} = \frac{V_4}{V_3}$$

$$V_3 \times \left(\frac{P_3}{P_4}\right)^{1/n} = V_4$$

$$V_c\left(\frac{P_2}{P_1}\right)^{1/n} = V_4 \; [\because V_3 = V_c \text{ and } P_3 = P_2 \text{ and } P_4 = P_1] \qquad ...(2)$$

As we know actual volume of air taken in,

$$V_a = V_1 - V_4$$

$$= V_1 - V_c\left(\frac{P_3}{P_4}\right)^{1/n} = (V_c + V_s) - V_c\left(\frac{P_3}{P_4}\right)^{1/n} \quad \text{as } V_1 = V_c + V_s$$

or $$= V_s + V_c - V_c\left(\frac{P_3}{P_4}\right)^{1/n} = V_s - V_c\left(\frac{P_3}{P_4}\right)^{1/n} + V_c$$

$$V_a = V_s - V_c\left\{\left(\frac{P_2}{P_1}\right)^{1/n} - 1\right\}$$

i.e., $$V_a = V_s - V_c\left\{\left(\frac{P_2}{P_1}\right)^{1/n} - 1\right\} \quad \ldots(3)$$

$\therefore$ From (1), $$\eta_{vol} = \frac{V_a}{V_s} = \frac{V_s - V_c\left\{\left(\frac{P_2}{P_1}\right)^{1/n} - 1\right\}}{V_s} = 1 - \frac{V_c}{V_s}\left\{\left(\frac{P_2}{P_1}\right)^{1/n} - 1\right\}$$

$$\boldsymbol{\eta_{vol} = 1 - C\left\{\left(\frac{P_2}{P_1}\right)^{1/n} - 1\right\}}$$

or $$\boldsymbol{\eta_{vol} = 1 - C \times \left(\frac{P_2}{P_1}\right)^{1/n} + C}$$

or $$\boldsymbol{\eta_{vol} = 1 + C - C\left(\frac{P_2}{P_1}\right)^{1/n}}$$

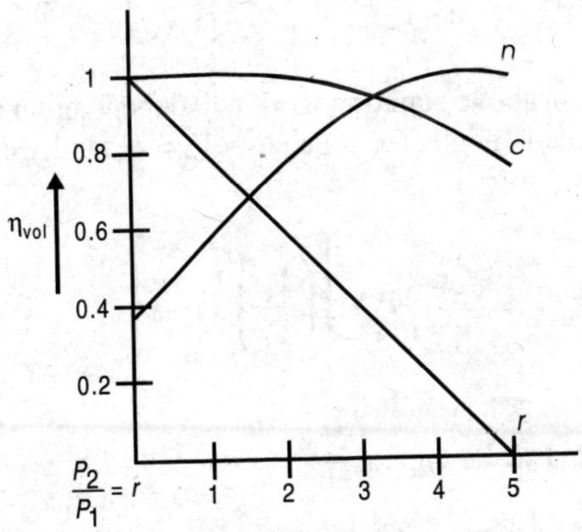

Fig. 19.10

Let the ratio $\frac{V_c}{V_s} = C$ = clearance ratio and $\frac{P_2}{P_1}$ = pressure ratio r.

The variation of volumetric efficiency with clearance volume and polytropic index is shown in Fig. 19.10. We note from this figure that

(i) As r increases η_{vol} decreases.
(ii) As n increases η_{vol} increases.
(iii) As c increases η_{vol} decreases.

Note.

(i) η_{vol} ranges from 60 to 85% and
(ii) Clearance ratio ranges from 4 to 10%
(iii) n ranges from 1.25 to 1.35 for air.

19.9 EXPRESSION FOR WORK INPUT WHEN CLEARANCE IS CONSIDERED

The work input required to compress the air with clearance is given by the area 1 – 2 – 3 – 4. (Ref. Fig. 19.9)

i.e.,
$$W_{input} = \text{Area } 1-2-3-4$$
$$= [\text{Area } 1-2-6-5] - [\text{Area } 3-4-5-6]$$

Assuming index of compression to expansion to be same as n.

$$W_{input} = \frac{n}{n-1}P_1V_1\left[\left(\frac{P_2}{P_1}\right)^{\frac{n-1}{n}} - 1\right] - \frac{n}{n-1}P_4V_4\left\{\left(\frac{P_3}{P_4}\right)^{\frac{n-1}{n}} - 1\right\}$$

But
$$P_4 = P_1,\ P_3 = P_2$$

$\therefore$
$$W_{input} = \frac{n}{n-1}P_1V_1\left\{\left(\frac{P_2}{P_1}\right)^{\frac{n-1}{n}} - 1\right\} - \frac{n}{n-1}P_1V_4\left\{\left(\frac{P_2}{P_1}\right)^{\frac{n-1}{n}} - 1\right\}$$

$$\mathbf{W_{input} = \frac{n}{n-1}P_1(V_1 - V_4)\left\{\left(\frac{P_2}{P_1}\right)^{\frac{n-1}{n}} - 1\right\}}$$

Therefore, it can be seen that from above equation work will depend upon volume of air taken/cycle and will not depend upon clearance. If clearance is neglected then $V_1 - V_4 = V_1$ and we get the expression for work without clearance as,

$$\mathbf{W_{input} = \frac{n}{n-1}P_1V_1\left[\left(\frac{P_2}{P_1}\right)^{\frac{n-1}{n}} - 1\right]}$$

$\therefore$
$$IP = \text{Indicated work/cycle} \times \frac{N}{60}$$

If V_1 is in m^3, then resultant unit will be kJ. Then multiply by $N/60$ to get power in kW.

19.10 ACTUAL INDICATOR DIAGRAM

In actual air compressor, some air is present in the clearance volume. This air expands during suction stroke.

At point 4, the inlet valve will not open in actual practice because of valve inertia and pressure differential required to open the valve. Thus the pressure drops away until the valve is forced to open. Some valve bounce will taken place and slowly intake will become nearly steady. This negative pressure, is known as *Intake depression.*

A similar situation occurs at point 2 i.e., at point 2 pressure has to increase otherwise valve will not open.

$\therefore$ Actual work required to compress and deliver the air will be greater than theoretical work.

And, $$\eta_{mech} = \frac{\text{Indicator work}}{\text{Actual work}} = \frac{\text{Indicator work}}{\text{Brake work}} = \frac{IP}{BP}$$

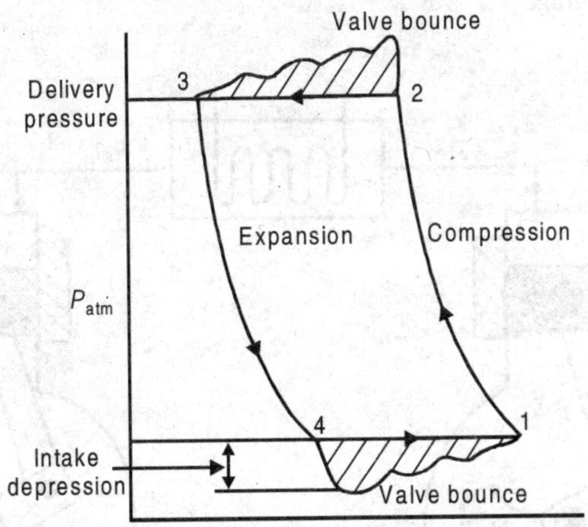

Fig. 19.11

19.11 FREE AIR DELIVERY (FAD)

It is the actual volume of air delivered by an air compressor reduced to either NTP or STP conditions or intake conditions. We know that from continuity equation.

Mass flow rate at the inlet of compressor = Mass flow rate at outlet of compressor.

Since $$PV = \dot{m}RT$$

$$\frac{PV}{RT} = \dot{m}$$

$\therefore$ $$\frac{P_1V_1}{RT_1} = \frac{P_2V_2}{RT_2} = \frac{P_fV_f}{T_f} \text{ neglecting clearance}$$

where P_fV_f and T_f are free air conditions and V_f will be FAD. For convenience take $P_f = 101.325$ kPa, $T_f = 288$ K if not given and V_f can be calculated.

Note. $\dfrac{P_1(V_1 - V_4)}{T_1} = \dfrac{P_2(V_2 - V_3)}{T_2}$ if clearance is considered.

19.12 MULTISTAGE AIR COMPRESSOR

A 2 stage air compressor with water jacketed cylinder and inter cooler is shown in Fig. 19.10 (i).

The suction in LP cylinder ends at 1 and the air is compressed polytropically to 2'. The LP cylinder then delivers the air along 2' – P_2 to the intercooler. Where air is cooled to initial intake temperature T_1 at constant pressure P_2 by circulating cold water in the inter cooler. When the air is cooled in the intercooler to initial temperature T_1, the cooling is perfect. The air when it is cooled at constant pressure, suffers reduction in volume from 2' – P_2 to P_2 – 2. The cooled air is drawn into the HP cylinder along P_2 –2 for II-stage compression, where it is compressed polytropically to final pressure P_3 along 2 – 3 and then delivered at constant pressure P_3 along 3 – P_3. (Ref. Fig. 19.10 (iii)).

Figure 19.10 (iv) shows the combined indicator diagram for LP and HP cylinder of a single acting 2-stage air compressor with perfect intercooling.

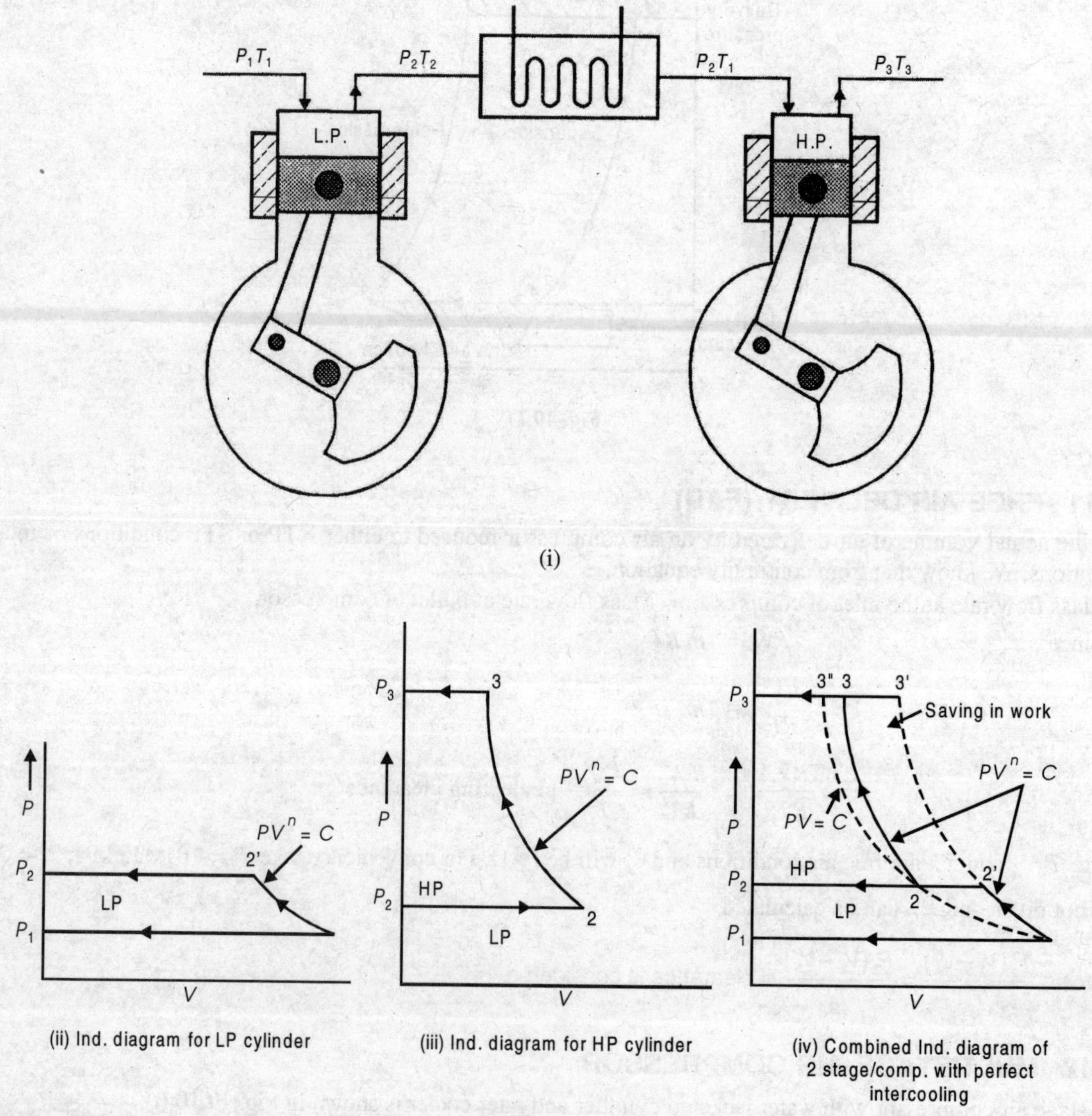

Fig. 19.12

LP diagram is shown as $P_1 - 1 - 2' - P_2$ and HP diagram is shown as $P_2 - 2 - 3 - P_3$. The reduction of work required due to intercooling is shown by shaded are $2 - 3 - 3' - 2'$, when the intercooling is perfect, the point 2 lies on the isothermal line $1 - 3''$.

Work required in LP cylinder/cycle.

$$= \frac{n}{n-1} P_1V_1 \left\{ \left(\frac{P_2}{P_1} \right)^{\frac{n-1}{n}} - 1 \right\}$$

And work required in HP cylinder/cycle

$$= \frac{n}{n-1} P_2V_2 \left\{ \left(\frac{P_3}{P_2} \right)^{\frac{n-1}{n}} - 1 \right\}$$

∴ Total work required /cycle,

$$W = \frac{n}{n-1} P_1V_1 \left\{ \left(\frac{P_2}{P_1} \right)^{\frac{n-1}{n}} - 1 \right\} + \frac{n}{n-1} P_2V_2 \left\{ \left(\frac{P_3}{P_2} \right)^{\frac{n-1}{n}} - 1 \right\} \quad \text{...(1)}$$

and

$$\text{IP of compressor} = \frac{WN}{60} \text{ J/sec or W.}$$

If the intercooling is perfect, then point 2 will lie on isothermal line, then $P_1V_1 = P_2V_2$ and Eq. (1) will be,

$$W = \frac{n}{n-1} P_1V_1 \left\{ \left(\frac{P_2}{P_1} \right)^{\frac{n-1}{n}} + \left(\frac{P_3}{P_2} \right)^{\frac{n-1}{n}} - 2 \right\} \text{J} \quad \text{...(2)}$$

and

$$\text{IP} = \frac{WN}{60} \text{ J/sec or W} \quad \text{...(3)}$$

Intercoolers and Aftercoolers. In the intercoolers air is cooled in-between the cylinders of a multistage compressor. And in the afterwords air is cooled after the last cylinder and before it enters the reservoir.

By aftercooling there is no saving in work as such, but by aftercooling volume of air decreases. So more amount of air can be stored in the reservoir. Hence capacity (mass of air to be stored) of air increases.

Ideal Intercooler Pressure. It may be noted that, saving in work increases as intercooling is increased. When the intercooling is perfect, saving of work is maximum and work input to compressor is given by Eq. (2).

It may further be noted that saving in work also varies with the chosen intercooler pressure P_2. When P_1 and P_3 are fixed, the best value, of intercooler pressure P_2 shall be fixed to give minimum work.

This value of P_2 can be found by differentiating the Eq. (2) w.r.t. P_2 and equating it to zero.

$$\therefore \quad W = \frac{n}{n-1} P_1V_1 \left[\left(\frac{P_2}{P_1} \right)^{\frac{n-1}{n}} + \left(\frac{P_3}{P_2} \right)^{\frac{n-1}{n}} - 2 \right] \quad \text{...(3)}$$

$$W = \frac{n}{n-1}P_1V_1\left[\left(\frac{P_2}{P_1}\right)^a + \left(\frac{P_3}{P_2}\right)^a - 2\right] \text{ where } a = \frac{n-1}{n}$$

$$W = \text{Constant}\times\left\{\left(\frac{P_2}{P_1}\right)^a + \left(\frac{P_3}{P_2}\right)^a - 2\right\}$$

Differentiating w.r.t. P_2 and equating it to zero for minimum work,

$$\frac{dW}{dP_2} = \frac{aP_2^{a-1}}{P_1^a} - \frac{aP_3^a}{P_2^{a+1}} = 0$$

Dividing by a throughout,

$$\frac{P_2^{a-1}}{P_1^a} = \frac{P_3^a}{P_2^{a+1}}$$

$$\therefore \qquad P_2^{2a} = (P_1P_3)^a$$

Taking a^{th} root throughout,

$$P_2^2 = P_1P_3 \Rightarrow P_2\sqrt{P_1P_3} \text{ or } \frac{P_2}{P_1} = \frac{P_3}{P_2}$$

This shows that for minimum work required or maximum efficiency, the intercooler pressure is the geometric mean of initial and final pressure, i.e., pressure ratio in each stage is same.

$\therefore$ Equation (3) will be,

$$W = \frac{n}{n-1}P_1V_1\left\{\left(\frac{P_2}{P_1}\right)^{\frac{n-1}{n}} + \left(\frac{P_3}{P_2}\right)^{\frac{n-1}{n}} - 2\right\}$$

$$W = \frac{2n}{n-1}P_1V_1\left\{\left(\frac{P_2}{P_1}\right)^{\frac{n-1}{n}} - 1\right\} \text{ since } \frac{P_2}{P_1} = \frac{P_3}{P_2}$$

$$= \frac{2n}{n-1}P_1V_1\left\{\left(\frac{(P_1P_3)^{1/2}}{P_1}\right)^{\frac{n-1}{n}} - 1\right\}$$

$$= \frac{2n}{n-1}P_1V_1\left\{\left[\left(\frac{P_3}{P_1}\right)^{1/2}\right]^{\frac{n-1}{n}} - 1\right\}$$

$$\mathbf{W = \frac{2n}{n-1}P_1V_1\left\{\left(\frac{P_3}{P_1}\right)^{\frac{n-1}{2n}} - 1\right\}}$$

and $\quad \min IP = \frac{WN}{60}$ J/sec

Similarly for 3 stage compressor

$$W = 3\frac{n}{n-1}\left\{\left(\frac{P_4}{P_1}\right)^{\frac{n-1}{3n}} - 1\right\}$$

And similarly for x-stage compress

$$W = \frac{xn}{n-1}P_1V_1\left\{\left(\frac{P_{x+1}}{P_1}\right)^{\frac{n-1}{xn}} - 1\right\}$$

Now in general,

$$\frac{P_2}{P_1} = \frac{P_3}{P_2} = \frac{P_4}{P_3} = \frac{P_{x+1}}{P_1} = k$$

where k is the pressure ratio is each stage.

From this,

$$P_2 = k P_1$$
$$P_3 = kP_2 = k\ k P_1 = k^2 P_1$$
$$P_4 = kP_3 = k.k^2 P_1 = k^3 P_1$$
$$P_{x+1} = kP_x = k^x P_1$$

$$\therefore \quad k^x = \left(\frac{P_{x+1}}{P_1}\right) \text{ or } k = \left(\frac{P_{x+1}}{P_1}\right)^{\frac{1}{x}}$$

i.e., $\quad k = (\text{Pressure Range})^{1/x}$

19.13 CAPACITY CONTROL OF COMPRESSORS

Generally compressors are not operated at full capacity. Depending upon the demand for high pressure air, they are operated.

Various types of controls for compressors are:

(i) Throttle control

(ii) Blowing air to atmosphere

(i) Throttle Control. When the demand for high pressure air is more and when less amount of air is present is the reservoir tank, then less pressure from the reservoir will act on the piston (1). Hence because of less pressure piston moves upwards and more amount of air is taken during suction stroke opposite operation will take place when the demand for high per air is less.

(ii) Blowing Air to Atmosphere. When the pressure in the receiver increases beyond the required limit, then because of high pressure air piston (1) of relay cylinder (2) in lifted and high pressure air enters cylinder (3) and presses the piston (4) downwards and high pressure air from the compressor is directly blown off to atmosphere.

And when the pressure comes below normal in the receiver, then piston (1) of relay cylinder moves down because of dead weights (5). Because of compressor air pressure check valve (6) opens and air enters receiver (7).

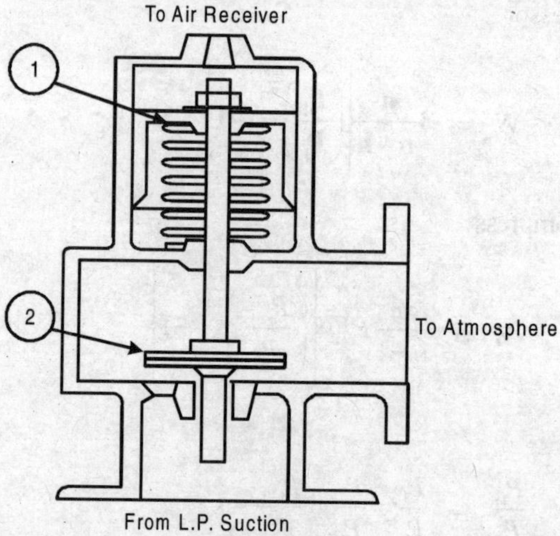

Fig. 19.13. *Throttle control.*

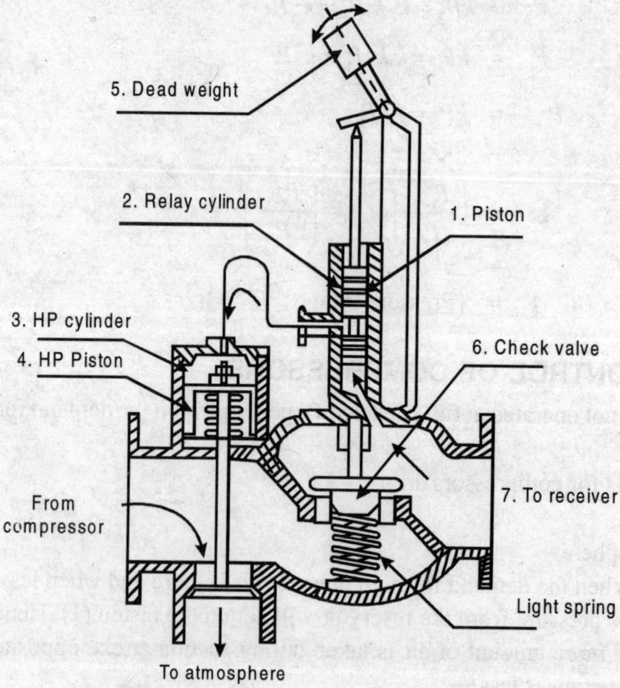

Fig. 19.14. *Blowing air to atmosphere.*

SOLVED EXAMPLES
(PROBLEMS ON SINGLE STAGE AIR COMPRESSOR)

Example 19.1. *An air compressor is required to compress air from a pressure of 1 bar to 10 bar. Calculate the amount of work required per kg of air, when n = 1.2 for compression. Assume initial temperature as 15°C and R for air is 0.287 kJ/kg-K.*

Data: P_1 = 1 bar 100 kPa P_2 = 10 bar = 1000 kPa

W.D/kg = ? n = 1.2

T_1 = 15 + 273 = 288 K R = 0.287 kJ/kg-K.

Solution

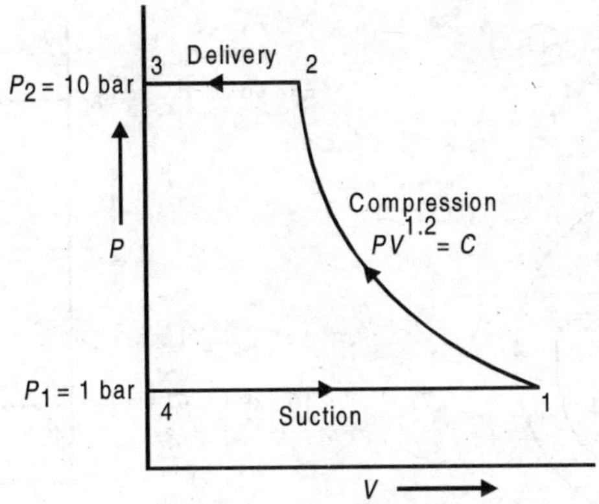

Fig. Ex. 19.1

We know that amount of work input required for compression and delivery of air polytropically/cycle,

$$W - \frac{n}{n-1} P_1 V\left\{\left(\frac{P_2}{P_1}\right)^{\frac{n-1}{n}} \;\; 1\right\}$$

$$W = \frac{n}{n-1} RT_1\left\{\left(\frac{P_2}{P_1}\right)^{\frac{n-1}{n}} - 1\right\} \quad \text{as } PV_1 = mRT_1$$

Note. m = 1 kg as work input/kg is required

$$W = \frac{1.2}{1.2-1} \times 0.287 \times 288\left\{\left(\frac{1000}{100}\right)^{\frac{1.2-1}{1.2}} - 1\right\}$$

$$= \frac{1.2}{0.2} \times 0.287 \times 288\{(10)^{0.2/1.2} - 1\}$$

$$= 495.936\ \{1.46779 - 1\}$$

$$= 495.936\ \{0.46779\}$$

W = 231.9985 kJ/kg

Example 19.2. *A single stage, single acting air compressor delivers air at 5 bar. The suction temperature is 20°C and suction pressure is 1 bar volume of air entering the compressor is 3 m³/min. Index of compression is 1.2. Calculate isothermal efficiency and power required to drive the compressor. Neglect clearance volume.*

Data: $T_1 = 20 + 273 = 293$ K $V_1 = 3$ m³/min

$n = 1.2$ $\eta_{iso} = ?$

W.D. = ?

Solution

We know that,

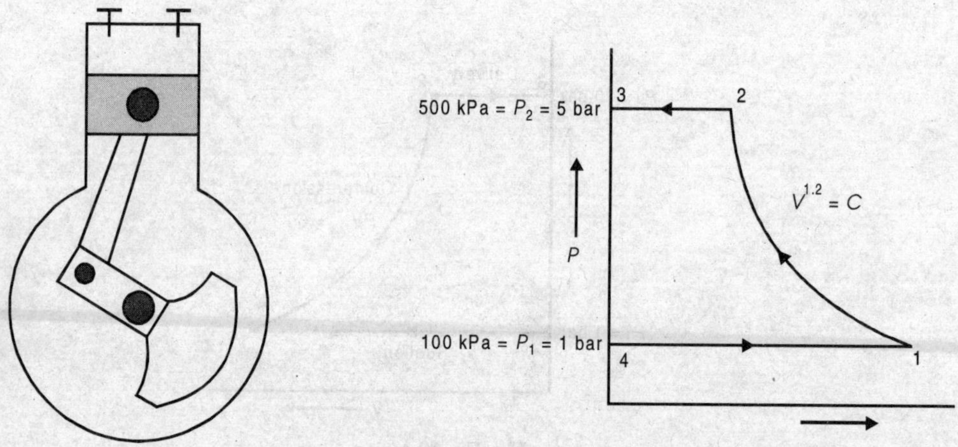

Fig. Ex. 19.2

$$\text{W.D. or I.P} = \frac{n}{n-1}P_1V_1\left\{\left(\frac{P_2}{P_1}\right)^{\frac{n-1}{n}} - 1\right\}$$

$$= \frac{1.2}{1.2-1}\times 100\times\frac{3}{60}\left[\left(\frac{500}{100}\right)^{\frac{1.2-1}{1.2}} - 1\right]$$

$$= \frac{\text{kN}}{\text{m}^2}\times\frac{\text{m}^3}{\text{sec}}[\text{ratio}-1] = \frac{\text{kJ}}{\text{sec}} = \text{kW}$$

$$\textbf{W.D. or I.P = 9.229 kW}$$

Also
$$\eta_{iso} = \frac{\text{Isothermal work of compression}}{\text{IP}}$$

$$= \frac{P_1V_1 \ln\frac{P_2}{P_1}}{9.229}$$

$$= \frac{100\times\frac{3}{60}\times\ln\frac{500}{100}}{9.229} = \frac{8.05}{9.229}$$

$$\eta_{iso} = \mathbf{87.2\ \%}$$

Example 19.3. *A single stage single acting/compressor with bore 15 cm and stroke 20 cm run at 200 rpm. Air is drawn in at 1 bar and 25°C. Air is delivered at 6 bar, law of compression is $PV^{1.3}$ = C. Determine FAD and indicated power required to drive the compressor in kW.*

Data: $D = 0.5\ m$ $\quad L = 0.2\ m$

$N = 200\ rpm$ $\quad P_1 = 100\ kPa$

$T_1 = 25 + 273 = 298\ K$

$FAD = ?$ $\quad IP = ?$

Solution

We know that,

$$\frac{P_1V_1}{T_1} = \frac{P_2V_2}{T_2} = \frac{P_fV_f}{T_f}$$

Now V_1 = volume of air handled/min

$$= \frac{\pi}{4}D^2LN$$

$$= \frac{\pi}{4}\times 0.15^2\times 0.2\times 200$$

$$= 0.707\ m^3/min$$

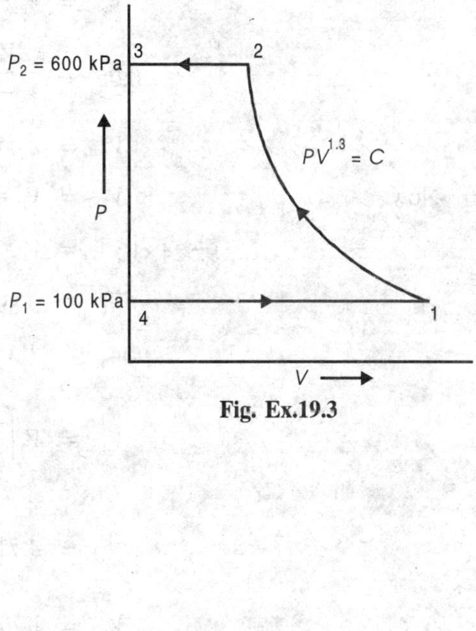

Fig. Ex.19.3

To find FAD i.e., V_f

$$\frac{P_1V_1}{T_1} = \frac{P_fV_f}{T_f}$$

Since nothing is mentioned take,

$$P_f = 101325\ kPa$$

$$T_f = 15 + 273 = 288\ K$$

$$\therefore \quad \frac{100\times 0.7087}{283} = \frac{101.325\times V_f}{288}$$

$$\mathbf{0.674\ m^3/min = V_f}$$

$$IP = \frac{n}{n-1}P_1V_1\left[\left(\frac{P_2}{P_1}\right)^{\frac{n-1}{n}} - 1\right]$$

$$= \frac{1.3}{1.3-1}\times 100\times\frac{0.707}{60}\left[\left(\frac{600}{100}\right)^{\frac{0.3}{1.3}} - 1\right]$$

$$\mathbf{IP = 2.615\ kW}$$

Example 19.4. *A single stage, single acting reciprocating air compressor has a bore of 200 mm and a stroke of 300 mm. It runs at a speed of 500 rpm. The clearance volume is 5% of swept volume and polytropic index is 1.3 throughout. Intake pressure and temperature are 97 kPa, 20°C and the compression pressure is 550 kPa. Detertermine:*

(i) FAD in m³/min
(ii) Air delivery temperature
(iii) Cycle power
(iv) η_{iso} neglecting clearance volume.

Solution

$$\frac{V_c}{V_s} = 0.05 = \frac{V_3}{V_1 - V_3} \qquad \text{...(1)}$$

$$V_s = \text{Stroke volume}$$

$$= \frac{\pi}{4} 0.2^2 \times 0.3$$

$$V_s = V_1 - V_3 = 9.425 \times 10^{-3}\,\text{m}^3$$

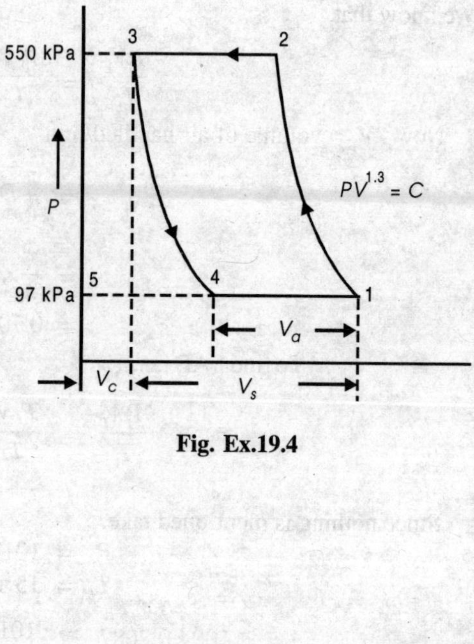

Fig. Ex.19.4

$$\therefore \quad 0.05 = \frac{V_3}{V_1 - V_3} = \frac{V_3}{9.425 \times 10^{-3}}$$

$$\Rightarrow \quad V_3 = 4.712 \times 10^{-4}\,\text{m}^3$$

Now $$V_s = V_1 - V_3$$

$$9.524 \times 10^{-3} = V_1 - 4.712 \times 10^{-4}$$

$$\therefore \quad V_1 = 9.896 \times 10^{-3}\,\text{m}^3$$

$$p_3 v_3^n = P_4 V_4^n$$

$$\therefore \quad V_4 = V_3 \left(\frac{P_3}{P_4}\right)^{1/n} = V_3 \left(\frac{P_2}{P_1}\right)^{1/n}$$

$$= 4.712 \times 10^{-4} \left(\frac{550}{97}\right)^{1/1.3}$$

$$V_4 = 1.79 \times 10^{-3}\,\text{m}^3$$

Then $$\frac{P_1(V_1 - V_4)}{T_1} = \frac{P_f V_f}{T_f}$$

$$\mathbf{V_f = 7.625 \times 10^{-3}\ m^3}$$

(i) $$\text{FAD} = V_f \times N = 3.814\,\text{m}^3/\text{min}$$

(ii) $$\eta_{\text{vol}} = \frac{V_1 - V_4}{V_1 - V_3} = 0.8656 \Rightarrow 86\%$$

(iii) $$\frac{T_1}{T_2} = \left(\frac{P_1}{T_2}\right)^{\frac{n-1}{n}} = \left(\frac{V_2}{V_1}\right)^{n-1}$$

$$\mathbf{T_2 = 437.29\,K}$$

(iv) $$\text{Cycle power} = \frac{n}{n-1}P_1(V_1 - V_4)\left[\left(\frac{P_2}{P_1}\right)^{\frac{n-1}{n}} - 1\right]\times\frac{N}{60}$$

$$= \frac{1.3}{0.3}\times 97\,(8.106\times10^{-3})\left[\left(\frac{550}{97}\right)^{\frac{0.3}{1.3}} - 1\right]\times\frac{500}{60}$$

$$\textbf{I.P.} = \mathbf{13.983\ kW}$$

Example 19.5. *A single stage, single acting air compressor works between 1 bar and 16 bar compression follows $PV^{1.3} = C$. The piston speed is 200 m/min. It runs at 350 rpm. It has an IP consumption of 30 kW and η_{vol} is 85%. Estimate the cylinder bore and stroke.*

Data: $P_1V_1^{1.3} = P_2V_2^{1.3}$ — *Piston speed* = *2 LN = 200 m/min*

N = *350 rpm* — *IP* = *30 kW*

η_{vol} = *85%* — *D* = *?*

L = *?*

Solution

Since Piston speed = 2 LN =200 m/min

$= 2 \times L \times 350 = 200$ m /min

$L = 0.286$ m

$\therefore$ Stroke L = 286 mm

$$\eta_{vol} = \frac{V_a}{V_s} = \frac{V_1 - V_4}{V_1 - V_3}$$

$$\text{IP} = \frac{n}{n-1}P_1(V_1 - V_4)\left[\left(\frac{P_2}{P_1}\right)^{\frac{n-1}{n}} - 1\right]\times\frac{N}{60}\frac{\text{kJ}}{\text{sec}}$$

$$30\,\text{kW} = \frac{1.3}{0.3}\times 100\,(V_1 - V_4)\left[\left(\frac{1600}{100}\right)^{\frac{0.3}{1.3}} - 1\right]\times\frac{350}{60}$$

$$(V_1 - V_4) = 0.01324\ \text{m}^3$$

Now from (1) $$0.85 = \frac{0.01324}{V_1 - V_3}$$

$$(V_1 - V_3) = 0.01558\ \text{m}^3 = V_s$$

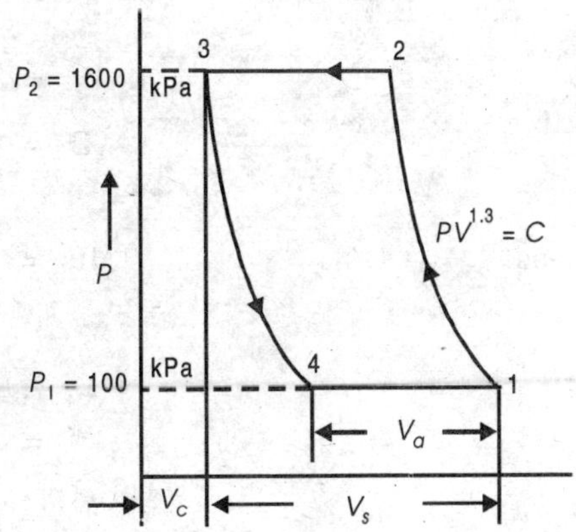

Fig. Ex. 19.5

$$\therefore \quad V_s = 0.01558\,\text{m}^3 = \frac{\pi}{4}D^2L$$

$$\therefore \quad D = 0.263\text{ m}$$

$$\mathbf{D = 263\ mm}$$

Example 19.6. *Actual volume of air taken in at 0.98 bar and 27°C by a single stage, acting reciprocating air compressor is 8.5 m³/min. The delivery conditions are 6.5 bar. The clearances volume is 5% of stroke volume compression and expansion follows $PV^{1.3} = C$. If the speed is 300 rpm. L/D ratio is 1:1. Find the cylinder dimensions by finding the volumetric efficiency required to run the compressor. Also find I.P.*

Data: $V_a = 8.5\ m^3/min$ $\quad P_2 = 6.5 \times 10^2\ kPa$

$T_1 = 27 + 273K$ $\quad P_1 = 0.98 \times 10^2\ kPa$

$\frac{V_c}{V_s} = 0.05$ $\quad n = 1.3, N = 300\ rpm$

Solution We know that,

$$\eta_{vol} = 1 - C\left\{\left(\frac{P_2}{P_1}\right)^{1/n} - 1\right\}$$

$$= 1 - 0.05\left\{\left(\frac{6.5\times10^2}{0.98\times10^2}\right)^{1/1.3} - 1\right\}$$

$$\boldsymbol{\eta_{vol} = 0.8352 = 83.57\%}$$

Also volumetric efficiency,

$$\eta_{vol} = \frac{V_a}{V_s}$$

$$\therefore \quad 0.8356 = \frac{8.5}{V_s}$$

$$V_s = 1.7712\ \text{m}^3/\text{min} = \frac{\pi}{4}D^2LN$$

$$\mathbf{D = L = 0.3508\ m}$$

and

$$\text{IP} = \frac{n}{n-1}P_1(V_1 - V_4)\left[\left(\frac{P_2}{P_1}\right)^{\frac{n-1}{n}} - 1\right]$$

$$= \frac{n}{n-1}P_1V_a = \left\{\left(\frac{P_2}{P_1}\right)^{\frac{n-1}{n}} - 1\right\}$$

$$= \frac{1.3}{1.3-1}\times\frac{0.98\times10^2\times8.5}{60}\left[\left(\frac{6.5}{0.98}\right)^{\frac{0.3}{1.3}} - 1\right]$$

$$\mathbf{IP = 33.6\ kW}$$

Example 19.7. *Atmospheric air at 1 bar and 27°C taken into a single stage single acting air compressor. It is compressed according to $PV^{1.3}$ = constant to the discharge pressure of 7 bar. The compressor takes 1 m^3 of air per min. Calculate:*

(1) IP

(2) Power of motor required to drive compressor.

(3) η_{iso}, and

(4) The cylinder dimensions, assuming speed of motor = 300 rpm, stroke to bore ratio = 1.5: 1; η_{mech} of compressor = 85%, Motor transmission efficiency 90%.

Data:

$$\eta = 90\,\% \qquad P_1 = 100\text{ kPa}$$

$$T_1 = 300\text{ K} \qquad PV^{1.3} = C$$

$$V_1 = 1\text{ m}^3/\text{min} \qquad N = 300\text{ rpm}$$

$$\frac{L}{D} = 1.5 \quad \Rightarrow \quad L = 1.5\,D$$

$$P_2 = 700\text{ kPa}$$

Solution Neglecting clearance

$$\text{IP} = \frac{n}{n-1} P_1V_1\left[\left(\frac{P_2}{P_1}\right)^{\frac{n-1}{n}} - 1\right]$$

$$= \frac{1.3}{0.3} \times 100 \times \frac{1}{60} \times \frac{1}{60}\left\{\left(\frac{7}{1}\right)^{\frac{0.3}{1.3}} - 1\right\}$$

$$\textbf{IP} = \textbf{4.09 kW}$$

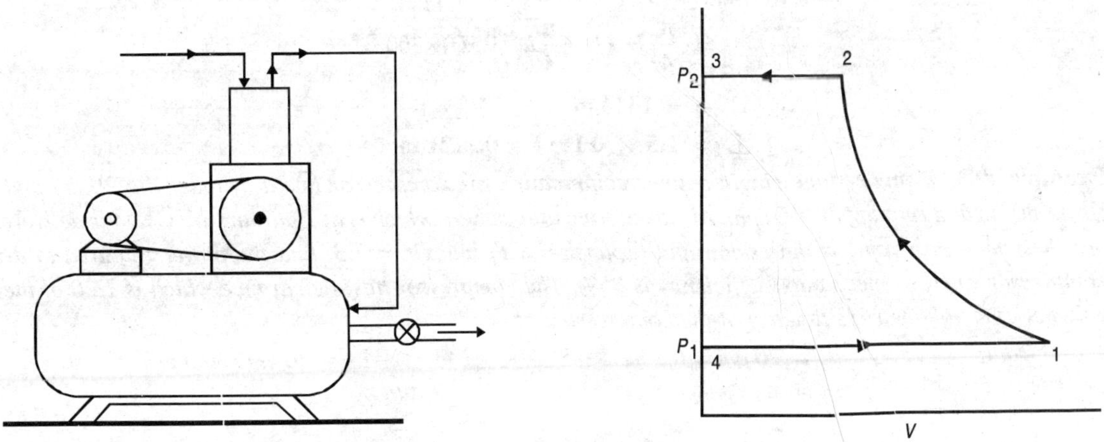

Fig. Ex. 19.7

Note. Actual work IP required to drive the compressor is more than IP, due to work required to overcome frictional lossess in compressor.

$$\therefore \quad \text{Shaft power} = \text{IP} + \text{FP for compressor}$$

And

$$\text{Compressor } \eta_{mech} = \frac{\text{IP}}{\text{shaft power}}$$

$$\therefore \quad \text{Shaft Power} = \frac{\text{IP}}{\eta_{mech} \text{ of compressor}} = \frac{4.09}{0.85}$$

$$= 4.81 \text{ kW}$$

And

$$\text{Power input to motor} = \frac{\text{Shaft power}}{\text{Transmission } \eta \text{ (inclusive of motor and drive } \eta)}$$

$$= \frac{4.81}{0.9} = 5.35 \text{ kW}$$

$$\text{Also Iso } \eta, \quad \eta_{iso} = \frac{\text{Iso. work}}{\text{Indi. work}}$$

$$\text{Iso Work} = P_1 V_1 \ln \frac{P_2}{P_1} = 150 \times \frac{1}{60} \times \ln \frac{7}{1} = 3.24 \text{ kW}$$

$$\therefore \quad \eta_{iso} = \frac{3.24}{4.09} = 0.792 \text{ i.e., } 79.2\%$$

Volume of air taken in/min

$$= \frac{\pi}{4} \cdot D^2 LN = \frac{\pi}{4} D^2 (1.5D) \times 300$$

$$\mathbf{D = 0.1414 \text{ m}}$$

$$\mathbf{L = 1.5 \times 0.1414 = 0.2121 \text{ m}}$$

Example 19.8. *A single stage single acting reciprocating air compressor has a cylinder dia 20 cm and piston stroke 30 cm and running at 650 rpm. Air taken from atmosphere which is at 1 bar and 27°C and is delivered at 11 bar. Assuming polytropic compression and expansion with index n = 1.3, find the power required to drive the compressor when its mechanical efficiency is 85%. The compressor has a clearance which is 1/20 of the stroke volume. Find volumetric efficiency of the compressor.*

Data:

$d = 20 \text{ cm} = 0.2 \text{ m}$ $\qquad L = 30 \text{ cm} = 0.3 \text{ m}$

$N = 650 \text{ rpm}$ $\qquad W/sec = ?$

$\eta_{mech} = 85\%$ $\qquad V_c = \frac{1}{20} V_c$

$\eta_{vol} = ?$

Solution

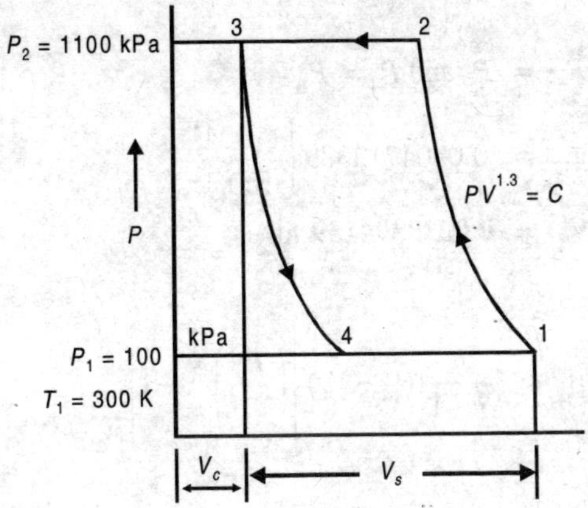

Fig. Ex. 19.8

We know that, work input to a compressor or I.P. of a compressor with clearance,

$$= \frac{n}{n-1} P_1(V_1 - V_4)\left\{\left(\frac{P_2}{P_1}\right)^{\frac{n-1}{n}} - 1\right\} \quad \ldots(1)$$

Now find V_1 and V_4. We also know that,

$$\text{Stroke volume} = V_s = V_1 - V_3 = \frac{\pi}{4} d^2 L$$

$$V_s = \frac{\pi}{4} \times (0.2)^2 \times 0.3$$

$$V_s = 0.094278 \text{ m}^3$$

As given $\quad V_c = \frac{1}{20} V_s$

$\therefore \quad V_c = \frac{1}{20} \times 0.0094278$

$\therefore \quad V_c = \frac{1}{20} \times 0.0094278$

$\therefore \quad V_c = 0.0004712389 \text{ m}^3$

From Fig. Ex. 19.8 $\quad V_1 = V_c + V_s = 0.0004712389 + 0.009427$

$$\mathbf{V_1 = 0.0098990 \text{ m}^3}$$

Now to find V_4 consider polytropic 3 – 4 for which

$$P_3 V_3^n = P_4 V_4^n$$

$$V_4 = V_3 \times \left(\frac{P_3}{P_4}\right)^{1/n}$$

$$V_4 = V_c \left\{\frac{P_2}{P_4}\right\}^{1/n}$$

As $\quad P_3 = P_2$ and $P_4 = P_1$

$$\therefore \quad V_4 = 0.0004712389 \left\{\frac{1100}{100}\right\}^{1/1.3}$$

$$\mathbf{V_4 = 0.0029894139\ m^3}$$

Now from Eq. (1) ,

$$\text{IP} = \frac{n}{n-1}((V_1 - V_4)\left\{\left(\frac{P_2}{P_1}\right)^{\frac{n-1}{n}} - 1\right\}$$

$$= \frac{1.3}{1.3-1} \times 100 \times (0.009890 - 0.0029804139) \times \left\{\left(\frac{1100}{100}\right)^{\frac{1.3-1}{1.3}} - 1\right\}$$

$$= 4.33 \times 100 \times (0.069186) \times \{1.739 - 1\}$$

$$\mathbf{IP = 2.2141\ kJ\ /cycle}$$

Since $\quad N = 650$ rpm

$$\text{IP} = 2.2141 \times \frac{650}{60}$$

$$\mathbf{IP = 23.9835\ kJ\ /\ sec\ or\ kW.}$$

Since Mechanical efficiency is given

$$\text{Actual power input} = \frac{\text{IP}}{\eta_{mech}} = \frac{23.98365}{0.85} = 28.2158\,\text{kW}$$

Also we know that, volumetric efficiency.

$$\eta_{vol} = \frac{V_a}{V_s} = \frac{V_1 - V_4}{V_1 - V_3}$$

$$= \frac{0.0098990 - 0.0029804139}{0.0098990 - 0.000471239} = \frac{0.0069186}{0.0094278}$$

$$\mathbf{\eta_{vol} = 0.73385\ \text{or}\ 73.385\%}$$

Example 19.9. *Calculate the volumetric efficiency of the compressor having a cylinder diameter 410 mm and stroke 610 mm. Compressor makes 420 rpm and delivers 30 kg/min of air at 1.01325 bar and 15° C.*

Data:

$\eta_{vol} = ?$ $\qquad d = 410$ mm $= 0.41$ m

$L = 610$ mm $= 0.61$ m

$N = 420$ rpm $\qquad m = 30$ kg/min

$P = 1.01325$ bar $= 101.325$ kPa

$T = 15 + 273 = 288$ K

Solution

We know that volumetric efficiency,

$$\eta_{vol} = \frac{V_a}{V_s}$$

and $$V_s = \text{Stroke volume} = \frac{\pi}{4} d^2 \times L \times N$$

Note. $m^2 \times m \times rmp = m^3/min$

$$= \frac{\pi}{4} \times 0.41^2 \times 0.61 \times 420$$

$$V_s = 33.824 \text{ m}^3/\text{min}$$

As 30 kg of air is delivered at 101.325 kPa of 288 K,

As $$PV = mRT$$

Actual volume of air handled

$$V_a = \frac{mRT}{P} = \frac{30\,\text{kg/min} \times 0.287 \times 288}{101.325}$$

Note. $$V = \frac{\frac{kg}{min} \times \frac{kJ}{kg\text{-}k} \times K}{k\frac{N}{m^2}} = \frac{k\text{-}Nm}{min} \times \frac{m^2}{kN}$$

$$V = m^3/min$$

$$V_a = 24.472539 \text{ m}^3/\text{min}$$

Now $$\eta_{vol} = \frac{V_a}{V_s} = \frac{24.472539}{38.8249} = 0.7235062$$

$$\boldsymbol{\eta_{vol} = 72.35\ \%}$$

Example 19.10. *A single stage single acting reciprocating air compressor delivers 10 m³ of free air per minute and it is running at 200 rpm. It takes air at a pressure of 1 bar and 303 K and delivers at 6 bar. Take n = 1.25 and clearance volume is 6% of stroke volume.*

Determine

(i) Delivery temperature

(ii) Volumetric efficiency

(iii) Stroke volume

(iv) Power required and

(v) Dimensions of cylinder if L = 1.2 D

Data:

$FAD = 10\ m^3/min = V_f$

$N = 200\ rpm$ $\quad V_c = 0.06\ V_s$

$T_2 = ?$ $\quad \eta_{vol} = ?$

$V_s = ?$ $\quad IP = ?$

$L = ?$ and $D = ?$

Solution

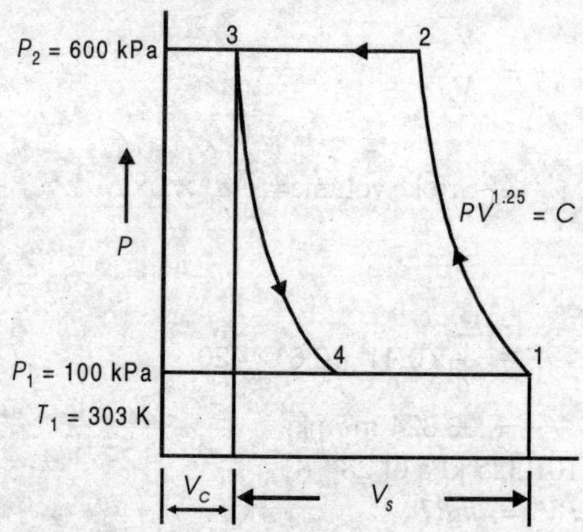

Fig. Ex. 19.10

(i) We know that,

$$\frac{T_2}{T_1} = \left(\frac{P_2}{T_1}\right)^{\frac{n-1}{n}}$$

$$T_2 = T_1\left(\frac{P_2}{P_1}\right)^{\frac{n-1}{n}} = 303\times\left(\frac{600}{100}\right)^{\frac{1.25-1}{1.25}}$$

$$T_2 = 303\times\left(\frac{6}{1}\right)^{\frac{0.25}{1.25}}$$

$$\mathbf{T_2 = 433.584\ K}$$

(ii) We also know that, volumetric efficiency,

$$\eta_{vol} = 1+C-C\times\left(\frac{P_2}{P_1}\right)^{1/n} = 1+0.06-0.06\times\left(\frac{6}{1}\right)^{\frac{1}{1.25}}$$

$$= 1 + 0.06 - 0.06 \times 4.19 = 0.8084$$

$$\mathbf{\eta_{vol} = 80.84\ \%}$$

(iii) To find stroke volume, we know that

$$\frac{P_f V_f}{T_f} = \frac{P_1(V_1 - V_4)}{T_1}$$

$$\frac{P_f V_f}{T_f}\times\frac{T_1}{P_1} = (V_1 - V_4)$$

$$\frac{101.325\times 10\,m^3/min}{288}\times\frac{303}{100} = (V_1 - V_4)$$

$$\mathbf{(V_1 - V_4) = 10.86 m^3/min = V_a}$$

Since the compressor is running at 200 rpm.

$$(V_1 - V_4) = V_a = \frac{10.66}{200}\text{m}^3$$

$$V_a = 0.0533 \text{ m}^3$$

Also since $$\eta_{vol} = \frac{V_a}{V_s}$$

$$0.8084 = \frac{0.533}{V_s}$$

$$\mathbf{V_s = 0.0659327m^3}$$

(iv) $$\text{I.P. required} = \frac{n}{n-1}P_1(V_1 - V_4)\left[\left(\frac{P_2}{P_1}\right)^{\frac{n-1}{n}} - 1\right]\frac{\text{rpm}}{60}$$

$$= \frac{1.25}{1.25-1}\times 100\times 0.533\left\{\left(\frac{6}{1}\right)^{\frac{0.25}{1.25}} - 1\right\}\times\frac{200}{60}$$

$$= 5 \times 100 \times 0.0533(1.43 - 1) \times\frac{200}{60}$$

$$\mathbf{I.P. = 38.284\ kW}$$

(v) To find cylinder dimensions :
We know that, stroke volume,

$$V_s = 0.0659327 = \frac{\pi}{4}D^2 \times L = \frac{\pi}{4}D^2 \times 1.2D$$

$$0.0659327 = D^3 \times 0.9425$$

$$\frac{0.0659327}{0.9425} = D^3$$

$$0.0699567 = D^3 \qquad \therefore\ \mathbf{0.4120\ m} = \boldsymbol{D}$$

As $$L = 1.2\text{ D} \qquad \therefore\ \text{L} = 1.2 \times 0.4120$$

$$\mathbf{L = 0.49445\ m}$$

Example 19.11. *The following data were recorded for a single acting single stage reciprocating air compressor; speed : 200 rpm.*

Free air delivered = 6 m³/min
Volume of intake air = 6 m³/min
n = 1.25
Clearance volume = 6% of stroke volume
Suction pressure = 1 bar
Delivery pressure = 7 bar
Temperature of air at inlet = 288 K
Determine:
(i) Temperature of air at delivery

(ii) Volumetric efficiency
(iii) Volume of air delivered at the outlet stroke
(iv) Compressor indicated power in kW neglecting clearance.
(v] Isothermal efficiency of the compressor.

Data: $N = 200$ rpm, $V_1 = 6$ m³/min, $V_c = 0.06\, V_s$, $P_2 = 700$ kPa, $FAD = V_f = 6$ m³/min, $PV^{1.25} = C$, $P_1 = 100$ kPa, $T_1 = 288$ K

Find: $T_2 = ?$, $V_s = ?$, $\eta_{vol} = ?$, $\eta_{iso} = ?$

Solution

(1) We know that, for a polytropic process,

$$\frac{T_1}{T_2} = \left(\frac{P_2}{P_1}\right)^{\frac{n-1}{n}}$$

$$\therefore \quad T_2 = T_1 \times \left(\frac{P_2}{P_1}\right)^{\frac{n-1}{n}} = 288 \times \left(\frac{7}{1}\right)^{\frac{1.25-1}{1.25}}$$

$$= 288 \times (7)\frac{0.25}{1.25} = 288 \times 1.475$$

$$\mathbf{T_2 = 425.022\ K}$$

(2) For volumetric efficiency, we know that,

$$\eta_{vol} = 1 + C - C\left(\frac{P_2}{P_1}\right)^{\frac{1}{n}} = 1 + 0.06 - 0.06(7)^{\frac{1}{1.25}}$$

$$= 1 + 0.06 - 0.06 \times 4.73 = 1 + 0.06 - 0.2859$$

$$= 0.7754034$$

$$\therefore \quad \boldsymbol{\eta_{vol} = 77.54\%}$$

(3) For volume of air delivered we know that,

$$\frac{T_2}{T_1} = \left(\frac{P_2}{P_1}\right)^{\frac{n-1}{n}} = \left(\frac{V_2}{V_1}\right)^{n-1}$$

$$\left(\frac{V_2}{V_1}\right)^{n-1} = \left(\frac{P_1}{P_2}\right)^{\frac{n-1}{n}}$$

or

$$\frac{V_2}{V_1} = \left(\frac{P_1}{P_2}\right)^{\frac{1}{n}}$$

$$V_2 = V_1 \times \left(\frac{P_1}{P_2}\right)^{\frac{1}{n}} = 6\,\text{m}^3/\text{min} \times \left(\frac{1}{7}\right)^{1/1.25}$$

$$= 6 \times 0.2108 = 1.2649\ \text{m}^3/\text{min}$$

$$= 1.2649 \text{ m}^3/\text{min} = \frac{1.2649}{200} \text{ m}^3/\text{stroke}$$

$$\mathbf{V_2 = 0.006324742 \ m^3/stroke}$$

(4) We know that, indicated power,

$$\text{I.P.} = \frac{n}{n-1} P_1 V_1 \left\{ \left(\frac{P_2}{P_1} \right)^{\frac{n-1}{n}} - 1 \right\}$$

$$= \frac{1.25}{1.25-1} \times 100 \times \frac{6}{10} \text{m}^3/\sec \left\{ (7)^{0.25/1.25} - 1 \right\}$$

$$= 5 \times 100 \times 0.1 \{1.4757 - 1\}$$

Note. $\frac{\text{kN}}{\text{m}^2} \times \frac{\text{m}^3}{\text{sec}} = \frac{\text{kNm}}{\text{sec}} = \frac{\text{kJ}}{\text{sec}} = \text{kW}$

$$\mathbf{IP = 23.785 \ kW}$$

(5) For isothermal efficiency, we know that,

$$\eta_{iso} = \frac{\text{Isothermal work}}{\text{Actual work (or IP)}}$$

$$\text{Isothermal work} = P_1 V_1 \ln \frac{P_2}{P_1}$$

$$= 100 \times \frac{6}{60} \times \ln \frac{7}{1}$$

$$\text{Isothermal work} = 19.459 \text{ kW}$$

Then Isothermal efficiency from Eqs (1) and (2)

$$\eta_{isothermal} = \frac{19.459}{23.785} = 0.8181294$$

$$\mathbf{\eta_{iso} = 81.8125 \ \%}$$

Example 19.12. *A single stage single acting air compressor is working between 1 bar and 7 bar, initial intake temperature is 27°C. Take FAD as 7.5 m³/min and clearance ratio as 0.04, L : D = 1.3 and N = 300 rpm and n = 1.3,*

Determine:

(a) Volumetric efficiency
(b) Bore and stroke dimensions
(c) Indicated power
(d) Isothermal efficiency

Data:

V_f = 7.5 m min　　　C = 0.04
LD = 1.3　　　N = 300 rpm
η_{vol} = ?　　　D = ?
L = ?　　　IP = ?
η_{iso} = ?

Solution

(a) We know that, volumetric efficiency of an air compressor is given by,

$$\eta_{vol} = 1 + C - C\left\{\frac{P_2}{P_1}\right\}^{1/n} = 1 + 0.04 - 0.04\left\{\frac{700}{100}\right\}^{1/1.3}$$

$$= 1 + 0.04 - 0.04 \times 4.467$$

$$\eta_{vol} = 0..8612955$$

$$\therefore \quad \boldsymbol{\eta_{vol} = 86.13\ \%}$$

(b) We also know that,

$$\frac{P_f V_f}{T_f} = \frac{P_1(V_1 - V_4)}{T_1}$$

$$\frac{P_f V_f}{T_f} \times \frac{T_1}{p_1} = V_1 - V_4 = V_a$$

$$\frac{101.325 \times 7.5\ \text{m}^3/\text{min}}{288} \times \frac{300}{100} = (V_1 - V_4) = V_a$$

$$\therefore \quad \mathbf{V_a = 7.9160\ m^3/min}$$

Since the compressor is running at 300 rpm

$$V_a = \frac{7.9160}{300}\ \text{m}^3$$

$$\mathbf{V_a = 0.0263866\ m^3}$$

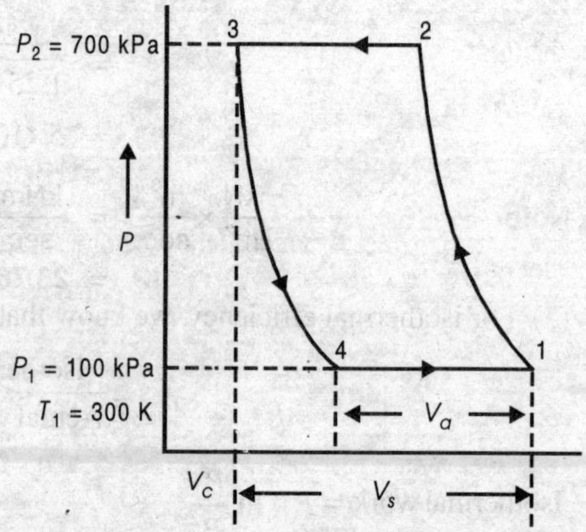

Fig. Ex. 19.12

Also $$\eta_{vol} = \frac{V_a}{V_s} = 0.8613$$

$$\therefore \quad 0.8613 = \frac{0.263867}{V_s}$$

$$\therefore \quad \mathbf{V_s = 0.0306358\ m^3}$$

and $$= 0.0306358 = \frac{\pi}{8}D^2 L = \frac{\pi}{4}D^2 \times 1.3D$$

$$0.0306358 = 1.0210\ D^3$$

$$0.0300051 = D^3$$

$$\mathbf{0.310741\ m = D}$$

and $$\mathbf{L = 1.3D = 0.4039634\ m}$$

(c) Now to find induced power,

$$\text{W or IP} = \frac{n}{n-1} P_1 V_a \left\{\left(\frac{P_2}{P_1}\right)^{\frac{n-1}{n}} - 1\right\}$$

$$= \frac{1.3}{1.3-1} \times 100 \times 0.26386 \left\{ \left(\frac{7}{1} \right)^{\frac{0.3}{1.3}} - 1 \right\}$$

Note. Unit kN/m² × m³ = kJ

$$= 4.33 \times 100 \times 0.26386 \{1.5668 - 1\}$$

$$\text{IP} = 6.476289 \text{ kJ}$$

As the compressor is running at 300 rpm.

$$\text{Power} = \frac{\text{WN}}{60} = 6.476289 \times \frac{300}{60}$$

$$\textbf{Power} = \textbf{32.38145 kW}$$

(d) Finally to find Isothermal efficiency

$$\eta_{vol} = \frac{\text{Isothermal work}}{\text{Actual work}} = \frac{P_1 V_1 \ln \frac{P_2}{P_1}}{\frac{n}{n-1} P_1 V_1 \left\{ \left(\frac{P_2}{P_1} \right)^{\frac{n-1}{n}} - 1 \right\}}$$

$$= \frac{\ln \frac{P_2}{P_1}}{\frac{n}{n-1} \left\{ \left(\frac{P_2}{P_1} \right)^{\frac{n-1}{n}} - 1 \right\}}$$

$$= \frac{\ln \left(\frac{7}{1} \right)}{\frac{1.3}{1.3-1} \left\{ \left(\frac{7}{1} \right)^{0.3-1.3} - 1 \right\}} = \frac{1.9459}{2.456275} = 0.7922$$

$$\eta_{vol} = \textbf{79.22\%}$$

Example 19.13. *For a single stage single acting air compressor, actual volume of air taken in is 10 m³/min, initial intake pressure 1.013 bar, initial temperature 27°C. Final pressure 900 kN/m². Clearance is 6% of stroke volume, L : D = 1.25 and compressor is running at 400 rpm, take n =1.3.*

Determine:

(i) *Volumetric efficiency*
(ii) *Cylinder dimensions*
(iii) *Indicated power*

Data:

V_a = 10 m³/mn P_1 = 1.013 bar = 101.3 kN/m²
T_1 = 27 + 273 = 300 K P_2 = 900 kN/m²
V_c = 0.06 V_s N = 400 rpm

Solution

(i) $\eta_{vol} = ?$ (ii) $L = ?$ $D = ?$ (iii) $IP = ?$

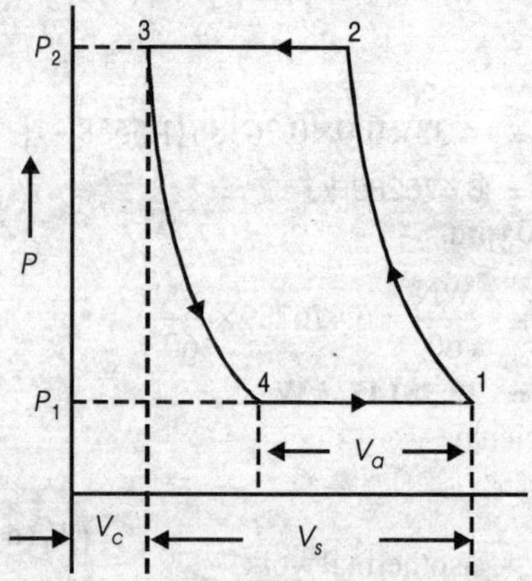

Fig. Ex. 19.13

(i) We know that the volumetric efficiency of a compressor is given by,

$$\eta_{vol} = 1 + C - C \times \left(\frac{P_2}{P_1}\right)^{1/n}$$

$$= 1 + 0.06 - 0.06 \times \left(\frac{900}{101.3}\right)^{1/1.3}$$

$$= 1 + 0.06 - 0.06 \times 5.366$$

$$\eta_{vol} = 0.73799 \cong 0.7380$$

$$\boldsymbol{\eta_{vol} = 73.80\%}$$

(ii) We also know that,

$$\eta_{vol} = \frac{\text{Actual volume of air taken in}}{\text{Stroke volume}}$$

$$\therefore \quad 0.7380 = \frac{10 \text{ m}^3/\text{min}}{V_s}$$

$$V_s = \frac{10 \text{ m}^3/\text{min}}{0.7380}$$

$$V_s = 13.550 \text{ m}^3/\text{min}$$

Since $N = 400$ rpm

$$V_s = \frac{13.550}{400}$$

$$V_s = 0.03387 = \frac{\pi}{4} D^2 \times L = \frac{\pi}{4} D^2 \times 1.25 D$$

$$0.03387 = 0.9817 \times D^3$$

$$D^3 = 0.0344996$$

$$\frac{0.3387}{0.9817} = D^3$$

$$(0.0344996)^{1/3} = D$$

$$\mathbf{D = 0.3255\ m}$$

and $\mathbf{L = 0.4069\ m}$

(iii) We also know that, *IP* of compressor is given by

$$= \frac{n}{n-1} P_1 V_a \left\{ \left(\frac{P_2}{P_1} \right)^{\frac{n-1}{n}} - 1 \right\}$$

$$= \frac{1.3}{1.3-1} \times 101.3 \times \frac{10}{60} \left\{ \left(\frac{900}{101.3} \right)^{\frac{1.3-1}{1.3}} - 1 \right\}$$

Note. $\left[\frac{kN}{m^2} \times \frac{m^3}{sec} = \frac{kNm}{sec} = \frac{kJ}{sec} = kW \right]$

$$= 4.33 \times 101.3 \times 0.166 \times \left\{ (8.88)^{0.3/1.3} - 1 \right\}$$

$$= 4.33 \times 101.3 \times 0.166 \times \{1.655 - 1\}$$

$$\mathbf{IP = 47.71 kW}$$

Example 19.14. *A single stage double acting compressor running at 120 rpm and power input = 75 kW, piston speed = 200 m/min. Suction pressure 1 bar and delivery pressure 10 bar, η_{vol} = 85%. Assuming $PV^{1.25} = C$ for expansion and compression find cylinder bore and V_c as a percent of V_s.*

Data: Double acting; N = 120 rpm $\quad P = 75\ kW$

Piston speed = 2 LN m/min – 200 m/min $\quad \eta_{vol} = 0.85$

$D = ?$ $\quad \frac{V_c}{V_s} = ?$

Solution

Note. N = Number of complete cycles/min
= RPM for single acting
= 2 × rpm for double acting

We know that,

$$\text{Piston speed} = 2LN = 200$$

$$L = \frac{200}{2 \times N} = \frac{200}{2 \times 120}$$

$$L = 0.8333\ \text{m}$$

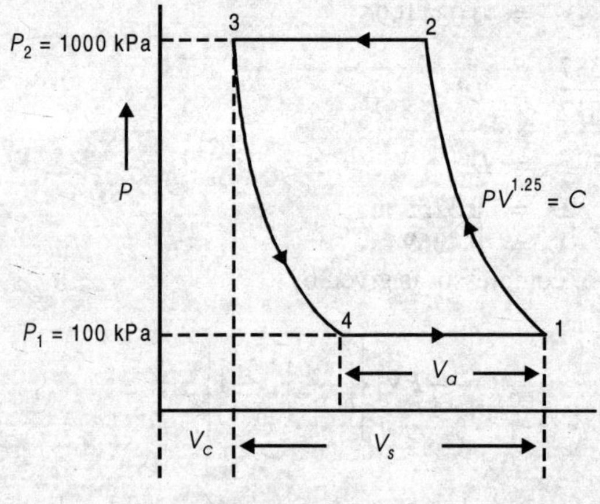

Fig. Ex. 19.14

Also $$\text{Power} = \frac{WN}{60}$$

$$75 = \frac{W \times (2 \times 120)}{60}$$

$$\mathbf{W = 18.75\ kJ/cycle}$$

Also work done per cycle is given by

$$W = \frac{n}{n-1} P_1 V_a \left\{ \left(\frac{P_2}{P_1} \right)^{\frac{n-1}{n}} - 1 \right\}$$

$$18.75 = \frac{1.25}{1.25-1} \times 100 \times V_a \left\{ \left(\frac{10}{1} \right)^{\frac{0.25}{1.25}} - 1 \right\}$$

$\Rightarrow$ $$\mathbf{V_a = 0.064\,m^3 = Actual\ vol.}$$

Then, we know that volumetric efficiency is given by,

$$\eta_{vol} = \frac{V_a}{V_s}$$

$$0.85 = \frac{0.064}{V_s}$$

$$V_s = 0.075\,\text{m}^3 = \frac{\pi}{4} D^2 \times L$$

$$0.075 = \frac{\pi}{4} \times D^2 \times 0.833$$

$$0.075 = D^2 \times 0.65447$$

$$0.11459 = D^2$$

$$\mathbf{0.3385\,m = D}$$

Now to find C, we know that,

$$\eta_{vol} = 1 + C - C\left\{\frac{P_2}{P_1}\right\}^{1/n}$$

or

$$= 1 - C\left\{\left(\frac{P_2}{P_1}\right)^{1/n} - 1\right\}$$

$$\therefore \quad 0.85 = 1 - C\left\{(10)^{1/0.25} - 1\right\}$$

$$0.85 = 1 - C\{5.31\}$$

$$0.85 = 1 - 5.31\,C$$

$$5.31\,C = 1 - 0.85$$

$$C = \frac{0.15}{5.31} = 0.0282$$

or

$$C = \frac{V_c}{V_s} = 0.0282$$

$$\mathbf{V_c = 2.82\,\%\,V_s}$$

Example 19.15. *A single stage, double acting air compressor has a cylinder diameter of 400 mm and a stroke of 300 mm. The clearance volume is 6% of V_s. The intake pressure is 0.95 bar and temperature is 25°C. If the delivery pressure is, 5 bar and speed is 250 rev/min, calculate the mass rate of air compressed. The index of compression and expansion is 1.3 and assume R = 0.287 kJ/kg-K.*

Data: Double Acting, so use N = 2 × rpm

$D = 0.4$ m; $L = 0.3$ m

$V_c = 0.06\,V_s$; $P_1 = 95$ kPa

$T_1 = 25 + 273 = 298$ K

$P_2 = 500$ kPa; $N = 250$ rpm

$m = ?$ kg/sec; $n = 1.3$

and $R = 0.287$ kJ/kg-K for air.

Solution

We know that, $P_1V_1 = mRT_1$

$$\dot{m} = \frac{P_1V}{RT_1} \quad \ldots(1)$$

Also, we know that, stroke volume

$$V_s = \frac{\pi}{4}D^2 \times L = \frac{\pi}{4} \times 0.4^2 \times 0.3$$

$$\mathbf{V_s = 0.037699\,m^3}$$

As $\quad V_c = 0.06\, V_s$

$\therefore \quad V_c = 0.06 \times 0.0377$

$\mathbf{V_c = 0.002262\ m^3}$

$\therefore \quad V_1 = V_c + V_s$

$V_1 = 0.002262 + 0.0377$

$V_1 = 0.0399619\ m^3$

$V_1 = 0.002262 + 0.0377$

$\mathbf{V_1 = 0.0399619\ m^3}$

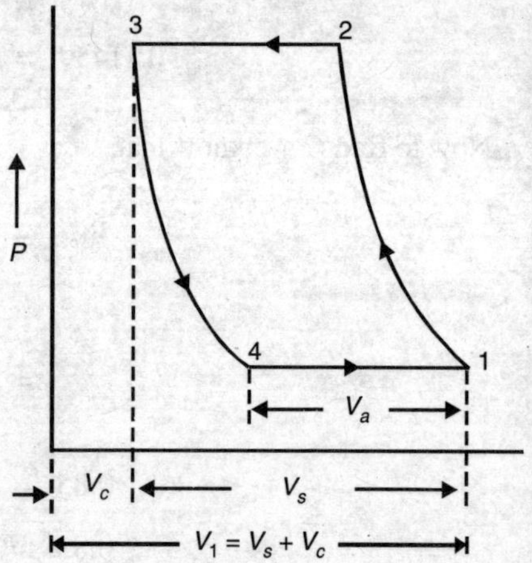

Fig. Ex. 19.15

Now from Eq. (1)

$$\dot{m} = \frac{P_1V_1}{RT_1} = \frac{95\times 0.0399619}{0.287\times 298}$$

$$\mathbf{\dot{m} = 0.0443887\,kg}$$

Now to find mass of air handled in kg/sec

$$\dot{m} = 0.0443887\times\frac{2N}{60}$$

(Double acting)

$$= 0.0443887\times\frac{2\times 250}{60}$$

$$\mathbf{\dot{m} = 0.3699\,kg/sec}$$

(PROBLEMS ON MULTISTAGE AIR COMPRESSORS)

Example 19.16. *A 2-stage reciprocating air compressor handles 2 kg/min of air corresponding to condition of 1 bar and 27°C. It is delivered of 25 bar. The polytropic index of compression is 1.25 and the intercooling is perfect. Neglect in clearance find,*

(1) Intermediate pressure for maximum η.

(2) Power required in kW.

(3) $\eta_{isothermal}$

(4) Saving in power required compared with that for single stage compression.

Data: $m = 2$ kg/min $\quad P_1 = 1$ bar $= 100$ kPa

$T_1 = 300$ K $\quad P_3 = 25$ bar $= 2500$ kPa

$n = 1.25$

Solution

Perfect intercooling

(I) Intermediate pressure P_2

$$= \sqrt{P_1\,P_3} = \sqrt{1\times 25} = 5\,\text{bar} = 500\text{ kPa}$$

(2) Power required with perfect intercooling

$$= \frac{2n}{n-1}\dot{m}RT_1\left\{\left(\frac{P_3}{P_1}\right)^{\frac{n-1}{2n}} - 1\right\}$$

$$= \frac{2\times1.25}{1.25-1}\times\frac{2}{60}\times \times0.287\times300\left\{\left(\frac{2500}{100}\right)^{\frac{0.25}{2.5}-1}\right\}$$

$= 10.89$ kW

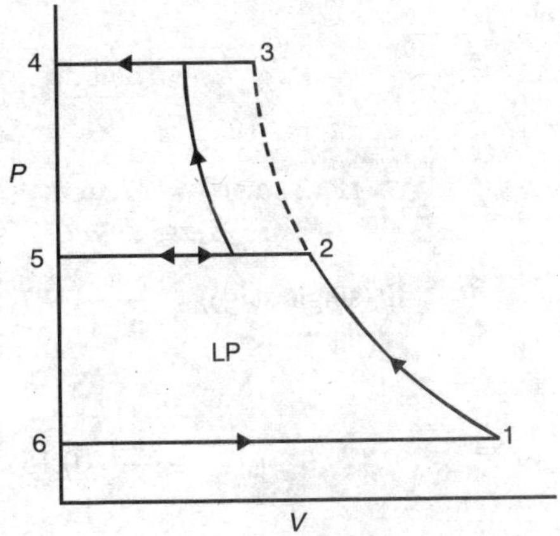

Fig. Ex. 19.16

(3) Isothermal Power input $= P_1V_1 \ln\frac{P_3}{P_1} = \dot{m}RT_1 \ln\frac{P_3}{P_1}$

$$= \frac{2}{60}\times0.287\times 300\ln\frac{2500}{100} = 9.24\,\text{kW}$$

$$\therefore \quad \eta_{iso} = \frac{\text{Isothermal power input}}{\text{Actual work input}} = \frac{9.24}{10.89}$$

$= 0.848$

η_{iso} **= 84.8%**

(4) Power input (single stage)

$$= \frac{n}{n-1}\times\dot{m}RT_1\left\{\left(\frac{P_3}{P_1}\right)^{\frac{n-1}{n}}-1\right\}$$

$$= \frac{1.25}{0.25}\times\frac{2}{60}\times0.287\times300\left\{\left(\frac{2500}{100}\right)^{\frac{0.25}{1.25}}-1\right\}$$

$= 12.96$ kW

Saving in power $= 12.96 - 10.89$

Saving in Power = 2.07 kW

Example 19.17. *A 2 stage single acting reciprocating air compressor draws in air at 1 bar and 300 K. The delivery pressure is 12 bar and intermediate pressure is ideal for minimum work and the intercooling is perfect. The compression follows the law $PV^{1.3} = C$. Flow rate of air through the compressor is 0.15 kg/sec.*

Determine:

(1) Power required to drive the compressor.

(2) Saving in power required to drive the compressor compared with single stage working.

(3) Isothermal η.

Data: P_1 = 1 bar = 100 kPa; T_1 = 300 K

P_3 = 12 bar = 1200 kPa; $PV^{1.3} = C$

m = 0.15 kg/sec

Solution

Since ideal intermediate pressure is there,

$$P_2 = \sqrt{P_1P_3} = 3.46\,\text{bar} = 346 \text{ kPa}$$

With ideal intermediate pressure and perfect intercooling,

$$\text{IP (2 stage)} = \frac{2n}{n-1} mRT_1 \left\{ \left(\frac{P_3}{P_1} \right)^{\frac{n-1}{2n}} - 1 \right\}$$

$$= \frac{2 \times 1.3}{0.3} \times 0.15 \times 0.287 \times 300 \left\{ \left(\frac{1200}{100} \right)^{\frac{0.3}{2.6}} - 1 \right\}$$

IP (2 stage) = 37.16 kW

$$\text{IP (single stage)} = \frac{n}{n-1} mRT \left\{ \left(\frac{P_3}{P_1} \right)^{\frac{n-1}{n}} - 1 \right\}$$

$$= \frac{1.3}{0.3} \times 0.15 \times 0.287 \times 300 \left\{ \left(\frac{1200}{100} \right)^{\frac{0.3}{1.3}} - 1 \right\}$$

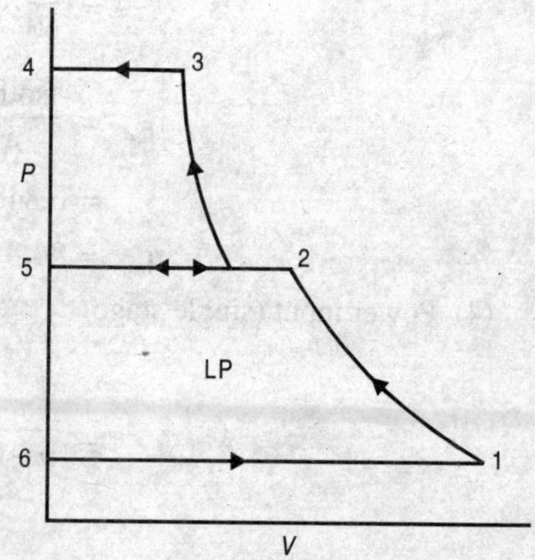

Fig. Ex. 19.17

IP (single stage) = 43.33 kW

Saving in power = 43.33 – 37.16

Saving in Power = 6.17 kW

And Isothermal work $= \dot{m}RT_1 \ln \frac{P_3}{P_1}$

$$= 0.15 \times 0.287 \times 300 \ln \frac{1200}{100} = 32.09 \text{ kW}$$

$$\therefore \quad \eta_{iso} = \frac{\text{Isothermal work}}{\text{IP}} = \frac{32.09}{37.16}$$

$$\boldsymbol{\eta_{iso} = 0.863 = 86.3\% \text{ (for multistage)}}$$

Example 19.18. *A 2 stage single acting air compressor takes in air at 1 bar and 300 K. Air is discharged at 10 bar. The intermediate pressure is ideal and intercooling is perfect. The law of compression is $PV^{1.3} = C$. The rate of discharge is 0.1 kg/sec. Find.*

1. *Power required to drive compressor.*
2. *Saving in work compared with single stage.*
3. *$\eta_{isothermal}$ for multistage and single stage.*
4. *Heat rejected in intercooler.*

Take: R = 0.287 and C_p = 1 kJ/kg-K

Solution

For ideal conditions,

$$P_2 = \sqrt{P_1 P_3} = \sqrt{1 \times 10} = 3.162 \text{ bar}$$

and

$$\text{IP (2 stage)} = \frac{2n}{n-1} \dot{m}RT_1 \left\{ \left(\frac{P_3}{P_1} \right)^{\frac{n-1}{2n}} - 1 \right\}$$

$$= \frac{2\times1.3}{1.3-1}\times0.1\times0.287\times300\left\{\left(\frac{10}{1}\right)^{\frac{1.3\times1}{2\times1.3}}-1\right\} = 22.62\,\text{kW}$$

$$\text{IP single in work} = \frac{n}{n-1}\dot{m}RT_1\left\{\left(\frac{P_3}{P_1}\right)^{\frac{n-1}{n}}-1\right\}$$

$$= \frac{1.3}{0.3}\times0.1\times0.28\times300\left\{\left(\frac{10}{1}\right)^{\frac{0.3}{1.3}}-1\right\} = 26.05\,\text{kW}$$

Saving in work = 26.05 – 22.62

Saving in work = 3.43 kW

(iii) Isothermal work $= m.RT_1 \ln\frac{P_3}{P_1} = 0.1\times0.287\times300\times\ln\frac{10}{1}$

= 19.82 kW

$$\eta_{iso} = \frac{19.82}{22.62} = 0.8765$$

η_{iso} = **87.65% with multistage**

$$\eta_{iso} = \frac{19.82}{26.05} = 0.7617$$

η_{iso} = **76.17% with single stage.**

(4) Heat rejected in intercooler.
Temperature rise after I stage

$$T_2 = T_1\left(\frac{P_2}{P_1}\right)^{\frac{n-1}{n}} = 300(3.162)^{0.3/1.3} = 390.95\,\text{K}$$

The heat lost in intercooler.

$$= \dot{m}C_p(T_2 - T_1) = 0.1\times1\times(390.95-300)$$

= 9.095 kJ/sec

Example 19.19. *A single acting two stage reciprocating air compressor with complete intercoooling delivers 10 kg/min of air at 16 bar pressure. The suction occurs at 1 bar and 15°C. The compression and expansion processes are reversible polytropic with polytropic index n = 1.25. Calculate*

(i) The power required
(ii) The isothermal efficiency
(iii) The free air delivered
(iv) Heat transferred in intercooler

Data: m = 10 kg/min, P_3 = 16 bar,
P_1 = 16 bar, T_1 = 15 + 273
n = 1.25.

Solution

For ideal condition

$$P_2 = \sqrt{P_1 P_3} = \sqrt{1\times 16} = 4\,\text{bar}$$

(i) Power required $= \dfrac{2n}{n-1}\dot{m}RT_1\left\{\left(\dfrac{P_2}{P_1}\right)^{\frac{n-1}{2n}}-1\right\}$

$$= \frac{2\times 1.25}{1.25-1}\times 10\times 0.287(15+273)\left\{(4)^{\frac{1.25-1}{2\times 1.25}}-1\right\}$$

Power = 44 kW

Isothermal work $= mRT_1 \ln\dfrac{P_3}{P_1}$

$$= \frac{10}{60}\times 0.287\times(15+273)\ln\left(\frac{16}{1}\right)$$

$= 38.2$ kW

(ii) Isothermal efficiency $= \dfrac{\text{Isothermal work}}{\text{Actual work}}$

$$\boldsymbol{\eta_{iso} = \frac{38.2}{44} = 0.868 = 86.8\%}$$

(iii) FAD $= \dfrac{mRT_1}{P_1} = \dfrac{10\times 0.287\times 288\times 1000}{1\times 10^5}$

$= 8.266\ \text{m}^3/\text{min}$

(iv) Temperature after end of compression upto intermediate pressure

$$\Rightarrow T_2 = T_1\times\left(\frac{P_2}{P_1}\right)^{\frac{n-1}{n}}$$

$$T_2 = 288\times(4)^{\frac{0.25}{1.25}}$$

$$T_2 = 380\ \text{K}$$

Heat transferred in the intercooler $Q = mC_p(T_2 - T_1)$

$$= \frac{10}{60}\times 1.005\times 1.005(380-288)$$

Q = 15.14kW

Example 19.20. *In a 3 stage compressor air is compressed from 98 kPa to 500 kPa. Calculate for 1m³ of air per sec.*

(*1*) *Work under ideal condition for n = 1.3 ,*
(*2*) *Isothermal work*
(*3*) *Saving in work due to multistaging*
(*4*) *Isothermal efficiency.*

Data: $P_1 = 98\ kPa$ $P_4 = 500\ kPa$
$V = 1\ m^3/sec$

Solution

Work input or IP for 3-stage

$$= \frac{3n}{n-1} P_1V_1 \left\{ \left(\frac{P_4}{P_1} \right)^{\frac{n-1}{n}} - 1 \right\}$$

$$= \frac{3\times1.3}{1.3-1} \times 98 \times 1 \left\{ \left(\frac{500}{98} \right)^{\frac{0.3}{3\times1.3}} - 1 \right\} = \mathbf{170\,kW}$$

For single stage

$$IP = \frac{n}{n-1} P_1V_1 \left\{ \left(\frac{P_4}{P_1} \right)^{\frac{n-1}{n}} - 1 \right\}$$

$$= \frac{1.3}{1.3-1} \times 98 \times 1 \left\{ \left(\frac{500}{98} \right)^{\frac{0.3}{1.3}} - 1 \right\}$$

$$= \mathbf{193.1}$$

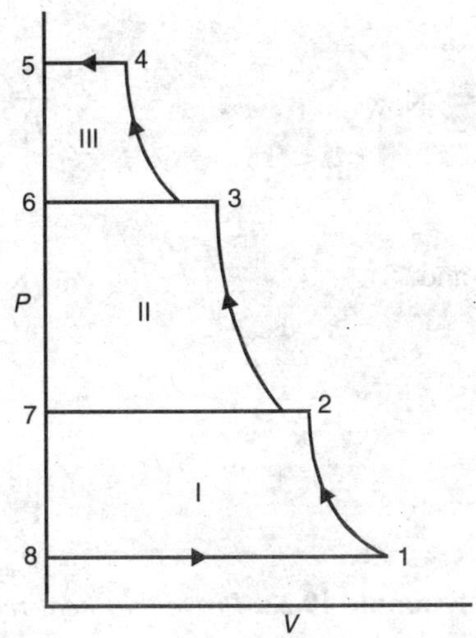

Fig. Ex. 19.20

$\therefore$ **Saving in work 193.1 – 170 = 23.1 kW**

$$\text{Isothermal work} = P_1 V_1 \ln \frac{P_{final}}{P_1} = 90 \times 1 \times \ln \frac{500}{98} = 159.7\,kW$$

$\therefore$ Isothermal efficiency for multistage,

$$\eta_{iso} = \frac{159.7}{170} = 93.89\%$$

$$\eta_{iso} \text{ for single stage} = \frac{159.7}{193} = 82.7\%$$

Example 19.21. *A 3 stage reciprocating air compressor compresses air from 1 bar and 17°C to 35 bar. The law for compression is $PV^{1.25} = C$ for all stages. Assuming perfect intercooling and neglecting clearance find the minimum work required to compress 15 m³/min of free air. Also find intermediate pressures.*

Data: 3-stage

$P_1 = 100\ kN/m^2$ $T_1 = 290\ K$
$P_4 = 3500\ kN/m^2$ $n = 1.25$
$V_1 = 15\ m^3/min$

Solution

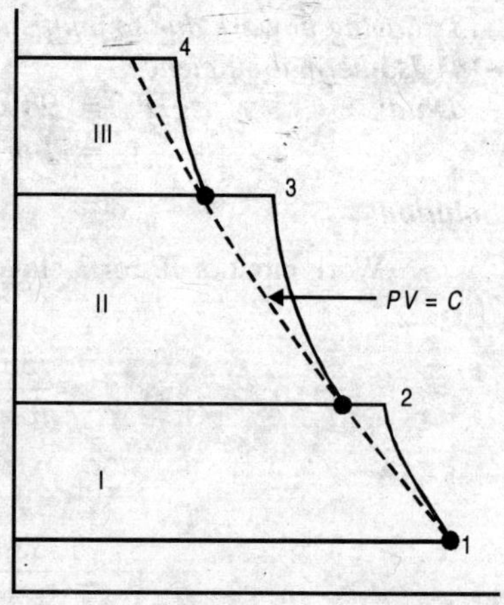

Fig. Ex. 19.21

$$\text{IP (3 stage)} = \frac{3n}{n-1} P_1 V_1 \left\{ \left(\frac{P_4}{P_1} \right)^{\frac{n-1}{3n}} - 1 \right\}$$

$$= \frac{3 \times 1.25}{1.25 - 1} \times 100 \times \frac{15}{60} \left\{ \left(\frac{3500}{100} \right)^{\frac{0.25}{3 \times 1.25}} - 1 \right\}$$

$$= 100.29 \text{ kJ/sec or kW}$$

Now $$\frac{P_2}{P_1} = \frac{P_3}{P_2} = \frac{P_n}{P_3} = k$$

and $$K = \left(\frac{P_{x+1}}{P_1} \right)^{1/x}$$

$$= (\text{Pressure Range})^{1/x} = \left(\frac{3500}{100} \right)^{\frac{1}{3}} = 3.27$$

$\therefore$ $\mathbf{P_2 = kP_1 = 3.27\ bar}$

$\mathbf{P_3 = P_1 k^2 = 10.674\ bar}$

Example 19.22. *Prove that heat rejected per stage per kg of air in a reciprocating air compressor with perfect intercooling is given by* $\left[C_p + C_v \left(\frac{\gamma - n}{n - 1} \right) \right] (T_2 - T_1)$ *where* $(T_2 - T_2)$ *is the temperature rise during each stage.*

Solution

Total heat rejected in a reciprocating compressor

= Heat rejected during compression + Heat rejected during intercooling ...(1)

Now heat rejected during compression per kg,

$$Q\ (\text{poly}). = \frac{\gamma - n}{\gamma - 1} \times W_{\text{poly}}/\text{kg (W.D. in a poly compression process)}$$

$$= \frac{\gamma - n}{\gamma - 1} \times \frac{P_2 V_2 - P_1 V_1}{n - 1}$$

$$= \frac{\gamma - n}{\gamma - 1} \times \frac{R(T_2 - T_1)}{n - 1} \quad ...(2)$$

And heat rejected during intercooling/kg of air

$$= C_p (T_2 - T_1)$$

Total heat rejected = (2) + (3)

$$= \frac{\gamma - n}{\gamma - 1} \times \frac{R(T_2 - T_1)}{n-1} + C_p (T_2 - T_1)$$

But $$R = C_p - C_v \text{ and } \gamma = \frac{C_p}{C_v}$$

$\therefore$ $$R = \gamma C_v - C_v$$

$$R = C_v(\gamma - 1)$$

$\therefore$ $$C_v = \frac{R}{\gamma - 1}$$

$\therefore$ Total heat rejected/kg/stage

$$= \frac{\gamma - n}{n-1} \times C_v (T_2 - T_1) + C_p (T_2 - T_1)$$

$$= \left\{ C_p + C_v \left(\frac{\gamma - n}{n-1} \right) \right\} (T_2 - T_1)$$

Example 19.23. *A 4 stage compressor works between limits of 1 bar and 115 bar. The index of compression in each stage is 1.29. The temperature at the start of compression in each stage is 35°C and the intermediate pressure are so chosen that the work is divided equal among the stages ($\therefore$ pressure ratio same). Neglecting clearance find,*

(1) Pressures P_2, P_3 and P_4
(2) Delivery temperature in each stage.
(3) Individual work/kg and
(4) Isothermal work and, $h_{isothermal}$

Data: $P_1 = 1$ bar, $n = 1.29$, $P_5 = 115$ bar, $T_1 = 35°C = 308$ K

Solution

For ideal intermediate pressure

$$\frac{P_2}{P_1} = \frac{P_3}{P_2} = \frac{P_4}{P_3} = \frac{P_5}{P_4} = k$$

$\therefore$ $$P_2 = kP_1$$

$$P_3 = kP_2 = k^2 P_1$$

$$P_{x+1} = k^x P_1$$

i.e., $$P_5 = k^4 P_1$$

$$\frac{P_5}{P_1} = k^4$$

$$\left(\frac{P_5}{P_1}\right)^{1/4} = k; \left(\frac{115}{1}\right)0.25 \quad \Rightarrow k = 3.27$$

$$P_2 = 327.47 \text{ kPa}$$

$$P_3 = 1072.38 \text{ kPa}$$

$$P_4 = 3511.74 \text{ kPa}$$

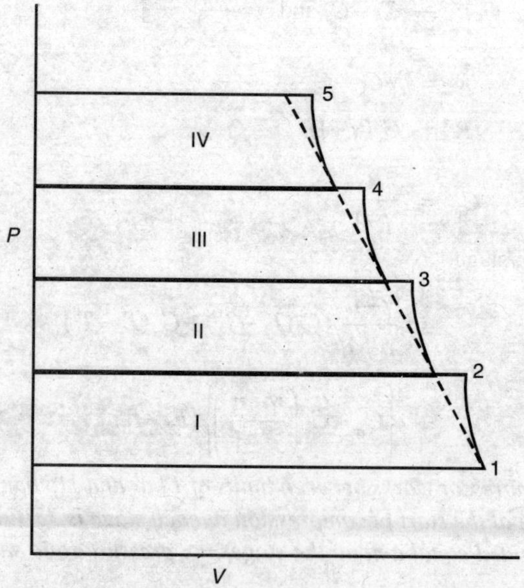

Fig. Ex. 19.23

Since compression is polytropic,

$$\frac{T_2}{T_1} = \left(\frac{P_2}{P_1}\right)^{\frac{n-1}{n}}$$

$$T_2 = 308(3.27)^{0.29/1.29}$$

$$= 401.39 = 402 \text{ K}$$

Delivery temperature from each stage.

$$\text{IP (4 stage)} = \frac{4n}{n-1} 1.RT_1 \left[\left(\frac{P_5}{P_1}\right)^{\frac{n-1}{4n}} - 1\right]$$

$$= \frac{4 \times 1.29}{0.29} \times 0.287 \times 308\{(115)^{0.29/5.26} - 1\}$$

$$= 480.7 \text{ kJ/kg}$$

$$\text{Isothermal work} = mRT_1 1\text{n} \frac{P_5}{P_1} \text{ and } m = 1 \text{ kg (for kJ/kg)}$$

$$= 0.287 \times 3081\text{n}\left(\frac{115}{1}\right)$$

$$= 419.43 \text{ kJ/kg}$$

$$\therefore \text{ Isothermal efficiency} \quad = \eta_{iso} = \frac{419.45}{480.7}$$

$$\boldsymbol{\eta_{iso} = 0.872 = 87.2\%}$$

Example 19.24. *A single acting, two stage RAC running at 280 rpm delivers air at a pressure of 18 bar, and the conditions at the commencement of compression being 0.98 bar and 305 K.*

The intermediate pressure is 4 bar and the clearance volume of LP cylinder is 5% of stroke volume (swept volume). Equation for compression and expansion for each cylinder is $PV^{1.25} = C$. The compressor has a free air capacity of 2.25 m³/min measured under free conditions of 1 bar and 290 K. Take mass of air handled as 2.7 kg/min.

Determine

(i) η_{vol} *of L.P. cylinder.*

(ii) *Theoretical work spent in driving the compressor.*

(iii) *Dimension of L.P. cylinder if L = D*

The given compressor has a perfect intercooling.

Data: $V_c = 0.05\, V_s$ *for L.P. cylinder*

$V_f = 2.25 \text{ m}^3/\text{min}$ $\qquad P_f = 100 \text{ kPa}$

$T_f = 290 \text{ K}$ $\qquad \dot{m} = 2.7 \text{ kg/min}$

Solution

(i) Now volumetric efficiency of L.P. cylinder

$$\eta_{vol}(\text{L.P.}) = 1 - C\left\{\left(\frac{P_2}{P_1}\right)^{\frac{1}{1.25}} - 1\right\}$$

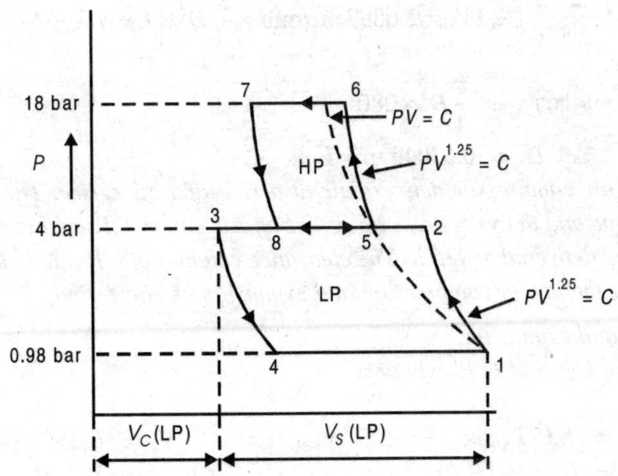

Fig. Ex.19.24

$$= 1-0.05\left\{\left(\frac{4}{0.98}\right)^{\frac{1}{1.25}}-1\right\} = 0.896 \text{ and } 89.6\%$$

(ii) Theoretical work input, for 2-stage air compressor, we know that,

$$\text{IP (2 stage)} = \frac{2n}{n-1}\dot{m}RT_1\left\{\left(\frac{P_3}{P_1}\right)^{\frac{n-1}{2n}}-1\right\}$$

$$= \frac{2\times1.25}{1.25-1}\times\frac{2.7}{60}\times0.287\times305\left\{\left(\frac{18}{0.98}\right)^{\frac{1.25-1}{2\times1.25}}-1\right\}$$

$$= 13.307 \text{ kW}$$

(iii) We also know that,

$$\frac{P_fV_f}{T_f} = \frac{P_1(V_1-V_4)}{T_1}$$

$$\frac{P_fV_f}{T_f}\times\frac{T_1}{P_1} = (V_1-V_4)$$

$$\frac{100\times2.25\,\text{m}^3/\text{min}}{210}\times\frac{305}{0.98\times100} = (V_1-V_4)$$

$$2.41\,\text{m}^3/\text{min} = (V_1-V_4)$$

$$\eta_{\text{vol LP}} = \frac{(V_1-V_4)}{V_s} = 0.896$$

$$\frac{2.41}{2.896} = V_s = 2.6897\,\text{m}^3/\text{min} = \frac{\pi}{4}D^2\times L\times N$$

$$2.6897 = \frac{\pi}{4}D^3\times280$$

$$\mathbf{D = 0.23040\ m = L}$$

Example 19.25. *A 2 stage air compressor delivers air at a rate of 1.35 kg/sec. The suction pressure is 1 bar, intermediate pressure is 7 bar and delivery pressure is 42 bar. Air enters L.P. cylinder at 290 K is intercooled to 305 K before inlet to H.P. and delivered at 415 K. The clearance volumes of L.P. and H.P. cylinders are 6% and 8% of stroke volumes. Assuming the law of compression and expansion be same, find,*

(i) Index of compression and expansion
(ii) Volumetric efficiency of L.P. and H.P. cylinders
(iii) Total power

Data:

$m = 1.35$ kg/sec $\qquad V_c = 0.06\ V_s$ (L.P)

$V_c = 0.08\ V_s$ (H.P) $\qquad V_{s\,LP} = V_1 - V_3$

$V_{s\,HP} = V_5 - V_7$

Solution

For the process 5–6

$$\frac{T_6}{T_5} = \left(\frac{P_6}{P_5}\right)^{\frac{n-1}{n}}$$

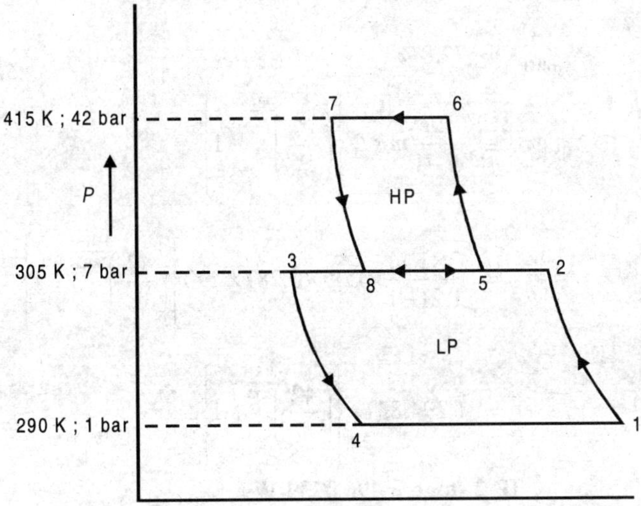

Fig. Ex. 19.25

$$\frac{415}{305} = \left(\frac{42}{7}\right)^{\frac{n-1}{n}}$$

$$1.36 = (6)^{n-1/n}$$

$$\ln 1.36 = \frac{n-1}{n}\ln 6$$

$$0.171 = \frac{n-1}{n}$$

$\therefore$ **n = 1.21** (We can get this value by considering process 1–2 also)

$$\eta_{vol}(\text{L.P.}) = 1 - C\left\{\left(\frac{P_2}{P_1}\right)^{1/n} - 1\right\}$$

$$= 1 - 0.06\left\{\left(\frac{7}{1}\right)^{1/n} - 1\right\}$$

$$= 1 - 0.06\left\{\left(\frac{7}{1}\right)^{1/n} - 1\right\} = 0.7603 \Rightarrow 76.03\%$$

$$\eta_{volHP} = 1 - C\left\{\left(\frac{P_3}{P_2}\right)^{1/n} - 1\right\}$$

$$= 1 - 0.08\left\{\left(\frac{42}{7}\right)^{\frac{1}{1.21}} - 1\right\} = 0.728$$

$$\eta_{volHP} = \mathbf{72.8\%}$$

$$\text{IP 2 stage} = \frac{2n}{n-1}\dot{m}RT_1\left\{\left(\frac{P_3}{P_2}\right)^{\frac{n-1}{2n}} - 1\right\}$$

$$= \frac{2\times1.21}{1.21-1}\times1.35\times0.287\times290\left\{\left(\frac{42}{1}\right)^{\frac{1.21-1}{2\times1.21}} - 1\right\}$$

$$= 1.294.831\times\left\{\left(\frac{42}{1}\right)^{\frac{0.21}{2.42}} - 1\right\}$$

IP 2 stage = 496.073 kW

THEORY QUESTIONS

19.1. State the uses of compressed air.

19.2. Explain classification of air compressor.

19.3. Explain: (i) Volumetric efficiency, (ii) Isothermal efficiency, (iii) FAD.

19.4. Obtain an expression for volumetric efficiency of an air compressor in terms of clearance ratio index of compression and pressure ratio.

19.5. Explain the reasons for low volumetric efficiency of a reciprocating air compressor.

19.6. Discuss the reasons for multistage compression. Also explain the additional benefits of multistage compression, if any.

19.7. Bring out the factors that influence the volumetric efficiency of a reciprocating. air compressor.

19.8. What are the advantages of multistage compression ?

19.9. Explain the working of a two stage reciprocating air compressor with intercooler.

19.10. Prove that the volumetric efficiency of air compressor is given by,

$$\eta_{vol} = 1 + C - C\left[\frac{P_d}{P_s}\right]^{1/n}$$

where C = Clearance ratio, P_d = Discharge pressure, P_s = Suction pressure.

19.11. Explain the effect of clearance on volumetric efficiency.

19.12. Explain the methods to improve the isothermal efficiency of air compressor.

19.13. Obtain an expression for IP of single stage air compressor.

19.14. Derive the expression for the volumetric efficiency of a reciprocating air comp. in terms of clearance ratio; pressure ratio; and index of compression.

19.15. Explain why staging is done in reciprocating compression.

19.16. Explain clearly how the actual indicator design for a single stage reciprocating compression in different than theoretical indicator design.

19.17. What are the advantage of multistage compressor ?

19.18. Explain the reasons for low volumetric efficiency of reciprocating air compressor.

19.19. Obtain an express for how volumetric efficiency of reciprocating air compressor.

19.20. State difference uses of compress air.

19.21. Why are multistage lamp are preferred in single stage ? Explain.

19.22. When is multistage compressor used for air ? What are its advantage ?

19.23. With the help of *PV* design, explain how work is seemed by a 2 stage compressor in the same pressure limit. Write expansion for amount of work seemed because of 2-state opm. State ending for major maximum work saving.

19.24. Explain clearly how actual *P–V* diagram for a single stage air compressor deviates from the theoretical diagram.

PROBLEMS FOR PRACTICE

1. A 2 stage single acting air compressor takes in air at 1 bar and 300 K. Air is discharged at 10 bar. The intermediate pressure is ideal and intercooling is perfect. The law of compression is $PV^{1.3} = C$. The rate of discharge is 0.1 kg/sec. Find

(i) Power required to drive the compressor. (**Ans.** 22.62 kW)

(ii) Saving in work compared with single stage. (**Ans.** Saving in work = 3.43 kW)

(iii) Isothermal efficiency for single and multistage.

(**Ans.** Iso η_{single} = 76.17°%, Iso η_{mult} = 87.65%)

2. A single acting single stage compressor is belt driven from an electric motor at 400 rpm. The cylinder diameter is 15 cm and stroke is 17.5 cm. The air is compressed from 1 bar to 7 bar and law of compression is $PV^{1.3} = C$. Find power of motor, if transmission efficiency is 97%, and η_{mech} = 90%. Neglect clearance effects.

CHAPTER 20

Rotary Air Compressors

20.1 INTRODUCTION

Compressor is a device which compresses air/gases or vapours from low pressure to high pressure. It needs external energy input in the form of work. Out of the total work input to the compressor, some work is utilised to compress the fluid while the remaining is lost in overcoming friction, some work is lost to cooling medium, etc.

Compressors are mainly grouped into two categories:

(i) Positive displacement compressor and

(ii) Dynamic compressors.

In a positive displacement compressor, the pressure of the gas is increased by reduction of its volume i.e., by positive displacement of the gas to the delivery side.

While in a dynamic compressor, the kinetic energy imparted to the gas by the rotation of the rotor (impeller) is converted into pressure energy partly in the rotor and the remaining in the diffuser. Thus, the pressure rise is developed due to the dynamic action of gas.

In the subsequent article, various types of compressors are listed. Compressors are extensively used for a variety of applications (Refer Sec. 20.15).

20.2 CLASSIFICATION OF COMPRESSORS

Compressors are classified on the basis of several criteria as follows.

(a) On the basis of design and principle of operation. Based on the design and principle of operation, compressors are classified into two main groups and further sub-divided into sub-groups as shown in flow chart given on next page.

(b) According to delivery pressure. Low pressure (upto 10 bar), medium pressure (10 – 80 bar), high pressure (80 – 1000 bar).

Hyper-compressors are multistage reciprocating compressors with delivery pressure upto 1000 bar.

(c) According to the pressure ratio (as per ASME code). Fans — Pressure ratio upto 1.1

Blowers — Pressure ratio between 1.1 to 2.3

Compressor — Pressure ratio above 2.3

(d) According to free air delivered (capacity). Small capacity — Upto 9 m^3/min

Medium capacity — Between 9 – 3000 m^3/min

Large capacity — Above 3000 m^3/min

(e) According to number of stages adopted. Single stage, Multi-stage

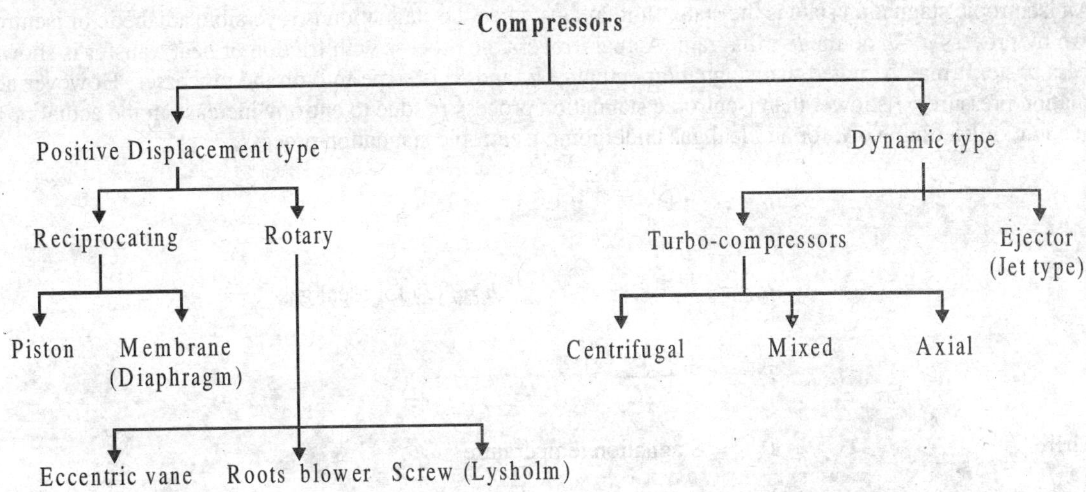

(f) According to the drive (prime mover). Electric motor driven, I.C. engine driven.

The present chapter deals with Rotary positive displacement compressors and dynamic compressors.

20.2.1 Stagnation Properties of a Flowing Fluid

When a flowing fluid having same velocity is brought to rest, it is said to have attained a stagnation state. The final stagnation state is governed by the manner in which it is attained. A reversible adiabetic or isentropic process has a great significance.

For an isentropic stagnation process, the steady flow energy equation simplifies to

$$h + \frac{V^2}{2} = h_o, \text{ kJ/kg}$$

where h_o is stagnation enthalpy and h is its initial enthalpy when fluid is flowing with a velocity of V m/s.

During the stagnation process, the kinetic energy of the fluid is converted to enthalpy, thereby pressure and temperature of the fluid increases as shown below.

The properties of the fluid at the stagnation state are referred as stagnation properties e.g., stagnation pressure, stagnation temperature, stagnation density, etc. The stagnation state and stagnation properties are denoted by lower suffix o.

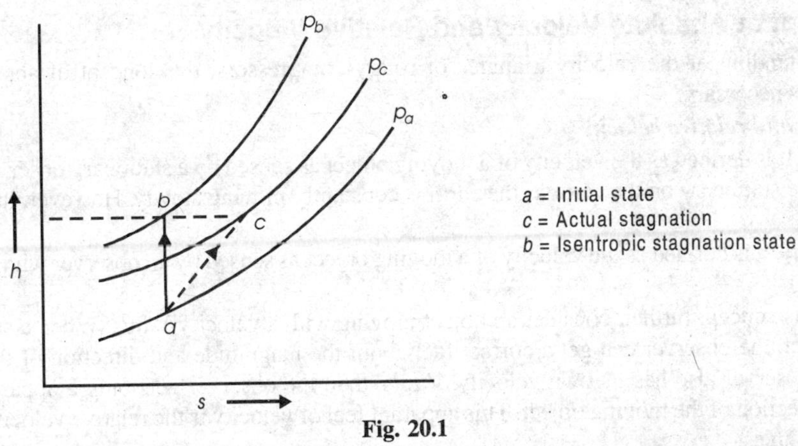

Fig. 20.1

An isentropic stagnation state is the state, fluid attains when the stagnation is reversible adiabetic or isentropic, shown by process $a - b$ on the h–s diagram. Actual irreversible process with friction or heat transfer is shown by process $a - c$. It may be noted that stagnation enthalpy h_b and h_c is same in both the processes. However actual stagnation pressure p_c is lower than isentropic stagnation process p_b, due to entropy increase in the actual process on account of friction. Now, for an ideal gas undergoing isentropic stagnation process,

$$h_o = h + \frac{V^2}{2}$$

$$\therefore \quad C_p T_o = C_p T + \frac{V^2}{2} \qquad h = f(T) \text{ for ideal gas}$$

$$\therefore \quad T_o = T + \frac{V^2}{2C_p}$$

where T_o = Stagnation temperature

$\frac{V^2}{2C_p}$ = Dynamic temperature

Now, stagnation pressure p_o can be found as

$$p_o = p_b = p_a \left(\frac{T_o}{T_1}\right)^{\frac{\gamma-1}{\gamma}}$$

Also, we have,

$$\frac{p_o}{p_a} = \frac{p_b}{p_a} = \left(\frac{T_o}{T_1}\right)^{\frac{1}{\gamma-1}}$$

Similarly, steady flow energy equation can be written as

$$q - w_s = \left(h_{o_b} - h_{o_a}\right) + g\left(z_b - z_a\right)$$

$$= C_p\left(T_{o_b} - T_{o_a}\right) + g\left(z_b - z_a\right)$$

where T_{o_b} and T_{o_a} are stagnation temperatures.

20.2.2 Concept of Absolute Velocity and Relative Velocity

For better understanding of the velocity triangles of rotary compressors, the concept of absolute velocity and relative velocity is necessary.

Absolute velocity and relative velocity

Absolute velocity. It is defined as the velocity of a moving object as sensed by a stationary observer. In a true sense, no observer can be stationary on the earth as the earth is constantly moving slowly. However, its slow motion can be ignored.

Relative velocity. It is defined as the velocity of a moving object as sensed by an observer who is moving with its own velocity.

To illustrate this concept further, consider an object moving with a velocity V m/s. When a stationary observer looks at this object, the observer can get a correct feel about the magnitude and direction of the moving object. However, if the observer also has its own velocity, V m/s, then the observer gets only a apparant feel about the magnitude and direction of the moving object. This apparant feel of velocity is the relative velocity. Let us consider the following examples.

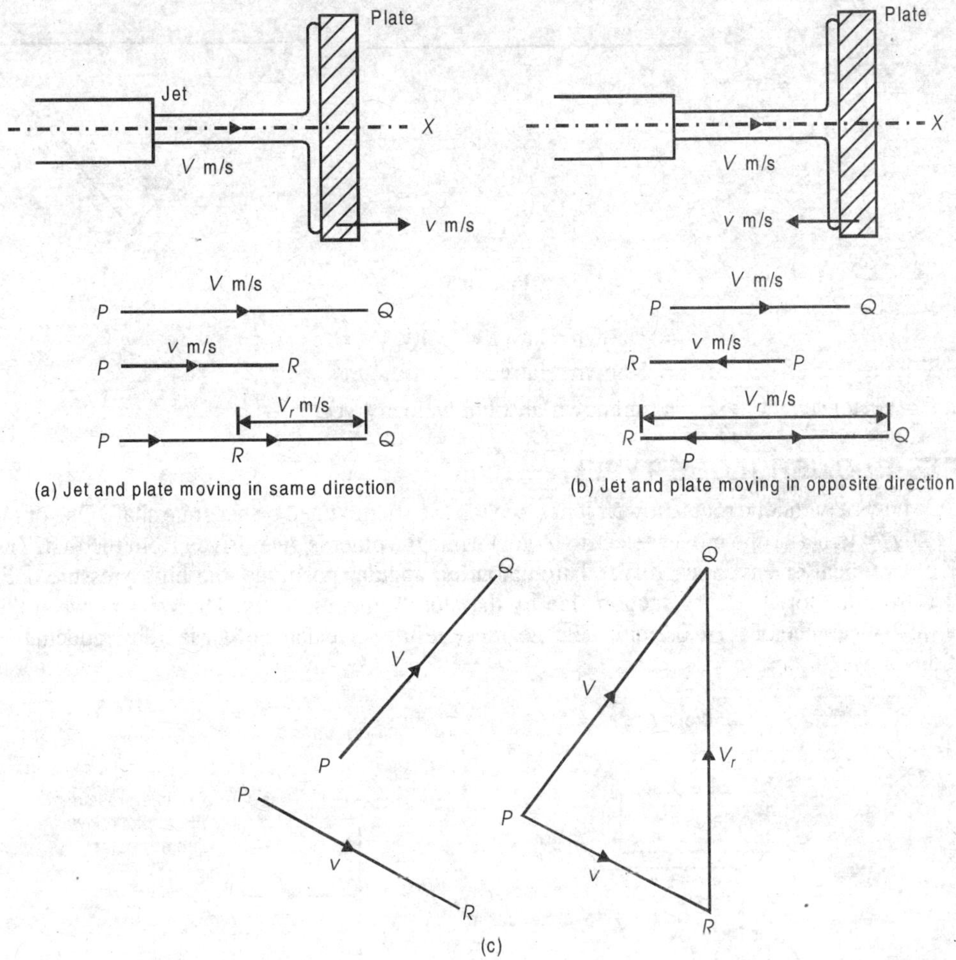

Fig. 20.2

Here, the relative velocity V_r is the vector difference of the two absolute velocity vectors. Figure 20.2 (b) shows the two moving objects moving in opposite direction. The relative velocity V_r is the vector sum of the two absolute velocity vectors. In general, the procedure to obtain the relative velocity between the two objects moving in their own directions can be stated as follows.

From a common starting point, set both absolute velocity vectors in their magnitude and direction. The line joining the ends of the two absolute velocity vectors represents the relative velocity which is the vector difference of the known absolute velocity vectors, see Fig. 20.2 (c).

Similarly, when a relative velocity and one of the absolute velocities are known, the unknown absolute velocity is given by the vector sum of the known relative velocity and known absolute velocity. The graphical procedure is as follows.

Draw the known relative velocity vector to some scale, in its magnitude and direction. In succession i.e., starting from the end of the relative velocity vector draw to same scale, the known absolute velocity vector in its magnitude and direction. Then the line joining the starting point to end point, representing vector sum, gives the unknown absolute velocity, as shown in Fig. 20.3.

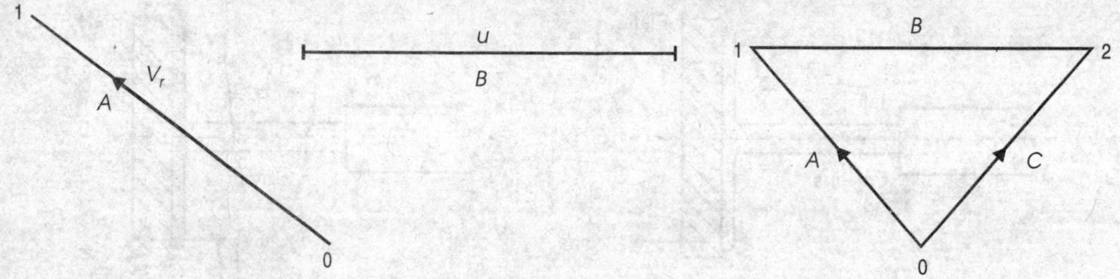

Fig. 20.3

$$A = \text{Known relative velocity, } V_r\text{, m/s}$$
$$B = \text{Known relative velocity, } u\text{, m/s}$$
$$\text{Then } C = \text{Vector sum } (V_r + u) = \text{unknown absolute velocity vector}$$

20.3 ROOTS BLOWER (LOBE TYPE)

Figure 20.4 (a) shows a twin lobe roots blower. It has two lobes each mounted on separate shaft. One of these shafts is connected to the external prime mover (electric motor) while the other is gear driven from the first. The lobes of the rotor are of cycloidal or envolute profile. Throughout all angular positions, the high pressure delivery side remains sealed from the low pressure suction side by the closely mating lobes. The wear between the lobes is avoided by a small clearance of 0.1 – 0.2 mm. The clearance results in leakage of air thereby reducing volumetric efficiency of the blower.

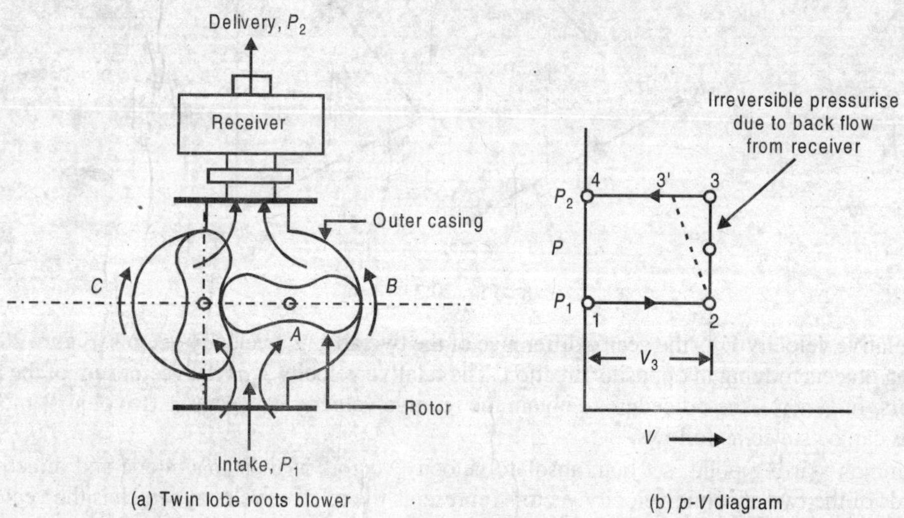

Fig. 20.4

Suction takes place through the intake port. The entrapped air between the lobes and casing is carried forward during the rotation and is finally discharged to the delivery port. There is no change in the flow area and no reduction of volume of air. However, when the delivery port is open, blower discharges air into the high pressure reservoir causing irreversible pressurerise as shown in the *p-V* diagram [Fig. 20.4 (b)]. The dotted line in *p-V* diagram shows compression process of a reciprocating compressor. The area represents excess work required due to irreversible pressure rise.

Let,

V = Volume of entrapped air in space A i.e., lobe face and casing, m^3

p_1 = Suction pressure, kPa

p_2 = Delivery pressure, kPa

N = Speed, r.p.m.

Volume of Free Air delivered/revolution = $4V$

$$\therefore \quad \text{Work required/rev.} = (p_2 - p_1) \times 4V \text{ kJ/rev.} \qquad \ldots(1)$$

$$\therefore \quad \text{Power required} = (p_2 - p_1) \times 4V \times \frac{N}{60} \text{ kW}$$

$$\text{Further isentropic work reqd./s} = \frac{\gamma}{\gamma - 1} p_1 \times 4V \left[\left(\frac{p_2}{p_1} \right)^{\frac{\gamma-1}{\gamma}} - 1 \right] \times \frac{N}{60} \text{ kW} \qquad \ldots(2)$$

Roots efficiency η_{root} is defined as the ratio of isentropic work required to the actual work required.

$$\therefore \quad \eta_{roots} = \frac{\frac{\gamma}{\gamma-1} \times p_1 4V \left[\left(\frac{p_2}{p_1} \right)^{\frac{\gamma-1}{\gamma}} - 1 \right]}{(p_2 - p_1) \times 4V}$$

$$= \frac{\frac{\gamma}{\gamma-1} \times p_1 \times 4V \left[\left(\frac{p_2}{p_1} \right)^{\frac{\gamma-1}{\gamma}} - 1 \right]}{p_1 (r_p - 1) \times 4V} = \frac{\frac{r}{r-1} \left[r_p^{\frac{\gamma-1}{\gamma}} - 1 \right]}{r_p - 1}$$

where

$$r_p = \text{pressure ratio} = \frac{p_2}{p_1}$$

$$\therefore \quad \eta_{roots} = \frac{C_p}{R} \frac{\left[r_p^{\frac{\gamma-1}{\gamma}} - 1 \right]}{r_p - 1} \qquad \ldots(3)$$

From Eq. (3), it is seen that with increase of pressure ratio r_p, roots efficiency η_{roots} decreases.

Roots blower have capacity ranging between 0.14 m^3/min to 1400 m^3/min and can develop pressure ratio of 2 in a single stage and of 3 in two stage.

The roots (lobes) are made of steel and are finished with close precision. Extremely fine clearances are required to be maintained during their manufacture and assembly.

20.4 ROTARY VANE COMPRESSOR

This compressor is also called as 'sliding vane compressor'. Figure 20.5 shows a multi-vane rotary compressor. It consists of a rotor eccentrically housed in the casing. Rotor has several radial slots in it, each housing a spring loaded vane. These vanes are made of steel or synthetic fibrous material. Larger the number of vanes, internal leakage of air decreases due to small pressure difference prevailing between the adjacent spaces around the rotor.

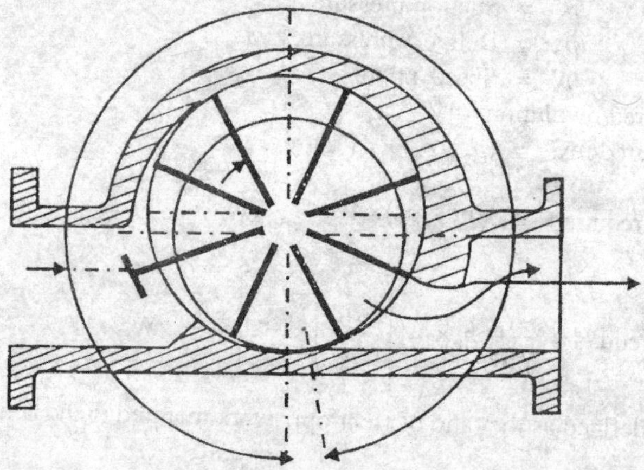

Fig. 20.5. *Vanes compressors.*

High pressure ratio requires large number of vanes (20 – 30). The casing has intake and delivery openings. These compressors are often used for capacities upto 150 m^3/min and for pressure ratios upto 8.5. For a given pressure ratio and FAD, vane compressor requires less work input than that for roots blower.

When the rotor rotates, vanes are driven out of the rotor towards the casing due to centrifugal force. The space between the two adjacent vanes, rotor and the casing increases creating vacuum. Thus, the gas is drawn in, from the suction opening. When the rotor crosses the point just opposite to its eccentricity, suction starts. As the rotor continues to rotate, the entrapped gas is compressed due to reduction in volume. The high pressure gas is then discharged through delivery opening. The *p-V* diagram of vane compressor is shown in the Fig. 20.6. Usually, half of the total pressure rise is developed during the internal reversible compression and the remaining pressure rise occurs irreversibly when the entrapped gas is released to the delivery side, due to back flow of high pressure air from the receiver.

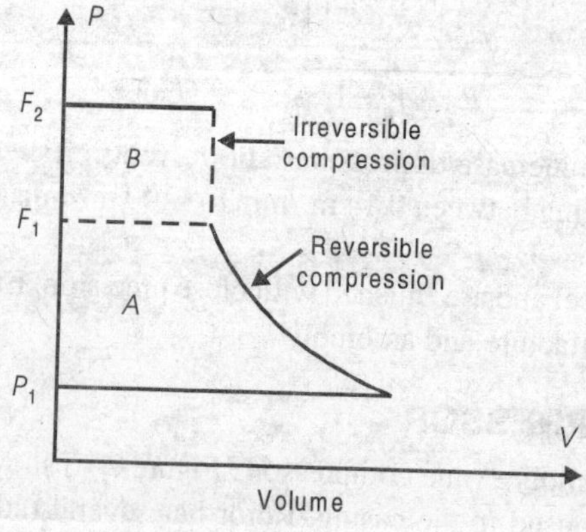

Fig. 20.6. *p-V diagram.*

Due to uniform torque, the compressor does not require any flywheel and is light weight.

Let, ε = eccentricity of rotor, m

L = axial length of rotor, m
t = thickness of each vane, m
D = diameter of rotor, m
n = number of vanes

$\therefore$ Swept volume/revolution $= 2L \times \varepsilon\,(\pi D - nt)$ m^3/rev. ...(4)

20.5 SCREW (LYSHOLM) COMPRESSOR

Figure 20.7 shows a screw (Lysholm) compressor. It consists of two rotors meshing with each other closely. Each rotor is helical with large pitch. The male rotor has three or four lobes while the female rotor has four, five or six recesses. The entrapped air is continuously compressed as it flow axially, due to narrowing of passages formed by the rotors. The inlet and discharge openings are somewhat oblique as shown in the Fig. 20.7.

The cycle of working of screw compressor consists of four processes namely:

(i) Suction. When the meshing rotors begins to disengage near the suction opening, gas is drawn in due to low pressure created. This is continuous so long as rotors unmesh.

(ii) Transportation. As the rotation of the rotor continues, the space filled with gas isolates from suction opening. This space is then gradually transported around the outer periphery at constant pressure until the male lobe starts meshing with the space being transported.

(iii) Compression. When rotors remesh from suction end gradually, the entrapped gas is compressed.

(iv) Delivery. Soon after the space filled with entrapped gas is exposed to discharge opening, the gas is driven out into the delivery line due to continuous penetration of male rotor into the space.

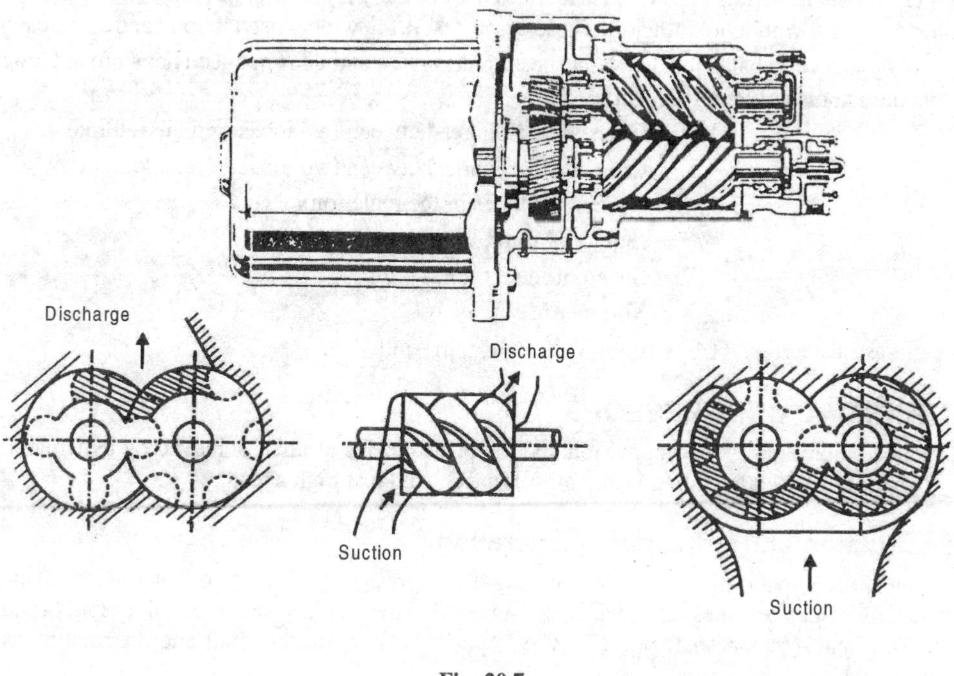

Fig. 20.7

The compressor is run at very high speeds, 3000 r.p.m. for large compressors and upto 32000 r.p.m. for small units. Rotors have peripheral velocity of 50 m/s to 100 m/s. The rotors made of Carbon steel need precision machining. The variation of volumetric efficiency and adiabetic efficiency versus pressure ratio is shown in Fig. 20.8. High pressure ratio increases internal leakage of gas back to suction opening and reduces volumetric efficiency.

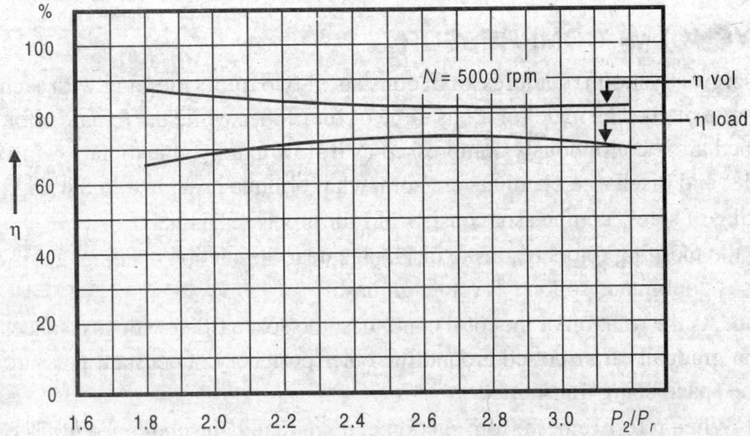

Fig. 20.8. *Variation of efficiencies.*

These compressors have capacity ranging between 3 m^3/min to 1000 m^3/min and pressure ratios of 4 (in single stage) and 8 – 11 (in double stage). The compressor does not need any lubrication, except for bearings. Thus, there is no contamination of gas with lubricating oil. These compressors are widely used in food and chemical industries.

The rotors are dynamically balanced, hence, it runs vibration free and does not need heavy foundation. Its main drawback is operating noise of high frequency.

Let,

A_A = Cross-sectional area between two lobes of male rotor, m^2

A_B = Cross-sectional area between two lobes of female rotor, m^2

n_A = Number of lobes in the male rotor

L = Length of rotor, m

N_A = Speed of male rotor, r.p.m.

η_{vol} = Volumetric efficiency

$\therefore$ Volume of gas sucked/rev. = $(A_A + A_B)\, n_A\, L\, N_A \times \eta_{vol}$, m^3/min. ...(5)

20.6 CENTRIFUGAL COMPRESSORS

This is a dynamic compressor. The compression and the pressure rise of air is achieved by dynamic action. The compressor is often directly coupled to a prime mover and is driven at high speed.

20.6.1 Construction and Principle of Operation

Figure 20.9 shows a schematic diagram of a centrifugal compressor. It consists of inlet (suction) pipe (1), an impeller (rotor) with blades or vanes (2), a vaned or vanless diffuser (3), a volute or scroll casing (4), an outlet or delivery pipe (5), a shaft (6) and shaft seal (7). The impeller is keyed to the shaft and the rotating assembly is dynamically balanced.

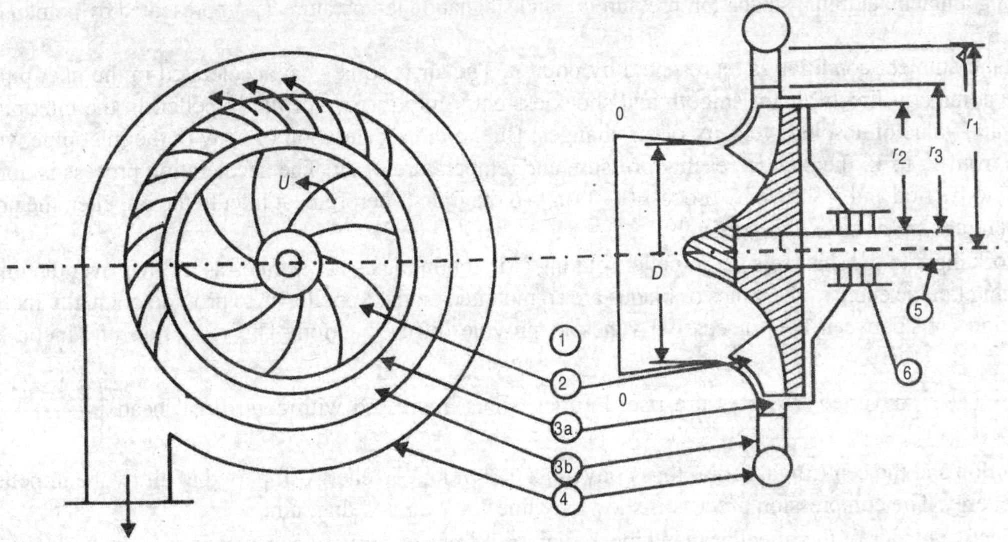

1. Inlet pipe, 2. Impeller, 3a. Vaneless diffuser, 3b. Vaned diffuser, 4. Volute casing, 5. Shaft, 6. Sealing.

Fig. 20.9. *Schematic diagram of a centrifugal compressor.*

The thermodynamic process of compression and pressure-rise over the rotor and in the diffuser is shown on *h–s* diagram in Fig. 20.10.

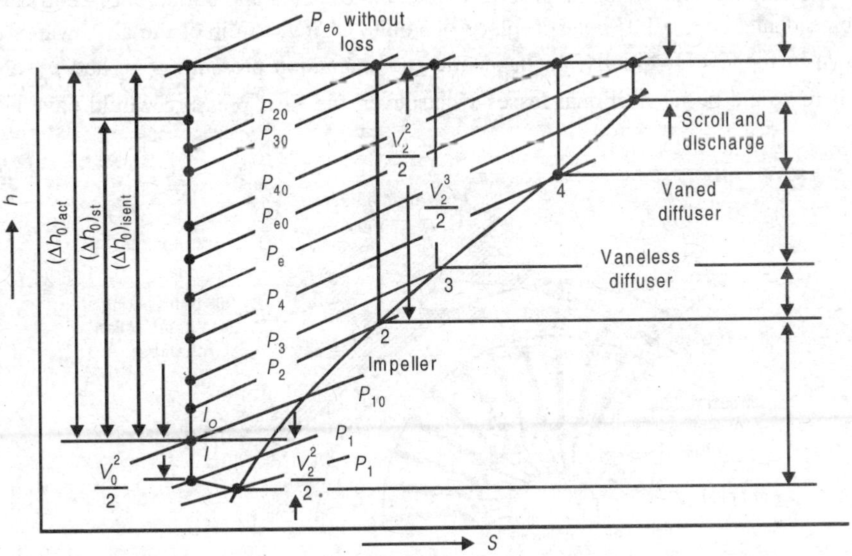

Fig. 20.10. *Centrifugal compressor process on h–S chart.*

Due to the rotation of impeller, vacuum is created at the eye of the impeller. Thus, air is sucked in from the surrounding ambient, at initial stagnation pressure p_{I_o} and stagnation temperature T_{I_o}, represented by point *a* on *h–S* diagram.

The static ambient condition is represented by point *o*. The air is somewhat accelerated in the inlet pipe. The inlet pipe admits or directs air for smooth and shockless entry into the eye of the impeller. In the inlet pipe, the direction and value of absolute velocity of air changes. Due to the acceleration of flow in the inlet pipe, velocity increases from V_o to V_1 thereby decreasing pressure and temperature of air. The acceleration process is non-isentropic due to friction and is shown by process 0 – 1 on *h–S* diagram. Therefore, at inlet to the impeller, the pressure and temperature are p_1 and T_1 respectively.

The work input to the shaft sets the impeller rotating. Air continuously accelerates as it flows over the impeller, thus, gaining kinetic energy. The vanes or blades are so mounted on the impeller as to provide gradually increasing cross-section area between two successive vanes to provide diffuser action. The reduction of kinetic energy $\left(\frac{V_{r_2}^2 - V_{r_1}^2}{2}\right)$ gets converted into pressure rise. Further, air is impressed with centrifugal head $\left(\frac{u_2^2 - u_1^2}{2}\right)$. The diffuser action and the centrifugal action thus converts a part of kinetic energy imparted to air by the impeller into pressure energy. The compression process is shown by line 1 – 2 on *h–S* diagram.

The vanes or blades of the impeller may have radial, backward or forward curvature.

Air leaves the impeller outlet at very high absolute velocity V_2 which is further converted into pressure in a vaneless or vaned diffuser and volute (scroll) casing. In the vaneless diffuser the absolute velocity of air decreases from V_2 to V_3, process 2–3 on *h–S* diagram. It also stabilises the flow leaving the vaned impeller tips for shockless entry into the vaned diffuser. The velocity of air further decreases from V_3 to V_4 in the vaned diffuser to increase pressure further. This process is represented by line 3–4 on *h–S* diagram.

The vaned diffuser discharges air into the volute (scroll) casing and finally air leaves from the outlet pipe. Process 4 – *e* on *h–S* diagram represents the process in the volute casing and outlet pipe. Centrifugal compressors used in gas turbine plants have a 90° bend in place of volute casing, to direct the air towards the combustion chamber as shown in Fig. 20.11. Finally, air has static and stagnation pressure of p_e and p_{e_o} at the exit of the compressor. If there would be no frictional losses whatsoever, the final pressure would have been together, as represented by $p_{e_{o(wl)}}$.

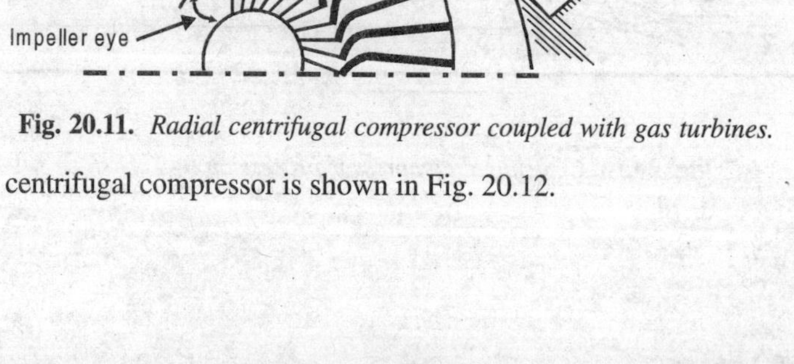

Fig. 20.11. *Radial centrifugal compressor coupled with gas turbines.*

A typical multistage centrifugal compressor is shown in Fig. 20.12.

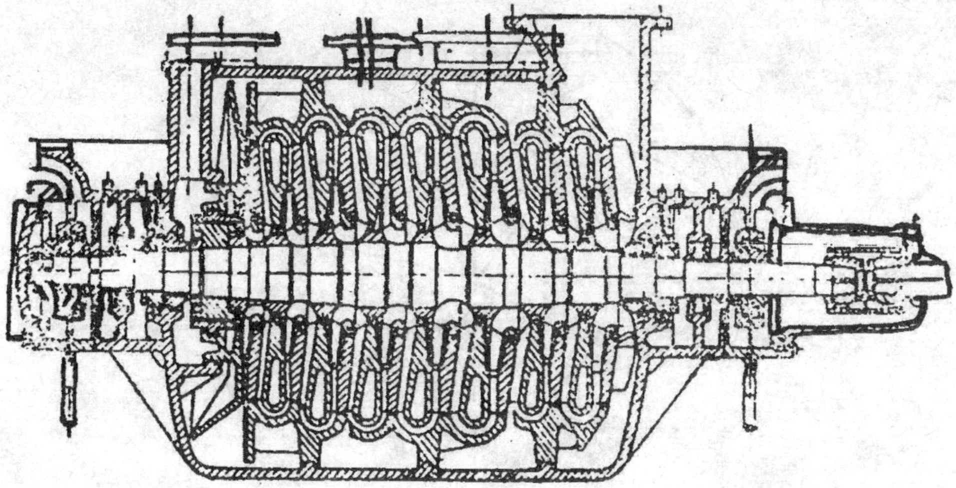

Fig. 20.12. *Multistage centrifugal compressor.*

20.6.2 Velocity Diagrams of a Centrifugal Compressor

Figure 20.13 shows velocity diagrams at the inlet and outlet of impeller and at inlet and outlet of diffusers.

Nomenclature. The following notations shall be adopted hereafter for energy analysis of the centrifugal compressor. Suffix '1' denotes parameters at inlet and '2' at outlet.

β_1 = Angle of the rotor blade at inlet
β_2 = Angle of the rotor blade at outlet
α_1 = Angle made by entering air or exit angle of guide blade
α_2 = Angle made by the outgoing air from rotor blade
V_1 and V_2 = Absolute velocity of air at inlet and outlet of rotor, m/s
V_{r_1} and V_{r_2} = Relative velocity of air at inlet and outlet of rotor, m/s
V_{f1} and V_{f2} = Velocity of flow at inlet and outlet of rotor, m/s
V_{w_1} and V_{w_2} = Velocity of whirl at inlet and outlet of rotor, m/s
u_1 and u_2 = Mean peripheral velocity of blade tip at inlet and outlet, m/s
r_1 and r_2 = Inner and outer radii of rotor, m
m = Mass flow rate of air, kg/s
α_3 = Vaned diffuser inlet angle or vaneless diffuser outlet angle

At the inlet to rotor, air enters with absolute velocity V_1 making an angle α_1 to the direction of motion of blade (usually $\alpha_1 = 90°$), without any shock and its whirl component $V_{w_1} = 0$.

Inlet triangle of velocities is now drawn to scale setting V_1 in the radial direction, u_1 in the tangential direction to inlet periphery. The vector joining end points of V_1 and u_1 represents relative velocity V_{r_1} at inlet. The blade tip at inlet has angle β_1 i.e., its curvature at inlet lies in the direction of V_{r_1}.

At the outlet from rotor, air leaves with a relative velocity V_{r_2} at an angle β_2 with the direction of motion. Now, as u_2 is known, V_2 can be found by vectorial addition.

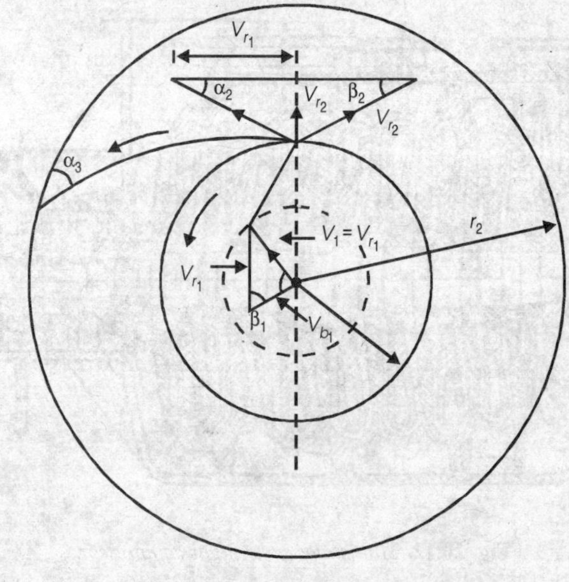

Fig. 20.13. *Velocity at inlet and outlet.*

In a vaneless diffuser, the flow is assumed to be logarithmic spiral and free-vortex. Air enters the vaned diffuser with velocity V_3 at an angle α_3 and leaves the diffuser with a velocity.

20.6.3 Work Requirement (Euler's work) for a Centrifugal Compressor

The work required/kg of air in a stage of a centrifugal compressor can be found by applying the moment of momentum theorem.

As per the Newtonian equation, force is given by rate of change of momentum.

Similarly, rate of change of moment of momentum about the centre of rotation gives torque.

Consider 1 kg of air flowing through the impeller (rotor). Theoretical torque supplied by impeller,

$$T = \left(V_{w_2} r_2 - V_{w_1} r_1\right) \text{ N-m/kg}$$

Now, for an angular velocity of impeller w rad/s

Theoretical work required/kg of air = $T \times w$ w/kg

$$\left(V_{w_2} r_2 - V_{w_1} r_1\right) w/\text{J/kg} \quad \ldots(6)$$

Equation (6) is known as Euler's equation and gives Euler's work.

Further, if air enters radially i.e., when $\alpha_1 = 90°$ and $V_{w1} = 0$,

Theoretical work reqd./kg = $V_{w2}\, u_2$ J/kg ...(7)

The theoretical work reqd./kg also represents theoretical or virtual head developed in the absence of all losses. Thus, virtual or theoretical head; $H_{\text{virt.}}$ is given by

$$H_{\text{virt.}} = \frac{V_{w_2} u_2}{g} \quad \ldots(8)$$

From inlet and outlet velocity triangles, we have

$$V_{r_1}^2 = V_1^2 + u_1^2 - 2\, V_{w_1}, u_1 \text{ vectorially}$$

and
$$V_{r_2}^2 = V_2^2 + u_2^2 - 2\, V_{w_2}, u_2 \text{ vectorially}$$

Substituting the value of $V_{w1}\,u_1$ and $V_{w2}\,u_2$ from the above equations in Eq. (6), we get theoretical work required/kg.

$$= \frac{V_{r_2}^2 - V_1^2}{2} + \frac{V_{r_1}^2 - V_{r_2}^2}{2} + \frac{u_2^2 - u_1^2}{2} \text{ w/kg} \quad \ldots(9)$$

In the Eq. (9), first term $\left(\frac{V_2^2 - V_1^2}{2}\right)$ represents the increase in kinetic energy per kg of fluid in the impeller which has to be converted into pressure energy in the diffuser.

The last two terms signify the pressure rise in the impeller due to diffusion action and centrifugal action respectively. The second term is known as diffusion action as the relative velocity of fluid decreases from inlet to outlet of impeller. The last term represents centrifugal action due to increase of peripheral velocity of the vane tips from inlet to outlet.

Therefore, out of kinetic energy imparted by the impeller to the fluid, the part of the kinetic energy converted into pressure energy in impeller is

$$\int_1^2 \frac{dp}{p} = \left(\frac{V_{r_1}^2 - V_{r_2}^2}{2}\right) + \left(\frac{u_2^2 - u_1^2}{2}\right) \quad \ldots(10)$$

Similarly, if velocity at outlet of the diffuser is V_4 then,

Decrease of kinetic energy in the diffuser

$$\frac{V_2^2 - V_4^2}{2} = \int_2^4 \frac{dp}{p} + \Delta h_{\text{diff-loss}}$$

Thus, decrease of the kinetic energy of fluid in the diffuser is partly converted into pressure and the remaining is lost irreversibly as heat losses in the diffuser.

20.6.4 Dimensionless Parameters of Centrifugal Compressors

For the design and performance analysis of centrifugal compressor, several dimensionless parameters are very useful. These are defined and explained in what follows.

(a) Flow coefficient (ϕ). It is defined as the actual mass flow rate to the mass flow rate referred to the tip of the impeller, corresponding to tip peripheral velocity.

$$\therefore \quad \text{Flow coefficient, } \phi = \frac{\dot{m}}{\rho_2 A_2 u_2} \quad \ldots(11)$$

where,
$\dot{m}$ = actual mass flow rate, kg/s
ρ_2 = density of fluid at the outlet tip of impeller, kg/m^3
A_2 = flow area at the exit tip of impeller, m^2

Applying continuity equation,

$$\text{Flow cofficient, } \phi = \frac{\rho_2 \pi D_2 b_2 V_{f_2}}{\rho_2 \pi D_2 b_2 u_2} = \frac{V_{f_2}}{u_2} \quad \ldots(12)$$

Value of flow coefficient often lies between 0.28 and 0.32, its optimum value being 0.3.

(b) Head coefficient 'λ'. It is defined as ratio of enthalpy increase in the stage to the kinetic energy corresponding to the tip peripheral velocity.

Thus,

$$\text{Head coefficient, } \lambda = \frac{\Delta h}{u_2^2/2} \qquad \ldots(13)$$

where Δh = Enthalpy increase in the stage, per kg of fluid. The enthalpy increase includes the work supplied /kg of fluid compressed, as given by the Euler's Eq. (6) plus half of the heat corresponding to rotor friction losses. So we get,

$$\text{Head coefficient, } \lambda = \frac{2\left[\left(V_{w_2}u_2 - V_{w_1}u_1\right) + \frac{1}{2}\alpha u_2^2\right]}{u_2^2} \qquad \ldots(14)$$

$$\lambda = 2\left[\mu \bar{V}w_{2\infty} - \left(\frac{D_1}{D_2}\right)^2 \bar{V}_{w_1}\right] + \alpha$$

where, μ = slip factor

$$\bar{V}w_{2\infty} = \frac{\bar{V}w_{2\infty}}{u_2} \text{ and } \bar{V}w_1 = \frac{\bar{V}w_1}{u_1}$$

For inlet velocity in axial direction,

$$V_1 = V_{f_1} \text{ and } V_{w_1} = 0$$

$$\therefore \quad \lambda = 2\mu \bar{V}w_{2\infty} + \alpha \qquad \ldots(15)$$

Now, if $\alpha = 0$, head coefficient for infinite number of blades is given by

$$\lambda = 2\mu \bar{V}_{w_{2\infty}} \qquad \ldots(16)$$

and head coefficient with finite number of blades is

$$\lambda = 2\mu \bar{V}_{w_{2\infty}} \qquad \ldots(16)$$

(c) Pressure coefficient (Ψ). It is defined as the ratio of isentropic enthalpy rise to the kinetic energy corresponding to tip peripheral velocity. Thus,

$$\text{Pressure coefficient, } \psi = \frac{\Delta h_{\text{isen}}}{u_2^2/2}$$

Considedring isentropic efficiency, η_{isent}, we get

$$\psi = \frac{\eta_{\text{isent}} \times \Delta h}{u_2^2/2} = \lambda \times \eta_{\text{isent}} \qquad \ldots(18)$$

where $$\eta_{\text{isent}} = \frac{\Delta h_{\text{isent}}}{\Delta h}$$

(d) Reaction (Ω). It is defined as the ratio of temperature rise in the rotor to the temperature rise in the stage. Thus,

$$\text{Reaction } \Omega = \frac{\Delta T_R}{\Delta T_{\text{stage}}} = 1 - \frac{C_p \Delta T_{\text{stator}}}{C_p \Delta T_{\text{stage}}}$$

$$= 1 - \frac{V_2^2 - V_1^2}{2C_p \Delta T_{stage}}$$

where it is assumed that $V_1 = V_4$ i.e., initial and final velocity is same. Now, for axial entry at inlet, $V_{w1} = 0$.

$$\text{Reaction } \Omega = 1 - \frac{V_{w_2}^2 + V_{f_2}^2 - V_{f_1}^2}{2u_2 V_{w_2}} \qquad \text{...(19)}$$

20.6.5 Slip Factor of a Centrifugal Compressors

Under ideal conditions, fluid particles follow exactly the same path of blade profile such that relative velocity at impeller outlet tip is inclined with the tangential direction at blade tip angle β_2 only, irrespective of mass flow rate, speed etc.

Such an ideal flow is possible when impeller has infinite number of blades of no thickness. Figure 20.14 shows velocity triangle for radial blades. Dotted lined diagram represents ideal conditions while full lined diagram represents actual conditions.

In actual practice, when impeller has finite number of blades, fluid is trapped between the impeller vanes due to its inertia and the fluid is reluctant to flow over the impeller.

This causes a pressure difference across the blades. There is a high pressure at the leading face and low pressure at the trailing face. This pressure difference generates a relative velocity gradient and formation of eddies.

The fluid is thus discharged at a certain average angle β'_2 which is less than β_2.

Therefore fluid is said to have slipped with respect to impeller during its flow across it.

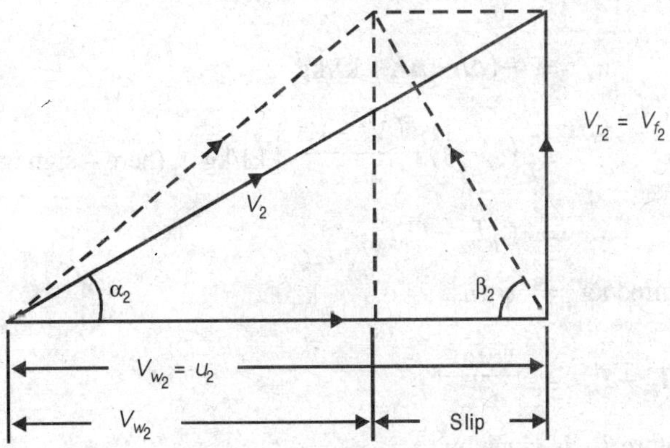

Fig. 20.14. *Slip over blackward curved bladed impller.*

The slip factor μ is defined as the *ratio of actual velocity of whirl to the ideal velocity of whirl.*

$$\text{Slip factor, } \mu = \frac{\text{Actual velocity of whirl}}{\text{Ideal velocity of whirl}}$$

$$\mu = \frac{V_{wa}}{V_{wi}} = \frac{V_{w_{2a}}}{u_2} \text{ (for radial blades)} \qquad \text{...(20)}$$

Slip factor depends upon number of blades and is usually 0.9. It does not reduces the efficiency but only reduces the head developed.

20.6.6 Pressure Ratio of Compression in a Centrifugal Compressor

Slip factor is also defined as the ratio of work required under actual conditions to the work required under ideal conditions.

Thus, $$\mu = \frac{V_{w_2}u_2 - V_{w_1}u_1}{V_{w_{2i}}u_2 - V_{w_{1i}}u_1}$$

But with radial entry, $V_{w_1} = V_{w_{1i}} = 0$

$$\mu = \frac{V_{w_2}}{V_{w_{2i}}}$$

$$\text{Work} = \mu \times V_{w_2} \times u_2$$

$$= \mu u_2^2 \qquad \text{...(21)}$$

Actual work required is greater than theoretical work given by Eq. (21) due to friction, windage etc. Power input factor Ψ is introduced now such that

$$\text{Actual Work req./kg} = \psi \mu u_2^2 \qquad \text{... (22)}$$

The value of power input factor Ψ is about 1.04.

As per steady flow energy equation applied at inlet and outlet of impeller and assuming no heat transfer,

$$Q = \Delta P + \Delta K + \Delta h + W_s \text{ kJ/kg}$$

With $Q = 0$, $\Delta P = 0$, we get

$$w_s = -(\Delta h + \Delta K) \text{ kJ/kg}$$

$$= (h_2 - h) + \left(\frac{V_2^2 - V_1^2}{2}\right) \text{ kJ/kg} \quad \text{(here – sign indicator work required)}$$

$$= C_p T_{o_2} - C_p T_{o_1} \qquad \text{...(23)}$$

But, $$\text{work required/kg} = \psi \mu u_2^2$$

$\therefore$ $$T_{o_2} - T_{o_1} = \frac{\psi \mu u_2^2}{C_p} \qquad \text{...(24)}$$

Actual stagnature temperature T_{o_1} is given by

$$T'_{o_2} - T_{o_1} = \frac{\psi \mu u_2^2}{C_p} \times \eta_{\text{isent}} \quad \text{where } \eta_{\text{isent}} = \text{isentropic efficiency}$$

Therefore, pressure ratio

$$\frac{p_{o_2}}{p_{o_1}} = \left(\frac{T'_{o_2}}{T_{o_1}}\right)^{\frac{\gamma}{\gamma-1}}$$

$$= \left[1 + \frac{T'_{o_2} - T_{o_1}}{T_{o_1}}\right]^{\frac{\gamma}{\gamma-1}}$$

$$= \left[1+\frac{\eta_{isent}\ \psi\mu u_2^2}{C_p T_{o_1}}\right]^{\frac{\gamma}{\gamma-1}} \quad ...(25)$$

If all factors are unity as in case of isentropic compression, then,

$$\frac{p_{o_2}}{p_{o_1}} = \left[1+\frac{u_2^2}{C_p T_{o_1}}\right]^{\frac{\gamma}{\gamma-1}} \quad ...(26)$$

The slip factor μ and the power input factor Ψ are different and independent from each other. Whereas the power input factor Ψ signifies the increase in the work input required which is lost in overcoming friction. This energy lost gets converted into heat and causes temperature rise. The temperature rise of compressed air is often not a loss as in gas turbine plants, the hot compressed air elevates the temperature in the combustion chamber without increase of fuel supply. Yet the power input factor should be low and compression should ideally be isentropic.

But the slip factor limits the work capacity of the compressor even with isentropic working. The slip factor should be as high as possible. High slip factor increases V_{w_2}. When $V_{w_2} = u_2$, work input is maximum which can be gainfully utilised.

Slip factor increases with greater number of vanes which in turn decreases area of flow passage for air. A compromise is often necessary. The present practice is to adopt about 20 vanes to obtain a slip factor of 0.9.

An impirical formula proposed by Stanitz is

$$\eta = 1-\frac{0.63\pi}{n}$$

where, n = number of blades

Higher tip speed increases the pressure ratio.

Maximum tip speed is restricted to 460 m/s for the common materials.

Also reducing the inlet temperature T_{o_1} raises the pressure ratio for a given work input.

20.6.7 Influence of Impeller Blade Geometry

The blades of the impeller have various shapes which can be described based on the blade angle at outlet β_2. Depending on the value of β_2, the blades have different curvatures compared with the forward direction of rotation of the impeller. The various shapes of blades are:

(i) Backward curved blades; $\beta_2 < 90°$

(ii) Radial blades; $\beta_2 = 90°$

(iii) Forward curved blades; $\beta_2 > 90°$

Figure 20.15 shows the velocity triangle for each case of the above.

In all cases, blade tip velocity at exit u_2 is kept same.

It is seen from Fig. 20.15 that for the backward curved blades, the tangential component V_{w_2} is least compared with those is other cases. Therefore, for a given speed work input required for the impeller is low, as given by Euler's work (Eq. 6).

In the case of forward curved blades, V_{w_2} is maximum, thus the impeller needs maximum work input. Too, the value of V_2 is highest.

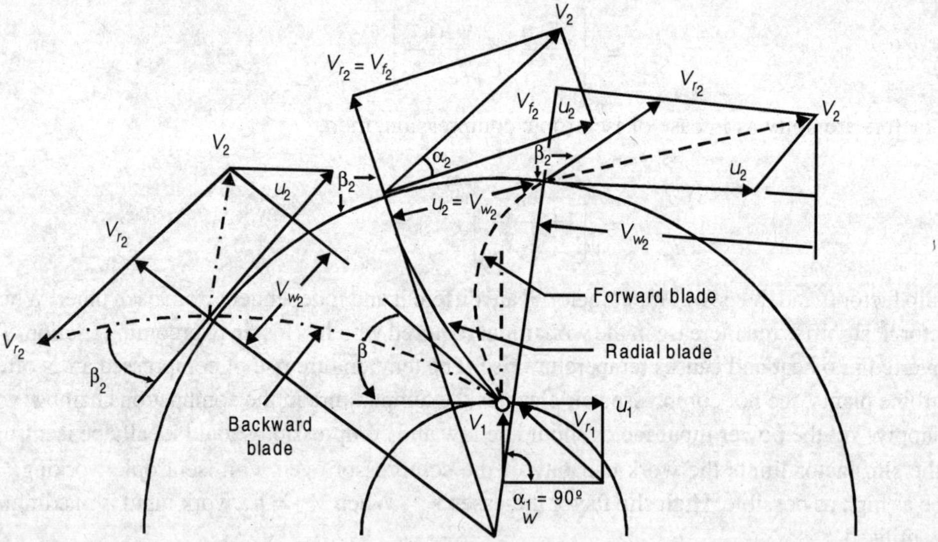

Fig. 20.15. *Velocity diagram for each blade shape.*

And in case of radial blades, V_{w_2} lies in between, that is, backward and forward curved blades.

Therefore, backward curved blades are best giving high efficiency.

The influence of blade geometry can also be explained by dimensionless parameters as follows.

The head coefficient is one dimensionless parameter which is used to express the work required per stage of centrifugal compressor.

For a working fluid treated as ideal gas, isentropic efficiency $\eta_{isent} = 1$. As such head coefficient λ and pressure coefficient Ψ are same, the work input to the fluid is then given by,

$$\psi_i = 2\overline{V_w} \quad = \quad 2\frac{V_{w_{2i}}}{u_2} \qquad \text{...(27)}$$

But,
$$\beta_2 \quad = \quad \frac{V_{f_2}}{u_2 - V_{w_{2i}}} = \frac{V_{f_2}/u_2}{1 - \dfrac{V_{w_{2i}}}{u_2}} \qquad \text{...(28)}$$

where,
$$\frac{V_{f_2}}{u_2} \quad = \quad \phi = \text{flow coefficient}$$

Therefore,
$$\psi_i \quad = \quad 2\left(1 - \frac{\phi}{\tan \beta_2}\right) \qquad \text{...(29)}$$

It is seen from Eq. (29) that for a given value of flow coefficient ϕ, the work input required per stage increases as blade angle at outlet β_2 increases.

For backward curved blades, $\beta_2 < 90°$ and with $\tan \beta_2 = \phi$, the $\Psi_i = 0$.

For radial blades, $\beta_2 = 90°$, $\Psi_i = 2$ and is independent of flow coefficient ϕ.

For forward curved blades, $\beta_2 > 90°$, $\tan \beta_2$ is negative and when $\phi = -\tan \beta_2$, $\Psi_i = 4$.

Further blade angle at outlet β_2 influences the reaction Ω of the stage. For an impeller with infinite number of blades,

$$\text{Reaction } \Omega = 1 - \frac{V_{w_2}}{u_2}$$

$$= 1 - \frac{\Psi_i}{4}$$

Thus, as pressure coefficient Ψ increases, the reaction of the stage decreases.

20.6.8 Influence of Compressor Geometry on the Performance of Centrifugal Compressor

Figure 20.20 (a) and (b) illustrates variation of rection of the stage Ω with pressure coefficient Ψ.

20.6.9 Pre-Whirl

The gas turbines and jet engines use centrifugal compressors running at high speed. There is often a possibility of development of a shock wave in the flow passage. This can be avoided if the Mach Number at any point in the flow passage is restricted below unity.

The maximum value of Mach Number referred to relative velocity is found at inlet.

$$\therefore \quad \text{Mach Number at inlet} = \frac{V_{r_1}}{\text{Velocity of sound}}$$

$$M_1 = \frac{V_{r_1}}{\sqrt{\gamma R T_1}} \quad \ldots(30)$$

where, T_1 = Static temperature at inlet, K

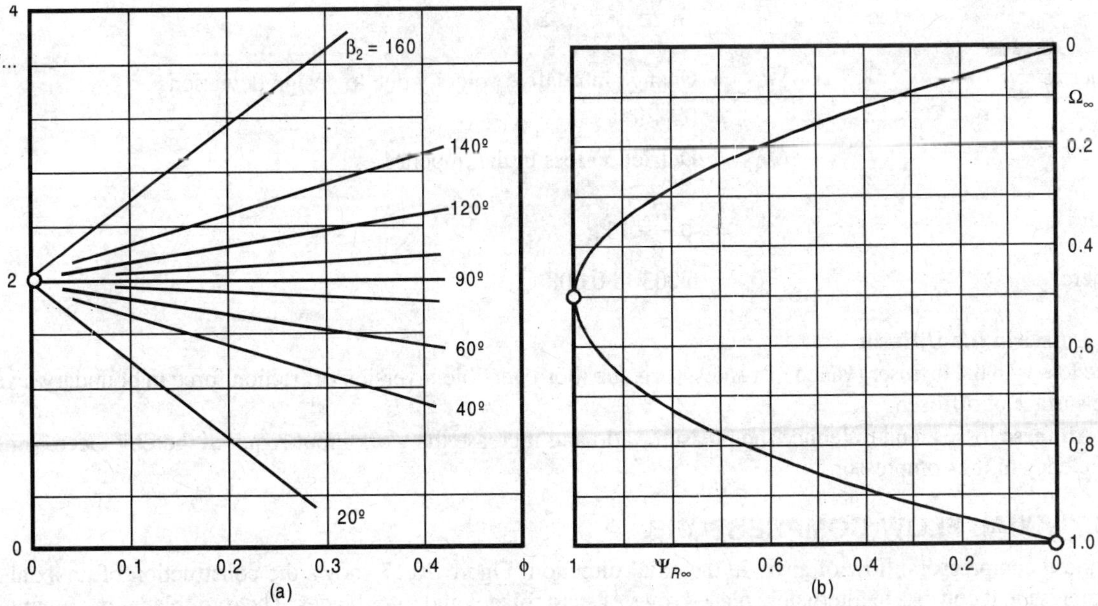

Fig. 20.20. *Influence of blade shape.*

For a given flow rate, if V_{r1} and u_1 is high for shockless operation of the compressor, the fluid is given an initial pre-rotation or pre-whirl with the help of fixed guide blades attached to the casing at inlet.

The entering air passing through the fixed guide blades is given pre-whirl. Thus V_{r1} is reduced without affecting V_{f1} or mass flow rate.

20.6.10 Losses in a Centrifugal Compressor

The frictional losses in a centrifugal compressor occur mainly at two places viz. over the impeller and in the diffuser.

I. Losses Over the Impeller

The fluid friction losses over the impeller are given by

$$\Delta h_{imp} = \Delta h_1 + \Delta h_2 + \Delta h_3 + \Delta h_4 + \Delta h_d$$

where Δh_1 = Loss in the inducer $= \xi_1 \dfrac{V_{r1}^2}{2}$ J/kg where, $= \xi_1 = 0.2$ to 0.3

Δh_2 = Loss due to change of direction from axial to radial at the impeller inlet

$= \xi_2 \dfrac{V_1^2}{2}$ J/kg where, ξ_2 = loss coefficient = 0.1 to 0.2

Δh_3 = Loss due to friction in the blade passages

$= \xi_3 \dfrac{Vr_1^2}{2}$ J/kg where, ξ_3 = loss coefficient 0.2 to 0.4

Δh_4 = Loss due to design deviations

$= \xi_4 \dfrac{(\Delta Vr_1)^2}{2}$ J/kg

where, ΔV_{r_1} = Change in relative velocity due to design deviation

= 0.54 to 0.60

Δh_d = Defriction loss in the impeller

$= \sigma \dfrac{u_2^2}{2}$ J/kg

where, σ = 0.003 to 0.008

II. Losses in the Diffuser

The losses in the diffuser (vaned or vanless) are due to irreversible inversion of friction force in boundary layer on the surface of diffuser.

All these losses tend to reduce the pressure ratio and increase the work input required thereby decreasing the efficiency of the compressor.

20.7 AXIAL FLOW COMPRESSORS

In these compressors, flow of air is in the axial direction. Figure 20.17 shows the construction of an axial flow compressor. It consists of alternately placed rows of stator blades and rotor blades. The rotor blades are mounted on the rotor while the stator blades are attached to the casing.

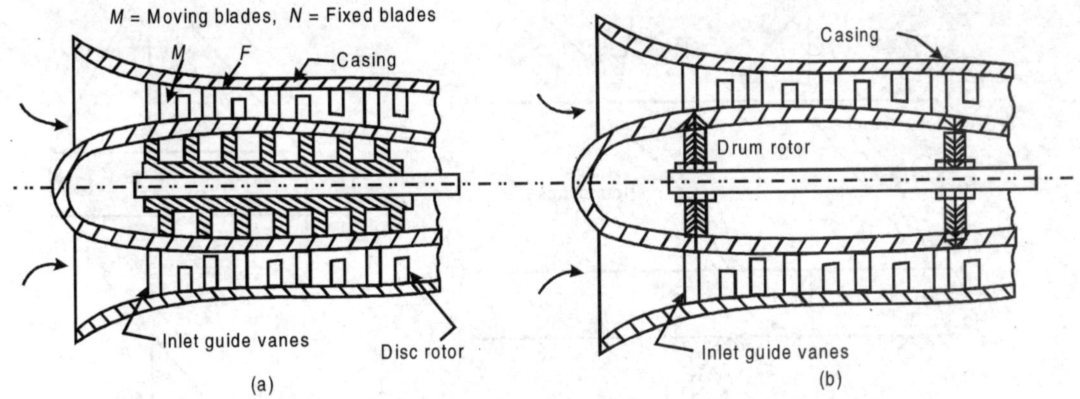

Fig. 20.17. *Axial flow compressor.*

The conversion of kinetic energy into pressure head i.e., the diffusion action takes place partly over the rotor blades and partly over the stator blades. The stator blades also direct air over the successive rotor blades for a shockless entry. The flow passages between rotor blades as well as stator blades provide diverging area. One row of rotor blades preceded by a row of stator blades is known as a stage of axial flow compressor. A single stage compressor can achieve a pressure ratio of 1.12 to 1.2 only. Therefore, for higher pressure ratio, often several stages of axial flow compressor are necessary.

20.7.1 Construction and Principle of Operation

Rotors are of two types viz. drum type and disc type (shown in Fig. 20.17). The disc type rotors being light weight are preferred for aircraft applications. The drum type rotors are used for static industrial applications. (Volume of casing converges from low pressure suction end to high pressure discharge end). It helps to maintain more or less a constant velocity throughout the length of compressor, though the density of fluid increases gradually. Constant axial velocity throughout the stage is an important design consideration of axial flow compressor. Contraction of the flow annulus is achieved by contraction of casing and diverging the rotor.

The materials used for various components of the compressor are listed below:

(i) **Rotor blades.** Fibrous composites, Aluminium, Titanium, Steel, Nickel alloy.

(ii) **Rotor.** Steel for shaft, disc. In aircrafts, first stage uses Titanium while the later stages use Nickel steel.

(iii) **Stator blades.** These are made of same materials as those for rotor blades.

(iv) **Casing.** It is either a casting of magnesium, aluminium, steel, iron or fabricated from Titanium or steel.

The blades of compressor are of aerofoil section. The axial flow compressors can be of impulse type or reaction type. The blade profile and variation of pressure and velocity in the stage are shown in the Fig. 20.17 (a). While in Fig. 20.17 (b), the difference in the shape of Axial flow turbines and Axial flow compressor is illustrated.

20.7.2 Velocity Diagrams of Axial Flow Compressors

As explained earlier, in the axial flow compressor, fluid is imparted with kinetic energy from the rotor. This kinetic energy of the fluid is converted into pressure energy by diffuser action. The process occurs over several stages.

The energy analysis is considered for one stage of the compressor. The flow is assumed to occur tangential to the mean blade speed of u.

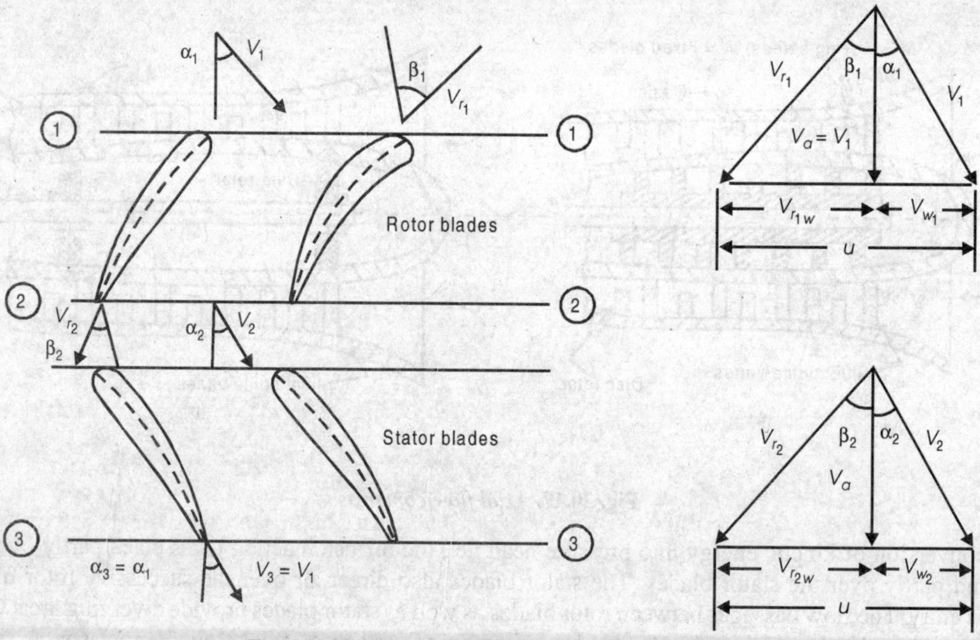

Fig. 20.18. *Velocity traingle for compressor stages.*

Figure 20.18 shows velocity triangles for one stage of the compressor. Air approaches rotor blade with absolute velocity V_1 at an angle α_1 to the axial direction. Combining V_1 with blade velocity u_1 gives relative velocity V_{r_1} at inlet, inclined at angle β_1 with the direction of motion.

While air flows through the diverging passage between two successive rotor blades, its absolute velocity increases and air leaves the rotor with relative humidity V_{r_2} at an angle β_2. The reduction in the relative velocity from V_{r_1} at inlet to V_{r_2} at exit causes some pressure rise over the rotor. The absolute velocity at exit V_2 can be found by vector difference between V_{r_2} and u_2. The absolute velocity at exit V_2 makes an angle α_2 with the direction of motion. The air then enters the next stage via stator blades. Over the stator blades, the pressure of air further increases due to diffusion action. Air leaves the stator blades with absolute velocity V_3 (less than V_2) making an angle α_3 with the axial direction. The stator blades are so designed that $V_3 = V_1$ and $\alpha_3 = \alpha_1$ for a shockless entry over the next stage.

It may particularly be noted that unlike in centrifugal compressors, in this case all angles are measured with axial direction and not from tangential direction.

Referring to the velocity triangles shown in Fig. 20.18 we have,

$$\frac{u}{V_f} = \tan\alpha_1 + \tan\beta_1 \qquad \text{...(31)}$$

and

$$\frac{u}{V_f} = \tan\beta_2 + \tan\alpha_2 \qquad \text{...(32)}$$

where, $V_{f_1} = V_{f_2} = V_f$ is axial velocity of flow assumed to be constant throughout the stage. Now, for 1 kg of air flowing through the stage

$$\text{Work required/kg of air} = u\left(V_{w_2} - V_{w_1}\right) = C_p\left(T_{o_2} - T_{o_1}\right) \text{ J/kg} \qquad \text{...(33)}$$

Equation (33) can also be written as

$$\text{Work required/kg of air} = uV_f(\tan\alpha_2 - \tan\alpha_1) \qquad \text{...(34)}$$

From Eqs. (31) and (32), we get,

$$\text{Work required/kg} = u \times V_f (\tan \beta_1 - \tan \beta_2) \qquad \text{...(35)}$$

Also, work required/kg can be obtained from Euler's equation,

$$\text{Work required/kg} = (u_2 V_{w_2} - u_1 V_{w_1}) \text{ J/kg} \qquad \text{...(36)}$$

This energy input is absorbed in raising pressure and velocity of air. Some energy is lost in overcoming frictional losses. However, this entire energy input appears as a rise in "stagnation" temperature of air, ΔT_{os}. If $V_3 = V_1$, then

$$\Delta T_{o_s} = \Delta T_{\text{static}} = \frac{uV_f}{C_p}(\tan\beta_1 - \tan\beta_2) \qquad \text{...(37)}$$

Considering work done factor λ, the actual stage temperature rise is

$$\text{Actual } \Delta T_{o_s} = \Delta T_{\text{static act}} = \frac{\lambda}{C_p} \times u \times V_f (\tan\beta_1 - \tan\beta_2)$$

And the pressure ratio R_s is given by,

$$R_s = \left[1 + \frac{\eta_s \Delta T_{o_s}}{T_{o_i}}\right]^{\frac{\gamma}{\gamma-1}} \qquad \text{...(38)}$$

where, T_{oi} = inlet stagnation temperature and η_s is isentropic efficiency of the stage

$$\eta_s = \frac{\text{Isentropic temperature rise}}{\text{Actual temperature rise}}$$

Using velocity triangles and cosine rule, Eq. (38) can be written as

$$\text{Work required/kg} = \frac{V_{r\,1}^2 - V_{r\,2}^2}{2} + \frac{V_2^2 - V_3^2}{2} \qquad \text{...(39)}$$

where, $V_3 = V_1$

Equation (39) is represented graphically on *h–S* diagram in Fig. 20.19.

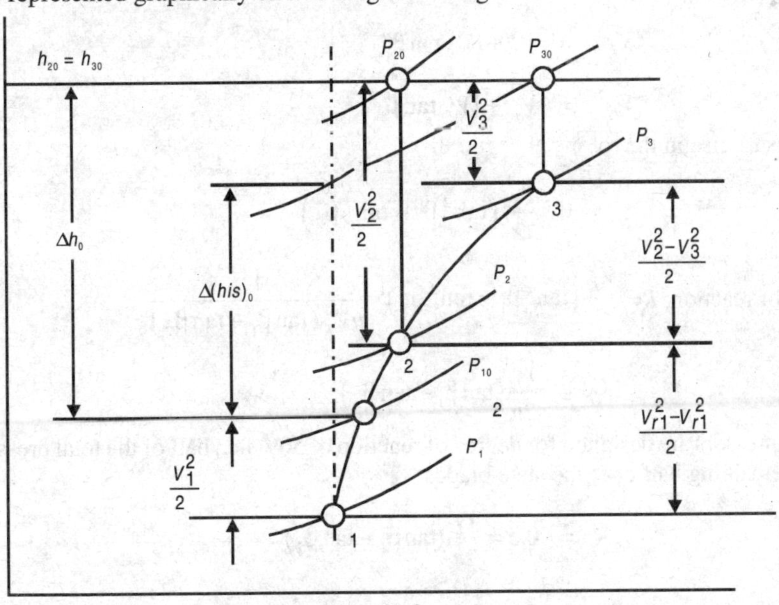

Fig. 20.19. *Process in axial flow compressor without GV on h-S diagram.*

The first term on the right hand side of Eq. (39) represents the increase of kinetic energy in the rotor which is converted into pressure due to diffusion action in the rotor itself while the second term represents increase of kinetic energy over the rotor that has to be converted into pressure energy in stator blades.

Comparing this Eq. (39) of work required/kg with Eq. (39) applicable for centrifugal compressor, it is seen that the term centrifugal head $\left(\frac{u_2^2 - u_2^2}{2}\right)$ is missing for axial flow compressor. Therefore, pressure ratio per stage in axial flow compressor is much less than that of centrifugal compressor.

20.7.3 Degree of Reaction of Axial Flow Compressor

The degree of reaction is defined as the ratio of pressure rise in the rotor to the pressure rise in the stage.

$$\text{Degree of reaction, } R = \frac{\text{Pressure rise in the door}}{\text{Pressure rise in the stage}} \quad \text{...(40)}$$

Total pressure rise in the stage of a compressor is given by work input per stage

$$= u\,(V_{w2} - V_{w1}) \quad \text{...(41)}$$

From velocity triangles,

$$V_{w_2} = u - V_f \tan\beta_2$$

and

$$V_{w_1} = u - V_f \tan\beta_1$$

Hence, pressure rise in the stage

$$= uV_f\left(\tan\beta_1 - \tan\beta_2\right) \quad \text{...(42)}$$

Further, pressure rise over the rotor is due to change in relative kinetic energy.

$$= \frac{V_{r_1}{}^2 - V_{r_2}{}^2}{2} \quad \text{...(43)}$$

From velocity traingles,

$$V_{r_1}{}^2 = V_f^2 + \left(V_f \tan\beta_1\right)^2$$

and

$$V_{r_2}{}^2 = V_f^2 + \left(V_f \tan\beta_2\right)^2$$

Therefore, pressure rise in the rotor

$$= \frac{V_f^2}{2}\left(\tan^2\beta_1 - \tan^2\beta_2\right) \quad \text{...(44)}$$

$$\text{Hence, degree of reaction, } R = \frac{V_f^2}{2}\left(\tan^2\beta_1 - \tan^2\beta_2\right) \times \frac{1}{uV_f\left(\tan\beta_1 - \tan\beta_2\right)}$$

$$= \frac{V_f}{2u}(\tan\beta_1 - \tan\beta_2) \quad \text{...(45)}$$

Axial flow compressors are designed for degree of reaction of 50% i.e., half of the total pressure rise occurs over the rotor and the remaining half over the state blades.

$$\therefore \quad R = 0.5 = \frac{V_f}{2u}(\tan\beta_1 - \tan\beta_2) \quad \text{... (46)}$$

$$\therefore \quad \frac{u}{V_f} = (\tan\beta_1 - \tan\beta_2) \quad \text{...(47)}$$

Also, from velocity traingles,

$$\frac{u}{V_f} = (\tan \alpha_1 - \tan \beta_1) \quad ...(48)$$

and

$$\frac{u}{V_f} = (\tan \alpha_2 - \tan \beta_2) \quad ...(49)$$

From Eqs (47), (48) and (49), it follows that

$$\alpha_1 = \beta_2 \text{ and } \alpha_2 = \beta_1$$

Also constant velocity of flow through the stage,

$$V_f = V_1 \cos \alpha_1 = V_3 \cos \alpha_3$$

But,

$$V_3 = V_1 \text{ (assumed)}$$

$\therefore$

$$\alpha_1 = \alpha_3$$

As $\alpha_1 = \beta_2 = \alpha_3$ and $\alpha_2 = \beta_1$, rotor blade and stator bade are symmetrical. In the above deductions, work done factor is assumed to be unity.

For work done factor lower than unity, the degree of reaction differs slightly.

20.7.4 Cascade or Diffuser Efficiency

It is defined as the ratio of actual pressure rise to the isentropic pressure rise. Thus,

$$\eta_c = \frac{P_2 - P_1}{P_{2_{is}} - P_1} \quad ...(50)$$

Friction in rotor and stator blade passages, shock losses, etc. reduce the efficiency of the cascade (stage).

20.7.5 Polytropic Efficiency

Consider the compression of air in a multistage axial flow rotary compressor shown on T–S diagram in Fig. 20.20.

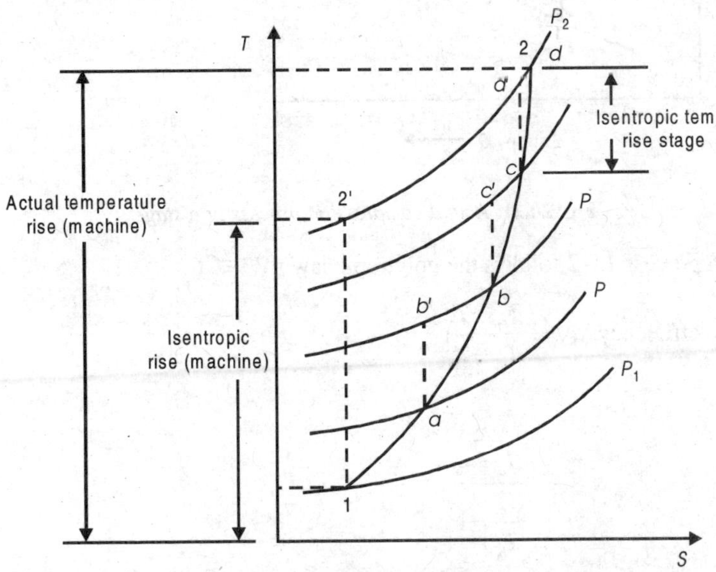

Fig. 20.20. *Multistage compression.*

It is customary to assume that all stages have equal isentropic efficiency where isentropic efficiency is given by

$$\eta_{is} = \text{stage isentropic efficiency} = \frac{\text{Isentropic temp. rise in the stage}}{\text{Actual temp. rise in the stage}}$$

Now, overall isentropic efficiency

$$\text{Overall} \qquad \eta_{is} = \frac{T_{o_{2'}} - T_{o_1}}{T_{o_2} - T_{o_1}}$$

$$\text{and,} \qquad \text{Polytropic efficiency} = \frac{\Sigma(\text{Isentropic temp. rise in the stage})}{T_{o_2} - T_{o_1}}$$

$$= \frac{(T_{o_{a'}} - T_{o_1}) + (T_{o_{b'}} - T_{o_a}) + (T_{o_{c'}} - T_{o_c}) + (T_{o_{d'}} - T_{o_d})}{T_{o_2} - T_{o_1}}$$

20.7.6 Polytropic Efficiency η Terms of *n*

Consider an actual compression process in a stage of an axial flow compressor, as shown in Fig. 20.21.

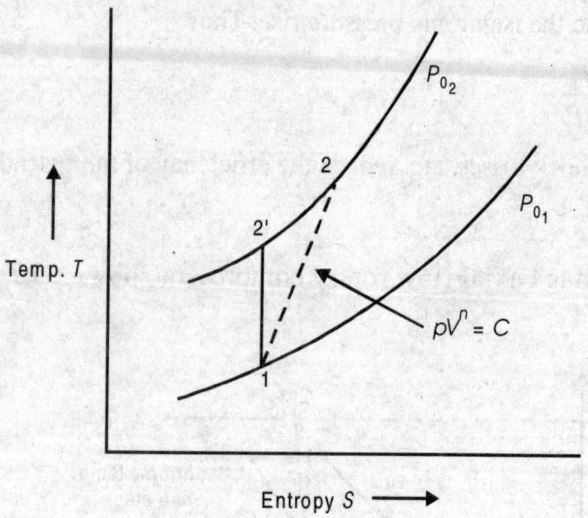

Fig. 20.21. *Actual compression process in a stage.*

Actual compression process 1 – 2 follows the polytropic law $pV^n \cong C$.

$$\therefore \qquad \text{Polytropic efficiency, } \eta = \frac{T_{o_{2'}} - T_{o_1}}{T_{o_2} - T_{o_1}}$$

$$= \frac{\dfrac{T_{o_{2'}}}{T_{o_1}} - 1}{\dfrac{T_{o_2}}{T_{o_1}} - 1}$$

$$= \frac{\left(\frac{P_{o_2}}{P_{o_1}}\right)^{\frac{\gamma-1}{\gamma}} - 1}{\left(\frac{P_{o_2}}{P_{o_1}}\right)^{\frac{n-1}{n}} - 1}$$

$$= \frac{\left(\frac{P_{o_1} + \Delta P_o}{P_{o_1}}\right)^{\frac{\gamma-1}{\gamma}} - 1}{\left(\frac{P_{o_1} + \Delta P_o}{P_{o_1}}\right)^{\frac{n-1}{n}} - 1}$$

$$= \frac{\left(1 + \frac{\Delta P_o}{P_{o_1}}\right)^{\frac{\gamma-1}{\gamma}} - 1}{\left(1 + \frac{\Delta P_o}{P_{o_1}}\right)^{\frac{n-1}{n}} - 1}$$

Expanding the above equation binomially and ignoring higher order terms, we may write,

$$\eta_{\text{poly}} = \frac{\left(\frac{\gamma-1}{\gamma}\right)\left(\frac{\Delta P_o}{P_{o_1}}\right)}{\left(\frac{\gamma-1}{\gamma}\right)\left(\frac{\Delta P_o}{P_{o_1}}\right)}$$

$$= \left(\frac{\gamma-1}{\gamma}\right)\left(\frac{n-1}{n}\right) \qquad \ldots(51)$$

The concept of isentropic efficiency cannot be used to compare the performance of two compressors having different pressure ratio or different number of stages. For such a comparison, polytropic efficiency is used.

Further, the work input required to a multistage compressor is given by

Work input required/kg of air $= C_p\left(T_{o_2} - T_{o_1}\right)$

$$= C_p \times \left(T_{o_2} - T_{o_1}\right) \times \eta_{o_{\text{isent}}}$$

$$= C_p \times \eta_{o_{\text{isent}}} \times T_{o_1}\left[\left(\frac{P_{o_2}}{P_{o_1}}\right)^{\frac{\gamma-1}{\gamma}} - 1\right] \qquad \ldots(52)$$

20.7.7 Work Done Factor

It is rather difficult to draw the actual velocity triangles at the root, mean and tip of the blades as accurate estimation of contraction of main stream cannot be done.

The axial velocity at the mean section is somewhat greater than the average axial velocity.

The actual temperature rise is somewhat lower than that estimated from the velocity triangles.

An empirical relation suggested by Hotwell, for stagnation enthalpy rise across a stage is

$$C_p \Delta T_o = \lambda u \left(V_{f_2} \tan \alpha_2 - V_{f_1} \tan \alpha_1\right)$$

where λ is work done factor which is approximately 0.85. Hotwell further suggested that the work done factor depends upon the number of stages. Greater the number of stages, the work done factor reduces.

20.7.8 Losses in Axial Flow Compressor Stage

During the flow of fluid through a stage of a compressor, various losses occur. The pressure loss is due to three factors, namely

(a) profile losses on the blade surface; (b) skin friction on the annulus walls and (c) secondary flow losses.

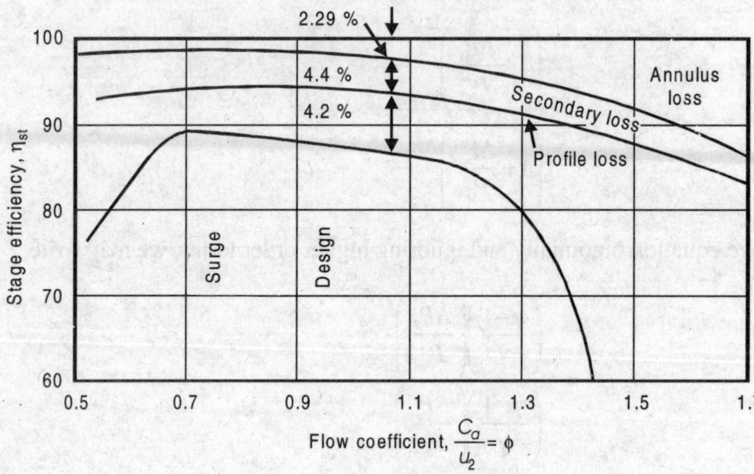

Fig. 20.22. *Losses in compressor stage.*

Figure 20.22 illustrates various losses influencing the stage efficiency.

(a) Profile losses. This is a pressure loss of two dimensional cascade arising from the skin friction on the blade surface and due to mixing of fluid particles after the blade.

(b) Skin friction losses on the annulus walls. The total pressure loss arise from the skin friction on the annulus walls and some secondary loss.

(c) Secondary flow losses. These losses occur due to combined effects of curvature and boundary layer.

20.7.9 Performance of Axial Flow Compressors

Figure 20.23 shows the performance characteristics of an axial flow compressor using dimensionless parameters namely stagnation pressure ratio $\frac{P_{o_2}}{P_{o_1}}$, dimensionless mass flow $\frac{m\sqrt{T_{o_1}}}{P_{o_1}}$ for dimensionless constant speeds $\frac{N}{\sqrt{T_{o_1}}}$ relative to design value.

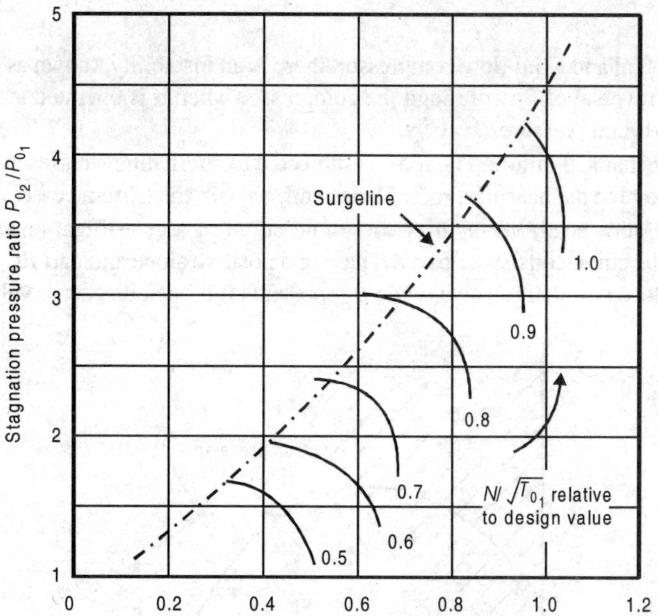

Fig. 20.23. *Axial compressor characteristics.*

It is seen that a fixed value of $\frac{N}{\sqrt{T_{o_1}}}$ covers only a narrow range of mass flow as compared with that in centrifugal compressors. As speed increases, the curves become steeper. Surging and choking impose limit at the upper end of $\frac{N}{\sqrt{T_{o_1}}}$.

The range of stable operation of axial flow compressor is narrow. Due care has to be taken to match the components for avoiding instability during operation when these compressors are used in gas turbine.

The variation of power, pressure ratio and efficiency of an axial flow compressor against flow rate is depicted in the Fig. 20.24.

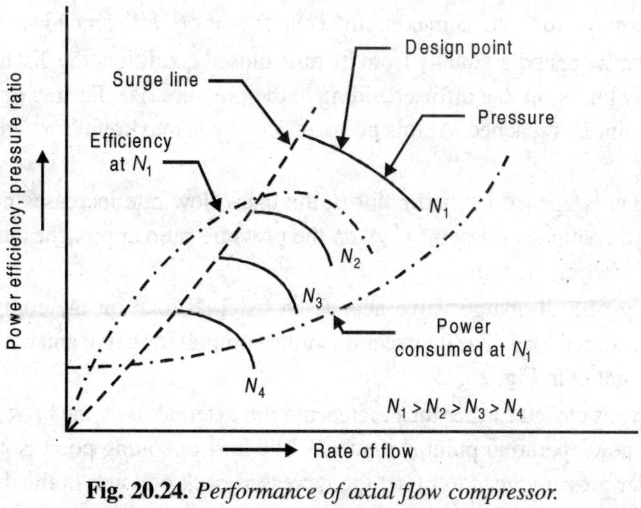

Fig. 20.24. *Performance of axial flow compressor.*

Figure 20.24 shows variation of pressure ratio versus volume flow rate for axial flow compressor.

20.8 SURGING

In the operation of centrifugal and axial flow compressor, there is an instability known as surging. It is caused due to unsteady, periodic and reversal of flow through the compressor when it is operated at less mass flow rate than that corresponding to maximum pressure.

Surging may lead to mechanical failure. The rotor is subjected to alternating stresses during this irregular working and may result in damage to the bearings, rotor blades and seals. In the extreme case, the rotor shaft may bend.

Consider the pressure ratio (head) versus mass flow rate curve of a centrifugal or axial flow compressor as shown in the Fig 20.25. The curve consists of part *AB* having a positive slope and part *BC* having a negative slope. Point *A* represents fully closed discharge valve while *B* represents full open discharge valve.

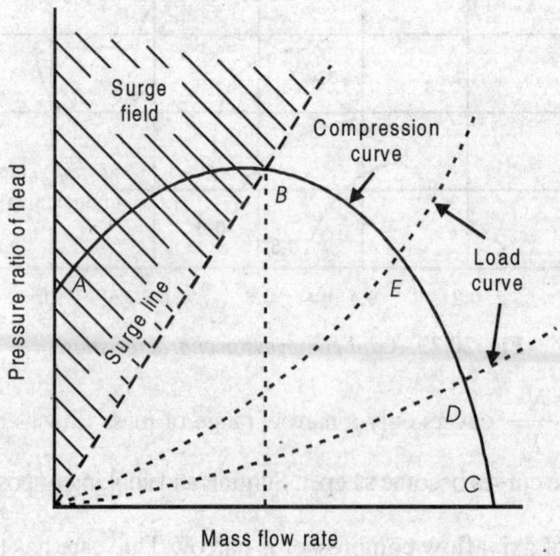

Fig. 20.25. *Surging.*

At point *A*, the flow rate is zero and pressure developed is called shut-off head. And at point *B*, pressure head developed is zero when the mass flow rate is maximum. Thus at point *B*, efficiency is zero.

When the discharge valve is opened gradually from its fully closed condition, the discharge of fluid commences and static pressure gradually builts-up due diffuser adding to the pressure rise. Further opening of discharge valve increases the pressure till point *B* is reached. At this point, efficiency is maximum for a given speed, inlet pressure and temperature.

Now, when delivery valve is opened beyond point *B*, the mass flow rate increases but pressure rise and efficiency decreases. This trend continues till point *C* when the pressure ratio approaches unity while the mass flow rate is maximum but the efficiency, is zero.

The opening and closing of a discharge valve acts as an external load on the compressor. The compressor operates in response to the external load. The intersection of the compressor curve and the load curve represents the operating point, say as at point *D* in Fig. 20.25.

When the discharge valve is closed further thus increasing the external load, the back pressure in the discharge line increases. As a result, new operating point shifts to *E*. The new operating point is possible and stable as the compressor develops greater pressure head to offset the increased back pressure in the discharge line.

With further closing of the discharge valve, external load increases, the operating point shifts in the region *A*–*B*.

During this operation, mass flow rate is less than the estimated value, say corresponding to point *B*. The compressor develops less head than that existing in the discharge line. As such, no discharge of fluid is possible. This leads to momentary flow reversal. A short while after, the fluid from the discharge line leaves and the back pressure would reduce. The delivery of fluid from the compressor would restart and the cycle would repeat with instability of operation again.

Therefore, when the flow rate from the compressor is less than the estimated value, a surge or pulsation commences. Air surges back and forth through the whole compressor rather than giving a unidirection steady stream of flow. The unstable operation of the compressor prevails in the region of positive slope of the curve shown in Fig. 20.25.

Various methods of avoiding and correcting surging conditions are listed below. It can be controlled,

(a) by speed control.
(b) by throttling inlet.
(c) by provision of flow control system.
(c–1) by venting discharge to the atmosphere.
(c–2) by passing fluid from discharge line to suction line.
(c–3) by providing integral expander on the by-pass line.
(d) by cut-off capacity control.
(e) by inlet guide vanes or adjustable diffuser vanes.

Figure 20.26 shows the surge fields of centrifugal compressor and axial flow compressor on pressure ratio $\frac{P_{o_2}}{P_{o_1}} - \frac{m\sqrt{T_{o_1}}}{P_{o_1}}$ plane. It is seen that the surge field of the centrifugal compressor is narrow as compressed with that of the axial flow compressor.

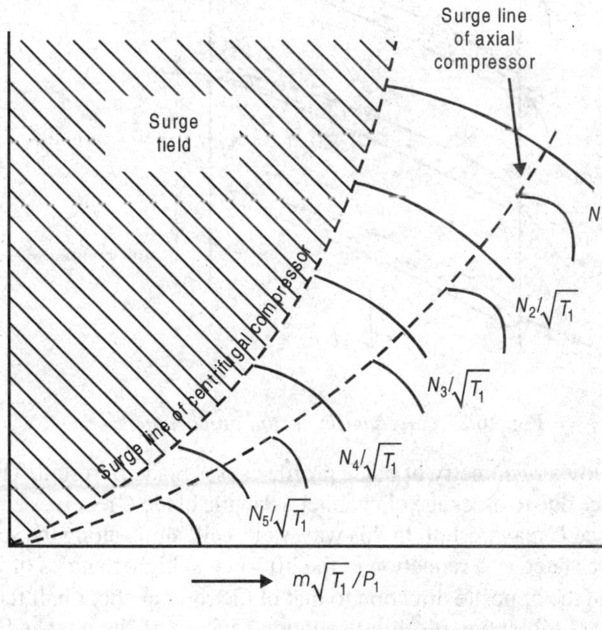

Fig. 20.26. *Surge line.*

20.9 CHOKING

Referring to the Fig. 20.25, when the mass flow rate is increased beyond B, the pressure ratio decreases and efficiency falls, as the air angle differs significantly from the vane angle causing the break-away of the air-stream.

This continues till point C where the pressure ratio becomes unity and efficiency zero. Entire power input is lost in overcoming internal friction. The point D on the curve BC, represents the maximum mass flow rate, known as 'choking mass flow'. The phenomena of choking limits the maximum mass flow rate.

As pressure ratio falls to unity, the theoretical mass flow rate is maximum. This occurs when the Mach number corresponding to inlet relative velocity becomes sonic.

20.10 STALLING

Stalling of a stage of axial flow compressor is defined as the *aerodynamic stall or breakaway of the flow stream from the suction side of the blade aerofoil*. It may be due to less mass flow rate than the designed value or due to non-uniformity in the blade profile.

The phenomena of stall preceds surging. A multistage compressor may operate in the stable unsurged region with some of its stages stalling. Thus, stalling is a local phenomena while surging is a complete system phenomena. The phenomena of stalling has been extensively researched by Smith and Fletcher.

A rotating stall is illustrated in Fig. 20.27.

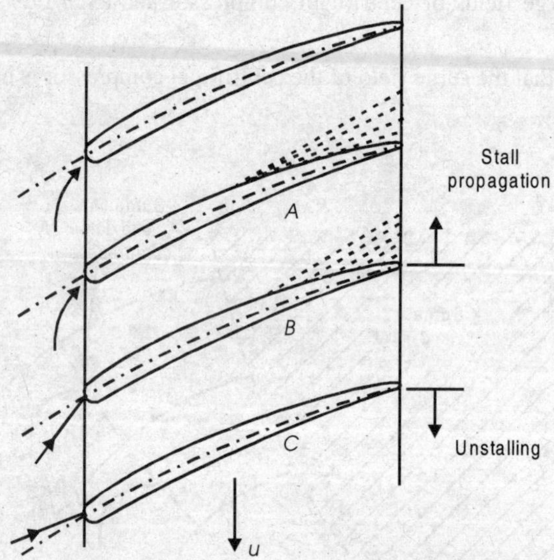

Fig. 20.27. *Mechanism of stall propagations.*

The non-uniformity of the flow or geometry of blade profile causes blade B to stall. The air now flows on to the blade A at an increased incidence due to blockage of channel AB while blade C receives air at reduced incidence. As a result blade A stalls while blade C may unstall. In this way, stall 'cell' shifts along the stage in the direction of lift of the blades. Thus, stall can be stated as a reduction in the lift force at higher angles of incidence.

A rotating stall may rotate in the opposite direction to that of the rotor at about half the rotational speed. It may lead to aerodynamically induced vibrations resulting in fatigue failure of the attached components. Axial flow compressors are more prone to stalling.

20.11 PERFORMANCE CHARACTERISTICS OF CENTRIFUGAL FLOW COMPRESSOR

The performance of centrifugal compressors is specified by illustrating the variation of delivery pressure and temperature with mass flow rate for various speeds. However, these characteristics are further influenced by other variables like inlet pressure and temperature.

To understand the interdependence of these variables, dimensionless parameters are often useful, such as

$$\frac{P_{o_2}}{P_{o_1}}, \frac{T_{o_2}}{T_{o_1}}, \frac{m\sqrt{T_{o_1}}}{P_{o_1}}, \frac{ND}{\sqrt{RT_{o_1}}}$$

Even isentropic efficiency is also expressed in a similar way as

$$\eta_{\text{isent}} = \frac{T_{o_2} - T_{o_1}}{T_{o_2} - T_{o_1}} = \frac{\left(\frac{P_{o_2}}{P_{o_1}}\right)^{\frac{\gamma-1}{\gamma}} - 1}{\left(\frac{T_{o_2}}{T_{o_1}} - 1\right)}$$

Figure 20.28 shows the plot of such parameters.

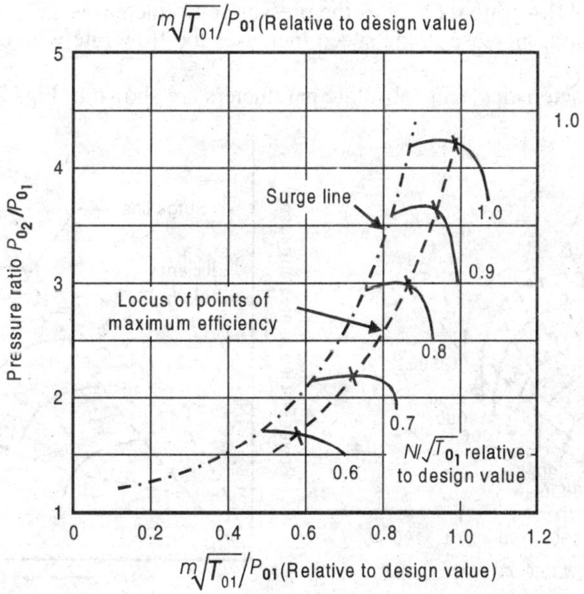

Fig. 20.28. *Centrifugal compressor charcuteristics.*

Points of low mass flow at the extremities of constant speed curves can be joined to obtain surge line. While the curve obtained by joining extremities on the right hand side of constant speed curves represents choking limit occurring at low pressure ratio. The compressor is operated only between these extreme limits.

Figure 20.29 shows the variation of isentropic efficiency with $\frac{m\sqrt{T_{o_1}}}{P_{o_1}}$ for various values of $\frac{N}{\sqrt{T_{o_1}}}$ relative to the design value. The maximum efficiency is almost same at all speeds.

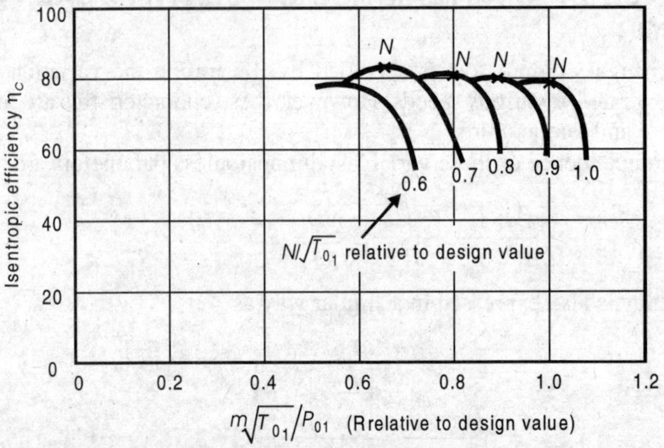

Fig. 20.29. *Efficiency and mass flow for centrifugal characteristics.*

From Figs 20.28 and 20.29, following deductions can be made:

1. The curves are fairly flat at lower speeds and are limited by surging. At high speed, the range is limited by surging at one end and choking at the other end.
2. At a given speed, mass flow rate reduces as the pressure ratio increases.
3. At a given pressure ratio, increase in the speed increases the flow rate with considerable reduction in the efficiency.

Similar performance characteristics using absolute parameters are shown in Figs 20.30 and 20.31.

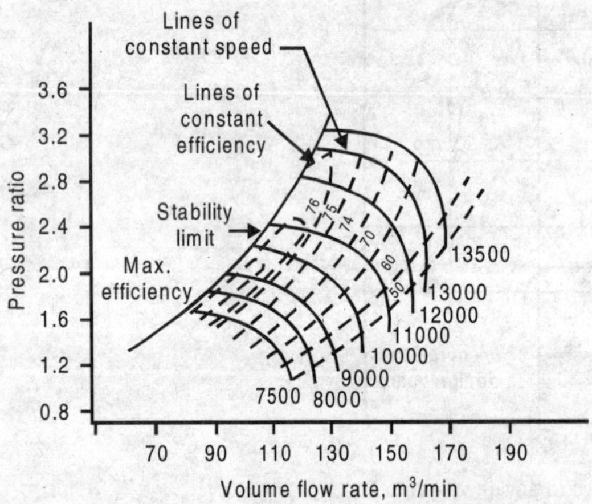

Fig. 20.30. *Performance charcateristics of centrifugal flow compressors.*

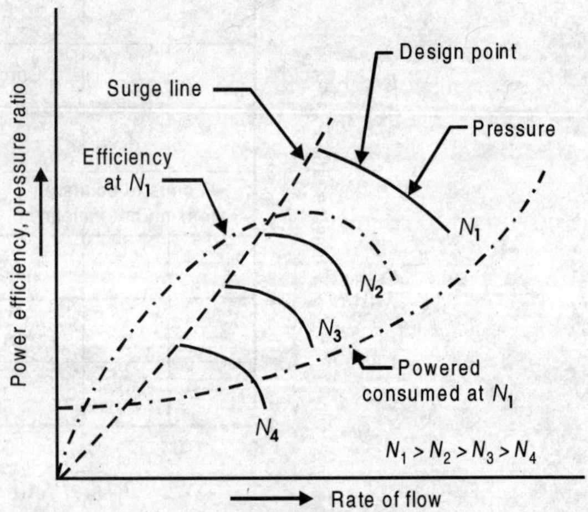

Fig. 20.31. *Performance curves of an centrifugal flow compressors.*

20.12 PERFORMANCE CHARACTERISTICS OF AXIAL FLOW COMPRESSORS

The variation of pressure ratio $\dfrac{P_{o_2}}{P_{o_1}}$ aganist $\dfrac{m\sqrt{T_{o_1}}}{P_{o_1}}$ relative to design value for various values of $\dfrac{N}{\sqrt{T_{o_1}}}$ relative to design value is shown in the Fig. 20.32.

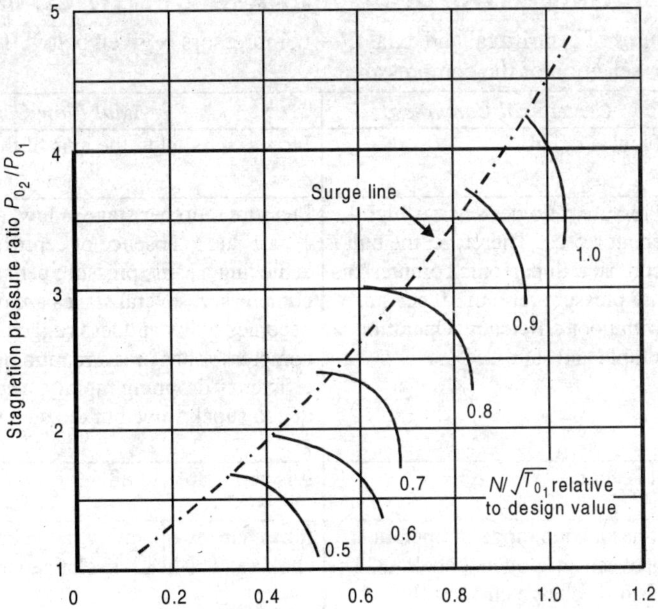

Fig. 20.32. *Axial compressor characterstics.*

It is seen that, for a given value of $\dfrac{N}{\sqrt{T_{o_1}}}$, these curves cover a much narrow range of mass flow rate as compared with those for centrifugal compressors. Also at higher rotational speeds, the curves become steep, nearly vertical. Therefore, the range of stable operation of axial flow compressor is much narrower. Hence, great care is necessary to match the components of a gas turbine plant for avoiding instability of operation. The phenomena of surging and stalling can hardly be distinguished as occurrence of one may lead to occurrence of the other. Stalling of these compressors results in severe vibrations of blade.

The performance characteristics using absolute parameters are also shown in Fig. 20.33.

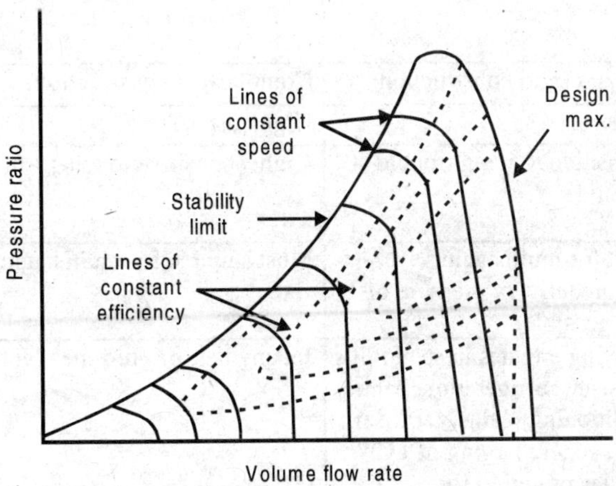

Fig. 20.33. *Performance curves of axial flow compressors.*

20.13 COMPARISON OF CENTRIFUGAL AND AXIAL FLOW COMPRESSORS

Comparison of main features of centrifugal and axial flow compressors is given below. The comparison enables to arrive at some criteria for selection of the compressor.

S. No.	*Item*	*Centrifugal Compressor*	*Axial Flow Compressor*
1.	Direction of flow	Flow is radial	Flow is parallel to the axis of the compressor.
2.	Pressure Ratio	Pressure ratio per stage is high about 4.5 : 1. Therefore, the unit is compact. Supersonic compressors use pressure ratio of 10 per stage with poor efficiency. Operation is simple and safe.	Pressure ratio per stage is low, about 1.2 : 1 due to absence of centrifugal action. Hence, for achieving a same pressure per stage of a centrifugal compressor, several stages are required. Thus, the unit becomes bulky and less rugged. The supersonic compressor has pressure ratio of 10 per stage with efficiency dropping rapidly. The compression is achieved due to shock wave but the operation is risky.
3.	Isentropic efficiency	It is about 80% to 82%.	It is about 86% to 88%.
4.	Operating range	It has a wide range of operation between surging and choking. The head-discharge curve is flat.	It has narrow operating range between surging and choking. The head-discharge curve is steep.
5.	Part load performance	It is better.	It is poor.
6.	Frontal area	It has a large frontal area.	It has smaller frontal area for a given mass floor rate and pressure than that of centrifugal compressor. Therefore, it is more suitable for jet engines.
7.	Effect of deposits on the blade surfaces	Accumulated deposits on the blade surfaces due to the flow of contaminated fluid has no adverse effect on the performance.	Performance is adversely affected due to accumulated deposits on the blade surfaces when contaminated fluid flows.
8.	Starting torque required	Need low starting torque.	Need high starting torque.
9.	Constructior	Simple and rigid construction.	Complicated construction.
10.	Cost	Low cost.	High cost.
11.	High altitude operation	Less sensitive to icing trouble at high altitude.	Highly sensitive to icing trouble at high altitude.
12.	Multistaging	Difficult for multistaging. Maximum delivery pressure of 400 bar possible.	Most suitable for multistaging.
13.	Field of application	For blowing engines in steel mills, low pressure refrigeration, central air conditioning plants, fertiliser industry, supercharging of I.C. engine, gas pumping for transportation, petrochemical industries.	Jet engines, gas turbine plants, steel mills, etc.

20.14 COMPARISON OF CENTRIFUGAL AND RECIPROCATING COMPRESSORS

The distinguishing features of these compressors are tabulated below.

Item	*Reciprocating Compressor*	*Centrifugal Compressor*
1. Mechanical balance	Poor mechanical balance and greater vibrations due to presence of reciprocating masses.	Absence of reciprocating masses provides better mechanical balance and vibration free operation.
2. Pressure ratio per stage	High upto 5–8.	Comparatively low upto 4.5.
3. Multistaging capability	Suitable for high delivery pressures with multistaging, delivery pressure upto 5000 bar possible.	Suitable for medium delivery pressure. Delivery pressure of 400 bar possible with multistaging.
4. Volume flow rate	Capable of delivering small volume flow rates.	Capable of delivering large volume of flow rates.
5. Operational flexibility	Greater flexibility in pressure range and capacity.	No flexibility in pressure range and capacity.
6. Maintenance cost	High.	Low
7. Effect of pressure ratio	For pressure ratios above 2, efficiency is high.	High efficiency for pressure ratio below 2.
8. Drive	Suitable for slow speed drive only.	Suitable for direct coupling to high speed drives like turbines.
9. Attention required	Greater attention required.	Less attention is adequate.
10. Contamination with lubricating oil	Greater chance of mixing working fluid with lubricating oil.	No chance of contamination of working fluid with lubricating oil.
11. Suitability	Suitable for low and medium and high delivery pressures with low/medium volume flow rates.	Suitable for low and medium pressures with large volume flow rates.

20.15 APPLICATIONS OF COMPRESSED AIR/GASES

The various applications of compressed air/gases are

1. For industrial furnaces for supporting combustion.
2. For infloating the vehicle tubes.
3. For actuating pneumatic tools such as hammers, chisels, drills, screw drivers, etc.
4. For industrial and household cleaning of floors, equipments, machine tools, components etc.
5. For pneumatic conveyor belts.
6. For pneumatic brakes of automobiles.
7. For air conditioning plants.
8. For ventilation of mines.
9. For production of aerated soft drinks.
10. For the production of synthetic ammonia, hydrogenation of brown coal, polymerisation of ethylene.
11. For air lift pumps, glass blowing.

12. For supercharging of IC engines.
13. For long distance transportation of industrial gases.
14. In gas turbine plants and jet engines.
15. For cabin pressurisation of air crafts.

SOLVED EXAMPLES

Example 20.1. *A centrifugal compressor has a total pressure head ratio 4:1 and discharges 18.2 kg of air per second. It runs at 15000 rpm. This inlet stagnation temperature is 15° C. The slip factor is 0.9 and power input factor is 1.4. It has isentropic efficiency of 80%.*

Determine the outer diameter of the impeller.

Solution

Let suffix *o* denote the stagnation temperature.

$$\frac{T_{O_2}}{T_{O_1}} = \left(\frac{P_{O_2}}{P_{O_1}}\right)^{\frac{\gamma-1}{\gamma}} = 4^{\frac{0.1}{1.4}}$$

$$\therefore \quad T_{O_2} = 288\times 4^{\frac{0.1}{1.4}} = 427.97 \text{ K}$$

$$\therefore \text{ Now, isentropic efficiency } \eta_{is} = \frac{T_{O_2}-T_{O_1}}{T_{O_{2'}}-T_{O_1}}$$

$$\therefore \text{ Actual temperature rise } \left(T_{O_2}-T_{O_1}\right) = \frac{427.97-288}{0.8} = 174.96 \text{ K}$$

$$\text{Work required/kg of air} = \frac{\sigma\mu u_2^2}{1000} = C_p\left(T_{O_2}-T_{O_1}\right)$$

$$\therefore \text{ Blade speed at outlet } u_2 = \sqrt{\frac{1.005\times1000\times174.96}{1.04\times0.9}} = 433.42 \text{ m/s}$$

But
$$u_2 = \frac{\pi D_2 N}{60}$$

$$D_2 = \frac{60\times433.42}{\pi\times15000} = 0.5518 = 55.18 \text{ cm}$$

Example 20.2. *A rotary vane compressor has pressure ratio of 1.4 and the intake pressure of 1.0 bar. The compressor discharges 00.6 m³/rev. of FDA.*

Calculate the work input required per revolution to drive the compressor when

(1) there is no internal compression (Roots blower)

(2) there is 50% pressure rise due to internal isentropic compression before back flow occurs (vane compressor). Also find the efficiency of compressor.

(i) When there is no internal compression (Roots blower):

The p–V diagram is shown below.

Data: $V_1 = 0.6$ m³/rev $\quad p_1 = 1.0$ bar $\quad p_2 = 1.4$ bar

Solution

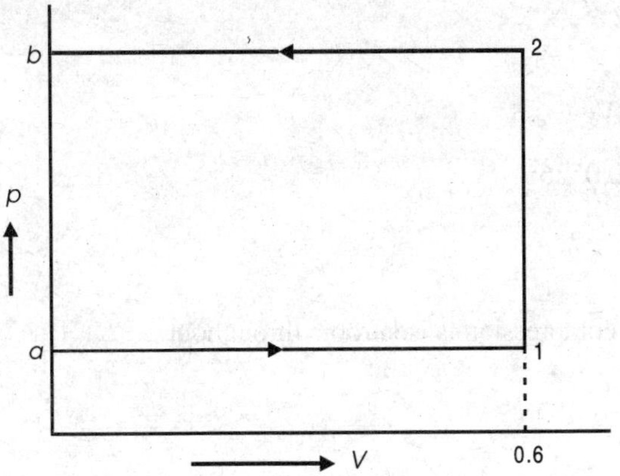

Fig. Ex. 20.1 (a)

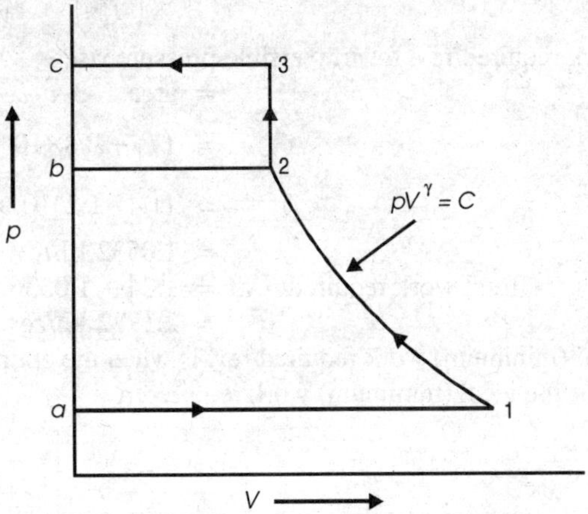

Fig. Ex. 20.1 (b)

$$\text{Work req./rev.} = (P_2 - P_1) \times V_1$$
$$= (1.4 - 1.0) \times 10^2 \times 0.06$$
$$= \mathbf{2.4\ kJ/rev.}$$

(ii) When half the pressure rise occurs due to internal compression (vane compressor). The *p–V* diagram is shown below.

Data: $p_1 = 1$ bar

$p_3 = ?$

$$\text{The intermediate pressure } P_2 = P_1 + \frac{(P_2 - P_1)}{2}$$
$$= 1 + \frac{1.4 - 1.0}{2}$$
$$= 1.2 \text{ bar}$$

Now, work required/rev. for isentropic compression 1 – 2

$$\text{Area } a-1-2-b = \frac{\gamma}{\gamma - 1} p_1 V_1 \left[\left(\frac{P_2}{P_1} \right)^{\frac{\gamma-1}{\gamma}} - 1 \right]$$
$$= \frac{1.4}{0.4} \times 1.0 \times 10^2 \times 0.06 \left[\left(\frac{1.2}{1.0} \right)^{\frac{r-1}{r}} - 1 \right]$$
$$= 1.24 \text{ kJ/rev.}$$

and $$V_2 = V_1\left(\frac{P_2}{P_1}\right)^{\frac{\gamma-1}{\gamma}} = \left(\frac{1.2}{1}\right)^{\frac{0.4}{1.4}} \times 0.06 = 0.05266\ \text{m}^3$$

Work required/rev. for irreversible pressure rise

$$= \text{area } 2\text{–}3\text{–}c\text{–}b\text{–}2$$

$$= (P_3 - P_2)\times 10^2 \times V_2$$

$$= (1.4-1.2)10^2 \times 0.05266$$

$$= 1.0532 \text{ kJ/rev.}$$

∴ Total work required/rev. $= 1.24 + 1.0536$

$$= \mathbf{2.1772\ kJ/rev.}$$

Ideal (minimum) work required/rev. is when the entire compression is isentropic throughout. Thus, the ideal (minimum) work required/rev.

$$= \frac{r-1}{r} p_1 V_1 \left[\left(\frac{P_2}{P_1}\right)^{\frac{\gamma-1}{\gamma}} - 1\right]$$

$$= \frac{1.4}{0.4} \times 1.0 \times 0.06 \times 10^2 \left[\left(\frac{1.4}{1}\right)^{\frac{0.4}{1}} - 1\right]$$

$$= 2.12 \text{ kJ/rev.}$$

Therefore, efficiency of Roots blower $= \frac{2.12}{2.4} \times 100$

$$= \mathbf{88.3\%}$$

Similarly, efficiency of vane compressor $= \frac{2.12}{2.1772} \times 100$

$$= \mathbf{97.3\%}$$

Example 20.3. *A centrifugal compressor delivers 10 m/s free air while running at 9000 r.p.m. The air is compressed from 1 bar and 20°C to 4 bar with an isentropic efficiency of 0.82.*

Blades are radial at outlet of the impeller and the flow velocity is 62 m/s which is constant throughout. The ratio of outer to inner radii of the impeller is 2 and the slip factor is 0.9. The blade area coefficient at inlet is 0.9.

Determine

(1) temperature of air discharged
(2) theoretical power required
(3) diameters of the impeller at inlet and outlet
(4) the breadth of impeller at inlet
(5) impeller blade angle at inlet
(6) diffuser blade angle at inlet

Data:

N = 9000 r. p. m.   F. A. D. = 10 m³/s

P_1 = 1.0 bar T_{o_2} = 293 K

P_2 = 4.0 bar η_{isent} = 0.82

Solution

Slip factor, $\mu = 0.9$

$$V_{f_1} = V_{f_2} = 62 \text{ m/s} \qquad \frac{R_2}{R_1} = 2.0$$

$$k_1 = 0.9$$

$$T_{o_2} = T_{o_1}\left(\frac{P_{o_2}}{P_{o_1}}\right)^{\frac{\gamma-1}{\gamma}} = 293\,(4)^{\frac{0.4}{1.4}}$$

$$= \mathbf{433.56\ K}$$

$$\text{Actual temp. rise} = \left(T_{O_2} - T_{O_1}\right) = \frac{T_{O_2} - T_{O_1}}{\eta_{isent}}$$

$$= \frac{433.56 - 293}{0.82}$$

$$= 173.85$$

$$\therefore \quad T_{o_2} = 293 + 173.85 = 466.85 \text{ K}$$

$$\text{Mass flow rate } \dot{m} = \frac{p_1 V}{RT_{o_1}} = \frac{1\times10^2\times10\times60}{0.287\times293}$$

$$= 713.5 \text{ kg/min.}$$

Now, work required/s = Theoretical power input required.

$$= \dot{m}\, C_P\left(T_{o_2} - T_{o_1}\right)$$

$$= \frac{713.5}{60}\times1.005\times173.85$$

$$= 2077.7 \text{ kW}$$

For radial blades, the work input required is given by

$$\text{Work input required/kg} = \frac{\sigma\mu u_2^2}{1000} = C_p\left(T'_{O_2} - T_{O_1}\right)$$

$$\therefore \quad u_2 = \sqrt{\frac{1000\times1.005\times173.85}{0.9}}$$

$$= 440.6 \text{ m/s} = \frac{\pi D_2 N}{60}$$

$$\therefore \quad D_2 = \frac{60\times440.6}{\pi\times9000} = 0.9349 \text{ cm}$$

$$= \mathbf{93.49\ cm}$$

and $$D_1 = \frac{D_2}{2} = \mathbf{46.745\ cm}$$

where k = blade area coefficient at inlet

$$b_1 = \frac{600}{60\times\pi\times0.46745\times62\times0.9}$$

$$= 0.122\text{ m} = 12.2\text{ cm}$$

Referring to inlet velocity triangle (Fig. 20.18)

$$\tan\beta_1 = \frac{V_{f_1}}{u_1} = \frac{62\times2}{440.0}\left[\because u_1 = \frac{1}{2}u_2\right]$$

$$= 0.2814$$

$$\therefore \quad \beta_1 = 15.7°$$

Also, $$\tan\alpha_2 = \frac{V_{f_2}}{\mu u_2} = \frac{62}{0.9\times440.6}$$

$$= 0.1563$$

$$\alpha_2 = 8.9°.$$

Example 20.4. *A centrifugal air compressor delivers air with a stagnation pressure ratio of 4 :1. Its speed is 15000 r.p.m. The stagnation temperature at inlet is 17°C and the power input factor is 1.04. It has overall diameter 0.5538 m and isentropic efficiency of 0.8.*

Determine the slip factor.

Data: N = *15000 r.p.m.* η_{isent} = *0.8*

ψ = *1.04* T_{o_1} = *280 K*

$\frac{P_{o_2}}{P_{o_1}}$ = *4* D_2 = *0.5538 m*

Solution

$$\therefore \quad T_{o_2} = T_{o_1}\left(\frac{P_{o_2}}{P_{o_1}}\right)^{\frac{\gamma-1}{\gamma}} = 290\times4^{\frac{0.4}{1.4}}$$

$$= 431\text{ K}$$

Actual stagnation temperature after compression T'_{o_2} can be calculated as

$$T'_{o_2} - T_{o_1} = \frac{T_{o_2} - T_{o_1}}{\eta_{isent}} = \frac{431-290}{0.8}$$

$$= 176.25\text{ K}$$

Now, Work required/kg $\dfrac{\Psi \mu u'^2_2}{1000} = C_p\left(T'_{O_2} - T_{O_1}\right)$

But, $u_2 = \dfrac{\pi D_2 N}{60} = \dfrac{\pi \times 0.5538 \times 15000}{60}$

$= 435$ m/s

Ship factor $\mu = \dfrac{1.005(176.25) \times 1000}{1.04 \times 435^2}$

$=$ **0.9**

Example 20.5. *An aeroengine uses centrifugal compressor for supercharging. The compressor delivers 3 kg/s of air. Air enters the compressor at 1 bar, 280 K and 90 m/s after compression in the impeller, air has pressure 1.5 bar temperature 335 K and 230 m/s.*

Determine:

(1) isentropic efficiency

(2) power required to drive the compressor

(3) overall efficiency.

Assume that all the kinetic energy of air gained in the impeller is entirely converted into pressure in the diffuser.

Solution

$$\dot{m} = 3 \text{ kg/s}$$

$p_1 = 1$ bar $\quad T_1 = 280$ K, $\quad V_1 = 90$ m/s

$p_2 = 1.5$ bar $\quad T_2 = 335$ K, $\quad V_2 = 230$ m/s

Referring to the *T–S* diagram,

Actual work required/kg of air in the impeller

$$= C_p(T_2 - T_1) + \left(\frac{V_2^2 - V_1^2}{2 \times 1000}\right)$$

$$= 1.005(335 - 280) + \frac{230^2 - 90^2}{2 \times 1000}$$

$$= 81.71 \text{ kJ/kg}$$

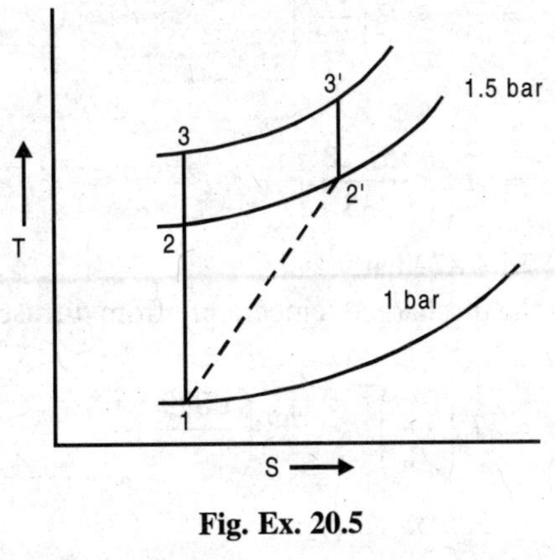

Fig. Ex. 20.5

$$\frac{T_2}{T_1} = \left(\frac{P_2}{P_1}\right)^{\frac{\gamma-1}{\gamma}} \qquad \therefore\ T_2 = 280\times1.5^{\frac{0.4}{1.4}} = 314.4\text{ K}$$

Isentropic work required/kg of air

$$= C_p(T_2 - T_1) + \frac{V_2^2 - V_1^2}{2\times1000}$$

$$= 1.005(314.4 - 280) + \frac{230^2 - 90^2}{2\times1000}$$

$$= 56.97\text{ kJ/kg}$$

$$\therefore \quad \text{Isentropic efficiency} = \frac{\text{Isentropic work reqd./kg}}{\text{Actual work reqd./kg}}$$

$$= \frac{56.97}{87.71}$$

$$= 69.72\%$$

$$\text{Power required} = \text{Actual work required/kg} \times m$$

$$= 81.71 \times 3 = \mathbf{245.13\ kW}$$

As kinetic energy gained in the impeller is converted into pressure, therefore,

$$C_p(T_3' - T_2') = \frac{V_2^2 - V_1^2}{2\times1000}$$

$$1.005\ (T_3' - 335) = \frac{230^2 - 90^2}{2000}$$

$$T_3' = \mathbf{357.278\ K}$$

Now, the pressure of air after leaving the diffuser

$$= P_3 = P_2\left(\frac{T_3'}{T_2'}\right)^{\frac{\gamma-1}{\gamma}}$$

$$= 1.5\left(\frac{357.28}{335}\right)^{\frac{1.4}{0.4}}$$

$$= 1.879\text{ bar}$$

In case of isentropic compression the discharged temperature from diffuser is

$$T_2' = T_1\left(\frac{P_2}{P_1}\right)^{\frac{r-1}{r}} = 280\left(\frac{1.879}{1}\right)^{\frac{0.4}{1.4}}$$

$$= 335.35\text{ K}$$

$$\text{Overall isentropic efficiency} = \frac{335.35-280}{357.28-280}$$

$$= \mathbf{71.62\%}$$

Example 20.6. *A single suction centrifugal compressor delivers 6 kg/s of air. The ambient air conditions are 1 bar and 20°C. It runs at 22000 r.p.m. with an isentropic efficiency 82%. Air is compressed from 1 bar to 4.2 bar total (stagnation) pressure. Air enters the eye of the impeller with a velocity of 150 m/s without any prewhirl. Assuming the ratio of whirl speed to tip speed as 0/9, calculate*

(*a*) *rise in stagnation temperature during compression if the change in kinetic energy is negligible*
(*b*) *tip diameter of the impeller*
(*c*) *power required*
(*d*) *diameter of the eye of the impeller if the hub diameter is 10 cm.*

Data: $T_1 = 293$ K, $V_1 = 150$ m/s

$N = 22000$ r. p.m., $\eta_{isent} = 82\%$

$\frac{V_{w_2}}{u_2} = 0.9$, $P_{o_2} = 4.2$ bar

$p_1 = 1.0$ bar, $\dot{m} = 6$ kg/s

$V_{w_1} = 0$

Solution

$$T_{o_1} = T_1 + \frac{V_1^2}{2C_P} = 293 + \frac{150^2}{2\times1.005\times1000}$$

$$= 304.19 \text{ K}$$

$$P_{o_1} = P_1\left(\frac{T_{o_1}}{T_1}\right)^{\frac{\gamma-1}{\gamma}}$$

and

$$T_{o_2} = T_{o_1}\left(\frac{P_{o_2}}{P_{o_1}}\right)^{\frac{\gamma-1}{\gamma}} = T_{o_1}\left[\frac{P_{o_2}}{P_1\left(\frac{T_{o_1}}{T_1}\right)^{\frac{\gamma-1}{\gamma}}}\right]^{\frac{\gamma-1}{\gamma}}$$

$$= T_{o_1}\left(\frac{p_{o_2}}{p_{o_1}}\right)^{\frac{\gamma-1}{\gamma}}$$

$$= 293\left(\frac{4.2}{1}\right)^{\frac{0.4}{1.4}} = 441.68 \text{ K}$$

Referring to the Fig. Ex. 20.6

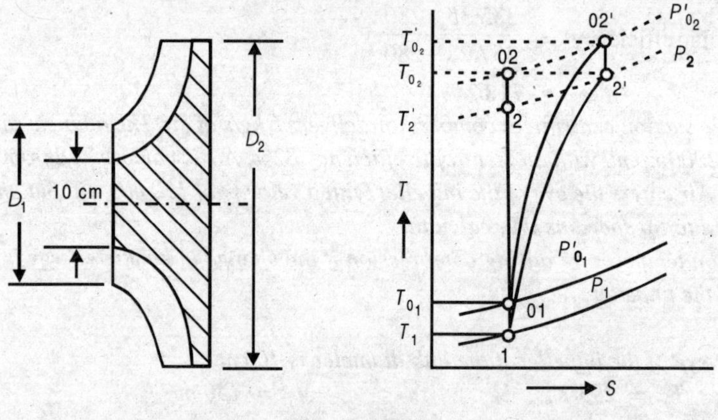

Fig. Ex. 20.6

But, $$\frac{V_{w_r}}{u_2} = 0.9$$

$$\therefore \quad 168.5 = \frac{u_2^2 \times 0.9}{1000}$$

$$\therefore \quad u_2 = 432.69 \text{ m/s}$$

and $$u_2 = 432.69 = \frac{\pi D_2 N}{60}$$

$$\therefore \quad D_2 = 0.3756 \text{ m}$$
$$= \mathbf{37.56 \text{ cm}}$$

Similarly, power required = Work required/kg × $\dot{m}$
$$= 168.5 \times 6$$
$$= \mathbf{1011.00 \text{ kW}}$$

Now, using continuity equation,

$$\dot{m} = \frac{\pi}{4}\left(D^2{}_1 - D^2{}_h\right)V_1 \times \rho_1$$

where, $$\rho_1 = \frac{P_1}{RT_1} = \frac{1 \times 10^5}{0.287 \times 293}$$
$$= 1.189 \text{ kg/m}^3$$

$$\therefore \quad \rho = \frac{\pi}{4}\left(D_1^2 - 0.1^2\right) \times 90 \times 1.189$$

$$\therefore \quad D_1 = 0.2298 \text{ m}$$
$$= \mathbf{22.98 \text{ cm}}$$

Example 20.7. *A centrifugal compressor is used as a supercharger for an aircraft engine developing 750 kW power and having specific fuel consumption of 0.273 kg/kW-h. The supercharger supplies fuel : air mixture at 1.25 bar. The air : fuel ratio is 17. Air enters the supercharger at 0.55 bar and 0°C if the adiabetic efficiency of the* *supercharger is 85%, find the volume of mixture to be supplied and the power required to drive the supercharger.*

Assume C_p = 1.0 kJ/kg-K and C_v = 0.277 kJ/kg-K.

Solution

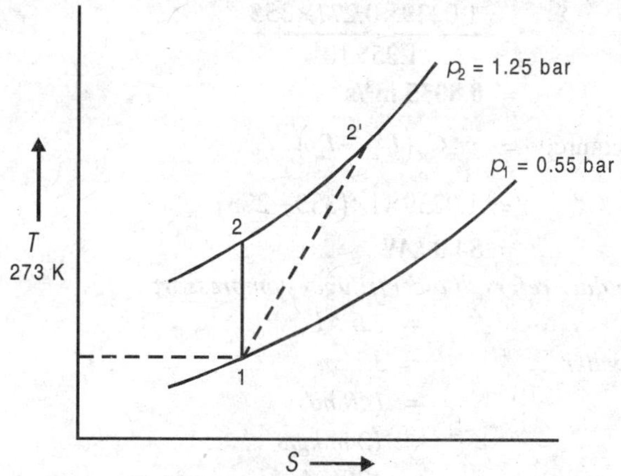

Fig. Ex. 20.7

Fuel consumption of the jet engine = Specific fuel comsumption × Power developed

$$= \frac{750 \times 0.273}{3600} \text{ kg/s}$$

$$\dot{m}_f = 0.05688 \text{ kg/s}$$

$$\text{Air supplied/s} = \dot{m}_f \times \text{air : fuel ratio}$$

$$= 0.5688 \times 17$$

$$\dot{m}_a = 0.967 \text{ kg/s}$$

Hence, mass of mixture = $\dot{m}_f + \dot{m}_a$

$$m_m = 1.0239 \text{ kg/s}$$

Now,
$$r = \frac{C_p}{C_v} = \frac{C_p}{C_p + R} = \frac{1.0}{1.0 - 0.277} = 1.383$$

$$T_{o_2} = T_{o_1}\left\{\frac{P_{o_2}}{P_{o_1}}\right\}^{\frac{\gamma-1}{\gamma}} = 273\left\{\frac{1.25}{0.55}\right\}^{\frac{0.4}{1.4}}$$

$$= 342.7 \text{ K}$$

Isentropic efficiency,
$$\eta_{isent} = \frac{T'_{o_2} - T_{o_1}}{T_{o_2} - T_{o_1}}$$

$$0.85 = \frac{342.7 - 273}{T'_{o_2} - 273}$$

$$\therefore \quad T_{o_2} = 355 \text{ K}$$

$$\therefore \text{ Volume of mixture to be supplied, } V_m = \frac{\dot{m}_m RT'_{o_{2'}}}{P_o}$$

$$= \frac{1.0239 \times 0.277 \times 355}{1.25 \times 10^2}$$

$$= \mathbf{0.8055\ m^3/s}$$

$$\text{And, Power required } = \dot{m}_m C_p \left(T'_{o_2} - T_{o_1}\right)$$

$$= 1.0239 \times 1 \times (355 - 273)$$

$$= \mathbf{84.0\ kW}$$

Example 20.8. *The following data refers to a centrifugal compressor:*

Ratio of total pressure	*= 3.6 : 1*
Diameter of the eye of impeller	*= 35 cm*
Axial velocity at inlet	*= 140 m/s*
Mass flow rate of air	*= 12 m kg/s*
Velocity in the delivery duct	*= 120 m/s*
Tip speed of impeller	*= 460 m/s*
Speed	*= 20000 r.p.m*
Total head isentropic efficiency	*= 80%*
Pressure coefficient	*= 0.73*
Ambient conditions	*= 1.013 bar and 15ºC*

Calculate:

(i) Static pressure ratio

(ii) Static pressure and temperature at inlet and outlet of compressor

(iii) Work reqd./kg of air

(iv) Theoretical power required.

Data: $r_p = 3.6$

$T_{o_1} = 288$ K, $\quad \eta_{isent} = 0.80$

$\dot{m} = 12$ kg/s $\quad \psi = 0.73$

Solution

$$T_{o_2} = T_{o_1}\left(r_p\right)h^{\frac{\gamma-1}{\gamma}} = 288(3.66)^{\frac{0.4}{1.4}} \text{ K}$$

$$\therefore \quad T'_{o_2} - T_{o_1} = \frac{T_{o_2} - T_{o_1}}{\eta_{isent}} = \frac{288\left[3.6^{\frac{0.4}{1.4}} - 1\right]}{0.80} = 159.7^\circ\text{C}$$

$$T_{o_2} = 288 + 159.7 = \mathbf{447.1\ K}$$

$$\text{Now static temperature at exit} = T'_{o_2} - \frac{120^\circ}{2C_p}$$

$$= 447.1 - \frac{120^{o}}{2\times1.005\times10^{3}}$$

$$T_2 = 440 \text{ K}$$

Also static pressure at exit

$$P_2 = P_{o_2} - \frac{\rho_2 \times 120^2}{2}$$

where

$$= 3.65 - \frac{P_2 \times 120^2}{0.287\times440\times2\times10^3} \qquad \rho_2 = \frac{p_2}{RT_2}$$

$$= 3.45 \text{ bar} \qquad \text{and} \qquad P_{o_2} = 1.013\times3.6 = 3.65 \text{ bar}$$

Now static condition at inlet

$$T_{o_1} = T_1 + \frac{140^2}{2\times C_p} = T_1 + \frac{140^2}{2\times1.005\times10^3} = T_1 + 9.75$$

$$T_1 = 288 - 9.5$$

$$= 278.25 \text{ K}$$

Also,

$$P_1 = P_{o_1} - \frac{140^2 \times P_1}{2\times0.287\times278.25\times10^3}$$

$$= 1.013 - 0.123\, P_1$$

∴ $\mathbf{P_1 = 0.9 \text{ bar}}$

∴ **Static pressure ratio** $= \mathbf{\frac{P_2}{P_1} = \frac{3.45}{0.9} = 3.83}$

Furthur, work required/kg of air = $C_p\left(T'_{o_2}, T_{o_1}\right)$

$$= 1.005\ (159 - 1)$$

$$= \mathbf{159.89 \text{ kJ/kg}}$$

$$= \dot{m}C_p\left(T'_{o_2}, T_{o_1}\right)$$

$$= 10 \times 159.89 \text{ kW}$$

$$= \mathbf{1598.9 \text{ kW}}$$

Example 20.9. *A centrifugal compressor running at 18000 r.p.m. admits at 300 K. The other data is as follows:*

Isentropic total head efficiency = 0.76

Outer diameter of blade tip = 550 mm

Slip factor = 0.82

Calculate:

(i) Temperature rise of air

(ii) Static pressure ratio.

Assume that absolute velocity of air at inlet and exit to be same.

Take C_p = 1.005 kJ/kg-K

Solution

T_1 = 300 K, $\qquad N$ = 18000 r. p. m.

D_2 = 0.55 m $\qquad \eta_{isent}$ = 0.76

μ = 0.82

$$u_2 = \frac{\pi D_2 N}{60} = \frac{\pi \times 0.55 \times 18000}{60}$$
$$= 518.36 \text{ m/s}$$

Work required/kg of air = $\left(V_{w_2} u_2 - V_{w_1} u_1\right)$ N-m/kg

But, $V_{w_1} = 0$ and $\mu = \dfrac{V_{w_2}}{u_2}$

$\therefore$ Work reqd./kg $= \mu u_2^2 = 0.82 \times 518.36^2$ N-m/kg

$= \mathbf{220.331}$ **kN/kg**

Also, work reqd./kg of air = $C_p \left(T_{o_2} - T_{o_1}\right)$

$$220.331 = 1.005\left(T'_{o_2} - T_{o_1}\right)$$

$$\therefore \quad \left(T'_{o_2} - T_{o_1}\right) = \frac{220.331}{1.005} = 219.23^\circ\text{C}$$

And Isentropic efficiency = $\dfrac{T_{o_2} - T_{o_1}}{T'_{o_2} - T_{o_1}}$

$$\therefore \quad \mathbf{T_{o_2} = T_{o_1} + 0.76(219.36) = 466.6\ K}$$

And, Static pressure ratio $\left(\dfrac{P_2}{P_1}\right) = \left(\dfrac{T_{o_2}}{T_{o_1}}\right)^{\frac{\gamma-1}{\gamma}} = \mathbf{4.69}$

Example 20.10. *An axial flow compressor has 8 stages and with 50 % degree of reaction compresses air with a pressure ratio of 4 : 1. Air enters the compressor at 20°C and flows through it with a constant speed of 90 m/s. The mean blade speed is 180 m/s. The isentropic efficiency is 82 %.*

Calculate:

(i) Work required/kg of air

(ii) Blade angles.

Assume $\gamma = 1.4$ and $C_p = 1.005$ kJ/kg-K

Solution

No. of stages = 8

D. R. = 0.50, $\therefore \alpha_1 = \beta_2$ and $\beta_1 = \alpha_2$

$r_p = 4.0$, $T_1 = 293$ K

$u = 180$ m/s

$$\eta_{isent} = \frac{T'_2 - T_1}{T'_2 - T_1} \qquad \text{But} \qquad T_2 = T_1 \left(r_p\right)^{\frac{\gamma-1}{\gamma}}$$

$$= 293 \times 4^{\frac{0.4}{1.4}} = 435.4 \text{ K}$$

$$0.82 = \frac{435.4 - 293}{T'_2 - 293}$$

$$T'_2 = 466.6\,\text{K}$$

$$\text{Total work reqd./kg} = C_p(T'_2 - T_1) = 1.005\,(466.6 - 293)\ \text{kJ/kg}$$

$$= \mathbf{174.47\ kJ/kg}$$

$$\text{Total work reqd./kg} = \text{No. of stages} \times u\left(V_{w_2} - V_{w_1}\right) = 8 \times u \times V_f\,(\tan\alpha_2 - \tan\alpha_1)$$

$$\therefore \quad 174.47 = 8 \times 180 \times 90\,(\tan\alpha_2 - \tan\alpha_1) \times 10^3$$

$$\therefore \quad \tan\alpha_2 - \tan\alpha_1 = 1.346 \qquad \text{...(i)}$$

With 50% degree of reaction,

$$\alpha_1 = \beta_2 \text{ and } \beta_2 = \alpha_1$$

Also,

$$\tan\alpha_1 + \tan\beta_1 = \frac{u}{V_f} = \frac{180}{90} = 2 \qquad \text{...(ii)}$$

Solving Eqs (i) and (ii)

$$\boldsymbol{\beta_1 = \alpha_2 = 59.1°}$$

$$\boldsymbol{\alpha_1 = \beta_2 = 18.1°}$$

Example 20.11. *Determine the pressure rise and work done by a rotary axial flow compressor for which the following data is available.*

$$u = 200\ \text{m/s},\ V_w = V_f = 186\ \text{m/s},$$

$$\alpha_1 = 45°,\ \alpha_2 = 14°,$$

and

$$\rho = 1.0\ \text{kg/m}^3.$$

Solution

Pressure rise through a ring of axial flow compressor

$$\Delta p = \rho V_w^2\left(\tan^2\alpha_1 - \tan^2\alpha_2\right)$$

$$= \frac{1 \times 186^2}{2 \times 10^5}\left(\tan^2 45 - \tan^2 14\right)$$

$$= \mathbf{0.2022\ bar}$$

Work required/kg of air on the blades

$$= uV_w\left(\tan\alpha_1 - \tan\alpha_2\right)$$

$$= \frac{200 \times 186}{10^3}(\tan 45 - \tan 14)$$

$$= \mathbf{27.92\ kW}$$

Example 20.12. *An eight stage axial flow compressor admits air at 1.0 bar and 20°C. It has an overall pressure ratio of 5:1 and overall isentropic efficiency of 90%. The work is equally divided between the stages. The mean blade speed is 175 m/s and 50% reaction design is used. The axial velocity of flow through the compressor is contant at 100 m/s.*

Calculate the power required and blade angles.

Data: *Degree of reaction* = 50%

P_1 = 1.0 bar T_1 = 293 K

η_{isent} = 0.90, r_p = 5.0

u = 175 m/s V_f = 100 m/s

$\alpha_1 = \beta_2$ *and* $\alpha_2 = \beta_1$

Solution

Let sufix z denote the number of stages.

$$T_{o_2} = T_{o_1} \times r_p^{\frac{\gamma-1}{\gamma}} = 293 \times 5^{\frac{1.4}{1.4}}$$

$$= 464 \text{ K}$$

But
$$\eta_{isent} = \frac{T_{o_2} - T_{o_1}}{T'_{o_2} - T_{o_1}}$$

$$\therefore \quad 0.90 = \frac{464 - 293}{T'_{o_2} - 293} \qquad \therefore \quad T'_{o_2} = 483 \text{ K}$$

$$\text{Work reqd./kg} = C_p\left(T'_{o_2} - T_{o_1}\right) = \left(V_{w_2} - V_{w_1}\right) u \times z$$

$$= V_f\left(\tan\alpha_2 - \tan\alpha_1\right) u \times z$$

$$\therefore \quad \left(\tan\alpha_2 - \tan\alpha_1\right) = \frac{1.004(483 - 293) \times 10^3}{100 \times 175 \times 8}$$

$$= 1.3639 \qquad \ldots\text{(i)}$$

Referring to velocity triangles,

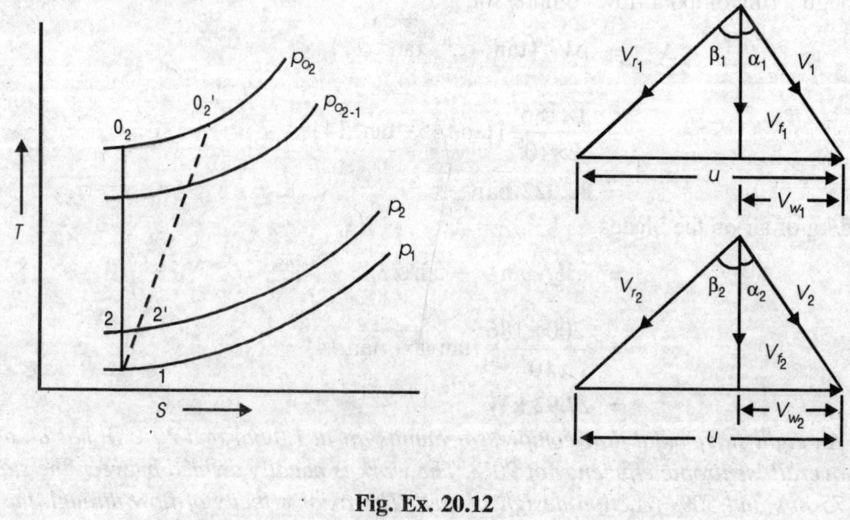

Fig. Ex. 20.12

$$\frac{u}{V_f} = \tan\alpha_1 - \tan\beta_1 = \frac{175}{100}$$

$$= 1.75$$

From Eqs (i) and (ii), we get

$$\tan\beta_1 = \frac{1.365+1.75}{2}$$

$$= 1.5569$$

$$\therefore \quad \beta_1 = \alpha_2 = 57.4^\circ$$

and substituting in Eq. (i),

$$\alpha_1 = \beta_2 = 10.95^\circ$$

And power required to drive the compressor

$$= \dot{m}C_p\left(T'_{o_2} - T_{o_1}\right)$$

$$= 1.005\ (483 - 293)\ /\ 1$$

$$= \mathbf{190.93\ kW}$$

Example 20.13. *An axial flow compressor operates with an overall stagnation pressure ratio of 4 and overall stagnation isentropic efficiency 86 %. The inlet stagnation pressure and temperature are 1.0 bar and 320 K. The mean blade speed is 190 m/s. The degree of reaction is 0.5 at the mean radius with relative air angles of 30º and 10º at inlet and outlet of rotor respectively. The work done factor is 0.88.*

Determine the stagnation temperature. If the hub : tib ratio is 0.4 and mass flow rate is 20 kg/s, calculate the blade height in the first stage and stagnation polytropic efficiency.

Data: $\frac{p_{o_2}}{P_{o_1}} = 4, \qquad \eta_{isent} = 0.86$

$D.R. = 0.50, \qquad \alpha_2 = \beta_2 = 10^\circ$

$\alpha_2 = \beta_1 = 30^\circ$

$p^1 = 1.0$ bar, $\qquad T_1 = 320$ K

$u = 190$ m/s $\qquad$ Work done factor $= 0.88$

$\dot{m} = 20$ kg/s

Solution

$$T_{o_2} = T_{o_1}\left(\frac{P_{o_2}}{P_{o_1}}\right)^{\frac{\gamma-1}{\gamma}} = 320/4^{\frac{0.4}{1.4}} = 475.7\ \text{K}$$

$$\eta_{isent} = 0.86 = \frac{T_{o_2} - T_{o_1}}{T'_{o_2} - T_{o_1}} = \frac{475.7-320}{T_{o_2} - 320}$$

$$T'_{o_2} = 501\ \text{K}$$

$$\text{Stagnation polytropic efficiency, } \eta_p = \frac{\ln\left(\frac{P_{o_2}}{P_{o_1}}\right)^{\frac{\gamma-1}{\gamma}}}{\ln\left(\frac{T_{o_2}}{T_{o_1}}\right)}$$

$$= \frac{\ln(4)^{\frac{0.4}{1.4}}}{\ln\left(\frac{501}{302}\right)}$$

$$= \mathbf{88.4\%}$$

Referring to velocity triangles shown in Fig. Ex. 20.12.

$$\frac{u}{V_f} = \tan\alpha_1 + \tan\beta_1 = \tan 10^\circ + \tan 30^\circ$$

$$= 0.7536$$

$$\therefore \quad V_f = \frac{190}{0.7536} = 252.12 \text{ m/s}$$

Also $V_{w_2} = V_f \tan\alpha_2 = 252.12 \tan 30^\circ = 145.56$ m/s

$$V_{w_1} = V_f \tan\alpha_1 = 252.12 \tan 10^\circ = 44.45 \text{ m/s}$$

∴ Work required/kg of air per stage

$$= u\left(V_{w_2} - V_{w_1}\right)\times\lambda\times 10^{-3} \text{ kJ/kg}$$

$$= \frac{190\times 0.88(145.56\text{-}44.45)}{1000}$$

$$= 16.906 \text{ kJ/kg per stage}$$

Total work required/kg for all stages

$$= C_p\left(T'_{o_2} - T_{o_1}\right)$$

$$= 1.005\ (501 - 320)$$

$$= 181.9 \text{ kJ/kg}$$

$$\therefore \quad \text{No. of stages } z = \frac{181.9}{16.906} = 10.76$$

$$= 11 \text{ stages}$$

Now, $V_1 = \dfrac{V_f}{\cos\alpha_1} = \dfrac{252.12}{\cos 10^\circ} = 256$ m/s

and $T_1 = T'_{o_1} - \dfrac{V^2}{2C_p} = 320 - \dfrac{190^2}{2\times 1.005\times 1000} = 302.03$ K

$$\frac{P_1}{P_{o_1}} = \left(\frac{T_1}{T_{o_1}}\right)^{\frac{\gamma}{\gamma-1}}$$

$$\therefore \quad \frac{P_1}{P_{o_1}} = \left(\frac{302.03}{320}\right)^{\frac{1.4}{0.4}}$$

$$\therefore \quad P_1 = 0.816 \text{ bar}$$

and $$\rho_1 = \frac{P_1}{RT_1} = \frac{0.816\times10^2}{0.287\times302.03}$$

$$= 0.941 \text{ kg/m}^3$$

From the continuity equation,

$$\rho_1 A_1 V = m = 20 \text{ kg/s}$$

$$\therefore 0.941\times\pi r_1^2\left[1-0.4^2\right]\times252.12 = 20$$

$$r_1 = 0.1787 \quad m = 17.87 \text{ cm}$$

But, $$\frac{r_2}{r_1} = 0.4$$

$$\therefore \quad r_1 = 17.87\times0.4 = 7.1 \text{ cm}$$

$\therefore$ Hence height of blade in the first stage

$$z = r_2 - r_1$$

$$= 17.87 - 7.148$$

$$= 10.722 \text{ cm}$$

PROBLEMS FOR PRACTICE

20.1. Air at 101.3 kPa and 288 K is to be compressed at a rate of 5.6 m³/min to 11.75 bar. Two compressors are considered (i) Roots blower and (ii) Sliding vane rotary compressor. Compare the power required assuming that for vane type compressor, 75% of the pressure rise occurs due to internal compression before the delivery takes place and that compressor is uncooled.

[**Ans.** (i) 6.88 kW, (ii) 5.75 kW]

20.2. Free air of 30 m³/min. is compressed from 101.3 kPa to 2.23 bar. Determine the power required (i) if the roots blower is used, (ii) if vane blower is used. Assume that there is 25% reduction in volume before the back flow occurs. Also find the isentropic efficiency in each case.

[**Ans.** (i) 60.85 kW, (ii) 48.46 kW, (iii) 73.69%, (iv) 92.53%]

20.3. The design pressure ratio of a centrifugal compressor is 3.5 : 1. The diameter of the impeller eye is 30 cm. The axial velocity of flow at inlet is 130 m/s and the mass flow rate of air is 5 kg/s. The velocity in the delivery duct is 115 m/s. The tip speed of the impeller is 450 m/s and runs at 20000 r.p.m. with a total head isentropic efficiency of 78%. The pressure coefficient is 0.72.

Determine:

(i) Static pressure ratio

(ii) Static pressure and temperature at the inlet and outlet of the compressor

(iii) Work required/kg of air

(iv) Theoretical power required.

Assume ambient conditions as 1.013 bar and 15°C.

[**Ans.** (i) 4.21, (ii) 0.917 bar, 279.6 K, 3.86 bar, 461.67 K, (iii) 180.29 kJ/kg, (iv) 901.45 kW]

20.4. A centrifugal compressor runs at 20000 r.p.m. draws in air at 290 K. It has slip factor 0.80, total head isentropic efficiency 0.75 and outer diameter of blade tip 50 cm.

Calculate,

(i) temperature rise of air flowing through the compression

(ii) static pressure ratio.

Assume that velocities of air at inlet and outlet to remain constant.

(**Ans.** (i) 218.62°C, (ii) 4.8)

20.5. Air at a temperature of 17°C flows axially into the eye of the impeller of an uncooled centrifugal compressor running at 18000 r.p.m. and leaves the impeller with a velocity of whirl 0.88 times the velocity of the blade tip. The outer diameter of the blade tip is 46 cm and the kinetic energy of air at the outlet from the impeller is equal to that at the inlet to the impeller. Assuming isentropic efficiency of 0.75, determine

(i) Rise in temperature of the air compressed

(ii) Static head pressure ratio.

(**Ans.** (i) Actual Δt = 205°C, (ii) Pressure ratio = 3.48)

20.6. A single sided centrifugal compressor of a gas turbine plant has to deliver 8.2 kg/s of air with a pressure ratio of 4.4 : 1 at 18000 r.p.m. The intake conditions are static pressure 1.013 bar and 15°C. Assuming isentropic efficiency of 78%, ratio of whirl speed to tip speed 0.94. Calculate rise in stagnation temperature, tip speed, external diameter of impeller, external diameter of the eye for which the internal diameter is 12.5 cm and the axial velocity is to be 140 m/s with no pre-whirl.

(**Ans.** Outer dia. of impeller D_2 = 48.5 cm, Eye external dia. = 28.6 cm, $\Delta T_{o'}$ = 195 °C)

20.7. An eight stage axial flow compressor takes in air at a temperature of 20°C at a rate of 3 kg/s. The pressure ratio is 6 and the isentropic efficiency is 0.89. The compression process is adiabetic. Ali stages of the compressor are similar and operate with 50 reaction. In each stage, the mean blade speed is 180 m/s. The constant axial velocity of flow of air is 105 m/s.

Determine the power required and the direction of air at the entry to and exit from the rotor and stator. Assume air as an ideal gas.

(**Ans.** Power = 660.3 kW, $\alpha_1 = \beta_2$ = 7.4°. $\alpha_2 = \beta_1$ = 57.8°)

20.8. The mean blade speed in the second stage of an uncooled axial flow compressor is 230 rn/s and the axial velocity of flow is 150 m/s. The shape of the fixed and moving blades is same and the difference between the tangents of the angles at inlet and outlet is 0.56 If the adiabetic efficiency of the stage is 85% and the air inlet temperature to the stage is 290 K, find the number of required for a pressure rise from 1 bar to 3.8 bar. Also find the power required to compress a mass flow of 1000 kg/min. Assume C_p = 1.005, mechanical efficiency 99%. Neglect the difference in the kinetic energy entering sriccessive stages is negligible.

(**Ans.** W = 19.32 kJ/kg, ΔT = 19.22°C, r_p = 1.211, No. of stages = 7, power = 2276 kW)

20.9. An axial flow compressor compresses air from an inlet condition of 1 bar and 290 K so a delivery pressure of 5 bar with an overall isentropic efficiency 87%. The degree of reaction is 0.5 and the blade angles at inlet and outlet are 44° and 13° respectively. The mean blade speed and axial velocity of flow are constant throughout the compressor. Assuming a blade velocityof 180 m/s and work done factor of 0.85,

Determine:

(i) Number of stages,

(ii) Change in entropy.

(**Ans.** (i) 12, (ii) 0.4898 kJ/kg-K])

20.10. The first stage of an axial flow compressor is to be designed for an axial velocity of flow 133 m/s, mass flow rate of air 22.7 kg/s and for ambient conditions of 1 bar an 10°C.

At the mean radius, both moving and fixed blade sections are to be suitable for relative air angles of 50° at inlet and 30° at outlet. A row of guide vanes giving an outlet air angle of 30° is to be fitted before each stage. Draw the triangles of velocity assuming the ratio of the hub to tip diameters as 0.68, free vortex conditions.

Calculate:

(i) Speed

(ii) Theoretical temperature rise across the stage

(iii) Number of similar stages theoretically required to give a total head delivery, pressure of 8.4 bar if the stage isentropic efficiency is 85%.

(**Ans.** (i) 9000 r.p.m., (ii) 20°C, (iii) 13)

20.11. A six stage axial flow compressor take air at a temperature of 25°C at a rate of 5 kg/s. The pressure ratio is 5 and isentropic efficiency 0.85. The compressor stag are similar and has 50% degree of reaction. In each stage, the mean blade speed is 190 m/s and axial velocity of flow is 100 m/s. Calculate the power required and blade angles.

(**Ans.** 1074.675 kW, $\beta_1 = 72.23°$, $\beta_2 = 62.72°$)

20.12. In an xial flow air compressor has a pressure ratio 6 : 1. Air enters at 20°, mean velocity of rotor blades is 200 m/s and the inlet and exit angles of both moving and fixed blades are 45° and 15° respectively. The degree of reaction is 50%, the work done factor is 0.86 throughout. There are 12 stages and axial velocity remains constant throughout.

Calculate the isentropic efficiency of the compressor. Explain why the overall pressure ratio is not equal to the stage pressure ratio raised to the power of *n*, is the number of stages.

(**Ans.** $\eta_{isent} = 0.83$)

20.13. Sketch the velocity diagram for an axial flow compressor and derive the relation,

$$\frac{p_{o_2}}{p_{o_1}} = \left[1 + \eta_c \frac{V_b(V_{w_2} - V_{w_1})}{C_p T_{o_1}}\right]^{\frac{\gamma}{\gamma-1}}$$

CHAPTER 21

Compressible Fluid Flow

21.1 INTRODUCTION

The science of gas dynamics is primarily concerned with the motion of gases and consequent effects. Gas dynamics differs from classical fluid dynamics in that, although both deal with fluid motion, fluid dynamics usually does not consider thermal or chemical effects and gas dynamics does.

The term gas dynamics is open to criticism as being too general and other names have been suggested. For example, supersonic flow, compressible flow, aerothermochemistry and aerothermodynamics have been considered. Of these, supersonic flow is too restricted because it deals with one particular domain, namely, velocities which are greater and sonic velocity or sound velocity. Compressible flow may be misleading because gas density does not necessarily vary in a flow process. The term aerothermochemistry implies at once that chemical phenomena enter the picture. This is true, for example, in the combustion chamber of a jet engine, but not true for the diffuser of the engine. The last term, aerothermodynamics, appears to be a more appropriate name for the field covered by gas dynamics, but as yet it has relatively few proponents.

21.2 SONIC VELOCITY (SOUND VELOCITY, ACOUSTIC VELOCITY)

The speed of sound is called acoustic or sonic speed. Since sound is a small pressure wave, some medium is essential for its transmission.

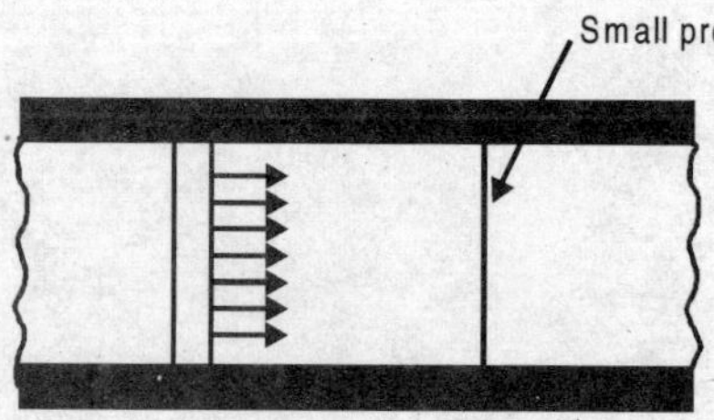

Fig. 21.1

The acoustic velocity or the velocity of sound in a fluid medium, is by definition the speed with which a small disturbance is transmitted through the fluid.

Consider a piston which may slide without friction in a tube shown in Fig. 21.1 and assume that the piston and gas in the tube are at rest originally. Let p be the pressure of the gas uniform throughout. If the small impulse is given to the piston, the gas immediately adjacent to the piston will experience a slight rise in pressure or, in other words, it will be compressed. The change in density takes place because the gas is compressible and there is, therefore, a lapse of time between the motion of the piston and the time this is observed at the far end of the tube. Thus it will take the pulse a certain time to reach the far end of the tube or, in other words, there is a finite velocity of propagation which by definition is the acoustic velocity.

Let the change in pressure be denoted by dp, and let the change of density of gas, due to compression of the latter be $d\rho$.

The bulk modulus E is defined by

$$E = \rho \cdot \frac{dp}{d\rho} \qquad \text{...(21.1)}$$

Dimensionally, $\sqrt{\frac{E}{\rho}}$ is a velocity and it can be shown that this is also the acoustic velocity.

$$a = \sqrt{E/\rho} \qquad \text{... (21.2)}$$

$$= \sqrt{\frac{dp}{d\rho}} \qquad \text{... (21.3)}$$

Equation 21.3 shows that the velocity with which a small pressure wave is propagated, that the velocity of sound, are related to the compressibility of the gas medium. It can be seen from Eq. 21.3 that greater the compressibility of gas, lower will be the speed of sound. The velocity of the pressure wave cannot always be defined by Eq. 1.3 as this is valid only when the pressure and velocity increments across the wave are so small that they can be approximated by their differentials. If the variations in pressure, velocity and temperature are vanishingly small, it can be concluded that the process takes place reversibly. Also, the process takes place very rapidly and no drastic temperature gradients are involved, so that the process is essentially adiabatic and a sound wave can be treated as an isentropic phenomenon. Hence the derivative in Eq. 21.3 must be taken along an isentropic process. It is, therefore, more correct to use partial differential notation, namely

$$a = \sqrt{\left(\frac{\partial p}{\partial \rho}\right)_S} \qquad \text{... (21.4)}$$

We have for perfect or ideal gas,

$$pv\gamma = \text{constant or } \frac{p}{\rho\gamma} = \text{constant} \qquad \text{... (21.5)}$$

$$\frac{\partial p}{\partial \rho} = \text{constant} \times \gamma . \rho^{\gamma-1} = \frac{C \times \gamma \cdot \rho^{\gamma}}{\rho}$$

$$= \frac{\gamma}{\rho}(C \times \rho^{\gamma}) = \frac{\gamma \cdot p}{\rho}$$

$$= \gamma \cdot RT$$

$$\boldsymbol{\frac{\partial p}{\partial \rho} = \gamma \cdot RT}$$

∴ For a perfect gas, the acoustic or sonic velocity is given by

$$a = \sqrt{\left(\frac{\partial p}{\partial \rho}\right)_s}$$

$$= \sqrt{\gamma R T} \quad ...(21.6)$$

$$= \sqrt{\frac{\gamma p}{\rho}} \quad ...(21.7)$$

$$\mathbf{a = \sqrt{\gamma RT}} \quad ...(21.8)$$

Velocity of sound as defined by Eqs. 21.6 and 21.7 is in reasonal agreement with the experimental data. Nevertheless the equation is developed on the premise that the gas under consideration obeys perfect gas laws.

The NACA standard velocity at 15°C is 340.7 m/sec. In comparison, the speed of sound in steel is 4944 m/s and in water it is 1438.2 m/s at NTP.

For air,

$$a = \sqrt{\gamma R T}$$

$$= \sqrt{1.4 \times 287\, T} \quad ...(21.9)$$

$$= 20.045\sqrt{T}$$

where a = Sonic velocity

T = Temperature in K.

21.3 MACH NUMBER, MACH LINE AND MACH ANGLE

A parameter that is fundamental in compressible flow theory is the speed of sound a. In a flowing fluid, the speed of sound is a significant measure of the effects of compressibility only when it is compared to the speed of the flow. This introduces a dimensionless parameter which is called the *Mach number* denoted by M.

$$\mathbf{M = \frac{V}{a} = \frac{V}{\sqrt{\gamma RT}}} \quad ...(21.10)$$

It is the most important parameter in compressible flow theory.

It is evident that M will vary from point to point in a given flow, not only because c changes but also because a depends on the local conditions.

Flow may be classified according to the value of the existing Mach number as :

Mach Number	*Classification*
$M < 1$	Subsonic flow
$M = 1$	Sonic flow
$M > 1$	Supersonic flow
$M > 5$	Hypersonic flow

Mach number is based on velocity and not on frequency. In acoustical work, the term *ultrasonic* is used in referring to the frequencies above the audible range. This usage has no place in Gas Dynamics.

It is often desirable to define variations of the Mach number in terms of changes in velocity and temperature.

$$M = \frac{V}{\sqrt{\gamma R T}} \text{ or } M^2 = \frac{V^2}{\sqrt{\gamma R T}} \quad ...(21.11)$$

$$2M \cdot dM = \frac{2V \cdot dV}{\gamma R T} - \frac{V^2}{\gamma R} \times \frac{1}{T^2} \cdot dT$$

$$M \cdot dM = \frac{V \cdot dV}{\gamma RT} - \frac{V^2 \cdot dT}{2\gamma RT^2}$$

Dividing by M^2 or $\frac{V^2}{\gamma RT}$, we get

$$\frac{dM}{M} = \frac{dV}{V} - \frac{dT}{2T} \quad ...(21.12)$$

This Eq. 21.12 is particularly useful in isothermal analysis.

Consider a stationary object omitting small pressure impulses as a fluid moves over it with various velocities or, in other words, in the stationary fluid the object is moving with various velocities.

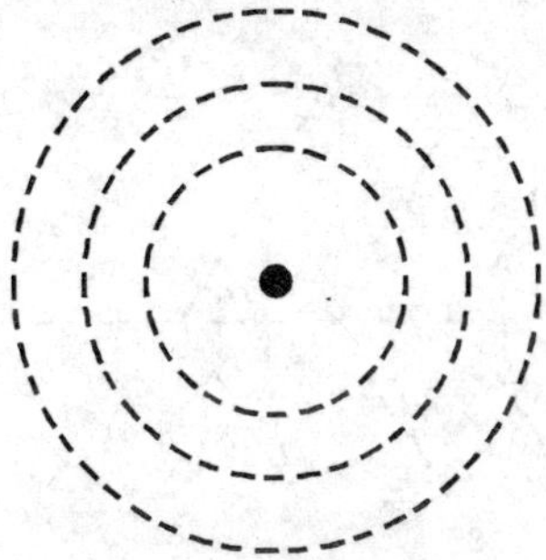

Fig. 21.2 (a)

Figure 21.2 (a) shows the situation with the fluid moving very slowly or not at all. In this the pressure impulses spread uniformly in all the directions. This is the case for any fluid with a very large sonic velocity, i.e., incompressible fluid. When the Mach number is small, either the velocity may be very small or the acoustic velocity may be large and for either case the fluid may be considered to be incompressible for many calculations.

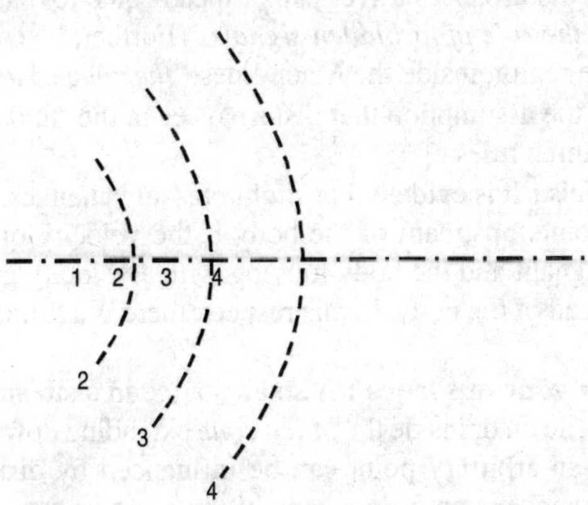

Fig. 21.2 (b)

Figure 21.2 (b) indicates that for Mach number is less than unity, but not negligible, the pressure impulses are felt in all directions but the intensity is not symmetrical.

Consider a point source travelling with a velocity $V < a$ as shown in Fig. 21.2 (b). At reference time zero the point disturbance is assumed to be at A. At unit time later (1t), the point source will have moved to B, a distance ct. At 2 time units later, 2t, the source will have moved to C, a distance 2Vt. Mean while, the spherical pressure wavefront which started out from A will have grown after time 2t to a radius 2at and the wavefront from B which could not start until the source reached B will have grown to a radius at. At time 2t, the source will have just reached point C. This step-by-step analysis of a point source moving with subsonic velocity indicates that each succeeding wavefront will always be contained within the initial or preceding wavefront.

If the point source moves at a speed V such that $V > a$, the wavefront circles will no longer contain the source. The condition is shown in Fig. 21.2 (c). The envelope to this family of circles is a straight line, known as *Mach line*. It is easy to see from Fig. 21.2 (c) that

$$\sin \alpha = \frac{a}{V} = \frac{1}{M}$$

or

$$\alpha = \sin^{-1} \frac{1}{M}$$

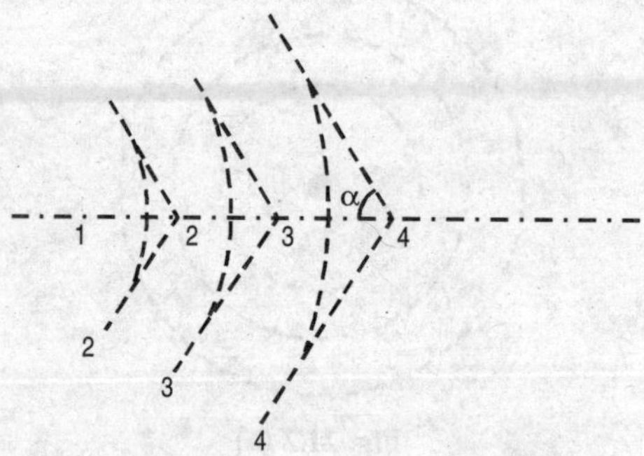

Fig. 21.2 (c)

The angle between the Mach line and the direction of flow is known as *Mach angle*.

The Mach line constitutes a demarcation. It is easy to see that the fluid outside the Mach line will not receive any signal from the source because the disturbances travel with a speed equal to that of sound. Von Karman has appropriately called this phenomenon *"the rule of forbidden signals."* Further, he called the region ahead of the Mach lines *"the zone of silence"* and the region inside the Mach lines *"the zone of action."*

Based on figures above and on the assumption that disturbances in the fluid are small i.e., acoustic waves, Von Karman postulated the following three rules :

(a) The rule of forbidden signals : It is evident that slight pressure changes, produced in the moving fluid by a stationary source, cannot reach points upstream of the body if the velocity of flow is greater than the speed of sound. Similarly, if the fluid is stagnant and the body moving with a velocity greater than the sound, the pressure fluctuation cannot reach points ahead of the body. In this respect, there is a fundamental contrast between subsonic and supersonic flow.

(b) The zone of action and the zone of silence : A small source in a stream of fluid with supersonic velocity produces effects only on points that lie on or inside the *Mach cone* extending downstream from the source. Similarly, in a stagnant fluid, conditions at an arbitrary point can be influenced by disturbances which are moving with supersonic speed only if the disturbance is on or in a cone of the same vertex angle extending upstream from the chosen arbitrary point.

(c) The rule of concentrated action : This rule concerns the distribution of the pressure effect in the fluid relative to the body. The distance between the circles (spheres in three-dimensions) which represent pressure impulses is an inverse measurement of the strength of the pressure disturbances in that region. The closer the circles are, the weaker the impulse becomes. For the case of imcompressible flow (or stationary fluid and source), the field is symmetrical about the source. For subsonic flow, the field is unsymmetrical; and for supersonic flow, the pressure disturbance is concentrated in the viscinity of the Mach cone which bound the zone of action.

From Fig. 21.2 (c) we see that all the impulses are limited to a cone. The pressure signals cannot be observed upstream since the cone has its apex at the stationary source of the impulses and the orientation is such that the apex points into the flow. This cone is called the *Mach cone*.

The lines which bound the Mach cone, having an inclination of α with the direction of flow, are called *Mach lines* and in some of the literature, *Mach waves*.

In a plane (two-dimensional) there are two lines emanating from any point in the flow. Each of the lines makes an angle α with the streamline and if we look down the streamline, it is evident that one line runs to the left and the other to the right. Consequently, it is possible to refer to left running Mach lines and right running Mach lines. These lines are *characteristics* of the flow, or of the mathematics describing the flow, and hence have been given the name characteristics.

21.4 STATIC AND STAGNATION PROPERTIES

When a flowing fluid having same velocity is brought to rest, it is said to have attained a stagnation state. The final stagnation state is governed by the manner in which it is attained. A reversible adiabatic or isentropic process has a great significance.

For an isentropic stagnation process, the steady flow energy equation simplifies to

$$h + \frac{V^2}{2} = h_o, \text{ kJ/kg}$$

where h_o is stagnation enthalpy and h is its initial enthalpy when fluid is flowing with a velocity of V m/s.

During the stagnation process, the kinetic energy of the fluid is converted to enthalpy, thereby pressure and temperature of the fluid increases as shown below.

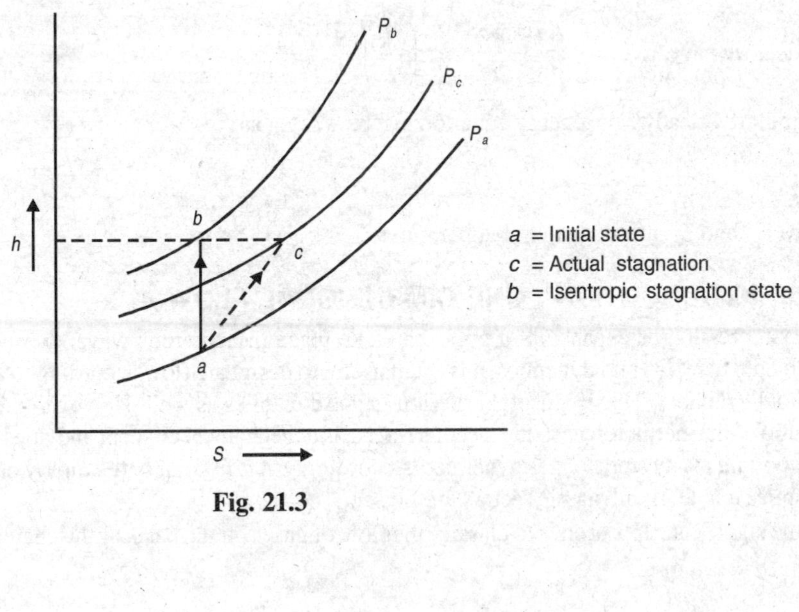

Fig. 21.3

The properties of the fluid at the stagnation state are referred as stagnation properties e.g., stagnation pressure, stagnation temperature, stagnation density, etc. The stagnation state and stagnation properties are denoted by lower suffix *o*.

An isentropic stagnation state is the state, fluid attains when the stagnation is reversible adiabatic or isentropic, shown by process $a - b$ on the h–s diagram. Actual irreversible process with friction or heat transfer is shown by process $a - c$. It may be noted that stagnation enthalpy h_b and h_c is same in both the processes. However actual stagnation pressure p_c is lower than isentropic stagnation process p_b, due to entropy increase in the actual process on account of friction. Now, for an ideal gas undergoing isentropic stagnation process,

$$h_o = h + \frac{V^2}{2}$$

$$\therefore \quad C_p T_o = C_p T + \frac{V^2}{2} \qquad \because h = f(T) \text{ for ideal gas}$$

$$\therefore \quad T_o = T + \frac{V^2}{2\,C_p}$$

where T_o = Stagnation temperature

$\frac{V^2}{2C_p}$ = Dynamic temperature

Now, stagnation, pressure p_o can be found as

$$p_o = pb = pa\left(\frac{T_o}{T_1}\right)^{\frac{\gamma-1}{\gamma}}$$

Also, we have,
$$\frac{\rho o}{\rho a} = \frac{\rho b}{\rho a} = \left(\frac{T_o}{T}\right)^{\frac{1}{\gamma-1}}$$

Similarly, steady flow energy equation can be written as

$$\begin{aligned} q - w_s &= (h_{ob} - h_{oa}) + g\,(z_b - z_a) \\ &= C_p\,(T_{ob} - T_{oa}) + g\,(z_b - z_a) \end{aligned}$$

where T_{ob} and T_{oa} are stagnation temperatures.

21.5 INTRODUCTION : ONE DIMENSIONAL FLOW

The compression or expansion of a gas can take place in numerous ways, the most important of which will be considered here. In fluid dynamics, it is often useful to designate flow according to a dimensionless parameter – the Reynold's number. This is a useful criterion in gas dynamics as well. However, the Reynold's number, defined as the ratio of the inertia forces to the viscous forces, is a mechanical concept and does not consider the thermal effects so important in any study of gas dynamics. Accordingly, it is necessary to employ other approaches when both fluid dynamics and thermodynamic factors are present.

One significant approach is to classify the flow of gases on the basis of the change of entropy accompanying the process. A flow process is *isentropic* if it proceeds reversibly and adiabatically ($Q = 0$). A different type of flow is one which proceeds without frictional losses, but during the flow the gas exchanges heat with its surroundings. Such a reversible flow is called *diabatic flow*. During diabatic flow, the entropy may increase or decrease depending on the direction of heat exchange between the gas and the surroundings.

A flow process is said to be irreversible if events take place which reduce the total available energy of the system and the surroundings. Such flows are accompanied by losses of the type associated with friction or shock waves. If irreversible processes take place adiabatically, the second law specifies that the entropy must increase.

21.6 THE FLOW SYSTEM

Essentially all the problems of gas dynamics may be considered to take place in a flow system which may be considered as represented in Fig. 21.4. This consists of an infinite reservoir and a flow section. The conditions in the reservoir are denoted by the subscript 0. The flow section may be of any shape. It may be a convergent section, a straight section or a convergent-divergent section. In the flow section there is a point where the Mach No. is equal to unit and this section is called the critical section. In a convergent-divergent flow section and with reversible adiabatic flow, the throat is always the critical section.

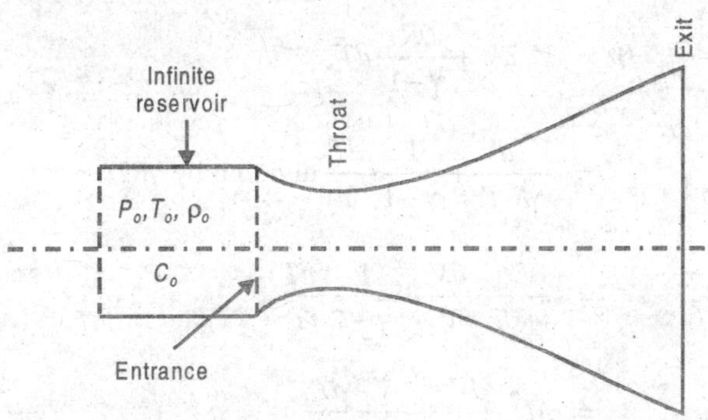

Fig. 21.4

At any section, we have,

$$h_o + \frac{V_o^2}{2} = h + \frac{V^2}{2}$$

or

$$h_o = h + \frac{V^2}{2} \text{ as } V_o = 0$$

$$(h_o - h) = \frac{V^2}{2} = C_p (T_o - T)$$

$$= \frac{\gamma R}{\gamma - 1}(T_o - T) \text{ for ideal gas}$$

$$dh = \frac{\gamma R}{\gamma - 1} \cdot dT$$

$$= du + d(p \cdot \upsilon)$$

$$= du + d\left(\frac{p}{\rho}\right)$$

$$= du + \frac{dp}{\rho} - \frac{p\,d\rho}{\rho^2} \qquad \text{... (21.13)}$$

With isentropic flow, $W_{sf} = 0, Q = 0$ and with $\Delta p = 0$, we have

$$0 = dK + \Delta h$$

$$= dK + dh$$

$$= V \cdot dV + C_p\ dT$$

$$= V \cdot dV + \frac{\gamma R}{\gamma - 1} \cdot dT \qquad \text{... (21.14)}$$

Equation 21.14 is basic to most derivations involving isentropic flow. From this many equations can be obtained.

$$0 = V \cdot dV + \frac{\gamma R}{\gamma - 1} \cdot dT$$

$$= \frac{V \cdot dV}{\gamma R\,T} + \frac{1}{\gamma - 1} \cdot \frac{dT}{T} \text{ dividing by } \gamma RT$$

$$= \frac{V^2}{\gamma R\,T} \cdot \frac{dV}{V} + \frac{1}{\gamma - 1} \cdot \frac{dT}{T}$$

$$= M^2 \cdot \frac{dV}{V} + \frac{1}{\gamma - 1} \frac{dT}{T}$$

$$\therefore \quad 0 = \frac{dT}{T} + (\gamma - 1) M^2 \cdot \frac{dV}{V} \qquad \text{... (21.15)}$$

A relation between temperature, velocity and Mach No.

Flow Parameters :

We have,
$$0 = V\,dV + \frac{\gamma R}{\gamma - 1} \cdot dT$$

$$\text{Constant} = \int_{T_0}^{V} V \cdot dV + \frac{\gamma R}{\gamma - 1} \int_{T_0}^{T} dT$$

$$= \frac{V^2}{2} + \frac{\gamma R}{\gamma - 1} \cdot T$$

$$\frac{V_0^2}{2} + \frac{\gamma R}{\gamma - 1} \cdot T_o = \frac{V^2}{2} + \frac{\gamma R}{\gamma - 1} \cdot T$$

$$\frac{\gamma R}{\gamma-1}\cdot T_o = \frac{V^2}{2}+\frac{\gamma R\,T}{\gamma-1} \qquad (\because V_o=0)$$

or $$\frac{a_0^2}{\gamma-1} = \frac{V^2}{2}+\frac{\gamma RT}{\gamma-1} \qquad ...(21.16)$$

or $$T_o = \frac{V^2}{2}\times\frac{\gamma-1}{\gamma R}+T$$

This equation implies that in a reversible adiabatic expansion process, the acoustic velocity based on the reservoir conditions is greater than the local acoustic velocity at any point of the flow section. On the basis of temperature, this means that the reservoir temperature is higher than the local temperature at each subsequent point. This can easily be understood if it is remembered that a gas is cooled when it is expanded reversibly and adiabatically.

Now, let the temperature at some point in the expansion section be reduced to absolute zero. (Theoretically).

$$\therefore \qquad \frac{\gamma R}{\gamma-1}T\to 0$$

∴ Local velocity V will be maximum.

$$\frac{a_0^2}{\gamma-1} = \frac{V_{max}^2}{2} \text{ or } V_{max}^2 = \frac{2\,a_0^2}{\gamma-1}$$

$$V_{max} = \sqrt{\frac{2\,a_0^2}{\gamma-1}} \qquad ...(21.17)$$

$$= a_o\sqrt{\frac{2}{\gamma-1}}$$

For air, $\gamma - 1.4$

$$V_{max} = a_o\sqrt{\frac{2}{0.4}=5}$$

$$= a_o\sqrt{5}$$

$$= 2.24\,a_o$$

$$= 2.24\sqrt{\gamma RT_o}$$

$$= 2.24\sqrt{1.4\times 287\;T_o}$$

$$= 44.8\sqrt{T_o}$$

$$\mathbf{V_{max} = 44.8\sqrt{T_o}} \qquad ...(21.18)$$

In practice V_{max} is never reached; nevertheless, Eq. 21.18 shows why the air entering supersonic wind tunnels must be heated if extremely high velocities such as those in hypersonic wind tunnels are desired in the test section.

At V_{max}, $M \to \infty$.

Let us now consider subsonic flow and imagine a gradual increase of the local velocity until the acoustic velocity is approached at a particular location. The expansion and increase in velocity results in a lowering of temperature and this also lowers the local acoustic velocity. Thus, while the local velocity increases, the local acoustic velocity decreases. With further expansion, the stream velocity could become equal to the acoustic velocity $V = a$ at this condition and this point is denoted by a^*.

From Eq. 21.16, we have

$$\frac{a_0^2}{\gamma-1} = \frac{a^2}{\gamma-1}+\frac{V^2}{2}=\frac{a^{*2}}{\gamma-1}+\frac{a^{*2}}{2}$$

$$= a^{*2}\left[\frac{1}{\gamma-1}+\frac{1}{2}\right]=a^{*2}\left[\frac{2+\gamma-1}{2(\gamma-1)}\right]$$

$$= \frac{\gamma+1}{2(\gamma-1)}a^{*2}$$

or

$$a^{*2} = \frac{2(\gamma-1)a_0^2}{(\gamma+1)(\gamma-1)}=\frac{2\,a_0^2}{\gamma+1} \quad \text{... (21.19)}$$

For air, the critical acoustic velocity and the acoustic velocity in the reservoir are related by

$$a^{*2} = \frac{2\,a_0^2}{1.4+1}$$

$$= \frac{2\times a_0^2}{2.4}=\frac{1}{1.2}\,a_0^2$$

$$a^* = a_o\sqrt{1/1.2} = 0.93\,a_o$$

$$\mathbf{a^* = 0.93\,a_o} \quad \text{... (21.20)}$$

It follows that for any particular reservoir, the maximum possible velocity is limited. However, it should not be inferred from this that the speed of the aircraft is limited by these same considerations. Theoretically, there is no limit on the maximum speed which air planes may reach.

The critical acoustic velocity serves a very useful purpose in defining a velocity ratio based on it, namely

$$M^* = \frac{V}{V^*}=\frac{V}{a^*} \quad \text{... (21.21)}$$

Although that is in the form of a Mach number, it pertains to no particular local conditions, but constitutes a reference ratio. There are two advantages in defining M^*. First, it is a parameter which for any particular section is a function of the velocity only. Second, at extremely high speeds the acoustic velocity decreases while the velocity itself increases. Hence, the Mach number tends towards infinity, complicating its use in equations. Defining the Mach number M based on a^* obviates this difficulty.

For stagnation temperature, we have

$$C_p\,(T_o - T) = \frac{V^2}{2}$$

or $$T_o = T + \frac{V^2}{2\,C_p \times 1000}$$

$$= T + \frac{V^2}{2010}$$

Also, $$T_o = T + \frac{V^2}{2\,C_p} = T + \frac{V^2}{2000 \times \frac{\gamma R}{\gamma - 1}}$$

$$= T + \frac{(\gamma - 1)V^2}{2000 \times \gamma R}$$

$$\therefore \quad \frac{T_o}{T} = 1 + \frac{\gamma - 1 \cdot V^2}{2000 \times \gamma RT}$$

$$\mathbf{\frac{T_o}{T} = 1 + \frac{\gamma - 1}{2} \times M^2} \qquad \text{...(21.22)}$$

From this equation, we get,

$$\frac{T_o}{T_1} = 1 + \frac{\gamma - 1}{2} M_1^2$$

$$\frac{T_o}{T_2} = 1 + \frac{\gamma - 1}{2} M_2^2$$

$$\therefore \quad \mathbf{\frac{T_1}{T_2} = \frac{1 + \frac{\gamma - 1}{2} M_2^2}{1 + \frac{\gamma - 1}{2} \cdot M_1^2}}$$

For stagnation pressure, we have,

$$\frac{T_o}{T} = \left(\frac{p_o}{p}\right)^{\frac{\gamma - 1}{\gamma}} = 1 + \frac{\gamma - 1}{2} M^2$$

$$\therefore \quad \mathbf{\frac{p_o}{p} = \left\{1 + \frac{\gamma - 1}{2} M^2\right\}^{\frac{\gamma}{\gamma - 1}}} \qquad \text{...(21.23)}$$

The dynamic pressure q is defined by

$$q = \frac{1}{2} \rho V^2$$

Also $$M = \frac{V}{\sqrt{\gamma R\,T}} \text{ or } M^2 = \frac{V^2}{\gamma R\,T}$$

$$\therefore \quad V^2 = M^2 \cdot \gamma RT = M^2 \cdot \gamma \cdot \frac{p}{\rho}$$

$$\therefore \quad q = \frac{1}{2} \cdot \rho \cdot \frac{M^2 \gamma p}{\rho}$$

$$\mathbf{q = \frac{M^2 \gamma p}{2}} \qquad \text{... (21.24)}$$

Also, we have

$$p_o = p\left\{1 + \frac{\gamma - 1}{2} M^2\right\}^{\frac{\gamma}{\gamma - 1}}$$

$$\therefore \quad p_o - p = \left[p\left\{1 + \frac{\gamma - 1}{2} M^2\right\}^{\frac{\gamma}{\gamma - 1}} - 1\right]$$

$$\therefore \quad \frac{p_o - p}{q} = p\left[\left\{1 + \frac{\gamma - 1}{2} M^2\right\}^{\frac{\gamma}{\gamma - 1}} - 1\right] \frac{2}{\gamma p\, M^2}$$

$$= \frac{2}{\gamma M^2}\left[\left\{1 + \frac{\gamma - 1}{2} M^2\right\}^{\frac{\gamma}{\gamma - 1}} - 1\right]$$

$$= \frac{2}{\gamma M^2}\left[1 + \frac{\gamma}{2} M^2 + \frac{\gamma}{8} M^4 + \frac{\gamma}{48}(2 - \gamma) M^6 + \ldots - 1\right]$$

$$\therefore \quad p_o - p = \frac{1}{2} \rho\, V^2 \frac{2}{\gamma M^2}\left[\frac{\gamma}{2} M^2 + \frac{\gamma}{8} M^4 + \frac{\gamma}{48}(2 - \gamma) M^6 + \ldots\right]$$

$$\therefore \quad p_o = p + \frac{\rho V^2}{2}\left[\frac{2}{\gamma M^2} + \frac{M^2}{4} + \frac{2 - \gamma}{12} \cdot \frac{M^4}{\gamma} + \ldots\right] \qquad \text{... (21.25)}$$

21.7 ISENTROPIC FLOW THROUGH A PASSAGE OR VARYING CROSS-SECTION : CHOKING OR NOZZLES

Let us consider any flow passage of varying section area in the direction of flow.

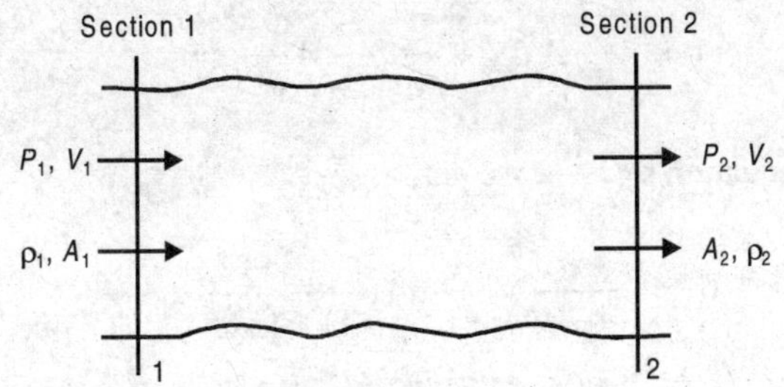

Fig. 21.5 *Flow through duct*

Consider section 1 – 1 and section 2 – 2 and applying continuity equation and using usual notation for P_1, V_1, ρ_1 and A_1, we get

$$\rho_1 A_1 V_1 = \rho_2 A_2 V_2$$

Similarly, applying energy equation to this flow,

$$\frac{V_1^2}{2} + z_1 \cdot g + h_1 + q = \frac{V_2^2}{2} + z_2 \cdot g + h_2 + w$$

As $z_1 = z_2$ and since no work is done, $w = 0$.

We get,
$$\frac{V_1^2}{2} + h_1 + q = \frac{V_2^2}{2} + h_2$$

$$\therefore \quad q = \frac{V_2^2}{2} - \frac{V_1^2}{2} + h_2 - h_1$$

$$\therefore \quad dq = dK + dh$$

where
$$\frac{V_2^2}{2} - \frac{V_1^2}{2} = dK \text{ and } h_2 - h_1 = dh$$

$$\therefore \quad dq = dK + du + d(Pv)$$

$$\therefore \quad dq = dK + du + Pdv + vdP$$

but
$$dq = du + Pdv$$

$$\therefore \quad du + Pdv = dK + du + Pdv + vdP$$

$$\therefore \quad 0 = dK + vdP$$

$$\left(\text{Change in kinetic energy} = \frac{V_2^2}{2} \text{ neglecting } \frac{V_1^2}{2}\right)$$

The differential form,
$$dK = \frac{2V}{2} \cdot dV = V\,dV$$

$$\therefore \quad 0 = V\,dV + v\,dP$$

$$\therefore \quad V\,dV = -v\,dP \qquad \text{... (21.26)}$$

Now, consider continuity equation which is

$$\rho_1 A_1 V_1 = \rho_2 A_2 V_2 = \rho AV = \text{constant}$$

where
ρ = density, kg/m³
A = area, m²
V = velocity in m/sec

By partial differentiation, we get,

$$0 = \frac{1}{V} \cdot A\, dV + \frac{V}{v} dA - A \frac{V}{v^2} \cdot dv$$

Dividing both sides of this equation by $\frac{Ac}{v}$, we get

$$0 = \frac{A\, dV}{v \cdot AV} v + \frac{V}{v} dA \cdot \frac{v}{AV} - \frac{AV}{v^2} \cdot dv \cdot \frac{v}{AV}$$

$$= \frac{dV}{V} + \frac{dA}{A} - \frac{dv}{v} \qquad \text{... (21.27)}$$

$$\frac{dA}{A} = \frac{v - dp}{V^2} - \frac{dv}{v}$$

$$= \frac{v - dp}{V^2}\left[1 - \frac{dV}{v} \cdot \frac{V}{vdp}\right]$$

$$\frac{dv}{V^2} \cdot \frac{V^2}{v^2} = -\frac{V - dp}{V - V} + \frac{dA}{A} - \frac{dv}{v}$$

$$1 - M^2 \cdot \frac{1}{v^2} \frac{dv}{dp} = \frac{1}{a^2}$$

Now, substituting the value of dV from Eq. 21.26, we get,

$$0 = -\frac{V\, dP}{V} \cdot \frac{1}{V} + \frac{dA}{A} - \frac{dV}{V}$$

$$\therefore \quad 0 = -\frac{v\, dP}{V^2} + \frac{dA}{A} - \frac{dV}{V}$$

$$\therefore \quad \frac{dA}{A} = \frac{dV}{V} + \frac{vdP}{V^2} = \frac{vdP}{V^2}\left[\frac{dV}{V} \cdot \frac{V^2}{VdP} + 1\right]$$

But
$$\frac{dV}{V^2 - dP} = -\frac{1}{(\text{sonic velocity})^2}$$

$$= -\frac{1}{a^2} \text{ (From sonic velocity derivation)}$$

where a is the sonic velocity.

$$\therefore \quad \frac{dA}{A} = \frac{VdP}{V^2}\left[-\frac{V^2}{a^2} + 1\right]$$

$$= \frac{vdP}{V^2}\left[1 - M^2\right]$$

Substituting for vdP from Eq. 21.26, we get

$$\frac{dA}{A} = -\frac{VdV}{V^2}\left[1 - M^2\right]$$

$$\therefore \quad \frac{dA}{A} = \frac{dV}{V}\left[M^2 - 1\right] \qquad \text{...(21.28)}$$

This is an important relation between velocity, area and *acoustic velocity* at any particular case.

Now, $M = \dfrac{\text{Velocity}}{\text{Velocity of sound}} = \text{Mach Number} = \dfrac{V}{a}$.

Case I :

If $M < 1$ i.e., velocity is less than one Mach number

$M^2 < 1$ and $M^2 - 1$ is negative.

i.e., $$\frac{dA}{A} = \frac{dV}{V}(-\text{ ve quantity})$$

$$\therefore \quad \frac{dV}{V} = -\frac{dA}{A}$$

Now, if we want velocity to increase, $\frac{dV}{V}$ will be +ve and $\frac{dA}{A}$ will be negative. Therefore sectional area is decreasing in the direction of flow. Such a flow passage is called nozzle and the velocity is then sonic velocity, the nozzle is called subsonic nozzle.

In the subsonic nozzle the flow is accelerated while the area is reducing and the limit is that the velocity equal to velocity of sound when $M = 1$.

Case II :

If $M > 1$, i.e., the velocity is greater than sonic velocity

$M^2 > 1$ and $M^2 - 1$ is positive,

i.e., $$\frac{dA}{A} = \frac{dV}{V}(+\text{ve quantity})$$

$$\therefore \quad \frac{dV}{V} = +\frac{dA}{A}$$

Now, if we want velocity to increase, $\frac{dV}{V}$ will be +ve and $\frac{dA}{A}$ will be positive. Therefore sectional area is increasing in the direction of flow and such a passage is called diverging nozzle. Here supersonic flow is further accelerated and the nozzle is called supersonic nozzle.

Therefore subsonic nozzle will be accelerated to sonic velocity and it will be further accelerated to supersonic velocity. Thus, if a fluid with subsonic velocity enters convergent-divergent nozzle we will get supersonic velocity at the outlet.

Hence, following conclusions can be drawn,

1. For subsonic velocities ($M < 1$) dA and dV must be opposite in sign i.e., an increase of cross-sectional area causes a reduction of velocity and vice versa.

2. For supersonic velocities ($M > 1$) dA and dV are of same sign. An increase of cross-sectional area then causes an increase of velocity and vice versa.

3. When $M = 1$, dA must be zero and since the second derivative is positive, A must be minimum. Thus, if the velocity of flow equals the sonic velocity anywhere, it must be at minimum area which is throat.

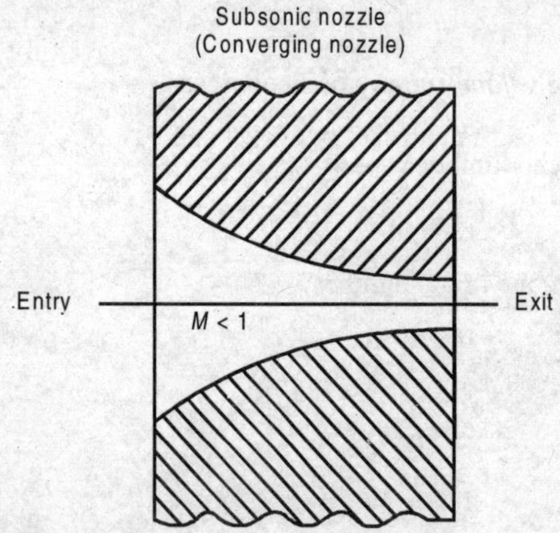

Fig. 21.6 *Subsonic nozzle*

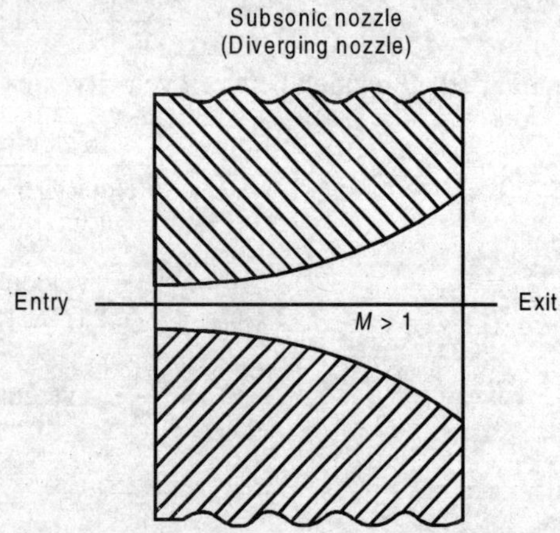

Fig. 21.7 *Supersonic nozzle*

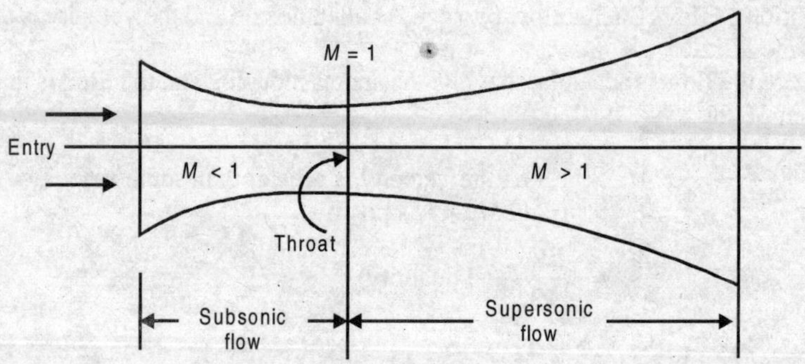

Fig. 21.8 *Convergent-divergent nozzle*

Nozzles are used for increasing the velocity of steam. At the exit of the boiler, velocity is negligible. Then steam passes through subsonic nozzle and its velocity increases to sonic velocity ($M = 1$) and it reaches throat. We still want higher velocity, hence divergent nozzle has to be used.

Flow characteristic of a nozzle and chocking of nozzle :

$$m = \text{Mass flow rate} = \frac{\text{Area} \times \text{velocity}}{\text{Specific volume}} = \frac{AV}{v}$$

Now,

$$\text{velocity, } V = \frac{V}{a} = M$$

$$= M\sqrt{\gamma RT}$$

and

$$\frac{1}{v} = \frac{P}{RT}$$

We get $$m = \frac{APM}{\sqrt{T}}\sqrt{\frac{\gamma}{R}}$$

For sonic velocity $M = 1$, pressure and temperature at the throat will have critical values. The critical values are

$$P_c = \left(\frac{2}{\gamma+1}\right)^{\frac{\gamma}{\gamma-1}} \cdot P_1$$

and $$T_c = \left(\frac{2}{\gamma+1}\right)T_1$$

Substituting these values in the above equation,

$$m = \frac{A\cdot\left(\frac{2}{\gamma+1}\right)^{\frac{\gamma}{\gamma-1}}\cdot P_1}{\sqrt{\frac{2}{\gamma+1}\cdot T_1}}\cdot\sqrt{\frac{\gamma}{R}}.$$

If R and γ are fixed for a gas, mass flow is a function of upstream pressure and temperature only.

The mass flow rate increases as ratio $\frac{P_c}{P_1}$ increases. At throat $\frac{P_c}{P_1} = 0.5283$ for perfect gas, 0.546 for superheated steam, and 0.578 for dry saturated steam. When this ratio increases beyond this value, the mass flow starts decreasing and when this ratio is 1, i.e., when throat pressure is equal to inlet pressure, the discharge is zero.

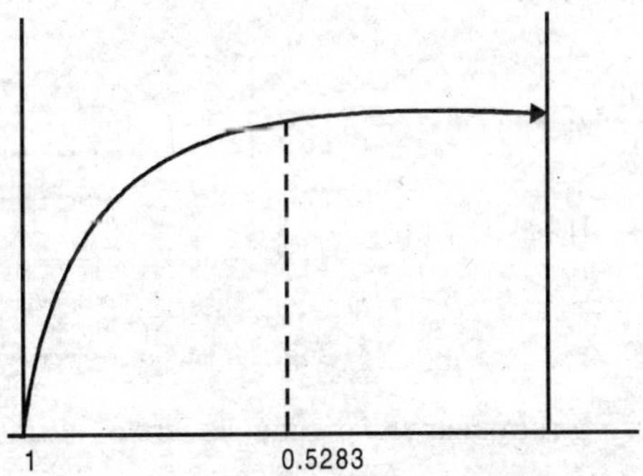

Fig. 21.9 *Pressure ratio,* $\frac{P_c}{P_1}$

The nozzles are always designed for maximum discharge and the discharge at point B is maximum and at this point the nozzle is said to be choked. The discharge of nozzle cannot be more than this value.

21.8 NOZZLES

A nozzle is a device which serves two purposes :

1. To convert thermal energy into kinetic energy.

2. To direct mass flow from it at a specified angle.

There are many applications which require high velocity jet of fluid.

Nozzles are used in steam and gas turbines, jet engines, etc. The other applications of the nozzle are the injectors for pumping feed water into the boiler and the ejectors for removing air from the condensers.

Steam Turbine Layout :

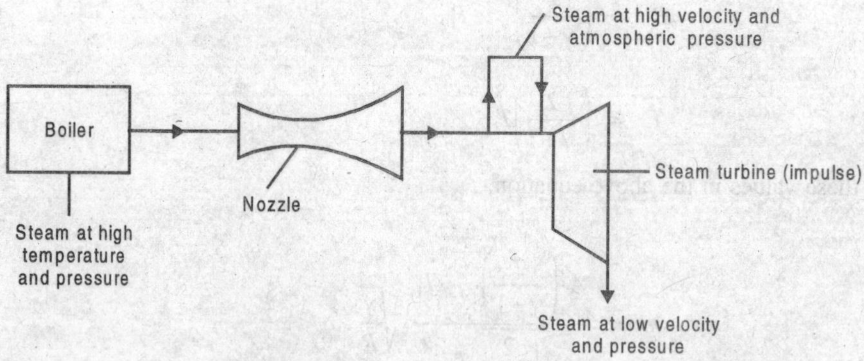

Fig. 21.10 *Steam Turbine Layout*

21.9 NOZZLE SHAPES

There are two types of nozzle shapes.

(i) **Convergent Nozzle :** (ii) **Convergent-divergent nozzle**

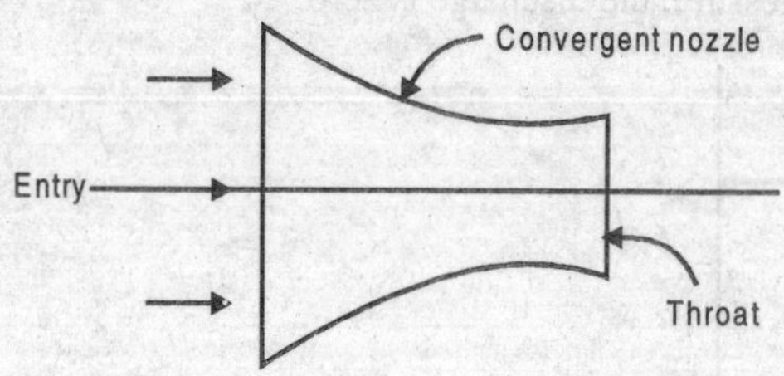

Fig. 21.11 *Convergent Nozzle*

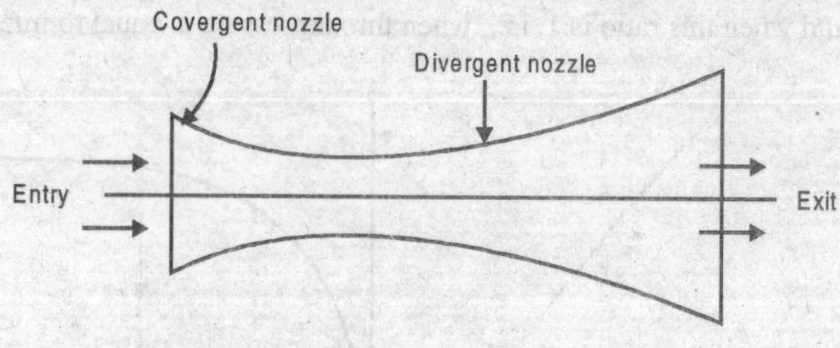

Fig. 21.12 *Convergent-divergent nozzle*

The cross-section of the nozzle can be circular, rectangular, square or elliptical.

21.10 NOZZLE FLOW CONDITIONS

The nozzle is an open system in which flow of fluid is a steady flow process.

Hence, the flow of fluid through the nozzle should satisfy the following conditions :

1. Mass flow rate into and out of system are equal and do not vary with time.
2. State and energy of the fluid at entrance and exit should not vary with time.
3. The rates of heat and work transfer into and out of the system should not change with time.

We will derive relations for exit velocity and discharge for ideal gas flowing through the nozzle and apply these relations to steam by using suitable value of index *n*.

21.11 FLOW OF AN IDEAL GAS THROUGH NOZZLE

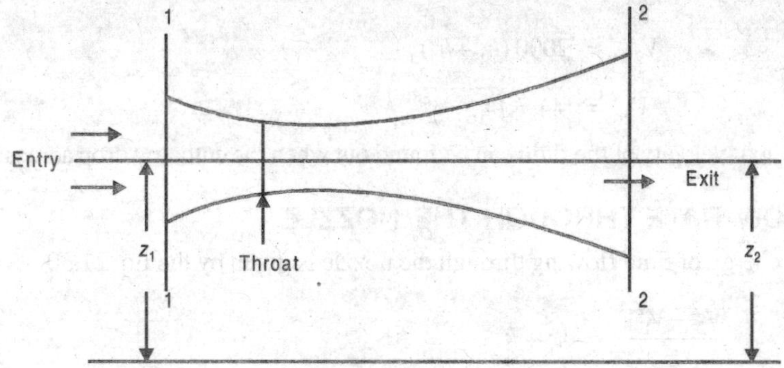

Fig. 21.13 *Flow of gas through nozzle*

Let,

$$
\begin{aligned}
A_1, A_2 &= \text{Cross-sectional area of nozzle at entrance and exit respectively}\\
m &= \text{Mass flow rate in kg/sec}\\
P_1, P_2 &= \text{Pressure absolute in N/m}^2\\
v_1, v_2 &= \text{Specific volume in m}^3\text{/kg}\\
u_1, u_2 &= \text{Specific internal energy in J/kg}\\
V_1, V_2 &= \text{Velocity in metres/sec}\\
z_1, z_2 &= \text{Elevation above arbitrary datum in metre}\\
q &= \text{Net rate of heat transfer in J/kg}\\
w &= \text{Net rate of work transfer in J/kg}
\end{aligned}
$$

Applying steady state flow energy equation to nozzle,

$$\frac{V_1^2}{2} + z_1 \cdot g + h_1 + q = \frac{V_2^2}{2} + z_2 \cdot g + h_2 + w \qquad \text{...(21.29)}$$

Assumptions :

(i) Normally nozzle is horizontal $\quad \therefore z_1 = z_2$

(ii) Flow is isentropic $\quad \therefore q = 0$

(iii) No work is done $\quad \therefore w = 0$

$\therefore$ Equation 21.28 reduces to,

$$\frac{V_1^2}{2} + h_1 = \frac{V_2^2}{2} + h_2$$

or

$$\frac{V_2^2 - V_1^2}{2} = h_1 - h_2 \qquad \text{...(21.30)}$$

Velocity of fluid when it enters the nozzle is known as velocity of approach (V_1). Normally, velocity of approach is very small as compared with exit velocity and hence can be neglected.

$$\therefore \qquad V_2 = \sqrt{h_1 - h_2}$$

In the above equation, unit of every term is J/kg. However, as the enthalpy for ideal gas is given in kJ/kg, the equation can be written with each term in kJ/kg.

$$\therefore \quad \frac{V_2^2}{1000} = 2(h_1 - h_2) \quad \text{...(21.31)}$$

$$\therefore \quad V_2^2 = 2000(h_1 - h_2)$$

$$\therefore \quad V_2 = 44.7\sqrt{h_1 - h_2} \quad \text{...(21.32)}$$

By using Eq. 21.32, exit velocity of the fluid can be found out when the enthalpy drop across the nozzle is known.

21.12 MASS FLOW RATE THROUGH THE NOZZLE

The change in kinetic energy of fluid flowing through the nozzle is given by the Eq. 21.30

$$\frac{V_2^2 - V_1^2}{2} = h_1 - h_2$$

$$\therefore \quad \frac{V_2^2 - V_1^2}{2} = u_1 + P_1 v_1 - u_2 - P_2 v_2$$

$$\therefore \quad \frac{V_2^2 - V_1^2}{2} = (P_1 v_1 - P_2 v_2) + (u_1 - u_2)$$

$$= (P_1 v_1 - P_2 v_2) + C_v (T_1 - T_2)$$

$$(\because du = C_v dt)$$

For ideal gas $P_1 v_1 = RT_1$ per kg and $C_p - C_v = R$.

$$\therefore \quad \frac{V_2^2 - V_1^2}{2} = (P_1 v_1 - P_2 v_2) + \frac{C_v \cdot R}{R}(T_1 - T_2)$$

$$\therefore \quad \frac{V_2^2 - V_1^2}{2} = (P_1 v_1 - P_2 v_2) + \frac{C_v}{C_p - C_v}(RT_1 - RT_2)$$

$$\therefore \quad \frac{V_2^2 - V_1^2}{2} = (P_1 v_1 - P_2 v_2) + \frac{1}{\gamma - 1}(P_1 v_1 - P_2 v_2)$$

(In expression $\frac{C_v}{C_p - C_v}$, dividing numerator and denominator by C_v).

$$\therefore \quad \frac{V_2^2 - V_1^2}{2} = (P_1 v_1 - P_2 v_2)\left[1 + \frac{1}{\gamma - 1}\right]$$

$$\therefore \quad \frac{V_2^2 - V_1^2}{2} = (P_1 v_1 - P_2 v_2)\left(\frac{\gamma}{\gamma - 1}\right)$$

If the velocity of approach V_1 is neglected,

$$\frac{V_2^2}{2} = (P_1 v_1 - P_2 v_2)\frac{\gamma}{\gamma - 1}$$

$$\therefore \quad V_2 = \sqrt{\frac{2\cdot\gamma}{\gamma-1}(P_1 v_1 - P_2 v_2)}$$

$$\therefore \quad V_2 = \sqrt{2\cdot\frac{\gamma}{\gamma-1}\cdot P_1 v_1\left(1-\frac{P_2 v_2}{P_1 v_1}\right)}$$

As $$P_1 v_1^{\gamma} = P_2 v_2^{\gamma}$$

$$\therefore \quad \frac{v_2}{v_1} = \left(\frac{P_2}{P_1}\right)^{-1/\gamma}$$

Substituting for $\frac{v_2}{v_1}$,

We get, $$V_2 = \sqrt{2\cdot\frac{\gamma}{\gamma-1}P_1 v_1\left\{1-\left(\frac{P_2}{P_1}\right)^{\frac{\gamma-1}{\gamma}}\right\}} \quad \text{... (21.33)}$$

$$\text{Mass flow rate in kg/sec.} = \frac{\text{Area} \times \text{Velocity}}{\text{Specific volume}}$$

Hence, the mass flow rate in kg/sec when the fluid flows through cross sectional area A_2 and where specific volume of fluid is v_2 is given by

$$m = \frac{A_2}{v_2}\sqrt{2\frac{\gamma}{\gamma-1}P_1 v_1\left\{1-\left(\frac{P_2}{P_1}\right)^{\frac{\gamma-1}{\gamma}}\right\}} \quad \text{... (21.34)}$$

We will convert this expression in terms of P_1, v_1, A_2 and ratio $\frac{P_1}{P_2}$ which are known as *quantities*.

Now, $$P_1 v_1^{\gamma} = P_2 v_2^{\gamma}$$

$$\therefore \quad \left(\frac{v_2}{v_1}\right)^{\gamma} = \frac{P_1}{P_2}$$

$$\therefore \quad \frac{v_2}{v_1} = \left(\frac{P_1}{P_2}\right)^{1/\gamma}$$

$$\therefore \quad v_2 = v_1\left(\frac{P_1}{P_2}\right)^{1/\gamma} = v_1\left(\frac{P_2}{P_1}\right)^{-1/\gamma}$$

Substituting this value of v_2 in Eq. 21.34, we get

$$m = \frac{A_2}{v_1}\left(\frac{P_2}{P_1}\right)^{1/\gamma} \sqrt{2\cdot\frac{\gamma}{\gamma-1}\cdot P_1 v_1 \left\{1-\left(\frac{P_2}{P_1}\right)^{\frac{\gamma-1}{\gamma}}\right\}}$$

$$= A_2\sqrt{2\cdot\frac{\gamma}{\gamma-1}\cdot P_1 v_1 \left(\frac{P_2}{P_1}\right)^{\frac{2}{\gamma}}\cdot\frac{1}{v_1^2}\left\{1-\left(\frac{P_2}{P_1}\right)^{\frac{\gamma-1}{\gamma}}\right\}}$$

$$= A_2\sqrt{2\cdot\frac{P_1}{v_1}\cdot\frac{\gamma}{\gamma-1}\left\{\left(\frac{P_2}{P_1}\right)^{\frac{2}{\gamma}}-\left(\frac{P_2}{P_1}\right)^{\frac{\gamma+1}{\gamma}}\right\}} \quad \text{...(21.35)}$$

This is a very important relation which gives mass flow rate in kg/sec in terms of known quantities P_1, P_2, v_1 and A_2.

21.13 CONDITIONS FOR MAXIMUM DISCHARGE THROUGH NOZZLE

Nozzles are always designed for maximum discharge conditions. The maximum discharge condition determines the pressure at throat and area at throat. The pressure and volume at entry are constant and considering maximum discharge conditions, pressure and volume at throat are determined.

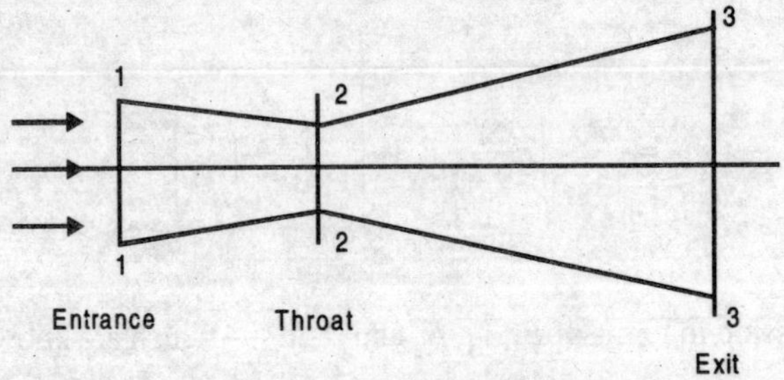

Fig. 21.14 *Nozzle*

In the expression

$$\dot{m} = A_2\sqrt{\frac{P_1}{v_1}\frac{\gamma}{\gamma-1}\cdot\left\{\left(\frac{P_2}{P_1}\right)^{\frac{2}{\gamma}}-\left(\frac{P_2}{P_1}\right)^{\frac{\gamma+1}{\gamma}}\right\}}$$

P_1, v_1 and γ are all constants so the mass flow rate per unit area is dependent on ratio $\frac{P_2}{P_1}$ where P_2 is the pressure at throat and P_1 is pressure at entrance.

Hence there is only one value of $\frac{P_2}{P_1}$ which will give maximum discharge.

Hence differentiating the term $\left\{\left(\frac{P_2}{P_1}\right)^{\frac{2}{\gamma}} - \left(\frac{P_2}{P_1}\right)^{\frac{\gamma+1}{\gamma}}\right\}$ with respect to $\frac{P_2}{P_1}$ and equating it to zero will give the value of $\frac{P_2}{P_1}$ at which discharge will be maximum.

$$\therefore \quad \frac{d}{d\left(\frac{P_2}{P_1}\right)}\left\{\left(\frac{P_2}{P_1}\right)^{\frac{2}{\gamma}} - \left(\frac{P_2}{P_1}\right)^{\frac{\gamma+1}{\gamma}}\right\} = 0$$

$$\therefore \quad \frac{2}{\gamma}\left(\frac{P_2}{P_1}\right)^{\frac{2}{\gamma}-1} - \frac{\gamma+1}{\gamma}\left(\frac{P_2}{P_1}\right)^{\frac{\gamma+1}{\gamma}-1} = 0$$

$$\therefore \text{ Simplifying} \quad \frac{P_c}{P_1} = \frac{P_2}{P_1} = \left(\frac{2}{\gamma+1}\right)^{\frac{\gamma}{\gamma-1}} \qquad \text{...(21.36)}$$

$$\therefore \quad P_2 = P_1\left(\frac{2}{\gamma+1}\right)^{\frac{\gamma}{\gamma-1}} = P_c \qquad \text{...(21.37)}$$

This is known as critical pressure which gives maximum discharge and $\frac{P_c}{P_1} = \left(\frac{2}{\gamma+1}\right)^{\frac{\gamma}{\gamma-1}}$ is called *critical pressure ratio.*

21.14 CRITICAL TEMPERATURE RATIO

This is the ratio of temperature at throat and at entrance for maximum discharge condition.

Let T_c be the temperature at throat and T_1 be the temperature at entrance.

$$\text{For maximum discharge,} \quad \frac{P_c}{P_1} = \left(\frac{2}{\gamma+1}\right)^{\frac{\gamma}{\gamma-1}}$$

Now, general relation between pressure and temperature for an ideal gas is

$$\frac{T_c}{T_1} = \left(\frac{P_c}{P_1}\right)^{\frac{\gamma-1}{\gamma}}.$$

Substituting $\frac{P_c}{P_1} = \left(\frac{2}{\gamma+1}\right)^{\frac{\gamma}{\gamma-1}}$ in the above equation, we get

$$\frac{T_c}{T_1} = \left[\left(\frac{2}{\gamma+1}\right)^{\frac{\gamma}{\gamma-1}}\right]^{\frac{\gamma-1}{\gamma}} = \frac{2}{\gamma+1} \quad \text{...(21.38)}$$

The ratio $\frac{T_c}{T_1}$ is called *critical temperature ratio.*

21.15 FLOW OF STEAM THROUGH NOZZLE

The expression for critical pressure and maximum discharge derived earlier apply to perfect gas and not vapours. This is due to the fact that in case of vapours the exponent of expansion is no longer constant but depends upon initial and final states of vapour.

Close approximation is achieved for expansion of steam through nozzle if the expansion is assumed to follow the law $Pv^n = C$.

1. For steam initially superheated, $n = 1.3$
2. For steam initially dry saturated, $n = 1.135$.
3. For steam initially wet with dryness fraction 'X', $n^* = 1.035 + X\,(0.10)$
4. For air, $n = \gamma = 1.4$

Using these values of n, we can write expression of critical pressure ratios as under

1. For superheated steam,

$$\frac{P_c}{P_1} = \left(\frac{2}{1.3+1}\right)^{\frac{1.3}{0.3}} = \mathbf{0.546} \quad \text{... (21.39)}$$

2. For dry saturated steam,

$$\frac{P_c}{P_1} = \left(\frac{2}{1.135+1}\right)^{\frac{1.135}{1.135-1}} = \mathbf{0.578} \quad \text{... (21.40)}$$

3. For wet steam, $n = 1.035 + (0.6)\,(0.1)$
(say with dryness fraction = 0.6) $= 1.095$

$$\frac{P_c}{P_1} = \left(\frac{2}{1.095+1}\right)^{\frac{1.095}{1.095-1}}$$

$$= (0.9546)^{11.52} = 0.585$$

$$\frac{P_c}{P_1} = \left(\frac{2}{1.4+1}\right)^{0.2857} = \mathbf{0.5283}$$

Now for 7 bar and 275°C temperature, entropy of steam from steam tables = 7.2031 kJ/kgK

* This is known as *Zeuner's equation.*

$$\therefore \quad s_1 = 7.2031 = s_3 = s_f + X_{s_{fg}}$$

$$\therefore \quad 7.2031 = 1.3027 + X \cdot 6.0571$$

$$\therefore \quad X = 0.974$$

$$\therefore \quad v_3 = X \cdot v_g = (0.974)(1.6937) = 1.649 \text{ m}^3/\text{kg}$$

For steady flow,
$$\frac{A_2 V_2}{v_2} = \frac{A_3 V_3}{v_3}$$

$$\therefore \quad A_3 = \frac{A_2 V_2}{v_2} \cdot \frac{v_3}{V_2}$$

$$= \frac{\left(\frac{\pi}{4}\right)(0.005)^2 \times 533 \times 1.649}{0.5112 \times 838} \times 10^6$$

$$A_3 = 40.28 \text{ mm}^2 = \frac{\pi}{4} d_3^2$$

$$\therefore \quad d_3 = \mathbf{7.16 \text{ mm}}$$

21.16 FLOW WITH FRICTION

This type of flow may simulate actual operation in technological devices, but more often than not friction must be taken into consideration. Hence, it is necessary to be able to account properly for the losses incurred by friction.

Certain assumptions are made which led to a type of flow which is called *Rayleigh flow*. In this chapter too, certain assumptions will be made, which when followed, lead to a pattern known as *Fanno flow*. The requirements for Fanno flow are that :

1. The flow must be steady.
2. The flow to be considered one-dimensional.
3. Area of the duct is constant throughout.
4. The flow is adiabatic.
5. No system which deliver or receive mechanical work.
6. Gas is perfect and hence has constant specific heat.

21.17 FRICTION LOSS

Although the first law of thermodynamics in the form of the steady flow energy equation is applicable to the flow with friction, it cannot predict the magnitude of loss incurred by friction. The second law of thermodynamics is more useful because the loss in available energy from frictional effects causes an increase in entropy which can be used to indicate the amount of loss. According to the second law of thermodynamics, the entropy of a system increases in any adiabatic but irreversible process.

Whether the flow is reversible or not, a drop in pressure will usually take place in the direction of flow. For a reversible flow, this pressure drop is expressed by *Euler equation*, namely

$$dp + \rho V \cdot dV = 0$$

or

$$dp = -\rho V \, dV$$

If friction accompanies the flow, there will be a further drop in pressure although nothing else has been changed in the system viz. pipe length etc. Denoting this pressure drop due to friction by dp_f, the total drop becomes

$$dp_T = dp + dp_f = -(\rho v\, dv) + (dp_f) \qquad ...(21.41)$$

Here the pressure loss from friction is considered as a separate item.

21.18 THE FANNING EQUATION

The reduction in pressure from friction is given with reasonable accuracy by **D' Arcy-Weisbach equation** or the **Fanning equation.**

The D' Arcy-Weisbach equation is

$$dp_f = \frac{f' \rho V^2}{2D} \cdot dx \qquad ...(21.42)$$

where, dp_f = Pressure drop due to friction (N/m²)

f' = Dimensionless friction factor for D' Arcy-Weisbach equation

ρ = Density (kg/m³)

V = Velocity (m/sec)

D = Diameter of the passage (metres)

dx = Length of the passage (metres)

The negative sign is inserted becaus the pressure drop results in the direction of positive flow → dx.

The Fanning equation is

$$dp_f = -\frac{f \times \rho V^2}{2r_n} \cdot dx \qquad ...(21.43)$$

where, r_h = Hydraulic radius of the passage usually in metres

The term hydraulic radius has particular significance when a duct or passage is not round in cross-section or the fluid does not fill the duct. The hydraulic radius r_h is defined as the ratio of the cross-sectional area to the wetted perimeter. Hence, for a circular duct or pipe,

$$r_h = \frac{\pi r^2}{2\pi r} = \frac{r}{2} = \frac{D}{4}$$

∴ Fanning equation for circular duct is

$$dp_f = -f \times \frac{\rho V^2}{2r_h} dx$$

$$= -f \times \frac{\rho V^2}{2 \times \frac{D}{4}} dx$$

$$\mathbf{dp_f = -\frac{4f\, \rho V^2 \cdot dx}{2D}} \qquad ...(21.44)$$

From this we see that the friction factor in D' Arcy-Weisbach equation is 4 times that in the Fanning equation.

21.19 FRICTION FACTOR

Regardless of whether the Fanning or D' Arcy-Weisbach equation is used, the friction coefficient is defined as the ratio of the shearing stress to the dynamic flow pressure. To illustrate, consider a constant area, insulated duct and a control volume of length *dx* as shown in Fig. 21.15.

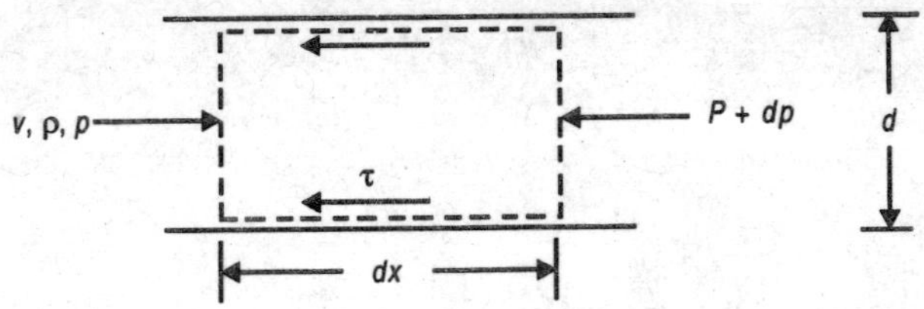

Fig. 21.15

Let τ be the shearing stress between the wall and flow. Then, by definition, the coefficient of friction is given as

$$f = \frac{\tau}{\frac{\rho V^2}{2}} \qquad \text{...(21.45)}$$

For incompressible flow, (Experiments by Moody) the friction factor is a function of the Reynold's number and of a dimensionless parameter $\frac{\varepsilon}{D}$, called the roughness ratio, in which ε is a linear measure of absolute roughness of the inside pipe surface and *D* is the inside diameter.

For compressible flow, (Experiments by Keenan and Neumann) which is supersonic flow of air in a tube, friction coefficient is influenced as usual by the Reynold's number and dimensionless parameter $\frac{L}{D}$. Because compressibility effects must be considered for supersonic flow, dimensional analysis indicates that at high velocities the friction coefficient will depend also on the Mach number.

When the ratio $\frac{L}{D} > 50$, the friction factor was approximately equal to friction factor for incompressible flow for the same value of Reynold's number. Also it was found that supersonic flow can rarely be maintained if the $\frac{L}{D} > 50$. In other words, the effect of the Mach number is to limit the range of values of the ratio $\frac{L}{D}$.

21.20 FRICTION PARAMETER

In the absence of gravity forces, the total pressure drop over a distance *dx* is found from Eq. 21.40 which may be combined with Eq. 21.44 to give

$$dp = -4f \cdot \frac{\rho V^2}{2D} dx - \rho V \cdot dV \qquad \text{...(21.46)}$$

$$\therefore \quad \frac{dp}{p} = -\frac{4f\,\rho V^2}{2D}\frac{dx}{p} - \frac{\rho V \cdot dV}{p}$$

We have,

$$V^2 = M^2 \gamma RT \text{ and } p = \rho RT$$

$$= M^2 \times \frac{\gamma p}{\rho}$$

$$\therefore \quad \frac{\rho V^2}{p} = \gamma M^2 \text{ and } \frac{\rho V}{p} = \frac{\rho V^2}{pV} = \frac{\gamma M^2}{V}$$

$$\therefore \quad \frac{dp}{p} = -4f\frac{\gamma M^2}{2D}dx - \gamma M^2 \cdot \frac{dV}{V}$$

$$\therefore \quad 4f\frac{\gamma M^2}{2D}dx = -\frac{dp}{p} - \gamma M^2 \cdot \frac{dV}{V} \quad \text{...(21.47)}$$

For a constant area duct, the perfect gas equation combined with continuity equation yields :

$$\rho_1 V_1 = \rho_2 V_2$$

and

$$p = \rho RT$$

or

$$\rho V = c$$

$$dp = \rho R\, dT + RT \cdot d\rho$$

$$\therefore \quad V \cdot d\rho + \rho dV = 0$$

$$\therefore \quad \frac{dp}{p} = \frac{dT}{T} + \frac{d\rho}{\rho}$$

$$\frac{d\rho}{\rho} + \frac{dV}{V} = 0$$

$$= \frac{dT}{T} - \frac{dV}{V}$$

Similarly,

$$M^2 \cdot \gamma RT = V^2$$

$$\therefore \quad 2M\,\gamma RT \cdot dM + M^2\,\gamma R\, dT = 2V \cdot dV$$

$$\therefore \quad 2\frac{dM}{M} + \frac{dT}{T} = \frac{2 \cdot dV}{V}$$

$$\therefore \quad \frac{dT}{T} = 2\frac{dV}{V} - 2\frac{dM}{M}$$

$$\therefore \quad \frac{dT}{T} = 2\frac{dV}{V} - 2\frac{dM}{M}$$

$$\therefore \quad 2\frac{dV}{V} - 2\frac{dM}{M} = -(\gamma-1)M^2 \cdot \frac{dV}{V}$$

and

$$\frac{dT}{T} = -(\gamma-1)M^2 \cdot \frac{dV}{V}$$

$$\therefore \quad 2\frac{dV}{V} + (\gamma-1)M^2 \cdot \frac{dV}{V} = 2\frac{dM}{M}$$

$$\therefore \quad \frac{dV}{V} + \frac{\gamma-1}{2}M^2 \cdot \frac{dV}{V} = \frac{dM}{M}$$

$$\therefore \quad \frac{dV}{V}\left[1+\frac{\gamma-1}{2}M^2\right] = \frac{dM}{M}$$

$$\therefore \quad \frac{dV}{V} = \frac{dM}{M\left[1+\frac{\gamma-1}{2}M^2\right]} \qquad \text{... (21.48)}$$

$\therefore$ From Eq. 21.47, we have

$$4f\frac{\gamma M^2}{2D}dx = -\frac{dp}{p} - \frac{\gamma M^2 \cdot dV}{V}$$

$$= -\left(\frac{dT}{T} - \frac{dV}{V}\right) - \frac{\gamma M^2 dV}{V}$$

$$= -\left(2\frac{dV}{V} - 2\frac{dM}{M} - \frac{dV}{V}\right) - \frac{\gamma M^2 \cdot dV}{V}$$

$$= -\frac{dV}{V} + \frac{2dM}{M} - \frac{\gamma M^2 dV}{V} = \frac{2dM}{M} - \frac{dV}{V}\left[1+\gamma M^2\right]$$

$$= \frac{2dM}{M} - \frac{dM}{M\left[1+\frac{\gamma-1}{2}M^2\right]}[1+\gamma M^2]$$

$$= \frac{dM}{M}\left(2 - \frac{1+\gamma M^2}{1+\frac{\gamma-1}{2}M^2}\right) = \frac{dM}{M}\left[\frac{2+(\gamma-1)M^2-1-\gamma M^2}{1+\frac{\gamma-1}{2}M^2}\right]$$

$$= \frac{dM}{M}\left[\frac{1+\gamma M^2 - M^2 - \gamma M^2}{1+\frac{\gamma-1}{2}M^2}\right]$$

$$= \frac{1-M^2}{1+\frac{\gamma-1}{2}M^2} \cdot \frac{dM}{M}$$

$$= \frac{2\left(1-M^2\right)}{2+(\gamma-1)M^2} \cdot \frac{dM}{M}$$

$$\therefore \quad \frac{4f \cdot dx}{D} = \frac{2\left(1-M^2\right)}{2+(\gamma-1)M^2} \cdot \frac{dM}{M} \cdot \frac{2}{\gamma M^2}$$

$$= \frac{2\left(1-M^2\right)dM}{\gamma M^3\left[1+\frac{\gamma-1}{2}M^2\right]} \quad \text{... (21.49)}$$

This equation relates the drop in pressure in relation to the Mach number in a constant area duct when the flow is accompanied by friction.

In the presence of friction it is impossible to cross from the supersonic regime into the subsonic and vice-a-versa. Accordingly, it becomes convenient to integrate Eq. 21.49 using a Mach number of unity as the reference. Thus,

$$\int_{x=0}^{x=L_{max}} \frac{4f\cdot dx}{D} = \frac{4f\cdot L_{max}}{D} = \int_{M}^{1} \frac{2\left(1-M^2\right)dM}{\gamma M^3\left[1+\frac{\gamma-1}{2}M^2\right]} \quad \text{... (21.50)}$$

The term $\frac{4f\ L_{max}}{D}$ is called the *friction parameter* and is given in the table.

21.21 FANNO LINES

Frictional, steady flow which takes place in constant area ducts in the absence of heat transport is generally termed as **Fanno flow**. Furthermore, equations which are applicable to such flow are called **Fanno equations**. It is evident that numerous Fanno equations could be derived, each of which finds its origin in the energy, continuity and momentum equations.

The stagnation enthalpy is defined by

$$h_o = h+\frac{V^2}{2}$$

For steady flow, the mass flow rate ρAV is a constant and for a pipe, the mass velocity ρV = constant.

$$\therefore \quad h_o = h+\frac{G^2}{2\rho^2} \quad \text{... (21.51)}$$

From this equation, when density increases, the enthalpy must increase also. For Fanno flow, curves of enthalpy are called **Fanno lines** or any curves on enthalpy-entropy co-ordinates are called **Fanno lines**.

For Fanno flow, which is adiabatic, the steady-flow energy equation is

$$0 = du+d\left(\frac{p}{\rho}\right)+V\cdot dV$$

Also entropy can be introduced by making use of the equation

$$T{\cdot}ds = du + p{\cdot}dV$$

$$= du - p\cdot\frac{d\rho}{\rho^2}$$

$$d\left(\frac{p}{\rho}\right) = -p\cdot\frac{d\rho}{\rho^2}+\frac{dp}{\rho}$$

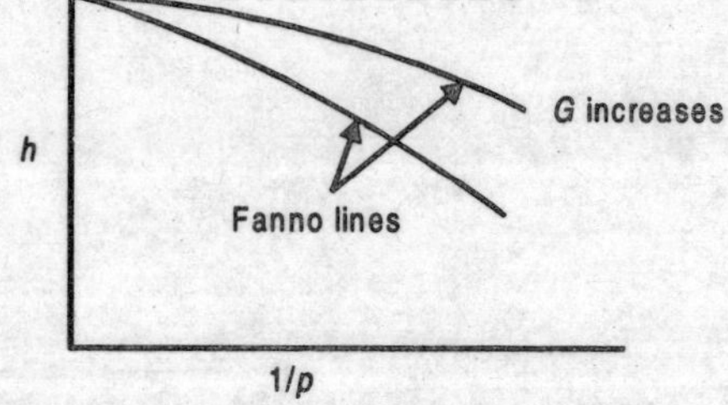

Fig. 21.16

$$d(pv) = p\,dv + v\,dp = pdv + \frac{dp}{\rho}$$

$$\therefore \quad p \cdot dv = -p.\frac{d\rho}{\rho^2}$$

which is the fundamental equation developed by combining the first and the second law of thermodynamics.

$$\therefore \quad Tds = -\left(\frac{dp}{\rho} + v \cdot dv\right)$$

$$\therefore \quad \rho T\,ds = -(dp + \rho v \cdot dv) \qquad \text{... (21.52)}$$

When this is applied to an adiabatic but irreversible process, the magnitude of ds is a measure of the entropy increase introduced by friction.

Equation 21.46 gives,

$$dp = -\frac{4f\,\rho v^2 \cdot dx}{2D} - \rho v\,dv$$

∴ Equation 21.52 gives

$$\rho T\,ds = +\frac{4f\rho v^2 \cdot dx}{2D} - \rho v\,dv + \rho v\,dv$$

$$= +\frac{4f\rho v^2 \cdot dx}{2D}$$

$$\therefore \quad ds = +\frac{4f\rho v^2 \cdot dx}{2D} \times \frac{1}{\rho T}$$

$$ds = +\frac{4f\,v^2 \cdot dx}{2TD} \qquad \text{... (21.53)}$$

But

$$v^2 = M^2 \gamma RT$$

$$\therefore \quad ds = +\frac{4f \cdot M^2 \gamma RT \cdot dx}{2TD}$$

$$= +\frac{4f\gamma R}{2D} \cdot M^2 \cdot dx \qquad \text{... (21.54)}$$

$$= +\frac{4f \cdot dx}{D} \cdot \frac{\gamma RM^2}{2}$$

$$\therefore \quad +\frac{4f\,dx}{D} = +\frac{2ds}{\gamma RM^2}$$

Equation 21.49 gives

$$\frac{4f\,dx}{D} = \frac{2(1-M^2)dM}{\gamma M^3\left[1+\frac{\gamma-1}{2}M^2\right]}$$

$$\therefore \quad +\frac{2ds}{\gamma RM^2} = \frac{2(1-M^2)dM}{\gamma M^3\left[1+\frac{\gamma-1}{2}M^2\right]}$$

$$\therefore \quad ds = +\frac{R(1-M^2)dM}{M\left(1+\frac{\gamma-1}{2}M^2\right)}$$

$$\therefore \quad \frac{ds}{dM} = \frac{R(1-M^2)}{M\left(1+\frac{\gamma-1}{2}M^2\right)} \quad \text{...(21.55)}$$

It can be seen from Eq. 21.55 that the entropy is maximum when $\frac{ds}{dM} = 0$ or when $M = 1$. In other words, if the flow is supersonic and there exists friction, the effect of friction will be for the flow Mach number to approach a value of unity. Whether or not, it will become unity, depends on the extent of the friction, which, for any particular pipe, will be determined by its length. Friction decelerates a supersonic flow and this deceleration may continue, if there is enough friction, until the Mach number becomes unity. If the length for the flow or friction exceeds a magnitude commensurate to a Mach number of unity, there will occur first shock-phenomenon and then finally choking.

If the flow is subsonic originally, the effect of friction will be to accelerate the flow. Again, the limiting condition will be a Mach number of unity, which will be reached when the entropy becomes a maximum. If friction continues to exist after the equilibrium state is reached, the flow will be choked and will have to read just. These phenomena are depicted qualitatively in Fig. 21.17.

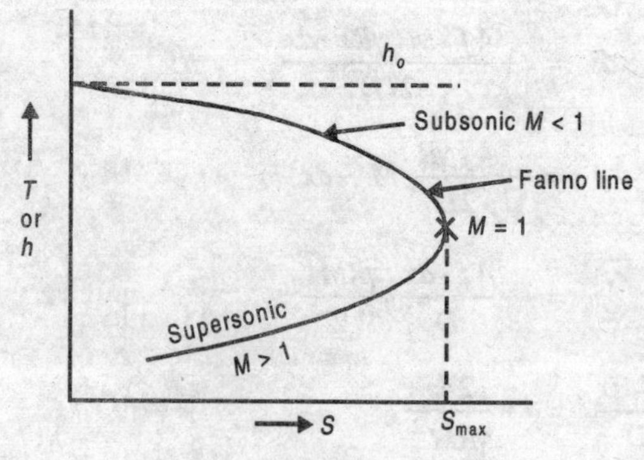

Fig. 21.17

Generation of shock-waves and/or choking will occur if the so-called theoretical length of the pipe is less than the actual given length of the pipe.

21.22 FANNO EQUATIONS

Several additional forms for the Fanno equation will now be derived to supplement those already given.

Under adiabatic conditions of flow,

$$\frac{\gamma R}{\gamma-1} dT + V \cdot dV = 0$$

or

$$\gamma R\, dT + (\gamma-1) V \cdot dV = 0$$

Integrating between the reservoir conditions and any local point we obtain

$$\gamma R(T - T_o) + (\gamma-1)\frac{V^2}{2} = 0$$

or

$$\frac{\gamma R}{\gamma-1}(T - T_o) + \frac{V^2}{2} = 0$$

$$\therefore \quad \frac{V^2}{2} = \frac{\gamma R}{\gamma-1}(T_o - T)$$

$$v = \sqrt{\frac{2\gamma R}{\gamma-1}(T_o - T)}$$

Also,

$$\frac{T_o}{T} = 1 + \frac{\gamma-1}{2}M^2$$

$$\therefore \quad T = \frac{T_o}{1 + \frac{\gamma-1}{2}M^2}$$

$$\therefore \quad v = \sqrt{\frac{2\gamma R}{\gamma-1}T_o\left\{1 - \frac{1}{1+\frac{\gamma-1}{2}M^2}\right\}}$$

21.23 DIABATIC FLOW — RAYLEIGH FLOW

Heat exchangers and combustion chambers are two devices in which heat interchanges occur. Whether the heat is added across the system boundaries as in a heat exchanger, or whether it is added internally by virtue of a chemical reaction as in a combustion chamber, does not alter the analysis in principle and the generic term diabatic process is applicable to both of these related types of phenomena. The equations describing non-adiabatic or diabatic process are complicated and consequently certain limiting assumptions are usually required in order to make solutions of the equations possible. The assumptions are that :

1. The flow takes place in a constant-area section.
2. There is no friction.
3. The gas is perfect and hence has constant specific heats.

4. The composition of the gas does not change.
5. There are no devices in the system which deliver or receive mechanical work.
6. The flow is steady.

21.24 STAGNATION TEMPERATURE CHANGE

We know that heat may be added to a gaseous system either by external or by internal means. The flow system in each case is slightly different. Thus, if the heat is added as in a heat exchanger, the system can be represented as shown in Fig. 21.18.

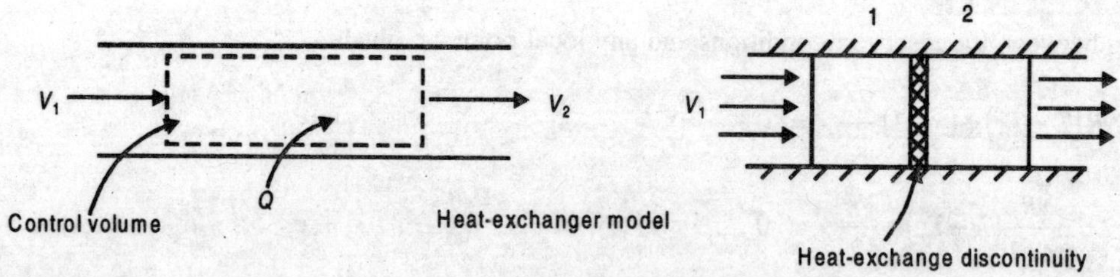

Fig. 21.18

It is also possible for a thermal manifestation of heat addition to occur in a thin discontinuity as shown. For this it is assumed that the duct is insulated so that theoretically there is no heat exchange between the fluid within the control boundaries and the surroundings and furthermore, that there is no heat transfer between parts of the fluid in the control volume due to convection, conduction and radiation. That the heat addition takes place by virtue of a thin discontinuity may seem to be an overly exaggerated simplification, but it actually can be strongly supported. A typical example would be a flame front travelling in a combustion tube and experiments show that such a flame front is actually very thin.

Neglecting ΔP and W_s,

$$dQ = dh + dk \text{ from steady flow energy equation}$$

$$= C_p \cdot dT + V \cdot dV$$

$$= \frac{\gamma R}{\gamma - 1} dT + V \cdot dV \quad \text{... (21.56)}$$

This equation can be integrated between the conditions indicated at 1 and 2.

$$\therefore \quad Q_{1 \to 2} = \frac{\gamma R}{\gamma - 1}(T_2 - T_1) + \frac{V_2^2 - V_1^2}{2}$$

$$= C_p (T_2 - T_1) + \frac{V_2^2 - V_1^2}{2}$$

$$= (h_2 - h_1) + \frac{V_2^2 - V_1^2}{2} \quad \text{... (21.57)}$$

$$= \left(h_2 + \frac{V_2^2}{2}\right) - \left(h_1 + \frac{V_1^2}{2}\right)$$

$$= h_{o_2} - h_{o_1}$$

$$= C_p\left(T_{o_2} - T_{o_1}\right) \quad \text{...(21.58)}$$

For diabatic flow, $Q_{1-2} \neq 0$ and as $C_p > 0$.

$$\mathbf{T_{o_2} \neq T_{o_1}}$$

This inequality states that in diabatic flow the stagnation temperature is not solely determined by the reservoir conditions, as in the case of adiabatic flow.

Heating raises the stagnation temperature and cooling lowers it.

21.25 RAYLEIGH LINE

The locus of points of properties during a constant area, frictionless flow with heat exchange is called the **Rayleigh line**. Along a Rayleigh line the continuity equation and the momentum equation must apply. Thus, for a steady flow in a constant area duct, the equation of continuity appears as

$$G' = \rho VA \text{ and the mass velocity}$$

$$G = \rho V \frac{\text{kgm}}{\text{m}^2\cdot\text{sec}}$$

and the momentum equation is

$$p + \rho V^2 = \text{Constant.}$$

Combining these equations, we have

$$p + \frac{G^2}{\rho} = \text{Constant.} \quad \text{... (21.59)}$$

which is one of the Rayleigh line equations.

Any equation expressing the relation among the properties of a system undergoing adiabatic process in conformity with Eq. 21.59 is a **Rayleigh equation.**

21.26 PRESSURE RATIO

An equation expressing the pressure ratio for a Rayleigh process is derived as follows :

$$p + \rho V^2 = \text{Constant}$$

$$\therefore \quad p_2 + \rho_2 V_2^2 = p_1 + \rho_1 V_1^2 \text{ for constant area section}$$

$$\therefore \quad p_2 - p_1 + \rho_2 V_2^2 - \rho_1 V_1^2 = 0$$

But we have, $\quad M^2 = \dfrac{V^2}{\gamma R\, T}$ and $RT = \dfrac{p}{\rho}$

$$\therefore \quad M^2 = \frac{V^2 \rho}{\gamma p}$$

$$\therefore \quad \rho V^2 = M^2 \gamma p$$

Substituting this in the above equation, we have

$$(p_2 - p_1) + \gamma p_2 M_2^2 - \gamma p_1 M_1^2 = 0$$

$$\therefore \left(p_2 + \gamma p_2 M_2^2\right) - \left(p_1 + \gamma p_1 M_1^2\right) = 0$$

$$p_2\left(1+\gamma M_2^2\right) - p_1\left(1+\gamma M_1^2\right) = 0$$

$$\therefore \quad \left(1+\gamma M_2^2\right) - \frac{p_1}{p_2}\left(1+\gamma M_1^2\right) = 0 \qquad \text{... (21.60)}$$

$$\therefore \quad \frac{p_2}{p_1} = \frac{1+\gamma M_1^2}{1+\gamma M_2^2}$$

$$\mathbf{\frac{p_2}{p_1} = \frac{1+\gamma M_1^2}{1+\gamma M_2^2}} \qquad \text{... (21.61)}$$

In diabatic flow the stations before and after the heat addition have different corresponding reservoir conditions because the stagnation temperature changes. Therefore, a new type of reference condition must be used and it is convenient to define this at the conditions where the Mach number is unity. In Eq. 21.61, the conditions at either point may be used as reference. Arbitrarily selecting this condition at 1 and denoting the reference by asterisk, we obtain

$$\frac{p}{p^*} = \frac{1+\gamma}{1+\gamma M^2} \qquad \text{... (21.62)}$$

We have, $$\frac{p_o}{p} = \left\{1+\frac{\gamma-1}{2}M^2\right\}^{\gamma/\gamma-1}$$

Writing this equation between two stations 1 and 2 dividing one by the other, and finally substituting from the Rayleigh line, Eq. 21.61, we obtain

$$\frac{p_{o_1}}{p_1} = \left\{1+\frac{\gamma-1}{2}M_1^2\right\}^{\gamma/\gamma-1} \quad \text{and} \quad \frac{p_{o_2}}{p_2} = \left\{1+\frac{\gamma-1}{2}M_2^2\right\}^{\gamma/\gamma-1}$$

$$\frac{p_{o_2}}{p_2}\times\frac{p_1}{p_{o_1}} = \frac{p_{o_2}}{p_{o_1}}\times\frac{p_1}{p_2} = \frac{\left\{1+\frac{\gamma-1}{2}M_2^2\right\}^{\gamma/\gamma-1}}{\left\{1+\frac{\gamma-1}{2}M_1^2\right\}^{\gamma/\gamma-1}}$$

$$\therefore \quad \mathbf{\frac{p_{o_2}}{p_{o_1}} = \left\{\frac{1+\frac{\gamma-1}{2}M_2^2}{1+\frac{\gamma-1}{2}M_1^2}\right\}^{\frac{\gamma}{\gamma-1}} \cdot \frac{1+\gamma M_1^2}{1+\gamma M_2^2}} \qquad \text{... (21.63)}$$

$$\frac{p_o}{p_o^*} = \left\{\frac{2\left(1+\frac{\gamma-1}{2}M^2\right)}{1+\gamma}\right\}^{\gamma/\gamma-1} \cdot \frac{1+\gamma}{1+\gamma M^2}$$

21.27 TEMPERATURE RATIO

Equations relating the temperature and the Mach number will now be developed.

At the upstream side of the heat addition, by the definition of the stagnation temperature, we have

$$\frac{T_{o_1}}{T_1} = 1+\frac{\gamma-1}{2}M_1^2$$

and

$$\frac{T_{o_2}}{T_2} = 1+\frac{\gamma-1}{2}M_2^2$$

$\therefore$

$$\frac{T_{o_2}}{T_{o_1}} = \frac{T_2}{T_1} \cdot \frac{1+\frac{\gamma-1}{2}M_2^2}{1+\frac{\gamma-1}{2}M_1^2} \quad \text{...(21.64)}$$

For perfect gas,

$$\frac{p}{\rho} = RT$$

$\therefore$

$$T_2 = \frac{p_2}{\rho_2 R}$$

$$T_1 = \frac{p_1}{\rho_1 R}$$

$\therefore$

$$\frac{T_2}{T_1} = \frac{p_2\rho_1}{p_1\rho_2} \quad \text{...(21.65)}$$

For a constant-area flow, from the equation of continuity,

$$e_1 V_1 = \rho_2 V_2$$

$\therefore$

$$\frac{\rho_1}{\rho_2} = \frac{V_2}{V_1} \quad \text{...(21.66)}$$

and

$$\frac{p_2}{p_1} = \frac{1+\gamma M_1^2}{1+\gamma M_2^2} \text{ from Eq. 21.61}$$

$\therefore$

$$\frac{T_2}{T_1} = \left\{\frac{1+\gamma M_1^2}{1+\gamma M_2^2}\right\} \cdot \frac{V_2}{V_1} \quad \text{...(21.67)}$$

But we have, $$M_1 = \frac{V_1}{\sqrt{\gamma R T_1}} \text{ and } M_2 = \frac{V_2}{\sqrt{\gamma R T_2}}$$

$$\therefore \quad \frac{V_2}{\sqrt{\gamma R T_2}} \cdot \frac{\sqrt{\gamma R T_1}}{v_1} = \frac{M_2}{M_1}$$

$$\therefore \quad \frac{V_2}{V_1} = \frac{M_2}{M_1} \cdot \sqrt{\frac{T_2}{T_1}}$$

$$\therefore \quad \frac{T_2}{T_1} = \left\{\frac{1+\gamma M_1^2}{1+\gamma M_2^2}\right\} \cdot \frac{M_2}{M_1} \sqrt{\frac{T_2}{T_1}}$$

$$\therefore \quad \mathbf{\frac{T_2}{T_1} = \left[\frac{M_2}{M_1}\left\{\frac{1+\gamma M_1^2}{1+\gamma M_2^2}\right\}\right]^2} \qquad \text{... (21.68)}$$

$$\frac{\rho^*}{\rho} = \frac{V}{V^*} = \frac{(\gamma+1)M}{1+\gamma M^2}$$

From Eq. 21.64,

$$\mathbf{\frac{T_{o_2}}{T_{o_1}} = \left[\frac{M_2}{M_1}\left\{\frac{1+\gamma M_1^2}{1+\gamma M_2^2}\right\}\right]^2 \cdot \left\{\frac{1+\frac{\gamma-1}{2}M_2^2}{1+\frac{\gamma-1}{2}M_1^2}\right\}} \qquad \text{... (21.69)}$$

This equation is convenient because it is in terms of quantities which can be measured with relative ease. However, it is possible to modify it slightly and make it even more convenient for analytical work. To do this, we again let $M_1 = 1$ and denote all properties at this point by subscript asterisk.

Thus for $M_1 = M^* = 1$, $T_1 = T^*$, $p_1 = p^*$ and $T_{o_1} = T_o^*$

$$\therefore \quad \frac{T}{T^*} = \frac{M^2(1+\gamma)^2}{\left(1+\gamma M_2^2\right)^2} \text{ from Eq. 21.68} \qquad \text{... (21.70)}$$

and $$\frac{T_o}{T_o^*} = \left[\frac{M_2}{1}\left\{\frac{1+\gamma}{1+\gamma M^2}\right\}\right]^2 \cdot \left\{\frac{1+\frac{\gamma-1}{2}M^2}{1+\frac{\gamma-1}{2}}\right\}$$

$$= \frac{2M^2(1+\gamma)^2}{\left(1+\gamma M^2\right)^2} \cdot \left\{\frac{1+\frac{\gamma-1}{2}M^2}{(1+\gamma)}\right\}$$

$$= \frac{2M^2(1+\gamma)}{\left(1+\gamma M^2\right)^2} \cdot \left\{1+\frac{\gamma-1}{2}M^2\right\} \qquad \text{... (21.71)}$$

These ratios are tabulated in tables for Mach numbers 0–10 for various *k*.

21.28 NORMAL SHOCK WAVES

Basically, a shock is said to be normal if the flow direction upstream as well as downstream from it have the same direction. Oblique and normal shocks are similar in nature in that both results in abrupt changes of thermodynamic properties across them. Geometrically, an oblique shock can be described by an observer who sees a normal shock while at the same time moving parallel to it with a velocity V_t. This is illustrated in Fig. 21.19.

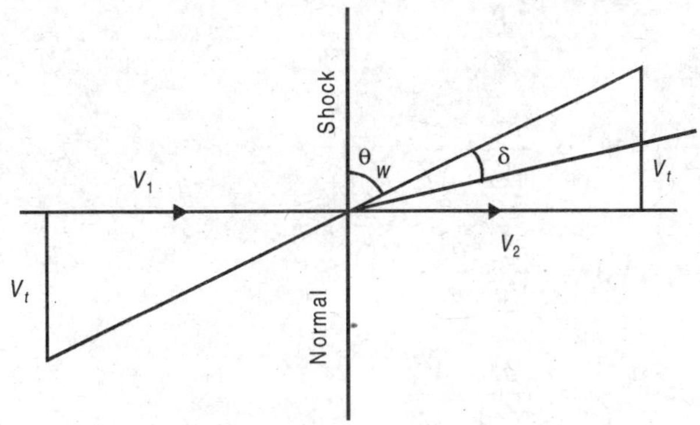

Fig. 21.19

If the observer is stationary, the shock will appear to be normal to the flow. If the observer is in motion, the shock will seem to occur at an angle θ_w with the flow. As the velocity of the observer increases, the shock will seem to become increasingly oblique to the flow.

Assume that we are considering only normal shock as shown in Fig. 21.20.

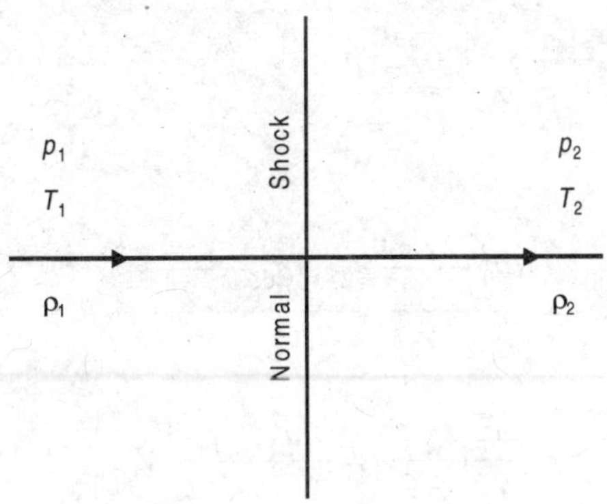

Fig. 21.20

We have,
$$V_1 = V_{n1}$$
$$V_2 = V_{n2}$$

$$\therefore \quad \rho_1 V_1 = \rho_2 V_{n2} \qquad \text{... (21.72)}$$

$$p_1 + \rho_1 V_1^2 = p_2 + \rho_2 V_2^2 \qquad \text{... (21.73)}$$

From Eq. (21.72), dividing by $\rho_1 V_1 = \rho_2 V_2$

$$\frac{p_1}{\rho_1 V_1} + V_1 = \frac{p_2}{\rho_2 V_2} + V_2 \qquad \text{... (21.74)}$$

$$\therefore \quad V_1 - V_2 = \frac{p_2}{\rho_2 V_2} - \frac{p_1}{\rho_1 V_1}$$

We know that, $a^2 = \gamma \dfrac{p}{\rho}$

$\therefore$ From Eq. 21.74,

$$V_1 - V_2 = \frac{a_2^2}{\gamma V_2} - \frac{a_1^2}{\gamma V_1}$$

or

$$\gamma (V_1 - V_2) = \frac{a_2^2}{V_2} - \frac{a_1^2}{V_1} \qquad \text{... (21.75)}$$

We know, $\dfrac{a_0^2}{\gamma - 1} = \dfrac{a_1^2}{\gamma - 1} + \dfrac{V_1^2}{2}$

$$\therefore \quad 2a_1^2 = 2a_0^2 - (\gamma - 1)V_1^2 \qquad \text{... (21.76)}$$

$$2a_2^2 = 2a_0^2 - (\gamma - 1)V_2^2 \qquad \text{... (21.77)}$$

$$\therefore \quad a_1^2 = a_0^2 - \frac{(\gamma - 1)V_1^2}{2} \qquad \text{... (21.78)}$$

$$a_2^2 = a_0^2 - \frac{(\gamma - 1)V_2^2}{2}$$

Substituting in Eq. 21.75, we get

$$\gamma (V_1 - V_2) = \frac{a_0^2 - \frac{\gamma - 1}{2} V_2^2}{V_2} - \frac{a_0^2 - \frac{\gamma - 1}{2} V_1^2}{V_1}$$

$$= \frac{a_0^2}{V_2} - \frac{\gamma - 1}{2} V_2 - \frac{a_0^2}{V_1} + \frac{\gamma - 1}{2} V_1$$

$$= a_0^2 \left[\frac{1}{V_2} - \frac{1}{V_1} \right] + \frac{\gamma - 1}{2} [V_1 - V_2]$$

$$\therefore \quad \gamma(V_1 - V_2) = a_0^2\left[\frac{V_1 - V_2}{V_1V_2}\right] + \frac{\gamma-1}{2}[V_1 - V_2]$$

$$\therefore \quad \gamma = \frac{a_0^2}{V_1\,V_2} + \frac{\gamma-1}{2}$$

$$\gamma\cdot V_1\,V_2 = a_0^2 + \frac{\gamma-1}{2}V_1\,V_2$$

$$(\gamma V_1V_2) - \left(\frac{\gamma-1}{2}V_1V_2\right) = a_0^2$$

$$V_1V_2\left(\gamma - \frac{\gamma-1}{2}\right) = a_0^2$$

$$V_1V_2\left(\frac{2\gamma-\gamma+1}{2}\right) = a_0^2$$

$$V_1V_2\left(\frac{\gamma+1}{2}\right) = a_0^2$$

$$\therefore \quad \mathbf{V_1V_2 = \frac{2}{\gamma+1}a_0^2} \qquad \text{....(21.79)}$$

But $$a_0^2\cdot\frac{2}{\gamma+1} = a^{*2}$$

$\therefore$ For critical conditions we have

$$\mathbf{V_1V_2 = a^{*2}} \qquad \text{... (21.79)}$$

This Eq. 21.80 gives the velocity change across the normal shock and is usually known as the **Rankine-Hugonlot equation for normal shock**. For a normal shock wave to occur, V_1 must be supersonic and a^* is the critical acoustic velocity. It is easy to see from Eq. 21.80 that the velocity behind the normal shock wave must be subsonic.

Frequently it is desirable to show explicitly the changes in density, pressure or temperature across a normal shock.

We have, $$\rho_1\,V_1 = \rho_2\,V_2 \text{ or } r_2 = \frac{\rho_1V_1}{V_2}$$

and $$p_1 + \rho_1\,V_1^2 = p_2 + \rho_2\,V_2^2 = p_2 + \rho_1V_1V_2$$

Dividing both sides of this equation by p_1, we have

$$1 + \frac{\rho_1V_1^2}{p_1} = \frac{p_2}{p_1} + \frac{\rho_1}{p_1}V_1V_2 \qquad \text{... (21.80)}$$

$$\therefore \quad \frac{p_2}{p_1} = 1+\frac{\rho_1 V_1^2}{p_1}-\frac{\rho_1 V_1 V_2}{p_1}$$

$$= 1+\frac{\rho_1 V_1}{p_1}(V_1 - V_2) \qquad \text{... (21.81)}$$

From Eq. 21.76 and 21.77

$$a_0^2 = a_1^2 + \frac{V_1^2}{2}(\gamma-1) = a_2^2 + (\gamma-1)\frac{V_2^2}{2}$$

Let, $a_2 = a^* = V_2$ and $a_1 = a$ and $M^2a^2 = V^2$

$$\therefore \quad a^2 + M^2a^2\frac{(\gamma-1)}{2} = a^{*2} + a^{*2}\cdot\frac{\gamma-1}{2} = a^{*2}\left(1+\frac{\gamma-1}{2}\right) = a^{*2}\left(\frac{\gamma+1}{2}\right)$$

$$\therefore \quad a^{*2} = \frac{2}{\gamma+1}\cdot\left[a^2 + M^2a^2\frac{(\gamma-1)}{2}\right]$$

$$= a^2\left[\frac{2}{\gamma+1}+\frac{\gamma-1}{\gamma+1}M^2\right] \qquad \text{... (21.82)}$$

$\therefore$ From Eq. 21.80,

$$V_1 V_2 = a^{*2} = a^2\left[\frac{2}{\gamma+1}+\frac{\gamma-1}{\gamma+1}M^2\right]$$

$\therefore$ From Eq. 21.81,

$$\frac{p_2}{p_1} = 1+\frac{\rho_1 V_1}{p_1}(V_1 - V_2)$$

$$= 1+\frac{\rho_1}{p_1}\left[V_1^2 - a_1^2\left(\frac{2}{\gamma+1}+\frac{\gamma-1}{\gamma+1}M_1^2\right)\right]$$

$$= 1+\frac{\gamma V_1^2}{a_1^2} - \gamma\left[\frac{2}{\gamma+1}+\frac{\gamma-1}{\gamma+1}M_1^2\right]$$

$$= 1+\gamma M_1^2 - \gamma\left[\frac{2+M_1^2(\gamma-1)}{\gamma+1}\right]$$

$$= 1+\gamma M_1^2 - \gamma\left[\frac{2+\gamma M_1^2 - M_1^2}{\gamma+1}\right]$$

$$= 1+\gamma\left\{\frac{M_1^2\gamma + M_1^2 - 2 - \gamma M_1^2 + M_1^2}{\gamma+1}\right\}$$

$$= 1+\frac{\gamma}{\gamma+1}\cdot 2\left(M_1^2-1\right)$$

$$\therefore \quad \frac{\mathbf{p_2}}{\mathbf{p_1}}=\mathbf{1}+\frac{\mathbf{2\gamma\left(M_1^2-1\right)}}{\mathbf{\gamma+1}} \qquad \text{...(21.83)}$$

An expression defining the density variation across a normal shock is developed as follows :

$$\frac{a_0^2}{\gamma-1} = \frac{a_1^2}{\gamma-1}+\frac{V_1^2}{2}$$

But

$$V_1 V_2 = \frac{2}{\gamma+1}a_0^2$$

$$\therefore \quad a_0^2 = \frac{\gamma+1}{2}\cdot V_1 V_2$$

$$\therefore \quad \frac{\gamma+1}{2(\gamma-1)}V_1\cdot V_2 = \frac{a_1^2}{\gamma-1}+\frac{V_1^2}{2}$$

$$\therefore \quad V_1 V_2 = \frac{2(\gamma-1)}{(\gamma+1)}\times\frac{a_1^2}{\gamma-1}+\frac{2(\gamma-1)}{(\gamma+1)}\cdot\frac{V_1^2}{2}$$

$$\frac{V_2}{V_1} = \frac{2}{\gamma+1}\frac{a_1^2}{V_1^2}+\frac{\gamma-1}{\gamma+1}$$

$$\frac{\rho_1}{\rho_2} = \frac{2}{(\gamma+1)M_1^2}+\frac{\gamma-1}{\gamma+1}$$

$$\therefore \quad \frac{\boldsymbol{\rho_2}}{\boldsymbol{\rho_1}}=\frac{\mathbf{1}}{\dfrac{\mathbf{2}}{\mathbf{(\gamma+1)M_1^2}}+\dfrac{\mathbf{\gamma-1}}{\mathbf{\gamma+1}}} \qquad \text{...(21.84)}$$

An expression for the temperature change across a normal shock may be obtained by the use of Eq. 21.83 and 21.84 in connection with the equation of state for a perfect gas.

$$\frac{p_1}{\rho_1 T_1} = \frac{p_2}{\rho_2 T_2}$$

$$\therefore \quad \frac{T_2}{T_1} = \frac{p_2}{p_1}\times\frac{\rho_1}{\rho_2}$$

$$\therefore \quad \frac{\mathbf{T_2}}{\mathbf{T_1}}=\left\{\mathbf{1}+\frac{\mathbf{2\gamma\left(M_1^2-1\right)}}{\mathbf{\gamma-1}}\right\}\times\left\{\frac{\mathbf{2}}{\mathbf{(\gamma+1)M_1^2}}+\frac{\mathbf{\gamma-1}}{\mathbf{\gamma+1}}\right\} \qquad \text{... (21.85)}$$

The change in the Mach number across a normal shock wave can also be easily obtained. For station (1), before the shock wave,

$$\frac{2}{\gamma+1}\cdot\frac{a_1^2}{V_1^2}+\frac{\gamma-1}{\gamma+1} = \frac{a^{*2}}{V_1^2} \text{ i.e., } \left(\frac{\frac{2}{\gamma+1}}{M_1^2}\right)+\frac{\gamma-1}{\gamma+1}=\frac{a^{*2}}{V_1^2}$$

Similarly, for station (2), after the shock wave,

$$\frac{2}{\gamma+1}\cdot\frac{a_2^2}{V_2^2}+\frac{\gamma-1}{\gamma+1} = \frac{a^{*2}}{V_2^2} \text{ i.e., } \left(\frac{\frac{2}{\gamma+1}}{M_2^2}\right)+\frac{\gamma-1}{\gamma+1}=\frac{a^{*2}}{V_2^2}$$

But $$a^{*2} = V_1V_2$$

$$\therefore \quad \frac{V_1V_2}{V_1^2} = \left(\frac{\frac{2}{\gamma+1}}{M_1^2}\right)+\frac{\gamma-1}{\gamma+1} \text{ and } \frac{V_1V_2}{V_2^2}=\left(\frac{\frac{2}{\gamma+1}}{M_2^2}\right)+\frac{\gamma-1}{\gamma+1}$$

$$\therefore \quad \frac{V_2}{V_1} = \left(\frac{\frac{2}{\gamma+1}}{M_1^2}\right)+\frac{\gamma-1}{\gamma+1} \text{ and } \frac{V_1}{V_2}=\left(\frac{\frac{2}{\gamma+1}}{M_2^2}\right)+\frac{\gamma-1}{\gamma+1}$$

$$\therefore \quad \mathbf{1}=\left\{\frac{2}{(\gamma+1)M_1^2}+\frac{\gamma-1}{\gamma+1}\right\}\times\left\{\frac{2}{M_2^2(\gamma+1)}+\frac{\gamma-1}{\gamma+1}\right\} \quad \text{... (21.86)}$$

SOLVED EXAMPLES

Example 21.1 *Find the lowest speed of sound which may be encountered in the standard atmosphere.*

Solution

$$\begin{aligned} a &= \text{acoustic velocity} = \text{sound velocity} \\ &= \sqrt{\gamma RT} = \sqrt{1.4\times 287\,T} \\ &= 20\sqrt{T} = 20\sqrt{288} \text{ (standard temperature = 15°C)} \\ &= \mathbf{339.4 \text{ m/sec}} \end{aligned}$$

Example 21.2 *The speed of sound in air in km/hr may be computed by the equation* $a = e\sqrt{T}$. *If T is given in °K, find the value and units of c.*

Solution

For air,

$$\begin{aligned} a &= \sqrt{\gamma RT} \\ &= 20\sqrt{T} \text{ m/s} \\ &= 20\sqrt{T}\times\frac{3600}{1000} \text{ km/hr.} \end{aligned}$$

$$= 72\sqrt{T} = c\sqrt{T}$$

$$c = 72\frac{\textbf{km}}{\textbf{hour}} \times \frac{1}{°\textbf{K}}$$

Example 21.3 *A model is placed in a wind tunnel operating at a test section where M = 1.2.*
(a) Find the velocity of air stream based on standard air.
(b) Find the Mach angle α.

Solution

(a) For standard air, $a = 339.4$ m/sec

and $M = 1.2 = \frac{V}{a}$

$\therefore$ V = Velocity of air stream
$= 1.2 \times 339.4$
= **407.28 m/sec**

(b) Mach angle is given by

$$\sin\alpha = \frac{1}{M} = \frac{1}{1.2} = 0.8333$$

$\therefore$ α = Mach angle
= **56°.27**

Example 21.4 *What is the dynamic pressure at 4500 m (ρ_w = 0.024 kg_f/m^3) when the velocity in 600 km/hr?*

Solution

$$\text{Dynamic pressure} = \frac{1}{2}\rho V^2$$

where V = 600 km/hr

$$= \frac{1}{2} \times 0.024 \times \frac{(500)^2}{9}$$

$$= 600 \times \frac{1000}{3600} \text{ m/s}$$

$$= 0.012 \times \frac{250000}{9}$$

$$= \frac{500}{3} \text{ km/sec}$$

= **0.0333 kg/cm²**

Example 21.5 *For a hypersonic wind tunnel having a test section Mach No. of 100, find*

(a) the pressure ratio $\frac{p_o}{p}$

(b) the temperature ratio $\frac{T_o}{T}$.

Solution

(a) Pressure ratio is given by

$$\frac{p_o}{p} = \left[1+\frac{\gamma-1}{2}\cdot M^2\right]^{\frac{\gamma}{\gamma-1}}$$

$$= \left[1+\frac{0.4}{2}\cdot 10^2\right]^{\frac{1.4}{0.4}}$$

$$= (1+20)^{3.5}$$

$$= (21)^{3.5}$$

$$= \mathbf{42430}$$

(b) Temperature ratio is given by

$$\frac{T_o}{T} = \left[1+\frac{\gamma-1}{2}M^2\right]$$

$$= \left[1+\frac{0.4\times 100}{2}\right]$$

$$= \mathbf{21}$$

Example 21.6 *Standard air at zero velocity expands to M = 0.9. Find (a) the final density and (b) density change.*

Solution

(a) Temperature of standard air is 288°K and its density before expansion = 1.222 kg/cu.m

Now we have, $$\frac{p_o}{p} = \left[1+\frac{\gamma-1}{2}M^2\right]^{\frac{\gamma}{\gamma-1}}$$

$$= \left[1+\frac{0.4}{2}\times 0.81\right]^{3.5}$$

$$= (1.162)^{3.5} = \mathbf{1.695}$$

Also, we have $$\frac{T_o}{T} = \left(1+\frac{\gamma-1}{2}M^2\right)$$

$$= \mathbf{1.162}$$

∴ We can write $$\frac{p_o}{\rho_o T_o} = \frac{p}{\rho T}$$

∴ $$\frac{\rho_o}{\rho} = \frac{p_o}{p}\times\frac{T}{T_o} = \frac{1.695}{1.162}$$

$$= \mathbf{1.46}$$

∴ $$\rho = \frac{1.222}{1.46} = \mathbf{0.837\ kg/m^2}$$

(b) Density change

$$\Delta\rho = -(\rho_o - \rho)$$
$$= -(1.222 - 0.837)$$
$$= \mathbf{-0.385}$$

Example 21.7 *Standard air at 48 km/hr is accelerated isentropically in a nozzle to 800 km/hr.*

Find : (a) the change in temperature,

(b) the change in pressure,

(c) change in density,

(d) change in stagnation temperature,

(e) the change in stagnation pressure.

Solution

(a) Change in temperature :

$$48 \text{ km/hr} = \frac{48 \times 1000}{3600} = \frac{40}{3} \text{ m/s}$$

$$800 \text{ km/hr} = \frac{800 \times 1000}{3600} = \frac{2000}{9}$$

For steady flow process and isentropic process,

$$0 = \Delta K + \Delta h \text{ as } Q = 0, W_{sf} = 0, \Delta P = 0$$

$$\therefore \quad K_1 + h_1 = K_2 + h_2$$

$$h_1 - h_2 = \frac{V_2^2 - V_1^2}{2} \text{ or } C_p(t_1 - t_2) = \frac{\frac{(2000)^2}{81} - \frac{(40)^2}{9}}{2}$$

$$\therefore \quad t_2 - t_1 = \text{Change in temperature}$$

$$= \frac{\left(\frac{1600}{9} - \frac{4 \times 10^6}{81}\right)}{2 \times 1000 \times 1.05}$$

$$= \frac{14400 - 4000000}{81 \times 2 \times 1000 \times 1.05}$$

$$= \frac{-39856}{1620 \times 1.05}$$

$$= \mathbf{-24.45°C}$$

– ve sign indicates that the temperature decreases.

(b) Change in pressure :

Temperature of standard air $= 15° = 288°$ K

$$\therefore \quad T_2 = 288 - 24.45 = 263.55° \text{ K}$$

For isentropic flow,

$$\frac{T_2}{T_1} = \left(\frac{p_2}{p_1}\right)^{\frac{\gamma-1}{\gamma}} \text{ or } \frac{T_1}{T_2} = \left(\frac{p_1}{p_2}\right)^{\frac{\gamma-1}{\gamma}} = \frac{288}{263.55} = 1.135$$

$$\therefore \qquad 1.135 = \left(\frac{p_1}{p_2}\right)^{2/7} \quad \therefore \frac{p_1}{p_2} = 1.56$$

$$\therefore \qquad \frac{p_2}{p_1} = 0.64 \quad \therefore \frac{p_2 - p_1}{p_1} = \frac{p_2}{p_1} - 1 = 0.64 - 1 = -0.36$$

p_1 = standard pressure = 1.01325 bar

$$\therefore \qquad p_2 - p_1 = \text{Change in pressure} = -0.36 \times 1.01325$$

$$= \mathbf{-0.3648\ bar}$$

– ve sign indicates that the pressure decreases.

(c) Change in density :

Initial density $\qquad \rho_1 = 1.222\ \text{kg/m}^3$

For final density, $\qquad \dfrac{p_2}{\rho_2} = RT_2$

$$\rho_2 \frac{p_2}{RT_a} = \frac{10^5 (1.01325 - 0.3648)}{287 \times 263.55}$$

$$= \frac{0.64845 \times 10^5}{287 \times 263.55} = 0.8573\ \text{kg/m}^3$$

$\therefore$ Change of density $= 0.8573 - 1.2220 = \mathbf{-0.3657}$

– ve sign indicates that the density decreases.

(d) Change in stagnation temperature :

Stagnation temperature is given by

$$T_o = T + \frac{V^2}{2010}$$

$$\therefore \qquad T_{o_1} = T_1 + \frac{V_1^2}{2010} = 15 + \frac{(40/3)^2}{2010}$$

$$= 15 + \frac{1600}{9 \times 2010} = 15 + 0.08825$$

$$= \mathbf{15.08825°C\ or\ 288.08825°K}$$

Similarly, $\qquad T_{o_2} = T_2 + \dfrac{V_2^2}{2010} = 263.55 + \dfrac{(2000/9)^2}{2010}$

$$= 263.55 + \frac{4 \times 10^6}{81 \times 2010}$$

$$= 263.55 + 24.57$$

$$= \mathbf{288.12°K}$$

$\therefore$ Change in stagnation temperature

$$= T_{o_2} - T_{o_1} = 288.12 - 288.08825$$

$$= \mathbf{0.03175°K}$$

(e) Change in stagnation pressure :

We have
$$\left(\frac{p_{o_1}}{p_1}\right)^{\frac{\gamma-1}{\gamma}} = \frac{T_o}{T_1} = \frac{288.08825}{288}$$
$$= 1$$

$$\therefore \quad \frac{p_{o_1}}{p_1} = 1 \text{ or } p_o = p_1 = \mathbf{1.01325\ bar}$$

Similarly,
$$\left(\frac{p_{o_2}}{p_2}\right)^{\frac{\gamma-1}{\gamma}} = \frac{T_{o_2}}{T_2} = \frac{288.12}{263.55} = 1.093$$

$$\therefore \quad \frac{p_{o_2}}{p_2} = (1.093)^{1.4/0.4} = (1.093)^{3.5}$$
$$= 1.366$$

$$\therefore \quad p_{o_2} = 1.366 \times (1.01325 - 0.3648)$$
$$= 1.366 \times 0.64845$$
$$= \mathbf{0.886}$$

∴ Change in stagnation pressure
$$= p_{o_2} - p_{o_1} = 0.886 - 1.01325$$
$$= \mathbf{-0.12725\ bar}$$

Example 21.8 *For air undergoing an isentropic process, derive the equation* $a^* = c\sqrt{T_o}$ *and determine the numerical value of the constant c when a* is given in m/sec and* T_o, *°K.*

Solution

We have for isentropic flow,

$$\frac{a_0^2}{\gamma-1} = \frac{a^2}{\gamma-1} + \frac{V^2}{2}$$

When $V = a$, $a = a^*$

$$\therefore \quad \frac{a_0^2}{\gamma-1} = \frac{a^2}{\gamma-1} + \frac{a^2}{2} = \frac{2a^2 + a^2(\gamma-1)}{2(\gamma-1)}$$
$$= \frac{a^2(\gamma+1)}{2(\gamma-1)}$$

$$\therefore \quad a_0^2 = \frac{a^2}{2} \times \frac{\gamma+1}{\gamma-1}$$

$$\therefore \quad a_0^2 = \frac{a^{*2}(\gamma+1)}{2}$$

$$\therefore \quad a^{*2} = \frac{2\,a_0^2}{\gamma+1}$$

$$\therefore \quad a^* = \frac{a_o}{\gamma+1}\sqrt{\left(\frac{2}{\gamma+1}\right)}\,a_o = 20\sqrt{T_o} \times \sqrt{\frac{2}{\gamma+1}}$$

$$= 20\sqrt{\frac{2}{2.4}} \times \sqrt{T_o} = 20\sqrt{\frac{1}{1.2}} \times \sqrt{T_o}$$

$$\therefore \quad a^* = 18.26\sqrt{T_o}$$

$$\therefore \quad c\sqrt{T_o} = 18.26\sqrt{T_o}$$

$$\therefore \quad \mathbf{c = 18.26}$$

Example 21.9 *Gas enters a convergent-divergent nozzle at a pressure of 5 bar and temperature of 800°C. The exit pressure is 1 bar. Gas flow rate is 20 kg/s. Find : (a) Velocity of gas at throat and exit, (b) Area of nozzle and throat and exit in mm². Assume gas properties as that of air.*

Solution

The flow is shown on *T–S* diagram in Fig. Ex. 21.9

Given : $p_1 = 5$ bar,

$p_3 = 1$ bar

$T_1 = 800 + 273 + 1073$ K

$m_a = 20$ kg/s

Assume the critical pressure ratio for gas as that for air i.e.,

$$\frac{p_2}{p_1} = \left(\frac{2}{\gamma+1}\right)^{\frac{\gamma}{\gamma+1}}$$

$$= \left(\frac{2}{2.4}\right)^{\frac{1.4}{0.4}} = \left(\frac{2}{2.4}\right)^{3.5}$$

$$= 0.5283$$

$$\therefore \quad p_2 = 5 \times 0.5283$$

$$= \mathbf{2.6414\ bar}$$

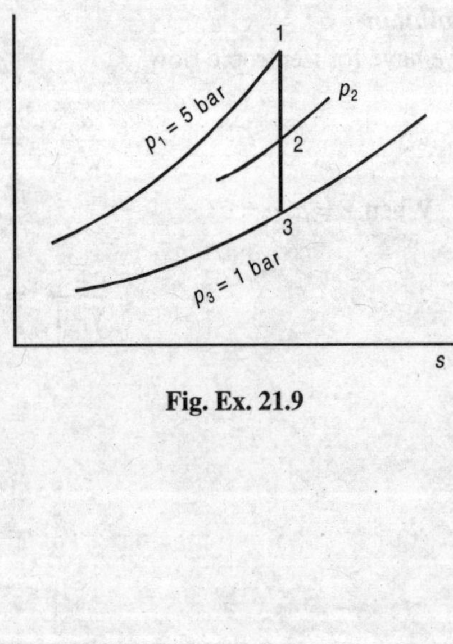

Fig. Ex. 21.9

$$\therefore \quad \frac{T_1}{T_2} = \left(\frac{p_1}{p_2}\right)^{\frac{\gamma-1}{\gamma}} = \left(\frac{5}{2.6414}\right)^{\frac{0.4}{1.4}}$$

$$= (1.893)^{0.2857}$$

$$= 1.2$$

$$\therefore \quad T_2 = \text{Temperature at the throat}$$

$$= \frac{T_1}{1.2} = \frac{1073}{1.2}$$

$$= \mathbf{894.2\ K}$$

Similarly, temperature at the exit is given by

$$\frac{T_1}{T_3} = \left(\frac{p_1}{p_3}\right)^{\frac{\gamma}{\gamma-1}}$$

$$= (5)^{0.2857}$$

$$= \mathbf{1.584}$$

$$\therefore \qquad T_3 = \frac{1073}{1.584} = \mathbf{683.7\ K}$$

(a) Throat and exit velocity :

Throat : Enthalpy drop upto throat

$$= h_1 - h_2$$

$$= C_p (T_1 - T_2) = 1.005\,(1073 - 894.2)$$

$$= 1.005 \times 178.8$$

$$= \mathbf{179.69\ \ kJ/kg}$$

$\therefore$ Velocity of gas at the throat

$$V_2 = 44.72\sqrt{179.69}$$

$$= \mathbf{599.46\ m/s}$$

Exit : Enthalpy drop between inlet and exit

$$= h_1 - h_3$$

$$= C_p (T_1 - T_3)$$

$$= 1.005\,(1073 - 683.7)$$

$$= 1.005 \times 389.3$$

$$= \mathbf{391.25\ \ kJ/kg}$$

$\therefore$ Velocity of gas at exit, $\quad V_3 = 44.72\sqrt{391.25}$

$$= \mathbf{884.56\ m/s}$$

(b) Area of throat and exit :

Mass flow rate = 20 kg/s

Specific volume of gas at throat conditions

$$v_2 = \frac{RT_2}{p_2} = \frac{287 \times 894.2}{2.6414 \times 10^5}$$

$$= \mathbf{0.9716\ m^3/kg}$$

Similarly, specific volume at exit conditions

$$v_3 = \frac{RT_3}{p_3} = \frac{287 \times 683.7}{1 \times 10^5}$$

$$= \mathbf{1.962\ m^3/kg}$$

$$\therefore \quad m = \frac{A_2V_2}{v_2} = \frac{A_3V_3}{v_3}$$

$$\therefore \quad A_2 = \frac{20 \times 0.9716}{599.46}$$

$$= 0.0324\,\text{m}^2$$

$$= \mathbf{32400\ mm^2}$$

Similarly,

$$A_3 = \frac{mv_3}{V_3}$$

$$= \frac{20 \times 1.962}{884.56}$$

$$= \mathbf{0.0444\ m^2}$$

$$= \mathbf{44400\ mm^2}$$

Example 21.10 *Air, which is at a pressure of 6 bar and 850°C, expands in a convergent nozzle to 3.5 bar. The area of throat is 18000 mm². Neglecting approach velocity, calculate the exit velocity and mass rate if*

(a) *reversible expansion takes place and*

(b) *velocity friction factor is 0.9.*

Solution

Given : *T–S* diagram shows the nozzle process.

1–2 Reversible expansion process (Isentropic process)

1–2′ Expansion process with friction

$$p_1 = 6 \text{ bar}$$

$$p_2 = 3.5 \text{ bar}$$

$$T_1 = 850 + 273 = \mathbf{1123\ K}$$

$$A_2 = 18000 \text{ mm}^2$$

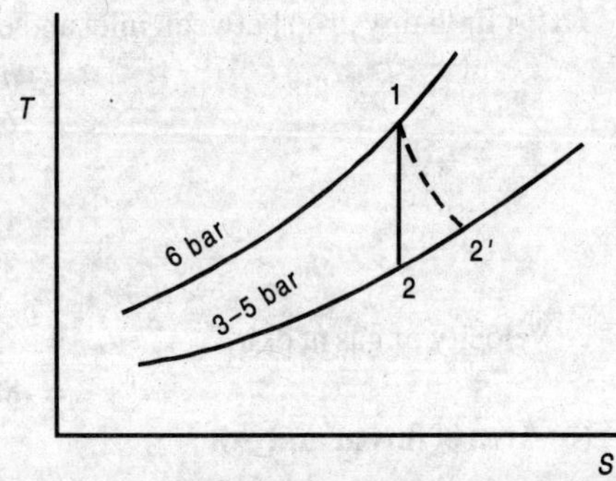

Fig. Ex. 21.10

(a) Reversible expansion : We have,

$$\frac{T_1}{T_2} = \left(\frac{p_1}{p_2}\right)^{\frac{\gamma-1}{\gamma}}$$

$$= \left(\frac{6}{3.5}\right)^{0.2857}$$

$$= 1.1665$$

$$\therefore \quad T_2 = \text{Temperature at throat section (exit)}$$

$$= \frac{1123}{1.1665}$$

$$= \mathbf{962.71\ K}$$

$$\therefore \quad \text{Enthalpy drop upto exit} = h_1 - h_2$$

$$= C_p(T_1 - T_2)$$

$$= 1.005\,(1123 - 962.71)$$

$$= 1.005 \times 160.29$$
$$= \mathbf{161.09\ kJ/kg}$$

$\therefore \quad V_2$ = Velocity at exit

$$= 44.72\sqrt{h_1 - h_2}$$
$$= 44.72\sqrt{161.09}$$
$$= \mathbf{567.6\ m/s}$$

Specific volume of exit, $v_2 = \dfrac{RT_2}{p_2}$

$$= \frac{287 \times 962.71}{3.5 \times 10^5}$$
$$= \mathbf{0.789\ m^3/kg}$$

$\therefore$ Mass flow rate $= m = \dfrac{A_2V_2}{v_2}$

$$= \frac{18000}{10^6} \times \frac{567.6}{0.789}$$
$$= \mathbf{12.95\ kg/s}$$

(b) Expansion with friction :

Actual velocity $= V_{2'} = 0.9 \times 567.6$

$$= \mathbf{510.84\ m/s}$$
$$= 44.72\sqrt{h_1 - h_{2'}}$$

$$\therefore \quad \sqrt{h_1 - h_{2'}} = \frac{510.84}{44.72}$$
$$= 11.423$$

$$\therefore \quad h_1 - h_{2'} = 130.49 \text{ kJ/kg}$$
$$= C_p(T_1 - T_{2'})$$

$$\therefore \quad 130.49 = 1.005\,(1123 - T_{2'})$$

$$\therefore \quad 1123 - T_{2'} = \frac{130.49}{1.005} = 129.84$$

$$\therefore \quad T_{2'} = 1123 - 129.84$$
$$= \mathbf{993.16\ K}$$

$$\therefore \quad v_{2'} = \frac{287 \times 993.16}{3.5 \times 10^5}$$
$$= \mathbf{0.814\ m^3/kg}$$

$\therefore$ Mass flow rate, $m = \dfrac{A_2V_{2'}}{v_{2'}}$

$$= \frac{18000}{10^6} \times \frac{510.84}{0.814}$$

$$= \mathbf{11.296\ m/s}$$

Example 21.11 *Gas nozzle in gas turbine is supplied with pressure of 6 bar and 900°C and the gas is discharged at 1 bar. If the diverging portion of the nozzle is 60 mm log and the throat diameter is 8 mm, determine the cone angle of the divergent portion. Assume 10% of the total available enthalpy drop to be lost in friction in the divergent portion. Also determine the velocity and temperature of gas at the throat.*

Assume C_p for gas as 1.1, R = 287, γ = 1.35 with usual units.

Solution

Assume that the nozzle is designed for maximum discharge conditions.

Given :

$$p_1 = 6 \text{ bar}$$
$$T_1 = 1173 \text{ K}$$
$$p_3 = 1 \text{ bar}$$
$$d_2 = 8 \text{ mm}$$
$$C_{pg} = 1.1 \text{ kJ/kg-K}$$
$$\gamma \text{ for gas} = 1.35$$

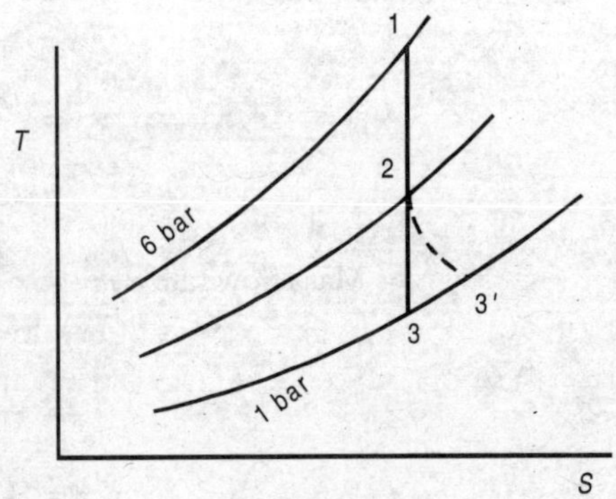

Fig. Ex. 21.11 (a)

10% loss in divergent nozzle.

Temperature and Velocity at Throat :

Pressure at throat is given by,

$$p_2 = 0.5283\, p_1 = 0.5283 \times 6$$
$$= \mathbf{3.1698\ bar}$$

$$\therefore \quad \frac{T_2}{T_1} = \left(\frac{p_2}{p_1}\right)^{\frac{\gamma-1}{\gamma}} = (0.5283)^{0.2857}$$
$$= 0.833$$

$$\therefore \quad T_2 = \text{Temperature at throat}$$
$$= 0.833 \times 1173$$
$$= \mathbf{977.52\ K}$$

$$\therefore \quad h_1 - h_2 = C_p (T_1 - T_2)$$
$$= 1.1\,(1173 - 977.52) = 1.1 \times 195.48$$
$$= \mathbf{215.028\ kJ/kg}$$

$$\therefore \quad V_2 = \text{Velocity of gas at throat}$$
$$= 44.72\sqrt{215.028}$$
$$= \mathbf{655.77\ m/s}$$

Cone angle of divergent portion : Specific volume of gas at throat is given by

$$v_2 = \frac{RT_2}{p_2} = \frac{287 \times 977.52}{3.1698 \times 10^5}$$
$$= \mathbf{0.885\ m^3/kg}$$

For divergent portion,

$$\frac{T_1}{T_3} = \left(\frac{p_1}{p_3}\right)^{\frac{\gamma-1}{\gamma}} = (6)^{0.2857}$$

$$= 1.668$$

$$\therefore \quad T_3 = \frac{1173}{1.668} = \mathbf{703.24\ K}$$

Isentropic enthalpy drop from inlet to exit

$$= h_1 - h_3 = C_{pg}(T_1 - T_3)$$
$$= 1.1\,(1173 - 703.24)$$
$$= 1.1 \times 469.76$$
$$= \mathbf{516.736\ \ kJ/kg}$$

$$\therefore \quad \text{Actual enthalpy drop} = 0.9 \times 516.736$$
$$= \mathbf{465.06\ \ kJ/kg}$$
$$= C_{pg}(T_1 - T_{3'}) = 1.1\,(1173 - T_{3'})$$

$$\therefore \quad T_{3'} = 1173 - \frac{465.06}{1.1} = 1173 - 422.78$$
$$= \mathbf{750.22\ K}$$

$$v_{3'} = \text{Specific volume at exit}$$
$$= \frac{RT_{3'}}{p_3} = \frac{287 \times 750.22}{1 \times 10^5}$$
$$= \mathbf{2.153\ m^3/kg}$$

Also, $\quad V_{3'} = \text{Velocity at exit}$

$$= 44.72\sqrt{465.06}$$
$$= \mathbf{964.4\ m/s}$$

As the mass flow is same, we have

$$\frac{A_2 V_2}{v_2} = \frac{A_3 V_{3'}}{v_3}$$

$$\therefore \quad \frac{d_2^2 \times 655.77}{0.885} = \frac{d_3^2 \times 964.4}{2.153}$$

$$\therefore \quad d_3^2 = d_2^2 \times \frac{655.77}{964.4} \times \frac{2.153}{0.885}$$
$$= d_2^2 \times 1.654$$

$$\therefore \quad d_3 = d_2 \times 1.286 = 8 \times 1.286$$
$$= \mathbf{10.288\ mm}$$

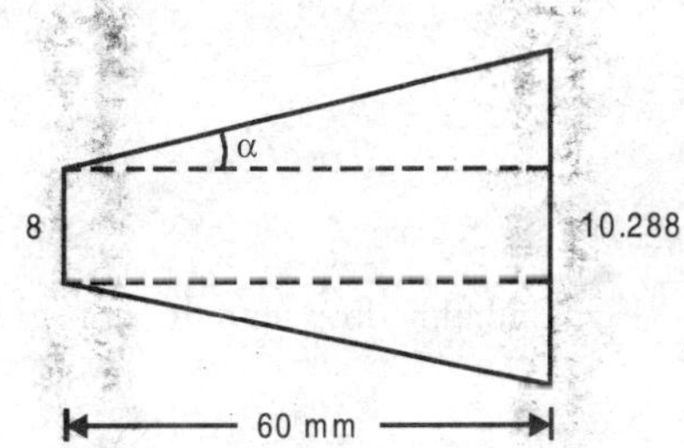

Fig. Ex. 21.11 (b)

$$\tan\alpha = \frac{10.288 - 8}{2 \times 60}$$
$$= 0.01906$$

$$\therefore \quad 2\alpha = \text{Cone angle of divergent portion}$$
$$= 2 \times 1.092 = \mathbf{2.184°}$$

Example 21.12 *Gas enters the nozzle of an impulse turbine at 6.5 bar and 900°C and leaves at 1 bar. The turbine develops 225 kW with specific gas consumption 16. If the throat diameter is 12 mm, find the number of nozzles and* *the exit diameter, assuming that 10% of the total enthalpy drop is lost in overcoming friction in the divergent portion only. Neglect velocity of approach and assume air properties for gas.*

Solution

Given : Nozzle is convergent-divergent.

$$\text{Gas consumption} = 225 \times 16 \text{ kg/hour} = 3600 \text{ kg/hour} = \mathbf{1 \text{ kg/sec}}$$

Number of nozzles :

$$p_1 = 6.5 \text{ bar},$$

$$T_1 = 900 + 273 = 1173 \text{ K}$$

$$p_3 = 1 \text{ bar}$$

$$d_2 = \text{Throat diameter} = 12 \text{ mm}$$

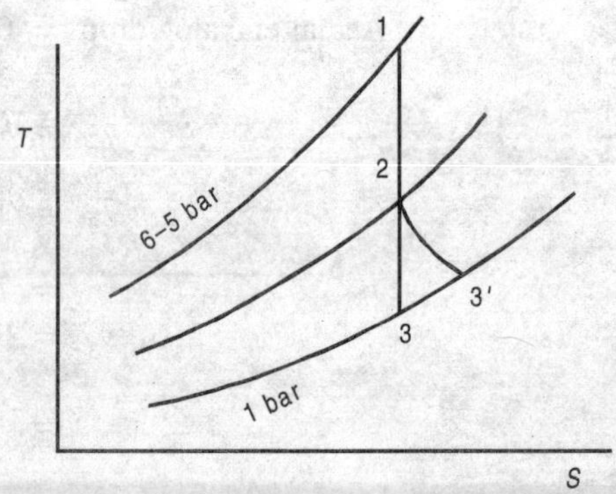

Fig. Ex. 21.12

Nozzle is designed for maximum discharge conditions.

$$\therefore \quad p_2 = \text{Throat pressure} = 0.5283\, p_1 = 0.5283 \times 6.5 = \mathbf{3.434 \text{ bar}}$$

We have,

$$\frac{T_1}{T_2} = \left(\frac{p_1}{p_2}\right)^{0.2857} = \left(\frac{6.5}{3.434}\right)^{0.2857} = 1.2$$

$$\therefore \quad T_2 = \text{Temperature at throat} = \frac{1173}{1.2} = \mathbf{977.5 \text{ K}}$$

$\therefore$ Enthalpy drop upto throat is given by,

$$h_1 - h_2 = C_p (T_1 - T_2) = 1.005\,(1173 - 977.5) = 1.005 \times 195.5 = \mathbf{196.48 \text{ kJ/kg}}$$

$\therefore$ Velocity at throat is given by

$$V_2 = 44.72\sqrt{196.48} = \mathbf{626.84 \text{ m/s}}$$

Again specific volume of gas is given by

$$v_2 = \frac{RT_2}{p_2} = \frac{287 \times 977.5}{3.434 \times 10^5} = \mathbf{0.817 \text{ m}^3\text{/kg}}$$

$$\therefore \quad m = \text{Discharge} = \frac{A_2 V_2}{v_2} = \frac{\pi}{4} \times \frac{144}{10^6} \times \frac{626.84}{0.817}$$

$$= \mathbf{0.08681\ kg/s}$$

$$\therefore \quad \text{Number of nozzles} = \frac{1}{0.08681}$$

$$= 11.52$$

i.e.,

$$\approx \mathbf{12}$$

Exit diameter : For isentropic flow through the nozzle, we have

$$\frac{T_1}{T_3} = \left(\frac{p_1}{p_3}\right)^{\frac{\gamma-1}{\gamma}} = \left(\frac{6.5}{1}\right)^{0.2857}$$

$$= 1.707$$

$$\therefore \quad T_3 = \frac{1173}{1.707}$$

$$= \mathbf{687.17\ K}$$

$$\therefore \quad \text{Isentropic heat drop} = h_1 - h_3 = C_p (T_1 - T_3)$$

$$= 1.005\,(1173 - 687.17)$$

$$= 1.005 \times 485.83$$

$$= \mathbf{488.26\ kJ/kg}$$

$$\therefore \quad \text{Actual enthalpy drop} = 0.9 \times 488.26 = \mathbf{439.43\ kJ/kg}$$

$$\therefore \quad \text{Velocity at exit} = 44.72\sqrt{439.43}$$

$$V_{3'} = 937.45 \text{ m/s}$$

$$\text{Actual enthalpy drop} = 439.43 = C_p(T_1 - T_{3'})$$

$$= 1.005\,(1173 - T_{3'})$$

$$\therefore \quad T_{3'} = 1173 - \frac{439.43}{1.005} = 1173 - 437.24$$

$$= \mathbf{735.76\ K}$$

$$v_{3'} = \frac{RT_{3'}}{p_3} = \frac{287 \times 735.76}{1 \times 10^5}$$

$$= \mathbf{2.112\ m^3/kg}$$

$$m = \frac{A_3 V_{3'}}{v_{3'}} = \frac{A_3 \times 937.45}{2.112} = A_3 \times 443.87$$

$$\therefore \quad A_3 = \frac{\pi d_3^2}{4} = \frac{0.08681}{443.87}$$

$$= 0.0001956 \text{ m}^2$$

$$\therefore \quad d_3^2 = \frac{0.0001956}{0.7857} = 0.000249$$

$$\therefore \quad d_3 = \text{Exit diameter}$$
$$= 0.016 \text{m}$$
$$= \mathbf{16\,mm}$$

Example 21.13 *An air nozzle admits air at 5 bar and 800°C. The throat diameter is 10 mm. Expansion of gas upto throat is frictionless. The velocities at throat and exit are 600 m/s and 900 m/s respectively. The gas leaves the nozzle at 1 bar. Determine : (1) throat pressure and (2) exit diameter for flow of air as 1 kg/s.*

Solution

Figure Ex. 21.13 shows the process taking place in the nozzle.

Given :

$$p_1 = 5 \text{ bar}$$
$$p_3 = 1 \text{ bar}$$
$$T_1 = 800 + 273$$
$$= 1073 \text{ K}$$
$$V_2 = \text{Throat velocity}$$
$$= 600 \text{ m/s}$$
$$V_{3'} = \text{Exit velocity}$$
$$= 900 \text{ m/s}$$

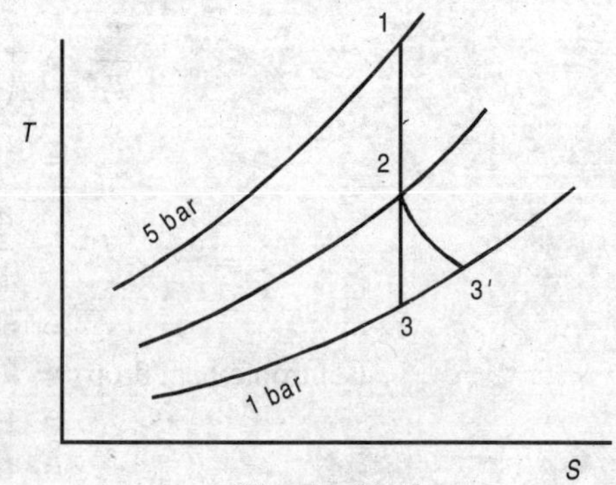

Fig. Ex. 21.13

Upto throat, the expansion is frictionless.

∴ For maximum discharge condition,

$$p_2 = \text{Pressure at throat}$$

$$\text{Exit velocity } v_{3'} = 44.72\sqrt{h_1 - h_{3'}}$$

$$900 = 44.72\sqrt{h_1 - h_{3'}}$$

$$\therefore \quad \sqrt{h_1 - h_{3'}} = \frac{900}{44.72} = 20.125$$

$$\therefore \quad h_1 - h_{3'} = \mathbf{405\ kJ/kg}$$
$$= C_p (T_1 - T_{3'}) = 1.005\,(1073 - T_{3'})$$

$$T_{3'} = 1073 - \frac{405}{1.005} = 1073 - 403 = 670 \text{ K}$$

Similarly, throat velocity is given by

$$V_2 = 600 = 44.72\sqrt{h_1 - h_2}$$

$$\therefore \quad h_1 - h_2 = \left(\frac{600}{44.72}\right)^2 = 180$$

$$\therefore \quad 180 = C_p (T_1 - T_2)$$
$$= 1.005\,(1073 - T_2)$$

$$\therefore \quad T_2 = 1073 - \frac{180}{1.005} = 1073 - 179.1$$
$$= 893.9 \text{ K}$$

But
$$\frac{T_1}{T_2} = \left(\frac{p_1}{p_2}\right)^{\frac{\gamma-1}{\gamma}}$$

$$\therefore \quad \frac{p_1}{p_2} = \left(\frac{T_1}{T_2}\right)^{3.5} = \left(\frac{1073}{893.9}\right)^{3.5}$$

$$= 1.895$$

$$\therefore \quad p_2 = \text{Throat pressure}$$

$$= \frac{5}{1.895} = \mathbf{2.638\ bar}$$

Also p_2 can be found by considering the maximum discharge condition.

$$\therefore \quad p_2 = 0.5283\,p_1 = 0.5283 \times 5 = 2.641 \text{ bar}$$

This is almost the same as calculated above.

(2) Exit diameter : Specific volume at exit

$$v_{3'} = \frac{RT_{3'}}{p_{3'}}$$

$$= \frac{287 \times 670}{1 \times 10^6}$$

$$= \mathbf{1.923\ m^3/kg}$$

$$\therefore \quad m = \frac{A_3 V_{3'}}{v_{3'}}$$

$$1 = \frac{A_3 \times 900}{1.923}$$

$$\therefore \quad A_3 = \frac{1.923}{900}\ \text{m}^2$$

$$= \frac{1.923}{900} \times 10^6\ \text{mm}^2$$

$$= 2136.7\ \text{mm}^2$$

$$\therefore \quad 2136.7 = \frac{\pi d^2}{4}$$

$$= 2719.5$$

$$\therefore \quad d = \text{Exit diameter}$$

$$= \mathbf{52.15\ mm \approx 52\ mm\ Approx.}$$

THEORY QUESTIONS

21.1. What are static and stagnation properties? Explain.

21.2. Define acoustic velocity and derive the relation for acousting velocity in terms of absolute temperature and adiabatic index.

21.3. Explain with sketches the following :
(a) Mach line
(b) Mach angle
(c) Mach cone
(d) Mach wave

21.4. What do you understand by 'the zone of silence' and 'zone of action'?

21.5. With usual notations derive the following relation for adiabatic flow

$$p_o = p + \frac{\rho V^2}{2}\left[\frac{2}{\gamma M^2} + \frac{M^2}{4} + \frac{2-\gamma}{12}.M^4 + ...\right]$$